Textbooks in Contemporary Dentistry

This textbook series presents the most recent advances in all fields of dentistry, with the aim of bridging the gap between basic science and clinical practice. It will equip readers with an excellent knowledge of how to provide optimal care reflecting current understanding and utilizing the latest materials and techniques. Each volume is written by internationally respected experts in the field who ensure that information is conveyed in a concise, consistent, and readily intelligible manner with the aid of a wealth of informative illustrations. *Textbooks in Contemporary Dentistry* will be especially valuable for advanced students, practitioners in the early stages of their career, and university instructors.

Donald J. Coluzzi • Steven P. A. Parker
Editors

Lasers in Dentistry—Current Concepts

Second Edition

Springer

Editors
Donald J. Coluzzi
Preventive and Restorative Dental Sciences
School of Dentistry, University of California
San Francisco, CA, USA

Steven P. A. Parker
Faculty of Health and Life Sciences
De Montfort University
Leicester, UK

ISSN 2524-4612 ISSN 2524-4620 (electronic)
Textbooks in Contemporary Dentistry
ISBN 978-3-031-43337-5 ISBN 978-3-031-43338-2 (eBook)
https://doi.org/10.1007/978-3-031-43338-2

This Springer imprint is published by the registered company Springer Nature Switzerland AG
The registered company address is: Gewerbestrasse 11, 6330 Cham, Switzerland

Paper in this product is recyclable.

Preface

The first laser specifically designed for dentistry was introduced in 1989 and used a crystal of neodymium-doped yttrium aluminum garnet (Nd:YAG) as its core active medium. Low average power photonic energy produced by this laser was delivered through a small-diameter optic fiber to target oral tissue. Such technology had been developed for use in medicine since 1975, and carbon dioxide (CO_2) lasers were commonly employed during the 1980s for general and oral surgery.

Nowadays, approximately 15% of dentists worldwide own lasers, and there are about 30 indications for their use in dental treatment. Whether used in addition to or instead of conventional instrumentation, lasers provide many unique patient benefits.

This textbook is intended to provide information about the basic science and tissue interactions of dental lasers and display the most current examples of clinical use in every dental discipline. The clinical cases were chosen to show the results of proper laser use for a particular procedure, and the accompanying text explains the rationale, advantages, and precautions of that use, documented with numerous citations.

Research studies continue to provide collaborative evidence demonstrating the efficacy of today's instrumentations. Furthermore, other investigations will enumerate novel clinical applications, and hopefully new laser wavelengths will be explored, developed to deliver highly specific power configurations to optimize laser-tissue interaction.

We are proud of the phenomenal success of the 2017 first edition of this textbook. Through the almost 7 years that have passed, the use of laser photonic energy, in all its power range applications, has continued to expand. Throughout the world, clinicians are discovering the wide array of clinical applications—diagnostics, photochemistry, sub-ablative photobiomodulation, and ablative surgical adjunctive use—and the ease with which these applications can deliver added benefits.

It is a testimony to the innovation of laser manufacturers, and to the worldwide research community and dental and oral clinicians, that both the number and variety of lasers continue to expand.

This book continues to be the product of those many highly respected dental clinicians, along with those in academia and involved in research throughout the world, and we are grateful for their efforts and their friendship. To those who have gone before, we say "thank you," and those new authors, we are appreciative of your effort in contributing.

Most importantly, we acknowledge the love, understanding, and support of our spouses, Catherine Coluzzi and Penny Parker.

We hope that you enjoy the book.

Donald J. Coluzzi
Steven P. A. Parker
Portola Valley, CA, USA
Harrogate, North Yorkshire, UK

Contents

Editors and Contributors

About the Editors

Donald J. Coluzzi, DDS
A 1970 graduate of the University of Southern California School of Dentistry, is a professor in the Department of Preventive and Restorative Dental Sciences at the University of California San Francisco School of Dentistry. He ran his own private practice of general dentistry in Redwood City, CA, and retired from it after 35 years. He is a life member of both the California Dental Association and the American Dental Association. He has served as a past president of the Academy of Laser Dentistry, received its Leon Goldman Award for Clinical Excellence, and is a past editor in chief of the *Journal of Laser Dentistry*. He has been using dental lasers since early 1991 and holds advanced proficiency in Nd:YAG and Er:YAG wavelengths. He is a fellow of the American College of Dentists as well as the International College of Dentists, is a University of California-certified dental laser educator, and is a member of Omicron Kappa Upsilon, the national dental honor society. He recently received the Outstanding Faculty Member Award from the American College of Dentists. Dr. Coluzzi serves as a reviewer for several journals and has presented about lasers worldwide, coauthored three textbooks, and published several peer-reviewed articles and book chapters.

Steven P. A. Parker, BDS, LDS RCS, FCGDent, PhD
Studied dentistry at the University College Hospital Medical School, University of London, UK, and graduated in 1974. He practiced clinical dentistry during a continuous period of 43 years. He is currently Honorary Professor, Faculty of Health and Life Sciences at De Montfort University, Leicester, UK, where he received his doctorate in laser photonics. Dr. Parker has been involved in the use of lasers in clinical dentistry since 1990. He is closely involved in the provision of education in laser use in dentistry. From 2010 to 2019, he served an appointment as *Professore a Contratto* in the Department of Surgical Sciences and Integrated Diagnostics, University of Genoa, Italy. During that period, he acted as International Coordinator and Lead Faculty of the Master of Science (*AMD Livello II*) degree program in laser dentistry at the University of Genoa.

Dr. Parker served as President of the Academy of Laser Dentistry (ALD) in 2005–2006. He holds advanced proficiency status in multiple laser wavelengths. He was awarded Mastership of the ALD in 2008. Awards gained with the Academy have been the Leon Goldman Award for Excellence in Clinical Laser Dentistry (1998) and Distinguished Service Award (2010). He was the dental consultant to the UK Medical Health Regulatory Agency (Dept. of Health) in the 2008 (Revised 2015) publication "*Guidance on the Safe Use of Lasers, Intense Light Source Systems and LEDs in Medical, Surgical, Dental and Aesthetic Practices*."

Dr. Parker has contributed chapters on aspects of laser use in dentistry in several textbooks and multimedia platforms. Additionally, he has received publication of over 80 peer-reviewed papers on the use of lasers in dentistry, including a 9-paper series "*The Use of Lasers in Dentistry*," published in the *British Dental Journal* in 2007 and later as a textbook. He serves as Associate Editor of *Lasers in Medical Science* and as referee for many peer-reviewed dental journals worldwide.

Contributors

Eugenia Anagnostaki, DDS, MSc Leicester School of Pharmacy, De Montfort University, Leicester, UK

Private Practice, Rethymno, Crete, Greece
e-mail: eugenia.anagnostakis@my365.dmu.ac.uk, eanagnostaki@densindente.de

Akira Aoki, DDS, PhD Department of Periodontology, Graduate School of Medical and Dental Sciences, Tokyo Medical and Dental University, Tokyo, Japan
e-mail: aoperi@tmd.ac.jp

Konstantinos Arapostathis, DDS, MSc, PhD Pediatric Dentistry Department, Aristotle University of Thessaloniki, Thessaloniki, Greece
e-mail: koarap@dent.auth.gr

Ali Borzabadi-Farahani, DDS, MScD, MOrth RCS (Ed) Craniofacial and Special Care Orthodontics, Children's Hospital Los Angeles, Los Angeles, CA, USA

University of Warwick, Warwick, UK
e-mail: faraortho@yahoo.com

Francesco Buoncristiani, DDS, MSc Private Practice, Milan, Italy
e-mail: info@dentistasanvincenzo.it

Nasim Chiniforush, DDS, PhD Total Core Academy, Dubai, UAE
e-mail: nasimch2002@yahoo.com

Donald J. Coluzzi, DDS Preventive and Restorative Dental Sciences, School of Dentistry, University of California, San Francisco, CA, USA
e-mail: doncoluzzi@gmail.com

Mark Cronshaw, BSc, BDS, LDS RCS (Eng), MSc Private Practice, Cowes, Isle of Wight, UK
e-mail: Drmarkcronshaw@outlook.com

Deepti Dua, DDS, MSc Royal Bahrain Hospital, Manama, Bahrain
e-mail: drdeeptidua@gmail.com

Daniel Fried, PhD Preventive and Restorative Dental Sciences, School of Dentistry, University of California, San Francisco, CA, USA
e-mail: daniel.fried@ucsf.edu

Roy George, BDS, MDS, PhD(QLD), ADC School of Dentistry and Oral Health, Griffith University, Nathan, QLD, Australia
e-mail: drroygeorge@gmail.com

Omar Hamadah, DDS, MSc, PhD Oral Medicine Department, The Faculty of Dental Medicine, Damascus University, Damascus, Syria
e-mail: Omar.hamadah@damascusuniversity.edu.sy

Kenneth Luk, BDS, DGDP(UK), MGD(HK), MSc, PhD Private Practice, Hong Kong, China
e-mail: drkluk@mac.com

Shally Mahajan, BDS, MDS Department of Dentistry, Dr Ram Manohar Lohia Institute of Medical Sciences, Lucknow, India
e-mail: drshally23@gmail.com

Robert J. Miller, DDS Department of Oral Implantology, Atlantic Coast Dental Research Clinic, Palm Beach, FL, USA

Private Practice, Delray Beach, FL, USA

e-mail: drrjmiller1@msn.com, drrjmiller1@aol.com

Alex Mathews Muruppel, BDS, MDS, Dipl. LAS. DENT., FPFA Department of Laser Surgery and Laser Therapy, Faculty of Medicine and Dentistry, University of Genoa, Genoa, Italy

Rajas Dental College and Hospital, Kanyakumari, Tamil Nadu, India

Private Practice, Trivandrum, Kerala, India
e-mail: alexmuruppel@gmail.com

Valina Mylona, DDS, MSc Private Practice, Athens, Greece
e-mail: val.mylona@yahoo.com

Penny J. Parker, DCP, RDN Cert. Dent. Rad. Harrogate, UK
e-mail: thewholetooth@gmail.com

Steven P.A. Parker, BDS, LDS RCS, FCGDent, PhD Faculty of Health and Social Sciences, De Montfort University, Leicester, UK
e-mail: thewholetooth@mail.com

Riccardo Poli, DDS, MSc Private Practice, Turin, Italy
e-mail: riccardo.poli.pro@gmail.com

Peter Rechmann, DMD, PhD, Prof. Dr. med. dent. Division of Prosthodontics, Preventive and Restorative Dental Sciences, School of Dentistry, University of California San Francisco, San Francisco, CA, USA
e-mail: Peter.Rechmann@ucsf.edu

Wayne Selting, DDS, BS, MS Private Practice, Colorado Springs, CO, USA
e-mail: WSelting@aol.com

Vipul Srivastava, BDS, MDS Private Practice, Lucknow, India
e-mail: vipul13@gmail.com

Dimitrios Velonis Private Practice, Larisa, Greece
e-mail: dimitrios.velonis@gmail.com

Laurence J. Walsh, BDSc, PhD, DDSc, GCEd University of Queensland, St. Lucia, QLD, Australia
e-mail: l.walsh@uq.edu.au

Joshua Weintraub, DDS Private Practice, Stevenson, MD, USA
e-mail: weintraubdds@gmail.com

Concepts of Laser Use

Contents

Lasers in Dentistry: Where to Begin?

Shally Mahajan, Vipul Srivastava, and Donald J. Coluzzi

Contents

D. J. Coluzzi, S. P. A. Parker (eds.), *Lasers in Dentistry—Current Concepts*, Textbooks in Contemporary Dentistry,
https://doi.org/10.1007/978-3-031-43338-2_1

Core Message

Over a few decades, lasers have not only emerged as high technology instrument but also as very helpful tool in all aspects of our daily lives. They have been slowly incorporated into the world of dentistry with their wide applications that have embraced all aspects of our profession. When given a choice, patients come to accept a treatment that is oriented at maximum tissue preservation, high quality, minimally invasive, and comfortable, with minimum postoperative care and complications. Fortunately, a practice that utilizes lasers can be termed patient friendly as it fulfills all of those goals. Moreover, the clinician can be proud of adhering to evidence-based techniques. The purpose of this chapter is to discuss some of the benefits of adopting lasers into a dental practice; what the clinician must know before purchasing a laser; and concepts of revenue generation. Moreover, a practitioner who is apprehensive about adopting this technology should also find information to help in making a decision to purchase.

1.1 Introduction

Light has been a major part of our ecosystem and an extremely important resource for our survival on this planet. It has fascinated mankind from many centuries while being used for regulation of body growth, sleep, and physiology.

There have been innumerable references to light being a source of healing and curing many diseases for ancient cultures. For example, many Roman homes featured solariums [1], while they and the neighboring Greeks took daily sunbaths. The use of light for photodynamic therapy enabled early civilizations to treat a variety of dermatologic conditions using photosynthesizer chemicals found in plants. Over 200 years ago, physicians in Europe offered similar therapy using both artificial and natural light [2, 3].

At present, laser technology has become associated within indispensable and diverse applications such as meteorology, science and engineering, medicine, communications, art and entertainment, research work, defense, and astronomy. It is impossible to even imagine state-of-the-art physics, chemistry, biology, and medicine research without the use of radiation from various laser systems.

In 1989, the first laser model specifically designed for the dental profession became available for treating oral soft tissue. Since then, many different wavelengths have been introduced and the practitioner can easily use them on both hard and soft tissue for both surgery and healing. This new technology greatly expands the scope of procedures while making them easier and more comfortable for patients. Encouraged by ever-increasing evidence of the safe and effective use of lasers, there is a growing number of practitioners embracing the technology and appreciating how their patients can benefit.

The question in this chapter's title may be properly expanded into, "why buy a laser and what do I need to know when I buy it?" The following sections should provide many details for that answer.

1.2 A Buyer's Guide for Choosing a Laser

While investigating a product for our personal use or a piece of equipment for our practice, several aspects should be considered to avoid any regret later. As a clinician, we should assess what value the equipment is going to add to our practice. A careful analysis should include all related features, benefits, assets, and liabilities. A well thought out decision leads to a better business operation and good management. Hasty decisions can lead to financial distress and instability in career and unnecessary emotional stress. Similarly, before investing in a laser, we can ask the following questions:

- **Is a laser worth the investment; in other words, is there value for the money?** The first and foremost thing before buying a laser is to identify your practice goals because that will help you optimally understand the demands of your patient and how you would meet their expectations. Thus, one response to the posed question is a multi-part one:
  1. Which procedures would I be able to perform with the laser that would produce beneficial results?
  2. Can I achieve a good return on my investment by an additional fee for the procedures that I already perform conventionally?
  3. Are there new procedures I can perform? Another section of this chapter will discuss these points in detail. After that analysis, the answer about value

should be very straightforward. In any case, and depending upon the various treatment applications, lasers are available in a variety of wavelengths, sizes, and competitive prices.

- **Where would I put the laser? What should be the room size for the laser unit to fit? We have personally experienced that many times, we buy an equipment out of our fascination and curiosity, but in reality, and due to a lack of space, it stays in a box and remains unused for longer periods due to the lack of time in unboxing before every procedure. This question is of prime importance and should be considered well before any investment. While considering this question, we should first be clear with what kind of procedures are we aiming at in our clinical practice.** For example, lasers for hard tissue—tooth preparation and osseous surgery—have a relatively large footprint, approximately the size of a standard dental cart. These lasers have air, water, and mains utility requirements similar to that dental cart so the room should accommodate those. Other lasers such as soft tissue diodes are smaller units and only need small plug-in adapters from AC power mains or are battery driven. They can be placed on any available small flat surface. In fact, some of those units are compact to the point of being shaped like a thick pencil and are self-contained. Technological sophistication continues to be developed, but each unit will have its unique space requirements. Clearly, we should be aware of how much open or extra space is available; or do we need to plan a remodel to gain the space required for the equipment.
- **What is the laser's portability and ease of set up?** As a corollary to the previous paragraph, all lasers have a degree of portability. In multiple chair clinics, portability is the feature which enables a clinician to roll the laser unit around wherever it is needed. The larger laser units have caster wheels attached, enabling mobility, and the smaller units are light weight enough to be lifted with one hand. Some have a wireless foot control pedal while the others have multiple cables to connect. Nonetheless, any unit can be moved between operatory rooms. Setting up the laser follows prescribed steps. Along with various safety features, the start-up protocol takes very little time. Manufacturers provide software that runs initial diagnostics and then displays error codes and prompts. The delivery systems have specific accessories that are simple to attach and the displays on main screen are easily readable. Many instruments have controllable presets so that the clinician can easily select the parameters that will enable the best efficiency for the proposed treatment. Protective eyewear is essential for the surgical team and for the patient and any observer in the treatment area. These should be stored close at hand. Each of these steps should become routine so that the laser use becomes seamlessly integrated into any patient care where it is needed.
- **What's the quality of construction?** All of the units are manufactured for patient care with necessary industrial standards that regulate not only electrically powered devices but also dictate infection control requirements. The quality of construction on every laser should be very high although some components will wear with normal use. The main concern of the practitioner is likely to be how comfortable the delivery system is to handle. Some devices have small flexible optical glass fibers, while other lasers have larger hollow tube or articulated arm assemblies. All terminate in a handpiece and some have small tips or tubes to direct the beam toward the target tissue. Tips are usually used in contact with the target tissue, whereas in some instruments, the photonic beam is simply aimed at the target tissue. Clearly, this is not the usual method for most dental procedures and this lack of tactile feedback will have to be practiced. Your hand should not fatigue while performing lengthy procedures and the handpiece should be able to reach in the remote vicinities of the mouth, thus enabling you to perform a wide range of clinical procedures with ease and precision.

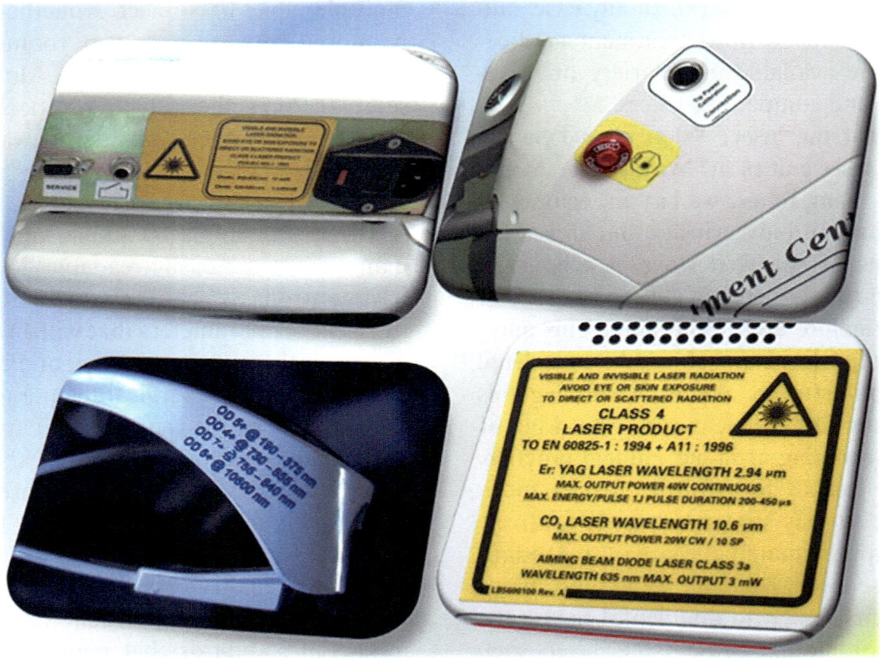

courtesy Dr D. Coluzzi / Dr S. Parker

- **What are the safety features?** Lasers can interact rapidly and precisely with tissue so safety should also be of universal concern. The operator must have a sound knowledge of all the aspects of laser safety. These are summarized here but will be covered in detail in ▶ Chap. 5. All dental lasers are well equipped with built in safety features subject to rigorous rules. Some examples of these features are: an emergency stop button, emission port shutters to prevent laser emission until the correct delivery system is attached, covered foot-switch to prevent accidental operation, a control panel that allows selection of the emission parameters such as average power, pulse details, amount of air and/or water needed, audible or visual signs of laser emission, locked unit panels to prevent unauthorized access to internal components, key or password protection, and remote inter-locks to minimize the risk of accidental exposure. A visible aiming beam for invisible wavelengths must be clear and bright. Clearly, the practitioner must be familiar with these protective items, and a laser safety officer must be appointed to supervise the laser's operation.
- **What is the cost of operation?** Aside from the initial investment of the device, each procedure will have a cost while performing a procedure. Some items or accessories are single use. An example is a tip for a diode laser; these tips are available in multiple diameters and lengths. One tip can generally be used for one patient visit although treatment of multiple areas may require more than one tip. Other components are designed as long lasting but could require replacement. An example is the delivery system itself. Optical fibers can lose some transmissive capability over time; some handpieces have mirrors or other components that degrade. Protective glasses can be scratched or damaged from repeated use. On the other hand, the active medium of the laser and other internal parts generally show little or no wear throughout the life of the laser. While the tip cost is a small percentage of the fee, other items can be a significant economic factor for the practice. In every case, the manufacturer should be able to service the unit, update any software, and offer replacement parts when necessary.
- **How are the parts sterilized or disinfected?** Sterilization and disinfection are essential aspects of every dental treatment and should never be compromised. It is extremely important to follow the manufacturer's instructions for infection control to prevent any cross contamination from patient to patient. Some components of a laser, especially those are in direct contact with oral tissues, are either autoclavable or disposable. The handpiece is an example of the former, and the single use tips are disposable. Other areas, like the control panel and the delivery system can be protected with barriers and subsequently disinfected with standard spray on liquids. The protective safety glasses can also be disinfected.

1.3 Integrating a Laser into Your Practice

Lasers have provided a new cutting-edge technology to the dental world. It is truly amazing to think about how such an investment like this could have such a huge impact on clinical practice. Incorporating lasers into conventional therapies helps in better prognosis and treatment outcomes. Lasers began as alternatives for soft tissue oral surgery and have expanded into all aspects of dentistry: orthodontics, endodontics, oral and maxillofacial surgery, periodontics, aesthetic dentistry, restorative dentistry, prosthodontics, dental implantology, and pediatric dentistry. In addition, low-level lasers can be used as adjuncts to treat chronic pathologies and within photodynamic therapy to treat infectious disease.

Several factors are presented for consideration about how a laser can be incorporated into a practice:

- **Identify your practice**. The first and perhaps foremost concept before buying a laser is to identify how you practice. Your treatment planning is based on the patient's oral health conditions and the goals of your care will help improve or maintain that health as well as meeting their expectations. Your clinical experience and scope of practice usually determines which procedures you perform, and a list of those should be studied so that you can begin to choose a laser instrument. Likewise, you may have thought about the addition of other newer treatments that will expand your services. Those could affect the type of laser you purchase.
- **Analyze what procedures do you currently perform that can be assisted with laser technology.** A dental laser can help you provide a higher level of care. In restorative dental procedures, management of soft tissue is simplified because the tedious and painful placement of retraction cord can be eliminated. Better impressions are possible for indirect restorations such as crowns and bridges, and clearer margins near the gingiva are revealed for optical scanning. Class V carious lesions can be prepared at or near the subgingival level with excellent hemostasis. This ensures an improved bond for composite materials and ultimately results in better aesthetics and a longer lasting restoration. Two minutes of disinfection treatment of an aphthous ulcer brings immediate relief to the patient who may have been suffering for days. Excellent hemostasis can be achieved during minor surgical procedures like an immediate loading implant or second-stage implant uncovering.

  Not only is there a clean dry operating site, but the improved visualization will save time for the other steps of the treatment. All this will save your time. Also, by differentiating your practice, you'll attract a more educated cliental. Patients associate laser procedures as less invasive leading to a better overall dental experience, and once treated will refer their families and friends. There is easy return of investment as the procedures are made simpler and easier.

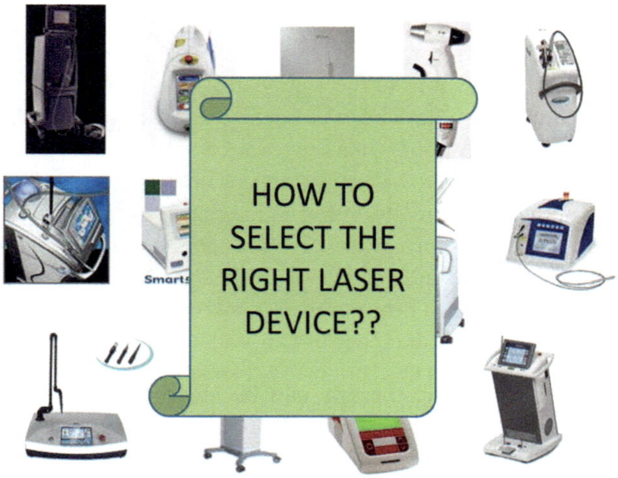

- **Think about which procedures that you do not perform that you would like to provide if you had a laser.** Within your scope of practice, there are procedures that can be accomplished with dental lasers in your office that you previously may have referred to a specialist, and/or did not offer your patients. Of course, proper training is necessary before you begin any procedure and is especially important when you are attempting a new one. However, understanding the fundamentals of the wavelength and watching the interaction as it happens will provide clinical experience and confidence for the clinician to continue offer additional treatment options at chairside.

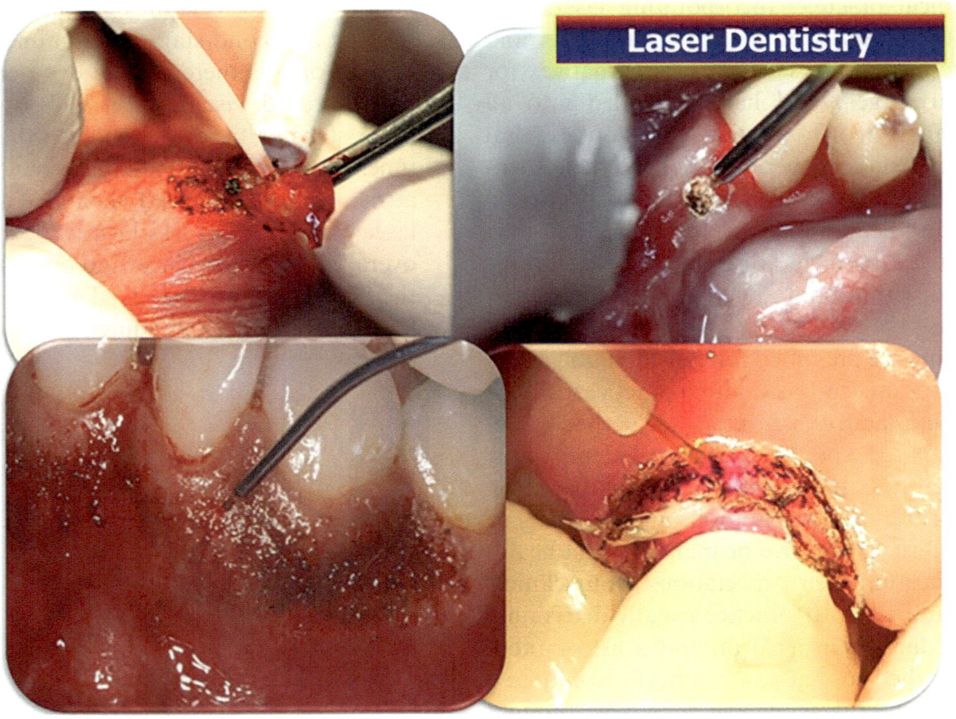

courtesy Dr S. Parker

Endodontic therapy can be aided by both laser debridement and pathogen reduction. Examples of laser soft tissue excisions are numerous: a removal fibrous tissue in an irritation fibroma, epulis in the soft tissues of patients wearing removal prosthodontic appliances, operculectomy treatment of an unerupted tooth, a frenectomy to prevent further adult periodontal problems, releasing a tongue tie in infants, and revising the frenum in a child's diastema to aid proper tooth positioning. Oral surgical procedures treatment of oral mucositis fibrosis, lichen planus, and leukoplakia can also be performed. Lasers can also be used to for aesthetic enhancement of the patient's smile by minor recontouring of gingival tissue, laser tooth whitening, and removal of depigmentation in the soft tissues.

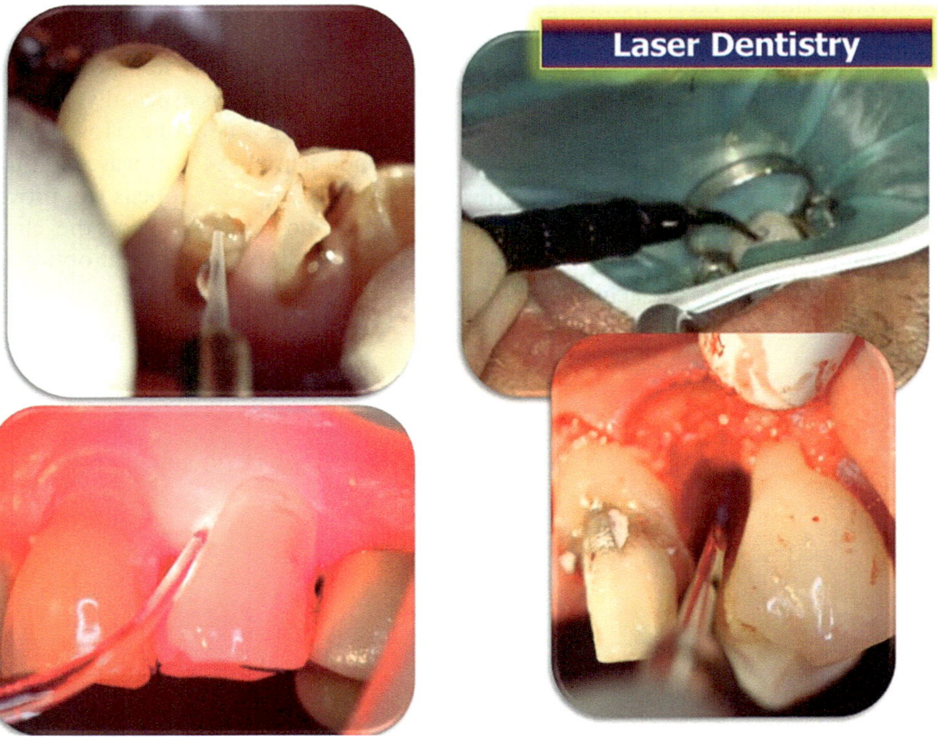

courtesy Dr S. Parker

Osseous crown lengthening for the treatment of altered passive eruption or to obtain adequate tooth structure for a restoration can proceed with the all-tissue lasers [4]. During the initial alignment phase of orthodontic treatment, photobiomodulation (PBM) formerly known as Low Level Laser Therapy (LLLT) can be used since it has shown to accelerate the tooth movement and also to relieve the discomfort occurred during the initial arch-wire changes [5]. That PBM effect can be used in patients with bruxism, temporomandibular joint disease, acute abscess areas, and many more applications [6]. Some of the biggest hurdles while taking diagnostic records like impression-making and intraoral radiographs, is gag reflex, which can be particularly strong in some patients. Low-level lasers are a boon in such cases, using lower doses of laser energy helps in minimizing the reflex [7]. When all these benefits explained in detail, there is no doubt the patient will accept the planned treatment. The increased revenue also helps to satisfy a further return on the initial cost of the laser.

- **There are several choices of laser instruments.** There are many lasers available for purchase. Their availability can be dependent on regional regulations of sales and clearances, along with support of service and training. There are world-wide standard and consistent classifications so that basic choices can be made. A generic division describes dental lasers as either surgical or non-surgical, and they are sometimes termed high level and low level, respectively. ◘ Figure 1.1 the basic categories of those classifications. After analyzing your practice's procedures, you can become familiar with how each available laser could be utilized.
- **Prioritize your clinical needs with respect to how a laser's use would be of benefit**. In a modern dental office, a patient has certain expectations: treatment should be less painful, more precise, less invasive with less bleeding, better healing, and fewer appointments. Fortunately, the practice of dentistry has been revolutionized and modernized so that our procedures have become more patient-friendly. With the incorporation of this device, an anxious patient feels more confident, there is noise free or no vibration of the drill or smell of conventional dental care, and with the fact that much of the treatment can be performed with "no anesthesia" or "needle-free" dentistry. These factors could transform patients who were resistant to conventional treatment into ones who readily accept treatment. Also in the future, we can expect more referrals to the practice, thus, lasers proving to be a safe investment and a true value for the money.

If your practice is focused on oral hygiene maintenance including sulcular debridement, prosthodontic or restorative soft tissue management such as gingival troughing, or aesthetic procedures involving

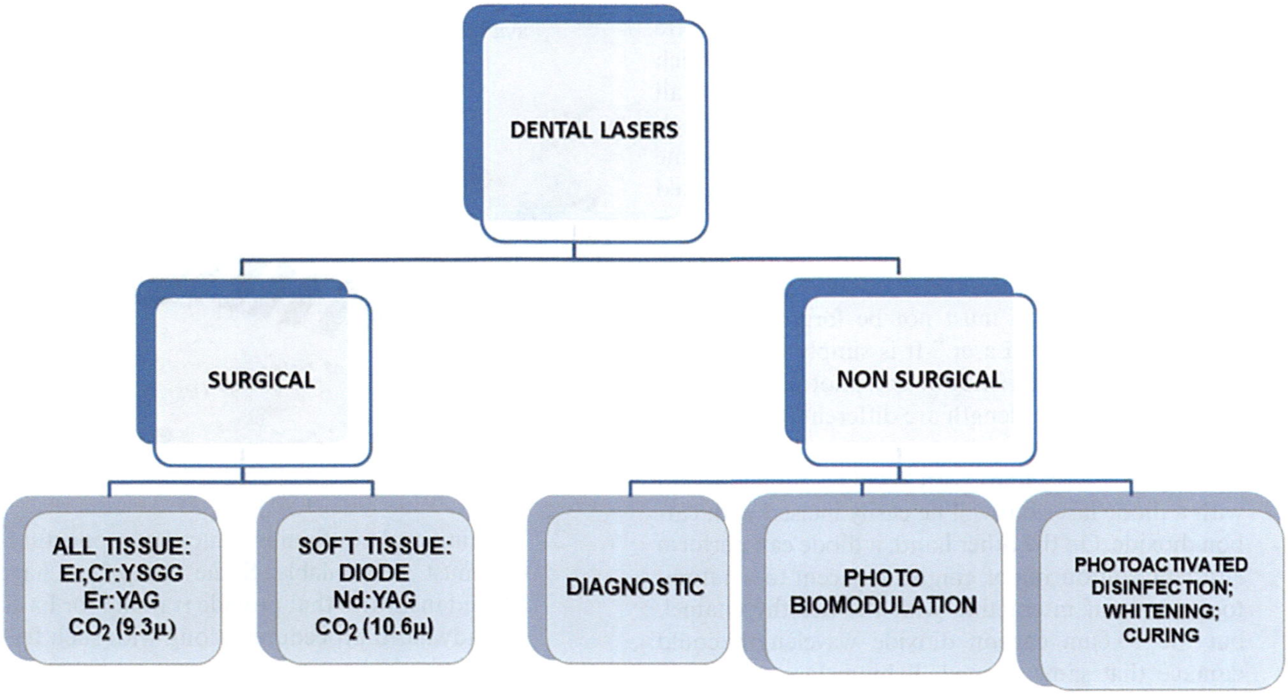

◘ **Fig. 1.1** A graphic representation of the type of dental lasers. The surgical lasers have output power sufficient for incision, excision, ablation, and coagulation of dental tissues, whereas the non-surgical lasers will not perform those procedures. Recent studies show that surgical devices do have some capabilities for photobiomodulation gingival recontouring, gingival depigmentation, and laser whitening), a diode or an Nd:YAG laser would be ideal. The small diameter fiber optic contact delivery can be safely used on soft tissue with minimal interaction with hard tissue. For the restorative practice and conservative dentistry, the erbium family (Er,Cr:YSGG, 2780 nm and Er:YAG, 2940 nm) and the 9300 nm carbon dioxide lasers offer a wonderful alternative and adjunct to the dental bur. A specialty practice that is mainly focused on oral and maxillofacial surgical procedures, or periodontal surgery, the afore mentioned all tissue lasers perform osseous surgery safely and rapidly while minimizing potential thermal damage to adjacent tissues and the blood supply. In addition to those instruments, the 10,600 nm carbon dioxide laser is often used for precise and rapid cutting during soft tissue surgery.

The clinician is constantly assessing and assimilating patient's needs and satisfaction while deciding on the proper treatment [8]. When choosing to add a laser to the procedure's protocol, certain wavelengths have advantages over others. To be clear, any of the available laser wavelengths suitable for soft tissue treatment. But if the dentist treats both soft and hard tissues, only the erbium or the 9300 nm carbon dioxide instruments will currently provide necessary energy and tissue interaction for those dental procedures.

- **There is no single perfect laser wavelength.** Currently, there are over two dozen indications for use of the various dental wavelengths, as listed in the different manufacturer's operating manuals. There are often many discussions in the profession about which laser is the best, as well as debates about how all lasers are the same. It should be clear that although similarities exist, every laser wavelength has some unique properties compared to another. When asked the question about which laser is best, a proper answer could be, "the one you know how to use in your practice!"

One thing which must not be forgotten is that there is "no perfect Laser." It is simply because the absorption characteristics of the photonic emission by a particular wavelength are different for the same tissues. Although every laser can be used for soft tissue surgery, a very fibrous area will be difficult to cut with a diode laser but will be easily incised with carbon dioxide. On the other hand, a diode can perform aesthetic contouring of gingiva adjacent to a natural tooth without interacting with the healthy enamel, but the 10.6μm carbon dioxide wavelength could damage that same enamel. Erbium lasers are very highly absorbed by water, which allows the easy removal of a carious lesion. However, highly fluoridated enamel can be more challenging to ablate because of the minimal water content. The 9.3μm carbon dioxide's photons react primarily with carbonated hydroxyapatite, so that tooth structure can be easily ablated. In addition, the laser's output can be a factor during treatment. Some procedures need only minimum energy levels; an example would be when desensitizing an aphthous ulcer. Likewise, photobiomodulation effects are performed at power densities well below any threshold of surgical cutting. In contrast, tooth preparation requires very high peak powers and very short pulses for efficient removal of the mineralized material without thermal damage.

Therefore, before investing, all the factors just discussed should help the clinician to identify what kind of laser is best suited for one's practice.

1.4 Sales, Training, and Company Support

1. The laser manufacturer is engaged in a highly competitive business with a limited market of purchasers. The sales team must be transparent and honest about their product's performance and avoid unrealistic assurances about everything from clinical efficacy to availability and shipping time for the device. The company's representatives should have a sound knowledge of the laser's operation so that they can initially demonstrate how the laser is set up along with knowing how to help in case of troubleshooting a problem. Customer support representatives should be available to answer questions and solve problems.

2. Training and continuing education opportunities must be available. Some companies have formed institutes that provide training for basic and advanced procedures, along with such features as educational resources, a discussion forum, examples of clinical cases, and other digital learning. Others sponsor courses and workshops during larger dental conferences.

3. It would be useful to know how long the particular device has been commercially available for purchase as well as to learn about the company's track record of efficiency, reliability, and service. Some companies have a global market, but local support in your country or state would be very desirable. Regional dental suppliers can also represent the company to provide sales and service. Since those suppliers already have a relationship with the dental practice, this could facilitate good support.
4. The operating manual enclosed with the laser is the guidebook for its use and describes the clinical procedures for which the device may be used. This is sometimes termed "indication for use," and simply means that there is solid evidence for safety and effectiveness, as opposed to "off label" treatment. All sections should be well written. Instructions should include the range of operating parameters for each procedure for the wavelength's use. Those settings are always guidelines and suggestions for modification should be listed. Factors such as beam diameter versus output power, approximate time of exposure, and varying tissue interaction must be considered. The steps necessary for the assembly and disassembly of the delivery system should describe every detail. The care and maintenance of each component of the laser should be illustrated. Warnings, precautions, and troubleshooting procedures should be explained, along with contact information for support.
5. As previously mentioned, there are various accessories necessary for using a laser. These include delivery system tips, foot control pedals, keys, interlocks, and protective eyewear. The initial and replacement cost of these items as well as any maintenance and availability should be noted. In some cases, accessories are optional and have additional costs. Those can negate an initial attractive initial price of the laser itself. Likewise, maintenance can be included for a period of time in the purchase, but a contract for service beyond that may incur a fee.
6. The warranty period should be clearly stated, and the dentist-purchaser should thoroughly understand the terms and conditions. Lasers are designed for precision delivery of photonic energy and the device is generally well constructed. However, any portion can be damaged with normal use and accidental breakage can occur. Warranty is a promise provided by the manufacturer to repair or replace the instrument within a specified time. That promise may stipulate what repairs are covered in specific circumstances.
7. The laser's operation is governed by software control of the internal components. Many companies offer updated versions of their software and may include them in the purchase price. Likewise, some of the hardware may undergo modification and it would be prudent to determine if any retrofitting or upgrades are appropriate and available for the model of laser purchased.

1.5 Education and Knowledge

1. A prudent question to ask is "how much training and do I need?" The simple answer is that you should continue to acquire knowledge all during your dental career. The elusive secret to success has always been to achieve better quality of patient care. That achievement can only be found with life-long learning. It starts with the sessions offered by the laser manufacturer after purchase. Unfortunately, some of these are simply didactic lectures available on playback media. Hands-on simulated exercises on animal tissue followed by over the shoulder supervised patient care are very superior learning methods. Whichever methods of initial training are taken, simple procedures performed with minimum power settings will help to overcome your fears and increase the level of your skills. Observing the rate of tissue interaction and the progress of reaching the treatment objective may appear to be at a slower pace than you first expected. Your patience will be rewarded; in fact, a slow sweeping motion for tissue removal is usually preferred. Moreover, you will avoid unnecessary thermal damage while precisely cutting and contouring tissue; and that will produce a successful outcome. That continuing journey toward mastering how a procedure is performed can bring you a lot of satisfaction. During that time, your range of comfort with all procedures will certainly increase.
2. For continuing education about the use of lasers in dentistry, several opportunities are available. Local study clubs and regional academies have regularly scheduled meetings where members can share information. Many major dental conferences feature presentations and workshop courses. There are university affiliated programs which offer both information and assess competency. Advanced programs,

Fellowship, Mastership, and MSc programs are offered in many countries. A document entitled "Curriculum Guidelines and Standards for Dental Laser Education," was developed in 1993 and is often used as a reference for these learning opportunities [9].

3. Finding a mentor would be a bonus for any laser clinician. There is no faster way to improve your skills and knowledge then to have someone to guide you as you work on your goals. That person should have the right attitude about teaching along with the experience to demonstrate the proper way to perform the procedure while correcting any of your deficiencies. Your confidence in delivering care will also increase. In addition, you can gain insights about new techniques and treatments.
4. Another question that can be asked is "what are the rules for laser use?" The response is that various regulatory agencies exist to ensure safe and efficacious use of lasers for the health and welfare of patients. The practitioner must have knowledge of those regulations and comply with their provisions. A review of those is presented here and will be detailed in the chapter on laser safety.
   - Regional or local bodies issue a license to practice dentistry to a properly qualified dentist. That allows the dentist to offer dental care according to the scope of practice—i.e., the general or specialty services that are provided. That care is delivered in a manner that is based on the practitioner's training, education, and clinical experience. It should be remembered that "laser dentistry" is not a recognized specialty; in contrast, it is a description of using an instrument during a procedure.
   - Certain agencies control the manufacturer and their products, but do not control the practice of dentistry. One example is the United States Food and Drug Administration through its Center for Devices and Radiologic Health regulates the construction of the laser to ensure compliance with medical device legislation. That same agency awards the manufacturer a marketing clearance for a procedure which states that the treatment where the laser is used will be safe and effective.

     The International Electrotechnical Commission prepares and publishes international standards for all electrical, electronic, and related technologies that includes regulations and conformity assessment for lasers in a similar manner to the Food and Drug Administration.

     Both organizations strongly influence regulatory agencies in other countries.
   - Currently, there is no common agreement about what defines a proper credential for a dental laser practitioner. Some local licensing jurisdictions have a course requirement. A small number of dental schools have introduced laser care into the pre-doctoral curriculum.
5. Evidence-based dental practice comprises an equal combination of the integration of clinically relevant scientific evidence, the clinician's experience, and the patient's treatment needs

and experience. Regarding dental lasers, the peer-reviewed literature offers an abundance of studies, clinical cases, and meta-analysis. Some reviews proclaim controversies that exist with regard to superiority of incorporating lasers into the treatment protocol. However, many manuscripts using controlled clinical studies do show effectiveness of these instruments. The laser practitioner should be familiar with as much of the literature, published articles, case reports, and scientific reviews that are readily available online or offline. Less reliable blogs and forums can offer information and networking about personal experiences. All of those resources contribute to evidence that has a place in the hierarchy of learning. The knowledge of how a particular wavelength would serve the purpose will be very beneficial to the success of your practice.

1.6 Investing in Your Team

Nurturing your employees is an important part of creating an engaged workforce. Invest in their personal and professional development and it will pay handsome dividends down the line by giving you a happy, capable, and productive team in an optimized practice. Your patients will immediately notice the professional and friendly atmosphere where you have created a healthy working environment with a caring and holistic approach toward their treatment.

Everyone on your staff from the receptionist and administrators at the front desk to the clinical team of assistants, hygienists, and other associated doctors must be educated about dental lasers.

With proper training and experience, they can answer any of the patient inquiries about how a laser might be used for treatment. They can increase the patients' awareness of the advantages and limitations of the technology. They can also address any apprehension about a procedure. Interestingly, many people are familiar with lasers because of previous medical procedures; and a few have expressed a misunderstanding about the word radiation as it represents the last letter in the laser acronym. Regarding the latter, a well-informed staff member can clarify the fact that dental lasers do indeed emit radiation in the thermal portion of the electromagnetic spectrum and not in the ionizing portion used for radiographs. The entire team would benefit by attending an introductory course about the use of lasers in dentistry as well as being actively interested in other continuing education offerings.

1.7 Marketing

One of the secrets to a successful practice lies in its marketing. Marketing is a process by which a product or service is introduced and promoted to potential customers [10]. It is the best means to make people aware of the quality of service being provided. The overall marketing umbrella covers advertising, public relations, promotions, and sales.

There are mainly three methods of marketing, namely: online, offline, and word-of-mouth. The latter is sometimes termed "internal marketing." Online marketing uses the Internet, e- mail, social networking websites, and blogs as online channels for delivering marketing content to the public. Offline marketing is disseminated through the "conventional" media: radio, television, and print ads. Word-of-mouth marketing is the best of the three approaches for any dental office. The starting point is that the staff or team must have knowledge of the practice. A written strategic plan composed of a vision, mission statement, goals, and objectives will certainly provide a framework for that knowledge. Each employee should be able to articulate the fact that the entire office constantly stays updated with current innovations and procures the latest technology to ensure the best treatment, care, in a comfortable environment. That in turn will clearly influence how the patient will speak about the practice, encouraging friends, neighbors, and relatives to seek dental care there.

For the best results, all three modalities should be incorporated in a marketing plan. For the first two portions, you also may consider promoters, publicists, and professional marketing outlets to support any of your large-scale promotion efforts. Obviously, a budget must be prepared so that the dollars necessary for any of the above are well spent. However, word of mouth marketing usually just involves spending some time inside the office to ensure a consistent and high standard of delivery of dental care.

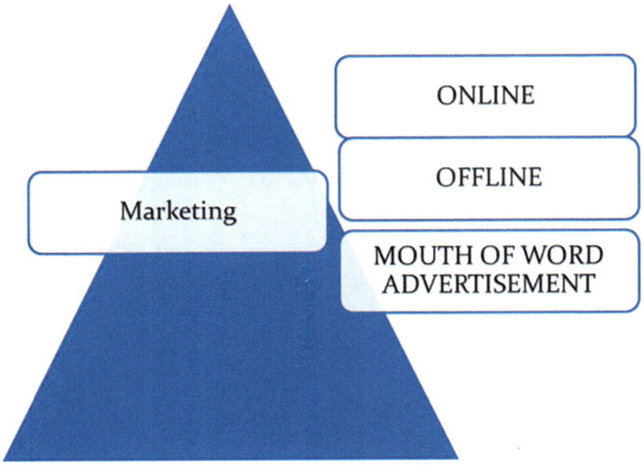

Lasers can be an excellent marketing tool. In a surgical case, the dentist who utilizes a laser is no longer bound by conventional treatment that always involves injectable anesthesia, along with bleeding, and sutures. Instead, the patient can be treated with the alternate laser technique that may require minimal or no anesthesia, with no bleeding, and minimal to no suturing. Similarly, for restorative dentistry, the traditional cavity preparation with rotary high- and low-speed drills and burs can give way to laser ablation of the carious lesion and preparation of the preparation margins. As an additional benefit, some of the treatment can also be performed with less anesthesia and more patient comfort.

Patients are becoming more techno-savvy these days, and because of that they can spend an inordinate amount of time researching dental treatment options on various online portals with varying degrees of opinion and education. Nonetheless, they gain knowledge about their options; and they rarely oppose treatments with laser if given a choice. They know and understand that the technology is up to date and it can provide faster, more comfortable dental care, while achieving those better results in less time.

1.8 Why Lasers in Dentistry

Lasers are in common use in every aspect of our lives, be it military, industrial, or medical. Now in its third decade, laser dentistry is no exception. The term laser itself evokes a positive response in patient's mind. Possibly because of prior experience with other medical procedures, the patient will associate the treatment performed with the instrument as very beneficial. Laser dental care can be quicker and more efficient along with markedly reduced pain, lack of bleeding, minimal need of anesthesia and last but not least minimal postoperative discomfort. The patients can resume their daily activities shortly after the treatment is rendered [11]

- **Why should I buy a laser?** The dental laser should become part of the practitioner's armamentarium. The photonic energy, with its unique properties of monochromatism and coherency, transmitted through an ergonomic delivery system, becomes a novel instrument for dental care. When used with proper knowledge, understanding and correct training, it can function as an integral part of any dental treatment appointment. The clinician can have assurance that each laser procedure is being safely and easily performed without some of the disadvantages that were present when the scalpel or electrosurgery were used. Two examples can be listed: disinfection during a laser incision versus a bleeding scalpel cut; and safe removal of tissue during laser implant fixture exposure versus certain damage with an electrosurgical tip. In addition, for those who challenge themselves to constantly better their skills, a laser is a must "have" for them—not as a "gadget," but as a surgical instrument.
- **What difference will it make for my patients and myself?** The incorporation of technology in dentistry has improved the way we serve our patients. Digitized radiographs are replacing traditional radiographs, diagnosis is done on 3D model of teeth and bone (CBCT), single sitting root canals, CAD / CAM technology is gaining popularity. All these advancements including lasers are being incorporated to improving the dental care provided to the patients on daily basis [12]. Dental pain is scored amongst the world's first ten phobias. In some patients, just the sight of dental chair, the whining noise of air rotor, or the white coat of a dentist can create panic attacks. Dental lasers make a huge difference in the life of such patients since they can reduce the level of stress and anxiety [13] and help a clinician to deliver best of dentistry. As an added benefit, it has been shown that lasers can help to provide neural blockage leading to analgesic effect and anti-inflammatory effects [14].
- **Will it be income generating?** Dental lasers can help the practitioner to formulate treatment plans for the benefit of the patients. As mentioned previously, the existing procedures will be improved and new or previously referred treatments can be offered. The dentist may necessarily increase the fee schedule to reflect the additional cost of the laser purchase, but that adjustment should be explained by simply enumerating the benefits of using the instrument.

The surgical procedures are generally shorter than traditional surgeries and are usually performed on an outpatient basis. Patients usually have less pain, swelling, and scarring than with traditional surgeries. This makes a huge difference in the quality of life of patient since there is usually no long recovery period. Just as important, the practitioner can be more efficient because the surgical appointment and necessary pre- and post-procedure protocol is less complex and time consuming. Thus, there could be more time available for other patients which will in turn generate more revenue and help to grow the practice. An additional advantage is that multiple procedures may be performed during one visit, thus increasing production. It naturally follows that more the patient's acceptance of a proposed treatment, coupled with a positive, comfortable, and healthy outcome, the more it will result in confident referrals of new patients to the practice.

- **When should a laser be used?** The clinician should know the indications for use of the laser. This textbook will describe all of those in detail. Continuing education will certainly provide suggested techniques and protocols. When reviewing the steps necessary for a procedure, there should be an analysis of how the laser could be used either as an adjunct or as monotherapy. Equally importantly, the instrument and all of the needed accessories should be easily obtained—within reach or stored close by—so that it can be inserted into the procedure. As the clinician continues to utilize the laser, it will become essential in the armamentarium. For some treatments, it can be substituted for other instruments; in other procedures, it can be used adjunctively. Likewise, the experience of repeated use will result in confidence in delivering excellent patient care. Indeed, the laser will become the smart investment that was hoped for during purchase. ◘ Figure 1.2 shows a small sampling of clinical procedures where a laser can be used. In every case depicted, the laser was used instead of conventional instrumentation.

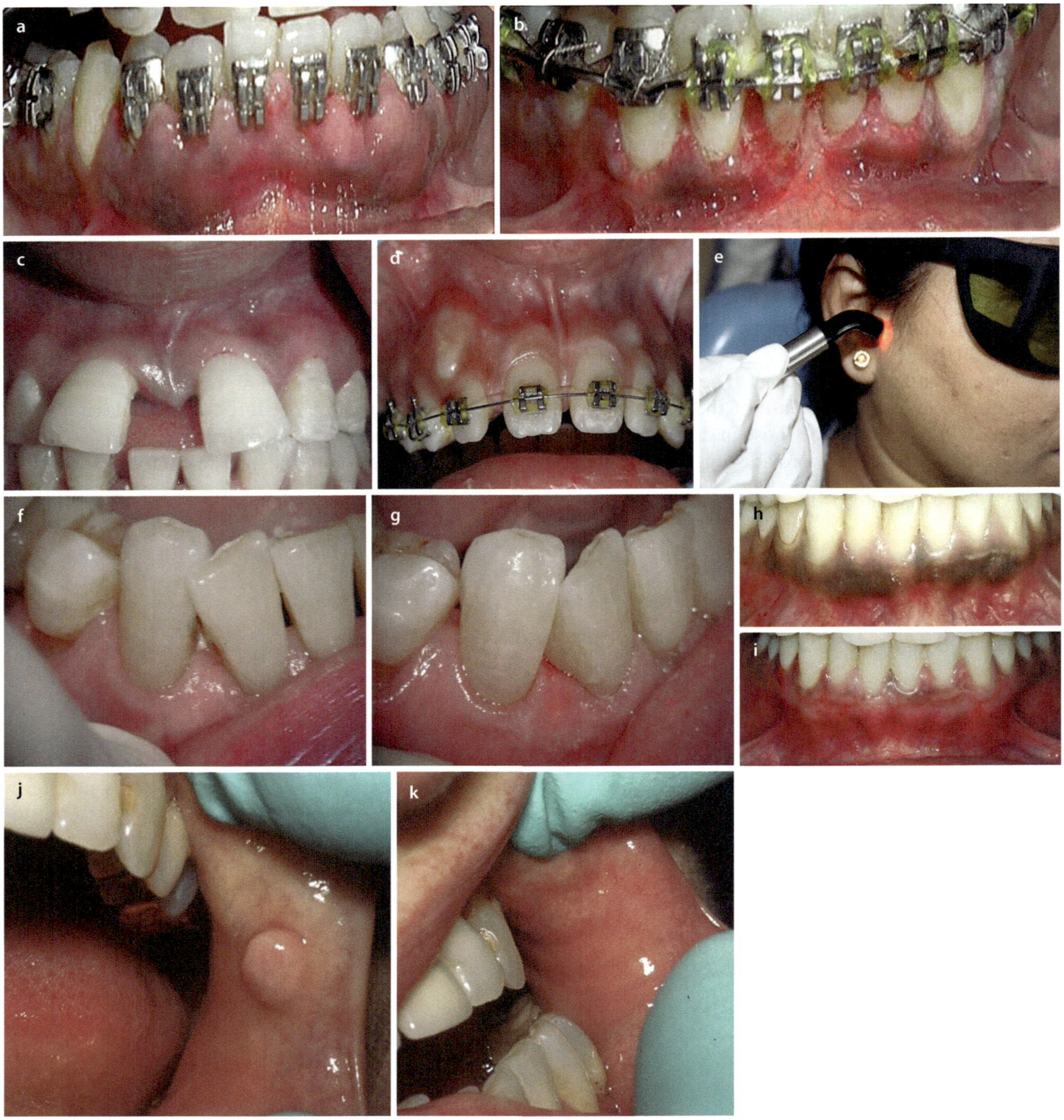

Fig. 1.2 **a** Pre-operative view of hyperplastic tissue present during orthodontic therapy. **b** Postoperative view showing tissue removal, with more normal periodontium. **c** A pre-operative view of a wide maxillary diastema with frenum involvement. **d** Photo depicting the healed frenum revision, gingivectomy, and good progression of orthodontic alignment. **e** A photobiomodulation laser is used for the treatment of temporomandibular joint inflammation. **f** Pre-operative view of interproximal carious lesions. **g** Immediate postoperative view of the new restorations. Both teeth were prepared with the laser instead of the dental handpiece. **h** Pre-operative view of pigmentation on the mucosa. **i** Postoperative view showing the pigment removed. **j** Pre-operative view of a benign irritation fibroma. **k** Postoperative view showing healed area

1.9 Limitations of Laser Dentistry

If you are a proficient clinician using a laser, then you can see the almost limitless and enormous possibilities of using them for treatment. However, as with any instrumentation, certain considerations apply. The clinician should be very well trained to judge the disease to be treated. After proper selection of the case, an appropriate decision is to be made on what wavelength, power, or energy density will be used and will be dependent on the absorptive pattern of the target tissue. This of course implies a very thorough understanding of the fundamentals of laser physics, tissue interactions, and the safe use of the device.

There are some disadvantages to the currently available dental laser instruments. They are relatively high cost and require training. Most of the laser emission tips are end-cutting although there are some radial firing ones available. Some instruments do not use a contact tip; rather the beam must be focused to a specific area on the target tissue. This lack of tactile sense can be initially challenging to clinicians since the majority of dental instruments are both side- and end-cutting and are used in contact. The laser practitioner will be necessarily required to employ a modification of clinical technique. A laser incision is not as sharp- edged as one made with a scalpel. Furthermore, since sutures are seldom used compared to one from a surgical blade, a laser wound heals by secondary intention. The patient must be given the appropriate postoperative instructions to correctly care for the area during healing. As mentioned, no single wavelength will optimally treat all dental disease. Accessibility to the surgical area can sometimes be a problem with some current delivery systems, and the clinician must prevent overheating the tissue while attempting to complete a procedure. One additional drawback of the erbium family and 9300 nm carbon dioxide lasers is the inability to remove defective metallic and cast porcelain restorations. Of course, this limitation in some cases could be quite beneficial when treating small areas of recurrent decay around otherwise sound restorations. Sometimes, the slower pace of laser soft tissue surgery can lead to tissue charring or carbonization during any surgical procedure. This can be due to a combination of too much average power or moving the laser beam too slowly. Both of those can be corrected with experience. One aspect that should not be ignored is the production of the laser plume which is by-product of vaporized water (steam), carbon and other harmful molecular particles, and possibly infectious cellular products, which combine to produce a malodorous scent. Maintaining the suction wand within 4 cm of the surgical site to remove as much of the plume as possible is recommended [15, 16].

1.10 Enjoying Benefits of Laser Dentistry

Over the time, the developments in the art and science of dentistry have provided us with the ability to allow the clinician to provide minimally invasive solutions to the patient's disease. From the incorporation of less invasive treatment of periodontitis to comprehensive cosmetic restorative treatment, the current standard is to conserve as much of the dentition and surrounding structure as possible. With the advancements in innovative materials, new and improved clinical techniques, that goal can be achieved. The rapid use of laser technology has gained popularity in various dental specialties and disciplines including endodontics, prosthodontics, oral and maxillofacial surgery, orthodontics, dental implantology, pediatric dentistry, aesthetic dentistry, and periodontics. It has revolutionized some treatment protocols and is certainly a practice building tool.

The benefits enumerated above can transform a patient who was previously resistant to conventional treatment plans into a more relaxed and certainly cooperative one. Moreover, the fact that dental practice can be very physically demanding and stressful during normal patient care. For more special needs patients such as those who are mentally and physically challenged, it is possible for the laser clinician to perform more procedures with efficiency and confidence, while conserving time and respecting the patient's tolerance. Lasers are especially helpful in geriatric patients as it makes the procedure more tolerable and help them overcome some of the barriers in providing dental care to them including severe dental complexity, multiple medical conditions, and diminished functional status. Similarly, laser-assisted pediatric dental treatment can result in a happy, healthy, and trusting child whose parents will appreciate the gentle and efficient care.

In today's digital world, patients interact almost instantly with their multimedia friends, share their experiences and concerns, and better understand diagnoses and treatment options. They are more likely to accept recommendations for treatment, and they certainly are willing to invest in a procedure that they value and that is as comfortable as possible. If a patient has a positive laser experience that will likely produce more referrals to the practice. In short, lasers can enable the dentist to render better quality dentistry [17].

1.11 Conclusion

We live in a fast-paced world. The practice of dentistry is constantly evolving and there are mainly two main reasons we change: one is that we want to strive to deliver the optimum treatment available for our patients; the other is that we want to keep abreast with the latest and best method to achieve that. Never stop learning or else we shall stop growing. In the present era, it is always important to improve your skills and abilities and we should continue to learn so that we can continue to grow in knowledge as a life-long pursuit. Willingness and openness to learn new things is the key to success. Whenever we think we are good, we can be even better.

The first step toward laser dentistry is to seek objective information on all aspects of the instrument and its uses. Eventually, the decision to purchase a laser should be based on sound scientific evidence, your own experience, knowledge, and training, and upon the patient's preference for treatment options.

References

1. Sognnaes RF, Stern RH. Laser effect on resistance of human dental enamel to demineralization in vitro. J South Calif State Dent Assoc. 1965;33:328–9.
2. Taylor R, Shklar G, Roeber F. The effects of laser radiation on teeth, dental pulp and oral mucosa of experimental animals. Oral Surg. 1965;19:786–95.
3. Lobene RR, Fine S. Interaction of laser radiation with oral hard tissues. J Prosthet Dent. 1966;16:16589–97.
4. Coluzzi DJ, Convissar RA. Atlas of laser applications in dentistry. Chicago, IL: Quintessence Publishing; 2007.
5. Farias R, Closs L, Miguens S. Evaluation of the use of low-level laser therapy in pain control in orthodontic patients: a randomized split-mouth clinical trial. Angle Orthod. 2016;86(2):193–8.
6. Sayed N, Murugavel C, Gnanam A. Management of temporomandibular disorders with low level laser therapy. J Maxillofac Oral Surg. 2014;13(4):444–50.
7. Elbay M, et al. The use of low-level laser therapy for controlling the gag reflex in children during intraoral radiography. Lasers Med Sci. 2016;31(2):355–61.
8. Weiner GP. Laser dentistry practice management. Dent Clin N Am. 2004;48:1105–1126, ix.
9. White JM, et al. Curriculum guidelines and standards for dental laser education. Proc SPIE San Francisco. 1998;3593:110–22.
10. Schwab D. What your staff needs to know about marketing your practice. Dent Econ. 1999;89:50–53, 95.
11. Cakart K. Evaluation of patient perceptions of frenectomy: a comparison of Nd:YAG laser and conventional techniques. Photomed Laser Surg. 2008;26(2):147–52.
12. Farman AG. Image guidance: the present future of dental care. Pract Proced Aesthet Dent. 2006;18(6):342–4.
13. Dundee JW, Yang J. Prolongation of the antiemetic action of P6 acupuncture by acupressure in patients having cancer chemotherapy. J R Soc Med. 1990;83:360–2.
14. Armida MM. Laser therapy and its applications in dentistry. Pract Odontol. 1989;10:9–16.
15. Srivastava V, Mahajan S. Practice management with dental lasers. J Laser Dent. 2011;19(2):209–11.
16. Myers TD, Sulewski JG. Evaluating dental lasers: what the clinician should know. Dent Clin N Am. 2004;48(4):1127–44.
17. Srivastava VK, Mahajan S. Diode lasers: a magical wand to an orthodontic practice. Indian J Dent Res. 2014;25(1):78–82.

Laser and Light Fundamentals

Donald J. Coluzzi

Contents

D. J. Coluzzi, S. P. A. Parker (eds.), *Lasers in Dentistry—Current Concepts*, Textbooks in Contemporary Dentistry,
https://doi.org/10.1007/978-3-031-43338-2_2

Core Message

The word LASER is an acronym for Light Amplification by Stimulated Emission of Radiation. The theory was postulated by Albert Einstein in 1916. A brief description of each of those five words will begin to explain the unique qualities of a laser instrument.

Once the laser beam is created, it is delivered to the target tissue. Furthermore, each device has certain controls that the clinician can operate during the procedure.

An understanding of these fundamentals will become the foundation for further elaboration of the basic concepts of how lasers are used in dentistry.

2.1 Light

2.1.1 Origins and Curiosities of Light

The word light has been used for many centuries, including biblical references such as in the beginning sentences of the Book of Genesis. Early civilization seemed to understand that the cycle of day and night with the sun, the moon, and the stars produced differences in ambient brightness. Historical investigations into the nature of light produced interesting and sometimes conflicting studies. Ancient people were curious about this brightness: the Greek philosopher, Pythagoras, began to develop wave equations about 400 BC. Over a century later, the Greek mathematician Euclid claimed that light is emitted in rays from the eye; he then proclaimed the law of reflection of those waves. It took until 1021 for a mathematician from Basra, Ibn al-Haytham, to correct the concept and prove that light enters rather than emanates from the eye. In addition, al-Haytham postulated that there are tiny particles of energy coming from the Sun that produce light. In 1672, British physicist Issac Newton was studying the laws of reflection and refraction and concluded that light was made of particles, which he called "corpuscles" [1]. He concluded that light is a combination of seven colored particles—violet, indigo, blue, green, yellow, orange, and red (in keeping with the belief that seven is a mystical number). Those particles combine to produce "white" light [2]. A few years later in 1678, the Dutch physicist Christiaan Huygens insisted that light was made up only of waves and published the "Huygens' Principle" [3]. As history would have it, both Newton and Huygens were at best half correct.

Over a hundred years later, new discoveries of light emerged. In 1800, William Herschel, a German born musician and astronomer, moved to England and investigated individual temperatures of the visible color. From those experiments, he discovered infrared light [4]. Johann Ritter, from a region of Eastern Europe now known as Poland, discovered ultraviolet light in 1801, by observing how the common chemical silver chloride changes color when exposed to sunlight [5]. The British physicist Michael Faraday produced evidence that light and electromagnetism were related [6]. In 1865, his Scottish colleague James Maxwell then explained electromagnetic radiation: that is electricity, magnetism, and light are in fact interrelated in the same phenomenon [7]. His discovery quantified the different wavelengths of radiation and thus helped to explain our current understanding of the existence of light in more than just the visible spectrum of Newton's colors. In 1895, Wilhelm Roentgen, a German professor of physics, added X radiation to the electromagnetic spectrum, after studying many experiments from colleagues such as Philipp Lenard and Nikola Tesla [8]. He used the terminology of "X" to signify an unknown quantity. A theoretical physicist Max Planck, also from Germany, proposed that light energy is emitted in packets he termed "quanta" in 1900 [9]. He formulated an equation that gave a relationship between energy and wavelength or frequency. In 1905, the German scientist Albert Einstein discovered what he termed the photoelectric effect. He observed that shining light on many metals causes them to emit electrons, and he termed them photoelectrons. He then deduced that the beam of light is not just a wave traveling through space but must also be composed of discrete packets of energy, as described by Planck. Einstein called these tiny particles photons [10] thus crystallizing the particle-wave dual nature of light.

2.1.2 The Duality of Light

Based on the discoveries and arguments over the last three millennia, it can now be stated that light is a form of electromagnetic energy with a dual nature. It behaves as a particle and travels in waves at a constant velocity. The basic packet or quantum of this particle of radiant energy is called a photon [11]; a photon is a stable particle that only exists when moving at the speed of light in a vacuum. By implication of the theory of relativity, it has no mass. When decelerated, it no longer exists, and its energy is transformed.

The wave of photons that travels at the speed of light can be defined by two basic properties, as shown in ◘ Fig. 2.1. The first is amplitude, which is defined as the vertical height of the wave oscillation from the zero axis to its peak. This correlates to the amount of energy carried in the wave: the larger the amplitude, the greater the amount of energy available that can do useful work. The second property of a wave is wavelength, which is the horizontal distance between any two corresponding points on the wave. This measurement is very important both in respect to how the laser energy is delivered to the tissue and what the interaction will be. Wavelength is

Amplitude is the height of the wave from the zero axis to the peak.
Wavelength is the horizontal distance between two adjacent parts of a wave.

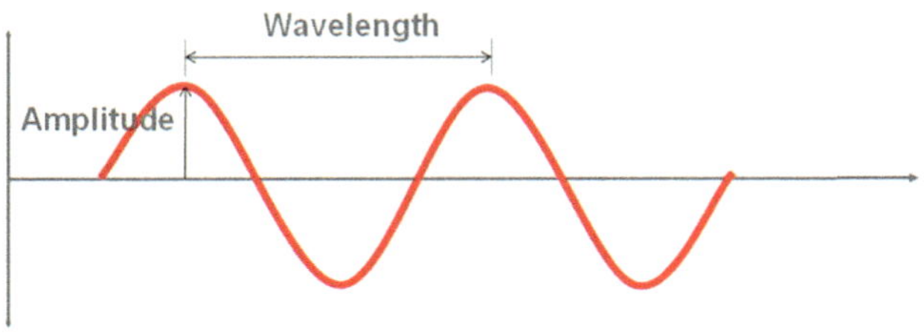

Fig. 2.1 A depiction of electromagnetic waves showing the two important quantities of amplitude and wavelength

measured in meters; and dental lasers have wavelengths on the order of much smaller units using terminology of either nanometers (10^{-9} m) or microns (10^{-6} m). As waves travel, they oscillate several times per second, which is termed frequency. Frequency is inversely proportional to wavelength: the shorter the wavelength, the higher the frequency and vice versa.

2.1.3 Properties of Light and Laser Energy

Ordinary light produced by a table lamp, as an example, is usually a white glow. The white color seen by the human eye is really a sum of the many colors of the visible spectrum—for example, red, orange, yellow, green, blue, and violet, as first described by Isaac Newton. The light is usually diffuse and not well focused.

Laser energy is distinguished from ordinary light by two properties. One is *monochromaticity* which means the generated light wave is a single specific color. For dental instruments, that color is usually invisible to our eyes. Secondly, each wave has *coherency*, identical in physical size and shape along its axis, producing a specific form of electromagnetic energy. This wave is characterized by spatial coherency—that is the beam can be well defined; the beam's intensity and amplitude follow the Gaussian beam's bell curve in that most of the energy is in the center, with rapid drop off at the edges. There is also temporal coherency, meaning that the single wavelength's emission has identical oscillations over a time period. The final laser beam begins in collimated form and can be emitted over a long distance in that fashion. However, beams emanating from optical fibers usually diverge at the tip; and beams from "tipless" handpieces will also lose their collimation. By using lenses, beams can be precisely focused; and this monochromatic, coherent beam of light energy can accomplish the treatment objective.

Using a household fixture as an example, a 100-W lamp will produce a moderate amount of light and proportionally more heat in a room. On the other hand, 2 W of laser power can be used for a precise excision of an irritation fibroma, providing adequate hemostasis on the surgical site without disturbing the surrounding tissue.

2.2 Emission

2.2.1 Spontaneous Emission

In 1913, Niels Bohr, a Danish physicist, developed his model of an atom, and applying the quantum principle of Planck. He proposed distinct energy "orbits" or levels of energy around the nucleus of that atom. Bohr found that an electron could "jump" to a higher (and unstable) level by absorbing a photon and then the electron would return to a lower (more stable) level while releasing a photon [12]. He termed this "spontaneous emission." The nuance to this emission is that, since there are several possible orbital levels in the atom, the wavelength of the photonic emission would be determined by the energy of the emitted photon, according to Planck's equation. It should also be noted that the emitted photon will likely have a random direction and phase. In more simple terms, spontaneous emission can be demonstrated when a conventional electric light bulb is switched on. The filament glows brightly emitting light and heat as the electrons are excited to higher energy states and then return to their ground conditions. Different broad groups of wavelengths (e.g., white light) will be produced during emission from the higher energy levels. A light emitting diode also produces spontaneous emitted light by using a flow of energized electrons recombining on the positive side of the wafer to produce luminescence. The color (wavelength) of the emitted light will depend on the chemical composition of the diode wafer [13].

2.2.2 Stimulated Emission

In 1916, Albert Einstein postulated the theory of lasers [14]. Using Bohr's model, he postulated that during process of spontaneous emission, an additional photon, if present in the field of the already excited atom with the same excitation level, would produce a release of two quanta. These would be identical in phase, direction, and wavelength. In addition, these emission photons would share monochromatic and coherent properties—thus a laser is born.

2.3 Amplification

Amplification is part of a process that occurs inside the laser. Once stimulated emission occurs, the process should theoretically continue as more photons enter the field to both excite the atoms as well as to interact with the excited photons returning to their ground state. One could imagine a geometric progression of the number of emitted photons; and, at some point, a population inversion occurs, meaning that a majority of atoms are in the elevated rather than the resting state. As Bohr implied, there can be several potential levels of energy available in most atoms. Having multiple levels (more than two) would aid in maintaining a population inversion because there would be no possibility of equal rates of absorption back into the ground state and stimulated emission. This amplification effect can only occur if there is a constant and sufficient source of energy, which is supplied by a pumping mechanism.

2.4 Radiation

The basic properties of a wave were discussed in ▶ Sect. 2.1.2. The entire array of wave energy is described by the electromagnetic spectrum (EM)—in other words, all frequencies and wavelengths of radiation [15]. The EM has several regions with rough boundaries of wavelength or frequency. There are seven general classes, with increasing order of wavelength to describe the radiation: Gamma rays, X rays, Ultraviolet radiation, Visible radiation, Infrared Radiation, Microwaves, and Radio frequency waves. These wavelengths range in size: gamma rays measure about 10^{-12} m; on the other end of the spectrum, radio waves have wavelengths up to thousands of meters. The EM can be generally divided into two divisions: Gamma rays, X-rays, and ultraviolet light are termed ionizing radiation, whille all others are termed non-ionizing. Ionizing simply means that the radiant wave has enough photon energy to remove an electron from an atom, and those wavelengths can cause mutagenic changes in cellular DNA. The human eye responds to wavelengths from approximately 380 to 750 nm, with those two numbers representing deep violet and dark red, respectively. That range is termed the visible spectrum. The term thermal radiation can be applied to many wavelengths. For example, an infrared lamp generates heat; the sun provides both light and heat; and the ionization present in plasma can also produce high temperatures.

The energy of a photon can be calculated using the equation from Max Planck. It states that the energy is directly related to the frequency of wave or inversely proportional to the wavelength. Thus, gamma or x-radiation with very short wavelengths (ranging from 10^{-12} to 10^{-10} m) has very high energy, while radio waves (approximately 3 m to 1 km) have significantly lower energy by comparison.

2.5 Components of a Laser

Identifying the components of a laser instrument is useful in understanding how the energy is produced. All dental lasers share common features—an active medium, a pumping mechanism, and a resonator. In addition, a cooling system, controls, and a delivery system complete the laser device.

All available dental laser devices have emission wavelengths of approximately 0.45 μm, or 450 nm to 10.6 μm or 10,600 nm. That places them in either the visible or the invisible portion non-ionizing portion of the electromagnetic spectrum. ◘ Figure 2.2 is a graphic depiction of those lasers on a portion of the electromagnetic spectrum.

2.5.1 Active Medium

Lasers are generically named for the material that is being stimulated; such material is called the active medium. As mentioned above, the atoms (or molecules) of that material absorb photonic energy, then begin to spontaneously emit. Subsequently under the right conditions, the process of stimulated emission will begin. Common materials for dental lasers can be broadly designated as one of three types: a container of gas, a solid-state crystal, or a semiconductor. The active medium is at the center or core of the laser, termed the optical "cavity."

Gas Lasers

The most common gas dental laser is Carbon Dioxide, which contains a gas mixture of Carbon Dioxide, Helium, and Nitrogen. Helium is not directly involved in the lasing process, but Nitrogen does interact with the excitation process and ultimate transfers that energy to the Carbon Dioxide molecules.

A second gaseous laser is the Argon ion instrument. A tube of this noble gas when excited can produce several radiant emissions, the most common being a visible blue and blue green beam of collimated light. The physical demands of power and cooling have rendered this laser to a very limited application in dentistry.

One of the first lasers developed was the Helium-Neon gas laser, which has a visible red color emission.

Solid-State Crystal Lasers

Various solid-state crystals are used in dental lasers. The host material is composed of Yttrium Aluminum Garnet (YAG), Yttrium Aluminum Perovskite (YAP), or Yttrium Scandium Gallium Garnet (YSGG). Any of these can then be "doped" with ions of Neodymium,

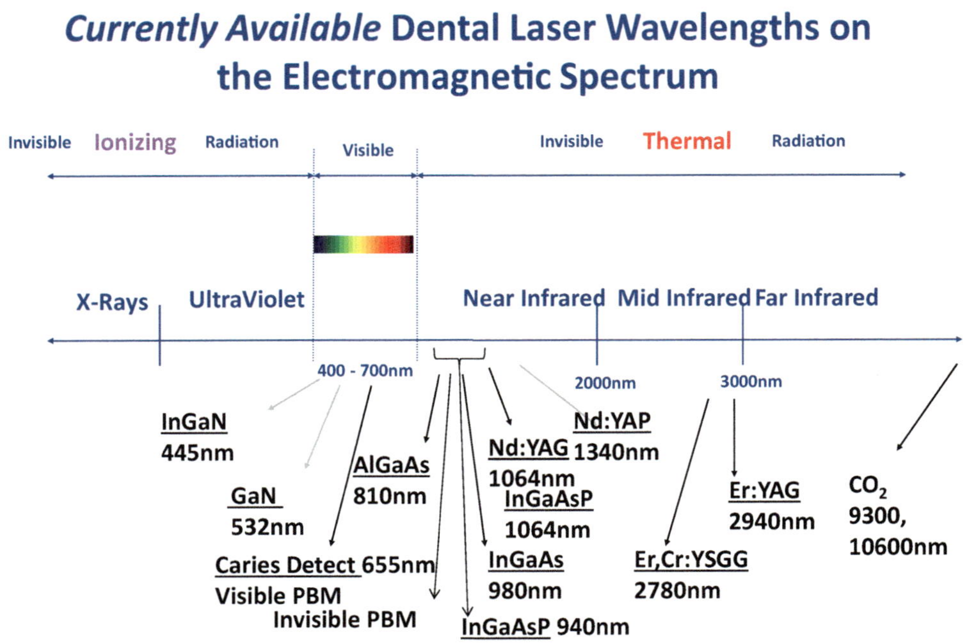

Fig. 2.2 A graphic showing the currently available dental wavelengths' position on the visible and invisible non-ionizing portion of the Electromagnetic Spectrum. Note that most of the wavelengths also include the composition of the active medium which produces that wavelength. PBM is an abbreviation for photobiomodulation, and those instruments use various active media

Erbium, and Chromium. The resulting designation would be written as Nd:YAG, for example, which would be a Neodymium-doped Yttrium Aluminum Garnet crystal.

Semiconductor Dental Lasers

A semiconductor laser utilizes the basic positive-negative (p-n) junction of everyday electronic circuits—the diode: that is, a two pole oppositely charged wafer. The flow of negatively charged electrons into the positively charged holes diffuses across the junction. The lasing action takes place between the charged layers, called the depletion region. This small rectangle will emit coherent and monochromatic light, but collimation must be performed by an external lens. Current diode lasers consist of various atomic elements in binary, ternary, or quaternary form arranged in a wafer-like structure. Examples would be Gallium Arsenide (GaAs), Aluminum Gallium Arsenide (AlGaAs), Indium Gallium Arsenide (AlGaAs), and Indium Gallium Arsenide Phosphate (InGaAsP). These elements provide a checkerboard-like crystalline structure to allow lasing to occur; the usual silicon-based semiconductor is not used because of its symmetry. The single diode wafer just described is then arranged in a linear array for cooling and the number of wafers determines the power output.

2.5.2 Pumping Mechanism

Surrounding this optical cavity with its active medium is an excitation source, known as the pumping mechanism. Pumping is the used to transfer energy into the optical cavity, and that energy must be of sufficient quantity and duration so that the occupation of a higher energy level exceeds that of a lower level. This condition is called a population inversion and it allows amplification to occur.

Although the above-described process occurs rapidly, it still takes some time. Most lasers are described as "three-level" or "four-level." A three-level system describes the basic concept: level 1 would be the stable, ground state, sometimes designated as Energy level zero; a pumped level (Energy level 2) and a lasing level (Energy level). Despite rapid decay from level 2, with enough pumped energy, there will be a population inversion between level 1 and 2. A four-level system is similar with the pumped level designated as 3, the "upper" lasing level as 2, and the "lower" lasing level as 1. The difference between these two lasing levels will aid in producing a population inversion. Certain active media operate as either three- or four-level systems.

In the laser industry, there are a wide variety of pumping mechanisms. The pumping of dental lasers is usually performed with optical devices—high power lamps or lasers—or by electricity—either with direct current mains or with electronic modulation of alternating current. Currently, diode lasers are electronically pumped; solid-state crystal lasers use high powered strobes (flash lamps); and Carbon Dioxide lasers can be operated with AC or DC current or radio frequency (RF) pumping methods. As a variation in pumping, one form of Carbon Dioxide technology uses very high-pressure gas and many electrodes along the length of the gas tube. This is known as a transversely excited atmosphere (TEA) laser.

2.5.3 Resonator

The resonator, sometimes known as the optical cavity or optical resonator, is the laser component surrounding the active medium. In most lasers, there are two mirrors one at each end of the optical cavity, placed parallel to each other; or in the case of a semiconductor, either a cleaved and polished surface exists at the end of the wafer, or there is reflection within the wafer. In all cases, these mirrored surfaces then produce constructive interference of the waves: that is, the incident wave and the reflected wave can superimpose on each other producing an increase in their combined amplitude. Clearly, some waves will not combine and will soon lose their intensity, but others will continue to be amplified in this "resonator." With the mirror system, this continued effect will help to collimate the developing beam. As mentioned previously, a diode laser collimation occurs externally.

2.5.4 Other Mechanical Components

A cooling system is necessary for all lasers, and higher output power requires increasing dissipation of the heat produced by pumping and stimulated emission. Air circulation around the active medium can control the heat, especially with diode lasers; the solid-state crystal lasers and some gaseous lasers require additional circulating water cooling.

Focusing lenses are employed for each beam; and in the case of diode lasers, for collimation. The delivery system will ultimately determine the diameter of the emitted wavelength.

The laser control panel allows the user to adjust the parameters of energy emission, along with a foot or finger switch for "on-off" or variable output operation on some devices.

2.5.5 Components Assembled

Laser energy is produced because the active medium is energized by the pumping mechanism. That energy in the form of photons is absorbed into the active medium, raising its atomic electrons to higher orbital levels. As the electrons return to their stable ground state, photons are emitted while other entering photons can produce stimulated emission. The resonator allows more numbers of these photonic interactions and will continue the amplification process.

The operation is temperature controlled, the beam is focused, and the clinician can control the laser used. ◘ Figure 2.3 shows a graphic of a solid-state laser such as an Nd:YAG or a gas laser such as Carbon Dioxide; and ◘ Fig. 2.4 depicts a schematic of a single semiconductor laser wafer. ◘ Tables 2.1 and 2.2 provide details of the currently available dental lasers with their active medium, common usage, and emission wavelength.

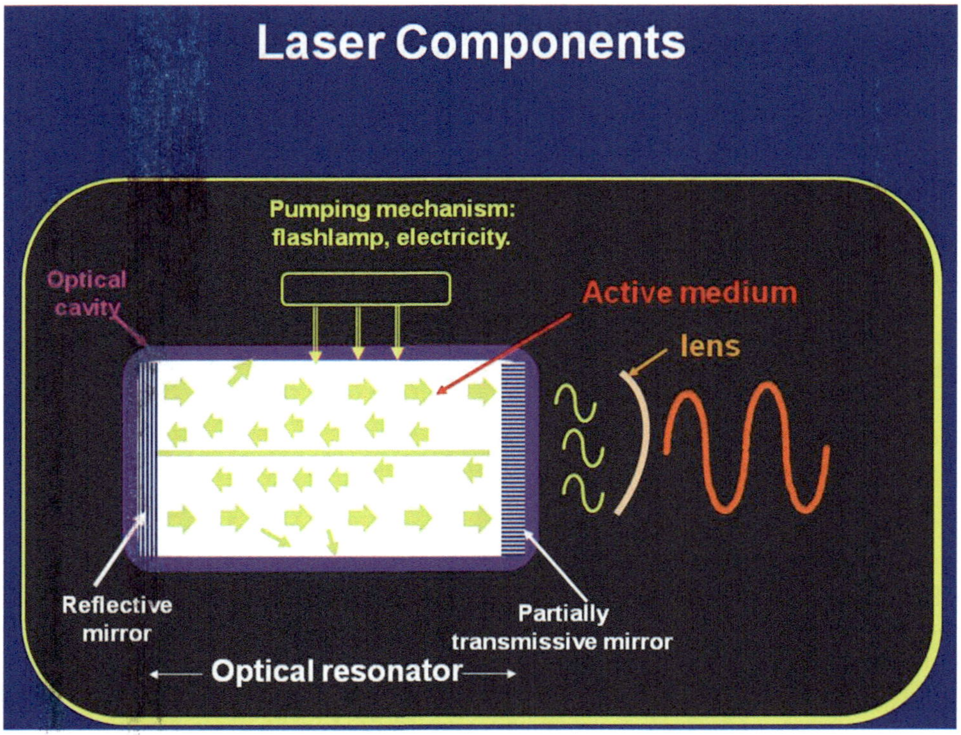

Fig. 2.3 General schematic of a laser. The active medium can be solid state (like Nd:YAG) or a gas (like Carbon Dioxide). The pumping mechanism provides the initial energy and the resonator consists of the active medium and axial mirrors. One mirror is totally reflective and the opposite one is partially transmissive. When a sufficient population inversion is present, laser photonic energy is produced and focused by lenses

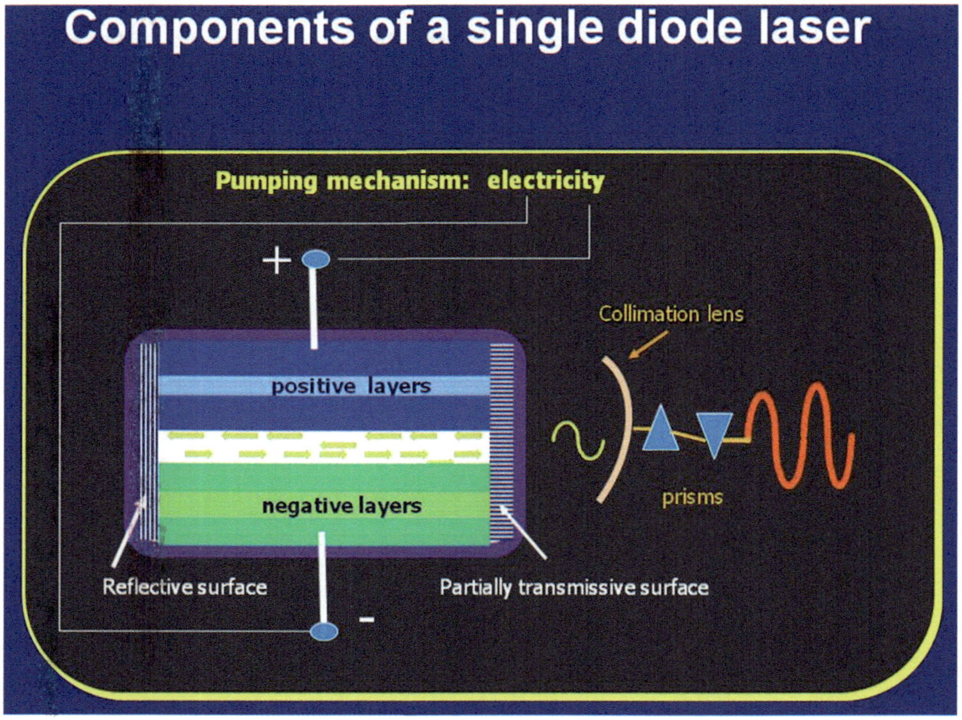

Fig. 2.4 Schematic of a single (individual) diode laser wafer. There are layers of positively and negatively charged compounds, pumped by electricity. The white layer with the yellow arrows represents the active layer where stimulated emission takes place. In this example, a reflective coating is applied to opposite ends of the wafer. In the right area are examples of lenses and prisms that would be placed at the emission end of an array of wafers to produce useful powers of diode laser photonic energy

Table 2.1 Currently available visible spectrum dental lasers. The type of laser and its emission spectrum is listed in column 1; column 2 indicates the general usage in dentistry; column 3 describes the active medium; column 4 shows the emission mode with the following abbreviations: *CW* continuous wave, *GP* acquired pulse

Type of laser and emission spectrum	General uses	Active medium	Wave-length (nm)	Emis-sion mode
Semiconductor diode, visible blue	Soft tissue surgical procedures, tooth whitening	Indium Gallium Nitride	445	CW, GP
Semiconductor diode, visible green	Soft tissue procedures, tooth whitening	Gallium Nitride	532	CW, GP
Low power output lasers, visible red light emission semiconductor or gas lasers	Photobiomodulation therapy (PBM), photodynamic therapy (PDT), or carious lesion detection	Variations of Gallium Arsenide or Indium Gallium Arsenide Phosphorus diodes	600–670	CW, GP

Table 2.2 Currently available invisible infrared dental lasers. Note that the term, "surgical" is applied to those lasers with sufficient output power. There is increasing evidence that any wavelength can produce a PBM effect. The type of laser and its emission spectrum is listed in column 1; column 2 indicates the general usage in dentistry; column 3 describes the active medium; column 4 shows the emission mode with the following abbreviations: *CW* continuous wave, *GP* acquired pulse, *FRP* free-running pulse

Type of laser and emission spectrum	General uses	Active medium	Wavelength (nm)	Emission mode
Low power output lasers, (invisible) near infrared	Photobiomodulation therapy (PBM), photodynamic therapy (PDT)	Variations of Aluminum Gallium Arsenide diodes	800–900	CW, GP
Semiconductor diode, near infrared	Soft tissue surgical procedures	Aluminum Gallium Arsenide	800–830	CW, GP
Semiconductor diode, near infrared	Soft tissue surgical procedures	Indium Gallium Arsenide Phosphorus	940	CW, GP
Semiconductor diode, near infrared	Soft tissue surgical procedures	Indium Gallium Arsenide	980	CW, GP
Semiconductor diode, near infrared	Soft tissue surgical procedures	Indium Gallium Arsenide Phosphorus	1064	CW, GP
Solid state, near infrared	Soft tissue surgical procedures	Neodymium-doped yttrium aluminum garnet (Nd:YAG)	1064	FRP
Solid state, near infrared	Soft tissue surgical procedures, endontic procedures	Neodymium-doped yttrium aluminum perovskite (Nd:YAP)	1340	FRP
Solid state, mid infrared	Soft tissue surgical procedures, hard tissue surgical procedures	Erbium, chromium-doped yttrium scandium gallium garnet (Er,Cr:YSGG)	2780	FRP
Solid state, mid infrared	Soft tissue surgical procedures, hard tissue surgical procedures	Erbium-doped yttrium aluminum garnet (Er:YAG)	2940	FRP
Gas, far infrared	Soft tissue surgical procedures, hard tissue surgical procedures	Carbon Dioxide (CO_2) laser, with an active medium isotopic gas	9300	GP
Gas, far infrared	Soft tissue surgical procedures	Carbon Dioxide (CO_2) laser with an active medium of a mixture of gases	10,600	CW, GP

2.6 History of Laser Development

After Einstein's laser theory was published, experiments to build a device didn't appear until the 1950s. Charles Townes of Columbia University in New York began working with a microwave amplification in 1951. In 1957, another Columbia graduate student Gordon Gould described in his laboratory notebook the basic idea of how to build a laser. That was considered the first time the term was used. The first laser was built by Dr. Theodore Maiman in 1960 at Hughes laboratory [16]. He used a 1 × 2 cm synthetic ruby cylinder for the active medium, photographic flash lamps for the pumping mechanism, and produced a brilliant red light pulsed emission. At the end of that year, three other scientists at Bell labs developed the helium-neon gas laser with a continuous output of red light. Other wavelength instruments were rapidly developed during that decade. Notable is the 1964 invention of the Carbon Dioxide laser, with a 10.6 μm wavelength, by Kumar Patel; and, in the same year, the Nd:YAG laser was built by Joseph Geusic and Richard Smith, all at Bell labs. In the spring of 1970, a team of Russian and American scientists independently developed a continuous wave room temperature semiconductor laser [17]. The first laser specifically designed for dentistry was marketed in 1989 [18].

2.7 Laser Delivery Systems

Laser energy can be delivered to the surgical site by various means that should be ergonomic and precise. There are three general modalities:

- An optical fiber
- A hollow waveguide
- An articulated arm

Shorter wavelength instruments, such as KTP, diode, and Nd:YAG lasers, have small, flexible fiber-optic systems with bare glass fibers or disposable tips that deliver the laser energy to the target tissue through a handpiece. A few low-powered diode lasers are offered as "handheld" units with disposable glass tips. Photobiomodulation (PBM) effects can be efficiently produced by using a handpiece with a much larger beam diameter than is used for surgical procedures.

Erbium devices are constructed with more rigid glass fibers, semi-flexible hollow waveguides, or articulated arms. Carbon Dioxide lasers use waveguides or articulated arms. Some of the erbium systems employ small quartz or sapphire tips and Carbon Dioxide instruments employ metal cylinders that attach to the handpiece. All of the tips are used in contact with target tissue although they can direct the beam toward the tissue when not directly touching it. Other lasers in these wavelengths use tip-less (and therefore non-contact) delivery systems. In addition, the erbium lasers and the 9.3 μm Carbon Dioxide laser employ a water spray for cooling hard tissue.

2.7.1 Optical Fiber

An optical glass fiber usually made of quartz-silica. This glass core conducts the laser beam along its length. A thin polyamide coating surrounds the core to contain the light, and a pliable thicker jacket covers both to protect the integrity of the system. A specific connector couples the fiber to the laser instrument; a handpiece and tip are added to the operative end. ◘ Figure 2.5a shows the components of a bare optical fober, the handpiece, and the canula. ◘ Figure 2.5b shows the assembled fiber-optic delivery system. In ◘ Fig. 2.5c, a permanently sealed fiber-optic cable and coupled handpiece are pictured. ◘ Figure 2.5d depicts disposable glass tips that are then coupled to the end of that permanently sealed fiber system, and the complete assembly is shown in ◘ Fig. 2.5e. The handpiece in ◘ Fig. 2.5f can be coupled to an optical fiber to produce a photobiomodulation effect. ◘ Figure 2.5g shows a hand-held low output power laser and tip that is used for photoactivated disinfection.

2.7.2 Hollow Waveguide

A hollow waveguide is a jacketed flexible tube. The internal surface has a reflective coating like silver iodide to allow the beam's transmission. A series of protective jackets complete the system. The waveguide is connected to the emission port on the laser, and a handpiece and optional tip is connected to the operative end. ◘ Figure 2.6 shows a typical hollow waveguide and handpiece assembly.

2.7.3 Articulated Arm

An articulated arm, consisting of a series of reflective hollow tubes with pivoting internally mirrored joints along its length. The arm has a counterweight to provide ease in movement. The laser emission port is coupled with the first tube, and a handpiece and optional tip is added to the operative end of the distal tube. ◘ Figure 2.7a depicts the basic arm assembly, and ◘ Fig. 2.7b shows the handpiece without a tip, typical for this delivery system.

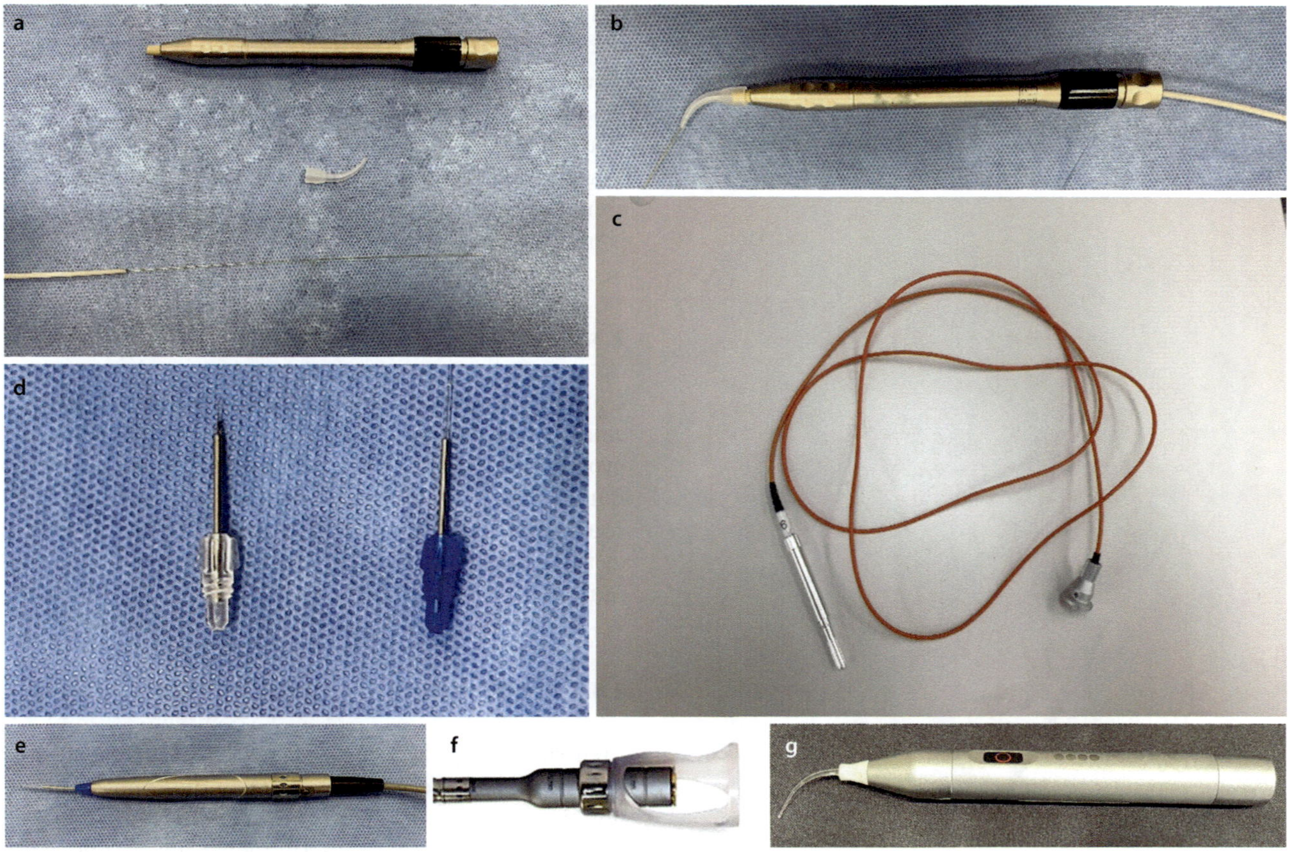

Fig. 2.5 **a** A bare optical fiber, handpiece, and canula. **b** Assembled bare fiber delivery system. **c** Permanently sealed fiber-optic cable and handpiece. **d** Disposable tips. **e** Assembled fiber-optic delivery system with a disposable tip. **f** A handpiece that can be used for PBM. **g** A hand-held laser and disposable tip that can be used for photodisinfection

2.7.4 Contact and Non-contact Procedures

All conventional dental instrumentation, either hand or rotary, must physically touch the tissue being treated, giving the operator instant feedback. As mentioned, dental lasers can be used either in contact or out of contact. Clinically, a laser used in contact can provide easy access to otherwise difficult to reach areas of tissue. The fiber tip can easily be inserted into a periodontal pocket to remove small amounts of granulation tissue, for example. In non-contact, the beam is aimed at the target at some distance away from it. This modality is useful for following various tissue contours, but the loss of tactile sensation demands that the surgeon pays close attention to the tissue interaction with the laser energy.

The active beam is focused by lenses. With the hollow waveguide or articulated arm, there will be a precise spot at the focal point where the energy is the greatest, and that spot should be used for incisional and excisional surgery. For the optic fiber, the focal point is at or near the tip of the fiber, which again has the greatest energy. When the handpiece is moved away from the tissue and away from the focal point, the beam is defocused, and becomes more divergent. At a small divergent distance, the beam can cover a wider area, which would be useful in achieving hemostasis. At a greater distance away, the beam will lose its effectiveness because the energy will dissipate. This concept will be further discussed in ► Sect. 2.8.

2.7.5 Aiming Beam

All the invisible dental lasers are equipped with a separate aiming beam, which can either be laser or conventional light. The aiming beam is delivered co-axially along the fiber or waveguide and shows the operator the exact spot where the laser energy will be focused.

Fig. 2.6 A hollow waveguide assembly

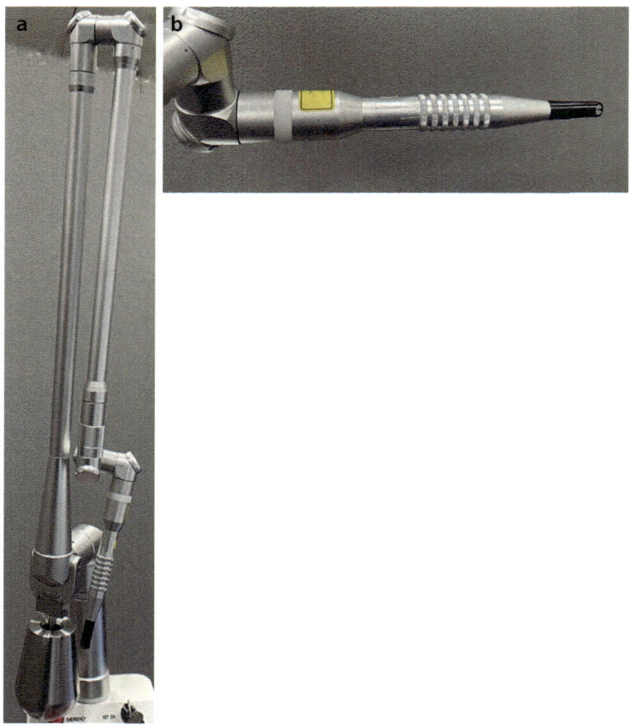

Fig. 2.7 **a** An articulated arm delivery system. **b** A handpiece typically used in an articulated arm delivery system

2.8 Emission Modes

There are two natural modes of wavelength emission for dental lasers, based on the excitation source: Continuous Wave and Free-Running Pulsed. A subset of Continuous Wave mode is a gated pulsed emission, where there is some means of modification performed after the beam is initially generated.

2.8.1 Continuous Wave

Continuous wave emission means that laser energy is emitted continuously when the laser is switched on and produces constant tissue interaction. These lasers are pumped with a constant direct current electrical field source. KTP, diode, and older model CO_2 lasers operate in this manner. The energy and/or power have a level output.

2.8.2 Free-Running Pulse

Free-running pulse emission occurs with very short bursts of laser energy due to a very rapid on/off pumping mechanism. Two examples are a high-powered strobing lamp or a radio frequency electronic field. The usual pulse durations of energy can be measured in microseconds, and there is a relatively long interval between pulses. The power produced has a high peak and low average level, which will be discussed in ▶ Sect. 2.9. Nd:YAG, Nd:YAP, Er:YAG, and Er,Cr:YSGG devices operate as free-running, direct pulsed lasers.

2.8.3 Gated Pulsed Mode

Some laser instruments equipped with a mechanical shutter with a time circuit or a digital mechanism to produce pulsed energy. Pulse durations can range from tenths of a second to microseconds. Some diode and Carbon Dioxide lasers have these gated pulses from their continuous wave emission. There can be high peak and low average power levels produced.

Another method to produce very short pulses is called Q switching (the Q indicates the quality factor of the optical resonator). An attenuating mechanism modulates the rate of stimulated emission, while the pumping mechanism continues to provide energy into the resonator. When the Q switch is turned off (opened) the result is a very short pulse of light, on the order of tens of nanoseconds. Peak powers can be very high.

Alternatively, an acousto-optic modulator can be placed in the laser cavity to ensure that the phases of emission all constructively interfere with each other. This is called mode locking and can produce pico- or femtosecond pulse durations with resulting extremely high peak powers.

Current dental lasers do not utilize Q switching or mode locking emission modes.

2.9 Terminology

The laser instrument's wavelength has a unique and unchangeable photon energy emission. However, the clinician can adjust various parameters of that emission from both the control panel and the handpiece's position on the target tissue. Throughout the remainder of this book, various terms will be used when describing laser procedures. The Glossary at the end of this chapter contains many of the terms and definitions that are standard for lasers. A few of those terms, listed in ◘ Table 2.3, will be described in more detail in this section.

2.9.1 Energy and Fluence

Energy is a fundamental physics term defined as the ability to do work. This energy is usually delivered in a pulse. A joule (J) is a unit of energy; a useful quantity for dentistry is a millijoule (mJ), which is one-one thousandth of a joule. Pulse energy is therefore the amount of energy in one pulse.

Fluence is a measurement of energy per area and is expressed as J/cm^2. This is also known as Energy Density. Procedures on different dental tissues will require various fluences for both efficiency and safety.

◘ **Table 2.3** Important terminology for laser use

Term	Definition	Abbreviation
Energy	The ability to do work	J (Joule) or mJ (millijoule)
Fluence	Energy per area	J/cm^2
Power	Work performed over time	W (Watt)
Power density	Power per area	W/cm^2
Beam size	The area of the projected laser beam on the tissue	(Usually measured in microns or millimeters)

2.9.2 Power and Power Density

Power is the measurement of work completed over a period of time and is measured in watts (W). One watt equals 1 J delivered for 1 s.

Power density is the measurement of power used per unit of area and is expressed as W/cm^2. Alternate terms are intensity or radiance.

2.9.3 Pulses

Except for Continuous Wave operation, all lasers can produce pulsed emission, that is, several bursts of energy can occur in a second. The number of pulses per second (pps) is the usual term applied, and an alternate word is hertz. That word could be confused with the description of the number of cycles per second of alternating electrical current.

Pulse Duration, Pulse Interval, Emission Cycle

The length of each pulse is called the pulse duration or sometimes pulse width and can be a short as 1 μs (10^{-6} s). The pulse interval is that time period between the pulses, when no laser energy is emitted. The emission cycle is the ratio, usually expressed as a percentage of the individual pulse duration to the total time of that pulse duration plus the subsequent pulse interval. In other words, if the pulse duration is 0.5 s and the pulse interval is 0.5 s, that is one pulse per second and the emission cycle is 50%. The emission cycle is sometimes referred to as the duty cycle. Similar to hertz, that similarity is unfortunate since the phrase duty cycle actually refers to how long on a device can remain on and working before it must be switched off for cooling.

2.9.4 Average and Peak Power

Average Power is what the tissue experiences during the duration of the procedure. Peak Power is the power of each pulse. Obviously, with continuous wave lasers, there is really no peak power. For any pulsed laser, the average power will be less than the peak power.

The calculation of peak power is the result of dividing the pulse energy by the pulse duration. For example, a 100 mJ pulse with a duration of 100 μs would have a peak power of 1000 W. This a common peak power achieved in free-running pulsed dental lasers. However, those same lasers are generally used with a low pulses per second parameter, which means that the pulse interval is relatively large. This results in a correspondingly

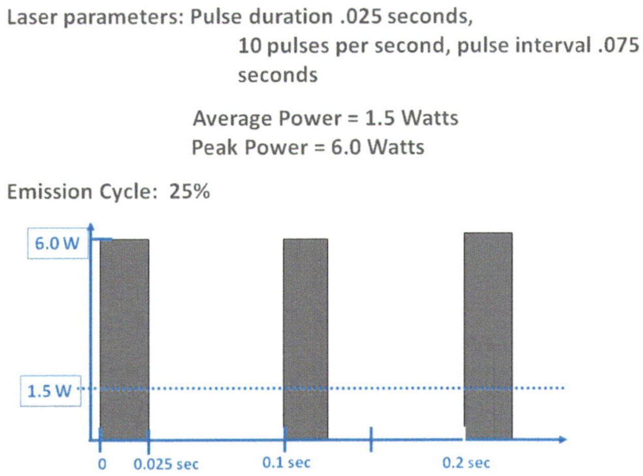

Fig. 2.8 This graphic shows the relationship between peak and average power along with the emission cycle. The pulses of laser energy in light blue bars. The individual pulse duration is 0.025 s and the pulse interval is 0.075 s. Each pulse has a peak power of 6 W, but the average power is 1.5 W, due to the emission cycle of 25%

low percentage emission cycle. Using the above example of a pulse duration of 100 μs at 50 pulses/s, the total emission time is 5/1000 of a second, which means the pulse total pulse interval is 995/1000 of a second. The duty cycle is then calculated at approximately 1%. Figure 2.8 shows a graphic depicting the relationship between Peak and Average power with basic laser parameters.

2.9.5 Beam Size

This is the area of the photonic emission that will interact with the target tissue. Lasers that employ tips have their nominal size indicated on the tip and non-contact lasers also have an area of focus. Laser tips are available in several diameters; typical sizes are 200, 300, 400, and 600 μm. Other tip less lasers can produce beam sizes with similar measurements. Clearly, the fluence and Power Density measurements will be based on that beam size. As mentioned previously, the laser beam will diverge at a prescribed angle from a quartz or sapphire tip, increasing its area. Likewise, a focused beam from a tip less delivery system will have a larger area when the beam is defocused. If the average power remains the same, both the fluence and the Power Density will be reduced.

Conversely, choosing a smaller diameter tip or producing a smaller focused area would increase the fluence or Power density with the same laser output setting. This could affect the tissue interaction. Figure 2.9 is a graphic showing how the difference in tip sizes would affect the Power Density.

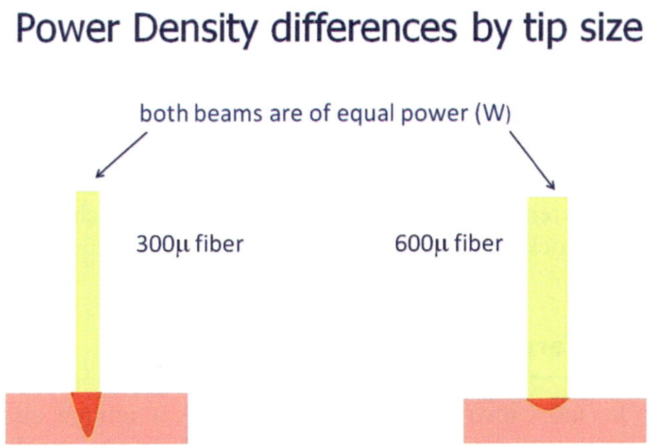

Fig. 2.9 This graphic shows the difference between Power Density areas using a 300 μm tip/beam size and a 600 μm tip/beam size. The smaller fiber has a larger area of interaction because of the larger Power Density calculation

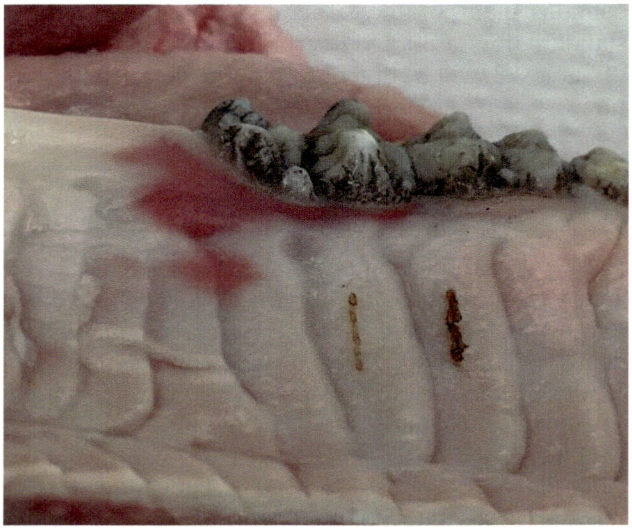

Fig. 2.10 An 810 nm diode laser with a 400 μm contact fiber was used at 1.0 W continuous wave for both incisions on a porcine maxilla specimen. The left incision was made with a faster vertical movement than the right incision. The left incision is narrower; the right incision is wider, more ragged, and produced a higher temperature in the tissue. Thus, the power density was larger for that incision

2.9.6 Hand Speed

In addition to the above parameter adjustments, an important principle of laser use is the speed at which the beam moves on the target tissue. A slower speed will increase the Power Density because of the longer time the energy remains in the tissue and could result in a larger area of interaction. This may or may not be a desirable effect, especially if the treatment objective is a minimally invasive procedure. Figure 2.10 shows a laboratory comparison of hand speed for soft tissue incision.

2.10 Conclusion

This chapter provided details of light and lasers. From basic experiments with light to the sophisticated development of different instruments, it should be clear that laser photonic energy can be precisely produced and controlled to be used for dental procedures.

References

1. Shapiro AE, editor. The optical papers of Isaac Newton, The optical lectures, 1670–72, vol. 1. Cambridge University Press; 1984.
2. Newton I. Opticks: or, a treatise of the reflexions, refractions, inflexions and colours of light. Also two treatises of the species and magnitude of curvilinear figures. London; 1704.
3. Huygens C. Traitė de la Lumiere (Treatise on light). Leyden; 1690.
4. Herschel W. Observations tending to investigate the nature of the Sun, in order to find the causes or symptoms of its variable emission of light and heat; With remarks on the use that may possibly be drawn from solar observation. Philos Trans R Soc Lond. 1801;91:265–318.
5. Berg H. Johann Wilhelm Ritter—the founder of scientific electrochemistry. Rev Polarogr. 2008;54(2):99–103.
6. Faraday M. In: Martin T, editor. Faraday's diary, Nov 12, 1839–Jun 26, 1847, vol. IV. London: George Bell and Sons; 1933.
7. Maxwell JC. A dynamical theory of the electromagnetic field. Philos Trans R Soc Lond. 1865;155:459–512.
8. Stanton A. Wilhelm Conrad Röntgen on a new kind of rays: translation of a paper read before the Würzburg Physical and Medical Society, 1895. Nature. 1806;53:1369, 274–276.
9. Planck M. Zur theorie des gesetzes der energievertielung im normal spektrum (On the theory of energy distribution law of the normal spectrum of radiation). Verhandlungen der Deutschen Physikalischen Gesellschaft im Jahre. 1900;2: 237–45.
10. Einstein A. Uber einen die erzeugung und verwandlung des lichtes betreffenden heuristischen gesichtspunkt (On a heuristic point of view about the creation and conversion of light). Ann Phys. 1905;17(6):132–48.
11. The photonics dictionary, 46th ed. Pittsfield, MA: Laurin Publishing; 2000.
12. Bohr N. On the constitution of atoms and molecules. Philos Mag. 1913;26:1–24.
13. Hummel R. Electronic properties of materials. 3rd ed. New York: Springer-Verlag; 2001. p. 263–264, 279–281.
14. Einstein A. Strahlungs-emission und-absorption nach der quantentheorie (Emission and absorption of radiation in quantum theory). Verhandlungen der Deutschen Physiklischen Gesellschaft. 1916;18:318–23.
15. Slater J, Frank N. Electromagnetism. New York: Dover Publications; 1969, Chapter 8.
16. Maiman T. Stimulated optical radiation in ruby. Nature. 1960;187:493–4.
17. Hecht J. Short history of laser development. Appl Opt. 2010;49(25):99–122.
18. Myers TD. Lasers in dentistry. J Am Dent Assoc. 1991;122(1):46–50.

Further Reading

Coluzzi DJ, Convissar RA. Atlas of laser applications in dentistry. Hanover Park, IL: Quintessence; 2007.

Convissar RA, editor. Principles and practice of laser dentistry. 2nd ed. St Louis, MO: Elsevier Mosby; 2015.

Manni JG. Dental applications of advanced lasers. Burlington, MA: JGM Associates; 1996.

Meserendio LJ, Pick RM. Lasers in dentistry. Chicago, IL: Quintessence; 1995.

Moritz A, editor. Oral laser application. Berlin: Quintessenz Verlags; 2006.

Parker S. Lasers in dentistry. London: British Dental Association; 2007.

Laser-Tissue Interaction

Steven P. A. Parker

Contents

D. J. Coluzzi, S. P. A. Parker (eds.), *Lasers in Dentistry—Current Concepts*, Textbooks in Contemporary Dentistry,
https://doi.org/10.1007/978-3-031-43338-2_3

3

Core Message

The potential for laser-tissue interaction forms the basis of the usefulness of predictable employment of laser photonic energy as an adjunct to clinical dental and oral therapy. Appreciation of the underlying mechanisms together with acknowledgment of limitations will help the clinician to provide laser therapy and minimize collateral damage.

There is an often-cited belief that in order to obtain benefit from laser photonic energy irradiation of target tissue, there must be absorption of the energy. Such understanding has merit but not the entire truth. Owing to the multi-structural nature of oral hard and soft tissue, the possibility of incident photonic energy reacting in a definite, predictable and exclusive manner with target tissue molecules is flawed through the anisotropic nature of the varying structures. Interaction may be a combination of surface, deeper, scattered and refracted energy distribution; true absorption of power values predicated through laser control panel selection may be impossible to achieve because of the varying interactive phenomena that may occur.

All oral tissues are receptive to laser treatment, but the biophysics governing laser-tissue interaction demands a knowledge of all factors involved in delivery of this modality. Through this knowledge, correct and appropriate treatment can be delivered in a predictable manner.

This chapter looks at the concepts of electromagnetic energy distribution within oral hard and soft tissue and examines the potential for true photonic energy ablation of target molecules. Prime concepts of photothermal action as a pathway to tissue change are explained, and adjunctive spatial and temporal components of the incident beam and the effects of such variance are explored.

The inconsistencies of laser-tissue interaction continue to pose some difficulty for the dental clinician; however, the development of many laser machines, amounting to a facility to produce laser photonic energy at several wavelengths between the visible and far-infrared areas of the electromagnetic spectrum, addresses many of the inconsistencies.

3.1 Introduction

Our understanding of the concepts of color helps to define the interaction of an incident beam of multi-wavelength (λ) electromagnetic (EM) energy—so-called white light—with a target structure. Human interpretation of "light" as a concept is limited to the ability of the retina to respond to this energy and the visual cortex to correlate stimulation in terms of a very limited range of the EM spectrum (λ 350–750 nm), termed the "visible spectrum." White light is seen in nature as the consequence of solar energy that filters through the many layers of the earth's atmosphere, or the emission of a man-made incandescent light source. "Light waves" that arise from such sources are multi-directional (not in phase—incoherent), and through the inverse proportionality relationship of wavelength with photonic energy, of consequent multiple energy values.

The fundamental theories on light of the latter nineteenth and early twentieth centuries—notably the works of Maxwell, Planck, Hertz, Einstein and Bohr—provided a coalescence of the prevailing opinions of light being composed of either particles or waves. Newton, through his "corpuscular" theorem [1], in which light traveled as discrete packets ("corpuscles"), was at variance with earlier work of Huygens. Popular acceptance of a predominant belief in light propagation by waves re-emerged in the early eighteenth-century England with the slit experiments of Thomas Young [2]. The confirmation that light energy was a form of electromagnetic (EM) radiation, capable of causing a photoelectric effect with certain metals, proposed a duality of existence for "packets" of light energy. Einstein is attributed [3] with providing the annotation "photon" (one quantum—smallest unit—of electromagnetic energy is called a photon (origin Greek "φως," meaning "light")) and with others listed before, provided that there is an understanding that photonic energy is a form of energized EM radiation, with each photon traveling at the speed of light (approx. 300×10^6 m/s) in a sinusoidal wave pattern. From this, it is fundamental to our understanding of so-called laser-tissue interaction that EM energy in its many forms is interrelated, and the energy contained therein is capable of conversion (subject to incident power value) to thermal and ablative equivalent within target tissue, through the law of conservation of energy (sic energy cannot be created or destroyed, just transformed from one form to another).

In determining a prescribed level of energy-derived physical change in target oral tissue, it is necessary to appreciate the quantity of incident energy, the degree of positive interaction and the potential for energy conversion. Inherent in every incident laser beam is the photonic energy.

Laser "light" is considered unique in that, unlike other forms of light (sunlight, incandescent, LED irradiation), there are two inherent properties—monochromaticity and coherence. The single wavelength concept is founded in the physics of laser EM propagation, and using the Einstein/Bohr postulations, the delivery of laser irradiation is in sinusoidal waveform with successive waves in phase—so-called wave coherence. Additional man-made configuration of the photonic energy produced can provide high-density beam spatial density over distance—this is termed "collimation." In terms of the benefits of laser-tissue interaction due to these properties, the monochromatic absorption of a chosen laser,

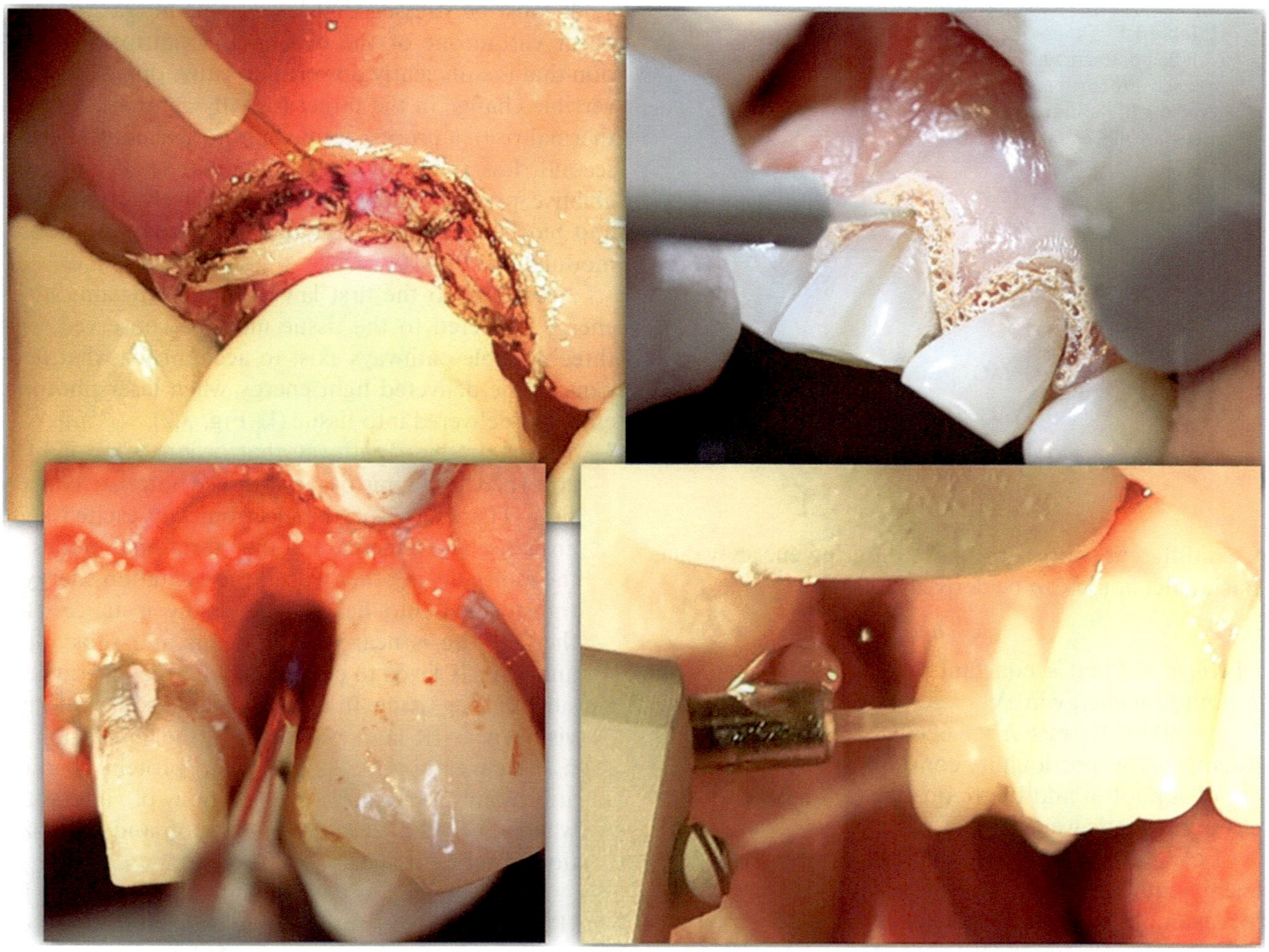

Fig. 3.1 Laser-assisted oral tissue surgery is photothermal in nature. Incident (photonic) energy is absorbed by target tissue elements (chromophores), relative to the laser wavelength. This leads to rapid temperature rise, protein denaturation and water vaporization. This constitutes an example of photoablation

together with the coherence of the beam, will offer selective tissue interaction of high quality; the collimation and ability to focus the beam will define a degree of accuracy and power density.

Figure 3.1 provides clinical examples of predictable laser-tissue interaction.

3.2 Photonic Energy

The emission of a single photon from an atom is the result of a shift in the energy status of that origin. Plank proposed that all matter existed in a state of energy relative to extremes of a lower (ground) state and a higher (energized) state, commensurate with entropic physical form [4]. Boltzmann, through his theories on thermodynamics [5], readily accounted a direct relationship between matter, energy and temperature. Put simply, an electric light filament at room temperature is a dull, inert wire but rapidly heats when energized by an electric circuit. The resistance of the wire leads to thermal conversion of the electrical (EM) energy. At this induced higher energy state of the light filament, the volatility of constituent electrons gives rise to higher thermal energy and emission of such energy as light.

Laser photonic energy assumes the production of high-energy photons from an energized source, whereby each photon scribes an identical waveform, and each photon has identical energy value. Plank and Einstein had established an inverse relation between wavelength and photonic energy, a direct proportional relationship between photonic energy and frequency, and Neils Bohr paved a way for the "quantum" (amount) nature of emitted photons to be calculated; thus, it provided a predictable base for the development of the maser and optical-maser, or laser.

The energy of emitted photons is expressed in Joules, or more conveniently, electron-volt eV (energy derived by acceleration through a PD of 1 V). Since photonic energy is related to wavelength (λ), photons emitted

3

Table 3.1 Commonly used laser wavelengths associated with dental treatment. Photonic energy and wavelength are inversely proportional. With ascending numerical value of wavelength, the corresponding photonic energy (expressed in electron volt—eV) is reduced

(eV)	Laser	λ (nm)
2.4	KTP	532
2.0	He–Ne	633
1.6	Diode	810
1.2	Nd:YAG	1064
0.4	Er:YAG	2940
0.1	CO_2	10,600

from different sources will have differing energy values. Basic calculation can be derived through:

$$\lambda = hc / E$$

where h = Plank's constant, c = speed of light and E = photon energy in eV.

$c = 300 \times 10^6$ m/s; $h = 6.626068 \times 10^{-34}$ m^2 kg/s represents the proportionality constant between the energy (E) of a photon and the frequency (ν) of its associated electromagnetic wave.

1 eV = 1.602×10^{-19} J, and it is possible to evaluate energy-equivalent values for the many laser wavelengths; for example, photons of wavelength 1240 nm (near infrared) equate to an eV value of 1.0, whereas an eV value of 2.0 is 621 nm (visible red) and eV 0.13 equates to 9600 nm.

Table 3.1 provides an overview of laser wavelengths commonly used in dentistry with corresponding photonic values.

3.3 Photonic Energy and Target Molecular Structures

A simplistic look at one of the many graphic representations of the relationship of target tissue elements, incident laser wavelengths and relative absorption potential would suggest that laser photonic energy is capable of ablative interaction with target tissue elements (chromophores). A chromophore is defined as "a chemical group capable of selective light absorption resulting in the coloration of certain organic compounds" [6]. For those compounds whose color is discernible within the visible spectrum, the definition may be sustained; however, given the concept that the chemical group confers a preferential ability to absorb (to a greater or lesser degree) photonic (EM) energy, the wavelength of that energy may fall within a spectrum of ultraviolet to far infrared, a narrow component of which may be visible to the human eye.

It must be stressed that laser-tissue interaction may occur within one of the two basic scenarios—interaction that is sufficiently powerful to cause direct and irreversible change in the target (usually achieved through thermal rise), a process termed photothermolysis, and a second, less powerful interaction that results in non-ablative, predominately (but not exclusively) stimulatory and biochemically mediated change, termed photobiomodulation (PBM).

According to the first law of thermodynamics, the energy delivered to the tissue must be conserved, and three possible pathways exist to account for what happens to the delivered light energy when laser photonic energy is delivered into tissue (Fig. 3.2):

1. The commonest pathway that occurs when light is absorbed by living tissue is called internal conversion. The energy of the electronically excited state gives rise to an increase in the vibrational modes of the molecule; in other words, the excitation energy is transformed into heat [7]. In many instances, the thermal rise is near instantaneous and substantial and quickly leads to conductive thermal energy into surrounding tissue. In the case of oral soft tissue and visible/near-IR laser wavelengths, the absorption by tissue chromophores gives rise to protein denaturation and secondary vaporization of interstitial water. The result is a visible ablation and vaporization of target tissue [8].

   With longer laser wavelengths, mid-IR and far IR, the prime absorptive tissue element in both soft and hard oral tissue is water. Ablation of tissue is achieved through the near-instantaneous vaporization of interstitial water, leading to an explosive fragmentation of tissue structure. With hard oral/dental tissue, this interaction can be quite dramatic [8].
2. With incident laser photonic energy values that fall below target tissue ablation, a second pathway can occur as fluorescence. Fluorescence is a luminescence or re-emission of light in which the molecular absorption of a photon triggers the emission of another photon with a longer wavelength. Such action provides the basis for optical scanning techniques used in caries detection in enamel and dentine and tomographic techniques in the scanning of soft tissue for neoplastic change.
3. The third pathway is broadly termed photochemistry [9]. Because of the energy of the photons involved, covalent bonds cannot be broken. However, the energy is sufficient for the first excited singlet state to be formed, and this can undergo intersystem crossing to the long-lived triplet state of the chromophore. The long life of this species allows reactions to occur, such as energy transfer to ground-state molecular oxygen to form the reactive species, singlet oxygen. Singlet or nascent oxygen is an ultra-short-lived form of the parent molecule that can

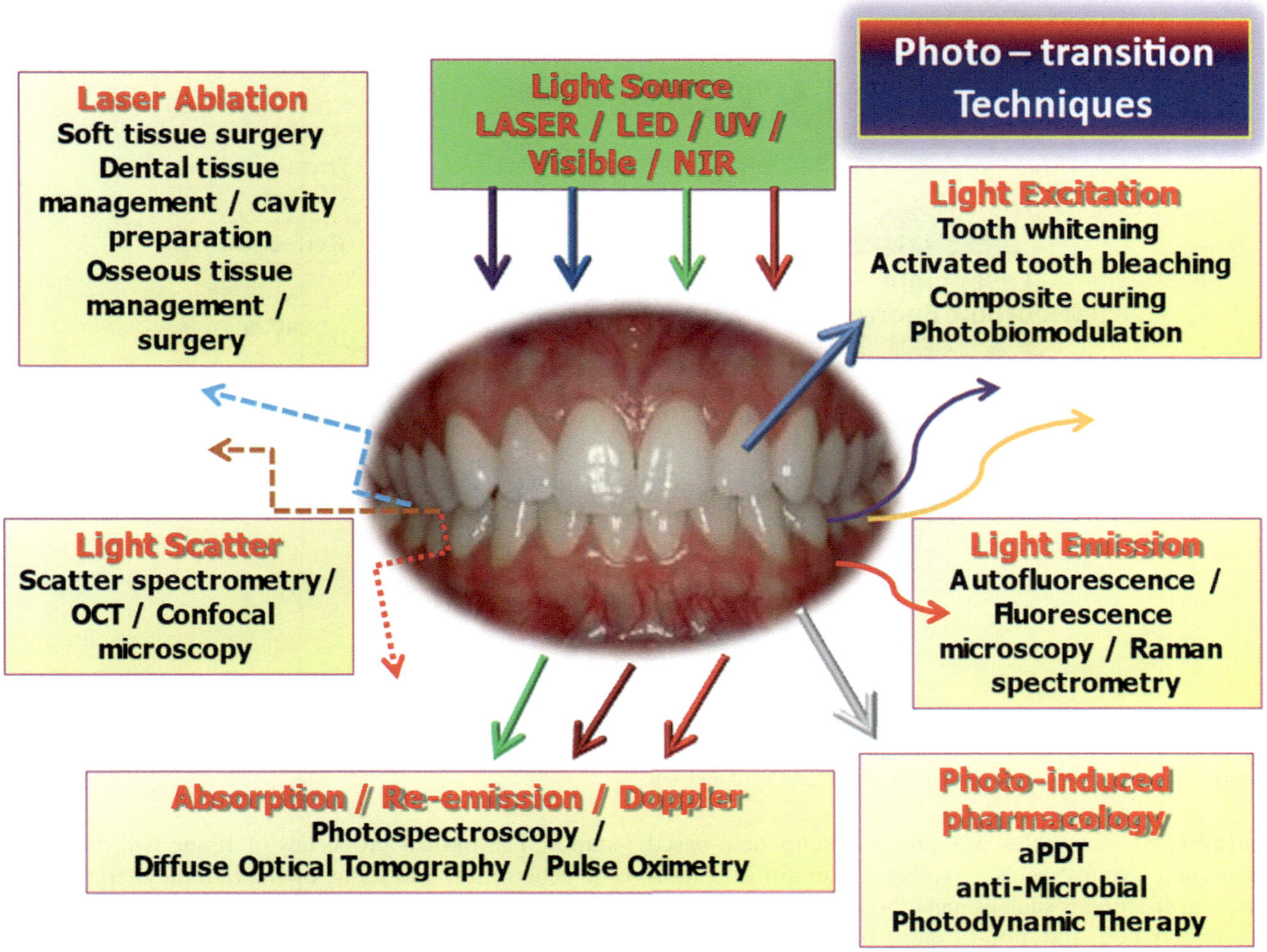

Fig. 3.2 An overview of the manipulation of incident photonic energy, such as laser light as an adjunct to screening, diagnostic and therapeutic clinical activity

cause cell apoptosis through oxidative stress. Such action can be commonly seen in photodynamic therapies where an intermediary chemical—photosensitiser—is employed to direct energy transfer to target tissue sites [10, 11].

Electron transfer reactions are highly important in the host cell mitochondrial respiratory chain [12], where the principal chromophores involved in laser therapy are thought to be situated. An additional photochemistry pathway that can occur after the absorption of a red or NIR photon within a host cell is the dissociation of a non-covalent bound ligand from a binding site on a metal containing co-factor in an enzyme. The most likely candidate for this pathway is the binding of nitric oxide to the iron-containing and copper-containing redox centers in unit IV of the mitochondrial respiratory chain, known as cytochrome c oxidase. Such action may induce an increase in cell pH and production of ATP and has been cited as basic cellular theory in photobiomodulation with low-level lasers.

3.4 Basics of Photothermolysis

Incident photonic irradiation directed onto target tissue will behave in one of the four main ways: transmission, reflection, scatter and absorption. The defining criteria can be simply summarized as dependent on the nature of the target tissue and wavelength of the incident beam (hence the predictability of absorption or transmission), nature of the tissue and its heterogeneity (hence the scope for scatter) and angle of the beam incident to the tissue surface (incident beam angle < total reflective angle) wherein reflection may have predominant effects (Fig. 3.3).

Oral hard and soft tissue is complex and heterogenous, anisotropic and of varying degrees of thickness, commensurate with structural anatomy. Within such tissue, component elements may be found that represent key molecules capable of selective absorption of photonic energy; within the visible EM spectrum, such molecules are termed chromophores, and with longer (IR) spectra, a terminology of absorptive tissue element may

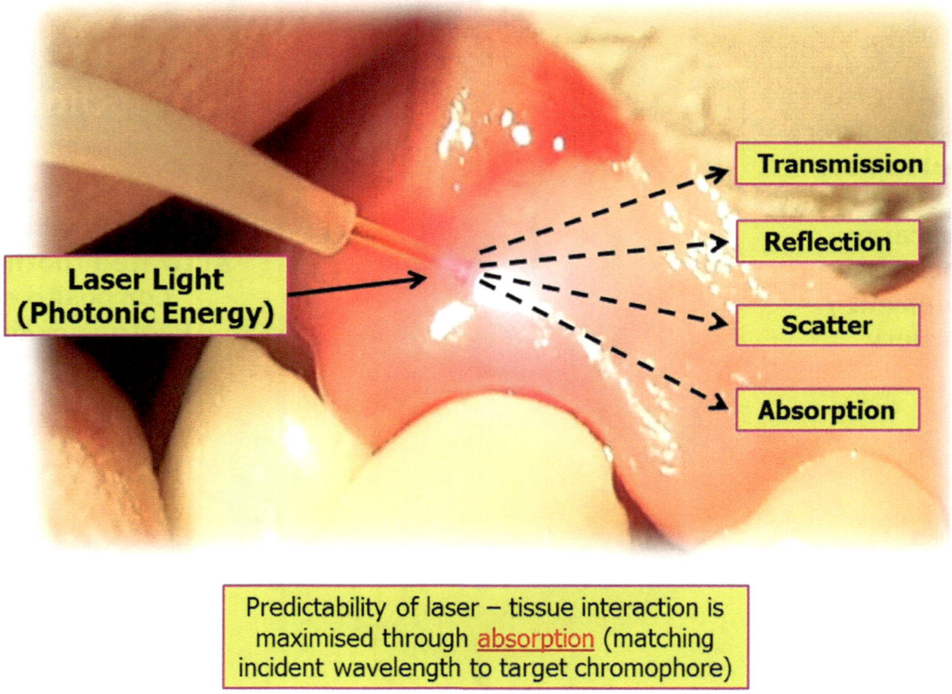

Fig. 3.3 Summary of the basic interactive phenomena of incident laser energy and target tissue. Such is the variance in tissue structure and heterogeneity commonly found in oral hard and soft tissue that there may be multiple and complex degrees of each interactive phenomenon

be adopted. Examples are protein/amino acid-based molecular groups, such as collagen, keratin and non-structural proteins such as melanin, hemoglobin in both its oxygenated and non-oxygenated (HbO_2 and Hb) forms. Dental and osseous tissues are based upon the calcium phosphate complex referred to as hydroxyapatite (HA) found as a structural crystal in bone and the carbonated lattice crystal as a mineral component of enamel and dentine (CHA). Water—as the intra-cellular medium base of cell cytoplasm, a component of circulating blood and plasma, a free molecule in interstitial tissue structure or as a hydroxyl (OH^-) radical as part of the hydroxyapatite molecule—represents a major ubiquitous molecular group of varying degrees of absorptive potential relative to incident photonic wavelength. Figure 3.4 demonstrates a graphical interpretation of the interaction of ascending photonic wavelengths with each of the major tissue chromophores that are found in oral and dental tissues. The term adopted as a measurement of the level of energy absorption by a chromophore is absorption coefficient—considered as a measure of the rate of decrease in the intensity of electromagnetic radiation (as light) as it passes through a given substance. The optical properties of tissue will determine the penetration into tissue of the radiant energy from a laser source. Absorption coefficient is inversely proportional to transmittance, and the depth of penetration of photons within a given chromophore will reduce as the absorption coefficient increases.

Each chromosome has molecular structure and for each there is a "ground state" which defines the structure, atomic configuration and interatomic binding energy at body temperature [13]. If external energy is applied, a point may be reached when molecular vibration is sufficient to overcome the forces binding atoms or molecules together. Examples include protein dissociation and water vaporization. True photonic ablation of a target molecule therefore represents incident energy sufficient to break interatomic binding forces and is termed dissociation energy. Table 3.2 provides examples of commonly found chromophore molecules and the dissociation energy value required to break the interatomic bond.

As is evident from data in Table 3.1, almost none of the popular laser photonic energies is capable of direct intra-molecular bond cleavage and one may be

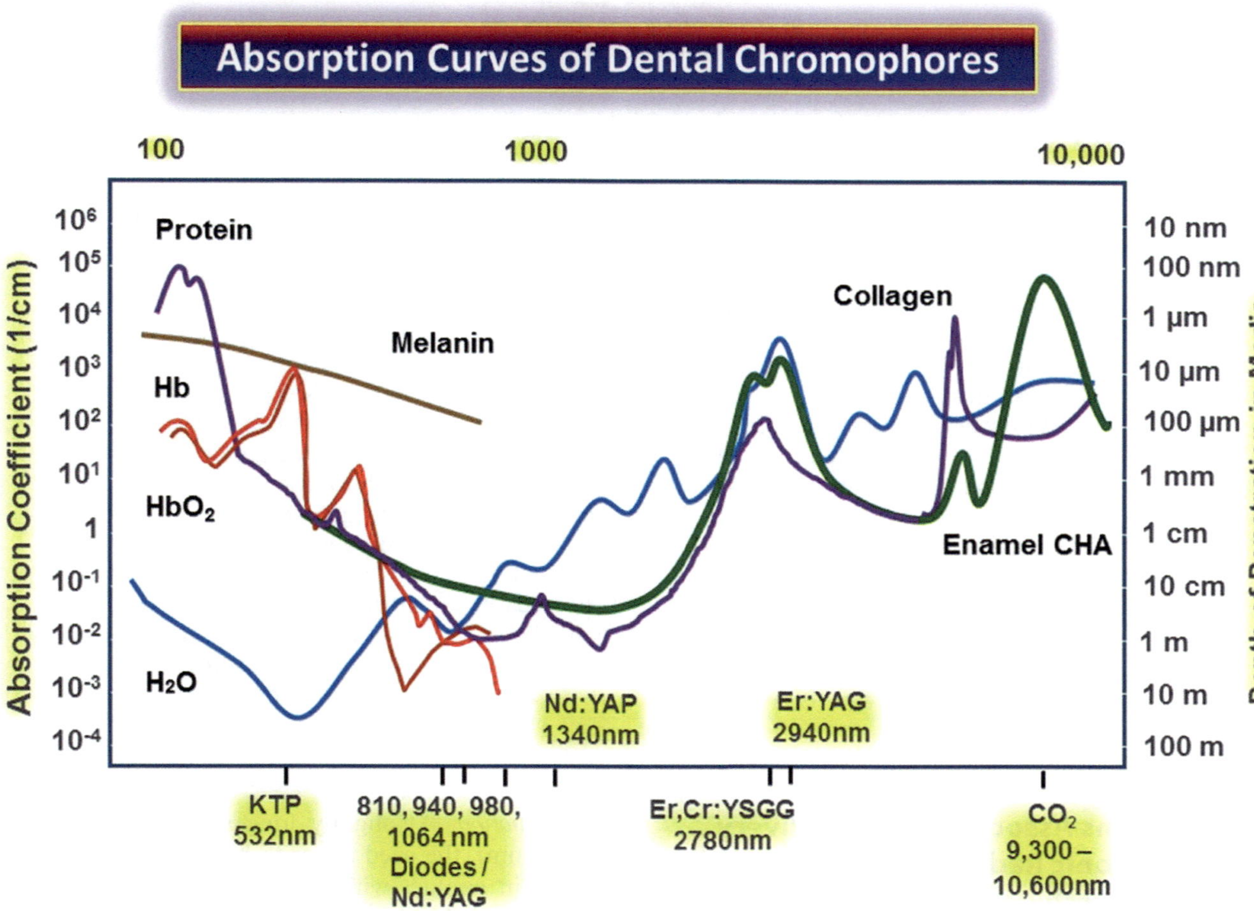

Fig. 3.4 Absorption coefficient curves for commonly found chromophores, relative to incident photonic wavelength

Table 3.2 Provides examples of commonly found chromophore molecules and the dissociation energy value required to break the interatomic bond

General concepts	
Dissociation energy of selected chemical bonds[a]	
Type of bond	**Dissociation energy (eV)**
C=O	7.1
C=C	6.4
O–H	4.8
N–H	4.1
C–C	3.6
C–N	3.0
C–S	2.7
Fe–OH	0.35
HA lattice	310

Dissociation energy, expressed in eV values, required to break the bonds (covalent, ionic, etc.) that bind atoms of common chromophores. Examples represent component molecules within tissue water, protein, blood and ionic forces within the crystal lattice of hydroxyapatite

Data taken from Mó O, Yáñez M, et al. J Phys Chem A. 2005;109(19):4359–65 [13]

forgiven for concluding that dental lasers cannot ablate target oral tissue through the use of empirical state photonic energy. Certainly, when the binding (ionic) lattice energies of crystalline carbonated hydroxyapatite are exposed to the mid-IR laser wavelengths (Er,Cr:YSGG, Er:YAG), the photonic energy value is pitiful compared to the dissociation energy of hard dental tissue [14].

Something else must be happening.

Figure 3.5 offers a summary of the interaction between a photon and target chromophore molecule, through successive stages of absorption, excitation and dissociation. Such predictive events might account for why certain laser wavelengths interact (are absorbed) with certain oral tissues.

Although individual photons possess insufficient energy to break apart target molecules, with each successive photon absorbed, the energy causes increasing molecular vibration up to a point where sufficiently high power density (energy density within ultrashort time) drives molecular fragmentation, or—more commonly seen with current dental lasers—molecular vibration converted into thermal rise leads to protein denaturation and water vaporization.

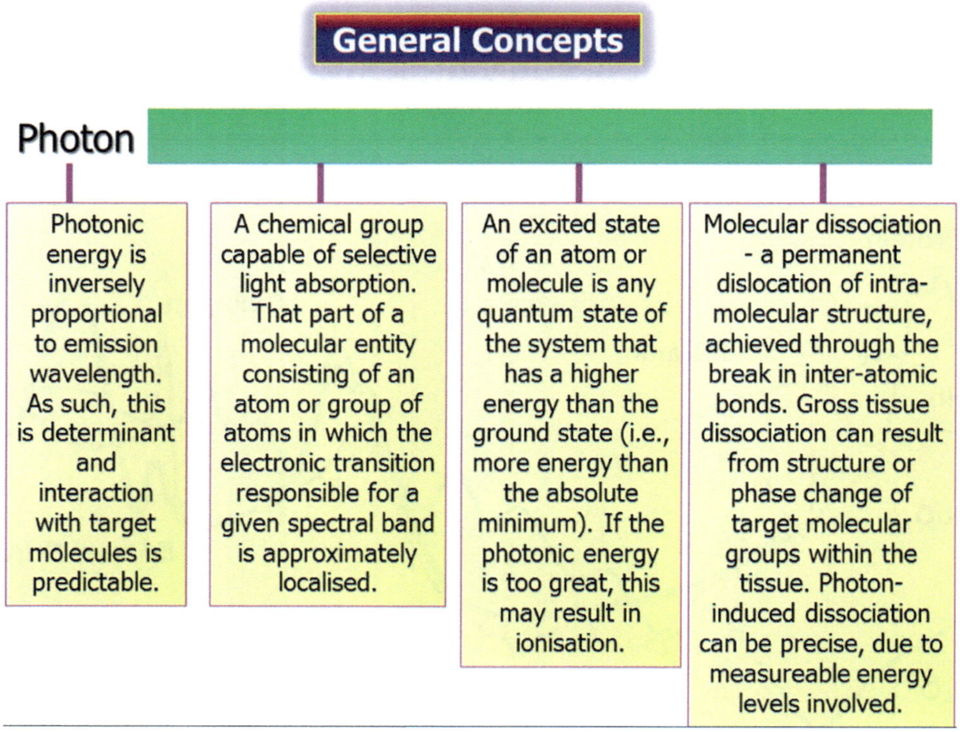

Fig. 3.5 Photonic energy interaction with target chromophore molecules

3.5 Problems Associated with Delivery of Photonic Radiation vs. Laser Wavelength

Considering a clinical application of high-intensity lasers, parameters such as wavelength, energy density, intensity, peak power, average power, repetition rate and pulse length are extremely important to heat generation due to irradiation on any biological tissue. The amount of heat inside the tissue is highly dependent on its optical properties, such as absorption and scattering coefficients [15].

The complex nature of oral soft tissue structure can pose some problems in delivering predictive laser-tissue interaction. For some wavelengths in the visible and near-infrared regions of the EM spectrum, the prime pigmented chromophores may be at some depth within the oral epithelium and covered with a thick keratinised layer. During photothermolysis, photonic energy may be delivered to the target and theoretically transferred (and undergo conversion) in one of the three [16] ways:

- *Radiation*: Non-contact laser waves are emitted from the delivery tip and absorbed by the target. Energy conversion occurs. This is referred to as the "real" laser effect. High-photonic-energy wavelengths such as the KTP (532 nm) may be delivered through a "non-contact" technique and direct photoablation may occur.
- *Conduction*: At slightly longer wavelengths (810–1064 nm), the need arises to both hold the delivery fiber in contact with the tissue and "initiate" the fiber tip with suitable absorbent material. A proportion of the incident laser energy is absorbed and gives rise to a "hot-tip" effect, whereby thermal energy is conducted to the tissue and aids the ablation of the tissue.

  Some concern is expressed as to possible disadvantages of this technique in that the effect of the hot tip on tissue is independent of the wavelength of laser radiation and that the heated fiber (sic) transmits only thermal energy, with no direct radiation energy [17].
- *Convection*: Moving heat within large volumes of liquid or gas either toward or away from the target. A similar effect of cooling may be seen through the effect of using a water spray, high-volume suction or air.

3.6 Concepts of "Power Density"

Figure 3.6 provides an example of simple manipulation of sunlight, using a magnification device such as a simple lens. Multi-wavelength cosmic radiation from the sun, although powerful and capable of tissue damage over time, can be brought to a focal area, and with the

Fig. 3.6 Manipulation of incident beam power over area to enable the power density effect to be a major factor in delivering predictable and powerful laser-tissue interaction

power in the beam concentrated to a small spot, the effects are much more dramatic. The use of a magnifying glass to concentrate the sun's rays is a simple example of power density. Power (energy per second) is an expression of a laser's ability to do work, and when measured over the area exposed to the beam, it will be readily acknowledged that the greater the concentration of photons, the greater the level of potential interaction. Consequently, for any given laser delivery system, reduce the spot size of the beam and one can expect to speed up the interaction—assuming that all other parameters remain constant [18].

"Power density" (PD) is the delivery of energy through time divided by the area of the exposed tissue; it is expressed in W/cm^2.

The output of any laser over time is expressed as average power and equates to the total number of Watts delivered per second. For a continuous wave emission laser, the average power will equal the maximum output; for a micro-pulsed free-running emission, the average power output may be of the order of a few Watts, but due to the active photonic emission only lasting a possible 20% of each second, there will be peaks of energy. A typical free-running emission laser, such as Nd:YAG or Er:YAG, may deliver an average power value of 3.0 W, but due to the pulse width of 150 μs, there will be peak power bursts of 1000+ W [19].

As has been seen elsewhere, the predominant temporal emission mode is a Gaussian distribution, and this lends itself readily to being brought to a focal spot. With control over the area of irradiation, the concept of power density as a prime factor in laser-tissue interaction becomes valid. Even in those conditions where there is little if any direct absorption, the concentration of laser power in ever-higher values over ever-shorter time periods gives rise to power density of such magnitude that photodisruption and photoionization of target molecules can occur.

Technology has much to deliver in terms of future developments, but already it is possible to see predictable laser-tissue interaction involving PD of values of 10^6+ W/cm^2 for microsecond periods and even shorter, enabling photovaporization of interstitial water in tooth tissue and consequent disruption of the crystalline solids.

Nowhere within this discussion does the influence of laser wavelength occur. Of course, the harmonization of incident wavelength, its inherent energy value and selective absorption within a suitable target chromosome will remain as empirical in our understanding of interaction concepts; but as we have already seen, the continuous bombardment of a target tissue with photons of a suitably absorbed wavelength will lead to thermal rise and eventually sufficient heat to effect physical change in the tissue. Such processes take time, and the risk of collateral thermal damage becomes an ever-present threat. To deliver sufficient energy in a form of concentration of both area of exposure and time must be seen as a distinct advantage.

As our understanding develops, there evolves the interaction of three components to predictive laser-tissue interaction: the absorptive potential of the target tissue, relative to the incident laser wavelength; secondly, the transfer of energy—from initial photonic energy and delivery (emission) mode of the laser, together with power density and time of exposure; and thirdly, the availability of thermal relaxation to enable the target tissue to avoid progressive overheating. Thermal relaxation is inherent with free-running pulsed emission modes and impossible to deliver with continuous-wave emission modes. The clinician would therefore need to be aware of the potential for thermal damage and allow sufficient time and respite periods to enable the tissue to recover.

In Fig. 3.7, the relationship between PD and exposure time is represented with reference to an ascending physical change in the target tissue. Very low irradiance over extended periods may give rise to subtle stimulation of biochemical pathways associated with tissue heath and reparative capabilities and is the core of understanding of photobiomodulation. With ever-shortening time and an ascending level of power density, not only do irreversible physical changes occur in tissue, but with exposure times of micro- and nanoseconds, even with relatively low average power values, the effects on the target can also be spectacularly rapid and without leaving a "thermal thumbprint."

For any given laser-tissue interaction, assuming that absorption can occur through the equation of power density with exposure time may enable the clinician to influence the type of interaction that occurs. It has already been established that at everyday levels of power delivery in dentistry, the predominant effect is tissue ablation through thermal rise—photothermolysis. By reducing the exposure time to milliseconds and microseconds, successively higher peak power density above 10^8 W/cm^2 can be obtained. At such powerful levels, the intensity of energy is so great that electromagnetic fields developed around the interaction are sufficient to tear target molecules apart—photoplasmolysis [20]. Reference to the work of Boulnois and the graphic representation of laser-tissue interaction can be seen as an ascending phenomenon and product of ultrashort exposure time and megawatt peak power [21].

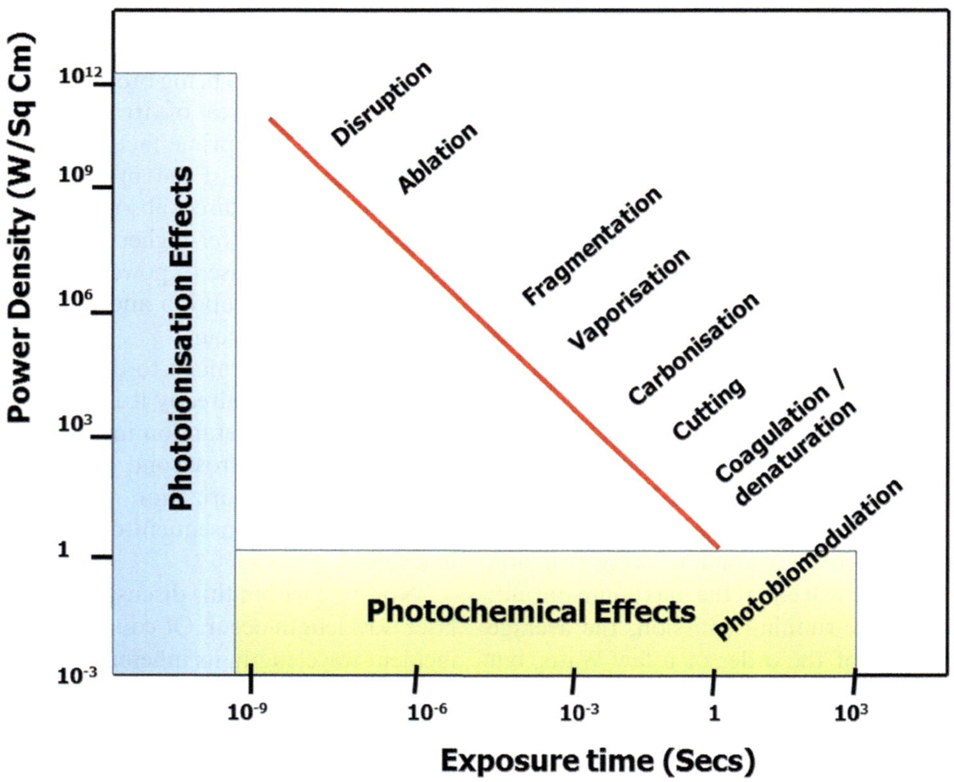

Fig. 3.7 Relationship of incident photonic power density and exposure time. (Source: Boulnois J-L. Laser Med. Sci. 1986;(1):47–66 [21])

3.7 Thermal Rise and Thermal Relaxation

In considering the broadest concepts of photothermal action and regardless of the laser system used for a soft tissue surgical application, the effects may be broadly classified as follows:

- *Tissue heating*: To a non-destructive level, the warming of tissue may be a desired total effect (as part of a biomodulation therapy) or may occur at some distance from an ablation site, along a thermal gradient within the tissue. This latter example is covered in greater detail later.
- *Tissue coagulation*: Over a period of time, the temperature rise in soft tissue at or above 45 °C will constitute "destructive heating," i.e., amounting to progressive irreversible change. At about 50 °C, bacteria can be demonstrated to achieve a state of deactivation, with tissue protein denaturation occurring at around 60 °C. Within this zone of thermal rise, the walls of small-diameter vessels (arterioles, venules and lymphatics) within the irradiated area will undergo structural change of vessel walls and lead to progressive blood and lymph coagulation. Dependant on the wavelength of the laser used and concentration of chromophores specific to that wavelength, a concept of "selective photothermolysis" can be considered, such as in the ablation of melanin in diode laser-assisted gingival de-pigmentation.
- *Vaporization*: At normal atmospheric pressure (1 Bar), vaporization of water occurs at 100 °C. Within soft tissue, the vaporization of water will accompany existing protein denaturation that may have occurred at a lower temperature. The phenomenon of water vaporization is accompanied by volumetric change and expansion in the ratio of 1:1600 as the liquid is vaporized to steam. With short wavelengths (visible and near IR), the structural change in soft tissue collagen scaffolding would allow water vaporization to occur as part of the overall ablation process. With mid-IR erbium family wavelengths, the scenario is often very different owing to the poor absorption of these wavelengths in pigmented tissue, the limitation of thermal rise in the tissue when a co-axial water spray is being used and the FRP emission mode of the lasers with associated high peak power values, and with the extremely high absorption of these wavelengths in water, the vaporization is often more dynamic and accompanied by audible "popping."

  When erbium lasers are used on soft tissue without water spray, or when CO_2 laser wavelengths are used (often without water spray), the vaporization is reflective of a more "thermal" exchange, where the heating of the tissue as well as the absorption of the photons in water occurs. Often, the result of such vaporization leads to visible signs of tissue desiccation, structural shrinkage and predisposition to rapid heating and carbonization.
- *Carbonization*: As laser-assisted surgery proceeds, the risk remains to be the potential for tissue heating that leads to the production of end-stage molecular destruction and residual carbon. It is generally considered to be at temperatures around 200 °C, although the actual rise may be considerably higher. Carbonization would only occur as a result of either an inappropriate high dose, relative to the parameters consistent with a desired surgical outcome, or the application of laser photonic energy over excessive time, resulting in opportunity for destructive collateral effects. The characteristic visual sign is the development of black residue associated with the soft tissue incision.

  By far, the consequence of such development is the preferential absorption in carbon residue of (in effect) all incident EM photonic wavelengths. This is the basis of the "black body" concepts of preferential absorption and characterized by the re-emission of multi-wavelength, incandescent near-IR thermal radiation; in consequence, the carbonized tissue continues to absorb incident laser energy and becomes the source of thermal conductive energy that significantly contributes to collateral damage during soft tissue surgery.
- *Photoacoustic phenomena*: As has been seen already, incident coherent photonic energy can be subject to conversion into other forms of energy, notably thermal in the dominant effect of photothermolysis. With the instantaneous phase change of water from stable liquid to vapor, the volume change can give rise to a cavitation phenomenon and consequent shock wave. Additionally, the energy may be changed to sound, and this may be witnessed with mid-IR interaction with tissue and the "popping" sound often heard. True photoacoustic effects are used elsewhere in medicine and surgery in procedures such as lithotripsy, where kidney and gallstones are fragmented using indirect shock waves; in maxilla-facial surgery, a similar approach has been reported to assist in the fragmentation and subsequent safe passage of sialoliths within the submandibular gland [22, 23].

Given the current limitations of laser emission parameters and the consequence that by far the greater consequence of laser-tissue interaction is photothermal shift and temperature rise, the ablation of target tissue can be severely compromised by excessive thermal rise and a build-up of the ablation residue that may rapidly overheat.

The effects of thermal rise can be both subtle and dramatic—depending on the rate of warming. ◘ Figure 3.8 provides a tabulated outline of the effects that temperature rise may have, relative to both the visual change and the biological change (the latter as may be applied to soft tissue). In addition, the varying stages of thermal rise have been investigated and provide opportunity to influence the structural changes in the tissue and the effect of heat on associated bacterial cells. Irradiated tissue should not be regarded as sterile, although there will be significant pathogen reduction at the site of maximum laser-tissue interaction.

Two concepts of ablation may be considered: a zone of tissue removal/permanent change preceded by an "ablation" front and a second advancing line denoting the permanent effect of change rendered by thermal rise—a "thermal" front. In an ideal situation, the "ablation" front will denote the predicted volume of ablated tissue and through the correct management of heat rise and debridement, the risk of unwanted thermal damage can be avoided. In hard tissue management, the concepts of ablation and thermal zones will be discussed with specific reference to tooth cavity preparation.

Thermal relaxation time can be deduced mathematically [24] as the time taken for the irradiated tissue to dissipate about 63% of the incident thermal energy. It is additionally related to the area of the irradiated tissue and thermal diffusivity and bulk of the tissue.

Thermal damage time is the time required, for the entire target, including the primary chromophore (e.g., melanin) and the surrounding target (e.g., gingiva), to cool by about 63%. It includes cooling of the primary chromophore as well as the entire target.

Extinction length is the thickness of material necessary to absorb 98% of incident energy.

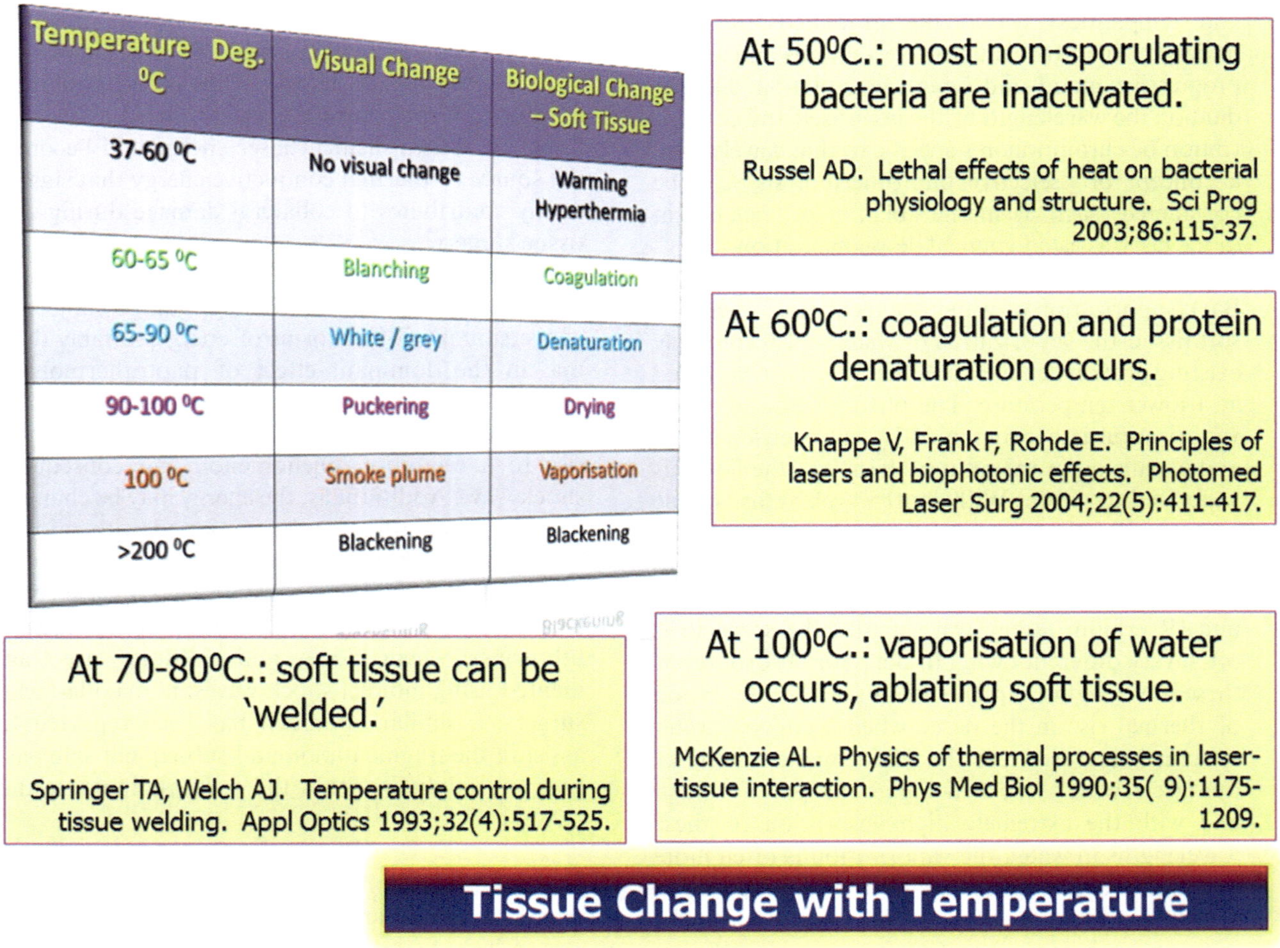

Temperature Deg. °C	Visual Change	Biological Change – Soft Tissue
37-60 °C	No visual change	Warming Hyperthermia
60-65 °C	Blanching	Coagulation
65-90 °C	White / grey	Denaturation
90-100 °C	Puckering	Drying
100 °C	Smoke plume	Vaporisation
>200 °C	Blackening	Blackening

◘ **Fig. 3.8** Effects of thermal rise on (soft) tissue

3.8 Laser Photonic Energy and Target Soft Tissue

Various laser wavelengths are available for clinical use with target oral soft tissue and span the visible (blue) EM spectrum through to the far infrared. Examples of laser wavelength currently available to the clinician are shown in Fig. 3.9.

With current configurations of emission modes, power limits and commercial technology application, all soft tissue ablation achievable in clinical dentistry is primarily and almost exclusively due to photothermolysis [25]; in general, chromophore absorption is by pigmented molecules (haem, melanin) with short wavelengths (532–1064 nm), whereas longer wavelengths experience greater interaction with tissue water components (H_2O and OH^- radicals), with peak absorption occurring at approximately 3000 and 10,600 nm. The emergence of commercially available laser units with wavelength emissions at 450–490 nm offers the opportunity of utilizing absorption in water with a diode-source active medium. Protein as a structural component of oral soft tissue appears to have moderate absorption of ultraviolet wavelengths, together with peaks at 3.0 and 7.0 μm. Visible and near-IR wavelengths have limited absorption in protein, but as has been discussed above, secondary thermal rise consequent upon time-related photonic energy exposure will give rise to conductive heat changes in proteinaceous material.

As such, oral soft tissue, high in water and protein with varying degrees of pigment and blood perfusion, remains a straightforward target tissue wherein low-dose irradiation can be configured to deliver predictable laser-tissue interaction with limited collateral damage.

Of practical interest to the clinician, the following factors (Table 3.3) will each and collectively affect the absorption of laser light by a chosen target tissue [26]:

Shorter wavelengths tend to penetrate soft tissue to depths of 2–6 mm [27], and scatter is a significant event, both back-scatter of photons and forward scatter into the tissue. Longer wavelengths are attenuated at or near the tissue surface, due to water content of cellular tissue. As tissue ablation proceeds, short-wavelength photonic energy causes protein denaturation and conductive effects as the tissue is heated. A typical soft tissue zone of near-IR laser ablation is surrounded by a zone of reversible edema and little evidence of acute inflammatory response. Classically, the progression of near-IR laser ablation of soft tissue is through a crater-shaped zone where depth and volume removed appear proportional [28].

In Fig. 3.10, the interaction and progression of near-IR irradiation in soft tissue are graphically represented. In an ideal fashion, the zone of ablation and

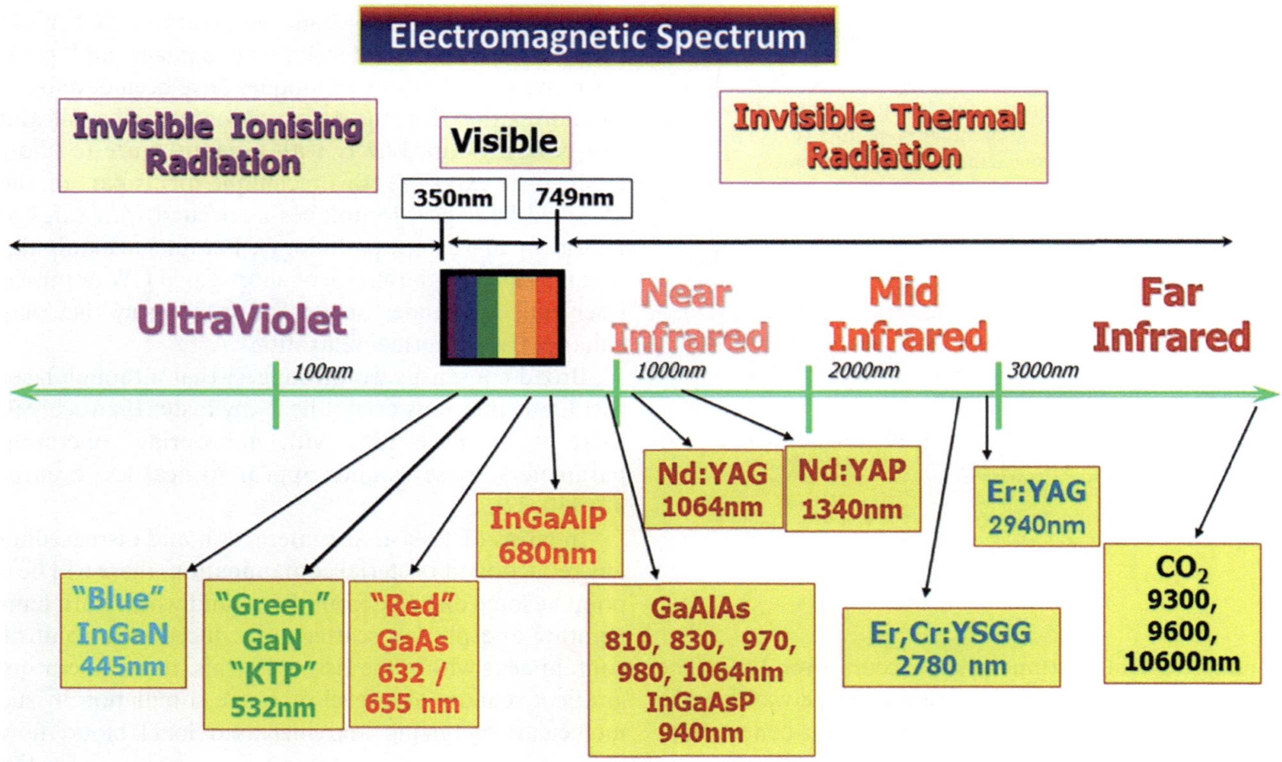

Fig. 3.9 Laser wavelengths commonly available for use in dentistry

Table 3.3 Individual factors associated with laser-tissue interaction that may affect the predictability of clinical use of a chosen laser

Factor	Comment
Laser wavelength	Individual wavelengths (visible extending to far infrared non-ionizing radiation) and inversely proportional to the photonic energy
Laser emission mode	Inherently continuous wave (CW) or free-running pulsed (FRP), due to the excitation source or additionally modified by the manufacturer to deliver gated CW and mode-locked CW or modification in pulse width (>10 µs) with FRP
Laser power value	With increasing power delivery, there is potential for thermal rise. Below the ablation threshold, this may be reversible (tissue warming/PBM)
Exposure time	Together with laser power, "spot size" and emission mode, this will affect power density and thermal relaxation
Tissue-type composition	All oral tissue is heterogenous, and the proportions of common chromophore content will alter the potential for individual laser wavelength absorption
Tissue thickness	Thicker tissue will take longer to incise/ablate. Additional factors may be thermal diffusivity and longer thermal relaxation times
Tissue surface wetness	Due to water or saliva—of note with longer wavelengths > approx. 1500 nm. Wetness will affect tissue reflection (below)
Incident angle of the laser beam	Incident angle of beam to tissue of 90° will define maximum potential for interaction. As angle approaches the reflection limit (TIR), this reduces the potential for interaction to zero
Contact vs. non-contact modes	Employed between laser delivery tip and tissue. With visible and near-IR wavelengths, contact technique may be essential to allow a "hot-tip" technique. Non-contact may have a focussed beam, and distance of tip/handpiece to tissue may be crucial to maximize laser-tissue interaction
Thermal relaxation factors	Exogenous (water spray, tissue pre-cooling, high-speed suction, pulsing/gating laser emission) Endogenous (tissue type and density, blood supply)

conductive temperature spread occurs over time. The predominant scatter phenomenon of these wavelengths gives rise to a complex pattern of photon penetration, wherein there may be indeterminate tissue effects, giving rise to the acronym WYDSCHY—"What You Don't See Can Harm You," coined by Fisher in his 1993 paper [29]. In essence, visible and near-IR laser wavelengths have deeper penetrating effects on oral soft tissue and demand that optimal and non-excessive operating parameters are used in order to avoid unwanted tissue damage.

Clinically, such interaction can be seen in Fig. 3.11, during the removal of a fibroma from the lateral tongue, using an 810 nm diode laser:

Certainly, even incisions will have a "U"-shaped cross-sectional appearance, and this is due in part to progression of photonic energy through scatter as well as some direct conductive thermal spread [29]. A simple in vitro example using pig mucosa with short and longer wavelengths provides an excellent example of the structure of the incision and difference in laser-tissue interaction between the two wavelengths (Fig. 3.12).

With longer laser wavelengths (mid-IR approx. 3.0 µm—Er,Cr:YSGG, Er:YAG—and far IR approx. 10 µm—CO_2), a more "V"-shaped cross-sectional appearance prevails (Figs. 3.13 and 3.14). The bulk of laser-tissue interaction occurs at or within the confines of the tissue surface, and as an incision is developed, the majority of excess energy (thermal) is released through the escape of vaporized tissue water [29]. This predominant effect reduces the conductive thermal rise into adjacent tissue. A risk exists with soft tissue in that desiccation of target tissue can predispose to the formation of carbonized tissue elements (surface char—termed eschar) and the preferential absorption of this material leading to very high temperatures that might cause conductive collateral tissue damage and post-operative pain. Various techniques have been developed to address this risk; the eschar is loosely adherent and can be easily wiped away with a damp gauze to allow fresh tissue exposure (this technique forms part of the so-called laser peel techniques associated with surgical treatment of surface pathology). Parameter manipulation may include the choice of short-gated CW or pulsed laser emission modes or coaxial water spray that may enhance tissue thermal relaxation.

Broad consensus would suggest that although laser soft tissue incisions do not heal any faster than scalpel, there is evidence that with appropriate operating parameters, these wounds appear to heal less eventually [30–32].

In terms of laser-tissue interaction and disregarding any reduction in bacterial contamination, there will be a point at some distance from the wound where both temperature and photon scatter are reduced to a point of containment within the tissue. By this, the temperature gradient reduces to a level of tissue stimulation, tissue molecular energizing and increased local blood flow [33]. In addition, a scatter gradient exists where the energy delivered is reduced to that point where biomodulation effects predominate [34]. For these reasons, it

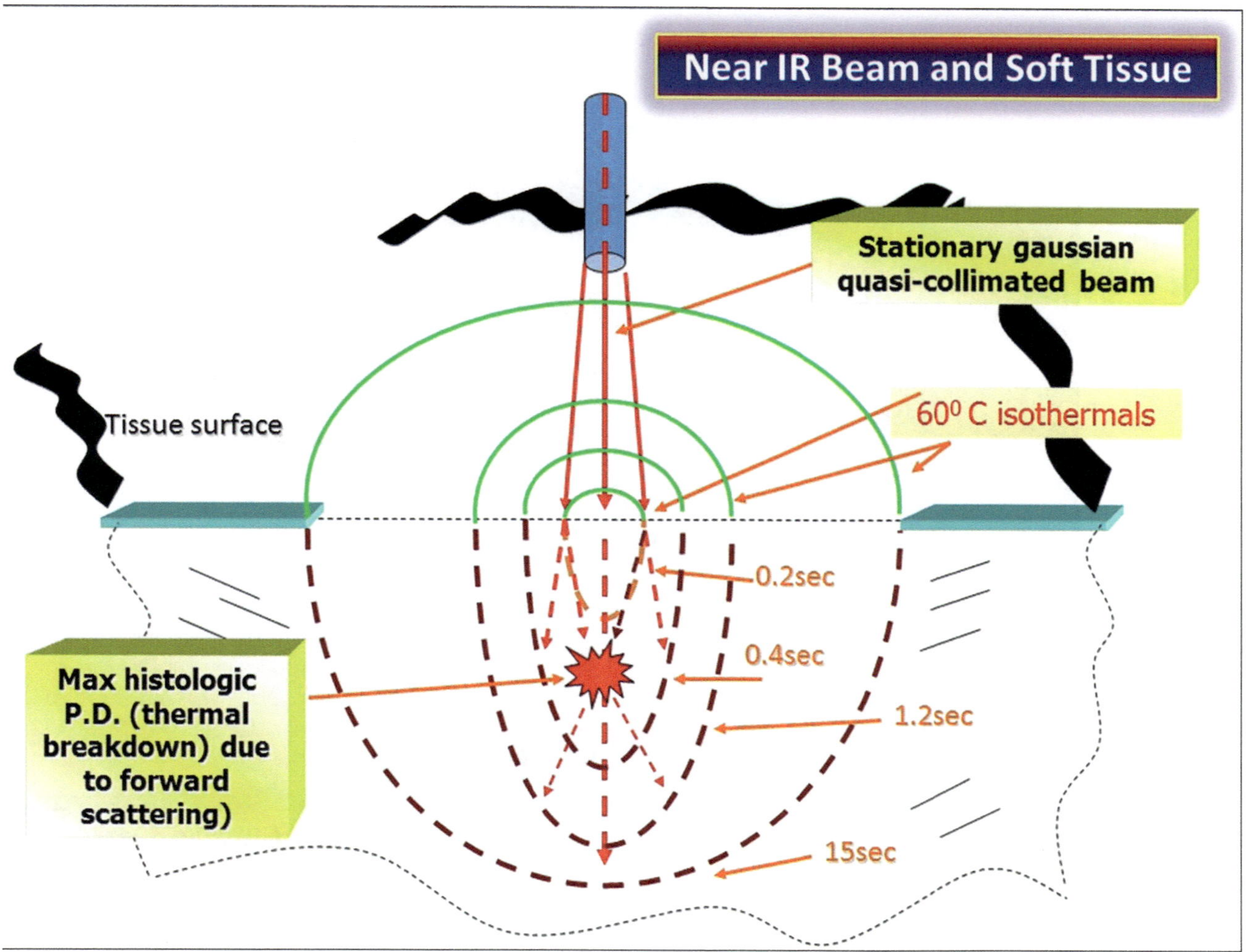

Fig. 3.10 Graphic representation of visible and near-IR laser photonic energy interaction with oral soft tissue. (Graphics: S. Parker after Fisher J.C. 1993)

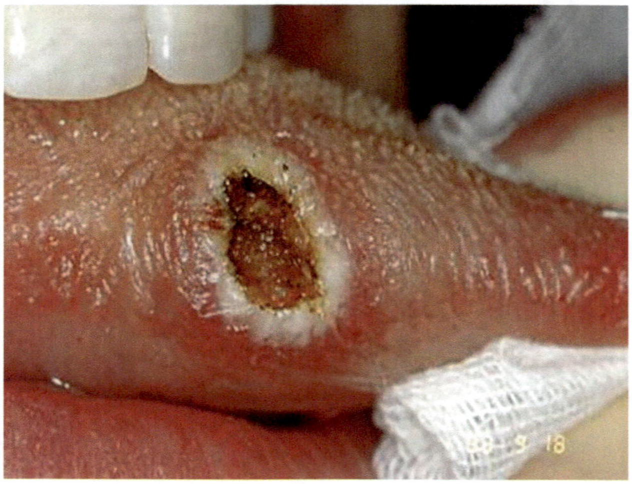

Fig. 3.11 Following removal of a fibroma lateral tongue, using an 810 nm diode laser. The central ablation zone is surrounded by an area of edema. The lack of carbon residue indicates a correct choice of laser power parameters

may be seen that laser-assisted surgical wounds respond in a positive and supportive framework that delivers less eventful healing (Fig. 3.15).

Fisher [35] defines a comprehensive understanding of photon scattering into deeper soft tissue areas that is seen with the use of visible and near-IR lasers. With successive interaction and as photons are absorbed, the possibility exists for a scenario whereby the ablation threshold of the host tissue at deeper sites is greater than the photonic energy. This "energy gradient" phenomenon might provide explanation as to how distant effects of (surgical) laser use may mimic essentially low-level (photobiomodulation) stimulation of cells and host tissue. Standard textbooks [36] provide authoritative and evidence-based explanations of how the host tissue may respond positively to low-level photonic energy, and the reader is directed to such references for further information.

Positive healing effects following laser surgery: One of the often-cited side effects of laser-assisted surgery is

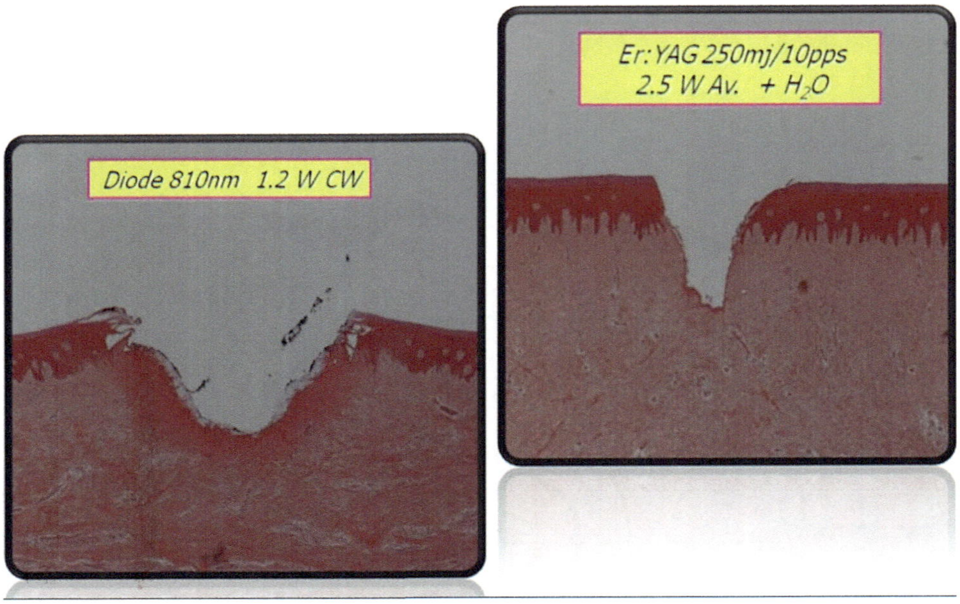

Fig. 3.12 Histological representation of two laser wavelengths (diode 810 nm and Er:YAG 2940 nm interaction with pig mucosa in vitro). This demonstrates the progressive crater-shaped incision with shorter wavelengths and a "V"-shaped incision with longer wavelengths

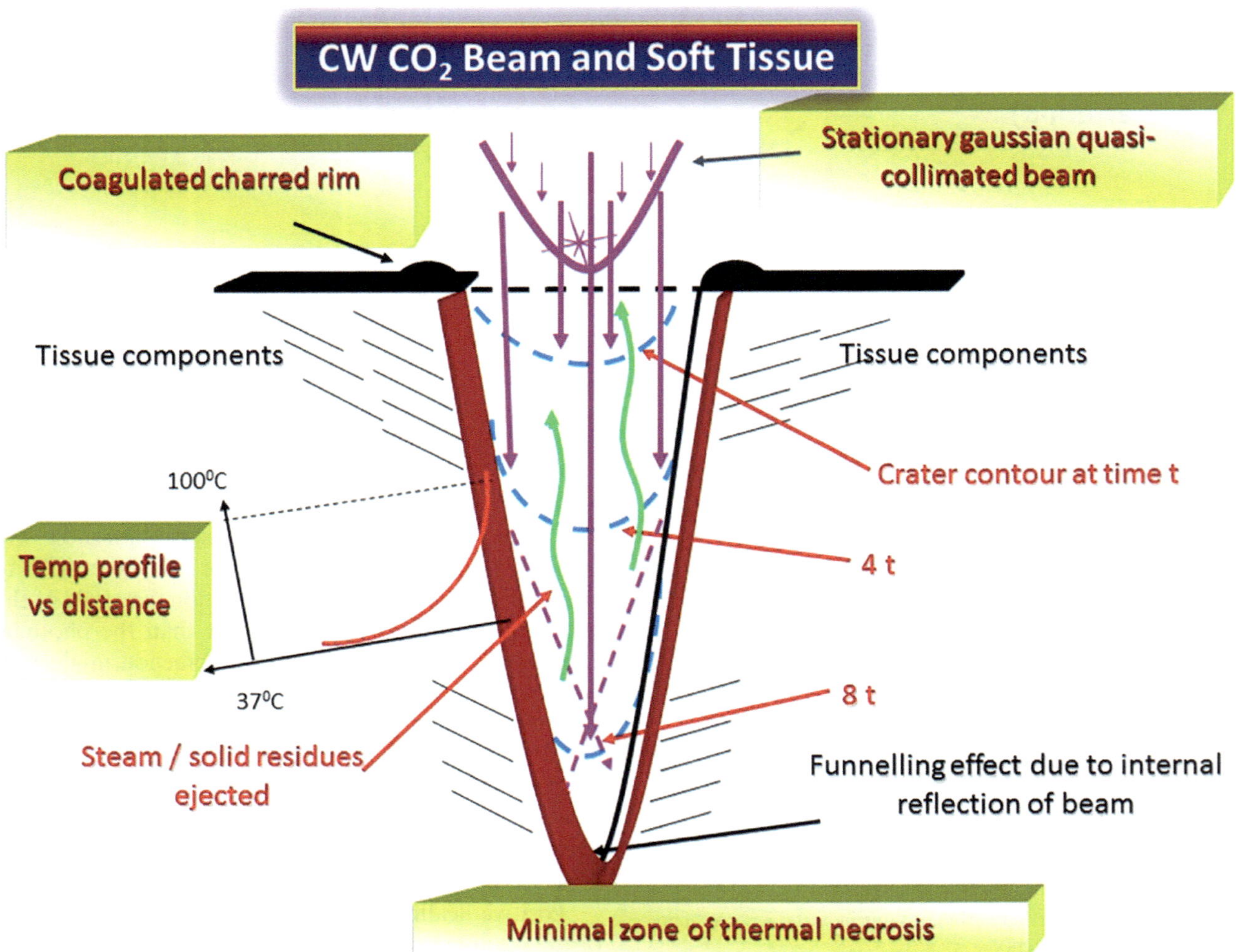

Fig. 3.13 Graphic representation of far-IR (and potentially mid-IR) laser photonic energy interaction with oral soft tissue. (Graphics: S. Parker after Fisher J.C. 1993)

the lack of post-operative inflammation and uneventful healing. Inasmuch as many claims are anecdotal, often if not always, the need for dressings or sutures can be avoided, and irrespective of the laser wavelength employed, all soft tissue healing will be by secondary intention in that it will be impossible to oppose the cut tissue edges to their original alignment. Of note, however, is the phenomenon of lack of post-incisional contamination by bacteria, due to a possible sterility of the cut surface [37] but certainly through the protective layer of coagulum of plasma and blood products—a tenacious film that allows early healing to take place underneath [38]. Additionally, studies with longer wavelengths show that there is a lack of fibroblast alignment associated with the incision line and consequent reduced tissue shrinkage through scarring [39]. Such findings are often borne out in the clinical setting.

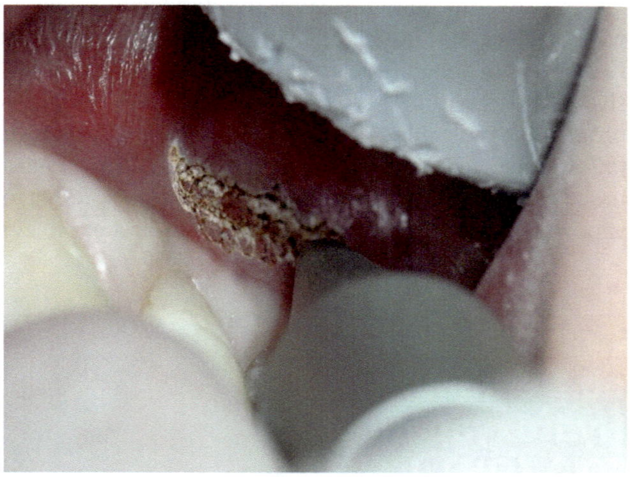

Fig. 3.14 Clinical example of a mucosal incision using a CO_2 laser. In the absence of a water spray, note the build-up of eschar, which can be easily removed with damp gauze to minimize thermal damage

3.9 Laser Photonic Energy and Target Oral Hard Tissue

Oral hard tissue includes cortical and trabecular (cancellous) bone and components of deciduous and permanent teeth (enamel, dentine, cementum). Within this group, it acknowledges the association of dental caries, being the predominant reason why teeth are subject to surgical intervention.

In common with laser-tissue interaction mechanisms described above, the current limitations of operating parameters of those lasers that are commercially available in dentistry center on the targeting of chromophores within the host tissue.

From the development of early lasers to ablate dental hard tissue, the predominant chromophore has been water—both interstitial "whole molecular" water and OH^- radicals forming part of the carbonated hydroxyapatite molecule ($[Ca_{10}(PO_4)_{6-Y}(CO_3)_Z(OH)_2] + H_2O$). Prime tissue groupings of oral hard tissue are listed by percentage of constituent structural elements [40, 41] in Table 3.4.

Absorption curves for both water and carbonated hydroxyapatite between wavelengths of approximately 3.0 and 10 μm are graphically listed in Fig. 3.16. The

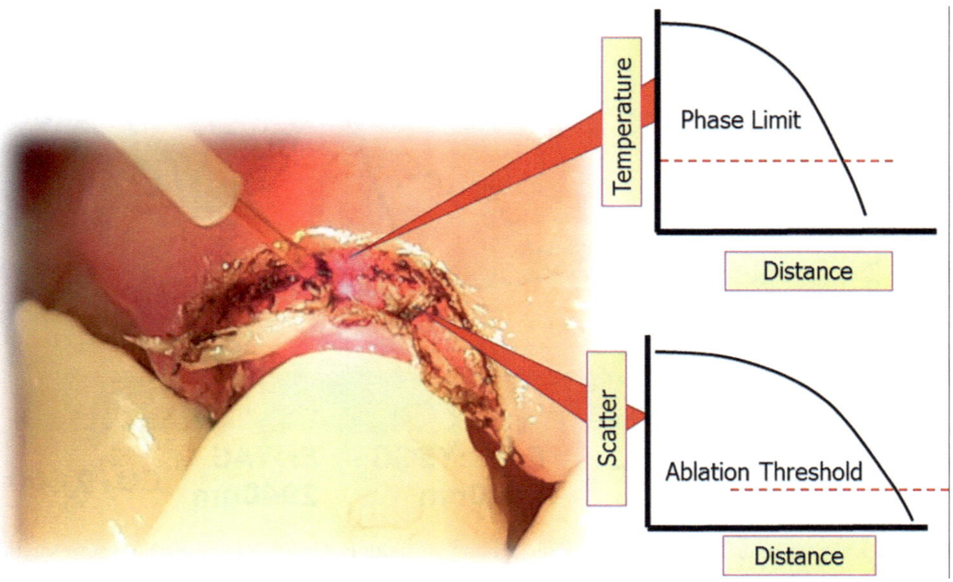

Fig. 3.15 At the point of surgical ablation of tissue, two intra-tissue gradients predominate. One is a thermal gradient, and with distance, a reduction in temperature will define a point where the temperature provides tissue stimulation. The other gradient—scatter—can produce a similar point-at-distance stimulation wherein biomodulation effects predominate

3

Table 3.4 Oral and dental hard tissues have structural components that may be viewed as potential chromophores—mineral, protein and water. For each tissue, the percentage of each of these chromophores will differ and define a level of laser-tissue interaction with a suitable laser wavelength

Tissue	Component chromophore as percentage		
	Mineral HA/CHA (%)	Protein collagens I and II (%)	Water (%)
Cortical bone	65	25	10
Cancellous (trabecular bone)	55	28	17
Tooth enamel	85–90	1–3	4–12
Tooth dentine	47	33	20
Tooth cementum	50	40	10
Dental caries	>5	70	25

carbonated hydroxyapatite molecule (CHA) is a relatively complex inorganic molecule with a parent calcium chain supporting radicals of phosphate, carbonate and hydroxyl subgroups. Additionally, within clinical specimens, there is whole-molecule free water; each radical is capable of preferential absorption, and peaks occur to indicate that laser interaction is possible, assuming that correct spatial and temporal operating parameters are used [42].

As was seen in the earlier table (Table 3.2), there is insufficient energy associated with an incident photon of erbium YAG (2940 nm) whose value is 0.42 eV, to break the atomic bond of a hydroxyl radical (value—4.8 eV). By the same measure, the dissociation energy within the crystal lattice of hydroxyapatite at 310 eV is two orders of magnitude greater [14]. Given that successive photons within a stream of irradiance will lead to progressive molecular vibration in the target structure, it follows that the chief goal of laser ablation of oral hard tissue would be the induced phase transition of water to vapor (steam), leading to the dislocation and explosive

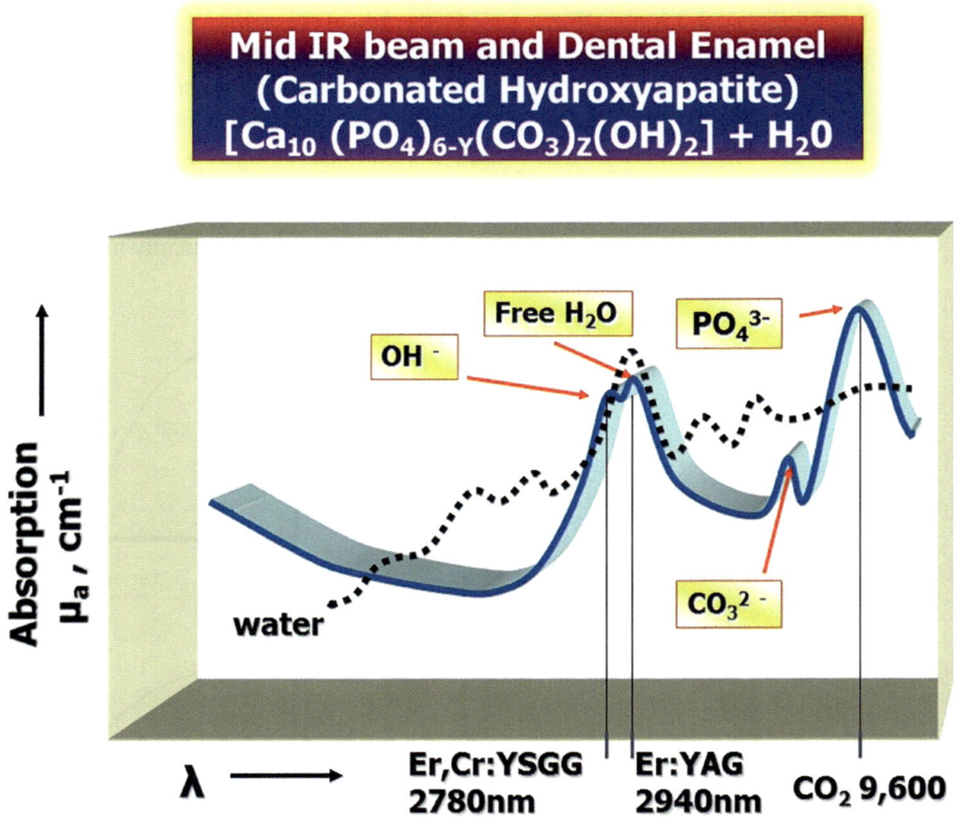

Fig. 3.16 Absorption peaks for water and CHA exist coincidentally for both Er,Cr:YSGG and Er:YAG wavelengths. High absorption exists in the phosphate group of CHA, coincident with CO_2 at 9300 and 9600 nm. (Source: Parker S. BDJ 2007;202(8);445–454)

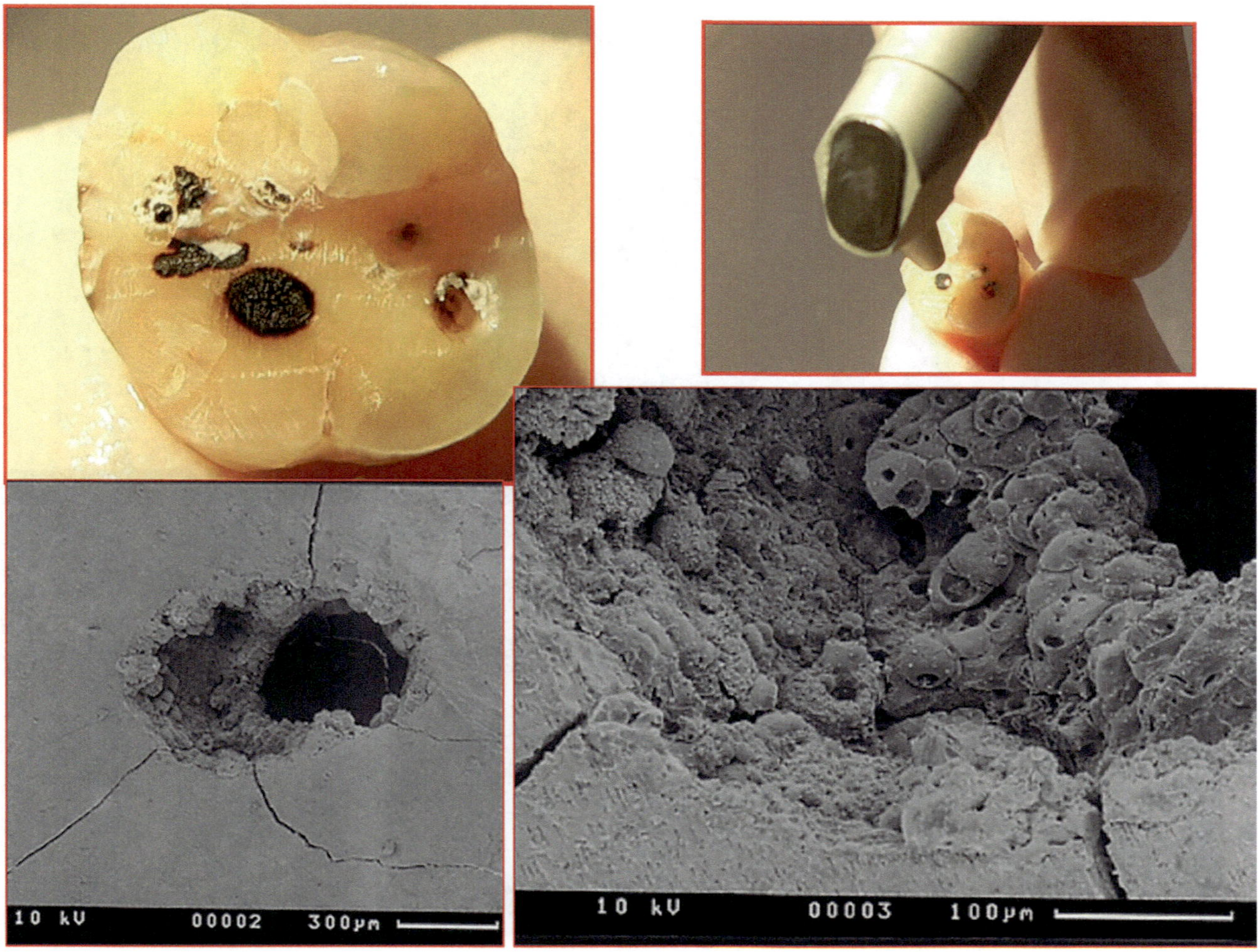

■ **Fig. 3.17** In vitro exposure of molar tooth to CO_2 laser irradiation and SEM examination. Globules of melted and re-solidified (amorphous) hydroxyapatite are present, with large voids and gross disruption of structure. Average power 1.5 W CW with no water spray

derangement of the surrounding crystal lattice. Both erbium (erbium YAG, erbium chromium YSGG) laser wavelengths have free-running pulsed emission modes (pulse width 50–150 µs), which give rise to high peak power levels (>1000 W). Such power levels result in an instantaneous, explosive vaporization of the water content of enamel and dentine, which leads to dissociation of the tissue and ejection of micro-fragments [43].

The increased water content of caries results in rapid and preferential ablation of such material compared to normal enamel and dentine; to some extent, this may allow cavity preparation to be accomplished with a more conservative preservation of intact dental tissue.

Sustained exposure of hydroxyapatite and carbonated hydroxyapatite to laser irradiance will quickly render the structure to overheating, first to drive off any residual water and then to rapidly melt the mineral and produce signs of carbonization (■ Fig. 3.17). It is evident that sufficient heat may be produced to cause the melting of hydroxyapatite (several hundred degrees Celsius) and associated thermal cracking. Of course, such temperatures would lead to direct pulpal damage through heat conduction.

In consequence, the interactions of high-intensity laser irradiation with bone and dental hard tissues are the result of a photothermal action [44] targeting both molecular and interstitial water.

With an appropriate laser wavelength such as the Er,Cr:YSGG (2780 nm) and Er:YAG (2940 nm) and operating parameters configured to maximize interaction, together with adequate coaxial water spray, the outcome is completely different. With both enamel and dentine, the outcome of the "explosive" vaporization and ejection of tooth fragments results in a clean-cut surface, without smear layer often associated with rotary instrumentation. Due to the outward dissipation of energy, there is minimal thermal rise within the structure of the tooth and conduction to the pulp.

The fragmented appearance of cut enamel (■ Fig. 3.18) especially was historically thought to enhance the facility for bonding of restorative resins and composites without the need for acid etching. However,

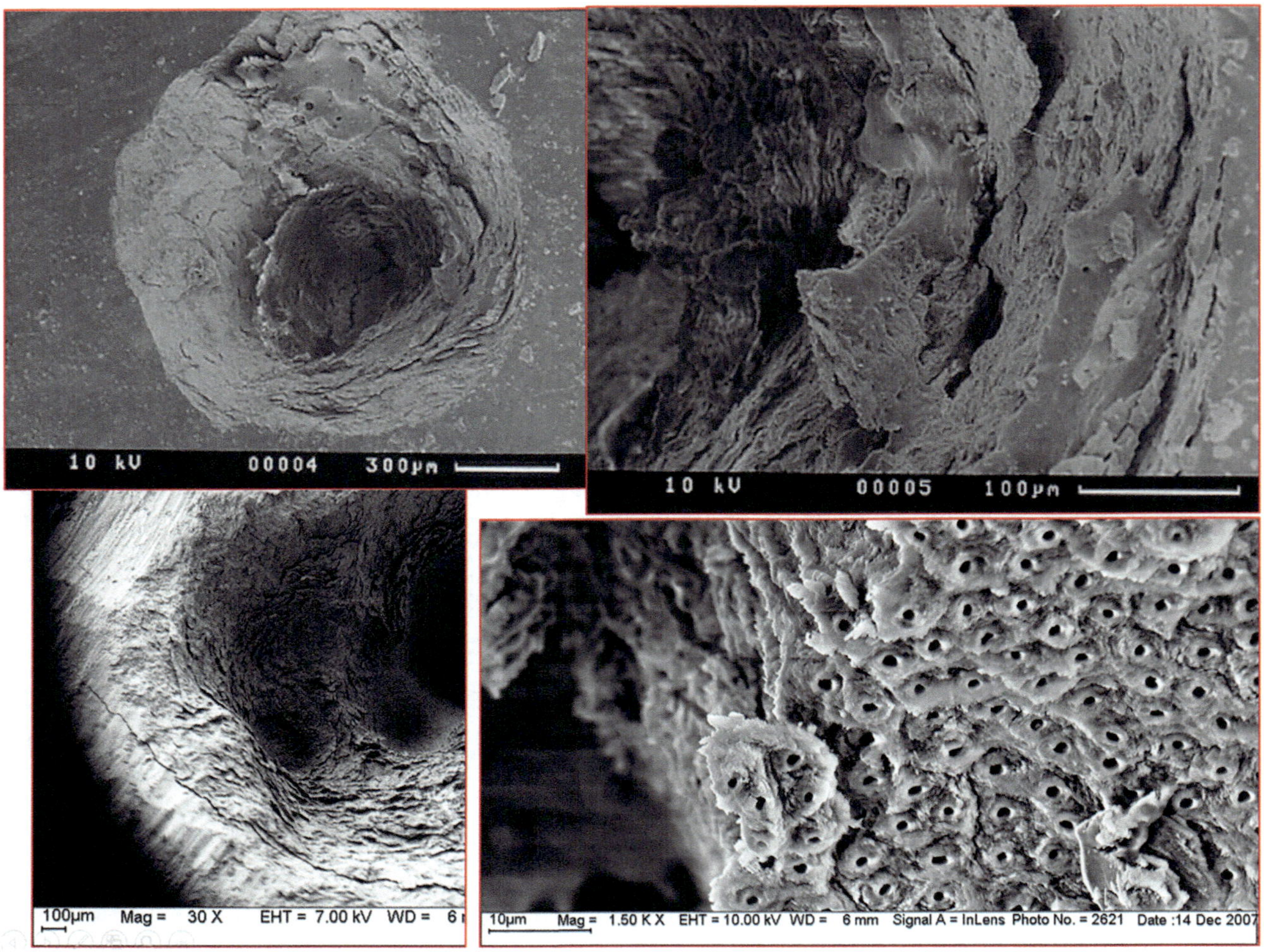

Fig. 3.18 SEM examination of tooth structure exposed to Er:YAG laser irradiation. Top images L and R—enamel structure showing evidence of dislocation and some fragmentation of the cut surface. Lower images L and R—similar cut surface of dentine, showing absence of smear layer and open tubules

many studies have highlighted the fragility of the cut margin in enamel and subsequent failure of the restoration margin as weakened tooth fragments gave way under tensile stress, with resulting failure and secondary caries risk [45, 46].

Mid-IR laser beam interaction with enamel (and to some extent also with dentine and bone) is a combination of temperature and pressure [47]. Both can be seen to rise rapidly during the pulse train of a clinical ablation procedure. The increase in volume as water vaporizes (1:1600) is significant and will give rise to significant rise in pressure just prior to the explosive dislocation of the enamel structure. As pressure rises, the continued vaporization leads to increase in temperature, resulting within the micro-confines of the interaction in "superheating" of the water and temperatures of several hundred degrees Celsius.

Furthermore, there are additional characteristics of laser ablation of hard tissue, surrounding the use of coaxial water, a necessary component to both aid excessive temperature rise and help wash away debris as a result of ablation. In addition, commercial models of both lasers use coaxial water spray to aid dispersal of ablated tissue and to cool the target [48], in a process called "water augmentation" [49]. According to this study, when dental hard tissues are irradiated with Er,Cr:YSGG and Er:YAG emission wavelengths with an additional thin water layer, the cutting efficiency increases at the same time that the pulp temperature decreases. However, the thickness of water layer should be well controlled to avoid a compromise in cutting efficiency and the blurring of the visual field.

As a liquid, water has a moderately high surface tension (72.8 mN/m at 20 °C), and this accounts for the intact film that may surround the tooth surface during water-augmented laser irradiation. In a further study [50], the influence of water thickness was investigated.

During ablation, the stream of photons is emitted in a free-running train of microsecond pulses (>100 μs), and any water between the laser delivery tip and the tar-

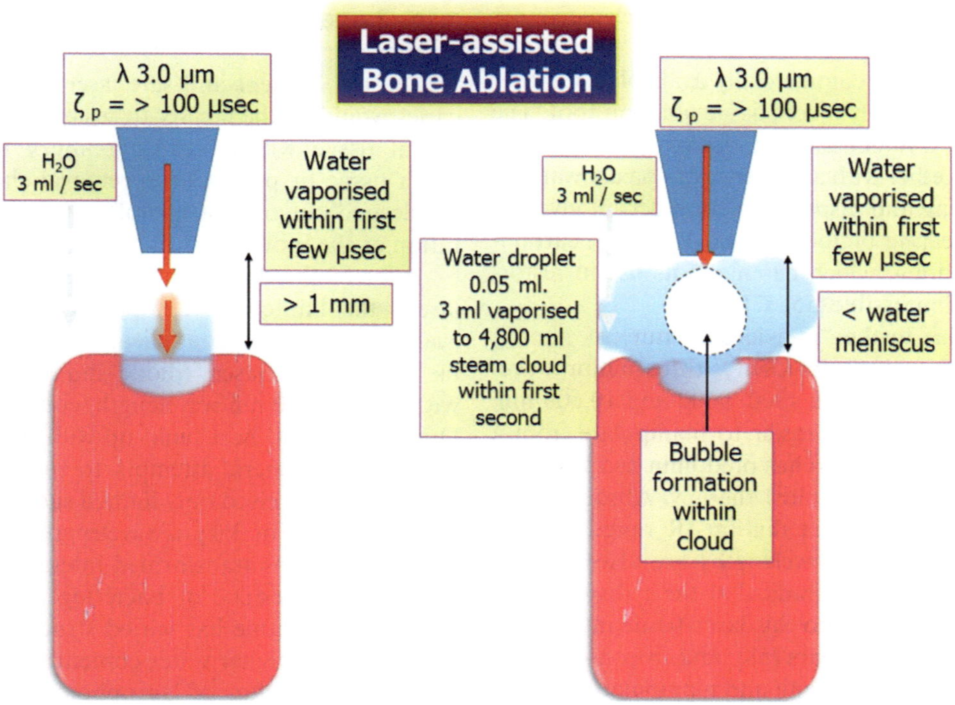

Fig. 3.19 Diagrammatic influence of the tip-to-tissue distance and influence of a contiguous water film, facilitating laser-induced cavitation phenomena. (Adapted S. Parker from: Mir M, Gutknecht N, et al. Lasers Med Sci (2009) 24:365–374 [50])

get would be vaporized during the first >30 µs of each pulse, allowing successive photons to interact with the target. If the tip-to-target distance is greater than the distance wherein the integrity of the water meniscus is maintained, the photon stream will pass through air before interacting with water at the surface of the tissue.

However, in circumstances where the delivery tip is close enough to the tissue surface and due to the surface tension of the water, a continuous envelopment of that distance occurs, the vaporization of the water film happens as before, but the vapor is contained within a rapidly expanding bubble; as it collapses, it gives rise to a cavitation phenomenon, and associated pressure waves may be sufficient (50–100 MPa) to initiate laser-induced "tripsy" (disintegration) of the tooth surface (Fig. 3.19).

This concept is exciting and in common with the similar phenomenon of laser-induced cavitation in water using wavelengths at 3.0 µm that may occur in endodontic and laser-assisted osteotomy procedures. With reference to the study by Mir et al. above [50], the staging can be summarized as follows:

- Energy intensity in the first pulses leads to absorption in the first µm layers of water opposite to the tip.
- Bubble formation with higher output energy density bubble dimensions were not clear, and a cloud-shaped appearance of laser-water interactions was recorded.
- In, for example, 140 µs pulses, after approximately 20–30 µs of the start of the pulse curve, a plume of vapor (comparable with a cloud) covers the tissue surface.
- The suction force exerted by the collapsing bubble and by the impact of the high-velocity jet generated during bubble collapse results in tissue ablation.

It remains to be seen to what extent this contributes to the "classic" understanding of 3.0 µm-mediated hard tissue ablation, but it is worthy of note that superheating of the vapor and hyperbaric pressure phenomena play a part.

In consideration of the ever-changing nature of available technology and its incorporation into laser wavelength choice, a precise irradiation parameter must be chosen in order to avoid collateral damage. This has importance no more than in terms of hard dental morphological damage, such as surface carbonization or cracking, which could produce structural and aesthetic damage and post-operative complications such as transient pulpitis. Moreover, the energy densities used must be safe with regard to pulp and periodontal tissue vitality [51]. Studies have indicated that temperature increments above 5.6 °C can be considered potentially threatening to the vitality of the pulp [52], and increments in excess of 16 °C can result in complete pulpal necrosis [53]. In comparison with rotary instrumentation, pulpal temperature rise is minimal when erbium laser wavelengths are employed in cavity preparation [54].

Until recently, the commercially available CO_2 laser has been predominately a soft tissue ablation tool. The CW and gated CW emission modes of the 10,600 nm wavelength, together with an absence of coaxial water to aid tissue cooling and disperse ablation debris, give rise to rapid overheating of tooth tissue, cracking, carbonization and melting, which has made its use in restorative dentistry impossible [55–57].

However, due to the "four-level" nature of photon generation within the laser cavity, three major wavelength emissions occur—at 9300, 9600 and 10,600 nm. The longer wavelength is easier to manipulate from a technical point of view and has predominated the availability of CO_2 laser in clinical therapy. Absorption in water is a strong feature at this far-IR range, but the shorter 9300 and 9600 wavelengths are also strongly absorbed in the phosphate radical of the hydroxyapatite molecule. Investigation into this laser-tissue interaction has spanned almost 20 years [58], and in consequence, with a shorter CO_2 wavelength and manipulation of the emission to allow microsecond bursts together with a coaxial water spray to minimize heat generation, the interaction is both more positive and clinically acceptable [59].

If pulse durations in the range of 5–20 μs are used, efficient ablation occurs with minimal peripheral thermal damage [60, 61], and this has now resulted in a (for instance) commercially available laser unit emitting at 9300 nm with a specimen pulse duration of 10–15 μs and repetition rate of 300 Hz, demonstrating that enamel and dentine surfaces can be rapidly ablated by such lasers with minimal peripheral thermal and mechanical damage and without excessive heat accumulation [62, 63].

Commercial pressures may dictate the direction and speed of investigation into laser-tissue interaction with dental hard tissue that is based upon concepts of power density and pulse width as predominant factors to minimize collateral thermal damage, as opposed to the pure selection of chromophore-related laser wavelengths. By way of an example, in a most recent published investigation, a diode-pumped, thin-disk femtosecond laser (wavelength 1025 nm, pulse width 400 fs) was used for the ablation of enamel and dentin. Laser fluence, scanning line spacing and ablation depth all significantly affected femtosecond laser ablation efficiency and were predominant in comparison with the intuitively inappropriate choice of a near-IR laser wavelength of 1025 nm [64].

As may be seen with the erbium family of lasers at 3.0 μm, there is also the potential for easy disruption and ablation of composite restorative materials, and this has been the subject of published data [65, 66].

3.10 Laser Interaction with Dental Caries

Key to the efficient and safe laser-assisted removal of caries would be the facility to selectively ablate decayed dental tissue without causing injury to surrounding tooth tissue or pulp. Caries removal should be within the desired outcome of complementary aesthetic restoration of the tooth.

Prior to the development of the erbium family of lasers, the limited laser wavelengths and technology available within dental application meant that only visible and near-IR lasers (diode and Nd:YAG), together with CO_2 10,600 nm wavelength, could be investigated. Due to the rapid heat build-up with the available laser delivery parameters, attempts to remove (pigmented) dental caries received only limited success [67–69].

With the use of 3.0 μm wavelengths, the high absorption in water has transformed interaction. Caries as a demineralised residue of bacteriogenic acid action on enamel and dentine has varied structure but predominately a much higher water content than normal tooth structure. Interaction of laser photonic energy with this material will allow some selective ablation relative to the tooth tissue, and this remains a major advantage of lasers over more conventional rotary instrumentation [70, 71]. Additional positive indications support the development of the new generation of micro-pulsed CO_2 lasers and the ability to utilize the absorption on water at this wavelength. This has been the subject of intense research and investigation by the group at UCSF in San Francisco, USA [72, 73].

3.11 Caries Prevention

Laser-tissue interaction with hard dental tissue may pose difficult challenges relative to the wavelength and operating parameters, and this has been outlined above. Peripheral to the blunt outcome of thermal damage potential has been the careful application of several wavelengths to achieve a thermally mediated change in the carbonated hydroxyapatite structure of enamel, to change the crystal lattice to a more acid-resistant amorphous "glass-like" state. This change has been shown to occur with injudicious laser use on hard tissue, but with care, a number of studies [74–78] have proposed that many laser wavelengths may be manipulated to provide caries resistance in non-diseased teeth.

3.12 Laser-Tissue Interaction with Bone

The structure of osseous tissue of the maxilla and mandible resembles that of dentine in terms of proportional ratios of mineral, protein and water (◘ Table 3.4). Bone

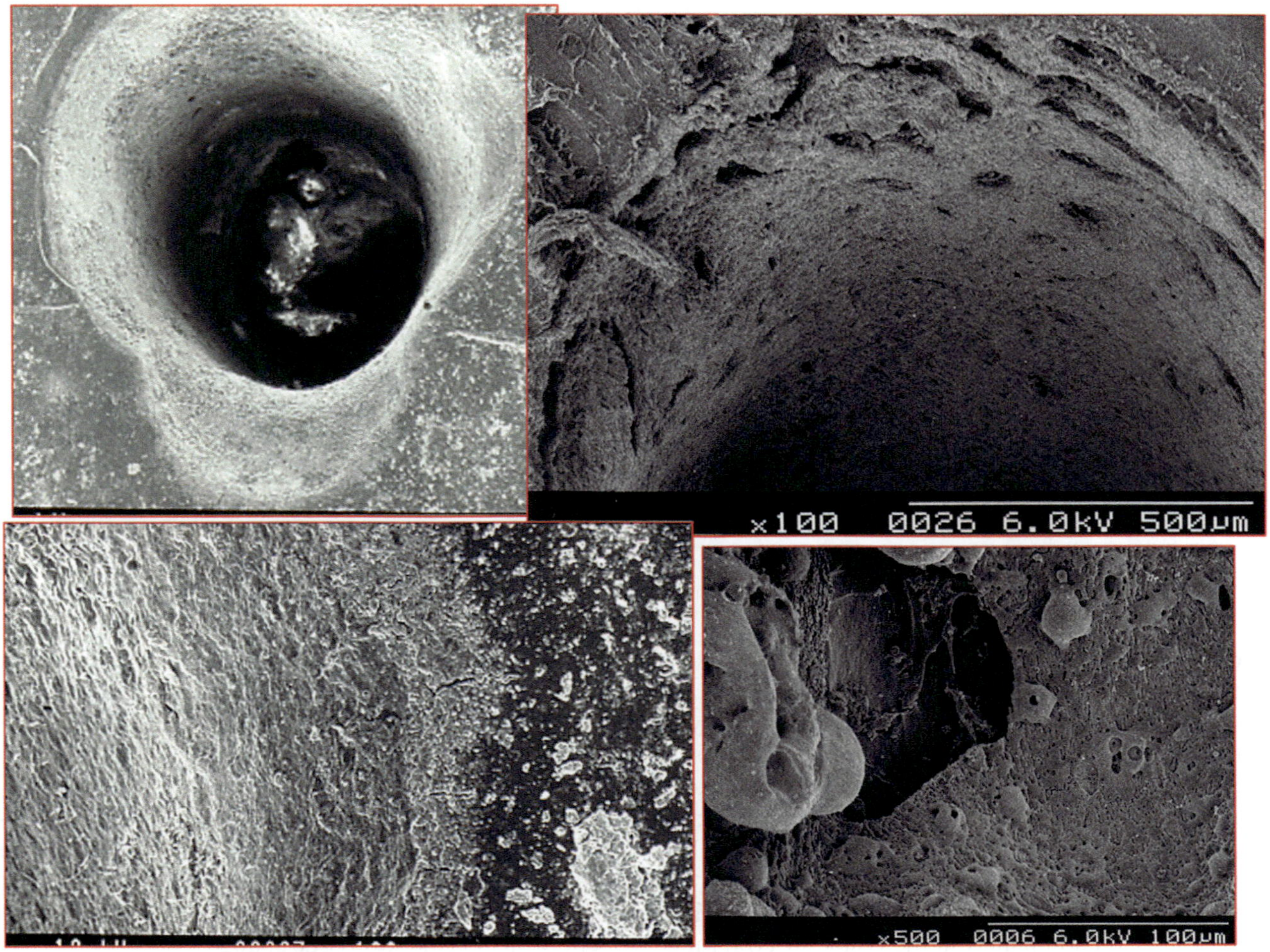

Fig. 3.20 SEM representation of laser interaction with osseous tissue. Top left, right and lower left images relate to ascending magnification of Er:YAG (2940 nm) with porcine rib bone in vitro. The cut is clean with minimal evidence of tissue disruption and thermal damage beyond the cut margin. Compare this to the image at lower right, using Nd:YAG (1064 nm), where apatite melting and large thermally induced voids are visible

is a much more dynamic tissue with reference to cell activity and turnover, compared to tooth tissue, and care must be observed to respect the potential for disruptive consequences of using inappropriate laser parameters. Although early reports of supportive use of CO_2 (10,600 nm) laser wavelength in surgical bone management are recorded [79], the potential for photothermolysis and collateral damage remains high. Laser ablation of bone with erbium laser wavelengths (2780 and 2940 nm) with high absorption in water defines a level of selective ablation through the vaporization of water and tissue structure disruption, and in this way, laser ablation of bone proceeds in a similar fashion to that seen in laser-mediated tooth tissue ablation. The higher water content and lower density of bone compared to enamel allow faster cutting, through dislocation of hydroxyapatite and cleavage of the collagen matrix (Fig. 3.20). This ease of cutting places the use of Er:YAG and Er,Cr:YSGG laser wavelengths as the preferred choice for laser bone ablation when compared to other wavelengths [80].

3.13 Laser-Tissue Photofluorescence

In an earlier section of this chapter, it was defined that fluorescence, as a form of sub-ablative laser-tissue interaction, is a luminescence or re-emission of light in which the molecular absorption of a photon triggers the emission of another photon with a longer wavelength. In absorbing incident photonic energy, some of that energy is expended and the difference between the absorbed and emitted photons ends up as molecular vibrations or heat. With re-emission, that energy loss is seen as a longer wavelength. This event is governed by the biophysical nature of tissue molecules involved (*termed fluorophores*) and as such can be the basis for optical scanning techniques used in caries detection in enamel

3

■ **Table 3.5** Common fluorophores found within the oral cavity and dentistry. Each fluorophore is capable of excitation at specific light wavelength, and corresponding re-emission measurement can prove helpful in differential diagnosis of tissue change

Autofluorescence wavelength?			
Fluorophore	**Excitation (nm)**	**Fluorescence peak**	**Comments**
Tryptophan	275	350	Protein
Collagen	335	390	Connective tissue (CT)
Elastin	360	410	CT
Keratin	370	505	Surface analysis
Porphyrins	405, 630	590, 625, 635, 705	Cell mitochondria/metallo-, copro-, proto-porphyrins
Healthy enamel	405	533	
Caries	405, 488, 655	580–700	
Inorganic composites	655	Mean fluorescence intensity closely matched to healthy enamel	
GI composites	655	Mean fluorescence intensity closely matched to carious enamel	

Source: Kim A, Roy M, et al. J Biomed Opt. 2010;15(6):066026 [81]

and dentine and tomographic techniques in the scanning of soft tissue for neoplastic change.

The oral cavity provides substantial opportunity for scanning and fluorescence techniques, due to the ease of access of oral structures and the database of the excitation and emission wavelengths of individual tissue elements as well as non-biologic materials that may have use in dentistry [81]. In ■ Table 3.5, it may be seen that through the choice of precise monochromatic laser wavelengths, predominately in the visible spectral range, the resultant re-emission would help to provide analysis of the target composition. A specific example might be that whereas healthy enamel exposed to a blue incident irradiation re-emits as a green color, the presence of porphyrin (pigment component of dental caries) re-emits at a longer red-brown color and would allow differentiation diagnosis of dental caries to be made.

Fluorescent and photodynamic diagnosis may provide screening facility or part of a hierarchical series of tissue investigation and must be regarded as an adjunct to a range of investigations—direct visual, microscopic and histologic examination, and genetic analysis—to provide support to the clinician, especially within the field of soft tissue health screening [82].

In oral soft tissue structure, disease changes the concentration of the fluorophores as well as the light scattering and absorption properties of the tissue, due to changes in blood concentration, collagen content and epithelial thickness. Such effects may be seen as follows:

- Recorded fluorescence signal will be lower in the case of hyperplasia—the epithelial layer shields the strongly fluorescent collagen layer.
- Excessive keratin production by lesions may produce an increase in autofluorescence intensity.
- Cell metabolism may increase with malignant changes, which changes the balance between fluorescent NADH (increase) and non-fluorescent NAD+ (decrease).

Many studies have been performed to investigate photodynamic diagnosis and fluorescence techniques in the oral cavity. These studies may be grouped in an attempt to address specific criteria of relevance in clinical assessment of neoplastic soft tissue change:

1. Whether autofluorescence imaging is capable of providing a higher contrast between a lesion and healthy tissue than white light or tactile and visual inspection? This is certainly the case for flat, early lesions [83, 84].
2. Whether autofluorescence imaging is helpful in differentiating between different lesion types, in particular between benign, dysplastic and malignant lesions? Overall, the specificity of autofluorescence imaging for distinguishing (pre) malignant from benign lesions does not seem to be very promising [85, 86].
3. The detection of unknown lesions and unknown extensions of known lesions, which would be useful for tumor demarcation: Indications have indeed been found that autofluorescence imaging is capable of detecting invisible lesions or invisible tumor extensions [87, 88].

Autofluorescence imaging might be appropriate as an easy-to-use, sensitive and inexpensive method for lesion detection, although further research is still necessary. In general, autofluorescence imaging may give good results for the distinction of lesions from normal mucosa. However, suspect lesions of the oral mucosa must be subjected to biopsy and other investigations, and certainly, it is inappropriate to place autofluorescence investigation in any role other than as an adjunctive scanning technique. If possible, autofluorescence spectroscopy could be used to find the optimal, most dysplastic location for biopsy, although the literature shows that autofluorescence is not specific enough for this purpose.

An allied area of laser-tissue interaction and spectroscopic analysis of re-emission is Raman scattering. This is a special, very weak form of light scattering in which energy is lost or gained to a molecule through a phenomenon known as inelastic scattering, where the frequency of photons in monochromatic light changes upon interaction with a tissue sample under investigation. The frequency (frequency and photon energy are inversely proportional) of the re-emitted photons is shifted up or down in comparison with original monochromatic frequency, and this is called the Raman effect. This shift provides information about vibrational, rotational and other low-frequency transitions in molecules.

Raman spectroscopy can be used to study solid, liquid and gaseous samples. Vibrational information is specific to the chemical bonds and symmetry of molecules. Therefore, it provides a fingerprint by which the molecule can be identified and has an important role within the area of tissue photo-analysis as it impacts disease and health.

3.14 Laser-Tissue Interaction and Photobiomodulation

Photobiomodulation (PBM) is the manipulation of cellular behavior using low-intensity light sources and the delivery of laser therapy (application of photonic energy at specific wavelengths) to induce a biological response through energy transfer. Sub-ablative photonic energy delivered into the tissue modulates biological processes within that tissue and within the biological system of which that tissue is a part. Key to the limits of benefit is the restriction of laser operating parameters to ensure that PBM has no appreciable thermal effect in irradiated tissue.

Phototherapy is characterized by its ability to induce photo-biological processes in cells [89]. This conforms to the first law of photobiology (light absorption by specific molecular chromophores). There is a so-called optical window in tissue (approx. 650–1100 nm), where the effective tissue penetration of light is maximized. The use of PBM therapy in patients almost exclusively involves red and near-infrared light (600–1100 nm) [90].

The absorption and scattering of light in tissue are both much higher in the blue region of the spectrum than the red. The principal tissue chromophores (hemoglobin and melanin) have high absorption bands at shorter wavelengths, tissue scattering of light is higher at shorter wavelengths and water strongly absorbs infrared light at wavelengths >1100 nm.

Wavelengths in the 600–700 nm range are chosen for treating superficial tissue, and those between 780 and 950 nm are chosen for deeper seated tissues, due to longer optical penetration distances through tissue with the latter group. Beam coherence is maintained as the laser beam penetrates the tissue and along with polarization may be an important factor in allowing the laser to effectively treat deeper tissues.

Incident photons of wavelengths as referenced above are absorbed into mitochondria and cell membranes of the target cells. Photonic energy is incorporated into a molecule to increase kinetic energy, activate or deactivate enzymes or alter physical or chemical properties of main macromolecules.

Growth factor response within cells and tissue may be seen as a result of increased ATP and protein synthesis [91], change in cell membrane permeability to Ca^{2+} uptake and cell proliferation, and overall, a cascade of metabolic "downstream" effects results in physiological changes resulting in improved tissue repair, faster resolution of the inflammatory response and a reduction in pain [92].

In summary, the results of laser-tissue interaction that promote PBM may be seen within three clinical areas of benefit—anti-inflammatory effects [93], analgesic and pain suppression effects [94] and effects that promote healing in the irradiated tissue [95].

Investigation into pain response during surgical laser use has revealed findings that are inconsistent with many anecdotal reports and may provide an opportunity for the essential subjective aspects of patient receptiveness to be accepted. Pain is a defence mechanism, and pain perception is innate and subjective. Equally, all stimuli applied to excess will result in pain.

Pain perception is multi-factorial and may be influenced through the following, either singularly or in combination:

- Emotion: fear, anxiety, stress syndrome, excitement
- Awareness: trust, previous experience, conditioning, e.g., hypnosis, activity subordination
- Threshold potential: age, infirmity, drugs, alcohol, social factors

The avoidance of pain during restorative and dental surgical procedures remains a strong factor in promoting patient acceptance of treatment, and many studies have been carried out to evaluate this in terms of laser-tissue interaction [96–98]. The use of the Nd:YAG (1064 nm) laser in developing pulpal analge-

3

sia, possibly through interference with the "gate theory" of neural stimulus propagation, was an early mainstay benefit of this laser following its launch into dental practice in 1990. However, investigation into the subjectivity or placebo effect has rendered its application inconsistent [99–101]. Chaiyavej et al. found that Er:YAG laser use, similar to rotary bur cutting of tooth tissue, caused neural response in both A and C intra-dental fibers [102].

Perhaps of greater significance in exploring this area may be the lack of tactile and thermal stimulation compared to rotary instrumentation during laser-assisted restorative dentistry. In seeking to understand the essentially anecdotal reports of soft tissue surgery using laser photonic energy in what is a thermally based interchange, there is the patient-centered factor of trust in the operator, together with a possible harmonization of micro-pulsed free-running emissions with regeneration potential of acetylcholine at synaptic junctions within the sensory neurone.

It is without question that when used correctly and with recommended operating parameters to maximize laser-tissue interaction, laser-assisted surgical procedures on soft and hard tissue are less physically injurious when compared to both scalpel and rotary bur. Patient acceptance, peer pressure and a general acceptance of "hi-tech" approach to treatment may all propose an enhancement of tolerance of sensory stimulation.

3.15 Conclusion

An overview has been presented to explore the physical and biological aspects of interaction of laser photonic energy with target oral hard and soft tissue. Any inconsistency may be viewed in terms of the precise mechanisms governing molecular energy dynamics but also the great diversity in tissue types and their close approximation within the oral cavity. The clinician is faced with technical challenges in manipulating any given laser wavelength to employ it as broadly as possible during clinical procedures; the additional facility of power density phenomena in helping to initiate an essentially photothermal event may help to deliver predictable and precise surgical outcomes. In ◘ Fig. 3.21, it is possible to appreciate the complex yet interconnected relationship that exists between incident laser choice and operating parameters, the consequent gradient of effects that result from such choice and how the laser-tissue effects so produced can represent a breadth of clinical application— in terms of both therapy and diagnosis. The photon distribution and concentration (fluence) delivered will influence the degree of (essentially) photothermal effect and consequent reaction of the target tissue. At low fluences, the benefits can be seen as diagnostic value and PBM (photochemical) effects; at higher fluences (with the same laser wavelength or another), the photothermal interaction predominates with ablative and per-

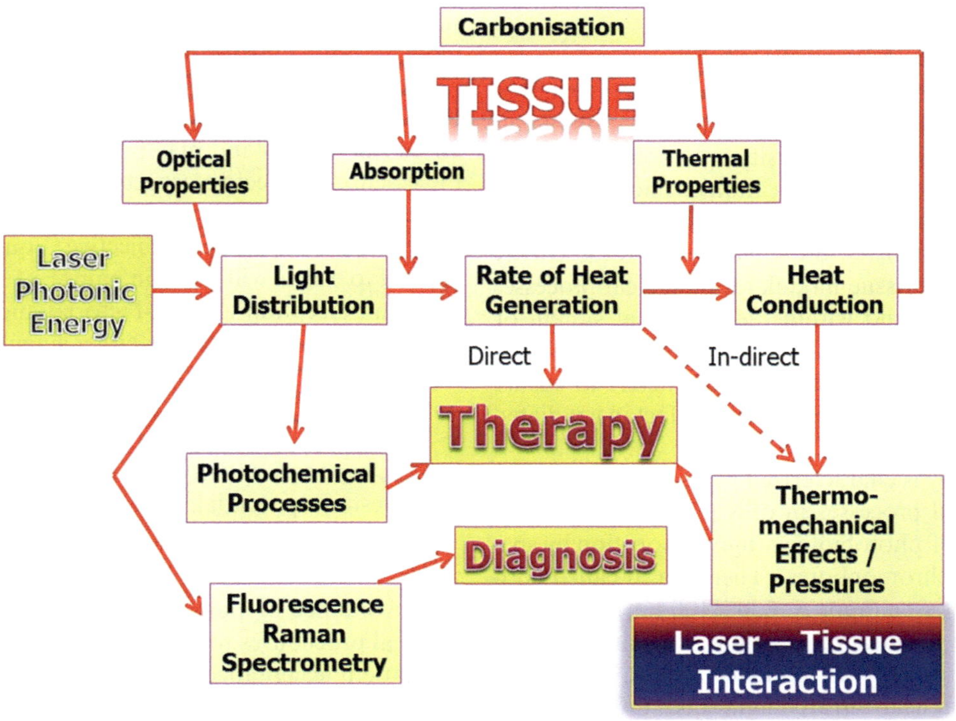

◘ **Fig. 3.21** Summary of laser-tissue interaction to demonstrate the breadth of application, relative to photon delivery manent structural tissue change. The overload of fluence values may give rise to destructive collateral effects such as carbonization and consequent changes in tissue optical properties and absorption potential.

In summary, laser photonic energy offers a degree of "purity" through empirical properties—wave coherence and single (mono) wavelength. Unique wavelength value confers predictable photonic EM energy value through an inverse proportional relationship.

All matter has constituent atomic and molecular energy, consistent with intra-, inter- and extra-atomic and molecular binding forces. For any system, the "resting" gross energy value determines a ground-state physical form (solid, gas, liquid) relative to temperature.

Predictable ("pure") interaction with tissue can only occur if incident energy is absorbed by the tissue although levels of laser photonic energy may be viewed as being largely insufficient to overcome interatomic covalent or intra-lattice ionic binding forces within target tissue.

A better explanation might emerge, based on chromophore absorption of photonic energy leading to temperature rise within the system, and based on such assumption, laser photothermolytic interaction with tissue is mainly due to the indirect consequences of the conversion of EM photonic energy into thermal energy.

Photoacoustic and photochemical effects may be viewed as further consequential effects of primary photothermolysis.

References

1. Newton I. Opticks: or, a treatise of the reflections, refractions, inflexions and colours of light. 1702.
2. Young T. Experimental demonstration of the general law of the interference of light. Philos Trans R Soc Lond. 1804;94:1–16.
3. Einstein A. Zur Quantentheorie der Strahlung. Physiol Z. 1917;18:121–8.
4. Planck M. On the law of distribution of energy in the normal spectrum. Ann Phys. 1901;4:553 ff.
5. Boltzmann L. Ableitung des Stefan'schen Gesetzes, betreffend die Abhängigkeit der Wärmestrahlung von der Temperatur aus der electromagnetischen Lichttheorie. Ann Phys Chem. 1884;22:291–4.
6. www.thefreedictionary.com/chromophore.
7. Knappe V, Frank F, Rohde E. Principles of lasers and biophotonic effects. Photomed Laser Surg. 2004;22:411–7.
8. Parker S. Laser tissue interaction. Br Dent J. 2007;202:73–81.
9. Karu T. Primary and secondary mechanisms of action of visible to near-IR radiation on cells. J Photochem Photobiol B. 1999;49(1):1–17.
10. Kwiatkowski S, Knap B, Przystupski D, et al. Photodynamic therapy—mechanisms, photosensitizers and combinations. Biomed Pharmacother. 2018;106:1098–107. https://doi.org/10.1016/j.biopha.2018.07.049.
11. Warrier A, Mazumder N, Prabhu S, Satyamoorthy K, Murali TS. Photodynamic therapy to control microbial biofilms. Photodiagn Photodyn Ther. 2021;33:102090. https://doi.org/10.1016/j.pdpdt.2020.102090.
12. Hamblin MR, Demidove TN. Mechanisms of low level light therapy. In: Hamblin MR, Waynant RW, Anders J, editors. Mechanisms for low-light therapy, January 22 and 24, 2006, San Jose, CA, Proc. SPIE, vol. 6140. Bellingham, WA: SPIE—The International Society for Optical Engineering; 2006. p. 614001-1–614001-12.
13. Mó O, Yáñez M, et al. Periodic trends in bond dissociation energies. A theoretical study. J Phys Chem A. 2005;109(19):4359–65.
14. Zhang D, Tamilselvan A. Lattice energy and mechanical stiffness of hydroxyapatite. J Mater Sci Mater Med. 2007;18(1):79–87.
15. Brown WS, Dewey WA, Jacobs HR. Thermal properties of teeth. J Dent Res. 1970;49(4):752755.
16. Parker S, Cronshaw M, Anagnostaki E, Mylona V, Lynch E, Grootveld M. Current concepts of laser-oral tissue interaction. Dent J (Basel). 2020;8(3):61. https://doi.org/10.3390/dj8030061.
17. Frank F, Hessel S, Kramp C. Ch. 1–4.3: Optical applications. In: Berlien HP, Müller G, editors. Applied laser medicine. Berlin: Springer-Verlag; 2003. p. 161.
18. Dompe C, Moncrieff L, Matys J, Grzech-Leśniak K, et al. Photobiomodulation-underlying mechanism and clinical applications. J Clin Med. 2020;9(6):1724. https://doi.org/10.3390/jcm9061724.
19. Dederich DN. Laser/tissue interaction. Alpha Omegan. 1991;84(4):33–6.
20. Fisher JC. Photons, physiatrics, and physicians: a practical guide to understanding laser light interaction with living tissue, part I. J Clin Laser Med Surg. 1992;10:419–26.
21. Boulnois J-L. Photophysical processes in recent medical laser developments: a review. Laser Med Sci. 1986;1:47–66.
22. Schrötzlmair F, Müller M, et al. Laser lithotripsy of salivary stones: correlation with physical and radiological parameters. Lasers Surg Med. 2015;47(4):342–9.
23. Rai V, Walvekar RR, Verma J, Monga U, Rai D, Munjal M. Laser-assisted sialolithotripsy: a correlation of objective and subjective outcomes. Laryngoscope. 2022;132(12):2344–9. https://doi.org/10.1002/lary.30106.
24. Kumar YR. Definitions in laser technology. J Cutan Aesthet Surg. 2009;2(1):45–6.
25. Partovi F, Izatt JA, et al. A model for thermal ablation of biological tissue using laser radiation. Lasers Surg Med. 1987;7(2):141–54.
26. Dederich DN. Laser/tissue interaction: what happens to laser light when it strikes tissue? J Am Dent Assoc. 1993;124:57–61.
27. Ball KA. Lasers: the perioperative challenge. 2nd ed. St Louis: Mosby-Year Book; 1995. p. 14–7.
28. Parker S. Laser tissue interaction. Br Dent J. 2007;202:76.
29. Fisher JC. Photons, physiatrics, and physicians: a practical guide to understanding interaction of laser light with living tissue: part II: basic mechanisms of tissue destruction by laser beams. J Clin Laser Med Surg. 1993;11(6):291–303.
30. Lingamaneni S, Mandadi LR, Pathakota KR. Assessment of healing following low-level laser irradiation after gingivectomy operations using a novel soft tissue healing index: a randomized, double-blind, split-mouth clinical pilot study. J Indian Soc Periodontol. 2019;23(1):53–7. https://doi.org/10.4103/jisp.jisp_226_18.
31. Inchingolo F, Tatullo M, et al. Comparison between traditional surgery, CO_2 and Nd:YAG laser treatment for generalized gingival hyperplasia in Sturge-Weber syndrome: a retrospective study. J Investig Clin Dent. 2010;1(2):85–9.

32. Tamarit-Borrás M, Delgado-Molina E, et al. Removal of hyperplastic lesions of the oral cavity. A retrospective study of 128 cases. Med Oral Patol Oral Cir Bucal. 2005;10(2):151–62.
33. Parrish JA, Deutsch TF. Laser photomedicine. IEEE J Quantum Electron. 1984;QE-20(12):1386–96.
34. Karu T. Mechanisms of interaction of monochromatic visible light with cells. In: Proc. SPIE, vol. 2639. Bellingham, WA: SPIE—The International Society for Optical Engineering; 1996. p. 2–9.
35. Fisher J. In: Shapshay SM, editor. Endoscopic laser surgery handbook. Dekker; 1987. p. 109, Fig. 29.
36. Vo-Dinh T, editor. Biomedical photonics handbook. Boca Raton, FL: CRC Press; 2014. ISBN: 0-8493-1116-0.
37. Kaminer R, Liebow C, Margarone JE 3rd, Zambon JJ. Bacteremia following laser and conventional surgery in hamsters. J Oral Maxillofac Surg. 1990;48:45–8.
38. Ratre MS, Chaudhari PA, Khetarpal S, Kumar P. Effective management of focal reactive gingival overgrowths by diode laser: a review and report of two cases. Laser Ther. 2019;28(4):291–7. https://doi.org/10.5978/islsm.19-CR-03.
39. Fisher SE, Frame JW, Browne RM, Tranter RM. A comparative histological study of wound healing following CO2 laser and conventional surgical excision of canine buccal mucosa. Arch Oral Biol. 1983;28:287–91.
40. Li B, Aspden R. Composition and mechanical properties of cancellous bone from the femoral head of patients with osteoporosis or osteoarthritis. J Bone Miner Res. 1997;12(4):641–51.
41. Schroeder HE. Oral structure biology: embryology, structure, and function of normal hard and soft tissues of the oral cavity and temporomandibular joints. G. Thieme Verlag; 1991.
42. Corrêa-Afonso A, Bachmann L, et al. FTIR and SEM analysis of CO2 laser irradiated human enamel. Arch Oral Biol. 2012;57(9):1153–8.
43. Wigdor H, Abt E, Ashrafi S, Walsh JT Jr. The effect of lasers on dental hard tissues. J Am Dent Assoc. 1993;124:65–70.
44. Seka W, Featherstone JDB, Fried D, Visuri SR, Walsh JT. Laser ablation of dental hard tissue: from explosive ablation to plasma-mediated ablation. In: Wigdor HA, Featherstone JDB, White JM, Neev J, editors. Lasers in dentistry II, vol. 2672. Bellingham, WA: SPIE; 1996. p. 144158.
45. Heyder M, Sigusch B, Hoder-Przyrembel C, Schuetze J, Kranz S, Reise M. Clinical effects of laser-based cavity preparation on class V resin-composite fillings. PLoS One. 2022;17(6):e0270312. https://doi.org/10.1371/journal.pone.0270312.
46. Tuna EB, Ozel E, Kasimoglu Y, Firatli E. Investigation of the Er: YAG laser and diamond bur cavity preparation on the marginal microleakage of class V cavities restored with different flowable composites. Microsc Res Tech. 2017;80(5):530–6. https://doi.org/10.1002/jemt.22827.
47. Walsh JT Jr, Cummings JP. Effect of the dynamic optical properties of water on mid-infrared laser ablation. Lasers Surg Med. 1994;15:295–305.
48. Hoke JA, Burkes EJ Jr, Gomes ED, Wolbarsht ML. Erbium:YAG (2.94 mum) laser effects on dental tissues. J Laser Appl. 1990;2:61–5.
49. Fried D, Ashouri N, Breunig T, Shori R. Mechanism of water augmentation during IR laser ablation of dental enamel. Lasers Surg Med. 2002;31:186–93.
50. Mir M, Gutknecht N, et al. Visualising the procedures in the influence of water on the ablation of dental hard tissue with erbium:yttrium–aluminium–garnet and erbium, chromium:yttrium–scandium–gallium-garnet laser pulses. Lasers Med Sci. 2009;24:365–74.
51. Ana PA, Blay A, Miyakawa W, Zezell DM. Thermal analysis of teeth irradiated with Er,C:YSGG laser at low fluences. Laser Phys Lett. 2007;4:827–30.
52. Zach L, Cohen G. Pulp response to externally applied heat. Oral Surg. 1965;19:515–30.
53. Baldissara P, Catapano S, Scotti R. Clinical and histological evaluation of thermal injury thresholds in human teeth: a preliminary study. J Oral Rehabil. 1997;24:791801.
54. Rizoiu I, Kohanghadosh F, Kimmel AI, Eversole LR. Pulpal thermal responses to an erbium,chromium:YSGG pulsed laser hydrokinetic system. Oral Surg Oral Med Oral Pathol Oral Radiol Endod. 1998;86:220–3.
55. Malmstrom HS, McCormack SM, Fried D, Featherstone JD. Effect of CO2 laser on pulpal temperature and surface morphology: an in vitro study. J Dent. 2001;29:521–9.
56. Watanabe I, Lopes RA, Brugnera A, Katayama AY, Gardini AE. Effect of CO2 laser on class V cavities of human molar teeth under a scanning electron microscope. Braz Dent J. 1996;7:27–31.
57. Friedman S, Liu M, Izawa T, Moynihan M, Dorscher-Kim J, Kim S. Effects of CO2 laser irradiation on pulpal blood flow. Proc Finn Dent Soc. 1992;88(Suppl 1):167–71.
58. Takahashi K, Kimura Y, Matsumoto K. Morphological and atomic analytical changes after CO_2 laser irradiation emitted at 9.3 microns on human dental hard tissues. J Clin Laser Med Surg. 1998;16(3):167–73.
59. Anton Y, Otero CI, Bortolotto T, DiBella E, Krejci I. Influence of 9.3 μm CO2 and Er:YAG laser preparations on marginal adaptation of adhesive mixed class V composite restorations with one component universal adhesive. Am J Dent. 2021;34(1):31–8.
60. Fan K, Bell P, Fried D. The rapid and conservative ablation and modification of enamel, dentin and alveolar bone using a high repetition rate TEA CO_2 laser operating at $\lambda = 9.3$ μm. J Biomed Opt. 2006;11(6):064008.
61. Mullejans R, Eyrich G, Raab WH, Frentzen M. Cavity preparation using a super-pulsed 9.6-microm CO2 laser—a histological investigation. Lasers Surg Med. 2002;30:331–6.
62. Nguyen D, Chang K, et al. High-speed scanning ablation of dental hard tissues with a $\lambda = 9.3$ μm CO2 laser: adhesion, mechanical strength, heat accumulation, and peripheral thermal damage. J Biomed Opt. 2011;16(7):071410.
63. Tom H, Chan K, Darling C, Fried D. Near-IR image-guided laser ablation of demineralization on tooth occlusal surfaces. Lasers Surg Med. 2016;48(1):52–61.
64. Chen H, Li H. Femtosecond laser for cavity preparation in enamel and dentin: ablation efficiency related factors. Sci Rep. 2016;6:20950.
65. Cardoso M, Coelho A, Lima R, Amaro I, Paula A, Marto CM, Sousa J, Spagnuolo G, Marques Ferreira M, Carrilho E. Efficacy and patient's acceptance of alternative methods for caries removal—a systematic review. J Clin Med. 2020;9(11):3407. https://doi.org/10.3390/jcm9113407.
66. Prabhakar A, Lokeshwari M, Naik SV, Yavagal C. Efficacy of caries removal by Carie-Care and erbium-doped yttrium aluminum garnet laser in primary molars: a scanning electron microscope study. Int J Clin Pediatr Dent. 2018;11(4):323–9. https://doi.org/10.5005/jp-journals-10005-1533.
67. Morrant GA. Lasers: an appraisal of their possible use in dentistry. Dent Pract Dent Rec. 1965;16(1):5–9.
68. Melcer J, Chaumette MT, Melcer F, et al. Treatment of dental decay by CO2 laser beam: preliminary results. Lasers Surg Med. 1984;4(4):311–21.
69. Myers TD, Myers WD. The use of a laser for debridement of incipient caries. J Prosthet Dent. 1985;53(6):776–9.

70. Aoki A, Ishikawa I, Yamada T, et al. Comparison between Er:YAG laser and conventional technique for root caries treatment in vitro. J Dent Res. 1998;77(6):1404–14.
71. Eberhard J, Eisenbeiss AK, Braun A, Hedderich J, Jepsen S. Evaluation of selective caries removal by a fluorescence feedback-controlled Er:YAG laser in vitro. Caries Res. 2005;39(6):496–504.
72. Chan KH, Tom H, Darling CL, Fried D. Serial removal of caries lesions from tooth occlusal surfaces using near-IR image-guided IR laser ablation. Proc SPIE Int Soc Opt Eng. 2015 Feb;24:9306.
73. Chung LC, Tom H, Chan KH, Simon JC, Fried D, Darling CL. Image-guided removal of occlusal caries lesions with a λ= 9.3-μm CO_2 laser using near-IR transillumination. Proc SPIE Int Soc Opt Eng. 2015;9306:93060N.
74. Myers TD, McDaniel JD. The pulsed Nd:YAG dental laser: review of clinical applications. J Calif Dent Assoc. 1991;19(11):25–30.
75. Featherstone JD. Lasers in dentistry 3. The use of lasers for the prevention of dental caries. Ned Tijdschr Tandheelkd. 2002;109(5):162–7.
76. Valério RA, Rocha CT, Galo R, Borsatto MC, Saraiva MC, Corona SA. CO2 laser and topical fluoride therapy in the control of caries lesions on demineralized primary enamel. Sci World J. 2015;2015:547569.
77. Mohan AG, Ebenezar AV, Ghani MF, Martina L, Narayanan A, Mony B. Surface and mineral changes of enamel with different remineralizing agents in conjunction with carbon-dioxide laser. Eur J Dent. 2014;8(1):118–23.
78. Vieira KA, Steiner-Oliveira C, Soares LE, et al. In vitro evaluation of enamel demineralization after several overlapping CO2 laser applications. Lasers Med Sci. 2015;30(2):901–7.
79. Panossian B, Lacau Saint-Guily J, et al. Effects of the CO2 laser on mandibular bone. Preliminary study. Ann Otolaryngol Chir Cervicofac. 1984;101(8):653–5.
80. Noba C, Mello-Moura ACV, Gimenez T, Tedesco TK, Moura-Netto C. Laser for bone healing after oral surgery: systematic review. Lasers Med Sci. 2018;33(3):667–74. https://doi.org/10.1007/s10103-017-2400-x.
81. Giovannacci I, Magnoni C, Vescovi P, Painelli A, Tarentini E, Meleti M. Which are the main fluorophores in skin and oral mucosa? A review with emphasis on clinical applications of tissue autofluorescence. Arch Oral Biol. 2019;105:89–98. https://doi.org/10.1016/j.archoralbio.2019.07.001.
82. Meleti M, Giovannacci I, Vescovi P, Pedrazzi G, Govoni P, Magnoni C. Histopathological determinants of autofluorescence patterns in oral carcinoma. Oral Dis. 2020;26:1185. https://doi.org/10.1111/odi.13304.
83. Betz C, et al. A comparative study of normal inspection, autofluorescence and 5-ALA-induced PPIX fluorescence for oral cancer diagnosis. Int J Cancer. 2002;97:245–52.
84. Ayoub HM, Newcomb TL, et al. The use of fluorescence technology versus visual and tactile examination in the detection of oral lesions: a pilot study. J Dent Hyg. 2015;89(1):63–71.
85. Kulapaditharom B, Boonkitticharoen V. Performance characteristics of fluorescence endoscope in detection of head and neck cancers. Ann Otol Rhinol Laryngol. 2001;110:45–52.
86. Kraft M, Betz CS, et al. Value of fluorescence endoscopy for the early diagnosis of laryngeal cancer and its precursor lesions. Head Neck. 2011;33(7):941–8.
87. Rück A, Hauser C, et al. Spectrally resolved fluorescence lifetime imaging to investigate cell metabolism in malignant and nonmalignant oral mucosa cells. J Biomed Opt. 2014;19(9):96005.
88. Kolli V, et al. Native cellular fluorescence can identify changes in epithelial thickness in-vivo in the upper aerodigestive tract. Am J Surg. 1995;170:495–8.
89. Karu TI, Kolyakov SF. Exact action spectra for cellular responses relevant to phototherapy. Photomed Laser Surg. 2005;23:355–61.
90. Lipko NB. Photobiomodulation: evolution and adaptation. Photobiomodul Photomed Laser Surg. 2022;40(4):213–33. https://doi.org/10.1089/photob.2021.0145.
91. Amaroli A, Parker S, Dorigo G, Benedicenti A, Benedicenti S. Paramecium: a promising non-animal bioassay to study the effect of 808 nm infrared diode laser photobiomodulation. Photomed Laser Surg. 2015;33(1):35–40.
92. Hamblin MR. Mechanisms and applications of the anti-inflammatory effects of photobiomodulation. AIMS Biophys. 2017;4(3):337–61. https://doi.org/10.3934/biophy.2017.3.337.
93. Glass GE. Photobiomodulation: a review of the molecular evidence for low level light therapy. J Plast Reconstr Aesthet Surg. 2021;74(5):1050–60. https://doi.org/10.1016/j.bjps.2020.12.059.
94. Cronshaw M, Parker S, Anagnostaki E, Mylona V, Lynch E, Grootveld M. Photobiomodulation and oral mucositis: a systematic review. Dent J (Basel). 2020;8(3):87. https://doi.org/10.3390/dj8030087.
95. Parker S, Cronshaw M, Anagnostaki E, Bordin-Aykroyd SR, Lynch E. Systematic review of delivery parameters used in dental photobiomodulation therapy. Photobiomodul Photomed Laser Surg. 2019;37(12):784–97. https://doi.org/10.1089/photob.2019.4694.
96. Poli R, Parker S, Anagnostaki E, Mylona V, Lynch E, Grootveld M. Laser analgesia associated with restorative dental care: a systematic review of the rationale, techniques, and energy dose considerations. Dent J (Basel). 2020;8(4):128. https://doi.org/10.3390/dj8040128.
97. Martins IP, Martins RP, Caldas SGFR, Dos Santos-Pinto A, Buschang PH, Pretel H. Low-level laser therapy (830 nm) on orthodontic pain: blinded randomized clinical trial. Lasers Med Sci. 2019;34(2):281–6. https://doi.org/10.1007/s10103-018-2583-9.
98. Vahdatinia F, Gholami L, Karkehabadi H, Fekrazad R. Photobiomodulation in endodontic, restorative, and prosthetic dentistry: a review of the literature. Photobiomodul Photomed Laser Surg. 2019;37(12):869–86. https://doi.org/10.1089/photob.2019.4707.
99. Whitters CJ, Hall A, Creanor SL, et al. A clinical study of pulsed Nd:YAG laser-induced pulpal analgesia. J Dent. 1995;23:145–50.
100. Orchardson R, Whitters CJ. Effect of HeNe and pulsed Nd:YAG laser irradiation on intradental nerve responses to mechanical stimulation of dentine. Lasers Surg Med. 2000;26:241–9.
101. Orchardson R, Peacock JM, Whitters CJ. Effect of pulsed Nd:YAG laser radiation on action potential conduction in isolated mammalian spinal nerves. Lasers Surg Med. 1997;21:142–8.
102. Chaiyavej S, Yamamoto H, Takeda A, Suda H. Response of feline intra-dental nerve fibers to tooth cutting by Er:YAG laser. Lasers Surg Med. 2000;27:341–9.

Laser Operating Parameters for Hard and Soft Tissue, Surgical and PBM Management

Wayne Selting

Contents

D. J. Coluzzi, S. P. A. Parker (eds.), *Lasers in Dentistry—Current Concepts*, Textbooks in Contemporary Dentistry,
https://doi.org/10.1007/978-3-031-43338-2_4

Core Message

New technologies introduced into clinical dentistry in recent years have added immeasurably to the quality of care that may be provided. Lasers, dental implants, CAD/CAM, and motorized endodontics have all improved clinical outcomes but require a significant investment in hardware and, most importantly, education to understand concepts and protocols. As with all medical instrumentation, it is not enough to follow basic guidelines or "preset" parameters in approaching each patient situation. A deep understanding of the technology, how it interacts with the patient's tissues, and what variables are important to consider are necessary for a successful clinical outcome.

Of the four technologies mentioned above, lasers are, perhaps, the easiest to abuse. Without basic education, it is nearly impossible for a clinician to accomplish the first successful case of implant placement, CAM/CAD crown fabrication, or motorized endodontic treatment. However, with a manufacturer-provided user's manual, a dentist with a new laser may push a button that says "gingivectomy" and attempt a procedure.

The purpose of this chapter is to explore the parameters that are important, and in some cases critical, to successful laser therapy. This discussion then leads to insights into specific lasers and tissues.

As has been presented, various lasers can be used for surgical treatment of both hard and soft tissues. In addition, photonic energy can be injected into tissue to affect cellular metabolism beneficially. In each of these therapies, applied energy density, its wavelength, and the time over which it is applied represent critical factors affecting the outcome.

4.1 Intrinsic Properties

A number of parameters are dictated by design decisions during the manufacture of any clinical laser. They cannot be chosen or controlled by the operator and must be accepted with their benefits and shortcomings.

4.1.1 The Photon

A photon is a quantum of electromagnetic energy and has been discussed extensively in previous chapters. All photons are certainly not alike. The amount of energy in each photon is determined by the wavelength or, more precisely, the wavelength is determined by the photonic energy released.

The number of photons generated by a laser is mind-boggling. If an 810 nm diode laser is set to deliver 1 W of output power, 4.08 quintillion photons are produced every second. That is, 4,080,000,000,000,000,000 photons are released each second. On the other hand, an Er:YAG laser at 2940 nm must produce 14.799 quintillion photons in a second [1]. Photons exert their influence on tissue through absorption by cellular elements, and that absorption is highly dependent on wavelength.

4.1.2 Wavelength

As an electron drops from one orbital to one of lower energy, a precise amount of photonic energy is released producing a specific wavelength. Every laser is capable of generating and does generate several wavelengths. The resonant cavity is designed to dampen unwanted wavelengths through destructive interference leaving only the desired output to be amplified and emitted.

4.1.3 The Laser

Every laser produces only one predominant wavelength of photons. The laser is usually identified by its active medium and host material. The active medium is a material that can absorb photons of energy and then release them when stimulated further by more photons. Many different materials have been shown to produce this unique "stimulated emission." Lasers are designated therefore as solid state, gas, semiconductor, and liquid, based on the material used.

Solid-state lasers are most commonly trivalent rare earth ions such as neodymium (Nd) and erbium (Er). The host material suspends the active ions and comprises the majority of the laser crystal. They are most commonly grown-crystal structures such as yttrium aluminum garnet abbreviated as YAG [2]. This combination of elements accounts for the naming convention of Er:YAG and Nd:YAG.

Gas lasers consist of a combination of gases. Carbon dioxide lasers are actually a mixture of approximately one part carbon dioxide combined with four parts nitrogen as the active medium distributed in approximately five parts helium as the host material.

Liquid or so called dye lasers use fluorescein, malachite green, coumarin, or rhodamine as the active medium suspended in water, alcohol, or glycol as a host material. They have the great advantage of being "tuneable" from approximately 365 to 1000 nm and, because of the inherent cooling provided by the liquid medium, can produce as much as 20 W of continuous wave output and 1.4 kW of pulsed output. Unfortunately, they are bulky, complex, and expensive and are not currently used in dental applications [3].

Finally, semiconductor lasers, represented by the diode laser, are the most common laser available in dentistry today. They are small, simple, and relatively inexpensive making them very attractive. Most diode lasers are based on combining group III and group V compounds from the periodic table. Those fabricated from gallium arsenide and its derivatives typically lase at wavelengths between 660 and 900 nm, while those utilizing indium phosphide-based compounds produce wavelengths between 1300 and 1550 nm [4]. Recent advances in technology allow laser emission at wavelengths ranging from as little as 370 nm to an amazing 15,000 nm. Only a few of these have practical application in dentistry and are limited to approximately 808–1064 nm [5].

All photons produced at a particular wavelength are the same. While a laser manufacturer would like you to believe that their photons are different and better than those of their competitor, it is simply not true. How they deliver those photons to the tissue and how you are able to control that process are unique to each product and could be a factor in your decision to purchase that particular laser.

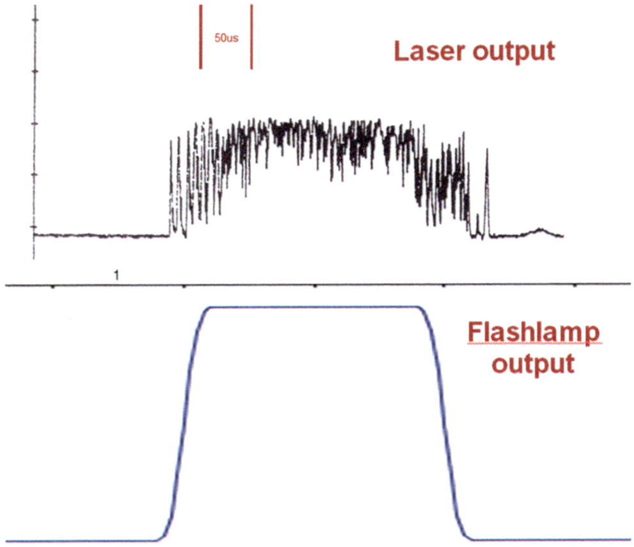

Fig. 4.1 As long as photons are injected into a resonator by a flashlamp or other sources, stimulated coherent photons will be emitted. The amount of lasing activity changes throughout the pulse

Free-Running Pulse Laser

Free-running pulse is a term applied to a laser, meaning that the laser emission lasts as long as the pumping process is sufficient to sustain lasing conditions. Lasers such as Nd:YAG, Er,Cr:YSGG, and Er:YAG release photons as a train of pulses. Most commonly, a flashlamp similar to a photographic flash injects a large number of photons into the active medium. This begins a cascade of events that have been described previously. As long as the flashlamp is injecting photons, lasing will continue. Therefore, the pulse width or length of time that photons will be emitted is controlled by how long the flash lasts. Usually, the duration of the flash in dental lasers and, therefore, the laser output of photons are in the range of 50–1000 μs. Complex control electronics usually allow the operator to select this vital parameter through a touch screen.

The amount of energy in each of these pulses is controlled by the intensity of the flashlamp. If the flashlamp injects twice as many photons to stimulate the active medium, approximately twice as many coherent laser photons will be produced in the resonant cavity and be emitted within the same selected time frame or pulse width.

Lasing is not a continuous activity during the duration of the pulse. It takes some time measured in microseconds for the process to build. As energy is released, the number of stimulated photons is depleted and drops below the state of population inversion. As more photons continue to be injected by the flashlamp, the critical level is again exceeded and laser photon production again occurs. This process repeats itself about every 5 μs. Figure 4.1 is a representation of actual laser activity.

Continuous-Wave Lasers

Lasers such as the KTP, diode, and many CO_2 emit photons on a continuous basis rather than in pulses. The flashlamp is replaced by an electric current, which injects electrical energy instead of photons into the active medium. However, the result is still the release of photons in the familiar stimulated emission mode. As long as the laser is energized, a continuous stream of photons will be emitted. A light-emitting diode (LED) is not capable of laser activity but is a common example of electrical energy being converted to photonic energy. All diode lasers rely on this direct energy conversion.

The CO_2 laser is a bit different. Electrons are passed through a gas starting the lasing process using a high-voltage transformer that can operate continuously. As long as the gas is energized, lasing will occur. Using control circuits, the output can be continuous wave, variable gated continuous wave, pulsed, super-pulsed, gain-switched, or Q-switched.

In super-pulsed mode, the pulse peak power driving the laser discharge can be several times the average continuous-wave power. A super-pulsed CO_2 laser is gated "on" for about 1/3 of the time. This allows it to be driven three times as hard when "on" to produce the same average power while still avoiding overheating. The pulse width is usually in the 5–1000 μs range with peak power reaching triple that of its continuous-wave counterpart [6]. Super-pulsed lasers available in dentistry usually operate at a wavelength of 9.3 or 10.6 μm.

Gain-switched CO_2 lasers can produce megawatts of peak power, while Q-switched lasers with pulse widths in the nanosecond range can produce peak powers that are several hundred times the average output. These lasers are not currently applied in dentistry.

Variable-Gated Continuous-Wave Lasers

Continuous-wave lasers may be turned on and off repeatedly by an electronic circuit generating what is commonly called a variable-gated continuous wave. The term "variable" suggests that the ratio of on-time to off-time is not fixed but each parameter can be varied independently. Currently, on-times and off-times can be selected in the range of approximately 10–1000 ms. One laser manufacturer has developed a unique system that allows on-times of as little as 18 μs with pulse repetition rates of up to 20,000 pulses per second.

Clarification of Pulsed Laser Concept

Confusion continues regarding very important differences in the nature of laser photon emission. Continuous-wave output is inherently different than pulsed output.

Variable-gated continuous-wave lasers are often incorrectly designated as "pulsed lasers." Continuous-wave lasers produce the same number of photons in each microsecond whether they are gated or not. Gating simply represents an on-off switch. If the on-time is twice as long, twice as many photons will be emitted. Of course, no photons are delivered to the target tissue during the off-time. This concept is illustrated in ◘ Fig. 4.8.

The most important characteristic of a pulsed laser is the ability to store and release energy very rapidly. This creates very high peak powers. Pulsed lasers, as described in ► Sect. 4.1.3.1, emit a predetermined amount of energy during the pulse. If a pulse is half as long, the same total number of photons will be emitted but all must be emitted in half the time. Therefore, twice as many photons will be emitted in each microsecond during the pulse. It should be evident that if the pulse width is made very short, very high peak powers will result—a very large number of photons will be emitted in each microsecond. This concept is illustrated in ◘ Fig. 4.9.

The best way to illustrate this dramatic difference between pulsed and continuous-wave emission is to compare two lasers that, on the surface, appear to be nearly the same. ◘ Table 4.1 shows the large difference in actual output.

◘ Table 4.1 Comparison of pulsed Nd:YAG and variable-gated continuous-wave lasers adjusted to what appear to be identical settings. The resulting pulse energy, average power, peak power, and peak power density are calculated to be dramatically different

Laser type	Nd:YAG	Diode
Wavelength	1064 nm	1064 nm
Mode	Pulsed	Variable-gated continuous wave
Delivery fiber diameter	600 μm	600 μm
Power displayed on screen	5 W	5 W
Pulse width	100 μs	100 μs
Pulse repetition rate	50 pps	50 pps
Pulse energy	100 mJ	0.5 mJ
Average power	5 W	0.025 W
Peak power	1000 W	5 W
Peak power density	353,678 W/cm^2	1768 W/cm^2

4.1.4 The Photon Beam

As photons leave the resonating chamber through the partially reflective mirror, they are gathered by a series of lenses. These lenses collimate the photons making them all travel parallel to each other in a common direction, or they focus the beam to the correct spot size to be fed into the delivery device.

Beam Profile

The spatial profile of the laser beam is often, erroneously, assumed to be homogeneous with energy output uniform across the entire beam. Most lasers emit in the "fundamental transverse mode" also called the "TEM_{00} mode." This output is Gaussian in cross section as it leaves the resonator as depicted in ◘ Fig. 4.2.

The density of photons is significantly higher at the center of the beam. Cells directly on the beam axis are irradiated at a very high fluence, while those cells on the periphery of the incident beam receive insufficient cellular energy to produce any surgical effect. Laboratory measurement of the actual output from a diode laser reveals that the power density at the beam center is more than twice the average calculated from the applied power while power density on the periphery is as low as 5% of applied power. Significant amounts of sub-ablative energy are deposited into tissue on the fringes causing heating and dehydration instead of the desired ablation.

By the time the light is emitted, imperfections in the laser tip and the fiber-optic bundle can cause further variation from this theoretical output as shown in ◘ Fig. 4.2c.

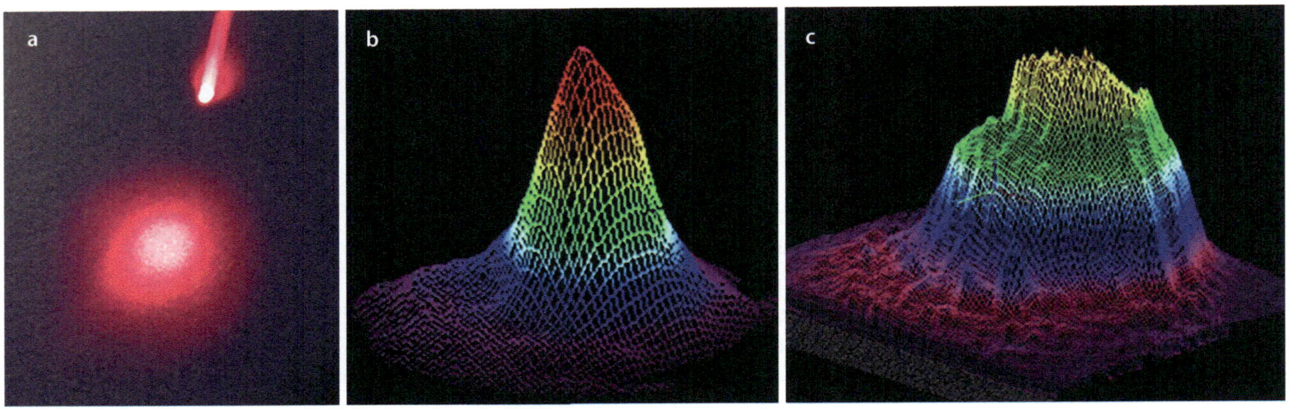

Fig. 4.2 The output from the laser fiber in **a** is obviously nonuniform. The plot in **b** depicts power density at different points across the beam and is approximately Gaussian in cross section. **c** Depicts the output profile of an Er:YAG quartz. The vertical axis indicates relative energy density. (Photo courtesy of Frank Yung DDS)

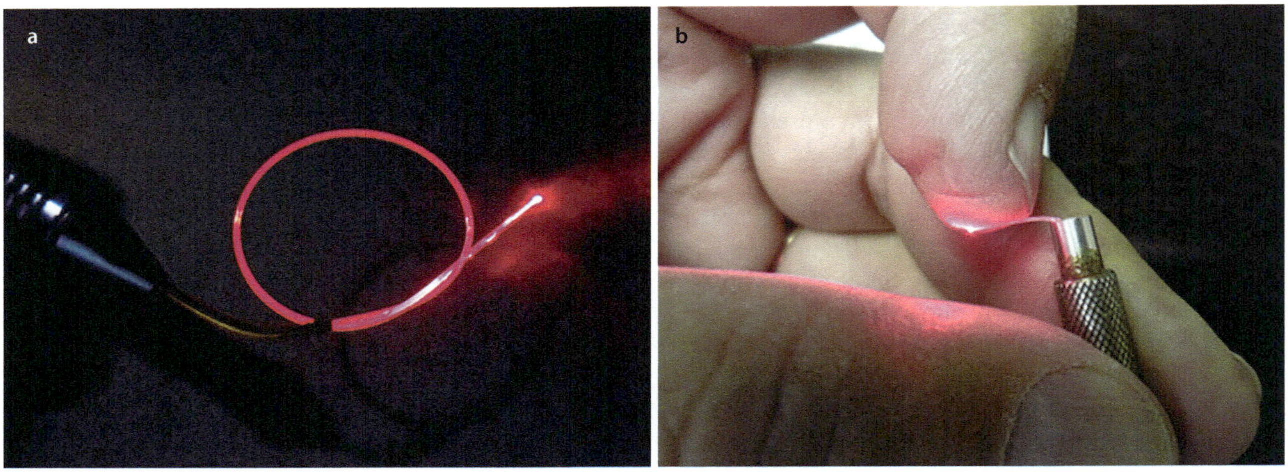

Fig. 4.3 **a** A 400 μm diameter optical fiber can be tied in a knot and still transmit most of the laser energy. The red color visible in the knot indicates that a small number of photons are escaping. **b** As long as the protective polyamide coating is present, the same fiber can be bent at nearly a 90° angle without breaking

4.1.5 Delivery System

Laser energy exiting the resonant chamber and passing through collimating and sizing lenses is delivered to one of the three devices for transportation to the tissue: optical fiber, semiflexible waveguide, or articulated arm.

Optical Fiber

The propagation of light in fibers depends on the principle of total internal reflection [7]. The core layer of the fiber, made of fused silica, has a larger refractive index than the outer cladding layer. The incident laser energy is reflected off this boundary layer and is trapped inside the core. The core and cladding are coated with a buffer material, such as polyamide which has a refractive index slightly greater than that of the cladding and has the additional advantage of mechanically protecting the fiber.

Optical fibers are exceptionally flexible as shown in Fig. 4.3a, b. Even with a severe bend, they continue to reflect photons off the boundary and continue to travel inside the fiber. Figure 4.3b illustrates this phenomenon. However, even a slight nick in the fiber will result in an instantaneous fracture. The polyamide plastic coating surrounds the entire fiber to prevent damage from accidental contact.

For dental use, these fibers are produced in diameters from 100 μm to more than 1200 μm. The significance of selecting a particular size will be discussed later in this chapter. These optical fibers are very attractive to use since their flexibility and lightweight make them easy to maneuver when accessing challenging areas in the oral cavity. However, they are only able to be used with short-wavelength lasers such as diode and Nd:YAG. Longer wavelength photons are strongly absorbed in all optical fibers.

Semirigid Hollow Waveguides

Photons of longer wavelength travel easily through an air medium. Most erbium wavelengths deliver energy through a hollow, semirigid waveguide with inner diameters of typically 300–1000 μm. The inner surface is coated with a silver mirror finish and then with silver halide, creating a very efficient dielectric reflector for infrared wavelengths.

Photons traveling down the lumen of this tube reflect off the sides and continue to the exit portal. However, while bouncing back and forth, photons lose any semblance of collimation and, also, assume a Gaussian distribution. Therefore, lenses are again needed to collimate the output and direct it into the final delivery handpiece. From there, it is delivered to the tissue, using a variety of end attachments, such as a tipless handpiece, a sapphire or quartz tip, and metal or ceramic "guides."

These waveguides have very limited flexibility and a large diameter making them somewhat unwieldy to maneuver within the oral cavity. There are also some limitations on the power level that may pass through without damaging the silver coating.

Articulated Arm Waveguides

For even longer wavelengths such as CO_2, rigid, hollow waveguides are used. Light is redirected using articulated arms and reflective mirrors. These usually consist of seven segments connected with joints containing carefully aligned mirrors. The mirrors move in such a fashion that the beam is always directed down the center, no matter how the joint is turned. Photons entering each segment are transmitted in a straight line down the center of the tube. When they reach the end, they strike a mirror and are reflected precisely down the center of the next segment.

At the end of the last segment, the photons are presented to mirrors in the handpiece and then on to the delivery tip. In many cases, they are delivered, instead, to a final lens to be focused at some distance away from the handpiece for delivery to the tissue. The distance is usually 6–10 mm to allow vision around the handpiece to the underlying hard or soft tissue. This distance creates some difficulty in precise application of the laser energy although aiming beams provide considerable guidance.

Semirigid hollow waveguides and articulated arm waveguides are used almost exclusively for longer wavelength lasers such as Er:YAG, Er,Cr:YSGG, and CO_2. While bulkier and more difficult to maneuver than optical fibers, they overcome the problem of transmission through a solid medium.

4.1.6 Emitting Device

The final tip or lens through which the laser energy passes has a very significant impact on interaction with target tissues. An optical fiber has a flat end and will emit in the pattern shown in ◘ Fig. 4.4a. The beam immediately diverges as will be discussed in ▶ Sect. 4.1.7 below. This pattern also applies when using a removeable flat-ended tip with an Er:YAG or Er,Cr:YSGG laser. These tips are constructed of either quartz or sapphire.

Erbium and CO_2 lasers using a tipless delivery will emit as shown in ◘ Fig. 4.4b. A lens is the last element in the path of the beam and focuses it. The normal distance between the lens and the focal point is 6–10 mm.

Maximum power density is achieved by positioning the handpiece so that the tissue surface is at the system focal point. In practical terms, the most efficient ablation occurs at this distance. A significant advantage of this system is that, as the tip is moved either nearer or further away than this distance, the beam diverges, spot diameter increases, and power density decreases. This technique is often referred to as "defocusing" the beam and is of value when the aim is to inject energy into the tissue without the consequences of high power density such as ablation. This point is further discussed in ▶ Sect. 4.2.7.

Laser pointers use a collimating lens as shown in ◘ Fig. 4.4c. This allows the beam to have a consistent diameter at a large range of distances. This system is not used in surgical lasers for safety reasons since a misaimed beam could accidently interact with a distant tissue with the same power density as at the intended target. Also, if the beam were to reflect off a mirror, it could ablate other tissues in random locations. Instead, by using the patterns in ◘ Fig. 4.4a or b, the beam safely loses its power density and, consequently, its surgical ability with distance.

The exception to this rule is seen in the newly developed "flattop" handpiece used in photobiomodulation. In this application, a predictable power density that is independent of tip-to-tissue distance is extremely desirable and important. However, the power levels used are very low, thus mitigating the risk of inadvertent tissue damage.

There are many instances when it is desirable to direct the laser energy laterally. Examples would be when irradiating into dentinal tubules or the threads of a failing implant. A tip has been developed by several manufacturers with a conical terminal end. Energy is internally reflected from this surface and then passes through the opposite side. Practical considerations allow redirection up to approximately 60° from the original forward axis as shown in ◘ Fig. 4.4d.

4.1.7 Beam Divergence

As laser energy exits any optical fiber or flat-ended handpiece tip, the beam diverges at a predictable angle. In practical applications, this is usually on the order of 8–15° per side angle. While this may seem to be a small

4

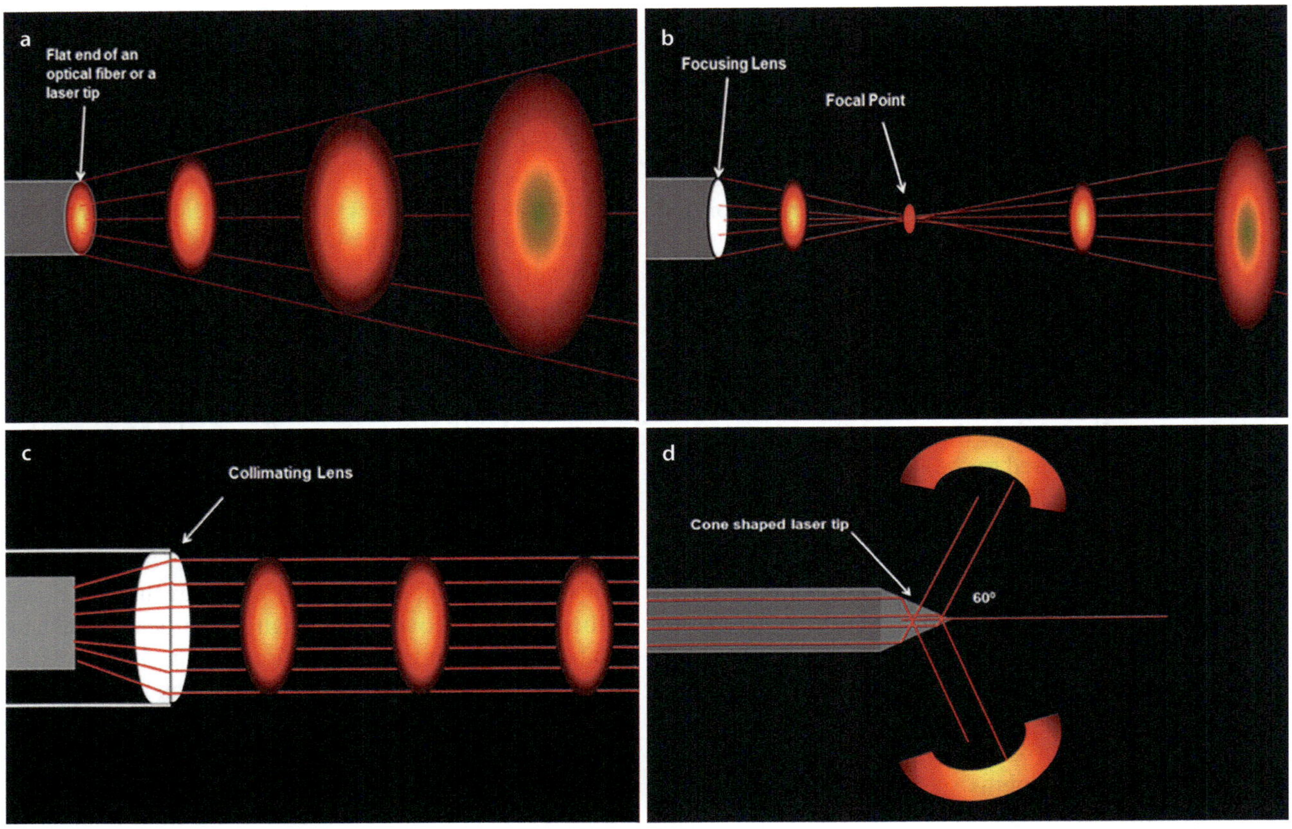

Fig. 4.4 The output from different laser devices is not the same. **a** Optical fiber or laser tip, **b** tipless delivery system, **c** laser pointer or flattop handpiece, **d** radial firing laser tip

amount, the area on which the delivered photons impact increases rapidly with distance. The influence of this divergence on power density is of great significance and will be discussed in detail later in this chapter.

4.2 Adjustable Parameters

The parameters discussed above are all intrinsic to the particular laser being used. They are determined by the laser manufacturer and begin the process of optimally matching the photon delivery to the tissue of interest. The operator has control of a number of parameters that profoundly affect the interaction and, therefore, the outcome of laser therapy. Fluence, irradiance, and peak power are parameters critical to effective laser treatment. However, determining and controlling these parameters are not as straightforward as it would at first seem.

In order to discuss the scientific basis of parameters, it is necessary to have an understanding of definitions:

- Power (measured in W or J/s) is the measure of how much energy is delivered in 1 s of time. One watt is an abbreviation and equivalent of 1 J/s.
- Power density (measured in W/cm^2) is the power delivered to a unit area of target tissue, usually here 1 cm^2.
- Irradiance is another term for power density.
- Energy density (measured in J/cm^2) is calculated as the power density in W/cm^2 multiplied by the total time of illumination.
- Fluence is another term for energy density.

Each laser manufacturer displays a different group of parameters on the control interface. In addition, some manufacturers do not provide direct control of every variable. Figures 4.5, 4.6, and 4.7 are representative of dental lasers currently produced by different manufacturers. Throughout the remainder of this chapter, these three lasers will be used as the basis for discussion and calculation. The parameters available to be directly controlled are different in each case. Therefore, a unique set of equations must be used in each instance to determine the critical parameters. Examples of all appropriate calculations will be provided in Appendix of this book.

4.2.1 Average Power

Power and energy are often confused. Energy is, classically, a measure of the ability to do work. It is a "quantity" and is measured in joules. It is important to remember that a joule of energy is produced by a vari-

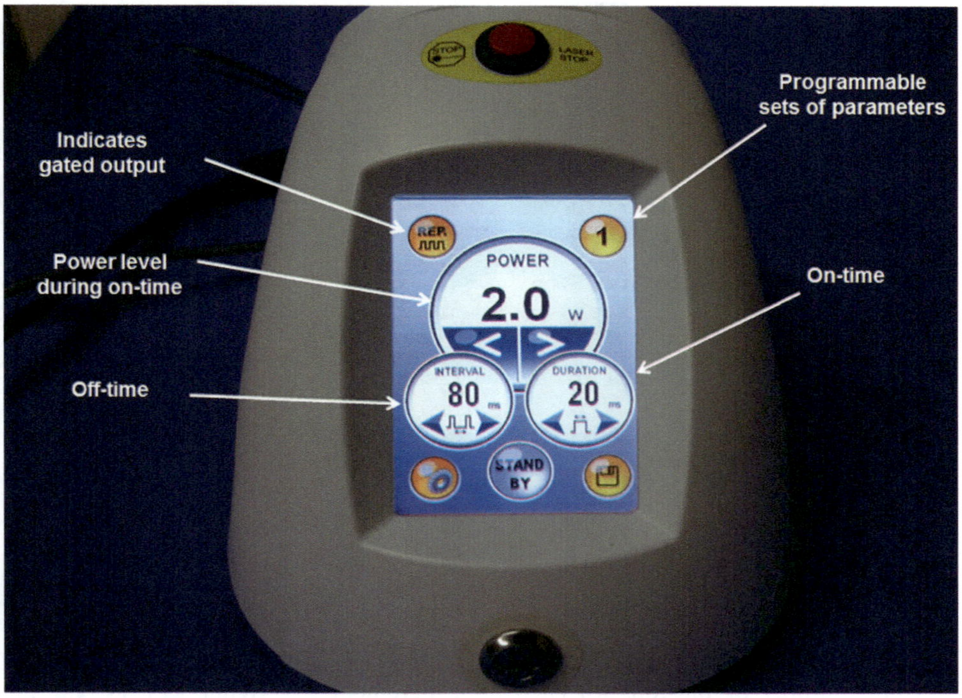

Fig. 4.5 This 810 nm diode laser has a limited number of variables that may be controlled directly

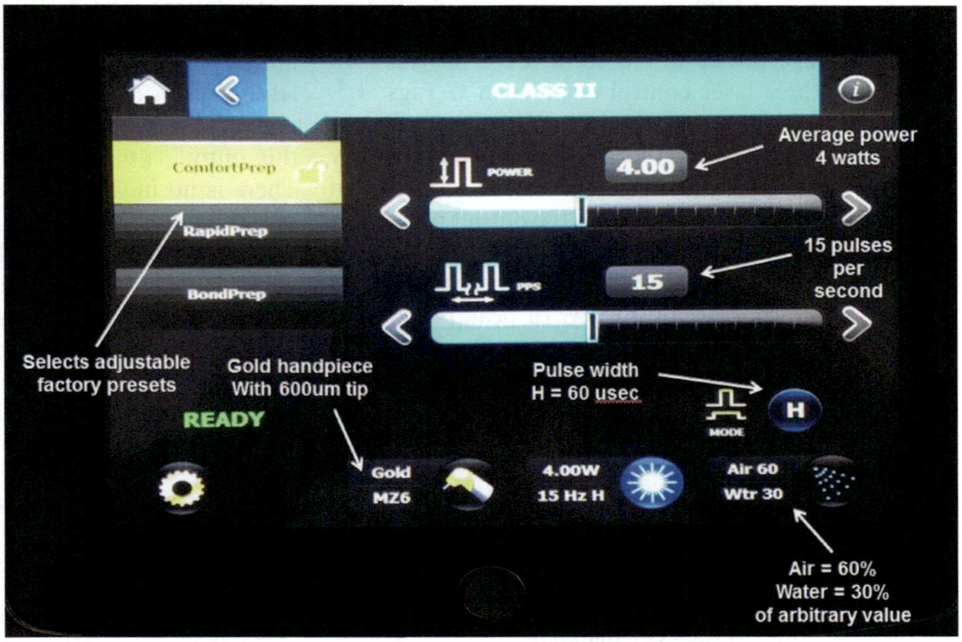

Fig. 4.6 This 2780 nm Er,Cr:YSGG laser has a choice of two pulse widths. It allows control of average power, but energy per pulse and peak power must be calculated. Levels of auxiliary air and water are set as percentage of arbitrary values determined by supply to the individual dental suite

able number of photons since each photon contains a different amount of energy based on its wavelength.

Power is the rate of producing energy. It is the number of joules created in each second of time and is designated in joules per second. A watt is simply an abbreviated word to mean "joule per second." If 1 J/s or W is produced and continues to be produced for 10 s, then 10 J of total energy will have been produced. If energy continues to be produced at this rate for 60 s, 60 J of total energy will have been produced. If instead the power (the rate of doing work) is increased to 6 J/s (W) and is applied for just 10 s, the same 60 J of total energy will have produced.

Since average power, peak power, energy per pulse, pulse width, and pulse frequency are so interrelated, they will be discussed as a parameter group interspersed with examples.

Average power represents the total energy produced in 1 s, no matter how uniformly that happens within that length of time. A continuous-wave laser produces a

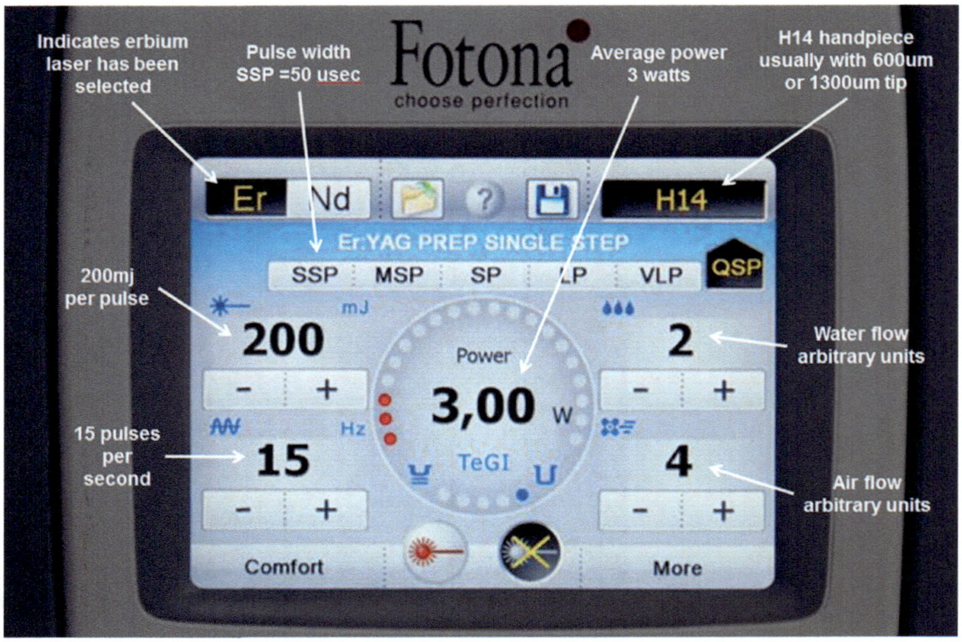

Fig. 4.7 This laser contains both a 2940 nm Er:YAG laser and a 1064 nm Nd:YAG laser. In erbium mode, it allows selection of six different pulse widths. However, the operator must refer to a manual to find the value. It allows control of pulse energy and pulse frequency. The resulting average power is displayed but not directly selectable

constant number of photons for each time increment that it is activated while a pulsed laser produces photons in short bursts ranging from femtoseconds to milliseconds.

The average power would seem to be provided on two of the three laser control displays in Figs. 4.5, 4.6, and 4.7, but a very important distinction must be made. With a diode laser, the power displayed is usually (but not always) the peak power. The diode laser in Fig. 4.5 allows direct control of the output power, but the power displayed is the "continuous wave" output rather than the average power. Any time that this laser is emitting photons, it is emitting 2 W of peak power.

In *variable-gated mode*, independent selection of off-time and on-time is provided and results in vastly different average power for a given indicated power. If the laser is "on" for 20 ms and "off" for 80 ms, then it is on for 20% of the time:

$$\begin{aligned}\text{Percentage of "on}-\text{time"} &= 20\,\text{ms}/\left(20\,\text{ms}+80\,\text{ms}\right)\\ &= 20\,\text{ms}/100\,\text{ms} = 0.2 = 20\%\end{aligned}$$

Average power is then calculated as:

$$\begin{aligned}\text{Power}_{\text{avg}} &= \text{power}_{\text{indicated}} \times \text{percentage of "on}-\text{time"}\\ &= 2\,\text{W} \times 20\% = 0.4\,\text{W}\end{aligned}$$

If the lengths of the on and off periods are changed but stay in the same ratio, average power will be the same. Figure 4.8 illustrates these concepts.

Peak power will always be the same as that indicated on this display, no matter what magnitude or ratio of "on" and "off" times are selected.

While Figs. 4.6 and 4.7 are both erbium lasers, their control features are distinctly different. The Er,Cr:YSGG laser shown in Fig. 4.6 allows control of average power. The pulse width is selected as "H" for hard tissue and represents a pulse width of 60 μs. The laser displays the pulse repetition rate (15 pulses per second). The peak power is adjusted by the laser to provide this output although it cannot be directly controlled. There is no indication of the energy produced in each pulse.

The Er:YAG laser in Fig. 4.7 allows control of energy per pulse and pulse frequency. While average power is indicated on the display, it is not directly controllable. If the amount of energy contained in each pulse and the number of pulses in each second are known, the average power is calculated as the total of all these pulses in 1 s:

$$\text{Power}_{\text{avg}} = \text{Energy per pulse} \times \text{Number of pulses per second}$$

4.2.2 Peak Power

Peak power represents the maximum instantaneous power that the laser produces at any single time. In a continuous-wave laser, the peak power does not change from that selected on the user interface, no matter what other parameters are changed. As has been discussed, it is the value displayed by the diode laser in Fig. 4.5.

With pulsed lasers such as all dental erbium and Nd:YAG lasers, peak power is dramatically affected by the interrelated control of average power, energy per

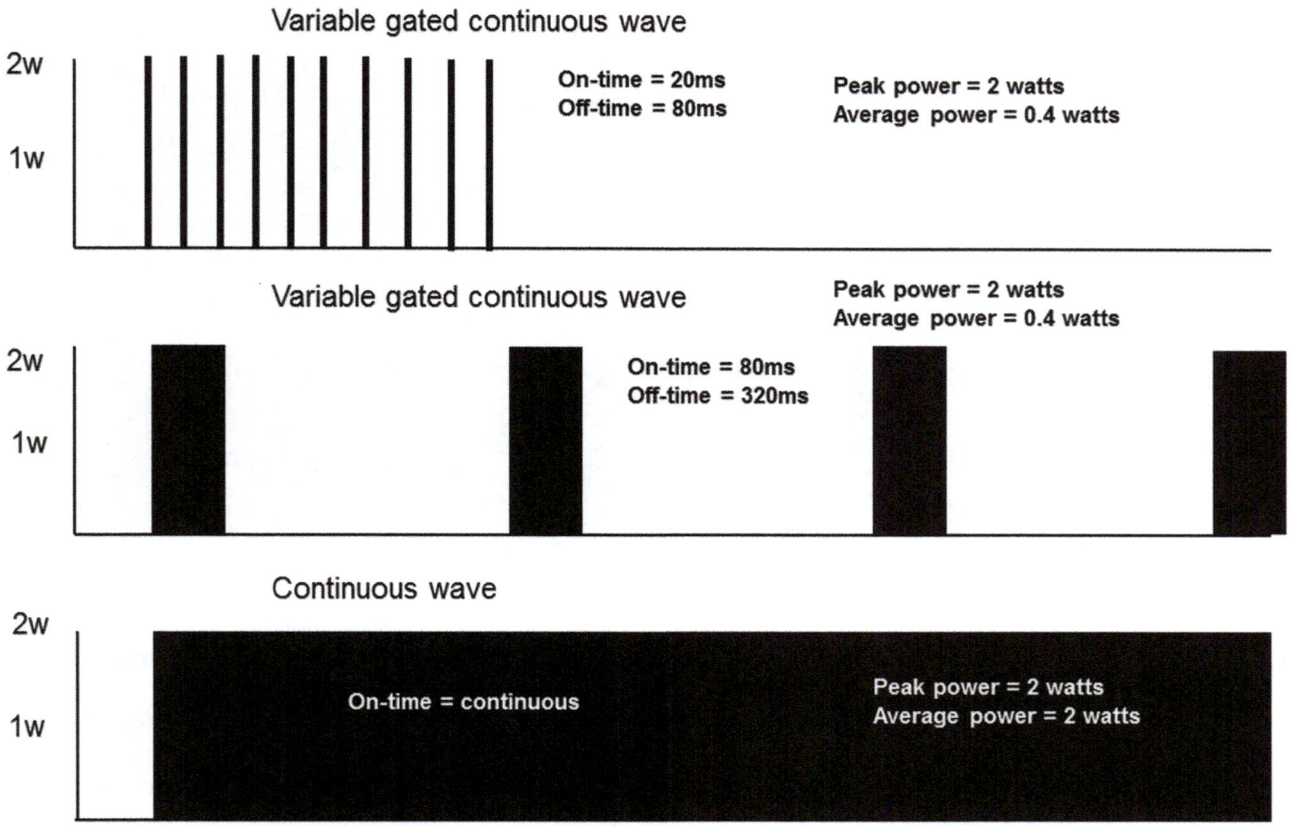

Fig. 4.8 Calculated peak and average power produced at the settings with the diode laser shown in Fig. 4.5

pulse, pulse width, and pulse frequency. If these parameters are not known, they can be calculated.

Peak power can be calculated as:

$$\text{Power}_{\text{peak}} = \text{Energy per pulse} / \text{pulse width}$$

4.2.3 Pulse Energy

Some manufacturers allow direct control of the total amount of energy in each pulse instead of controlling the average power. By necessity, if the pulse energy is held constant, altering the pulse width will change the peak power. Figure 4.9 illustrates this concept. It is based on the Er:YAG laser shown in Fig. 4.7. In each case, the energy per pulse is the same. If that energy is delivered in one-half the time, the power or rate of delivery must be doubled.

Some pulsed lasers do not allow direct control of pulse energy but, instead, provide selection of average output power. This output results in vastly different pulse energy and peak power as illustrated in the following example.

In Fig. 4.6, if average power is set to 4 W and there are 15 pulses per second:

$$\text{Energy / pulse} = \text{Total energy per second} / \text{number of pulses containing that energy}$$

$$\begin{aligned}\text{Energy / pulse} &= 4\,\text{J/s} / 15\,\text{pulses per second} \\ &= 0.267\,\text{J/pulse} = 267\,\text{mJ/pulse}\end{aligned}$$

At the same time, if average power were to be set to 8 W and there were 30 pulses per second:

$$\begin{aligned}\text{Energy / pulse} &= 8\,\text{J/s} / 30\,\text{pulses per second} \\ &= 0.267\,\text{J/pulse} = 267\,\text{mJ/pulse}\end{aligned}$$

The energy per pulse is the same even though the average power is doubled.

4.2.4 Pulse Width

In most pulsed lasers, selection of several different pulse widths is provided. As an example:

For the control panel shown in Fig. 4.6:

- "H" = "hard tissue" = 60 µs
- "S" = "soft tissue" = 700 µs

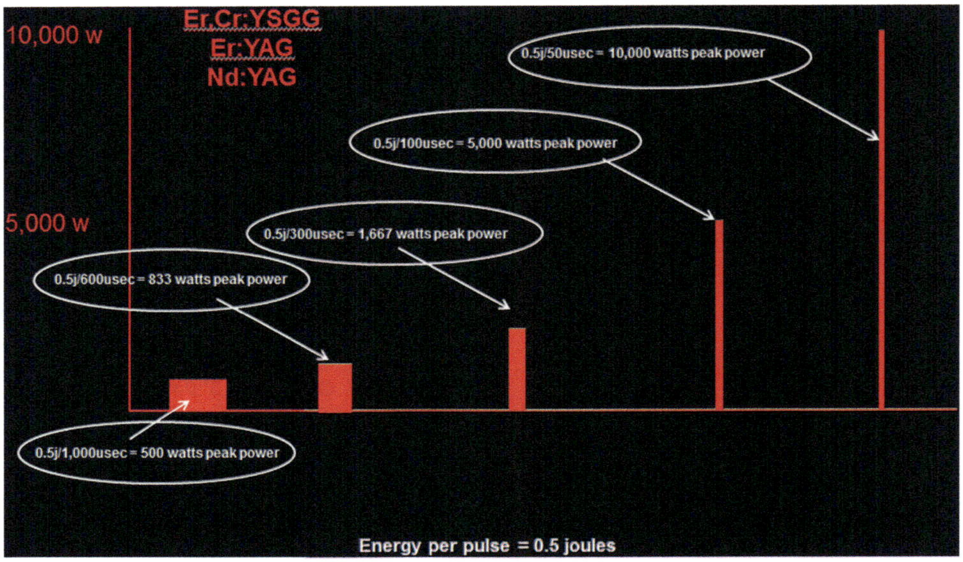

Fig. 4.9 The peak power emitted by a laser change with changes in pulse width if the energy per pulse is constant. This graph shows examples of free-running pulsed lasers, but some variable-gated pulsed lasers use a similar concept

For the control panel shown in Fig. 4.7:

- "SSP" = "super-short pulse" = 50 μs
- "MSP" = "medium short pulse" = 100 μs
- "SP" = "short pulse" = 300 μs
- "LP" = "long pulse" = 600 μs
- "VLP" = "very long pulse" = 1000 μs
- "QSP" = "quantum square pulse" = a unique pulse train not relevant to this discussion

4.2.5 Pulse Repetition Rate

Doubling the number of pulses that are delivered in a second will double the total amount of energy delivered to the target tissue. If the pulse width and energy per pulse are left unchanged, the peak power will be unchanged. Since there are twice as many pulses applied in a second, the average power will double.

In the Er:YAG laser example of Fig. 4.7:

$$\begin{aligned}\text{Power}_{\text{avg}} &= 0.200\,\text{J}/\text{pulse}\times 15\,\text{pulses per second}\\ &= 3.0\,\text{W}\,(\text{J}/\text{s})\end{aligned}$$

If the pulse repetition rate were doubled to 30 pulses per second:

$$\begin{aligned}\text{Power}_{\text{avg}} &= 0.200\,\text{J}/\text{pulse}\times 30\,\text{pulses per second}\\ &= 6.0\,\text{W}\,(\text{J}/\text{s})\end{aligned}$$

In each case, the peak power would be 4000 W, if the laser were set to SSP mode.

4.2.6 Diameter of the Final Emitting Device

The number of photons that strike an individual cell and their associated energy is one of the primary determinants of tissue effect. Power density or fluence is a critical factor in determining laser-tissue interaction.

Power density is, by definition, the number of photons passing through a specific area in each second of time. A 400 μm laser fiber has a cross-sectional area of:

$$\text{Area} = \pi \times r^2 = 3.14159 \times (0.02\,\text{cm})^2 = 0.0013\,\text{cm}^2$$

A seemingly small output of 1 W, if passed through this fiber, results in a power density striking an individual cell of 796 W/cm^2.

$$\begin{aligned}\text{Power density} &= \text{Power}/\text{area} = 1\,\text{W}/0.0013\,\text{cm}^2\\ &= 796\,\text{W}/\text{cm}^2\end{aligned}$$

The same 1 W passing through a 200 μm diameter fiber would have a power density of 3185 W/cm^2.

Each of these outputs will have a profoundly different effect on a target cell. Table 4.2 shows the dramatic range of power densities that occur with different optical fiber diameters at average power outputs routinely available in dentistry. It is evident that the choice of fiber diameter has a critical effect on the delivered power density. Simply choosing a 200 μm diameter fiber instead of a 1200 μm fiber makes a 36-fold difference.

The combined influence of pulse width and fiber or tip diameter looms as an overwhelming determinant of

Table 4.2 Average power density in W/cm^2 with different fiber size and power setting

Power density (W/cm^2)						
Fiber diameter		**200 µm**	**400 µm**	**800 µm**	**1200 µm**	**1.15 cm**
Power (W)	0.5	1591	397	99	44	0.5
	1	3185	795	198	88	1
	2	6370	1591	397	176	2

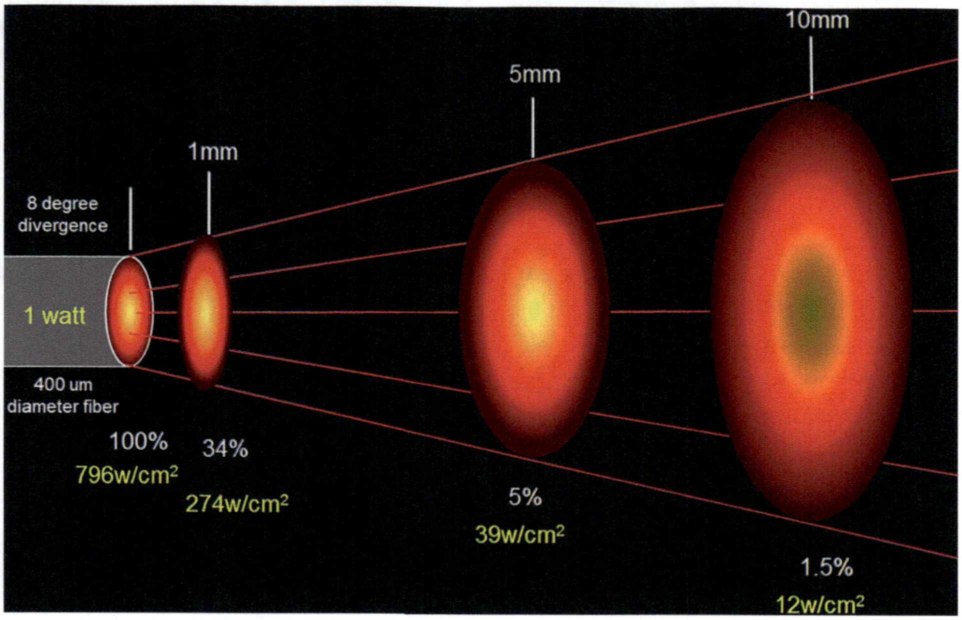

Fig. 4.10 The effect of tip-to-tissue distance on spot diameter and, therefore, on power density. These calculations are based on a beam divergence of 8° per side angle

laser-tissue interaction. In the example of the Er:YAG laser in Fig. 4.7, using an 800 µm diameter tip, producing 1000 mJ of energy per pulse with a pulse width of 50 µs and 50 pps will deposit a mind-boggling peak power density of 4,000,000 W/cm^2 into the target tissue with an average power density of 10,000 W/cm^2. These calculations are illustrated in Appendix.

4.2.7 Tip-to-Tissue or Focus-to-Tissue Distance

Under many circumstances, the optical fiber tip, quartz tip, or sapphire tip is held at some distance away from the tissue. In addition, there are a number of erbium and carbon dioxide lasers that are routinely used without a tip but rely on focused laser energy as discussed in ▶ Sect. 4.1.6 above. With a tipless system, the smallest spot diameter and, consequently, highest power density exist at the point of focus. All measurements of distance are referenced to that point.

Tip-to-tissue distance, or the comparable focus-to-tissue distance, provides yet another parameter that dramatically affects the density of photons striking an individual target cell. As laser energy exits an optical fiber or a quartz or sapphire tip, it diverges quite significantly.

The applied energy is distributed over an increasing area as the tip-to-tissue distance increases, dramatically affecting power density at a cellular level as shown in Fig. 4.10. Power density can be diminished by 95% with only 5 mm of tip-to-tissue distance. This is further demonstrated in Fig. 4.11. When using this standard delivery system, the repeatable application of an appropriate energy density is extremely technique and operator sensitive.

Table 4.3 illustrates how significantly the power density changes with just minor changes in the tip-to-tissue distance. It is also apparent that the effect of beam divergence is far greater with smaller diameter emitting devices. A collimated beam eliminates this destructive divergence but is only possible to produce if a lens is the final element in the path of the laser energy as it is emitted

All concepts associated with beam divergence from a flat-ended tip apply to tipless systems with distances being measured from the focal point as demonstrated in Fig. 4.12.

4

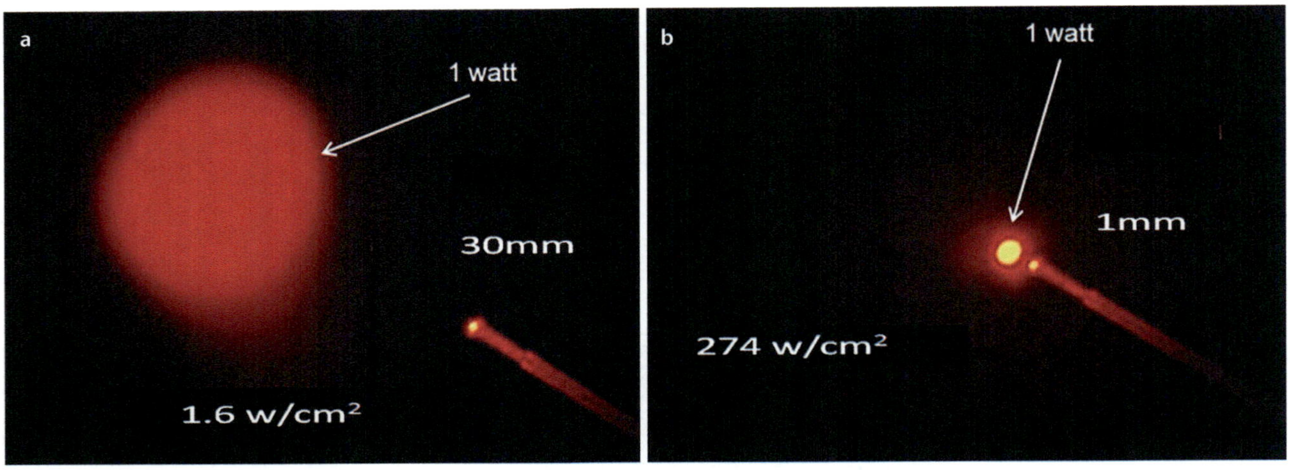

Fig. 4.11 Power density at tip-to-tissue distances of 30 mm **a** and 1 mm **b** with an identical 1 W of applied diode laser power

Table 4.3 Average power density at 1 W laser output for different tip diameters and tip-to-tissue distances assuming an 8° per side beam divergence

Power density (W/cm²)						
T–T distance		**Contact**	**1 mm**	**2 mm**	**5 mm**	**10 mm**
Tip diameter	400 μm	796	275	138	39	12
	800 μm	199	109	69	26	10
	8 mm	2	1.9	1.7	1.4	1.1
	1.15 cm	1	0.92	0.88	0.76	0.62
	1.15 cm (collimated)	1	1	1	1	1

4.2.8 Auxiliary Water

All erbium lasers, together with the 9300 nm CO_2 laser, have the ability to spray water on the tissue during treatment. In the case of hard tissue, this spray is used to cool the tooth or bone and remove debris from the ablation site. Without extrinsic water, the tooth will overheat under repeated pulses and ablation may cease.

Er:YAG laser energy is absorbed in approximately 0.8 μm of water [8–10]. Logic suggests that, if auxiliary water is used, all energy will be absorbed by overlying water before it can reach the tissue surface and no ablation will occur. On the other hand, without extrinsic water, charring will rapidly occur and ablation will cease with significant damage to the tooth.

In practice, other principles resolve this issue [11]. When laser energy strikes the surface of a water film, it is rapidly absorbed causing vaporization. This rapid phase change occurring in a few microseconds creates a shock wave of very significant magnitude generating pressures of a few hundred atmospheres. This shock wave displaces water at the point of contact creating an open channel to the underlying tooth surface [12] as shown in Fig. 4.13.

If the laser tip is completely submerged, water is evacuated from the laser path by a slightly different phenomenon. When laser energy strikes the water directly in front of the tip, it once again causes vaporization. An air bubble is created, as shown in Fig. 4.14, which pushes the water out of the way allowing energy from the remainder of the pulse to strike the tooth surface and cause ablation.

Nahen and Vogel [13] showed that such a bubble forms in the first 10–20 μs of the laser pulse. The balance of the pulse energy is then able to ablate as though no water is even present. Nearly all of the laser energy strikes the tooth surface unimpeded allowing excellent ablation. As the bubble collapses between each pulse, the newly ablated surface is bathed in water, thus cooling, cleaning, and rehydrating the structure.

Within limits, the thickness of the water layer is of little consequence. The bubble formed has a finite size estimated by the author from ultra-speed photographs, such as in Fig. 4.15, to be approximately 3 mm in length.

When only a water mist is used, any droplets in the path of the laser energy are rapidly vaporized allowing passage of subsequent energy.

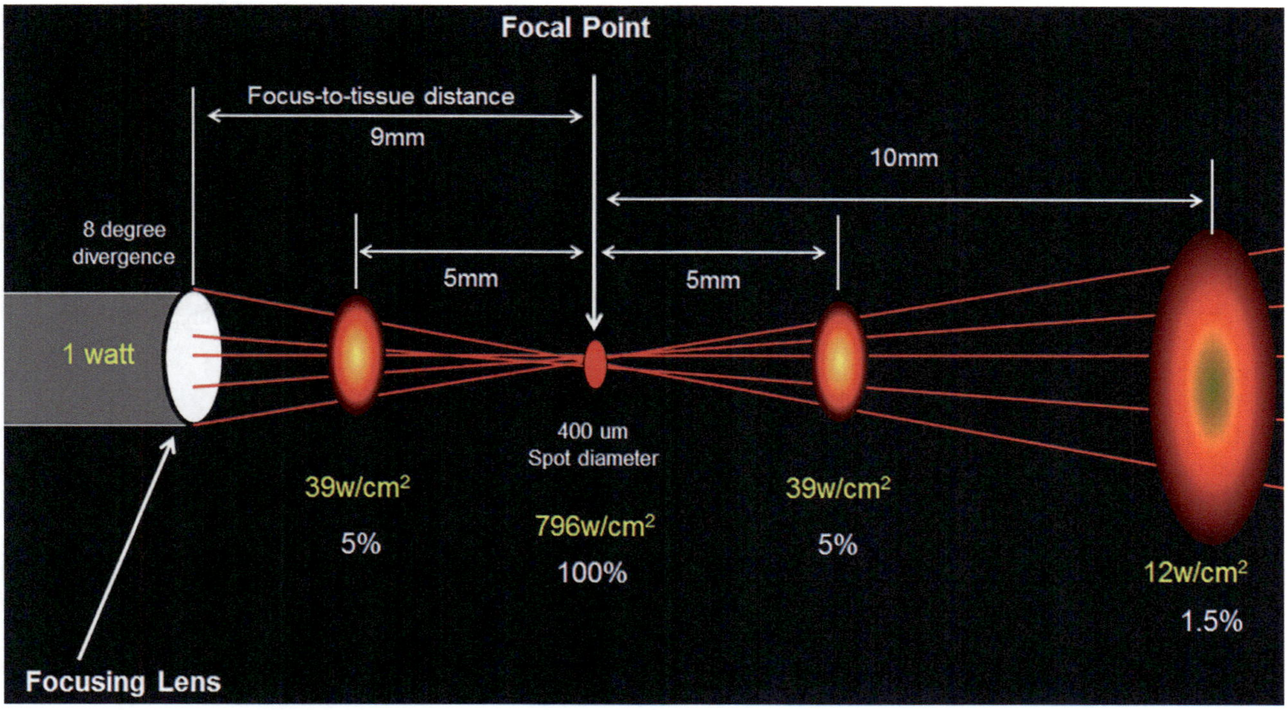

Fig. 4.12 The effect of focus-to-tissue distance on spot diameter and, therefore, on power density. These calculations are based on a beam divergence of 8° per side angle. It is evident that all of the distances from the focus point are the same as those from the tip in Fig. 4.10. Consequently, the calculated values are identical

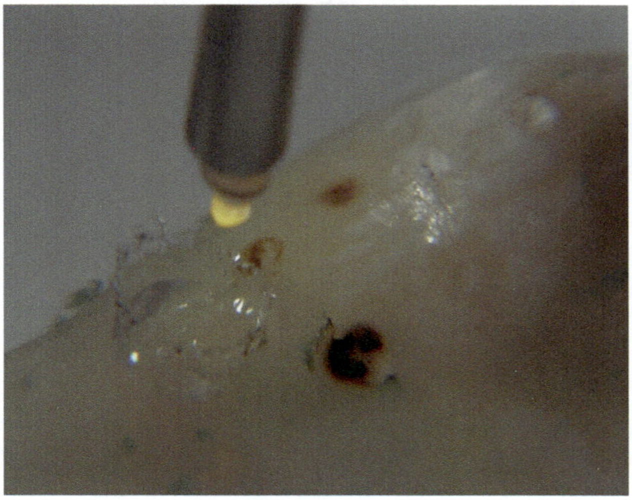

Fig. 4.13 Ultra-speed macrophotograph of Er:YAG laser energy interaction with a superficial water layer. The shock wave generated by the laser pulse displaces the water on the tooth surface

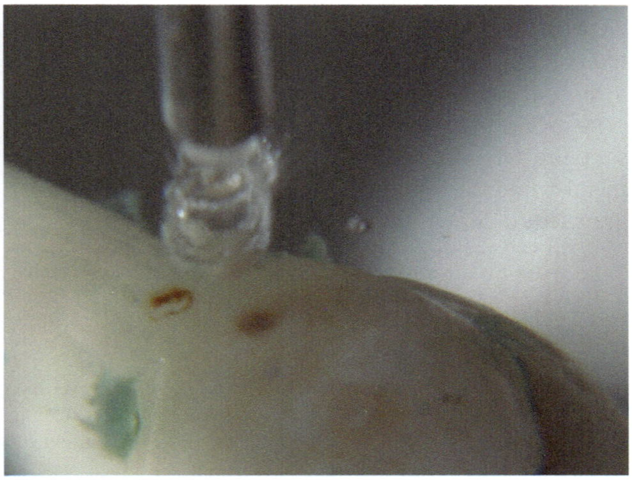

Fig. 4.14 Ultra-speed macrophotography of Er:YAG laser energy interaction when the tip is completely submerged in water. Vaporization of the water in front of the tip creates a bubble forming a channel for ablation

In an experiment to verify the theoretical concepts discussed above, the flow rate of irrigation water was controlled at 1, 2, 4, 8, 16, and 24 mL/min while the tip-to-tissue distance was held at 0.5 mm. Other samples were ablated completely submerged in water at the same controlled distance [11].

Figure 4.16 shows that Er:YAG ablation efficiency is essentially independent of irrigation water flow rate. In addition, applying energy with the tip completely submerged in water has no detrimental effects on ablation efficiency as long as the laser tip is within about 2 mm of the tissue surface.

At low irrigation water flow rates, the possibility of dehydration and charring is significant. Without water coolant, residual heat quickly dehydrates the enamel making further ablation inefficient or impossible. At the

4

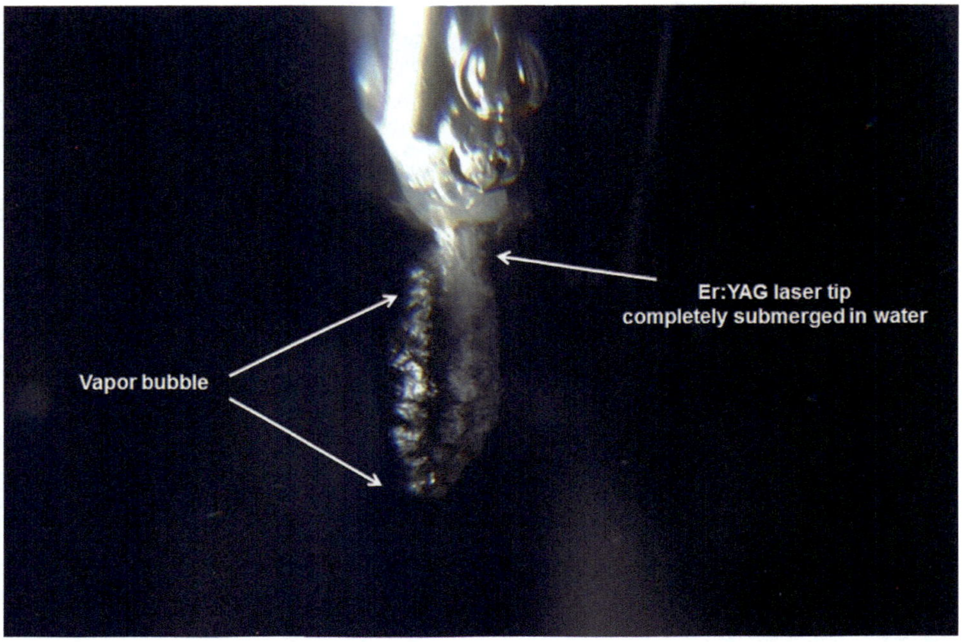

Fig. 4.15 Ultra-speed macrophotography of the bubble formed surrounding an Er:YAG laser tip when it is completely submerged in water. The vaporization bubble in this picture has a diameter of approximately 1.5 mm and a length of more than 3 mm beyond the tip

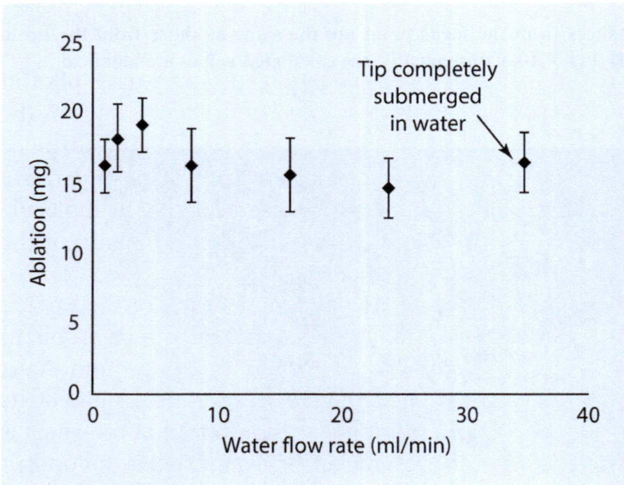

Fig. 4.16 Effect of irrigation water flow rate on Er:YAG laser ablation of enamel. Tip-to-tissue distance was constant at 0.5 mm. The volume of irrigation water has a minimal effect on ablation efficiency

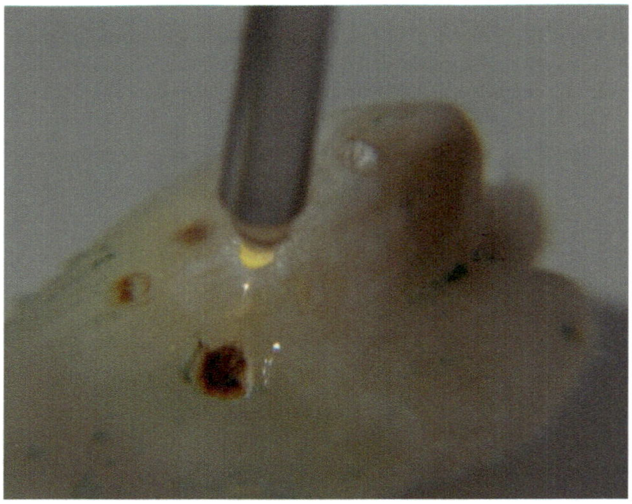

Fig. 4.17 Effect of irrigation water flow rate on Er:YAG laser ablation of enamel. Without irrigation, hydroxyapatite absorbs laser energy causing progressive desiccation, overheating, and carbonization

same time, a layer of exploded hydroxyapatite particles covers the enamel surface and absorbs some of the energy of the next energy pulse, heating this unwanted layer significantly.

Figure 4.17 shows enamel ablated without water. Thermal damage results in carbonization, cracking, and a loss of laser ablation effectiveness. Water spray both cools the tooth and washes away the debris. Irrigation also serves to enhance ablation rate and efficiency, improve surface morphology, alter chemical composition, and enhance adhesion to restorative materials.

A flow of at least 8 mL/min is needed to minimize adverse effects. Since flows of up to 24 mL/min have been shown to have no negative effect on laser ablation efficiency, it is prudent to use 8–24 mL/min to ensure adequate hydration. In fact, it can be argued that using the maximum amount of water possible is an appropriate strategy. However, ultimately, site visibility by the dentist as well as patient tolerance for

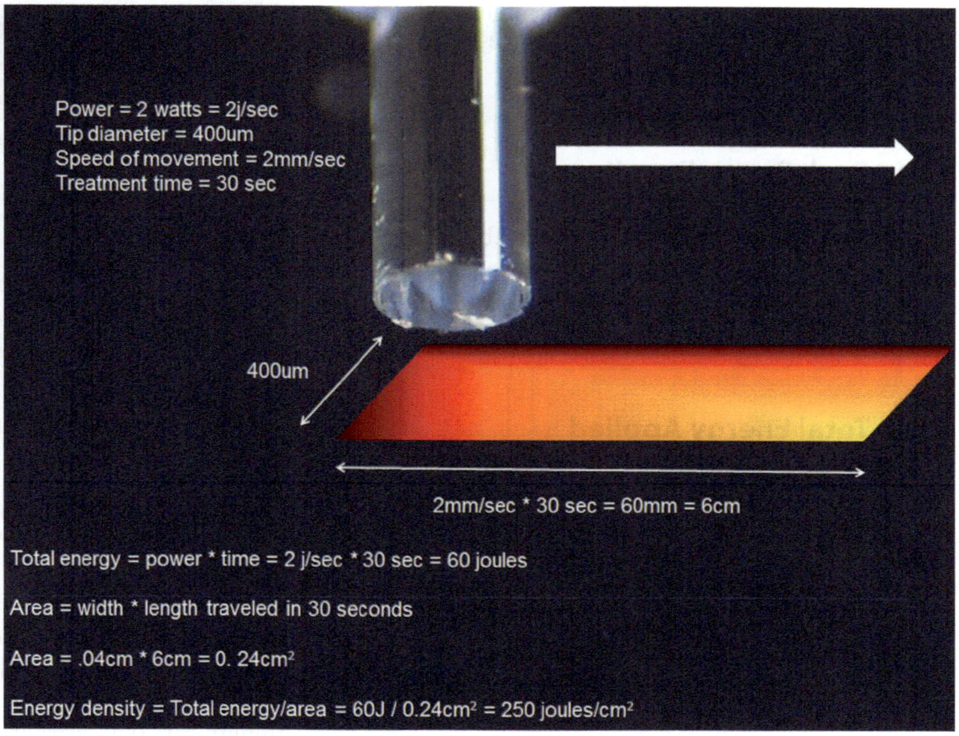

Fig. 4.18 The energy density striking individual cells is calculated by dividing the total energy delivered by the total area irradiated

large water volumes will dictate the water flow rate used.

Super-pulsed CO_2 lasers also use auxiliary water to cool the tooth or bone and remove debris from the ablation site, and it is vital to effective use of this modality. Water is not used with continuous-wave CO_2 lasers since it would absorb all the energy in the scenario discussed in the second paragraph of this section.

4.2.9 Auxiliary Air

Auxiliary air is also provided by all erbium lasers. However, its value and effect are not as well understood. Certainly, air may help dissipate accumulated thermal energy, but its primary function is to remove debris from the path of the laser energy.

4.2.10 Speed of Movement

Energy is being dispensed as long as the laser is activated. The speed with which the tip is moved across the target tissue will dictate how much energy strikes each individual cell. Moving too quickly will not allow sufficient energy absorption to initiate ablation, and the tissue will be detrimentally heated. Moving too slowly or maintaining the tip in one position will ablate the tissue but also causing overheating. A method of calculating energy density with movement is illustrated in Fig. 4.18.

4.3 Further Calculated Parameters

Having adjusted laser parameters in accordance with an established protocol, it is instructive to understand the nature of the resulting energy applied to the target. Average power, peak power, fluence (energy density), and irradiance (power density) all have a very significant effect on the target tissue. Further parameters are also important.

4.3.1 Spot Area at the Tissue

The spot area at the tissue surface must be determined before fluence and irradiance can be calculated. Incorporating the effect of beam divergence and tip-to-tissue distance can often make this a confusing calculation. Further confusion arises regarding even the definition of spot size.

In physics and engineering, spot size refers very specifically to the radius of the laser beam [14], while in many medical publications, it is used to describe the beam diameter or even the area. While the exact definition is of minor importance, knowledge of the author's intent is critical. It is preferable to use the more specific terms of "spot diameter at tissue" and "spot area at tissue."

4.3.2 Average Power Density

Average power density is determined by dividing the average power by the spot area at the tissue surface.

Since spot area is generally very small as shown in ▶ Sect. 4.2.6 above, average power densities can be surprisingly large.

4.3.3 Peak Power Density

4

Peak power density is simply determined by dividing the peak power by the spot area at the tissue surface. Once again with a small spot area, it is possible to get a stunning 4,000,000 W/cm^2 as shown in ▶ Sect. 4.2.6.

4.3.4 Total Energy Applied

The total energy applied to the tissue is simply the average power multiplied by the total number of seconds of treatment.

The discussion in this chapter highlights the importance of understanding and controlling the exact nature of the parameters selected in any clinical protocol in order to effectively interact with either hard or soft tissue. While the necessary calculations are not complicated, they need to be precise. Appendix (Sect. 4.7) assembles all relevant information and provides calculations based on the settings shown in ◘ Figs. 4.5, 4.6, and 4.7.

4.4 Hard Tissue Considerations

4.4.1 Most Appropriate Laser for Hard Tissue Procedures

In the past, erbium lasers were the only devices appropriate for ablation of hard tissues. In recent years, super-pulsed carbon dioxide lasers have been developed that are capable of ablating dentin, enamel, and bone effectively and safely. In addition, research has shown that almost any wavelength of laser energy can be used if the pulse width is short enough and the peak power high enough. The next decade will, no doubt, see the development of practical nanosecond and femtosecond pulsed devices.

4.4.2 Mechanism of Hard Tissue Ablation

Mechanism of Ablation for Erbium Lasers

While a hydrokinetic theory was originally proposed to explain tissue interaction with Er,Cr:YSGG laser energy, a different mechanism is now most widely accepted [15–17]. All erbium lasers interact with tissue in a similar fashion. Ablation of enamel, dentin, and bone occurs through the explosive removal of tissue in a thermome-

◘ Fig. 4.19 Ultra-speed microphotography of Er:YAG laser energy interaction with an enamel surface

chanical event. Because laser energy at erbium wavelengths is very highly absorbed by water, molecules in the target tissue are rapidly superheated [18]. When the steam pressure within the tissue exceeds the structural strength of the overlying material, micro-explosions occur, ejecting particles of fractured material as shown in ◘ Fig. 4.19. This explosive phenomenon is initiated long before melting of the carbonated hydroxyapatite occurs.

Considerable pressure is needed to fracture enamel and dentin, so temperatures much higher than the intuitive 100 °C must be generated. A study suggests that ablation actually occurs at superheated temperatures of approximately 600 °C for enamel and 500 °C for dentin [19].

Mechanism of Ablation for Carbon Dioxide Lasers

Ablation at CO_2 wavelengths is predominantly a photothermal phenomenon [19]. Laser energy is directly absorbed by the carbonated hydroxyapatite. Melting of the mineral is nearly instantaneous, occurring in less than a picosecond, and is followed rapidly, with increased temperature, by vaporization and expulsion of molten mineral droplets [20]. This liquid mineral quickly resolidifies into characteristic globules routinely seen in SEMs. Lesser absorption by water and protein at these wavelengths only has a minor effect on the ablation process.

If laser energy at CO_2 wavelengths is applied in a continuous-wave mode, the principal interaction consists of melting of carbonated hydroxyapatite and massive transfer of thermal energy to the surrounding tooth structure. Charring and pulp death are the inevitable outcomes. However, applying this energy in super-pulsed mode with very short pulses, ablation as described above still occurs, but most of the thermal energy is ejected with the molten carbonated hydroxyapatite. Water spray directly removes further amounts of diffused energy.

Mechanism of Ablation for Femtosecond Lasers

Femtosecond lasers emit energy in extremely short pulses with durations of less than a nanosecond. Even with pulse energy as low as 20 μJ/pulse, a 400 fs pulse width produces peak power of 50 million W. This is sufficient to ionize carbonated hydroxyapatite directly to plasma. The plasma then quickly dissipates, carrying virtually all energy away and transferring almost none to surrounding tissues. Residual thermal energy and collateral damage are almost nonexistent. The wavelength of the applied laser energy is of little consequence as a study of efficient enamel ablation using a 1025 nm laser shows [21]. Ablation simply relies on plasma formation. It should be noted that this wavelength is nearly identical to that of the Nd:YAG laser (1064 nm). Peak ablation efficiency occurs at approximately 5 J/cm^2. Clinical femtosecond lasers are not currently available in dentistry but show great promise.

4.4.3 Role of Hydroxyapatite in Ablation

To further clarify the discussion above, carbonated hydroxyapatite is a major absorber of carbon dioxide laser energy and is the primary vehicle of ablation. It absorbs energy, injecting heat into the enamel or dentin, promoting dehydration and structural change. If sufficient energy is absorbed to raise the temperature to about 1280 °C, melting followed by vaporization of the enamel will occur. Melting is a short-lived change in physical state, and there is a new solid created upon cooling. This new solid has a modified structure and composition and will interact with subsequent laser energy differently.

This is a completely different phenomenon than that seen with erbium lasers where water is the primary vehicle of ablation. With these lasers, ablation is the explosive removal of tissue using thermal energy to vaporize internal water. In either case, if the enamel temperature is sufficiently elevated, some of the thermal energy will diffuse into the interior of the tooth compromising pulp vitality. Femtosecond lasers have the potential to nearly eliminate this problem.

A simple experiment illustrates this point. Directing erbium laser energy at a dried extracted tooth with no water spray using 240 mJ at 25 pps (commonly recommended parameters for enamel ablation) from a distance of 10 mm results in sub-ablative irradiation. After about 10 s, the tooth becomes too hot to hold having reached about 200 °C. No ablation has occurred, but a significant amount of energy has been absorbed [18]. Moving even closer (5 mm) causes melting, charring, and overheating of the enamel as depicted in ◘ Fig. 4.16. No true explosive ablation occurs under these conditions.

4.4.4 Cautionary Considerations

Interaction with Non-biological Materials in the Oral Cavity

In the course of applying clinical lasers to intraoral structures, a number of non-biological restorative materials will be encountered. Each of these may lead to reflection, absorption, scattering, refraction, or transmission of energy. Research on the potential negative results of accidental or intentional interaction has been sparse.

Many publications have asserted that laser energy is reflected off the surface of metal restorations and so is of little consequence except to use caution with reflective surfaces. In fact, laser energy of most wavelengths is absorbed by metals. Lasers are routinely used in industrial situations to both cut and weld steel, copper, and aluminum.

◘ Figure 4.20 illustrates the interaction with existing amalgam restorations. Some might suggest that the interaction is due to adsorbed water, but ◘ Fig. 4.20b shows that the interaction with dental amalgam powder in the absence of water displays the destructive effects of the interaction on the laser tip.

All of these photographs represent the effect of commonly used laser parameters. This suggests that these negative side effects will likely occur any time restorative materials are inadvertently exposed to laser irradiation.

Laser tips and fibers can become coated with exploded or melted debris making them ineffective for future use. Techniques have been developed to alleviate this problem [21] (◘ Fig. 4.21).

Other Considerations

As discussed earlier in this chapter, applying erbium energy to enamel, dentin, or bone without adequate irrigation will almost always be disastrous. The significant thermal energy absorbed by the target will cause irreversible changes. A minimum of 8 mL/min is the recommended level.

Majaron [22] showed that maintaining a gap of 0.3–0.7 mm between the tip and tissue increased enamel ablation significantly. Theoretically, this would allow more water in to flush accumulated debris, allow debris to exit the gap, and remove thermal energy that had been absorbed by the hydroxyapatite. As the distance exceeds 1 mm, decreases in energy density as well as absorption and refraction from interposed water will minimize these advantages.

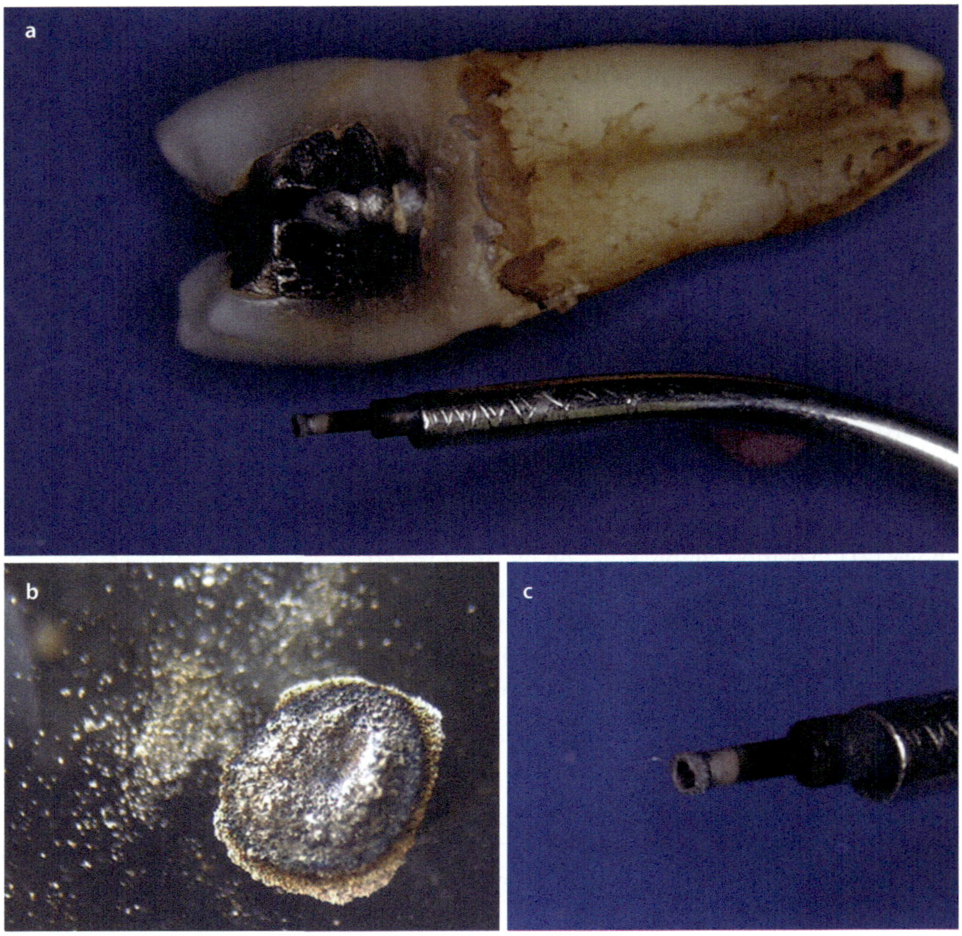

Fig. 4.20 **a** The erbium laser energy produces a significant interaction with amalgam restorations. **b** Interaction with amalgam powder suggests that it is not interaction with adsorbed water. Any such interaction will severely damage the laser tip. **c** An erbium laser was applied at 240 mJ and 25 pulses per second

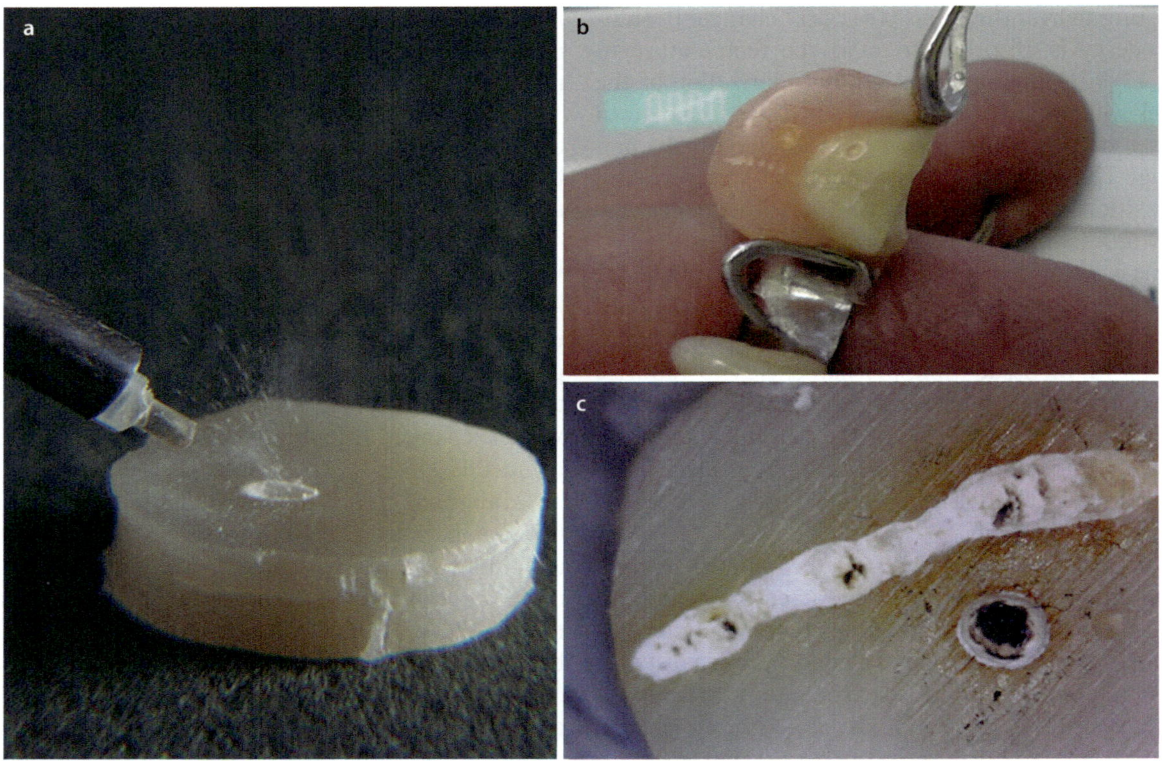

Fig. 4.21 Energy from both erbium lasers **a** and **b** as well as diode lasers **c** interacts with composite resins and dental acrylic materials at commonly used power levels

4.5 Soft Tissue Surgery Considerations

4.5.1 Most Appropriate Laser for Soft Tissue Procedures

All of the existing dental lasers can surgically incise soft tissue.

The diode laser is the most popular dental laser with thousands in use in dental offices around the world. They are attractive because they are small, lightweight, low maintenance, durable, and relatively inexpensive.

While a valuable tool, these devices are plagued by a very significant shortcoming. They are very poorly absorbed in soft tissue. With diode lasers ranging from approximately 808 to 1064 nm, melanin, hemoglobin, and protein are the primary chromophores. However, water which makes up more than 70% of most tissues is not a chromophore and produces almost no interaction at diode wavelengths. This suggests that the diode is a poor laser for incising soft tissue.

Since the laser energy spreads through a large area of tissue, the risk of significant collateral damage is high. On the other hand, this energy is highly absorbed in blood so causes excellent coagulation and hemostasis.

The Nd:YAG laser at 1064 nm suffers from the same poor absorption as all diode lasers, again suggesting that this is not a good laser for incising soft tissue. Its advantage lies in its very high peak power with short pulse width providing thermal containment and moderate ablation with very good hemostasis. Nd:YAG lasers are more complex and considerably more expensive than diode lasers.

Erbium lasers have a primary chromophore of water, which is the most abundant constituent of biological tissue making them an excellent choice to perform surgical soft tissue procedures. Both the Er:YAG and the Er,Cr:YSGG lasers are absorbed efficiently in a few microns of tissue. Since little energy is transferred to the surrounding area, little collateral tissue damage occurs. Because of this same property, these lasers do not produce significant hemostasis from thermal absorption at the incision margins.

Finally, the CO_2 laser has primary chromophores of water and collagen, making it a useful soft tissue laser. Its action is very superficial but produces melting of collagen as well as explosion of tissue through steam generation. Since this energy is delivered via a noncontact handpiece, precision is more difficult to achieve.

4.5.2 Fiber Initiation Consideration

While diode lasers are poor soft tissue lasers, they are small and relatively inexpensive. The optical fiber delivery system is flexible and very maneuverable allowing access to hard-to-reach areas of the oral cavity. Fiber initiation provides a technique to circumvent this laser's shortcomings.

An energy-absorbing material is applied to the end of the fiber as shown in ◘ Fig. 4.22 in a process called initiation or activation. Material is transferred to the fiber end creating a thin layer that absorbs subsequent laser energy.

When the laser is activated, this thin layer absorbs the emitted photons and is rapidly heated to several hundred degrees. This "hot tip" is used to melt proteins, thus separating the tissue. The argument is often made that this technique is similar to electrosurgery or applying a heated instrument. However, because only the thin layer of absorbent material is heated, the thermal energy is concentrated and there is significantly less collateral thermal damage than with these other modalities.

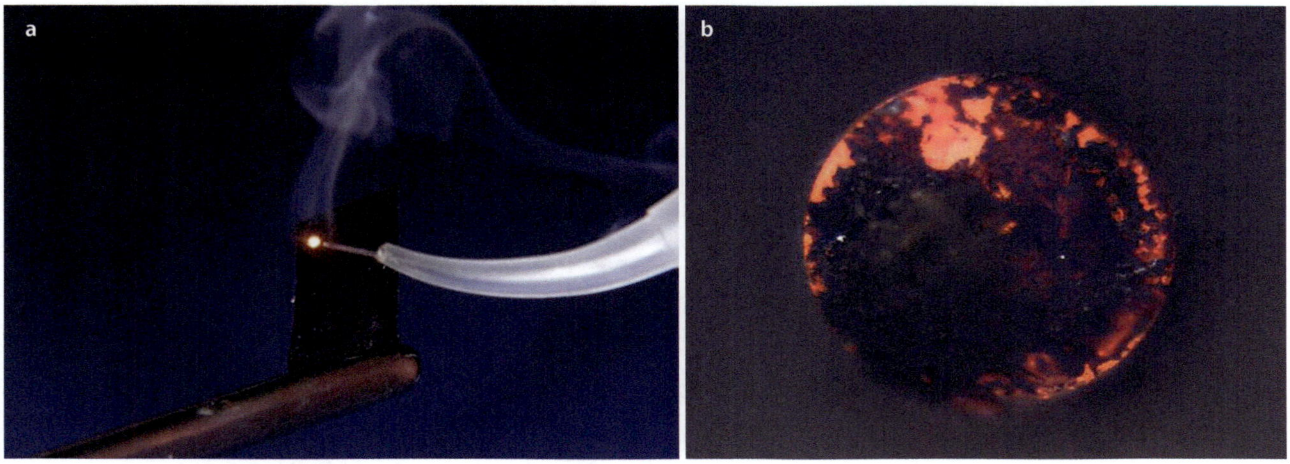

◘ **Fig. 4.22** The technical process of initiation is shown in **a** while the resulting coat on the fiber end is shown in **b**

While each laser manufacturer has suggestions for proper fiber initiation, the following technique has been shown to provide superior results:

- Adjust the laser to a very low power setting of less than 0.5 W.
- Touch the fiber tip to articulating paper without activating the laser.
- Activate the laser momentarily until a perforation is seen in the articulating paper.
- Repeat at least eight times.
- Observe the aiming beam against a surface. If it is still clearly visible, repeat more times.

4.5.3 Effect of Fiber Size

As has been discussed in ► Sect. 4.2.6 above and ◘ Table 4.2, smaller fiber diameters significantly increase power density and positively affect the ability of the laser to incise tissue. However, fibers with a diameter smaller than 400 μm are very flexible affecting the ability to cut precisely and are at higher risk of spontaneous fracture.

4.5.4 Effect of Debris Accumulation

In the course of clinical treatment, cellular debris adheres to the warm fiber tip as shown in ◘ Fig. 4.23. This material absorbs laser energy and is progressively heated, increasing the risk of collateral thermal damage. It is prudent to remove this debris repeatedly during a surgical procedure. This is best accomplished with wet gauze.

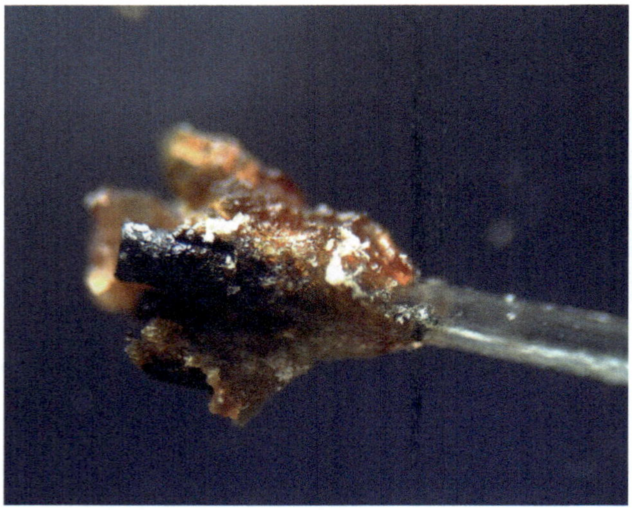

◘ **Fig. 4.23** During clinical application, tissue adheres to the fiber tip. It in turn is heated causing collateral tissue damage

4.5.5 Precooling of Tissue

Collateral damage occurs through the prolonged heating of cells. A study by Simanovskii et al. [23] asserted that mammalian cells can survive a temperature increase to 42–47 °C for prolonged periods of time. Above this point, the length of time at higher temperatures becomes a critical variable. Cells can only survive at 70 °C for about 1 s and 130 °C for 300 μs. Since optical fiber tip temperature can exceed 800 °C, it becomes apparent that this is the source of damage.

The margins of a surgical excision will be rapidly heated although not to this level. It has been shown that approximately 350 μm of collateral tissue death may occur [24]. Precooling the tissue with an air/water spray can reduce, but certainly not eliminate, collateral damage. Frequent cooling during procedures is also considered to be of benefit. However, using a constant flow of water during surgery is not recommended. Vision will be compromised, and, more importantly, thermal energy will be diffused over the tissue surface scalding a superficial layer as shown in ◘ Fig. 4.24.

4.5.6 Cautionary Considerations

The preceding discussions illustrate the need for careful attention to tissue response during even the most rudimentary procedure.

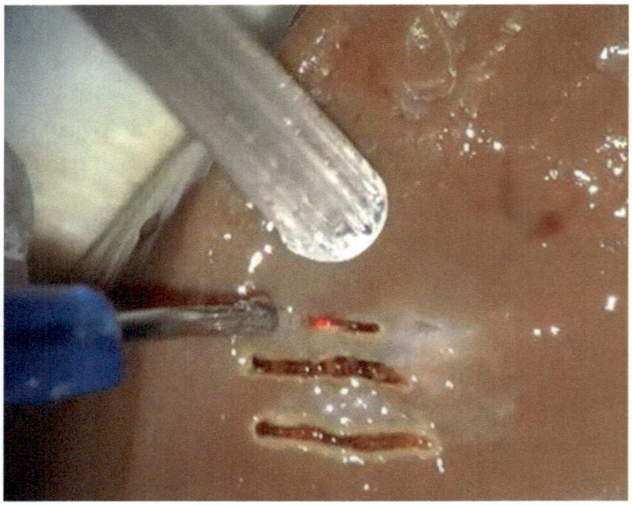

◘ **Fig. 4.24** Three cuts all performed with a 940 nm diode laser at 1 W, CW. The bottom cut done with no precooling and air or water irrigation. Note the significant collateral damage. The middle cut was performed with air directed over the surface. Speed of cut was slower but as effective with much less collateral damage. Applying an air/water spray during treatment in the top cut caused significant surface cell death. (Courtesy of Dr. Mark Cronshaw)

4.6 Photobiomodulation Considerations

It is well established that, in general, photobiomodulation can have a positive effect on tissue [25–27]. However, photonic energy must reach target cells at the appropriate intensity to be effective. This energy density is widely accepted to be between 3 and 10 J/cm^2, relative to the individual cell surface, in accordance with the Arndt-Schultz law [28, 29]. At less than 3 J/cm^2, there is insufficient energy absorbed by the cell to increase its metabolism, while densities of more than 10 J/cm^2 appear to inhibit cellular function.

While energy is applied on a macroscopic level, the interaction with tissue occurs on a subcellular level [30]. Ideally, each individual cell in the path of the applied energy should be illuminated with the same number of photons within this range. The positive effects of photobiomodulation are very strongly associated with the density of photons striking the cells. Unfortunately, a method of delivering photons to a group of individual cells, often deep within a tissue mass, in a uniform and predictable manner has been lacking.

4.6.1 Most Appropriate Laser for Photobiomodulation

Between approximately 600 and 1200 nm, very few absorbing chromophores exist allowing laser energy to penetrate deeply into the tissue mass. This range of wavelengths has been labeled the "optical window" [31] and is exploited in most photobiomodulation protocols. In general, diode lasers are the most appropriate laser for this application.

4.6.2 Recommended Parameters

Further chapters will explore the intricacies of dosage in photobiomodulation. The energy density at the cellular level is a key determinant of biostimulation and appears to be between 3 and 10 J/cm^2 depending on the cell type and should be applied at a rate of approximately 100 mW/cm^2.

For lesions on the surface of tissue, calculation of the appropriate dose is rudimentary. However, as energy enters tissue, it is moderately absorbed and highly scattered reducing the effective dose.

4.6.3 Distribution of Applied Energy in a Tissue Volume

As energy penetrates a tissue mass, it is absorbed and widely scattered. Bashkatov [32] asserted that photons are 100 times more likely to be scattered than to be absorbed. When treating a muscle or a temporomandibular joint, energy reaching the target cells is predictably and drastically reduced. ◘ Figure 4.25 is a presentation of energy reaching cells deep within the tissue.

4.6.4 Effect of Lesion Depth and Tissue Type on Parameter Calculations

Many of the anatomic structures that are purported to benefit from low-level laser therapy such as the TMJ or joint bursa are located at some distance within the surrounding tissue. ◘ Figure 4.26 suggests that, at 5 mm into representative tissue ranging from beef muscle (similar to human muscle) to beef liver (similar to highly vascularized tissue), energy density has decreased to about 10% of its surface value.

4.6.5 Cautionary Considerations

Arany et al. [33] state that a skin temperature increase to greater than 45 °C is phototoxic. While it is tempting to apply large amounts of energy in a short period of time in order to stimulate deep structures and to speed treatment, very negative unintended consequences may occur in overlying tissues.

Of all the protocols for laser use in dentistry, photobiomodulation is at the greatest risk of misunderstood technique and unintended consequences. Unfortunately, there is no direct visual feedback of effectiveness leaving the clinician to rely on an understanding of parameters.

While not settled science, the currently accepted guidelines recommend applying a fluence of 100 mW/cm^2 and a total cellular dose of 3–10 J/cm^2.

There are several unique devices available for applying therapeutic doses including large-diameter (8–10 mm) optical fibers, rectangular therapy/bleaching handpieces, and flattop handpieces. Many clinicians use the expedient of a surgical optical fiber held at a distance from the tissue surface as a perceived equivalent. However, these devices can produce very different tissue exposure.

If 1 W of power is applied through a 400 μm fiber from a distance of 10 mm for 30 s as it is moved over a tissue area of 2 cm^2, the result is:

$$\text{Fluence} = 1\,\text{W} / 2\,\text{cm}^2 = 0.5\,\text{W} / \text{cm}^2 = 500\,\text{mW} / \text{cm}^2$$

$$\text{Irradiance} = 0.5\,\text{W} / \text{cm}^2 \times 30\,\text{s} = 15\,\text{J} / \text{cm}^2$$

Applying the same parameters through an 8 mm diameter PBM handpiece in contact without movement results in:

4

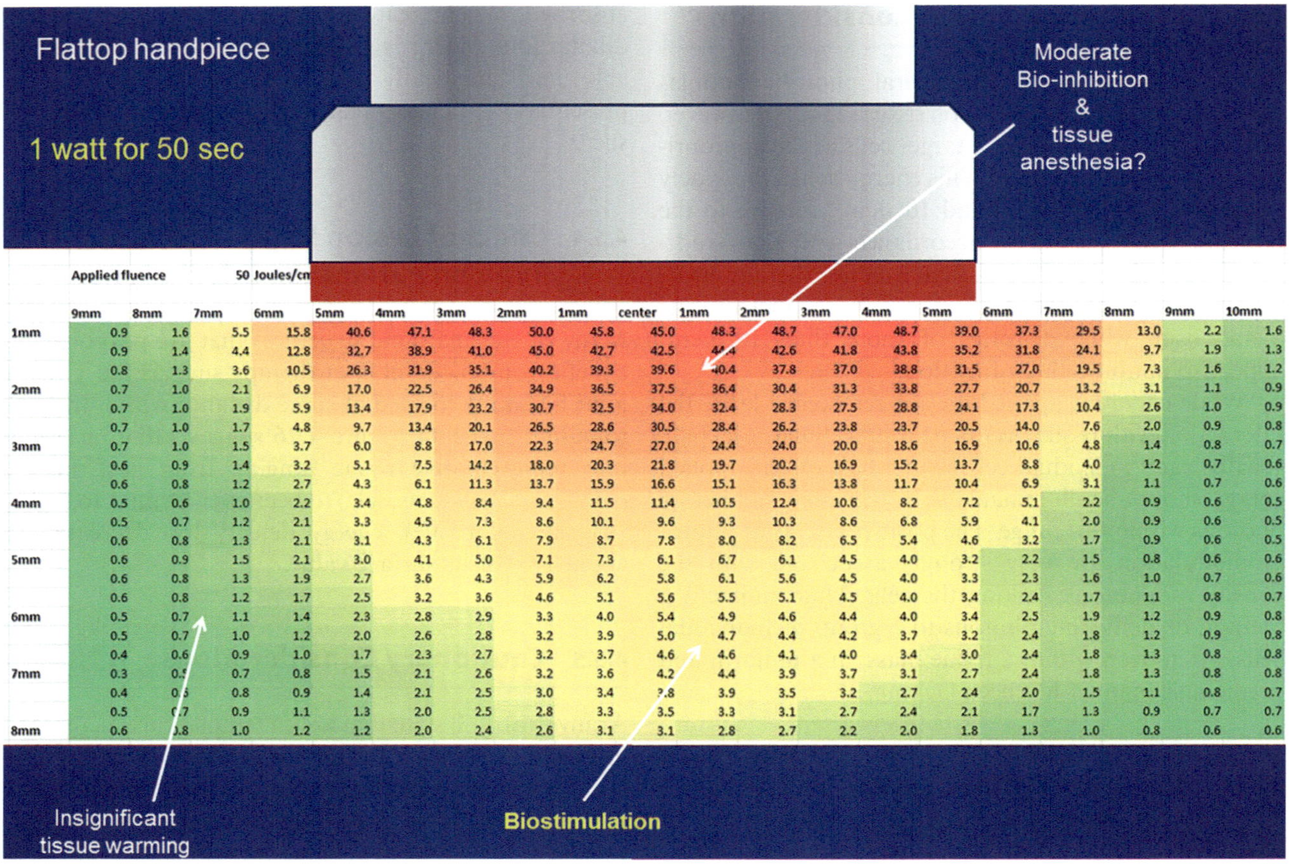

Fig. 4.25 All tissue areas in red exceed the optimal range and experience bioinhibition and perhaps even irreversible cellular damage. This zone extends about 3.5 mm into the tissue. Biostimulation would occur from 10 J/cm² to about 3 J/cm² as depicted in yellow. This semicircular band is about 4 mm wide directly in front of the handpiece tapering down to about 1 mm wide on the periphery. Beyond that band, the tissue cells would be perhaps warmed a little bit with no discernable photobiomodulation effect

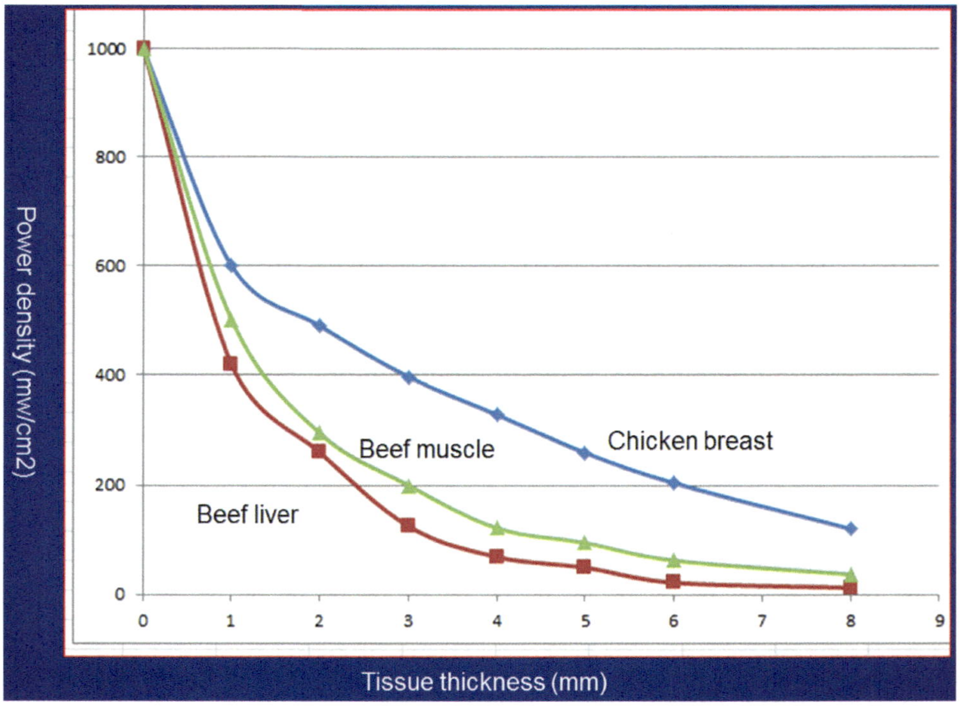

Fig. 4.26 Power density diminishes rapidly as the laser energy penetrates the tissue. Beef muscle and liver are representative of the range of tissues seen in the oral cavity

$$\text{Fluence} = 1\,\text{W} / 0.5\,\text{cm}^2 = 2\,\text{W} / \text{cm}^2 = 2000\,\text{mW} / \text{cm}^2$$

$$\text{Irradiance} = 2\,\text{W} / \text{cm}^2 \times 30\,\text{s} = 60\,\text{J} / \text{cm}^2$$

Finally, applying the same parameters through a rectangular therapy/bleaching handpiece with dimensions of 8 mm by 35 mm results in:

$$\text{Fluence} = 1\,\text{W} / 2.8\,\text{cm}^2 = 0.36\,\text{W} / \text{cm}^2 = 360\,\text{mW} / \text{cm}^2$$

$$\text{Irradiance} = 0.36\,\text{W} / \text{cm}^2 \times 30\,\text{s} = 10.7\,\text{J} / \text{cm}^2$$

It is easy to understand that, while using identical laser parameters, each of these common delivery systems will have a vastly different effect on the target tissue.

Whichever means of energy delivery is chosen, it is incumbent on the clinician to carefully calculate the tissue dose and to adjust parameters appropriately. To achieve the recommended protocol:

- The 400 μm fiber should be applied at 0.2 W for 50 s from a distance of 10 mm.
- The PBM handpiece should be applied at 0.05 W (an impossibility) for 100 s.
- The therapy handpiece should be applied at 0.3 W for 90 s.

Summary

Knowledge of actual laser energy being applied to the target tissue with any laser is vital to achieving positive results. This chapter has focused on understanding the meaning and the effect of each of the variable parameters available to the clinician when making a decision to apply laser technology for the benefit of a patient.

4.7 Appendix – Laser Photonic Energy – Mathematical Quantification and Calculation

Calculation of parameters is provided for each of the lasers shown below. While some readers may find this rudimentary, others will appreciate the review of physics and mathematical calculations.

Please note that, for calculations to be valid, it is necessary to convert all dimensions to centimeters, all times to seconds, all energies to joules, and all powers to watts before computation. This standardized system allows the statement of parameters as W/cm^2, etc.

Because of basic differences in variable-gated continuous-wave and true pulsed lasers, each needs a different set of equations to calculate the same set of variables (Fig. 4.27).

4.7.1 Variable Gated Continuous Wave Laser (Fig. 4.27)

Emission Cycle

If the laser is “on” for 20 ms and “off” for 80 ms, then it is on for 20% of the time:

$$\text{Percentage of "on – time"} = 20\,\text{ms} / (20\,\text{ms} + 80\,\text{ms}) = 20\,\text{ms} / 100\,\text{ms} = 0.2 = 20\%$$

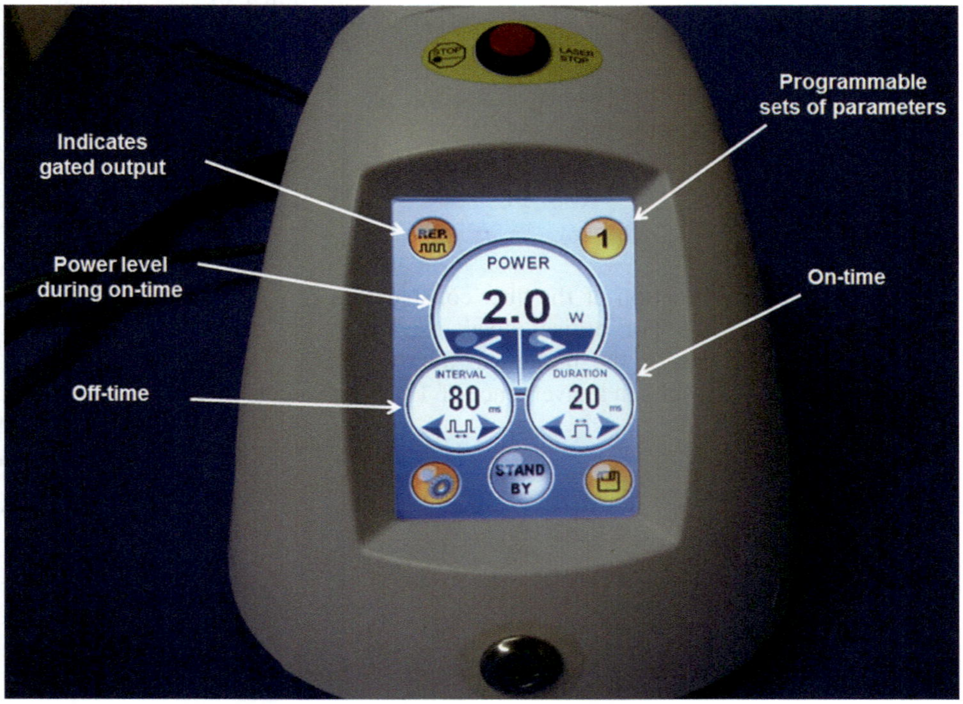

Fig. 4.27 This 810 nm diode laser has limited variables that may be controlled directly. It is used here with a 400 μm diameter optical fiber

Average Power

This diode laser allows direct control of the output power, but the power displayed is the "continuous wave" output rather than the average power. Any time that this laser is emitting photons, it is emitting 2 W. Average power is calculated as:

$$\begin{aligned}\text{Power}_{\text{avg}} &= \text{power}_{\text{indicated}} \times \text{percentage of "on} - \text{time"} \\ &= 2\,\text{W} \times \left(20\,\text{ms} / \left(20\,\text{ms} + 80\,\text{ms}\right)\right) = 0.4\,\text{W}\end{aligned}$$

If the lengths of the on and off periods are changed but stay in the same ratio, average power will be the same.

$$\begin{aligned}\text{Power}_{\text{avg}} &= \text{power}_{\text{indicated}} \times \text{percentage of "on} - \text{time"} \\ &= 2\,\text{W} \times \left(80\,\text{ms} / \left(80\,\text{ms} + 320\,\text{ms}\right)\right) = 0.4\,\text{W}\end{aligned}$$

Peak Power

For this diode laser, peak power will always be the same as that indicated on the display, no matter what magnitude or ratio of "on" and "off" times is selected.

$$\text{Power}_{\text{peak}} = \text{power}_{\text{indicated}} = 2\,\text{W}$$

Pulse Width

Pulse width, in this case, is simply the "on" time:

$$\text{Pulse width} = 20\,\text{ms}$$

Pulse Repetition Rate

One repetition is the same as one emission cycle or the sum of one "on" time and one "off" time. So:

$$\text{Emission cycle} = 20\,\text{ms}\left(\text{"on"}\right) + 80\,\text{ms}\left(\text{"off"}\right) = 100\,\text{ms}$$

Then, the number of these events that occur in 1 s represents the pulse repetition rate.

$$\text{Pulse repetition rate} = 1\,\text{s} / 100\,\text{ms} = 10\,\text{pulses per second}$$

Beam Divergence

Beam divergence is a function of the fiber construction and numerical aperture. For optical fibers used with diode and Nd:YAG lasers, the numerical aperture is commonly 0.22. This results in a divergence of 12.7° per side angle.

Tip Area

Tip area is the radius squared of the tip multiplied by π:

$$\begin{aligned}\text{Tip area} &= 200\,\mu m \times 200\,\mu m \times 3.14159 \\ &= 0.02\,\text{cm} \times 0.02\,\text{cm} \times 3.14159 = 0.00126\,\text{cm}^2\end{aligned}$$

Spot Diameter at Tissue

Diode lasers are used in contact for surgical procedures. Therefore, the spot diameter at the tissue surface is the same as the tip diameter:

$$\text{Spot diameter at tissue} = 400\,\mu m = 0.04\,\text{cm}$$

Spot Area at Tissue

Similarly, the spot area at the tissue surface is the same as the area of the optical fiber tip.

$$\text{Spot area at tissue} = 0.00126\,\text{cm}^2$$

Power Density

Power density is the total energy in joules passing through a specific area in 1 s of time. With the attached 400 μm laser fiber, this diode laser produces:

$$\begin{aligned}\text{Power density}_{\text{avg}} &= \text{power}_{\text{avg}} / \text{area} = 0.4\,\text{W} / 0.00126\,\text{cm}^2 \\ &= 318\,\text{W} / \text{cm}^2\end{aligned}$$

$$\begin{aligned}\text{Power density}_{\text{peak}} &= \text{power}_{\text{peak}} / \text{area} = 2\,\text{W} / 0.00126\,\text{cm}^2 \\ &= 1587\,\text{W} / \text{cm}^2\end{aligned}$$

Total Energy

The total amount of energy delivered to the tissue during a 30-s treatment is:

$$\text{Total energy} = \text{power}_{\text{avg}} \times \text{time} = 0.4\,\text{W} \times 30\,\text{s} = 12\,\text{J}$$

Energy Density

If the fiber tip is held stationary in one location, the energy density is calculated as:

$$\begin{aligned}\text{Energy density}_{\text{avg}} &= \text{power}_{\text{avg}} \times \text{time} / \text{area} \\ &= 0.4\,\text{W} \times 30\,\text{s} / 0.00126\,\text{cm}^2 \\ &= 9540\,\text{J} / \text{cm}^2\end{aligned}$$

$$\begin{aligned}\text{Energy density}_{\text{peak}} &= \text{power}_{\text{peak}} \times \text{time} / \text{area} \\ &= 2\,\text{W} \times 30\,\text{s} / 0.00126\,\text{cm}^2 \\ &= 47{,}610\,\text{J} / \text{cm}^2\end{aligned}$$

Movement of the Tip

If the fiber tip is moved through the tissue at a speed of 2 mm/s, the applied energy is distributed over a much larger area affecting energy density. This may be calculated as shown below (Fig. 4.28):

Energy density with movement:

$$\text{Area} = \text{width} \times \text{length traveled in } 30\,\text{s}$$

$$\text{Area} = 0.04\,\text{cm} \times 6\,\text{cm} = 0.24\,\text{cm}^2$$

$$\text{Energy density} = 0.4\,\text{W} \times 30\,\text{s} / 0.24\,\text{cm}^2 = 50\,\text{J}/\text{cm}^2$$

4.7.2 True Pulsed Laser with Control of Average Power (Fig. 4.30)

Average Power

This erbium laser allows direct control of the average output power. In this case, it has been adjusted to 4 W.

Pulse Width

Pulse width is established by choosing one of the two values available—"H" = "hard tissue" = 60 µs or "S" = "soft tissue" = 700 µs. In this case, 60 µs has been selected.

Pulse Repetition Rate

Pulse repetition rate is selected as 15 pulses per second (pps) Fig. 4.29

Emission Cycle

The emission cycle indicates the percentage of time that the laser is actually emitting energy. With any true pulsed laser, this is a very small amount. It is calculated as the length of time that the laser is emitting during each second or the sum of the width of all pulses:

$$\text{Emission cycle} = \begin{pmatrix} \text{number of pulses per second} \\ \times \text{ width of each pulse} / 1\,\text{s} \end{pmatrix} \times 100\%$$

$$\begin{aligned}\text{Emission cycle} &= 15 \times 60\,\mu s = 15 \times 0.0006\,\text{s} \\ &= 0.009 \times 100 = 0.9\%\end{aligned}$$

Energy Per Pulse

If 4 W is produced in 1 s, then 4 J/s is produced. If this energy is divided equally between 15 pulses, then:

$$\text{Energy per pulse} = \text{energy per second} / \text{number of pulses per second}$$

$$\begin{aligned}\text{Energy per pulse} &= 4\,\text{J}/\text{s}/15\,\text{pps} \\ &= 0.267\,\text{J}/\text{pulse} \\ &= 267\,\text{mJ}/\text{pulse}\end{aligned}$$

A key point must be considered here. With any pulsed laser that allows control of average power, the energy per pulse changes if the pulse repetition rate changes. As an example, if the laser in Fig. 4.30 has the pulse repetition rate changed to 5 pulses per second, while the average power is unchanged, the energy per pulse changes to:

$$\begin{aligned}\text{Energy per pulse} &= 4\,\text{J}/\text{s}/5\,\text{pps} \\ &= 0.8\,\text{J}/\text{pulse} \\ &= 800\,\text{mJ}/\text{pulse}\end{aligned}$$

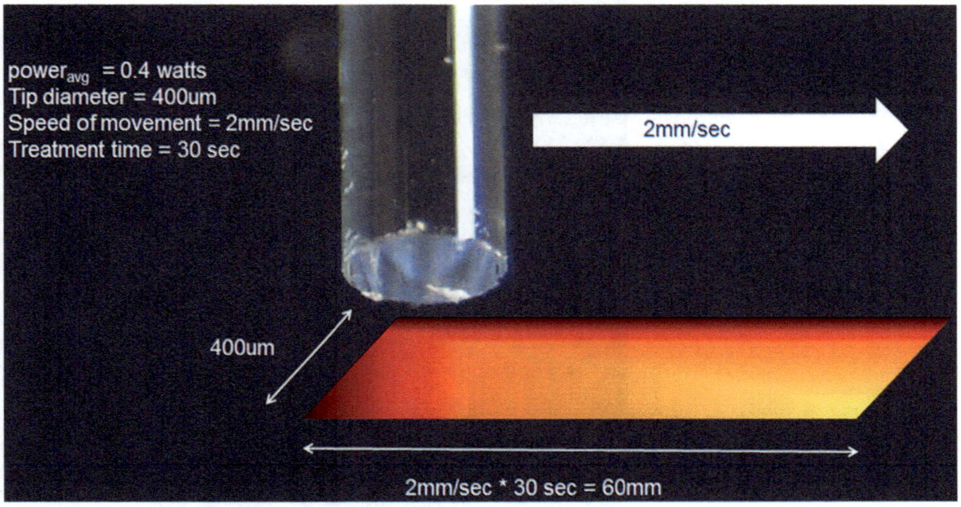

Fig. 4.28 When the optical fiber or emitting tip is moved across the tissue during treatment, the energy produced is distributed over a larger area based on the speed of movement. When determining energy density received by an individual group of cells, it is necessary to calculate the entire area that is irradiated

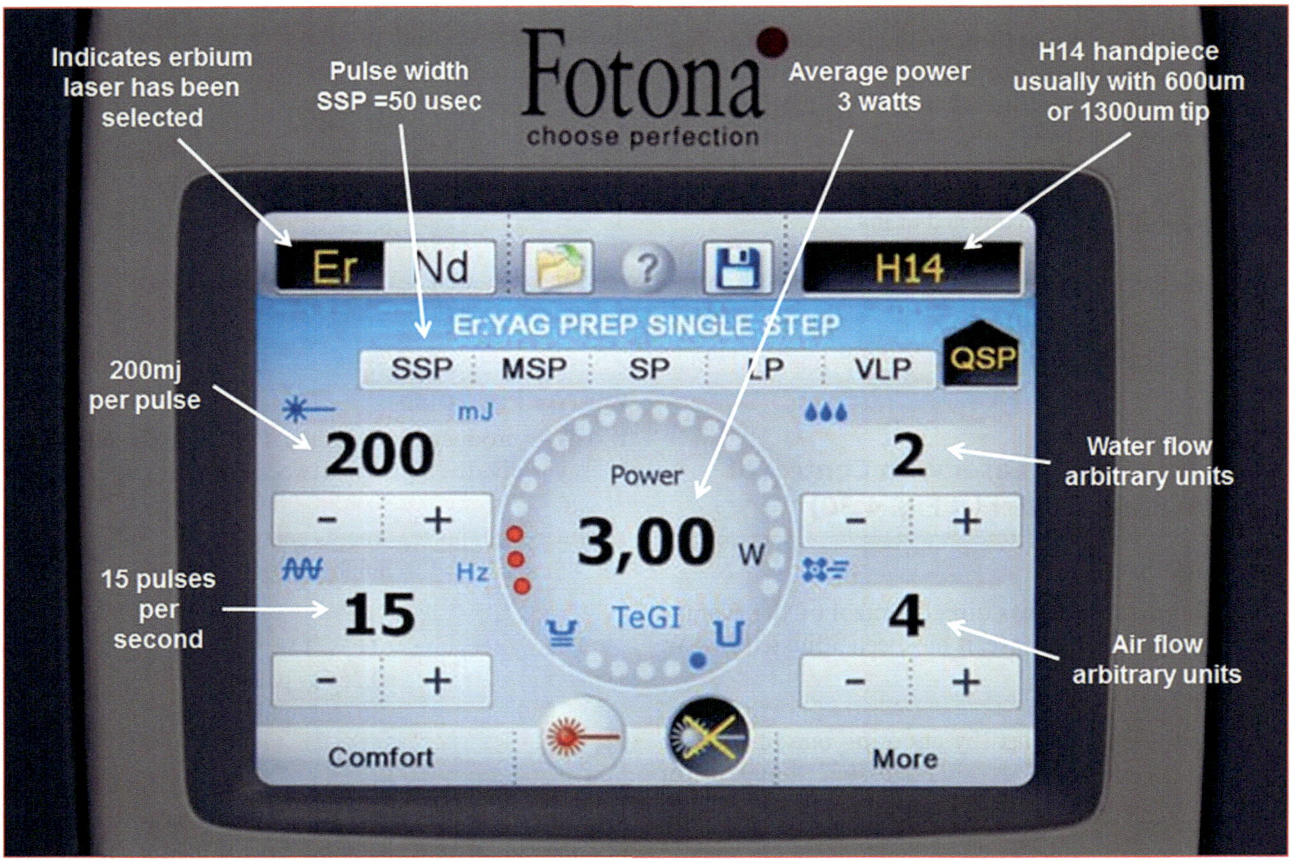

Fig. 4.29 This laser contains both a 2940 nm Er:YAG laser and a 1064 nm Nd:YAG laser. In erbium mode, it allows selection of six different pulse widths. However, the operator must refer to a manual to find the value. It allows control of pulse energy and pulse frequency. The resulting average power is displayed but not directly selectable

Fig. 4.30 This 2780 nm Er,Cr:YSGG laser has a choice of two pulse widths. It allows control of average power, but energy per pulse and peak power must be calculated. Levels of auxiliary air and water are set as percentage of an arbitrary value

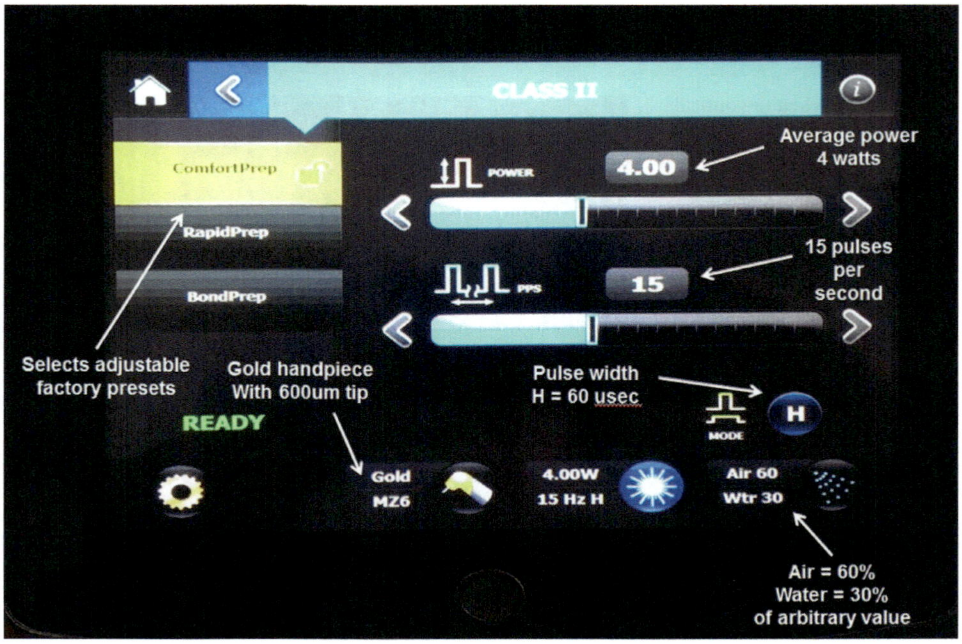

This concept applies to any Er:YAG, Er,Cr:YSGG, or Nd:YAG laser that provides control of average power instead of energy per pulse.

Peak Power

Peak power represents the maximum instantaneous power that the laser produces.

Peak power can be calculated as:

$$\text{Power}_{\text{peak}} = \text{energy per pulse} / \text{pulse width}$$

$$\text{Power}_{\text{peak}} = 0.267\,\text{J} / 0.00006\,\text{s} = 4444\,\text{W}$$

Tip Area

Commonly available tips for this laser are cylindrical, tapered, and chisel shaped. While calculating the area, divergence and spot area for chisel tips are beyond this discussion, and the diameter of the output end of a tapered tip can be treated the same as a cylindrical tip as calculated below.

Assuming the use of a 600 μm diameter cylindrical tip allows the following calculations:

Tip area is the radius squared of the tip multiplied by π:

$$\begin{aligned}\text{Tip area} &= 300\,\mu m \times 300\,\mu m \times 3.14159\\ &= 0.03\,\text{cm} \times 0.03\,\text{cm} \times 3.14159\\ &= 0.00283\,\text{cm}^2\end{aligned}$$

A tip that tapers from 1200 to 600 μm at the output end would have the same tip area of 0.00283 cm^2.

Spot Diameter at Tissue

Erbium lasers are usually used in contact for soft tissue surgical procedures. Therefore, the spot diameter at the tissue surface is the same as the tip diameter:

$$\text{Spot diameter at tissue} = 600\,\mu m = 0.06\,\text{cm}$$

Spot Area at Tissue

For this scenario, the spot area at the tissue surface is the same as the area of the optical fiber tip.

$$\text{Spot area at tissue} = 0.00126\,\text{cm}^2$$

Beam Divergence

When this or any laser is used out of contact, the effect of beam divergence has a critical effect on spot diameter at the tissue surface and therefore energy and power density.

Beam divergence is a function of the particular tip chosen. For tips used with this laser, it is commonly 12° per side angle. For a tip-to-tissue distance

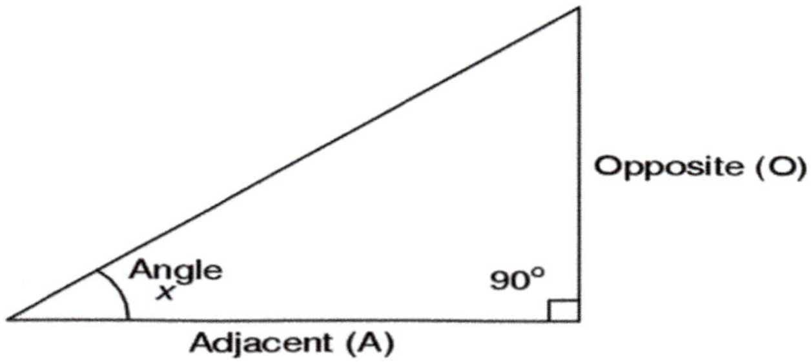

Fig. 4.31 Basic geometry establishes the relationship between different variables in a right triangle

(O) of 2 mm and an angle (x), from geometry (Fig. 4.31):

$$O = A \times \tan(x)$$

$$O = 2\,\text{mm} \times \tan(12°)$$

$$O = 2\,\text{mm} \times 0.2126 = 0.425\,\text{mm}$$

$$\begin{aligned}\text{Spot radius at tissue} &= \text{tip radius} + \text{opposite}\\ &= 0.03 + 0.0425 = 0.0725\,\text{cm}\end{aligned}$$

$$\text{Spot diameter at tissue} = 0.145\,\text{cm}$$

As shown in Fig. 4.32, at a tip-to-tissue distance of 2 mm with a 12° beam divergence per side angle:

$$\begin{aligned}\text{Spot area at tissue} &= \pi \times r^2\\ &= 3.14159 \times 0.0725\,\text{cm} \times 0.0725\,\text{cm}\\ &= 0.0165\,\text{cm}^2\end{aligned}$$

The effect of divergence on parameters is similar to that with an optical fiber.

Power Density

Power density is the total energy in joules passing through a specific area in 1 s of time. With the attached 600 μm laser fiber and a tip-to-tissue distance of 2 mm, this erbium laser produces:

$$\begin{aligned}\text{Power density}_{\text{avg}} &= \text{power}_{\text{avg}} / \text{area}\\ &= 4\,\text{W} / 0.0165\,\text{cm}^2 = 242\,\text{W} / \text{cm}^2\end{aligned}$$

$$\begin{aligned}\text{Power density}_{\text{peak}} &= \text{power}_{\text{peak}} / \text{area}\\ &= 4444\,\text{W} / 0.0165\,\text{cm}^2\\ &= 269{,}333\,\text{W} / \text{cm}^2\end{aligned}$$

This peak power calculation should make it clear how a pulsed laser can ablate enamel.

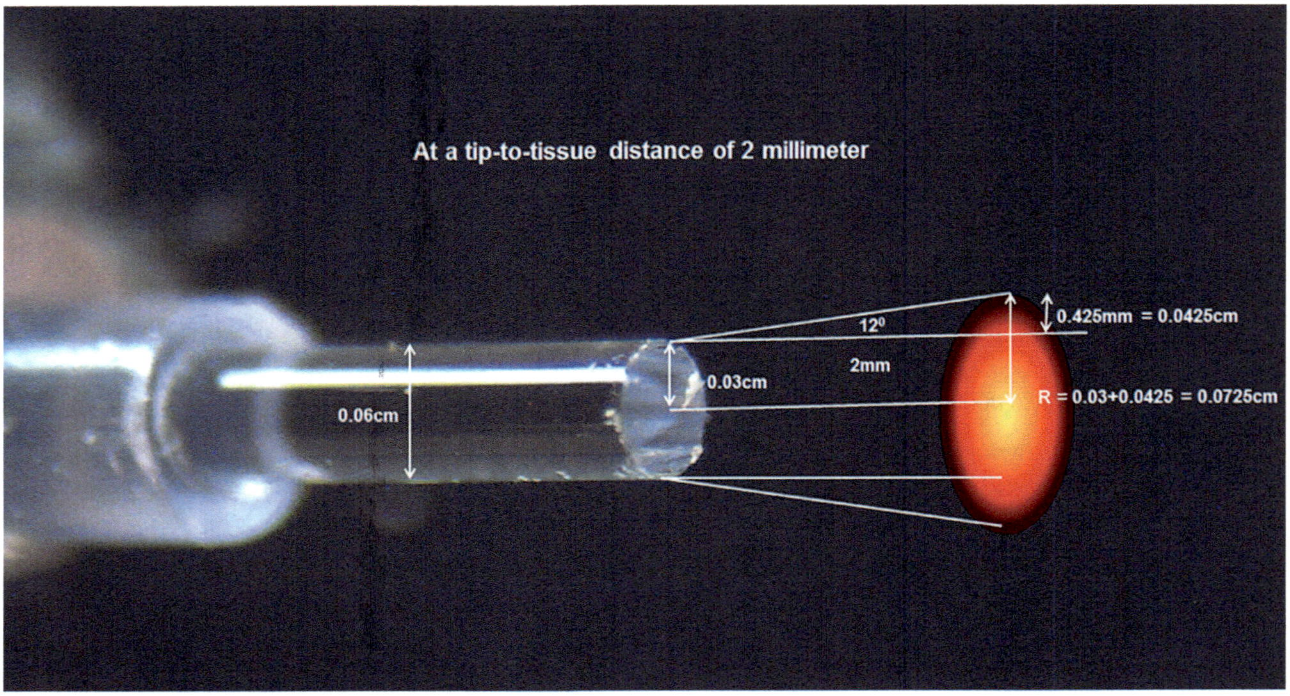

Fig. 4.32 This is a depiction of the various values calculated to determine the spot area at the tissue with a divergent beam

Total Energy

The total amount of energy delivered to the tissue during a 30-s treatment is:

$$\text{Total energy} = \text{power}_{\text{avg}} \times \text{time} = 4\,\text{W} \times 30\,\text{s} = 120\,\text{J}$$

Energy Density

If the fiber tip is held stationary in one location, the energy density is calculated as:

$$\begin{aligned}\text{Energy density}_{\text{avg}} &= \text{power}_{\text{avg}} \times \text{time} / \text{area} \\ &= 4\,\text{W} \times 30\,\text{s} / 0.0165\,\text{cm}^2 \\ &= 7273\,\text{J} / \text{cm}^2\end{aligned}$$

$$\begin{aligned}\text{Energy density}_{\text{peak}} &= \text{power}_{\text{peak}} \times \text{time} / \text{area} \\ &= 4444\,\text{W} \times 30\,\text{s} / 0.0165\,\text{cm}^2 \\ &= 8{,}080{,}000\,\text{J} / \text{cm}^2\end{aligned}$$

4.7.3 True Pulsed Laser with Control of Pulse Energy (Fig. 4.29)

Pulse Width

Pulse width is established by choosing one of the six values available—choosing the SSP mode selects a pulse width of 50 μs.

Pulse Repetition Rate

Pulse repetition rate is selected as 15 pulses per second (pps).

Emission Cycle

The emission cycle indicates the percentage of time that the laser is actually emitting energy. With any true pulsed laser, this is a very small amount. It is calculated as the length of time that the laser is emitting during each second or the sum of the width of all pulses:

$$\text{Emission cycle} = \left(\begin{array}{l}\text{number of pulses per second} \\ \times \text{ width of each pulse} / 1\,\text{s}\end{array}\right) \times 100\%$$

$$\begin{aligned}\text{Emission cycle} &= 15 \times 50\,\mu s = 15 \times 0.0005\,\text{s} \\ &= 0.0075 \times 100 = 0.75\%\end{aligned}$$

Energy Per Pulse

The energy per pulse is directly selected as 200 mJ.

Average Power

This erbium laser indicates an average output power of 3 W. In this case, it is not directly adjustable but is internally calculated by the laser. It can be calculated as shown below:

$$\begin{aligned}\text{Power}_{\text{avg}} &= \text{energy per pulse} \\ &\quad \times \text{number of pulses per second}\end{aligned}$$

$$\text{Power}_{\text{avg}} = 0.2\,\text{J} / \text{pulse} \times 15\,\text{pulses per second} = 3.0\,\text{W}$$

Peak Power

Peak power represents the maximum instantaneous power that the laser produces.

Peak power can be calculated as:

$$\text{Power}_{\text{peak}} = \text{energy per pulse} / \text{pulse width}$$

$$\text{Power}_{\text{peak}} = 0.200\,\text{J} / 0.00005\,\text{s} = 4000\,\text{W}$$

Tip Area

As with the Er,Cr:YSGG laser discussed above, commonly available tips for this laser are cylindrical, tapered, and chisel shaped. All calculations below are based on the use of a 600 mm diameter tip at a tip-to-tissue distance of 2 mm. The resulting calculations will be identical to those above.

Assuming the use of a 600 μm diameter cylindrical tip allows the following calculations:

Tip area is the radius squared of the tip multiplied by π:

$$\begin{aligned}\text{Tip area} &= 300\,\mu m \times 300\,\mu m \times 3.14159 \\ &= 0.03\,\text{cm} \times 0.03\,\text{cm} \times 3.14159 = 0.00283\,\text{cm}^2\end{aligned}$$

A tip that tapers from 1200 to 600 μm at the output end would have the same tip area of 0.00283 cm^2.

As before, at a tip-to-tissue distance of 2 mm with a 12° beam divergence per side angle:

$$\begin{aligned}\text{Spot area at tissue} &= \pi \times r^2 \\ &= 3.14159 \times 0.0725\,\text{cm} \times 0.0725\,\text{cm} \\ &= 0.0165\,\text{cm}^2\end{aligned}$$

Power Density

Power density is the total energy in joules passing through a specific area in 1 s of time. With the attached 600 μm laser fiber and a tip-to-tissue distance of 12°, this erbium laser produces:

$$\begin{aligned}\text{Power density}_{\text{avg}} &= \text{power}_{\text{avg}} / \text{area} \\ &= 3\,\text{W} / 0.0165\,\text{cm}^2 = 182\,\text{W} / \text{cm}^2\end{aligned}$$

$$\begin{aligned}\text{Power density}_{\text{peak}} &= \text{power}_{\text{peak}} / \text{area} \\ &= 4000\,\text{W} / 0.0165\,\text{cm}^2 \\ &= 242{,}424\,\text{W} / \text{cm}^2\end{aligned}$$

Total Energy

The total amount of energy delivered to the tissue during a 30-s treatment is:

$$\text{Total energy} = \text{power}_{\text{avg}} \times \text{time} = 3\,\text{W} \times 30\,\text{s} = 90\,\text{J}$$

Energy Density

If the fiber tip is held stationary in one location, the energy density is calculated as:

$$\begin{aligned}\text{Energy density}_{\text{avg}} &= \text{power}_{\text{avg}} \times \text{time} / \text{area} \\ &= 3\,\text{W} \times 30\,\text{s} / 0.0165\,\text{cm}^2 \\ &= 5455\,\text{J} / \text{cm}^2\end{aligned}$$

$$\begin{aligned}\text{Energy density}_{\text{peak}} &= \text{power}_{\text{peak}} \times \text{time} / \text{area} \\ &= 4000\,\text{W} \times 30\,\text{s} / 0.0165\,\text{cm}^2 \\ &= 8{,}080{,}000\,\text{J} / \text{cm}^2\end{aligned}$$

Assumptions

Several assumptions are made in these calculations:

- Output is assumed to be uniform over the radiated area. In practice, this is rarely true since a Gaussian output is normally produced.
- Output is assumed to be uniform over time. That is, each pulse is assumed to rise instantaneously from zero to the peak power being produced. Again, this is rarely true.
- Output is assumed to be constant over the duration of the pulse. In fact, as shown in this chapter, ◘ Fig. 4.27 of this text, the true output varies significantly and must be averaged for these calculations.
- The output is assumed to be the same as that displayed on the interface screen. Laser manufacturers are careful to point out in their technical specifications that most parameters are only guaranteed to be within about 20% of the stated value.
- Actual spot diameter is as calculated. Because of the Gaussian nature of the beam cross section, a portion of the photons will strike tissue outside of the calculated area.

Summary

The calculations presented here are based on representative values as shown on actual dental lasers for all parameters. By substituting actual known values for the laser being used into the equations, it is possible to determine appropriate parameters. In general, all continuous-wave lasers can be assessed using the set of calculations in Sect. 4.7.1, while true pulsed lasers can be analyzed using calculations in Sects. 4.7.2 and 4.7.3 depending on which parameters are displayed or can be selected.

4

References

1. http://www.pveducation.org/equations/photon-energy-ev. Accessed 17 Feb 2011.
2. https://www.rp-photonics.com/rare_earth_doped_gain_media.html. Accessed 17 Feb 2011.
3. https://www.rp-photonics.com/dye_lasers.html?s=ak. Accessed 30 Jul 2016.
4. http://www.laserdiodesource.com/. Accessed 30 Jul 2016.
5. http://www.olympusmicro.com/primer/java/lasers/diodelasers/index.html. Accessed 30 Jul 2016.
6. https://spie.org/membership/spie-professional-magazine/spie-professional-archives-and-special-content/jan2010-spie-professional/co2-laser. Accessed 30 Jun 2016.
7. http://hyperphysics.phy-astr.gsu.edu/hbase/phyopt/totint.html. Accessed 20 Sept 2015.
8. Hale G, Querry M. Optical constants of water in the 200 nm to 20 um wavelength region. Appl Opt. 1973;12:555–63.
9. Vogel A, Venugopalan V. Mechanisms of pulsed laser ablation of biological tissues. Chem Rev. 2003;103:577–644.
10. Niemz M. Laser-tissue interactions. 3rd ed. Berlin: Springer-Verlag; 2007.
11. Selting W. Fundamental erbium laser concepts; part II. J Laser Dent. 2010;18(3):116–22.
12. Ith M, Pratisto H, Altermatt HJ, Frenz M, Weber HP. Dynamics of laser-induced channel formation in water and influence of pulse duration on the ablation of biotissue under water with pulsed erbium-laser radiation. Appl Phys B Lasers Opt. 1994;59:621–9.
13. Nahen K, Vogel A. Plume dynamics and shielding of the ablation plume during Er:YAG laser ablation. J Biomed Opt. 2002;7(2):165–78.
14. http://vlab.amrita.edu/?sub=1&brch=189&sim=342&cnt=1. Accessed 14 Jun 2015.
15. Hibst R, Keller U. The mechanism of Er:YAG laser induced ablation of dental hard substances. Proc SPIE. 1993;1880:165–2.
16. Farrar SR, Attril DC, Dickinson MR, King TA, Blinkhorn AS. Etch rate and spectroscopic ablation studies of Er:YAG laser-irradiated dentine. Appl Opt. 1997;36(22):5641–6.
17. Niemz M. Investigation and spectral analysis of the plasma-induced ablation mechanism of dental hydroxyapatite. Appl Phys B Lasers Opt. 1994;58:273–81.
18. Selting W. Fundamental erbium laser concepts; part 1. J Laser Dent. 2009;17(2):89–95.
19. Fried D, Zuerlein M, Featherstone J, Seka W, Duhn C, McCormack S. IR laser ablation of dental enamel: mechanistic dependence on the primary absorber. Appl Surf Sci. 1998;127:852–6.
20. Majaron B, Sustersic D, Lukac M, Skaleric U, Funduk N. Heat diffusion and debris screening in Er:YAG laser ablation of hard biological tissues. Appl Phys B Lasers Opt. 1998;66:1–9.
21. Selting W. The effect of tip wear on Er:YAG laser ablation efficiency. J Laser Dent. 2007;15(2):74–7.
22. Majaron B, Prosen T, Sustercic D, Lukac M. Fiber-tip drilling of hard dental tissue with Er:YAG laser. In: Featherstone JBD, Rechmann P, Fried DS, editors. Lasers in dentistry IV, January 25–26, 1998, San Jose, CA, Proc. SPIE, vol. 3248. Bellingham, WA: SPIE—The International Society of Optical Engineering; 1998. p. 69–76.
23. Simanovskii D, Mackanos M, Irani A, O'Connell-Rodwell C, Contag C, Schwettman H, Palanker D. Cellular tolerance to pulsed hyperthermia. Phys Rev. 2006;74(011915):1–7.
24. Angiero F, Parma L, Crippa R, Benedicenti S. Diode laser (808 nm) applied to oral soft tissue lesions: a retrospective study to assess histopathological diagnosis and evaluate physical damage. LIMS. 2012;27(2):383–8.
25. Turner J, Hode L. Low level laser therapy, clinical particle and scientific background. Grangesberg: Prima Books; 1999.
26. Karu T. The science of low-power laser therapy. Gordon & Breach Science Publishers; 1998.
27. Karu T. Is it time to consider photobiomodulation as a drug equivalent? Photomed Laser Surg. 2013;31(5):189–91.
28. Woodruff L, Bounkeo J, Brannon W, Dawes K, Barham C, Waddell D, Enwemeka C. The efficacy of laser therapy in wound repair: a meta-analysis of the literature. Photomed Laser Surg. 2004;22(3):241–7.
29. Huang YY, Sharma SK, Carroll J, Hamblin MR. Biphasic dose response in low level light therapy-an update. Dose Response. 2011;9(4):602–18.
30. Jacques S. Optical properties of biological tissues: a review. Phys Med Biol. 2013;58:R37–61.
31. Huang Y-Y, Hamblin M. Biphasic dose response in low level light therapy. Dose Response. 2009;7:358–83.
32. Bashkatov A, Genina E, Kochubey V, Tuchin V. Optical properties of human skin, subcutaneous and mucous tissues in the wavelength range from 400 to 2000nm. J Phys D Appl Phys. 2005;38:2543–55.
33. Khan I, Tang E, Arany P. Molecular pathway of near-infrared laser phototoxicity involves ATF-4 orchestrated ER stress. Sci Rep. 2015;5:10581.

Laser Safety in Dentistry

Penny J. Parker and Steven P. A. Parker

Contents

D. J. Coluzzi, S. P. A. Parker (eds.), *Lasers in Dentistry—Current Concepts*, Textbooks in Contemporary Dentistry,
https://doi.org/10.1007/978-3-031-43338-2_5

Core Message

Laser use in general dental practice has grown considerably since first introduced in 1989, reflected both in numbers of machines and the scope of usage. As with other areas of primary healthcare, general and specific measures must be employed to ensure the safe use of lasers in dentistry. Regulations—specific to lasers or within the licenced scope of practice—apply a duty of care to all dental healthcare professionals in the application of lasers in clinical practice. Such laser regulations may exist through the adoption of international standards and/or through National or Regional legislation. The duty of care extends to all staff as well as patients. The registered laser owner or lead clinician is responsible for ensuring that all staff personnel have a thorough knowledge of laser safety. Additional devolved responsibility may be applied to the laser safety officer (LSO)—also referred to as laser safety supervisor (LSS).

Laser safety measures relate primarily according to the Class of laser being used. Within the range of laser photonic devices, there is an ascending classification from I to IV, to denote an ascending level of power output. Sub-classes relevant to laser operation may be applied to those instances where adjunctive optical magnification (loupes, microscope, etc.) are employed. Specific irradiation safety risks apply to the unprotected eye and skin, according to the laser class and emission wavelength. Non-beam risks also exist, including those posed by the laser plume.

It is considered therefore—from advisory to mandatory, depending on prevailing regulations—that those personnel involved in the delivery of clinical laser therapy should undergo training sufficient to alert them to their responsibility. Prime among this responsibility is to any patient receiving such therapy to receive adequate measures to safeguard risk from photonic energy that is capable of risk to unprotected ocular and non-target tissue.

5.1 Introduction

LASER—coherent, single wavelength, non-ionizing electromagnetic radiation is a comparatively intense form of light energy; a potential may exist for such energy, directed at biological tissue to be absorbed—in part or totally—and give rise to structural change within the tissue. Lasers used in dentistry are configured to deliver photonic energy that would mostly convert to thermal change within tissue, and it is this thermal rise that may be sufficient to progress through tissue warming (seen with sub-ablative photobiomodulation, PBM) to a level and above, where irreversible change such as protein denaturation and water vaporization may occur (seen with surgical hard and soft tissue lasers). Additional laser photonic sources may be directed to initiate intermediate photochemical action within an applied photosensitive dye or bleaching gel. The fundamental safety aspect is defined not by the specific applied photonic dose associated with a chosen procedure, but by the risk associated with the Class of laser being employed; thus, the application of, for example, PBM using a Class IV laser source would nevertheless mandate full safety protection measures.

To deliver controlled, predictable, and positive laser-assisted change represents the ideal of the dental clinician in providing such treatment. However, inappropriate energy levels and/or delivery of laser photonic beam to non-target tissue can represent a risk, in turn maybe sufficient to cause irreparable and permanent damage. For dental professionals, this may occur within the oral cavity, to circumoral and facial skin structure or, most importantly to the structures of the unprotected eye.

As the purchase and use of lasers in dentistry continues to grow, so must concern for laser safety. Failing to wear available eye protection is one of the most frequent contributing factors to laser injuries [1]. Within such risk and pertinent to the use of lasers in general, laser safety measures are delineated for interpretation and application in the workplace, such as within the dental operatory.

In a systematic review of papers relevant to beam and non-beam medical laser hazards, together with access to the Rockwell Laser Industries Laser Accident Database, it was concluded that occupational hazards associated with medical laser applications remain poorly understood and uncharacterized. There are relatively few published accounts of laser accidents; they tend to suffer from the problem of reliance on self-reporting. Eye injuries, skin burns, injuries related to the onset of fires, and electric shock have been reported in relation to medical laser use. It is probable that both acute and chronic health effects have been experienced by medical personnel as a result of exposure to laser generated air contaminants (plume) [2].

A more substantial study of laser-related injuries reported on risk factors but was not limited to medical laser application [3]. Four database sources within the United States—Center for Devices and Radiological Health of the US Food and Drug Administration (FDA), Rockwell Laser Industries (RLI), the US Army's Laser Accidents and Incidents Registry (LAIR) and the Federal Aviation Administration (FAA) safety reporting system—revealed a total of 869 injuries and deaths. Of these where injury or death occurred, a total of 663 (52%) were located in some type of medical facility.

It may seem innocuous to the dental professional new laser user that safety is a key objective to protect the patient and clinical team. Useful lessons can often be learned from studying laser accidents and although most countries require such accidents to be reported to the relevant authority, often such information is not accessible to laser users. The risk of such accidents may be reduced by performing a risk assessment and this is often a legal requirement, such as in the countries of the European Union and United Kingdom [4, 5].

In a paper that reported laser-associated accidents within medical practice [6], 12 incidents were reported. Simple errors relating to the operation of laser equipment accounted for five of the reported incidents; in four incidents, there was temporary or permanent damage to the operator's eye, but of greater significance was the reporting of three incidents where the laser emission caused either direct inflammation of gauze and drapes or damage to endotracheal tubing that led to potentially fatal airway fires in an anesthetized patient. The paper suggested that operator error was at fault in 67% of the incidents reported and that equipment fault (25%), laser-induced fire (25%), and broken delivery fiber (17%) constituted the (avoidable) risks in these unfortunate accidents.

It is only when viewed in the context of permanent and possibly life-threatening risk to the clinician and attendant staff but more so, the patient that the importance of knowledge and appreciation of laser safety issues must be acknowledged.

5.2 Regulatory Framework

The regulatory framework governing the safe use of lasers may be seen as a hierarchical devolvement, most referenced through the I.E.C. (International Electrotechnical Commission) and A.N.S.I. (American National Standards Institute). From these organizations and their representation, National Regulations may apply—either as specific statutory instruments or more often as part of laws and regulations that might apply within the workplace (◘ Fig. 5.1) [7–9].

IEC, formed in 1906 is responsible for the development of world standards for the electrical and electronics area and is composed of the national committees from countries around the world. In 1930, the IEC established electrical units under a system commonly known as the "Système International," or SI for short. In 1974, the IEC created Technical Committee 76, to address standards relating to lasers, with a particular focus on safety. This committee developed a four-class system for lasers that was the global reference, later modified in 2002.

The American National Standards Institute (ANSI) developed in parallel although initially unconnected to the IEC (◘ Fig. 5.2). ANSI was originally established in 1919 as the American Engineering Standards Committee and created the International Standards Association (ISA), an organization that would eventually become the International Organization for Standardization (ISO). In 1969, ANSI adopted its present name.

The ANSI Federation also initiated what has now become an annual series of discussions with the European Committee for Standardization (CEN), the European Committee for Electro-technical Standardization (CENELEC), and the European Telecommunications Standards Institute (ETSI).

The United States has always had its own regulation on lasers (known as FDA CFR 21 1040.10). This is a US government regulation and is written into US law. As an interpretation of recommendations, the ANSI Z136 series is recognized by the Occupational Safety and Health Administration (OSHA) as the authoritative series of laser safety in United States.

An important ANSI Federation member and accredited standards developer, the Laser Institute of America (LIA) is the professional society dedicated to fostering lasers, laser applications, and laser safety worldwide. In

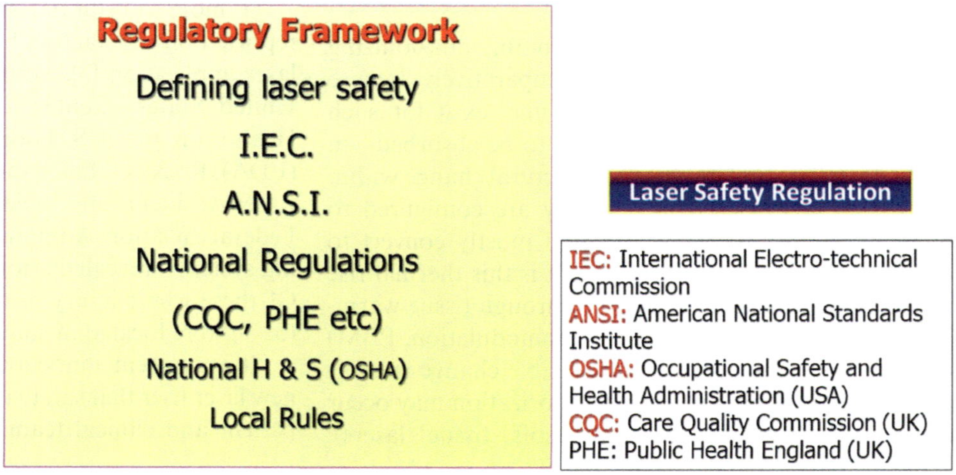

◘ **Fig. 5.1** International and National regulatory bodies associated with laser safety

Fig. 5.2 International Electro-technical Committee and American National Standards Institute: a near-contemporary developmental history

2005, the LIA published ANSI Z136.4—Recommended Practice for Laser Safety Measurements for Hazard Evaluation. This provides guidance for optical measurements associated with laser safety requirements.

In addition, the US Food and Drug Administration (FDA) informed laser product manufacturers that the US FDA would henceforth accept IEC classification and labeling [10].

In addition to other activities, IEC publishes International Standards, Technical Specifications, Technical Reports, Publicly Available Specifications (PAS) and Guides. Most pertinent to laser and (some aspects of) light emitting diode (LED) safety is the publication IEC 60825 (European variant EN 60825) which has been published in several formats with revisions from 1994 to 2017 [11].

The Standard IEC (EN) 60825-1 "Safety of laser products Part 1: Equipment classification, requirements and user's guide" sets out regulation governing the following:

- To introduce a system of classification of lasers and laser products emitting radiation in the wavelength range 180 nm to 1 mm according to their degree of optical radiation hazard in order to aid hazard evaluation and to aid the determination of user control measures.
- To establish requirements for the manufacturer to supply information so that proper precautions can be adopted.
- To ensure, through labels and instructions, adequate warning to individuals of hazards associated with accessible radiation from laser products.
- To reduce the possibility of injury by minimizing unnecessary accessible radiation and to give improved control of the laser radiation hazards through protective features.

From such broad regulation, safety and risk assessment when using lasers and LED equipment in clinical dental surgery can be devolved to include [12]:

- Suitability for use, clinical parameters
- Administrative code, record keeping
- Safety features of laser, laser maintenance
- Environment safety, patient safety
- Laser Safety Officer, Laser Protection Advisor

Table 5.1 provides a glossary of terminology associated with aspects of laser safety.

5

Table 5.1 Glossary of terminology associated with aspects of laser safety

Laser safety term	Explanation
Laser classification I–IV	Laser/LED unit capability of ascending laser beam emission power, relative to the safety risk posed. I—low; IV—high.
Aversion response	Blinking of the eye, or movement of the head to avoid exposure to a bright light. Approximately 0.25 s.
Accessible emission limits (AEL)	Maximum accessible level of laser radiation permitted within a particular laser class.
Maximum permissible exposure (MPE)	The level of laser radiation to which a person may be exposed without hazardous effect or adverse biological changes in the eye or skin.
Nominal hazard zone (NHZ)	Aka Nominal Ocular Hazard Zone. Area within which the level of the direct, reflected, or scattered radiation during normal operation exceeds the applicable MPE.
Controlled area	Area within which occupancy and activity is subject to control and supervision for the purpose of protection from laser radiation hazards.
Intra-beam viewing	Exposure of the eye to all or part of a laser beam.
Specular reflection	A mirror-like reflection of a laser beam, with or without loss of beam fluence.
Diffuse reflection	Change in the spatial (area) distribution of a laser beam when it is reflected in multiple directions by a rough or matt surface.
American National Standards Institute (ANSI) ANSI Z 136.1,2,3,4,7	US Regulations governing areas of laser safety pertinent to clinical laser dentistry. Z 136.1—Safe use of lasers Z 136.2—Safe use of optical fiber communication systems utilizing laser diode and LED sources Z 136.3—Safe use of lasers in healthcare Z 136.4—Recommended practice for laser safety measurements for Hazard evaluation Z 136.7—Testing and labeling of laser protective equipment
International Electro-technical Commission (IEC) IEC 60825 + updates	Broad range specifications relating to laser manufacture and use

5.3 Laser Classification

Lasers and laser systems are divided into four major classifications according to their potential to cause biological damage to the eye or skin. The purpose of these classifications is to warn users of the hazards associated with the laser and LED relative to the Accessible Emission Limits (AEL). These limits are based on laser output energy or power, radiation wavelengths, exposure duration, and cross-sectional area of the laser beam at the point of interest.

Prior to 2002, the classification of lasers ran from I to IV, with Class IV to include surgical lasers.

- Class I: Laser products are generally exempt from radiation hazard controls during operation and as such do not pose any specific risk in normal usage.
- Class II: Low-power visible lasers that emit above Class I levels but at a radiant power not above 1 mW. The concept is that the human aversion reaction (blink response) to bright light will protect a person. Only limited controls are specified.
- Class IIIA: Intermediate power lasers. Some limited controls are usually recommended.
- Class IIIB: Moderate power lasers. In general, Class IIIB lasers will not be a fire hazard, nor are they generally capable of producing a hazardous diffuse reflection. However, specific controls are recommended.
- Class IV: High-power lasers are hazardous to view under any condition (directly or diffusely scattered) and are a potential fire hazard and a skin hazard. Significant controls are required of Class IV laser facilities.

From 2002, this classification has been refined and adopted by the IEC. The revision was prompted by increasing sophistication in laser technology, together with the increased adoption of magnification devices as adjuncts to operative medicine and dentistry—operating microscopes, loupes, etc. In general, LEDs would be in the lower Classes (I, IM, II, IIM, IIIR), but very exceptionally may be Class IIIB. As such, this classification will be examined in greater detail.

The IEC Laser classifications are summarized as follows:

Fundamental to the classification and inherent risk of tissue damage of a laser, the power density (irradiance—W/m^2) of an incident laser beam and its relation to a notional maximum tissue tolerance, may be expressed in terms of Maximum Permissible Exposure (MPE). However, the computed MPE limit may be exceeded for a given laser irrespective of its AEL—for example, the possibility of visual damage associated with extended direct laser pointer irradiation.

- **Class I**: Class I lasers are safe under all operating conditions. There is no risk to eyes or skin. This means the maximum permissible exposure (MPE) cannot be exceeded when viewing a laser with the naked eye or with the aid of typical magnifying optics (e.g., telescope or microscope). To verify compliance, the standard specifies the aperture and distance corresponding to the naked eye, a typical telescope viewing a collimated beam, and a typical microscope viewing a divergent beam. Class I lasers may consist of a higher power laser housed within an enclosure.
- **Class IM**: Class IM lasers are not capable of producing hazardous exposure under normal operating conditions but may be hazardous if viewed with the aid of optical instruments—magnifying optics such as microscopes and telescopes.

  Class IM lasers produce large-diameter beams, or beams that are divergent. The MPE for a Class IM laser cannot normally be exceeded unless focusing or imaging optics are used to narrow the beam. Given the nominal higher photonic energy of shorter wavelengths, the maximum power delivery of Class I and IM must not exceed 40 μW (blue) or 400 μW (red).
- **Class IC**: Under IEC 60825:2014, a new Class IC is defined where C stands for "contact" but in some interpretations also stands for "conditional" [13]. Currently, the class is limited to products intended for the treatment of the skin or internal tissue in contact or close to the skin where the product is designed to be safe for the eye.
- **Class II**: Class II lasers are limited to 1 mW continuous wave (CW), or more if the emission time is less than 0.25 s or if the light is not spatially coherent. A Class II laser is safe because the blink reflex will limit the exposure to no more than 0.25 s. It only applies to visible-light lasers (400–700 nm). Intentional suppression of the blink reflex could lead to eye injury. Many laser pointers and measuring instruments are Class II. There is no hazard from exposure to diffuse radiation.
- **Class IIM**: Low-power lasers (CW: up to 1 mW) in visible wavelength range (400–700 nm). Class IIM lasers are not hazardous under normal operating conditions because of the aversion reaction (blink reflex). Class IIM lasers may be hazardous if viewed with the aid of optical instruments. As with Class IM, this applies to laser beams with a large diameter or large divergence, for which the amount of light passing through the pupil cannot exceed the limits for Class II.
- **Class IIIR**: Moderate power lasers (CW: up to 5 mW) for visible wavelengths (400–700 nm). Up to a factor of five over maximum allowable exposure of Class II lasers for other wavelengths. A Class IIIR laser is considered safe if handled carefully, with restricted beam viewing. With a Class IIIR laser, the MPE can be exceeded, but with a low risk of injury. Visible continuous lasers in Class IIIR are limited to 5 mW. For other wavelengths and for pulsed lasers, other limits apply.

  Since the exposure limits (MPEs) for the eye are the direct basis for the AEL (accessible emission limits) for the laser product safety classes Class I, IM, II, IIM, and IIIR, any changes in the MPE will also result in equivalent changes of the AEL values and thus in the permitted output powers for these classes [14].
- **Class IIIB**: Moderate power lasers (CW: up to 500 mW, Pulsed up to 30 mJ) in wavelength range of 300 nm to far infrared. Direct eye exposure to Class IIIB lasers is hazardous; however, diffusely scattered radiation is generally safe. Direct exposure to skin is a potential hazard.

  The AEL for continuous lasers in the wavelength range from 300 nm to far infrared is 500 mW. For pulsed lasers between 400 and 700 nm, the limit is 30 mJ. Other limits apply to other wavelengths and to ultrashort pulsed lasers. Protective eyewear is typically required where direct viewing of a class IIIB laser beam may occur. Class IIIB lasers must be equipped with a key switch and a safety interlock.
- **Class IV**: High-power lasers (CW: above 500 mW). Class IV is the highest and most dangerous class of laser, including all lasers that exceed the Class IIIB AEL. By definition, a Class IV laser can burn the skin, or cause devastating and permanent eye damage as a result of direct, diffuse or indirect beam viewing. Class IV lasers are also a potential fire hazard.

  These hazards may also apply to indirect or non-specular reflections of the beam, even from apparently matt surfaces and therefore great care must be taken to control the beam path. Class IV lasers must be equipped with a key switch and a safety interlock.

A summary of laser classes commonly used in clinical dentistry is shown in Table 5.2.

Table 5.2 A summary of laser classes commonly used in clinical dentistry

Laser class	Maximum output	Use in dentistry	Possible hazard	Safety measures
Class I	40 μW (blue)	Integral scanning Laser caries detector	No implicit risk	Blink response
Class IM	400 μW (red)		Possible risk with magnified beam	Laser safety labels
Class II	1.0 milli-Watt (mW)	Laser caries detection	Possible risk with direct viewing	Sight aversion response
Class IIM		Aiming beams	Significant risk with magnified beam (Class IIM)	Laser safety labels
Class IIIR	Visible: 5.0 mW Invisible: 2.0 mW	Some low-level lasers. Aiming beams.	Eye damage	Safety eyewear Safety personnel
Class III	500 mW (0.5 W)	Low-level lasers	Eye damage—direct or specular Maximum output may pose slight fire/skin risk	Training for Class IIIR and IIIB lasers
Class IV	No upper limit	All surgical lasers	Eye/skin damage Non-target tissue Fire hazard Plume hazard Possible ionizing effects with UV lasers	Safety eyewear Safety personnel Training/local rules Possible registration with National regulations

5

5.4 Hazards of Laser Beams

A laser injury can be defined as an event causing (a) physical harm or damage, (b) physiological dysfunction, (c) an adverse surgical outcome, resulting directly from the laser energy, or a consequence of the device's inherent technology, or (d) a second treatment procedure to correct the first procedure.

Discussion as to all aspects of laser hazard and risk assessment when using laser devices, will acknowledge the over-riding importance and relevance of those more-powerful lasers classes of IIIB and IV.

The risk posed by coherent light irradiation may be assessed with regard to the following:

1. The laser wavelength
2. The power intensity of the incident laser irradiation
3. Optical risks
4. Non-target oral tissue
5. Non-target skin
6. Inhalation and laser plume risks
7. Other Associated Risks: Hazards—mechanical, chemical, fire, sterilization

5.4.1 The Laser Wavelength

In clinical dentistry, the range of laser wavelengths falls within 370 and 10,600 nm, i.e., from the blue visible limit to the far infrared, non-ionizing spectrum. All laser tissue interaction may be viewed as photothermal in that incident photonic energy is absorbed by chromophore molecules, raising the molecular energy level and leading to disruption, relative to ascending power density and target tissue ablation threshold [15]. With regard to laser safety, it must be appreciated that absorption is but one of the basic physical phenomena associated with laser–tissue interaction. Absorption confers the maximal interaction, with scatter as an associated, quantitatively less-precise energy transfer. Additional phenomena are not associated with energy transfer to the exposed tissue are laser beam reflection and transmission. All four interactions may occur simultaneously in non-homogenized tissue and pose a mixture of threats that must be nullified through adequate safety measures.

5.4.2 The Power of the Irradiation

This will be related to the amount of energy delivered over time, together with consideration of the area of tissue that is exposed to the beam. In consequence, the power density of the beam may represent a threshold above which irreversible change may occur in tissue exposed to the beam. As was seen in Table 5.1, the classification of lasers sets out the upper limit of power delivered. In addition, the potential for the laser to deliver pulsed irradiation may influence the maximum power achievable (peak power).

5.4.3 Optical Risks

The unprotected eye is generally regarded as the organ at greatest risk from accidental laser exposure. Several cases of laser-induced eye injuries have been documented [16–19]. For any given laser beam that exceeds an output value in excess of the MPE of ocular tissue, a risk pertains that must be anticipated, and safety protocols of eye protection employed.

An important natural reflex in limiting the potential for contact and injury to the eye is blinking. The eyelid response time is of the order of 0.2 and 0.25 s and is cited as a significant natural aversion response to visible lasers within Class I and II. Of course, key to this protective mechanism is the word "visible." It will have no beneficial effect with beams of wavelengths outside the visible spectrum and in addition, some laser beam intensities are so great that injury can occur faster than the protective action of the eyelid reflex [20].

Laser hazards to the eye depend most predominantly upon wavelength [21, 22] as shown in ◘ Fig. 5.3.

As described above, laser photonic energy cannot damage tissue unless the light energy is able to be absorbed within that structure. Visible and near-infrared wavelengths, which can be transmitted through clear ocular media can, subject to MPE levels be absorbed in the retina and cause significant damage. Lasers operating between 400 and 1400 nm are particularly dangerous; the high collimation of the incident beam permits the rays to be focused to an extremely small spot on the retina with most of the light being absorbed by melanin pigments in the pigment epithelium just behind the photoreceptors [23], causing burns in the retina. This spectral band of wavelengths is often referred to as the retinal hazard region since the increased concentration of light after entering the eye and falling on the retina is of the order of 100,000 times. Hence, a collimated beam of 1 W/cm^2 at the cornea will focus to an area on the retina with an irradiance of 100 kW/cm^2.

Due to the action of the lens, incident beams will be focused on the macula and its fovea; if these areas are damaged by laser radiation, substantial loss of vision can result.

Laser wavelengths outside the retinal hazard region (1400–10,600 nm) may give rise to injury to the anterior region of the unprotected eye. Injury to the cornea is normally very superficial, involving only the corneal epithelium and with the cornea's high metabolic rate, repair occurs within a day or two and total recovery of vision will occur. However, if significant injury occurs in deeper corneal layers, in the stroma or endothelium, corneal scars can result, leading to possible permanent loss of vision.

The types of damage sustained by non-protected ocular structures exposed to laser beams of varying wavelengths are listed in ◘ Table 5.3.

Optical risks therefore assume a "worse case" scenario and the choice of maximum permitted exposure values defines objective quantifiable levels of risk, expressed in irradiance terms (W/cm^2) that would be measured at the cornea [24]. ◘ Table 5.4 provides a range of values by wavelength. In order to provide perspective, the values are expressed within three possible time-related scenarios that may be applicable to laser use in dentistry:

- *0.25 second*: The human aversion time for bright-light stimuli (the blink reflex).
- *10 seconds*: The time period chosen by the ANSI Z 136.1 committees represents the optimum "worst-case" time period for ocular exposures to infrared (principally near-infrared) laser sources.
- *30,000 seconds*: The time period that represents a full 1-day (8-h) occupational exposure.

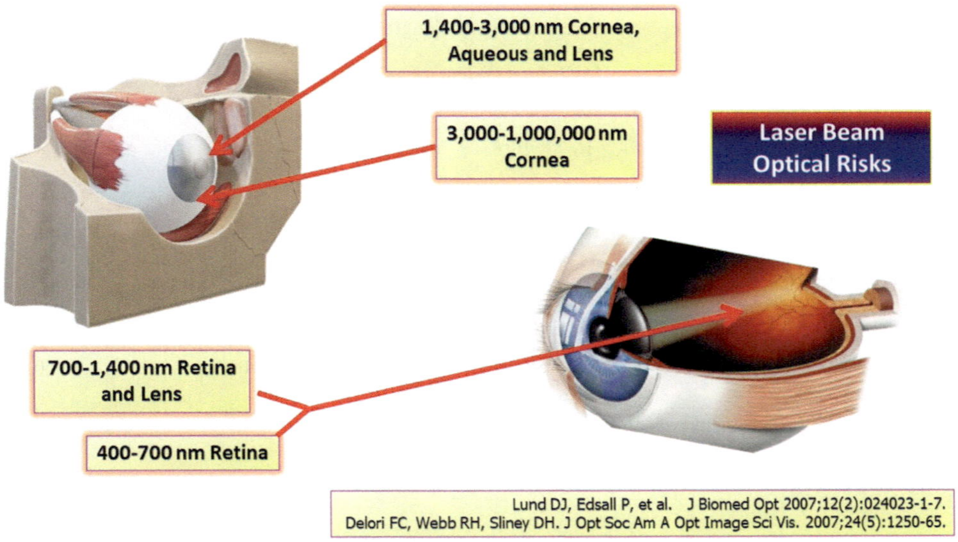

◘ **Fig. 5.3** Graphic representation of potential risk to regions of the unprotected eye, relative to incident laser wavelength in nanometers

According to regulations, for lasers operating in a pulsed emission mode, the MPE value was calculated by considering the most restrictive among the following conditions: (a) the MPE from a single pulse within a train of pulses does not exceed the MPE for a single pulse; (b) the average exposure for a complete pulse train (of constant amplitude) during a time interval shall not exceed the MPE values for a single pulse as given in the standard [25].

Of particular note is the substantially higher value of MPE in the region of 3.0 μm—Er:YAG laser. This may be accounted for through the intensely high absorption of this wavelength in water—<1.0 μm in corneal surface depth [26].

In summary, coherent laser photonic beams pose significant risk to the unprotected eye and to some extent non-target tissue and skin (Fig. 5.4).

Table 5.3 Types of damage sustained by non-protected ocular structures exposed to laser beams of varying incident wavelengths

Incident laser wavelength	Pathological effect
315–400 nm (UV-A)	Photochemical cataract (clouding of the eye lens)
400–780 nm (visible)	Photochemical damage to the retina, retinal burn
780–1400 nm (near-IR)	Cataract, retinal burn
1.4–3.0 μm (mid-IR)	Aqueous flare (protein in the aqueous humor), cataract, corneal burn
3.0–10.6 μm (far-IR)	Corneal burn

The length of exposure may be instantaneous or may be cumulative over a period of time. Direct exposure to the laser beam would maximize the risk but of concern would be those instances where the beam may be reflected. Figure 5.5 provides an overview of risk in these events.

5.4.4 Non-target Oral Tissue

The oral cavity may present significant challenge in terms of access and non-intentional/function-related movements of structural components in the conscious dental patient. Often the visualization and access for the treatment of individual teeth and areas of supporting tissues is compromised by restriction in space together with involuntary movement of the cheeks, lips, and tongue. Care should always be taken to ensure free and easy identification of and access to the treatment site, together with planned manipulation of the laser delivery handpiece and tip prior to operation and firing of the laser.

Oral tissue is anisotropic and non-homogeneous, and this may impact the absorption characteristics when using a laser; certainly, this may pose a risk where adjacent non-target oral tissue has a higher absorption coefficient than the target tissue. An example may be where high peak-power mid IR irradiation of a cervical tooth cavity encroaching and causing collateral damage to adjacent gingival tissue, or unintentional transmission of near-IR and visible wavelengths through tooth tissue to the pulp.

Additional risk may be posed to non-target oral tissue through reflection phenomena. Many instruments used in dentistry are metal and in many instances the clinician will operate using a mouth mirror. Direct reflection or specular (diffuse) reflection of the incident

Table 5.4 MPE values for unprotected eye exposure to laser wavelengths

Laser type	Wavelength (μm)	MPE level (W/cm^2)		
		0.25 s	10 s	30,000 s
KTP (CW)	0.532	16.7×10^{-6}		1.0×10^{-6}
HeNe (CW)	0.633	2.5×10^{-3}		17.6×10^{-6}
GaAs (CW)	0.810		1.9×10^{-3}	610.0×10^{-6}
Nd:YAG (FRP)	1.064		17.0×10^{-6}	2.3×10^{-6}
Er:YAG[a]	2.94		1.0×10^{-2}	1.0×10^{-2}
CO_2 (CW)	10.6		100.0×10^{-3}	100.0×10^{-3}

Note: Blink reflex applies only to visible wavelengths
Source: ANZI Z 136.1
[a] Schulmeister K, Sliney D, et al. J. Laser Appl. 2008;20(2):98–105 [26]

Photobiological Spectral Domain (CIE Band)	Eye Effects	Skin Effects
Ultraviolet C (200–280 nm)	Photokeratitis	Erythema (Sunburn) Skin Cancer
Ultraviolet B (280–315 nm)	Photokeratitis	Erythema (Sunburn) Accelerated Skin Aging Increased Pigmentation
Ultraviolet A (315–400 nm)	Photochemical UV Cataract	Pigment Darkening Skin Burn
Visible (400–780 nm)	Photochemical and Thermal Retinal Injury Color and Night Vision Degradation	Skin Burn Photosensitive Reactions
Infrared A (780–1400 nm)	Retinal Burns Cataract	Skin Burn
Infrared B (1400–3000 nm)	Corneal Burn Aqueous Flare IR Cataract	Skin Burn
Infrared C (3000–1 million nm)	Corneal Burn	Skin Burn

Fig. 5.4 Potential damaging effects to irradiated unprotected ocular and dermal tissue, relative to photobiological spectral domain. (Source: IEC (EN) 60825-1 "Safety of laser products Part 1: Equipment classification, requirements and user's guide")

laser beam may expose non-target oral tissue to ablative beam fluences and may not be detected at the time of the treatment. Moist oral tissue may lead to specular reflection.

Care should be taken when using hollow metal delivery tips where such may come to lie against non-target structures such as the lips or tongue. It is likely that such tips may become extremely hot when no co-axial water spray is used and may cause direct thermal burns to the tissue being retracted.

5.4.5 Non-target Skin

The potential for risk to skin from laser photonic energy is considered low in terms of dental treatment. There may be specific scope of practice limits to a dentist delivering clinical (surgical) treatment beyond the vermillion border of the patient's lips. As such, the need for protective measures is not as high as would be applicable to facial cosmetic and dermatological laser procedures.

Notwithstanding, the varying absorption coefficients of pigmented and non-pigmented structures will predispose to varying depths of penetration, as shown in Fig. 5.6. The greater level of depth is seen with near-IR wavelengths and the additional phenomenon of high photon scatter at these wavelengths may compound the risk.

The most common trans-dermal exposure using lasers that may affect dentists may occur during sub-ablative photobiomodulation techniques as adjunct to TMJ and associated muscular dysfunction conditions or facial pain syndromes. In view of the use of near-IR wavelengths as preferred in these treatments, some recommendations suggest relatively high fluences to allow the deep penetration of the energy into the TMJ area. Care should therefore be taken to ensure surface irradiance of the skin does not lead to structural damage and dosage modified to accommodate differing skin types, racial differences, and those who may have drug-associated skin hypersensitivity.

5.4.6 Inhalation and Laser Plume Risks

A by-product of the surgical ablation of target oral tissue is the production of the so-called laser plume. This represents a significant hazard in breathing airborne contaminants produced during the laser vaporization of tissues. Studies of the production of both the chemical toxicity of photothermolysis products and the potential viability of infectious particulates (e.g., viral fragments)

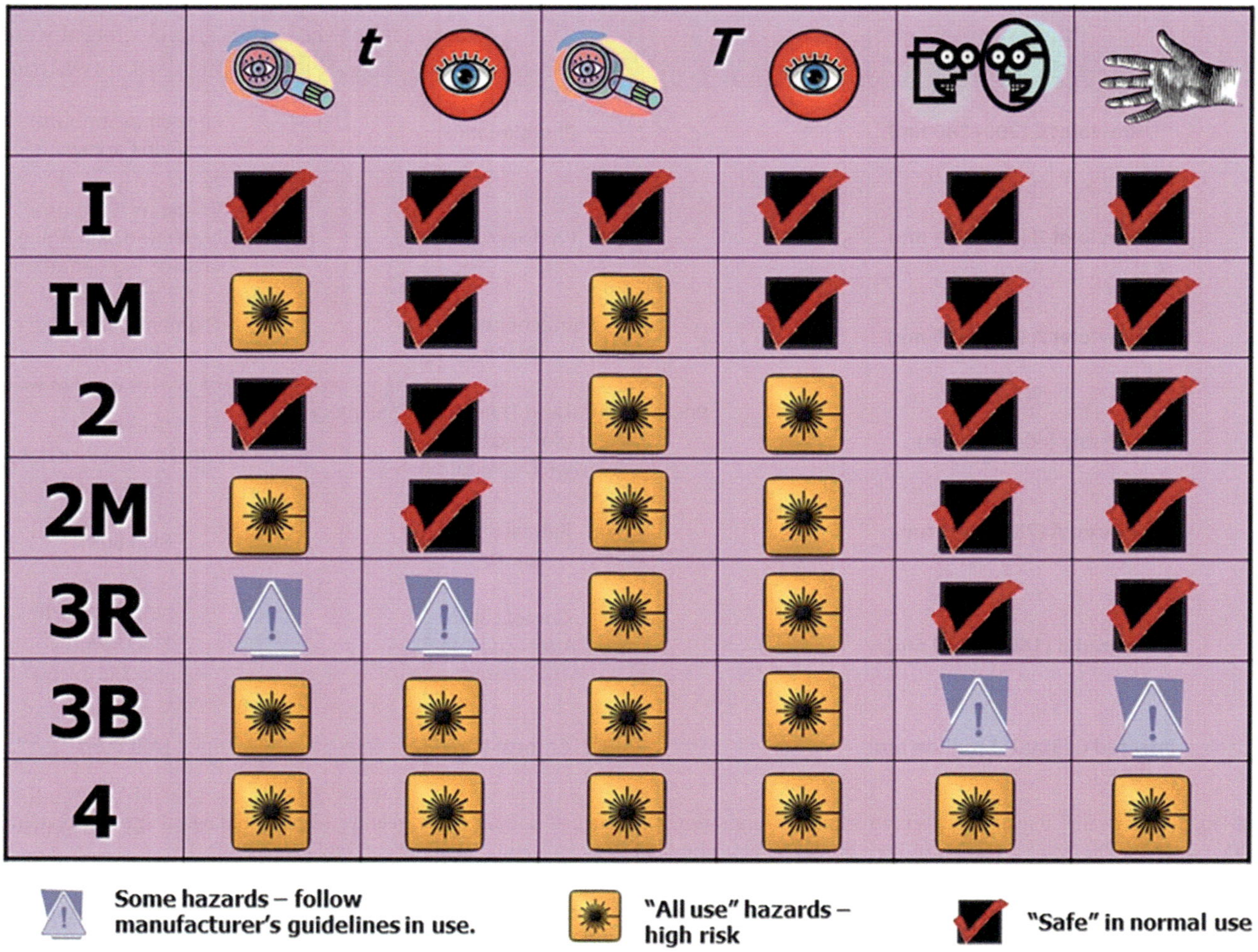

Fig. 5.5 List of laser Classes (IEC post-2002) with attendant risk that each may pose. "t" denotes instantaneous risk—both unprotected eye and when using magnification devices. "T" denotes longer exposure. Risk posed during reflection scenario and risk to unprotected skin are also represented

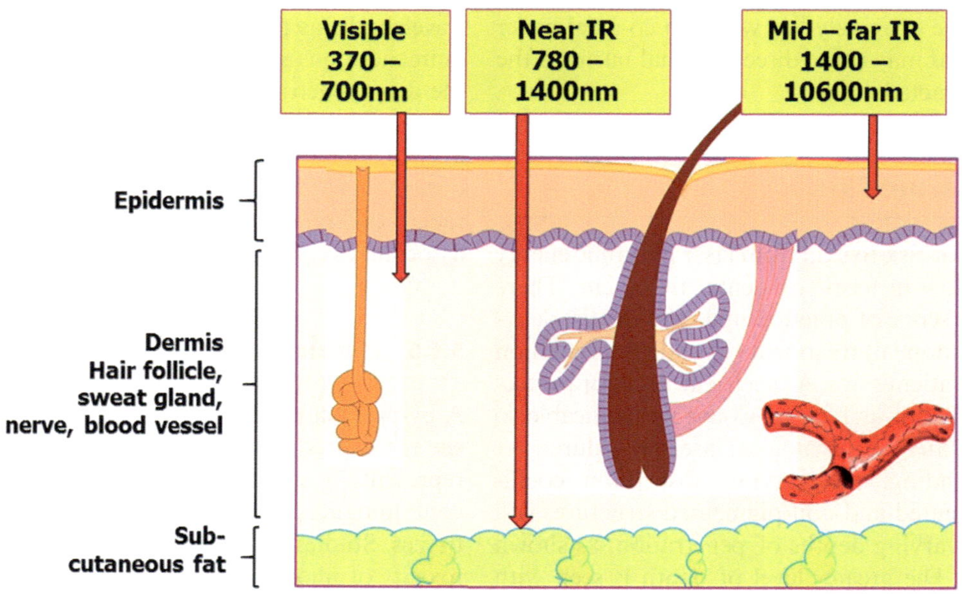

Fig. 5.6 Schematic showing superficial skin structure and penetration of laser photonic energy relative to wavelength

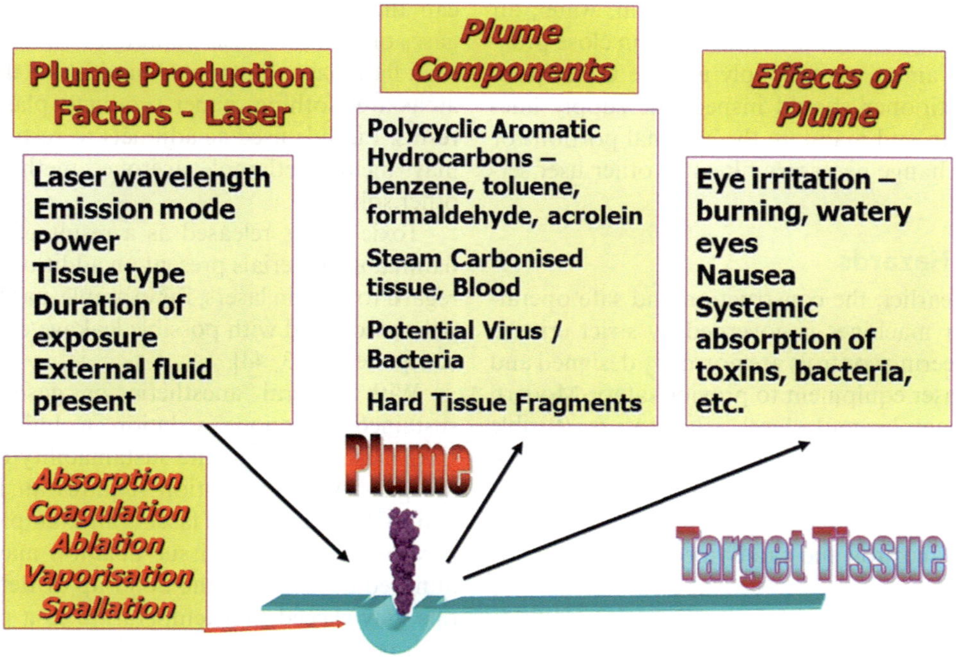

Fig. 5.7 Laser plume production, composition components and effects

have shown cause for concern unless efficient aspiration, local exhaust ventilation, and operating staff facial protection measures are employed. Although many associated factors may provide some influence, the amount of plume produced and volume of contaminants will be related to incident laser power, nature of target tissue, laser emission mode (CW, FRP), and co-axial supplies such as water and air spray [27–30].

As summarized in Fig. 5.7, investigation into the components of the laser plume has shown a number of chemicals—water vapor, hydrocarbon gases, carbon monoxide and dioxide, together with metal fumes, particulate organic and inorganic matter, bacteria and viral bodies [31–34]. The hazard presented by the laser plume may include eye irritation, nausea, and breathing difficulties, together with the possibility of transfer of infective bacteria and viruses [35, 36].

In dentistry, additional aspects of plume production and control may be seen in terms of the site of interaction and potential for airborne chemicals and bacteria [37], the use or otherwise of co-axial water spray [38] and the potential for using near-IR lasers that could be operated on soft tissue within a film of water that acts as a shield [39]. In any event, to combat the risks associated with the laser plume, eye protection, specific fine-mesh face masks capable of filtering 0.1-μm particles should be worn and the spread of the plume minimized through the use of high-speed suction aspiration that is capable of close-circuit filtration [40]. In addition, normal surgical protective clothing must also be employed.

Figure 5.7 provides a summary of the mechanisms associated with surgical ablation of target tissue and variables that might influence the degree of laser-tissue interaction. The components of the plume produced will depend upon the type and constituent structure of the target tissue. Effects on the operator, support staff, and patient will be minimized through adequate protective measures.

5.4.7 Other Associated Hazards: Service Hazards

Most laser systems involve high potential, high current electrical supplies. Early lasers used three-phase mains electrical supply, but this is no longer necessary for units used in dentistry. However, even with compact units the risk from electrocution is significant and the most serious accidents reported with lasers have been due to electrocution [41]. Safe manufacturing practices offer adequate protection from these hazards and insulation, shielding, grounding, and housing of high voltage electrical components provide adequate protection under most circumstances from electrical injury. No attempt should be made to access internal parts of the machine during use.

Installation of laser equipment should always be performed by qualified personnel and not by the dentist. The laser should be serviced regularly according to manufacturer's recommendations and only by qualified personnel.

Many surgical lasers will have electrical, water, air and supply cabling, connectors, and filters in close proximity. Co-axial air or water supply may be under pressure. The practitioner should inspect the supply lines and cables, clean and maintain the external portions of the laser, and change necessary filters or other user serviceable items.

Mechanical Hazards

As referenced earlier, the construction and safe operating of all laser machines is governed by strict criteria [11, 42]. Engineering controls are normally designed and built into the laser equipment to provide safety. Modern laser machines employ multi-level safety features (fusible plugs, interlocks, pressure relief valves, warning lights, etc.) to inactivate the machine in the event of a component failure.

Mechanical hazards and safety mechanisms may be listed as follows [12]:

Laser device hardware:

- Locked unit panels to prevent unauthorized access to internal machinery.
- Control panel to ensure correct emission parameters.
- Emission port shutters to prevent laser emission until the correct delivery system is attached.
- Covered footswitch to prevent accidental operation.
- Remote interlocks to govern against accidental access to the operatory by unauthorized and non-protected personnel.
- Casters, if present, must be lockable.

During laser use:

- Key or password protection. When disabled (key or code removed), the laser cannot be operated.
- Laser software diagnostics and error messages.
- Display of parameters.
- Audible or visual signs of laser emission—recommended as an area control for Class IIIB laser operation. Such a warning system is mandatory for Class IV lasers.
- Specific stand-by and laser-emission modes.
- Time-lapsed default to stand-by mode Class IV lasers require a permanently attached beam stop or attenuator which can reduce the output emission to a level at or below the appropriate MPE level when the laser system is on "standby."
- Emergency "Stop" button.
- Additional hazards may exist however, due to heavy articulated arm delivery systems or the risk of needle-stick injury with fine quartz optic fiber cables.

Chemical and Fire Hazards

In the presence of flammable materials, lasers may pose significant hazards. The high temperatures that are possible in the use of Class IV and certain Class IIIB lasers can themselves either cause ignition of material and gases or promote flash-point ignition. Some of the common flammable materials found in the dental treatment areas are clothing, paper products, plastic, waxes, and resins. Liquids used as adjuncts to restorative materials may include ethanol, acetone, methyl methacrylate, other solvents.

Toxic fumes released as a result of combustion of flammable materials present an additional hazard. With regard to certain lasers, for example, gas lasers, there are risks associated with possible leakage of active medium components [43, 44].

With general anesthetic or gaseous anesthetic-sustained conscious sedation techniques, significant risks exist relative to the sustainability of explosion or fire [45]. Any combustion requires an ignition source and a Class IV surgical laser is an example; the field temperature during soft tissue ablation may reach several hundred degrees and the build-up of debris on delivery tips may exceed that temperature to a point above the flash-point of chemicals and equipment. Such high-temperature ignition may burn with a blue flame, and this may be very difficult to spot under intense operating lights. Additionally, and more dangerously the ignition may occur within the respiratory airway and be delivered deeper through forced inhalation.

With a laser as an ignition source, the presence of fuel sources (gauze, drapes, prep fluids, alcohol, and anesthetic gases) may be commonplace items within a dental operatory. Greater risk is further posed when an oxygen-enriched atmosphere (above 20%) is used in anesthetic and sedation techniques.

Rubber-based endotracheal tubes should be avoided to prevent the possibility of combustion of the material and subsequent airway burns, either through coatings of non-reflective, non-absorbent material or cuffed tubes to prevent leakage of anesthetic gases. With gaseous conscious sedation procedures, such as the use of a nosepiece to deliver oxygen/nitrous oxide mixtures, it is recommended to use a closed-circuit delivery system and scavenging system. A summary of risk and hazards associated with laser use is provided in ◘ Fig. 5.8.

Other Hazards: Sterilization

There exists a risk of damage posed by those personnel working in close proximity to the laser. Care should be shown to the possibility of contamination of all laser hardware and protective sleeves and screens used where possible, any components that may contact the surgical site subject to bagging and autoclaving or disposal in a safe container. Especially, care must be exercised in the manipulation of quartz optic fibers and details of safety measures should be entered in the local rules.

It is considered a "best practice" approach to utilize a tray system for instruments used during laser-assisted procedures and to rotate such through a "sterile" to

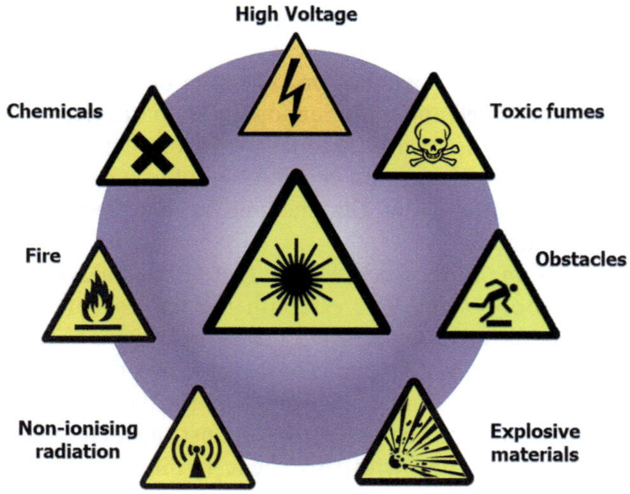

Fig. 5.8 Areas of risk and hazards associated with laser use

"operation" to "unsterile" status before disposal or re-sterilization.

Staff should consider the wearing of disposable gowns during surgical procedures.

5.5 Laser Safety Within the Dental Operatory

By far the provision of primary dental healthcare occurs within privately-run bespoke practices. These small-to-medium businesses are run as self-sufficient establishments and, it will be well-known to those involved, the breadth of regulation and compliance that allows the provision of dentistry to be such an efficient and safe health profession available to the patient. Such efficiency mirrors those public healthcare establishments that are bigger clinics and hospitals. Within such a framework of efficiency and safety, a protocol is necessary to allow all laser instruments, from low-powered units up to those high-powered Class III and IV lasers, to be used with predictable regard for acknowledgment of risks, planning to anticipate and avoid such risk and in a way that is readily applicable to all staff and patients who may be exposed to laser irradiation.

For all operators using lasers within the delivery of dental care, the concept of this protocol may be seen within four areas:

1. The physical dimensions of the care delivery area—the "controlled area" and "nominal ocular hazard zone." "Laser protection advisor" role.
2. The duties of the "laser safety officer" in determining and enforcing safety within the controlled area during delivery of dental care to patients.
3. Physical laser protection of eyes, skin, and general sterility within the controlled area.
4. Local rules, record keeping, and adverse reaction protocols.

5.5.1 Controlled Area

The "controlled area" is any location or area where there are one or more lasers and where activity of personnel is subject to control and supervision. In many cases, such an area will be a dental operatory, with physical barriers—walls, doors, windows by/through which any laser beam shall be attenuated. Control of such an area can be achieved through display of notices, remote interlocks, etc.

The controlled area must be indicated and marked by laser warning signs that specify the risk and conform to national regulations. Within the controlled area, all surfaces should be non-reflective and suitable measures should be employed to ensure that all laser supply cables, and delivery systems (optic fibers) are protected from inadvertent damage. A secure designated place for the (applicable) laser operating key should be assigned together with a designated place for all laser accessories. A suitable fire extinguisher should be sited for easy access.

Those dental clinics that operate a multi-chair, open-plan environment would need to address the physical dimensions and administration of the controlled area in greater detail.

The defining goal is to provide an area for the treatment of dimension beyond which laser irradiation falls below MPE value. This may be a primary laser beam or reflected, scattered, and diffuse photonic energy that may pose a danger to the unprotected eye.

Having regard to permissible MPE levels, the controlled area may also be referenced as the "nominal ocular hazard zone" (NOHZ). Calculation of the MPE has been discussed earlier and appropriate values for exposure of the unprotected eye at 0.25, 10, and 30×10^3 s periods determined.

The NOHZ is a complex calculation that can be done by a medical physicist, but for practical purposes a concept of a controlled area should be applied whereby a combination of the NOHD, and physical barriers can minimize risk [46]. The following factors are required in NOHZ computations:

- Wavelength, maximum/minimum laser energy output.
- Emission mode (CW and pulse repetition if applicable). Maximum exposure duration.
- Beam diameter, beam optics and beam path, beam divergence.
- Lens: focal length.

Subject to Federal or National Regulations that may apply, dental practices offering Class IIIB and IV laser treatment must appoint a laser protection advisor (LPA) and/or a laser safety officer (LSO). The LPA is usually a medical physicist who will advise on the protective measures required, MPE and NOHZ calculations, additional measures to provide attenuation of lasers being used and generally provide assistance and advice regarding all laser, LED, and intense pulsed light units within the practice of dentistry.

5.5.2 Laser Safety Officer (LSO) (aka Laser Safety Supervisor: LSS)

The LSO is appointed to ensure that all safety aspects of laser use are identified and enforced (see also ▶ Sect. 5.6). Ideally, the LSO could be a suitably trained and qualified dental surgery assistant.

The LSO standing duties refer to the minimum level of responsibility:

- Read the manufacturers' instructions concerning installation and use of the laser equipment. Confirm the Class of the laser.
- Be familiar with and oversee maintenance protocols for laser equipment.
- Train other staff in the safe use of lasers.
- Maintain an adverse effects reporting system.

LSO responsibilities during laser use include:

- Define and oversee the controlled area specific to the laser being used and limit unauthorized access.
- Post appropriate warning signs at all points of access to the controlled area.
- Make sure that laser equipment is properly assembled for use, together with all disposables. Carry out or supervise a "test fire" of all laser equipment before the patient enters the controlled area.
- Recommend appropriate personal protective equipment such as eye wear and protective clothing (suitable face masks, gowns, etc.).
- Maintain a log of all laser procedures carried out to include the patient details, the procedure performed, and laser operating parameters employed.
- Assume overall control for laser use and interrupt the procedure if any safety measure is infringed.

5.5.3 Laser Protection and Sterility

Earlier sections of this chapter have dealt with the beam hazards that exist to unprotected biological tissue. Key measures to be adopted should include the following:

1. The test fire procedure is designed to allow all working components of the laser to be checked before attempting the clinical procedure on the patient. It is obligatory for the LSO or dentist to carry this out before admitting the patient to the operatory and all safety measures—warning signs door/window protection and interlocks employed, to prevent unauthorized access as though a laser procedure was being performed. All personnel within the controlled area shall wear appropriate eye protection before the laser device is switched on. The laser delivery system is assembled and minimum operating parameters are chosen. The laser is directed away from the eyes and a suitable absorption medium should be used. The laser is fired within that medium so that the beam is attenuated.

   Some authorities believe that this would be an ideal opportunity to establish the level (if any) of power loss through the laser delivery system. The use of an approved laser power meter allows the LSO to check the parameter display on the laser control panel and to check the value against the meter reading. Associated with this check is the opportunity with many laser units to use a calibration port on the laser as part of the set-up procedure.

   For visible and near-infrared wavelengths, a suitable medium would be pigmented/dark articulating paper and for longer mid and far-infrared wavelengths (Erbium family and Carbon Dioxide) the suitable medium should be water. See ◘ Fig. 5.9.

   The objectives of the laser test fire are to check the following:

- Test operational ability of the machine
- Test the cleave of delivery quartz fiber (where appropriate)
- Demonstrate patency of delivery mechanism
- Demonstrate patency of aiming beam/co-axial air/water

In this way, it can be satisfied that the laser is in operating mode for the chosen clinical procedure.

2. Skin protection may not be a prime risk factor for the dentist working within the oral cavity. For those ablative procedures within the vermilion border of the lips, care should be observed to avoid encroaching beyond the target tissue, especially with mid and far-infrared wavelengths and shorter wavelengths operating at high power. Shielding of skin can be employed, using damp gauze to minimize the possibility of collateral damage.

   ◘ Figure 5.10 provides an example of laser handpiece selection (Lumenis Corp., Israel). Metal delivery conduits are potentially liable to overheating and care should be taken when using these delivery mechanisms for intraoral soft tissue surgery and the metal conduit rests against the lips and peri-oral skin.
3. Eye protection for all persons within the controlled area is mandatory with all Class IV lasers and any other laser Classes as advised by individual manufacturers and regulatory agencies (see ◘ Fig. 5.5). Regulations that specify the nature and suitability of laser protective eyewear are contained in ANSI Z 136.1 and IEC 60825 (EN 207/208) documentation.

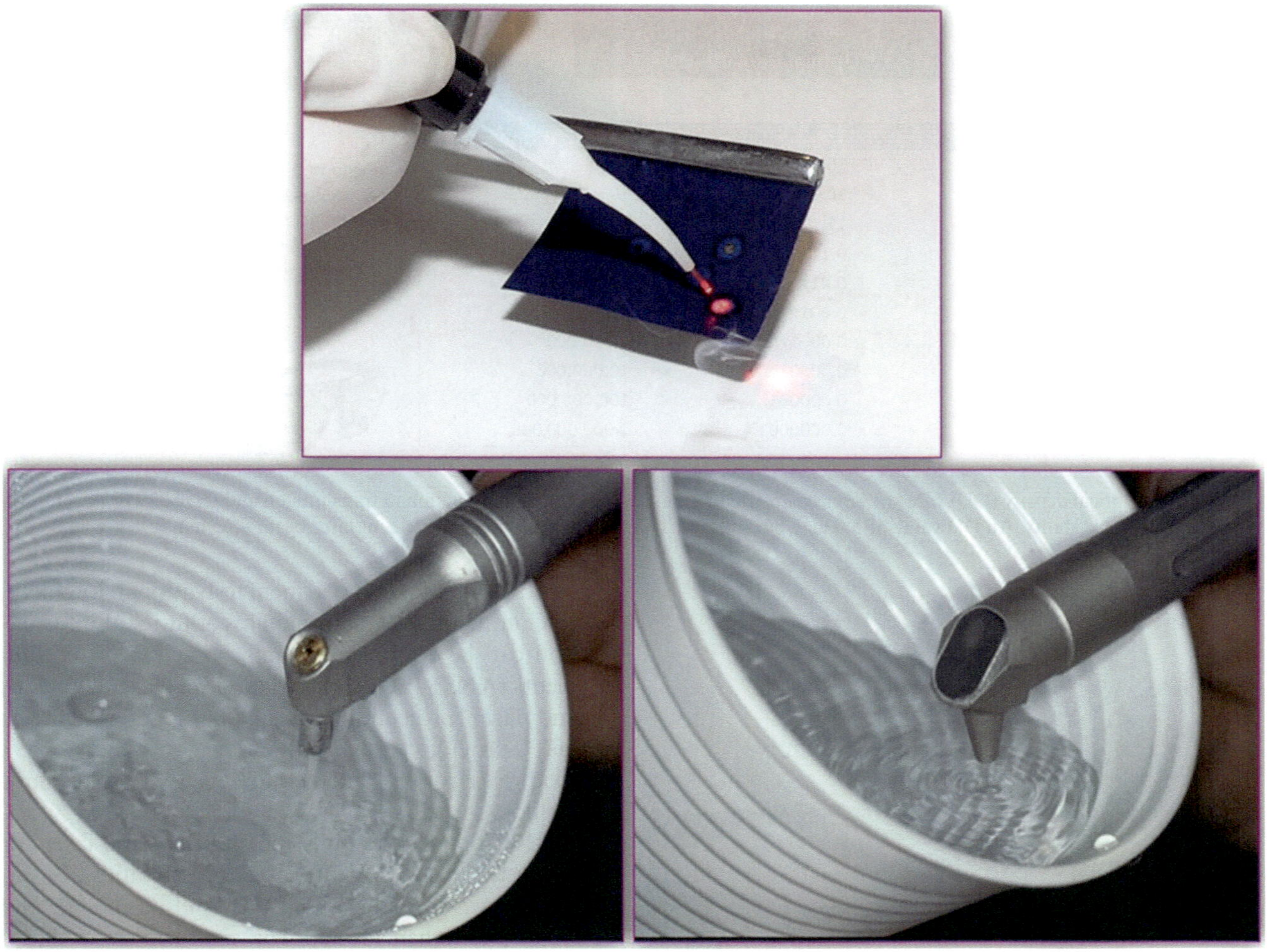

Fig. 5.9 Laser test fire. Using minimum laser operating parameters, the beam is directed into a suitable attenuation medium. Top center: for visible and NIR wavelengths pigmented articulating paper. Bottom left: water for (mid IR) Erbium YAG and bottom right: water for CO_2 wavelengths

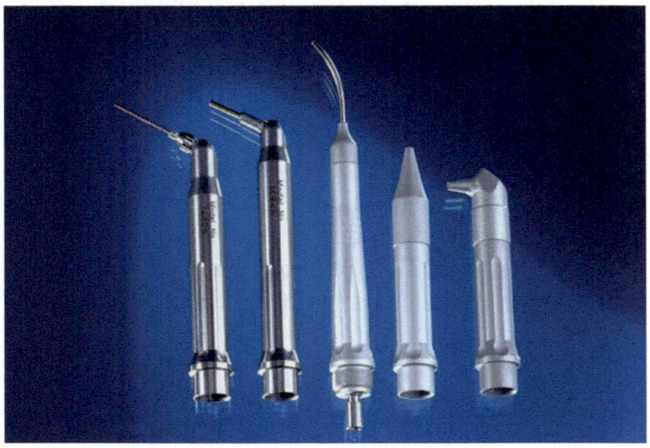

Fig. 5.10 Examples of dental laser handpieces. Center: this metal CO_2 delivery tip is liable to overheating and care should be exercised to avoid contact with peri-oral skin tissue

Eye protection during laser use can be summarized as follows:

- "Protocols as to use" is mandatory, including "patient on first-off last" as a maxim to represent the safety of the patient as paramount. With the patient considered first, each member of the clinical team within the controlled area would be obliged to similarly wear eye protection.
- Glasses/goggles must cover the entire peri-orbital region, be free of any surface scratches or damage and be fitted with suitable side-panels to prevent diffuse laser beam entry.
- Eyewear should be constructed of wavelength-specific material to attenuate the laser energy or to contain the energy within MPE values.
- Glasses or goggles should be marked with the wavelength for which protection is given, either as a spe-

Eye Protection Measures
Optical Density (OD)

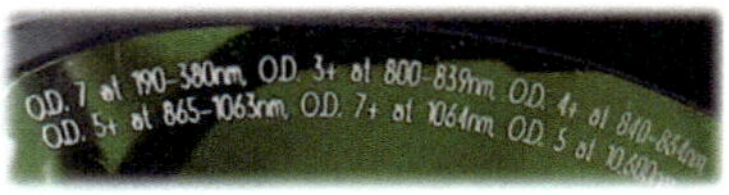

OD	Transmission	Attenuation factor
0	100%	1
1	10%	10
2	1%	100
3	0.1%	1,000
4	0.01%	10,000
5	0.001%	100,000
6	0.0001%	1,000,000
7	0.00001%	10,000,000
8	0.000001%	100,000,000
9	0.0000001%	1,000,000,000

Ability of a material to reduce laser energy of a specific wavelength to a safe level equal or below the MPE

Fig. 5.11 Optical density (OD) value of laser beam attenuation, corresponding % reduction in transmission and relevance to wavelength-specific eyewear suitability. An OD value of 5.0 may be considered a minimum safe level of incident laser beam attenuation

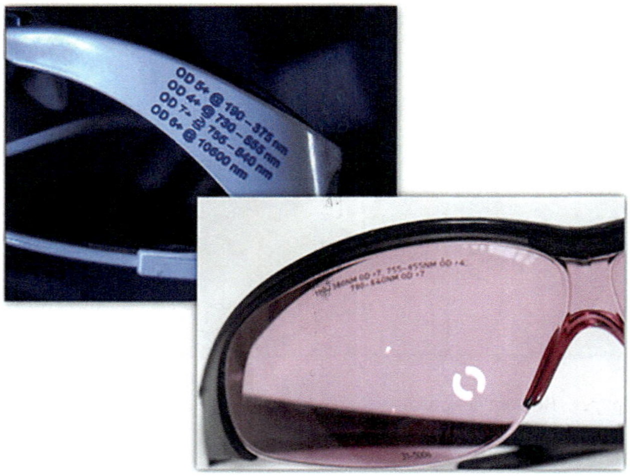

Fig. 5.12 Examples of laser safety eyewear, with relevant optical density (OD) values applicable to specific wavelengths. This information may be found on the eyewear lens or on the sidebars

cific wavelength value or range of wavelengths within which protection may be afforded by the chosen eyewear.

The level of protection is specifically expressed in terms of Optical Density (OD). This quantitative measurement represents the ability of the eyewear lens material to reduce laser energy of a specific wavelength to a safe level below the MPE. The OD value is measured as a log10 scale of laser beam attenuation and should be "5.0" or above for adequate protection [47]. See Figs. 5.11 and 5.12.

Worldwide, minimum information that must be present on protective eyewear (laser wavelength or a range of wavelengths that are covered by the material and OD value) may be further enhanced according to the nature of the eyewear, its specific intended use and other factors pertinent to National, Regional, or Continental regulations. This additional information can be referenced against laser safety requirements (ANSI/IEC/EN) for individual laser clinicians and laser protection advisors/laser safety officers may provide additional advice (Fig. 5.13). Examples of supplemental eyewear data may include:

- "DIR": defines the emission mode of the laser for which the eyewear is intended. "D" signifies Continuous Wave emission (CW), "I" (pulsed mode), and "R" (Q-switched mode). The significance of this is further enhanced in respect of the following:
- "DIN": Direct Impact Number—this is the ability of the eyewear material to attenuate direct laser beam energy to within MPE limits. This ability may be applied to exposure to a train of 100 pulses when the emission mode is gated or free-running pulsed, or to withstand direct exposure to a continuous-wave emission for 10 s.
- "L6A": a protective grade which defines a suitability for use of the protective eyewear within an intended clinical, industrial, or research condition.
- "CE" logo: "Conformité Européenne" will indicate a licence approved for distribution and use within the countries of the European Community.
- Manufacturer's identification mark.

An increasing number of dental practitioners use loupes during clinical work and such magnification may significantly increase the risk to the unprotected eye. For those procedures and laser classes of risk to the eye when using loupes (laser Classes IM, II, IIM, IIIR, IIIB, and Class IV), the clinician must wear the appropriate protective insert or shield that is specific to the wavelength being used. See Fig. 5.14.

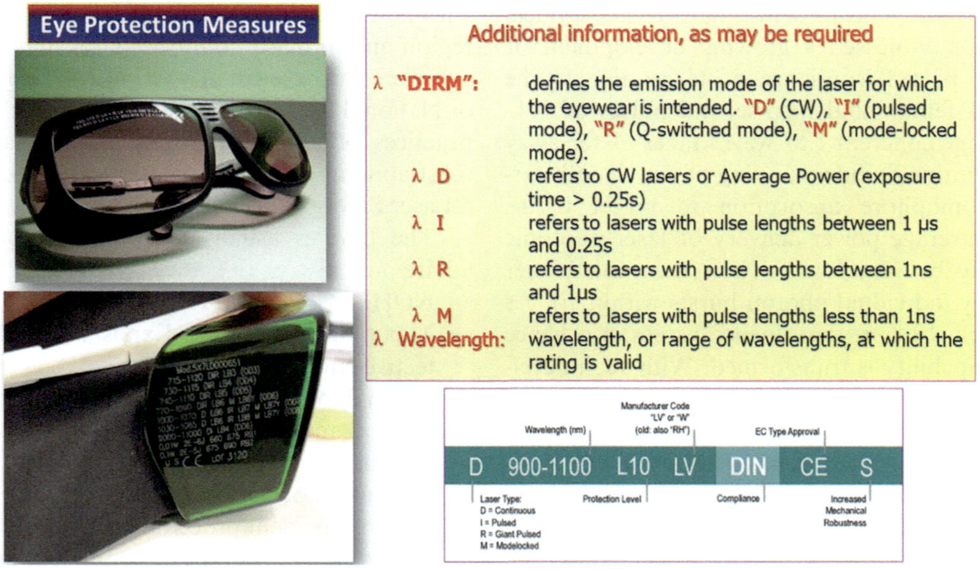

Fig. 5.13 Additional data pertaining to eyewear suitability, as may be found relative to intended level of protection and specific use

Fig. 5.14 Example of surgical loupes with wavelength-specific inserts, designed to fit on the inner aspect of the eyewear between the loupe lens and the operator's eye. Surgical microscope laser protection may benefit from liaison with equipment manufacturer and estimation of through-transmission of the laser beam within an SLR prismatic microscope

In the same manner and with the same breadth of laser classes, dental practitioners using an operating microscope to facilitate high-definition observation of laser surgery must fit the appropriate filters and maintain close eye contact with the oculars. The following is taken from a published paper (Saegusa H, Watanabe S, et al. 2010) to investigate the effectiveness of wavelength-specific ocular filters with an operating microscope [48]:

> *The aim of this study was to investigate the safety of laser use under the dental microscope. Nd:YAG, Er:YAG and diode lasers were used. The end of the tips was positioned at a distance of 5 cm from the objective lens of a dental microscope. Each eye protector was made into a flat disc, which was fixed on the lens of the microscope. The filters were placed in front of the objective lens or behind the eye lens. Transmitted energy through the microscope with or without the filters was measured. No transmitted laser energy was detected when using matched eye protectors. Mismatched eye protectors were not effective for shutting out laser energy, especially for Nd:YAG and diode lasers. None or very little laser energy was detected through the microscope even without any laser filter. Matched filters shut out all laser energy irrespective of their positions.*

The current trend in dental hard tissue surgical ablation has witnessed a growing development of ultrashort, pulsed irradiance of target tissue. The delivery of femto- and pico-second pulses of 0.4–3.0 μm range coherent EM waves takes us far away from conventional understanding of wavelength-targeted chromophore absorption in target tissue. Although average power delivery of laser photonic energy may be very low, such is the peak power achieved by individual photon bursts within 10^{-12} s and possibly 10^{-15} s, that fluence values and hence ablation capability is transformed. With such developments, laser-tissue interaction becomes ever-more plasma-mediated as opposed to photothermal in nature. Some investigation has shown that the modern laser eye protection seems to be robust except for the irradiance possible with ultrashort laser pulse exposure [49]. As always, it is incumbent upon the laser protection advisor and laser safety officer to ensure that protective eyewear is appropriate to the laser being used and the procedure being carried out.

5.6 Training of Staff Using Lasers

As has been seen throughout this chapter, in accordance with all Federal, National, and local regulations that may pertain, the responsibility of the lead clinician shall be to ensure the protection of the patient and operatory personnel and prevent inadvertent exposure of others during laser emission. In some countries, this responsibility is part-devolved through the senior personnel in charge of the dental clinic, as part of broader over-reaching statute governing care standards in healthcare [50]. Despite the nature of assumed responsibility for patient safety, the day-to-day approach to its observance must be measurable in qualitative and quantitative terms. Through this approach not only should a written protocol exist but also staff performance within such protocol should continue to develop a "best practice" approach to laser safety with audit-driven refinement at regular intervals. The prime tenet of any structured laser safety program should be the protection of the patient from inadvertent harm during laser-mediated clinical dental care. It is vital that all staff involved in the patient/client's treatment are aware of each other's role during treatment. Good communication between staff is essential.

Regulatory agencies recognize the essential nature of appropriate training in laser use and there is an implied necessity that clinicians should receive training as part of their duty of care and dental licencing [51]. It would be incumbent upon the laser user to acquaint themselves with how their laser use is regulated in their country or region and in many countries, laser manufacturers and suppliers have a legal duty to inform. Subject to Federal or National Regulations that may apply, clinical/medical practices offering Class IIIB and IV laser treatment must appoint a Laser Protection Advisor (LPA) and/or a Laser Safety Officer (LSO) [52, 53].

The LPA is usually a medical physicist who will advise on:

- NOHD, Controlled Area/Safe Access
- Maximum Permissible Exposure (MPE) levels; Protective equipment
- Normal operating procedures; safety checks; safe working methods
- Authorized user's responsibilities
- Prevention of use by unauthorized persons
- Adverse incident procedures
- Procedure in the event of equipment failure

The support of a laser protection advisor (LPA) would overcome any doubts that may arise through interpretation of any regulations as applied to a given country or region.

All healthcare establishments should have written local rules specific to each clinical application and for each laser, IPL (Intense Pulsed Light) and LED device. All staff involved in the use of these devices should read the local rules and sign them to indicate that they have been understood. This should be undertaken before staff use the equipment.

The following summarizes a "best practice" approach to laser safety in dental practice:

Appointment of a laser safety officer (LSO), suitably trained and aware of responsibilities (◘ Fig. 5.15). As referenced earlier, general responsibilities of the LSO shall include the reading and understanding of the laser operating manual, with specific regard to suitability as to use, operating settings by procedure, "set up" and "set down" procedures as applied to a specific laser and laser safety features.

The Laser Safety Officer/Laser Safety Supervisor 1 is defined as:

> One who has the authority to monitor and enforce the control of laser hazards and effect the knowledgeable evaluation and control of laser hazards.

The LPO/LSS [52] is usually an employee of the laser establishment. Drawing upon the recommendations referenced in Ref. [52], the following provides an overview of applicable duty and responsibility:

- If a service is performed on a laser product with an enclosed Class IIIB or Class IV laser, it is necessary to designate an LSO.

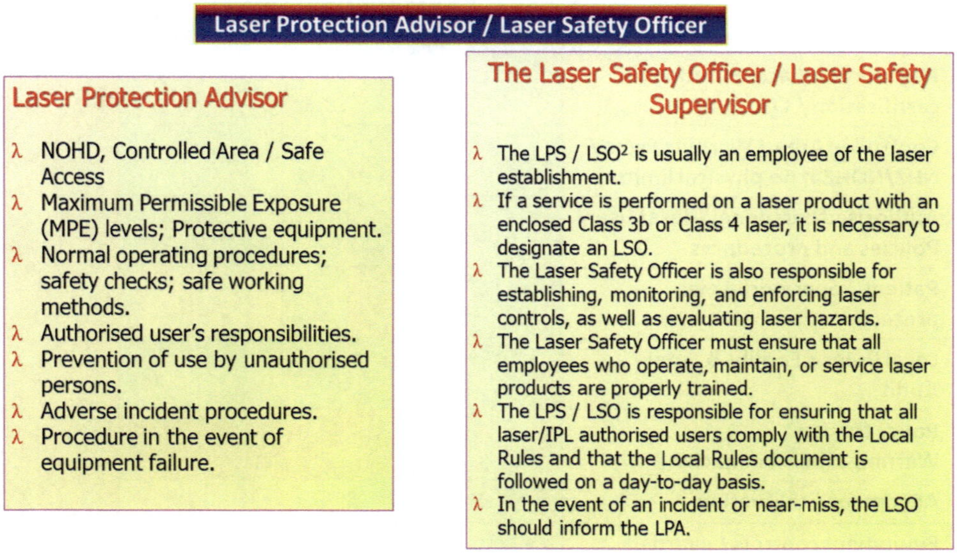

Fig. 5.15 Summary of responsibility and duties of a Laser Protection Advisor (LPA) and Laser Safety Officer/Supervisor (LSO/LSS). Subject to Federal or National Regulations that may apply

- The Laser Safety Officer is also responsible for establishing, monitoring, and enforcing laser controls, as well as evaluating laser hazards.
- The Laser Safety Officer must ensure that all employees who operate, maintain, or service laser products are properly trained.
- The LPS/LSO is responsible for ensuring that all laser/IPL authorized users comply with the Local Rules and that the Local Rules document is followed on a day-to-day basis.
- In the event of an incident or near-miss, the LSO should inform the LPA.

Above all, the Class of laser being used must be established, together with the nature of visual operating enhancement (loupes/operating microscope) as they may impact upon the safe non-risk (relative to MPE) levels and need for specific laser safety measures.

The LSO (under direction of the LPA if required by National regulation) must define the Nominal Hazard (safe ocular) Zone (NHZ). In practice, any operatory confined by (non-transparent, physically intact) walls and doors would satisfy this requirement, but any "open plan" operatory should adopt a protocol whereby adjacent areas are marked by signs to limit traffic and alert others to the need for caution.

The LSO should ensure the laser is appropriate to the proposed clinical procedure, is properly maintained and assembled. Accessories (optical fibers, tips, connectors, etc.) must be suitable for both the make and model of device that they are to be used with. Only use the accessories in accordance with the manufacturer's instructions. Additionally, equipment error messages or fault should be recorded in the equipment fault log. The log should be regularly reviewed, and appropriate action taken to inform the laser protection advisor or laser protection supervisor of all issues. Figure 5.16 provides a summary of protocols, applicable to successive laser class and within the scope of a trained laser safety officer.

Identify and eliminate all environmental risks. Apply laser warning signs, at the boundaries of the controlled area. Such laser warning signs must meet approved standards as applied through IEC/ANSI regulations and might include the laser wavelength being used, the need for eye protection, as well as over-riding precaution against unauthorized access. Arm all remote interlocks if applicable.

Test fire the laser prior to admission of the patient within the controlled area. Following this, the laser is deactivated, and the patient admitted.

The surgeon/practitioner shall choose the laser operating parameters appropriate for the intended treatment, commensurate with a policy of minimal power values to achieve the desired clinical outcome. The LSO shall closely monitor and assess the procedure and advise or adjust those operating parameters.

Commensurate with need defined by laser Class, the patient and personnel within the controlled area (NHZ) shall wear appropriate eyewear. Adjunctive safety measures—suitable filtration masks, gloves, and high-speed suction shall be employed. Non-target tissue shall be suitably protected (tissue retraction, use of non-reflective instruments, wet gauze).

The LSO shall be authorized to abort the procedure in the event of detected risk. The prime responsibility shall be the safety of the patient.

An adverse effect is defined as one that causes injury or death through direct use of the laser and will involve

Protocol ↓	Laser Class →	1	1C	1M	2	2M	3R	3B	4
Registration / accreditation / certification / CQC (UK)								X	X
Controlled Area / Warning Signs / NHZ/NOHZ if no physical limits								X LPA	X LPA
Authorised personnel – laser user/ Policies and procedures								X	X
Patient / operator(s) eye protection.				Care!		Care!	Care!	X	X
Local Rules / Facility & Safety Audit								X	X
Policy for use Warning labels on machine				X	X	X	X	X	X
Alignment / calibration					Advised			X	X
Equipment controls / guarded switchgear / accessory equipment								X	X
Maintenance and service as per manufacturer's recommendations		X	X	X	X	X	X	X	X

Fig. 5.16 A summary of action protocols relating to laser ownership and clinical use. "Red" areas would be mandatory, relative to applicable IEC/ANSI regulations and regions that apply such. Frequency and date of each protocol should be recorded in the Practice Local Rules

regulatory agency notification, according to national use. Additional effects reporting may require access to emergency clinical services in the event of eye or skin damage.

Suitable sterilization control measures should be employed. Reasonable measures should be employed to minimize risk of cross-infection, but minimally to include the use of protective barriers, chemical and autoclaving sterilization procedures.

The Standard of Care dictates that any part of the dental laser that contacts the oral tissues and/or the bloodstream must either be heat sterilized or if a single-use device be properly disposed. Portions of the laser that can contact the oral tissues must be disinfected with a suitable chemical agent.

Optimally, instruments should be employed through a tray system. Most clinicians will be used to employing specific protocols to recognize "dirty," "clean," and "sterile" in relation to dental instrumentation. Steam autoclaves are mandatory as part of a re-cycling of adjunctive non-consumable instruments; most will use metal trays which can be used as part of the correct storage of sterile elements. Additional to some recommended sterilization protocols, "dishwasher" style disinfecting units may deliver pre-autoclave cycles capable of destroying prions and removing proteinaceous debris. It should be the responsibility of the lead clinician to avail of all supporting advice as would apply to which parts of the laser may be disposable and if re-usable, what cleansing, disinfecting, and sterilization treatments are recommended and applicable.

With those delivery systems that use quartz optic fibers, appropriate cleaving techniques should be employed to remove damaged or contaminated elements and thorough cleaning before autoclaving. The cleaved piece must be disposed into the "sharps" container and regard given to disposal of plastic delivery tips. Many fiber delivery units have optic cabling as part of the assembly and the outer sheath will require regular checking to ensure sufficient quartz fiber is available to pass through the handpiece. This checking should be done prior to "bagging" (if applicable) and autoclaving of the delivery cable. Other laser units have fiber tip inserts which may be deemed as single-use or re-usable. In these cases, the manufacturer's recommendations must be followed to minimize the risk of cross-contamination between successive patient treatment sessions.

Smoke plumes—minimizing harmful effects [9]. The LSO shall take precautionary action to reduce, if not remove the plume. The amount of smoke plume and other deleterious matter generated varies with the procedure being undertaken, nature and type of target tissue, technique employed, duration of energy applied to tissue and laser emission mode used to vaporize the tissue. Staff involved in procedures resulting in smoke plumes should be educated on how they are produced and how to reduce or eliminate exposure.

Training by suitably qualified in-house personnel or smoke evacuator manufacturers should be considered. The most effective way of protecting clinical personnel and patients from inhaling the constituents of the smoke plume is to use either a stand-alone smoke evacuator or an evacuation system that is incorporated into the laser system. All smoke evacuators should have a high-efficiency filter that collects all smoke generated during the procedure.

Medical vacuum systems (operating theater wall suction systems) are not suitable for smoke plume removal. The accumulation of particles over time eventually decreases suction capability in theater evacuation systems. All evacuated airborne particles are deposited into a central vacuum system, which can become blocked and bacteria can then multiply.

All clinical staff within the controlled area and especially involved in the clinical procedure should wear well-fitting, high filtration-efficiency face masks (e.g., particulate respirators that filter particles of 0.1 μm in size) during all laser procedures. Standard surgical face masks are not sufficient to act as the primary method of particle filtration.

Record the laser-assisted procedure in the patient's notes. It is advisable and may be mandatory to keep an additional log of laser use, to record each clinical procedure according to laser, wavelength, operating parameters, and clinical outcome.

5.7 Local Rules

The following is taken from the 2015 publication "Lasers, intense light source systems and LEDs in medical, surgical, dental and aesthetic practices" in which the author acted as dental consultant to the United Kingdom Medicines and Healthcare products Regulatory Agency [54].

Local rules should be related to a risk assessment. They should contain the working practices, procedures, and information required to address the hazards and risks identified in that assessment.

The local rules should be prominently displayed in the laser/IPL room or the theater office.

All authorized users, assisting staff, or other individuals who work within the delivery of laser treatment should read the local rules then sign the associated form to show that they have understood them and agree to follow them.

Example of local rules:

5.7.1 Local Rules for the Use of the "X" Laser in "X" Dental Practice/Dental Clinic Premises

Dental Surgery: NAME, ADDRESS.

5.7.2 Nature of Hazards to Persons

The laser can injure the skin and eyes from both the direct and scattered beams. The aiming beam may also be hazardous. Safe use of the laser depends on people strictly following the rules:

The operating beam(s) is (are): X Laser Type, at operating wavelength(s) of X nm. It (they) constitutes an appreciable hazard.

The device(s) aiming beam is a low-powered diode laser and operates in the visible region of the EMS (X nm). It is designated as a Class I laser.

(Other wavelengths to be specified).

5.7.3 Laser Device(s)

1. Name of the laser
2. Any other Class IV laser (unspecified)

5.7.4 Laser Authorized Users

1. Dr. NAME
2. Manufacturer's/Supplier's Representative(s)

5.7.5 Laser Protection Advisor

1. As appointed by Regulatory Authority

5.7.6 Laser Safety Officer

1. NAME

5.7.7 Personnel Authorized to Assist in the Operation of the Laser(s)

1. NAME
2. Any other assistant, following training by Laser Protection Supervisor(s)

5.7.8 Controlled Area Designation and Access

The room in which the laser is used is designated a "controlled area," and the laser should only be used in this area. Approved warning signs should be fitted to the door.

A notice should be fixed to the laser indicating that its use is subject to the local rules.

The controlled area is the dental surgery in which the laser(s) is (are) installed. The controlled area shall be that area as designated by the Laser Protection Adviser, in accordance with the safe use of laser(s) and respective nominal ocular hazard distance(s). The controlled area shall be enforced only during such times as the laser is in clinical use. A sign shall be placed on the outside of the surgery door during the periods when the laser(s) is (are) in use.

Protective spectacles or goggles must be worn by operator, assistant, and patient and visitors whose presence is required. Eyewear must be wavelength specific to attenuate emitted laser light of laser in use. The laser safety officer shall be responsible for ensuring that protective eyewear shall be worn prior to activation of the laser and during all emission periods.

5.7.9 Restriction of Use to Authorized Persons

The equipment should only be used by an authorized user. Responsibilities and duties of laser protection supervisors shall be to:

- Ensure that the local rules are followed.
- Inform the laser protection adviser if they consider that the existing rules require amending.
- Ensure that the register of authorized users is maintained and that the correct procedure for authorization has been undertaken.
- Obtain written statements from each authorized user that they have read and understood the local rules and send copies of statements to the laser protection adviser.
- Ensure that only authorized users operate the laser.
- Inform the laser protection adviser as soon as possible in the event of an incident occurring.
- Seek assistance from the laser protection adviser on the safety implication when a change in operating procedure is envisaged.

The laser protection supervisor, in consultation with the laser protection adviser, can decide if another person is suitable to use the equipment. Their name can be added to the register, provided the person has signed a statement that they have read and understood the local rules. A copy of the signed statement should be sent to the laser protection adviser.

5.7.10 Operation of Equipment

1. The laser(s) may only be operated by an authorized user (see above).
2. Keys required to activate laser equipment may only be held by authorized user(s). Keys should be marked "LASER—for authorized use only" and should be stored in the safe.
3. The operating beam should never be activated unless accurately aimed at the operating site. Exception to this is during the "test fire" of the laser to be used, which shall be under the strict supervision of the laser safety officer and in accordance with safe practice regulations.
4. Neither the aiming nor operating beam(s) shall be directed toward the eyes of the operator, assistant(s), or patient.
5. When not in use for a treatment procedure, the laser shall be switched to "stand-by" mode.
6. When not in use for treatment, the laser shall be switched off, by use of the operating key or screen command.
7. The laser(s) shall not be used in the presence of anesthetic gases, or other explosive gases or liquids.
8. The operating beam shall not be directed toward amalgam restorations, since these are prone to vaporize, or other shiny surfaces, especially metal.
9. The operator/laser safety officer shall be responsible for ensuring that other individuals present at operation are sufficiently trained in laser safety and is responsible for the safety of any visitors within the controlled area.
10. When the laser is to be used, anyone who does not need to be present should leave the controlled area.
11. All operators must sign statements that they have read and understood the local rules. Assistants must sign statements that they have read and understood the local rules. These will be filed by the Laser Protection Supervisor.

5.7.11 Accidents Involving Eyes

Should a person's unprotected eyes be exposed to radiation from a laser beam, an arrangement exists between us (Dental Practice NAME) and the consultant ophthalmologist at NAME Hospital, who will carry out an examination within 24 h of the accident occurring.

Consultant: NAME. Tel: X.

Copy of report to be sent to: Dentist defence indemnifier/manufacturer/H & S contact as regulated.

All accidents arising from inadvertent and non-target laser radiation must be reported in the first instance to the Laser Protection Advisor.

5.7.12 Record Keeping

Full operating details are to be logged in the patient records. A separate book shall be kept, in which a record is made for the inspection of the Registering Authority. Details shall include the following:

- The name of the operator.
- Date and nature of procedure carried out.
- Identification of patient.
- Statement of machine maintenance.
- Appropriate information from the manufacturer, enabling the Registering Authority Laser Protection Advisor to be adequately informed in respect of his/her duties in advising the Registering Authority on the control of the laser hazards that may be potentially involved.
- The name and contact number of the appointed service agent.

5.7.13 Maintenance of Laser(s)

The authorized user is responsible for ensuring that the manufacturer's recommended planned preventative maintenance schedule is followed and that records are kept of service attendance (Figs. 5.17 and 5.18).

Date

Signed: ____________________ **Dr NAME Dentist**

____________________ **LSO/LSS as appointed**

____________________ **LPA as appointed**

Review date + 2 years

NAME Dental Practice / Local Rules / Safety / Risk Assessment /2023
Compiled by Dr NAME, Registered Owner.

Risk	Possible Adverse Effect	Risk Assessment	Precautions Applicable	Risk Assessment with Precaution
Direct Beam Hazard				
Skin	Laser burn	Low	Only use laser within oral cavity LSO to oversee "laser in use"	Minimal within dental setting
Eye	Cornea / Retina	Cornea – High with Mid – Far IR λ Retina – Low with Mid to Far IR λ	Correct safety eyewear / λ LSO to oversee eye protection. "Patient on-first, off-last"	Minimal
Oral Tissue	Collateral / distant exposure to laser energy	Moderate, made greater with use of mirror, with loose tissue, or with transactional excision. Intra-dentally, risk of trans-dental energy to pulp tissues	Correct use of power parameters / procedure. Use of attenuating barriers, e.g. wet gauze.	Low
Specular Reflection Hazard	Use of mirrors intra-orally. Wet tissue surfaces	Moderate, assuming low regard to risk	Use of anodised surfaces where possible. Protection of distal targets.	Low
Indirect Hazard				
Cables, Traffic	Damage to cables, waveforms	Low within surgery configuration	Proper positioning of laser relative to operator	Low
Anaesthetics	Explosive / combustion by anaesthetic agents	Non-existent. No GA administered.	N/A	N/A

Laser Risk Assessment. *DATE*

Fig. 5.17 a, b Example of Local Rule Risk Assessment (Page #1)

NAME Dental Practice / Local Rules / Safety / Risk Assessment /2023
Compiled by Dr NAME, Registered Owner.

Risk	Possible Adverse Effect	Risk Assessment	Precautions Applicable	Risk Assessment with Precaution
Indirect Hazard				
Liquids	Combustion of alcohol-based liquids	Moderate to high.	No alcohol-based liquids to be used in preparatory or laser-based treatments.	Low
Safety Hazards				
Access	Un-authorised access to controlled area	Given location of surgery - low	Location …, all staff instructed in precautions, use of laser warning signs posted at only door access	Low
Personnel	Operator / Assistant / Patient exposure to laser beam	LSO always in attendance. Risk is low	LSO governance	Low
Operating Parameters	Use of inappropriate power parameters during clinical procedure	Low	Operator / LSO choose operating parameters	Low
Distant Risks	Possible exposure to laser beams outside controlled area	Minimal	Controlled area is …. and window ……Window is glazed with heavy lead glass. Vertical blinds hung at window. Attenuation of possible beam exposure to glass considered maximal. Distal targets at and above 1st floor level > 50 metres.	Minimal

Laser Risk Assessment. *DATE*

Fig. 5.17 (continued)

NAME Dental Practice | Clinical Protocols and Monitoring | Compiled: Dr NAME

Clinical Protocols for Laser Treatment

Preliminary to patient treatment:

- Secure operating room-place proper laser warning signs
- Set up laser and test proper laser operation.
- Laser test fire in suitable attenuating medium, using all safety measures and minimum power.
- Supplies dispensed, equipment and sterile instruments arranged.
- Patients' information: review charting, x-rays, etc
- Patients seated: review treatment plan and informed consent.
- Safety: eye protection placed; patients first followed by operating personnel.

Treatment Protocol:

- Local anaesthetic as required or requested.
- LSO clearance
- Choice of laser operating parameters chosen emission mode, power level
- Define non-target tissue protection.
- LSO clearance that laser in operating (ready) mode
- Laser procedure carried out.
- LSO clearance that laser in stand-by mode
- Removal of protective measures
- Treatment procedure completed.
- Post-operative instructions issued.
- Patient leaves surgery
- LSO Warning signs de-activated / controlled area removed.

Treatment quality Monitoring:

*Outline of laser treatment provided entered in patients notes.
* Specific laser parameters and procedure entered in laser treatment schedule sheet in patient's notes.
* Minimum parameters: laser wavelength, emission mode, – delivery handpiece / fibre tip, average power, time taken.
* Review of procedures and outcomes undertaken yearly, with next review due DATE
* Treatment notes recorded

Clinical Protocols. *DATE*

Fig. 5.18 Example of Local Rules / Clinical Protocol

5.8 Conclusion

The development of laser photonic technology within primary healthcare has highlighted the absolute need for safety considerations. In general, a multi-tiered administrative structure has been developed and applied internationally, drawing upon various national regulations to form an over-riding set of guidance and rules that may govern and enhance the safe use of lasers. Within this framework, it has been possible to explore where and to what extent the use of laser photonic energy can be safely employed by the dental clinician.

The various classes of lasers have been demonstrated and their impact on risk explored and demonstrated. Of prime consideration has been the risk to the unprotected eye and all aspects of environmental, operatory, and personnel aspects of this important consideration, explored, together with guidance on the use of correct protective eyewear for all personnel within a designated "at risk" zone.

Secondary issues of non-target protection, post-ablation laser plume management, and the impact of lasers within general health and safety at work policy have been explained. Through the presentation of specimen local rules, guidance has been offered to support the adoption of appropriate and compliant measures that may be employed within the Practice/Office setting and for the use of all staff.

The prime concern is to safeguard the patient receiving laser dentistry procedures, and it is hoped that the foregoing chapter can provide both the background and applicable measures to implement the highest levels of safe practice when performing laser-assisted therapy.

References

1. Sweeney C. Laser safety in dentistry. Gen Dent. 2008;56(7):653–9.
2. Pierce J, Lacey S, et al. An assessment of the occupational hazards related to medical lasers. J Occup Environ Med. 2011;53(11):1302–9.
3. Clark KR, Johnson TE. Trends in laser injury reporting. Lasers Med Sci. 2003;18:S64.
4. The management of health and safety at work regulations. London: HMSO; 1999.
5. Directive 2013/35/EU of The European Parliament and of The Council of 26 June 2013 on the minimum health and safety requirements regarding the exposure of workers to the risks arising from physical agents (electromagnetic fields) (20th individual Directive within the meaning of Article 16(1) of Directive 89/391/EEC) and repealing Directive 2004/40/EC.
6. Moseley H. Operator error is the key factor contributing to medical laser accidents. Lasers Med Sci. 2004;19:105–11.
7. ANSI Z 136.1-9. American National Standard for Safe Use of Lasers. 2014.
8. Occupational Safety and Health Administration.
9. Lasers, intense light source systems and LEDs—guidance for safe use in medical, surgical, dental and aesthetic practices. Crown copyright 2015. Published by the UK Medicines and Healthcare Products Regulatory Agency.
10. FDA Laser Notice 50. Laser products—conformance with IEC 60825-1, Am. 2 and IEC 60601-2-22; final guidance for industry and FDA (laser notice no. 50), 26 Jul 2001.
11. International Electrotechnical Commission. IEC 60825-1 Safety of laser products—part 1: equipment classification and requirements. Edition 3.0. May 2014. ISBN: 978-2-8322-1499-2.
12. Parker S. Laser regulation and safety in general dental practice. Br Dent J. 2007;202(9):523–32.
13. Sliney DH, Dennis J. Smart skin treatment lasers—the need for a new class; ILSC 2009 paper #702. 2009.
14. Schulmeister K. The upcoming new editions of IEC 60825-1 and ANSI Z136.1—examples on impact for classification and exposure limits. In: Proceedings of the international laser safety conference, Orlando, FL, Mar 18–21, 2013, p. 330–7.
15. Parker S, Cronshaw M, Anagnostaki E, Lynch E. Laser essentials for the dental practitioner: foundation knowledge—construction, modes of operation and safety. EC Dent Sci. 2019;18(9):2020–7.
16. Boldray E, et al. Eye injuries from lasers. Ophthalmologica. 1981;88:101.
17. Burnett WD. Evaluation of laser hazards to the eye and the skin. Am Ind Hyg Assoc J. 1969;30(6):582–7.
18. Lund D, Stuck B, Edsall P. Retinal injury thresholds for blue wavelength lasers. Health Phys. 2006;90(5):477–84.
19. Zuclich J, Connolly J. Ocular damage induced by near-ultraviolet laser radiation. Invest Ophthalmol Vis Sci. 1976;15(9):760–4.
20. Sliney DH, Wolbarsht ML. Safety with lasers and other optical sources. New York: Plenum; 1980.
21. Lund D, Edsall P, et al. J Biomed Opt. 2007;12(2):024023-1–7.
22. Delori F, Webb R, Sliney D. J Opt Soc Am A Opt Image Sci Vis. 2007;24(5):1250–65.
23. Barkana Y, Belkin M. Laser eye injuries. Surv Ophthalmol. 2000;44(6):459–78.
24. American National Standards Institute (ANSI) ANSI Z 136.1-2014. Safe Use of Lasers.
25. De Luca D, Delfino I, Lepore M. Laser safety standards and measurements of hazard parameters for medical lasers. Int J Opt Appl. 2012;2(6):80–6.
26. Schulmeister K, Sliney D, et al. Review of exposure limits and experimental data for corneal and lenticular damage from short pulsed UV and IR laser radiation. J Laser Appl. 2008;20(2):98–105.
27. Lopez R, Lacey S. Characterization of size-specific particulate matter emission rates for a simulated medical laser procedure—a pilot study. Ann Occup Hyg. 2015;59(4):514–24.
28. Mowbray N, Ansell J, et al. Is surgical smoke harmful to theater staff? A systematic review. Surg Endosc. 2013;27(9):3100–7.
29. Pierce J, Lacey S, et al. Laser-generated air contaminants from medical laser applications: a state-of-the-science review of exposure characterization, health effects, and control. J Occup Environ Hyg. 2011;8(7):447–66.
30. Smith J, Moss C, Bryant C, Fleeger A. Evaluation of a smoke evacuator used for laser surgery. Lasers Surg Med. 1989;9(3):276–81.
31. Manson L, Damrose E. Does exposure to laser plume place the surgeon at high risk for acquiring clinical human papillomavirus infection? Laryngoscope. 2013;123(6):1319–20.
32. Garden JM, O'Banion MK, Bakus AD, Olson C. Viral disease transmitted by laser-generated plume (aerosol). Arch Dermatol. 2002;138(10):1303–7.
33. Wisniewski PM, Warhol MJ, Rando RF, Sedlacek TV, Kemp JE, Fisher JC. Studies on the transmission of viral disease via the CO_2 laser plume and ejecta. J Reprod Med. 1990;35(12):1117–23.
34. Ziegler BL, Thomas CA, Meier T, Müller R, Fliedner TM, Weber L. Generation of infectious retrovirus aerosol through medical laser irradiation. Lasers Surg Med. 1998;22(1):37–41.
35. Bigony L. Risks associated with exposure to surgical smoke plume: a review of the literature. AORN J. 2007;86(6):1013–24.
36. Mowbray N, Ansell J, Warren N, Wall P, Torkington J. Is surgical smoke harmful to theater staff? A systematic review. Surg Endosc. 2013;27(9):3100–7.
37. McKinley IB Jr, Ludlow MO. Hazards of laser smoke during endodontic therapy. J Endod. 1994;20(11):558–9.
38. Israel M, Cobb CM, Rossmann JA, Spencer P. The effects of CO_2, Nd:YAG and Er:YAG lasers with and without surface coolant on tooth root surfaces. An in vitro study. J Clin Periodontol. 1997;24(9 Pt 1):595–602.
39. Yanyan H, Zhengjia L, Chuyun H, Yuncheng Y. KTP green laser vaporization of biologic tissue under water and its clinical application. Photomed Laser Surg. 2008;26(4):337–41.
40. Li PT, Tang SP, Gomersall CD. Protecting staff against airborne viral particles: in vivo efficiency of laser masks. J Hosp Infect. 2006;64(3):278–81.
41. Dudelzak J, Goldberg DJ. Laser safety. Curr Probl Dermatol. 2011;42:35–9.
42. American National Standards Institute. American National Standard for the safe use of lasers. New York: Laser Institute of America; ANSI Z; 1993 + updates. ANSI Z Publication Z 136.1-7.
43. Davis RK, Simpson GT. Safety with the carbon dioxide laser. Otolaryngol Clin N Am. 1983;16:801–14.
44. Lim RY, Kenney CL. Precaution and safety in carbon dioxide laser surgery. Otolaryngol Head Neck Surg. 1986;95:239–41.
45. Daane SP, Toth BA. Fire in the operating room: principles and prevention. Plast Reconstr Surg. 2005;11(5):73e–5e.
46. Marshall WJ, Conner PW. Field laser hazard calculations. Health Phys. 1987;52(1):27–37.
47. Laser safety eyewear. Health Devices. 1993;22(4):159–204.
48. Saegusa H, Watanabe S, Anjo T, Ebihara A, Suda H. Safety of laser use under the dental microscope. Aust Endod J. 2010;36(1):6–11.

49. Stolarski D, Stolarski J, et al. Reduction of protection from laser eyewear with ultra-short exposure. In: Laser-tissue interaction XII: photochemical, photothermal, and photomechanical, Proc. SPIE, vol. 4257. SPIE—The International Society of Optical Engineering; 2001. p. 125.
50. UK Care Quality Commission. www.cqc.org.uk/dentist/laser.
51. Toon S. Doctors using lasers risk problems without proper training, protection. Occup Health Safety. 1988;57(7):30.
52. American National Standards Institute. American National Standard for the safe use of lasers: ANSI Z-136.1-2000. Orlando, FL: LIA. p. 2000.
53. Essential standards regarding class 3B and class 4 lasers and intense light sources in non-surgical applications. BMLA. May 2017.
54. Parker S. (Dental Consultant in) United Kingdom Medicines and Healthcare products Regulatory Agency. Lasers, intense light source systems and LEDs—guidance for safe use in medical, surgical, dental and aesthetic practices. Appendix A—Example of Local Rules. 2015. p. 61–5.

Laser Assisted Diagnostics

Alex Mathews Muruppel and Daniel Fried

Contents

D. J. Coluzzi, S. P. A. Parker (eds.), *Lasers in Dentistry—Current Concepts*, Textbooks in Contemporary Dentistry,
https://doi.org/10.1007/978-3-031-43338-2_6

Core Message

An objective and accurate diagnosis is an essential and key component in the formulation of safe, comprehensive management, and treatment of dental patients. The framework of such diagnosis should be based on clear criteria, applied with sound diagnostic methodologies which can assess, grade, and detect the presenting symptoms of any individual case. Various diagnostic approaches therefore play an essential part in developing a provisional and final diagnosis, from which treatment modalities and strategy can then be planned and implemented.

The dental clinician plays a pivotal role in being a skilled examiner, a physician, and a surgeon, all of which is dependent upon each patient's needs and desired treatment outcome. The oral cavity and its varied range of both hard and soft tissues coupled with the concomitant host of microbial flora is easily subject to pathology that may be a simple, single tissue disease or extending to anatomical, regional, or may render multi-structural tissue change. Hence, the role of an appropriate and accurate diagnosis cannot be understated.

Specifically, diagnostic techniques should ideally be simple, cost-effective, non-invasive, reproducible, measurable, and most importantly should be in tandem with advancing scientific research and technology. The diagnostic technique should have application within hard dental disease as well as soft tissue pathology and should ideally be available within a general dental practice setting.

Laser photonic energy has been shown to interact with oral tissues and within a sub-ablative power envelope may provide measurable data to assist the clinician in distinguishing between healthy and diseased tissue and furthermore can quantify or assay the disease process and can even provide for time related monitoring of the potential disease progress.

This chapter sets out the underlying science of laser fluorescence and its integration and application within a hierarchy of diagnostic measures in the assessment of oral disease.

6.1 Introduction

Lasers when applied to the realm of diagnostic sciences signify minimally invasive techniques that provide unparalleled precision and accuracy. An illustration would be the case of optical biopsies that provide elaborate and miniscule details of tissue without actually having to physically injure the patient by having to take a sample to do so. The varied plethora of techniques available today is testament to the concerted and dedicated research work, in tandem with the growth of the field of lasers and their use in medicine and dentistry. These techniques could be applicable to soft tissue and hard tissue in various clinical scenarios and even branch out to the treatment of specific conditions as in the case of photodynamic therapy (PDT). However, some techniques continue to be expensive and hence not all of the benefits of these methods have been universally available yet.

It is intriguing to note that though there is indeed a veritable list of laser assisted techniques in diagnosis the scientific basis of all these boils down to light-related phenomena like Fluorescence, Phosphorescence, and Spectroscopy. Such physical outcomes of incident irradiation of target material have been the subject of pioneering work by investigators such as George Gabriel Stokes (1852) [1] and Sir Chandrasekhara Venkata Raman (1930) [2, 3].

Stokes in his treatise, "On the Change of Refrangibility of Light" [1], reported the ability of mineral calcium difluoride, CaF_2 (fluorspar), and uranium glass to convert incident invisible (UV) light beyond the violet end of the visible spectrum into re-emitted blue light. He coined the word fluorescence (being light from fluorite) taking a leaf out of the term opalescence being derived from the color change of hydrated silica, SiO_2.

Similarly, Sir C.V. Raman's seminal work [2, 3] on the inelastic scattering of light led to the development of spectroscopic techniques which could detect vibrational and rotational changes caused in a molecule by photonic energy and therefore even lend a description to the molecular framework of a particular material.

Photochemical changes induced by lasers and the subsequent vibrational and radiational relaxation evoked in molecules would give distinctive information on their constitutional make up and led to the development of a myriad of diagnostic techniques [4]. The high intensity energy of laser light allowed easy excitation of any molecule and furthermore differentiation from spontaneous radiation induced by absorption of background low intensity ordinary sources of light was simple and unambiguous. Thus the phenomenon of Florescence and Phosphorescence and spectroscopic techniques are the basis and foundational fundamental in these diagnostic methods. Hence to enrich our understanding of these methods, we would first dwell on the theoretical processes of these phenomena.

6.2 Basics of Fluorescence and Phosphorescence

All matter around us is in a state of vibration, such vibration is the resultant effect of electrons within constituent atoms in a molecule being in constant transla-

6

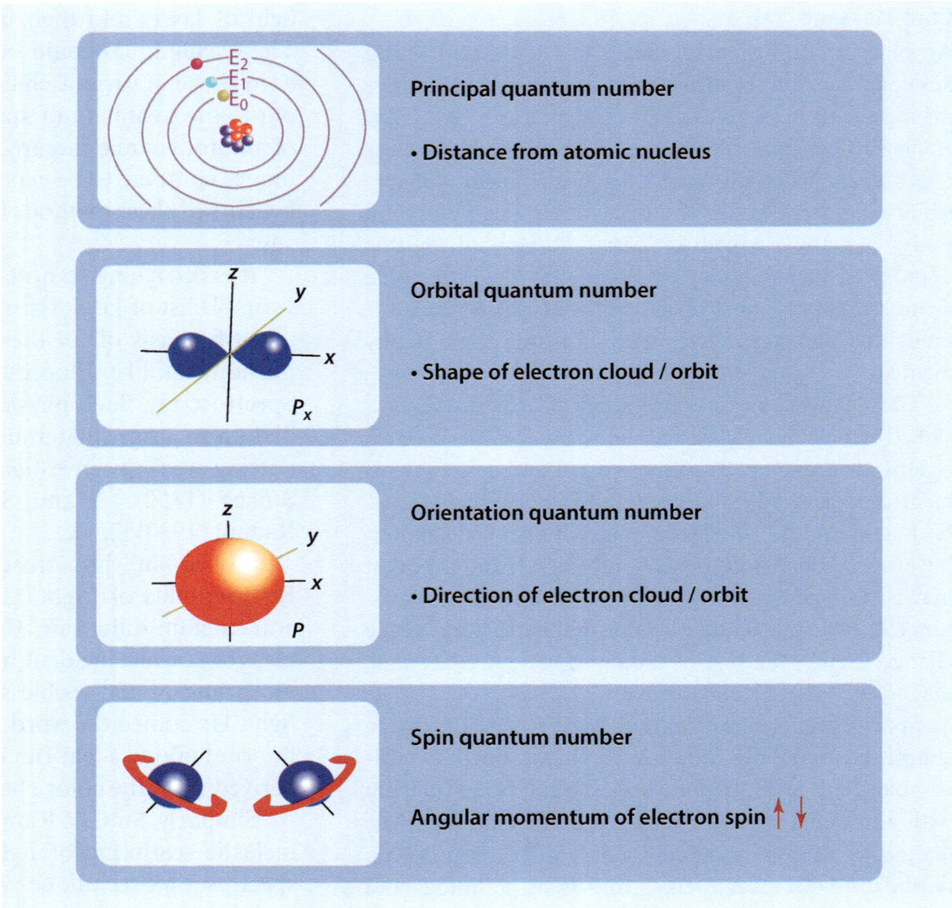

Fig. 6.1 Description of quantum numbers. Quantum numbers indicate the energy level of an individual electron in terms of both electrical and magnetic (spin and orientation) energy

tional and rotational motion. Blank [5] writes that all organisms have an endogenous electromagnetic field (EEMF) albeit of very low intensity (below 1 Hz up to 1015 Hz). Cellular processes like enzymatic peroxidation, ATP production, the Krebs cycle, and natural luminophores in nucleic acids and proteins generate an electromagnetic field [6].

The energy levels that dictate such translational and rotational motion is "quantized" (M. Planck, N. Bohr) and are described according to discrete quantities of energy that the electron has according to the four quantum numbers that characterize the electron according to (1) radius of distance from the atomic nucleus—Principal Quantum Number, (2) orbital angular momentum of the electron as *s*, *p*, *d*, or *f*—Orbital Quantum Number, (3) direction of the electron cloud vector relative to an electric field—Orientation Quantum number, (4) electron spin or angular momentum (the electron has an intrinsic magnetic moment directed along its spin axis)—Spin Quantum Number (Fig. 6.1).

Note: Electrons normally are configured in atomic orbitals as pairs with opposite spins within the same orbital.

Such molecular vibration could be increased by excitation caused by absorbing a quantum of energy, E, corresponding to the vibration's frequency, ν, according to the relation, $E = h\nu$ (where h is Planck's constant) (Fig. 6.2).

The term "Ground state" applies to the normal electronic state of an atom or molecule where the electron has stable and paired spins. However, it will still have a normal rotational and vibrational energy characteristic to the element or molecular bond.

The term "Excited state" on the other hand refers to a higher energy level to which an electron has climbed subsequent to absorption of a quantum of energy. The "lifetime" of the excited state would vary according to the quantum of energy gained or mode of excitation. Eventually de-excitation can occur by radiating a quantum of energy (fluorescence or phosphorescence), by expending the gained quantum of energy as vibration and heat, internal conversion, or intersystem crossing.

6.2.1 Light as an Oscillating Electric Field

Thus, a quantum of energy, for example, from an incoming "green" photon (for illustration sake) could excite an electron of a target atom that is in a ground state if the frequency of the light wave equals the natural frequency

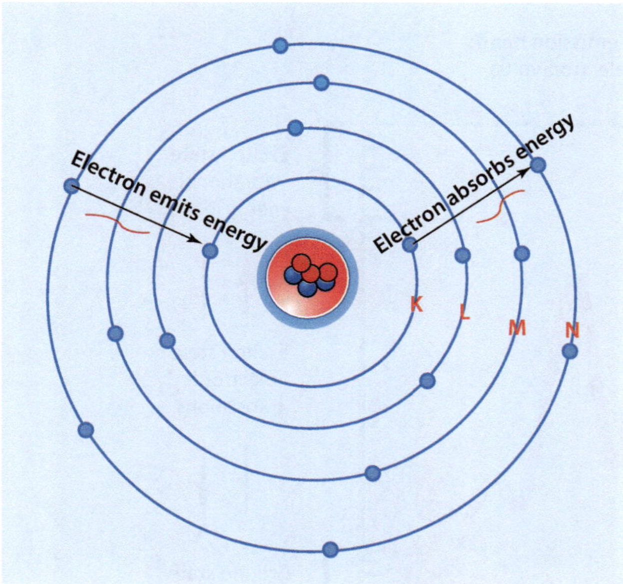

Fig. 6.2 Excitation and de-excitation of electrons—absorption occurs when an incident photon's energy couples with the electron in the ground state (non-excited state) and goes to an excited state (higher energy level). Emission occurs when the electron "decays" from the excited state or sheds the energy it gained to return to the ground state

of free vibrations of the atom. This excitation can lead to the electron absorbing the energy of the green photon, leaving its paired electron (having opposite spin) in the ground state, and forming an excited singlet state (Fig. 6.3).

Interaction strength between the incoming photon and the molecule depends on various factors such as the field dipole, induced dipole strength, and the distance between them, and most importantly it is imperative that the frequency (ν) of oscillating field of the molecule must "match" the electronic oscillation frequency of the incoming photon [7, 8].

It is termed as a singlet state as even in the excited state the electron still has the paired and opposite spin to its counterpart electron which is in the ground state. Thereafter, the excited singlet electron loses some of its absorbed energy through vibrational relaxation which is dissipated as a small packet of heat called a "phonon." It now only has a lesser amount of energy left and hence emits a longer wavelength, yellow photon and returns to the ground state by internal conversion. Internal conversion is a term used to describe the transition of an excited singlet state electron back to its paired counterpart in the ground state dissipating some of the energy

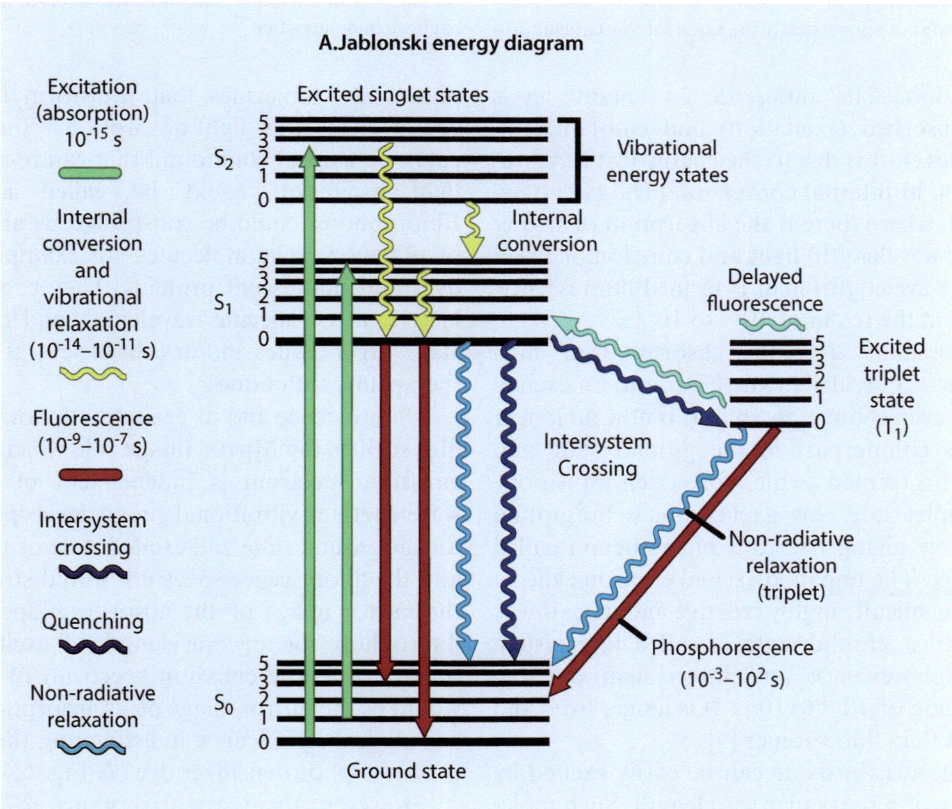

Fig. 6.3 Jablonski energy diagram—excitation → $S_0 + h\nu_{ex} \rightarrow S_2$, de excitation → $S_2 - S_0$ [Stokes Shift (non-radiative relaxation – heat (vibrations)] + $h\nu_{em}$ → fluorescence + HEAT. Radiational relaxation of the electron occurs from an excited state (light is liberated) by fluorescence-radiative decay from an excited singlet state, or phosphorescence-radiative decay from an excited triplet state. Non-radiative relaxation occurs by internal conversion or intersystem crossing, but could also occur by intermolecular energy transfer as quenching when the energy is consumed (collisional or complex formation) or when a new excited species is created it is called photosensitization

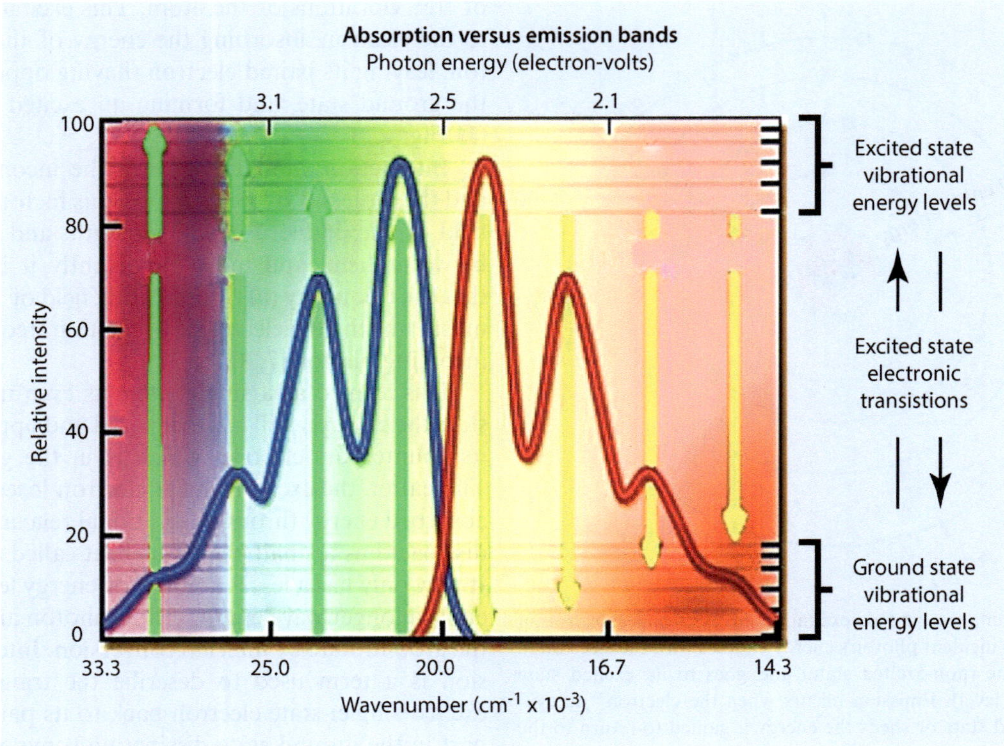

Fig. 6.4 Mirror image rule where though the spectrum changes according to the wavelength of the excitation light the absorption and emission spectrum is diametrically the same for the same fluorophore. Adaptation of this rule is useful in estimating the radiative lifetime, excitation levels and absorptive levels of any particular fluorophore in diagnostics

through vibration. The difference in energy levels between the absorbed (excitation) and emitted light, termed as Stokes shift is due to the energy lost as vibrational relaxation in internal conversion (The phenomenon, in general, where there is the absorption of higher energy (shorter wavelength) light and emission of lower energy (longer wavelength) light is termed fluorescence. It is short lived in the realm of 10^{-9} to 10^{-7} s.

Another possibility is when absorption of high energy photonic energy drives an electron to an excited state wherein it even changes its spin. It is now no longer paired with its counterpart in the ground state and undergoes what is termed as the "forbidden transition" to an excited triplet state. Now a relaxation to the ground state is very slow giving rise to a phenomenon called phosphorescence. The unpaired excited electron called a free radical is chemically highly reactive and its pathway of relaxation to a ground state is called intersystem crossing. Phosphorescence and intersystem crossing occur in the region of 10^{-3} to 10^{2} s. It is longer lived and more persistent than fluorescence [9].

Specific molecules in tissue can be easily excited by photonic energy of a particular wavelength. Such molecules or substances are termed as Chromophores. Waynant [10] defines a chromophore as "a substance or specific target tissue that serves as an attractant for a laser photon." Arnat and Rigau [11] terms chromophores as "molecules that transform their electronic energy levels after light absorption." Similarly, a molecule or chemical compound that can re-emit light upon light excitation could be called a fluorophore. Fluorophores could be constituted by aromatic groups, or plane or cyclic molecules, for example—conjugated dyes and fluorescent proteins. They can be excited by laser light of specific wavelength and can be used to stain target tissues and/or cells in several diagnostic and therapeutic techniques.

Fluorescence and its excitation follow certain principles such as the Mirror Image rule which states that the emission spectrum is independent of the excitation wavelength as vibrational energy level spacing is similar for the ground state and excited state of the fluorophore and the fluorescence spectrum would strongly resemble the mirror image of the absorption spectrum for that fluorophore for any wavelength of excitation light. In other words, the emission spectrum of a fluorophore would be the mirror image of its absorption spectrum. It has clinical significance in estimating the radiative lifetime of a photosensitizer dye (Fig. 6.4).

However, there are exceptions to this rule too. Resonance fluorescence (electron absorbs two photons instead of one) leads to the emission of a shorter wavelength of higher energy light than what was absorbed. In terms of spectroscopic principles such a decrease in

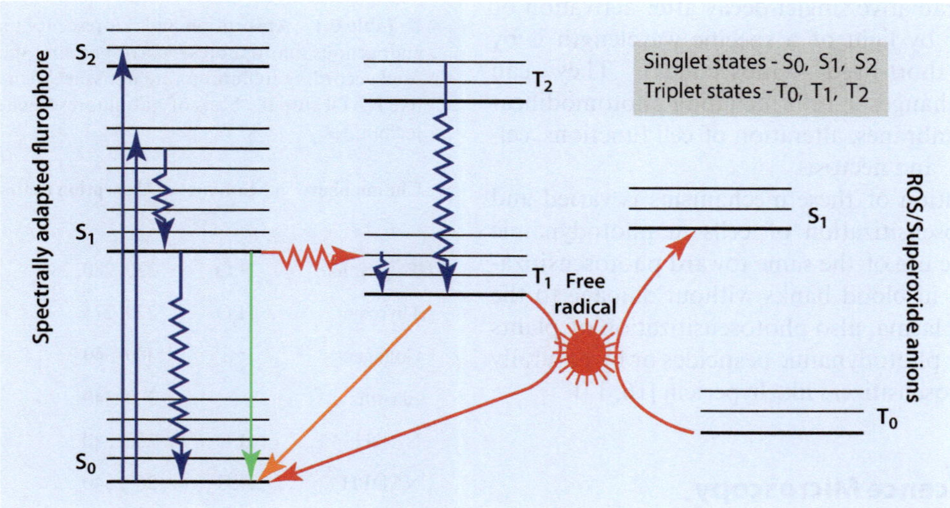

Fig. 6.5 Pathways of de-excitation of a photosensitizer through fluorescence (singlet) or phosphorescence (triplet) in photodynamic therapy (PDT). Excited singlet states or triplet states give rise to cytotoxic species which are used therapeutically against bacteria or even cancer cells

wavelength of emitted light is called blue shift. Whereas, usually in fluorescence, there is an increase in the wavelength (lower energy) of the emitted light, virtually demonstrating a shift to the red end of the spectrum and is called a red shift. In practical terms, a redshift would entail a lower frequency and lower photon energy whereas a blue shift means a higher frequency and higher photon energy of the emitted light. In summation, redshift or blue shift describes the relative difference between the absorbed and emitted wavelengths (or frequency) of a fluorophore.

Michael Kasha's rule states that photon emission (from fluorescence or phosphorescence) would occur in appreciable yield only from the lowest excited state. Sergei I. Vavilov modified this dictum and stated that the quantum yield of luminescence is generally independent of the excitation wavelength. This has clinical significance in that the fluorescence yield is proportional to the fluorescence lifetime Φ = photons emitted/photon absorbed.

Understanding of these basic phenomena is important as they would serve as the basic mechanisms for the formation of excited singlet states such as singlet oxygen in photobiomodulation and even the formation of long-lived excited triplet states such as superoxide formation in photodynamic therapy (PDT) or photo-activated chemotherapy (PACT).

Biological tissues are composed of dipole structures like enzymes, ionic pumps, nuclear material and nucleotide molecules, polar molecules like water and bound electrons which can be stimulated chiefly by the electric field of light (as biologic tissues are not magnetic, they are not much affected by the magnetic field of light) [12, 13]. Tissues are composed of 70% or more of water which is highly polar and have energy carriers such as ATP and also macromolecules of proteins having transition metal complexes. The polar nature of water and dissolved tissue oxygen can make it highly reactive with fluorophores in the triplet state, leading to the formation of free radicals (singlet oxygen) that are toxic to cells. Fluorophores in the triplet state can also react directly with the above-mentioned biological molecules too through sequential decay and effect oxidation of cell structures [14]. Such a compound which is capable of causing light-induced reactions in other sensitive and receptive molecules is called a photosensitizer. A photosensitizer has also been termed as a "spectrally adapted chromophore" as it can be activated only by light of a specific wavelength [14].

Foote [15] (though first proposed by, H. Kautsky in 1939) classified the subsequent chemical reactions that ensue after light activation of the photosensitizer as either Type 1 or Type 2. Typically, Type I reactions involve a direct interaction of the photosensitizer and the surrounding molecule and the creation of a free radical. The commission of the photosensitizer into a long-lived triplet state (milliseconds to several seconds) allows it to then react with other biological molecules (other than oxygen) leading to the production of hydroxyl and superoxide anions. On the other hand, Type II reaction entails transfer of energy from the photosensitizer to oxygen producing a singlet state of oxygen [14] (Fig. 6.5).

Singlet oxygen reacts readily with cellular constituents like amino acids in proteins such as tryptophan, tyrosine, histidine, cysteine, and methionine and guanine bases of DNA and RNA, and also in unsaturated lipids, including cholesterol and unsaturated fatty acids

[16, 17]. Such radiative singlet decay after activation of photosensitizer by light of a specific wavelength is by fluorescence (short-lived—nanoseconds). They can effect various changes in cells including photomodification of cell membranes, alteration of cell functions, cellular oxidation, and necrosis.

The application of these mechanisms is varied and explains photosensitization of cells in photodynamic therapy and the use of the same toward photosensitization of viruses in blood banks without damage to the blood cells or plasma, also photosensitization of plants and animals by photodynamic pesticides or by naturally occurring photosensitizers like hypericin [16, 17].

6

6.3 Fluorescence Microscopy

6.3.1 Soft Tissue Applications

Fluorescence microscopy has become ubiquitous and invaluable in almost every branch of medical and biologic sciences. Various techniques and a host of fluorescent proteins which serve as optical probes to investigate cellular processes non-invasively have been developed and lend further clarity to diagnostic processes. A fluorescence microscope utilizes emitted light from fluorescence and phosphorescence to develop the image; hence, it would have an excitation source and an optical element that is receptive to longer wavelength, i.e., fluorescent light, and filters or blocks out all other light including the exciting light and autofluorescence sources too.

There are a wide range of techniques used for assaying fluorescence such as fluorescence-lifetime imaging microscopy (FLIM) which investigates the interactions between fluorescent proteins and cellular processes involved in its immediate proximate environ. Fluorescence resonance energy transfer (FRET) refers to the transfer of energy from the excited fluorophore to another molecule in the near proximity, and this leads to emitting light of a longer wavelength. This allows resolution of a very high detail to the realm of molecules within cellular processes and their interactions thereof. Other techniques such as total internal reflection fluorescence (TIRF) (which allows imaging within hundreds of nanometers and selective imaging of molecules such as of a cell membrane) and stimulated emission depletion (STED) which gives a high degree of selectivity by allowing illumination and fluorescence of only a particular area and thereby providing precise resolution are also being researched [16, 17].

Fluorophores employed in these microscopic techniques can be Endogenous or Intrinsic fluorophores such as the amino acid Tryptophan (which fluoresce in UV light) or nicotinamide adenine dinucleotide (NADH) which fluoresce in the blue region. They form the basis of autofluorescence techniques (Table 6.1). Synthetic organic fluorophores such as tissue dyes like Rhodamine, Hoechst, 4-6-diamidino-2-phenylindole (DAPI) (UV and Blue), or Quantum dots form a second class. Whereas fluorescent proteins capable of forming an intrinsic fluorophore by being genetically encoded like avGP from Aequorea Victoria jellyfish (fluoresce in green light) or even a hybrid of a synthetic dye bound by a covalent bond to a genetically encoded protein forms a third class of fluorophores [17].

Fluorescent proteins can even be in a dormant stage and can then be "photoactivable," like PA-GFP, "photoconvertible" meaning changed from one band of fluorescence to another like Kaede (518–582) or

Table 6.1 Absorption and fluorescence maxima of endogenous fluorophores—corresponding spectral light to be used according to fluorophore in tissue. Such fluorophores like NADH are the basis of autofluorescence diagnostic techniques

Chromophore	Solvent	Absorption (nm)	Fluorescence (nm)
Tryptophan	H_2O	220, 280, 288	320–350
Thyrosin	H_2O	220, 275	305
Collagen		300–340	420–460
Elastin		300–340	420–460
NADH	H_2O	260, 340	470
NADPH	H_2O	260, 340	470
Flavins	H_2O	260, 370, 450	530
Zn-coproporphyrin	DMSO	411, 539	580
Zn-protoporphyrin	DMSO	421, 548, 585	592
Uroporphyrin	DMSO	404, 501, 533, 568, 622	624
Coproporphyrin	DMSO	398, 497, 531, 565, 620	622
Protoporphyrin	DMSO	406, 505, 540, 575, 630	633
Chlorophyll a	Ether	425, 670	685
Chlorophyll b	Ether	455, 642	660

Reproduced from Koenig and Schneckenburger Journal of Fluorescence, Vol. 4, No. 1, 1994 with permission [17]
DMSO dimethyl sulfoxide

photoswitchable like Dronpa by being able to be activated when illuminated at 488 nm or not [16].

Autofluorescence of endogenous fluorophores in tissues can be assayed by laser-induced autofluorescence and can have a range of diagnostic applications such as differentiation between types of tissues, detection of infections by microorganisms and metabolic states of tissues and even detection of metabolic defects. Mitochondrial intracellular oxidation rates and intracellular oxygen concentration were evaluated by Chance et al. [18] in 1962 in vivo (on rat brain and kidney) by studying NADH fluorescence at 460 nm (higher NADH concentrations as titered by its fluorescence, means reduced intracellular oxygen levels). This work is in fact a continuation of Chance's work with Jobsis [19] in 1959 on frog Sartorius muscle, demonstrating increased fluorescence at 443 nm of cytoplasmic and mitochondrial pyridine nucleotides on muscle contraction. Mayevsky [20] continued on the same line of work (1972 and 1988) by surface fluorometry-reflectometry using flexible optical fibers and a corrected fluorescent signal (at 450 nm) devoid of other background influences of tissue absorption or blood volume changes (which reflected light at 336 nm) in the puppy or adult dog brain in vivo and also on Mongolian gerbil [21]. They demonstrated the effects of ischemia, hypoxia, and anoxia with a correlated increase in fluorescence of NADH.

Autofluorescence when employed for differentiation of tissue could be effective for detection of tumors. Lohmann and Paul in 1988 [22] showed that melanomas could be detected in situ at 475 nm when excited at 365 nm. Lohmann (1990) [23, 24] used the same technique to describe and correlate the detection of cancer and also cellular dysplasia in vitro cryosections of the uterine cervix and in ex vivo sections of lung tumors as compared to fluorescence of normal tissue and found that the fluorescence in the areas around the tumor had markedly higher intensity. However, autofluorescence of skin becomes slightly more complicated as in addition to the established diagnostic autofluorescence of the reduced pyridine coenzymes NADH and NADPH, the presence of various other fluorophores (such as collagen, elastin, keratin extracellularly in addition to, tryptophan and hemoglobin found intracellularly) can confound the resultant spectrum. Furthermore in vivo redox states may make quantification of NADH more difficult. Non-melanoma skin cancers were detected in vivo by Ra et al. [25] using a indomethacin-based fluorescent probe called fluorocoxib (developed by Mannet et al.) administered systemically on genetically engineered mouse with an accuracy of up to 88% for macroscopic tumors and up to 85% for microscopic tumors. In this study, fluorescence signals from 500 to 800 nm were monitored at 10 nm intervals after excitation from 503 to 555 nm.

Understanding the basis of autofluorescent detection of cancer and the biologic fundamentals is a foundational prerequisite before practical evaluations and clinical modalities are discussed. Much of our present perception of these processes can be attributed to the work of Pavlova [26] and her team (from 2003 through 2008). In her study in 2008, autofluorescence of 49 biopsies of normal, benign and neoplastic samples were captured by confocal microscopy and the images proved that the autofluorescence patterns, from UV light (351 and 364 nm) and 488 nm Argon laser, depended not just on pathologic or normal nature, but also according to the site in the oral cavity. While UV light was intended for autofluorescence of NADH in the epithelium and collagen in the connective tissue, the 488 nm was directed at autofluorescence of flavin adenine dinucleotide (FAD) of the epithelium and the connective tissue. The study found that epithelium of the palate and gingival were highly fluorescent as compared to the epithelium from the buccal mucosa, floor of the mouth and tongue. Notably, benign lesions of the epithelial mucosa had only weak fluorescence whereas neoplastic versions had increased fluorescence to UV light as compared to normal samples. Intriguingly connective tissue of both benign and neoplastic samples (irrespective of anatomic site) had decreased fluorescence to both 488 nm and UV light [27, 28]. They surmised that this decreased fluorescence of the connective tissue in neoplastic samples is because of the loss of collagen cross links due to inflammation, lymphocytic activity, and matrix degrading proteases [29].

The epithelium of the normal palate and gingiva being highly fluorescent was attributed to the highly keratinized masticatory mucosa which limits the penetration and increases the scattering of light, thereby having only an NADPH contribution to fluorescence and not from collagen. In a neoplastic condition, however, the biochemical pathways of cell signaling between the epithelium and connective tissue is changed, this coupled with neoangiogenesis of the cancerous tissue and loss of collagen in the connective tissue is thought to be the reason for the decreased fluorescence of neoplastic epithelial samples [28].

Roblyer et al. [30] confirmed the above-mentioned facts in their study designed on 56 patients and 11 volunteers through quantitative autofluorescence imaging using the Multispectral Digital Microscope (MDM, a wide-field optical microscope, color CCD camera, variable range of 1–7 cm) at 365, 380, 405, and 450 nm. The focal region to be analyzed was selected by another practitioner who was blinded to the grouping of the samples and thereafter diagnostic algorithms related to the red-to-green fluorescence intensity ratio, etc. were applied. The results showed that this diagnostic modality of quantitative autofluorescence could discriminate between neoplastic and non-neoplastic tissue with a sensitivity and specificity of 96%.

Shin et al. [31] in 2010 described the various diagnostic techniques in vogue today like ViziLite (Zila Pharmaceuticals, USA) which has chemiluminescent blue light source and can detect pathologic changes however studies are yet undecided about its accuracy.

The VELscope or Visually Enhanced Lesion Scope (LED Dental, USA) invented by M. Suyama is a handheld device that is designed for the diagnosis of oral precancerous, cancerous, or pathologic lesions using fluorescent light of 400–600 nm. It works on the notion that normal mucosa would appear green or pale green due to autofluorescence and a pathologic or neoplastic lesion would appear brownish to black. It is based on the concept of *visual* autofluorescence as opposed to the autofluorescence imaging techniques described earlier in this text. It can however aid a practitioner in identifying and delineating cancerous or precancerous lesions. However, it is quite subjective as it relies heavily on the judgment, training, and experience of the practitioner, false positives with benign lesions may also occur as they may also have connective tissue changes leading to a change in autofluorescence of tissues. A range of studies have been designed to evaluate the efficacy of this device and the results have been ambiguous.

6

Lane et al. [32] was the first group to investigate the VElscope device in 2006 and reported a sensitivity of 98% and specificity of 100%. Subsequently, Poh et al. [33] did their evaluations on 20 patients from whom they derived 122 oral mucosa biopsies and reported a sensitivity of 97% and specificity of 94%. Poh [34] designed a similar study a year later in 2007 with another team and published a report that out of a total of 60 patients who underwent surgery for oral cancer, 7 patients (25%) of the 22 control-group patients experienced recurrence, while none of the 38 patients where the surgeon was guided by the visual fluorescence (VELscope) to place a 10 mm margin of excision experienced recurrence. However, studies done by McNamara et al. [35] in 2012 on 42 patients do not concur with the previous studies and stated that visual examination was better than visual fluorescence by VELscope.

Cancer diagnosis through autofluorescence could also be based on the detection of porphyrin or its derivatives in neoplastic tissues. There are various porphyrin-related fluorophores in the tissue such as protoporphyrin IX (PP IX), coproporphyrin III (CP III), uroporphyrin III (UP III), and hematoporphyrin IX (HP IX). Strong red fluorescence of porphyrins when exposed to UV light was reported as early as 1924 by Policard in sarcomas of rats. This would later on develop into the foundational work of photodynamic therapy (PDT). It was not until 1942 when Auler and Banzer demonstrated hematoporphyrin in malignant cells and later in 1960 when Ghadially and Neish studying squamous cell carcinoma in rabbits that established endogenous PP IX as the main fluorophore responsible for red autofluorescence [36].

Yuanlong et al. (1987) [37] showed that squamous cell cancer tissues when exposed to 365 nm Xenon ion-pulsed laser showed fluorescence at 630 and 690 nm at 89% correlation with the traditional biopsy method. Koenig and Schneckenburger [38] continued these investigations using the 364-nm argon laser with a fiber-optical sensor and elicited autofluorescence at 673 nm in the skin and in subcutaneously transplanted solid Ehrlich carcinoma of mice at 638 and 680 nm, and at 635 nm in patients with squamous cell carcinoma.

6.4 Optical Coherence Tomography

6.4.1 Optical Biopsy

Optical Coherence Tomography (OCT), a modality adapted from ophthalmology (Naohiro Tanno and David Huang 1991) is a non-invasive, live, imaging technique that provides three-dimensional high-resolution (10–15μm) images to a depth of ~2 mm. Literally, it allows the diagnosis of pathology in tissue by optical methods without actually physically harvesting it from a patient surgically [39]. It is based on the principle of back scattered light analyzed through low coherence interferometry, which in simple terms means that the light scattering back off a surface is superimposed (as constructive interference and destructive interference) to develop useful diagnostic data. The back scattered (reflected) light from the substrate generates a low coherence beam from differing depth levels (depending on tissue structures in the volume being imaged) to the interferometer (◘ Fig. 6.6). The optical data from each single scan point is interpreted by the interferometer as an interference pattern and recorded as a depth profile (A-scan) whereas in a linear scan across the sample delivers cross-sectional (B-scan) data. The images can be based on spectral domain scans of a single focus or swept laser light sources where light of varying frequencies can be emitted sequentially.

The light is directed along two arms, a reference arm (mirror) and a sample/substrate arm using a handheld (X-Y) scanning device similar to an endoscopic probe (◘ Fig. 6.7). The source of light typically is a near-infrared diode lasers operating in continuous wave or a short pulsed femtosecond laser and is used in tandem with time domain or frequency domain interferometers (◘ Fig. 6.8).

Normal and pathologic tissue can be visualized in three-dimensional images with differing contrast as the spatial differences in refractive index of different tissue

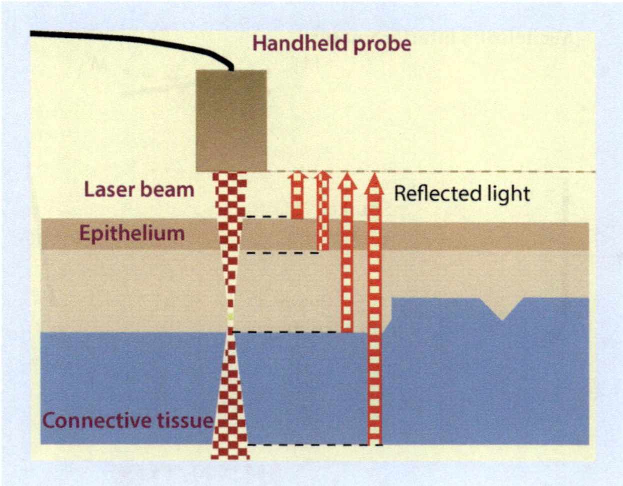

Fig. 6.6 Light from a laser source generates backscatter depending on tissue components at various depth levels. This is analyzed by a low coherence interferometer that generates depth-wise information or a series of cross-sectional images according to tissue constituents that cause the back scatter at different levels

constituents changes with depth and hence this variation lends the depth and difference to images. Furthermore, it allows qualitative morphological imaging of skin in vivo and can help in diagnosis of precancerous or cancerous lesions of the skin [40]. Optical biopsies from this modality has led to the diagnosis of neoplastic lesions of both non-melanoma skin cancers (NMSC) like basal cell carcinoma and squamous cell carcinoma [41–45] and also melanomas [46, 47].

OCT has also been used in the diagnosis of mucosal lesions of the oral cavity. Adegun et al. (2012) [48] applied OCT for the diagnosis of epithelial dysplasia and compared B-Scan images to histological sections. The relevance and significance of this study, they stated was that the choice of determining the anatomic site for obtaining the biopsy is subjective to the clinician and hence the diagnosis also would vary, and hence OCT provided non-invasive technique that could in the future direct the practitioner to selecting the appropriate site for taking the biopsy only if necessary and prompted by

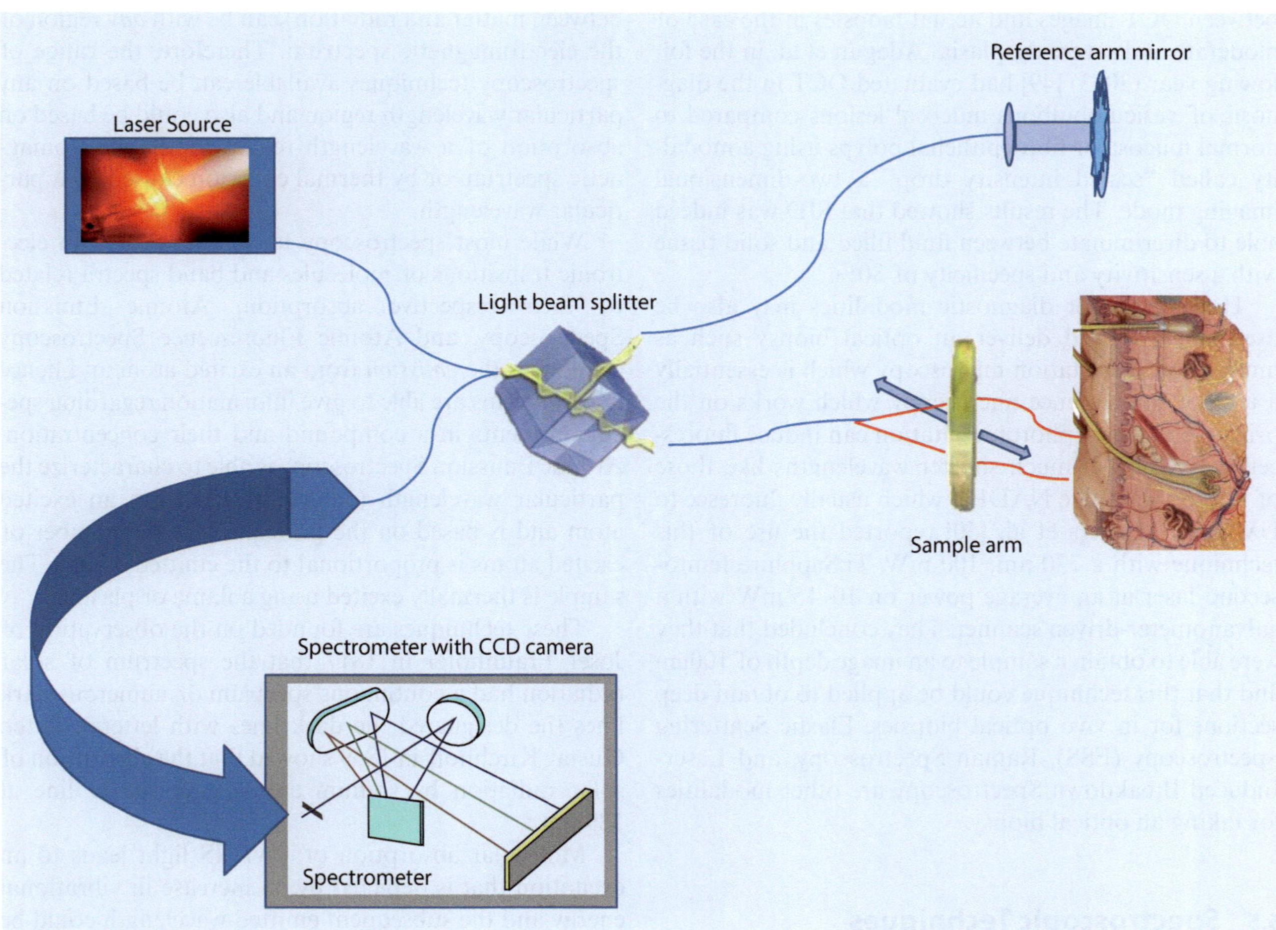

Fig. 6.7 The components of optical coherence tomography—which consists of a scanner probe which has the laser light source is directed along two arms by a beam splitter, one from the sample (recorded by CCD detectors or CMOS imaging) and from a reference mirror. The light from both arms are recombined and the spectrum analyzed by computer. *CCD* charge coupled device, *CMOS* complementary-symmetry metal-oxide semiconductor

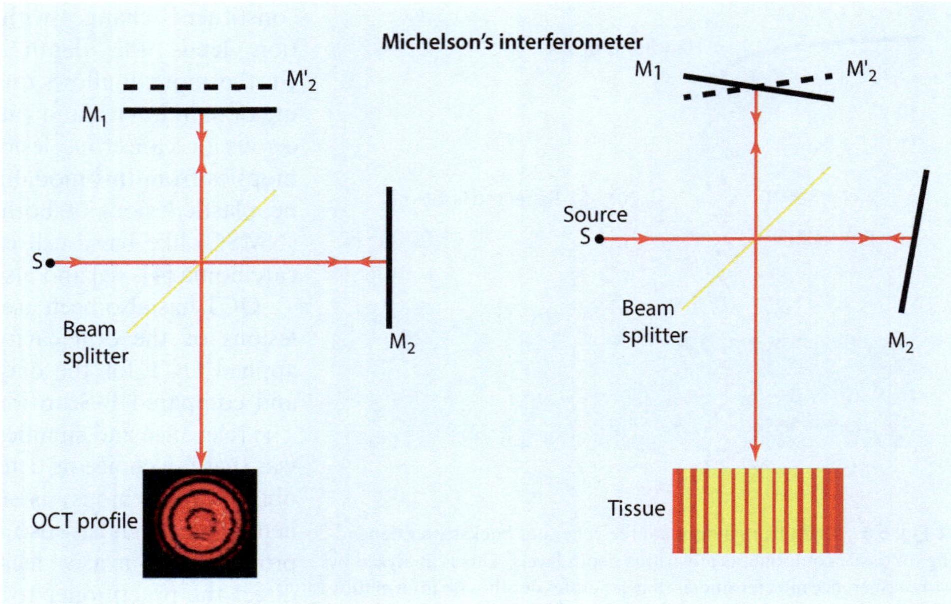

Fig. 6.8 Michelson's interferometer consists of a beam splitter that splits the light from the source into two arms, the reflected light from both arms (sample and reference mirror) is then recombined to produce the OCT image

the OCT image. The results showed a poor correlation between OCT images and actual biopsies in the case of moderate and severe dysplasia. Adegun et al. in the following year (2013) [49] had evaluated OCT in the diagnosis of vesiculobulbous mucosal lesions compared to normal mucosa or fibroepithelial polyps using a modality called "scaled intensity drop" a two-dimensional imaging mode. The results showed that SID was indeed able to discriminate between fluid filled and solid tissue with a sensitivity and specificity of 80%.

However, other diagnostic modalities may also be used similarly and deliver an optical biopsy such as multi-photon excitation microscopy which is essentially a type of fluorescence microscopy which works on the principle that two photon excitation can induce fluorescence emission at much shorter wavelengths like those of fluorophores like NADPH which usually fluoresce to UV light. Masters et al. [50] reported the use of this technique with a 730 nm, 100 mW, Ti:Sapphire femtosecond laser at an average power on 10–15 mW with a galvanometer-driven scanner. They concluded that they were able to obtain a sample to an image depth of 100μm and that this technique could be applied to obtain deep sections for in vivo optical biopsies. Elastic Scattering Spectroscopy (ESS), Raman Spectroscopy, and Laser-Induced Breakdown Spectroscopy are other modalities for taking an optical biopsy.

6.5 Spectroscopic Techniques

Skoog and Holler [51] define spectroscopy as the study of the interaction between matter and electromagnetic radiation. Helmenstine [52] refines this definition further by stating that this analysis (of the interaction between matter and radiation) can be with *any* region of the electromagnetic spectrum. Therefore, the range of spectroscopy techniques available can be based on any particular wavelength region and also could be based on absorption of a wavelength region of the electromagnetic spectrum or by thermal emission caused by a particular wavelength.

While most spectroscopy techniques deal with electronic transitions of molecules and band spectra related to their respective absorption, Atomic Emission Spectroscopy, and Atomic Fluorescence Spectroscopy catalogues the *emission* from an excited atom and hence in plain terms are able to give information regarding specific elements in a compound and their concentration. Atomic Emission Spectroscopy is able to characterize the particular wavelength (color) emitted from an excited atom and is based on the principle that the number of excited atoms is proportional to the emitted energy. The sample is thermally excited using a flame or plasma.

These techniques are founded on the observation of Josef Fraunhofer in 1817 that the spectrum of solar radiation had a continuous spectrum of numerous dark lines (he designated the dark lines with letters). Later, Gustav Kirchhoff in 1859 showed that the absorption of solar radiation by sodium atoms gave the D line at 589 nm.

Molecular absorption of UV/VIS light leads to an excitation that is depicted by an increase in vibrational energy and the subsequent emitted wavelength could be measured as against the absorbed wavelength in techniques of absorption spectrometry such as Atomic Absorption Spectroscopy (displays these as black lines on a white background). Laser-induced breakdown spectroscopy (LIBS) is a type of atomic emission spectroscopy where a highly energetic focused laser pulse is the excitation source which atomizes the substrate to form plasma. This plasma would have a characteristic atomic or molecular signature of the substrate.

Kumar et al. (2004) [53] had demonstrated the use of LIBS in vivo in canine hemangiomas and showed that calcium and potassium and copper and potassium levels were different in tumor cells compared to the normal cells.

Kanawade et al. (2015) [54] applied LIBS for the differentiation of tissues by analyzing their basic atomic composition as against the National Institute of Standards and Technology (NIST) [55] data base using and Excimer Laser 193 nm, 28 ns pulses at 10 Hz, 0.6 × 0.4 mm spot size, and energy/pulse 38 mJ on fat, muscle, nerve, and skin tissue samples from pig source. They proved that emission intensity ratios of Na to C, K to Na, and O to C ratios could be used to differentiate between tissues and that this information could be used for future application in minimally invasive laser-guided surgeries.

Photoluminescence spectroscopy specifically assays the fluorescent (fluorescent quantum yield) or phosphorescent (phosphorescent quantum yield) pathways of relaxation and dates back to the mid-1800s. The spectra depict the intensity of the emitted radiation as a function of either the excitation (excitation spectra) wavelength or the emission (emission spectra) wavelength. Excitation spectra shows the emission at a given fixed wavelength while varying the excitation wavelength. Whereas the emission spectra show the intensity of the emitted radiation when a fixed wavelength is used to excite the sample.

Spectroscopic measurements of scattered light are applied in the case of techniques such as Raman Spectroscopy.

6.6 Inelastic Scattering of Light vs Elastic (Rayleigh) Scattering

Elastic scattering suggests that the scattered light has the same energy as the incident light and the particle on which it was incident is much smaller than the wavelength of the light. This is based on Rayleigh's law which states that scattering is inversely proportional to the fourth power of wavelength, which means that shorter the wavelength (violet, blue) the greater will be the scatter [14].

Conversely inelastic scattering would cause a change in the frequency and wavelength of the incident light in that there is a transfer of energy toward vibrational states of the molecule. This transfer of energy enables the molecule to reach higher (excited) "virtual" vibrational states and as a result there would an equal and proportional decrease in energy from the scattered light than its incident energy. This would increase the wavelength (redshift) of the scattered light and the difference in energy would correspond to the difference between the vibrational states of the molecule. This is called Stokes shift.

However, in such a situation, when the incident photon was to hit a molecule which is already in its excited vibrational state then the scattered photon would have more energy than the incident and hence it would have a shorter wavelength (blue shift) thereafter. This is termed as anti-Stokes shift (◘ Figs. 6.9 and ◘ 6.10).

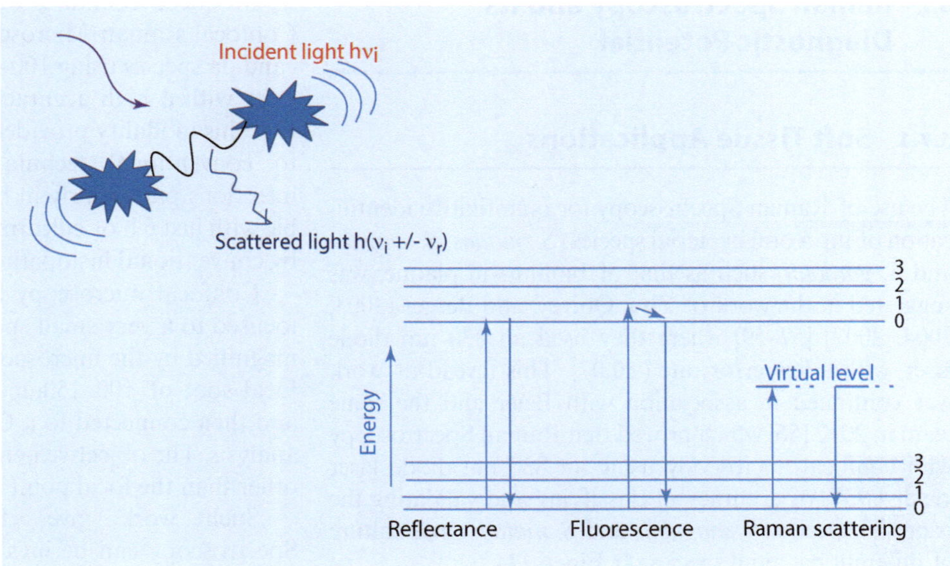

◘ **Fig. 6.9** Inelastic scattering of incident light ($h\nu_i$) that induces the molecule to a higher "virtual" excited state with subsequent loss (or gain) of energy of scattered light ($h\nu_R$). This loss (or gain) of energy of the scattered light can be detected spectroscopic means

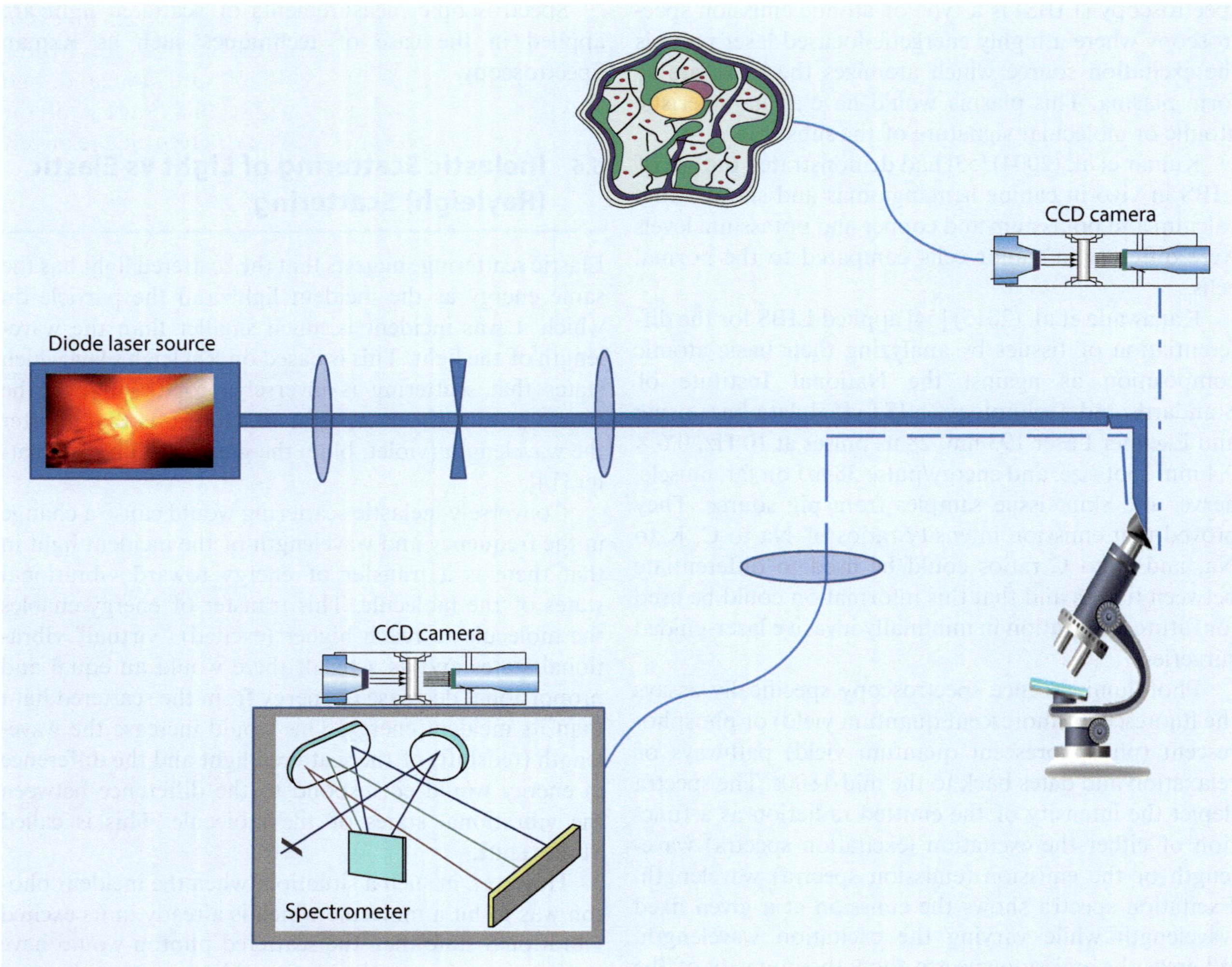

Fig. 6.10 The diagnostic setup for Raman Spectrometry. The excitation source can be a Diode Laser, a fiber-optic sampling probe (which directs the laser light to the source and also channelizes the scattered light from the source to the spectrometer), and a spectrometer detector. (Reproduced from Beier B.D. et al. (2012), AMB Express 2012, 2:35, © 2012, Beier et al. [56]; licensee Springer)

6.7 Raman Spectroscopy and Its Diagnostic Potential

6.7.1 Soft Tissue Applications

The use of Raman Spectroscopy for quantitative identification of intra oral bacterial species (*S. mutans*, *S. sanguis*, and *S. gordonii*) such as that of biofilms in plaque, was suggested in the work of Zhu, Quivey, and Berger (2003, 2004, 2007) [57–59] where they used an 830 nm diode laser, with a low error rate (±0.07). This thread of work was continued in association with Beier and the same team in 2012 [56] which proved that Raman Spectroscopy with confocal microscopy using an 830 nm diode laser yielded 93% of accuracy in classifying and validating the bacterial species (*S. sanguinis* and *S. mutans*) in a culture of different bacterial species (Fig. 6.11).

Maquelin et al. in 2002 [60] first reported the use of Confocal Raman Microscopy for the identification of candida species using 100–150 mW, 830 nm Ti:Sapphire laser with a high accuracy of 97–100%. They advised that this modality provided a viable and fast alternative to conventional techniques in identifying candida infections particularly in hospital settings as it is possible with just 6 h of culturing as compared to the 24–48 h by conventional histopathological processes.

Confocal microscopy allows the laser beam to be focused to a very small spot size (~250 nm in diameter) magnified by the microscope and projected onto a confocal spot of 100–150μm diameter or an optical fiber and then connected to a CCD camera for spectral data analysis. The objective here is that all other signals from other than the focal point are simply blocked out.

Such work gives the promise that Raman Spectroscopy can be invaluable in the early identifica-

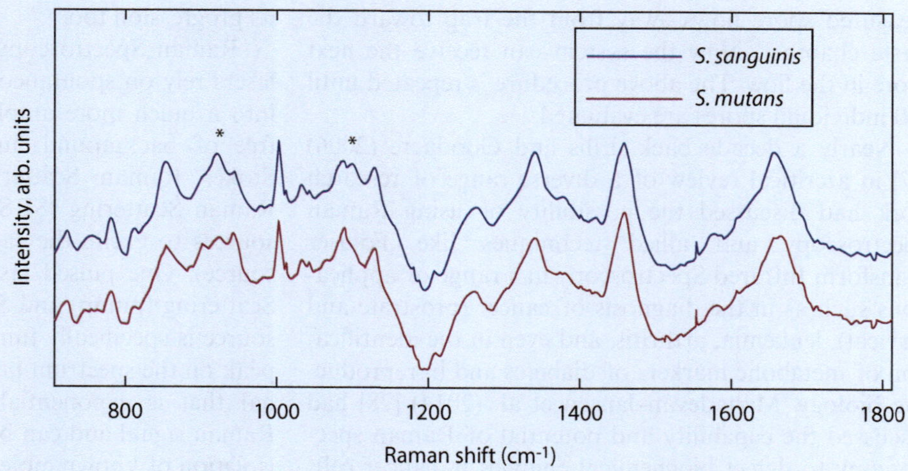

Fig. 6.11 Raman spectra of bacteria. (Reproduced from Beier B.D. et al. (2012) AMB Express 2012, 2:35 © 2012 Beier et al. [56]; licensee Springer)

tion of caries as well as susceptibility of the patient and designation of preventive measures. The inspiration for such work though dates back to the work of Puppels [61–63], Nelson [64], Sperry [64], Manoharan and Ghiamati [65–67] in the early 1900s who first used this technique for the identification of bacteria. The choice of bacteria used in these studies are indeed obvious as *S. mutans* is implicated as one of the primary causative organisms and *S. sanguis*, and *S. gordonii* are secondary causative agents in dental caries.

Chan et al. in 2006 [68] had used the same principle with over 98.3% accuracy in the identification of neoplastic cells and differentiation of live healthy hematopoietic cells and later with 97% accuracy using Laser Tweezers Raman spectroscopy (LTRS) with a 30 mW 633-nm He–Ne laser as an excitation source [69].

LTRS utilizes a single beam of infrared lasers to analyze a cell several microns away from any other substrate, virtually as an optical tweezer, thereby precluding any other background signals. This is based on the principle of "optical trapping or optical levitation" described by Ashkin and Dziedzic [70] (using a 120 mW argon laser) where viruses, bacteria, or cells in gaseous or liquid media are virtually held at a focal point of the laser beam where the scattering and intensity of the laser radiation is at a balance. Prerequisite to this process is that the biologic (or otherwise) particle being "trapped" should have a refractive index that is higher than the surrounding medium. "Optical trapping" works on the principle that a gradient force proportional to the intensity of the laser beam draws high refractive index particles toward the beam axis but repels low refractive index particles away from its axis. The laser bean could virtually suspend a particle as the radiational field counterbalances the gravitational forces on a particle. (Magneto optical trap, Dipole trap) [71]

Previously in 2004, Chan et al. [72] had demonstrated the successful use of Confocal Laser Tweezers Raman Spectroscopy in identification of single bacillus spores using a 50 mW Argon Laser. Later, the same research team headed by Chan in 2008 [73] demonstrated the application of LTRS using 10 mW, 633 nm He–Ne laser (both as excitation and optical trapping) in differentiating between T and B cells of young leukemic patients with an accuracy of 95% in the case of normal cells and 90% in classifying cells according to their respective types. Huang et al. [74] employed a similar technique of microfluidic Raman tweezers using 785 nm diode laser source (this wavelength is preferred as there is very less water absorption at this range and hence damage to optically trapped cells is prevented) in quantifying the levels of pyridine-2,6-dicarboxylic acid (dipicolinic acid or DPA) chelate with calcium ion (Ca-DPA) in bacillus spores. Ca-DPA levels have been correlated with spore resistance and stability in a medium [75]. The study showed that identification of individual bacillus spores was possible without altering individual spores with a laser exposure time up to 20 s beyond 20 s there was the release of DPA as all bands disappeared from the Raman spectrum [76].

The microfluidic technique is a flow cytometry like patented technique developed by the researchers themselves which comprised a square quartz capillary tube (50µm by 50µm; 5 cm long) with two connecting chambers, one with the spore sample and the other with water. A precise flow pump controlled the flow that would deliver individual spores to the focal point of the laser where optical trapping of the spore would give the Raman spectrum and then the flow was immediately turned off so that the spore is held for Raman acquisition. Once the Raman spectrum is obtained, the flow is turned on again, while the laser beam is blocked for a short time (~10 ms) and when it is turned back on the measured spore flows away from the trap toward the waste chamber. Now the system can receive the next spore in the flow. The above procedure is repeated until 200 individual spores are evaluated.

Nearly a decade back, Ellis and Goodacre (2006) [77] in a critical review of a diverse range of research work had discussed the possibility of using Raman Spectroscopy and allied techniques like Fourier Transform Infrared Spectroscopy in a range of applications such as in the diagnosis of cancer (prostrate and cervical), leukemia, arthritis, and even in the identification of metabolic markers of diabetes and in reproductive biology. Mahadevan-Jansen et al. (2014) [78] had discussed the capability and potential of Raman spectroscopy to detect biochemical changes in cancer cells such as the increase in nucleic acid content, glycogen, and collagen.

Nguyen et al. (2016) [79] successfully employed Raman Spectroscopy in the in vivo verification of tissue margins between normal and neoplastic tissues in soft tissue sarcomas with a sensitivity of 89.5% and specificity of 96.4%. They used a portable Raman spectroscopy device with a handheld probe, 400μm excitation fiber and seven 300μm collection fibers excited at 785 nm, providing information to a depth of 700μm. This technique, they reported, allows for more accurate appraisal of designing margins in surgical resection of tumors.

Raman spectroscopy when combined with intuitive computer-aided statistical analysis of the spectra by Multivariate Analyses (MVA) such as Principal Component Analysis (PCA), Hierarchical Cluster Analysis (HCA), Discrimination Function Analysis (DFA), or geometric based Vertex Component Analysis (VCA) give a clearer and composite picture leading to the interpretation, classification, and diagnosis of the sample under study.

In related work by Escoriza et al. in 2000 [80], Raman Spectroscopy was successfully employed to discriminate and differentiate between viable and non-viable cells of *Escherichia coli* and *Staphylococcus epidermidis*. The study demonstrated a sensitivity of Raman Spectroscopy to differentiate viable and non-viable cells up to 86% and was able to discriminate between species with accuracy of 87%.

Raman spectroscopy provides a unique fingerprint that differentiates the biochemical molecular alterations between the samples. Most interestingly, Beier et al. in 2012 [56] showed that this technique lends further depth to the diagnosis by giving actual information on the difference in spatial positioning of the different bacterial species in the three-dimensional network of the biofilm. Such essential constitutive differences can build up a three-dimensional mosaic of the complex microbial structure such as of a biofilm and could give unique information and inference regarding cariogenicity and its progression too.

Raman Spectroscopy (which uses continuous wave lasers rely on spontaneous emission) could be clarified into a much more amplified, efficient, coherent signal free of background fluorescence by Coherent Anti-Stokes Raman Scattering (CARS) and Stimulated Raman Scattering (SRS) which uses two pulsed laser sources to excite the sample (instead of a single CW source). One pulsed laser source initiates the Raman Scattering (pump) and Stokes shift whereas the second source is specifically tuned to the frequency of a single peak on the spectrum bringing about a blue shifted signal that is exponentially stronger than the original Raman signal and can be used for specific imaging and isolation of known molecules.

CARS has been used for the specific and quantitative intracellular identification of lipid molecules in living cells. Nan et al. (2006) [81] have been able to correlate this to the tracking of Hepatitis C virus whose pathogenicity is closely linked to its insidious ability to be concealed in triglyceride-rich lipoprotein particles. They used two pulsed Ti:Sapphire lasers with a repetition rate of 80 MHz, 2 ps pulse width and pump wavelength at 711 nm and Raman resonance (Stokes shift) wavelength at 892 nm. CARS had been employed using fiber-optic delivery by Buschman et al. in 2000 [82] and thereafter by subsequently Mostaco-Guidolin et al. in 2010 [83] in animal studies of myocardial infarction prone, Watanabe heritable, hyperlipidaemic rabbits for detection of atherosclerosis in ex vivo samples (dissected arteries) using Ti:Sapphire oscillator at 800 nm (for Raman resonance) 100 fs pulse duration and a 532 nm 7.25 W green laser provided the pumping. They were able to establish a correlation between the severity of atherosclerosis and age of the animals. Significantly, Motz et al. in 2006 [84] demonstrated (830 nm diode laser) in vivo usage of Raman Spectroscopy on diagnosis of atherosclerosis in femoral bypass and breast lumpectomy surgeries. CARS has also been used for the identification and imaging of bacterial spores.

Another technique for amplification of the Raman signal (by 14 or 15 orders of magnitude) is Surface-enhanced Raman scattering (SERS) spectroscopy. Zhang et al. in 2003 [85] used a 632 nm He–Ne laser to detect low concentration of dipicolinic acid (DPA).

6.7.2 Choice of Wavelength in Raman Spectroscopy

These studies mentioned illustrate the range of differing excitation wavelengths used by the various research groups using differing techniques across the decades. However, the common feature is that the wavelength chosen for Raman spectroscopy are mostly in the red or near-infrared region. This is because shorter wavelength particularly below 600 nm will elicit considerable background autofluorescence from cells which can confound and confuse results and also because cells typically have chromophores (proteins, amino acids, nucleic acids, etc.) which will absorb shorter wavelengths and cause cellular alteration or even cell damage (opticution). A key feature of Raman spectroscopy is that is non-invasive and gives real-time, vital information of living cells. Wavelengths between 600 and 900 nm have low water absorption and elicit more scattering. Blue or green wavelengths have also been used by some studies, but notably these are on fixed cells or ex vivo tissues and not in live homeostasis state of tissues.

6.8 Hard Tissue Applications

Dental caries is often overlooked or continues undetected owing to the insidious nature of the lesion. Radiographic detection often requires at least around 30% of demineralization to warrant detection and uses ionizing radiation [86]. Early detection of incipient caries (non-cavitated—"white spot") avoids the use of invasive restorative modalities. Incipient caries has a sub-surface second layer which is very porous and can extend to a depth of 100–250μm whereas a cavitated lesion can extend up to 1.5 mm or deeper.

6.8.1 Fiber-Optic Transillumination (FOTI) and Digital Imaging Fiber-Optic Transillumination (DIFOTI)

Fiber-Optic Transillumination (FOTI) and Digital Imaging Fiber-Optic Transillumination (DIFOTI) are transillumination techniques. DIFOTI uses high intensity light from a fiber-optic device source to illuminate a tooth and the backscattered light is captured by a digital camera and analyzed by a computer. Carious enamel and dentin would scatter light more than sound tooth structure and appear darker (Fig. 6.12). This technique has the advantage that it precludes the use of ionizing X-radiation and also provides diagnosis in real-time and studies have shown that DIFOTI has a higher sensitivity than conventional radiography. The technique relies heavily on the judgment of the practitioner and suffers from a lot of variability. It does not describe the lesion depth or severity and this has been reported in a study by Young and Featherstone (2005) [87].

Bin-Shuwaish et al. (2008) [88] evaluated the accuracy of DIFOTI as against Digital Radiography using complementary metal-oxide silicon (CMOS) sensor in

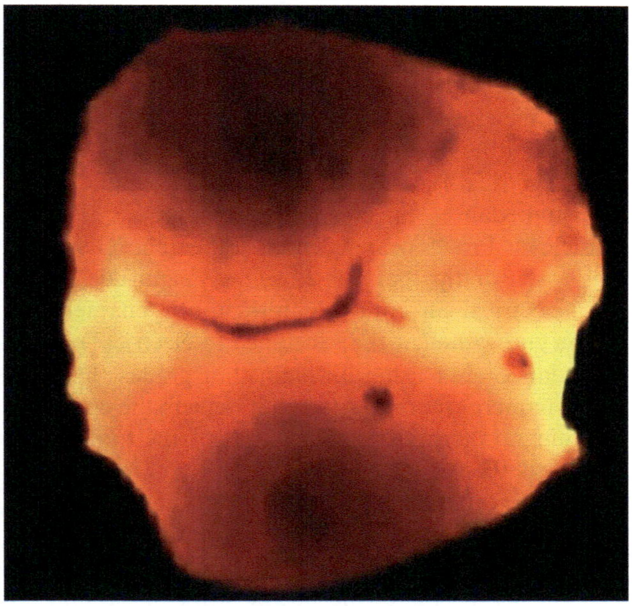

Fig. 6.12 FOTI image—transillumination depicts the carious areas as darker areas contrasted against sound enamel. (Reproduced from J. Gomez [90], BMC Oral Health 2015, 15(Suppl 1):S3 under the terms of the Creative Commons Attribution License (http://creativecommons.org/licenses/by/4.0), © 2015)

52 patients having Class II carious lesions. They concluded that though DIFOTI does significantly detect the lesion depth (particularly smaller lesions, in comparison to radiographs the technique was less accurate. They add that as DIFOTI had a sensitivity and specificity lower than visual examination, however they remarked that it can be a useful adjunct to radiographic diagnosis [89].

Electroconductivity measurement (ECM) is based on the principle that sound tooth surfaces will have little or no conductivity whereas areas which have demineralization due to caries would have conductivity proportional to the extent of demineralization. It involves covering the tooth surface with a conducting medium and measuring conductivity using a probe. But this technique had a lot of false positives and lack of specificity. Reports from literature average its accuracy to around 80% [91]. The ECM could be used to predict the probability that a sealant or a sealant restoration would be required within 18–24 months after eruption [92].

A related technique is electrochemical impedance spectroscopy (EIS) which is based on the application of electric currents of differing frequencies to detect carious lesions. Dental hard tissues have their own characteristic electrical signature, and these reference values can be used to detect carious tooth structure as compared to normal enamel and dentin. The demineralization caused by the carious process renders the tooth structure to be porous and subsequently permeated by fluids replete with ions, which in turn therefore would increase the electrical conductivity of carious teeth. Simply stated the carious teeth structure would have a higher electrical conductivity than sound and healthy teeth. Likewise sound dentin which is naturally more porous and has dentinal fluid within itself would have more electrical conductivity than sound enamel itself. AC Impedance Spectroscopy Technique (ACIST, CarieScan Pro™ orangedental, Biberach, Germany) uses a sensor tip that passes a low magnitude (micro amp) electrical current through the tooth (enamel, dentin, or pulp) to elicit the resistance to electrical current (impedance) and using an algorithm grades that data into a numeral score (0–100) that depicts the comparative mineral density or caries status of the tooth being examined [93–96].

On the other hand, laser assisted diagnostic techniques lend the advantage and accuracy of early detection by assessing biochemical and fluorescent changes in tooth structure. Following on the work of Bommer in 1927 using UV light to detect plaque on tooth, Benedict in 1928 and Hartles and Leaver in 1953 and later Armstrong in 1963 examined the fluorescence of healthy and carious dentin samples under UV light. However, it was Alfano who reported the fluorescence of carious teeth at 550 nm on excitation at 488 nm, these foundational works led on to the development of newer and still emerging diagnostic techniques in hard tissues [17].

6.8.2 Bioluminescence

Initial studies by Jablonski-Momineni, Moos et al. in 2016–2018 proposed that the bioluminescence (480 nm, Blue light Phosphorescence) emitted by a photoprotein (calcium-sensitive photoprotein aequorin—CSP—aequorin WO/2008/075081) on reacting with the higher level of free calcium ions released as a result of the demineralization process in an active caries site can be used to detect progressing carious lesions. The authors investigated the effectiveness of this system (Calcivis, Ltd., UK) in an in vitro study on extracted teeth and found that this novel method demonstrated high diagnostic credibility over visual or tactile methods [97].

Longbottom, Vernon, Pitts (who are the original inventors of this system) et al. further investigated bioluminescence as a diagnostic modality of caries in 2020–2021 through a couple of in vitro studies and they reported that bioluminescence is not affected by stains or other factors that can affect fluorescent signals and can therefore effectively detect an active carious lesion and that the bioluminescence thus detected is also directly proportional to the intensity of the caries lesion [98].

A blinded study was conducted by Pitts, Shank, Longbottom et al. in 2020 that confirmed, through FDA approval, that Calcivis is a relevant and efficient method of caries diagnosis, however further research through elegant, peer-reviewed and randomized controlled trials is still wanting before this diagnostic modality could become fully accepted by practitioners and the profession in general [99].

6.8.3 Quantitative Light Fluorescence and Laser Fluorescence (DIAGNOdent, KaVo)

Based on the technique developed by Bjelkhagen and Sundström (1981) [100] and later applied clinically in vivo in conjunction with de Jong (1995) [101], quantitative light-induced fluorescence (QLF) traditionally used Argon laser light at 488 nm to induce fluorescence in a tooth. More recently, QLF devices use an arc lamp or xenon lamp that generates light at 370 nm light (290–488, violet-blue) that is then transmitted through optical fiber into a handpiece and captured using a charge coupled device (CCD) camera [102]. The display of natural fluorescence of the tooth would contrast against dark areas of caries where the tooth's natural fluorescence is impeded. The fluorescence generated from the tooth by the exciting light would then be filtered by a high pass filter which allows only wavelengths greater than 520 nm to be detected. Karlsson and Tranaeus stated that QLF can detect carious lesions on occlusal and smooth surfaces to a depth of 500μm [103].

Saliva, plaque and hypoplastic areas would skew the results; nonetheless, an application of QLF that is not so widely researched is that it can detect red fluorescence from bacteria in plaque [103]. But the technique was not able to specifically differentiate between caries, hypoplasia, and similar conditions. Another serious limitation of this technique is that it is not able to differentiate or detect lesions extending into dentin [104].

An application based on the same concept of autofluorescence is Soprocare (Acteon, France) which uses light at 450 nm from three diode lasers in a clinical handpiece and elicits autofluorescence from the tooth that is recorded by a high magnification camera to detect caries or even from plaque and can even detect gingival inflammation. Rechmann et al. (2014) [105] reported the use of this device recently. The device has a period mode that detects plaque in an orange to red gradient and gingival inflammation in a magenta color and a caries mode that could detect carious areas as red fluorescent regions. In their study on 55 subjects and 638 teeth, they compared the Soprocare autofluorescence grading of plaque and gingival inflammation, to conventional grading methods such as Loe and Silness index for gingival inflammation and a plaque index, which was a

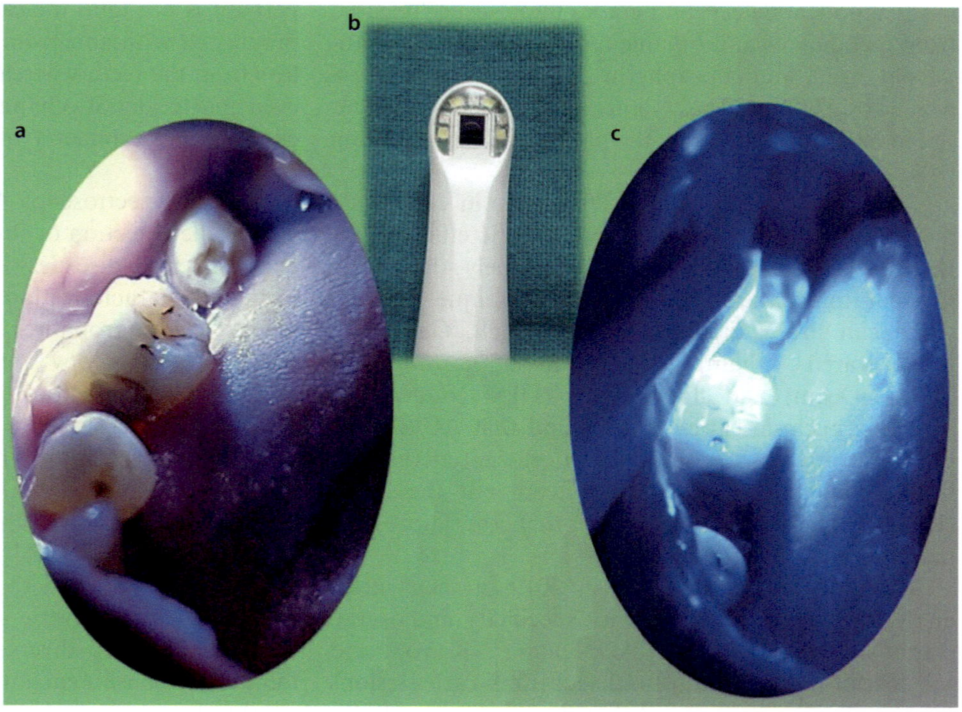

Fig. 6.13 Soprocare autofluorescence showing **a** Carious tooth in daylight mode. **b** The Soprocare handpiece with three blue light emitting diode lasers. **c** Carious tooth in blue light autofluorescence. (Courtesy Dr. Niladri Maiti)

Turesky Modification of the Quigley Hein Plaque Index. They concluded that the autofluorescence method of Soprocare correlated well to the clinical evaluation methods for plaque and gingival inflammation (Fig. 6.13).

Drancourt, Roger-Leroi et al. compared the efficacy (sensitivity and specificity) of autofluorescence (Soprocare) and bioluminescence (Calcivis) to detect caries in occlusal surfaces of extracted teeth, the study showed both methodologies being nearly equally efficient wherein Sporolife showed a higher sensitivity whereas Calcivis showed a better specificity in detecting caries [106].

In the technique of laser fluorescence (DIAGNOdent, KaVo), longitudinal laser light at 655 nm is directed at the tooth substrate and the reflected fluorescence from bacterially produced porphyrins is detected and measured (much like the detection of sarcomas described by P Policard as early as 1924). This technique was developed based on the work of Hibst and Gall. The amount of fluorescence produced can be graded (0–99) and correlated to the extent of caries (5–25 enamel caries, dentin caries greater than 35). Dental plaque, stains, calculus, and deposits could influence and interfere with the diagnostic results of this modality. Various studies have given different sets of validity and accuracy to the Diagnodent's assessment and specificity, but most of them are in vitro and cannot perhaps be directly extrapolated to clinical practice as such studies do not simulate clinical scenarios. In a clinical situation, Diagnodent readings may be adversely affected by stains, calculus, or even developmental or mineralization anomalies of teeth [107].

Barberia et al. (2008) [108] reported a specificity of 89% in a convenience sample of 320 molar teeth in children aged 6–14 years where the Diagnodent was able to successfully diagnose carious lesions with a high sensitivity of 0.89 in primary molars but 0.40 in permanent molar teeth and a specificity of 0.87 overall. Lussi et al. (2001) [109] in a previous clinical study had showed that Diagnodent was not able to clearly differentiate between superficial and deep dentinal caries.

6.8.4 Differences Between QLF and Diagnodent

QLF uses a much shorter wavelength at 488 nm and is designed to detect and receive 540 nm autofluorescent light from the enamel (it filters out the shorter wavelength scattered light suing a 520 nm filter).

The Diagnodent uses a longer wavelength 655 nm light and measures fluorescence intensity (uses a 680 nm filter).

Diagnodent can be related to dentinal decay but QLF results cannot correlate with dentinal decay.

QLF measures the amount of induced natural (autofluorescence) fluorescence from the tooth structure, particularly enamel; hence, it is better suited for detection of early lesions and more so on smooth surfaces rather than occlusal or in fissures [110]. This autofluorescence decreases according to the severity (demineralization) of the carious lesion.

Diagnodent detects and grades fluorescence released from metabolites and organic molecules such as protoporphyrin IX and coporphyrin IX released by bacteria which absorbs the red laser light (655 nm) of Diagnodent. It can hence even monitor the progression of the lesion [104, 109].

Sürme, Kara et al. [111] in 2019 in an in vitro study on extracted teeth assessed the diagnostic efficiency of three different diagnostic modalities, namely Alternating Current Impedance Spectroscopy Technique (ACIST, CarieScan Pro™ orangedental, Biberach, Germany), near-infrared light transillumination (NILT, DIAGNOcam) and fluorescence (DIAGNOdent) in detecting occlusal caries. They found that ACIST was comparatively not as effective as fluorescent methods in both permanent and deciduous teeth but the NILT (Diagnocam) was more accurate than the fluorescent methods of DIAGNOdent.

Iranzo-Cortés et al. [112] in 2019 in their meta-analysis compared the sensitivity, specificity, and overall diagnostic ability of DIAGNOdent (KaVo) and VistaProof (Durr Dental) to detect frank caries lesions (cavitated). They concluded that though both methodologies are efficacious and cogent with "moderate to high" sensitivity and specificity; however, they recommended that both methods should be adjunct to traditional methods of caries diagnosis. Furthermore, the results also show VistaProof to have a higher sensitivity and specificity than DIAGNOdent.

Mitchell, Zaku et al. [113] in 2021 evaluated the efficiency of DIAGNOdent (KaVo) in diagnosis of root caries through a prospective in vivo study. They reported that DIAGNOdent was most accurate in detecting active root caries with a 95% confidence interval. They recommended that DIAGNOdent could be reliably used for detecting root caries with a high sensitivity and specificity.

American Association of Dental Consultants (AADC) position statement [114] on laser fluorescence in caries diagnosis based on the studies and clinical trials conducted states that, "laser fluorescence can be used as an adjunct to traditional caries detection methods and not used as a primary diagnostic tool."

6.8.5 Raman Spectroscopy in Hard Tissues

Raman spectroscopy has varied non-invasive diagnostic applications for oral hard tissues. The precision of such techniques makes it invaluable in the early diagnosis of carious lesions. Tactile, visual, or radiographic detection perhaps is not until the carious lesion has considerably progressed with an accompanying quantum of mineral loss from the teeth whereas this diagnostic technique is even able to identify changes in crystallinity such as orientation and symmetry of crystals of the tooth structure.

Raman spectroscopy has found useful indication also in the detection of fluorosis and manifestations of developmental disorders such as amelogenesis imperfecta and characterization of the mineral phases in calculus. The fluorophores that could be detected by Raman Spectroscopy are also produced by various groups of bacteria too, hence detection of plaque or incipient carious lesions is also possible.

Raman spectroscopy has been applied experimentally for the compositional assessment of bone by evaluating bone mineral characteristics and collagen. Makowski et al. in 2013 [115] used phase matching (for polarization control) with Confocal Raman microscope at 785 nm on cadaveric specimens of femur bone of both genders in a diverse age group of 48–96. They found that phase matching (to quantify the phase and amplitude Raman peaks of bone) to reduce polarization bias could give specific diagnostic information related to peak rations of bone composition like mineral to collagen ratios; however, they added that optimizing polarization is necessary for specific discrimination between bone characteristics.

Raman spectroscopy is unique and unparalleled in caries detection as it allows for wide spatial resolution and thus early diagnosis of caries. A range of studies have been conducted in this area in the past decade comparing this technique to conventional methods and other spectroscopic methods and has lent a lot of clarity to this technique-sensitive area of diagnosis. Ko et al. (2005) [116] discussed a technique combining the high resolution, (20μm) [121], morphological imaging of OCT with the biochemical and molecular specificity of Raman spectroscopy in the diagnosis of dental caries. The OCT images were acquired using a diode laser 850 nm, spot size 10–20μm at 750 μmW and Raman spectroscopic data was acquired using an 830 nm diode laser at 24–52 mW from tooth samples extracted for orthodontics reasons (caries free and with incipient caries). While OCT images lend morphological depth as to the extent and spread of the carious lesion, Raman Spectroscopy could diagnose even early and incipient caries through biochemical changes (Table 6.2).

De Carvalho et al. (2013) [117] compared Raman spectroscopic techniques to Laser Fluorescence (Diagnodent) measurements on teeth having smooth surface carious lesions and non-carious lesions. Raman Spectra were obtained using a 785 nm, 500 mW diode laser with a 20 s exposure time. They concluded that the Diagnodent could not detect subtle changes in mineral

Table 6.2 Raman spectral bands and assignments

Spectral bands	Assignments	Author
431, 446 cm^{-1}	Phosphate PO_4^{3-} (Symmetric bending ν_2)	Ko et al. (2005) [116] Coello (2015) [108]
579, 590, 608, 614 cm^{-1}	Phosphate PO_4^{3-} (Asymmetric bending ν_4)	Ko et al. (2005) [116]
~575 cm^{-1}	Fluoridated apatite	de Carvalho et al. (2013) [117]
~960 $cm^{-1}{}_4$	Symmetric stretching (ν_1) Phosphate hydroxyapatite	Ko et al. (2005) [116] de Carvalho et al. (2013) [117]
1023, 1043, 1046 cm^{-1}, 1052, 1069 cm–1071 cm^{-1}	Asymmetric stretching vibration PO_4^{3-} (ν_3)	Ko et al. (2005) [116] Coello (2015) [118]
1069 cm^{-1}	Symmetric stretching mode of CO_3-type B (ν_1)	Boskey et al. (2005) [119]
1104 cm^{-1}	Symmetric stretching mode of CO_3-type A (ν_1)	Boskey et al. (2005) [119]
1200–1400 cm^{-1}	Amide III	Coello (2015) [118]
~1450 cm^{-1}	Organic matrix	de Carvalho et al. (2013) [117]
1670 cm^{-1}	Amide I	Xu and Wang (2012) [120]
2941 cm^{-1}	Lipids and proteins (C–H and C–H_2 groups)	Coello et al. (2015) [118]
2874 cm^{-1}	Unsaturated bonds of lipids	Coello et al. (2015) [118]
2926 cm^{-1}	Saturated bonds of lipids	Coello et al. (2015) [118]

content such as in an incipient carious lesion and had a lower sensitivity as compared to the specific detection of mineral changes by Raman Spectroscopy. The Diagnodent was able to detect organic changes much more than inorganic changes and this they felt corroborates the finding of earlier studies regarding the suitability of Diagnodent for initial detection of caries.

Coello et al. (2015) [118] proposed diagnostic quantitative mineralization indices of teeth through Raman spectroscopy with the objective of establishing a diagnostic scale with regard to demineralization. Raman Spectra was obtained using an Nd:YAG laser at 1500 mW, from different zones of extracted teeth. They concluded that the ability of Raman Spectroscopy to assess and differentiate the organic and inorganic areas of the tooth qualifies it to assess demineralization of the tooth. The MIb (Mineralization Index bending) and MIs (Mineralization Index stretching) indices, which they suggested, were able to diagnose even initial demineralization and demonstrated the value of Raman Spectroscopy in this area. Thus Raman Spectroscopy can be a convenient and effective modality in diagnosis and monitoring of diseases like fluorosis and amelogenesis imperfecta.

Similarly, Raman Spectroscopy was able to deepen our understanding of the chemical composition and make up of dentine, particularly in laying to rest the speculation about the composition and make up (organic and inorganic) of peritubular and intertubular dentine. Xu and Wang (2012) [120] cleared the confusion regarding the collagenous or non-collagenous nature of the peritubular dentine and composition of the intertubular dentin using micro-Raman spectroscopy (μmRs) with a He–Ne (632.8 nm) laser, 60 s, and atomic force microscopy (AFM). They proved that the peritubular dentine was hypermineralized with respect to the intertubular dentin with a mineral: matrix ratio three times higher than the intertubular dentin. The study showed that the peritubular dentin had an inorganic content of 96% (organic content of 4%) as compared to the inorganic content of 88% (organic content of 12%) for intertubular dentine though the crystalline nature of the peritubular dentin was much similar to the intertubular dentine.

6.8.6 SWIR and NIR Imaging of Tooth Surfaces

Short wavelength infrared (SWIR) imaging is ideally suited for screening for dental caries, due to the high transparency of enamel, the lack of interference of stains and the high contrast between sound and demineralized tissues. It is also well suited for imaging cracks in teeth, identifying composite restorations, and assessing lesion activity. Light scattering in dental enamel decreases

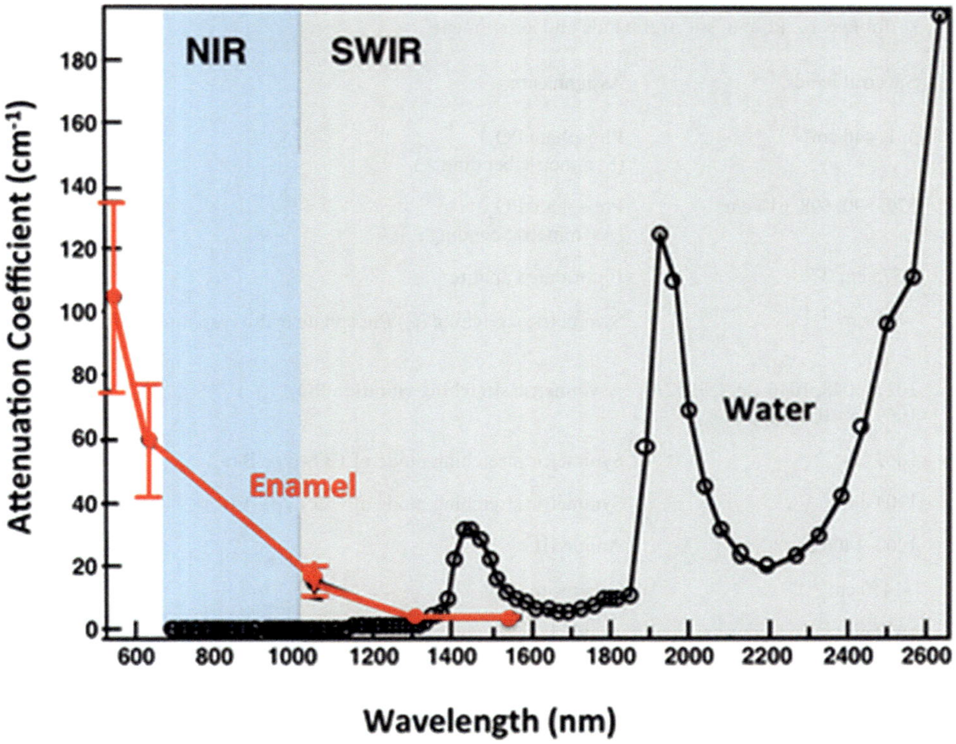

Fig. 6.14 The attenuation coefficient for dental enamel (red) and the absorption coefficient of water (black) in the visible to IR [123–125]. NIR range is 700–1000 nm (blue) and SWIR range is 1000–2500 nm (gray)

markedly with increasing wavelength and is orders of magnitude lower at SWIR wavelengths. It is common practice today to define the near-infrared (NIR) as 700–1000 nm and the short wavelength infrared (SWIR) as 1000–2600 nm; however, the entire range has been referred to as the NIR in the past. Light scattering in dental enamel is high at UV wavelengths and decreases by two orders of magnitude from 543 to 1300 nm and the wavelength dependence suggests that scattering is dominated by Rayleigh scattering due to the small crystals of enamel that are much smaller than the wavelength [122]. A plot of the attenuation of light in enamel and water is shown in Fig. 6.14 as a function of wavelength from 400 to 2000 nm [122, 124, 125]. Light attenuation increases beyond 1300 nm due to a rise in light absorption by water. Further studies suggest that the scattering coefficient of enamel continues to decrease beyond 1300 nm [126]. Light scattering increases by orders of magnitude upon demineralization due to the formation of small pores in the enamel that highly scatter light [127]. This change in light scattering produces large differences in the contrast between sound and demineralized enamel particularly at longer wavelengths where the light scattering in sound enamel is low. Light scattering remains high in dentin with increasing wavelength due to the strong scattering by the dentinal tubules that are larger than the wavelength of light [122]. Light scattering in dentin does increase with demineralization producing measurable contrast between sound and demineralized dentin but the contrast is not as high as it is for enamel.

An important advantage of imaging in the SWIR is that the organic molecules responsible for stains on teeth do not absorb light at longer wavelengths [128, 129]. This is of particular importance in the stained pits and fissures of the tooth occlusal surface. It is necessary to use wavelengths greater than 1200 nm to avoid significant interference from stains on tooth surfaces [130]. Therefore, stains can be easily differentiated from actual demineralization at SWIR wavelengths. At visible and NIR wavelengths stains interfere to such a degree that absorption due to stains contributes more to the lesion contrast than increased scattering due to demineralization [131]. Lesion contrast increases from the visible range where it is negative to extremely high positive values beyond 1700 nm [130]. Since it is impractical to remove stains from the deep grooves and fissures on tooth occlusal surfaces, lack of interference from stains at longer SWIR wavelengths is a significant advantage. Figure 6.15 shows three reflectance images of a tooth with occlusal decay at 400–700, 1300, and 1450 nm. Sound areas of the tooth get darker with increasing wavelength yielding extremely high contrast of demineralization at wavelengths beyond 1400 nm [129]. Also notable is that the stains in the visible light image dominate the lesion contrast, producing negative contrast, i.e., the demineralized areas appear darker as opposed to appearing lighter. A plot of the mean lesion contrast as a function of wavelength is shown in Fig. 6.15 for occlusal lesions on 55 extracted teeth. The mean contrast was negative at visible wavelengths (400–700 nm) and at 850 nm and increased significantly for each wavelength range [130].

Reflectance and multiple geometries of transillumination are practical for imaging teeth for the detection of caries lesions on tooth proximal and occlusal surfaces. SWIR imaging offers many advantages over radiography and conventional visual and tactile methods for the detection of dental caries in addition to reduced exposure to ionizing radiation. SWIR clinical imaging studies were first carried out more than a dozen years ago. Interproximal lesions that appeared on radiographs could be detected with SWIR proximal and occlusal transillumination with similar sensitivity [132]. Even though the sensitivity of radiographs is not very high [133–137], the specificity of radiographs is above 90%, therefore it is a suitable standard for comparison. In addition to proximal transillumination, a novel occlusal imaging geometry, occlusal transillumination, was also employed to detect interproximal lesions in vivo [132, 138]. Further clinical studies showed the usefulness of SWIR reflectance imaging and that the diagnostic performance of SWIR imaging was better than radiographs for both occlusal and interproximal lesions [138, 139]. Lesions on proximal surfaces located well below the tooth surface can be imaged with high contrast in SWIR reflectance as can be seen in ▫ Fig. 6.16. A tooth with a large interproximal lesion that is not visible in the color image is visible in both reflectance and transillumination SWIR images with high contrast. This tooth was also diagnosed with occlusal decay in the stained central groove based on visual and tactile examination but that appears to be a false positive. No demineralization was visible with either SWIR imaging or optical coherence tomography [139]. The reflectance and transillumination images in ▫ Fig. 6.16 were taken with two separate SWIR reflectance and transillumination devices. It is feasible to combine both devices in a single multispectral SWIR reflectance and transillumination system and the images shown in ▫ Fig. 6.17 were taken with a device that acquires simultaneous SWIR reflectance and transillumination images [140].

Even though studies indicate the contrast of demineralization is much higher at SWIR wavelengths than at NIR wavelengths and that the sound enamel is more transparent [141], all the currently available commercial devices utilize shorter NIR wavelengths. This is most likely due to the higher cost of InGaAs imaging arrays compared to Si imaging arrays. The first commercially available NIR imaging device called the Diagnocam from Kavo (Biberach, Germany) uses an occlusal transillumination probe with 780 nm light [142, 143]. The shorter wavelength allows the use of a less expensive silicon detector as opposed to InGaAs-based detectors. It is also important to point out that stains still interfere

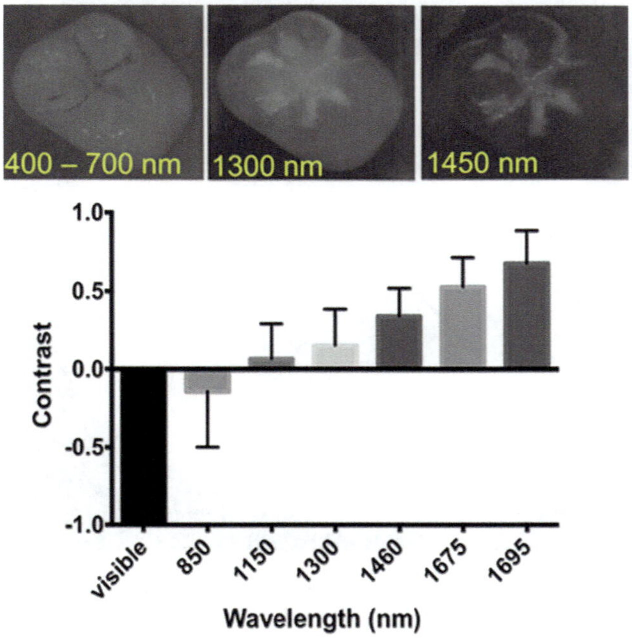

▫ **Fig. 6.15** Visible (400–700 nm), NIR and SWIR reflectance images of an extracted tooth with occlusal decay. Plot of the lesion contrast as a function of wavelength for 55 teeth with occlusal decay. The contrast of demineralization should be positive in reflectance. All bars are significantly different ($p < 0.05$) [130]

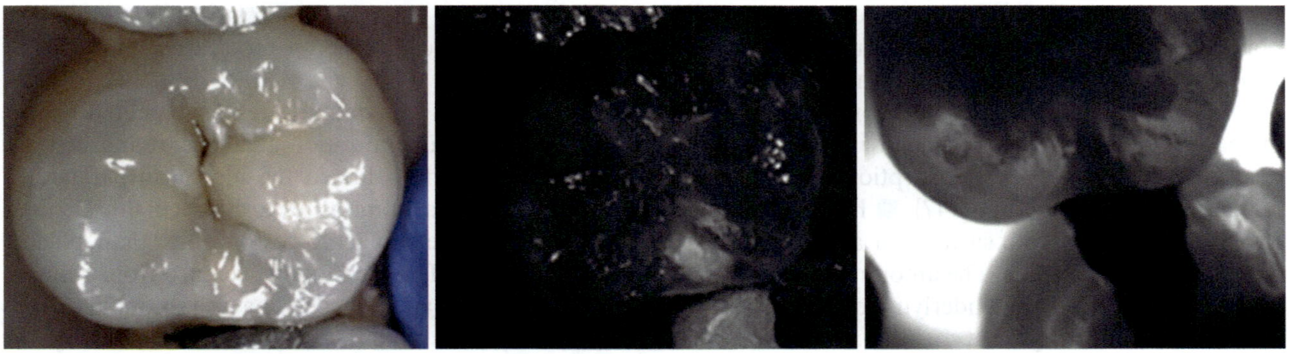

▫ **Fig. 6.16** Color, SWIR reflectance at 1600 nm, SWIR transillumination at 1300 nm images for one tooth with suspected occlusal and proximal decay [139]

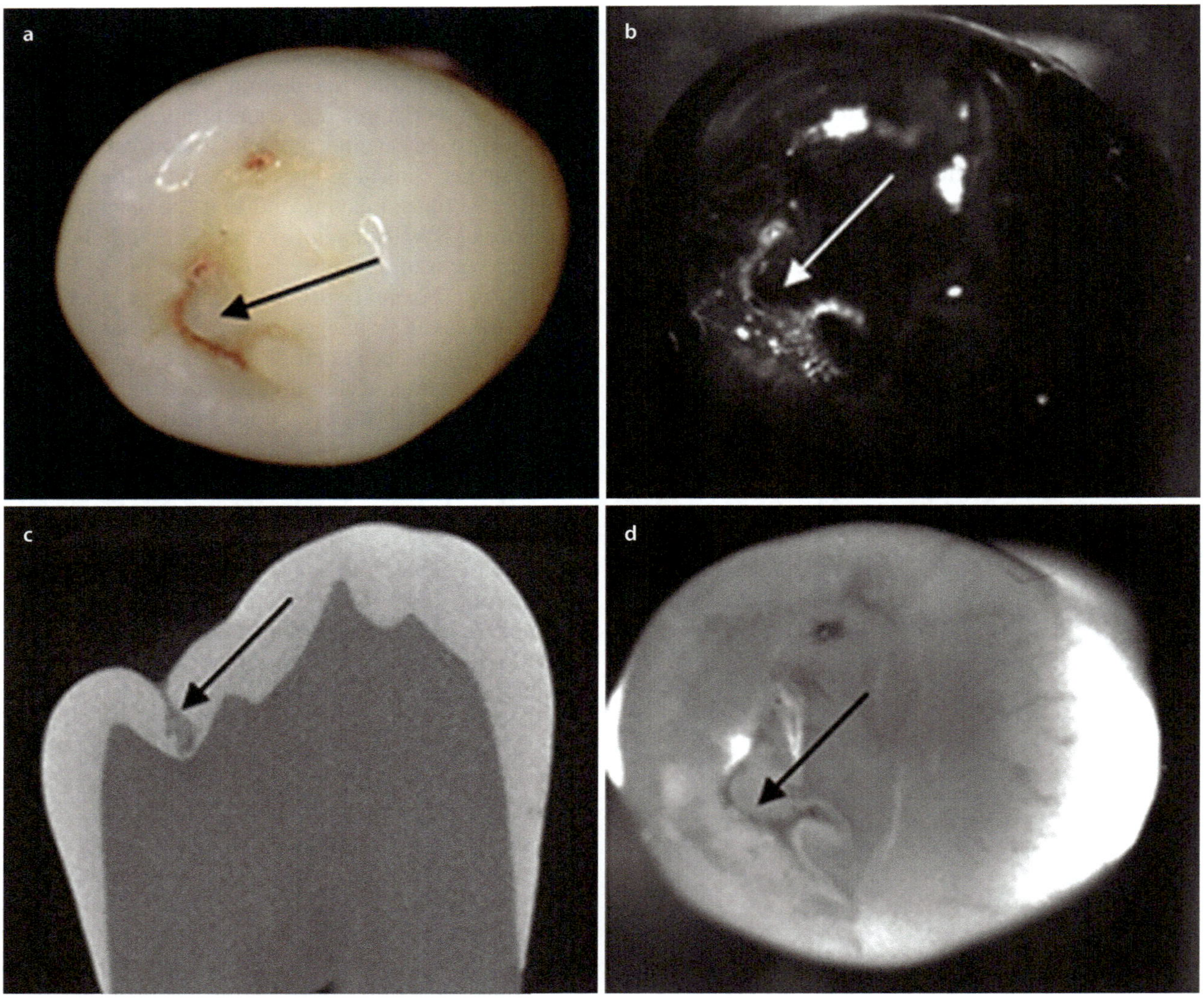

Fig. 6.17 **a** Color, **b** SWIR reflectance at 1600 nm, **c** microCT and **d** transillumination at 1300 nm images for one tooth with occlusal decay. Arrows point to lesion. SWIR images acquired with dual SWIR transillumination/reflectance device [140]

at 780 nm so stains are still a potential problem with shorter wavelength NIR devices [130]. The first commercial NIR reflectance system operating at 850 nm was the Vistacam iX Proxi from Durr Dental (Bietigheim-Bissingen, Germany) [144].

Root caries is an increasing problem with our aging population [145, 146]. Imaging root caries at SWIR wavelengths may be equally advantageous due to the lack of interference of stains beyond 1200 nm and increased suppression of the reflectivity from the sound dentin due to higher water absorption which increases markedly beyond 1400 nm [147]. Figure 6.18 shows SWIR images of the root surfaces of an extracted tooth with lesions and calculus. The amount of light reflected (backscattered) from the underlying dentin depends on the ratio of scattering to absorption. Therefore, wavelengths with higher water absorption can yield high lesion contrast even though the light scattering in sound dentin is still relatively high. Better visualization can lead to more precise methods for removal, preserving healthy tissues. One possible approach is to use lasers for the selective removal of calculus without damaging the cementum [148–150]. SWIR images at wavelengths beyond 1400 nm may be ideally suited for the image-guided laser removal of both root caries and calculus [151, 152].

SWIR imaging modalities are ideally suited for the detection of demineralization adjacent to and beneath restorative materials due to the high transparency of enamel and dental composites to SWIR light. Dentists spend more time replacing restorations than placing new ones [153]. The lack of interference from stains is also a significant advantage in imaging secondary caries since composites often become stained masking the peripheral demineralization. Composite restorative materials have unique spectral signatures in the SWIR

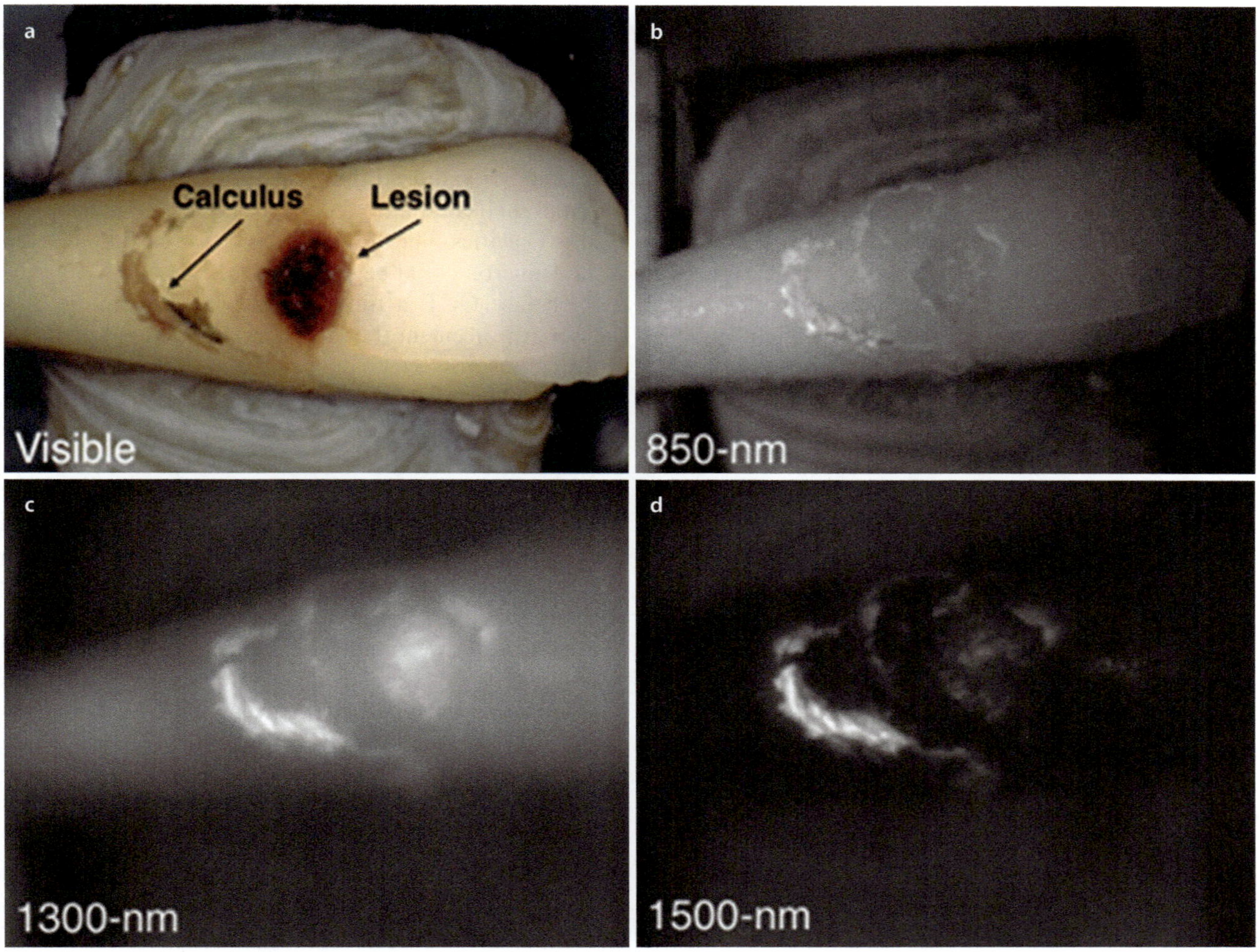

Fig. 6.18 **a** Color, **b** SWIR reflectance at 1600 nm, **c** microCT and **d** transillumination at 1300 nm images for one tooth with occlusal decay. Arrows point to lesion. SWIR images acquired with dual SWIR transillumination/reflectance device [147]

resulting from combination absorption bands that can be exploited for differentiating tooth structure and other type of composites. The most prominent dental resin absorption bands lie at 1171, 1400, 1440, 1620, and 1700 nm and result from overtones and combinations of the fundamental mid-IR vibrational bands from C–H, N–H, and O–H groups found in both resin and water [154, 155]. The overall visualization of the restoration boundaries can be enhanced when viewed at different SWIR wavelengths [156, 157]. The ability to distinguish between composite, sound enamel and demineralized enamel is necessary for both the detection of secondary decay and the replacement of restorations with minimal loss of sound tissue structure. Tooth-color-matched restorations manifest very high contrast with sound tooth structure at SWIR wavelengths coincident with high water absorption. SWIR light is attenuated to a lesser degree in composite due to the lower water content of composite compared to enamel and dentin. This can also enable the selective removal of composite by lasers [158, 159].

Cracks in teeth are an important problem in dentistry; they may cause a large range of symptoms and lead to the loss of healthy tooth structure [160]. Cracks can originate from coronal tooth structure or from within the root, and they may occur in both horizontal and vertical directions on the crown and/or root. Ultrasound can also potentially be used to locate cracks [161]. Since tooth enamel is most transparent at 1300 nm, SWIR reflectance and transillumination imaging is ideally suited for imaging cracks in teeth [162]. Crack formation and propagation can be imaged in real-time during laser irradiation using SWIR transillumination [162–164]. Sapra, Darbar et al. used an 810 nm diode laser (1 W, CW, 200µm, 5 mm distance, 1 mm/s) in an in vivo prospective study over 4 years in 12 patients who had sharp dull or aching pain on biting and release, they concluded that the laser can be complementary to conventional diagnostic methods in identifying and allowing early treatment of teeth with cracks [165].

Many caries lesions that are detected have been arrested or are undergoing remineralization and do not require intervention. As lesions become arrested by mineral deposition or remineralization, the permeability of the mineralized surface layer decreases. Studies have shown that permeability changes can be monitored using quantitative light fluorescence (QLF) [166–168], thermal and SWIR reflectance imaging [169–172] during lesion dehydration. In vivo studies using QLF [169] and thermal imaging have been reported [172]. A small increase in the surface layer thickness can lead in large changes in permeability [173]. Changes in the lesion contrast during drying are particularly high at wavelengths coincident with water absorption bands near 1450 and 1950 nm suggesting that those wavelengths are optimal for the assessment of lesion activity on tooth coronal surface [174, 175].

6.8.7 Optical Coherence Tomography in Hard tissues

OCT or its newer adapted advanced versions provides for varied and versatile applications in hard tissues such as the detection of caries and even marginal gaps of restorations or cracks in teeth. Shimada (2012, 2015) [176] illustrated this in his review article.

Polarization-sensitive OCT (PS-OCT), a functional OCT technique, with near-infrared excitation could give structural and positional information based on birefringence of the sample and polarization of back scattered light [177]. PS-OCT can detail dental carious lesions and give images which potentially can even monitor the progression of lesions over time [178].

Wang et al. (1999) [179] had shown that PS-OCT using 856 nm, 0.8 W diode laser, could be applied toward the investigations of enamel and dentin by analyzing their intrinsic birefringence. They were able to determine refractive index values of enamel and dentin but though banded birefringence of enamel crystals was seen up to the enamel dentin junction, considerable scattering and anisotropy in dentine was observed with no birefringence and resolution of finer features [180]. Gossage et al. (2003) [181] had stated that this was due to speckle interference (in dentinal structures such as collagen—form birefringence) and that speckle noise prevents resolution of structures and gives only sub-resolution features.

However, dentin of attrited teeth were investigated successfully by Mandurah et al. (2015) [182] using Swept Source-OCT in extracted teeth, where a 1310 nm laser (use of longer wavelength gives greater penetration and more resolution of the deeper layers) with higher frequency scanning speeds of 20 kHz provides better axial (11μm) and lateral resolution (17μm). Shimada et al. (2010) [183] had shown that this technique could deliver an imaging depth of up to 2–3 mm. The dentin showed different back scattering patterns between carious and sound dentin and in attrited dentin there was a low attenuation coefficient because of the sclerosed dentinal tubules. Makishi et al. (2011) [184] and Bakhsh et al. (2011) [185] had used the same modality for evaluation of marginal adaptation of resin-based restorations in Class I cavities. The authors report that gaps up to even a half micron could be detected by SS OCT.

En face OCT, a time-domain OCT technique, is a newer advancement that provides the ability to precisely localize lesions by giving transverse (cross sectional) images of layers at a specified even microscopic depth. Todea et al. (2010) [186] used en face OCT to evaluate and compare the quality of endodontic treatment between diode laser assisted (980 nm), Nd:YAG laser assisted (1064 nm), and conventional groups of extracted teeth. They found that though all the groups had obturation defect the laser groups had significantly lesser defects compared to the conventional group and that en face OCT allowed a precise, non-invasive, diagnostic evaluation.

Most papers published regarding OCT imaging dental hard tissues involves OCT imaging at 1300 nm where there is deeper optical penetration due to the higher transparency of enamel. Longer wavelengths beyond 1300 nm have been investigated to achieve higher performance for imaging hard tissues, but higher performance has not yet been established at these wavelengths [187, 188]. There have been several review articles regarding the use of OCT imaging to image hard and soft tissues in the oral cavity [189–194]. The first images of the soft and hard tissue structures of the oral cavity were acquired by Colston et al. [195, 196]. Feldchtein et al. [197] presented the first clinical images of dental hard tissues and restorations. OCT can be used to measure the reflectivity within dental hard tissues to a depth of up to 3–4 mm in enamel and 1–2 mm in dentin. OCT is valuable for obtaining high-resolution images of lesion structure and severity to aid diagnosis (◘ Fig. 6.19).

Baumgartner et al. [199, 200] presented the first polarization-sensitive (PS-OCT) images of dental caries. PS-OCT images are typically processed in the form of phase and intensity images [201, 202], such images best show variations in the birefringence of the tissues. Later PS-OCT measurements demonstrated the advantage of using the cross polarization images (CP-OCT) to quantify lesion severity and track changes in lesion severity over time [203]. Cross polarization is useful for enhancing the contrast of demineralization and for reducing the interference of the strong surface reflection that can inhibit the resolution of structures near the tooth surface [200, 203–204]. This is demonstrated in ◘ Fig. 6.20. CP-OCT and PS-OCT are particularly useful for moni-

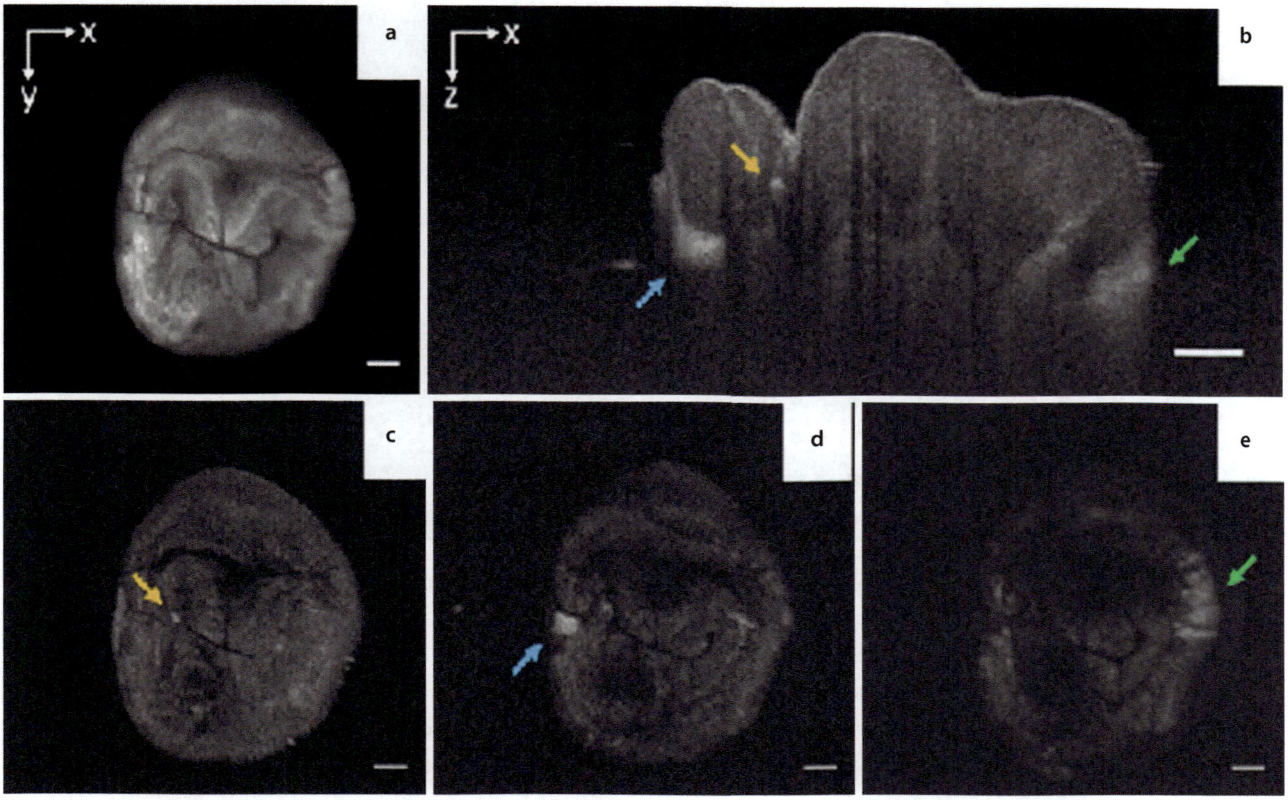

■ **Fig. 6.19** OCT tomographic images of a premolar with interproximal lesions and occlusal decay at 1310 nm showing: **a** the overall reflectivity, **b** optical cross section or b-scan image (*xz* plane) from mesial to distal surfaces and **c–e** en face images (*xy* plane) at different depths (*z*) from the tooth surface. Yellow arrow is a small occlusal lesion that penetrates to the DEJ and the blue + green arrows are interproximal lesions. White bar is 1 mm [198]

toring the repair or remineralization of caries lesions by resolving the formation of a distinct highly mineralized transparent surface zone indicative of a reduction in lesion activity and lesion arrest [205–207].

High-speed Fourier domain OCT systems (FD-OCT) have long since replaced the early time-domain OCT systems (TD-OCT). For TD-OCT, the sensitivity (signal to noise) decreases markedly with increased scan rate, this limits the maximum scanning rate to 1–2 kHz. This is not a problem with the FD-OCT systems and very high scanning rates can be achieved exceeding 100 kHz. Galvanometers or Microelectromechanical (MEMS) scanning mirrors can be used to scan the beam in two dimensions to acquire 3D images of entire teeth in a 1–2 s as can be seen in ■ Fig. 6.19 [198]. Almost all OCT systems available today utilize either swept-source (SS-OCT) or spectral domain (SD-OCT) systems [209–212]. SD-OCT systems are very popular for Ophthalmology and for Dermatology, but they have had limited applicability to dentistry since it has been difficult to achieve high axial resolution while maintaining the higher scanning ranges needed for dentistry, 7 mm of depth is needed to scan tooth occlusal surfaces and the high refractive index of enamel (1.63) further restricts the scanning range.

Many clinicians are primarily interested in the lesion depth so that they can decide whether a restoration is necessary. The ability of radiographs to detect occlusal lesions is extremely poor with an accuracy of ~50% [213, 214]. The failure to correctly diagnose occlusal lesions that don't appear on radiographs is a major problem and likely results in many unnecessary cavity preparations. OCT is ideally suited for monitoring and improving the diagnosis of occlusal lesions. Lesions spread laterally under the enamel upon contacting the more soluble softer dentin. Therefore, OCT can be used to determine if occlusal lesions have penetrated to the underlying dentin by detecting the lateral spread across the dentinal-enamel junction (DEJ), and this approach has been successfully demonstrated clinically [139, 215]. Interproximal lesions can also be imaged from the occlusal surface as can be seen in ■ Fig. 6.19 and the depth from the proximal surface can be measured to assess the penetration depth.

OCT has also been used successfully to measure demineralization on dentin and on root surfaces (cementum) [216–218]. ■ Figure 6.21 shows an OCT scan of the root surface of a tooth with a root caries lesion and with dental calculus. Cementum has lower reflectivity than dentin in OCT images, making it possible to iden-

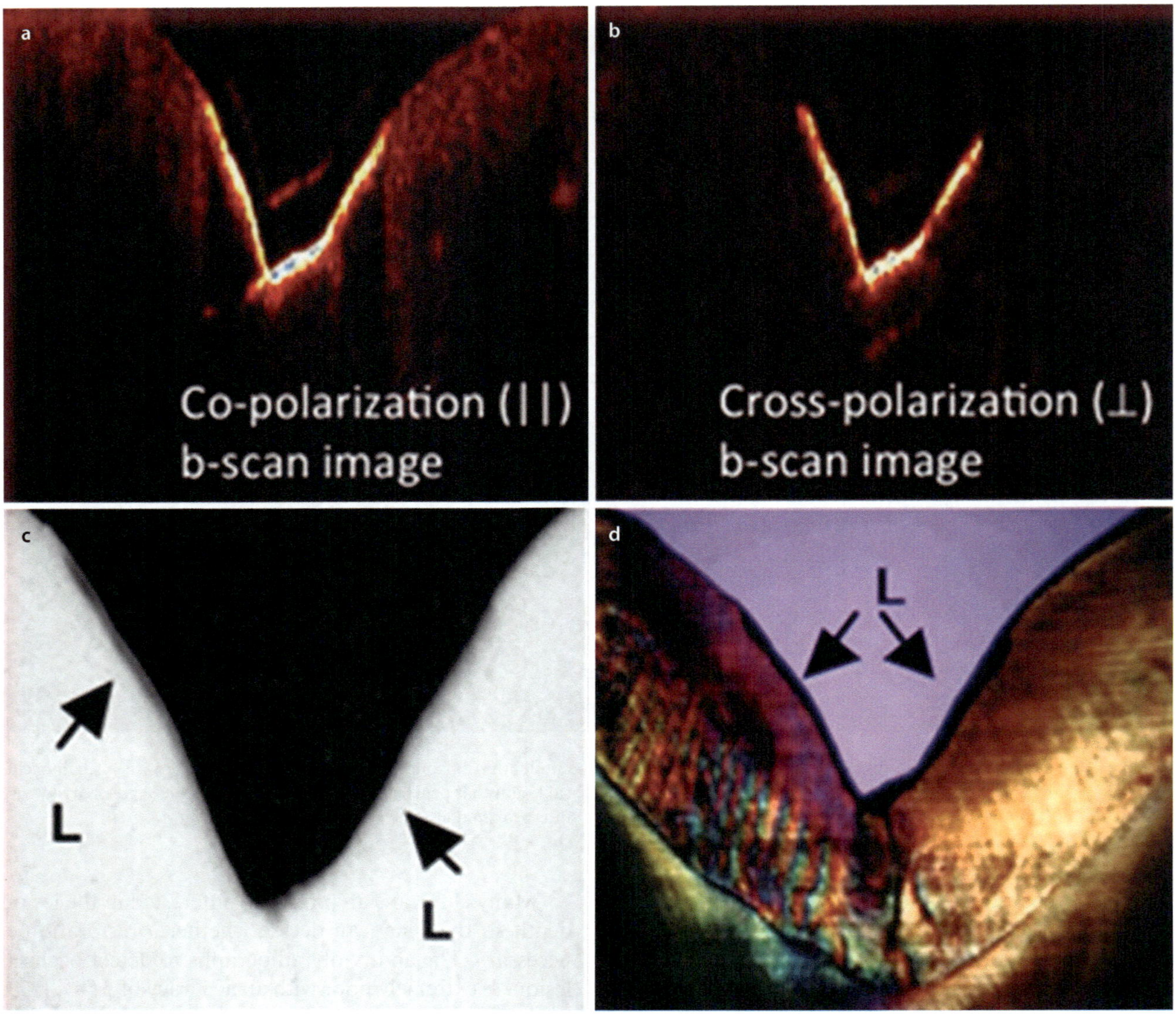

Fig. 6.20 **a** Co-polarization ||—PS-OCT, **b** ⊥—PS-OCT images (red-yellow color table), **c** TMR, and **d** PLM images of an artificial lesion produced in an occlusal surface. Mineral loss is black and sound enamel is white in the TMR image, mineral loss appears black in PLM. L indicates lesion position [208]

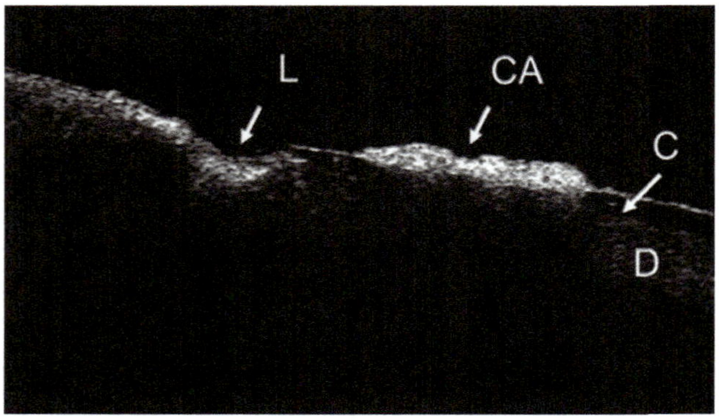

Fig. 6.21 OCT b-scan images of the root surface of an extracted tooth showing a slightly cavitated arrested root caries lesion (L), calculus on the surface (CA), and the more transparent layer of intact cementum (C) over the dentin (D)

tify the remaining cementum thickness [216, 218]. Shrinkage occurs in demineralized dentin due to the loss of water from the collagen with dehydration. Active root caries lesions manifest shrinkage while arrested lesions that have been successfully remineralized have reduced shrinkage [218]. OCT [219, 220] has also been investigated for imaging dental calculus. Root fractures are a major concern and endodontists lack a means of detection. Mineral fills the tubules of dentin with age and the tooth root becomes brittle and may fracture. Shemesh et al. [221] showed that OCT can potentially be used to detect vertical root fractures using an inter-catheter rotating—pullback scanning probe that can be inserted into a root canal. The tooth root becomes increasingly transparent with increased mineral deposition with age at SWIR wavelengths [222].

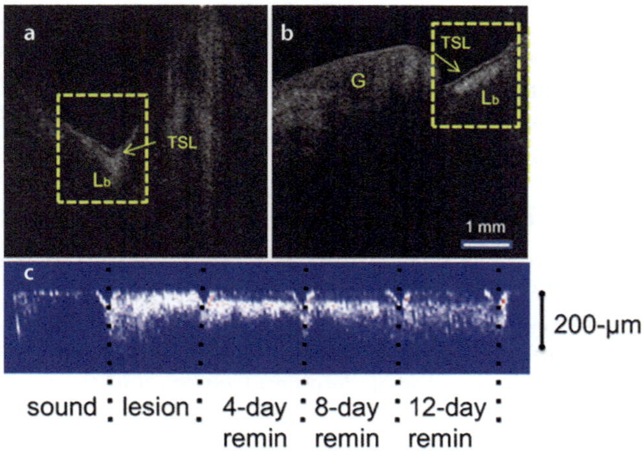

Fig. 6.22 Clinical CP-OCT b-scans of **a** an arrested occlusal lesion on a primary tooth and **b** an arrested root caries lesion. Both lesions have a transparent surface layer (TSL) over the lesion body (Lb). A b-scan image of artificial lesions produced on a block of bovine enamel **c** that were exposed to a remineralization solution showing the formation of the TSL at the surface of the lesion. At baseline before remineralization all four treatment windows looked similar to the lesion window [205]

Since OCT is capable of acquiring high-resolution images of lesion structure, it is well suited for monitoring the severity of demineralization and assessing whether remineralization has taken place. PS-OCT and CP-OCT are particularly well suited for this purpose [171, 203, 204, 207, 208, 223–225]. The ability of CP-OCT to monitor remineralization and the formation of a distinct transparent surface zone has also been demonstrated [205–207, 226–229]. Figure 6.22c shows CP-OCT images of a bovine enamel sample with 5 windows showing sound, lesion and lesion areas that have been exposed for 4, 8, and 12 days to a remineralization solution. There is minimal reflectivity in the sound regions outside the four windows, while the lesion areas have much higher reflectivity. Although there was a high degree of remineralization, there was still incomplete remineralization of the body of the lesion. The most obvious change was the formation of a distinct transparent outer surface layer (TSL) ~50 µmm thick. Clinical images of arrested lesions on the occlusal and root surfaces are also shown in Fig. 6.22 indicating that the TSL can be measured in vivo and can serve as a key indicator of lesion arrest.

In the first clinical study using OCT to monitor demineralization, PS-OCT was able to non-destructively measure significant increases in demineralization on both the buccal and occlusal surfaces over time [230]. In a subsequent study, a high-speed swept-source CP-OCT system with an integrated MEMS scanner from Santec (Komaki Aichi, Japan) was used to acquire 3D volumetric images of the area at the base of orthodontic brackets over a period of 12 months after placement. Even though an increase in demineralization was not visibly apparent in images taken before and after 12 months, CP-OCT was successful in measuring a significant ($p < 0.05$) increase in the mean lesion depth (Ld) and integrated reflectivity with time over the area cervical to the brackets further validating the utility of CP-OCT for monitoring early demineralization [231]. Therefore, changes in the internal microstructure of caries lesions can be monitored overtime during preventative intervention.

OCT has also been used to look at different restorative materials and identify pit and fissure sealants [133, 204, 232]. Other studies have investigated the use of polarization-sensitive optical coherence tomography (PS-OCT), at ~1300 nm, for the detection of demineralization beneath sealants and composites in addition to primary lesions [233–237]. OCT has also been found to be valuable for assessing the interfaces between tooth structure and resin restorations [184, 238, 239].

6.8.8 Laser-Induced Breakdown Spectroscopy in Hard Tissue and Soft Tissue

Laser-Induced breakdown Spectroscopy (LIBS) works on the principle that highly intense but very short laser pulses of ultraviolet (UV), visible or infrared range (IR) would ablate target tissue into an expanding plasma plume of electrons, ions, and atoms. This luminous plasma plume would emit characteristic qualitative and quantitative structural and molecular information about the tissue being examined. The creation of such plasma and theoretical basis of its breakdown was first suggested by Ambartsumyan and Basov et al. [240] as early as 1965 (unpublished work by W.S. Boyle 1962); however, it was J. Maxwell who gave the first description of the equipment which used a Q-switched Ruby laser at the Jarrell –Ash laboratory. Later, Rosan [240] presented the first spectra on biologic samples using this equipment at the First Annual Conference on Biologic Effects of Laser Radiation at Washington, USA, in 1964.

The phenomenon of plasma formation, "optical breakdown" through shock wave generation, cavitation, and jet formation by the Nd.YLF laser (1053 nm) with 30 ps pulses, 30µm spot size, and 1 mJ pulse energy was discussed in detail by Niemz in his textbook [14]. Further and sustained laser pulses can lead to a cascade of laser-induced plasma. This was proven further in Niemz's work (1994) [241] on LIBS for diagnosis of caries. Ca and Na spectral lines were observed in healthy and carious teeth. The spectra from carious tooth mineral were "weaker" (in intensity and line width) than those from healthy tooth structure.

Samek et al. (2001) [242] used a Q-switched Nd:YAG laser, 1064 nm at 20 Hz, 4–8 ns pulses, with pulse energy 10–30 mJ on 159 carious and healthy extracted teeth and even in vivo on the molar tooth of a volunteer. They confirmed that LIBS along with pattern recognition algorithms like discriminant analysis is able to positively identify between carious and healthy tooth structure to a precision of 100–200μm laterally and 10μm in depth. They inferred that this is possible through analysis of the spectra of matrix elements such as Ca or P and non-matrix elements like Li, Sr, Ba, Na, Mg, Zn, and C using pattern recognition algorithms. They stated that in carious tooth structure there would be a decrease in matrix elements and an increase in non-matrix elements.

LIBS has found varied applications [243] in the breakdown (Fang et al. 2005) [244], qualitative and quantitative characterization of urinary and kidney stones [245–247], or gall bladder stones and even the analysis of fluids [248] like those containing glucose or organic matter.

6.9 Guided Selective Laser Ablation

New sensors and imaging devices have been developed over the past three decades for acoustic, spectral, and optical feedback that are well suited for guiding laser systems for the selective removal of dental caries, composite restorative materials, and calculus from tooth surfaces. New and more sensitive optical methods of caries imaging have recently become available that allow acquisition of 2 and 3-dimensional images of lesion position, depth, activity, and severity on tooth surfaces [194, 249–251]. Such imaging systems are ideally suited for integration with computer-controlled laser scanning systems.

Over the past 50 years, the nature of dental caries has changed with the majority of new lesions occurring in the pits and fissures of the occlusal surfaces [252, 253]. Demineralization in these areas may harbor an active biofilm [254], and it can reduce the adhesive properties to sealants [255–257]. Lasers are ideally suited for the selective removal of demineralized tissue in the pits and fissures of the occlusal surfaces and for the thermal sterilization of such surfaces [258]. If needed lasers can also be used to selectively remove composite sealants from the pits and fissures with minimal loss of sound enamel [259, 260].

Selective ablation can be achieved by simply tuning the laser to a wavelength that has a lower ablation threshold for specific tissue components [261, 262]. Demineralized dental hard tissues are typically removed at higher rates than sound tissues due to the higher ratio of water, protein, and lipid to mineral, offering some level of selectivity without additional feedback. In the visible range, stains in caries lesions also have increased absorption. Several studies report the selective removal of caries lesions using various lasers including the Nd:YAG [263, 264], Er:YAG [265], frequency-doubled alexandrite (377-nm), frequency-tripled Nd:YAG (355-nm) [266, 267], and excimer [268] based on increased absorption of stains. The same principle can be applied for the removal of calculus from tooth surfaces using UV and visible laser pulses [149, 269]. However, stains are not a good indicator of demineralization, and the mineral gradients in caries lesions are highly variable, and it is often necessary to remove sound tissues to reach the underlying body of the lesion. In addition, the laser should be able to remove both sound and demineralized enamel and dentin efficiently in order to be commercially viable. Therefore, a more practical approach is to choose a laser with a wavelength and pulse duration that efficiently removes sound enamel, and employ some form of sensor, either optical, spectral, or acoustic in nature that can be used to discriminate between sound and demineralized tooth structure and composite materials.

A major advantage of lasers is that they can be integrated with computer-controlled scanning systems for high precision and selectivity. Lasers have been used for many years for computer-aided design/machining (CAD/CAM). It is clinically feasible to use high-speed computer-controlled scanning systems to selectively remove dental caries and composite restorative materials using spectral and image guidance. Robotic procedures in medicine are now commonplace, however few if any robotic procedures have been developed for dental applications. Lasers are ideally suited for automated control and after more than two decades of research, carbon dioxide, and erbium lasers are now both cleared by the FDA for caries removal. However, these lasers are still used clinically in a similar fashion to the conventional dental handpiece used for mechanical removal in which the laser handpiece is handheld and visually guided. The potential for integration with spectral, image, and acoustic guidance systems and automated control systems for precision and selective removal of caries and restorative materials has not yet been realized.

6.9.1 Acoustic Feedback

One of the first feedback mechanisms investigated was acoustic feedback. A complaint most often voiced by dentists with respect to hard tissue ablation is the loss of tactile feedback. Tactile contact provides the dentist with important feedback with respect to discrimination of carious vs. non-carious tissue and control of cutting rate. Basic acoustic feedback can be used to determine if the laser is efficiently cutting tissue, whether it is cutting enamel, dentin, or carious tissue and if there is excessive risk of acoustic damage due to too high of a fluence [270–272]. This mechanism can be implemented just by training the clinician to recognize the appropriate acoustic signatures. Measurement of the photoacoustic response (acoustic response to pulsed laser impact) provides a simple tool for measuring the relative absorption of carious and non-carious dental hard tissue [266, 273, 276]. Additionally, it is a sensitive method for detecting the respective ablation thresholds for various tissues [272].

Pigmented caries has been selectively removed by visible and near-IR (NIR) lasers [275, 276]. Erbium laser wavelengths are highly absorbed by water and its water-mediated ablation is a highly explosive process. Monitoring the magnitude of the photoacoustic signal as a function of the number of laser pulses provides important information about the evolving absorption characteristics of the tissue. During ablation of graphite with the Nd:YAG laser, the signal is initially high due to the strong absorption. Photoacoustic magnitude falls of gradually as a crater is formed and the surface topography changes markedly. For dentin, there is initially a weak signal due to the poor absorption, followed by an increase in absorption as the protein chars creating a highly absorbing target. After reaching maximum intensity, the signal drops after the char have been removed and an ablation crater has formed. Thermal modification of the tissue changes its absorption characteristics and crater formation increases the irradiated surface area (decreases incident fluence). Therefore, the photoacoustic response is an excellent tool for determining both the onset of ablation and subsequent "stallout" after thermal modification of the tissue. Even though this method of feedback was developed more than 30 years ago, it does not appear that there have been any successful attempts in dentistry to couple acoustic feedback with laser scanning for selective ablation.

6.9.2 Spectral Guided Feedback

LIBS, plume emission spectroscopy LIBS, plume emission spectroscopy, or laser microprobe emission spectroscopy described in the previous section can be used for the identification of many materials. The method was first used to examine teeth over 40 years ago [277] and has been used to identify calcified plaque in arteries [278]. Over the past 20 years, very compact fiber-optic spectrometers have become available that are relatively inexpensive and can be readily interfaced with laser ablation systems. This method has been proposed to differentiate between caries and sound tooth structure [242, 279, 280]. However, there are high concentrations of calcium in both sound and demineralized dental hard tissues and spectral guidance does not work for the removal of caries lesions [281], and an alternative image-guided approach is required to remove caries lesions.

Dental composites are used as restorative materials for filling cavities, shaping, and covering teeth for aesthetic purposes, and as adhesives. Dentists spend more time replacing existing restorations that fail due to microleakage and secondary caries than they do placing new restorations. Tooth-colored restorations are difficult to differentiate from the surrounding tooth structure and adhere strongly to the underlying enamel and dentin making them challenging to remove without damaging tooth structure. Excessive amounts of healthy tooth structure are often removed to ensure complete removal of the composite. Multiple studies have shown that spectral feedback can be used effectively to discriminate between the ablation of dental composite adhesive/restorative materials and dental hard tissues [282–286]. Lasers are ideally suited for selective ablation of failed restorations and sealants, removing composite adhesives such as residual composite left after debonding orthodontic brackets [283], and for the repair of aesthetic bonding [287].

Spectral analysis of the plume has revealed that the distinctive high intensity calcium emission line at 605 nm found in dental hard tissues can be used to differentiate tissues rich in hydroxyapatite (i.e., dentin and enamel) from composite [259, 260, 266] (◘ Fig. 6.23). Composite restorative materials lack calcium and thus lack the calcium emission lines that are very strong between 580 and 650 nm which gives the plume a distinct red appearance.

Similar results can be achieved faster and cheaper by substituting the spectrometer with a pair of filtered photodiodes. Spectral guided selective removal of dental composite was demonstrated in vivo in a small clinical study [288, 289] using a CO_2 laser with an articulating arm, galvanometer, and two diodes with narrow bandpass filters at 620 and 540 nm (◘ Fig. 6.23).

6.9.3 Image-Guided Ablation

Fluorescence has been proposed as a means of guiding caries removal by both laser and mechanical means [290–293]. Increased levels of porphyrins accumulate in dentinal caries lesions due to the high porosity and that

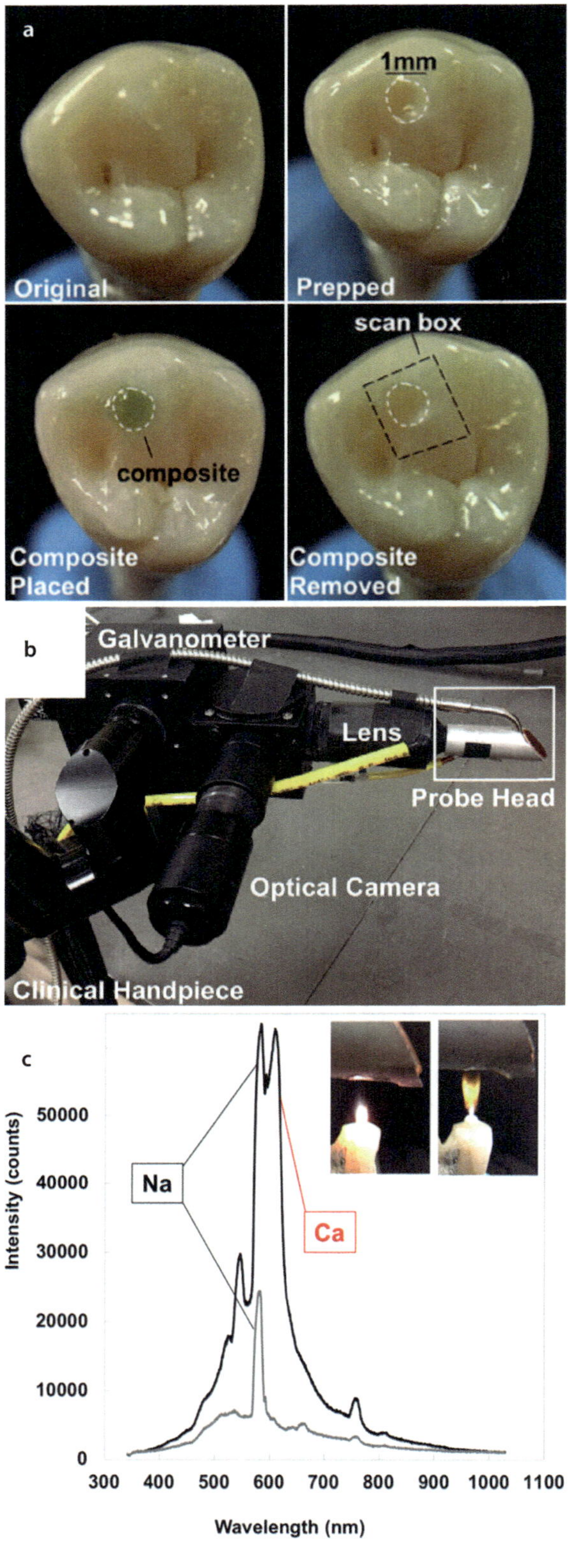

Fig. 6.23 **a** Composite placed and removed by spectral guided laser ablation (SGLA). **b** Clinical handpiece used for SGLA with 9.3 µm CO_2 laser, galvanometer-based scanning system, video camera, spectral feedback system with two diode detectors at 550 and 650 nm, and air and water cooling. **c** Spectrum of enamel (black) and composite (gray), inset shows plumes generated when ablating enamel and composite that appear colored red and yellow due to the Na and Ca atoms [288, 289]

fluorescence has been employed for caries detection. However, the fluorescence is not specifically associated with cariogenic bacteria since they do not contain porphyrins nor is it correlated with the degree of demineralization. Moreover, the fluorescence is weak and diffuse and poorly localized and is not well suited for precise guidance of the laser. In addition, stains can lead to false-positive readings, which is a major problem for fluorescence imaging. However, the highly conjugated organic molecules associated with staining do not absorb light at wavelengths longer than 1200 nm [139, 294–296].

Multiple in vitro and in vivo studies evaluating the coronal surfaces of teeth have demonstrated that longer wavelengths yield much higher contrast between demineralized and sound tooth structure than visible reflectance and fluorescence. In addition, stains do not interfere at wavelengths longer than 1100 nm [128, 131, 132, 138, 293, 297]. Studies show that stains completely mask demineralization on tooth occlusal surfaces [130, 131]. Therefore, longer wavelengths beyond 1200 nm are best suited for image-guided ablation on tooth surfaces. Imaging root caries at longer short wavelength IR (SWIR) wavelengths appears to be equally advantageous due to the lack of interference of stains beyond 1200 nm. Root surfaces are often heavily stained, and it can be difficult to differentiate between stained eroded areas (affected dentin) and root caries (demineralized dentin/cementum). In addition, there is increased suppression of the reflectivity from the sound dentin due to the higher water absorption that increases markedly beyond 1400 nm [147]. The amount of reflected light (backscattered) from the underlying dentin depends on the ratio of scattering to absorption. Therefore, wavelengths with higher water absorption can yield high lesion contrast even though the light scattering in sound dentin is still relatively high.

Several ablation studies have been carried out using SWIR images for guidance [151, 152, 259, 298, 299]. CO_2 lasers are well suited for selective ablation due to the strong absorption by the mineral in hard tissues and the high pulse repetition rates that are easily achievable with these laser systems. Figure 6.24 shows an example of a small occlusal lesion that was removed by a CO_2 laser using serial 1450 nm SWIR images to guide removal. OCT images acquired before and after removal shows that the removal was complete and highly selective. Since the ablation laser is being scanned over the tooth surface to remove all the decay, it is possible to scan a second SWIR laser to acquire the SWIR images. In a recent study, coaxial CO_2 and SWIR lasers were used to acquire images of demineralization on the tooth surface and remove it selectively. This greatly reduces the cost of the system since an expensive InGaAs cam-

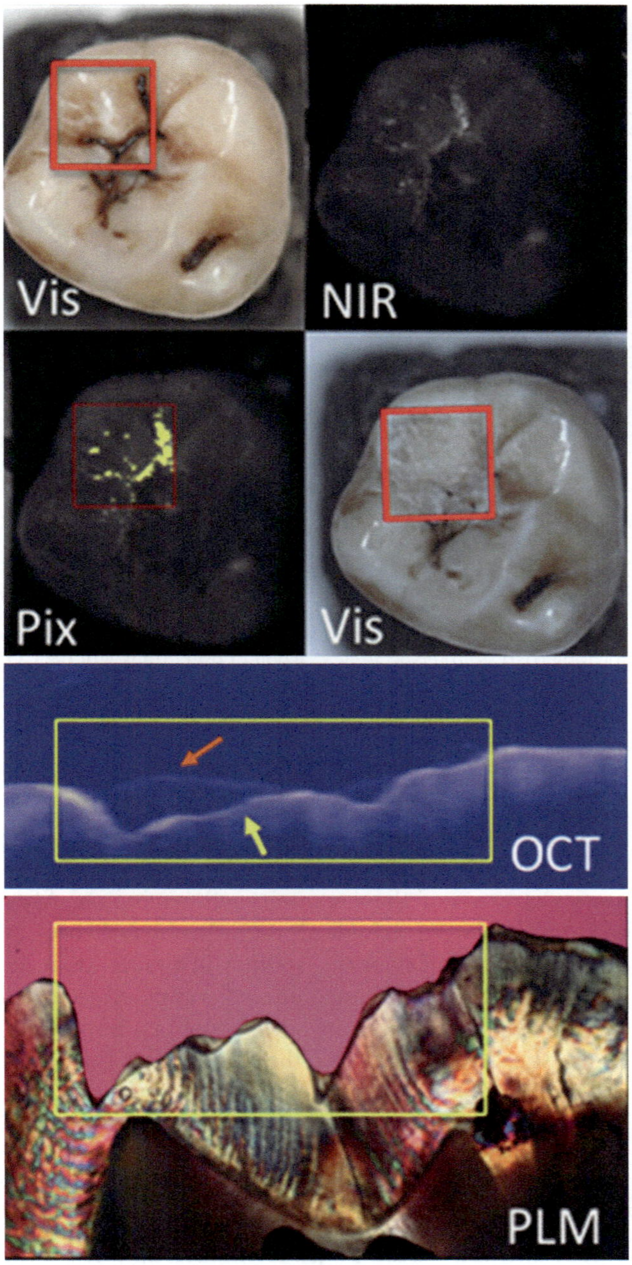

Fig. 6.24 (Vis) Color images before and after removal with a CO_2 laser. (NIR/SWIR) 1450 nm image before removal (Pix) area selected for removal by the laser in the area of the red box in the NIR image. (OCT) OCT b-scans before (red arrow) and after (yellow) removal. (PLM) PLM image of a tooth cross section taken after removal [300]

era is no longer needed. Moreover, light can be used at wavelengths longer than the sensitivity of conventional InGaAs cameras ($\lambda > 1700$ nm) where higher contrast is attainable such as 1880 or 1950 nm [151].

Calculus can form sub- and supragingivally and can contribute to irritation and inflammation of the gingiva that may lead to gingivitis and periodontitis. Scaling and root planning are often recommended to remove the calculus. Improved visualization of that calculus is important for both monitoring and complete removal. Several approaches have been used to image calculus including fluorescence-based methods [301–308], Raman spectroscopy [220, 309], and optical coherence tomography (OCT) [147, 219, 220, 310]. In addition to investigating the selective removal of caries lesions with a fluorescence-based feedback system, Krause et al. [148] investigated the removal of calculus. SWIR reflectance imaging at longer wavelengths is also advantageous for imaging dental calculus or mineralized dental plaque on tooth surfaces due to the lack of interference of stains and the higher light scattering of mineralized plaque [147].

Diode-pumped solid-state (DPSS) Er:YAG lasers with high pulse repetition rates are more suitable for the selective removal of dental caries, composites, and calculus than existing Er:YAG lasers [311]. The flash-lamp pumped erbium solid-state lasers presently being used for dental hard tissue ablation are poorly suited for selective laser ablation since they utilize high energy pulses and relatively low pulse repetition rates. DPSS Er:YAG lasers are now available operating with pulse repetition rates as high as 1–2 kHz and studies have been carried out demonstrating their utility for the ablation of dental hard tissues and bone [312–314] and the selective removal of composite from tooth surfaces [159]. In a recent study, a DPSS Er:YAG laser was combined with SWIR imaging for the selective removal of calculus from root surfaces [152].

In addition to two-dimensional imaging, 3D imaging systems are promising for image-guided ablation such as optical coherence tomography and cone beam computed tomography (CBCT). Simon et al. was able to use CBCT to guide a CO_2 laser to cut precise minimally invasive access holes for root canal therapy (Fig. 6.25).

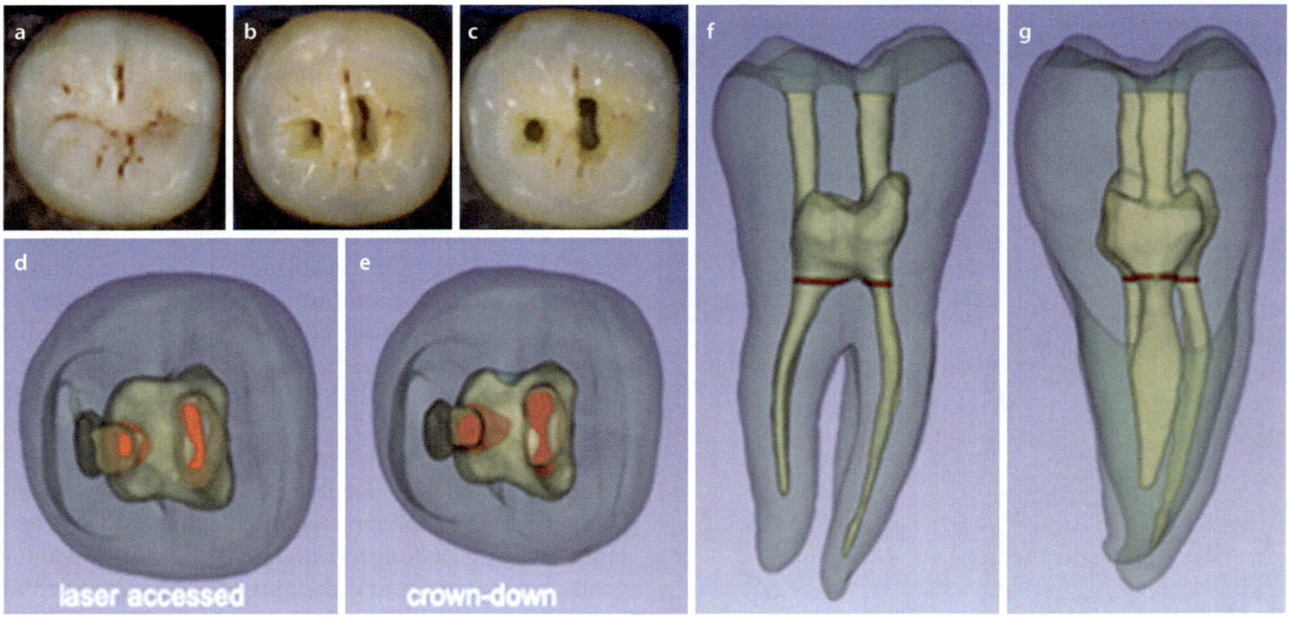

Fig. 6.25 Mandibular molar accessed with truss design. **a–c** Visible microscope images showing pre-ablation **a**, post laser access **b**, and post crown down **c**. **d**, **e** CBCT volumetric renderings indicating positioning of the laser preparations (yellow) in relation to orifice location (red) from the occlusal viewpoint post cutting with the CO_2 laser **d** and post crown down procedure **e**. **f**, **g** Volumetric renderings showing mesial-distal **f** and buccal-lingual **g** views [315]

6.10 Photodynamic Diagnosis

Photodynamic therapy has manifold applications in dentistry, from invaluable and unparalleled use in implantology, periodontal disease, dental caries, endodontic infections, oral mucosal lesions, mucosal infections such as in candidiasis and also in diagnosis and treatment of oral neoplasias. Photodynamic therapy involves the administration of a photosensitizer (topically, intravenously, or locally), which is a spectrally adapted chromophore (mostly organic dyes), which means it can be activated by a light of a particular wavelength(s) of the electromagnetically spectrum only. It accumulates by having a selective uptake at areas of diseased or pathologic tissues (areas of rapid cell turn over). It then undergoes certain molecular or chemical transformations on absorption of the light (of a particular wavelength) and is excited from the ground state to an excited triplet state. The triplet state can interact directly with biomolecules producing free radicals or with molecular oxygen-producing reactive oxygen species like singlet oxygen, superoxide, and hydroxyl radical. These can prove to be cytotoxic by causing disruption of phospholipid molecules of cellular membranes and even DNA.

Migita et al. (2010) [316] demonstrated the use of Talaporfin sodium (mono-L-aspartyl chlorin e6/NPe6) in conjunction with a hyperspectral imaging system in detecting oral squamous cell carcinoma in 69 Sprague Dawley rats to which a carcinogen was administered. 5 mg/kg of Talaporfin sodium was given intravenously in 0.1 mL saline and the subjects were kept in the dark. Talaporfin is expected to have selective uptake and accumulate at areas of the squamous cell carcinoma. After the animals were sacrificed and the absorbance of 664 nm light by Talaporfin sodium in tissue was measured by hyperspectral imaging and compared to histological samples classifying the tissue as normal or neoplastic. The study proved the efficiency of Talaporfin sodium in photodynamic diagnosis in that it had higher absorbance signifying higher accumulation at areas of neoplastic tissue than normal tissue. The authors surmise that this is because there is a higher rate of intra nuclear cell division in the areas of squamous cell carcinoma, and this would engender higher uptake of the photosensitizer.

Chang et al. (2005) [317] evaluated the efficacy of Photofrin (Porfimer sodium) in photodynamic diagnosis in hamsters. Carcinoma was induced in cheek pouches and the photosensitizer was applied topically. They used Xenon lamp (380–420 nm) for excitation of the photosensitizer after 3 h. The digital images were analyzed using Photoshop 5.0 software under RGB gradient and gray scale modes. The biopsies from the cheek tissues were taken and studied histologically. They concluded that Photofrin indeed had clearly made identification of the neoplastic areas better with a high degree of sensitivity (93.27%) and specificity (97.17%). They believed that topical application was better than the intravenous administration of photosensitizers as there could be possibility of other areas being affected other than the tumor site. They stated that the fluorescence emitted from Photofrin is because of its interaction with endogenous protoporphyrin IX (PP IX) in the tumor areas.

6.11 Laser Doppler flowmetry

Laser Doppler Flowmetry (LDF), originally in development, termed as laser Doppler velocimetry, uses laser light (632, 780–820 nm) to measure pulpal blood flow and thus tooth vitality by measuring the backscatter (change in frequency or collimation) on the same principle of interferometry as used in OCT. Though the first reports of clinical use of this technique in ophthalmology came in the early 1970s [318], the adaptation of this technique to dentistry was only in the late 1980s [319, 320].

Laser Doppler Flowmetry has since then found a range of applications in dentistry, though chiefly in endodontics and pediatric dentistry, (in assessing the pulpal vitality of traumatized or avulsed teeth [321, 322] and age-related changes in pulp [323]) but also in other fields like in orthodontics (measuring pulpal blood flow during tooth movement and treatment such as Rapid Maxillary Expansion [324, 325] (RME)), in maxillofacial surgery (to confirm the vitality of teeth after orthognathic surgery [326] and revascularization of reimplanted teeth [327, 328]), and even in implantology (as an aid in evaluating implant stability [329, 330]).

A range of review articles describe the reliability, role, and value of this diagnostic measure [331]. Unlike other biomedical applications where LDF is limited or affected by artifacts produced by movement of the imaged structure (such as the heart), in dentistry the tooth, alveolar bone, or periodontal structures are static and the only movement is that of the red blood cells in the pulp and their mean velocity per second. Polat et al. 2005 [332] have shown that LDF could penetrate up to 6–13 mm (depending on the density) in teeth.

However, it does have limitations LDF may not be very effective in teeth with restorations, the signal being affected by adjacent tissues like periodontal tissues and requiring isolation of the teeth being examined by a rubber dam or cotton rolls. Variations between the anatomy and position of the tooth in the arch, the laser wavelength, frequency of bandwidth and filter, probe design, and fiber separation (250–500μm) all can affect the reliability of LDF. But studies have attributed 80–90% accuracy for pulp vitality assessments [333–336]. It is definitely a non-invasive method, but incumbent costs and initial costs have yet to make it a ubiquitous diagnostic methodology [337].

6.12 Conclusion

Lasers have virtually ushered in a newer, accurate, and non-invasive era in diagnosis. Laser-assisted diagnostic techniques have indeed shed light on the intricacies of disease processes and lent better modalities to treat them. In developing a theory for the mechanism of homeopathy, del Giudice [338], suggests that the electromagnetic information of a substance can be transduced into the surrounding water molecules (for instance, by affecting the field produced by large clusters of molecules). Light and its electromagnetic energy is similarly transferred to molecules such as fluorophores or photosensitizers.

Adey [339] elaborates on this notion by stating "Biological effects of oscillating environmental electric fields are related to the electric gradient which they introduce in the tissue [340]. This will be determined by the degree of coupling between the field and the tissues." Such coupling of energy (or loss of energy) can be detected by spectroscopic processes to the extent that they elucidate molecular mechanisms of functioning in cells and their biochemical structure. These techniques do give us the early start over disease processes by diagnosis of incipient lesions or neoplasia, yet translating research into painless, non-intrusive treatment methodologies still remains a challenge and is at a pioneering stage.

Laser-assisted diagnosis has indeed given us the winning edge over disease and infection, yet today's understanding is but at the verge and threshold of greater discoveries which need to be proved by future research work.

Acknowledgment The contributor gratefully acknowledges the support of Dr. Steven Parker, Professor (a.c) University of Genova, Italy, Dr. Donald Coluzzi, Professor, University of San Francisco at California, U.S.A., Dr. Stefano Benedicenti, Dean DiSC, University of Genova, Mirza Hasanuzzaman, Associate Professor, Sher-e-Bangla Agricultural University, Dhaka, Bangladesh, and Daniel Mathews Muruppel, Project Leader, Kuwait Airways Corporation, Kuwait, toward his work.

References

1. Stokes GG. On the change of refrangibility of light. Phil Trans R Soc Lond. 1852;142:463–562.
2. Raman CV. A new radiation. Indian J Phys. 1928;2:387–98.
3. Raman CV. A new radiation. Indian J Phys. 1928;2:399–419.
4. Letokhov VS. Laser-induced chemistry. Nature. 1983;305:103–8.

5. Blank M. Biological effects of electromagnetic fields. Bioelectrochem Bioenerg. 1993;32:203–10.
6. Lakhovsky G. The secret of life: electricity, radiation and your body. Costa Mesa: Noontide Press; 1992, 214 p.
7. Warnke U. Influence of light on cellular respiration. In: Popp F-A, et al., editors. Electromagnetic bio-information. Munchen: Urban & Schwarzenberg; 1989. p. 213–20.
8. Pohl RW. Optik und atomphysik. Springer Verlag; 1976.
9. Ditchburn RW. Light. New York: Dover Publications; 1961. p. 407–8.
10. Waynant RW. Lasers in medicine. Boca Raton: CRC Press; 2011.
11. Arnat A, et al. J Photochem Photobiol B Biol. 2006;82: 152–60.
12. Grant EH, Sheppard RJ, South GP. Dielectric behaviour of biological molecules in solution. Oxford: Clarendon Press; 1978, Chapter I.
13. Quaglino D, Capri M, Zecca L, Franceschi C, Ronchetti IP. The effect on rat thymocytes of the simultaneous in vivo exposure to 50-Hz electric and magnetic field and to continuous light. Sci World J. 2004;4(Suppl 2):91–9.
14. Niemz MH. Laser-tissue interactions fundamentals and applications. 3rd ed. Springer; 2007. p. 47–9.
15. Foote CS. Mechanisms of photosensitized oxidation. Science. 1968;162:963–70.
16. Lichtman JW, Conchello JW. Fluorescence microscopy. Nat Methods. 2005;2(12):910.
17. Koenig K, Schneckenburger H. Laser-induced autofluorescence for medical diagnosis. J Fluoresc. 1994;4(1):17.
18. Chance B, et al. Intracellular oxidation-reduction states in vivo. Science. 1962;137:499–508.
19. Chance B, Jobsis FF. Changes in fluorescence in a frog sartorius muscle following a twitch. Nature. 1959;184:195–6.
20. Mayevski A. Microcirculation in circulatory disorders. In: Microcirculatory and ionic responses to ischemia in the Mongolian Gerbil. Springer; 1998. p. 273–6.
21. Mayevsky A, Nioka S, Chance B. Fiber optic surface fluorometry/reflectometry and 31-p-NMR for monitoring the intracellular energy state in vivo. Adv Exp Med Biol. 1988;222:365–74.
22. Lohmann W, Paul E. In situ detection of melanomas by fluorescence measurements. Naturwissenschaften. 1988;75:201–2.
23. Lohmann W, Mussmann J, Lohmann C, Künzel W. Native fluorescence of the cervix uteri as a marker for dysplasia and invasive carcinoma. Eur J Obstet Gynecol Reprod Biol. 1989;31:249–53.
24. Lohmann W, Hirzinger B, Braun J, Schwemmle K, Muhrer K-H, Schulz A. Fluorescence studies on lung tumors. Z Naturforsch. 1990;45c:1063–6.
25. Ra H, et al. Detection of non-melanoma skin cancer by in vivo fluorescence imaging with fluorocoxib A. Neoplasia. 2015;17(2):201–7.
26. Pavlova I, Williams M, El-Naggar A, Richards-Kortum R, Gillenwater A. Understanding the biological basis of autofluorescence imaging for oral cancer detection: high-resolution fluorescence microscopy in viable tissue. Clin Cancer Res. 2008;14(8):2396–404. https://doi.org/10.1158/1078-0432.CCR-07-1609.
27. Arifler D, Pavlova I, Gillenwater A, Richards-Kortum R. Light scattering from collagen fiber networks: micro-optical properties of normal and neoplastic stroma. Biophys J. 2007;92:3260–74.
28. Pavlova I. Monte Carlo model to describe depth selective fluorescence spectra of epithelial tissue applications for diagnosis of oral precancer. J Biomed Opt. 2008;13(6):064012. https://doi.org/10.1117/1.3006066.
29. Pavlova I, Sokolov K, Drezek R, Malpica A, Follen M, Richards-Kortum R. Microanatomical and biochemical origins of normal and precancerous cervical autofluorescence using laser-scanning fluorescence confocal microscopy. Photochem Photobiol. 2003;77(5):550–5.
30. Roblyer D, Kurachi C, Stepanek V, Williams MD, El-Naggar AK, Lee JJ, Gillenwater AM, Richards-Kortum R. Objective detection and delineation of oral neoplasia using autofluorescence imaging. Cancer Prev Res (Phila). 2009;2(5):423–31. Published online 2009 Apr 28. https://doi.org/10.1158/1940-6207.CAPR-08-0229.
31. Shin D, Vigneswaran N, Gillenwater A, Richards-Kortum R. Advances in fluorescence imaging techniques to detect oral cancer and its precursors. Future Oncol. 2010;6(7):1143–54. https://doi.org/10.2217/fon.10.79.
32. Lane PM, Gilhuly T, Whitehead P, et al. Simple device for the direct visualization of oral-cavity tissue fluorescence. J Biomed Opt. 2006;11(2):024006.
33. Poh CF, Ng SP, Williams PM, et al. Direct fluorescence visualization of clinically occult high-risk oral premalignant disease using a simple hand-held device. Head Neck. 2007;29(1):71–6.
34. Poh CF, MacAulay CE, Zhang L, Rosin MP. Tracing the "at-risk" oral mucosa field with autofluorescence: steps toward clinical impact. Cancer Prev Res. 2009;2(5):401–4.
35. McNamara KK, Martin BD, Evans EW, Kalmar JR. The role of direct visual fluorescent examination (VELscope) in routine screening for potentially malignant oral mucosal lesions. Oral Surg Oral Med Oral Pathol Oral Radiol. 2012;114:636–43.
36. Balasubramaniam AM, Sriraman R, Sindhuja P, Mohideen K, Parameswar RA, Haris KTM. Autofluorescence based diagnostic techniques for oral cancer. J Pharm Bioallied Sci. 2015;7(Suppl 2):S374–7. https://doi.org/10.4103/0975-7406.163456.
37. Yuanlong Y, Yanming Y, Fuming L, Yufen L, Paozhong M. Characteristic autofluorescence for cancer diagnosis and its origin. Lasers Surg Med. 1987;7(6):528–32.
38. Koenig K, Hemmer J, Schneckenburger H. In: Spinelli P, DalFante M, Marchesini R, editors. Photodynamic therapy and biomedical lasers. Amsterdam: Elsevier; 1992. p. 903–6.
39. Huang D, Swanson EA, Lin CP, Schuman JS, Stinson WG, Chang W, Hee MR, Flotte T, Gregory K, Puliafito CA, et al. Optical coherence tomography. Science. 1991;254(5035):1178–81.
40. Wessels R, De Bruin DM, Faber DJ, Van Leeuwen TG, Van Beurden M, Ruers TJM. Optical biopsy of epithelial cancers by optical coherence tomography (OCT). Lasers Med Sci. 2014;29:1297–305.
41. Strasswimmer J, Pierce MC, Park BH, Neel V, de Boer JF. Polarization-sensitive optical coherence tomography of invasive basal cell carcinoma. J Biomed Opt. 2004;9(2):292–8.
42. Olmedo JM, Warschaw KE, Schmitt JM, Swanson DL. Optical coherence tomography for the characterization of basal cell carcinoma in vivo: a pilot study. J Am Acad Dermatol. 2006;55(3):408–12.
43. Gambichler T, Orlikov A, Vasa R, Moussa G, Hoffmann K, Stucker M, Altmeyer P, Bechara FG. In vivo optical coherence tomography of basal cell carcinoma. J Dermatol Sci. 2007;45(3):167–73.
44. Mogensen M, Nurnberg BM, Forman JL, Thomsen JB, Thrane L, Jemec GB. In vivo thickness measurement of basal cell carcinoma and actinic keratosis with optical coherence tomography and 20-MHz ultrasound. Br J Dermatol. 2009;160(5):1026–33.

45. Mogensen M, Joergensen TM, Nurnberg BM, Morsy HA, Thomsen JB, Thrane L, Jemec GB. Assessment of optical coherence tomography imaging in the diagnosis of non-melanoma skin cancer and benign lesions versus normal skin: observer-blinded evaluation by dermatologists and pathologists. Dermatol Surg. 2009;35(6):965–72.
46. de Giorgi SM, Massi D, Mavilia L, Cappugi P, Carli P. Possible histopathologic correlates of dermoscopic features in pigmented melanocytic lesions identified by means of optical coherence tomography. Exp Dermatol. 2005;14(1):56–9.
47. Gambichler T, Regeniter P, Bechara FG, Orlikov A, Vasa R, Moussa G, Stucker M, Altmeyer P, Hoffmann K. Characterization of benign and malignant melanocytic skin lesions using optical coherence tomography in vivo. J Am Acad Dermatol. 2007;57(4):629–37.
48. Adegun OK, Tomlins PH, Hagi-Pavli E, Mckenzie G, Piper K, Bader DL, Fortune F. Quantitative analysis of optical coherence tomography and histopathology images of normal and dysplastic oral mucosal tissues. Lasers Med Sci. 2012;27:795–804. https://doi.org/10.1007/s10103-011-0975-1.
49. Adegun OK, Tomlins PH, Hagi-Pavli E, Mckenzie G, Piper K, Bader DL, Fortune F. Quantitative optical coherence tomography of fluid-filled oral mucosal lesions. Lasers Med Sci. 2013;28:1249–55. https://doi.org/10.1007/s10103-012-1208-y.
50. Masterrs BR, So PTC, Gratton E. Optical biopsy of in vivo human skin : multiphoton excitation microscopy. Lasers Med Sci. 1998;13:196–203.
51. Skoog DA, Holler FJ, Nieman TA. Principles of instrumental analysis. Saunders College Pub.; 1998.
52. Helmenstine AM. Analytical chemistry definition. http://chemistry.about.com/od/chemistryglossary/a/analyticaldef.htm. Accessed Feb 2023.
53. Kumar A, Yueh FY, Singh JP, Burgess S. Characterization of malignant tissue cells by laser-induced breakdown spectroscopy. Appl Opt. 2004;43:5399–403.
54. Kanawade R, Mahari F, Klampfl F, Rohde M, Knipfer C, Tangermann-Gerk K, Adler W, Schmidt M, Stelzle F. Qualitative tissue differentiation by analysing the intensity ratios of atomic emission lines using laser induced breakdown spectroscopy (LIBS): prospects for a feedback mechanism for surgical laser systems. J Biophotonics. 2015;8(1–2):153–61.
55. Kramida A, Ralchenko Y, Reader J. NIST atomic spectra database (ver. 5.0). Gaithersburg, MD: National Institute of Standards and Technology; 2012. http://physics.nist.gov/asd.
56. Beier BD, Quivey RG, Berger AJ. Raman microspectroscopy for species identification and mapping within bacterial biofilms. AMB Express. 2012;2:35.
57. Berger AJ, Zhu Q, Quivey RG. Raman spectroscopy of oral bacteria. In: European conference on biomedical optics, Munich, Germany, 22 Jun 2003.
58. Zhu Q, Quivey RG Jr, Berger AJ. Measurement of bacterial concentration fractions in polymicrobial mixtures by Raman microspectroscopy. J Biomed Optic. 2004;9(6):1182–6.
59. Zhu Q, Quivey RG Jr, Berger AJ. Raman spectroscopic measurement of relative concentrations in mixtures of oral bacteria. Appl Spectrosc. 2007;61:1233–7.
60. Maquelin K, Choo-Smith L-P, Endtz HP, Bruining HA, Puppels GJ. Rapid identification of Candida species by confocal Raman microspectroscopy. J Clin Microbiol. 2002;40(2):594–600.
61. Puppels GJ, Demul FFM, Otto C, Greve J, Robertnicoud M, et al. Studying single living cells and chromosomes by confocal Raman microspectroscopy. Nature V. 1990;347:301–3.
62. Puppels GJ, Colier W, Olminkhof JHF, Otto C, Demul FFM, et al. Description and performance of a highly sensitive confocal Raman microspectrometer. J Raman Spectrosc V. 1991;22:217–25.
63. Puppels GJ, Olminkhof JHF, Segersnolten GMJ, Otto C, Demul FM, et al. Laser irradiation and Raman spectroscopy of single living cells and chromosomes—sample degradation occurs with 514.5nm but not with 660nm laser light. Exp Cell Res. 1991;V195:361–7.
64. Nelson WH, Manoharan R, Sperry JF. Appl Spectrosc Rev. 1992;27:67–124.
65. Manoharan R, Ghiamati E, Britton KA, Nelson WH, Sperry JF. Appl Spectrosc. 1991;45:307–11.
66. Ghiamati E, Manoharan R, Nelson WH, Sperry JF. Appl Spectrosc. 1992;46:357–64.
67. Manoharan R, Ghiamati E, Sperry JF, Nelson WH. Abstr Pap Am Chem Soc. 1990;200:138.
68. Chan JW, Taylor DS, Zwerdling T, Lane SM, Ihara K, et al. Micro-Raman spectroscopy detects individual neoplastic and normal hematopoietic cells. Biophys J. 2006;90:648–56.
69. Chan JW, Taylor DS, Lane S, Zwerdling T, Tuscano J, et al. Non-destructive identification of individual leukemia cells by laser tweezers Raman spectroscopy. Anal Chem. 2008;80:2180–7.
70. Ashkin A, Dziedzic JM. Optical trapping and manipulation of viruses and bacteria. Science. 1987;235:1517–20.
71. Padgett MJ, Molloy J, McGloin D. Optical tweezers: methods and applications. CRC Press; 2010. p. 1–3.
72. Chan JW, Esposito AP, Talley CE, Hollars CW, Lane SM, Huser T. Reagentless identification of single bacterial spores in aqueous solution by confocal laser tweezers Raman spectroscopy. Anal Chem. 2004;76:599–603.
73. Chan JW, Esposito AP, Talley CE, Hollars CW, Lane SM, Huser T. Nondestructive identification of individual leukemia cells by laser trapping Raman spectroscopy. Anal Chem. 2008;80:2180–7.
74. Huang SS, Chen D, Pelczar PL, Vepachedu VR, Setlow P, Li YQ. Levels of Ca2-dipicolinic acid in individual bacillus spores determined using microfluidic Raman tweezers. J Bacteriol. 2007;189(13):4681–7.
75. Paidhungat M, Setlow B, Driks A, Setlow P. Characterization of spores of Bacillus subtilis which lack dipicolinic acid. J Bacteriol. 2000;182:5505–12.
76. Setlow B, Atluri S, Kitchel R, Koziol-Dube K, Setlow P. Role of dipicolinic acid in resistance and stability of spores of Bacillus subtilis with or without DNA-protective α/β type small acid-soluble proteins. J Bacteriol. 2006;188:3740–7.
77. Ellis DI, Goodacre R. Metabolic fingerprinting in disease diagnosis: biomedical applications of infrared and Raman spectroscopy. Analyst. 2006;131:875–85.
78. Mahadevan-Jansen A, Patil C, Pence I. Raman spectroscopy: from benchtop to bedside. In: Vo-Dinh T, editor. Biomedical photonics handbook. 2nd ed. Boca Raton, FL: CRC Press; 2014. p. 759–802.
79. Nguyen JQ, Gowani ZS, O'Connor M, Pence IJ, Nguyen T-Q, Holt GE, Schwartz HS, Halpern JL, Mahadevan-Jansen A. Intraoperative Raman spectroscopy of soft tissue sarcomas. Lasers Surg Med. 2016 https://doi.org/10.1002/lsm.22564.
80. Escoriza MF, Vanbriesen JM, Stewart S, Maier J. Appl Spectrosc. 2000;61:8.
81. Nan XL, Tonary AM, Stolow A, Xie XS, Pezacki JP. Intracellular imaging of HCV RNA and cellular lipids by using simultaneous two-photon fluorescence and coherent anti-stokes Raman scattering microscopies. Chembiochem. 2006;7:1895–7.

82. Buschman HP, Marple ET, Wach ML, Bennett B, Schut TCB, et al. In vivo determination of the molecular composition of artery wall by intravascular Raman spectroscopy. Anal Chem. 2000;72:3771–5.
83. Mostaco-Guidolin LB, Sowa MG, Ridsdale A, Pegoraro AF, Smith MS, et al. Differentiating atherosclerotic plaque burden in arterial tissues using femtosecond CARS-based multimodal nonlinear optical imaging. Biomed Opt Express. 2010;1:59–73.
84. Motz JT, Fitzmaurice M, Miller A, Gandhi SJ, Haka AS, et al. In vivo Raman spectral pathology of human atherosclerosis and vulnerable plaque. J Biomed Opt. 2006;11:021003.
85. Zhang X, Yonzon CR, Van Duyne RP. Proc SPIE. 2003;5221:82–91.
86. White DJ. The application of in vitro models to research on demineralization and remineralization of the teeth. Adv Dent Res. 1995;9:175–93.
87. Young DA, Featherstone JDB. Digital imaging fiber-optic transillumination, F-speed radiographic film and depth of approximal lesions. J Am Dent Assoc. 2005;136(12):1682–7.
88. Bin-Shuwaish M, Yaman P, Dennison J, Neiva G. The correlation of DIFOTI to clinical and radiographic images in Class II carious lesions. J Am Dent Assoc. 2008;139(10):1374–81.
89. Schneiderman A, Elbaum M, Shultz T, Keem S, Greenebaum M, Driller J. Assessment of dental caries with Digital Imaging Fiber-Optic Trans Illumination (DIFOTI): in vitro study. Caries Res. 1997;31(2):103–10.
90. Gomez J. Detection and diagnosis of the early caries lesion. BMC Oral Health. 2015;15(Suppl 1):S3.
91. Tam LE, McComb D. Diagnosis of occlusal caries: part II. Recent diagnostic technologies. J Can Dent Assoc. 2001;67(8):459–63.
92. Fejerskov O, Kidd E. Dental caries: the disease and its clinical management. 2nd ed. Wiley-Blackwell; 2008.
93. Longbottom C, Huysmans MCDNJM. Electrical measurements for use in caries trials. J Dent Res. 2004;83:C76–9.
94. Ashley PF, Blinkhorn AS, Davies RM. Occlusal caries diagnosis: an in vitro histological validation of the Electronic Caries Monitor (ECM) and other methods. J Dent. 1998;26:83–8.
95. Guimerà A, Calderón E, Los P, Christie AM. Method and device for bio-impedance measurement with hard-tissue applications. Physiol Meas. 2008;29:S279–90.
96. Jablonski-Momeni A, Heinzel-Gutenbrunner M, Haak R, et al. Use of AC impedance spectroscopy for monitoring sound teeth and incipient carious lesions. Clin Oral Invest. 2017;21:2421–7. https://doi.org/10.1007/s00784-016-2038-2.
97. Jablonski-Momeni A, Kneib L. Assessment of caries activity using the CALCIVIS caries activity imaging system. Open Access J Sci Technol. 2016; https://doi.org/10.11131/2016/101241.
98. Longbottom C, Vernon B, Perfect E, Haughey AM, Christie A, Pitts N. Initial investigations of a novel bioluminescence method for imaging dental demineralization. Clin Exp Dent Res. 2021;7(5):786–94. https://doi.org/10.1002/cre2.402. Epub 2021 Jan 28.
99. Pitts N, Shanks N, Longbottom C, Willins M, Vernon B. Clinical validation of a novel bioluminescence imaging technology for aiding the assessment of carious lesion activity status. Clin Exp Dent Res. 2021;7(5):772–85. https://doi.org/10.1002/cre2.400.
100. Bjelkhagen H, Sundström F. A clinically applicable laser luminescence method for the early detection of dental caries. IEEE J Quant Electron. 1981;17(12):2580–2.
101. de Josselin de Jong E, Sundström F, Westerling H, Tranæus S, ten Bosch JJ, Angmar-Månsson B. A new method for in vivo quantification of changes in initial enamel caries with laser fluorescence. Caries Res. 1995;29(1):2–7.
102. Karlsson L, Tranæus S. Supplementary methods for detection and quantification of dental caries. J Laser Dent. 2008;16(1):6–14.
103. van der Veen MH, Thomas RZ, Huysmans MC, de Soet JJ. Red autofluorescence of dental plaque bacteria. Caries Res. 2006;40(6):542–5.
104. Banerjee A, Boyde A. Autofluorescence and mineral content of carious dentine: scanning optical and backscattered electron microscopic studies. Caries Res. 1998;32(3):219–26.
105. Rechmann P, Liou SW, Rechmann BMT, Featherstone JDB. Soprocare—450 nm wavelength detection tool for microbial plaque and gingival inflammation—a clinical study. In: Rechmann P, Fried D, editors. Lasers in dentistry XX, Proc SPIE, vol. 8929; 2014. p. 892906. https://doi.org/10.1117/12.2047275.
106. Drancourt N, Roger-Leroi V, Pereira B, et al. Validity of Soprolife camera and Calcivis device in caries lesion activity assessment. Br Dent J. 2020; https://doi.org/10.1038/s41415-020-2316-x.
107. Akarsu S, Köprülü H. In vivo comparison of the efficacy of DIAGNOdent by visual inspection and radiographic diagnostic techniques in the diagnosis of occlusal caries. J Clin Dent. 2006;17(3):53–8.
108. Barbería E, Maroto M, Arenas M, Silva CC. A clinical study of caries diagnosis with a laser fluorescence system. JADA. 2008;139(5):572–9.
109. Lussi A, Megert B, Longbottom C, Reich E, Francescut P. Clinical performance of a laser fluorescence device for detection of occlusal caries lesions. Eur J Oral Sci. 2001;109(1):14–9.
110. Konig K, Flemming G, Hibst R. Laser-induced autofluorescence spectroscopy of dental caries. Cell Mol Biol. 1999;44(8):1293–300.
111. Sürme K, Kara NB, Yilmaz Y. In vitro evaluation of occlusal caries detection methods in primary and permanent teeth: a comparison of CarieScan PRO, DIAGNOdent Pen, and DIAGNOcam methods. Photobiomodul Photomed Laser Surg. 2020;38(2):105–11. https://doi.org/10.1089/photob.2019.4686. Epub 2019 Oct 7.
112. Iranzo-Cortés JE, Montiel-Company JM, Almerich-Torres T, Bellot-Arcís C, Almerich-Silla JM. Use of DIAGNOdent and VistaProof in diagnostic of pre-cavitated caries lesions-a systematic review and meta-analysis. J Clin Med. 2019;9(1):20. https://doi.org/10.3390/jcm9010020.
113. Mitchell C, Zaku H, Milgrom P, et al. The accuracy of laser fluorescence (DIAGNOdent) in assessing caries lesion activity on root surfaces, around crown margins, and in furcations in older adults. BDJ Open. 2021;7:14. https://doi.org/10.1038/s41405-021-00069-2.
114. Beauchamp J, et al. Evidence-based clinical recommendations for the use of pit-and-fissure sealants: a report of the American Dental Association Council on Scientific Affairs. JADA. 2008;139:257. http://jada.ada.org.
115. Makowski AJ, Patil CA, Mahadevan-Jansen A, Nyman JS. Polarization control of Raman spectroscopy optimizes the assessment of bone tissue. J Biomed Opt. 2013;18(5):055005.
116. Ko AC-T, Choo-Smith L-P, Hewko M, Leonardi L, Sowa MG. Ex vivo detection and characterization of early dental caries by optical coherence tomography and Raman spectroscopy. J Biomed Opt. 2005;10(3):031118.
117. de Carvalho FB, Barbosa AFS, Zanin FAA, Júnior AB, Júnior LS, Pinheiro ALB. Use of laser fluorescence in dental caries diagnosis: a fluorescence x bio molecular vibrational

spectroscopic comparative study. Braz Dent J. 2013;24(1):59–63.
118. Coello B, López-Álvarez M, Rodríguez-Domínguez M, Serra J, González P. Quantitative evaluation of the mineralization level of dental tissues by Raman spectroscopy. Biomed Phys Eng Express. 2015;1:045204.
119. Boskey AL, Mendelsohn R. Infrared spectroscopic characterization of mineralized tissues. Vib Spectrosc. 2005;38(1–2):107–14.
120. Xu C, Wang Y. Chemical composition and structure of peritubular and intertubular human dentine revisited. Arch Oral Biol. 2012;57:383–91.
121. Feldchtein FI, Gelikonov GV, Gelikonov VM, Iksanov RR, Kuranov RV, Sergeev AM, Gladkova ND, Ourutina MN, Warren JA Jr, Reitze DH. In vivo OCT imaging of hard and soft tissue of the oral cavity. Opt Express. 1998;3:239–50.
122. Fried D, Glena RE, Featherstone JD, Seka W. Nature of light scattering in dental enamel and dentin at visible and near-infrared wavelengths. Appl Opt. 1995;34(7):1278–85.
123. Fried D, Featherstone JDB, Glena RE, Seka W. The nature of light scattering in dental enamel and dentin at visible and near-IR wavelengths. Appl Opt. 1995;34(7):1278–85.
124. Hale GM, Querry MR. Optical constants of water in the 200-nm to 200-μm wavelength region. Appl Opt. 1973;12:555–63.
125. Jones RS, Fried D. Attenuation of 1310-nm and 1550-nm laser light through sound dental enamel. In: Lasers in dentistry VIII, Proc SPIE, vol. 4610. Bellingham, WA: SPIE; 2002. p. 187–90.
126. Chan KH, Fried D. Multispectral cross-polarization reflectance measurements suggest high contrast of demineralization on tooth surfaces at wavelengths beyond 1300-nm due to reduced light scattering in sound enamel. J Biomed Opt. 2018;23(6):060501.
127. Darling CL, Huynh GD, Fried D. Light scattering properties of natural and artificially demineralized dental enamel at 1310-nm. J Biomed Opt. 2006;11(3):034023.
128. Buhler C, Ngaotheppitak P, Fried D. Imaging of occlusal dental caries (decay) with near-IR light at 1310-nm. Opt Express. 2005;13(2):573–82.
129. Zakian C, Pretty I, Ellwood R. Near-infrared hyperspectral imaging of teeth for dental caries detection. J Biomed Opt. 2009;14(6):064047.
130. Ng C, Almaz EC, Simon JC, Fried D, Darling CL. Near-infrared imaging of demineralization on the occlusal surfaces of teeth without the interference of stains. J Biomed Opt. 2019;24(3):036002.
131. Chung S, Fried D, Staninec M, Darling CL. Multispectral near-IR reflectance and transillumination imaging of teeth. Biomed Opt Express. 2011;2(10):2804–14.
132. Staninec M, Lee C, Darling CL, Fried D. In vivo near-IR imaging of approximal dental decay at 1,310 nm. Lasers Surg Med. 2010;42(4):292–8.
133. Peers A, Hill FJ, Mitropoulos CM, Holloway PJ. Validity and reproducibility of clinical examination, fibre-optic transillumination, and bite-wing radiology for the diagnosis of small approximal carious lesions. Caries Res. 1993;27:307–11.
134. Pine CM, ten Bosch JJ. Dynamics of and diagnostic methods for detecting small carious lesions. Caries Res. 1996;30(6):381–8.
135. Purdell-Lewis DJ, Pot T. A comparison of radiographic and fibre-optic diagnoses of approximal caries lesions. J Dent. 1974;2(4):143–8.
136. Vaarkamp J, ten Bosch JJ, Verdonschot EH, Bronkhoorst EM. The real performance of bitewing radiography and fiber-optic transillumination in approximal caries diagnosis. J Dent Res. 2000;79(10):1747–51.
137. Stephen KW, Russell JI, Creanor SL, Burchell CK. Comparison of fibre optic transillumination with clinical and radiographic caries diagnosis. Community Dent Oral Epidemiol. 1987;15(2):90–4.
138. Simon JC, Lucas SA, Lee RC, Staninec M, Tom H, Chan KH, Darling CL, Fried D. Near-IR transillumination and reflectance imaging at 1300-nm and 1500-1700-nm for in vivo caries detection. Lasers Surg Med. 2016;48(6):828–36.
139. Simon JC, Kang H, Staninec M, Jang AT, Chan KH, Darling CL, Lee RC, Fried D. Near-IR and CP-OCT imaging of suspected occlusal caries lesions. Lasers Surg Med. 2017;49(3):215–24.
140. Zhu Y, Abdelaziz M, Simon J, Le O, Fried D. Dual short wavelength infrared transillumination/reflectance mode imaging for caries detection. J Biomed Opt. 2021;26(4):043004.
141. Jones G, Jones RS, Fried D. Transillumination of interproximal caries lesions with 830-nm light. In: Rechmann P, editor. Lasers in dentistry X, Proc SPIE, vol. 5313. Bellingham, WA: SPIE; 2004. p. 17–22.
142. Kuhnisch J, Sochtig F, Pitchika V, Laubender R, Neuhaus KW, Lussi A, Hickel R. In vivo validation of near-infrared light transillumination for interproximal dentin caries detection. Clin Oral Investig. 2015;20(4):821–9.
143. Sochtig F, Hickel R, Kuhnisch J. Caries detection and diagnostics with near-infrared light transillumination: clinical experiences. Quintessence Int. 2014;45(6):531–8.
144. Jablonski-Momeni A, Jablonski B, Lippe N. Clinical performance of the near-infrared imaging system VistaCam iX Proxi for detection of approximal enamel lesions. BDJ Open. 2017;3:17012.
145. Dye BA, Tan S, Lewis BG, Barker LK, Thornton-Evans TG, Eke PI, Beltrán-Aguilar ED, Horowitz AM, Li CH. Trends in oral health status, United States, 1988–1994 and 1999–2004. Vital Health Stat 11. 2007;248:1–92.
146. Dye BA, Thornton-Evans T, Li X, Iafolla TJ. Dental caries and tooth loss in adults in the United States, 2011–2012. In: National Center for Health Statistics, editor. NCHS data brief, #197. National Center for Health Statistics; 2015.
147. Yang VB, Curtis DA, Fried D. Cross-polarization reflectance imaging of root caries and dental calculus at wavelengths from 400-2350-nm. J Biophotonics. 2018;11:e201800113.
148. Krause F, Braun A, Brede O, Eberhard J, Frentzen M, Jepsen S. Evaluation of selective calculus removal by a fluorescence feedback-controlled Er:YAG laser in vitro. J Clin Periodontol. 2007;34(1):66–71.
149. Rechmann P. Dental laser research: selective ablation of caries, calculus, and microbial plaque: from the idea to the first in vivo investigation. Dent Clin N Am. 2004;48(4):1077–1104, ix.
150. Schoenly JE, Seka W, Rechmann P. Investigation into the optimum beam shape and fluence for selective ablation of dental calculus at lambda = 400 nm. Lasers Surg Med. 2010;42(1):51–61.
151. Chan KH, Fried D. Selective ablation of dental caries using coaxial CO2 (9.3-μm) and near-IR (1880-nm) lasers. Lasers Surg Med. 2019;51:176–84.
152. Fried WA, Chan KH, Darling CL, Curtis DA, Fried D. Image-guided ablation of dental calculus from root surfaces using a DPSS Er:YAG laser. Lasers Surg Med. 2020;52(3):247–58.
153. Mjor I, Moorehead J, Dahl J. Reasons for replacement of restorations in permanent teeth in general dental practice. Int Dent J. 2000;50:361–6.

154. Li X, King TA. Microstructure and optical properties of PMMA/gel silica glass composites. J Sol-Gel Sci Technol. 1995;4:75–82.
155. Stansbury JW, Dickens SH. Determination of double bond conversion in dental resins by near infrared spectroscopy. Dent Mater. 2001;17:71–9.
156. Logan CM, Co KU, Fried WA, Simon JC, Staninec M, Fried D, Darling CL. Multispectral near-infrared imaging of composite restorations in extracted teeth. In: Lasers in dentistry XX, Proc SPIE, vol. 8929. Bellingham, WA: SPIE; 2014. p. 89290R.
157. Fried WA, Simon JC, Darling CL, Le O, Fried D. High-contrast reflectance imaging of composite restorations color-matched to tooth structure at 1000–2300-nm. In: Lasers in dentistry XXIII, Proc SPIE, vol. 10044. Bellingham, WA: SPIE; 2017. p. 100440J.
158. Alexander R, Fried D. Selective removal of orthodontic composite using 355-nm Q-switched laser pulses. Lasers Surg Med. 2001;30:240–5.
159. Fried WA, Chan KH, Darling CL, Fried D. Use of a DPSS Er:YAG laser for the selective removal of composite from tooth surfaces. Biomed Opt Express. 2016;9(10):5026–36.
160. Kahler W. The cracked tooth conundrum: terminology, classification, diagnosis, and management. Am J Dent. 2008;21(5):275–82.
161. Culjat MO, Singh RS, Brown ER, Neurgaonkar RR, Yoon DC, White SN. Ultrasound crack detection in a simulated human tooth. Dentomaxillofac Radiol. 2005;34(2):80–5.
162. Fried WA, Simon JC, Lucas S, Chan KH, Darling CL, Staninec M, Fried D. Near-IR imaging of cracks in teeth. In: Lasers in dentistry XX, Proc SPIE, vol. 8929. Bellingham, WA: SPIE; 2014. p. 89290Q.
163. Darling CL, Fried D. Real-time near IR (1310 nm) imaging of CO2 laser ablation of enamel. Opt Express. 2008;16(4):2685–93.
164. Maung LH, Lee C, Fried D. Near-IR imaging of thermal changes in enamel during laser ablation. In: Lasers in dentistry XVI, Proc SPIE, vol. 7549. Bellingham, WA: SPIE; 2010. p. 754902.
165. Sapra A, Darbar A, George R. Laser-assisted diagnosis of symptomatic cracks in teeth with cracked tooth: a 4-year in-vivo follow-up study. Aust Endod J. 2020;46(2):197–203. https://doi.org/10.1111/aej.12391. Epub 2019 Dec 9.
166. Stookey GK. Quantitative light fluorescence: a technology for early monitoring of the caries process. Dent Clin N Am. 2005;49(4):753–70.
167. Ando M, Stookey GK, Zero DT. Ability of quantitative light-induced fluorescence (QLF) to assess the activity of white spot lesions during dehydration. Am J Dent. 2006;19(1):15–8.
168. Ando M, Ferreira-Zandona AG, Eckert GJ, Zero DT, Stookey GK. Pilot clinical study to assess caries lesion activity using quantitative light-induced fluorescence during dehydration. J Biomed Opt. 2017;22(3):35005.
169. Lee RC, Darling CL, Fried D. Assessment of remineralization via measurement of dehydration rates with thermal and near-IR reflectance imaging. J Dent. 2015;43:1032–42.
170. Lee RC, Darling CL, Fried D. Activity assessment of root caries lesions with thermal and near-infrared imaging methods. J Biophotonics. 2016;10(3):433–45.
171. Lee RC, Kang H, Darling CL, Fried D. Automated assessment of the remineralization of artificial enamel lesions with polarization-sensitive optical coherence tomography. Biomed Opt Express. 2014;5(9):2950–62.
172. Yang V, Zhu Y, Curtis D, Le O, Chang N, Fried W, Simon JC, Banan P, Darling C, Fried D. Thermal imaging of root caries in vivo. J Dent Res. 2020;99(13):1502–8.
173. Chang NN, Jew JM, Fried D. Lesion dehydration rate changes with the surface layer thickness during enamel remineralization. In: Lasers in dentistry XXIV, Proc SPIE, vol. 10473. Bellingham, WA: SPIE; 2018. p. 104730D.
174. Fried WA, Abdelaziz M, Darling CL, Fried D. High contrast reflectance imaging of enamel demineralization and remineralization at 1950-nm for the assessment of lesion activity. Lasers Surg Med. 2021;53(7):968–77.
175. Tressel J, Abdelaziz M, Fried D. Dynamic SWIR imaging near the 1950 nm water absorption band for caries lesion diagnosis. J Biomed Opt. 2021;26(5):056006.
176. Shimada Y, Sadr A, Sumi Y, Tagami J. Application of optical coherence tomography (OCT) for diagnosis of caries, cracks, and defects of restorations. Curr Oral Health Rep. 2015;2(2):73–80.
177. Hitzenberger CK, Gotzinger E, Sticker M, Pircher M, Fercher AF. Measurement and imaging of birefringence and optic axis orientation by phase resolved polarization sensitive optical coherence tomography. Opt Express. 2001;9:780–90.
178. Fried D, Xie J, Shafi S, Featherstone JDB, Breunig TM, Lee C. Imaging carious lesions and lesion progression with polarization sensitive optical coherence tomography. J Biomed Opt. 2002;7:618–27.
179. Wang XJ, Milner TE, de Boer JF, Zhang Y, Pashley DH, Nelson JS. Characterization of dentin and enamel by use of optical coherence tomography. Appl Opt. 1999;38(10):2092.
180. Amaechi BT, Higham SM, Podoleanu AG, Rogers JA, Jackson DA. Use of optical coherence tomography for assessment of dental caries: quantitative procedure. J Oral Rehabil. 2001;28:1092–3.
181. Gossage KW, Tkaczyk TS, Rodriguez JJ, Barton JK. Texture analysis of optical coherence tomography images: feasibility for tissue classification. J Biomed Opt. 2003;8:570–5. https://doi.org/10.1117/1.1577575.
182. Mandurah MM, Sadr A, Bakhsh TA, Shimada Y, Sumi Y, Tagami J. Characterization of transparent dentin in attrited teeth using optical coherence tomography. Lasers Med Sci. 2015;30(4):1189–96. https://doi.org/10.1007/s10103-014-1541-4.
183. Shimada Y, Sadr A, Burrow MF, Tagami J, Ozawa N, Sumi Y. Validation of swept-source optical coherence tomography (SS-OCT) for the diagnosis of occlusal caries. J Dent. 2010;38(8):655–65. https://doi.org/10.1016/j.jdent.2010.05.004. Epub 2010 May 12.
184. Makishi P, Shimada Y, Sadr A, Tagami J, Sumi Y. Non-destructive 3D imaging of composite restorations using optical coherence tomography: marginal adaptation of self-etch adhesives. J Dent. 2011;39(4):316–25. https://doi.org/10.1016/j.jdent.2011.01.011.
185. Bakhsh TA, Sadr A, Shimada Y, Tagami J, Sumi Y. Non-invasive quantification of resin-dentin interfacial gaps using optical coherence tomography: validation against confocal microscopy. Dent Mater. 2011;27(9):915–25.
186. Todea C, Balabuc C, Sinescu C, et al. En face optical coherence tomography investigation of apical microleakage after laser assisted endodontic treatment. Lasers Med Sci. 2010;25:629. https://doi.org/10.1007/s10103-009-0680-5.
187. Sordillo LA, Pu Y, Pratavieira S, Budansky Y, Alfano RR. Deep optical imaging of tissue using the second and third near-infrared spectral windows. J Biomed Opt. 2014;19(5):056004.

188. Weber JR, Baribeau F, Grenier P, Emond F, Dubois S, Duchesne F, Girard M, Pope T, Gallant P, Mermut O, Moghadam HG. Towards a bimodal proximity sensor for in situ neurovascular bundle detection during dental implant surgery. Biomed Opt Express. 2013;5(1):16–30.
189. Bouma BE, Tearney GJ. Handbook of optical coherence tomography. New York: Marcel Dekker; 2002.
190. Drexler W, Fujimoto JG, editors. Optical coherence tomography technology and applications. New York: Springer; 2008.
191. Otis LL, Everett MJ, Sathyam US, Colston BW Jr. Optical coherence tomography: a new imaging technology for dentistry. J Am Dent Assoc. 2000;131(4):511–4.
192. Fried D, Featherstone JDB, Darling CL, Jones RS, Ngaotheppitak P, Buehler CM. Early caries imaging and monitoring with near-IR light. Dent Clin North Am. 2005;49(4):771–94.
193. Fried D. Ch. 68: Dentistry: diagnostics and spectroscopy. In: Popp J, editor. Handbook of biophotonics, Photonics for health care, vol. 2. New York: Wiley; 2012.
194. Katkar RA, Tadinada SA, Amaechi BT, Fried D. Optical coherence tomography. Dent Clin N Am. 2018;62(3):421–34.
195. Colston B, Everett M, Da Silva L, Otis L, Stroeve P, Nathel H. Imaging of hard and soft tissue structure in the oral cavity by optical coherence tomography. Appl Opt. 1998;37(19):3582–5.
196. Colston BW, Sathyam US, DaSilva LB, Everett MJ, Stroeve P. Dental OCT. Opt Express. 1998;3(3):230–8.
197. Feldchtein FI, Gelikonov GV, Gelikonov VM, Iksanov RR, Kuranov RV, Sergeev AM, Gladkova ND, Ourutina MN, Warren JA, Reitze DH. In vivo OCT imaging of hard and soft tissue of the oral cavity. Opt Express. 1998;3(3):239–51.
198. Zuluaga AF, Yang V, Jabbour J, Ford T, Kemp N, Fried D. Real-time visualization of hidden occlusal and approximal lesions with an OCT dental handpiece. In: Lasers in dentistry XXV, Proc SPIE, vol. 10857. Bellingham, WA: SPIE; 2019. p. 108570E.
199. Baumgartner A, Hitzenberger CK, Dicht S, Sattmann H, Moritz A, Sperr W, Fercher AF. Optical coherence tomography for dental structures. In: Lasers in dentistry IV, Proc SPIE, vol. 3248. Bellingham, WA: SPIE; 1998. p. 130–6.
200. Baumgartner A, Dicht S, Hitzenberger CK, Sattmann H, Robi B, Moritz A, Sperr W, Fercher AF. Polarization-sensitive optical coherence tomography of dental structures. Caries Res. 2000;34:59–69.
201. Everett MJ, Colston BW, Sathyam US, Silva LBD, Fried D, Featherstone JDB. Non-invasive diagnosis of early caries with polarization sensitive optical coherence tomography (PS-OCT). In: Featherstone R, Fried D, editors. Lasers in dentistry V, Proc SPIE, vol. 3593. Bellingham, WA: SPIE; 1999. p. 177–83.
202. Wang XJ, Zhang JY, Milner TE, de Boer JF, Zhang Y, Pashley DH, Nelson JS. Characterization of dentin and enamel by use of optical coherence tomography. Appl Opt. 1999;38(10):585–90.
203. Fried D, Xie J, Shafi S, Featherstone JDB, Breunig T, Lee CQ. Early detection of dental caries and lesion progression with polarization sensitive optical coherence tomography. J Biomed Opt. 2002;7(4):618–27.
204. Jones RS, Staninec M, Fried D. Imaging artificial caries under composite sealants and restorations. J Biomed Opt. 2004;9(6):1297–304.
205. Kang H, Darling CL, Fried D. Nondestructive monitoring of the repair of enamel artificial lesions by an acidic remineralization model using polarization-sensitive optical coherence tomography. Dent Mater. 2012;28(5):488–94.
206. Jones RS, Fried D. Remineralization of enamel caries can decrease optical reflectivity. J Dent Res. 2006;85(9):804–8.
207. Jones RS, Darling CL, Featherstone JD, Fried D. Remineralization of in vitro dental caries assessed with polarization-sensitive optical coherence tomography. J Biomed Opt. 2006;11(1):014016.
208. Jones RS, Darling CL, Featherstone JD, Fried D. Imaging artificial caries on the occlusal surfaces with polarization-sensitive optical coherence tomography. Caries Res. 2006;40(2):81–9.
209. Madjarova VD, Yasuno Y, Makita S, Hori Y, Voeffray JB, Itoh M, Yatagai T, Tamura M, Nanbu T. Investigations of soft and hard tissues in oral cavity by spectral domain optical coherence tomography. In: Coherence domain optical methods and optical coherence tomography in biomedicine X, Proc SPIE, vol. 6079. Bellingham, WA: SPIE; 2006. p. 60790N.
210. Seon YR, Jihoon N, Hae YC, Woo JC, Byeong HL, Gil-Ho Y. Realization of fiber-based OCT system with broadband photonic crystal fiber coupler. In: Coherence domain optical methods and optical coherence tomography in biomedicine X, Proc SPIE, vol. 6079. Bellingham, WA: SPIE; 2006. p. 60790I.
211. Yamanari M, Makita S, Violeta DM, Yatagai T, Yasuno Y. Fiber-based polarization-sensitive Fourier domain optical coherence tomography using B-scan-oriented polarization modulation method. Opt Express. 2006;14(14):6502.
212. Furukawa H, Hiro-Oka H, Amano T, DongHak C, Miyazawa T, Yoshimura R, Shimizu K, Ohbayashi K. Reconstruction of three-dimensional structure of an extracted tooth by OFDR-OCT. In: Coherence domain optical methods and optical coherence tomography in biomedicine X, Proc SPIE, vol. 6079. Bellingham, WA: SPIE; 2006. p. 60790T.
213. Bader JD, Shugars DA. The evidence supporting alternative management strategies for early occlusal caries and suspected occlusal dentinal caries. J Evid Based Dent Pract. 2006;6(1):91–100.
214. Bader JD, Shugars DA, Bonito AJ. A systematic review of the performance of methods for identifying carious lesions. J Public Health Dent. 2002;62(4):201–13.
215. Staninec M, Douglas SM, Darling CL, Chan K, Kang H, Lee RC, Fried D. Nondestructive clinical assessment of occlusal caries lesions using near-IR imaging methods. Lasers Surg Med. 2011;43(10):951–9.
216. Lee C, Darling C, Fried D. Polarization sensitive optical coherence tomographic imaging of artificial demineralization on exposed surfaces of tooth roots. Dent Mater. 2009;25(6):721–8.
217. Manesh SK, Darling CL, Fried D. Nondestructive assessment of dentin demineralization using polarization-sensitive optical coherence tomography after exposure to fluoride and laser irradiation. J Biomed Mater Res B Appl Biomater. 2009;90(2):802–12.
218. Manesh SK, Darling CL, Fried D. Polarization-sensitive optical coherence tomography for the nondestructive assessment of the remineralization of dentin. J Biomed Opt. 2009;14(4):044002.
219. Hsieh YS, Ho YC, Lee SY, Lu CW, Jiang CP, Chuang CC, Wang CY, Sun CW. Subgingival calculus imaging based on swept-source optical coherence tomography. J Biomed Opt. 2011;16(7):071409.

220. Kao MC, Lin CL, Kung CY, Huang YF, Kuo WC. Miniature endoscopic optical coherence tomography for calculus detection. Appl Opt. 2015;54(24):7419–23.
221. Shemesh H, van Soest G, Wu MK, Wesselink PR. Diagnosis of vertical root fractures with optical coherence tomography. J Endod. 2008;34(6):739–42.
222. Yang VB, Curtis DA, Fried D. Use of optical clearing agents for imaging root surfaces with optical coherence tomography. IEEE J Sel Topics Quant Electron. 2018;25(1): 1–7.
223. Chong SL, Darling CL, Fried D. Nondestructive measurement of the inhibition of demineralization on smooth surfaces using polarization-sensitive optical coherence tomography. Lasers Surg Med. 2007;39(5):422–7.
224. Hirasuna K, Fried D, Darling CL. Near-IR imaging of developmental defects in dental enamel. J Biomed Opt. 2008;13(4):044011.
225. Chan KH, Chan AC, Fried WA, Simon JC, Darling CL, Fried D. Use of 2D images of depth and integrated reflectivity to represent the severity of demineralization in cross-polarization optical coherence tomography. J Biophotonics. 2015;8(1–2):36–45.
226. Fried D, Ngaotheppitak P, Darling CL, Ho CM. Polarization sensitive optical coherence tomography for quantifying the severity of natural caries lesions on occlusal surfaces. In: Lasers in dentistry XIII, Proc SPIE, vol. 6425. Bellingham, WA: SPIE; 2007. p. 64250U.
227. Ngaotheppitak P, Darling CL, Fried D. Polarization optical coherence tomography for the measuring the severity of caries lesions. Lasers Surg Med. 2005;37(1):78–88.
228. Ngaotheppitak P, Darling CL, Fried D, Bush J, Bell S. PS-OCT of occlusal and interproximal caries lesions viewed from occlusal surfaces. In: Lasers in dentistry XII, Proc SPIE, vol. 6137. Bellingham, WA: SPIE; 2006. p. 61370L.
229. Ngaotheppitak P, Darling CL, Fried D. PS-OCT of natural pigmented and non-pigmented interproximal caries lesions. In: Lasers in dentistry XI, Proc SPIE, vol. 5687. Bellingham, WA: SPIE; 2006. p. 25–33.
230. Louie T, Lee C, Hsu D, Hirasuna K, Manesh S, Staninec M, Darling CL, Fried D. Clinical assessment of early tooth demineralization using polarization sensitive optical coherence tomography. Lasers Surg Med. 2010;42:738–45.
231. Nee A, Chan K, Kang H, Staninec M, Darling CL, Fried D. Longitudinal monitoring of demineralization peripheral to orthodontic brackets using cross polarization optical coherence tomography. J Dent. 2014;42(5):547–55.
232. Otis LL, Al-Sadhan RI, Meiers J, Redford-Badwal D. Identification of occlusal sealants using optical coherence tomography. J Clin Dent. 2000;14(1):7–10.
233. Lenton P, Rudney J, Chen R, Fok A, Aparicio C, Jones RS. Imaging in vivo secondary caries and ex vivo dental biofilms using cross-polarization optical coherence tomography. Dent Mater. 2012;28(7):792–800.
234. Holtzman JS, Osann K, Pharar J, Lee K, Ahn YC, Tucker T, Sabet S, Chen Z, Gukasyan R, Wilder-Smith P. Ability of optical coherence tomography to detect caries beneath commonly used dental sealants. Lasers Surg Med. 2010;42(8):752–9.
235. Stahl J, Kang H, Fried D. Imaging simulated secondary caries lesions with cross polarization OCT. In: Lasers in dentistry XVI, Proc SPIE, vol. 7549. Bellingham, WA: SPIE; 2010. p. 754905.
236. Lammeier C, Li Y, Lunos S, Fok A, Rudney J, Jones RS. Influence of dental resin material composition on cross-polarization-optical coherence tomography imaging. J Biomed Opt. 2012;17(10):106002.
237. Tom H, Simon JC, Chan KH, Darling CL, Fried D. Near-infrared imaging of demineralization under sealants. J Biomed Opt. 2014;19(7):77003.
238. de Melo LS, de Araujo RE, Freitas AZ, Zezell D, Vieira ND, Girkin J, Hall A, Carvalho MT, Gomes AS. Evaluation of enamel dental restoration interface by optical coherence tomography. J Biomed Opt. 2005;10(6):064027.
239. Nazari A, Sadr A, Shimada Y, Tagami J, Sumi Y. 3D assessment of void and gap formation in flowable resin composites using optical coherence tomography. J Adhes Dent. 2013;15(3):237–43.
240. Baudelet M, Smith BW. The first years of laser-induced breakdown spectroscopy. J Anal At Spectrom. 2013; https://doi.org/10.1039/C3JA50027F.
241. Niemz MH. Diagnosis of caries by spectral analysis of laser induced plasma sparks. Proc SPIE. 1994;2327:56.
242. Samek O, Telle HH, Beddows DCS. Laser-induced breakdown spectroscopy: a tool for real-time, in vitro and in vivo identification of carious teeth. BMC Oral Health. 2001;1:1–9.
243. Singh VK, Kumar V, Sharma J. Importance of laser-induced breakdown spectroscopy for hard tissues (bone, teeth) and other calcified tissue materials. Lasers Med Sci. 2015;30:1763–78.
244. Fang X, Ahmad SR, Mayo M, Iqbal S. Elemental analysis of urinary calculi by laser-induced plasma spectroscopy. Lasers Med Sci. 2005;20:132–7.
245. Singh VK, Rai AK, Rai PK, Jindal PK. Cross-sectional study of kidney stones by laser-induced breakdown spectroscopy. Lasers Med Sci. 2009;24:749–59.
246. Anzano J, Lasheras RJ. Strategies for the identification of urinary calculus by laser induced breakdown spectroscopy. Talanta. 2009;79:352–60.
247. Pathak AK, Singh VK, Rai NK, Rai AK, Rai PK, Rai PK, Rai S, Baruah GD. Study of different concentric rings inside gallstones with LIBS. Lasers Med Sci. 2011; https://doi.org/10.1007/s10103-011-0886-1.
248. Wu J, Zhang W, Shao X, Lin Z, Liu X. Simulated body fluid by laser-induced breakdown spectroscopy. Chin J Laser B. 2008;35:445–7.
249. Amaechi BT, Owosho AA, Fried D. Fluorescence and near-infrared light transillumination. Dent Clin N Am. 2018;62(3):435–52.
250. Wilder-Smith P, Ajdaharian J, editors. Oral diagnosis: minimally invasive imaging approaches. Springer; 2020.
251. Zandona AF, Longbottom C. Detection and assessment of dental caries. Springer; 2019.
252. Harris N, Garcia-Godoy F. Primary preventive dentistry. Stamford, CT: Appleton & Lange; 1999.
253. Mertz-Fairhurst EJ. Pit-and-fissure sealants: a global lack of scientific transfer? J Dent Res. 1992;115:1543–4.
254. Fejerskov O, Nyvad B, Kidd E, editors. Dental caries: the disease and its clinical management. Wiley Blackwell; 2015.
255. Hevinga MA, Opdam NJ, Frencken JE, Bronkhorst EM, Truin GJ. Microleakage and sealant penetration in contaminated carious fissures. J Dent. 2007;35(12):909–14.
256. Hevinga MA, Opdam NJ, Frencken JE, Bronkhorst EM, Truin GJ. Can caries fissures be sealed as adequately as sound fissures? J Dent Res. 2008;87(5):495–8.
257. Kidd EA. How 'clean' must a cavity be before restoration? Caries Res. 2004;38(3):305–13.
258. Hibst R, Graser R, Udart M, Stock K. Mechanism of high-power NIR laser bacteria inactivation. J Biophotonics. 2010;3(5–6):296–303.

259. Chan KH, Hirasuna K, Fried D. Rapid and selective removal of composite from tooth surfaces with a 9.3-μm CO2 laser using spectral feedback. Lasers Surg Med. 2011;43(8):824–32.
260. Louie TM, Jones RS, Sarma AV, Fried D. Selective removal of composite sealants with near-ultraviolet laser pulses of nanosecond duration. J Biomed Opt. 2005;10(1):14001.
261. Alexander R, Xie J, Fried D. Selective removal of residual composite from dental enamel surfaces using the third harmonic of a Q-switched Nd:YAG laser. Lasers Surg Med. 2002;30(3):240–5.
262. Wheeler CR, Fried D, Featherstone JD, Watanabe LG, Le CQ. Irradiation of dental enamel with Q-switched lambda = 355-nm laser pulses: surface morphology, fluoride adsorption, and adhesion to composite resin. Lasers Surg Med. 2003;32(4):310–7.
263. Myers TD, Myers WD. The use of a laser for debridement of incipient caries. J Prosthet Dent. 1985;53:776–9.
264. Harris DM, White JM, Goodis H, Arcoria CJ, Simon J, Carpenter WM, Fried D, Burkart J, Yessik M, Myers T. Selective ablation of surface enamel caries with a pulsed Nd:YAG dental laser. Lasers Surg Med. 2002;30(5):342–50.
265. Hibst R, Keller U. Experimental studies of the application of the Er:YAG laser on dental hard substances: I. Measurement of the ablation rate. Lasers Surg Med. 1989;9:338–44.
266. Hennig T, Rechmann P, Jeitner P, Kaufmann R. Caries-selective ablation: the second threshold. In: Lasers in orthopedic, dental, and veterinary medicine II, Proc SPIE, vol. 1880. Bellingham, WA: SPIE; 1993. p. 117.
267. Fan K, Fried D. Scanning ablation of root caries with acoustic feedback control. In: Lasers in dentistry XIII, Proc SPIE, vol. 6425. Bellingham, WA: SPIE; 2007. p. 64250J.
268. Arima MK, Matsumoto K. Effects of ARF: excimer laser irradiation on human enamel and dentin. Lasers Surg Med. 1993;13:97–105.
269. Schoenly JE, Seka W, Rechmann P. Pulsed laser ablation of dental calculus in the near ultraviolet. J Biomed Opt. 2014;19(2):028003.
270. Grad L, Mozina J, Susteric D, Fundek N, Skaleric U, Lukac M, Cencic M, Nemes K. Optoacoustic studies of Er:YAG laser ablation in hard tissue. In: Lasers in surgery: advanced characterization, therapeutics, and systems IV, Proc SPIE, vol. 2128. Bellingham, WA: SPIE; 1994. p. 456–65.
271. Esenaliev RO, Oraevsky AA, Letokhov VS, Karabutov AA, Malinsky TV. Studies of acoustical shock waves in the pulsed laser ablation of biotissue. Lasers Surg Med. 1993;13:470–84.
272. Hennig T, Rechmann P, Pilgrim C, Kaufmann R. Basic principles of caries selective ablation by pulsed lasers. In: Proceedings of the third international congress on lasers in dentistry. 1992. p. 119–20.
273. Hennig T, Rechmann P, Jeitner P. Effects of a second harmonic Alexandrite laser on human dentin. In: Advanced laser dentistry, Proc SPIE, vol. 1984. Bellingham, WA: SPIE; 1995. p. 24.
274. Hennig T, Rechmann P, Pilgrim CG, Schwarzmaier H-J, Kaufmann R. Caries selective ablation by pulsed lasers. In: Lasers in orthopedic, dental, and veterinary medicine, Proc SPIE, vol. 1424. Bellingham, WA: SPIE; 1991. p. 99.
275. Rechmann P, Hennig T. Caries selective ablation: first histological examinations. In: Laser surgery: advanced characterization, therapeutics, and systems IV, Proc SPIE, vol. 2128. Bellingham, WA: SPIE; 1994. p. 389.
276. Harris DM, Goodis HE, White JM, Arcoria CJ, Simon J, Carpenter WM, Fried D, Burkart J, Yessik M, Myers T. Selective ablation of surface enamel caries with a pulsed Nd:YAG laser. Lasers Surg Med. 2002;30(5):342–50.
277. Sherman DB, Ruben MP, Goldman HM. The application of laser for the spectrochemical analysis of calcified tissues. Ann N Y Acad Sci. 1965;122:767–72.
278. Oraevsky AA, Jacques SL, Pettit GH, Tittel FK, Henry PD. XeCl laser ablation of atherosclerotic aorta: luminescence spectroscopy of ablation products. Lasers Surg Med. 1993;13:168–78.
279. Niemz MH. Investigation and spectral analysis of the plasma-induced ablation mechanism of dental hydroxyapatite. Appl Phys B Lasers Opt. 1994;58:273–81.
280. Samek, Liska M, Kaiser J, Beddows DC, Telle HH, Kukhlevsky SV. Clinical application of laser-induced breakdown spectroscopy to the analysis of teeth and dental materials. J Clin Laser Med Surg. 2000;18(6):281–9.
281. Cheng JY, Fan K, Fried D. Use of a compact fiber optic spectrometer for spectral feedback during the laser ablation of dental hard tissues and restorative materials. In: Peter R, Daniel F, editors. Lasers in dentistry XII, Proc SPIE, vol. 6137. Bellingham, WA: SPIE; 2006. p. 61370F.
282. Alexander R, Fried D. Selective removal of orthodontic composite using 355-nm Q-switched laser pulses. Lasers Surg Med. 2002;30:240–5.
283. Dumore T, Fried D. Selective ablation of orthodontic composite using sub-microsecond IR laser pulses with optical feedback. Lasers Surg Med. 2000;27(2):103–10.
284. Lizarelli RFZ, Moriyama LT, Bagnato VS. Ablation of composite resins using Er:YAG laser—comparison with enamel and dentin. Lasers Surg Med. 2003;33(2):132–9.
285. Louie TM, Sarma AV, Fried D. Selective removal of composite restorative materials using Q-switched 355-nm laser pulses. J Biomed Opt. 2005;10(1):014001.
286. Smith SC, Walsh LJ, Taverne AA. Removal of orthodontic bonding resin residues by CO2 laser radiation: surface effects. J Clin Laser Med Surg. 1999;17(1):13–8.
287. Yi I, Chan KH, Tsuji GH, Staninec M, Darling CL, Fried D. Selective removal of esthetic composite restorations with spectral guided laser ablation. In: Lasers in dentistry XXII, Proc SPIE, vol. 9692. Bellingham, WA: SPIE; 2016. p. 96920U.
288. Simon JC, Choi JH, Jang A, Fried D. In vivo spectral guided removal of composite from tooth surfaces with a CO2 laser. Proc SPIE. 2020;11217:112170K.
289. Jang AT, Chan KH, Fried D. Automated ablation of dental composite using an IR pulsed laser coupled to a plume emission spectral feedback system. Lasers Surg Med. 2017;49(7):658–65.
290. Eberhard J, Bode K, Hedderich J, Jepsen S. Cavity size difference after caries removal by a fluorescence-controlled Er:YAG laser and by conventional bur treatment. Clin Oral Investig. 2008;12(4):311–8.
291. Eberhard J, Eisenbeiss AK, Braun A, Hedderich J, Jepsen S. Evaluation of selective caries removal by a fluorescence feedback-controlled Er:YAG laser in vitro. Caries Res. 2005;39(6):496–504.
292. Jepsen S, Acil Y, Peschel T, Kargas K, Eberhard J. Biochemical and morphological analysis of dentin following selective caries removal with a fluorescence-controlled Er:YAG laser. Lasers Surg Med. 2008;40(5):350–7.
293. Fried WA, Chan KH, Fried D, Darling CL. High contrast reflectance imaging of simulated lesions on tooth occlusal surfaces at near-IR wavelengths. Lasers Surg Med. 2013;45(8):533–41.
294. Kleter GA. Discoloration of dental carious lesions (a review). Arch Oral Biol. 1998;43:629–32.

295. Sarna T, Sealy RC. Photoinduced oxygen consumption in melanin systems. Action spectra and quantum yields for eumelanin and synthetic melanin. Photochem Photobiol. 1984;39:69–74.
296. Fu D, Ye T, Matthews TE, Yurtsever G, Warren WS. Two-color, two-photon, and excited-state absorption microscopy. J Biomed Opt. 2007;12(5):054004.
297. Jones RS, Huynh GD, Jones GC, Fried D. Near-IR transillumination at 1310-nm for the imaging of early dental caries. Opt Express. 2003;11(18):2259–65.
298. Tao YC, Fried D. Near-infrared image-guided laser ablation of dental decay. J Biomed Opt. 2009;14(5):054045.
299. Chan KH, Hirasuna K, Fried D. Analysis of enamel surface damage after selective laser ablation of composite from tooth surfaces. Photon Lasers Med. 2014;3(1):37–45.
300. Chan KH, Fried D. Selective removal of demineralization using near infrared cross polarization reflectance and a carbon dioxide laser. Proc SPIE. 2012;8208:82080U.
301. Badran Z, Demoersman J, Struillou X, Boutigny H, Weiss P, Soueidan A. Laser-induced fluorescence for subgingival calculus detection: scientific rational and clinical application in periodontology. Photomed Laser Surg. 2011;29(9):593–6.
302. Folwaczny M, Heym R, Mehl A, Hickel R. The effectiveness of InGaAsP diode laser radiation to detect subgingival calculus as compared to an explorer. J Periodontol. 2004;75(5):744–9.
303. Krause F, Braun A, Jepsen S, Frentzen M. Detection of subgingival calculus with a novel LED-based optical probe. J Periodontol. 2005;76(7):1202–6.
304. Kurihara E, Koseki T, Gohara K, Nishihara T, Ansai T, Takehara T. Detection of subgingival calculus and dentine caries by laser fluorescence. J Periodontal Res. 2004;39(1):59–65.
305. Qin YL, Luan XL, Bi LJ, Lu Z, Sheng YQ, Somesfalean G, Zhou CN, Zhang ZG. Real-time detection of dental calculus by blue-LED-induced fluorescence spectroscopy. J Photochem Photobiol B. 2007;87(2):88–94.
306. Rams TE, Alwaqyan AY. In vitro performance of DIAGNOdent laser fluorescence device for dental calculus detection on human tooth root surfaces. Saudi Dent J. 2017;29(4):171–8.
307. Shakibaie F, Walsh LJ. Laser fluorescence detection of subgingival calculus using the DIAGNOdent Classic versus periodontal probing. Lasers Med Sci. 2016;31(8):1621–6.
308. Tung OH, Lee SY, Lai YL, Chen HF. Characteristics of subgingival calculus detection by multiphoton fluorescence microscopy. J Biomed Opt. 2011;16(6):066017.
309. Tsuda H, Jongebloed WL, Stokroos I, Arends J. A micro-Raman spectroscopic study of hydrazine-treated human dental calculus. Scanning Microsc. 1996;10(4):1015–23; discussion 1023–4.
310. Huminicki A, Dong C, Cleghorn B, Sowa M, Hewko M, Choo-Smith LP. Determining the effect of calculus, hypocalcification, and stain on using optical coherence tomography and polarized Raman spectroscopy for detecting white spot lesions. Int J Dent. 2010;2010:879252.
311. Yan R, Chan KH, Tom H, Simon JC, Darling CL, Fried D. Selective removal of dental caries with a diode-pumped Er:YAG laser. In: Lasers in dentistry XX, Proc SPIE, vol. 9306. Bellingham, WA: SPIE; 2015. p. 93060O.
312. Stock K, Diebolder R, Hausladen F, Hibst R. Efficient bone cutting with the novel diode pumped Er:YAG laser system: in vitro investigation and optimization of the treatment parameters. Proc SPIE. 2014;8926:89260P.
313. Stock K, Diebolder R, Hausladen F, Wurm H, Lorenz S, Hibst R. Primary investigations on the potential of a novel diode pumped Er:YAG laser system for bone surgery. In: Photonic therapeutics and diagnostics IX, Proc SPIE, vol. 8565. Bellingham, WA: SPIE; 2013. p. 85650D.
314. Stock K, Hausladen F, Hibst R. Investigations on the potential of a novel diode pumped Er:YAG laser system for dental applications. Proc SPIE. 2012;8208:82080D.
315. Simon JC, Kwok JW, Vinculado F, Fried D. Computer-controlled CO2 laser ablation system for cone-beam computed tomography and digital image guided endodontic access: a pilot study. J Endod. 2021;47:1445–52.
316. Migitaa M, Kamiyama I, Matsuzaka K, Nakamura A, Souta T, Aizawa K, Shibahara T. Photodynamic diagnosis of oral carcinoma using talaporfin sodium and a hyperspectral imaging system: an animal study. Asian J Oral Maxillofac Surg. 2010;22(3):126–32.
317. Chang CJ, Lin MS, Hwang PS, Cheng SMH. Topical application of Photofrin® for oral neoplasms in animal. Opt Quant Electron. 2005;37:1353–65.
318. Riva C, Ross B, Benedek GB. Laser Doppler measurements of blood flow in capillary tubes and retinal arteries. Invest Ophthalmol. 1972;11:936–44.
319. Gazelius B, Olgart L, Edwall B, Edwall L. Non-invasive recording of blood flow in human dental pulp. Endod Dent Traumatol. 1986;2:219–21.
320. Olgart L, Gazelius B, Lindh-Stromberg U. Laser Doppler flowmetry in assessing vitality in luxated permanent teeth. Int Endod J. 1988;21:300–6.
321. Andreasen FM, Andreasen JO. Luxation injuries. In: Andreasen JO, Andreasen FM, editors. Textbook and color atlas of traumatic injuries to the teeth. 3rd ed. Copenhagen: Munksgaard; 1994. p. 353–4.
322. Ebihara A, Tokita Y, Izawa T, Suda H. Pulpal blood flow assessed by laser Doppler flowmetry in a tooth with a horizontal root fracture. Oral Surg Oral Med Oral Pathol Oral Radiol Endod. 1996;81:229–33.
323. Ikawa M, Komatsu H, Ikawa K, Mayanagi H, Shimauchi H. Age-related changes in the human pulpal blood flow measured by laser Doppler flowmetry. Dent Traumatol. 2003;19:36–40.
324. Babacan H, Doruk C, Bicakci AA. Pulpal blood flow changes due to rapid maxillary expansion. Angle Orthod. 2010;80:1136–40.
325. Cho JJ, Efstratiadis S, Hasselgren G. Pulp vitality after rapid palatal expansion. Am J Orthod Dentofac Orthop. 2010;137:254–8.
326. Chen E, Goonewardene M, Abbott P. Monitoring dental pulp sensibility and blood flow in patients receiving mandibular orthognathic surgery. Int Endod J. 2012;45:215–23.
327. Emshoff R, Kranewitter R, Norer B. Effect of Le Fort I osteotomy on maxillary tooth-type-related pulpal blood-flow characteristics. Oral Surg Oral Med Oral Pathol Oral Radiol Endod. 2000;89:88–90.
328. Emshoff R, Kranewitter R, Gerhard S, Norer B, Hell B. Effect of segmental Le Fort I osteotomy on maxillary tooth type-related pulpal blood-flow characteristics. Oral Surg Oral Med Oral Pathol Oral Radiol Endod. 2000;89:749–52.
329. Meredith N. Assessment of implant stability as a prognostic determinant. Int J Prosthodont. 1998;11:491–501.
330. Riecke B, Heiland M, Hothan A, Morlock M, Amling M, Blake FA. Primary implant stability after maxillary sinus augmentation with autogenous mesenchymal stem cells: a biomechanical evaluation in rabbits. Clin Oral Implants Res. 2011;22:1242–6.

331. Jafarzadeh H. Laser Doppler flowmetry in endodontics: a review. Int Endod J. 2009;42:476–90.
332. Polat S, Er K, Polat NT. Penetration depth of laser Doppler flowmetry beam in teeth. Oral Surg Oral Med Oral Pathol Oral Radiol Endod. 2005;100:125–9.
333. Wilder-Smith PEEB. A new method for the non-invasive measurement of pulpal blood flow. Int Endod J. 1988;21:307–12.
334. Evans D, Reid J, Strang R, Stirrups D. A comparison of laser Doppler flowmetry with other methods of assessing the vitality of traumatised anterior teeth. Endod Dent Traumatol. 1999;15:284–90.
335. Roeykens HJJ, Van Maele GOG, De Moor RJC, Martens LCM. Reliability of laser Doppler flowmetry in a 2-probe assessment of pulpal blood flow. Oral Surg Oral Med Oral Pathol Oral Radiol Endod. 1999;87:742–8.
336. Roebuck EM, Evans DJP, Stirrups D, Strang R. The effect of wavelength, bandwidth, and probe design and position on assessing the vitality of anterior teeth with laser Doppler flowmetry. Int J Paediatr Dent. 2000;10:213–20.
337. Limjeerajarus C. Laser Doppler flowmetry: basic principle, current clinical and research applications in dentistry. CU Dent J. 2014;37:123–36.
338. Del Giudice E, Doglia S, Milani M. Order and structures in living systems. In: Ross Adey W, Lawrence AF, editors. Nonlinear electrodynamics in biological systems. New York: Plenum Press; 1983. p. 477–88.
339. Adey WR. Frequency and power windowing in tissue interactions with weak electromagnetic fields. Proc IEEE. 1980;63(1):119–25.
340. Venz S, Dickens B. NIR-spectroscopic investigation of water sorption characteristics of dental resins and composites. J Biomed Mater Res. 1991;25:1231–48.

Further Reading

Fejerskov O, Kidd E. Dental caries: the disease and its clinical management. John Wiley & Sons; 2009.

How a Raman instrument works. http://nicolet.com/theory.html#Raman.

Kincade K. Raman spectroscopy enhances in vivo diagnosis. Laser Focus World, Jul 1998. p. 83–91.

Minet O, Müller GJ, Beuthan J. Selected papers on optical tomography: fundamentals and applications in medicine, SPIE milestone series, vol. MS147. SPIE Optical Engineering Press; 1998. ISBN: 0819428779, 9780819428776.

Photobiomodulation Therapy Within Clinical Dentistry: Theoretical and Applied Concepts

Mark Cronshaw and Valina Mylona

Contents

D. J. Coluzzi, S. P. A. Parker (eds.), *Lasers in Dentistry—Current Concepts*, Textbooks in Contemporary Dentistry,
https://doi.org/10.1007/978-3-031-43338-2_7

Core Message

Photobiomodulation therapy (PBMT), formerly known as low-level laser therapy, is the targeted application of light for therapeutic purposes at a low level below that associated with damage to structural proteins. This may be a laser or a broader spectrum light source such as a light-emitting diode (LED), which may selectively achieve changes in subcellular, local, regional and centrally mediated systemic processes associated with healing, repair and the regeneration of tissues [1, 2]. In addition to the potential benefits of the promotion of an increase in cellular numbers by mitosis and a corresponding increase in the production of important matrix materials such as collagen and bone along with an enhanced vasculature, there can also be seen resolution of inflammation and the mitigation, up to inhibition of pain [3, 4]. Furthermore, therapeutic photonic conditioning of tissues as a prequel or synchronous to exposure to a noxious stimulus such as chemotherapy or radiotherapy can increase cellular stress resistance against apoptosis [5, 6]. Given the very wide range of possible applications, PBMT has excited considerable interest across a multitude of possible applications, as an adjunct to augment the normal procedures in clinical dentistry [7, 8]. Following over 50 years of research at the level of cellular, animal and more recently human clinical trials, there is a considerable published evidence base. To date the broader integration of this approach as a tool to assist the dental team to optimize a clinical intervention has been slow to be adopted. Given the extent of the current knowledge base which is rapidly expanding due to the heightened awareness of the potential value of this approach to enhanced healing and tissue management, it is inevitable that PBMT will in time become a standard evidence-based procedure in daily clinical practice. In this chapter we provide a contemporary overview of this important subject with a particular view to help guide the clinician with an interest in learning the processes involved to safely and successfully explore the potential merits of this approach to patient care.

Key clinical points are drawn from the many wavelengths in the visible to far infra-red that have been associated with positive healing events. Additionally, PBM may be best viewed as a multiphase therapy where at lower end dosimetry there is an increase in cellular activity whereas at a higher level there can be inhibition of cellular function which may be beneficial in pain relief [9]. The interest in regard to the mechanisms of action of PBM is relevant in respect of the understanding required to more deliberately direct the delivered photonic energy to achieve the desired beneficial effects. This may impact on the choices made in respect to parameters as well as the clinician's choice of clinical device.

7.1 Principles of Action: Historical Perspectives

Radiant solar energy is essential to sustain the majority of living organisms as without light there is no source of photonic energy to drive photosynthesis nor is there the primordial photothermal source to enable the enzyme processes of all but the hardiest of life forms. The capture by plant chlorophyll of incident photonic exposure in the visible to near-infra-red spectrum of light enables the fixation of gaseous CO_2 into carbohydrates which then in turn become the substrate for the controlled release of the captured solar energy to form key cellular biochemicals via the Krebs cycle and the mitochondrial electron transport chain. Aside from the Calvin reaction which is a magnesium-based carbon capture process associated with photosynthesis, there are also many other photon transduction mechanisms which can result in energy capture. These include other photochemical processes as well as photoelectric, photothermal, photomagnetic and photomechanical effects and photofluorescence [10–13].

The history of using light and color for health and healing (heliotherapy) can be traced back thousands of years in human history. Virtually all ancient civilizations (China, India, Egypt, Greece, Meso-America) implemented the use of sunlight or colored light in healing.

Indian medical literature dating back to 1500 BC describes a treatment combining herbs with natural sunlight to treat non-pigmented skin areas. Buddhist literature from about 200 AD and tenth-century Chinese documents made similar references. The nineteenth century saw a resurgence of this concept with the use of colored light sources for the promotion of wound healing and pain reduction [7, 14, 15].

Niels Finsen (Denmark) developed the first artificial light source for the purpose of treating lupus vulgaris. He received the Nobel Prize in Physiology or Medicine in 1903 [16–18].

The latter half of the twentieth century witnessed a rediscovery of the science by researchers including Gurwitsch (Russia), Mester (Hungary), Karu (Russia), Fröhlich and Popp (Germany) [10, 19–21].

In the early 1800s, Christian von Grotthus and John Draper formulated the first law of photochemistry, which states that only the light absorbed by a substance can produce a photochemical change. This has been adapted and adopted as the first law of photobiology which requires that there is a cellular receiver for the transfer of energy from the incident photons to effect change [22, 23]. The site of absorption has been referred to as a chromophore which is the part of a molecule responsible for its color within the visible spectrum. There are molecules such as water which are colorless yet are high absorbers with some wavelengths. The use of the term chromophore in the context of energy transfer in PBM is not strictly accurate; however it has been broadly adopted within the PBM literature.

Professor Endre Mester is credited with the discovery of the biological effects of low-power lasers. He performed early science experiments on the biological effects of laser irradiation. While applying lasers to the backs of shaven mice, Mester noticed that the shaved hair grew back more quickly on the treated group than the untreated group [24].

PBM therapy (PBMT) through the application of photonic energy works on the principle of inducing a biological response through energy transfer. Such non-ablative photonic energy delivered into tissues modulates biological processes within that tissue and within the biological system of which that tissue is a component part. It is a source of some debate, however, which concerns the thermal rise in irradiated cells or tissue, bearing in mind that because of absorption, photonic energy will impart increased target molecular activity. However, within a correct incident dose, PBM has no detrimental thermal effects in irradiated tissue [25].

The medical benefits of the therapeutic application of photons to patients initially may seem a strange idea. However, human physiology is highly responsive to light as evidenced by the circadian rhythm which is entrained through bright light. Specialized retinal receptors are activated by bright light which via axonal stimulation of the suprachiasmatic nucleus results in the inhibition of the production of melatonin in the pineal gland. Disruption to the normal diurnal rhythm of a daytime reduction and a nighttime rise in CNS melatonin levels results in disturbance to sleep patterns as may be experienced by long haul travelers with so called jet lag. Also, irregularity in the normal daytime sympathetic as opposed to nighttime parasympathetic nervous system physiological activity cycle has been associated with an increased risk for night workers of developing metabolic syndrome and possibly an increased risk of developing some types of cancer. A further example of human photo-reactivity is lack of exposure to ultraviolet "B" (UVB) 280–315 nm light, which is absent during the winter months in the northern hemisphere. This results in vitamin D deficiency with a consequence of negative impacts on the immune system as well as on bone metabolism. Furthermore, light can impact on mood as the production of some key neurochemicals including serotonin, dopamine and adrenaline is depleted in the shorter daylight periods of winter predisposing some individuals to clinical depression [26–30].

7.2 Principles of Action: Cellular Perspectives

To optimize success in PBMT, it is essential to have an understanding of optical transport processes in biological tissues, plus some knowledge of the optical properties of different wavelengths and an appreciation of photonic energy transduction pathways [31–33]. On a cellular basis, all eukaryote cells are photoreactive as indeed are many unicellular organisms such as bacteria and Archaeans. Eukaryote cells compose the essential base units of all multicellular organisms and are distinguished by the presence of a nucleus as well as by cellular organelles, including mitochondria. The mitochondria are generally regarded as one of the more important recipients for the applied photons in PBMT [34, 35]. Mitochondria are the active sites for the manufacture of adenosine triphosphate (ATP) which is the essential energy currency used by the cell to drive metabolic processes associated with protein and fatty acid manufacture and catabolism. Mitochondria are also pivotal to the nitrogen-based urea cycle, as well as being an important center for the regulation of calcium ions plus the production of nitric oxide (NO) and the collateral manufacture as a side product of aerobic and anaerobic metabolism of reactive oxygen species (ROS) [36, 37]. Furthermore, mitochondria are sites of high molecular activity from which the cells derive a source of heat, enabling a carefully regulated system of endothermic metabolism [9, 38–40]. Given the central role played by mitochondria in the anabolism and catabolism of proteins and fatty acids as well as the many other key aspects of eukaryote metabolism, the capacity to influence mitochondrial activity by exposure to light has attracted considerable attention in the PBM literature. The wider recognition of the importance of mitochondria in health and disease has resulted in very considerable cross discipline investigations from scientists including cell biologists, pharmacologists, biochemists, oncologists and researchers into degenerative CNS conditions, including Alzheimer's and Parkinson's disease along with many other disorders [9, 41].

The intracellular aspects of cellular regulation represent a heavily researched area which has revealed an extraordinarily complex system of checks and balances. In respect of PBMT, a good deal of attention has focused on the inner mitochondrial membrane and location of an array of five principal units (I–V) known as the electron transport chain (ETC). The functions of the ETC include the manufacture at unit V of ATP by means of a "molecular turbine" driven by a stream of protons, built up across the mitochondrial inner membrane from the core central matrix by the biochemical activities of the preceding four units. In the literature there is a considerable emphasis on the prime importance of unit 1 V of the ETC, which is a cupric–ferrous (Cu–Fe) proteinaceous enzyme cluster called cytochrome c oxidase (CCO). The selective absorption of incoming red to near-infra-red (NIR) light by the alpha 1-haem and haem-3 units of this 13-subunit complex was proposed in 1998 as the central step in the increase in activity of the ETC observed with the production of more ATP, along with increases in the mitochondrial membrane potential and oxygen consumption [4, 42, 43]. Furthermore, it was proposed that nitric oxide is a key cellular regulator of CCO due to the NO competitively binding to oxygen ion sites within CCO in hypoxic and stressed cells, thus reducing CCO activity. It has been shown that the inhibitory NO can be displaced from the oxygen binding sites by light-induced photodissociation, enabling an increase in aerobic metabolism with a resultant increase in ATP output [44–48].

Unit IV is one of the sites of manufacture of NO by nitrite reductase. PBM using a 590 nm LED was found to promote NO synthesis catalyzed by CCO and NO is a potent vasodilator. NO acts as a secondary messenger, able to modulate the activity of enzymes such as guanylyl cyclase that synthesizes cyclic guanosine monophosphate (cGMP) starting from guanosine-5′-triphosphate (GTP). cGMP relaxes smooth muscle of tissues and blood vessels, inducing vasodilation that leads to an increased blood flow as well as improved lymphatic drainage [49–51].

However, some of the concepts of this neat theoretical encapsulation of PBM activation have been subject to re-evaluation following more recent research. Lima et al., for example, characterized two different cell lines that lack the fully assembled CCO enzyme. PBM at 660 nm was still found to be effective in promoting mitosis along with an increase in ATP production. The long-standing concept that the optimal wavelengths for PBM are in the red to near-infra-red range due to their affinity for CCO has been challenged as the peak absorption of CCO corresponds to 82% for the 415 nm and 26.7% for the 540 nm whereas the 660 nm absorbs around 8.7% and the 810 nm 16.2%. A contrary effect of a reduction in ATP production by irradiation with the 415 nm blue wavelength as opposed to the increase observed with red to NIR sources implies that strong absorption by CCO may not be the intrinsic cause of the ATP upregulation. In respect of Lane's hypothesis, PBM has been identified as a trigger for new enzymic activity at CCO to produce NO. However, this process depends on the presence of a sufficient supply of nitrite which may not be present in significant amounts in all cells and tissues [52, 53].

Newer theories of PBM action have moved away from the earlier concepts of optimal photochemical pathways of specific applied wavelengths in the red to NIR. Current thinking proposes that there are very many potential targets for photon transduction including water. Biological tissues have a high water content with an average adult male body comprising around 55% by volume water, a female at a slightly higher water content of around 60% and a newborn baby who can have as high a water content as 85% [54]. Bulk water is not generally a barrier to onward transmission of light irradiation to wavelengths in the blue to NIR range as water has both low absorption and is a low scattering medium to visible to NIR wavelengths of light. The exception to this is the 980 nm wavelength which has a small but significant absorption peak for water [55, 56].

In close proximity to hydrophobic molecules, water has been proposed to form a semirigid nanostructure microlayer which may be photoreactive. Sommer has suggested an interesting hypothesis that combines the idea of nanostructured water acting as a photon receiver, with the well-established response to irradiation of increased production of ATP observed in mitochondria [57, 58]. ATP synthase (unit V of the ETC) is a membrane protein complex composed of two interconnected rotary molecular motors powered by the flow of protons generated by the electron transport chain driving the primary F0 rotor across the mitochondrial membrane. The F0 rotor is connected to the second F1 rotor, which produces ATP from ADP and phosphate. The hypothesis is that a photoinduced reduction in the viscosity of nanostructure water surrounding the mitochondrial membrane will allow these rotors to turn faster and therefore create more ATP (◘ Fig. 7.1).

There are many potential subcellular sites for the absorption of light including flavins, porphyrins, nuclear chromatin, cytochromes, opsins, cryptochromes and transient reactive vanilloid potential membrane ion gates (TRPVs). Examples of endogenous chromophores are melanin, hemoglobin (oxyhemoglobin, deoxyhemoglobin and methemoglobin), protein, peptide bonds, aromatic amino acids, urocanic acid and bilirubin [34]. One common thread to these sites of energy transference is the presence of pigmented materials, which by virtue of their molecular density as well as the content of transition metals such as iron/copper containing compounds are broad band receivers of visible to NIR light. Should the resonant frequency of the incoming photons coincide with the valence band width of the target, this can drive electrons into a higher valence band. In addition, any energy in excess of this value results in an increase in molecular vibration which is expressed in the form of heat [32]. Many transition metals are broad band receivers of photonic energy resulting in photoelectrical and photothermal effects as

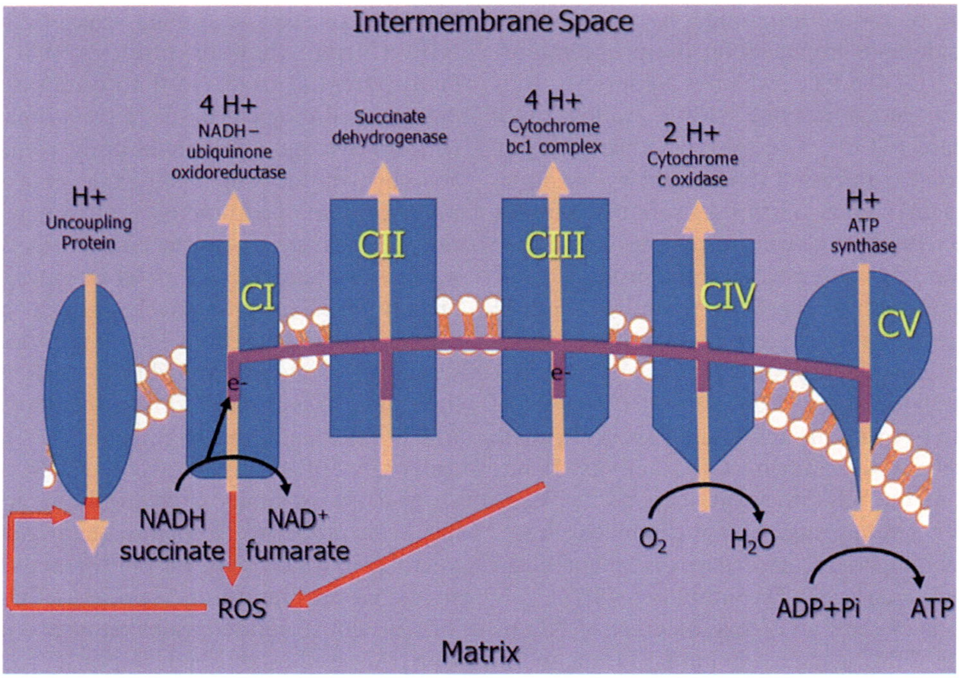

Fig. 7.1 The electron transport chain is a five-unit complex found in the inner mitochondrial membrane plates (cristae) separating the inner core of the mitochondrion (matrix) from the intermembrane space. A proton gradient is built up across the two compartments of the mitochondrion with an electron flow to unit V which permits the flow of protons back into the matrix which drives the twin molecular rotors culminating the phosphorylation of ADP into ATP. A by-product of this process is reactive oxygen species (ROS) which are carefully regulated to avoid excessive build up which can prove damaging to essential elements of the cell. Protective elements include key antioxidant enzymes, plus in cells with a high rate of metabolic activity, there may be a category of transmembrane proton transporters called uncoupling proteins (UCPs) which can act as a safety mechanism to reduce mitochondrial production of ATP. Each of the elements of the ETC is photosensitive and as such is regarded as key operand in the cytochemistry associated with PBM. (Graphics by S. Parker)

well as photomagnetic, photofluorescent and photomechanical events.

In addition to the potential selective photochemical stimulus of CCO and any lower-level photothermal effects, transition metals are electron acceptors as well as donors. Also, there can be changes in the valence band of electrons depending on the resonant frequency of the incoming photon wavelength coinciding with the chromophore. These effects can result in electron flow, or alternatively the activated state of the higher energized electrons may result in the release of a further photon as it decays to its ground state. In the case of a metastable valence band, this may result in phosphorescence rather than the more rapid decay observed in fluorescence. These effects may seem esoteric; however this type of induced secondary photon release is recognized as being a trigger for synchronized mitosis [59]. A further effect on metal containing compounds of photonic exposure is the realignment of dipoles, and this photomagnetic effect may change the configuration of enzymes' opening and closing binding sites, hence changing the activity of the enzyme as well as possibly opening membrane-bound ion gates [60].

In the event that the photon stream is delivered as a gated pulse, this can result in photomechanical (photoacoustic) induced changes in cell membrane permeability. It has been postulated that a low hertz gated pulse may result in a photomechanical pressure wave that can temporarily produce a cell membrane conformational change that can open membrane-bound ion gates. In nerves this could interfere with the sodium–potassium pump, and this mechanism has been proposed to be a contributory factor to a reduction of transmission along an axon by a wave of depolarization [2, 61].

The potential energy of a photon is expressed by the term electron volt which is the amount of energy necessary to move an electron through a 1 V electrical field. There is an inverse relation between energy and wavelength as shorter wavelengths have a higher cycle frequency and an associated higher-energy electromagnetic field. In consequence, the effects of a shorter-wavelength photon striking a molecular target are markedly different to a longer wavelength as the shorter wavelengths on absorption can have a more pronounced electromagnetic effect. To break molecular bonds requires an amount of energy which if sufficient can result in a pho-

Dissociation energy of selected chemical bonds	
Type of Bond	Dissociation energy (eV)
C=O	7.1
C=C	6.4
O-H	4.8
N-H	4.1
C-O	3.6
C-C	3.6
S-H	3.5
C-N	3.0
C-S	2.7
Fe-OH	0.35
HA Lattice	310

Laser wavelength and photonic energy (1000 nm = 1.24 eV)		
Laser	Wavelength (nm)	Photon energy (eV)
KTP (diode)	532	2.4
He-Ne	633	2.0
Diode	810	1.6
Nd:YAG	1064	1.2
Er:YAG	2940	0.4
CO_2	10600	0.1

Fig. 7.2 The dissociation energy required to break chemical bonds (eV) is higher than the eV available from single photons from laser wavelengths commonly used for PBMT. Except on the occasion of the simultaneous collision of two photons of the same wavelength at an absorbing molecule, there is insufficient energy to break molecular bonds. Hence visible to far-infra-red wavelengths are not capable of ionization of DNA base pairs within chromosomes and are not considered mutagenic. (Source: Mó, O., Yáñez, M., Eckert-Maksić, M., Maksić, Z.B., Alkorta, I., Elguero, J. (2005). Periodic trends in bond dissociation energies. A theoretical study. The Journal of Physical Chemistry A109, pp. 4359–4365)

tochemical change. As can be seen in the table below in the red to NIR wavelengths, there is insufficient energy to break the chemical bonds of many common tissue constituents [62]. However, the repeated and accumulated energy of a stream of photons can result in photothermal effects due to an indirect kinetic effect consequent to an increase in elastic and inelastic molecular vibration. It is recognized that although the red to near-infra-red wavelengths are less than the threshold eV necessary to break chemical bonds, indirect photochemical-induced effects such as an increase in the production of NO are observed [63] (Fig. 7.2).

One of the by-products of aerobic and anaerobic metabolism is reactive oxygen species (ROS). Persistently elevated higher levels of ROS are damaging to cells and tissues and are considered a major cause of the symptoms of many diseases and conditions as well as being associated with some of the degenerative changes seen in aging. However, this does not mean that ROS are entirely detrimental to cells and tissues. It has been found that lower-level short-duration increases in ROS production can be beneficial [37, 64, 65].

ROS have a number of positive roles one of which is to act as important cellular signalers. On short bursts and cumulative lower levels of ROS production, a signaling cascade is instigated which ultimately leads to the phosphorylation of IκB, an inhibitor of the pro-inflammatory transcription factor nuclear factor kappa B (NFκB). In its inactive state, IκB is bound to NFκB in the cytoplasm; however, once phosphorylated, IκB dissociates from NFκB and is targeted to the proteasome for degradation. This then allows the translocation of free NFκB to the nucleus binding to DNA and initiates a series of gene transcription changes, mRNA production and the downstream expression of key cytokines, chemokines and growth factors including interleukin 8 (IL-8), IL-6 and vascular endothelial growth factor (VEGF) which promotes the growth of new blood vessels. Another gene directly regulated by NFκB activation is cyclooxygenase-2 (COX-2). Its main role is to catalyze the conversion of cell membrane-bound arachidonic acid to prostaglandins including PGE2. It is interesting to note that some of these factors such as IL-6 and PGE2 are pro-inflammatory which in respect of the observed effects of PBM to promote good-quality healing and repair with the good resolution of inflammation may at first sight seem contradictory. However, the absence of inflammation is associated with poor healing as the acute inflammatory response is the primary trigger for the processes associated with cellular repair. Any short-term uptick in inflammation is however mitigated by the photoinduced generation of pro-healing growth factors such as transforming growth factors beta (TGFs-β) and VEGF, an increase in oxidative metabolism, mitosis and the increased expression of genes associated with the resolution of inflammation. In an in vivo animal and in vitro tissue culture investigation of cellular photon transduction pathways, Arany et al. employed a laser emitting a wavelength of 904 nm with radiant exposure outputs ranging from 0.1 to 6 J/cm^2 and concluded that PBM was able to activate latent TGF-β1. TGFs-β have a crucial role in tissue healing and can be activated by a range of stimuli including heat and pH changes [66, 67].

Intra-mitochondrial ROS arise primarily at units I and III of the ETC, and there are multiple systems to regulate excess ROS levels which unchecked can cause irreversible tissue damage. ROS are highly chemically active and can result in the formation of toxic reactive nitrosyl species. Unregulated this results in damage to proteins and fatty acids associated with cellular membranes with consequences in the progressive degradation of cellular function. Production of ROS is associated with both low and high levels of activation of the ETC. Manganese superoxide dismutase (SOD) within the mitochondria, as well as cupric–zinc SOD in the cytoplasm, converts the potentially harmful ROS ions into hydrogen peroxide, and there are other highly active protective antioxidants, including catalase, glutathione reductase, vitamin C, coenzyme Q10, resveratrol and melatonin, amongst others. ROS levels can however rise beyond the immediate shielding capacity of the cellular systems resulting in elevated levels building in the cytosol and the mitochondrial matrix. Should the levels of ROS continue to increase, this can potentially prove highly toxic, and an initial regulatory response results in an adaptive stress reaction mediated by the endoplasmic reticulum (ER) via a stress-induced protein called activated transforming factor 4 (ATF-4) [68]. The potential for phototoxicity was investigated by Khan and Arany in a murine model from which a pathway was characterized involving a system of chaperone proteins referred to as heat stress proteins (HSPs) which act in concert with ATF-4. The outcome of higher-dose photo-irradiation was identified to be temperature related where even a mild increase in temperature quickly inactivated the ROS-buffering enzymes catalase and glutathione reductase and also resulted in strong activation of the heat stress protein (HSP) chaperone system. In response to photon-induced temperature elevations, an increase in ROS levels was noted which when in excess of the capacity of the cells' normal antioxidant systems to reduce the ROS activates a sequence mediated by ATF-4 resulting in calcium ions being released from deposits in the endoplasmic reticulum. Elevated cytoplasmic calcium ions activate opening of the mitochondrial transport pore which in parallel with the actions of the HSP cascade effectively closes down mitochondrial activity and places the cell into a temporary form of stasis. There is an increase in the production and activity of ATPase with a corresponding reduction in the availability of ATP, plus there are changes in the actinic cellular skeleton which result in the translocation of mitochondria away from sites of key activity such as synaptic junctions. Temporary reversible changes are also seen in the axonal membranes with the appearance of structures referred to as axonal varices which may impede the progression of an axonal wave of depolarization [69]. This photon-induced inhibition of cellular metabolism is a protective response to the stimulus, and these pathways have been proposed to be in part some of the underlying mechanisms associated with photon-induced bio-inhibition of axons resulting in analgesia.

ATF-4 and HSP70 act to protect cellular and tissue integrity enabling a cell repair and tissue protection response which increases cellular resistance to stress that could otherwise result in apoptosis.

The ETC is not 100% efficient as there is some leakage of electrons and protons back across from the inner membrane space into the mitochondrial matrix. This occurs by virtue of two mechanisms: One is basal mediated by the catalytic activity of adenine nucleotide translocation associated with free fatty acids in the inner mitochondrial membrane. The other is inducible courtesy of a group of proton transporters related to the superfamily of transmembrane proteins associated with the transfer and carriage of metabolites. Brown adipose tissue was the first to be identified as possessing these inducible proton carriers, which are referred to as uncoupling proteins (UCPs). The presence of UCPs has subsequently been identified in many organs of high metabolic activity as well as muscle tissue sites. The UCPs open passage for protons away from the inner mitochondrial membrane back into the matrix, which results in an elevation of temperature and a reduction in the inner membrane electrical gradient. The reduced availability of inner membrane protons and the associated reduced electrical membrane gradient result in a reduced efficiency in production of ATP by the ETC. It is the current consensus opinion that elevated levels of ROS act as an initiator for the manufacture and activity of the UCPs [70–73]. The UCP response to elevated ROS levels has been suggested to be a further contributory factor to the mechanisms of photoinduced analgesia [9] (◘ Fig. 7.3).

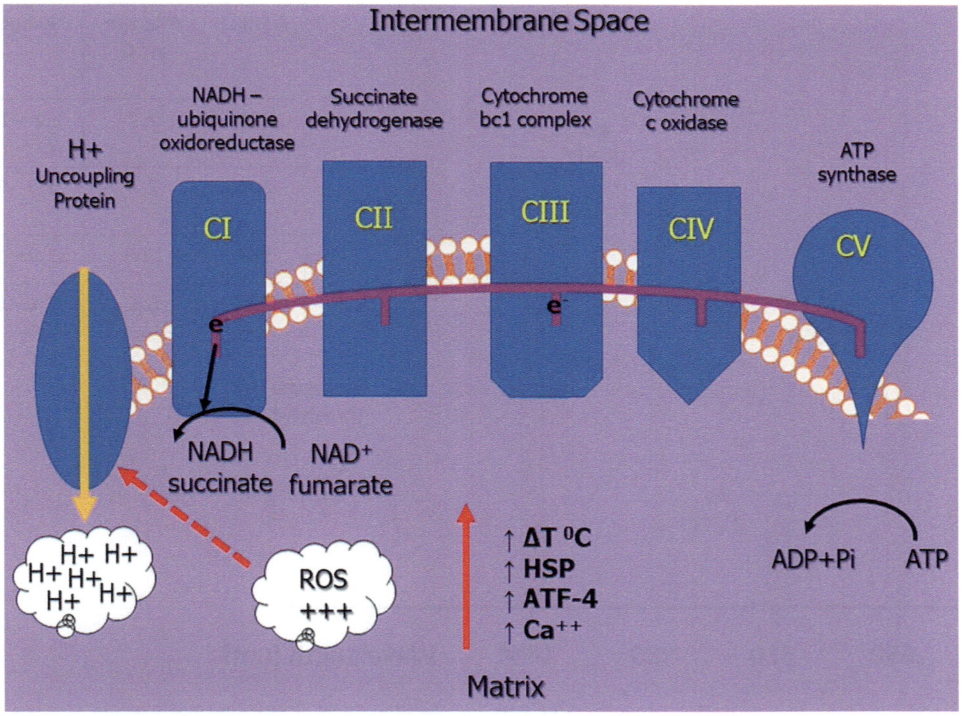

Fig. 7.3 Reverse electron transfer can occur on elevation of ROS levels resulting in re-formation of ubiquinone (unit 1 of the ETC) from ubiquinol in response to an increase in ROS. Also, in the presence of UCPs, the proton gradient can move back into the mitochondrial matrix bypassing unit V. This results in an elevation of the matrix temperature with the activation of HSPs, a depletion of ATP and decoupling of the ETC resulting in anaerobic metabolism. The cytoplasmic levels of ROS rise with the resultant activation of ATF-4 which in concert with the HSPs offers some protection of the cell from metabolic stress. These processes have been proposed to be contributory elements to improving cellular resistance to recurrent stressors (pre-conditioning) as well possibly as contributory factors to the mechanisms associated with photoinduced analgesia (Graphics S. Parker Reproduced from Scialò F, Fernández-Ayala DJ, Sanz A. Role of mitochondrial reverse electron transport in ROS signaling: potential roles in health and disease. Frontiers in physiology. 2017 Jun 27;8:428)

7.3 Principles of Action: Regional (Tissue) Perspectives

Although PBM-like effects are always associated with surgical laser usage, there is an awareness that to optimize the observed clinical results, there may be merit in the selective application of dual or multiple wavelengths. The bioavailability of photons to deeper structures varies markedly according to wavelength; hence a philosophy of care may involve the predominant surface absorption of high-intensity irradiation, from an ultrashort pulse duration of a mid- to far-infra-red wavelength laser to effect surgery with the minimal amount of collateral tissue damage. This may then be followed by a dedicated PBM therapy with a low-intensity application of a shorter wavelength in the visible to near-infra-red range, which is relatively poorly absorbed and enabling deep delivery of photons to an extended area of the wound periphery.

7.4 Wavelength

The choice of wavelength influences the depth of penetration of the photonic target tissues to depth due to substantial differences in optical penetration; this is in turn consequent upon absorption and scattering by tissue elements such as pigments, proteins, fats and water. There is some increased absorption of the 980 nm by water which can lead to a significant surface and subsurface thermal rise, as opposed to a 800–810 nm wavelength. The increased absorption of water by the 980 nm is relatively small compared to other wavelengths; however, due to the high prevalence of water in biological tissues, the excitation of the water can rapidly generate a localized thermal rise. The table below is taken from a recent in vitro study which indicates the limited utility of the 980 nm for extended use, due to an increase in surface and sub-surface temperature, in comparison to an 810 nm source for sub-surface dose delivery [56] (Fig. 7.4).

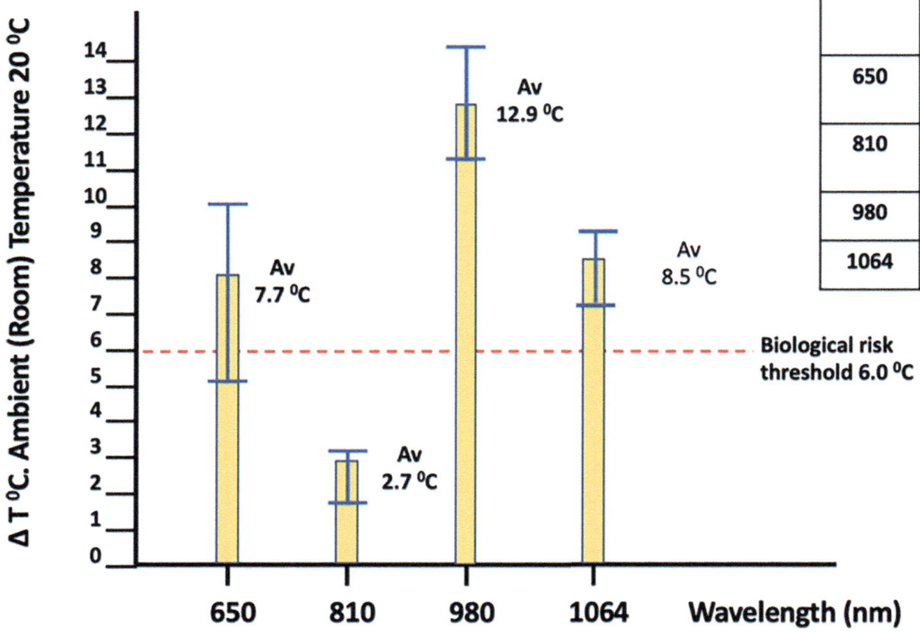

λ (nm)	Average Δ °C	Range °C	Number of readings
650	7.7	5.2 – 10.0	17
810	2.7	1.9 – 3.2	10
980	12.9	11.3 – 14.6	11
1064	8.5	7.8 – 8.7	7

Fig. 7.4 In vitro porcine muscle comparison of surface ΔT values as a function of wavelength (λ). Average power 2.0 W/cm^2 (continuous wave). Surface thermal rises exceed the 6.0 °C threshold limit for all wavelengths except 810 nm. (Reproduced from Cronshaw et al. Photothermal aspects of high energy photobiomodulation therapies: an in vitro investigation. Biomedicines 2023, 11, 1634 [56])

To a lesser degree, there is caution attached to the use of the 650 and 1064 nm sources, as again these wavelengths can be associated with a significant rise in tissue temperature with extended use. However, all these wavelengths can be of use subject to training, as the operator needs to be aware of the methods to avoid surface and sub-surface tissue overheating by means of a good technique of application, as discussed later in this chapter.

To avoid a rapid rise in surface temperature, the use of the shorter wavelengths such as the 445 nm (blue) laser requires low output power <100 mW and a corresponding low irradiance 50–100 mW/cm^2 for brief periods of exposure. Due to the high energy of the blue end of the spectrum, the high superficial absorption and rapid thermal rise seen with the shorter-wavelength sources, the utility of this end of the spectrum is at present limited for PBMT.

By convention PBMT is defined as not being primarily a significant photothermal effect resultant in adverse irreversible changes to structural proteins. However, some localized heat is generated, and a small highly localized thermal rise within the cell and subcellular structures can result in a range of interesting responses. At an elevation of temperature of 1–2 °C, slight warming of tissues has been found by the dermatology community to be associated with good-quality healing [74–77]. All enzyme functions are promoted by a small increment in temperature, and this may reasonably be anticipated to have a positive effect on cellular productivity and metabolism. However, above a 2 °C increase in temperature, a protective response is activated by an evolutionary primordial system common to all eukaryotes of heat stress proteins (HSPs) [78]. Primordial pre-eukaryote life was based on a variety of cycles of metabolism, including aerobic and anaerobic respiration, as well as sulfur metabolism in single-cell Archaeans and bacteria. The evolution or adoption of the mitochondrial Krebs cycle and ETC by the early eukaryotes provided a lot of additional adenosine triphosphate (ATP)-generating capacity. In addition, as this is an exothermic reaction, the essential platform was built that provided the added energy that could sustain the building of more complex organisms, resulting in multicellular species. Control over cellular temperature is a highly conserved feature of eukaryote cytochemistry. The thermal aspects of mitochondrial metabolism are conventionally denied a role in PBM, on the grounds that by definition PBM is a nonthermal event. This premise, however, may be true in respect of distinguishing the physiological effects of heat compared to light. However, the selective absorption by elements of the ETC may produce a localized increase in the intracellular temperature [38, 79, 80]. The HSP system protects against thermal damage to part formed proteins and enables the cell to conserve the partially built materials. A cellular response is elicited which reduces cellular activity, including the downregulation of the activity of the ETC. Once the adverse thermal stimulus has been removed, the cell can then revert to normal functional physiology without harm, providing the duration was not too long and the peak of the thermal insult not too high [68, 81].

The microthermal aspects of PBM have not yet been comprehensively investigated. Beckham et al. found that a thermal rise to 43 °C was a safe and effective primer for the stimulation of the production of HSPs. This preconditioning of the cells resulted in an increase in viability and a corresponding reduction in apoptosis on later laser exposure to much higher temperatures. We explore the issues around the safe administration of PBMT later in this chapter [6].

7.5 Optical Transport Pathways and the Relation to Wavelength

Light striking the surface of tissues can be transmitted, reflected, refracted, absorbed or scattered. Optical transmission into the tissues depends on a number of factors including the angle of incidence which with a more oblique angle results in an increase in reflection ranging from 2% to 7% [31, 32] (◘ Fig. 7.5).

On entering the tissues, the increased optical density of the medium changes the direction of the incident beam and there can be internal reflection and optical loss to deeper tissues (remission) at the epidermal–dermal boundary.

On entering the deeper layers of the dermis, some photons can travel straight through the tissues (so called ballistic photons). Some ricochet off and are deflected by tissue sub-constituents such as tissue boundaries, organelles and collagen which results in the phenomenon of scatter. With a coherent (laser) light source where all the photons are in the same phase and wavelength, should photons collide, this results in localized amplification and the phenomenon of speckling where bright spots of light can be seen which are a characteristic of lasers observable in the visible range. Some of the scattered photons can then travel onward (so called snake photons). Scattered photons can be directed in different directions once they hit particles within the tissues. The coincidence of the incoming photon stream from the surface source plus the photons deflected back by scatter to the dermal–epidermal boundary result in a higher fluence in the immediate sub-surface layers of the tissues than that applied at the surface [83]. Depending on wavelength absorption, this can result in a significant reduction in optical delivery to depth and may cumulatively lead to a higher temperature at this immediate sub-surface tissue level compared to the surface.

The anisotropy factor (*g*) is a function of the angle at which photons are scattered. In most tissues this is mostly a forward direction in the red to NIR wavelength range, although there can also be some scatter back toward the source; this results in energy losses to deeper tissue layers and radiant exitance. Shorter wavelengths 400–600 nm are highly absorbed by proteins and pig-

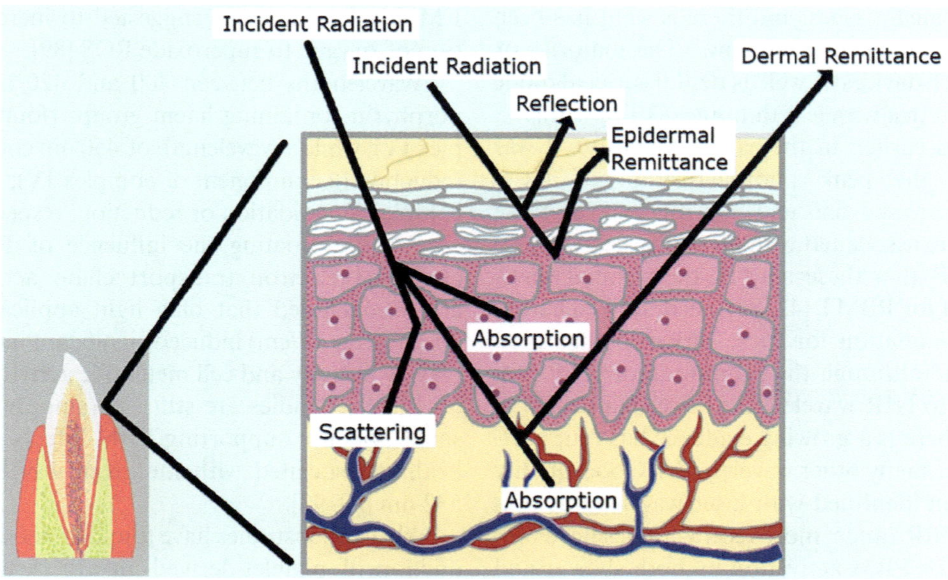

◘ **Fig. 7.5** Optical transport through biological tissues results in high power losses to deeper layers within the first few millimeters of application at the surface. Outcome variables include the thickness of the epithelium, skin pigmentation, wavelength applied, angle of incidence at the surface and the consistency of the tissues. In general it can reasonably be expected to deliver 2–10% of the surface applied to a depth of 1 cm using wavelengths in the red to NIR. (Reproduced from Cronshaw et al. Photobiomodulation dose parameters in dentistry: a systematic review and meta-analysis. Dentistry Journal. 2020 Oct 6;8(4):114 [82])

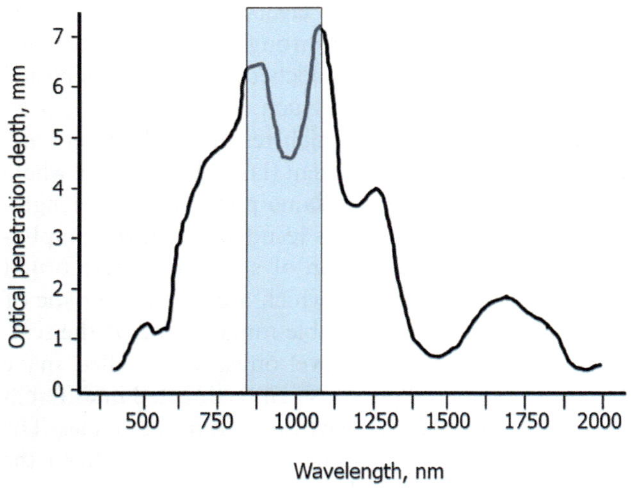

7

Fig. 7.6 The optical penetration depth δ of light into human mucous tissue over the wavelength range from 400 to 2000 nm. (Graphics by S. Parker taken from Bashkatov A, Genina E, Kochubey V, Tuchin V. Optical properties of human skin, subcutaneous and mucous tissues in the wavelength range 400 to 2000 nm. J. Phys. D: Appl. Phys. 2005;38:2543–2555 [84])

mented materials including melanin. Longer wavelengths of 1500–3000 nm are predominantly absorbed by water. If there is high absorption, the effects of optical scatter are not significant in predicting depth of optical penetration; this is relatively superficial in the shorter and longer wavelength ranges, in contrast to the 600–1100 nm commonly in use for PBMT [84].

The red to NIR wavelengths by comparison to shorter and longer light sources have relatively low absorption enabling optical penetration of 5 mm or more into biological tissues, and there is what has been described as a "therapeutic window." The majority of dedicated PBMT devices as well as dental surgical diode lasers fall within this wavelength range (Fig. 7.6).

As discussed earlier, in the earlier literature, it was proposed that the peak photoactivation of CCO occurred at or around 650 and 800 nm and, as these wavelengths were associated with an increase in the production of ATP, that these may be the optimal wavelengths to select for PBMT [42, 85]. As described earlier the scientific foundation for these proposals has been reappraised, and although the promotion of ATP production by red to NIR wavelengths is well supported in the literature, there is a growing evidence base supporting the merits of many other wavelengths. Good-quality healing has been identified with laser wavelengths outside the red to NIR range; more recent studies have provided evidence of PBM activation by both shorter and longer wavelengths than those originally believed to be the ones most likely to be effective [86]. The potential to reach deeper tissues is enhanced by the choice of a wavelength which is less highly absorbed, and a peak depth of penetration is associated with wavelengths at or around 800 nm. However, pending further research defining optimal photon transduction mechanisms for surface and sub-surface target tissues at the time of writing, there is no current consensus, and the published guidelines on wavelength are best viewed as indicative rather than strictly prescriptive.

A recent systematic review found evidence supporting PBM actions in many wavelengths outside of the visible to NIR wavelength ranges previously viewed as the essential choice for therapy. As there may be many pathways associated with the positive influence of incident light on biological tissues, the term quasi-photobiomodulation (Q-PBM) has been proposed in order to distinguish the effects as being related to but possibly operating on different sites of action to those characterized from light sources within the red to NIR spectrum (Fig. 7.7).

Opländer et al. found that blue wavelengths were highly active in generating NO in the epidermis. Irradiation of human skin with moderate doses of blue light caused a significant increase in enzyme-independent cutaneous NO formation as well as NO-dependent local biological responses, i.e., increased blood flow. The effects were attributed to blue light-induced release of NO from cutaneous photolabile NO derivates [88]. Serrage et al. demonstrated that blue LED light (400–450 nm, 5.76 J/cm^2) was as effective in inducing increases in mitochondrial activity as NIR LED light (810 nm, 5.76 J/cm^2). Blue light activates flavins and flavoproteins associated with units I and II of the ETC resulting in an increase in the activation of flavin mononucleotide FMN which has been suggested to increase the reduction of oxygen to superoxide ROS [89].

Wavelengths between 400 and 420 nm can oxidize porphyrin-containing haem groups (found within complex IV), while a wavelength of 450 nm could induce CuB reduction (a component of complex IV), hence inducing complex IV oxidation or reduction, respectively.

While evaluating the influence of PBM on mitochondrial electron transport chain activity, Buravlev et al. concluded that blue light application (442 nm, 30 mW/cm, 3 J/cm) induced significant increases in complex IV activity and cell metabolic activity [90].

Further studies are still emerging; however there is some evidence supporting PBM effects with enhanced healing associated with the green wavelengths around 532 nm [89–93].

Also, other studies have found an increase in the production of platelet-derived growth factor (PDGF) by the 2940 nm Er:YAG laser [94]. PDGF is an important chemokine to attract stem cells to sites of injury and elevated levels are associated with enhanced osseous repair. Furthermore, there is evidence supporting the

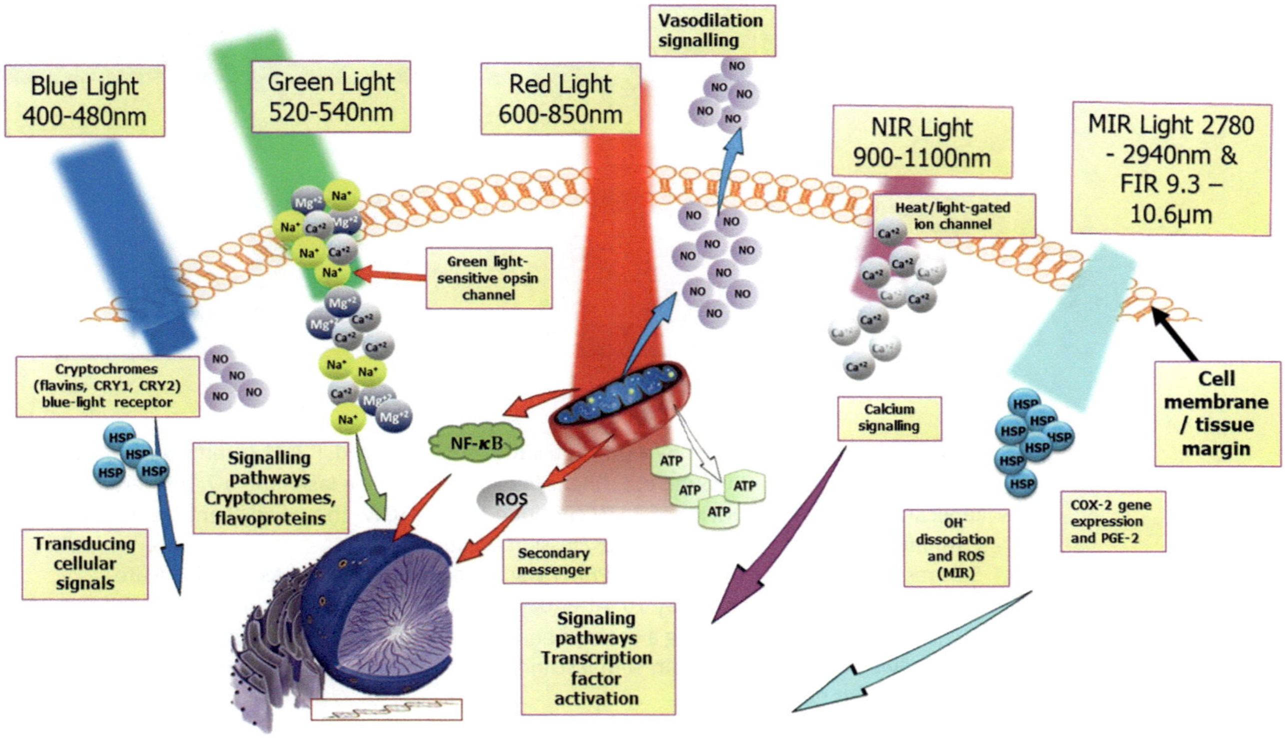

Fig. 7.7 Q-PBM tissue/cellular signaling with visible/IR λ. (Graphics by S. Parker taken from Salehpour F, Mahmoudi J, Kamari F, Sadigh-Eteghad S, Rasta SH, Hamblin MR. Brain Photobiomodulation Therapy: a Narrative Review. Mol Neurobiol. 2018 Aug;55(8):6601–6636. ► https://doi.org/10.1007/s12035-017-0852-4 [87])

analgesic effects of the 2780 nm Er,Cr:YSGG laser, as well as clinical and histological evidence of good-quality healing associated with the 10,600 nm CO_2 laser [86, 95, 96]. As a collateral effect of using a surgical diode, erbium or CO_2 laser, it is likely that there are PBM effects despite the limited optical penetration possible consequent to the very high remission of these wavelengths. Contrary effects have been described for many wavelengths and there are many apparent conflicts in reported outcomes. Resolution of optimized parameters for PBMT is reliant on high standards of experimental design and reportage which require a thorough appreciation of the optical physics associated with dosimetry, as discussed below.

7.6 Principles of Action: Clinical Perspectives

All clinical dental lasers have the capacity to demonstrate an associated positive tissue outcome subject to the knowledge of the operator to deliver the energy in a deliberate and safe fashion. All surgical lasers are recognized as being a good choice of tool to use subject to adequate training, as consistent good-quality healing can be observed across the full spectrum of clinical surgical lasers currently adopted in contemporary clinics. The collateral benefits of surgical laser integration into routine clinical practice include a dual advantage of precise controlled tissue manipulation along with excellent quality healing, repair and the capacity to promote regeneration of damaged or lost tissues. This "two for one" benefit is seen as being in part a consequence of the action of photons below the dosimetry threshold to damage tissues; this effects positive cellular physiological pathways which include an increase in stress resistance and the promotion of highly productive aerobic metabolism, angiogenesis and mitosis [86]. The observed tissue healing response outside the immediate surgical zone of destruction is marked by a reduction in acute inflammation with an associated amelioration of postsurgical pain and swelling [97]. The enhanced quality of the wound resolution in contrast to the conventional scalpel, drill and suture surgical approaches is entirely remarkable and to the suitably trained clinician represents a major gain toward the desired clinical outcome as well as facilitating the patient journey (Fig. 7.8).

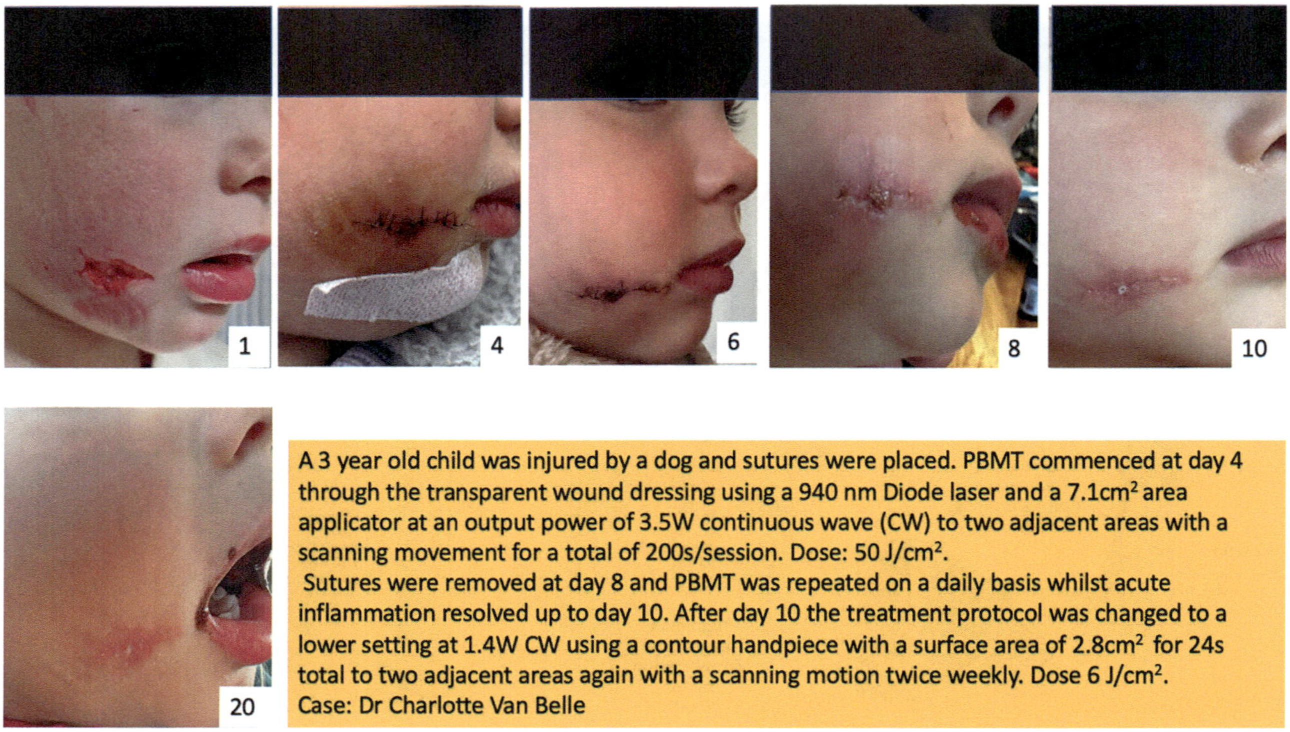

Fig. 7.8 Clinical case: childhood trauma is a regular cause of urgent care attention. PBMT can assist in the mitigation of the acute inflammatory phase associated with primary wound healing. Also an early intervention can improve the viability of the damaged tissues as well as reduce the peak and duration of the physiological response to injury. The reduction in pain and swelling plus the enhanced quality of the final repair make PBMT a highly useful clinical modality. (Images: days 1–20)

7.7 Dosimetry

The application of therapeutic light has been compared to the dosimetry associated with the prescription of a drug. The prescription of a drug requires the knowledge of its active principle, pharmacokinetics, toxicity, side effects, interactions, dosage, duration of treatment and modalities of administration. Indeed, a number of the pathways characterized as PBMT actions are identical to those targeted by drugs such as steroids and nonsteroidal anti-inflammatory medications [98–101].

The same applies for light, with the significant difference that undesirable side effects, toxicity and interactions are virtually nil or comparable to a placebo, as confirmed by results from the current available scientific literature. As with a drug, it has been recognized that to achieve the desired effects, certain parameters are required to trigger the anticipated positive clinical outcome of enhanced healing or perhaps analgesia.

In relation to PBMT when we refer to dose, this corresponds to the time accumulated radiant exposure which is the energy (J) over a standard unit area (1 cm^2) with the variable of the rate of delivery (W) and the duration of exposure (s). In many publications on PBM, there are referrals to a biphasic response as by underdosing the tissues at cellular levels, nothing happens and then, at a higher dose, there is a zone of positive activity followed at higher end dosimetry by a zone of inhibition of cellular physiology. By analogy to the contrary effects of drugs dependent on pharmaceutical dose observed by Hugo Schulz and Rudolf Arndt, this has been enshrined in the PBM literature as the Arndt–Schulz law [102–104]. The premise of Arndt–Schulz states that "for every substance, small doses stimulate, moderate doses inhibit and large doses kill." In modern pharmacology texts in recognition of the many exceptions to this rule, it has been supplanted by the concept of a hormetic response. Hormesis is a stress-induced characteristic of many biological processes manifesting as a biphasic or triphasic response to exposure to increasing amounts of a substance, and within the hormetic zone, the biological responses to toxins and other stressors are generally favorable.

7.8 Output Power, Radiant Exposure and Irradiance

In simple terms, to deliver the correct dose, the operator must provide sufficient energy (J) to an area (cm^2) at a rate (W) and an intensity (W/cm^2) consistent with the biological capacity of the cells at a subcellular level to respond favorably to the photonic stimulus. The prob-

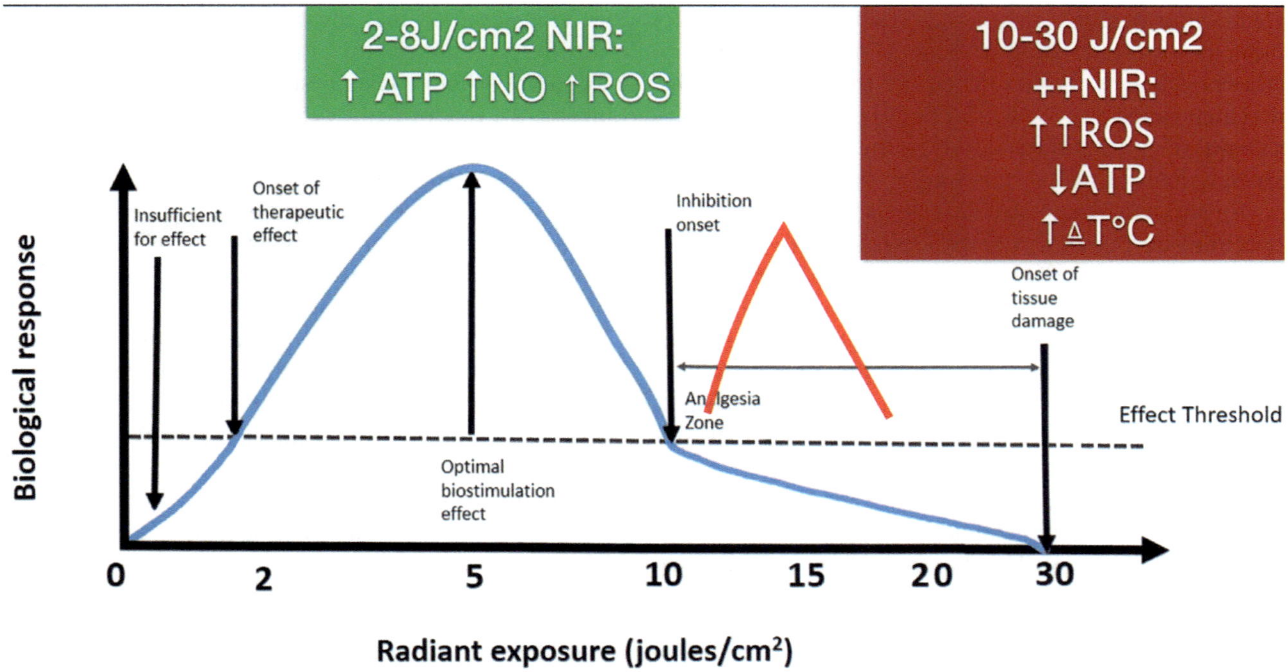

Fig. 7.9 Two at tissue-level dose ranges with the effects of stimulation or inhibition have been identified. This multiphasic response to dose is seen frequently in pharmacy and PBMT has been compared to being a drug equivalent. (Adapted from Cronshaw et al. Feeling the heat: evolutionary and microbial basis for the analgesic mechanisms of photobiomodulation therapy. Photobiomodulation, Photomedicine, and Laser Surgery. 2019 Sep 1;37(9):517–26 [9])

lem of calculating the correct dose has been compounded by confusion on what constitutes the delivered dose. The correct scientific term for the 2D planar energy/cm^2 is radiant exposure and for the 2D power/cm^2 is irradiance. The irradiance is in essence the intensity of the light, and as discussed below, along with the accumulated delivery of the correct dose (J/cm^2), the rate of delivery (W) and the intensity of the source (W/cm^2) are of prime importance. The delivered dose is time dependent on the power output to the area exposed, so, for example, 0.5 W/cm^2 represents 0.5 J/cm^2 delivered in 1 s. The dose (fluence) is the accumulated radiant exposure J/cm^2 × time; hence to deliver 5 J/cm^2 at an output power of 0.5 W to an area of 1 cm^2, it takes 10 × 0.5 = 10 s [105].

In PBMT it is recognized that in the red to NIR wavelengths, the optimal accumulated dose at target tissue level to stimulate cellular activity is in the range of 2–8 J/cm^2. Less than this, dose very little happens and beyond this in the range of 10–30 J/cm^2, a contrary effect of bio-inhibition has been identified. Although at higher dosimetry the positive aspects of PBM of an increase in cellular activity are absent, higher range treatments have been associated with pain relief as well as an increase in cellular stress resistance [6, 9, 106]. As this may be the desired clinical objective, it has been proposed that rather than viewing PBM dosimetry as a binary all-or-none biphasic response, it is more useful practically to consider it as a multiphasic response (see Fig. 7.9). There is a commonality of the stimulatory or inhibitory/protective PBM effects on all cellular types where the contrary effects of positive or negative responses can be observed in the two therapeutic dose windows in the red to NIR wavelength range.

In respect of radiant exposure (J/cm^2), a joule is a joule regardless of wavelength. However, in respect of spectral energy (J/nm), there is a considerable difference between wavelengths. For example, a 405 nm source has an eV of 3.06, whereas a 810 nm source has an eV of around 1.53. The molecular effects of the absorption of a more highly energetic 405 nm photon are highly likely to be quite different to that achieved by the more pedestrian 810 nm photon. In consequence, is it necessary to take into consideration wavelength when we are selecting a unit dose? A recent in vitro cellular study by Young et al. investigated comparative PBM dosing using a variety of wavelengths [107]. By amending the treatment exposure time to allow for the differences in individual photon energy, a comparable positive stimulatory effect could be seen across wavelengths in the range of 447–1064 nm. A proposal was made to use a new unified unit of dose using the 810 nm as a baseline reference unit from which a calibration factor could be applied to calculate the appropriate photon eV adjusted radiant exposure. At the time of writing, the majority of PBMT devices in use adopt the 600–1100 nm waveband, and the difference in dosimetry within this range is of the order of ±10%. At target cell level, given this relatively small order of difference, it is advisable with the red to NIR to adopt a median of 5 J/cm^2 for biostimulatory effects, which is approximately in the middle of the accepted range of 2–8 J/cm^2. For pain relief it is recommended to apply at target level around 15 J/cm^2 as a safe median dose for red to NIR wavelengths [9, 108].

In the case of sub-surface targets, it is necessary to take into account the energy losses due to reflection, refraction, radiant exitance, absorption and scatter. In an in vivo study by Alvarenga et al. of gingival tissues transilluminated by a 660 nm laser source, it was found that at 5 mm in depth, there was approximately a 50% attenuation (energy loss) to that applied at the surface [109]. Across the red to NIR range, it is reasonable to expect 2–10% of the surface applied energy to reach 1 cm in depth [83, 110–112]. As a result, to achieve the required dose at depth demands the application of more energy at the tissue surface to compensate for this high energy loss. At deeper depths beyond 1 cm below the surface, it becomes progressively more difficult to deliver the required dose either without an excessively long treatment time or with higher output power device adverse photothermal effects in the surface tissues. Fortunately, in dentistry the majority of tissue targets are in the surface to 1 cm in depth range.

7.9 Dose Delivery: Global Issues

There are a number of important considerations to be addressed prior to the practical aspects of administering the dose. First of all, the operator must decide the objective: stimulation of cellular activity or pain relief. Next is the location and approximate size of the target: for example, pain relief associated with a small aphthous ulcer will require a much-reduced overall dose compared to the objective of pain relief for a temporomandibular disorder (TMD). An aphthous ulcer is a small and superficial target, whereas myalgia associated with TMD may involve larger areas of sub-surface muscle and multiple muscle groups. Clearly it takes longer to treat a larger target, particularly with the added complication of compensating for the energy losses expected with delivery of the dose to depth.

Having determined the treatment objective, location and approximate area of the treatment sites, the next decision is to choose the optimum tool to deliver the required dose. This may be dictated by the restrictions imposed on the equipment available for the operator to use. This can be critically evaluated according to wavelength, range of output power and size of the applicator probes (◘ Fig. 7.10).

It is essential to consider the issue of time-dependent energy delivery and power output as the rate (power = W) and intensity (irradiance = W/cm^2) are critical parameters. It is accepted that the optimum cellular levels for photo-biostimulation of time accumulated radiant exposure is 5 J/cm^2. The question arises as to the importance of the rate of delivery which is expressed as watts (power) and the concentration of the watts to a unit area which is described as irradiance (W/cm^2). As 1 W is 1 J delivered in 1 s and the required dose is 5 J/cm^2, can the dose be delivered rapidly as say 5 W for 1 s? In short, the answer is no, as the rate of energy delivery has a considerable influence on cellular response. Also, at an output power greater than 250 mW, if the area of exposure is smaller than 0.5 cm^2, the outcome may be harmful. Short-duration high-dose delivery or the inappropriate use of a fine diameter probe can easily overload the biological capacity of the tissues to respond favorably to the stimulus to the extent of triggering irreversible cellular damage as opposed to the benefits associated with PBM [113, 114].

Many first-generation dedicated PBMT devices have a low power output range typically in the vicinity of 1–300 mW. A typical set of delivery probes offer glass optical guides of 4–6 mm in diameter with a corresponding surface area of 0.04–0.3 cm^2. At an output power of 40 mW, a 4-mm-diameter probe can deliver 40 mW/0.04 cm^2 = 1 W/cm^2. This would at first sight appear consistent with the dosimetry recommendations, as in 5 s this would apparently deliver 5 × 1 J = 5 J/cm^2. However, in reality at a power output of 40 mW in 5 s, the device delivers 5 × 40 mW = 200 mJ to a very small area of 0.04 cm^2. As the tip surface area is 0.04 cm^2, to cover an area of 1 cm^2 requires either multiple point applications or a scanning movement to treat an area of 1 cm^2 which is 25 × the tip area of 0.04 cm^2. As a result, the total treatment area time to deliver 5 J/cm^2 becomes 125 s.

Regrettably confusion over the relation between output power (W), irradiance (W/cm^2) and total delivered radiant exposure (J/cm^2) has been the source of many errors in clinical trials resulting in an outcome of a null or very low-level beneficial effects [82, 115].

It is of great importance that there is recognition of the many sources of potential error in past research to permit the evidence base to progress. A recent systematic review found that the majority of studies do not indicate the use of a power meter to confirm that the power output of the device matches that on the device display. There are many possible sources of power losses, and the use of a power meter must become a standard approach to validate the experimental outcome. In addition, it is essential to provide details of the applied parameters to permit critical evaluation and possible acceptance at the level of peer review as well as reproduction of the outcome by later researchers. The diagram below summarizes the suggested reportage of parameters in future PBM studies (◘ Fig. 7.11):

Fig. 7.10 The choice of dose delivery may depend on access, availability and size and depth of the target tissues. A representative selection more commonly available is shown in the images. Other devices may be provided by the manufacturer, and the key parameters to prescribe the correct dose are the optical spot size at the target, the output power of the device, the spectral beam profile and the emission mode (gated vs. continuous mode)

Fig. 7.11 Laser emission and interactive parameters. (Source: Parker S. Laser Photonic Energy Delivery in Clinical Dentistry: Scrutiny of Parameter Variables. PhD Thesis 2023, pp. 152 [116])

Type of study—such as RCT	Sample size/groups/control/randomization	Blinding (single / double blinding)
Laser used—emission wavelength (nm)	Delivery system Fibre, Waveguide, Articulated Arm	Gaussian / ''Flat top'' beam x-section
Emission mode (CW, Gated-CW, FRP)	Power meter used and output calibrated	Mean, median, max / min power (W)
Target tissue / lesion dimensions (mm) / Target tissue depth (mm)	Tip to tissue distance (mm)	Irradiated ''spot'' size (cm^2)
Fluence / Radiant exposure (J/cm^2)	Irradiation / Power density (W/cm^2)	Total energy delivered (J)
Irradiation PBM therapy time (sec)	Irradiation frequency (n x days)	Tip movement / area covered (mm/area)

7.10 Optical Spot Size of Applicators

Given the inherent difficulties in dose delivery associated with low power output devices and small diameter optical delivery probes, it is noteworthy that this type of delivery equipment has been used in many successful clinical trials. However, in a recent systematic review, a comparative analysis of reported outcomes and the relation to optic surface spot sizes applied found clear evidence demonstrating that the use of small diameter probes <5 mm in diameter is associated with less consistent reported clinical success, in the treatment of both superficial and deeper-seated pathologies [56, 82, 108].

In recognition of the potential added clinical outcome gain of PBMT, many surgical diode lasers have been re-tasked for the purpose as an additional nonsurgical treatment modality. The standard configuration of surgical lasers is designed to confine the energy to a very small point of application typically of 300–400 μm in diameter, to focus the output power (high irradiance) to a level sufficient to generate localized heat at temperatures above 100 °C, resulting in the separation of the tissue by photothermolysis. As biological tissues show signs of damage at sustained tissue temperatures above 45 °C, in contrast to surgery, it is important for PBMT to keep the irradiance (W/cm^2) low [68, 81, 117]. This can be achieved by setting the power output to a larger area of exposure to keep the irradiance at between 250 and 500 mW/cm^2. It is important to avoid confining the energy emission to a small point at the site of application, and this can be achieved by use of a surgical handpiece with or without a narrow diameter surgical tip at a distance from the treatment site.

The surgical handpiece with or without a tip can be a convenient way to deliver radiant energy to small areas as well as offering good ergonomics in confined intraoral areas. The beam emission is divergent by around 16–18°. The visible spectrum aiming beam can be used to assist the operator in creating the required sized spot at the tissue surface. With an area of approximately 1 cm^2, the delivered dose can then be calculated by multiplying the output power by the time of exposure. To take an example, with a 250 mW output power delivered over 20 s to an area of 1 cm^2, the total energy delivered to the area is 20 × 250 mW = 5 J/cm^2.

Some devices are provided with custom glass optic guides for PBM. The diameter of these tips is typically between 4 and 8 mm, and if held in contact with the tissues, this delivery device covers an area at the surface of between 0.04 and 0.5 cm^2. This type of tip can be held in close proximity to the treatment area; however as discussed earlier, if the area of the lesion is much larger than the tip, the treatment time has to be multiplied to ensure sufficient energy delivery. Furthermore, the practical operator facility to deliver sufficient energy to larger targets at depth is significantly reduced with a smaller optical spot size by comparison to the usage of a larger surface area delivery tool. A minimum surface spot size of 1 cm^2 is advisable, and for larger target areas, a greater than 1 cm^2 surface spot size applicator can be useful to spare precious time as well as help deliver sufficient energy to the required depth. However, although larger surface spot applicators 3–7 cm^2 are available, these require a higher output power to achieve the same irradiance, and the operator must have an appreciation of the significance of the spectral beam profile of the source as well as an understanding of the importance of a good application technique [56].

Many surgical lasers are supplied by the manufacturers with a handpiece designed to be used in conjunction with a high strength peroxide gel for laser-assisted tooth whitening. The surface area of these tools varies between manufacturers and typically falls into the range of 2.8–4.0 cm^2. By measuring the optic window area, the operator can set the output power of the laser to deliver an average irradiance in close proximity to the surface of the target of 250–500 mW/cm^2. For example, with a surface area of 2.8 cm^2, by setting the output power at 1.4 W, the energy is delivered to a total treatment area of 2.8 cm^2 at an average irradiance of 0.5 W/cm^2. At this setting the device will deliver in 10 s 10 × 0.5 W = 5 J/cm^2 which corresponds to a tissue surface delivered dose of 14 J to an area of 2.8 cm^2 = 5 J/cm^2.

In recognition of the rising use of surgical diode lasers for PBMT, a number of manufacturers have produced some custom PBM peripheral delivery tools. There are some interesting designs to assist dose delivery including larger surface spot sizes of up to 7 cm^2+. Larger applicators can be of considerable value should it be necessary to treat larger areas and volumes of tissue, for example, in the management of myalgia associated with temporomandibular disorders (TMD). Because this condition may involve multiple larger target muscle sites, it is possible to reduce the total treatment times as with a larger applicator more tissue is treated in the same unit of time. An added bonus to using medium- to larger-sized optic delivery devices is that as previously mentioned, the evidence base supports a more consistent positive treatment outcome by comparison to smaller optic probes, especially with subsurface conditions. This may be due to the delivery of photons to a greater volume of target tissue at depth (◘ Fig. 7.12).

Aside from re-tasked surgical diode lasers, there are many custom PBMT devices. Many of these devices are designed for sports physiotherapy applications to treat deep tissue injuries over large treatment areas; however, they can be easily adapted to dentistry as well as other

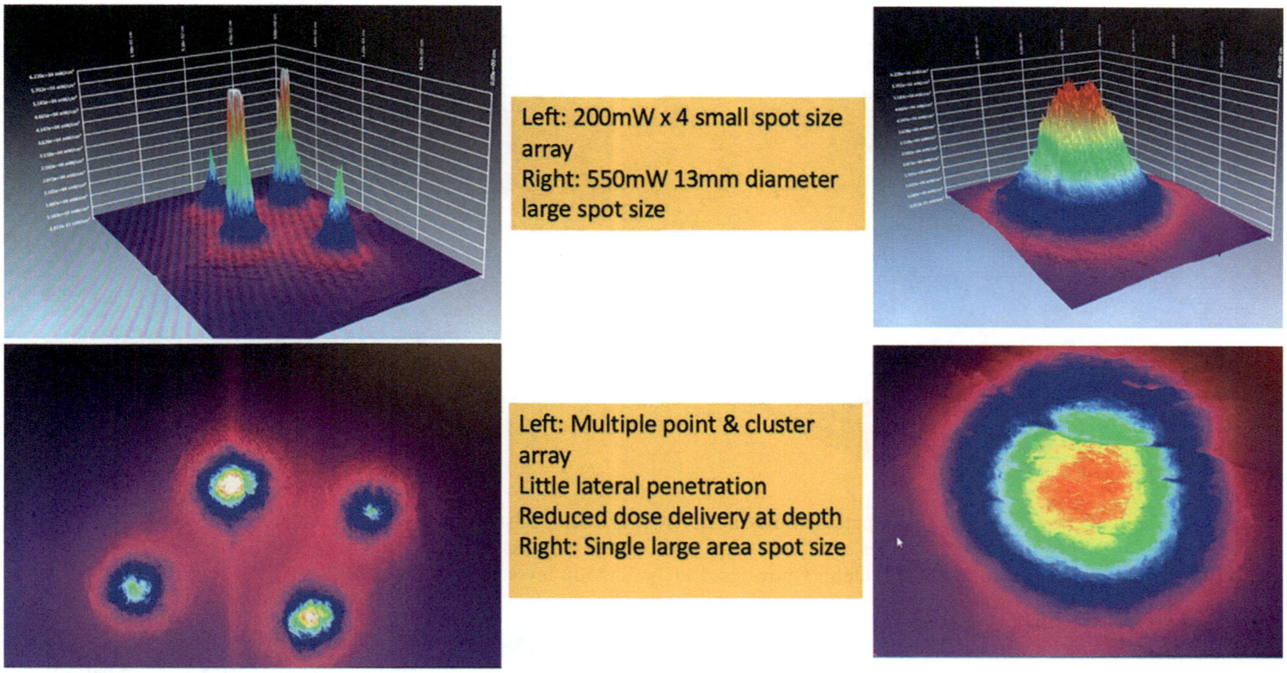

Fig. 7.12 Beam profilometer images of the spectral power distribution on transillumination through lean muscle porcine tissue samples. On the left upper and lower: 3D render and 2D images of a four-probe small spot size array. On the right upper and lower: 3D render and 2D image of a single large spot size device to the same area of exposure. (Adapted from Parker S et al. Photobiomodulation delivery parameters in dentistry: an evidence-based approach. Photobiomodulation, Photomedicine, and Laser Surgery. 2022 Jan 1;40(1):42–50 [108])

medical applications. Given the considerable rise in interest amongst practitioners of the value of PBMT, this is a rapidly expanding range of equipment. There are a variety of interesting design features available intended by manufacturers to assist the operator. Amongst the original designs are clusters of multiple low output diode lasers which may include LEDs as well as simultaneous irradiation with lasers of mixed wavelengths. There are also high output power devices associated with multiple simultaneous wavelength delivery lasers and applicators which can vary in size for a single spot size of 12 cm^2 or more. The more sophisticated delivery peripherals have an inbuilt thermal camera which enables the operator to avoid overheating the surface overlying a deeper placed target. The user interface can assist the operator in selecting the manufacturers' recommended parameters and offers guidance in delivery techniques. The relative merits of the very many designs available are, at the time of writing, a matter of opinion rather than proven notwithstanding the many claims made by competing manufacturers. For the more experienced practitioner seeking an advanced PBMT machine given the current evidence base at the time of writing, there is no consensus on an ideal configuration. This is an area of vigorous continued research efforts, and we discuss this further at the end of this chapter (see ► Sect. 7.38–7.40).

7.11 Spectral Beam Profile

The majority of optical spot size applicators result in a Gaussian optical beam spectral profile, and this can complicate calculations of safe delivery parameters, especially with larger surface spot sizes 2–7 cm^2+. With a Gaussian device, the power distribution typically generates a peak in the mid-third of the beam (Fig. 7.13).

Most therapeutic laser systems use a base transmission emission mode (TEM_{00}) with a typical beam divergence angle of around 16°. Aside from the uneven energy distribution across the beam, beam divergence can complicate dosimetry if care is not taken by the clinician to achieve the desired surface spot size, and in this respect the visible aiming beam can be helpful, or a spacer may be used to assist the clinician to deliver the dose to a predetermined surface spot size.

It is possible to modify the optics of the output spectral beam profile to create a collimated and even beam across the beam diameter. These so called "flat-top" devices contain internal optic prisms which convert the inherent Gaussian shape of the beam to an even profile. Also, by the use of internal lenses, the beam is corrected to prevent beam divergence and to create an even energy distribution with a beam to an area maintained even at a distance from the target (see images below). The merits of the "flat-top" concept in respect of dose delivery are

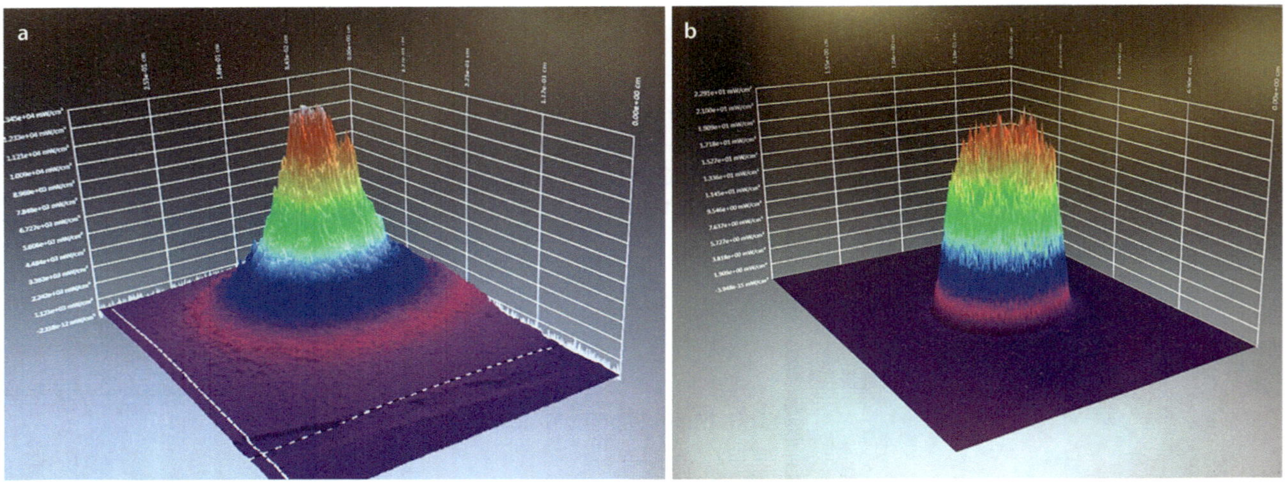

7

Fig. 7.13 3D rendering of optical beam spatial profiles. **a** Gaussian beam. **b** Flat top. Spot size: 1 cm^2. (Images of sources taken with an Ophir beam profilometer, Beamgage v5.5 Ophir Optronics, LLC, Jerusalem, Israel. Reproduced from Cronshaw M et al. Photothermal aspects of high energy photobiomodulation therapies: an in vitro investigation. Biomedicines 2023, 11, 1634 [56])

to enable consistent predictable dose delivery along with a number of other possible benefits as discussed here in relation to dose delivery techniques [118–120].

In the case of a typical Gaussian beam, the spectral beam power profile peaks in the center, with approximately 68% of the total power in the central third of the area exposed. With a small probe and a correspondingly lower output power device, clinically this is not a major issue. However, with larger surface optic applicators with higher power output devices, this effect can be highly significant as it can result in a level of irradiance which may pose a biological challenge to surface and immediate sub-surface tissues, resulting in a photoinduced dermatitis [56].

Current guidelines on irradiance recommend safe limits according to wavelength as follows (Fig. 7.14):

WALT Dose Recommendation			
Wavelength (nm)		p Factor	*Adjustment*
Blue	(400-499 nm)	1.8	*80% less*
Green	(500-599 nm)	1.5	*50% less*
Yellow	(600-699 nm)	1.3	*30 % less*
Red	(700-799 nm)	1.1	*10% less*
Near Infrared	(800-899 nm)	1	*Einstein*
Near Infrared	(900-999 nm)	0.9	*10% more**
Near Infrared	(1000-1100 nm)	0.8	*20% more**
Near Infrared	(1100-1200 nm)	0.7	*30% more**
Proposed irradiance limits: 800-900 nm: 750mW/cm2 600-700 nm: 300mW/cm2 400-500 nm: 100mW/cm2			

Fig. 7.14 The current consensus therapeutic objective when applying a Gaussian profile device is to avoid exceeding a limit of 750 mW/cm^2 using NIR, 300 mW/cm^2 for 600–700 nm and 100 mW/cm^2 for 400–500 nm wavelengths. (Adapted from Young NC et al. Thermodynamic basis for comparative photobiomodulation dosing with multiple wavelengths to direct odontoblast differentiation. Journal of Biophotonics. 2022 Jun;15(6):e202100398 [107])

These recommendations are based on cellular and animal outcome studies as well as the expert consensus opinions of the World Association for Photobiomodulation Therapy (WALT) [121]. As can be seen from the table, there is recognition of the differences in photonic energy between wavelengths. These represent safe guidelines which will no doubt be subject to refinement and revision as the evidence base further evolves.

Based on a recent thermography study [56] which investigated high power output devices along with large surface applicators, a research recommendation has been made that a calculation be made of the irradiance of the mid-third of the beam, with a further calculation for very large applicators of the central core zone of 1 cm^2. As illustrated in the example shown in Fig. 7.15, even at an average level of irradiance far below the upper limits recommended for PBM, it is possible to significantly exceed the suggested NIR peak irradiance value of 750 mW/cm^2 to a significant area of tissue (Fig. 7.16).

When large applicators with a correspondingly higher power output are applied, there can be a resultant marked temperature rise associated within the proposed average parameters. The surface and sub-surface temperature rises can exceed the advisable in vivo physiological limits of a near 6 °C rise above the normal body temperature of 37.5 °C. Beyond this level, there is the potential for permanent deformation of structural tissue proteins, and perhaps other biomolecules, subject to protracted exposure to temperatures above 45 °C [6]. The higher the thermal rise beyond 45 °C, the shorter the duration that the cells can tolerate the thermal stress; the higher the rise, the more it provides a corresponding risk of permanent harm to tissues both at the surface and in the near-surface layers overlying a more deeply

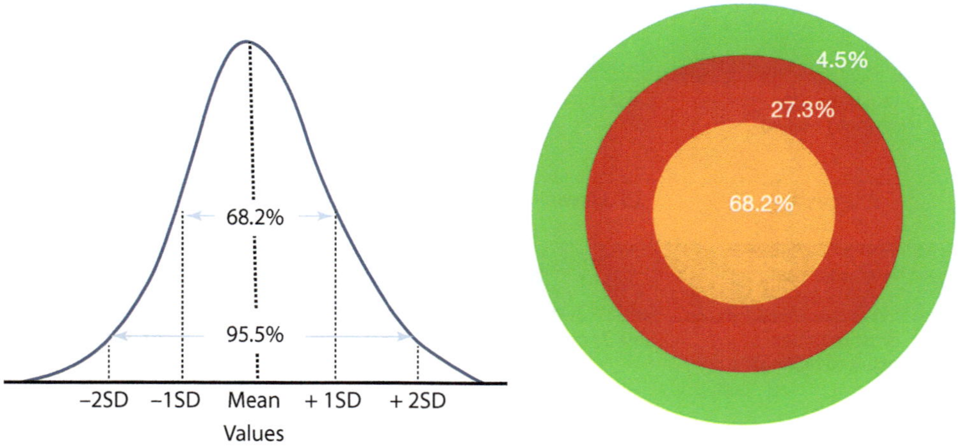

Fig. 7.15 Vertical and cross section of a Gaussian beam source showing mean values of power distribution across the beam. The power distribution of a Gaussian beam device is typically around 68% in the mid-third of the beam, whereas in the next third of the beam, around 27% of the power is available, and only ca. 5% of the beam is available in the peripheral third. (Reproduced from Cronshaw M et al. Photothermal aspects of high energy photobiomodulation therapies: an in vitro investigation. Biomedicines 2023, 11, 1634 [56])

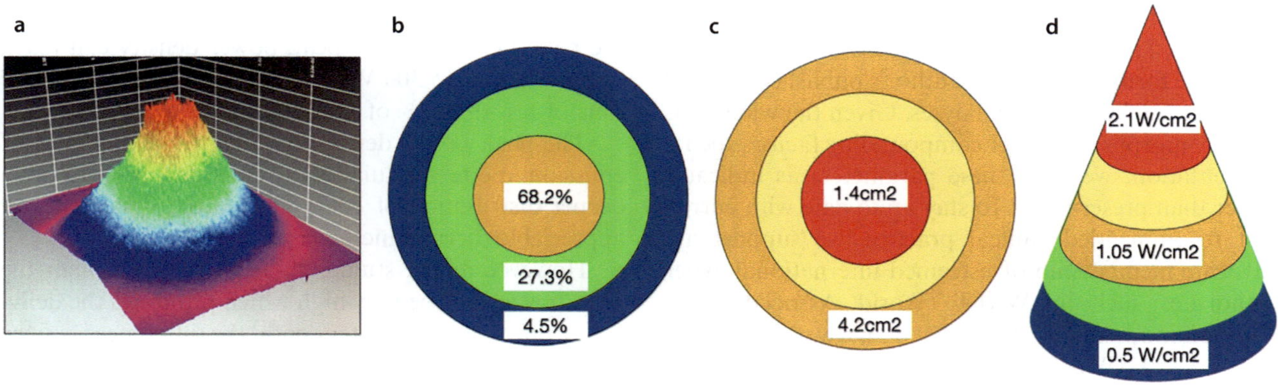

Fig. 7.16 **a** Gaussian beam profilometer 3D render of a 4-cm-diameter applicator with a surface beam area of 12.5 cm^2. **b** Approximately 68% of the power is concentrated in the mid-third of the beam, with around 27% in the mid-third and under 5% in the outer third. **c** In the mid-third of the beam, the area is around 4.2 cm^2 and the power further progresses to a peak in the mid-third of this central area. **d** At a power output of 6.25 W, the average irradiance across the entire area of 12.5 cm^2 is 0.5 W/cm^2. However, as indicated above the average irradiance in the central area is 1.05 W/cm^2 with a further peak within the innermost central zone in excess of 2.1 W/cm^2. (Reproduced from Cronshaw M et al. Photothermal aspects of high energy photobiomodulation therapies: an in vitro investigation. Biomedicines 2023, 11, 1634 [56])

placed target. Progressively steeper temperature rises above this threshold can culminate in cellular dysfunction, a process that can ultimately result in apoptosis with associated tissue damage.

By contrast the "flat-top" devices maintain an even power distribution across the target tissue. This avoids the inherent dosimetry problems associated with larger Gaussian beam devices. In addition, the evidence base suggests the use of a higher output power of 1 W/cm^2 may be possible with an 810 nm source. There is some limited evidence from in vitro studies that show a reduction in peak surface temperatures on this higher level of irradiance in contrast to a Gaussian device [56]. Also, the optical footprint up to a depth of around 1 cm may be wider as well as a little deeper [122].

The disadvantages of the "flat-top" devices are that the currently available optic spot sizes are restricted to 1 cm^2, so in contrast to a larger spot size Gaussian applicator, it can require extended treatment times to cover large target areas. Newer and larger 5-cm-diameter flat-top applicators have been developed although there is limited manufacturer fiber attachment compatibility of this type of peripheral attachment [123]. Pending further research, the merits of the flat-top concept in contrast to a Gaussian beam device have yet to be proven clinically, although the authors view this as an emergent technology with some supportive in vitro and animal studies.

Larger spot size Gaussian and flat-top devices are increasingly being adopted in clinical practice, and there

Surface Targets	Optical Spot ≈/= 1.0 cm²	Optical Spot > 3.0 cm²	Optical Spot > 5.0 cm²
Biostimulation (enhanced healing)	5 J/cm²	3 J/cm²	2 J/cm²
Analgesia	15 J/cm²	10 J/cm²	8 J/cm²
Sub-surface Targets (to 1.0 cm depth)			
Biostimulation (enhanced healing)	50 J/cm²	25 J/cm²	20 J/cm²
Analgesia	150 J/cm²	100 J/cm²	80 J/cm²

7

Fig. 7.17 Larger optical spot size devices have a significant-sized core central third area. As the average irradiance in this central zone is an average of 2 × the overall average irradiance, this may lead to over-dosimetry of the core area. This is particularly relevant to very large applicators as in the central area due to the Gaussian effect the beam continues to peak. The suggested dose tables above are the author's preferred best current recommendations

is an evident need for safe evidence-based guidelines on dosimetry. The parameters recommended in Fig. 7.17 below are formulated on the author's published research and our own laboratory analyses. Given the wide range of body morphs and other compounding factors such as pigmentation, we offer these parameters as indicative rather than prescriptive. To stay up to date with current best recommended clinical practice, we support and advocate membership of esteemed international expert authorities such as WALT (World Association for Photobiomodulation Therapy, ▶ www.waltpbm.org) and WFLD (World Federation for Laser Dentistry, ▶ https://wfldlaser.com).

7.12 Emission Modes

The laser beam can be emitted with different modes (Fig. 7.18) that are substantially "pulsed" (FRP lasers) and "gated" or "chopped," a mechanical or electrical interruption of a continuous wave with a given frequency (e.g., diode lasers).

However, it is common to find that the term "pulse" is erroneously and indifferently ascribed to both the modes. The diode lasers used in dentistry cannot achieve high peak power when used in continuous wave mode or gated mode, due to the intrinsic characteristics of the semiconductors (too much heat generation destroys the semiconductors).

Therefore, in some devices the electrical pumping system is modulated to produce very short T_{on} (micro- and nanoseconds), with a frequency up to 40,000 Hz. In this way, although the average power is relatively low, the peak power and the peak power density can reach high values up to 300 W although with an accumulated time of less than 1% of the total time.

The peak power density deriving from this kind of emission mode modulation is intended to lead to a deeper distribution of energy inside the tissue without appreciable thermal increase.

However recent studies indicate that although the peak power may be very high which promotes the delivery of photons to a greater depth, the difference is of the order of a few millimeters compared to lower hertz or continuous delivery emissions. Also, the accumulated time required to deliver the required dose at depth may need extended periods of surface application.

The lasers operating in this manner are called "superpulsed," and some devices offer the possibility to deliver packages of "pulses" called "pulse trains."

In 2010 Hashmi et al. published a meta-analysis of peer-reviewed literature about the effects of pulsing in PBM [61]. The conclusion was that, all being equal, pulsed light seems to be superior to CW light for wound healing, while CW light seems to be superior for nerve regeneration. However due to inconsistency in the reportage of parameters in the literature, there is no consensus about the clinical relevance of pulsed or gated emission compared to continuous emission to determine accepted protocols [117, 124, 125]. There is some limited evidence to support the superiority of a low hertz gated pulse train which is thought to create a photomechanical shock wave and which opens membrane-bound ion gates, reducing axonal activity. This mechanism has

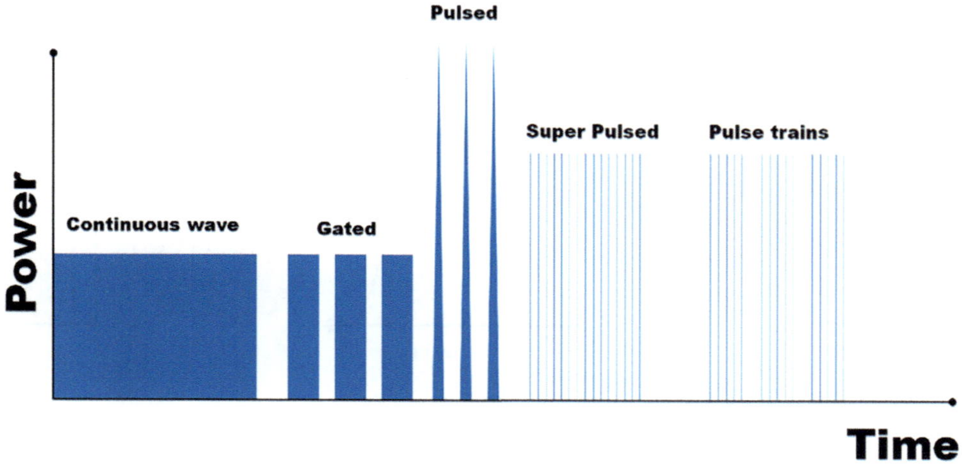

Fig. 7.18 Laser emission modes define the photonic source as being delivered as a continuous stream; as an interrupted "gated" pattern (on and off); as high peak power short-duration pulses which may be ultrashort "superpulsed" (<50 µs); as intermittent "pulse trains" which may be of variable peak power (not shown)

been proposed to explain the apparent benefits of this gated approach in pain management, as axonal transmission of a wave of depolarization is impeded by a loss of cross membrane electrical potential. A further suggestion is that a pulsed approach at around 5 kHz may coincide with the natural resonance frequency of the axon triggering blockade of axonal gates (gate control theory) [125–127].

7.13 Optical Delivery Techniques

A number of approaches to dose administration have been advocated including the application of a static spot applicator in contact with the surface as well as noncontact methods [114]. Some authors recommend applying physical pressure to indent the tissues again with a static contact. This is intended to shorten the optical path length as well as reduce the presence of blood which has been suggested as a high absorbent of 600–1100 nm light. Also, optical scanning techniques are described whereby the treatment area is painted with a moving beam [56, 128, 129]. Furthermore, some applicator devices as previously mentioned have multiple small lasers which may be interspersed with some LED devices in a cluster handpiece. Also, as previously mentioned there are many shapes and sizes of peripheral applicators (see Fig. 7.10).

In respect of recommended technique, it is advisable to use a surface optic spot of around 1 cm^2+ for anything other than very small lesions (Fig. 7.19).

The adoption of a scanning method is recommended for a surface spot size of 1 cm^2+, as this can dissipate excess heat build up in the tissues exposed to the central third of the beam [56]. The rate of movement depends on the size of the optic footprint and the corresponding irradiance, and, as a general rule, high output larger surface spot size devices (3 cm^2+) require an ascending more rapid rate of movement, compared to a slower sweeping action with smaller devices. As movement covers larger areas of tissue, it is an important calculation to determine the size of the target as well as the area treated, as with a smaller lesion such as a herpetic eruption or an aphthous ulcer, the use of a scanning movement can involve covering a larger area than the size of the lesion. It is beneficial to cover the wider area as there may be an associated peripheral area of inflammatory swelling beyond the boundaries of the lesion, and at the dosimetry recommended here, no adverse effects can be expected from exposed normal collateral tissues. Patients may notice some gentle warmth associated with the treatment, and it is our observation that at surface temperatures of around 44 °C+, patients will complain that the tissues can feel a little hot. Clearly at that point, it is appropriate to let the tissues cool a little before further proceeding. A further consideration for deeper sub-surface targets is that intervening subcutaneous fat deposits are a high scattering medium and can be associated with heat accumulation as well as a diminution of dose delivery to depth. Incremental treatments may be necessary in the presence of thick keratinised tissues especially in association with dense skin pigmentation [33, 55, 130, 131]. To help overcome these potential difficulties in the presence of thick tissues and dense deposits of melanin, the use of wavelengths longer than 800 nm combined with a larger optical surface spot size of 1 cm^2+ is recommended (Fig. 7.20).

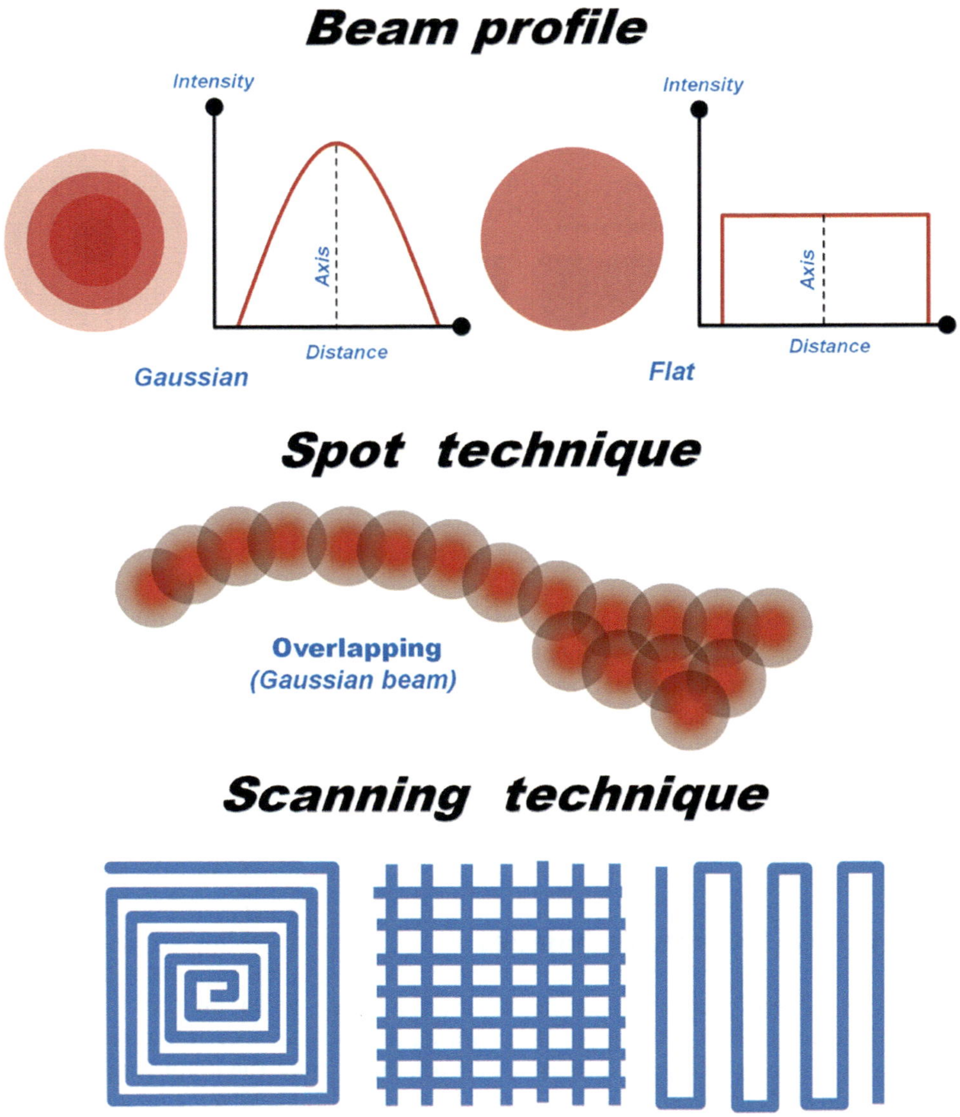

Fig. 7.19 A static beam may not cover the treatment area. Also with higher dosimetry devices with larger applicators, there is a potential risk of excessive dosimetry in the mid-third of the beam associated with a static spot method. Consequently a scanning technique can be the best approach

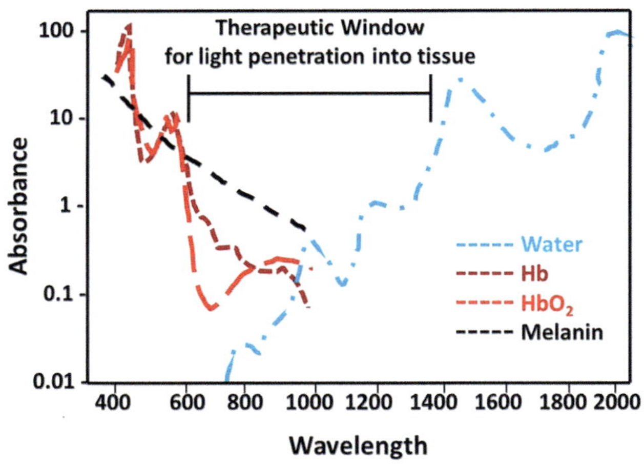

Fig. 7.20 For Caucasian skin, the remittance is about 0.1 at 300 nm, 0.2 at 360 nm and about 0.5 at 600 nm. The corresponding numbers for dark, Negroid skin are 0.02 at 300 nm, 0.09 at 360 nm and 0.2 at 600 nm. (Source: Richter AL, Barrera J, Markus RF, Brissett A. Laser skin treatment in non-Caucasian patients. Facial Plast Surg Clin North Am. 2014 Aug;22(3):439–46. ▸ https://doi.org/10.1016/j.fsc.2014.04.006 [55])

7.14 Timing of the Intervention and Frequency of Reapplication

Given the many pathways available associated with an increase in cellular viability on later exposure to a stressor, the idea of pre-conditioning tissues has attracted interest in the literature. There is clinical evidence supporting this concept from the oncology community where the prior application of PBM as a prequel, or application at the time of radio- or chemotherapy, has found to be a highly significant factor in reducing the incidence of clinical complications such as oral mucositis or radiodermatitis [132]. A further example is found in orthodontics where the application of PBMT at the time of placement of orthodontic separators or bands has been found to markedly reduce the incidence of post treatment pain [133–136]. An additional example is the application of PBMT at the time of the surgical extraction of lower wisdom teeth, which has an effect in the reduction of trismus and postoperative angular swelling equivalent to the administration of steroids [137, 138].

In respect of primary research using in vitro cellular models and animal studies, there is an extensive literature which identifies an increase in stress resistance in PBM test groups to chemical and bacteriological toxins, as well as a corresponding reduction in apoptosis. For example, Choi et al. exposed human gingival fibroblast cultures to bacterial-derived lipopolysaccharides and found a marked reduction in the PBM-treated test groups in the production of some key pro-inflammatory interleukins [139]. PBM has anti-inflammatory properties associated with the selective uptake of pro-inflammatory mediators and the inhibition of production of key pro-inflammatory prostaglandins such as PGE2 [140–142]. Furthermore, cells in a productive cycle of oxygenated phosphorylation are more robust and better able to withstand physiological stress [143, 144]. PBM can also increase the beneficial activity of superoxide dismutase to convert potentially harmful ROS into peroxide, which can then be safely converted into water and O_2 by catalase and glutathione. PBM is also known to increase the production of NO, which results in improved lymphatic drainage and the increased availability of oxygenated blood to the tissues consequent to vasodilatation [89, 90]. Also, it is recognized that PBM can regulate histone acetylation signaling of human dental papilla stem cells, which influence gene expression to increase stem cell viability. At higher levels of fluence (10 J/cm^2+), there can be a small but significant rise in intracellular temperature, and at a ΔT increase of 2 °C, heat stress proteins are activated. Furthermore, higher dosimetry of PBM can trigger protective (hormetic) pathways leading to a temporary reduction of cellular metabolism; subject to the duration and the extent of the physiological stressor, this improves cellular viability without damage to part formed proteins. Also, via gene transcription the production of HSPs and other antiapoptotic factors is increased, which increases stress resistance [145].

In sum, there appears to be substantial clinical gain to be made by applying PBM at the time of or as a prequel to tissue stressor events. Although this possibility has yet to be properly investigated, the added value in medically compromised or indeed all patients to reduce or mitigate postoperative complications seems highly attractive. Some common clinical procedures are recognized as being more prone to triggering postoperative pain and discomfort. The routine immediate postoperative application of PBMT may go far to enhance the patient experience and improve treatment compliance. This proposal is discussed further below in relation to wound healing, as well as in relation to specific conditions later in this chapter.

7.15 Wound Healing

Based on tissue culture, animal and now an increasing number of human studies, there is considerable interest reflected in the published peer-reviewed literature in the potential to apply PBMT as a tool to mitigate inflammation as well as to promote good-quality healing [66, 98, 146, 147].

Photobiomodulation as an adjunctive therapeutic intervention has been subject to some clinical trials as a device to mitigate postsurgical trauma with interesting results [137, 138].

In response to trauma whether accidental or that associated with a surgical intervention, a multiphasic sequence is initiated. This involves an acute inflammatory response which is characterized by pain, tissue swelling, erythema and localized heat which may persist for 4–10 days after the injury. A blood clot may be associated with the site of the injury which is progressively remodeled from the periphery and the base of the wound as the initial fibrin clot is replaced with primary wound collagen. Subsequently there is revascularization and continued deposition of new and replacement wound matrix which can in favorable circumstances be associated with regeneration of tissues such as muscle and bone along with the regrowth of sensory and motor axons. The primary phases of wound healing take around 30 days in normal circumstances with a progressive continued reorganization of the damaged tissues for a year or more [148]. The diagram below distills this process into the essential phases of the acute response, the initial matrix deposition phase followed by primary wound closure and progressive remodeling (◻ Fig. 7.21):

The acute inflammatory response is mediated by the immune system in tandem with an adrenergic response of the sympathetic nervous system to acute vascular injury, which results in vasodilatation and bleeding at the site of the wound. Arachidonic acid is released from the damaged cell walls of the traumatized tissues, and this is one of the primary activators for the production of prostaglandin E2 (PGE2) by cyclooxygenase [150]. PGE2 is in essence the acute inflammatory master switch which when released attracts stem cells to the site along with neutrophils, as well later on with pro-inflammatory macrophages. The local capillaries become porous releasing tissue fluids resulting in localized edema as well as a drop in oxygen tension. The wound

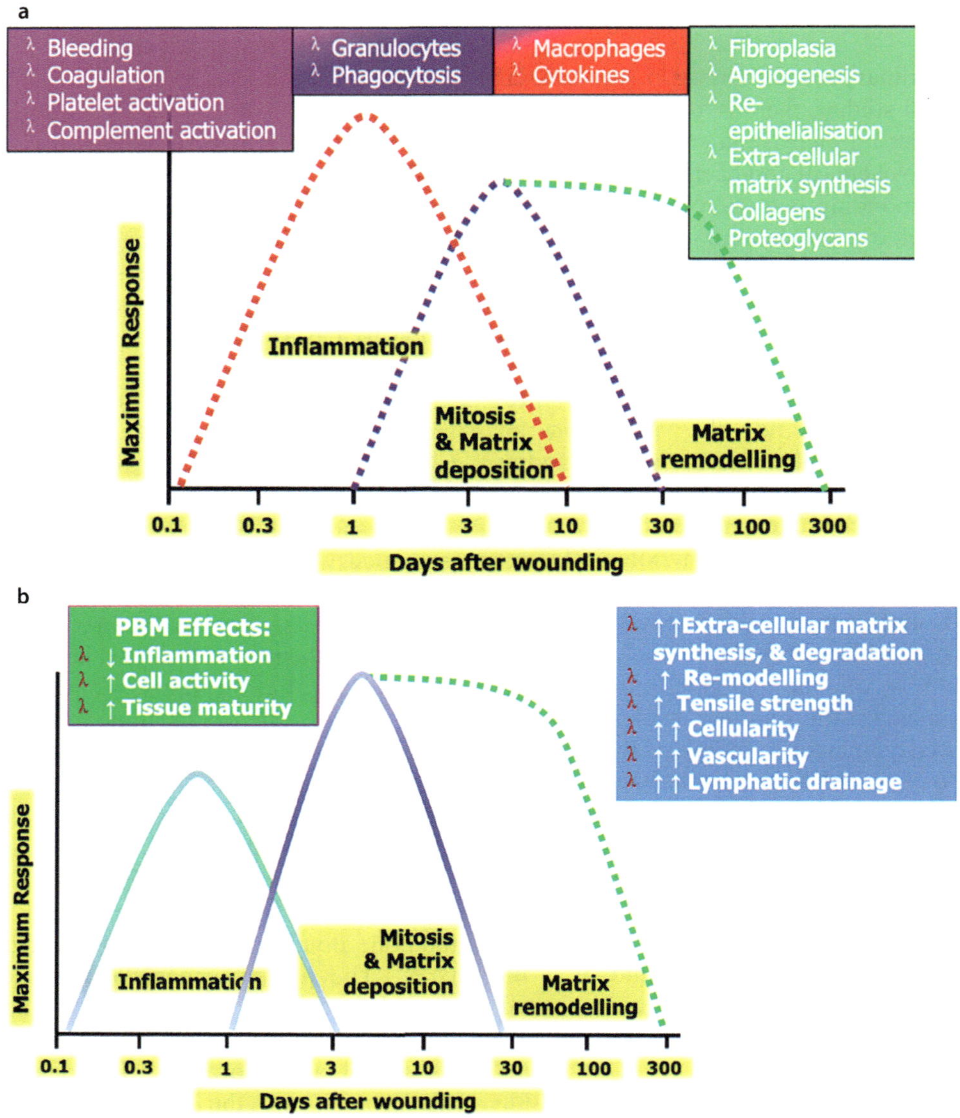

Fig. 7.21 Graphic **a**, standard healing. Graphic **b**, healing with adjunctive PBMT. PBM affects the peak and duration of the acute inflammatory tissue response to trauma or infection. There is also an increase in cellular volumes with an associated increase in matrix production. This positively impacts on the quality of the final repair optimizing the biological capacity for regeneration. (Graphics source: Adapted S. Parker from Enoch S., Price P. Cellular, molecular and biochemical differences in the pathophysiology of healing between acute wounds, chronic wounds and wounds in the aged. ▸ www.worldwidewounds.com, Aug 2004 [149])

becomes relatively anaerobic, and the tissues become more acidic due to the cells entering into a glycolytic anaerobic cycle of metabolism associated with the increased production of lactic acid. Histamine is released from mast cells, and there is a step increase in production of a number of pro-inflammatory cytokines including IL-1β, IL-6 and TNF-alpha, along with bradykinin, substance P and CGRP. Along with the classic signs of acute inflammation, there is also an associated reduction in the threshold required to trigger sensory nerve axonal depolarization. Mediated by the spine and the CNS, there are amplification circuits such that a local trigger can increase to involve regional axons affecting the segment, as well as on occasion resulting in activation and stimulation of other branches of the same cranial nerve and producing cross arch effects. In addition, motor nerves may become involved producing an increase in muscular tonicity which can manifest itself as trismus as the muscle may lock into spasm (Fig. 7.22).

Mediator	Molecules	Action and/or effects
Growth factors	BNF, GDNF, FGF, β- FGF, IGF-1 KGF, PDGF, TGF- β, VEGF	Proliferation, Differentiation, Angiogenesis, Migration, Chemotaxis
Anti-inflammatory cytokines	IL-2, IL-4, IL-8, IL-10	Differentiation, Proliferation, Immune activation, Chemotaxis, Angiogenesis
Pro-inflammatory cytokines	IL-1a, IL-1b, IL-6, TNF- α, PGE2, COX2	Stimulate and accelerate inflammation, Proliferation, Angiogenesis, Promote migration, Anti- and pro-apoptosis
Heat shock proteins	HSP90, HSP70, HSP25	Chaperone protein, Enhanced cell survival
Matrix Metalloproteinases	MMP2, MMP9	Cell survival, Prevent terminal differentiation, Tissue re-modeling
Small molecules	ATP, GSH, ROS, Ca^{++}, NO, H^{+}	Normalisation of cell function, Migration, Angiogenesis, Proliferation

Fig. 7.22 Biological mediators associated with PBM exposure. (Source: Adapted S. Parker from Kim WS, Calderhead RG. Is light-emitting diode phototherapy (LED-LLLT) really effective? Laser Ther 2011;20:205–15 [151])

Healing cannot occur in the absence of the acute inflammatory response, and it is a common observation that immunocompromised patients, including those on systemic steroids, have a tendency to heal poorly. The acute inflammatory response is primarily a defensive protective response intended to prevent trauma to damaged tissues, as well as to mitigate the potential for infection to spread beyond the primary wound site. Furthermore, as mentioned earlier it is the essential primary trigger to initiate the sequences essential to healing, repair and regeneration of damaged tissues. However, an excessive inflammatory response is debilitating to patients, for example, the marked angular swelling associated with a traumatic procedure such as a third molar extraction or an apicoectomy. On the other hand, an impaired response due to immune suppression can result in failed wound healing which may be associated with chronic inflammation and the progressive loss of functional tissues. Also failed wound healing may be associated with persistent infection. For example, the progressive loss of bone and supporting periodontal tissues seen in untreated or poorly managed adult chronic periodontal disease is a consequence of chronic inflammation perpetuated by a persistent pathogenic biofilm (Fig. 7.23).

As an adjunctive tool to surgery, PBMT supports good-quality healing and repair and represents an opportunity for the clinician to optimize the clinical outcome of an intervention. PBMT may be used as part of an integrated approach to tissue management based on sound clinical principles of the need to eliminate infection, to remove failed wound healing matrices, to stabilize the replacement primary wound matrix as well as to mitigate the acute inflammatory response while stimulating the vasculature to provide good drainage along with an enriched blood supply [152, 153]. As a supportive therapy, PBM improves cellular viability to surgical trauma and other tissue stressors such as cancer radiotherapy or cytotoxic chemotherapy and also stimulates a powerful immune response directed toward wound resolution. At re-care visits, further increments of PBM may be administered in order to increase the volume of matrix production by the cellular optimization of aerobic metabolism as well as an increase in cell volumes [154–157]. Also, PBM pre-conditioning and re-care PBM applications can improve the biological capacity of tissues to repair in medically compromised patients with systemic illnesses as well as the elderly who may have a reduced immune response along with underlying multiple age-related inflammatory disorders such as diabetes, cardiovascular disease, chronic respiratory diseases, etc. [158, 159] (Fig. 7.24).

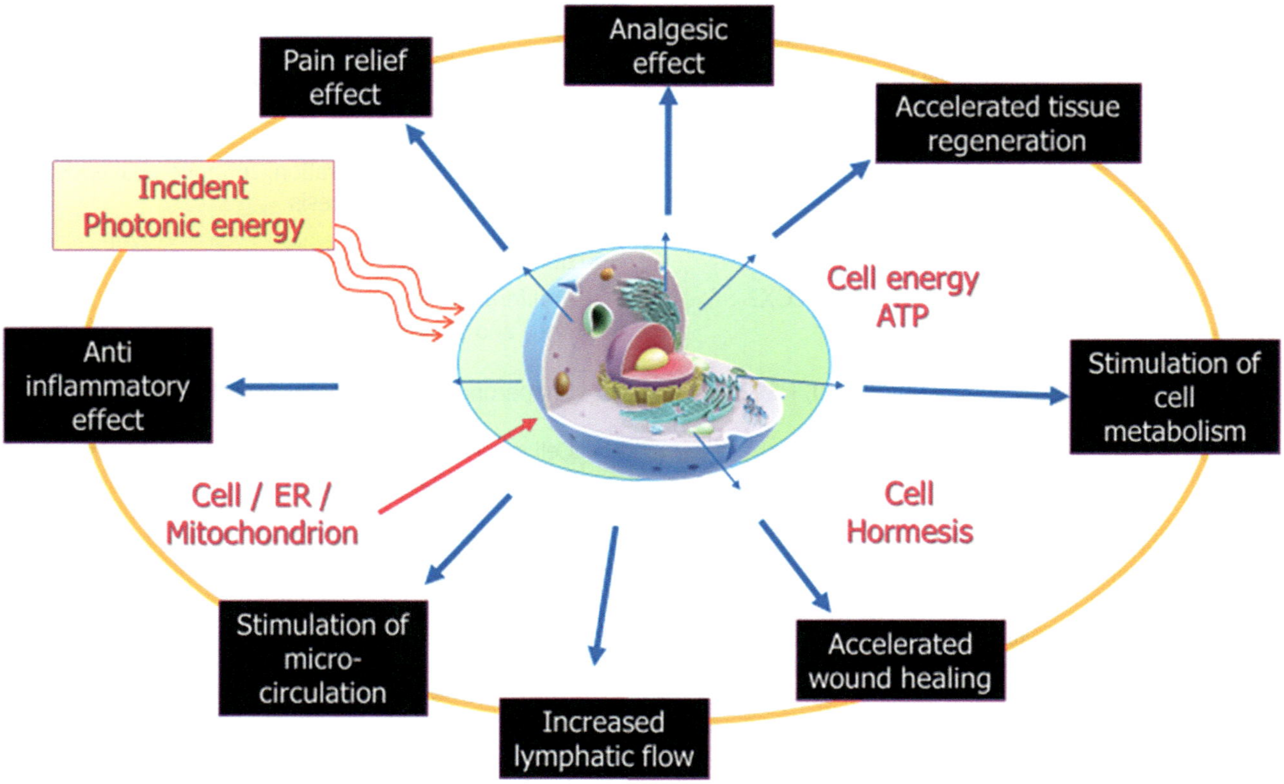

Fig. 7.23 Summary of PBM effects derived from wavelength-specific incident photonic energy. (Graphic: S. Parker)

7

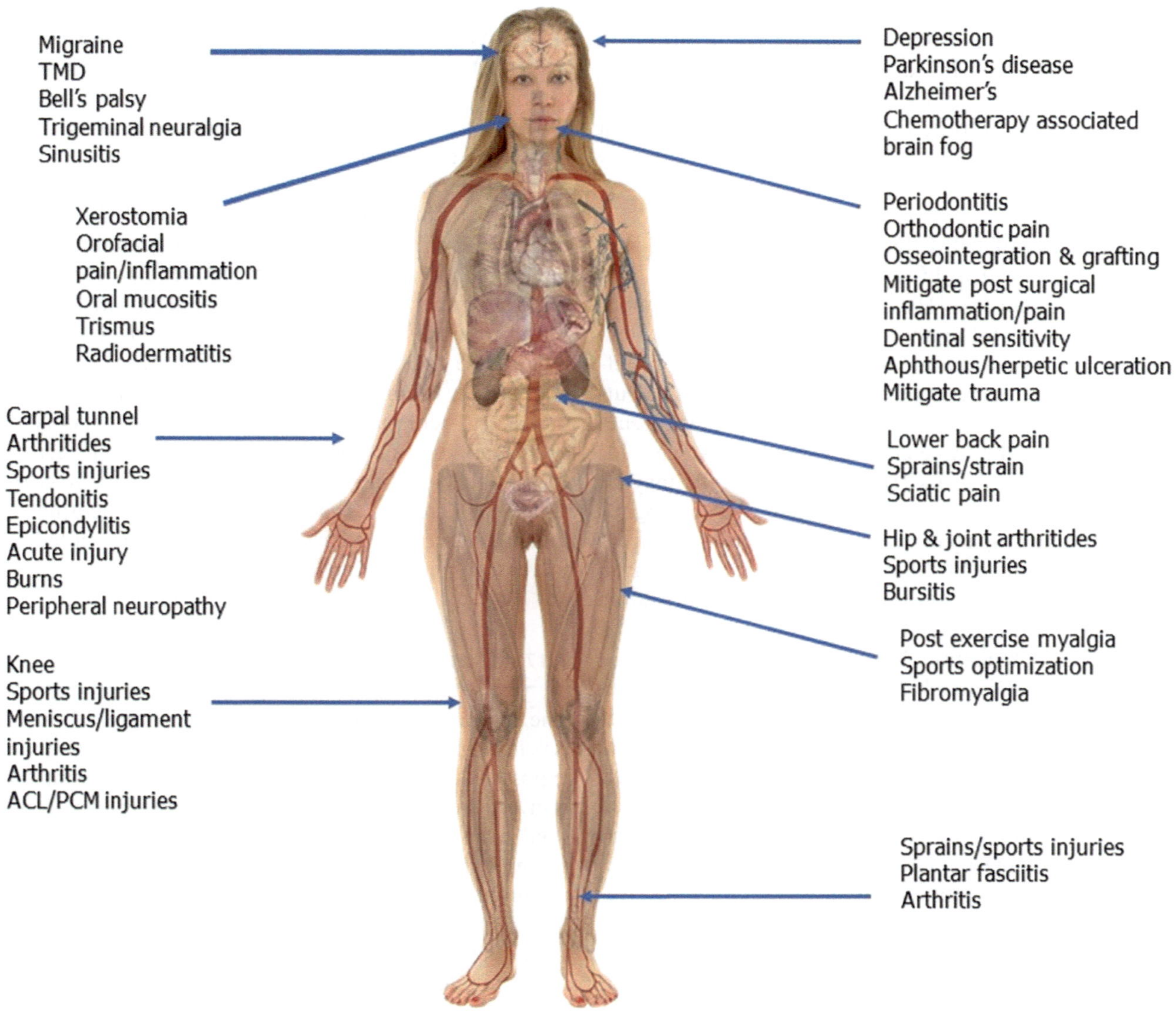

Fig. 7.24 The range of medical applications under investigation for PBM is very extensive. Many conditions are inflammatory in nature, and as such PBM offers great potential as a safe, highly localized and nontoxic therapeutic modality. (Graphic: S. Parker)

7.16 Photobiomodulation Therapy, Stem Cells and Regenerative Therapies

Arising from the disparate specialties of oral surgery, endodontics, periodontics, orthodontics and restorative dentistry, there is considerable interest in regenerative procedures to enhance the quality of healing and repair as well as to supplement tissues previously lost due to disease or trauma. A number of approaches have been adopted including the utilization of autologous-derived growth factors and natural tissue matrices such as platelet-rich fibrin and tissue grafting. In addition, there are many exogenous human- and animal-derived osseoinductive grafting materials, synthetic collagen matrices plus grafting materials such as beta tri-calcium phosphate, calcium sulfate and natural growth promoters such as enamel matrix derivative.

Given the properties identified with PBMT of an increase in cellular viability, stimulation of cellular activities and mitosis along with an increase in productive matrix manufacture of collagen and bone along with the promotion of angiogenesis and axonal regrowth, there is considerable interest within the literature to explore the potential of the addition of PBMT as an adjunct to established regenerative procedures. Coupled with the anti-inflammatory, analgesic and enhanced viability effects observed with PBM, there are very many potential clinical advantages that may be gained by the integration of this nontoxic therapy to very many aspects of routine and specialist dental care. Notwithstanding extensive published peer-reviewed cellular, animal and increasing numbers of human clinical trials, PBM is still at the stage of continued evaluation; despite the very many positive reports, there is as yet no agreement on consistent clinical procedures, although current best practice evidence-based guidelines have been issued by the World Association for Photobiomodulation Therapy (WALT).

One aspect of PBM that has attracted interest is in the potential of PBMT to enhance the number and activities of stem cells [160–163]. PBM has been demonstrated in vitro and in animal studies to stimulate stem cells with a resultant increase in their migration, proliferation, differentiation and protein expression. The outcome of tissue engineering to repair or regenerate tissues can be affected by three critical factors: cells, scaffolds and signaling mediators [164–166]. Stem cells may be derived from a number of sources including autologous human dental pulp stem cells (hDPSCs), stem cells from the apical papilla (SCAPs) and bone marrow-derived stem cells (BMSCs) amongst other sources including blood-derived stem cells separated from whole blood by fractional centrifugation. Various types of mesenchymal stem cells may be immediately harvested after the extraction of permanent teeth or the natural shedding of deciduous teeth, and oral tissues represent an excellent source of stem cells for therapeutic purposes [167, 168].

Stem cells can differentiate into the specialized cell types required for regeneration of tissues such as odontoblasts, osteoblasts, chondroblasts, fibroblasts, etc. However, cellular losses during the extraction and placement along with variability in the differentiation abilities and activities of the stem cell grafts have led to a lack of predictability and a compromised efficacy of these techniques [169]. Various factors have been demonstrated to regulate the stem cell properties, such as tissue origin (inflamed or not), donor age (the younger, the better), culture conditions (culture media, pH, temperature levels and available oxygen) and the presence or absence of growth factors [170]. Stem cell transplantations may lead to a very low engraftment rate, as their survival after placement into inflamed tissues is short, with a 24 h half-life [9]. The therapeutic benefits of stem cell therapies are associated with the paracrine secretion of soluble mediators, cytokines, secretomes, peptides, proteins, metabolites and micro-RNA, along with extracellular vesicles such as exosomes [167, 171].

Also, regulation of the immune system is a critical factor required in order to induce tissue regeneration [172, 173]. Elevated levels of reactive oxygen species (ROS) are produced as a response to the oxidative stress associated with an inflamed area, which may have detrimental effects on the surrounding tissues. Chronic inflammation is recognized to be associated with persistent infection, and senescent reparative cells may be present with a diminished capacity to heal and repair [174, 175]. Hence wound management is a critical factor in any attempt to regenerate tissues as this relies on the essential requirements to eliminate infection, remove failed wound healing matrices and create a new bio-scaffold with a good collateral blood supply.

As previously described above, PBM has very many positive attributes which in combination with a well-designed and well-executed surgical technique offer great potential to enhanced healing, repair and regeneration. In particular regard to stem cells, PBM has been found to increase the proliferation of stem cells as well as increase their viability. In addition, there is also an increase in stem cell metabolism with a continued increase in activities of the derived cellular phenotypes. For example, Arany found PBMT promoted the increase in numbers of pre-odontoblasts producing `a corresponding increase in the production of new dentine in a murine model. Similarly, Fekrazad et al. noted an increase in new dentine formation [164].

7.17 Clinical Aspects

The application of PBM in human regenerative therapies is being explored as an aid to improve the viability of vital exposed pulpal tissues. Also, the endodontic community is investigating the possibilities of promoting apex formation in immature nonvital teeth with open apices. For example, Moreira et al. noted the formation of a new odontoblast-like area in proximity to dentinal tubules in a model of regenerative apexification [176]. Zaccara et al. found that PBM can regulate histone acetylation signaling of human dental pulp stem cells (hDPSCs) [177]. Via an increase in nuclear-expressed modifying factors, this results in an increased expression of genes associated with the improved viability of hDPSCs. Matsui et al. associated PBMT with an increase in mineralization with a step increase in the levels of bone morphogenic proteins, osteocalcin and alkaline phosphatase [178–180]. Similar outcomes were noted by Suzuki et al. in an animal model of orthodontic tooth movement. PBMT was found to increase the number of osteoclasts present on the pressure side, while there was a corresponding increase in the number of osteoblasts on the tension side. Aside from the histomorphometric analysis, this interesting study applied an immunohistochemistry analysis of RANKL/OPG and tartrate-resistant acid phosphatase (TRAP) activity. This further correlated the histological findings of increased osteoclast activity in the test group on the pressure side (elevated levels of RANKL and TRAP) and increased bone apposition and osteoblast activity on the tension side (an increase in OPG). Furthermore, the rate of orthodontic tooth movement was found to be increased by around 40% compared to the control [181] (■ Fig. 7.25).

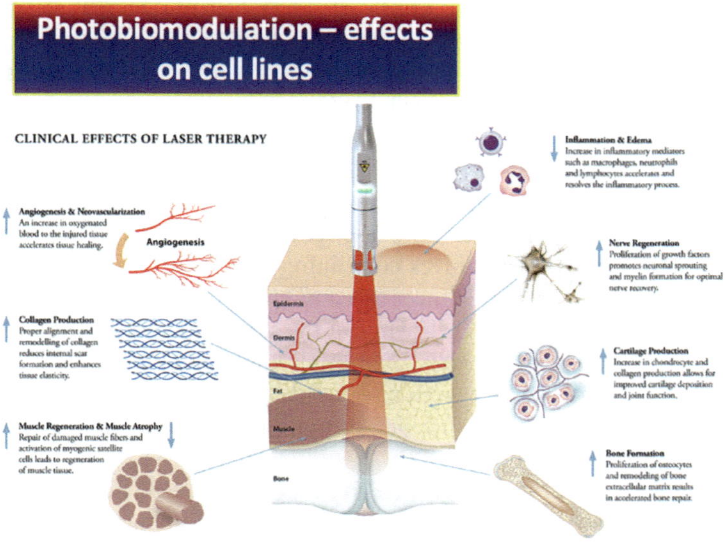

Cell type	Observed effect
Fibroblasts	Proliferation, maturation, locomotion. Transformation into myofibroblasts. ↓ secretion of PGE2 and IL-1 Enhanced secretion of βFGF
Macrophages	Phagocytosis Secretion of fibroblast growth factors. Fibrin resorption
Lymphocytes	Activation / ↑ proliferation
Epithelial cells	Motility
Endothelium	Increased granulation tissue Relaxation of vascular smooth muscle
Neural tissue	Reduced synthesis of inflammatory mediators maturation and regeneration of axonal growth
Osteoblasts	Enhanced proliferation

Fig. 7.25 PBM affects all cellular types and promotes increases in cellular actions including motility, matrix production and cellular volumes via mitosis. (Graphic: S. Parker)

As PBMT can impact on many aspects of wound healing, the integration with approaches to enhanced tissue repair and regeneration including stem cells is anticipated to become an established adjunctive procedure to optimize the benefits and overcome the above-mentioned drawbacks of the current standard techniques.

7.18 Safety Issues Associated with PBMT

Any intense light source can cause optical damage, and most particularly with the usage of wavelengths outside the visible range, wavelength-appropriate protective eyewear is a mandatory requirement. In regard to broader band low output power LED devices, the biohazard potential is not nil, and caution is advised particularly with higher output power devices >50 mW outside of the normal visual spectrum. The operator must not be lulled into a false sense of security in the mistaken belief that optical protection is not required with lower output power laser or LED devices, and the reader is directed to the section on optical safety earlier in this text where this topic is comprehensively covered (see Chap. 5).

Most of the contraindications and precautions (listed below) arise from the possibility that the biostimulating and immunomodulating effects of PBMT could interfere with some physiological and pathological conditions, or from the possibility of electromagnetic interferences with some obsolete implantable devices, such as pacemakers or ICD (implantable cardiac defibrillators). However, there is no evidence of an effective risk for all the situations listed below:

- Thyroid, gonads
- Hemorrhage
- Pregnancy (abdomen)
- Immunosuppressant treatments
- Epilepsy (subject to adequate optical protection in gated modes)
- Bone growth plates in children
- Neuropathies
- Hematological disorders

Concerns have however been expressed at the wisdom of applying a therapy recognized to stimulate cellular division, increase cellular resistance to stress and promote lymphatic drainage in patients with a current suspicious lesion [182, 183]. Similarly, is it safe to treat patients with a history of a prior malignancy or those with higher risk factors, for example, those with a high alcohol intake, smokers or those with signs of extant human papilloma viral infections? Regrettably cancers in one form or another affect around one in three of the population at one time or another. Given the high incidence of malignancies, is PBMT an additional risk factor and do special precautions need to be taken?

It is accepted that shorter wavelengths in the ultraviolet range 10–400 nm are an environmental risk factor for squamous cell carcinomas, melanomas and basal cell carcinomas. However, sun avoiders in contrast to those with the highest sun exposure have a reduced life expectancy of 0.6–2.1 years which is a comparable risk to smoking [184]. Visible to NIR wavelengths and beyond have insufficient eV to break chemical bonds in DNA. Direct damage to DNA can be caused by elevated levels of ROS; however phototoxic doses of laser wavelengths longer than UV do not cause plasmid DNA cleavage. This subject was investigated in a study of phototoxicity by Khan et al. using a murine model applying an 810 nm laser source at sublethal doses. Both the cellular and the animal research in this high-quality study showed no evidence of genotoxic damage. Although high NIR doses could be phototoxic, there was no evidence of a genotoxic or mutagenic harm, and it was concluded that the 810 nm laser is safe for clinical applications [68].

A number of tissue culture and animal studies have demonstrated a PBM stimulatory effect on tumor activity. However, in contrast to these few in vitro cellular studies and transgenic animal studies that show an increase in tumor cell growth, there are other studies which show precisely the opposite. In the absence of heterogenous tissues including a viable immune system in a monoculture, the biostimulation of the few tumor cell lines that show this effect in vitro would be subject to an in vivo invigorated host immune response. There are other associated effects of PBM such as vasodilatation which change the chemistry of the tissues away from an acidic pro-inflammatory area to an aerobic zone surrounding the cancerous tissues. For example, Ottaviani et al. found that PBMT inhibits tumor growth by promoting immune surveillance and vessel normalization. In a mouse model of melanoma, three different protocols of PBM were applied (660 nm, 50 mW/cm^2, 3 J/cm^2; 800 nm or 970 nm, 200 mW/cm^2, 6 J/cm^2, once a day for 4 days). It was found that all could reduce tumor growth and increase the recruitment of immune cells, in particular T lymphocytes and dendritic cells secreting type I interferons. PBM also reduced the number of highly angiogenic macrophages within the tumor mass which may be of benefit to control tumor progression [185].

By contrast, the clinical outcome value of animal models using transgenic species with a xenograft or a non-syngeneic immunodeficient murine model is open to question.

A further approach to eliminate cancerous cells by high-intensity laser irradiation has been investigated by use of a He–Ne 632 nm laser delivering 1200 J/cm^2 at 500 mW/cm^2 using an in vitro cell culture model an in vivo murine study of mice bearing breast cancer tumors. A single application of PBMT resulted in complete regression of the tumors as a response to overdose of the cancerous cells [186].

Another approach to provide adjunctive support using PBM relies on taking advantage of a differential effect of photon transduction between malignant cancer cells compared to the effects seen on healthy normal cells. This involves combining PBM with an additional cytotoxic anticancer therapy, so that it increases the killing of cancer cells while at the same time protecting normal healthy cells. Although this remains an area of continued research at present, there are some published articles that support this approach as an anticancer strategy.

For example, there are reports that PBM can potentiate the killing of cancer cells by photodynamic therapy and also by radiation therapy. Within the confines of their experimental studies to date, PBM increases cell death in cancer cells in response to cytotoxic stimuli. As an added benefit in normal cells, PBM exerts a protective effect as is well known in the case of neurotoxins, for example, as well as increasing stress resistance to thermal- and radiation-induced damage as discussed earlier in relation to the concept of PBM pre-conditioning.

As for human clinical studies, patients receiving PBMT have been found to have a significant improved complete response to treatment in comparison to those in a placebo group. Also, there is no evidence of field effects or an increase in recurrence in patients treated with PBMT.

In conclusion, the evidence supports the safety of PBMT as an adjunctive measure to assist patients undergoing chemoradiation therapies for some forms of cancers. There appear to be regional benefits to this approach which may improve the prognosis and reduce morbidity. Direct irradiation may be of benefit at high doses, and the current evidence base supports the use of PBMT as an adjunct to standard cancer therapies [187–190]. Due to theoretical concerns at the possibility of promoting a tumor to grow, PBMT is not advised as a stand-alone measure on a known or suspect lesion.

7.19 Clinical Management by Condition

PBM affects all cellular types, and as discussed earlier, dependent on dosimetry, effects achievable can vary from photoinduced stimulation of cellular physiology to photoinduced analgesia. In considering the processes attached to clinical care, it is recommended to adopt a decision process as exemplified for the condition of chemotherapy-associated oral mucositis below (▫ Fig. 7.26).

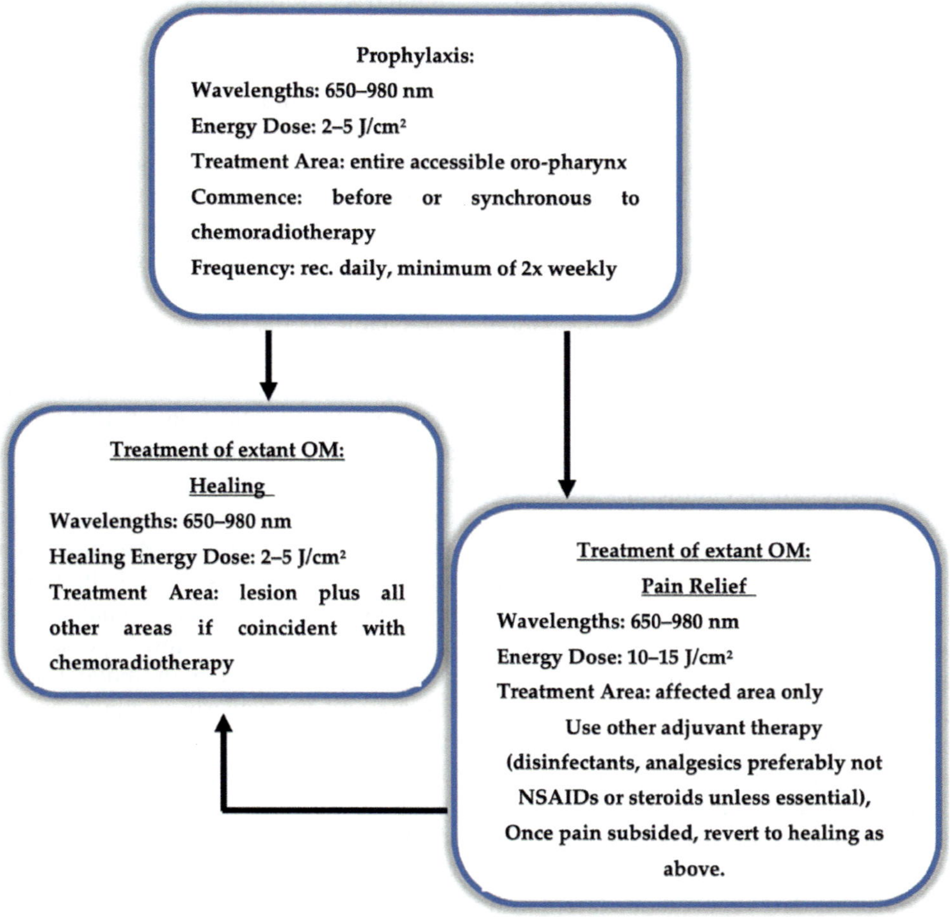

Fig. 7.26 Processes associated with PBM are the same for all conditions. The example demonstrates the treatment flow chart for chemotherapy-related oral mucositis. The dosimetry shown could equally apply to any superficial condition, for instance, aphthous ulceration or a herpetic eruption. If the condition is associated with a sub-surface target, it is advisable to increase the dose at the surface to compensate for energy losses at depth by a factor of 10× for a spot size of around 1 cm^2 and 5× for larger spot sizes greater than 3 cm^2. (Reproduced from Cronshaw, M.; Parker, S.; Anagnostaki, E.; Mylona, V.; Lynch, E.; Grootveld, M. Photobiomodulation and oral mucositis: A systematic review. Dentistry 2020. 8:87 [191])

As there is an essential similarity of the cellular response to equivalent radiant exposure, rather than the operator concentrating on a specific condition, it is more pragmatic to consider the desired outcome. The essential issues to decide are the treatment objectives, the location of the target tissues and the area to treat. Next, it is a matter of deciding which peripheral applicator may be best, and in this respect the operator needs to consider the issue of the size of the treatment area, as a larger surface area applicator may save a significant amount of treatment time but a smaller intraoral target may be best dealt with employing a smaller tool such as a surgical handpiece or a custom glass guide. The critical parameter is the area of tissue exposed as in the absence of a collimating lens system, the beam is inherently divergent. For this reason, some devices have a spacer attached to the handpiece which permits some standardization of area dosimetry. With smaller applicators such as a surgical handpiece, it may be possible to use the visible aiming beam as a visual guide to creating a spot size of around 1 cm^2.

Having determined the treatment objective of biostimulation or bio-inhibition (pain relief) and the issue of target location of surface or sub-surface, the operator can then consider the required output power. In accordance with current guidance, it is recommended to keep the average irradiance at or below 0.5 W/cm^2. As discussed earlier larger applicators require a higher overall power output to sustain the irradiance required. The power output of the device is easily determined for any optic spot area as the power required is half of the area. So, for instance, a 4-cm-diameter applicator has an area of 12.5 cm^2, and the required device power output for 0.5 W/cm^2 is 12.5/2 = 6.25 W. As 0.5 W/cm^2 delivers in 1 s 0.5 J/cm^2, to deliver 5 J, it takes 10 × 0.5 = 10 s.

7.20 Temporomandibular Joint Disorders

Temporomandibular disorders are a group of musculoskeletal disorders that cause facial pains with a high prevalence in both sexes. Symptoms are variable: pain, joint noises including clicking or popping and crackling, functional limitations up to articular disc displacement without reduction or dislocation. There are many causes—parafunctions, joint overload caused by impaired occlusion, stress, musculoskeletal diseases, trauma or acute and chronic inflammatory joint diseases [192].

The standard treatment of these disorders is multidisciplinary and is aimed at the control of pain symptoms through drug therapy, as well as at treating the functional limitation through physiotherapy treatment of the masticatory muscles, stress control, using cognitive behavioral therapy, massage and relaxation techniques, occlusal splints up to complex surgical and prosthetic rehabilitation.

PBM has been used to relieve pain for temporomandibular disorders, alone or in combination with conventional therapies such as physiotherapy exercises and occlusal splints [193–195]. Pre-auricular, intra-auricular and intraoral and extraoral maxillary points in the rear maxillary area are potential therapeutic targets, along with associated muscle groups affected by the symptoms which may be treated identified on palpation as trigger points. Published outcome data report diode lasers with a wavelength of between 650 and 1000 nm, with fluences of between 1.5 and 140 J/cm^2. Treatments are usually repeated two or three times a week until the symptoms have disappeared [196].

The term temporomandibular disorder (TMD) covers a wide range of disorders of the temporomandibular joint with or without associated issues manifest in the muscles of mastication. TMD can be classified into three types: myofascial pain, internal derangement of the joint and degenerative joint disease. Of the three types, patients most commonly present to their primary care dentist with acute or chronic pain associated with muscle groups including the lateral pterygoid, trapezius, sternocleidomastoid, masseter, temporalis and the posterior belly of the digastric muscles [197]. The pain may be severe and can be associated with trismus limiting the patient's ability to chew or open normally. This type of neuromuscular pain can be limited to one or two muscle groups or can involve multiple points which can be identified by digital palpation to identify trigger points. This type of pain is called myalgia and is caused by hypertonicity of the affected muscle which contracts into spasm. This restricts blood flow into the affected area resulting in anaerobic tissues associated with lactic acid build up [198]. Axonal nociceptor amplification can occur via the spinal tract resulting in radiating pain behind the eye on the affected side which typically starts in and around the zygoma along the line of the maxilla to the temporomandibular joint [146]. Diagnosis is by exclusion of radiographic or visual evidence of infection, as well as bimanual palpation of the joints and muscles in search of trigger points. It can be useful to measure the interincisal maximum opening which may be reduced significantly from the normal range of 32–35 mm+. It can be difficult to discriminate between odontogenic and myogenic pain, as the patient may have been subject in the past to multiple root canal treatments and extractions in attempts to relieve the patient of previous acute episodes of pain. The application of PBM which successfully ameliorates pain clearly indicates a myogenic rather than odontogenic origin and may be viewed as a useful diagnostic measure. There are other clinical signs that can support a provisional diagnosis of myalgia associated with TMD. These include flattened cusps, cervical abfractions and associated enamel cracks, abnormal temperature sensitivity and hyperacuity particularly to cold, the presence of mandibular tori plus a scalloped tongue and particularly in males enlarged muscles of mastication. In addition, there may be signs of premature incisal wear with obvious shiny wear facets, some highly localized periodontal bone loss and increased mobility of teeth trapped in a traumatic constricted envelope of masticatory function. Patients may or may not be aware of nocturnal parafunction which may be a contributory cause along with life stress and some prescription medications such as SSRI antidepressants to a pattern of nighttime bruxism, muscular hypertrophy and disturbance of sleep. Patients may be aware of frequently awakening with headaches and stiff muscles around the head and neck. There may also be a history of frequent fracture of both restored and previously intact teeth. The clinical manifestations of these disorders can include pain, malocclusion, midline deviation, joint noises and locking of the joint due to a displaced disc limiting mandibular movements.

Given a provisional diagnosis of myalgia-related TMD, PBM can be a highly useful measure as it can act as a palliative to the pain as well as relieve the trismus. Furthermore, on successfully reducing the acute discomfort, patients are receptive to proposals toward occlusal support and rehabilitation which may involve the use of various types of intraoral splint as well as, in some more complex cases, restoration of functional capacity involving a multidisciplinary approach (orthodontics, implants, etc.).

As a noninvasive and harmless intervention, PBMT has attracted increased interest in recent years [199–202]. The application of a coherent laser to extra- or intraorally transilluminate the affected muscle groups can

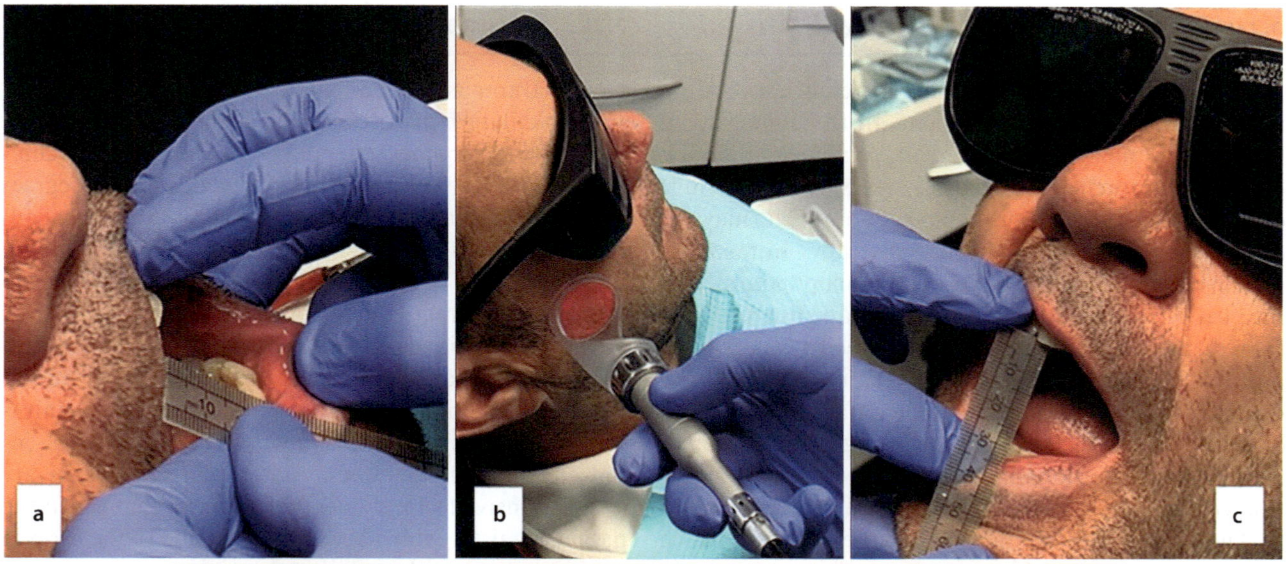

Case 1: Trismus associated with pericoronitis UR8.

Image A: Initial inter-incisal opening: 22mm

Image B: PBM with 940 nm 3.5W CW. 7.1cm² large applicator applied to three adjacent areas (total = 21cm²) with slow movement along the line of the underlying lateral pterygoid muscle. Treatment time 300 seconds. Average dose delivered: 50J/cm².

Image C: Final inter-incisal opening: 28mm

Fig. 7.27 A 38-year-old patient presented with pain and trismus associated with a pericoronitis at UR8. The treatment objectives were to relieve pain, improve the oxygenated blood supply to the affected area, improve lymphatic drainage and reduce the muscle spasm associated with the right lateral pterygoid muscle. Initial opening was limited at the measured inter-incisal tip distance at 22 mm. A 7.1 cm^2 area applicator was applied to the zygomatic arch region adjacent to the underlying lateral pterygoid muscle. An output power of 3.5 W was set using a 940 nm diode laser (Epic X, Biolase, CA). A slow sweeping scanning motion was adopted to an area encompassing the length of the zygomatic arch and the area immediately overlying the condylar head. This corresponded to approximately 3 × the area of the chosen applicator. A surface dose of 50 J/cm^2 was applied to each area which amounted to a total delivered surface dose of 1050 J to an area of 21 cm^2. The delivery applicator has an adjustable plastic spacer which when set at the maximum range results in a 7.1 cm^2 area of exposure. (Deep Tissue Handpiece, Biolase, CA, USA)

produce vasodilation, as well as at high doses of radiant exposure to a temporary reversible disruption of neuronal physiology. This contributes toward laser-assisted analgesia plus an anti-inflammatory effect with positive effects on painful muscles and joints. Mechanical injury to the TMJ, inclusive of ischemia–reperfusion, generates reactive oxygen species (ROS) in the articular TMJ tissues. PBMT decreases ROS generation with an increase in the production of adenosine triphosphate (ATP) in the mitochondria. Karic et al. in a recent in vitro study investigated the potential for promoting the differentiation of chondrocytes and fibroblasts on a biodegradable polylactide disc with human adipose-derived stem cells. This novel approach is intended to address the potential for the management of disc degenerative changes associated with TMD by use of a graft consisting of a PBM-treated stem cell-impregnated bio-scaffold. A considerable increase in both chondrocytes and fibroblasts plus an increase in the production of ATP was found after irradiation using 940 nm diode laser three times weekly over a period of 3 weeks [203].

As for clinical PBM irradiation protocols, there is at present no consensus, and a heterogeneity is observed concerning the radiant exposure, time of irradiation, emission mode, the number and frequency of sessions and the target application areas.

Based on a synthesis of systematic reviews and meta-analyses, the following clinical suggestions are proposed (Fig. 7.27):

- Radiant exposure: in the range of 35–100 J/cm^2
- Two sessions per week
- Eight to 12 sessions
- Exposure time: variable depending on the size of applicator used and the number of muscle and TMJ sites treated
- Continuous wave (CW) mode
- Multiple points of irradiation on the affected TMJ and muscles

7.21 Typical and Atypical Facial Pains

Facial pain is a real challenge for the clinician, both in terms of diagnosis and treatment. In the management of persistent moderate to severe atypical facial pain following a comprehensive medical assessment to exclude other causes, a multidisciplinary approach may be required. This may include cryotherapy, ultrasound, shock wave therapy, physiotherapy, acupuncture plus medications such as amitriptyline, gabapentin and low-dose diazepam. Pains secondary to specific causes, such as inflammation, infection, trauma, cancer, neurological degenerative lesions or abnormal contact of nerve and vascular structures (trigeminal neuralgia), are different from atypical or idiopathic facial pain, whose classification is diagnosis by exclusion and PBMT offers an additional option for treatment of difficult cases which otherwise are managed by symptoms often in combination with psychological support.

Referred pain may be excruciating with secondary symptoms such as paresthesia, paroxysmal response to minimal stimuli (allodynia) and depressive syndromes. These conditions have a high social cost both for the number of working days lost and for the cost of drugs and imaging techniques needed to arrive to a diagnosis; for this reason, the search for effective treatments with few adverse effects is important. PBMT's analgesic effect has been well studied and can be applied to various forms of facial pain in association or not with drug and/or conventional instrumental therapies.

Experimental studies on animals have investigated different fluences, and it has been noted that low fluences around 4.5 J/cm^2 are beneficial at a cellular level on reducing inflammatory mediators. High fluences can generate a temporary axonal block with an associated extended area of vasodilatation, plus there can be a reduction in the levels of important spinal and CNS mediators of pain, such as substance P and CGRP, plus there can be an increase in the manufacture of endogenous opiates such as β-endorphin [204, 205].

7.22 Trauma

Dental traumas consequent to falls, bicycle accidents, sports activities, playground accidents, etc. are frequent causes for emergency attendance to primary care clinicians. The consequential injuries include varying degrees of fractured maxillary central incisors, luxation, avulsion and lacerations. There may be marked swelling and bruising, pain and traumatic occlusion due to subluxation plus fractured restorations including anterior crowns and bridgework. The immediate management challenge to the clinician may be compounded due to soft tissue swelling and trismus limiting access for intraoral radiography as well as any urgently required essential stabilization therapies. Acute inflamed tissues are less responsive to local anesthesia due to changes in the tissue chemistry which become acidic due to the drop in oxygen tension with a resultant reduction in conversion of the local anesthetic to its active form. Many of the pro-inflammatory factors such as PGE2, IL-1β, IL-6, TNF-alpha, histamine, bradykinin, substance P, etc. are nocio-inductive resulting in increased axonal activity which by mediation via the spine and the CNS may result in regional pain and on occasion trismus due to conscription and activation of motor neurons [66, 98, 146, 147].

PBMT offers considerable near-immediate benefits as it is a vasodilator resulting in an increase in oxygenated blood in the tissues as well as an increase in lymphatic drainage. Also, at higher end dosimetry, it is possible to inhibit axonal transmission helping to palliate pain. Faced with an acute injury, this offers obvious clinical gains to both patient and clinician as aside from the mitigation of pain and acute swelling, it becomes easier to take intraoral radiographs plus provides any essential immediate stabilization treatments. Furthermore, an early intervention with PBM can shorten the acute inflammatory phase and assist in a good quality of healing, repair and regeneration of damaged tissues [206–208] (Figs. 7.28, 7.29 and 7.30).

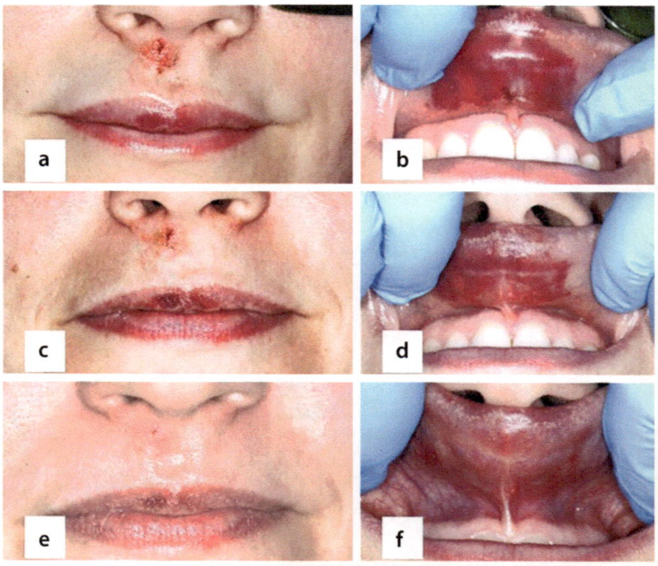

Case: 33yr old female with unremarkable medical history. She presented on a Monday morning having traumatised her lip and fraenum by jumping a fence and landing on her face. There were no hard tissue injuries.

She was treated for 300 seconds 940 nm 3.5W CW with a 7.1cm² area handpiece to the outside of her upper lip with a slow scanning motion and 120 seconds with 1.4W CW and a 2.8cm² handpiece in a scanning motion to the inside of her upper lip and fraenum. This was performed on the day of presentation and on days #3 and #5. Given the healing on day #5 it was agreed that no further PBMT would be required.

Images A/B: 24 hours post accident; B/C: Day 3 (after 2nd treatment);
E/F: Day 5 (after 3rd treatment); Case: Courtesy Dr Peter Orme, UK.

Fig. 7.28 Facial trauma results in a marked acute inflammatory response which can be disfiguring as well as being associated with pain and reduced function. As illustrated PBMT offers considerable benefits to ameliorate the trauma and promotes good wound resolution

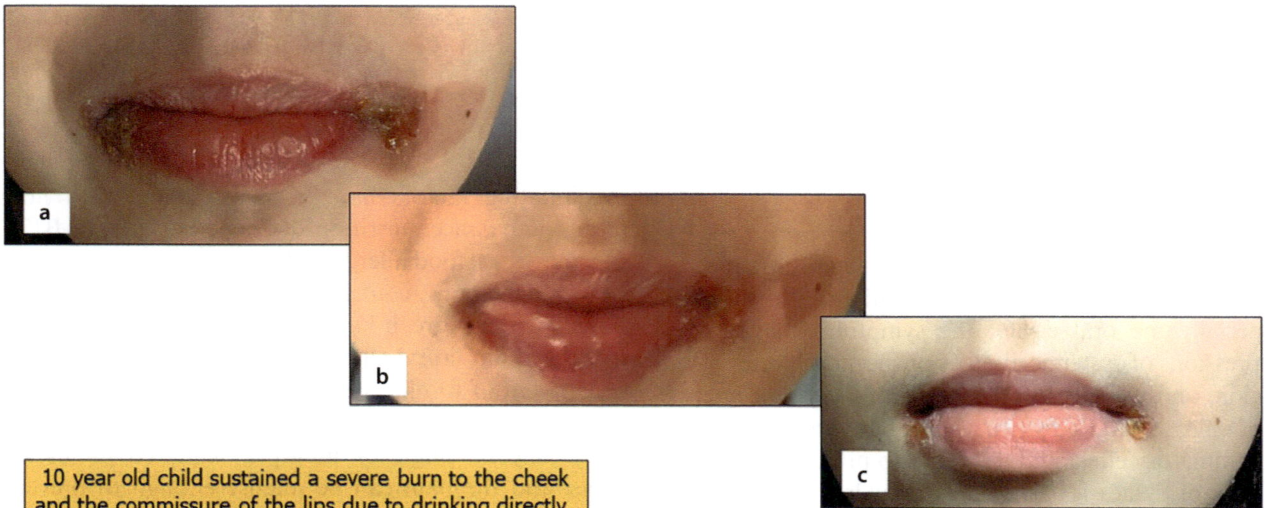

10 year old child sustained a severe burn to the cheek and the commissure of the lips due to drinking directly from a hot water tap.
This was treated over a period of six days on a daily basis by application with a scanning motion of a 940 nm laser 2.8cm² handpiece set at 1.4W CW.
The treatment time on each day was 100 seconds to each side which delivered 50J/cm² to the affected areas.

Image A: 24 hours after incident (day one of treatment);
B: Day three (after third PBM application);
C: Day six (after six daily applications).

Case: Courtesy Dr Mark Becker, UK.

Fig. 7.29 Burns may be deeply damaging to sensitive tissues and a severe burn often may require complex surgical procedures such as tissue grafting. PBMT improves the viability of the damaged tissues, stimulates repair and can greatly simplify patient management

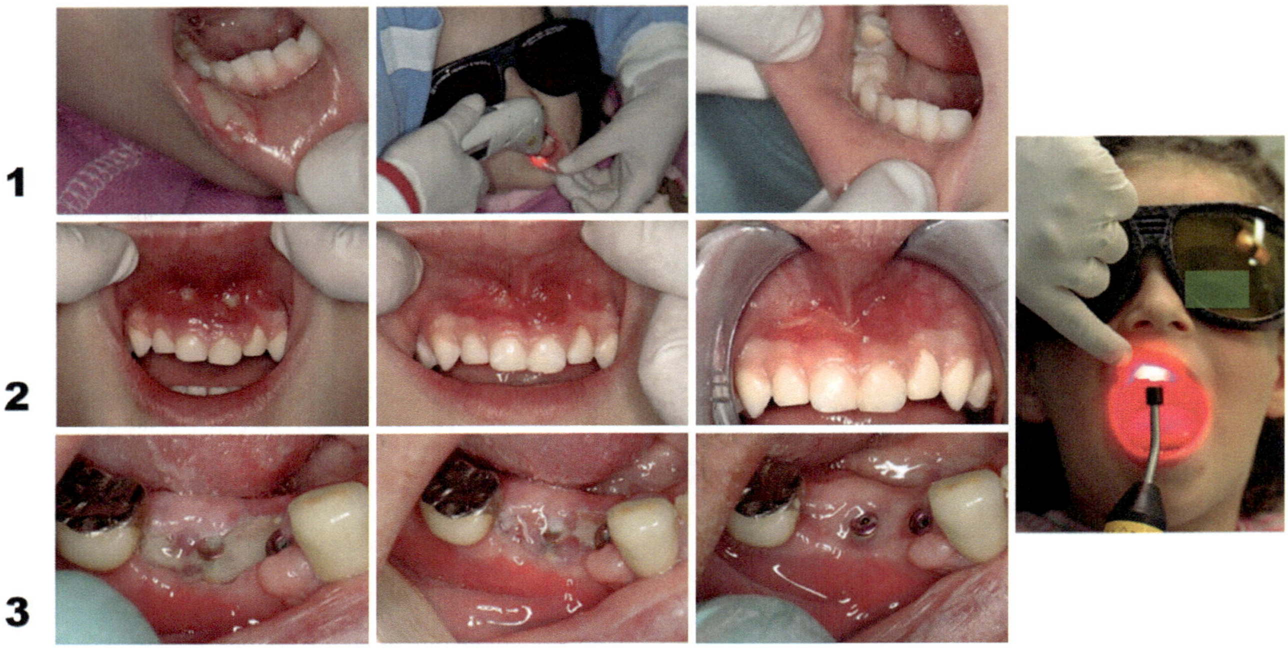

Fig. 7.30 (1) Four-year-old patient, with a traumatic painful lesion: biting following local anesthesia—treatment performed with diode laser 645 nm, fluence 4 J/cm^2, average power 5 mW, spot diameter 2 mm, power density 0.16 W/cm^2, CW, spot technique, 25 s per point, irradiations every other day. Immediate analgesic action, complete healing after five sessions. (2) Five-year-old patient, with a traumatic lesion with bone exposure (impact against the wooden edge of a pencil)—treatment performed with diode laser 904 nm, fluence 4 J/cm^2, average power 0.28 W, spot diameter 5 mm, power density 1.4 W/cm^2, superpulsed, spot technique, 3 s per point, irradiations every other day. Immediate analgesic action, complete healing after eight sessions. (3) Fifty-year-old patient: delayed healing after implant positioning and GBR, with fixture exposure—treatment performed with diode laser 810 nm, fluence 4 J/cm^2, average power 0.3 W, spot diameter 6 mm, power density 1.07 W/cm^2, CW, spot technique, 4 s per point, irradiations every other day, five sessions to trigger and to speed up the healing process. Complete healing after a month. (Cases courtesy of Dr E. Romagnoli and Dr A Cafaro, Italy)

7.23 PBMT in Oral Medicine

Oral medicine is concerned with the diagnosis and treatment of diseases affecting the maxillofacial region and especially the oral cavity; the symptoms maybe local expressions of systemic diseases or specific illnesses of the region, and it is the link between dentistry and internal medicine. PBMT, by virtue of its analgesic, anti-inflammatory effect and its ability to promote the regeneration of tissues, is particularly useful in the treatment of many pathological conditions of the oral cavity, especially in erosive manifestations of the oral mucosa [209].

7.24 Recurrent Aphthous Stomatitis (RAS)

Aphthae are ulcerations that vary in size (from a few millimeters to over a centimeter), affecting the oral non-keratinised mucosa; they are painful, have a rounded appearance, are gray/white in color and have an erythematous rim. The etiology remains unknown; however, predisposing factors have been proposed as consequential to an immune system dysfunction. Precipitating factors have been associated with hormonal imbalance, nutrition deficiencies (vitamin B12), specific drugs, food allergies, stress and mechanical injuries which may contribute to the occurrence of RAS. Furthermore, it has been reported that a sensitivity to sodium lauryl sulfate (SLS) which is a common constituent of toothpastes may be an important cause.

Aphthae tend to heal spontaneously in 10–15 days. They may be single or multiple, scattered in several areas, and the pain can be intense enough to prevent feeding, speech and oral hygiene.

There are many aids designed to reduce the symptoms and speed the healing of these injuries, such as painkillers, herbal or hyaluronic acid gels, topical corticosteroids and oral vitamin B12. Also a SLS-free dentifrice is advisable.

In literature there are many studies that have investigated the efficacy of PBMT in the treatment of RAS, and there is consensus that PBM laser treatment provides a good pain control. This can be achieved from a single application and there is better wound resolution compared to other therapies or placebo groups [210–216].

Diode lasers of various wavelengths (450–980 nm) with fluences of between 2 and 10 J/cm^2 with different treatment times according to the setting have been used. A suggested protocol can be adopted from the recommendations made for oral mucositis (see ◘ Fig. 7.26).

7.25 Herpes Simplex

Herpes simplex virus (HSV) is a common infection and has two types. Type 1 (HSV-1) is the most frequent in the oral cavity, where its prevalence is from 60% to 90%.

The virus can travel along the nerve sensory axons and cause chronic latent infection. Following primary infection it can remain latent in the trigeminal ganglion, and visible eruptions can recur triggered by fluctuations in the immune system, bright sunlight and physical stress. The progression of this lesion includes five stages: prodromal, vesicle, pustules or ulcers, crust and healing. Herpes labialis can be characterized as a benign, self-limiting lesion, which is spontaneously healed after 10 days after the first vesicles present [6].

Concerning treatment, topical and systemic antiviral medications, such as acyclovir and valacyclovir, are widely used. However, although these may alleviate the progress of the infection, there is a variable incidence of inevitable recurrence.

Until the stage of formation of a surface crust, the area is painful, and treatment is usually symptomatic with antiviral drugs (for large lesions or immunosuppressed patients), emollient creams, anesthetic ointments, painkillers and anti-inflammatory drugs which may be required, according to the gravity of the case.

The choice of PBM as a treatment method has been recommended to mitigate pain and reduce inflammation. Also, there can be stimulation of wound resolution and it is possible to arrest the progress of an early attack with a prompt PBM intervention.

Regarding the management by PBM treatment on herpes labialis lesions, it has been suggested to employ wavelengths in the range of 632–980 nm with an applied fluence of 4–10 J/cm^2. There are many suggested protocols some with multiple sessions and irradiation points during the different phases from the incipient prodromal to the crusty lesion stage. The recommended protocols are drawn from a limited number of studies in the literature [217–226]. The author's preferred approach based on clinical experience is to employ a high-intensity therapy. Employing a 1 cm^2 optical spot size, the affected area can be irradiated with a scanning motion to avoid excess heat accumulation. An output power of 0.5 W with wavelengths in the range 800–980 nm for 100 s delivers a fluence of around 50 J/cm^2. This high-energy treatment arrests the viral progress and at sub-surface levels promotes vasodilatation and stimulation of the immune system. Furthermore, patients experience good pain relief with prompt resolution of the condition. The validation of this approach however requires formal studies and evaluation under controlled circumstances, and as with so many aspects of dental science, more studies are required!

A further promising treatment approach combines the beneficial effects of aPDT (methylene blue + 660 nm) and PBM (660 nm). The former is performed in the vesicular phase, while the latter may offer benefits in the healing phase (◘ Fig. 7.31).

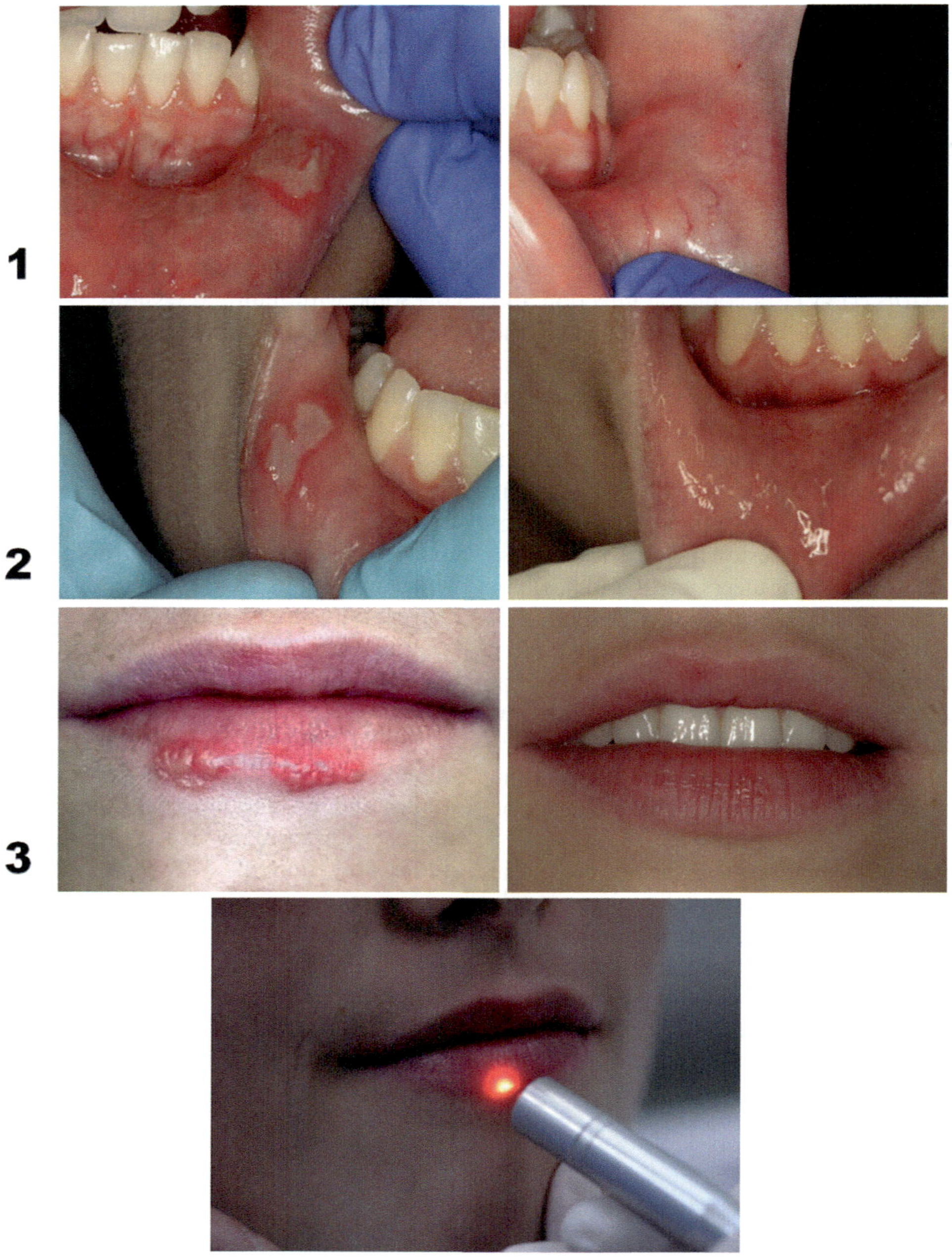

▪ **Fig. 7.31** (1) Fifteen-year-old patient, with a painful major aphthae—treatment performed with diode laser 904 nm, fluence 4 J/cm^2, average power 0.28 W, spot diameter 5 mm, power density 1.4 W/cm^2, superpulsed, spot technique, 3 s per point, irradiations every other day, immediate analgesic action and complete healing after four sessions. (2) Seventeen-year-old patient, with a painful major aphthae—treatment performed with diode laser 904 nm, fluence 4 J/cm^2, average power 0.28 W, spot diameter 5 mm, power density 1.4 W/cm^2, superpulsed, spot technique, 3 s per point, irradiations every other day, immediate analgesic action and complete healing after four sessions. (3) Twenty-six-year-old patient, with herpes simplex labialis—treatment performed with diode laser 810 nm, fluence 4 J/cm^2, average power 0.3 W, spot diameter 6 mm, power density 1.07 W/cm^2, CW, spot technique, 4 s per point, irradiations every other day. Complete healing after five sessions. (Cases courtesy of Dr E. Romagnoli and Dr A. Cafaro, Italy)

7.26 Mucosal Chronic Inflammatory and Autoimmune Diseases

The most well-known diseases included in this category are the vesiculobullous diseases and oral lichen planus.

7.26.1 Vesiculobullous Diseases (Pemphigus, Mucous Membrane Pemphigoid)

Pemphigoid vulgaris (PV) and bullous pemphigoid (BP) are defined as autoimmune mucocutaneous diseases, which clinically appear with blisters that can affect the skin and mucous membranes. Clinically these diseases appear with blisters that can affect the skin and mucous membranes. Intact blisters are rarely found and it is more common to see the result of their bursting. These present as erosions or ulcers that can be extended to large areas and can be very painful to the point of limiting the normal functions of the stomatognathic apparatus (hygiene, nutrition, speech). Diagnosis requires a specialist biopsy, and direct immunofluorescence staining of biopsied skin shows immunoglobulin and C3 deposits on the keratinocyte surface, and histology reveals keratinocyte acantholysis.

The treatment of choice is based on topical and/or systemic corticosteroids and other immunosuppressive drugs depending on the severity of the disease expression, with all the adverse effects that these substances have. There are few studies that have evaluated the effects of PBMT on these pathologies, and in each case a small population was studied, both because these diseases are not frequent and because of the involvement of other districts in addition to the mouth and because PBMT has almost always been used in addition to conventional therapy.

Laser diodes with a wavelength between the red (660 nm) and near infra-red (810–980 nm) have been tested, and two studies also used CO_2 lasers in defocused mode, with different fluences (4, -30, -60 J/cm^2).

Parameters employed for diode lasers are as follows:

- 660 nm, applied fluence of 35 J/cm^2, one session performed [2]
- 810 nm, applied fluence of 5 J/cm^2, seven sessions performed (every day for 1 week) [4]
- 980 nm, applied fluence of 4 J/cm^2, two sessions performed per week until the complete resolution of clinical signs

The results indicate a short- and long-term reduction in pain, faster healing of injuries and a minor recurrence of the disease [99, 227–230].

7.26.2 Oral Lichen Planus

Oral lichen planus (OLP) is a common mucocutaneous chronic disorder, and its prevalence is estimated to be between 0.5% and 2%. The correct initial diagnosis by histopathology is critical, as similar lesions are dysplastic and may be associated with an evolution into oral squamous cell carcinoma (OSCC). The main etiology remains unknown; however there is consensus that this is primarily of an immune-mediated origin. In view of this regular review, evaluations are advisable, which should be under the care and guidance of an appropriate specialist in oral medicine.

The various forms of this disease fall into two broad categories, white and/or red lesions. While "white" lichen is asymptomatic, the "red" ones (the atrophic–erosive and bullous variants) are very painful and have significant functional limitations. Erosions can hit multiple locations simultaneously; the most affected are the buccal mucosa, tongue, gums and more rarely the lips and the hard palate.

Therapy is based on topical and/or systemic corticosteroids depending on the clinical severity; a variable percentage of subjects are not responsive to this therapy, and therefore alternative treatments were considered, and PBMT is one of them. The ability to modulate the inflammation, the analgesic effect and the increase of the regenerative capacity of laser light are at the core of the excellent results obtained by the studies in literature [99, 227–234]. Diode lasers (630, 904, 980 nm) are reported to have been used, on average two treatments per week with fluence of up to 6 J/cm^2.

7.27 Medication-Related Osteonecrosis

Osteonecrosis of the jaw is an adverse effect following the use of drugs used for various metabolic and oncological diseases concerning the skeletal system (bone metastases, malignant hypercalcemia, Paget's disease, osteogenesis imperfecta, osteoporosis). This category includes bisphosphonates and antiangiogenic drugs. Bisphosphonates are drugs that have the ability to modulate bone turnover and reduce the process of resorption; they tend to be deposited in the tissue where they have a time-cumulative effect which seems to be at the basis of the adverse events.

The pathogenesis of osteonecrosis to date is not yet fully established. Amongst the most reliable theories are the inhibition of osteoclast activity, antiangiogenic action and the negative effect on circulating endothelial cells. Clinically, surgical therapy and tissue trauma are considered predisposing factors in the development of osteonecrosis, pre-existing periodontal disease and poor oral health conditions which can represent a negative prognostic factor.

Osteonecrosis treatment options are aimed both to the treatment of overt injury and especially to the prevention of them, even if there are not yet any specific guidelines based on scientific evidence. PBMT in recent years has been particularly studied for control of the osteo-mucous lesions, in addition to conventional surgical therapy or laser surgery (pain-relieving action, mucosal healing, support after resective surgery), but also to improve the healing of tissues in patients that underwent oral surgery to prevent exposure of necrotic bone and especially to induce formation of healthy bone [209, 235–248].

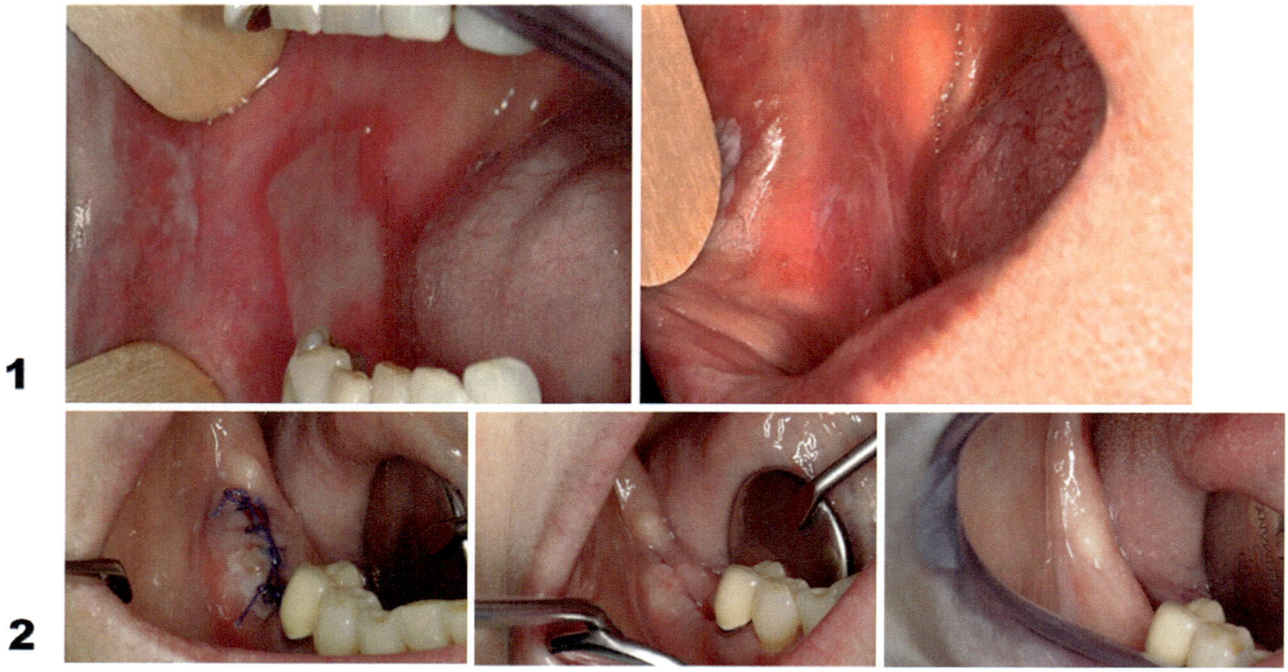

Fig. 7.32 (1) Fifty-eight-year-old patient with oral lichen planus (erosive form, unresponsive to topical steroids)—treatment performed with diode laser 980 nm, fluence 4 J/cm^2, average power 0.3 W, spot diameter 6 mm, power density 1.07 W/cm^2, CW, spot technique, 4 s per point, irradiations every other day, appreciable remission after 16 sessions (cases courtesy of Dr E. Romagnoli and Dr A Cafaro, Italy). (2) Sixty-year-old patient, assuming high doses of steroids and immunosuppressive drugs for a severe form of autoimmune hepatitis, with consequent osteoporosis and spontaneous bone fractures. The patient is undergoing a therapy with bisphosphonates by injection, with a high risk of oral necrosis of the jaw Preventive treatment, following dental extractions, performed with diode laser 904 nm, fluence 4 J/cm^2, average power 0.28 W, spot diameter 5 mm, power density 1.4 W/cm^2, superpulsed, spot technique, 3 s per point, irradiations every other day, six sessions. Complete healing after a month (cases courtesy of Dr E. Romagnoli and Dr A Cafaro, Italy)

The action of PBMT on bone, studied both in vitro and in vivo, is performed through the proliferation and differentiation of osteoblasts with increased calcium salt deposits which accelerate calcification, through the activation of MCM genes (minichromosome maintenance proteins) regulated by DNA replication and through type I collagen formation. The most commonly used wavelengths are in the range between 650 and 1064 nm, and power density is between 5 and 150 mW/cm^2, 30–60 s per point, and fluence between 0.3 and 9 J/cm^2. For 1064 nm wavelength, the parameters used are as follows: power 1.25 W, frequency 15 Hz, fiber 320 μm, PD 1562.5 W/cm^2 and fluence 7 J/cm^2 [111]. On average, laser sessions are performed two or three times per week (Fig. 7.32).

7.28 Burning Mouth Syndrome (BMS)

Burning mouth syndrome (BMS) has been classified as a distinct disease in 2004 by the International Headache Society, which defined the primary form as "a feeling of intraoral burning sensation for which no medical or dental cause can be found." The diagnosis is mainly by exclusion; it is chronic clinical entity characterized by burning or itching that affects the oral mucosa and perioral regions with a generally bilateral and symmetric distribution. Symptoms such as dysgeusia and xerostomia may accompany the burning sensation but without clinical and laboratory data that may suggest the combination of an organic disease.

The physiopathology of BMS is not entirely clear; it is believed that dysfunctions in the central nervous system, such as different processing by the brain of nociceptive and thermal stimuli and dysregulation of the dopaminergic nigrostriatal system, may represent plausible causes of oral burning. There is growing evidence in the most recent scientific literature that links the BMS to a peripheral neuropathic mechanism. Histopathological studies show a lower density of epithelial and subpapillary nerve fibers with axonal degeneration and an increase in the level of receptor mediators in charge of the processing of nociceptive response in patients with BMS compared to a control population. The tongue is the most frequently affected site.

There is no effective treatment plan; patients get some benefit by the use of anxiolytic drugs, anticonvulsants, herbal remedies, acupuncture and psycho-behavioral techniques. In recent years PBMT has also been used. The use of PBM has been explored for the management of BMS, due to its analgesic, anti-inflammatory and biostimulatory effects. Also, PBM can affect the microcirculation by vasodilation. Also, it has been shown that PBM can decrease the level of stress and salivary cortisol in BMS patients. The positive results in the treatment of BMS reported after laser treatments are due to the action that laser radiation has on pain control, through the release of endorphins, as well as preventing the arrival of ascending nociceptive stimulus to the higher cortical centers; a reduction of TNF-α and of IL-6 levels in saliva of patients after PBMT has also been reported.

The wide diversity of the parameters used varied from large numbers of in-contact small probe applications ("points") to scanning techniques with large diameter spot sizes 1 cm^2+. The lasers applied have a wavelength between 630 and 980 nm, power set from 40 to 300 mW and fluences from 0.4 to 176 J/cm^2. The sessions are performed one or two times a week for up to ten total treatments [249–267].

Recent systematic reviews do support the benefits of PBMT for this condition although the order of difference is relatively small as expressed as a statistical measure. Given the heterogeneity of approaches, it is not possible to make an evidence-based recommendation. However, the stability of the results over time and the effective reduction of inflammation mediators also suggest that there is a real photobiomodulating action.

7.29 Chemoradiation-Induced Mucositis

Oral mucositis (OM) is a common consequence of chemotherapeutic drug infusion, as well as head and neck radiotherapy (HNRT). A few days after such treatment has commenced, the patient complains of soreness, and the oral tissues appear red and smooth. Rapidly, the integrity of the mucosa breaks down, and ulceration occurs at sites on the buccal mucosa, ventral lingual mucosa, soft palate and the inner aspects of the lips and the floor of the mouth. Unfortunately, associated pain is intense, and OM can severely impair oral functions including speech and feeding [99]. This may lead to a requirement to discontinue cancer therapy, despite the increased risk of treatment failure. The serious morbidity that can arise from this clinical problem is associated with extended hospital-based care needs, together with a high degree of patient pain. Following extended multicenter clinical trials and multiple systematic reviews, the joint task force of the Multinational Association of Supportive Care in Cancer/International Society of Oral Oncology (MASCC/ISOO) has agreed new updated guidelines [268–270].

OM occurs in over 80% of patients undergoing hematopoietic stem cell therapy (HSCT), in addition to a high incidence in head and neck radiotherapy (HNRT). OM is also associated as a common complication (40%) of the management of many other types of cancer chemotherapy treatment. The pathology of OM is associated with pro-inflammatory pathways consequent upon the chemotherapeutically and radiotherapy-induced disruption of basal cells of the oral surface epithelium. Subsequent infection of the exposed dermal layers elicits an acute inflammatory response via increased cyclooxygenase production of prostaglandin E2, an upregulation of nuclear factor kappa B and interleukin 6, histamine release and increases in the production of bradykinin and substance P. This results in severe pain and swelling, which can be difficult to manage in already debilitated patients.

There is a large body of evidence demonstrating the efficacy of PBM for preventing OM in certain cancer patient populations, as recently outlined by the Multinational Association of Supportive Care in Cancer/International Society of Oral Oncology (MASCC/ISOO). Building on these, the WALT group outlines evidence and prescribed PBM treatment parameters for prophylactic and therapeutic use in supportive care for radiodermatitis, dysphagia, xerostomia, dysgeusia, trismus, mucosal and bone necrosis, lymphedema, hand–foot syndrome, alopecia, oral and dermatologic chronic graft-versus-host disease, voice/speech alterations, peripheral neuropathy and late fibrosis amongst cancer survivors [270, 271].

PBM parameters using low-level lasers or light-emitting diodes (LEDs) in cancer supportive care are usually within the red and near-infra-red (NIR) wavelength range between 600 nanometers (nm) and around 1000 nm, with an irradiance from 5 to 150 mW/cm^2 [7]. The duration of application varies according to the site, usually within 30–60 s per point. The therapeutic dosage varies between 4 and 12 J/cm^2. Higher end dosimetry is generally reserved for pain management as opposed to the prophylaxis (reduces incidence) and the promotion of healing of extant lesions. In recent years, LEDs with wavelengths in the red or NIR regions have been adopted as a prophylactic measure due to their safety, relatively low cost and suitability for home use.

In regard to management protocols, best evidence-based practice recommends prophylactic application as a prequel or on the same day as the medical intervention (chemo- or radiotherapy). This is repeated as required to coincide with further active intervention treatments. The area of exposure is the oropharynx, and administration of the dose is amenable either to multiple small point applications or to a larger optic beam device. There is some limited evidence supporting the benefits of a larger optical spot size, and indeed from a practical aspect, it is simpler and quicker to irradiate larger areas of tissues in contrast to a small probe tool [191].

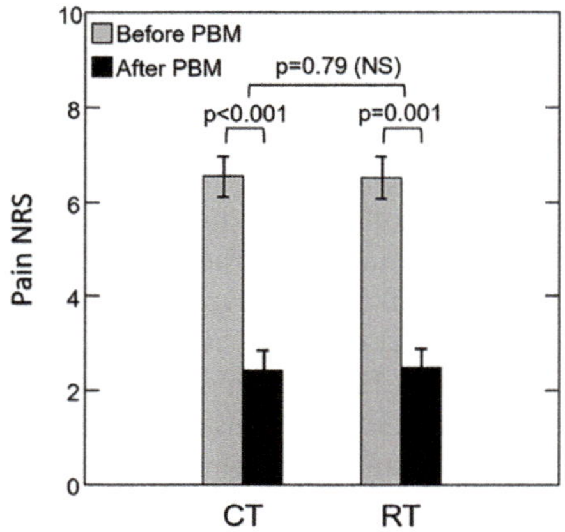

Severe pain associated with large areas of oral ulceration is a common occurrence in some forms of oncotherapy. This can complicate management as parenteral feeding may be required and can lead to an interruption in therapy. The resultant increase in morbidity and possible risk to mortality is a major clinical problem. The application of high dosimetry PBMT for pain relief can achieve an outcome which is of considerable clinical significance.

Fig. 7.33 A marked reduction following high-energy PBMT in a 10-point numerical rating scale (NRS) after chemotherapy (CT) and/or radiotherapy (RT). (Reproduced from Finfter O et al. High-power laser photobiomodulation therapy for immediate pain relief of refractory oral mucositis. Oral Diseases. 2023 May 19 [272])

The management of any extant lesions poses a dilemma as higher dosimetry PBM is effective at suppressing pain. However, the promotion of healing is best served by a lower dose window. As a possible clinical solution to this, a dosimetry flow chart has been suggested and this is reproduced above (Fig. 7.26). Higher-level dosimetry has been subject to some contemporary studies, and the research outcome offers support to the benefits for analgesia of a higher-intensity therapy for good pain relief.

In a study by Finfter et al., PBMT was administered intraorally using a 940 nm Indium–Gallium–Arsenide–Phosphorus (InGaAsP) semiconductor diode laser (Epic X, Biolase, Irvine, California, United States) [272]. A deep tissue handpiece with a defocused spot size of 1 cm^2 was applied with rapid rotatory movements over the ulcerated areas (but not over the tumor bed in cases of oral cancer) at a distance of 1 cm from the tissue. Each lesion was treated for 90 s. The parameters were as follows: 3 W with a duty cycle of 50% and pulse duration = 20 ms with a pulse interval = 20 ms. The irradiance was measured 1 cm from the laser aperture as 1.42 W/cm^2. Patients reported an immediate decrease in pain following 94% (74 of 79) of the PBM sessions, and after 61% (48 sessions), the pain reduction was over 50%. No occurrences of increased pain post-PBM were documented (Fig. 7.33).

7.30 Peripheral Neurological Lesions (Paresthesia, Anesthesia, Hyperesthesia)

Injuries of the inferior alveolar nerve, its branches and the lingual nerve are the major peripheral neurological lesions of oral interest. It is rare in everyday clinical practice to come across lesions to other nerve branches, as they are mainly due to complications following maxillofacial surgery.

The causes of injury of the IAN and the lingual nerve are third molar surgery, implantology, endodontics, orthodontic surgery and regional anesthesia (dental causes) but also operations for the removal of benign or malignant growths or operations concerning the salivary glands. The resulting symptomatology is variable, and complete absence of sensitivity (anesthesia) of the innervation district can be observed, or a decrease in sensitivity (hypoesthesia), sometimes accompanied by disabling pain (hyperesthesia) with a significant decrease in quality of life. Some lesions tend to resolve spontaneously in 2–3 months, especially if the cause is compressive and the cause (e.g., pressure on the mandibular canal or edema) is removed; if the damage is partial in an estimated 6–8 months' time, a partial recovery of the nerve due to spontaneous nerve regeneration from the proximal to the distal stump can be expected. In the case of complete rescission, a reconstructive microsurgical intervention can be attempted.

Drug therapy (high-dose corticosteroids with gastric coverage and neuroprotective drugs) should be started as early as possible to get a quick resolution of edema and allow recovery of cell function; if pain relieving pharmaceuticals therapies are required potent analgesics (clonazepam, gabapentin, carbamazepine) can be used.

In the literature, there are studies showing that the use of PBMT can be helpful in the management of peripheral nerve injury both in speeding up the regeneration of the nerve fiber and in control of pain [13, 273–276]. In vitro studies and animal model studies have shown an increase in the number of axons, Schwann cells and myelin, in the groups subjected to PBMT compared to control groups. An increase in metabolism in neurons was also observed, with increased production of basic fibroblast growth factor (bFGF) and neuronal growth factor (NGF) eGAP-43 (protein associated with peripheral axonal regeneration) [277].

The laser used is included in the range between 660 and 980 nm; the choice of the most suitable wavelength depends on the depth of penetration in relation

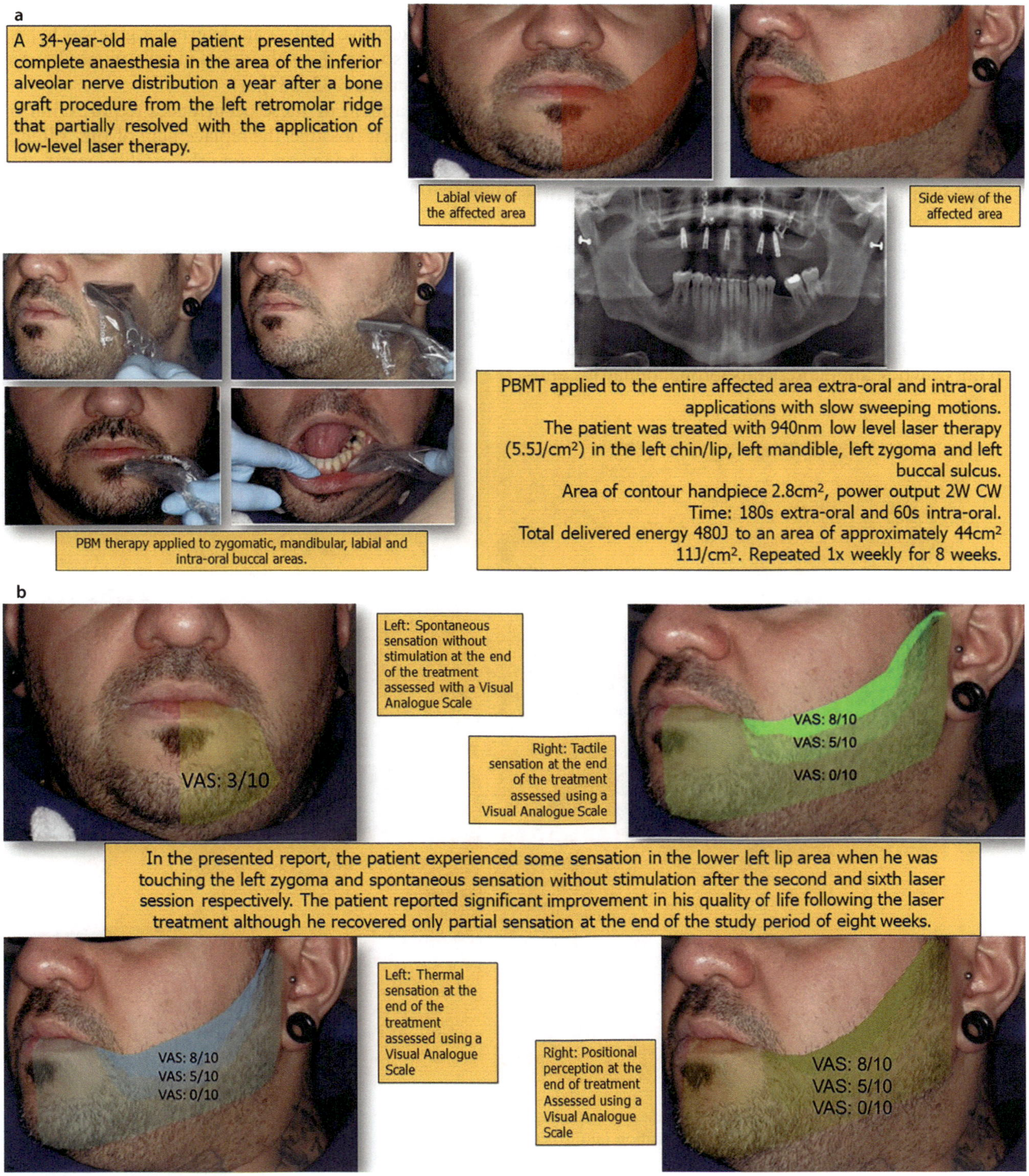

Fig. 7.34 **a**, **b** The neurotrophic effects of PBMT can be highly useful in efforts to promote axonal regrowth. (Case: Dr Despoina Chatzistavrianou and Dr Lochana Nanayakkara, UK)

to the anatomic site where the damage occurred. Fluences that have given the best results are between 0.2 J/cm^2 and 12 J/cm^2, treatment with repeatable cycles (ten sessions), irradiating the area affected by the nerve injury, after mapping, both intraorally and extraorally [278–281] (Fig. 7.34).

7.31 Adverse Effects of Drug Therapy on the Oral Mucosa

Pharmacological treatments can cause side effects even when used according to correct therapeutic schemes and dosages, and oral mucosa is one of the most affected places. Excluding systemic allergic reactions or effects caused by anticancer or antiangiogenic drugs, in the oral cavity it is possible to observe erosions/ulcerations, gingival hyperplasia, lichenoid reactions, salivation disorders, candidiasis, burning and redness, angioedema and multiform erythema up to Stevens–Johnson syndrome. Many drugs can cause effects on the oral mucosa; there are about 500 molecules capable of causing hyposalivation (anxiolytics, antidepressants, bronchodilators, antimigraine drugs, etc.); antihypertensives, anticonvulsants and contraceptive drugs can determine gingival hyperplasia, but so can also substances more commonly used as mouthwashes [282, 283].

The treatment of these occurrences requires first of all, when this is possible, the replacement of the "offending" drug and at the same time the establishment of a new therapy to promote healing of the tissues and counter the associated symptoms, with the possibility of experiencing new adverse effects. PBMT has been used successfully in the treatment of many drug-induced events, such as gingival hyperplasia [284], or to promote faster healing of wounds from the most diverse origins [285, 286] with no contraindications or side effects (◘ Fig. 7.35).

1

2

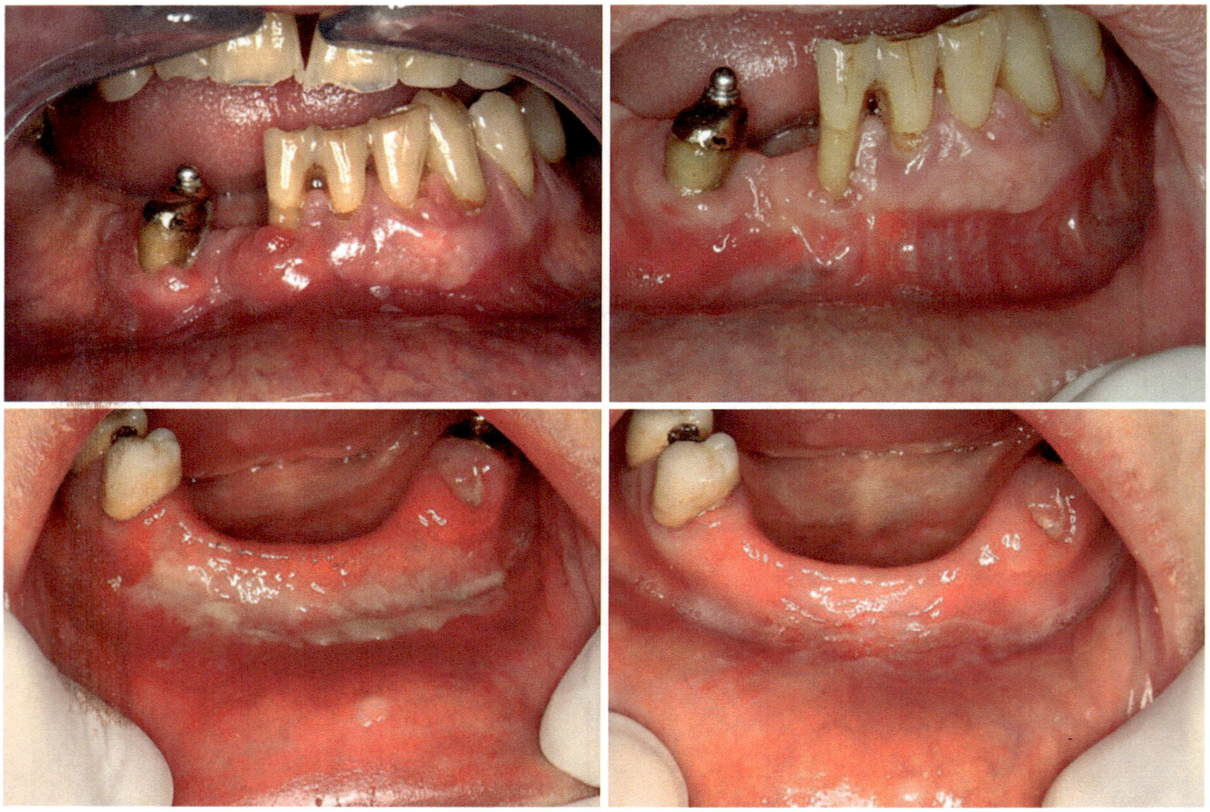

◘ **Fig. 7.35** (1) Seventy-year-old cardiopathic patient with a drug-induced hyperplasia and gingival painful inflammation (antihypertensive drug, non-replaceable). The patient was unable to wear the removable prosthesis. Treatment performed with diode laser 810 nm, fluence 4 J/cm^2, average power 0.3 W, spot diameter 6 mm, power density 1.07 W/cm^2, CW, spot technique, 4 s per point, irradiations every other day, appreciable improvement after six sessions. (2) Seventy-six-year-old patient, with Sjögren's syndrome, assuming systemic steroids. Allergic reaction to chlorhexidine. Treatment performed with diode laser 980 nm, fluence 4 J/cm^2, average power 0.3 W, spot diameter 6 mm, power density 1.07 W/cm^2, CW, spot technique, 4 s per point, irradiations every other day. Almost complete healing after 2 weeks (six sessions). (Cases courtesy of Dr E. Romagnoli and Dr A Cafaro, Italy)

7.32 PBMT and Bone

The bone tissue is a dynamic and plastic biological tissue characterized by considerable hardness and strength. It modulates its structure as a result of organic and mechanical stimuli, has the function of support and protection for the body and the internal organs, is a mineral salts reserve (calcium, 95%), contains the bone marrow and serves as insertion for the muscles. It is composed of an organic component rich in cells (osteoprogenitor cells, osteoblasts, osteocytes, osteoclasts), of the extracellular matrix rich in amorphous substance and collagen type I, which gives strength and elasticity, and of a part rich in minerals and inorganic salts (calcium phosphate and magnesium, sodium nitrate, potassium and manganese) which determine the hardness. The bone tissue is subject to a number of structural and functional changes due to age, nutrition and subjective conditions such as drug therapies.

Bone is a tissue with a high regenerative potential: healing of fractures is a clear example of this characteristic. The physiological mechanism of regeneration occurs through sequences activated by molecular and cellular factors and is similar to the healing of other types of tissue: there is an initial inflammatory phase, followed by reparative phase and the remodeling phase.

In the bone wound, a hematoma is formed that favors the supply of inflammatory cells (macrophages, monocytes, lymphocytes and nucleated polymorphic) that induce the production of bone morphogenetic proteins (BMPs) and growth factors. These in turn infiltrate the surrounding bone resulting in the formation of granulation tissue rich in newly formed vessels in which mesenchymal progenitor cells are recalled. During the reparative phase, the derived mesenchymal cells (monocytes and fibroblasts) begin to differentiate into bone cells and osteoblasts, which secrete the collagen matrix rich in fiber that creates a bridge between the various edges of the wound and leads to the formation of osteoid tissue, on which the mineral component is deposited. In the phase of tissue remodeling, the newly formed bone assumes the characteristics of the native bone in terms of shape, structure and mechanical strength; this process occurs in 3–6 months [287].

It has been observed that the use of anti-inflammatory drugs in the first phase can impede bone healing, and the use of tobacco can inhibit the formation of the supporting stroma of newly formed vessels.

The effect of PBMT on bone tissue has been studied both in vitro and in vivo demonstrating an increase in the synthesis of osteoblasts, modulation of inflammation and production of TGF-β which includes the BMP, one of the important factors that regulate the proliferation and differentiation of new bone. A decrease in osteoclast activity [288–291] has also been noted.

To validate the effectiveness of PBMT on osteoblasts, the activity of mitochondria was measured; after biostimulation with a 830 nm laser and fluence of 3 J/cm^2, an increase of cell proliferation by 30–50% was observed [291]. The wavelengths commonly used are in the range of near infra-red (deep penetration) with fluences between 2 and 5 J/cm^2.

The possibility to obtain an enhanced bone healing offers advantages in the treatment of surgical wounds in which the bone tissue is involved, after dental extractions, after oral surgery in general, for the osseointegration of implants and in intrabony defects subject to periodontal surgery.

7.33 PBMT and Implantology

The success of implant therapy is the result of many factors and depends both on the health of the soft tissue, the requirement for good osseous support and the connective/implants interface. PBMT has many potential applications as an adjunct to all aspects of surgical management. This ranges from the preservation of bone at extraction sites, the optimization of bone and soft tissue grafting plus the mitigation of postoperative pain and swelling. Furthermore, there may be benefits to increase success in osseointegration and improved primary implant stability. With an aging population, more patients in a mature age bracket are requesting implants as a restorative option, and as over 50% of patients over the age of 50 have a systemic disorder, any measure that may improve the prognosis deserves to be welcomed. Regrettably as in many areas of progressive dental clinical practice, the evidence base is far behind the clinical experience. Hence, we are at present largely but not entirely at the lower levels of evidence base pending more well-designed clinical RCTs.

At the level of histomorphometric, genomic and proteomic analyses, there is a considerable volume of evidence-based data on the effects of PBMT on cellular metabolism. Given the extensive knowledge base into PBMT, its safety and efficacy, it is useful to share clinical experiences pending the further development of the evidence base (see ◘ Figs. 7.36, 7.37 and 7.38).

Many of the studies available in the literature on implants have been performed on animal models [292, 293]. These indicate a much increased early resistance to implant removal, and the histology clearly shows good early osseointegration which implies improved primary stability.

The wavelengths used vary in the range of red and near infra-red with fluences ranging from 2 to 92 J/cm^2, mainly with values between 3 and 8 J/cm^2 with daily applications or every 48 h during first weeks after implant

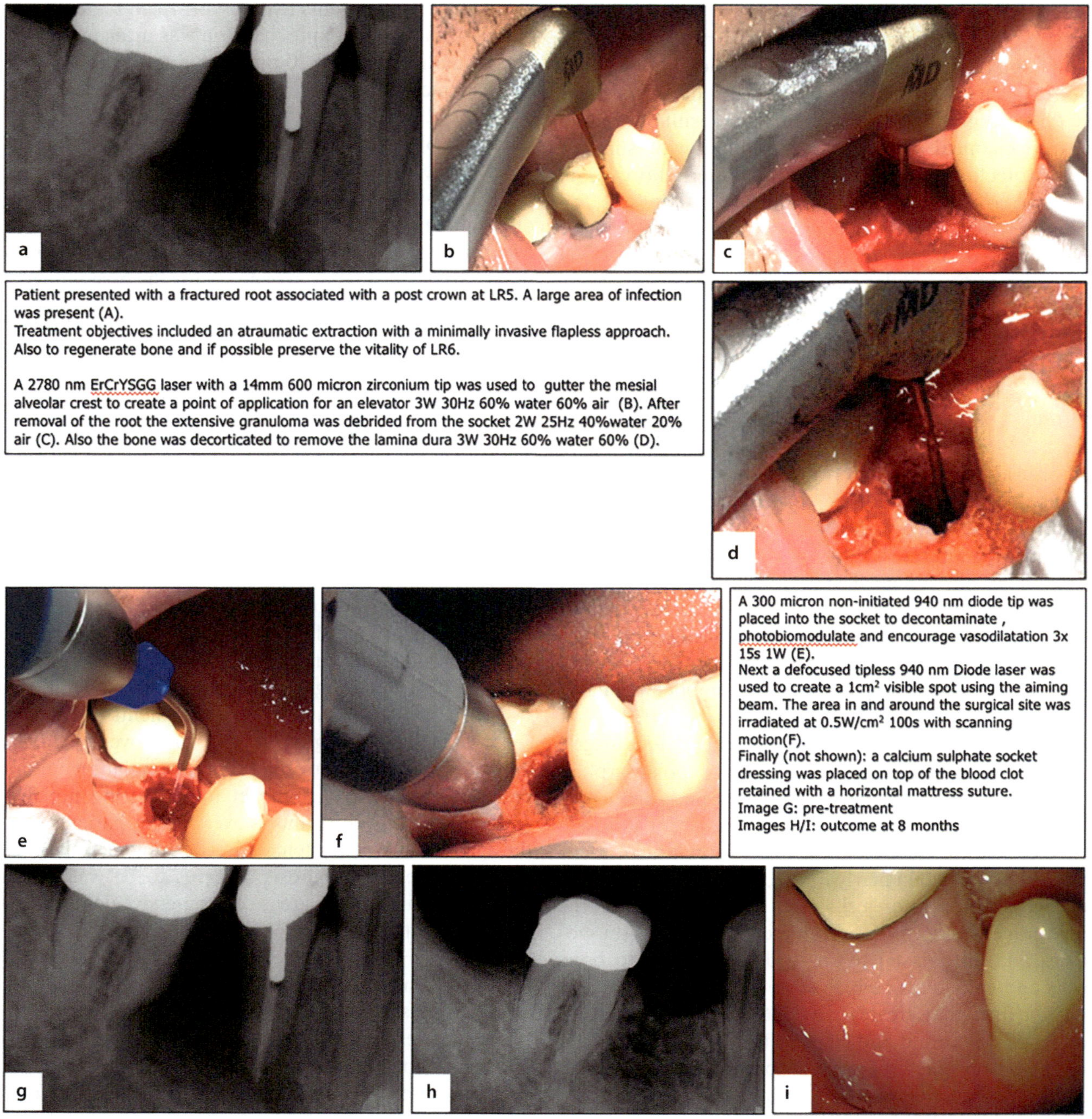

Fig. 7.36 An emergent treatment modality is to apply two different wavelengths. Due to high absorption in water, the erbium wavelengths are highly useful for minimally invasive hard and soft tissue surgery. Also, they decontaminate the surgical site and erbium lasers are recognized to have PBM benefits in promoting osseous regeneration. The diode wavelengths are used to decontaminate the site as well as to photobiomodulate the socket area. Also, there is an anti-inflammatory action and there is vasodilatation which promotes a good oxygenated blood supply to the surgical wound site

placement. The caveat to the dosimetry used in these animal studies is that due to differences in human tissue thicknesses and a corresponding increase in attenuation of the beam, it is recommended in patient care to increase the surface dosimetry to compensate (see Fig. 7.17). However, the possibility to speed up the early phases of osseointegration may be relevant for immediate loading of implants or for implants with a poor initial stability [293–297]. A further application relates to the management of peri-implantitis where for sure the added bonus of enhanced wound healing capacity is a benefit in an otherwise difficult to manage clinical situation [298–301].

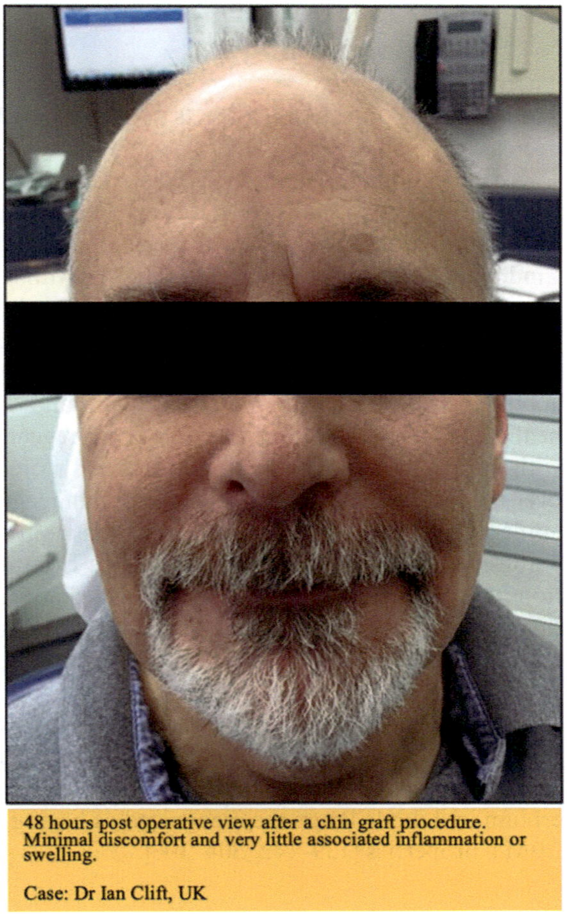

Fig. 7.37 No swelling or pain at 48 h post-operation having removed a bone graft from the chin. PBMT conducted with a 2.8 cm^2 contour handpiece with a scanning motion at 2 W CW for 100 s to an area of approximately 6 cm^2

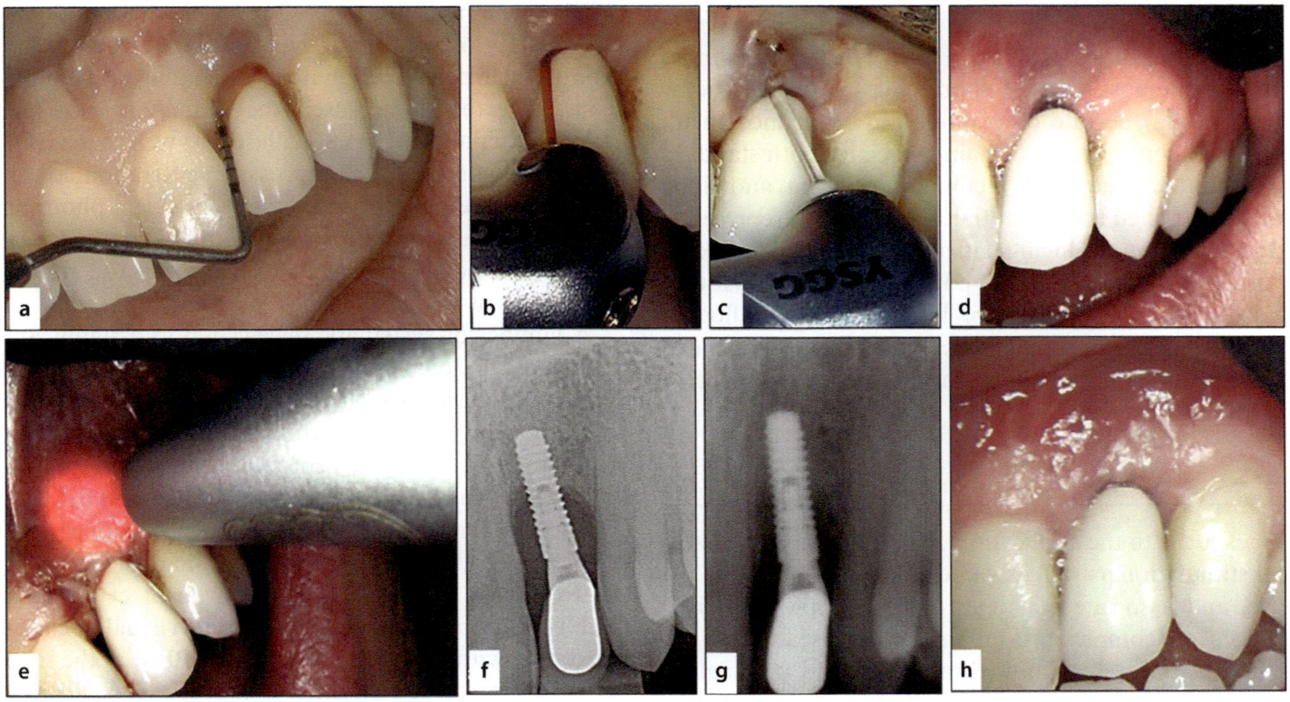

Fig. 7.38 Peri-implantitis at UL2 associated with a 9 mm pocket, pus and bleeding on probing **a**; debridement with a zirconium radial firing tip to remove peri-implant granulation tissue and disrupt biofilm **b**; a side firing sapphire tip debrides the implant threads **c**; healing at 8 months **d**; a tissue graft was placed and PBMT performed using a 940 nm 1 cm^2 spot size at 0.5 W CW for 100 s with a scanning motion. This procedure was also conducted after the flapless surgery 8 months prior **e**; pre-op radiograph **f**; post-op view at 8 months: new bone formation **g**; final view: healing at 1 month after the tissue graft **h**. (Case: Dr R Al-Falaki, UK)

7

7.34 PBMT in Intrabony Defects in Periodontology

The alveolar intrabony defects responsible for a decreased tooth stability are part of a framework of severe periodontal disease. The gold standard treatment is represented by specialist resective/regenerative surgery in combination with growth promoters such as Emdogain and platelet-rich fraction (PRF) with the aim of correcting the anatomy of the sites and promoting bone regeneration. As an adjunct to in addition to conventional treatments laser applications result in already-accepted effects on inflammation, pain and wound healing, as well as in an increased bone regeneration with improved clinical periodontal indices, more stable over the long term [301]. Laser-assisted management of adult chronic periodontitis is gaining support in the literature as a useful added measure. Aside from indications that the laser-assisted outcome may be more stable, there are patient factors related to a reduction in postoperative pain and swelling [302, 303]. The multifactorial aspects of laser care may require the use of an erbium laser for selective surgery of diseased tissues followed by a diode laser for the added depth of tissue penetration of the optic source into the target tissues. Both the erbium and diode lasers are recognized as having useful PBMT properties [139–142, 304–306].

7.35 PBMT in Orthodontics

Applications of lasers and LED PBMT for orthodontics provide the possibility of shortening treatment time by accelerating the rate of orthodontic tooth movement. Also, the complication of pain associated with standard orthodontic treatment is a common problem, and along with protracted treatment times, these issues represent a significant issue reducing patient compliance and acceptance of treatment [307–309]. Initial investigations studied the observable effects of laser devices to influence pain and discomfort experienced by patients associated with the forces applied to teeth to achieve movement.

7.35.1 Pain

In response to the application of load, there is mechanical stimulation as well as some damage of cells and tissues and associated changes in blood flow. This is a trigger to a complex pro-inflammatory cascade of cytokines including histamine, bradykinin and prostaglandins amongst others [310]. The nervous system contributes to the physiology of the resultant peripheral inflammation mediated via neuropeptides such as substance P, neurokinin A and calcitonin gene-related peptide. These potent mediators induce vasodilatation, increased vascular permeability and the activation of nuclear factor kappa B. The overall resulting biological response is to produce so called aseptic inflammation which results in the stimulation of C-nerve and A-delta nerve fibers producing pain symptoms. These symptoms can vary in intensity and duration and are most normally seen during the first hours after the application of forces. Pain usually reaches a peak after around 18–36 h with a gradual decline over the following week. This pain/discomfort experience is commonly associated with fixed and removable appliances, separator and band placement, bracket debonding and wire displacement. The consequent deterioration in patient comfort can in prevalence affect the majority of patients and is recognized as a key barrier to the completion of orthodontic treatment. The most common option has been to prescribe nonsteroidal anti-inflammatory drugs (NSAIDs). These are effective in pain relief; however they are associated with the hindering of osteoclastic activity due to the inhibition of the production of prostaglandins via COX2 suppression [311]. Also, NSAIDs can be associated with serious adverse effects such as gastric bleeding, ulcers and allergy. Studies in experimental animals have demonstrated a reduction in the rate of OTM in conditions where inflammation has been suppressed, and it is apparent that induced acute inflammation is a necessary component associated with OTM bone remodeling.

A variety of wavelengths of lasers have been found to be useful in producing analgesia including the helium–neon laser, the diode laser, the Nd:YAG laser, the Er:YAG laser and ErCr:YSGG lasers as well as the CO_2. Due to the potential for deep tissue penetration consequent upon the low absorption of the incident photonic energy by tissue chromophores such as water, the wavelengths that may be best suited for this purpose are the diode lasers as well as the Nd:YAG.

A number of systematic reviews report a significant reduction in pain amongst the treated patients. Successful treatment strategies in the literature describe applications at the cervical third of the gingivae on the buccal and palatal/lingual aspects. Dosimetry is reported to be in the range of 12–18 J/tooth as a divided delivery [132–135]. From a practical viewpoint, this appears to offer considerable benefits without adding much time to the treatment visit. For example, Marini et al. in a double-blind placebo-controlled trial virtually eliminated pain with a single application of PBM at the time of placement of orthodontic separators. Total PBMT treatment time is 360 s (◘ Fig. 7.39).

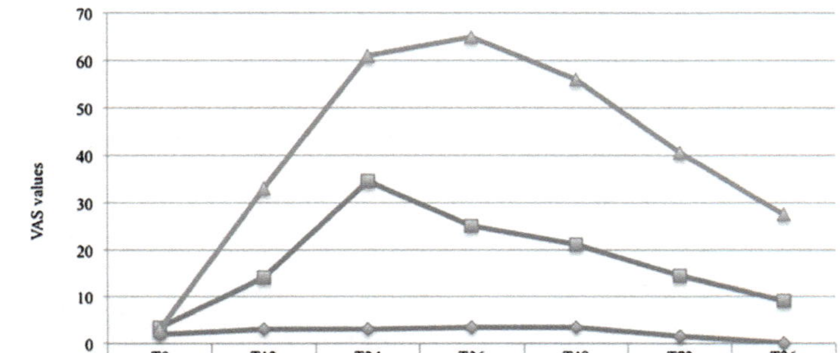

	T0	T12	T24	T36	T48	T72	T96
Laser	2	3	3	3,5	3,5	1,5	0,15
Placebo	3,5	14	34,5	25	21	14,5	9,15
Control	3	33	61	65	56	40,5	27,4

Parameters:
910 nm 160mW output power
Spot size: 0.5cm^2.
Applied buccal and palatal to the teeth with separators in place.
Delivered energy: 18J/tooth as divided doses.
Total treatment time (6 points of application on 3 teeth) 360s.

Fig. 7.39 A single application of PBMT at the time of orthodontic separator placement eliminated pain in the test group. (Reproduced from Marini I et al. The effect of diode superpulsed low-level laser therapy on experimental orthodontic pain caused by elastomeric separators: a randomized controlled clinical trial. Lasers in Medical Science. 2015 Jan;30:35–41 [312])

7.36 Acceleration of Orthodontic Tooth Movement (OTM)

On the application of an orthodontic force, a rapid acute inflammatory tissue response is elicited. The subsequent application of phototherapy apparently optimizes the cellular response permitting an increase in bone metabolism. Studies conducted in animal models using rats, dogs and rabbits have shown promise that laser and LED phototherapy can improve OTM [313]. Measures applied in the animal studies have included histology assessing bone density and volume, the proliferation of osteoclasts and osteoblasts, the number of capillaries and changes in the number of inflammatory cells. Also, by usage of monoclonal antibodies, there have been immunohistochemical measures for important cytokines involved in bone remodeling such as osteoprotegerin (OPG) and the receptor activator of nuclear factor-KB ligand (RANKL). The majority of the animal studies have used metrics on movement of the adult first molar in rats and dogs although perhaps controversially a few of the studies used movement of the rat incisor as the experimental model. Also, there are studies looking at the effects on the mid maxillary suture in rapid maxillary expansion augmented with laser or LED phototherapy. The majority of the animal studies showed that the application of lasers in the wavelengths of 650–940 nm increased the rate of tooth movement 2–3× compared to control groups. In addition this outcome was supported by histological evidence of increased cellular activity and significant signs of an increase in bone remodeling compared to control [181].

In an animal study by Suzuki et al. [181], laser phototherapy was found to increase the number of osteoclasts present on the pressure side, while there was a corresponding increase in the number of osteoblasts on the tension side. Aside from the histomorphometric analysis, this interesting study applied an immunohistochemistry analysis of RANKL/OPG and tartrate-resistant acid phosphatase (TRAP) activity which further correlated the histological findings of increased osteoclast activity in the test group on the pressure side (elevated levels of RANKL and TRAP) and increased bone apposition and osteoblast activity on the tension side (an increase in OPG). Furthermore, the rate of OTM was found to be increased by around 40% compared to the control.

The animal studies represent a substantial body of evidence-based research [2]; however there is no agreement on laser or LED wavelength, duration of treatment, frequency of treatment, irradiance (W/cm^2) applied or total dose (fluence). Faced with a heterogeneous set of experimental studies based on a variety of animal models ranging from rats, dogs to rabbits, it is not possible to extrapolate the results to human subjects. The animal studies are however highly supportive of a possible future role for photobiomodulation as an effective tool in accelerated bone metabolism in relation to OTM [327–333].

There is some limited clinical evidence supporting PBM to accelerate OTM, and an increase of 20–40% has been identified in a number of systematic reviews [314–317]. However due to the heterogeneity of the parameters plus issues related to the design of the studies, it is not at this stage possible to give evidence-based guidelines. One key problem is the logistical aspect of the need for frequent visits as this calls on good patient compliance plus added clinical time. In an attempt to overcome this problem, home administration patient LED devices have been developed. However, regrettably there is a lack of supporting good-quality blinded RCTs to evaluate the benefits of this novel approach, and the clinical value of this type of appliance has yet to be established (Fig. 7.40).

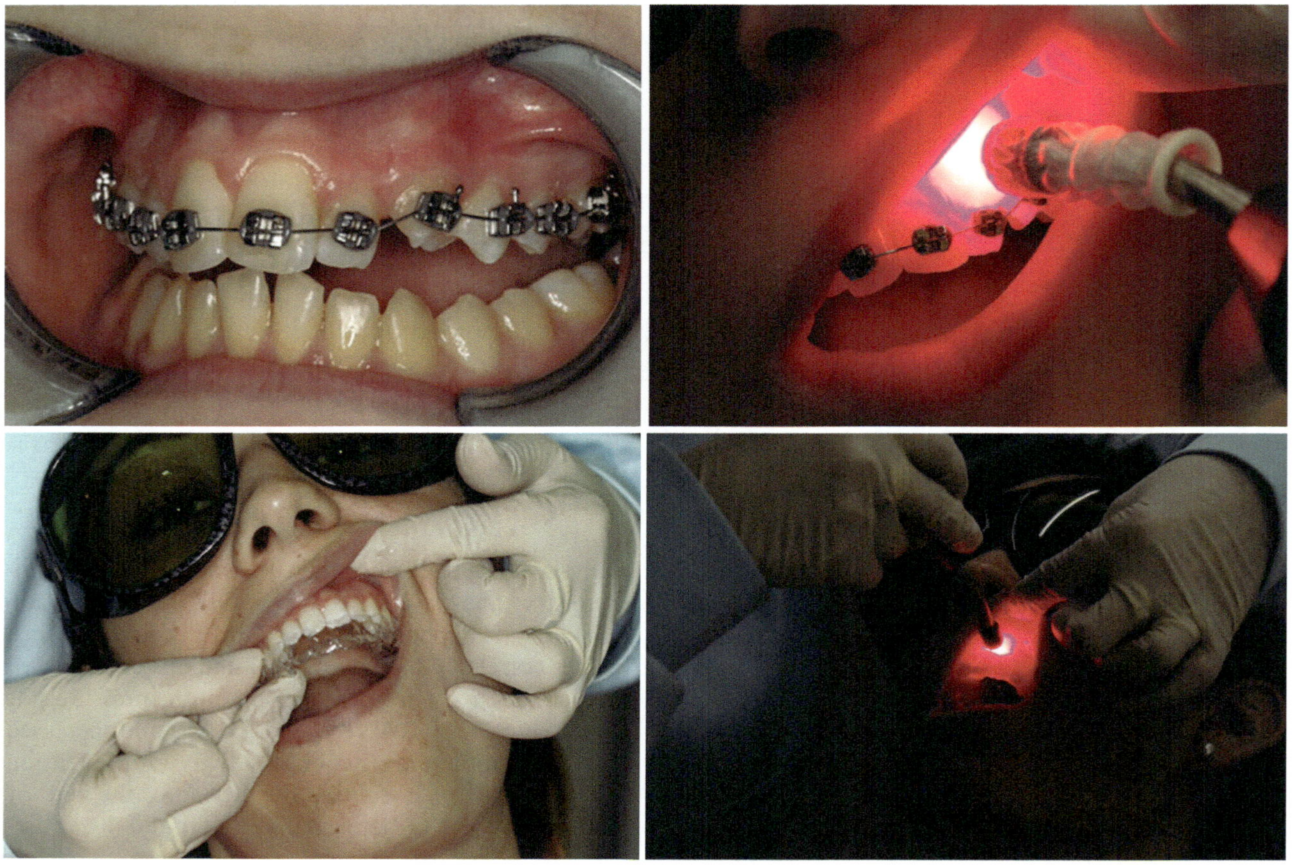

Fig. 7.40 Laser application in orthodontics (fixed appliance and removable aligner). Treatments performed with diode laser 904 nm, fluence 5.6 J/cm^2, average power 0.28 W, spot diameter 5 mm, power density 1.4 W/cm^2, superpulsed, spot technique, 4 s per point. Six points per tooth (mesial, distal, apical, vestibular side and lingual or palatal side). Irradiations every 15 days. (Case courtesy of Dr E. Romagnoli and Dr A. Cafaro, Italy)

7.37 PBMT and Dentin Hypersensitivity

One of the most common requests for dental examination concerns increased tooth sensitivity, mainly perceived as painful sensation due to thermal, tactile, osmotic or chemical stimuli. Excluding the most typical disease of the teeth (caries and its complications), dental hypersensitivity is established over time for dentine exposure due to factors such as gingival retraction or enamel erosion. The treatment of this condition involves the removal of the predisposing factors (incongruous brushing, saliva hyperacidity, eating disorders, parafunctions—teeth grinding and clenching—diseases of dental hard tissue, gastrooesophageal reflux disease, etc.).

Several agents are used with the purpose of sealing the open dentinal tubules responsible for the painful sensations, administered as home therapy or in the office. The patient should be properly instructed on proper oral hygiene techniques, on the limitation of acidic foods and drinks and on the use of toothpastes and mouthwashes containing fluoride or specific for dentin sensitivity. In-office professional therapies involve the use of fluoride varnishes, pastes containing oxalates, remineralising agents (calcium phosphate + casein), fillings with composite resins or glass ionomer cements and laser applications.

The action of PBMT on dentin hypersensitivity acts mainly on the control of the associated pain modifying the transmission of the nociceptive stimulus to the pulp, and by stimulating the normal cellular functions, the laser can promote the production of tertiary dentin, obliterating the dentinal tubules from the inside [318–323].

In a recent systematic review (2021), Mendes et al. included 14 studies employing wavelengths in the range of 780–2780 nm with a delivered fluence/tooth of 10–60 J/cm^2. Given that low-power diode lasers were included in this review with high-power free-running neodymium and erbium lasers, it is notable that the dosimetry employed falls within the suggested range for photoinduced analgesia [325].

After cleansing the area to be treated, the irradiation is performed perpendicular to the sensitive area, in three points (mesial, distal, central), repeating the sessions every 48–72 h until symptoms disappear. It is important to apply any surface desensitiser after the laser therapy

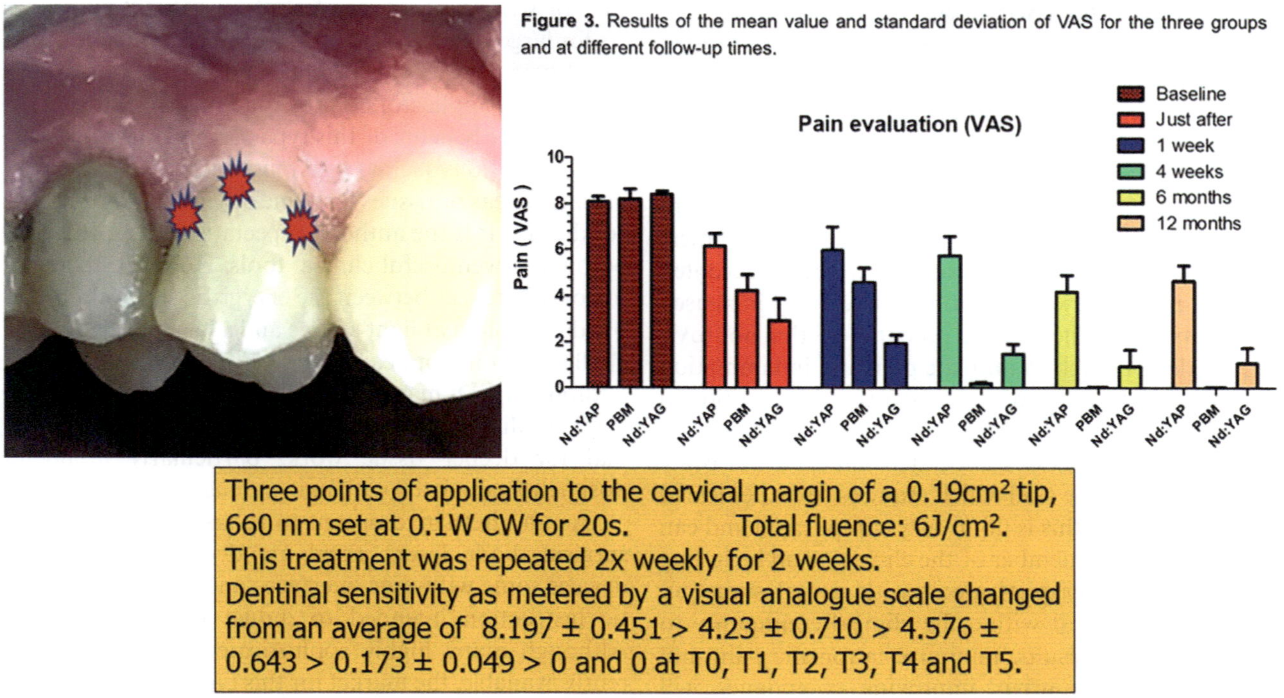

Fig. 7.41 PBMT compared favorably in outcome with the dentinal heat fusion technique employed. (Reproduced from Nammour S, et al.. Twelve-Month Follow-Up of Different Dentinal Hypersensitivity Treatments by Photobiomodulation Therapy, Nd:YAG and Nd:YAP Lasers. Life. 2022 Dec;12(12):1996 [324])

to optimize the adjunctive benefits (Dr E. Anagnostaki, personal communication).

In respect of the literature, there are articles that find PBMT is comparable in outcome to applications of fluoride gel and gel containing potassium and PBMT. A recent (2022) randomized controlled trial (RCT) showed that the combination of potassium nitrate gel and PBM with 808 nm and 35 J/cm^2 fluence for three sessions every 72 h could significantly reduce sensitivity [325].

Similarly, a RCT compared the usage of a 660 nm diode laser to both Nd:YAG and Nd:YAP high peak power free-running pulsed lasers. The neodymium lasers were both used to fuse and seal the open dentinal tubules in a single visit, whereas the 660 nm diode laser was used for repeated PBM 2 × weekly for 2 weeks. Both approaches resulted in a significant reduction in dentinal sensitivity; however the best outcome was found in the PBM red laser group [324] (Fig. 7.41).

There may be different modes of operation between the low-power and high-power devices. High-power lasers with a small spot size can generate significant heat which may melt the superficial apertures of the open dentinal tubules. There will also most likely be a dehydrating effect which will render the dentinal-treated surfaces more absorbent of any subsequently applied topical dentine desensitisers. It is accepted however that Er,Cr:YSGG lasers can have an analgesic effect and there may be a dual effect with high-power laser systems of surface modification and pulpal photobiomodulation. The details of the optimal evidence-based approach are elusive at present. However, in the absence of a consensus approach to treatment, it is apparent that many different wavelengths and parameters can prove highly effective [95, 96, 322, 326].

Also, the properties associated with PBM may be able to attenuate the damage and inflammation caused in pulp tissue by in-office bleaching agents. A systematic review by Carneiro et al. (2022) identified six studies investigating the potential benefits of PBMT to post-bleaching sensitivity [326]. All reported success; however there was no consensus on the parameters or the timing of the intervention, i.e., whether before, afterwards or both. The majority of the studies advocate the adjunctive post-PBMT use of high-fluoride dentifrices. The wavelengths employed ranged from 660 to 810 nm with a fluence in the range of 12–60 J/cm^2. Given the heterogeneity of the studies, it did not prove possible to conduct a meta-analysis. In respect of other PBM treatment modalities, pre-conditioning has been found to be beneficial in improving cellular viability and reducing the production of pro-inflammatory cytokines. The weight of current available evidence is insufficient to offer guidelines on the prophylaxis of post-bleaching sensitivity. However, this approach is likely to further develop as the outcome of the current studies is encouraging.

7.38 Future Perspectives

There is a considerable body of evidence base in respect of the PBM effects on biochemistry, immunology, gene transcription pathways and the histopathology associated with damaged and diseased tissues. However, PBMT has not as yet achieved a high level of acceptance as a useful adjunctive process in routine clinical care. As more well-designed studies are being accepted into the peer-reviewed literature, there is an increased awareness of the therapeutic potential of this nontoxic and gentle approach to mitigate pain and inflammation and to take advantage of the promotion of the optimal biological capacity of the tissues to heal, repair and regenerate. The processes associated clinical with PBMT entail a transferable skill which requires some dedicated training; however this is not overly complicated and can be learned by any member of the clinical team including hygienists and therapists along with dental surgeons. The level of interest within the clinical community to enhance clinical results, minimize iatrogenic pain and ease the patient journey improving compliance will inevitably result in PBMT being adopted as a routine procedure. In respect of standardization of equipment including wavelengths and peripheral applicators, most existing surgical diode lasers can already be adapted to the task dependent on the operator's knowledge and ability to work with what equipment is on hand. Due to variations in wavelengths and other inherent properties of the device, this may pose a challenge to a new adopter of PBMT.

The opportunity for discerning clinicians to adopt PBMT to good effect within the current knowledge base is limited only by the need for practical and didactic training. Given the ascending rise in awareness and interest in the already established effects of PBM on cellular and tissue metabolism, there are challenges ahead to satisfy the requirements of the evidence-based clinical community to offer standardized protocols. However, at the time of writing, PBM is the most active topic of scientific research papers on lasers in dentistry with over 5000 publications accepted in the peer-reviewed literature in the past 5 years alone. In combination with the use of regenerative materials such as autologous stem cells, the value of PBMT to clinical dental care is anticipated to be a key topic which will have a considerable impact on the patient outcome of many common treatment modalities.

LED appliances for home administration of PBMT as well as larger clinic-based units are currently available. For sure the attraction for patients to be able to self-administer low dosimetry PBMT at home is a great concept as it solves a logistical issue. However at the time of writing, this type of appliance is still unproven. As for larger clinic-based units, these are available as flat panel-type devices which can be multiwavelength as well as intraoral applicators. There is active current research assessing the benefits of this type of appliances for managing conditions such as oral mucositis as well as exposing large areas of tissues for some other oncology-related conditions. It is the authors' expectation that LEDs will become proven useful clinical tools. However, there are major differences between the energistic state of a broad band noncoherent light source and a laser which poses a challenge to the optical transport of photons to depth. Although LEDs may prove highly useful for some superficial conditions, it is probable LEDs may not be optimal for deeper tissue work, particularly if higher dosimetry associated with pain management is required.

As an alternative option to re-tasked extant dental surgical diode lasers, there are dedicated to PBMT devices some with user interfaces and supported didactic instruction intended to induct the new adopter. Although many highly sophisticated devices are currently available, the market for this type of more specialized equipment is directed primarily toward the physiotherapy community rather than in dentistry. One trend amongst the more advanced manufacturers for the physiotherapy community is to provide the ability to deliver multiple wavelengths simultaneously along a coaxial fiber. Also, the range of peripheral applicators is considerable, plus there are added built-in devices such as thermal cameras to help prevent hyperthermia of superficial tissues exposed to high output power sources. As the dental applications become more widely adopted, it is to be expected that the manufacturers will respond by building units dedicated to dentistry. Aside from thermography the range of wavelengths currently adopted for PBMT may include visible blue to green which offer some added benefits to the discerning practitioner. Shorter wavelengths carry more energy and may offer the ability to combine an antiviral, antifungal or antibacterial action with PBM as part of a mixture of wavelengths. This may be a direct photonic response or perhaps could be used in conjunction with a photosensitiser. Furthermore, as an adjunct to specialist care, PBMT has great potential in the management of medically compromised patients as well as in the care of conditions otherwise difficult to manage such as xerostomia. As a multifunctional clinical appliance with applications as diverse as disinfection, aPDT, the coagulation of venous anomalies, tooth whitening as well as the many healing, regenerative, anti-inflammatory and analgesic PBM treatment modalities, this type of specialized equipment will become a commonplace element in all modern dental clinics.

References

1. Hamblin MR. Low-level light therapy: photobiomodulation. Society of Photo-Optical Instrumentation Engineers (SPIE); 2017.
2. Chow RT, Armati PJ. Photobiomodulation: implications for anesthesia and pain relief. Photomed Laser Surg. 2016;34(12):599–609.
3. Anders JJ, Lanzafame RJ, Arany PR. Low-level light/laser therapy versus photobiomodulation therapy. Photomed Laser Surg. 2015;33(4):183.
4. Chung H, Dai T, Sharma SK, Huang YY, Carroll JD, Hamblin MR. The nuts and bolts of low-level laser (light) therapy. Ann Biomed Eng. 2012;40:516–33.
5. Agrawal T, Gupta GK, Rai V, Carroll JD, Hamblin MR. Pre-conditioning with low-level laser (light) therapy: light before the storm. Dose Response. 2014;12(4):619.
6. Beckham JT. The role of heat shock protein 70 in laser irradiation and thermal preconditioning. Doctoral dissertation, Vanderbilt University, Nashville, TN, USA, 18 Jul 2008.
7. Romagnoli E, Cafaro A. Theoretical and applied concepts of adjunctive use of LLLT/PBM within clinical dentistry. In: Coluzzi D, Parker S, editors. Lasers in dentistry-current concepts. Cham: Springer Nature; 2017. p. 125–57, Ch. 7. ISBN: 978-3-319-51943-2.
8. Fekrazad R, Arany P. Photobiomodulation therapy in clinical dentistry. Photobiomodul Photomed Laser Surg. 2019;37(12):737–8.
9. Cronshaw M, Parker S, Arany P. Feeling the heat: evolutionary and microbial basis for the analgesic mechanisms of photobiomodulation therapy. Photobiomodul Photomed Laser Surg. 2019;37(9):517–26.
10. Karu TI. Biophysical basis of low-power-laser effects. In: Laser chemistry, biophysics, and biomedicine, vol. 2802. SPIE; 1996. p. 142–51.
11. Hamblin MR, de Sousa MV, Arany PR, Carroll JD, Patthoff D. Low level laser (light) therapy and photobiomodulation: the path forward. In: Mechanisms for low-light therapy X, vol. 9309. SPIE; 2015. p. 930902.
12. Sharma SK, Sardana S, Hamblin MR. Role of opsins and light or heat activated transient receptor potential ion channels in the mechanisms of photobiomodulation and infrared therapy. J Photochem Photobiol. 2023;13:100160.
13. Liebert A, Capon W, Pang V, Vila D, Bicknell B, McLachlan C, Kiat H. Photophysical mechanisms of photobiomodulation therapy as precision medicine. Biomedicine. 2023;11(2):237.
14. Hönigsmann H. History of phototherapy in dermatology. Photochem Photobiol Sci. 2013;12(1):16–21.
15. Downes A, Blunt T. Researches on the effect of light upon bacteria and other organisms. Proc R Soc Lond B Biol Sci. 1877;26(179–184):488–500.
16. Finsen NR. Om anvendelse i medicinen af koncentrerede kemiske lysstraaler. Gyldendal; 1896.
17. Gotzsche P. Niels Finsen's treatment for lupus vulgaris. JRSM. 2011;104(1):41–2.
18. Howson CR. Heliotherapy in pulmonary tuberculosis—its possibilities and dangers. Cal West Med. 1928;29(1):25.
19. Gaspar L. Professor Endre Mester, the father of photobiomodulation. J Laser Dent. 2009;17:146–8.
20. Mester AF, Mester A. Mester's method of laser biostimulation. In: Laser/Optoelectronics in medicine/Laser/Optoelektronik in der Medizin. Berlin: Springer; 1986. p. 103–9.
21. Mester E, Ludány G, Sellyei M, et al. The stimulating effect of low power laser rays on biological systems. Laser Rev. 1968;1:3.
22. De Ment J. The first law of fluorescence. Science. 1953;118(3056):117.
23. Bolton JR, Mayor-Smith I, Linden KG. Rethinking the concepts of fluence (UV dose) and fluence rate: the importance of photon-based units—a systemic review. Photochem Photobiol. 2015;91(6):1252–62.
24. Mester E, Spiry T, Szende B, Tota JG. Effect of laser rays on wound healing. Am J Surg. 1971;122(4):532–5. https://doi.org/10.1016/0002-9610(71)90482-x.
25. Hamblin MR. Mechanisms of photobiomodulation in the brain. In: Photobiomodulation in the brain. Academic Press; 2019. p. 97–110.
26. Iyengar B. The melanocyte photosensory system in the human skin. Springerplus. 2013;2(1):1–8.
27. Fonken L, et al. Light at night increases body mass by shifting the time of food intake. PNAS. 2010;107(43):18664–9.
28. Scheer F, et al. Adverse metabolic and cardiovascular consequences of circadian misalignment. PNAS. 2009;106:4453–8.
29. McFadden E, et al. The relationship between obesity and exposure to light at night. Am J Epidemiol. 2014;180(3):245–50.
30. Blask D, et al. Circadian regulation of metabolic signaling mechanisms of human breast cancer growth by the nocturnal melatonin signal and the consequences of its disruption by light at night. J Pineal Res. 2011;51(3):259–69.
31. Niemz M. Laser tissue interactions. 3rd ed. New York: Springer; 2007.
32. Steiner R. Laser tissue interactions. In: Laser and IPL technology in dermatology and aesthetic medicine. New York: Springer; 2011.
33. Jacques SL. Optical properties of biological tissues: a review. Phys Med Biol. 2013;58(11):R37–61.
34. Hamblin MR. Mechanisms and mitochondrial redox signaling in photobiomodulation. Photochem Photobiol. 2018;94(2):199–212.
35. Avci P, Gupta A, Sadasivam M, Vecchio D, Pam Z, Pam N, Hamblin MR. Low-level laser (light) therapy (LLLT) in skin: stimulating, healing, restoring. Semin Cutaneous Med Surg. 2013;32(1):41.
36. Amaroli A, Pasquale C, Zekiy A, Utyuzh A, Benedicenti S, Signore A, Ravera S. Photobiomodulation and oxidative stress: 980 nm diode laser light regulates mitochondrial activity and reactive oxygen species production. Oxidative Med Cell Longev. 2021;2021:1–1.
37. Hamblin MR. Mechanisms and applications of the anti-inflammatory effects of photobiomodulation. AIMS Biophys. 2017;4(3):337.
38. Cretien B, Benit P, Ha H, et al. Mitochondria are physiologically maintained at close to 50C. PLoS Biol. 2018;16:e2003992.
39. Birceanu O. Mitochondria are too hot to handle. J Exp Biol. 2018;221:jeb170027.
40. Lane N. Hot mitochondria. PLoS Biol. 2018;16:e2005113.
41. Yang M, Yang Z, Wang P, Sun Z. Current application and future directions of photobiomodulation in central nervous diseases. Neural Regen Res. 2021;16(6):1177.
42. Karu T. Photobiology of low-power laser effects. Health Phys. 1989;56(5):691–704.
43. Karu T. Primary and secondary mechanisms of action of visible to near-IR radiation on cells. J Photochem Photobiol B Biol. 1999;49(1):1–7.

44. Moncada S, Erusalimsky JD. Does nitric oxide modulate mitochondrial energy generation and apoptosis? Nat Rev Mol Cell Biol. 2002;3(3):214–20.
45. Boelens R, Rademaker H, Pel R, Wever R. EPR studies of the photodissociation reactions of cytochrome c oxidase-nitric oxide complexes. Biochim Biophys Acta. 1982;679:84–94.
46. Boelens R, Wever R, Van Gelder BF, Rademaker H. An EPR study of the photodissociation reactions of oxidised cytochrome C oxidase-nitric oxide complexes. Biochim Biophys Acta. 1983;724:176–83.
47. Sarti P, Giuffre A, Forte E, Mastronicola D, Barone MC, Brunori M. Nitric oxide and cytochrome c oxidase: mechanisms of inhibition and NO degradation. Biochem Biophys Res Commun. 2000;274:183–7.
48. Brunori M, Giuffrè A, Forte E, Mastronicola D, Barone MC, Sarti P. Control of cytochrome c oxidase activity by nitric oxide. Biochim Biophys Acta Bioenerg. 2004;1655:365–71.
49. Quirk BJ, Whelan HT. What lies at the heart of photobiomodulation: light, cytochrome C oxidase, and nitric oxide—review of the evidence. Photobiomodul Photomed Laser Surg. 2020;38(9):527–30.
50. Vladimirov YA, Osipov AN, Klebanov GI. Photobiological principles of therapeutic applications of laser radiation. Biochem Mosc. 2004;69:81–90.
51. Hamblin MR, Demidova TN. Cellular chromophores and signaling in low level light therapy. In: Biomedical optics (BiOS). International Society for Optics and Photonics; 2007. p. 642802-1–642802-14.
52. Sommer AP. Revisiting the photon/cell interaction mechanism in low-level light therapy. Photobiomodul Photomed Laser Surg. 2019;37(6):336–41.
53. Hamblin M, Liebert A. Photobiomodulation therapy mechanisms beyond cytochrome c oxidase. Photobiomodul Photomed Laser Surg. 2022;40(2):75–7. https://doi.org/10.1089/photob.2021.0119.
54. Forbes RM, Cooper AR, Mitchell HH. The composition of the adult human body as determined by chemical analysis. J Biol Chem. 1953;203:359–66.
55. Richter AL, Barrera J, Markus RF, Brissett A. Laser skin treatment in non-Caucasian patients. Facial Plast Surg Clin North Am. 2014;22(3):439–46. https://doi.org/10.1016/j.fsc.2014.04.006.
56. Cronshaw M, Parker S, Grootveld M. Photothermal aspects of high energy photobiomodulation therapies: an in vitro investigation. Biomedicine. 2023;11:1634.
57. Sommer AP, Haddad MK, Fecht HJ. Light effect on water viscosity: implication for ATP biosynthesis. Sci Rep. 2015;5(1):12029.
58. Sommer AP. Mitochondrial cytochrome c oxidase is not the primary acceptor for near infrared light—it is mitochondrial bound water: the principles of low-level light therapy. Ann Transl Med. 2019;7(Suppl 1):S13.
59. Cifra M, Pospíšil P. Ultra-weak photon emission from biological samples: definition, mechanisms, properties, detection and applications. J Photochem Photobiol B Biol. 2014;139:2–10.
60. Simonova-Pushkar LI, Gertman VZ, Bilogurova LV. Application of photo-magnetic therapy for treatment of skin radiation damage in rats. Problemy Radiatsiinoi Medytsyny ta Radiobiolohii. 2014;19:458–70.
61. Hashmi JT, Huang YY, Sharma SK, Kurup DB, De Taboada L, Carroll JD, Hamblin MR. Effect of pulsing in low-level light therapy. Lasers Surg Med. 2010;42(6):450–66.
62. Mó O, Yáñez M, Eckert-Maksić M, Maksić ZB, Alkorta I, Elguero J. Periodic trends in bond dissociation energies. A theoretical study. J Phys Chem. 2005;A109:4359–65.
63. Niemz M. Laser tissue interactions. 3rd ed. Berlin: Springer; 2007. p. 92–3.
64. Chen AC, Arany PR, Huang YY, Tomkinson EM, Sharma SK, Kharkwal GB, Saleem T, Mooney D, Yull FE, Blackwell TS, Hamblin MR. Low-level laser therapy activates NF-kB via generation of reactive oxygen species in mouse embryonic fibroblasts. PLoS One. 2011;6(7):e22453.
65. Chen AC, Huang YY, Arany PR, Hamblin MR. Role of reactive oxygen species in low level light therapy. In: Mechanisms for low-light therapy IV, vol. 7165. SPIE; 2009. p. 9–19.
66. Khan I, Arany P. Biophysical approaches for oral wound healing: emphasis on photobiomodulation. Adv Wound Care. 2015;4(12):724–37.
67. Arany PR. Photobiomodulation-activated latent transforming growth factor-β1: a critical clinical therapeutic pathway and an endogenous optogenetic tool for discovery. Photobiomodul Photomed Laser Surg. 2022;40(2):136–47.
68. Khan I, Tang E, Arany P. Molecular pathway of near infrared laser phototoxicity involves ATF-4 orchestrated ER stress. Sci Rep. 2015;5:srep10581.
69. Chow R, David M, Armati P. 830 nm laser irradiation induces varicosity formation, reduces mitochondrial membrane potential and blocks fast axonal flow in small and medium diameter rat dorsal root ganglion neurons: implications for the analgesic effects of 830 nm laser. Photomed Laser Surg. 2007;12:28–39.
70. Cardoso S, Correia S, Carvalho C, et al. Perspectives on mitochondrial uncoupling proteins-mediated neuroprotection. J Bioenerg Biomembr. 2015;47:119–31.
71. Jovaisaite V, Mouchiroud L, Auwerx J. The mitochondrial unfolded protein response, a conserved stress pathway with implications in health and disease. J Exp Biol. 2014;217:137–43.
72. Esteves TC, Brand MD. The reactions catalyzed by the mitochondrial uncoupling proteins UCP2 and UCP3. Biochim Biophys. 2005;1709:25–44.
73. Busiello RA, Savarese S, Lombardi A. Mitochondrial uncoupling proteins and energy metabolism. Front Physiol. 2015; https://doi.org/10.3389/fphys.2015.00036.
74. Pereira FL, Ferreira MV, da Silva Mendes P, Rossi FM, Alves MP, Alves BL. Use of a high-power laser for wound healing: a case report. J Lasers Med Sci. 2020;11:112.
75. Lin YH, Chen YC, Cheng KS, Yu PJ, Wang JL, Ko NY. Higher periwound temperature associated with wound healing of pressure ulcers detected by infrared thermography. J Clin Med. 2021;10(13):2883.
76. Whitney JD, Salvadalena G, Higa L, Mich M. Treatment of pressure ulcers with noncontact normothermic wound therapy: healing and warming effects. J Wound Ostomy Continence Nurs. 2001;28:244–52.
77. Tobalem M, Harder Y, Tschanz E, Speidel V, Pittet-Cuénod B, Wettstein R. First-aid with warm water delays burn progression and increases skin survival. J Plast Reconstr Aesthet Surg. 2013;66:260–6.
78. Hu C, Yang J, Qi Z, Wu H, Wang B, Zou F, Mei H, Liu J, Wang W, Liu Q. Heat shock proteins: Biological functions, pathological roles, and therapeutic opportunities. MedComm. 2022;3(3):e161.
79. Chung DJ, Schulte PM. Mitochondria and the thermal limits of ectotherms. J Exp Biol. 2020;223(20):jeb227801.
80. Janowska MK, Baughman HE, Woods CN, Klevit RE. Mechanisms of small heat shock proteins. Cold Spring Harb Perspect Biol. 2019;11(10):a034025.
81. Niemz M. Laser tissue interactions. 3rd ed. Berlin: Springer; 2007. p. 78.

82. Cronshaw M, Parker S, Anagnostaki E, Mylona V, Lynch E, Grootveld M. Photobiomodulation dose parameters in dentistry: a systematic review and meta-analysis. Dent J. 2020;8(4):114.
83. Kaub L, Schmitz C. More than ninety percent of the light energy emitted by near-infrared laser therapy devices used to treat musculoskeletal disorders is absorbed within the first ten millimeters of biological tissue. Biomedicine. 2022;10:3204.
84. Bashkatov A, Genina E, Kochubey V, Tuchin V. Optical properties of human skin, subcutaneous and mucous tissues in the wavelength range 400 to 2000nm. J Phys D Appl Phys. 2005;38:2543–55.
85. Karu TI. Cellular mechanisms of low-power laser therapy. Laser Appl Med Biol Environ Sci. 2003;22:60–6.
86. Parker S, Anagnostaki E, Mylona V, Cronshaw M, Lynch E, Grootveld M. Systematic review of post-surgical laser-assisted oral soft tissue outcomes using surgical wavelengths outside the 650–1350 nm optical window. Photobiomodul Photomed Laser Surg. 2020;38(10):591–606.
87. Salehpour F, Mahmoudi J, Kamari F, Sadigh-Eteghad S, Rasta SH, Hamblin MR. Brain photobiomodulation therapy: a narrative review. Mol Neurobiol. 2018;55(8):6601–36. https://doi.org/10.1007/s12035-017-0852-4.
88. Opländer C, Deck A, Volkmar CM, Kirsch M, Liebmann J, Born M, Van Abeelen F, Van Faassen EE, Kröncke KD, Windolf J, Suschek CV. Mechanism and biological relevance of blue-light (420–453 nm)-induced nonenzymatic nitric oxide generation from photolabile nitric oxide derivates in human skin in vitro and in vivo. Free Radic Biol Med. 2013;65: 1363–77.
89. Serrage H, Heiskanen V, Palin WM, Cooper PR, Milward MR, Hadis M, Hamblin MR. Under the spotlight: mechanisms of photobiomodulation concentrating on blue and green light. Photochem Photobiol Sci. 2019;18(8):1877–909.
90. Buravlev EA, Zhidkova TV, Osipov AN, Vladimirov YA. Are the mitochondrial respiratory complexes blocked by NO the targets for the laser and LED therapy? Lasers Med Sci. 2015;30(1):173–80.
91. Wang Y, Huang YY, Wang Y, Lyu P, Hamblin MR. Red (660 nm) or near-infrared (810 nm) photobiomodulation stimulates, while blue (415 nm), green (540 nm) light inhibits proliferation in human adipose-derived stem cells. Sci Rep. 2017;7(1):7781.
92. Cheon MW, Kim TG, Lee YS, Kim SH. Low level light therapy by Red-Green-Blue LEDs improves healing in an excision model of Sprague-Dawley rats. Pers Ubiquit Comput. 2013;17(7):1421–8.
93. Fekrazad R, Mirmoezzi A, Kalhori KA, Arany P. The effect of red, green and blue lasers on healing of oral wounds in diabetic rats. J Photochem Photobiol B. 2015;148:242–5.
94. Kesler G, Shvero DK, Tov YS, Romanos G. Platelet derived growth factor secretion and bone healing after Er:YAG laser bone irradiation. J Oral Implantol. 2011;37(Sp 1): 195–204.
95. Poli R, Parker S. Achieving dental analgesia with the erbium chromium yttrium scandium gallium garnet laser (2780 nm): a protocol for painless conservative treatment. Photomed Laser Surg. 2015;33(7):364–71.
96. Poli R, Parker S, Anagnostaki E, Mylona V, Lynch E, Grootveld M. Laser analgesia associated with restorative dental care: a systematic review of the rationale, techniques, and energy dose considerations. Dent J. 2020;8(4):128.
97. Parker S, Cronshaw M, Anagnostaki E, Mylona V, Lynch E, Grootveld M. Current concepts of laser–oral tissue interaction. Dent J. 2020;8(3):61.
98. Mosca RC, Ong AA, Albasha O, Bass K, Arany P. Photobiomodulation therapy for wound care: a potent, noninvasive, photoceutical approach. Adv Skin Wound Care. 2019;32(4):157–67.
99. Merigo E, Rocca JP, Pinheiro AL, Fornaini C. Photobiomodulation therapy in oral medicine: a guide for the practitioner with focus on new possible protocols. Photobiomodul Photomed Laser Surg. 2019;37(11):669–80.
100. Mussttaf RA, Jenkins DF, Jha AN. Assessing the impact of low level laser therapy (LLLT) on biological systems: a review. Int J Radiat Biol. 2019;95(2):120–43.
101. Nie F, Hao S, Ji Y, Zhang Y, Sun H, Will M, Han W, Ding Y. Biphasic dose response in the anti-inflammation experiment of PBM. Lasers Med Sci. 2023;38(1):66.
102. Huang YY, Chen ACH, Carroll JD, Hamblin MR. Biphasic dose response in low level light therapy. Dose Response. 2009;7:358–83.
103. Huang YY, Sharma SK, Carroll J, Hamblin MR. Biphasic dose response in low level light therapy—an update. Dose Response. 2011;9:602–18.
104. Calabrese EJ, Agathokleous E. Theodosius Dobzhansky's view on biology and evolution v. 2.0: "Nothing in biology makes sense except in light of evolution and evolution's dependence on hormesis-mediated acquired resilience that optimizes biological performance and numerous diverse short and longer term protective strategies". Environ Res. 2020;186:109559.
105. Hadis MA, Zainal SA, Holder MJ, Carroll JD, Cooper PR, Milward MR, Palin WM. The dark art of light measurement: accurate radiometry for low-level light therapy. Lasers Med Sci. 2016;31:789–809.
106. Kate RJ, Rubatt S, Enwemeka CS, Huddleston WE. Optimal laser phototherapy parameters for pain relief. Photomed Laser Surg. 2018;36(7):354–62.
107. Young NC, Maximiano V, Arany PR. Thermodynamic basis for comparative photobiomodulation dosing with multiple wavelengths to direct odontoblast differentiation. J Biophotonics. 2022;15(6):e202100398.
108. Parker S, Cronshaw M, Grootveld M. Photobiomodulation delivery parameters in dentistry: an evidence-based approach. Photobiomodul Photomed Laser Surg. 2022;40(1):42–50.
109. Alvarenga LH, Ribeiro MS, Kato IT, Núñez SC, Prates RA. Evaluation of red light scattering in gingival tissue–in vivo study. Photodiagn Photodyn Ther. 2018;23:32–4.
110. Melnik IS, Steiner RW, Kienle A. Light penetration in human skin: in-vivo measurements using isotropic detector. In: Optical methods for tumor treatment and detection: mechanisms and techniques in photodynamic therapy II, vol. 1881. Bellingham, WA: SPIE; 1993. p. 222–30.
111. Henderson TA, Morries LD. Near-infrared photonic energy penetration: can infrared phototherapy effectively reach the human brain? Neuropsychiatr Dis Treat. 2015;11:2191–208.
112. Henderson TA, Morries LD. Infrared light cannot be doing what you think it is doing (re: DOI: 10.1089/photob.2018.4489). Photobiomodulation Photomed Laser Surg. 2019;37:124–5. https://doi.org/10.1089/photob.2018.4603.
113. Tunér J, Hode L. It's all in the parameters: a critical analysis of some well-known negative studies on low-level laser therapy. J Clin Laser Med Surg. 1998;16(5):245–8.
114. Tunér J, Hode L. Low level laser therapy. Graengesberg: Prima Books in Sweden AB; 1999. p. 326.
115. Parker S, Cronshaw M, Anagnostaki E, Bordin-Aykroyd SR, Lynch E. Systematic review of delivery parameters used in dental photobiomodulation therapy. Photobiomodul Photomed Laser Surg. 2019;37(12):784–97.

116. Parker S. Laser photonic energy delivery in clinical dentistry: scrutiny of parameter variables. PhD thesis. 2023. p. 152.
117. Zein R, Selting W, Hamblin MR. Review of light parameters and photobiomodulation efficacy: dive into complexity. J Biomed Opt. 2018;23(12):120901.
118. Amaroli A, Ravera S, Parker S, Panfoli I, Benedicenti A, Benedicenti S. An 808-nm diode laser with a flat-top handpiece positively photobiomodulates mitochondria activities. Photomed Laser Surg. 2016;34:564–71.
119. Amaroli A, Arany P, Pasquale C, Benedicenti S, Bosco A, Ravera S. Improving consistency of photobiomodulation therapy: a novel flat-top beam hand-piece versus standard gaussian probes on mitochondrial activity. Int J Mol Sci. 2021;22:7788.
120. Amaroli A, Ravera S, Parker S, Panfoli I, Benedicenti A, Benedicenti S. 808-nm laser therapy with a flat-top handpiece photobiomodulates mitochondria activities of Paramecium primaurelia (Protozoa). Lasers Med Sci. 2016;31:741–7.
121. Recommended treatment doses for low level laser therapies. Available online: https://waltpbm.org/wpcontent/uploads/2021/08/Dose_table_780–860nm_for_Low_Level_Laser_Therapy_WALT-2010.pdf. Accessed on 29 Mar 2023.
122. Wehner M, Betz P, Aden M. Influence of laser wavelength and beam profile on the coagulation depth in a soft tissue phantom model. Lasers Med Sci. 2019;34:335–41.
123. Lukač M, Hoefferle ŠL, Terlep S, Hreljac I, Vampelj U, Krisper MG, Vižintin Z. Characteristics of piano level laser therapy (PLLTTM) using novel 1064 nm laser handpiece technology. J Laser Health Acad. 2022;2022(1):1–6.
124. Al-Watban FA, Zhang XY. The comparison of effects between pulsed and CW lasers on wound healing. J Clin Laser Med Surg. 2004;22(1):15–8.
125. Chow R, David M, Armati P. 830 nm laser irradiation induces varicosity formation, reduces mitochondrial membrane potential and blocks fast axonal flow in small and medium diameter rat dorsal root ganglion neurons: Implications for the analgesic effects of 830 nm laser. J Peripher Nerv Syst. 2007;12:28–39.
126. Hosseinpour S, Tunér J, Fekrazad R. Photobiomodulation in oral surgery: a review. Photobiomodul Photomed Laser Surg. 2019;37(12):814–25.
127. Cheng K, Martin LF, Slepian MJ, Patwardhan AM, Ibrahim MM. Mechanisms and pathways of pain photobiomodulation: a narrative review. J Pain. 2021;22(7):763–77.
128. Zielińska P, Soroko M, Howell K, Godlewska M, Hildebrand W, Dudek K. Comparison of the effect of high-intensity laser therapy (HILT) on skin surface temperature and vein diameter in pigmented and non-pigmented skin in healthy racehorses. Animals. 1965;2021:11.
129. Zielińska P, Soroko-Dubrovina M, Dudek K, Ruzhanova-Gospodinova IS. A preliminary study of the influence of high intensity laser therapy (HILT) on skin surface temperature and longissimus dorsi muscle tone changes in thoroughbred racehorses with back pain. Animals. 2023;13:794.
130. Bashkatov AN, Genina EA, Kochubey VI, Tuchin VV. Optical properties of human skin, subcutaneous and mucous tissues in the wavelength range from 400 to 2000 nm. J Phys D Appl Phys. 2005;38:2543–55. https://doi.org/10.1088/0022-3727/38/15/004.
131. Bashkatov AN, Genina EA, Tuchin VV. Optical Properties of skin, subcutaneous, and muscle tissues: a review. J Innov Opt Health Sci. 2011;4:9–38. https://doi.org/10.1142/s1793545811001319.
132. Shi Q, Yang S, Jia F, Xu J. Does low level laser therapy relieve the pain caused by the placement of the orthodontic separators? A meta-analysis. Head Face Med. 2015;11:28.
133. Li FJ, Zhang JY, Zeng XT, Guo Y. Low-level laser therapy for orthodontic pain: a systematic review. Lasers Med Sci. 2015;30:1789–803.
134. Ren C, McGrath C, Yang Y. The effectiveness of low-level diode laser therapy on orthodontic pain management: a systematic review and meta-analysis. Lasers Med Sci. 2015;30:1881–93.
135. Sonesson M, et al. Efficacy of low-level laser therapy in accelerating tooth movement, preventing relapse and managing acute pain during orthodontic treatment in humans: a systematic review. BMC Oral Health. 2017;17:11. https://doi.org/10.1186/s12903-016-0242-8.
136. Cronshaw M. Ch. 12: Photobiomodulation concepts within orthodontics. In: Coluzzi D, Parker S, editors. Lasers in dentistry-current concepts. Cham: Springer Nature; 2017. p. 262–75. ISBN: 978-3-319-51943-2.
137. Ferrante M, Petrini M, Trentini P, Perfetti G, Spoto G. Effect of low-level laser therapy after extraction of impacted lower third molars. Lasers Med Sci. 2013;28(3):845–9.
138. Landucci A, Wosny AC, Uetanabaro LC, Moro A, Araujo MR. Efficacy of a single dose of low-level laser therapy in reducing pain, swelling, and trismus following third molar extraction surgery. Int J Oral Maxillofac Surg. 2016;45(3):392–8.
139. Choi H, Lim W, Kim I, Kim J, Ko Y, Kwon H, Kim S, Kabir KA, Li X, Kim O, Lee Y. Inflammatory cytokines are suppressed by light-emitting diode irradiation of P. gingivalis LPS-treated human gingival fibroblasts: inflammatory cytokine changes by LED irradiation. Lasers Med Sci. 2012;27:459–67.
140. Pesevska S, Nakova M, et al. Effect of laser on TNF-alpha expression in inflamed human gingival tissue. Lasers Med Sci. 2012;27:377–81.
141. Sakurai Y, et al. Inhibitory effects of low level laser irradiation on LPS stimulated prostaglandin E2 production and cyclooxygenase 2 in human gingival fibroblasts. Eur J Oral Sci. 2000;108(1):29–34.
142. Pesevska S, et al. The effect of lowlevel diode lasers on COX-2 gene expression in chronic periodontitis patients. Lasers Med Sci. 2017;32(7):1463–8.
143. Agrawal T, Gupta G, Carroll J, et al. Pre-conditioning with low level laser therapy: light before the storm. Dose Response. 2014;12:619–49.
144. Richter K, Haslbeck M, Buchner J. The heat shock response: life on the verge of death. Mol Cell. 2010;40:253–66.
145. Bakthisaran R, Tangirala R, Rao C. Heat shock proteins: role in cellular functions and pathology. Biochem Biophys Acta. 2015;1854:291–319.
146. Hahm E, Kulhari S, Arany P. Targeting the pain, inflammation and immune (PII) axis: plausible rationale for LLLT. Photon Lasers Med. 2012; https://doi.org/10.1515/plm-2012-0033.
147. Arany PR. Craniofacial wound healing with photobiomodulation therapy: new insights and current challenges. J Dent Res. 2016;95(9):977–84.
148. Azizi A, Osgouie KG. Dermal wound healing-remodeling phase: a biological review. In: 2010 The second international conference on computer and automation engineering (ICCAE), vol. 2. IEEE; 2010. p. 88–90.
149. Enoch S, Price P. Cellular, molecular and biochemical differences in the pathophysiology of healing between acute

7

wounds, chronic wounds and wounds in the aged. Aug 2004. www.worldwidewounds.com.
150. Ricciotti E, Fitzgerald GA. Prostaglandins and inflammation. Arterioscler Thromb Vasc Biol. 2011;31:986–1000.
151. Kim WS, Calderhead RG. Is light-emitting diode phototherapy (LED-LLLT) really effective? Laser Ther. 2011;20:205–15.
152. Ozcelik O, et al. Improved wound healing by low level laser irradiation after gingivectomy operations: a controlled clinical pilot study. J Clin Periodontol. 2008;35(3):250–4.
153. Fortuna T, Gonzalez AC, Ferreira MS, Reis SR, Medrado AA. Biomodulatory potential of low-level laser on neoangiogenesis and remodeling tissue. A literature review. J Dent Public Health. 2018;9(1):95–103.
154. Pagin MT, de Oliveira FA, Oliveira RC, Sant'Ana AC, de Rezende ML, Greghi SL, Damante CA. Laser and light-emitting diode effects on pre-osteoblast growth and differentiation. Lasers Med Sci. 2014;29:55–9.
155. Tripodi N, Corcoran D, Antonello P, Balic N, Caddy D, Knight A, Meehan C, Sidiroglou F, Fraser S, Kiatos D, Husaric M. The effects of photobiomodulation on human dermal fibroblasts in vitro: a systematic review. J Photochem Photobiol B Biol. 2021;214:112100.
156. Bakshi PV, Setty SB, Kulkarni MR. Photobiomodulation of human gingival fibroblasts with diode laser—a systematic review. J Indian Soc Periodontol. 2022;26(1):5.
157. Rossi F, Magni G, Tatini F, Banchelli M, Cherchi F, Rossi M, Coppi E, Pugliese AM, Rossi degl'Innocenti D, Alfieri D, Pavone FS. Photobiomodulation of human fibroblasts and keratinocytes with blue light: implications in wound healing. Biomedicine. 2021;9(1):41.
158. Ayuk SM, Abrahamse H, Houreld NN. The role of photobiomodulation on gene expression of cell adhesion molecules in diabetic wounded fibroblasts in vitro. J Photochem Photobiol B Biol. 2016;161:368–74.
159. Rajendran NK, Houreld NN, Abrahamse H. Photobiomodulation reduces oxidative stress in diabetic wounded fibroblast cells by inhibiting the FOXO1 signaling pathway. J Cell Commun Signal. 2021;15:195–206.
160. Trubiani O, Pizzicannella J, Caputi S, Marchisio M, Mazzon E, Paganelli R, et al. Periodontal ligament stem cells: current knowledge and future perspectives. Stem Cells Dev. 2019;28(15):995–1003.
161. Abrahamse H. Regenerative medicine, stem cells, and low-level laser therapy: future directives. Photomed Laser Surg. 2012;30(12):681–2.
162. Li B, Ouchi T, Cao Y, Zhao Z, Men Y. Dental-derived mesenchymal stem cells: state of the art. Front Cell Dev Biol. 2021;9:654559. https://doi.org/10.3389/fcell.2021.654559.
163. Cho H, Tarafder S, Fogge M, Kao K, Lee CH. Periodontal ligament stem/progenitor cells with protein-releasing scaffolds for cementum formation and integration on dentin surface. Connect Tissue Res. 2016;57(6):488–95. https://doi.org/10.1080/03008207.2016.1191478.
164. Fekrazad R, Asefi S, Eslaminejad MB, Taghiar L, Bordbar S, Hamblin MR. Photobiomodulation with single and combination laser wavelengths on bone marrow mesenchymal stem cells: proliferation and differentiation to bone or cartilage. Lasers Med Sci. 2019;34(1):115–26. https://doi.org/10.1007/s10103-018-2620-8.
165. Otsu K, Kumakami-Sakano M, Fujiwara N, Kikuchi K, Keller L, Lesot H, Harada H. Stem cell sources for tooth regeneration: current status and future prospects. Front Physiol. 2014;5:36. https://doi.org/10.3389/fphys.2014.00036.
166. Egusa H, Sonoyama W, Nishimura M, Atsuta I, Akiyama K. Stem cells in dentistry—part I: stem cell sources. J Prosthodont Res. 2012;56(3):151–65. https://doi.org/10.1016/j.jpor.2012.06.001.
167. Roato I, Chinigò G, Genova T, Munaron L, Mussano F. Oral cavity as a source of mesenchymal stem cells useful for regenerative medicine in dentistry. Biomedicine. 2021;9(9):1085. https://doi.org/10.3390/biomedicines9091085.
168. Hollands P, Aboyeji D, Orcharton M. Dental pulp stem cells in regenerative medicine. Br Dent J. 2018;224(9):747–50. https://doi.org/10.1038/sj.bdj.2018.348.
169. Zeng WY, Ning Y, Huang X. Advanced technologies in periodontal tissue regeneration based on stem cells: current status and future perspectives. J Dent Sci. 2021;16(1):501–7. https://doi.org/10.1016/j.jds.2020.07.008.
170. Zhu W, Liang M. Periodontal ligament stem cells: current status, concerns, and future prospects. Stem Cells Int. 2015;2015:972313. https://doi.org/10.1155/2015/972313.
171. Prockop DJ. The exciting prospects of new therapies with mesenchymal stromal cells. Cytotherapy. 2017;19(1):1–8. https://doi.org/10.1016/j.jcyt.2016.09.008.
172. Wang Y, Chen X, Cao W, Shi Y. Plasticity of mesenchymal stem cells in immunomodulation: pathological and therapeutic implications. Nat Immunol. 2014;15(11):1009–16. https://doi.org/10.1038/ni.3002.
173. Li W, Ren G, Huang Y, Su J, Han Y, Li J, Chen X, Cao K, Chen Q, Shou P, Zhang L, Yuan ZR, Roberts AI, Shi S, Le AD, Shi Y. Mesenchymal stem cells: a double-edged sword in regulating immune responses. Cell Death Differ. 2012;19(9):1505–13. https://doi.org/10.1038/cdd.2012.26.
174. Bunte K, Beikler T. Th17 cells and the IL-23/IL-17 axis in the pathogenesis of periodontitis and immune-mediated inflammatory diseases. Int J Mol Sci. 2019;20(14):3394. https://doi.org/10.3390/ijms20143394.
175. Paiva CN, Bozza MT. Are reactive oxygen species always detrimental to pathogens? Antioxid Redox Signal. 2014;20(6):1000–37. https://doi.org/10.1089/ars.2013.5447.
176. Moreira MS, Diniz IM, Rodrigues MF, de Carvalho RA, de Almeida Carrer FC, Neves II, Gavini G, Marques MM. In vivo experimental model of orthotopic dental pulp regeneration under the influence of photobiomodulation therapy. J Photochem Photobiol B. 2017;166:180–6.
177. Zaccara IM, Mestieri LB, Pilar EFS, Moreira MS, Grecca FS, Martins MD, Kopper PMP. Photobiomodulation therapy improves human dental pulp stem cell viability and migration in vitro associated to upregulation of histone acetylation. Lasers Med Sci. 2020;35:741–9.
178. Matsui S, Tsujimoto Y, Matsushima K. Stimulatory effects of hydroxyl radical generation by Ga-Al-As laser irradiation on mineralization ability of human dental pulp cells. Biol Pharm Bull. 2007;30:27–31.
179. Matsui S, Takeuchi H, Tsujimoto Y, Matsushima K. Effects of Smads and BMPs induced by Ga-Al-As laser irradiation on calcification ability of human dental pulp cells. J Oral Sci. 2008;50:75–81.
180. Ohbayashi E, Matsushima K, Hosoya S, Abiko Y, Yamazaki M. Stimulatory effect of laser irradiation on calcified nodule formation in human dental pulp fibroblasts. J Endod. 1999;25:30–3.
181. Suzuki SS, Garcez AS, Suzuki H, et al. Low level laser therapy stimulates bone metabolism and inhibits root resorption during tooth movement in a rodent model. J Biophotonics. 2016; https://doi.org/10.1002/jbio.201600016.

182. Hamblin MR, Nelson ST, Strahan JR. Photobiomodulation and cancer: what is the truth? Photomed Laser Surg. 2018;36(5):241–5.
183. Bensadoun RJ, Epstein JB, Nair RG, Barasch A, Raber-Durlacher JE, Migliorati C, Genot-Klastersky MT, Treister N, Arany P, Lodewijckx J, Robijns J. Safety and efficacy of photobiomodulation therapy in oncology: a systematic review. Cancer Med. 2020;9(22):8279–300.
184. Hoel DG, Berwick M, de Gruijl FR, Holick MF. The risks and benefits of sun exposure 2016. Dermato-endocrinology. 2016;8(1):e1248325.
185. Ottaviani G, Martinelli V, Rupel K, et al. Laser therapy inhibits tumor growth in mice by promoting immune surveillance and vessel normalization. EBioMedicine. 2016;11:165–72.
186. Arany PR. Healing tumors with light: science fiction or the future of medicine? Photomed Laser Surg. 2018;36(5):227–9.
187. Antunes HS, Herchenhorn D, Small IA, et al. Long-term survival of a randomized phase III trial of head and neck cancer patients receiving concurrent chemoradiation therapy with or without low-level laser therapy (LLLT) to prevent oral mucositis. Oral Oncol. 2017;71:11–5.
188. Wu S, Zhou F, Wei Y, Chen WR, Chen Q, Xing D. Cancer phototherapy via selective photoinactivation of respiratory chain oxidase to trigger a fatal superoxide anion burst. Antioxid Redox Signal. 2014;20:733–46.
189. Lu C, Zhou F, Wu S, Liu L, Xing D. Phototherapy-induced antitumor immunity: long-term tumor suppression effects via photoinactivation of respiratory chain oxidase-triggered superoxide anion burst. Antioxid Redox Signal. 2016;24: 249–62.
190. Kalyanaraman B. Teaching the basics of cancer metabolism: developing antitumor strategies by exploiting the differences between normal and cancer cell metabolism. Redox Biol. 2017;12:833–42.
191. Cronshaw M, Parker S, Anagnostaki E, Mylona V, Lynch E, Grootveld M. Photobiomodulation and oral mucositis: a systematic review. Dentistry. 2020;8:87.
192. Murphy MK, MacBarb RF, Wong ME, Athanasiou KA. Temporomandibular disorders: a review of etiology, clinical management, and tissue engineering strategies. Int J Oral Maxillofac Implants. 2013;28(6):e393–414. https://doi.org/10.11607/jomi.te20.
193. Ouanounou A, Goldberg M, Haas DA. Pharmacotherapy in temporomandibular disorders: a review. J Can Dent Assoc. 2017;83:1–8.
194. Khairnar S, Bhate K, Kumar SNS, Kshirsagar K, Jagtap B, Kakodkar P. Comparative evaluation of low-level laser therapy and ultrasound heat therapy in reducing temporomandibular joint disorder pain. J Dent Anesth Pain Med. 2019;19(5):289–94. https://doi.org/10.17245/jdapm.2019.19.5.289.
195. Chisnoiu AM, Picos AM, Popa S, Chisnoiu PD, Lascu L, Picos A, Chisnoiu R. Factors involved in the etiology of temporomandibular disorders—a literature review. Clujul Med. 2015;88(4):473–8. https://doi.org/10.15386/cjmed-485.
196. Xu GZ, Jia J, Jin L, Li JH, Wang ZY, Cao DY. Low-level laser therapy for temporomandibular disorders: a systematic review with meta-analysis. Pain Res Manag. 2018;2018:4230583. Published 2018 May 10. https://doi.org/10.1155/2018/4230583.
197. Liu F, Steinkeler A. Epidemiology, diagnosis, and treatment of temporomandibular disorders. Dent Clin N Am. 2013;57(3):465–79. https://doi.org/10.1016/j.cden.2013.04.006.
198. Maia ML, Bonjardim LR, Quintans Jde S, Ribeiro MA, Maia LG, Conti PC. Effect of low-level laser therapy on pain levels in patients with temporomandibular disorders: a systematic review. J Appl Oral Sci. 2012;20(6):594–602. https://doi.org/10.1590/s1678-77572012000600002.
199. Munguia FM, Jang J, Salem M, Clark GT, Enciso R. Efficacy of low-level laser therapy in the treatment of temporomandibular myofascial pain: a systematic review and meta-analysis. J Oral Facial Pain Headache. 2018;32(3):287–97. https://doi.org/10.11607/ofph.2032.
200. Tunér J, Hosseinpour S, Fekrazad R. Photobiomodulation in temporomandibular disorders. Photobiomodul Photomed Laser Surg. 2019;37(12):826–36.
201. Alsarhan J, El Feghali R, Alkhudari T, Benedicenti S, Pasquale C. Can Photobiomodulation support the management of temporomandibular joint pain? Molecular mechanisms and a systematic review of human clinical trials. Photonics. 2022;9(6):420.
202. da Silveira RB, Ferreira I, Botelho AL, Dos Reis AC. Effect of photobiomodulation treatment on pain control in patients with temporomandibular dysfunction disorder: systematic review. Cranio. 2022:1–1. https://doi.org/10.1080/08869634.2022.2086599.
203. Karic V, Chandran R, Abrahamse H. Photobiomodulation and stem cell therapy for temporomandibular joint disc disorders. Photobiomodul Photomed Laser Surg. 2020;38(7): 398–408.
204. de Pedro M, López-Pintor RM, de la Hoz-Aizpurua JL, Casanas E, Hernández DG. Efficacy of low-level laser therapy for the therapeutic management of neuropathic orofacial pain: a systematic review. J Oral Facial Pain Headache. 2020;34(1):13.
205. de Oliveira-Souza AI, Mohamad N, de Castro Carletti EM, Müggenborg F, Dennett L, de Oliveira DA, Armijo-Olivo S. What are the best parameters of low-level laser therapy to reduce pain intensity and improve mandibular function in orofacial pain? A systematic review and meta-analysis. Disabil Rehabil. 2023;45:3219–37.
206. de Paula EF, Gobbi MF, Bergamin LG, Migliorati CA, Bezinelli LM. Oral care and photobiomodulation protocol for the prevention of traumatic injuries and lip necrosis in critically ill patients with COVID-19: an observational study. Lasers Dent Sci. 2021;5(4):239–45.
207. Jana Neto FC, Martimbianco AL, Mesquita-Ferrari RA, Bussadori SK, Alves GP, Almeida PV, Delgado FG, Fonseca LR, Gama MZ, Jorge MD, Hamblin MR. Effects of multi-wavelength photobiomodulation for the treatment of traumatic soft tissue injuries associated with bone fractures: a double-blind, randomized controlled clinical trial. J Biophotonics. 2023;16(5):e202200299.
208. Khan I, Rahman SU, Tang E, Engel K, Hall B, Kulkarni AB, Arany PR. Accelerated burn wound healing with photobiomodulation therapy involves activation of endogenous latent TGF-β1. Sci Rep. 2021;11:1–5.
209. Kalhori KAM, Vahdatinia F, Jamalpour MR, Vescovi P, Fornaini C, Merigo E, Fekrazad R. Photobiomodulation in oral medicine. Photobiomodul Photomed Laser Surg. 2019;37:837–61. https://doi.org/10.1089/photob.2019.4706.
210. Chavan M, Jain H, Diwan N, Khedkar S, Shete A, Durkar S. Recurrent aphthous stomatitis: a review. J Oral Pathol Med. 2012;41(8):577–83.
211. Alli BY, Erinoso OA, Olawuyi AB. Effect of sodium lauryl sulfate on recurrent aphthous stomatitis: a systematic review. J Oral Pathol Med. 2019;48(5):358–64.
212. Huo X, Han N, Liu L. Effect of different treatments on recurrent aphthous stomatitis: laser versus medication. Lasers Med Sci. 2021;36:1095–100. https://doi.org/10.1007/s10103-020-03166-0.

7

213. Slebioda Z, Szponar E, Kowalska A. Etiopathogenesis of recurrent aphthous stomatitis and the role of immunologic aspects: literature review. Arch Immunol Ther Exp. 2013;62(3):205–15.
214. Suter VG, Sjolund S, Bornstein MM. Effect of laser on pain relief and wound healing of recurrent aphthous stomatitis: a systematic review. Lasers Med Sci. 2017;32:953–63.
215. Tarakji B, Gazal G, Al-Maweri SA, Azzeghaiby SN, Alaizari N. Guideline for the diagnosis and treatment of recurrent aphthous stomatitis for dental practitioners. J Int Oral Health. 2015;7(5):74–80.
216. Belenguer-Guallar I, Jiménez-soriano Y, Claramunt-lozano A. Treatment of recurrent aphthous stomatitis. A literature review. J Clin Exp Dent. 2014;6(2):e168–74.
217. Khalil M, Hamadah O. Association of photodynamic therapy and photobiomodulation as a promising treatment of herpes labialis: a systematic review. Photobiomodul Photomed Laser Surg. 2022;40:299–307. https://doi.org/10.1089/photob.2021.0186.
218. Cernik C, Gallina K, Brodell RT. The treatment of herpes simplex infections: an evidence-based review. Arch Intern Med. 2008;168:1137–44.
219. Griffin K. What is the best treatment for herpes labialis? Evid Based Pract. 2020;23:11–2.
220. Ahmadi M, Shirani AM, Farhad SZ, Khosravi M, Mir M. Comparison of 660-nm low-level and defocused 810-nm high-power laser for treatment of herpes labialis. Lasers Dent Sci. 2022;6:211–8. https://doi.org/10.1007/s41547-022-00164-z.
221. Eduardo CP, Aranha ACC, Simões A, Bello-Silva MS, Ramalho KM, Esteves-Oliveira M, Freitas PM, Marotti J, Tunér J. Laser treatment of recurrent herpes labialis: a literature review. Lasers Med Sci. 2014;29(4):1517–29. https://doi.org/10.1007/s10103-013-1311-8.
222. Barros AWP, da Hora Sales PH, de Barros Silva PG, Gomes ACA, Carvalho AAT, Leão JC. Is low-level laser therapy effective in the treatment of herpes labialis? Systematic review and meta-analysis. Lasers Med Sci. 2022;37:3393–402. https://doi.org/10.1007/s10103-022-03653-6.
223. Muñoz Sanchez PJ, et al. The effect of 670-nm low laser therapy on herpes simplex type 1. Photomed Laser Surg. 2012;30(1):37–40.
224. Stona P, et al. Recurrent labial herpes simplex in pediatric dentistry: low-level laser therapy as a treatment option. Int J Clin Pediatr Dent. 2014;7(2):140–3.
225. Donnarumma G, De Gregorio V, Fusco A, et al. Inhibition of HSV-1 replication by laser diode-irradiation: possible mechanism of action. Int J Immunopathol Pharmacol. 2010;23: 1167–76.
226. Dougal G, Lee SY. Evaluation of the efficacy of low-level light therapy using 1072 nm infrared light for the treatment of herpes simplex labialis. Clin Exp Dermatol 2013;38(7):713–718.
227. Dompe C, Moncrieff L, Matys J, Grzech-Leśniak K, Kocherova I, Bryja A, Bruska M, Dominiak M, Mozdziak P, Skiba TH, Shibli JA. Photobiomodulation—underlying mechanism and clinical applications. J Clin Med. 2020;9(6):1724.
228. Cafaro A, Broccoletti R, Arduino PG. Low-level laser therapy for oral mucous membrane pemphigoid. Lasers Med Sci. 2012;27:1247–50.
229. Minicucci EM, Miot HA, Barraviera SR, Almeida-Lopes L. Low-level laser therapy on the treatment of oral and cutaneous pemphigus vulgaris: case report. Lasers Med Sci. 2012;27:1103–6.
230. Pavlić V, Aleksić VV, Zubović N, Veselinović V. Pemphigus vulgaris and laser therapy: crucial role of dentists. Med Pregl. 2014;67(1–2):38–42.
231. Yilmaz HG, Kusakci-Seker B, Bayindir H, Tozum TF. Low-level laser therapy in the treatment of mucous membrane pemphigoid: a promising procedure. J Periodontol. 2010;81:1226–30.
232. Al-Maweri SA, Kalakonda B, Al-Soneidar WA, Al-Shamiri HM, Alakhali MS, Alaizari N. Efficacy of low-level laser therapy in management of symptomatic oral lichen planus: a systematic review. Lasers Med Sci. 2017;32:1429–37.
233. Payeras MR, Cherubini K, Figueiredo MA, Salum FG. Oral lichen planus: focus on etiopathogenesis. Arch Oral Biol. 2013;58:1057–69.
234. Yang H, Wu Y, Ma H, et al. Possible alternative therapies for oral lichen planus cases refractory to steroid therapies. Oral Surg Oral Med Oral Pathol Oral Radiol. 2016;121:496–509.
235. Del Vecchio A, Palaia G, Grassotti B, Tenore G, Ciolfi C, Podda G, Impellizzeri A, Mohsen A, Galluccio G, Romeo U. Effects of laser photobiomodulation in the management of oral lichen planus: a literature review. Clin Ter. 2021;172:467–83. https://doi.org/10.7417/CT.2021.2360.
236. Dodson TB. The frequency of medication-related osteonecrosis of the jaw and its associated risk factors. Oral Maxillofac Surg Clin North Am. 2015;27:509–16.
237. Al Dhalaan NA, BaQais A, Al-Omar A. Medication-related osteonecrosis of the jaw: a review. Cureus. 2020;12:e6944.
238. El Mobadder M, Grzech-Lesniak Z, El Mobadder W, Rifai M, Ghandour M, Nammour S. Management of medication-related osteonecrosis of the jaw with photobiomodulation and minimal surgical intervention. Dent J. 2023;11:1–9. https://doi.org/10.3390/dj11050127.
239. Rosella D, Papi P, Giardino R, Cicalini E, Piccoli L, Pompa G. Medication-related osteonecrosis of the jaw: clinical and practical guidelines. J Int Soc Prev Community Dent. 2016;6:97.
240. Merigo E, Cella L, Oppici A, et al. Combined approach to treat medication-related osteonecrosis of the jaws. J Lasers Med Sci. 2018;9:92.
241. Ruggiero SL, Dodson TB, Fantasia J, Goodday R, Aghaloo T, Mehrotra B, O'Ryan F. American Association of Oral and Maxillofacial Surgeons position paper on medication-related osteonecrosis of the jaw—2014 update. J Oral Maxillofac Surg. 2014;72:1938–56.
242. Sacco R, Leeson R, Nissan J, Olate S, de Castro CHBC, Acocella A, Lalli A. A systematic review of oxygen therapy for the management of medication-related osteonecrosis of the jaw (MRONJ). Appl Sci. 2019;9:1026.
243. Vanpoecke J, Verstraete L, Smeets M, Ferri J, Nicot R, Politis C. Medication-related osteonecrosis of the jaw (MRONJ) stage III: conservative and conservative surgical approaches versus an aggressive surgical intervention: a systematic review. J Cranio-Maxillofac Surg. 2020;48:435–43.
244. Weber JB, Camilotti RS, Ponte ME. Efficacy of laser therapy in the management of bisphosphonate-related osteonecrosis of the jaw (BRONJ): a systematic review. Lasers Med Sci. 2016;31:1261–72.
245. Razavi P, Jafari A, Vescovi P, Fekrazad R. Efficacy of adjunctive photobiomodulation in the management of medication-related osteonecrosis of the jaw: a systematic review. Photobiomodul Photomed Laser Surg. 2022;40:777–91.
246. De Freitas LF, Hamblin MR. Proposed mechanisms of photobiomodulation or low-level light therapy. IEEE J Sel Top Quantum Electron. 2016;22:348–64.

247. Tenore G, Mohsen A, Del Vecchio A, et al. Supportive pain management with super-pulsed low-level laser therapy of patients with medication related osteonecrosis of the jaw: clinical trial. Senses Sci. 2017;4:386–94.
248. Tenore G, Zimbalatti A, Rocchetti F, Graniero F, Gaglioti D, Mohsen A, Caputo M, Lollobrigida M, Lamazza L, De Biase A, et al. Management of medication-related osteonecrosis of the jaw (MRONJ) using leukocyte-and platelet-rich fibrin (l-PRF) and photobiomodulation: a retrospective study. J Clin Med. 2020;9:1–16. https://doi.org/10.3390/jcm9113505.
249. Li FL, Wu CB, Sun HJ, Zhou Q. Effectiveness of laser-assisted treatments for medication-related osteonecrosis of the jaw: a systematic review. Br J Oral Maxillofac Surg. 2020;58:256–67. https://doi.org/10.1016/j.bjoms.2019.12.001.
250. Andabak Rogulj A, Loncar Brzak B, Lončar Brzak B, Vučićević Boras V, Vidović Juras D, Škrinjar I. Burning mouth syndrome—a burning enigma. Med Flum. 2021; 57:4–16.
251. Imamura Y, Shinozaki T, Okada-Ogawa A, et al. An updated review on pathophysiology and management of burning mouth syndrome with endocrinological, psychological and neuropathic perspectives. J Oral Rehabil. 2019;46:574–87. https://doi.org/10.1111/joor.12795.
252. Ślebioda Z, Lukaszewska-Kuska M, Dorocka-Bobkowska B. Evaluation of the efficacy of treatment modalities in burning mouth syndrome-a systematic review. J Oral Rehabil. 2020;47:1435–47. https://doi.org/10.1111/joor.13102.
253. Périer J-M, Boucher Y. History of burning mouth syndrome (1800–1950): a review. Oral Dis. 2019;25:425–38. https://doi.org/10.1111/odi.12860.
254. Zakrzewska J, Buchanan JA. Burning mouth syndrome. BMJ Clin Evid. 2016;2016:1301.
255. Scala A, Checchi L, Montevecchi M, Marini I. Update on burning mouth syndrome: overview and patient management. Crit Rev Oral Biol Med. 2003;14:275–91.
256. Campello CP, Pellizzer EP, Vasconcelos BCDE, et al. Evaluation of IL-6 levels and +3954 polymorphism of IL-1β in burning mouth syndrome: a systematic review and meta-analysis. J Oral Pathol Med. 2020;49:961–8. https://doi.org/10.1111/jop.13018.
257. Kim M-J, Kim J, Kho H-S. Treatment outcomes and related clinical characteristics in patients with burning mouth syndrome. Oral Dis. 2020; https://doi.org/10.1111/odi.13693.
258. Stein J, Geisel J, Obeid R. Association between neuropathy and B-vitamins: a systematic review and meta-analysis. Eur J Neurol. 2021;28:2054–64.
259. Al-Maweri SA, Javed F, Kalakonda B, AlAizari NA, Al-Soneidar W, Al-Akwa A. Efficacy of low level laser therapy in the treatment of burning mouth syndrome: a systematic review. Photodiagn Photodyn Ther. 2017;17:188–93.
260. Lončar-Brzak B, Škrinjar I, Brailo V, Vidović-Juras D, Šumilin L, Andabak-Rogulj A. Burning mouth syndrome (BMS)— treatment with verbal and written information, B vitamins, probiotics, and low-level laser therapy: a randomized clinical trial. Dent J. 2022;10 https://doi.org/10.3390/dj10030044.
261. Bardellini E, Amadori F, Conti G, Majorana A. Efficacy of the photobiomodulation therapy in the treatment of the burning mouth syndrome. Med Oral Patol Oral Cir Bucal. 2019;24:e787–91.
262. Scardina GA, Casella S, Bilello G, Messina P. Photobiomodulation therapy in the management of burning mouth syndrome: morphological variations in the capillary bed. Dent J. 2020;8:99.
263. Škrinjar I, Lon˘car Brzak B, Vidranski V, Vučićević Boras V, Andabak Rogulj A, Pavelić B. Salivary cortisol levels and burning symptoms in patients with burning mouth syndrome before and after low level laser therapy: a double blind controlled randomized clinical trial. Acta Stomatol Croat. 2020;54:44–50.
264. Camolesi GCV, Marichalar-Mendía X, Padín-Iruegas ME, Spanemberg JC, López-López J, Blanco-Carrión A, Gándara-Vila P, Gallas-Torreira M, Pérez-Sayáns M. Efficacy of photobiomodulation in reducing pain and improving the quality of life in patients with idiopathic burning mouth syndrome. A systematic review and meta-analysis. Lasers Med Sci. 2022;37:2123–33. https://doi.org/10.1007/s10103-022-03518-y.
265. Hanna R, Dalvi S, Bensadoun RJ, Raber-Durlacher JE, Benedicenti S. Role of photobiomodulation therapy in neurological primary burning mouth syndrome. A systematic review and meta-analysis of human randomised controlled clinical trials. Pharmaceutics. 2021;13 https://doi.org/10.3390/pharmaceutics13111838.
266. de Matos ALP, Silva PUJ, Paranhos LR, Santana ITS, de Matos FR. Efficacy of the laser at low intensity on primary burning oral syndrome: a systematic review. Med Oral Patol Oral Cir Bucal. 2021;26:e216–25. https://doi.org/10.4317/medoral.24144.
267. Pasquale C, Utyuzh A, Mikhailova MV, Colombo E, Amaroli A. Recovery from idiopathic facial paralysis (Bell's palsy) using photobiomodulation in patients non-responsive to standard treatment: a case series study. Photonics. 2021;8(8):341.
268. Sonis ST. The pathobiology of oral mucositis. Nat Rev Cancer. 2004;4:277–84.
269. Zadik Y, Arany PR, Fregnani ER, Bossi P, Antunes H, Bensadoun R, Gueiros L, Majorana A, Nair R, Ranna V, et al. Systematic review of photobiomodulation for the management of oral mucositis in cancer patients and clinical practice guidelines. Support Care Cancer. 2019;27:3969–83.
270. Elad S, Cheng KKF, Lalla RV, et al. MASCC/ISOO clinical practice guidelines for the management of mucositis secondary to cancer therapy. Cancer. 2020;126(19):4423–31. https://doi.org/10.1002/cncr.33100.
271. Robijns J, Nair RG, Lodewijckx J, Arany P, Barasch A, Bjordal JM, Bossi P, Chilles A, Corby PM, Epstein JB, Elad S. Photobiomodulation therapy in management of cancer therapy-induced side effects: WALT position paper 2022. Front Oncol. 2022;12:927685.
272. Lalla RV, Brennan M, Schubert M. Oral complications of cancer therapy. In: Yagiela J, editor. Pharmacology and therapeutics for dentistry, vol. III. 6th ed. St Louis, MI: Mosby Elsevier; 2011. p. 782–98. ISBN: 978-0-323-05593-2.
273. Finfter O, Cohen R, Hanut A, Gavish L, Zadik Y. High-power laser photobiomodulation therapy for immediate pain relief of refractory oral mucositis. Oral Dis. 2023; https://doi.org/10.1111/odi.14618.
274. Saeed N, Morteza G, Negar P. The effect of photobiomodulation on regeneration of crushed inferior alveolar nerve caused by iatrogenic injuries in oral and maxillofacial surgeries: a systematic review. Adv Oral Maxillofac Surg. 2022;7:100155.
275. Cardoso FD, Salehpour F, Coimbra NC, Gonzalez-Lima F, Gomes da Silva S. Photobiomodulation for the treatment of neuroinflammation: a systematic review of controlled laboratory animal studies. Front Neurosci. 2022;16:1006031.
276. Biglioli F, et al. Surgical treatment of painful lesions of the inferior alveolar nerve. J Craniomaxillofac Surg. 2015;43(8):1541–5.
277. Biglioli F, et al. Lingual nerve lesion during ranula surgical treatment: case report. Minerva Stomatol. 2010;59(10):561–9.

278. Câmara CNDS, et al. Histological analysis of low-intensity laser therapy effects in peripheral nerve regeneration in Wistar rats. Acta Cir Bras. 2011;26(1):12–8.
279. Anders JJ, et al. In vitro and in vivo optimization of infrared laser treatment for injured peripheral nerves. Lasers Surg Med. 2014;46(1):34–45.
280. de Oliveira RF, et al. Laser therapy in the treatment of paresthesia: a retrospective study of 125 clinical cases. Photomed Laser Surg. 2015;33(8):415–23.
281. Falaki F, et al. The effect of low-level laser therapy on trigeminal neuralgia: a review of literature. J Dent Res Dent Clin Dent Prospects. 2014;8(1):1.
282. Ozen T, et al. Efficacy of low level laser therapy on neurosensory recovery after injury to the inferior alveolar nerve. Head Face Med. 2006;2(1):1.
283. Vučićević BV, et al. Adverse drug reactions in the oral cavity. Acta Clin Croat. 2015;54(2):208–15.
284. Nakonechna A, et al. Immediate hypersensitivity to chlorhexidine is increasingly recognised in the United Kingdom. Allergol Immunopathol. 2014;42(1):44–9.
285. Cafaro A, et al. Low level laser therapy (LLLT) as adjuvant in the management of drug induced gingival hyperplasia: a case report. Ann Stomatol. 2013;4(Suppl 2):8.
286. Kathuria V, et al. Low level laser therapy: a panacea for oral maladies. Laser Ther. 2015;24(3):215.
287. Cafaro A, et al. Low level laser therapy in oral graft-versus-host disease: a case report. Ann Stomatol. 2014;5(2):14.
288. Kalfas IH. Principles of bone healing. Neurosurg Focus. 2001;10(4):1–4.
289. Saracino S, et al. Superpulsed laser irradiation increases osteoblast activity via modulation of bone morphogenetic factors. Lasers Surg Med. 2009;41(4):298.
290. Aras MH, et al. Effects of low-level laser therapy on changes in inflammation and in the activity of osteoblasts in the expanded premaxillary suture in an ovariectomized rat model. Photomed Laser Surg. 2015;33(3):136–44.
291. Maman FAM, et al. Low-level laser therapy on bone repair of rat tibiae exposed to ionizing radiation. Photomed Laser Surg. 2014;32(11):618–26.
292. Tim CR, et al. Effects of low level laser therapy on inflammatory and angiogenic gene expression during the process of bone healing: a microarray analysis. J Photochem Photobiol B. 2016;154:8–15.
293. Prados-Frutos JC, et al. Lack of clinical evidence on low-level laser therapy (LLLT) on dental titanium implant: a systematic review. Lasers Med Sci. 2016;31(2):383–92.
294. Vande A, Sanyal PK, Nilesh K. Effectiveness of the photobiomodulation therapy using low-level laser around dental implants: a systematic review and meta-analysis. Dent Med Problems. 2022;59(2):281–9.
295. Tang E, et al. Photobiomodulation and implants: implications for dentistry. J Periodontal Implant Sci. 2013;43(6):262–8.
296. Campanha BP, et al. Low-level laser therapy for implants without initial stability. Photomed Laser Surg. 2010;28(3): 365–9.
297. Razaghi P, Haghgou JM, Khazaei S, Farhadian N, Fekrazad R, Gholami L. The effect of photobiomodulation therapy on the stability of orthodontic mini-implants in human and animal studies: a systematic review and meta-analysis. J Lasers Med Sci. 2022;13:e27.
298. Sourvanos D, Poon J, Lander B, Sarmiento H, Carroll J, Zhu TC, Fiorellini JP. Improving titanium implant stability with photobiomodulation: a review and meta-analysis of irradiation parameters. Photobiomodul Photomed Laser Surg. 2023;41(3):93–103.
299. Ting M, Alluri LS, Sulewski JG, Suzuki JB, Batista P, da Silva A. Laser treatment of peri-implantitis: a systematic review of radiographic outcomes. Dent J. 2022;10(2):20.
300. Santonocito S, Polizzi A, Cavalcanti R, Ronsivalle V, Chaurasia A, Spagnuolo G, Isola G. Impact of laser therapy on periodontal and peri-implant diseases. Photobiomodul Photomed Laser Surg. 2022;40(7):454–62.
301. Louvrou MK, Fragkioudakis I, Batas L. The use of lasers in peri-implant disease treatment and their efficacy in inflammation reduction: a narrative review. Lasers Dent Sci. 2022;6(1):15–26.
302. Coluzzi DJ, Mizutani K, Yukna R, Al-Falaki R, Lin T, Aoki A. Surgical laser therapy for periodontal and peri-implant disease. Clin Dent Rev. 2022;6(1):7.
303. Al-Falaki R, Hughes F, Wadia R, Eastman C, Kontogiorgos E, Low S. The effect of an Er,Cr:YSGG laser in the management of intrabony defects associated with chronic periodontitis using minimally invasive closed flap surgery. A case series. Laser Ther. 2016;25(2):131–9.
304. Clem D, Heard R, McGuire M, Scheyer ET, Richardson C, Toback G, Gwaltney C, Gunsolley JC. Comparison of Er,Cr:YSGG laser to minimally invasive surgical technique in the treatment of intrabony defects: six-month results of a multicenter, randomized, controlled study. J Periodontol. 2021;92(4):496–506.
305. Haxsen V, Schikora D, Sommer U, Remppis A, Greten J, Kasperk C. Relevance of laser irradiance threshold in the induction of alkaline phosphatase in human osteoblast cultures. Lasers Med Sci. 2008;23(4):381–4. https://doi.org/10.1007/s10103-007-0511-5.
306. Kesler G, et al. Platelet derived growth factor secretion and bone healing after Er:YAG laser bone irradiation. J Oral Implantol. 2011;37:195–204.
307. Pagin M, et al. Laser and light emitting diode effects on pre-osteoblast growth and differentiation. Lasers Med Sci. 2014;29:55–9.
308. Polat O. Pain and discomfort after orthodontic appointments. Semin Orthod. 2007;13(4):292–300.
309. Ngan P, Kess B, Wilson S. Perception of discomfort by patients undergoing orthodontic treatment. Am J Orthod Dentofac Orthop. 1989;96(1):47–53.
310. Royko A, Denes Z, Razouk G. The relationship between the length of orthodontic treatment and patient compliance. Fogorz Sz. 1999;92:79–86.
311. Yamaguchi M, Garlet G. The role of inflammation in defining the type and pattern of tissue response in orthodontic tooth movement. In: Krishna V, Davidovich Z, editors. Biological mechanisms of tooth movement. Chichester: Wiley; 2015. Ch. 9. ISBN: 978-1-118-68887-8.
312. Walker JB, Buring SM. NSAID impairment of orthodontic tooth movement. Ann Pharmacother. 2001;35(1):113–5.
313. Marini I, Bartolucci ML, Bortolotti F, Innocenti G, Gatto MR, Alessandri BG. The effect of diode superpulsed low-level laser therapy on experimental orthodontic pain caused by elastomeric separators: a randomized controlled clinical trial. Lasers Med Sci. 2015;30:35–41.
314. Fonseca P, de Lima FM, Higashi DT, et al. Effects of light emitting diode (LED) therapy at 940 nm on inflammatory root resorption in rats. Lasers Med Sci. 2013;28(1):49–55.
315. Yavagal CM, Matondkar SP, Yavagal PC. Efficacy of laser photobiomodulation in accelerating orthodontic tooth movement in children: a systematic review with meta-analysis. Int J Clin Pediatr Dent. 2021;14(Suppl 1):S94.
316. Reis CL, de Souza Furtado TC, Mendes WD, Matsumoto MA, Alves SY, Stuani MB, Borsatto MC, Corona

SA. Photobiomodulation impacts the levels of inflammatory mediators during orthodontic tooth movement? A systematic review with meta-analysis. Lasers Med Sci. 2022;37:1–7.
317. Huang T, Wang Z, Li J. Efficiency of photobiomodulation on accelerating the tooth movement in the alignment phase of orthodontic treatment—a systematic review and meta-analysis. Heliyon. 2023;9:e13220.
318. Fini MB, Olyaee P, Homayouni A. The effect of low-level laser therapy on the acceleration of orthodontic tooth movement. J Lasers Med Sci. 2020;11(2):204.
319. Douglas-De-Oliveira DW, Vitor GP, Silveira JO, Martins CC, Costa FO, Cota LOM. Effect of dentin hypersensitivity treatment on oral health related quality of life—a systematic review and meta-analysis. J Dent. 2017;71:1–8.
320. Canadian Advisory Board on Dentin Hypersensitivity. Consensus-based recommendations for the diagnosis and management of dentin hypersensitivity. J Can Dent Assoc. 2003;69:221–6.
321. Lopes AO, de Paula EC, Aranha ACC. Evaluation of different treatment protocols for dentin hypersensitivity: an 18-month randomized clinical trial. Lasers Med Sci. 2017;32:1023–30. https://doi.org/10.1007/S10103-017-2203-0.
322. Machado AC, Viana ÍEL, Farias-Neto AM, et al. Is photobiomodulation (PBM) effective for the treatment of dentin hypersensitivity? A systematic review. Lasers Med Sci. 2018;33:745–53. https://doi.org/10.1007/S10103-017-2403-7.
323. Cattoni F, Ferrante L, Mandile S, Tetè G, Polizzi EM, Gastaldi G. Comparison of lasers and desensitizing agents in dentinal hypersensitivity therapy. Dent J. 2023;11(3):63.
324. Tolentino AB, Zeola LF, Fernandes MRU, Pannuti CM, Soares PV, Aranha ACC. Photobiomodulation therapy and 3% potassium nitrate gel as treatment of cervical dentin hypersensitivity: a randomized clinical trial. Clin Oral Investig. 2022;26:6985–93. https://doi.org/10.1007/s00784-022-04652-1.
325. Mendes ST, Pereira CS, Oliveira JL, Santos VC, Gonçalves BB, Mendes DC. Treatment of dentin hypersensitivity with laser: systematic review. Brazil J Pain. 2021;4:152–60.
326. Rezazadeh F, Dehghanian P, Jafarpour D. Laser effects on the prevention and treatment of dentinal hypersensitivity: a systematic review. J Lasers Med Sci. 2019;10(1):1.
327. Torri S, Weber JBB. Influence of low level laser therapy on the rate of orthodontic tooth movement: a literature review. Photomed Laser Surg. 2013;31(9):411–21.
328. Altan A, Sokucu O, Ozkut M, et al. Metrical and histological effects of low level laser therapy on orthodontic tooth movement. Lasers Med Sci. 2012;27(1):131–40.
329. Rosa CB, Habib FA, de Araujo TM, et al. Effect of the laser and light emitting diode (LED) phototherapy on mid palatal suture bone formation after rapid maxillary expansion: a Raman spectroscopy analysis. Lasers Med Sci. 2014;29(3):859–67.
330. Shirazi M, Akhoundi A, Javadi E, et al. The effects of diode laser (660 nm) on the rate of tooth movement an animal study. Lasers Med Sci. 2015;30(2):713–8.
331. Cossetin E, Janson G, de Carvalho MG, et al. Influence of low level laser on bone remodelling during induced tooth movement in rats. Angle Orthod. 2013;83(6):1015–21.
332. Ekizer A, Uysal T, Yuksel Y. Light emitting diode photobiomodulation: effect on bone formation in orthopedically expanded suture in rats—early bone changes. Lasers Med Sci. 2013;28:1263–70.
333. Ekizer A, Uysal T, Guray E, et al. Effect of LED mediated photobiomodulation therapy on orthodontic tooth movement and root resorption in rats. Lasers Med Sci. 2015;30:779–85.
334. Nammour S, et al. Twelve-month follow-up of different dentinal hypersensitivity treatments by photobiomodulation therapy, Nd:YAG and Nd:YAP lasers. Life. 2022;12(12):1996.
335. Carneiro AM, Barros AP, de Oliveira RP, de Paula BL, Silva AM, de Melo AC, Silva CM. The effect of photobiomodulation using low-level laser therapy on tooth sensitivity after dental bleaching: a systematic review. Lasers Med Sci. 2022;37(7):2791–804.

Laser-Assisted Oral Hard Tissue Management

Contents

Laser Use in Dental Caries Management

Riccardo Poli, Francesco Buoncristiani, Deepti Dua, and Joshua Weintraub

Contents

D. J. Coluzzi, S. P. A. Parker (eds.), *Lasers in Dentistry—Current Concepts*, Textbooks in Contemporary Dentistry,
https://doi.org/10.1007/978-3-031-43338-2_8

Core Message

This chapter explores the range of benefits that relate to laser-assisted oral hard tissue management and details aspects of each wavelength in delivering adjunctive therapy. Of the currently available wavelengths of dental lasers, at the moment, only three can be used for hard tissue.

- "Erbium" lasers available on the market can have two different wavelengths: 2940 nm erbium yttrium-aluminum-garnet (Er:YAG) and 2780 nm erbium, chromium: yttrium-scandium-gallium-garnet (Er,Cr:YSGG), and their use is gradually increasing in dental practices as an alternative or as a complementary tool versus traditional dental treatments.
- During the last 15 years, researchers have developed a variation of the carbon dioxide CO_2 laser, traditionally used for soft tissue surgery, into a powerful hard tissue laser. The emission wavelength is 9300 nm, and the performance within clinical use in restorative dentistry is very promising.

These innovative properties can be easily perceived by comparing the use of these wavelengths and traditional techniques, envisaging the use of a high-speed handpiece and of a diamond bur, or alternative ones, for example, techniques such as air-abrasion or the use of decayed material dissolving gels. Thanks to its clear advantage, in restorative dentistry, every dentist can easily exploit the important characteristics that are revolutionizing dentistry.

This type of laser perfectly fits the Minimally Invasive Dentistry (MID) philosophy. The experience reported by the patient during its use for cavity preparation is completely different from the one when a rotary dental drill is used to prepare a decay lesion. ◘ Table 8.1 outlines the many advantages claimed during the use of a laser suitable for both hard and soft tissue surgical management. In most cases, local anesthesia through injection is not required, because the erbium laser triggers an analgesic effect in just a few seconds. This laser allows pain-free ablation of hard tissues. Furthermore, no vibration is felt as there is no bur working in contact with the surface, and thus the patient does not hear the traditional noise of the dental drill. However, the operator has to accept a learning curve, because the use of these wavelengths is neither intuitive nor immediate. A period of time is required to develop the optimal distance of the handpiece vis-à-vis the dental surface to be treated. By working "contactless," it is optimal to place the laser tip at about 1 mm in order to maximize ablation. Furthermore, the operator must have an in-depth knowledge on how to set and modify the various parameters (among which are energy output per pulse, frequency of pulses, and the air/water ratio for coaxial cooling irrigation) [1]. However, high- or low-speed burs used with the dental drill are still more efficient and fast in removing dental tissues. Preparing a cavity using a dental drill is much quicker. Burs ensure optimal control, and their use is more intuitive as all dentists are trained in their use. However, burs are more aggressive, non-selective, they generate intense vibration which may be harmful to the tooth structure, and they may cause cracking and pulp heat damage. Furthermore, during

◘ **Table 8.1** Advantages associated with hard/soft tissue laser use

Hard tissue laser advantages
• To be used on the hard tissue components of the tooth, on the bone and on soft tissues
• Possibility to cut soft tissues at the same time during cavity preparation (i.e., gingivoplasty or pulp exposure treatment during conservative therapy)
• Minimally invasive
• Reduced or no need for local anesthesia
• Suitable for preparation of very small cavities, without unnecessary removal of sound tissues
• Precision and accuracy in ablation on hard tissues
• Less noise associated with ablation compared to dental drill, no vibration, non-contact
• Ablation/excision selectivity of decayed hard tissues
• Limited risk of iatrogenic damage
• No smear layer in hard tissues
• Tissue decontamination due to bacteria removal from dentin tubules
• Less noise associated with ablation compared to dental drill, no vibration, non-contact
• Ablation/excision selectivity of decayed hard tissues
• No tissue/pulp heating
• No hard tissue cracking
• Increased useful surface for bonding (micro-retentive surface)
• Inhibition of dental demineralization with lower cario-receptivity
• Tissue decontamination due to bacteria removal from dentin tubules
• Biomodulation effect
• No tissue/pulp heating
• No hard tissue cracking
• Limited coagulation effect on soft tissues
• Working area on soft tissues remains clean
• No smear layer in hard tissues

Table 8.2 Aspects of performance between rotary bur and laser in dental restorative procedures

Restorative procedure	Handpiece and bur	Laser
Cutting enamel/dentine	Yes	Yes
Selective removal of caries	No	Yes
Precision	Precise >1–2 mm	Precise <300 μm
Smear layer	Smear layer produced	No smear layer
Thermal rise	Thermal rise >15 °C	Thermal rise <15 °C
Risk of iatrogenic damage	Greater	Less
Noise/vibration	>120 dB/vibration	<120 dB/no vibration
Bactericidal action	No	Surface decontamination
Speed of cutting enamel	Fast	<30% bur speed
Speed of cutting dentine	Fast	Comparable
Pain response	High	Less pain/no pain

Source: Parker S. [5]

8

their use, a large amount of smear layer is produced requiring acid etching for its removal before the application of the chosen adhesive system. Very frequently, it is necessary to use local anesthesia by injection in order to avoid pain. This traditional technique is the basis of the intense fear and phobia that patients feel when they undergo conservative dental therapy.

Table 8.2 shows a comparison among the characteristics resulting from the use of the traditional high-speed handpiece with diamond bur vis-à-vis the use of the erbium laser [2]. An alternative, more delicate, and less aggressive method is represented by air-abrasion, exploiting aluminum oxide particles (Al_2O_3) to remove carious tissue. With this method, the risk of cracking is lower than with the one using the diamond bur, and no smear layer is produced. Adhesion of cotmposite seems to have increased thanks to the micro-irregularities created and, as a consequence, there is less microleakage. The main disadvantage is represented by the abrasion particle layer that is deposited on the entire working area, which must be accurately removed before any adhesive technique. Decay chemical and mechanical removal systems envisage the use of sodium hypochlorite type of chemical substances (usually in the form of gel) or of enzymatic type. These substances can selectively dissolve decayed tissues, which are then removed through excavating tools.

8.1 Effect on Hard and Soft Tissues

At the end of laser preparation (provided correct parameters are used), the tooth surface is clean, without smear layer, with scaly and wavy appearance, with no thermal injury and no cracking and with open dentinal tubules. It is relevant for clinical purpose that at the end of erbium irradiation, the tooth surface, particularly in cervical areas shows a significant increase in roughness, proportional to applied energy density (up to seven times the roughness index of sound enamel). If areas of irradiated enamel or dentin remain exposed to saliva and not hybridized and covered by composite, they can significantly promote plaque retention and cariogenic bacteria adhesion with a greater caries risk [3, 4]. For this reason, it is fundamental to accurately avoid irradiation areas of the sound tooth surrounding the carious lesion. An additional advantage resulting from the use of the erbium laser compared to other methods is represented by the intra-operative possibility of performing ablation/excision of hard tissues and at the same time of treating surrounding soft tissues; as dictated by the surgical needs, adjunctive gingivoplasty, a more extensive gingivectomy and/or clinical crown lengthening which simultaneously modifies gum levels and possibly the bone margin by restoring the lost biological width may be considered. If during decay preparation, it is required to expose the healthy edge of the cavity covered by gingival encroachment, it is very easy to perform a light gingivoplasty in order to remove the superfluous keratinized tissue. Should it be required to move the excessive gum margin due to the size of the decayed cavity, it is possible to perform such procedure during the same conservative therapy session by simultaneously removing decayed and soft tissues. If the correct biological width is lost, performing the lengthening of the clinical crown with regard to soft tissues is very quick, and possibly also includes the underlying bone tissue. The entire procedure can be completed in a single session by considerably reducing operative times. Erbium lasers have moderate control over bleeding when used at low energy values, high frequency (30–40 Hz), with a high pulse width (i.e., 700 μs/pulse), without water and little cooling air to facilitate thermal interaction with tissues. This is a characteristic that can be used in conservative dentistry, as well as in oral surgery and dental prosthetics to facilitate hemostasis. If necessary, should the pulp be exposed following a trauma or penetrating decay, it can be decontaminated and coagulated before performing pulp capping or selective pulpotomy.

Figure 8.1a–f depicts a laser-adjunctive restorative procedure in treating an occlusal carious lesion.

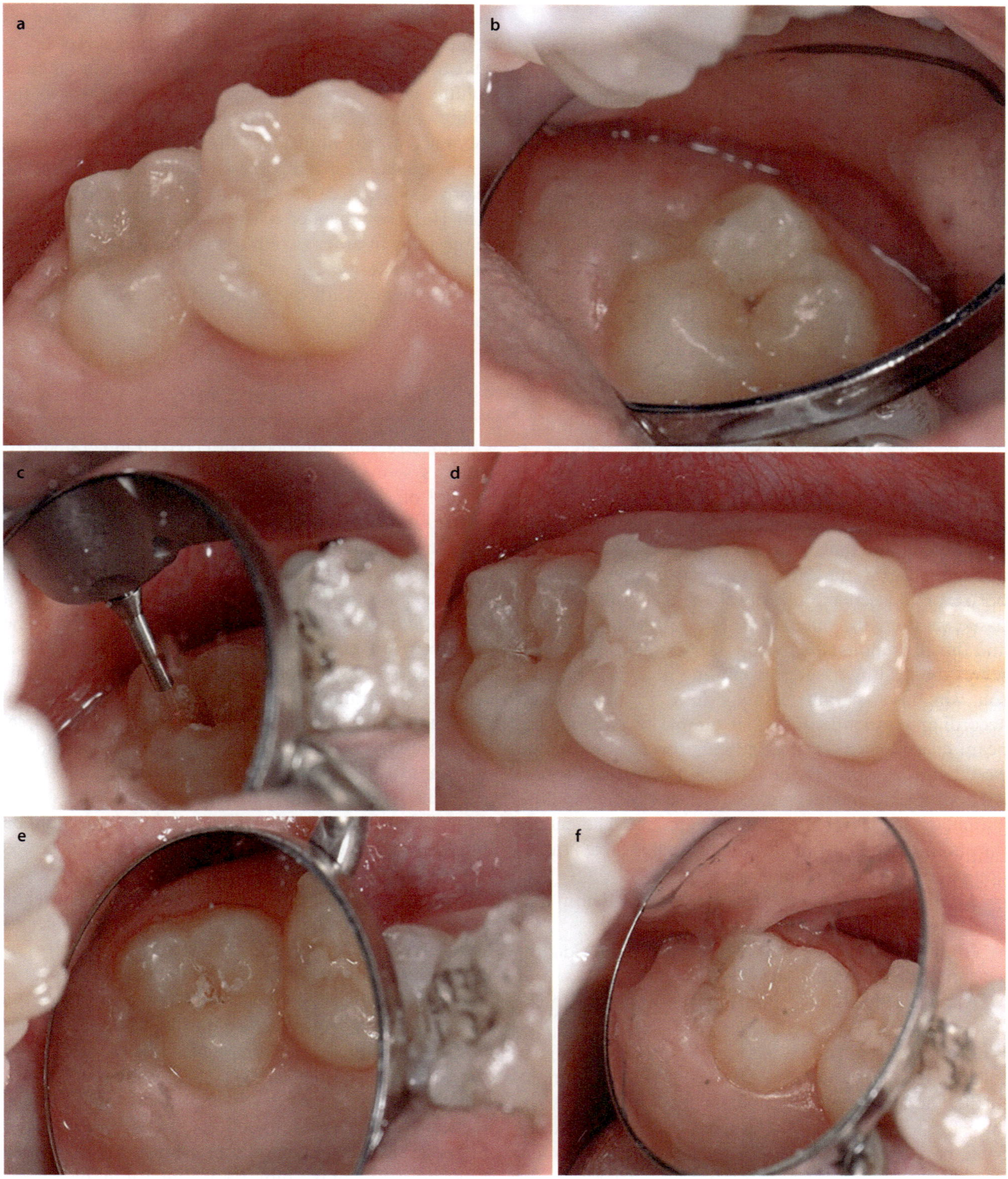

■ **Fig. 8.1** **a** Small cavity on occlusal side of tooth #15. **b** Detail of cavity on tooth #15. **c** Preparation with Er:YAG laser (Pulsar with wavelength of 2940 nm by LAMBDA Dr. Smile, Italy). Enamel settings: angulated handpiece, sapphire tip diameter 600 μm, length 12 mm, 2.6 W, 10 Hz, 260 mJ E/pulse, peak power 1625 W, average power density 426 W/cm^2, peak power density 266.521 W/cm^2, total energy 468 J, pulse width 160 μs, 1 mm tip-to-tissue distance, 60% water (20 mL/min), 100% air, total treatment time 180 s. Dentin and smear layer settings: sapphire tip diameter 600 μm, length 12 mm, 2 W, 15 Hz, 133 mJ E/pulse, peak power 831 W, average power density 323 W/cm^2, peak power density 136.336 W/cm^2, total energy 239 J, pulse width 160 μs, 1 mm tip-to-tissue distance, 50% water (18 mL/min), 80% air, total treatment time 120 s. **d** Detail of cavity after enamel preparation. **e** Detail of cavity after dentin preparation. **f** Completed composite restoration on tooth #15 (acid etching with orthophosphoric acid 37%), OptiBond FL total-etch adhesive system (Kerr, Orange, CA, USA), composite material (IPS Empress Direct Ivoclar Vivadent). (Procedure by Dr. Riccardo Poli)

8.2 Affinity with Water

Both erbium laser wavelengths (2780 and 2940 nm) have high affinity with water [5, 6]. Given the ubiquitous presence of water in human tissue and where water is a component of dental restorative materials, it may be seen that the greater the percentage content, the higher rate of absorption. Considering that the laser beam penetration is inversely related to absorption, incident laser radiation does not spread much in depth, thus the beam can penetrate dental tissues by only a few microns (for the wavelength of 2940 nm it is 7 μm into the enamel and 5 μm into the dentin, while for the wavelength of 2780 nm it is 21 μm into the enamel and 15 μm into the dentin) [7–11]. If the tissue has high water content (i.e., soft tissues vs hard tissues, dentin vs enamel, deciduous dentin vs permanent one, decayed dentin vs health dentin), the energy of erbium lasers could more easily cause explosive ablation at lower energy levels. The average threshold level at which ablation of hard tissues occurs is about 8–11 J/cm^2 for the Er:YAG laser and about 10–14 J/cm^2 for the Er,Cr:YSGG laser [10]. This phenomenon forms the basis of selective laser ablation; both erbium lasers more easily ablate the more hydrated tissue vis-à-vis the one with the lowest water content, and thus it is more effective on decayed dentin and preserves the healthy tissue surrounding it. This is the reason why laser parameters will have to be adjusted according to water content, for example, by reducing the beam energy used for a deciduous tooth vs what one would do to perform the ablation of a permanent tooth. An appropriate maxim would be to use a minimum dose parameter to deliver an optimal outcome. The energy threshold value that can allow a clinically efficient ablation of hard dental tissues is as follows:

- About 125 mJ (100–150 mJ) for primary dentin and decayed tissues
- About 150 mJ (100–200 mJ) for permanent dentin and primary enamel
- About 225 mJ (200–250 mJ) for permanent enamel

With regard to posterior teeth, or if tissues are highly calcified with less water content, it could be necessary to further increase energy parameters (up to about 350 mJ/pulse for healthy enamel). The speed of the laser cutting could be affected by numerous laser irradiation parameters such as energy density, frequency, water-ratio, and air-ratio in spray. Shinkai et al. [12] have shown that the depth of cut with 50% air is significantly deeper than when cut with 70% air. When the value of the air/water ratio was between 50% and 70%, a lower percentage of air translated into better cutting efficiency of the dentin compared to a higher percentage. Although it is not possible to determine from the article the exact amount and volume of air corresponding to 50% and 70% in the spray, a possible explanation of this phenomenon is that too much air might displace the cooling water, so causing a reduction in ablation efficiency. In the case of removal of deep carious dentin near the pulp, Shinkai et al. [12] recommended to use a higher percentage of air in the spray in order to reduce the ablation and preserve the pulp integrity. The volume of ablation per impulse depends significantly on the variability of the amount of water contained.

8.3 The Level of Laser Energy

Erbium lasers are equipped with coaxial integrated irrigation systems through an air/water spray. This allows to cool off targeted tissues, to keep the working area clean as it is key to prevent damages and thermal alterations on the cavity surface and on the tooth pulp. The operator should be able to accurately choose laser parameters in order to efficiently perform the ablation without damaging the surrounding healthy tissues. The first decisive factor is represented by the level of laser energy. It is always good practice to use minimum efficient value to obtain adequate excision. Excessive energy may damage the dental surface by altering, for example, the possibility of performing a good adhesive technique of composites [13]. The removal of a tissue occurs with a specific level of energy starting from the "threshold" value. Below it, no excision will occur, but there could be important structural or micro-structural modifications [14–18]. The chosen energy level can also be addressed toward a smaller or bigger surface. If the chosen level of energy is spread on a small surface, it will be easier to obtain the ablation effect vis-à-vis when the energy is spread on a bigger surface [14]. In fact, if the same amount of energy is spread on a bigger surface, the amount of energy per surface unit will be smaller and may be unsuitable to achieve the ablation threshold energy density (fluence) level capable of interacting with target tissue. By placing the tip in contact with the tissue, fluence will be maximum; delivery tip withdrawal from the tissue surface and increasing the distance we reduce it by about 70% at 0.5 mm, by 52% at 1 mm, by 32% at 2 mm, by 22% at 3 mm, and so forth. Obviously, the greater the energy density, the greater the interaction between laser and target tissue. Furthermore, this parameter can change by using, for example, a fiber or a tip with a different diameter. If the tip diameter is larger, the energy is spread over a larger target surface, compared to a tip with a smaller diameter, and thus, the subsequent effect will be less. On the other hand, above the threshold of 150–200 mJ, there would be a proportional increase of the excision, but also an increase in the risk of structural thermal alterations, especially if the air/water spray cooling is insufficient [19–21]. In such event, these alterations concern a surface depth of a few tenths of microns. Photonic energy is measured in Joules (J) and its density in J/cm^2. The energy density (fluence), measured in J/cm^2, is an important factor that determines the ablation selectivity of hard tissues. As previously stated, each hard tissue type has an ablation threshold related to the amount of water that is contained in it. The greater the water percentage, the easier the ablation will be. In an empirical research, we can set the Er:YAG laser using a tip of diameter of 600 μm, at 20 Hz, and coaxial coolant/dispersant water of 60% and air at 100%. By just changing the energy per pulse, we can demonstrate that we need increasing energy/pulse in order to ablate harder and harder tissues of the same individual tooth. With 0.5–0.8 W of average power (25–40 mJ/pulse, fluence 246–394 J/cm^2, peak power 156–250 W), there is no ablation of any tissue, because with such power levels, the amount of energy is insufficient to vaporize water. At 1 W of average power (50 mJ/pulse, fluence 492 J/cm^2, peak power 313 W), there is no ablation of tooth tissue but sufficient for the ablation of calculus, which is richer in water than cementum. This can be particularly useful in periodontology, as it can help the root debridement and scaling. At 1.3 W of average power (65 mJ/pulse, fluence 640 J/cm^2, peak power 406 W), harder calculus is ablated. At 1.7 W of average power (85 mJ/pulse, fluence 836 J/cm^2, peak power 531 W), the ablation of cementum on root surface is possible. This can be considered the average ablation threshold level at which the restorative dentistry takes place. At 3 W of average power (150 mJ/pulse, fluence 1476 J/cm^2, peak power 938 W), the ablation threshold of cervical enamel is reached; however, only at 4–5 W (200–250 mJ/pulse, fluence 1968–2460 J/cm^2, peak power 1250–1563 W) is the ablation of harder occlusal enamel possible, which contains less water than any other dental tissue and so needs more energy to be removed. The ablation threshold of dentin is reached at 2–2.5 W of average power (100–125 mJ/pulse, fluence 984–1230 J/cm^2, peak power 625–781 W). Additionally, decayed tissues are much more hydrated than healthy tissue counterparts, and similarly, primary tissues are less structured, less organized and richer in water than permanent ones, so both categories require lower parameters.

◘ Figure 8.2a–g demonstrates a laser-assisted treatment of a carious lesion involving enamel and dentin of tooth #13 on distal occlusal sides.

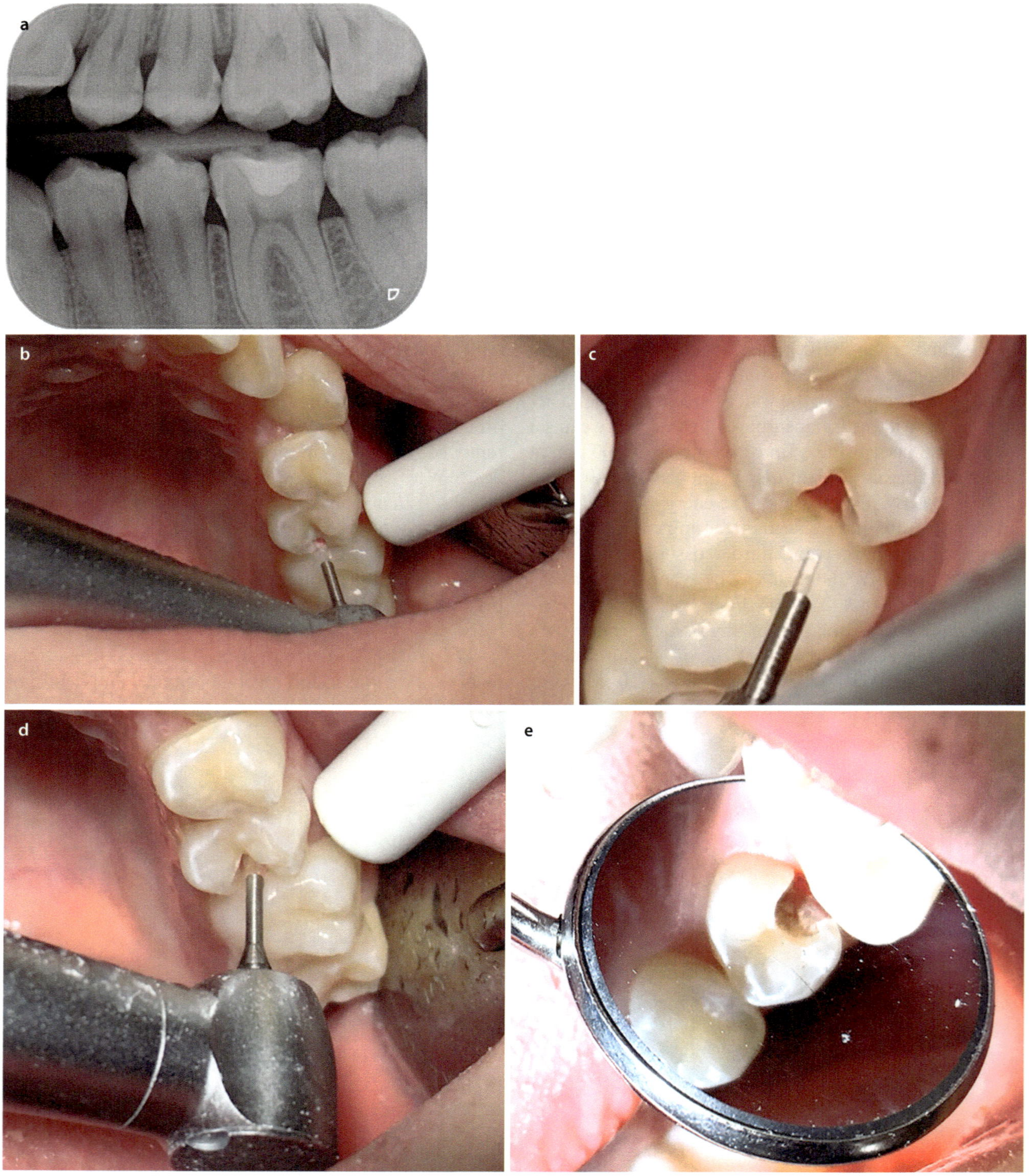

Fig. 8.2 **a** Intra-oral X-ray taken before the cavity preparation. Carious lesion involving enamel and dentin of tooth #13 on distal occlusal sides. **b–d** Cavity preparation with Er:YAG laser (Pulsar with wavelength of 2940 nm by LAMBDA Dr. Smile, Italy). Enamel settings: angulated handpiece, sapphire tip diameter 600 μm, length 12 mm, 4 W, 15 Hz, 267 mJ E/pulse, peak power 1669 W, average power density 657 W/cm^2, peak power density 273.696 W/cm^2, total energy 721 J, pulse width 160 μs, 1 mm tip-to-tissue distance, 40% water (16 mL/min), 100% air, total treatment time 180 s. Dentin and smear layer settings: sapphire tip diameter 600 μm, length 12 mm, 2 W, 15 Hz, 133 mJ E/pulse, peak power 831 W, average power density 327 W/cm^2, peak power density 136.336 W/cm^2, total energy 359 J, pulse width 160 μs, 1 mm tip-to-tissue distance, 50% water (18 mL/min), 80% air, total treatment time 120 s. **e** Detail of cavity after completed preparation. It is necessary to rectify and bevel enamel margins and following with etching, adhesive, and composite stratification steps (acid etching with orthophosphoric acid 37%, OptiBond FL total-etch adhesive system (Kerr, Orange, CA, USA), composite material (IPS Empress Direct Ivoclar Vivadent). **f** Composite reconstruction completed. **g** Final X-ray after restoration on tooth #13. (Procedure by Dr. Riccardo Poli)

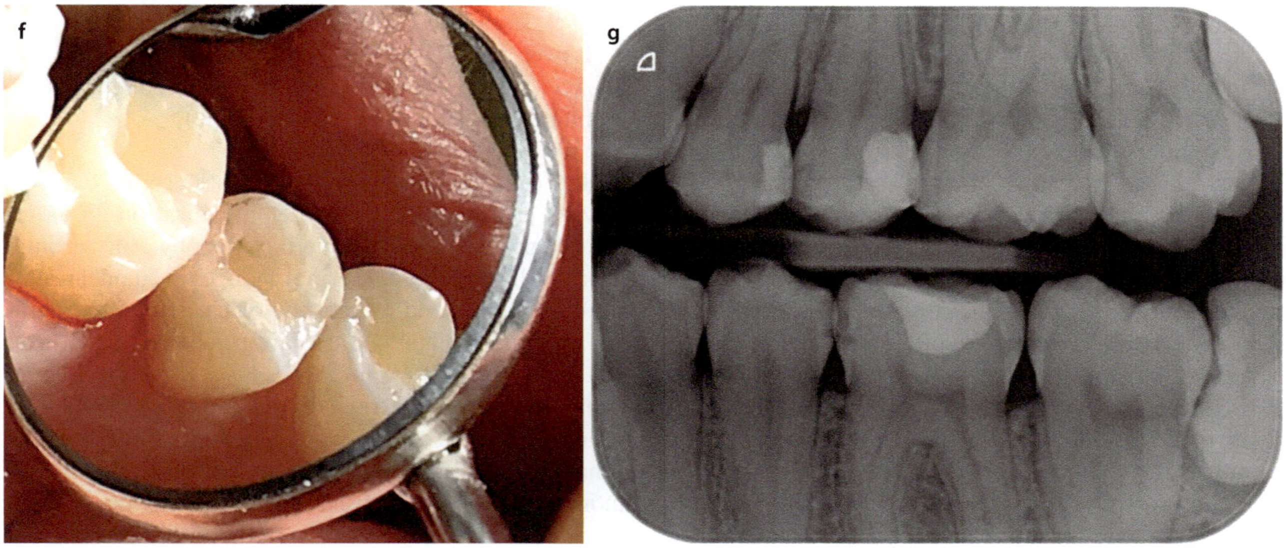

Fig. 8.2 (continued)

8.4 Pulse Dynamics and Frequency

Erbium laser radiation is through a free running pulse (FRP) train, inherently derived from the high intensity frequency of excitation energy applied to the laser active medium. The number of pulses released per second is termed frequency (or pulse frequency). This value is expressed as Hertz (Hz or pps, i.e., pulses per second). The larger the number of pulses per unit time, the greater and quicker the interaction with the target tissue will be, because a larger amount of energy is transferred to it. The amount of energy that is released per unit time identifies the power, i.e., the energy of each pulse times the number of pulses per second. It is measured in Watts (W). Thus, power depends on the ratio between energy and the number of pulses per second (W = J/pulse × Hz). When energy is provided through a short pulse (a pulse duration of about 50–150 μs), a high amount of energy interacts with the tissue in a fraction of a second and this means achieving huge power value. Each pulse can, however, achieve a maximum power (peak power) which has a major impact on the tissue. Interaction with a target is greater if the peak power is high. The shorter the pulse duration, the lower the degree of energy converted into heat. As a consequence, thermal interaction and the damage to tooth tissues following temperature increase will be reduced. The irradiated enamel and dentin surface appearance is very similar to that of etched tooth tissue: without smear layer, clean, wavy, micro-rough, irregular valleys and peaks, more pronounced when the applied energy is higher. Irradiated dentin results in open tubules; greater ablation occurs within the intertubular area as it is very hydrated, leaving peritubular areas more elevated and pronounced. The interaction between erbium energy and tooth tissue can create a series of chemical and structural modifications that are important to consider as there are clinical aspects involved in them. All these surface changes can affect the adhesion process and hamper the reconstruction duration but can also be useful because the changes can increase micro-hardness, the acid resistance and reduce the caries incidence.

8.5 Remineralization Effect

Loss of enamel can be caused by different processes in the oral cavity, namely attrition, erosion, abrasion, and tooth decay. Dental caries occurs due to the presence of acids, which are a result of bacterial carbohydrate metabolism, mineral loss occurs, and eventual cavity development. Once a cavity is formed, enamel cannot be repaired in a natural way and tooth restoration is necessary; enamel resistance against acid attack is viewed as a significant preventive measure. The effectiveness of fluoride for the prevention of dental caries has been clearly established in several studies as it enhances resistance to acid attack through the transformation of hydroxyapatite (HA) to fluorapatite, and at the same time, it reduces the ability of bacteria to produce acid, having joint bactericide and remineralizing effects. Fluoride is added to several remineralizing agents available on the market in order to improve the ability of fluoride to restore the balance of minerals of the dental structure, such as casein phosphopeptide-amorphous calcium phosphate (CPP-ACP) and HA, among others. The anti-cariogenic properties of such remineralizing agents are explained by their mineral contribution to the dental structure—

calcium (Ca) and phosphorus (P) ions, which inhibit demineralization and promote remineralization, resulting in the restoration of the dental surface. Some studies have tested Er:YAG laser irradiation at lower energy densities and have shown significant inhibition of enamel demineralization. It has been reported that lasers can increase resistance to enamel acids, and when it is associated with fluoride, both are reported to work in synergy, achieving a reduction of the solubility of enamel. Currently, other remineralizing agents have been shown to effectively inhibit enamel demineralization [22, 23]. The chemical modifications of the irradiated tooth surface happen during irradiation with ablative and also sub-ablative parameters. According to literature, the chemical analysis of post-irradiated dental structure with Er,Cr:YSGG lasers demonstrated a calcium ion increase and hydroxyapatite crystal reorganization after laser irradiation, due to enamel thermal effects [24]. This was later confirmed by Geraldo-Martins et al. (erbium increases enamel acid resistance by +23% with Er,Cr:YSGG using 0.25 W, 62.5 J/cm^2, no cooling with air/water) [16]. The exact mechanism leading to enamel gain acid resistance is not entirely explained; some authors attributed this to heating the surface (so causing structural and chemical changes), to vaporization of water, organic matrix, and inorganic components, to other interactions (photochemical effects or non-linear interactions), leading to less-soluble enamel, structural changes that reduced permeability, or dental enamel melting. Depending on the temperature following laser irradiation, different effects on dental enamel are produced that could result in decreased enamel solubility and increased acid resistance [10, 17, 18, 25, 26]. The above-mentioned changes happen even without an objective surface melting, but it is not surprising that even at low energy densities, the superficial layers temperature of enamel can reach values above 400 °C [19].

The effect of heat on hard tissues can be summarized in:

Decrease of water content and condensation of H phosphate ions (HPO_4^{2-}) with formation of pyrophosphate [20, 27, 28].

- At 200–400 °C inhibition of HA dissolution
- At 300–350 °C progressive increase of structural OH
- At 350 °C protein decomposition [29]
- At 400–650 °C reduction of carbonate ions (CO_3^{2-})
- At 400–800 °C enamel organic matrix evaporation
- At 800–1200 °C HA melting [30]

The main chemical alterations due to laser energy that are responsible of micro-hardness increase and greater acid resistance are [31] as follows:

- Increase of Ca and P ions
- Different ratio Ca/P
- Decrease of CO_3
- Lower ratio CO_3/PO_4
- Acid phosphate ions condense to form pyrophosphate ions (more stable, less soluble).

Ulusoy et al. [32] have proposed to use erbium lasers to prevent enamel and dentin demineralization by laser-induced prevention of demineralization, by which erbium laser irradiation causes thermal changes in enamel resulting in chemical and/or morphological structure alterations without ablation. The main mechanisms that lead to a less-soluble and more caries-resistant enamel structure are decreased carbonate and hydroxyl groups and organic matrix decomposition. The loss of calcium carbonates increases the degree of enamel crystallinity with an improvement in its structural proprieties. Lasers used at sub-ablative energy level in combination with traditional prophylaxis resulted in an increased caries prevention effectiveness compared with prophylaxis alone or to untreated teeth [33]. According to Abbasi et al. [34], the mechanisms behind increased enamel resistance against acid after laser application include:

- Decreased enamel permeability via melting and reforming enamel surface crystals.
- Decreased enamel solubility by forming compounds with more resistance against solubility like tetra calcium diphosphate monoxide.
- Decreased enamel solubility by changing enamel structure like reducing water and carbonate content and increasing hydroxyl ions, forming pyrophosphates, and breaking up protein chains [35].
- It was also reported that small spaces are created inside the enamel by laser irradiation. Calcium, phosphor, and fluoride ions are trapped inside them and deposit in enamel [36].

Apart from the mineral content changes, in literature there are also other mechanisms advocated to explain the lower carious lesion incidence following the laser irradiation of the tooth. Ying et al. demonstrated a partial decomposition of the organic matrix of the enamel during irradiation, with blockage of the inter- and intra-prismatic spaces and tubular blockage. They considered this as a possible reason of ion diffusion compromise in enamel with retardation of demineralization [21]. The ideal erbium laser parameters to inhibit demineralization have not been clearly determined. The laser parameters vary among studies and there is no consensus in literature regarding which is the best [16]; low energy densities applied for a short time can cause only water heating, vaporization, and pressure increase inside the enamel, without promoting melting or recrystallization on enamel structure. The laser effects on dental enamel become more evident as power increases and indeed a high power leads to undesired alterations, if not unwanted ablation. Parameters should be tested because irradiation conditions have already been shown to cause greater protective effects against erosion, and the increase in the resistance of enamel to acid demineralization is highly dependent on laser parameters such as pulse duration, energy density, and the number of overlapped pulses [37]. Thus, it is necessary to establish the ideal parameter to the clinical use of the laser that promotes the highest acid resistance with the lowest possible energy settings [38]. According to literature, laser-assisted remineralization is a further mechanism able to reduce cario-receptivity. Remineralization is the process through which Ca and PO_4 ions, delivered from an external source to the tooth, are deposited inside crystalline structure gaps of demineralized enamel, producing an increase of crystalline lattice [39]. This is the best approach for noncavitated white spot lesions due to caries process. Several chemical agents have been using for this purpose: fluoride, arginine, hydroxyapatite, bio glasses, casein phosphopeptide ACP (amorphous calcium phosphate), and citrate ACP. All these agents improve caries prevention, remineralization, inhibit demineralization, attenuate dentinal sensitivity, contrast erosion, and have antibacterial activity. The combination laser-chemical agent seems promising in augmenting the tooth intake of minerals. Every laser wavelength has been investigated with variable results, including CO_2, Nd:YAG, argon, diodes, and Er:YAG lasers. Zezell et al. [40] in a year—long double-blind crossover study evaluated the effects of Nd:YAG laser used with topical application of acidulated phosphate fluoride (APF) to prevent enamel demineralization in vivo. They obtained a 39% reduction in caries in laser group compared with control. The formation of white spot lesions was significantly less and they concluded that the combination of Nd:YAG and topical APF was effective for reducing the incidence of caries in vivo. Also Delbem et al. [41] found that Er:YAG laser use following the application of APF influenced the deposition of calcium fluoride on the enamel and exhibited a superficial anti-cariogenic action. Similar conclusions have been reached by Subramaniam and Pandey [42] who stated that Erbium Laser irradiation of primary teeth followed by CPP-ACP application increased surface micro-hardness of enamel (compared to negative and positive controls and CPP-ACP only) making it more resistant to acid challenge than normal enamel. At the moment, research into the combination of erbium and CPP-ACP is still conflictual, and their results are not uniform [43]. More recently, Yassaei et al. [44] confirmed the previously referred studies by irradiating tested teeth with Er:YAG 2.94 nm, 80 mJ/pulse, 4 Hz, 10 s, 20 mm tip-to-tissue, sweeping motion, with air and water cooling. They obtained a mineral loss due to in vitro acid challenge of—68% in the control group,—36% in CPP-ACP without fluoride + Er irradiation, and—32% in CPP-ACP with fluoride + Er irradiation. Conversely, in 2019 Yilmaz et al. [45] had promising remineralization results from the use of several chemical substances or from the Er:YAG irradiation alone, but the combination among them didn't show a notable impact in terms of eroded enamel improvement. For this reason, it is still sharable the affirmation of Ramalho et al. [46] in their literature review that "long-term clinical studies are necessary to validate this application because literature continues to be controversial." Areas at the restored cavity—healthy surface margins can become demineralized in case of a shift to acidic conditions. Evidence shows that laser irradiation can prevent mineral loss and demineralization of the tooth structure due to the effect of acids present in the oral environment [47]. The CO_2 and Er:YAG lasers alone seem to have no significant effect when used on hard tissues exposed to cariogenic solution. However, lasers can exert a synergistic effect when used with NaF varnish; fluoride varnish applied prior to laser irradiation is supposed to confer further resistance to the tooth structure and positively affect its hardness.

▣ Figure 8.3a–d provides detail of laser-assisted restorative treatment and placement of a Class I composite restoration at tooth #31.

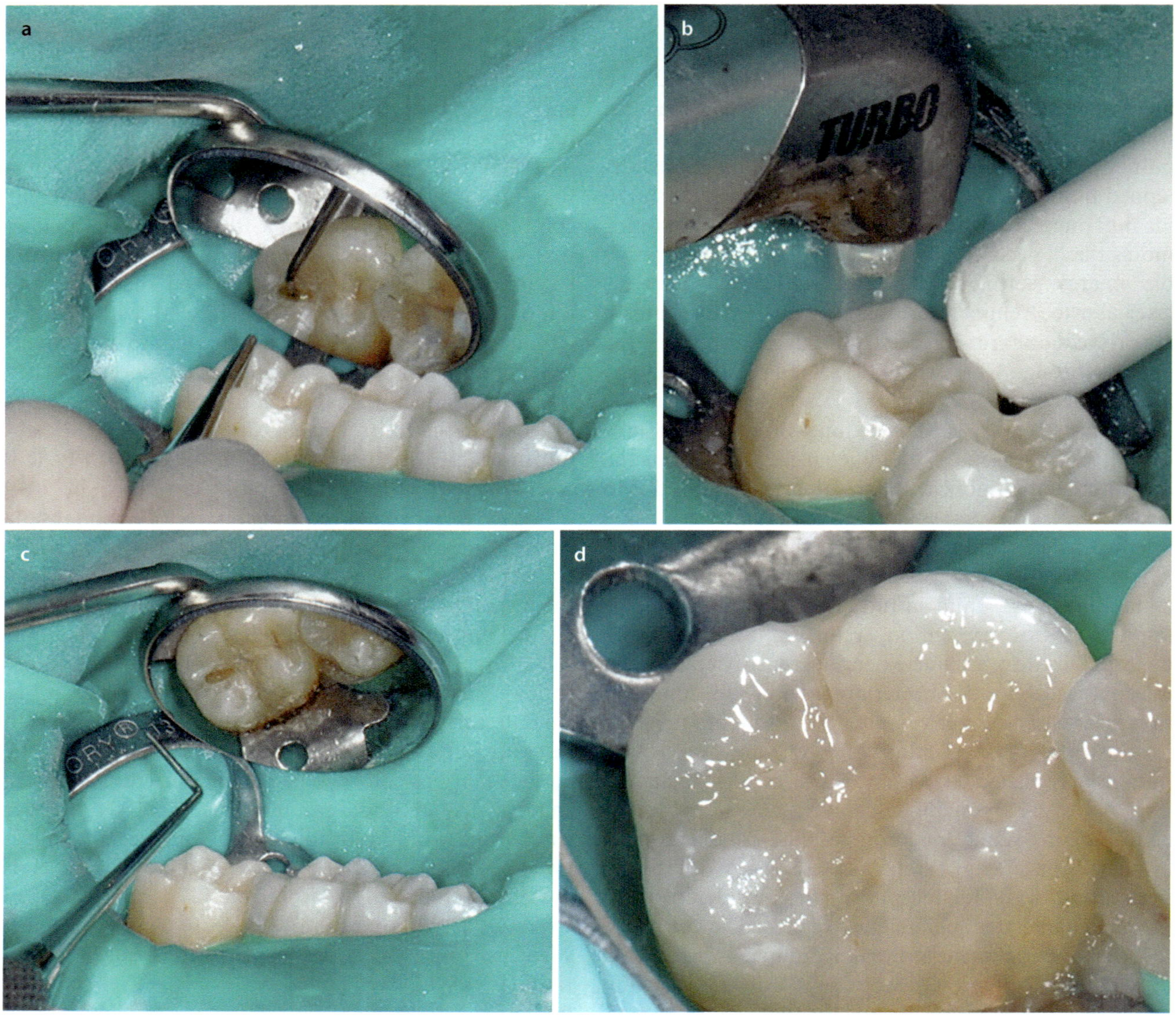

Fig. 8.3 **a** Carious lesion on occlusal side of tooth #31. **b** Preparation of a class I cavity in tooth #31 (Er,Cr:YSGG Waterlase iPlus laser with wavelength of 2780 nm by Biolase Technology, Irvine, CA, USA). Turbo handpiece, diameter 900 μm. Enamel settings: 3 W, 15 Hz, 200 mJ E/pulse, peak power 3333 W, average power density 492 W/cm^2, peak power density 546.710 W/cm^2, total energy 540 J, pulse width 60 μs, 7 mm tip-to-tissue distance, 60% water (18 mL/min), 60% air. Dentin and smear layer settings: 2.25 W, 15 Hz, 150 mJ E/pulse, peak power 2500 W, average power density 369 W/cm^2, peak power density 410.032 W/cm^2, total energy 511 J, pulse width 60 μs, 7 mm tip-to-tissue distance, 50% water (14 mL/min), 80% air. **c** At the end of the cavity preparation, enamel margins are rectified and beveled. Then, etching, adhesion, and composite stratification are performed (acid etching with orthophosphoric acid 37%, OptiBond FL total-etch adhesive system (Kerr, Orange, CA, USA), composite material (IPS Empress Direct Ivoclar Vivadent). **d** Composite reconstruction is completed. (Procedure by Dr. Riccardo Poli)

8.6 Distance to the Target

It is key that the operator works by holding the laser handpiece at the correct distance from the target, in order to optimize excision and treatment duration. Since the laser is contactless, excessive distance prevents efficient interaction with tissues. Furthermore, the hand piece must move slowly to allow the laser energy to interact with the tissue; the speed of the movement must be slower than the speed normally used with the dental drill. During cavity preparation work, the tip is gradually inclined on the one side to obtain widening [48]. As preparation proceeds, the tip must be positioned deeper to maintain the ideal distance vis-à-vis the target, and thus obtain a high and continuous energy density.

8.7 The Problem of Laser Etching

Several authors have suggested the use of erbium lasers instead of acid as a pre-bonding conditioner. The use of energy values below the ablation threshold (sub-ablative) allowing a micro-structural modification in dentin and enamel, creates a very similar surface to that obtained with orthophosphoric acid. Erroneously, this effect has been long defined "laser etching" in literature [36, 49–55]. As an example, recently Dilip et al. [56] proposed laser etching at 1.5 W/20 Hz because it achieved similar surface roughness and shear bond strength compared to conventional acid etching and could be a viable alternative for surface preparation of enamel. Moreso, they proposed lower power outputs in those clinical situations which demand lesser bond strength. The differences between acid etching and surface etching by erbium laser are numerous. More precisely, it would be appropriate to use the term "laser conditioning" [57, 58]. Acid etching is a process that has been used for decades to facilitate composite adhesion. Even though there are some issues associated with it (excessive decalcification with alteration of the enamel-dentin ideal architecture for adhesion, higher susceptibility to secondary decay, tooth sensitivity, excessive demineralization compared to the penetration ability of adhesive system monomers), the results obtainable through orthophosphoric acid are widely predictable [28, 59]. With regard to the use of orthophosphoric acid at 34–38%, erbium lasers generate a more irregular surface; the greater the energy and the lesser the frequency we respectively get deeper and more far-apart craters. Even if the final surface is very similar to the etched one, composite adhesion process to irradiated tooth hard tissues is a controversial phenomenon and its outcome should be further investigated as many authors deemed it of lower quality. In literature, data are quite contrasting and, often times, adhesion values came out much lower than those obtainable with the acid [53, 54, 60–64].

8.8 Microleakage

Difficulties encountered by operators during the bonding procedure between the composite and hard tooth tissues often translate into microleakages or the complete detachment of the reconstruction from its seat [65]. *Microleakage* can be defined as "*the clinically undetectable passage* of *bacteria and bacterial products, fluids, molecules or ions from the oral environment along the various gaps present in the cavity restoration interface*" [66]. It creates a loss of marginal seal between the restorative material used for tooth filling and the tooth cavity wall with the subsequent infiltration [67–70].

Among the causes of microleakage, we can mainly take the following elements into account:

1. Incomplete penetration of the bonding resin in the area that was decalcified by the etching acid or following the erbium laser effect. This gives rise to the formation of a weaker bonding area, which will be more sensitive to hydrolysis and infiltration.
2. Stress generated at tooth/reconstruction interface level following polymerization shrinkage, or due to oral environment temperature fluctuations [71], or due to cyclical phenomena of mechanical fatigue that are repeated during the masticatory load.

This poses a major concern since it may lead to microleakage and ultimately failure of adhesive bond between tooth substrate and restorative material [72]. Infiltration of bacteria or of fluids along the interface can cause hydrolytic collapse, both of the adhesive resin and of the collagen present in the hybrid layer, jeopardizing bond stability between the resin and the dentin surface. The minimum bond strength required at tooth-adhesive interface to sustain the forces in oral environment is 20 MPa [73]. Microleakage is the main factor of secondary decay and of reconstruction failure [74–76], and it is at the basis of dentin hypersensitivity, discoloration, and pulp damages. Composite as a restorative material has evolved to micro-filled, micro-hybrid, nanohybrid; and from conventional methacrylate-based to silorane-based systems to decrease the inherent problem of polymerization shrinkage associated with it [77]. Despite various advances in formulation of composite resin, the problem of polymerization shrinkage has not been completely overcome. When the gingival margins of a tooth preparation are in dentin or cementum or both, and the resin is firmly anchored to the etched enamel at the other margins, the material tends to pull away from the gingival margins during curing, because of polymerization shrinkage, resulting in formation of gap at that interface. Various factors affecting the polymerization shrinkage include degree of conversion, light-curing techniques, technique of composite placement, type of matrices and wedges used, direct or indirect method of restoration, and beveling of margins [78]. An additional cause of detachment between composite and tooth wall is related to the shape of the prepared cavity. The greater the number of walls (i.e., box-shaped cavities typical of Class I of Black's classification), the greater the relationship between bonded surfaces vis-à-vis non-bonded ones. This principle is defined as C factor [79, 80]. C factor is the ratio of bonded to non-bonded surface areas and is related to the cavity preparation geometry. The residual polymerization stress increases directly with this ratio. During curing, shrinkage leaves the bonded cavity surfaces in state of stress and the non-bonded surfaces in relaxed state. Incremental build up reduces C factor and hence the polymerization shrinkage. The horizontal placement technique has been reported to increase the C factor. Therefore, following types of incremental layering techniques are suggested [81]:

- Oblique technique
- Successive cusp build-up technique
- Stratified layering technique
- Split incremental horizontal layering

If the entire composite simultaneously adheres to the walls, as it happens in the occlusal cavities of molars, there will be many more cases of shrinkage-related stress because the composite adhering to many walls at the same time, by contacting, generates even greater stresses. On the other hand, if cavity tooth walls are just a few (i.e., interproximal preparations of Class II premolars and molars), we would assist to reduced stress since the part of non-adhering materials can compensate for polymerization shrinkage, releasing the effects toward the part free from constraints and, thus, there will be lesser risk of reconstruction detachment. Insufficient compensation of stresses resulting from polymerization shrinkage reduces the efficiency of the seal due to the reduced initial strength of the composite-cervical dentin bond. Another type of leakage, termed "*nanoleakage*," was described by Sano et al. (1995) [70]. Concerns have been raised that aggressive etching of dentin may cause demineralization to a depth that might be inaccessible to complete resin impregnation. A collagenous band at the base of the hybrid layer would weaken the resin-dentin bond resulting in a pathway for nanoleakage. This leakage occurs within the nanometer-sized spaces around the collagen fibrils within the hybrid layer that have not been completely infiltrated by resin. It has been shown to occur both at the bottom of the hybrid layer and/or scattered along its entire width, depending upon the bonding system employed [82]. This emphasizes the importance of adhesive systems, their selection based on composition plays an important role (◘ Table 8.3). It appears that bonding systems that etch

8

◘ **Table 8.3** Composition of adhesive systems

Classification	Adhesive	pH	Classification (based on pH)	Bond strength reference study	Microleakage reference study	Composition
Etch-n-rinse 3 step	Scotch bond MP	1		33, 48	56, 68	Etchant: 35% phosphoric acid, silica thickener Primer: HEMA, polyalkenoic acid copolymer, water Adhesive: HEMA, Bis-GMA
Etch-n-rinse 3 step	OptiBond FL	1		–	68	Etchant: 37.5% phosphoric acid, silica thickener Primer: HEMA, GPDM, PAMM, ethanol, water, photoinitiator Adhesive: TEGDMA, UDMA, GPDM, HEMA, Bis-GMA, filler, photoinitiator
Etch-n-rinse 2 step	Adper Single Bond 2/Single Bond 2	1		50, 51	55, 63, 65	HEMA, bis-GMA, DMA, methacrylate functional copolymer of polyacrylic and polyitaconic acids, water, ethanol, nanofiller, photoinitiator
Etch-n-rinse 2 step	Excite	1		39	–	HEMA, DMA, phosphonic acid acrylate, highly dispersed silicon dioxide, alcohol solution
Etch-n-rinse 2 step?	Adper single bond/ single bond	1		12, 36, 49, 52, 54	57, 61, 69	Bis-GMA, HEMA, DMA, Vitrebond copolymer, water/ethanol solv., silane-treated silica fillers
Etch-n-rinse 2 step	XP Bond	1		36	64	Carboxylic acid modified dimethacrylate (TCB resin), PENTA, UDMA, TEG-DMA, HEMA, butylated benzenediol (stabilizer), ethyl-4-dimethylaminobenzoate, camphorquinone, functionalized amorphous silica, *t*-butanol
Etch-n-rinse 2 step	Prime& Bond NT	1		25	25, 60	Di-and tri-methacrylate resins, functionalized amorphous silica, PENTA, photoinitiators, stabilizers, cetylamine hydrofluoride, acetone

Table 8.3 (continued)

Classification	Adhesive	pH	Classification (based on pH)	Bond strength reference study	Microleakage reference study	Composition
Etch-n-rinse 2 step	Optibond	1		–	58	Ethyl alcohol, Alkyl dimethacrylate resins, Barium aluminoborosilicate glass, Fumed silica (silicon dioxide), Sodium hexafluorosilicate
Etch-n-rinse 2 step	OptiBond Solo Plus	1		–	67	Ethyl alcohol, Alkyl dimethacrylate resins, Barium aluminoborosilicate glass, fumed silica (silicon dioxide), sodium hexafluorosilicate
Etch-n-rinse 2 step	Adper Single Bond Plus/Single Bond Plus	1		47, 53a	53b	BisGMA, HEMA, DMA, ethanol, water, nanofiller, a novel photoinitiator system, a methacrylate functional copolymer of polyacrylic and polyitaconic acids
Etch-n-rinse 2 step	One Coat Bond	1		38	–	HEMA, hydroxypropyl methacrylate, glycerol dimethacrylate, polyalkenoate methacrylate, UDMA, amorphous silica
Self-etch 2 step	Clearfil SE bond	2.0	Moderate	37, 52	57, 59	Self-etching primer: 10-MDP, HEMA, hydrophilic DMA, di-camphorquinone, *N*,*N*-diethanol-*p*-toluidine, water Adhesive resin: MDP, Bis-GMA, HEMA, hydrophobic DMA, di-camphorquinone, *N*,*N*-diethanol-*p*-toluidine, silanated colloidal silica
Self-etch 2 step	AdheSE	2.5–3.0	Mild	39	57	Self-etching primer: 10-MDP, HEMA, hydrophilic DMA, di-camphorquinone, *N*,*N*-diethanol-*p*-toluidine, water Adhesive resin: water, alcohol, HEMA, BisGMA, DMA, copolymers of polyacrylic acids and polyitaconic acid
Self-etch-1 step	One-Up Bond F	1.3	Moderate	–	55	Water, methyl methacrylate, HEMA, coumarin dye, methacryloyloxyalkyl acid phosphate, MAC-10, multifunctional methacrylic monomer, fluoroaluminosilicate glass, photoinitiator (aryl borate catalyst)
Self-etch 2 step	Prompt L-Pop	0.9–1.0	Aggressive	36	70	Methacrylated phosphoric esters, Bis-GMA, initiator, stabilizer, 2-HEMA, polyalkenoic acid, water
Self-etch 2 step	Adper SE Plus Bond (SE)	1.5	Moderate	51	64	Liquid A: water, HEMA, surfactant, pink colorant. Liquid B: UDMA, TEGDMA, TMPTMA, HEMA phosphates, MHP, bonded zirconia nanofiller, initiator system based on camphorquinone
Self-etch 2 step	Clearfil Protect Bond	1.9	Moderate	50	64	Primer: 5% MDPB, MDP, hydrophilic DMA, HEMA, water, photoinitiators Bond: MDP, bis-GMA, HEMA, DL-camphorquinone, NaF, silanated colloidal silica
Self-etch 2step	Xeno III	<1.0	Aggressive	36, 37	–	Liquid A: HEMA Purified water, ethanol, BHT, highly dispersed silicon dioxide Liquid B: Pyro-EMA, PEM-F, UDMA, BHT, camphorquinone, ethyl-4-dimethylamini benzonate

(continued)

8

Table 8.3 (continued)

Classification	Adhesive	pH	Classification (based on pH)	Bond strength reference study	Microleakage reference study	Composition
Self-etch 2 step	Silorane Adhesive System	2.7	Mild	49, 53a	53b, 64	Primer: phosphorylated methacrylates, Vitrebond copolymer, bis-GMA, HEMA, water/ethanol solv., silane-treated silica fillers Bond: phosphorylated methacrylates, hydrophobic DMA, TEGDMA, silane-treated silica fillers
Self-etch 2 step	MegaBond system	2.5	Mild	–	69	Primer: Water, MDP, HEMA, hydrophilic DMA, CQ, DET Adhesive: Bis-GMA, MDP, HEMA, Hydrophobic DMA, CQ, DET, silanated colloidal silica
Self-etch 1 step	Xeno V	<2.0	Moderate	–	62 (enamel), (dentin)	Bifunctional acrylic amides, acidic acrylic amide, functionalized phosphoric acid ester, acrylic acid, water, tertiary butanol, initiator, stabilizer
Self-etch 1 step	AdheSE One	1.5	Moderate	–	62 (enamel, dentin)	Derivatives of bis-acrylamide, water, bis-methacrylamide dihydrogen phosphate, amino acid acrylamide, hydroxyalkyl methacrylamide, highly dispersed silicon dioxide, catalysts, and stabilizers
Self-etch 1 step	Clearfil S3 Bond/Tri-S Bond (S3)	2.7	Mild	50, 51	62, 66	MDP, HEMA, Bis-GMA, hydrophobic DMA, DL-camphorquinone, ethanol, colloidal silica, water, initiators, accelerators
Self-etch 1 step	One Coat 7.0	2.0	Moderate	–	64	UDMA, TEGDMA, HEMA phosphates, Initiator system based on camphorquinone
Self-etch 1 step	Xeno Select	<2.0	Moderate	–	66	Bifunctional acryl resin with amide functions, acryloylamino alkylsulfonic acid, "inverse" functionalized phosphoric acid ester, camphorquinone, coinitiator, butylated benzenediol, water, tert-butanol
Self-etch 1 step	Clearfil s3 Bond Plus	2.7	Mild	–	63	Bis-GMA, HEMA, ethanol, sodium fluoride, 10-MDP, hydrophilic aliphatic DMA, hydrophobic aliphatic methacrylate, colloidal silica, DL-camphorquinone, accelerators, initiators, water
Self-etch 1 step	Futurabond	2	Moderate	–	66	Organic acids, bis-GMA, HEMA, TMPTMA, camphorquinone, amines, BHT, fluorides, nanofillers, ethanol
Self-etch 2 bottle-1 step	Etch and Prime 3.0	0.76	Aggressive	48	–	Primer/catalyst: pyrophosphate, HEMA, initiators, stabilizers Bond: HEMA, ethanol, distilled water, stabilizer
Self-etch 1 step	iBond	2.2	Mild	37	–	Acetone/water based, fillers <1%, methacrylates

Mild >2, moderate 1–2, aggressive <1, *HEMA* hydroxyethylmethacrylate, *GPDM* glycero-phosphate dimethacrylate, *PAMM*mono (2-metacryloxy ethyl) phthalate, *TEGDMA* triethylene glycol dimethacrylate, *UDMA* urethane dimethacrylate, *Bis-GMA* bisphenol A-glycidyl methacrylate, *PENTA* dipentaerythritol pentaacrylate phosphate, *DMA* dimethacrylate, *MDP* methacryloyloxydecyl dihydrogen phosphate, *TMPTMA* trimethylolpropane trimethacrylate, *MHP* methacrylated phosphates, *MDPB* methacryloyloxydodecylpyridinium bromide, *Pyro-EMA* phosphoric acid modified methacrylate, *BHT* butylated hydroxytoluene, *PEM-F* monofluoro phosphazene modified methacrylate, *CQ* camphorquinone, *DET* diethyl tartrate deeper into dentin are likely to show higher degrees of nanoleakage. The clinical significance of nanoleakage definitely indicates the inability of the adhesive resin to completely infiltrate the demineralized dentin, thereby leaving behind pores and spaces. A poorly infiltrated hybrid layer contains voids that predispose to accumulation of water and oral fluids which may accelerate the degradation of the bond. When demineralized dentin is fully infiltrated with resin, its modulus of elasticity is much higher than the values of the original demineralized dentin. If the hybrid layer fails to be penetrated completely, its modulus of elasticity might fall in between that of adhesive resin and the un-infiltrated dentin collagen [83]. The larger marginal gap is usually located on Class V gingival side and on the external edge of the Class II gingival margin (V-shaped gap). This is due to a lesser capacity of the dentin sublayer and of cement at tooth neck to favor strong bonding with the resin by means of an adhesive system [60, 84, 85]. A width gap below about 1 μm does not allow bacterial infiltration, but it may allow the spreading of toxins and of other tooth potentially dangerous bacteria-related substances (nanoleakage). When the cervical margin is located on the limit line between root dentin and cement, the leakage problem becomes more relevant because adhesive systems become less efficient at the level of these substrates vis-à-vis when they are used on the enamel. The bonding process to dentin is much more technique-sensitive and substrate-sensitive. The dentin substrate varies in microstructure in superficial and deep layers; there is difference in micromorphology of crown and root dentin with respect to number of dentinal tubules; altogether these factors influence adhesion [86, 87]. The ability of adhesive systems to bind to hybridized cementum must be discussed. "Cervical margin leakage can be correlated to the absence of dentin tubules in 100 μm within the cervical border itself, to the relatively reduced number of tubules in the first 200–300 μm of the gingival floor in the cavity, and to the mainly organic nature of the gingival substrate" [88]. When present in the cervical margin, enamel is usually thin, a-prismatic, and less receptive to bonding. When polymerized, composite resin shrinks toward the upper adhesion site of the occlusal cavity margin, while it gets far apart from the weakest adhesion placed at the gingival margin level. Further, the micromorphology of the prepared tooth surface after cavity preparation plays a key role in predicting the adhesive bond [89]. The bond strength of the adhesive system and microleakage are the major factors for the success of restorations. Shear stresses are the main reason for resin-tooth bond failure in clinical situations, therefore researchers have suggested shear bond strength as a viable method of assessment of clinical performance. Seal at restoration margins is essential for longevity of restorations and is assessed by testing microleakage or marginal gaps and cracks. Stresses are generated at tooth-resin interface due to polymerization shrinkage, repetitive masticatory load, or temperature fluctuations, thus it is imperative to include a thermocycling protocol in in vitro studies to simulate closely the in vivo condition [90]. Most of the studies in the current literature have used two-dimensional dye penetration test to assess microleakage which is the most common method used in in vitro studies due to low cost and ease of use. Three-dimensional method used in a study does not require sectioning of sample and provides objective reading. Variations were seen in types of dyes (methylene blue, silver nitrate, basic fuchsin, propylene glycol, acid red) used in various microleakage studies but this does not affect the results of studies [90].

▣ Figure 8.4a–g provides details of laser-assisted restorative treatment and placement of a Class III composite restoration at tooth #10.

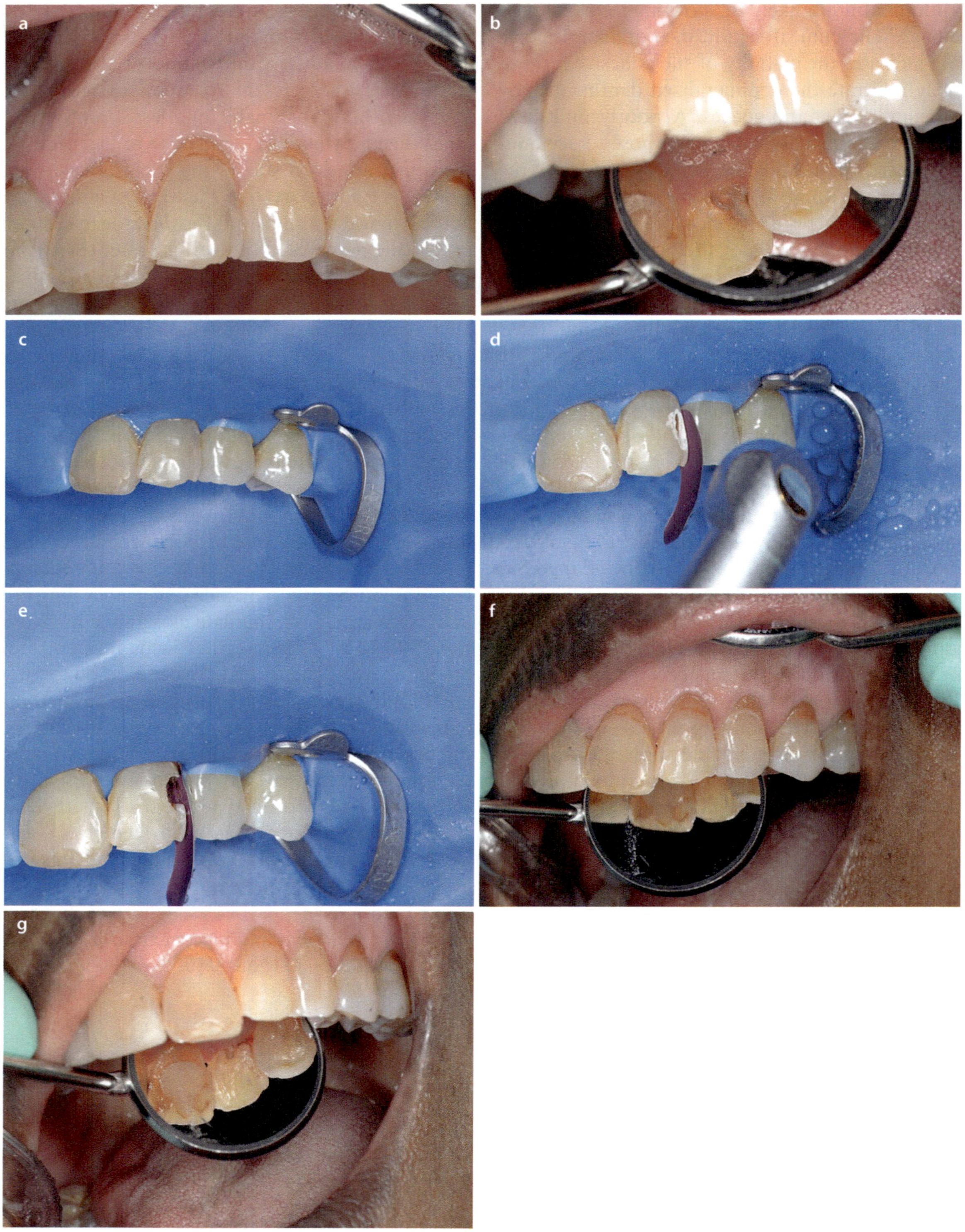

Fig. 8.4 **a** Carious lesion on distal side of tooth #10. **b** Palatal side view of carious lesion on tooth #10. **c** Dental dam positioned. **d** Preparation of a class III cavity in tooth #10 with Er:YAG laser (Pulsar with wavelength of 2940 nm by LAMBDA Dr. Smile, Italy). Enamel settings: tipless (Boost) handpiece, 900 μm, 5.5 W, 20 Hz, 275 mJ E/pulse, peak power 1719 W, average power density 902 W/cm^2, peak power density 281.897 W/cm^2, total energy 990 J, pulse width 160 μs, 7 mm tip-to-tissue distance, 70% water (24 mL/min), 100% air, total treatment time 180 s. Dentin and smear layer settings: 900 μm, 2 W, 15 Hz, 133 mJ E/pulse, peak power 831 W, average power density 323 W/cm^2, peak power density 136.336 W/cm^2, total energy 239 J, pulse width 160 μs, 7 mm tip-to-tissue distance, 50% water (18 mL/min), 80% air, total treatment time 120 s. **e** At the end of the cavity preparation, enamel margins are rectified and beveled. Then, etching, adhesion and composite stratification are performed (acid etching with orthophosphoric acid 37%, OptiBond FL total-etch adhesive system (Kerr, Orange, CA, USA), composite material (IPS Empress Direct Ivoclar Vivadent). **f** Composite reconstruction is completed (buccal view). **g** Palatal view. (Procedure by Dr. Riccardo Poli)

8.9 How to Increase Adhesion

The adhesion and microleakage values depend on the type of adhesive system and etching protocol used. There have been continuous efforts made to improve the adhesion of restorative material to enamel and dentin, making it more user-friendly and less technique-sensitive through alterations in chemical composition of adhesives and composites. Enamel etching results in three different micro morphologic patterns [91]. The type I pattern involves the dissolution of prism cores without dissolution of prism peripheries. The type II etching pattern is the opposite of type I: the peripheral enamel is dissolved, but the cores are left intact. Type III etching is less distinct than the other two patterns. It includes areas that resemble the other patterns and areas whose topography is not related to enamel prism morphology. An etching time of 60 s originally was recommended for permanent enamel using 30–40% phosphoric acid. Although a few studies have concluded that shorter etch times resulted in lower bond strengths, other studies using scanning electron microscopy (SEM) showed that a 15-s etch resulted in a similar surface roughness as that provided by a 60-s etch [89]. Other in vitro studies have shown similar bond strengths and leakage for etching times of 15 and 60 s [92, 93]. Most recent clinical recommendations advise enamel etching not exceeding 30 s and a very limited acid treatment on dentin (up to 15 s). As measured in the laboratory, shear bond strengths of composite to phosphoric acid-etched enamel usually exceed 20 megapascals (MPa) and can range up to over 50 MPa, depending on the test method used. Such bond strengths provide adequate retention for a broad variety of procedures and prevent leakage around enamel margins of restorations [94]. Studies of Nakabayashi et al. [95] have shown the importance of hybrid layer in achieving a sufficient bond to dentin. This involves exposure of collagen fibrils of dentin by dissolution of its mineral phase followed by infiltration with resin monomers and polymerization in situ. The formation of hybrid layer and its characteristics depend on the substrate morphological features [96]. Thus, bonding systems have evolved over years to have a better control over formation of hybrid layer according to the substrate and can be broadly classified into etch-n-rinse and self-etch adhesives which are further classified as: Etch-and-rinse 3-step (fourth generation), etch-and-rinse 2-step (fifth generation), self-etch adhesives 2-step (fifth generation), self-etch adhesives-1 step-2 bottle (sixth generation), and self-etch adhesives-1 step-1 bottle (seventh generation) [97]. Two strategies are used to overcome the low attachment strengths of the smear layer:

- Removal of the smear layer prior to bonding (***etch-and-rinse approach***)
- Use of bonding agents that can penetrate the smear layer and incorporate it into the hybrid layer (***self-etch approach***)

Conventional etching technique uses 35% phosphoric acid (pH 1.3) on both enamel and dentin in a total-etch approach, while self-etch technique involves a primer which has a less acidic (pH 1.9) resin monomer and does not require rinsing-off of the tooth with water as done in total-etch technique. The presence of thick smear layers may buffer the acidic monomers of self-etch adhesives, thus limiting the depth of resin infiltration in dentin. The manufacturers have attempted to alter the concentration of acidic resin monomers to create deeper hybrid layers and better bond at the tooth-adhesive interface. The self-etch adhesives are classified based upon the acidity of their monomer as strong (pH <1), moderate (pH 1–2), and mild (pH >2) (◘ Table 8.3). Some authors have shown a formation of thinner hybrid layer with etch and rinse adhesives on erbium laser ablated dentin when compared to acid-etched bur-cut dentin. This may be due to the possible increase of calcium and phosphorus since organic components are selectively removed, denaturation of the collagen network, and decrease of the dentin permeability, or due to the formation of a layer which has collagen fibers that are poorly attached to the underlying dentin substrate. Further, the thermal effects of laser irradiation may also extend into the dentin subsurface, thus impairing the inter-diffusion zone formation [98, 99]. Lately the focus has drifted toward investigating different types of self-etch adhesives and their effects on hybrid layer formation. In order to obtain better bonding conditions and facilitate monomers' spreading within the demineralized intertubular dentin, which was altered by laser irradiation, different post-irradiation dentin pre-treatments have been suggested for adhesion procedure.

Among them, we point out the use of:

- Sodium hypochlorite at a concentration ranging between 5% and 10%
- Orthophosphoric acid at 33–38% with an extended etching time [100]
- Polyacrylic acid (for glass ionomer material)
- Chlorhexidine gluconate
- Propolis
- Peroxide
- Ozone gas

Sodium hypochlorite can be used to remove collagen fiber frustules and dentin fragments modified by laser interaction. In such a way, following its use, we obtain a clean surface, free from the alterations produced during laser use (even if, thanks to the erbium laser, as we have already flagged out, there is no smear layer). The extension of the etching time by orthophosphoric acid apparently does not promote better adhesion, but on the contrary, it can generate an excessively etched tooth surface. It is appropriate to consider that the irradiated tooth surface does not have smear layer, because the erbium laser does not produce it, unlike what happens when using the high-speed hand piece and the diamond bur. For this reason, by performing the etching on hard tooth tissues, we obtain an immediate contact between acid and intra- and peritubular dentin. An excessive contact between acid and tubules could, on the contrary of what we would desire, completely destroy the dentin architecture favorable adhesion. There is no certain clinical proof that the different pre-treatments listed here could improve the action of adhesive systems for composites. According to Arslan et al. [101], "No adverse effect of different cavity disinfectants on microleakage were found when etch-and-rinse adhesive system was used."

8.10 Why Adhesion Can Be Impaired

There are different possible explanations on why the composite adhesion strength to the irradiated dentin could be lower than the one achieved through phosphoric acid. The surface of cavity created by Erbium laser is different from that created by conventional method of rotary burs. Different researchers believe that the main mechanism causing insufficient bonding between irradiated dentin and composite is the collapse and/or melting of collagen fiber network during laser excision [102]. In fact, the considerable increase of temperature following irradiation causing the instantaneous vaporization of the water component of the mineralized tooth matrix and of collagen fibers, initially spread and supported in this framework, tends to collapse because they are no longer supported by the crystalline structure. The consequence will be a reduction of bonding spreading within the network because the interfibrillar structure is reduced [102]. The substrate is free of debris and smear layer with micro-irregularities. There is loss of prismatic structure of enamel while dentinal surface exhibits open dentinal tubules with more prominent peritubular dentin than intertubular dentin [54, 103]. Thus, the hybrid layer will not be of optimal quality for adhesive procedures [102]. The main cause of reduced bond strengths with laser cavities in earlier literature has been due to the use of single component adhesive without etching with phosphoric acid [36, 53, 78, 104]. Erbium lasers are more effective on peritubular dentin due to higher water content but do not enlarge the dentinal tubules. Transverse electron microscopic analysis of erbium laser-irradiated dentin has shown that thermal effects extend till deeper layers of dentin resulting in altered dentin subsurface of approximately 3–4 μm thickness. There is a reduction in carbon to phosphorus ratio and formation of more stable compounds. The collagen fibrils beneath this layer are fused and do not demonstrate cross-banding which causes reduction or complete elimination of interfibrillar spaces [60]. Ablation of dentin melts collagen fibrils together, resulting in a lack of interfibrillar space that restricts resin diffusion into the subsurface of intertubular dentin, causing a lack of penetration of the resin and even a possible peeling off of the resin layer from the ablated dentin surface [105–107]. Erbium lasers used with excessive parameters can furthermore have a harmful effect on hard tissues. The strength of resin-dentin bond in erbium laser prepared cavities is due to the micromechanical retention through formation of resin tags into demineralized dentin [108]. The use of phosphoric acid etching allows partial demineralization of highly mineralized peritubular dentin and enlargement of dentinal tubules orifices and removes the weak laser modified layer. This promotes resin infiltration by increasing the surface area and formation of funnel shaped resin tags where enlarged part is present at the tubule orifices [53]. Bertrand et al. [53] have shown in their study the formation of surface irregularities and open dentinal tubules on Er:YAG irradiated dentin without subsequent acid etching which could be favorable to composite resin bonding. On the contrary, the erbium laser is less effective on peritubular dentin forming a crater-like surface which may indicate a necessity for additional etching procedure to expose collagen fibrils for resin infiltration [103]. There are several parameters which may influence laser interaction with the dental hard tissues. Corona et al. [109] have described the influence of energy density, number of pulses per second, and pulse energy on dentin ablation. Besides this, time of irradiation, air/water cooling, tip distance from tissue also play an important role in determining the micromorphology following ablation with laser and subsequent effect on bonding to adhesives [110]. Excessive laser energy values can cause cracking in tooth dental tissues, surface melting, surface scaling and flacking, marked loss of intertubular dentin, and collagen melting [102, 105]. It has also been thought that pulses could generate intense elastic waves inside tooth hard tissues during excision as a result of the interaction with the laser beam and due to alternate thermal expansion and shrinkage. By occurring inside a hard and stiff tissue, stress waves could cause micro-cracking and fractures in the dentin thickness and at the dentin/composite interface level, negatively affecting adhesion strength [1, 107, 111]. Energy threshold of approximately 150–200 or 200–250 mJ is required for clinically effective ablation of permanent dentin and enamel, respectively [110]. The erbium laser is more widely studied for the surface treatments at sub ablative energy thresholds as an alternative to conventional etching protocol where it causes modification in microstructure of enamel and dentin similar to that achieved with application of orthophosphoric acid. The term "laser etching/laser conditioning" has been used for this phenomenon [112, 113]. It has to be mentioned that laser etching is an insufficient term; laser conditioning is a more accurate term. Laser conditioning is a term used to describe a reduced ablative effect obtained with energy just above the threshold of ablation (15 J/cm^2 for Er:YAG and 20 J/cm^2 for Er;Cr:YSGG) [114]. While most of the literature shows a poor bond obtained after laser conditioning as compared to acid etching [60–62, 115], the results are contradictory for the adhesive protocol for dentin ablated with erbium lasers for cavity preparation. The authors have reported formation of fissured surface with water deficient non-apatite phases that may be loosely adherent to underlying unaltered dentin surface which necessitates etching before restoration. Koiliniotou-Koumpia et al. [116] and Brulat et al. [54] found no improvement in adhesion values of self-etch adhesives on erbium laser ablated dentin while Lee et al. [102] and Gurgan et al. [117] have stated comparable bond strengths in erbium laser prepared cavities in relation with conventional bur-prepared cavities. An additional explanation of the weaker bond between composite and dentin is represented by the deep craters that are created when laser energy is high: these valleys/hollows can prevent the optimal adaptation of the reconstruction material to the cavity walls since the resin would not be able to fill deeper concavities [118]. Furthermore, there could be an uneven distribution of the masticatory stress at adhesive-dentin interface [102, 119, 120]. Additionally, Dunn et al. [61] underlined that "laser irradiation of enamel surfaces produced surface fissures and a union or blending of a distinctive etch pattern normally seen in acid-etched enamel." This blending effect likely prevented the penetration of resin into enamel, resulting in lower enamel bond strength values. Er:YAG laser irradiation can obstruct the dentinal tubules and increase the dentin surface roughness which can cause microbial plaque retention and increase the risk of caries and periodontal disease [4]. It is very important that at the end of cavity preparation, unsupported enamel margins are removed, and margins are smoothed. This operation can be done at low power and high-speed Er:YAG or Er,Cr:YSGG laser, or with hand tools (enamel cutter, excavators) or low or high-speed tools with diamond or lamellar burs, or rubber tips [121]. Omitting this step may result in a diminished wall strength; it may cause incomplete adaptation of the composite to the preparation margin, and a subsequent chipping of the reconstruction margin and/or the enamel subjected to the masticatory load with subsequent microleakage. Recent research took into account the possibility that during irradiation by erbium laser, calcium phosphate insoluble molecules could be formed, which would prevent optimal composite adhesion [102]. On the other hand, authors believe that collagen denaturation during ablation causes an acid-resistant surface containing charred granular structures or structures covered with melted dentin particles. This denaturation could jeopardize infiltration of the adhesive system into the dentin structure, and it could prevent the creation of the hybrid layer [107, 122]. Such phenomena could concern cement during Class II and V cavity preparation, because "when cement is reached by erbium irradiation, it is altered, and a thin layer (5.7 μm) is formed. This can hamper hybridization because it becomes less affected by acid etching" [123]. If the resin cannot efficiently infiltrate into intra- and peritubular dentin, only shorter resin tags will be developed, without funnel shaped morphology and lateral resin projections and this would entail a damage to the resulting adhesion [53, 124, 125]. When using this type of laser, the absence of the smear layer, which is instead inevitably created during cavity preparation with burs, allows the immediate exposure of dentinal tubules and accentuates their permeability to dentin adhesives. Furthermore, the absence of smear plugs allows the passage of intra-tubular fluids to and from the pulp [125, 126]. "Loss of smear layer due to laser irradiation exposes the dentinal tubules and enhances the permeability of dentin adhesives. Intrinsic dentin wetness, as affected by pulpal pressure, could also affect the hydration state of dentin and the bond strength to dentin adhesives. Laser affects fluid perfusion of dentin more than bur" [126]. It is important to take into account the fact that greater perfusion could make dentin moister and for that reason it may interfere with some adhesive systems, especially water-based ones which could end up being more diluted. The best way to avoid or minimize the impact of these surface alterations which may cause difficulties to achieve optimal bonding is to reduce the laser energy for ablation to the lowest efficient level, compatibly with the time required to completely remove the decay. Many authors [127] agree on the fact that after irradiation it is preferable to perform enamel acid etching by orthophosphoric acid in order to obtain an even micro-rough surface. Enamel laser conditioning, on the other hand, would not be useful. The use of an acid on the dentin could be positive as it would allow the removal of the top layer altered by the erbium laser and exposes the network of collagen fibers which make up the ideal matrix required to create the hybrid layer for the bonding process. However, researchers' opinions are quite contrasting. Thus, although different acid application times are suggested, it would be appropriate not to exceed 30 s of contact time on the enamel and 15 s on irradiated dentin. It is advisable to remember that since the smear layer is absent, phosphoric acid acts more rapidly on the mineralized crystalline structure of hard tooth tissues, in particular on peritubular and intertubular dentin, and on collagen fibers. Other authors [128] advise to etch irradiated enamel for max 15 s and to avoid acid treatment on dentin at all.

8.11 Adhesive Systems for Irradiated Hard Tissues

An extensive discussion has prompted the opportunity of completely eliminating the smear layer (etch-and-rinse technique, total etch) or to modify it through suitable self-etch adhesive systems which would only remove one part of it and then maintain and exploit the remaining part of it to create suitable substrate for bonding [111, 129–141]. With regard to adhesion between composite materials and irradiated dentin, research is still debating if it would be possible to obtain optimal bonding through an Etch and Rinse or through self-etch adhesive systems. The results described in the most recent literature are extremely contrasting and contradictory. These adhesive systems were developed keeping in mind the surface micromorphology created by conventional rotary instruments. De Moor and Delmé [1] concluded that acid etching with phosphoric acid cannot be ignored on laser-irradiated enamel or dentin with respect to bond strength and marginal seal. There was no value-added benefit of laser conditioning of the enamel after laser preparation. However, the data on dentin bonding protocol with currently available adhesives systems on the cavity surfaces created by erbium lasers is still inconclusive. Truly diverse results have been found in the existing literature regarding the adhesive protocols used with erbium laser to achieve bond strength and marginal integrity comparable to the conventional method. This could be due to the difference in pH and compositions of adhesives, difference in substrate (enamel or dentin, superficial or deep dentin), laser parameters, type of wavelength, type of composite used and the method of bond tests (shear bond or tensile bond or microtensile bond), and methods of microleakage assessment. pH of adhesives plays a critical role in achieving predictable bond strength and marginal integrity if laser parameters are used within a particular range [142]. Varying laser parameters have been used in systematic review by Dua et al. [142]. Some studies have used very short pulse 100 μs with energy/pulse of 160 or 200 or 250 or 300 mJ [85, 104, 143]; some have used longer pulse durations 400 μs with energy/pulse of 200 or 350 or 500 mJ [53, 54, 144], while others have used very long pulse durations of up to 500 μs with energy/pulse of 250 or 450 mJ [136, 145]. Nishimoto et al. [146] demonstrated that increasing the pulse duration increases the depth of penetration of laser into dentin and decreases the radius of cavities. It has been shown that very long pulses have similar effect on surface morphology of dentin as that of conventional rotary drill demonstrating thick and regular smear layer, which makes a relatively stronger etching protocol imperative to remove or modify smear layer [147]. Long pulse duration of 400 μs or up to 500 μs has shown bond strength and marginal integrity in laser prepared cavities comparable or better than conventional cavities when only etch and rinse or moderate self-etch adhesives have been used. Roebuck et al. [118] have demonstrated that a deeper crater pattern is seen in erbium laser prepared cavities with high energy per pulse. These valleys/hollows can prevent the optimal adaptation of the reconstruction material to the cavity walls since the resin would not be able to fill deeper concavities [118]. Er:YAG and Er,Cr:YSGG with wavelengths of 2940 and 2780 nm, respectively, have high absorption in water and hydroxyapatite. The Er:YAG has its peak absorption in water while the Er,Cr:YSGG wavelength has its absorption slightly more in apatite mineral than in water. Both the lasers create similar surface characteristics, however since Er,Cr:YSGG has more absorption in hydroxyapatite, it can cause modification in intertubular dentin making it more acid resistant. Also, this can produce higher surface temperature despite the external water cooling which may affect the adhesion [148–150]. The evaluation of bonding of the FLD to dentin showed that the combined use of Er:YAG and Er,Cr:YSGG lasers with SBU-Single Bond Universal and AEO-Adper Easy One on dentin surfaces improved the dentinal bond strength of the FLD-Fuso Liquid Dentin [150]. Most of the mild, moderate, and aggressive adhesives have shown comparable or less microleakage in laser group [142]. 1 mild adhesive and 2 moderate adhesives have shown microleakage higher in laser group [151, 152]. The adhesives used were Clearfil tri-S (water and ethanol based), AdheSE One (water based), and Xeno V (water and ethanol based), respectively. The self-etch adhesives combine etching and priming steps. The etching solubilized calcium and phosphate ions from mineral component of dentin and suspends them in primer. The primer may be ethanol/acetone based or water and ethanol based. The low pH of self-etch adhesives causes enlargement of dentinal tubules by dissolution of calcium and phosphates from peritubular dentin, and the air drying of primer causes evaporation of volatile solvents, but there is no step of rinsing, thus increasing the concentration of calcium and phosphates which may limit the infiltration depth of the resin accounting for lower bond strengths with some self-etch adhesives (Xeno III—water and ethanol). The limited effectiveness of mild acid monomers is due to the insufficient demineralization of the superficial laser modified dentin that contain more stable surface and the subsurface which exhibit fused collagen fibrils [148, 153]. iBond is a mild adhesive but has low filler content which is said to affect the bond strength to dentin especially in laser prepared surfaces due to the absence of smear layer. Tay et al. [154] have suggested that inclusion of globular units of smear layer residues in polymerized resin acts as true mineral filler fillers. 10-MDP containing moderate self-etch adhesives Clearfil SE Bond and Clearfil Protect Bond have shown comparable or better bond strengths and microleakage in laser prepared cavities in all the studies where they were assessed [54, 104, 136, 155, 156]. This functional monomer 10-methacryloyloxydecyl dihydrogen phosphate causes minimal dissolution of smear plugs and limits the opening of tubules. This reduces dentin permeability, facilitates penetration, impregnation, polymerization, and entanglement of monomers with demineralized dentin to form a relatively thick layer. It has hydroxyl groups which may form ionic bond with calcium in enamel and dentin [157]. The all-in-one nano-filled self-etch adhesive Clearfil tri-S Bond creates a nano-interaction zone through ionic bond with the apatite of tooth structure. They do not diffuse into the deeper layers of dentin which exhibit fused collagen fibrils and are loosely attached to the underlying dentin surface. This prevents the negative effect on the shear bond strength in laser prepared cavities [136]. Some studies have shown no significant difference between etch and rinse and self-etch adhesives on bond strength and marginal integrity of laser prepared cavities. The aggressive self-etch adhesive and etch-n-rinse adhesive create a similar surface for micromechanical retention and form a thicker hybrid layer in lased dentin [158]. One study [159] in the systematic review by Dua et al. compared the effect of self-etch adhesive and selective etching on laser prepared cavities but found no significant difference. The varying results in different studies with same adhesives could be due to difference in laser parameters used or the difference in bonding substrate. Several studies [54, 147] have used Scotchbond MP for bonding laser and bur-prepared cavities but in one research [113] the microleakage was more in laser group. On the contrary, another study [147] had shown microleakage similar in laser and bur group. This may be because of additional beveling of enamel margins in study [147]. One study [160] has mechanically excavated laser prepared cavity before etching and bonding with an etch and rinse adhesive. It was suggested that this may modify enamel and dentin surface by eliminating excessive roughness of enamel margins and remove the loosely attached, poorly crystallized dentin subsurface and create some smear layer for improved adhesion. A different research [116] has analyzed cavities prepared in both superficial and deep dentin. In deep dentin, the bond strengths were significantly lower in laser treated group with both etch-and-rinse and self-etch adhesives as compared to bur-cut dentin. In superficial dentin—the bond strengths were significantly lower in laser-treated group bonded with etch-and-rinse adhesives while it was similar to that in bur-cut dentin when bonded with self-etch adhesives.

▣ Figure 8.5a–g depicts a laser-adjunctive restorative procedure in treating a disto-occlusal carious lesion at tooth #12 and placement of a composite restoration.

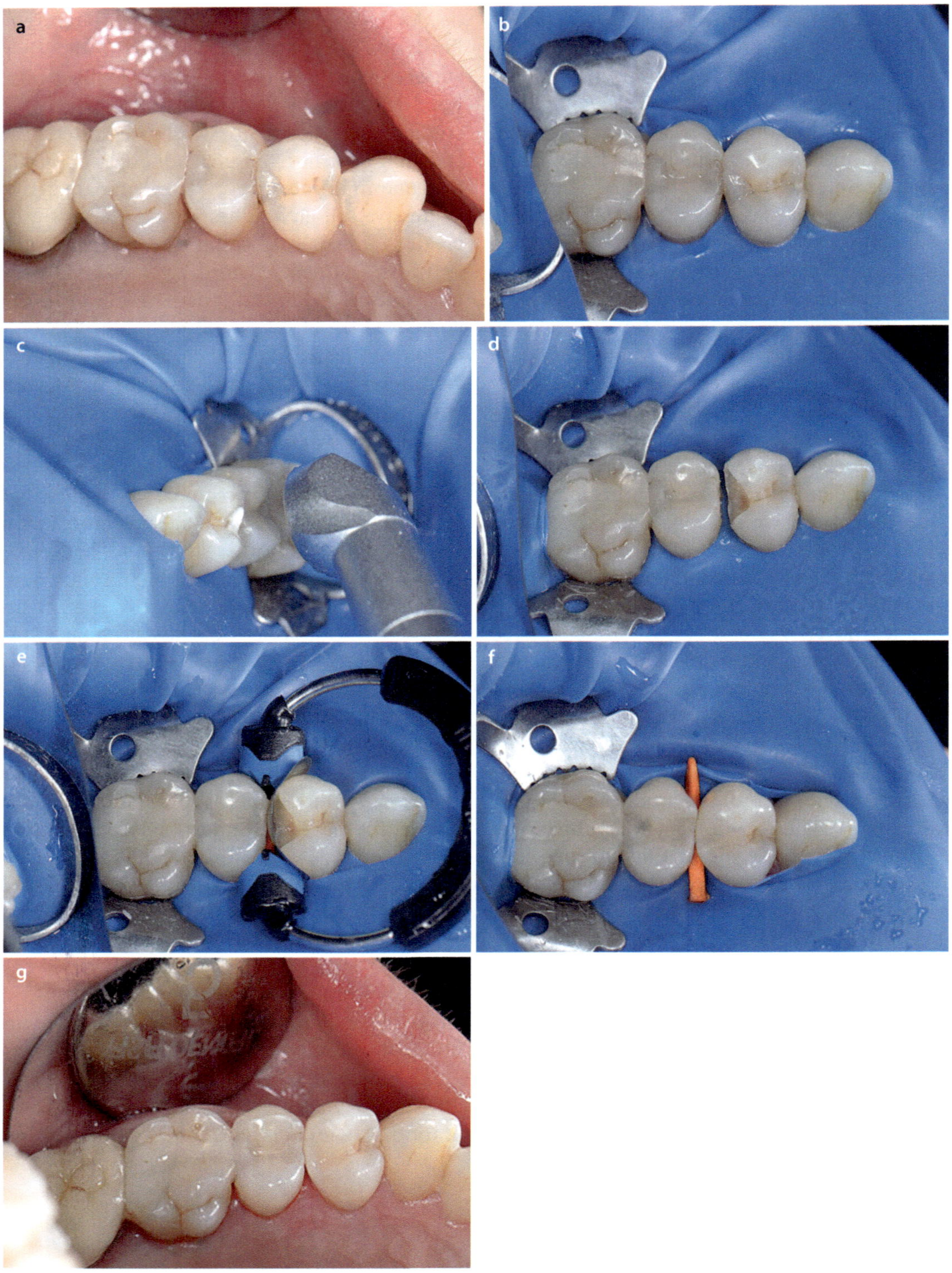

Fig. 8.5 **a** Occlusal view with distal carious lesion involving enamel and dentin on tooth #12. **b** Dental rubber dam isolation. **c** Cavity preparation with Er:YAG laser (Pulsar with wavelength of 2940 nm by LAMBDA Dr. Smile, Italy). Enamel settings: tipless (Boost) handpiece, 900 μm, 3.8 W, 15 Hz, 250 mJ E/pulse, peak power 1563 W, average power density 589 W/cm^2, peak power density 245.610 W/cm^2, total energy 675 J, pulse width 160 μs, 7 mm tip-to-tissue distance, 70% water (24 mL/min), 100% air, total treatment time 180 s. Dentin and smear layer settings: 900 μm, 2 W, 15 Hz, 133 mJ E/pulse, peak power 831 W, average power density 314 W/cm^2, peak power density 130.664 W/cm^2, total energy 359 J, pulse width 160 μs, 7 mm tip-to-tissue distance, 50% water (18 mL/min), 80% air, total treatment time 120 s. **d** Detail of cavity after completed preparation. **e** Rectification and beveling of enamel margins. Positioning of matrix, wedge, and hook. Acid etching, adhesive, and composite stratification steps (acid etching with orthophosphoric acid 37%, OptiBond FL total-etch adhesive system (Kerr, Orange, CA, USA), composite material (IPS Empress Direct Ivoclar Vivadent). **f** Composite reconstruction and polishing completed. **g** View after dental dam removal of composite reconstruction on tooth #12. (Procedure by Dr. Riccardo Poli)

8.12 Decontamination Effect

One big advantage in laser dentistry is represented by the decontamination effect of tissues. Also, during a restorative treatment, the operator performs dentin disinfection by vaporizing water of bacteria (bactericidal action), thus decontaminating the cavity [161–164]. As affirmed by Hibst et al., "*Bacteria below surface are killed during laser cavity preparation to a depth of 300–400 μm*" [165]. This means that the hard tissues treated with this wavelength achieve an important cariogenic microorganism count reduction in the irradiated layers, which is statistically significant when compared to the conventional bur excavation. This is due to Er:YAG laser irradiation itself which causes structural changes in bacteria and the following cavitation effect can cause cellular wall disruption. Considering microbial growth in permanent teeth affected by caries, it was found that microbial reduction following Er:YAG treatments was 91%, while a microbial clearance rate of 80.6% was achieved following conventional preparation [166].

8.13 Effect on Tissue Temperature

The energy generated by the erbium laser can be so powerful to breakdown the crystalline structure of hard tissues of the human body; however, if the energy levels do not excessively exceed ablation threshold limits, they can be much less aggressive than the diamond bur used on a high-speed drill. The vibration and pressure developed by using a traditional technique can very easily create micro fractures that branch out on the prepared decayed cavity walls. These will give rise to sensitivity, pain to heat/cold stimulation, risk of pulp infiltration damages, and secondary decay, causing reconstruction failure. If the erbium laser is used with the adequate selected parameters (i.e., low energy and frequency), sufficient enough to obtain decay ablation without creating any trauma inside the tissues, cracking will not occur. The very high water absorption coefficient for the two erbium laser wavelengths allows to limit the penetration of the beam by just a few microns (7 μm in the enamel and 5 μm in the dentin for Er:YAG 2940 nm, 21 μm in the enamel, and 15 μm in the dentin for the Er,Cr:YSGG 2780 nm) [7–11]. This limited penetration, especially if combined to a very short pulse width, allows a very limited transfer of heat into tissues. With optimal cooling made by an integrated air/water spray, temperature increase at pulp level will be below 5 °C [103, 167–169]. On average, there is a temperature increase by 1–2 °C in the pulp chamber, while the use of a high-speed bur entails a more frequent potential heat damage, especially in cavities where the floor is in close proximity with a pulp horn. Amasyali et al. during erbium use on the contrary had no temperature increase but obtained its reduction by −2 °C in the Er:YAG laser group [170]. It is obviously indispensable to use energy levels compatible with efficient excision and without being excessively traumatic or harmful for the tooth architecture.

8.14 The Cooling

It is also equally important to use a cooling spray with a water amount and an air volume sufficient to remove the fragments created during irradiation and cool the treated surface quickly. The minimum amount of water which should be used is 8 mL/min, but it would be better if it could be doubled. Not all erbium lasers available on the market accurately show on the display the amount of used water. Often times the display only shows a percentage vis-à-vis the 100% capability that can hold in the hand piece. However, the maximum value depends on the pressure present in the local aqueduct water network or in the building where the dental practice is, depending on manufacturer's settings, and also depending on the setting of the individual laser entered by the installer. In order not to run the risk of using an insufficient amount, it is advisable that the operator personally measures how much water per minute is delivered by the hand piece in percentages of 10–20–50–100%. In this way, we can be aware of how much water is used, obviating overheating tissues and avoiding thermal damages; these in turn could cause tooth pulp necrosis, a phenomenon of dental hypersensitivity, or alter the tooth surface with subsequent worsening of composite adhesion.

8.15 The Welding Effect

It is also possible to select a cooling spray containing a reduced amount of water when more thermal interaction is required. This type of use, which can modify the tooth surface, is called "welding" and it may reduce dentinal sensitivity and can transform the outer tooth wall, especially at tooth neck level or in case of preparation of a fixed prosthesis, in order to be more resistant to acids produced by decay-inducing bacteria and less permeable. The micro-structural effect is represented by the obliteration of dentinal tubules by the melting of the dentin outer layer and coagulation of collagen fibers. This procedure must be performed with low energy levels and for very short treatment periods, otherwise it is possible to cause severe pulp damage due to temperature increase.

8.16 Laser Analgesia

Use at low energy level and power allows one of the most important advantages that can be obtained through these lasers in conservative dentistry: laser analgesia; erbium wavelengths allow cavity preparation also in deep dentin, without the need to perform local anesthesia by injection and without causing pain to the patient. This is possible in a wide variety of cases [171], and it is also very useful in pediatric dentistry, for phobic patients, for all those patients who don't like injections, and for those who are allergic to local anesthetics. Any dentist knows that the fear for needles, also called trypanophobia, discourages many patients from going to see a dentist [172]. Vibrations, pain, and noise perceived when using the bur or the drill contribute to worsening the fear which is very frequently associated with dental care. All of this may trigger anxiety before dental treatments. Apart from fear, the patient can report correlated psychosomatic symptoms (dyspnea, tachycardia, sense of suffocation or light head, etc.) which may involve the possibility of not treating the patient or cause real discomfort and emergencies in the patient chair. Furthermore, anxious patients counteract the treatment by refusing it or by not collaborating. The clinical situation and the symptomatology become further complicated if the subject is "dental phobic," as extreme anxiety toward dental care will grow exponentially. It is believed that dental phobia affects 4–16% of adults and 6.7–20% of children [173, 174]. Its incidence tends to lower with age, but it may persist among the elderly. Thus, anxious patients are treated with extreme difficulty. The absence of rotating instruments, with associated discomfort due to vibration and noise, and of local anesthesia can facilitate the interaction between patient and dentist. In this way, two important factors contributing to off anxiety are removed. Dentists must be able to identify and treat such patients in order to lower their anxiety level [175, 176].

8.17 Alternatives to Local Anesthesia for Cavity Preparation

What possible alternatives are available for a clinician to avoid the use of the two therapeutic options so much opposed by patients? The methods that can somehow substitute local anesthesia for pain control during dental care include techniques with different degrees of probability of success and different abilities of anxiety and pain attenuation or suppression. Possible therapeutic alternatives designed to minimize fear and anxiety toward traditional dental treatments include hypnosis, conscious sedation with a mixture of nitrous oxide and oxygen, electronic anesthesia or electro-stimulation, high absorption coefficient topical anesthesia, general anesthesia, conscious sedation with oral drugs or by intravenous injection, and by using the erbium laser. Each one of the above listed techniques has pros and cons. None of them has a 100% success rate to eliminate anxiety and to facilitate patient compliance. Unfortunately, none of them allows painless treatment, free from discomfort for all patients; furthermore, some of them could have side effects and/or potential risks. It has been known for decades (or better, for hundreds of years) that achieving hypnosis status may permit medical therapies, even very invasive ones (delivery, endoscopy, surgery) without any pain. In dentistry, for example, it is possible to perform wisdom teeth extraction without any pain whatsoever. Hypnosis can be considered more helpful for its calming potential and to improve patient compliance. However, not every patient reaches a sufficiently deep level of trance able to obtain hypnotic analgesia. Conscious sedation with nitrous oxide and oxygen is based on the inhalation of a mixture of nitrous oxide and oxygen gases in variable proportions, using a nose mask. This mix reduces anxiety; it has a euphoric effect, it is lightly analgesic, reduces tissue sensitivity, it gives a mild retroactive amnesia, a feeling of well-being and reduces the perception of time. With a customizable proportion of the two gases (on average: 20–50% nitrous oxide and 80–50% oxygen), and after 3–5 min it is possible to obtain the desired effect and maintain it for all the time needed. Discontinuation of the mixture administration and the delivery of 100% oxygen permit recovery within a few seconds. This system, however, cannot be used to obtain a true and complete analgesia. It can be used as an aid to traditional local anesthesia or for laser analgesia support [177–180]. Some other therapeutic options are not completely proven or verified (for example, different brands of electro-stimulation or electronic anesthesia) or have unpleasant side effects like some topical anesthetics with very high absorption coefficient, i.e., EMLA 5%, cream containing prilocaine and lidocaine (which give a feeling of numbness, need 15 min of waiting time before they take effect and are quite distasteful) or potentially harmful (general anesthesia, conscious sedation with drugs and/or intravenous injection) [181]. Dental lasers are not completely able to replace the traditional bur and it is not always possible to avoid injected anesthesia, but this technology is particularly useful in pediatric dentistry (above all for primary dentition), for phobic patients and for those who do not like traditional anesthesia due to the feeling of numbness it causes, or because they are intolerant to it. This can explain why the use of this "no shot" modality can be highly appreciated by patients, especially by the youngest.

8.18 How Laser Analgesia Works

The mechanism by which the laser analgesia can take place is not completely known [141, 175, 182–190]. Laser pulses may hamper the possibility for neurotransmission to reach the central nervous system, since the former lasts only microseconds while it needs milliseconds to be modulated by the brain (Gate Theory). This over-loading of the peripheral and of the central nervous systems can be due to a physiological saturation caused by the laser beam. It has been assumed also that laser irradiation on pulp C fibers may cause a reduction of the Na^+–K^+ pump action. Temporary nervous transmission suppression could occur. Actually, the opinion of researchers converges on the role played by *Low Level Laser Therapy* (LLLT) in preconditioning tissues, and this is likely to be responsible for the onset of the analgesic effect. It is highly plausible that the laser phototherapy action on pain is a combination of several factors [191, 192]. In general, lasers and particularly free running pulsed (FRP) lasers reduce the annoyance factor which is a combination of the pressure applied to the tooth, the vibrations and noise recorded through the bones of the skull, the heat and smell generated at the interface between tooth and cutting instrument, and the time taken to perform the task. The annoyance factor of FRP lasers is lower compared with rotary instruments because the former produces less vibration than turbine; there is a lack of tactile physical force applied to the tooth, a different noise, less heat, and less smell (if enough cooling is applied). The limited pulp temperature increase [193–196] seems to be a further element that limits the pain during and after the treatment. All these factors can lower patients' anxiety and elevate pain threshold, because they are able to limit psychological distress.

8.18.1 Photo-Acoustic Effect of Pulsed Lasers

According to recent researches [194, 197], the real, direct laser influence on pain neurotransmission seems to be the short pulse duration, responsible to create shock waves and considered able to induce a gate control of painful sensations at peripheral and central levels of nervous system. All wavelengths are also able to reduce in a direct way the nerve transmission acting on the mechanism of depolarization of nervous fiber. In hard tissues, the generation of laser-induced shock waves (or stress waves) is a well-recognized phenomenon [197]. Erbium-generated shock waves during hard tissue ablation arise from an explosion with volume expansion and subsequent ablative recoil. These shock waves are influenced by several factors including the speed of sound in the tissue, the tissue density, the pulse duration and its shape. These waves upon reaching the pulp cause pressure fluctuations within the pulp chamber. This proprioceptive input to the nervous system, if of sufficient intensity, could potentially induce a dampening of responses through the gate control pathway.

8.18.2 The Gate Control Pathway

FRP lasers have an effect on pain but researchers have yet to reach a definitive conclusion over its mechanism. There are several theories proposed to explain them. The most accredited seems to be the gate control. It is well known that touching a painful area alleviates the perception of pain. Pain sensation is controlled by the balance of activity in small-diameter, slow-conducting fibers (*C and A delta*), and large-diameter, fast-conducting fibers (*A beta*) that enter the spinal cord. In particular, low level activity in *C* and *A delta*, that carry input from the nociceptors, can be blocked by *A beta* at the first synapse [198]. This is the so-called gate control mechanism, by which the sensory nervous system at its lower levels senses tactile stimuli (vibrations) first and then cannot perceive feelings of pain (nociception). Thus, a barrage of vibrations or laser-induced shockwaves could essentially mask any underlying discomfort. The inhibition of pain by interaction between painful and non-painful stimuli may be related to cortical processing. The results by Testani et al. (2015) [199] do not support the hypothesis of a spinal inhibition of pain and extend previous evidence, showing that non-painful stimuli can inhibit the arrival of the nociceptive input at the cerebral cortex.

8.18.3 Indirect Influence on Nerves

As a further support to this theory, it is necessary to consider that erbium radiation, differently from the near infrared wavelengths, is not able to reach the pulp tissue through overlying hard tissue in a direct way due to the very limited penetration. Very interesting are the studies on laser-induced analgesia by Orchardson et al. (1997–2000) with Nd:YAG laser, and by Zeredo et al. (2003–2007) using Er:YAG laser on rodents [200–207]. They showed that a dramatic blockage of neuronal activity and a corresponding increase in pain threshold of teeth occurred after laser irradiation. Furthermore, the laser-induced analgesic effect in these researches had a clear dose-response for its onset. The design of these animal studies clearly removes all possibility of placebo effects and psychogenic influences.

8

8.18.4 Influence on Na^+–K^+ Pump and Bio-Resonance

Apart from the gate control pathway, it is possible that infrared lasers influence the Na^+–K^+ pump action. This pump is located on the surface of the neuron axon, in a transmembrane position with an external part, one in cytoplasmic position and a controlling beta domain and a "nose" in the internal side of the membrane [208]. The pump is a hollow tubular protein which spans the cell membrane of nerves and establishes the gradients in Na^+ and K^+ which are essential for propagating an impulse down a nerve through membrane depolarization and repolarization. During each enzymatic cycle, the pump transports three Na^+ ions outside the cell and two K^+ ions into it, while hydrolyzing one molecule of ATP. This pump allows the generation of Na^+ and K^+ gradients across the plasma membrane in virtually all animal cells and is particularly important in nerves for re-establishing membrane polarity of impulses [209, 210]. Partial spatial dissociation of the beta domain of this ion channel protein, which controls the function of the pump, closes the lumen of the "pipe" and thus prevents the repolarization for a short period of time. Vibrational effects of laser energy are thought to act at the interface of the beta domain and the cytoplasmic "nose," closing the pump's inner channel and thereby blocking membrane repolarization. FRP lasers at 10–30 Hz seem able to provoke this spatial change thanks to their vibrational energy. This phenomenon has been termed "bio-resonance."

8.18.5 Direct Inhibition of Nerves

Another plausible mechanism that has been identified as a possible explanation in analgesia onset is the direct inhibition of neural activity. Transcutaneous visible and infrared lasers, with continuous wave or complimentary gated emission, applied at several points along the course of peripheral nerves, can cause conduction velocity slowing and decreased amplitudes potentials of *A delta* and *C* fibers [211–215]. In animals, studies have been proven a specific laser inhibition of these fibers which transmit nociceptive stimuli. This means that a laser beam is able to reduce the nociceptors response to a variety noxious stimuli, including pro-inflammatory substances [216, 217]. The exact mechanism by which photons can inhibit or at least slow or partially block nerve conduction causing a functional impairment remains unclear. According to Chow et al. [218], a possible explanation on how lasers reduce pain by producing an alteration in the conduction of action potentials in peripheral nerves is the generation of varicosities along the axons more than in cell bodies. These enlargements could reduce the speed of fast axonal flow and the mitochondrial membrane potentials, resulting in neurotransmission failure in *A delta* and *C* nociceptors fibers [219, 220]. In these experiments, after the end of irradiation, the neurons were able to return to original function because were not irrevocably damaged and the varicosities resolved after 24 h.

8.18.6 Which Energy Dose to Use?

The settings necessary to induce laser analgesia have been already investigated through several studies [48, 171], but further research is needed in order to determine the exact amount of medium infrared energy and the correct parametry necessary to produce a more predictable protocol and a reliable level of laser analgesia. Different to medium infrared irradiation, near infrared energy dose has been extensively studied and determined to obtain an analgesic effect. It is known that there is an irradiance threshold required to exert maximal suppression of nerve activity and control acute and/or chronic pain. According to Bjordal et al. [221], the optimal dose on acute pain in the first 72 h after the injury should be 7.5 J/cm^2. The majority of authors proposed greater levels of energy, such as Lizarelli (cited by Angelieri et al. in 2011) [222] who indicated a dose included in the range 5–20 J/cm^2 for severe pain. Other authors suggested a fluence of 35 J/cm^2 to reduce orthodontic pain, because 5 J/cm^2 were not effective [222]. In current literature, similar protocols reported conflicting results [223].

8.18.7 Effects of Laser Energy on Biochemistry

A final further possible explanation of the laser action on pain is represented by the biochemical action of low level laser therapy through the production and/or the inhibition of endogenous substances that create neuropharmacological effects locally and systemically [221, 224, 225]. It is well known that the wavelengths included in the so-called optical window (approximately between the range 600 and 1300 nm) are able to promote the synthesis, the modification, the release and to influence the metabolism or the inhibition of numerous biochemical substances. As an example, near infrared irradiation can increase beta endorphin, endogenous opioids, sustain the peripheral release of histamine and prostaglandins, facilitate central release of serotonin and acetylcholine, and allow cellular oxidation and the production of ATP [226–229]. Similarly, the possible inhibitory effects are represented by the reduction of several interleukins secretion (IL-1b, IL-6, IL-10) and other factors involved in inflammation process or nociception (COX2, nitric oxide, PGE2, TNFα, bradykinin) [230–232]. Additionally, in general there is a direct effect of laser energy on the increase of local hemodynamics with higher removal of pain stimulating substances [191, 220, 233]. Until now there is no incontestable demonstration that erbium family lasers have a similar biochemical range of effects. Laser-induced analgesia can be considered a possibly effective alternative treatment in affecting pain perception. Non-standardized study design, lack of consistent laser operating parameters, and lack of statistical analysis do not allow firm conclusions as to which laser wavelength is superior or what parameters might be employed to achieve predictability, and have failed to underline a strong protocol of how the mid-IR wavelengths may be applied in achieving laser analgesia [234]. In conclusion, to obtain successful analgesia it is highly plausible that it is necessary to apply low level energy (and power) or, more precisely, it is indispensable low energy density and low power density. Furthermore, initially it is useful to use low levels of air and water spray which can induce dental sensitivity due to the cooling effect of air and/or water.

8.19 Techniques for Laser Analgesia on Teeth

Two different techniques have been proposed in order to obtain the laser analgesia:

- Rabbit Technique (also called *Hare Technique*): The laser is immediately set on high power levels, able to perform hard tissues ablation and this is maintained during the whole treatment. At the beginning, however the beam is kept defocused at 6–10 mm from the tooth. So, the energy density is low and it takes advantage of the low level laser therapy. The tip is moved all around the tooth, at its neck level. Then, the tip is gradually brought closer up to 1 mm from the dental surface and so the ablation effect can start. At this point, if the patient feels some discomfort, it is possible to move aside the tip again. As soon as the beam gets through the enamel and reaches the dentin, the tip is again placed farther away, the laser irradiation becomes defocused (thus reducing the energy density), and cavity preparation is complete.
- Turtle Technique (also called *Tortoise T.*): The tip is immediately placed at 1 mm from the tooth and kept at this distance for the preparation procedure. Low power is then set in order to obtain pulp analgesia and has a lower risk of discomfort for the patient. Then, the energy is gradually increased up to a sufficient level able to obtain tissue ablation and this is carried on till enamel ablation is completed. When the dentin is reached, the power is reduced, and cavity preparation is completed. This latter technique is considered the most reliable to avoid patient's dental sensitivity during the restorative treatment. It is regarded as the most satisfactory, delicate, and effective to obtain dental analgesia [48].

It has been scientifically demonstrated that permanent teeth are more sensitive to pain than deciduous ones, and that laser analgesia is easier for the latter [186]. According to Moritz (2006) [164], the laser analgesic effect on the tooth should last approximately 15 min and after its disappearance no histological alteration of the pulp occurs. On the contrary, according to Whitters et al. [192], the pain threshold after laser analgesia obtained by the means of a Nd:YAG laser returned to baseline approximately after 60 min.

8.20 Protocol for Tooth Analgesia with the Erbium Laser

In order to study erbium laser analgesia, we recently [171] studied a protocol in order to propose a systematic painless restorative treatment of the teeth. We used the Er,Cr:YSGG (2780 nm) laser applying a combination between *Rabbit* and modified *Turtle technique*. Before starting cavity preparation, a laser-induced analgesia phase was always performed by initially using very low levels of energy, and then by gradually increasing them, without using any air/water spray. In this way, the dental pulp had the possibility to adapt to laser irradiation without triggering a mechanism of annoying sensitivity, but gradually performing analgesia. Then, it was possible to obtain a gradual painless ablation of tooth hard tissues. The analgesia phase was therefore started with power values of 0.1 W (consequently the energy had values of only 10 mJ) at a pulse repetition rate of 10 Hertz and afterwards these levels were gradually increased to 0.2 W, then to 0.5 W with a repetition rate of 15 Hz, and then, finally to 1 W and to 2 W with the same pulse repetition rate. Overall, this stage always lasted 3′30″ (210 s).

The study of laser-induced dental analgesia with regard to cavity preparation was performed by adopting the following sequence:

A. Preliminary pulp test using the electric pulp tester to evaluate dental vitality and to establish the baseline threshold of dental sensitivity.
B. Beginning of the dental analgesia induction phase by using power settings of 0.1 and then 0.2 W (energy per pulse of 10 and 20 mJ) at a pulse repetition rate of 10 Hz, for 30 s each (without using any air/water cooling spray), keeping the tip at 10 mm from the tooth using a spacer. Subsequently, the power was increased to 0.5 and then to 1 W (energy per pulse of 33 and 67 mJ) for 60 s each, with a spray composed by 15% of water (for our laser this means approximately 10 mL/min) and 20% of air, at a pulse repetition rate of 15 Hz, keeping the tip at the same distance from the tooth neck.
C. Preconditioning of hard tissues with 2 W of Power for 30 s with a cooling spray of 50% water (approximately 20 mL/min) and 80% air, at 15 Hz of pulse repetition rate, with the tip at approximately 1 mm from the tooth. The laser beam was kept in focus or, if the patient felt discomfort, it was defocused according to sensitivity.
D. Pulp test (EPT) performed again to evaluate the presence of analgesia and establish how the threshold value of dental sensitivity had changed.
E. Preconditioning and beginning of enamel ablation with a 3 W Power for 30 s (same previous settings as for pulse repetition rate, distance, and cooling spray).
F. Enamel ablation with 4 W of power (same previous settings).
G. Possible enamel ablation with 5–6 W of power (same previous settings).
H. Possible dentin ablation with 3–3.5 W of power (same previous settings).
I. Preparation completion and smear layer removal with a power of 2 W (same previous settings).
J. Pulp test at the end of the preparation. To assess if the threshold value of dental sensitivity had further changed after ablative laser irradiation.
K. Pulp test after 15′–20′ from the end of cavity preparation to assess if analgesia was over.

The entire period of laser analgesia induction had an overall duration of 3′30″ (210 s), and it was performed on all patients.

At the end, as specified, cavity preparation proceeded.

To correctly perform laser-induced analgesia in our protocol, we suggest to maintain the tip at a distance of 10 mm from the tooth from the start. In this way, it is possible to obtain a very low energy density from the initial stage (only 6 J/cm^2 with movement) and average power density (1 W/cm^2), thus allowing the pulp to progressively adapt to laser irradiation and achieve analgesia without risking painful or annoying sensations. With regard to discomfort felt by patients, the factors that seem to have a higher tendency to promote the shift to greater discomfort categories are as follows: posterior teeth compared to a superficial one, the time needed for ablation of hard tissues, and the use of laser at high power levels. One of the most important factors that influenced pain perception was age. In this study, all patients that felt greater discomfort or pain were in age brackets 20–29, 30–39, and 40–49. We think that younger patients could obtain analgesia more easily and quickly as their dental hard tissues are richer in water and they have wider dentinal tubules. This could facilitate ablation and progression of laser beam effect on pulp nerves. With regard to older patients, they could be less sensitive to irradiation for the opposite reasons: their dental tissues are more sclerotic and calcified, they have narrow dentinal tubules, and even if they are more difficult to ablate, they protect the pulp more and they are less influenced by stimuli. When a restorative treatment with erbium laser is planned, without resorting to any local injected anesthesia, it should be considered that cuspids and incisors may be more sensitive, especially if the decay is deep. For these teeth, the energy can rapidly affect nerve fibers of the pulp because of the limited thickness and cause pain. Actually, in our research we saw that the opposite was true: premolars and molars were more sensitive than front teeth. By using our protocol, initially applying very low energy levels and gradual irradiation, we obtained a better and quicker laser analgesia for anterior teeth. The possible explanation of this is connected to the greater thickness of hard tissues for posterior teeth compared to incisors and cuspids. It is also our opinion that the depth of the decay is important sensitivity-wise, but the time of preparation is more relevant in affecting it. This is due to the effect of erbium on the dentin. This kind of laser, combined with water, opens the dentinal tubules. The more the laser is used, the more the tubules will be opened, and the higher the patient's sensitivity will be. Besides, if dental hard tissues are not easily laser ablated (for example, if they include a lower water percentage) and if, for this reason, it is necessary to extend its use for a longer period of time or to increase the energy levels in order to facilitate ablation, then the risk of pain further raises. One additional element to consider is the fact that laser analgesia could not be completely effective to also achieve periodontal tissues analgesia. During this research, even if the tooth was completely insensitive to carious hard tissue ablation, we have quite frequently noticed that the patient could feel the discomfort provoked by the positioning of the dental dam clamp, the matrix, or the wedge. Thanks to the proposed protocol it was possible to perform a restorative treatment by using the Er,Cr:YSGG (2780 nm) in 24 patients out of 30 (80% of the sample) without resorting to any kind of local anesthesia and without the traditional hand piece and bur. These patients didn't feel any pain (in 57% of cases) or they felt only a very light sensitivity (in 23% of patients). The equipment we used was very likely to produce laser-induced analgesia, which allowed us to remove all the carious tissues and to complete the composite reconstruction without any pain for the patient. In relevant literature, the comparison between traditional hand piece with bur and erbium laser showed that with the former the dentist can obtain painless treatment in only 20–50% of patients [235]. The use of Er,Cr:YSGG laser precludes the administration of local anesthesia by injection and avoids the use of the traditional handpiece and bur. Thus, we can obtain reduced anxiety in patients, something that is frequently associated to dental therapies. Following these considerations, is it possible to draw some firm conclusions on the strong correlation between anxiety and discomfort? We believe it is possible to affirm that groups of patients reporting higher levels of anxiety before attending a dental session are the same who felt greater discomfort during therapy. So, it is likely that the anxiety factor contributes to generate a higher subjective evaluation of discomfort. In adult patients, we noticed that the level of anxiety felt during a dental session had more influence on the possibility to obtain complete leaser analgesia; this is probably due to the patient's individual difficulties related to dental care past experiences. On the other hand, with regard to pediatric patients, if they never had dental experiences in the past, if they were not very anxious by nature, but rather calm and happy, and if they did not have a negative influence from parents and/or relatives, they may be more incline to accept the dental treatment. This will mostly and more easily occur if the dentist adopts a psychological, positive, delicate, and serene approach. All of this will be facilitated by adopting the erbium laser; if neither needles nor local anesthetics are used, especially if the operator avoids noises and vibrations which are typical of the traditional handpiece combined with the bur. This approach will reinforce and maintain the patient trust toward the dentist.

◘ Figure 8.6a–f shows a cervical carious lesion involving enamel and dentin on tooth #4. Enamel decay removal using erbium:YAG laser.

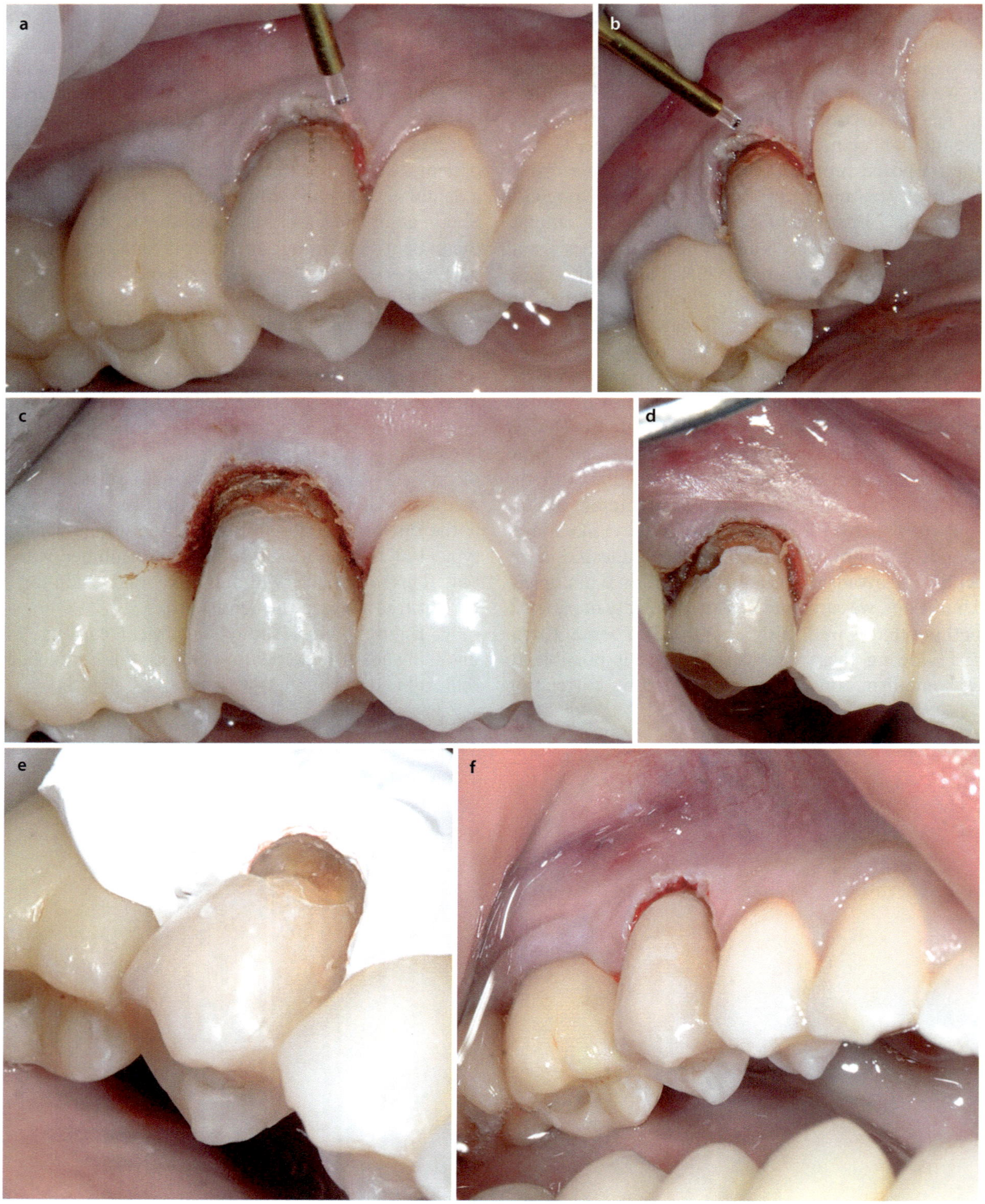

Fig. 8.6 **a** Buccal view with cervical carious lesion involving enamel and dentin on tooth #4. Enamel decay removal using erbium:YAG laser. **b** Detail of cavity on tooth #4. **c** Preparation with Er:YAG laser (Pulsar with wavelength of 2940 nm by LAMBDA Dr. Smile, Italy). Gingivectomy: angulated handpiece, sapphire tip diameter 400 μm, length 12 mm, 2.4 W, 20 Hz, 120 mJ E/pulse, peak power 750 W, average power density 659 W/cm^2, peak power density 205.860 W/cm^2, total energy 432 J, pulse width 160 μs, 1 mm tip-to-tissue distance, 50% water, 50% air, total treatment time 180 s. Enamel settings: angulated handpiece, sapphire tip diameter 800 μm, length 12 mm, 4.5 W, 20 Hz, 225 mJ E/pulse, peak power 1406 W, average power density 490 W/cm^2, peak power density 153.199 W/cm^2, total energy 810 J, pulse width 160 μs, 1 mm tip-to-tissue distance, 100% water, 70% air, total treatment time 180 s. Dentin and smear layer settings: sapphire tip diameter 800 μm, length 12 mm, 3.3 W, 20 Hz, 165 mJ, peak power 1031 W, average power density 360 W/cm^2, peak power density 112.346 W/cm^2, total energy 594 J, pulse width 160 μs, 1 mm tip-to-tissue distance, 100% water (18 mL/min), 70% air, total treatment time 120 s. **d** Detail of cavity on tooth #4 after gingivectomy. **e** Detail of completed cavity after dentin preparation. **f** Completed composite restoration. (Procedure by Dr. Francesco Buoncristiani)

8.21 The Laser Handpiece and Tips

Most modern erbium lasers are provided with two types of handpiece. One uses interchangeable tips of various lengths, diameter, and materials. On the market, there are tips with lengths ranging between 3 and 28 mm. With the shorter tips, it is possible to access small areas within the teeth arch or in difficult spots (i.e., the upper second and third molars, buccal areas of posterior teeth) or perform treatments when the patient has limited mouth opening (i.e., pedodontic patients). With longer tips and with a small diameter of 200–300 μm, it is possible to perform endodontic and periodontal laser-assisted treatments. They are made of quartz or sapphire. Often it is possible to differentiate them from one another because of their color, yellow for the former and white for the latter. Usually, the tips most used in restorative dentistry are the ones measuring 4–10 mm. The tip-to-tissue working distance should constantly be kept at 0.5–1 mm in order to obtain optimal energy density. In some scientific articles, this operational modality is wrongly defined as "in contact" since the operator is working at close proximity with the tooth. Nevertheless, there should never be contact, and the tooth surface should never be touched in order to avoid the creation of enamel-dentinal microfractures and to avoid damage to the delicate laser tips. This working modality is also called "focused," even if in reality the laser beam is not convergent, i.e., it is not focused on the target. The laser energy delivered by the tip is in fact immediately divergent with approximately an 8° divergence angle per side. The reason to keep the tip at 0.5–1 mm is due to the fact that at that distance, energy density (fluence) is optimal, and it is the one that allows a more efficient ablation. By increasing the distance, fluence will drop dramatically, preventing adequate excisional interaction with tissues. On the other hand, if the working distance is below 0.5 mm, the operator may run the risk of causing dental damage following contact, tip deterioration after accidental crash against the cavity surface may occur, reduced effectiveness of the air/water spray to cool and eliminate residues, and a very limited visibility on the working area. The diameter of tips used for hard tissue ablation usually ranges between 400 and 1000 μm. The smaller the diameter, the smaller the cavity the operator can prepare and save healthy tissues. Tips with larger diameter produce a spot size which will inevitably create a bigger size cavity; thus, it is not possible to do small or very conservative preparations. For minimally invasive dentistry including only occlusal grooves, it is preferable to use a tip with the smallest diameter possible. By doing so, preparation will be quicker because there will be more energy density since all the energy will be focused on a smaller surface. Usually, tips used for this type of therapy are cylindrical with a circular cross-section. Additionally, there are truncated-cone shaped tips, with rectangular section and chisel-shaped. Each one of them creates a different beam emission which in its turn provides an ablation zone and produces a different cavity shape. The second type of handpiece is also called *tipless* because it does not have the previously described tips, but a lens which can also be interchangeable, and it focuses the beam at a distance of about 5–10 mm from the surface. Operators work at a greater distance compared to the previous hand piece. At times in the scientific literature, this type of use is defined as "defocused," but this is not the correct term. In reality, the beam is focused at a few millimeters (usually 5–7 mm) away from the surface of the hand piece from which the beam is delivered. Such use is defined as "contactless." This greater distance from the target improves visibility, but it provides a less favorable and uncomfortable perspective in poorly accessible areas such as the upper molars. The reason is that it is more difficult to position at that distance and keep accuracy at 5–10 mm from target. Furthermore, it is very difficult to accurately irradiate a small target since the beam is wider. The target area (spot size) covered by the beam tends to be larger than when the operator works almost in contact; thus, it is very difficult to prepare very small cavities and, if the hand piece is not kept steady on the ablation target zone, the effect is often dispersed on a wider area, resulting in an unintended widening of the cavity and elimination of healthy tissue. This hand piece, however, is more efficient since it allows the removal of a larger amount of decayed tissue in less time.

8.22 The "Erbium Noise"

All types of hand piece produce a similar and characteristic noise. It is often defined as a "popcorn" type of noise, as it reminds corn popping in the pan. It is completely different from the noise produced by a traditional turbine and by the bur, thus patients tend not to associate it to the fear for the dentist. Noise intensity is directly proportional to the employed energy, which creates micro-explosions in the water present in tissues and in the one used for cooling. Also, irradiation frequency impacts noise. The higher the number of pulses, the lower the number of "explosions" heard as they will "merge" with one another, and they will sound as one noise. At about 30–40 Hz, the noise is continuous, without interruptions between pops.

8.23 Approach According to Cavity Classification

Depending on the place where the decay lesion is, the removal approach will be different.

8.23.1 Class I

Occlusal decay on posterior teeth (Class I) is obviously easier to treat, but the enamel is quite thick. Thus, it may be necessary additional time to fully remove this type of lesion, especially in the case they extend under the occlusal plane and when the decay opening is limited, and a lot of healthy tissue covers the entire lesion. In order to avoid extended laser ablation, and to reduce operating times, the operator can also open the decayed enamel grooves by using a small diamond bur, and only use laser irradiation later on. Small size cavities are more difficult to treat because they are less accessible. The combination of small high-speed diamond bur and lasers with minimum diameter tips is certainly advantageous.

Usually, ablation starts by placing the tip perpendicular to the tooth surface, by making small, very slow continuous movements and by keeping that position from the beginning of the creation of a small cavity. Later on, the beam should be gradually oriented toward the cavity walls outwardly (up to a maximum of 45° per side) to complete preparation [48]. For larger cavities inside the dentin and with large geometries, it is very difficult to reach each side of the walls. In this case, it could be necessary to eliminate much healthy tissue in order to be able to complete the full decay removal. The use of low-speed burs and manual excavators may allow to remove the residual decayed tissue and avoid the elimination of healthy tissue.

8

8.23.2 Class II

Cavities concerning premolar and moral interproximal areas (Black's class II), when they need preparation from the occlusal surface, they normally require more execution time by using the laser rather than the bur because the volume of tissue to be removed is considerable and the enamel wall can be large. Visibility is often reduced, as well as accessibility. To be able to reach any area concerned with the carious lesion, it is key that the tip is placed inside the cavity that is being created, in order to keep the correct tip-to-distance constant. The most difficult aspect is represented by the difficulty of sufficiently inclining the tip toward the walls to be prepared. It should be borne in mind that the angle formed by the tip vis-à-vis the decayed cavity wall should not exceed 45° otherwise ablation becomes ineffective.

8.23.3 Class III and IV

In the case of class III and class IV cavities, accessibility is much better, thus no major difficulties are noted. Enamel thickness is limited and, as such preparations normally require short times and reduced laser energy, compared to those to be used for posterior teeth. Laser preparations can also be more conservative than the ones obtained with traditional techniques. In order to protect nearby teeth and avoid damage to their surfaces, it is advisable to place a cellulose strip inter-proximally to prevent the laser beam to affect the healthy walls of nearby teeth. It would be better instead not to use metal matrixes for the same purpose as they could reflect the laser beam and thus become a source of potential risk for operator's eye safety.

8.23.4 Class V

Cervical cavities (Black's class V) can be prepared easily and rapidly, due to the limited thickness of the enamel and to the presence of root cement nearby. For anterior teeth, it is also possible to use a straight hand piece instead of the angled one to facilitate access to cervical tooth decay. With the cervical area of posterior teeth, it is usually easier to use an angled hand piece and 3–4 mm length tips which facilitate access to vestibular or lingual areas, despite the interference of cheek and/or tongue.

8.24 Interaction with Dental Materials

It is always important to pay attention to dental materials present in adjacent teeth. The erbium laser beam can very easily interact with amalgam, composites, and dental metal alloys. Irradiated silver-based amalgam can rapidly absorb energy, increase temperature, and create thermal problems to pulp and periodontium. Should the temperature further increase, amalgam melting may occur with subsequent damage to reconstruction and release of mercury vapor. In case of secondary decay or reoccurring decay under the amalgam filling, it is always necessary to remove the metal reconstruction through traditional methods (by high-speed bur) and only after this operation, carious tissue can be removed by laser. Interaction with composite occurs very easily when irradiated [236]. Their ablation is very easy thanks to the water content. However, the composite matrix undergoes explosive fragmentation as a result of the interaction with the laser energy; it re-solidifies and quickly aggregates around the tip and jeopardizes the integrity of the laser fiber. Tips altered by resin fragments must be cleaned and polished rapidly to avoid this risk. Tip inspection can be done by wearing amplifying glasses or by using a jeweler's lens with 30 magnifications. Polishing can be done by using rotating discs to polish composites mounted on a low-speed hand piece. It is then possible to gradually move on to rougher discs to the smoother one, and this allows to remove residues, burn marks, correct possible nicks and polish tips. Such procedure is much easier for quartz tips than for sapphire tips because the latter are harder. However, polishing may be more complex when there are composite fragments attached to the tip end. These materials, in fact, are difficult to remove when melted by laser energy and then they re-solidify on it, thus is advisable to avoid resin-based materials ablation during conservative therapy. In case composite reconstructions need to be redone, it is preferable to remove the old material by diamond bur; only later should the laser beam be used to ablate decayed tissue and to extend enamel preparation laterally, to complete dentin excision and condition the final surface before adhesive techniques are performed. If the tooth near the decayed element has a metal alloy crown, irradiation of the latter can involve thermal interaction quickly leading to temperature increase with potential risks of trauma to the tooth pulp and to the periodontium. Ceramic crowns, as well as temporary resin crowns, may sustain damage by erbium laser energy, the former can be fractured following rapid thermal expansion. If interaction is limited, damage may be limited to scratches or nicks. Temporary/provisional crowns may suffer damage similar to that seen with composite resin. Nonprecious metal alloys used for removable prostheses are easily damaged through temperature increase, while pink resins and artificial teeth can be similarly damaged. For all of these reasons, it is advisable to pay special attention to all surrounding dental materials during the entire conservative therapy.

8.25 Clinical Considerations

Clinical considerations for laser-assisted conservative dentistry are as follows:

- At the end of carious cavity preparation, eliminate unsupported enamel prisms by using manual instruments (enamel cutters, excavators) and/or high or low-speed hand piece fitted with diamond fine burs or rubber tips to bevel cavity margins. Alternative to this process, it is possible to use the erbium laser to bevel cavity margins. It is advisable to set limited energy values (40–80 mJ) and frequency of about 25–50 Hz.
- Perform acid etching by using orthophosphoric acid at 34–38% (maximum timing 30 s on enamel, 15 s on dentin) to optimize and make the treated surface uniform to regularize the areas affected by the erbium laser beam.
- Use an adequate adhesive system that takes patient's characteristics into account, such as age, teeth conditions (deciduous or permanent), decay depth, if the tooth has been subjected or not to the endodontic therapy. For superficial cavities, permanent teeth, and teeth treated with root canal therapy, it is advisable to use the *Etch and Rinse* adhesive system. In the other cases, it would be preferable to use a *Self-Etch* system, in particular if the cavity is deep and if the patient is very young (the same applies to permanent teeth).
- For deep cavities, it may be advisable to use glass ionomer cement or a flowable composite as liner on the cavity floor, in close proximity of the pulp, in order to obtain good protection of the pulp and of deep dentin, and to position in such area a low elasticity material and limited polymerization shrinkage; these characteristics facilitate optimal reconstruction adaptation to the cavity internal surface.
- Use a photo-polymerization lamp with controlled light irradiation (i.e., soft start or pulse delay technique) to limit polymerization shrinkage of composites and evenly reach all stratified areas.
- Always use an incremental technique to stratify the composite, in order to have max 1–2 mm layers of material and compensate for and minimize its polymerization shrinkage.
- Use low polymerization shrinkage composite resins for reconstruction (i.e., silane-based composites, even if their clinical use should still be further studied and validated).

8.26 Erbium Laser in Reconstruction with Post in Endodontically Treated Teeth

Teeth subjected to endodontic treatments can benefit from the erbium laser use during composite reconstruction performed in combination with a post. Thanks to irradiation, both canal walls following endodontic treatment, and the post can be optimized for the adhesive process. Most modern posts are made up of carbon, quartz, silica, or glass fibers, embedded in an epoxy matrix or in a methacrylic resin. They have an elasticity modulus similar to that of dentin, so that under mastication the material behaves similarly to tooth tissues and forces are discharged in an equivalent way. By combining this property to the possibility of obtaining an adhesive bond among the various materials (root and crown dentin, adhesive system, fiber post and core material in composite for cementing and reconstruction), it is possible to reduce the risk of fracture [237, 238]. These posts have high biocompatibility, they are easy to use, they have high mechanical resistance, good corrosion resistance, they are easy to be removed, and they have a very high appearance value (for quartz and glass posts). Post retention by the root depends on the chemical interaction and micromechanical strength among post, dentin, and resin-based cement. Should there be insufficient bonds between resin and dentin or at interface level between composite and post, restorative rehabilitation will fail, in association with the partial or total detachment of the reconstruction and of the post embedded in it. The bond strength is influenced by the degree of hydration/dehydration of the inter-canal dentin wall. If the inside of the canal is dehydrated, hydrophile monomers of the adhesive system will not be able to penetrate dentinal tubules resulting in a lack of hybrid layer. On the contrary, if the water content is excessive, monomers will be excessively diluted, and they will not perform. Other factors that contribute to determine a higher or lower retention strength between post and root are represented by physical property of the composite cement, unfavorable canal configuration (accentuated curvature, root with very thin walls not allowing a wider preparation) or due to insufficient canal length which does not allow the positioning of a sufficiently long retaining post, from adverse effects of canal sealing cements which, by containing eugenol they combat resin polymerization used for cementing, and due to anatomic or histological characteristics of dental tissues (i.e., number of tubules at the different levels inside the canal) [239]. The post fiber polymeric matrix is highly cross-linked, thus bonding phenomena with composite monomers do not easily occur. The bond between the reconstruction composite and the post occurs only partially, and the resin acts as a bond with glass or quartz fibers. To improve the odds of obtaining such link, different types of posts and canal walls pre-treatments have been proposed. For example, several authors frequently proposed the roughening of post surfaces in view of increasing retention. However, this exposes glass or quartz fibers, and it may give rise to their weakening. Sand blasting with A_2O_3 powders in 50 µm particles or the use of hydrofluoric acid must be performed with extreme attention to avoid too aggressive alteration of fibers. For quartz posts, it has been underlined [240, 241] that it would be useful to use the HF acid at a concentration below 9%. In such a way, higher tensile strength is obtained. However, the same treatment can be risky for glass fiber posts because it would induce corrosion. Some solvents could increase the adhesion strength between quartz or glass fiber posts and the resin core material. In particular, they have been tested with hydrogen peroxide (H_2O_2) at 24% for 1 min [242, 243] and dichloromethane (CH_2Cl_2) for 1 min [244]. Both solutions showed promising results, but they should still be tested on larger samples. Apart from its decontaminating effect, the erbium laser produces a smear layer free surface and a micro-rough texture which can facilitate retention. Following post space preparation, canal walls cleaning is a critical procedure, however indispensable, because there is a lot of smear layer inside the canal, as well as gutta-percha residues and endodontic cement on dentinal walls. All of this represents a contamination that may negatively affect adhesive procedures. Intratubular moisture and residues of irrigation liquids inside the tubules can furthermore complicate or impair adhesion process steps.

The erbium laser used to provide thorough cleaning and decontamination of the endodontic space allows an extremely accurate cleaning of the dentinal surface and the elimination of the smear layer, but, on the other hand, it is advisable to avoid its excessive use since it may induce dentin dehydration. It is very important to use limited energy values (from 100 to 125 mJ) and frequency of 10–20 Hz in order not to negatively impact the dental surface and not to create micro-structural damages to the dentin and to its hybrid layer [243]. The main difficulty in obtaining efficient laser conditioning on canal walls is represented by the fact that the tips inside the canal irradiate toward the apex with a divergence of 8° per side, thus the beam reaches dentinal walls with a very marked inclination and that, thus, does not interact much with the surface. For this reason, it is advisable to avoid excessive irradiation and excessively high parameters since they could damage hard dentinal tissues and their organic portion. The same parameters mentioned above can be used on the post surface to facilitate the formation of micro-roughness which may have a retention effect on the resin core material and increase post resistance and reconstruction.

8.27 The Use of the Dental Rubber Dam

As with all conservative dental treatment, laser-assisted procedures must be completed with dental dam in place, in order to avoid contamination in the operating area. Placement can be achieved before the preparation of decayed cavities, but after the step in which laser analgesia is attempted. Absence of the dental dam allows better irradiation in the tooth cervical area and of the gum surrounding the tooth. After formal analgesia, it will be possible to place the dam in place and proceed with cavity preparation by removing carious tissue. In this way, mild analgesia will be achieved in soft periodontal tissues as well, allowing the positioning of the dam clamp, the matrix, and the wedge, that will be perceived by the patient with minimal discomfort.

8.28 The Use of the CO_2 Laser with Hard Dental Tissues

The carbon dioxide laser (CO_2 10.6 μm) has been extensively used in the last 40 years for oral surgery. The continuous wave (CW) emission mode and complementary gated mode allow an efficient and quick vaporization and ablation of soft tissues, also obtaining a very good hemostasis. Early studies using a CW 10.600 nm CO_2 reported extensive cracking and charring of enamel, dentin, and bone [245, 246]. During the last 10 years, researchers have modified the native 10.6 μm CO_2 laser to emit at 9.3–9.6 μm wavelength, which is the peak of absorption for the molecule of phosphate in hydroxyapatite, by replacing the normal $^{12}C^{16}O_2$ molecule with an isotopic $^{12}C^{18}O_2$ [247]. The absorption is particularly important because in this case enamel, which is mainly composed of calcium and phosphate, absorption is 5–6 times higher at 9.3–9.6 μm than at the more commonly used 10.6 μm wavelength and it allows more efficient heating and ablation of dental hard tissue [248]. Transversely excited atmospheric-pressure (TEA) CO_2 lasers and other radio frequency excited (RF) pulsed lasers combined with 3D computer controlled programmable scanning systems are now available on the market from several manufacturers and seem even more versatile and efficient when compared to erbium family lasers [245, 246]. Using such lasers it is possible to perform a wide range of procedures. Such as modification of the pulse duration of new carbon dioxide laser in order to obtain an efficient removal of dental hard tissues (carious lesion ablation, caries prevention, removal of composite restorations) and bone, as well as retaining the surgical effect on soft tissue [249]. The most important feature regarding modern carbon dioxide devices is that they can be operated at high pulse repetition rates—in the order of kHz—and this allows a very practical removal rate of hard tissues and an incomparable ability of gingival and mucosal ablation [248]. The erbium lasers presently used for hard tissue ablation operate most efficiently at very low repetition rates (10–25 Hz). Therefore, in order to achieve higher cutting rate erbium lasers must deliver a larger amount of energy per pulse, in the range of 100–500 mJ. CO_2 lasers can be operated with very low single-pulse energies (in the order of μJ up to mJ) and fluence, while frequency can be increased for higher cutting rates [250]. The laser beam can also be scanned to minimize heat accumulation in the treatment area [245]. The wavelength of 9300–9600 nm is coincident with the strongest absorption of dental hard tissues due to phosphate ions in hydroxyapatite. Therefore, the energy necessary for ablation of tooth hard tissues is lower at these wavelengths versus others and this allows a reduced accumulation of heat in the tooth. Moreover, due to this very high absorption, the penetration is limited to under 1–2 μm. With erbium lasers, the shortest pulse width is 50–60 μs, while the CO_2 allows efficient ablation of enamel and dentin with laser pulses of 10–15 μs [247, 249], so it is possible to obtain high peak power values with lower energy levels, and this can be less aggressive and has a lower possibility of collateral damage of healthy dental structure [248, 251]. According to Staninec et al. [245], the thermal relaxation time of the energy deposited in enamel at these wavelengths is on the order of 1–2 μs for enamel and 5.5 μs for dentin [250], so the use of a laser with pulses of 10–20 μs width reduces the threshold for plasma shielding in the plume of ablated material, which would shield the surface and reduce the efficiency of irradiation, allowing the ablation of enamel and dentin at rates of 10–20 and 20–40 μm/pulse, respectively [245, 247]. The use of longer CO_2 laser pulses has the advantage of raising the plasma-shielding threshold allowing higher ablation rates per pulse, however the longer pulses are more likely to produce a larger zone of peripheral thermal damage. The practitioner should remember that, although ablation rates are higher for longer pulses, the peripheral thermal damage caused by these longer pulses may be too extensive for practical use. Such thermal damage may result in thermal stress cracking, accumulation of non-apatitic calcium phosphate (CaP) phases on the surface and excessive damage to the collagen matrix [245, 252], so it is advisable to limit the length of laser pulses.

Figure 8.7a–f shows carious tissue removal in tooth #3 MO using the CO_2 9.3 μm laser:

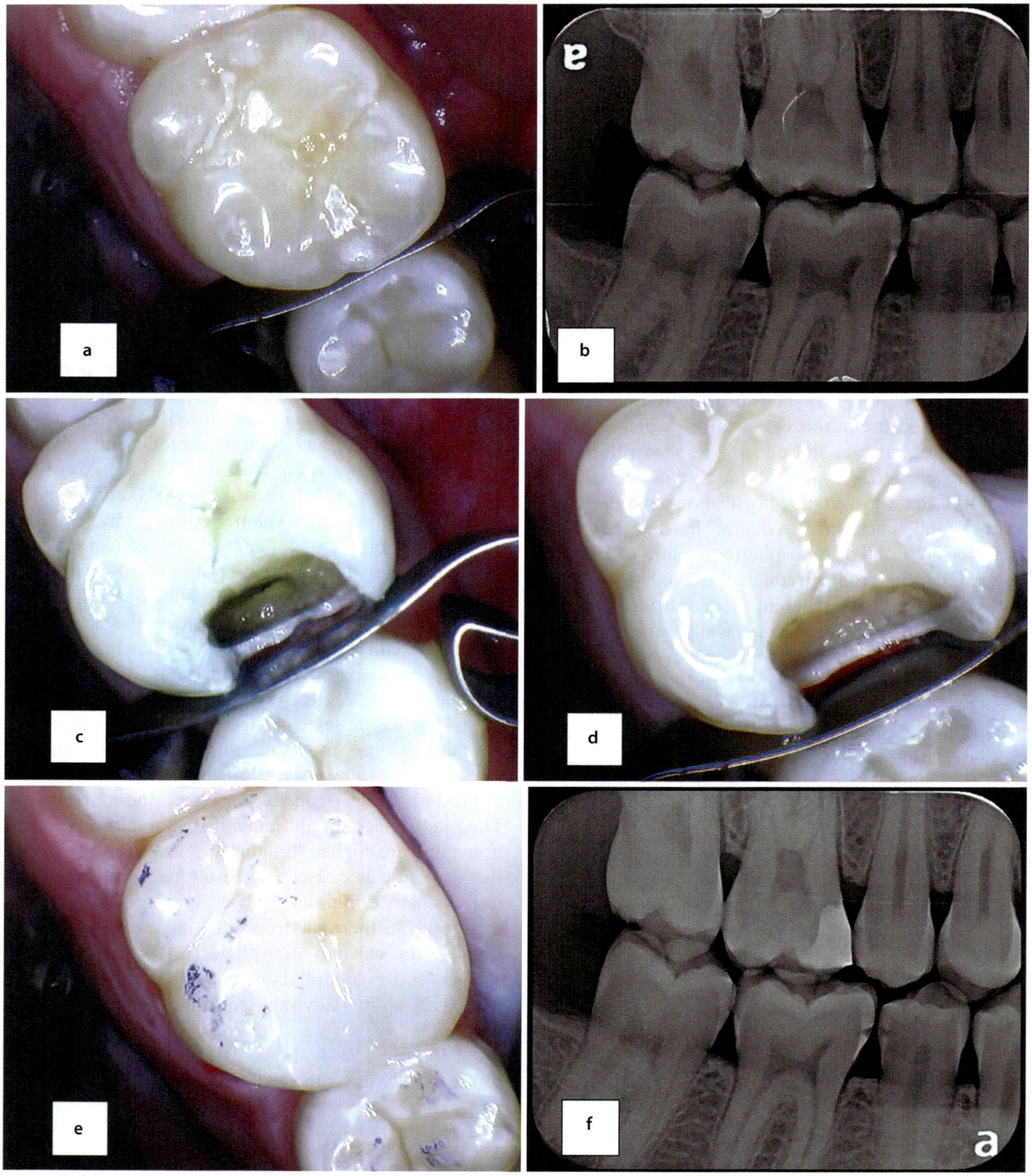

Fig. 8.7 **a**–**f** Carious tissue removal in tooth #3 MO using Solea CO_2 9.3 μm laser (Convergent Dental, Natick, MA, USA). Image key: **a** Preoperative occlusal view of the upper right molar with a carious lesion on the mesial surface. **b** Preoperative radiograph. **c** A photo of the partially completed carious lesion excavation. The 9300 nm laser with a 1.25 mm spot size was used with a cutting speed between 20% and 60%. Caries indicating solution was used to verify the progress of the preparation. Subsequently the laser was used with a 1.0 mm spot size at 20–40% cutting speed. **d** Photo of the completed preparation. **e** Immediate postoperative view of the restoration in place. **f** Postoperative radiograph showing the competed restoration. The entire procedure was performed without injected anesthesia using the "hard and soft tissue setting" and 100% mist. Caries indicator was used once more before restoring the tooth. The total laser time was 12 min. (Procedure by Dr. Joshua Weintraub)

8.29 Resistance to Acid

Another important advantage of CO_2 laser, in particular 9.3 μm CO_2, is the chemical and structural modification of enamel surface obtainable during irradiation [248]. This laser irradiation vaporizes water and carbonate and changes the chemical composition of the remaining mineral content of enamel, thus decreasing the solubility to acids with an enhanced resistance to secondary caries [253].

This allows to increase the acid resistance and consequently to reduce the incidence of carious lesions [254–258]. The occlusal pits and fissures are the areas of the tooth in which dental caries is most frequent. Thermal modifications of these surfaces with the 9.3 μm CO_2 laser have been shown to obtain a greater resistance to acid dissolution. In a clinical study, Rechmann et al. demonstrated over 12 months with 60 participants that the use of 9.3 μm CO_2 short-pulsed laser increases the caries resistance of occlusal pit and fissure surfaces in patients in addition to fluoride therapy. A total of 22% of the participants in the control group developed caries, while 0% of the participants in the test group (treated with Solea) developed caries [259].

One possible therapeutical approach is to irradiate the grooves of occlusal surfaces with this laser prior to placing sealants to further enhance the resistance to decay. Should the practitioner need to remove a sealant due to its failure, the same laser can be used for this purpose [258]. It is also important to underline that irradiation with this wavelength reduces the sensibility of dental tissues to acid etching.

◘ Figure 8.8a–e demonstrates a gingivectomy and Class V restoration, tooth #20, using the 9300 nm CO_2 laser.

8

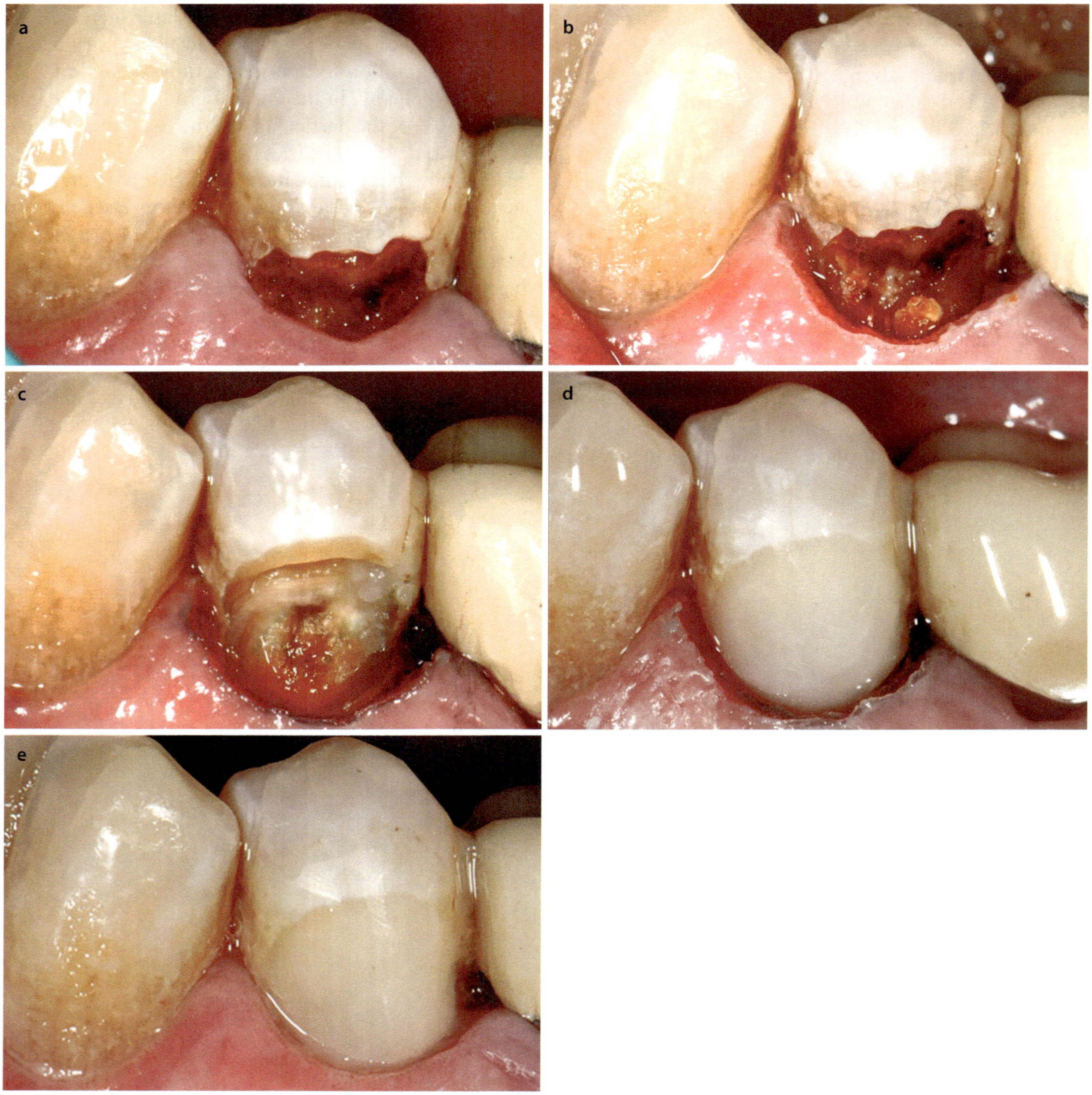

Fig. 8.8 **a** Gingivectomy and restoration, deep Class V decay tooth #20 B (Solea laser (Convergent Dental) 9300 nm CO_2. Pre-op. **b** Only topical anesthetic (TAC 20 alternative) was used. First, gingivectomy was performed to expose the apical extent of the deep subgingival decay, utilizing the 1.00 mm spot size, 1% mist, and cutting speed of ~30–40%. The procedure was performed utilizing the tipless contra angle handpiece. **c** Decay was then removed using the same settings except mist was increased from 1% to 80%. Precise control of speed of ablation was achieved by varying the working distance, hand speed, and rheostatically controlled cutting speed. Note, clean exposure of apical margin for best restoration. This is an advantage due to CO_2 ablating soft tissue more precisely than other wavelengths. **d** The tooth was then restored with Scotchbond Universal (3M) and TPH Spectra ST (Dentsply). **e** At 3 days post-op, note remarkably fast healing and excellent tissue appearance. (Procedure by Dr. Joshua Weintraub)

8.30 Pulpal Temperature Considerations

Enamel, dentin, and bone can be rapidly removed with the 9.3 μm CO_2 laser without peripheral thermal damage by mechanically scanning the laser beam and also with the aid of cooling water spray [248].

If compared to a high-speed traditional handpiece, this laser allows to avoid excessive peripheral thermal or mechanical damage [251], provided the use of enough water cooling is guaranteed, otherwise some desiccation of tissues might occur. Thermocouple measurements showed an increase in temperature of 3.3 ± 1.4 °C without water cooling versus 1.7 ± 1.6 °C with water cooling [245, 247]. Even though the tooth temperature rise was less than 5 °C without water cooling during an irradiation at 50 Hz, it is still necessary to use a water spray to produce the desired effect. This is advisable for the possible formation of non-apatitic calcium phosphate phases that are produced without a proper cooling due to excessive overheating of the mineral phase [245]. One disadvantage related to the use of this powerful laser is the formation of highly conical and deep ablation craters created when the irradiation is performed in the same spot by repeated laser pulses. This is also at the base of stalling phenomenon (cessation of ablation after penetration of 2–3 mm) and of excessive heat accumulation. To avoid this and obtain a more efficient ablation, it is necessary to use small spot sizes (<0.3 mm) and the laser should be scanned in two dimensions to expose a new area for each pulse [245, 251]. Scanning and positioning of the beam are now feasible due to recent advances in compact high-speed scanning technology such as the miniature galvanometer "galvo" based scanners [251, 260].

Figure 8.9a–g shows a gingivectomy and restoration, fractured and decayed tooth #8, using the CO_2 9.3 μm laser.

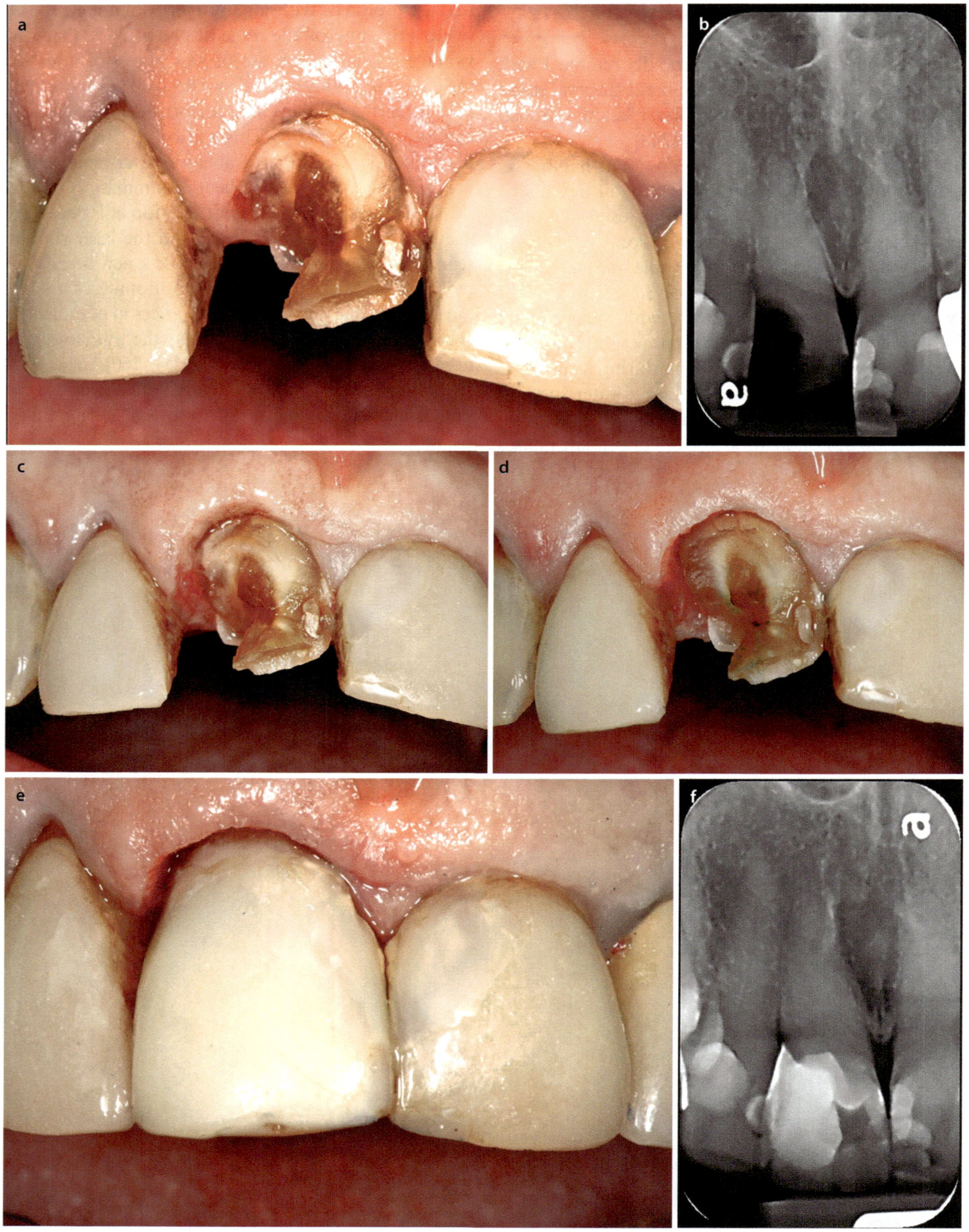

Fig. 8.9 **a** Preoperative appearance tooth #8 (Solea laser (Convergent Dental) 9300 nm CO_2). **b** Pre-op radiograph. **c** Only topical anesthetic (TAC 20 alternative) was used. First, gingivectomy was performed to expose the apical extent of the fracture and decay, utilizing the (Solea laser (Convergent Dental) 9300 nm CO_2) 1.00 mm spot size, 1% mist and cutting speed of ~30–40%. The procedure was performed utilizing the tipless contra angle handpiece. **d** Next, bone was precisely ablated utilizing the same handpiece, the 0.5 mm spot size, 80% mist, and cutting speed between ~60 and 80%. Decay was then removed utilizing the 1.00 mm spot size, 1% mist, and cutting speed of ~30–40%. Note, clean exposure of apical margin for best restoration. This is an advantage due to CO_2 ablating soft tissue and bone more precisely than other wavelengths. **e** The tooth was then restored temporarily (as the tooth was to be crowned) with Scotchbond Universal (3M) and TPH Spectra ST (Dentsply). **f** Post op radiograph. **g** At 1 week post-op, tissue has rapidly healed with excellent tissue tone. (Procedure by Dr. Joshua Weintraub)

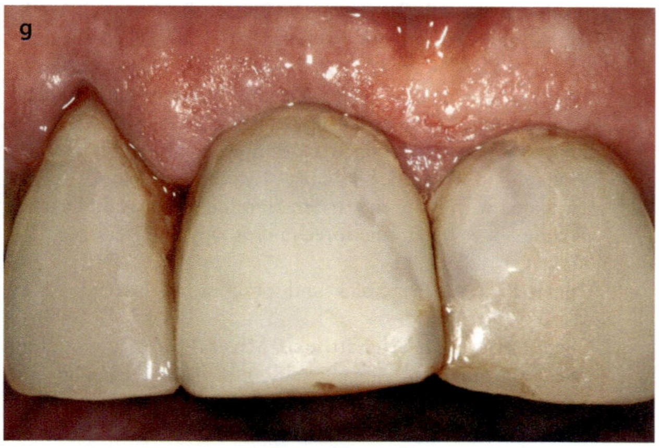

Fig. 8.9 (continued)

8.31 Composite Removal

Lasers can also be used for selective ablation of composite when replacing failed restorations or removing residual composite after debonding of orthodontic brackets [258]. The composite material can be easily and quickly removed without damage of dental surface, without any charring and limiting the removal of sound enamel. It can be effortlessly obtained at a reduced fluence, cleaning the operative area with water spray, to avoid discoloration and thermal damage [258]. For this purpose, the pulse duration should be comprised between 10 and 20 μs and the pulse repetition rate should be 200 Hz. The area of localized damage to enamel can reach a depth of less than 10 μm with a fluence of 3.2 J/cm^2 [257] or below 20 μm with a fluence in the range 5–10 J/cm^2 [258]. If the energy density exceeds the value of 4–5 J/cm^2, there will be a greater removal of enamel but this is considered unacceptable on buccal tooth surface [257]. This very limited amount of healthy enamel loss, due to the high degree of selectivity and minimal deposition of heat in the tooth, appears to be less than what it is obtainable with the conventional means of removal using dental low-speed and/or high-speed handpiece. Moreover, measurements of the enamel loss during a routine brush and prophylaxis reported average values ranging from 6 to 17 μm, depending on the material employed.

Conversely, the Er:YAG and Er,Cr:YSGG lasers, which are usually employed for this purpose, adopt higher single-pulse energy levels (100–500 mJ/pulse) and greater energy densities (20–100 J/cm^2) to remove hard tissue, orthodontic cements, and resin materials. These pulses can remove up to 50 μm of enamel and up to 200 μm of dentin each, possibly causing a severe damage to the underlying tooth structure. The temperature rise at the pulp level during composite ablation has an average maximum value of 1.9 ± 1.5 °C [137], below the critical limit of 5.5 °C [258] that it is considered dangerous for tooth vitality according to Zach and Cohen [261, 262]. In conclusion, dental hard tissues can be rapidly ablated with a mechanically scanned computer guided CO_2 laser at high pulse repetition rates without excessive heat accumulation in the tooth or peripheral thermal damage that produce no significant reduction in the tissue's mechanical strength or a major reduction of adhesive strength to restorative material [247, 250].

8.32 Air Quality in a Dental Clinic During Er:YAG Laser Use

Some of the gap phase contaminants generated and emitted from various dental procedures are volatile organic compounds. These particles have been associated with various effects on human health but mainly with respiratory disorders. Most dental aerosols have a maximum diameter of 5 μm, and long-term exposure is potentially dangerous not only for patients but, mainly, for dental staff. In 1987, Freitag and others carried out an in vivo study that highlighted the danger behind the smoke inhalation of the Nd:YAG laser; in the light of their findings, it was concluded that the mucociliary lung function was significantly depressed and that this depression was dose-dependent [263]. In 1991, a report confirmed the transmission of human papillomavirus DNA through smoke from the Nd:YAG laser [261, 264]. Though the results were not conclusive if the contamination happened due to instruments used or purely as a result of bacteria carried in the plume. McKinley and Ludow also reported that the smoke produced after argon laser irradiation of the root canals of extracted teeth inoculated with *E. coli* was positive for the growth of the bacterium used [264]. On the other hand, studies on aerosols generated with use of the 9.3 μm CO_2 laser show little to no living bacteria in the splatter as the vaporization temperatures reach above 1000 °C resulting in killing any living viruses or bacteria. [265]

Erbium ablation is thought to create an airborne particulate matter; any dentist who has used this device certainly has perceived the particular odor that spreads into the surrounding environment. Er:YAG laser irradiation on hard dental tissues produces a laser plume (the by-products of procedures such as vaporization, coagulation, and ablation include vapor, smoke, and particulate debris) due to dehydration of the tissue, cavitation of the irradiated water content, and subsequent heating of the residual solid matter to temperatures sufficient for combustion. Subsequently, the oxygen present in the ambient air will combine with the tissue elements to form a variety of by-products, many of which are unsafe. Therefore, the ablation of infected tissue can create susceptibility to cross infections due to the possible presence of infectious agents in the laser plume. A significant portion of the particles present in the laser plume is in the size range of 0.5–5.0 µm, and these particles seem too small to be effectively filtered by currently available surgical masks, while further studies are needed to evaluate the power of dental aspirators in eliminating smoke plumes. The isolation of the rubber dam in clinical practice helps to reduce contamination in the air, but its influence on volatile organic compounds and total PM, together with high power suction, remains to be evaluated. Physical ventilation in the environment is a key factor in improving air quality since both the concentrations of PM and those of volatile organic compounds decrease significantly [266].

8.33 Conclusion

For as long as laser photonic technology has been available within dentistry, there has been demand for laser-assisted hard dental and osseous tissue management. Notwithstanding the early adoption of the CO_2 soft tissue laser to offer bone ablation, much of the progress in developing clinically appropriate therapy occurred only with the development of the mid infrared wavelengths, commonly and collectively termed the "erbium family." Latterly, the emergence of a suitably tailored emission of 9300 nm CO_2 laser has broadened the options available to the restorative dentist and oral surgeon. Through this chapter, the multiple variants in energy manipulation necessary to provide sufficient power to ablate target hard oral tissue have been explored and their underlying association with the need to cause as little collateral damage to adjacent non-target tissue, especially the vital pulp, determined. Associated concepts of pain management through laser use have been evaluated together with appropriate techniques to allow the novice clinician to adopt these valuable added benefits. With a thorough understanding of the concepts of laser–tissue interaction, the biophysics involved and appreciation of the laser instruments available, the restorative clinician may easily and predictably incorporate laser photonic technology as a prime treatment adjunctive in the delivery of dental care.

References

1. De Moor RJ, Delmé KI. Laser-assisted cavity preparation and adhesion to erbium-lased tooth structure: part 1. Laser-assisted cavity preparation. J Adhes Dent. 2009;11:427–38.
2. Parker S. Lasers in restorative dentistry. In: Convissar R, editor. Principles and practice of laser dentistry, vol. 12. St. Louis: Mosby Elsevier; 2011. p. 181–202.
3. Teutle-Coyotecatl B, Contreras-Bulnes R, Scougall-Vilchis RJ, Almaguer-Flores A, Rodríguez-Vilchis LE, Velazquez-Enriquez U, Alatorre JÁA. Effect of Er:YAG laser irradiation on deciduous enamel roughness and bacterial adhesion: an in vitro study. Microsc Res Tech. 2019;82(11):1869–77.
4. Saberi S, Seyed Jabbari Doshanlo S, Bagheri H, Mir Mohammad Rezaei S, Shahabi S. Evaluation of tooth surface irradiated with erbium: yttrium aluminum garnet and carbon dioxide lasers by atomic force microscopy. J Lasers Med Sci. 2018;9(3):188–93.
5. Convissar RA. Principles and practice of laser dentistry. Mosby; 2011.
6. Chen P, Toroian D, McKittrick J. Minerals form a continuum phase in mature cancellous bone. Calcif Tissue Int. 2011;88(5):351–61.
7. Majaron B, Sustersic B, Lukac M, Skaleric U, Funduk N. Heat diffusion and debris screening. Er:YAG laser ablation of hard biological tissues. Appl Phys B Lasers Opt. 1998; 66:1–9.
8. Ivanov B, Hakimian AM, Peavy GM, Haglund RF. Mid-infrared laser ablation of hard biocomposite material: mechanistic studies of pulse duration and interface effects. Appl Surf Sci. 2003;208-9:77–84.
9. Perhavec T, Diaci J. Comparison of Er:YAG and Er,Cr:YSGG dental lasers. J Oral Laser Appl. 2008;8:87–94.
10. Apel C, Meister J, Ioana RS, Franzen R, Hering P, Gutknecht N. The ablation threshold of Er:YAG and Er,Cr:YSGG laser radiation in dental enamel. Lasers Med Sci. 2002;17:246–52.
11. Apel C, Franzen R, Meister J, Sarrafzadegan H, Thelen S, Gutknecht N. Influence of the pulse duration of an Er:YAG laser system on the ablation threshold of dental enamel. Lasers Med Sci. 2002;17:253–7.
12. Shinkai K, Takada M, Kawashima S, Suzuki M, Suzuki S. Effects of the percentage of air/water in spray on the efficiency of tooth ablation with erbium, chromium: yttrium-scandium-gallium-garnet (Er,Cr:YSGG) laser irradiation. Lasers Med Sci. 2019 Feb;34(1):99–105.
13. Gökçe B. Chap. 10: Effects of Er:YAG laser irradiation on dental hard tissues and all-ceramic materials: SEM evaluation. In: Kazmiruk V, editor. Scanning electron microscopy. IntechOpen; 2012.
14. Selting W. Fundamental erbium laser concepts: part I. J Lasers Dent. 2009;17:87–93.
15. Başaran G, Hamamcı N, Akkurt A. Shear bond strength of bonding to enamel with different laser irradiation distances. Lasers Med Sci. 2011;26:149–56.
16. Geraldo-Martins VR, Lepri CP, Palma-Dibb RG. Influence of Er,Cr:YSGG laser irradiation on enamel caries prevention. Lasers Med Sci. 2013;28:1056–9.
17. de Freitas PM, Rapozo-Hilo M, de Paula Eduardo C, Featherstone JD. In vitro evaluation of erbium, chromium:yttrium-scandium-gallium-garnet laser-treated enamel demineralization. Lasers Med Sci. 2010;25(2):165–70.
18. Ana PA, Zezell DM, Blay CC, Blay A, Eduardo CP, Miyazawa W. Thermal analysis of dental enamel following Er,Cr:YSGG laser irradiation at low fluencies. Lasers Surg Med. 2004;34(16):53–8.
19. Perhavec T, Diaci J. Comparison of heat deposition of Er:YAG and Er,Cr:YSGG lasers in hard dental tissues. J Laser Health Acad. 2009;2:1–6.
20. Featherstone JDB, Fried D. Fundamental interactions of lasers with dental hard tissues. Med Laser Appl. 2001;16: 181–94.
21. Ying D, Chuah GK, Hsu CS. Effect of Er:YAG laser and organic matrix on porosity changes in human enamel. J Dent. 2004;32:41–6.

22. Ceballos-Jiménez AY, Rodríguez-Vilchis LE, Contreras-Bulnes R, Alatorre JÁA, Velazquez-Enriquez U, García-Fabila MM. Acid resistance of dental enamel treated with remineralizing agents, Er:YAG laser and combined treatments. Dent Med Probl. 2018;55(3):255–9.
23. Mollabashi V, Rezaei-Soufi L, Farhadian M, Noorani AR. Effect of erbium, chromium-doped: yttrium, scandium, gallium, and garnet and erbium: yttrium-aluminum-garnet laser etching on enamel demineralization and shear bond strength of orthodontic brackets. Contemp Clin Dent. 2019;10(2):263–8.
24. Moslemi M, Fekrazad R, Tadayon N, Ghorbani M, Torabzadeh H, Shadkar MM. Effects of Er,Cr:YSGG laser irradiation and fluoride treatment on acid resistance of the enamel. Pediatr Dent. 2009;31:409–13.
25. Apel C, Schafer C, Gutknecht N. Demineralization of Er:YAG and Er,Cr:YSGG laser prepared enamel cavities in vitro. Caries Res. 2003;37:34–7.
26. Ana PA, Tabchoury CP, Cury JA, Zezell DM. Effect of Er,Cr:YSGG laser and professional fluoride application on enamel demineralization and on fluoride retention. Caries Res. 2012;46:441–51.
27. Fowler BO, Kuroda S. Changes in heated and in laser irradiated human tooth enamel and their probable effects on solubility. Calcif Tissue Int. 1986;38:198–208.
28. Keller U, Hibst R. Ultrastructural changes of enamel and dentin following Er:YAG laser radiation on teeth. Proc SPIE. 1990;1200:408–12.
29. Colucci V, de Souza Gabriel AE, Scatolin RS, et al. Effect of Er:YAG laser on enamel demineralization around restorations. Lasers Med Sci. 2015;30:1175–8.
30. Simsek H, Gurbuz T, Buyuk SK, Ozdemir Y. Evaluation of mineral content and photon interaction parameters of dental enamel after phosphoric acid and Er:YAG laser treatment. Photomed Laser Surg. 2017;35(5):270–7.
31. Moosavi H, Ghorbanzadeh S, Ahrari F. Structural and morphological changes in human and dentin after ablative and sub ablative Er:YAG laser irradiation. J Lasers Med Sci. 2016;7(2):86–91.
32. Ulusoy NB, Akbay Oba A, Cehreli ZC. Effect of Er,Cr:YSGG laser on the prevention of primary and permanent teeth enamel demineralization: SEM and EDS evaluation. Photobiomodul Photomed Laser Surg. 2020;38(5):308–15.
33. Pagano S, Lombardo G, Orso M, Abraha I, Capobianco B, Cianetti S. Lasers to prevent dental caries: a systematic review. BMJ Open. 2020;10(10):e038638.
34. Abbasi M, Nakhostin A, Namdar F, Chiniforush N, Hasani TM. The rate of demineralization in the teeth prepared by bur and Er:YAG laser. J Lasers Med Sci. 2018;9(2):82–6.
35. Takeda FH, Harashima T, Kimura Y, Matsumoto K. Efficacy of Er: YAG laser irradiation in removing debris and smear layer on root canal walls. J Endod. 1998;24(8):548–51.
36. Gutknecht N, Apel C, Schäfer C, Lampert F. Microleakage of composite fillings in Er, Cr: YSGG laser-prepared class II cavities. Lasers Surg Med. 2001;28(4):371–4.
37. Zamataro CB, Ana PA, Benetti C, Zezell DM. Influence of Er,Cr:YSGG laser on CaF2-like products formation because of professional acidulated fluoride or to domestic dentifrice application. Microsc Res Tech. 2013;76:704–13.
38. de Oliveira RM, de Souza VM, Esteves CM, de Oliveira Lima-Arsati YB, Cassoni A, Rodrigues JA, Brugnera JA. Er,Cr:YSGG laser energy delivery: pulse and power effects on enamel surface and erosive resistance. Photomed Laser Surg. 2017;35(11):639–46.
39. Cochrane NJ, Cai F, Huq NL, Burrow MF, Reynolds EC. New approaches to enhanced remineralization of tooth enamel. J Dent Res. 2010;89(11):1187–97.
40. Zezell DM, Boari HG, Ana PA, de Paula EC, Powell GL. Nd:YAG laser in caries prevention: a clinical trial. Lasers Surg Med. 2009;41:31–5.
41. Delbem AC, Cury JA, Nakassima CK, Gouveia VG, Theodoro LH. Effect of Er:YAG laser on CaF2 formation and its anticariogenic action on human enamel: an in vitro study. J Clin Laser Med Surg. 2003;21:197–201.
42. Subramaniam P, Pandey A. Effect of erbium, chromium: yttrium, scandium, gallium, garnet laser and casein phosphopeptide—amorphous calcium phosphate on surface microhardness of primary tooth enamel. Eur J Dent. 2014;8(3):402–6.
43. Ghelejkhani A, Nadalizadeh S, Rajabi M. Effect of casein-phosphopeptide amorphous calcium phosphate and fluoride with/without erbium, chromium-doped yttrium, scandium, gallium, and garnet laser irradiation on enamel microhardness of permanent teeth. Dent Res J (Isfahan). 2021;18:20.
44. Yassaei S, Aghili H, Shahraki N, Shahraki N, Safari I. Efficacy of erbium-doped yttrium aluminium garnet laser with casein phosphopeptide amorphous calcium phosphate with and without fluoride for remineralization of white spot lesions around orthodontic brackets. Eur J Dent. 2018;12(2):210–6.
45. Yilmaz N, Balaci E, Baygin O, Tüzüner T, Özkaya S, Canakci A. Effect of the usage of Er,Cr:YSGG laser with and without different remineralization agents on the enamel erosion of primary teeth. Lasers Med Sci. 2020;35(7):1607–20.
46. Ramalho KM, Hsu CY, de Freitas PM, Aranha AC, Esteves-Oliveira M, Rocha RG, de Paula Eduardo C. Erbium lasers for the prevention of enamel and dentin demineralization: a literature review. Photomed Laser Surg. 2015;33(6):301–19.
47. Valizadeh S, Rahimi Khub M, Chiniforush N, Kharazifard MJ, Hashemikamangar SS. Effect of laser irradiance and fluoride varnish on demineralization around dental composite restorations. J Lasers Med Sci. 2020;11(4):450–5.
48. Chen W. The clinical applications for the Er,Cr:YSGG laser system. Chen Laser Institute; 2011.
49. Niu W, Eto JN, Kimura Y, Takeda FH, Matsumoto K. A study on microleakage after resin filling of Class V cavities prepared by Er:YAG laser. J Clin Laser Med Surg. 1998;16(4):227–31.
50. Corona SA, Borsatto M, Dibb RG, Ramos RP, Brugnera A, Pécora JD. Microleakage of class V resin composite restorations after bur, air-abrasion or Er:YAG laser preparation. Oper Dent. 2001;26(5):491–7.
51. Kohara EK, Hossain M, Kimura Y, Matsumoto K, Inoue M, Sasa R. Morphological and microleakage studies of the cavities prepared by Er:YAG laser irradiation in primary teeth. J Clin Laser Med Surg. 2002;20(3):141–7.
52. Corona SA, Borsatto MC, Pecora JD, De SA Rocha RAS, Ramos TS, Palma-Dibb RG. Assessing microleakage of different class V restorations after Er:YAG laser and bur preparation. J Oral Rehabil. 2003;30(10):1008–14.
53. Bertrand MF, Semez G, Leforestier E, Muller-Bolla M, Nammour S, Rocca JP. Er:YAG laser cavity preparation and composite resin bonding with a single-component adhesive system: relationship between shear bond strength and microleakage. Lasers Surg Med. 2006;38:615–23.
54. Brulat N, Rocca JP, Leforestier E, Fiorucci G, Nammour S, Bertrand MF. Shear bond strength on self-etching adhesive systems to Er:YAG-laser-prepared dentin. Lasers Med Sci. 2009;24:53–7.

8

55. Delmé K, Meire M, De Bruyne M, Nammour S, De Moor R. Cavity preparation using an Er:YAG laser in the adult dentition. Rev Belg Med Dent. 2009;64:71–80.
56. Dilip S, Srinivas S, Mohammed Noufal MN, Ravi K, Krishnaraj R, Charles A. Comparison of surface roughness of enamel and shear bond strength, between conventional acid etching and erbium, chromium-doped: Yttrium scandium-gallium-garnet laser etching—an in vitro study. Dent Res J (Isfahan). 2018;15(4):248–55.
57. Hoshing UA, Patil S, Medha A, Bandekar SD. Comparison of shear bond strength of composite resin to enamel surface with laser etching versus acid etching: an in vitro evaluation. J Conserv Dent. 2014;17(4):320–4.
58. Jaberi Ansari Z, Fekrazad R, Felzi S, Younessian F, Kalhori KA, Gutknecht N. The effect of an Er,Cr:YSGG laser on the micro-shear bond strength of composite to the enamel and dentin of human permanent teeth. Lasers Med Sci. 2012;27:761–5.
59. Usumez S, Orhan M, Usumez A. Laser etching of enamel for direct bonding with an Er,Cr:YSGG hydrokinetic laser system. Am J Orthod Dentofac Orthop. 2002;122:649–56.
60. Ceballos L, Toledano M, Osorio R, Tay FR, Marshall GW. Bonding to Er:YAG-laser-treated dentin. J Dent Res. 2002;81:119–22.
61. Dunn WJ, Davis JT, Bush AC. Shear bond strength and SEM evaluation of composite bonded to Er:YAG laser prepared dentin and enamel. Dent Mater. 2005;21:616–24.
62. De Moor RJG, Delmè KIM. Erbium laser adhesion to tooth structure. J Oral Laser Appl. 2006;6:7–21.
63. Monghini EM, Wanderley RL, Pécora JD, Palma Dibb RG, Corona SA, Borsatto MC. Bond strength to dentin on primary teeth irradiated with varying Er:YAG laser energies and SEM examination of the surface morphology. Lasers Surg Med. 2004;34:254–9.
64. Sung EC, Lin CN, Harada V. Composite bond strength to primary dentin prepared with Er,Cr:YSGG laser. In: IADR 84th general session, Brisbane, Australia, June 28–July 1. J Dent Res. 2006;85(Special issue B).
65. Arbabzadeh Zavareh F, Samimi P, Birang R, Eskini M, Bouraima SA. Assessment of microleakage of class V composite resin restoration following erbium-doped yttrium aluminium garnet (Er:YAG) laser conditioning and acid etching with two different bonding systems. Lasers Med Sci. 2013 Winter;4(1):39–47.
66. Kidd EAM. Microleakage: a review. J Dent. 1976;4:199–205.
67. Araujo RM, Eduardo CP, Duarte JSL, Araujo MA, Loffredo LC. Microleakage and nanoleakage: influence of laser in cavity preparation and dentin pretreatment. J Clin Laser Med Surg. 2001;19(6):325–32.
68. Sano H, Shono T, Takatsu T, Hosoda H. Microporous dentin zone beneath resin-impregnated layer. Oper Dent. 1994;19(2):59–64.
69. Sano H, Takatsu T, Ciucchi B, Horner JA, Matthews WG, Pashley DH. Nanoleakage: leakage within the hybrid layer. Oper Dent. 1995;20(1):18–25.
70. Sano H, Yoshiyama M, Ebisu S, Burrow MF, Takatsu T, Ciucchi B, Carvalho R, Pashley DH. Comparative SEM and TEM observations of nanoleakage within the hybrid layer. Oper Dent. 1995;20(4):160–7.
71. Asmussen E. Composite restorative resins. Composition versus wall-to-wall polymerization contraction. Acta Odontol Scand. 1975;33:337–44.
72. Sun J, Fang R, Lin N, Eidelman N, Lin-Gibson S. Nondestructive quantification of leakage at the tooth-composite interface and its correlation with material performance parameters. Biomaterials. 2009;30:4457–62.
73. Anusavice KJ. Phillips' science of dental materials. 11th ed. St. Louis: Elsevier Science; 2003. p. 394.
74. Öznurhan F, Olmec A. Nanoleakage in primary teeth prepared by laser irradiation or bur. Lasers Med Sci. 2013;28:1099–105.
75. Li H, Burrow MF, Tyas MJ. Nanoleakage patterns of four dentin bonding systems. Dent Mater. 2000;16(1):48–56.
76. Dorfer CE, Staehle HJ, Wurst MW, Duschner H, Ploch T. The nanoleakage phenomenon: influence of different dentin bonding agents, thermocycling and etching time. Eur J Oral Sci. 2000;108(4):346–51.
77. Ilie N, Hickel R. Silorane-based dental composite: behavior and abilities. Dent Mater. 2006;25:445–54.
78. Braga RR, Ballester RY, Ferracane JL. Factors involved in the development of polymerization shrinkage stress in resin-composites: a systematic review. Dent Mater. 2005;21(10):962–70.
79. Gorucu J, Gurgan S, Cakir FY, Bicer CO, Gorucu H. The effect of different preparation and etching procedures on the microleakage of direct composite veneer restorations. Photomed Laser Surg. 2011;29(3):205–11.
80. Carvalho RM, Pereira JC, Yoshiyama M, Ashley DH. A review of polymerization contraction: the influence of stress development versus stress relief. Oper Dent. 1996;21:17–24.
81. Chandrasekhar V, Rudrapati L, Badami V, Tummala M. Incremental techniques in direct composite restoration. J Conserv Dent. 2017;20(6):386–91.
82. Eick JD, Gwinnett AJ, Pashley DH, Robinson SJ. Current concepts on adhesion to dentin. Crit Rev Oral Biol Med. 1997;8(3):306–35.
83. Labib LM, Nabih SM, Baroudi K. Nanoleakage evaluation of posterior teeth restored with low shrinkable resin composite—an invitro study. J Clin Diagn Res. 2016;10(7):ZC102–4.
84. Marshall GW Jr, Marshall SJ, Kinney JH, Balooch M. The dentin substrate: structure and properties related to bonding. J Dent. 1997;25:441–58.
85. Krmek SJ, Bogdan I, Simeon P, Mehicić GP, Katanec D, Anić I. A three-dimensional evaluation of microleakage of class V cavities prepared by the very short pulse mode of the erbium:yttrium-aluminium-garnet laser. Lasers Med Sci. 2010;25:823–8.
86. Pashley DH. Clinical correlations of dentin structure and function. J Prosthet Dent. 1991;66:777–81.
87. Perdigão J. Dentin bonding-variables related to the clinical situation and the substrate treatment. Dent Mater. 2010;26(2):e24–37.
88. Wendt SL, Mcinnes PM, Dickinson GL. The effect of thermocycling in microleakage analysis. Dent Mater. 1992;8:181–94.
89. Van Meerbeek B, De Munck J, Mattar D, Van Landuyt K, Lambrechts P. Microtensile bond strengths of an etch & rinse and self-etch adhesive to enamel and dentin as a function of surface treatment. Oper Dent. 2003;28:647–60.
90. Hilton TJ. Can modern restorative procedures and materials reliably seal cavities? In vitro investigations. Part 2. Am J Dent. 2002;15:279–89.
91. Silverstone LM, Saxton CA, Dogon IL, Fejerskov O. Variation in the pattern of acid etching of human dental enamel examined by scanning electron microscopy. Caries Res. 1975;9:373–87.
92. Barkmeier WW, Erickson RL, Kimmes NS, Latta MA, Wilwerding TM. Effect of enamel etching time on roughness and bond strength. Oper Dent. 2009;34(2):217–22.
93. Daher R, Krejci I, Mekki M, Marin C, Di Bella E, Ardu S. Effect of multiple enamel surface treatments on micro-shear bond strength. Polymers (Basel). 2021;13(20):3589.

94. Zhang QF, Yao H, Li ZY, Jin L, Wang HM. Optimal enamel conditioning strategy for rebonding orthodontic brackets: a laboratory study. Int J Clin Exp Med. 2014;7(9):2705–11.
95. Nakabayashi N, Kojima K, Masuhara E. The promotion of adhesion by the infiltration of monomers into tooth substances. J Biomed Mater Res. 1982;16:265–73.
96. Perdigao J, Swift EJ Jr, Denehy GE, Wefel JS, Donly KJ. In vitro bonds strengths and SEM evaluation of dentin bonding systems to different dentin substrates. J Dent Res. 1994;73(1):44–55.
97. Van Meerbeek B, Vargas M, Inoue S, Yoshida Y, Peumans M, Lambrechts P, Vanherle G. Adhesives and cements to promote preservation dentistry. Oper Dent. 2001;Suppl 6:119–44.
98. Rechmann P, Glodin DS, Henniing T. Changes in surface morphology of enamel after Er:YAG radiation. Lasers Dent. 1998;IV (3248):62–8.
99. Ceballos L, Osorio R, Toledano M, Marshall GW. Microleakage of composite restorations after acid or Er:YAG laser cavity treatments. Dent Mater. 2001;17:340–6.
100. Carvalho AO, Reis AF, de Oliveira MT, de Freitas PC, Aranha AC, Eduardo Cde P, Giannini M. Bond strength of adhesive systems to Er,Cr:YSGG laser-irradiated dentin. Photomed Laser Surg. 2011;29(11):747–52.
101. Arslan S, Yazici AR, Görücü J, Pala K, Antonson DE, Antonson SA, Silici S. Comparison of the effect of Er,Cr:YSGG laser and different cavity disinfection agents on microleakage of current adhesives. Lasers Med Sci. 2012;27:805–11.
102. Lee BS, Lin PY, Chen MH, Hsieh TT, Lin CP, Lai JY, Lan WH. Tensile bond strength of Er,Cr:YSGG laser-irradiated human dentin and analysis of dentin-resin interface. Dent Mater. 2007;23:570–8.
103. Hoke JA, Burkes EJ Jr, Gomes ED, Wolbarsht ML. Erbium:YAG (2.94 µm) laser effects on dental tissues. J Laser Appl. 1990;2:61.
104. Shirani F, Birang R, Malekipur MR, Zeilabi A, Shahmoradi M, Kazemi S, Khazaei S. Adhesion to Er:YAG laser and bur prepared root and crown dentine. Aust Dent J. 2012;57(2):138–43.
105. Lima DM, Tonetto MR, de Mendonça AA, Elossais AA, Saad JR, de Andrade MF, Pinto SC, Bandéca MC. Human dental enamel and dentin structural effects after Er:YAG laser irradiation. J Contemp Dent Pract. 2014;15(3):283–7.
106. Sennou HE, Lobule AA, Grégoire GL. X-ray photoelectron spectroscopy of the dentin-glass ionomer cement interface. Dent Mater. 1999;15(4):229–37.
107. Moretto SG, Azambuja N Jr, Arana-Chavez VE, Reis AF, Giannini M, Eduardo Cde P, De Freitas PM. Effects of ultramorphological changes on adhesion to lased dentin—scanning electron microscopy and transmission electron microscopy analysis. Microsc Res Tech. 2011;74:720–6.
108. Bertrand MF, Hessleyer D, Muller-Bolla M, Nammour S, Rocca JP. Scanning electron microscopic evaluation of resin–dentin interface after Er:YAG laser preparation. Lasers Surg Med. 2004;35:51–7.
109. Corona SA, de Souza AE, Chinelatti MA, Borsatto MC, Pecora JD, Palma-Dibb RG. Effect of energy density and pulse repetition rate of Er:YAG laser on dentin ablation ability and morphological analysis of the laser-irradiated surface. Photomed Laser Surg. 2007;25(1):26–33.
110. Coluzzi DJ, Parker SPA. In: Coluzzi DJ, Parker SP, editors. Textbooks in contemporary dentistry lasers in dentistry—current concepts. Springer; 2017, 397 p.
111. Oznurhan F. Morphological analysis of the resin-dentin interface in cavities prepared with Er,Cr:YSGG laser or bur in primary teeth. Photomed Laser Surg. 2013;31(8):386–91.
112. Hibst R, Keller U, Stainer R. The effect of pulsed Er:YAG laser radiation on dental hard tissues. Laser Med Surg. 1988;4:163–5.
113. Lin S, Caputo AA, Eversole LR, Rizoiu I. Topofraphical characteristics and shear bond strength of tooth surface cut with a laser-powered hydrokinetic system. J Prosthet Dent. 1999;82(4):451–5.
114. Olivi G, Olivi M. In: Olivi G, Olivi M, editors. Lasers in restorative dentistry—a practical guide. Springer; 2015. p. 95, Chapter 5.
115. Esteves-Oliveira M, Carvalho WL, de Eduardo C, Zezell DM. Influence of additional Er: YAG laser conditioning step on the microleakage of class V restorations. J Biomed Mater Res B Appl Biomater. 2008;87(2):538–43.
116. Koliniotou-Koumpia E, Kouros P, Zafiriadis L, Koumpia E, Dionysopoulos P, Karagiannis V. Bonding of adhesives to Er: YAG laser-treated dentin. Eur J Dent. 2012;6:16–23.
117. Gurgan S, Kiremitci A, Cakir FY, Yazici E, Gorucu J, Gutknecht N. Shear bond strength of composite bonded to erbium: yttrium-aluminium-garnet laser-prepared dentin. Laser Med Sci. 2009;24:117–22.
118. Roebuck EM, Sauders WP, Whitters CJ. Influence of erbium:YAG laser energies on the microleakage of class V resin-based composite restorations. Am J Dent. 2000;13:280–4.
119. Cardoso MV, Coutinho E, Ermis RB, Poitevin A, Van Landuyt K, De Munck J, Carvalho RC, Van Meerbeek B. Influence of dentin cavity surface finishing on micro-tensile bond strength of adhesives. Dent Mater. 2008;24:492–501.
120. Lopes RM, Trevelin LT, da Cunha SR, de Oliveira RF, de Andrade Salgado DM, de Freitas PM, de Paula EC, Aranha AC. Dental adhesion to erbium-lased tooth structure: a review of the literature. Photomed Laser Surg. 2015;33(8):393–403.
121. Shinoki T, Kato J, Otsuki M, Tatami J. Effect of cavity preparation with Er:YAG laser on marginal integrity of resin composite restorations. Asian Pac J Dent. 2011;11:19–25.
122. Ferreira LS, Apel C, Francci C, Simons A, Eduardo CP, Gutknecht N. Influence of etching time on bond strength in dentin irradiated with erbium lasers. Lasers Med Sci. 2010;25:849–54.
123. Maruyama H, Aoki A, Sasaki KM, Takasaki AA, Iwasaki K, Ichinose S, Oda S, Ishikawa I, Izumi Y. The effect of chemical and/or mechanical conditioning on the Er:YAG laser-treated root-cementum: analysis of surface morphology and periodontal ligament fibroblast attachment. Lasers Surg Med. 2008;40:211–22.
124. de Oliveira MT, Arrais CA, Aranha AC, de Paula EC, Miyake K, Rueggeberg FA, Giannini M. Micro-morphology of resin dentin interfaces using one-bottle etch & rinse and self-etching adhesive systems on laser-treated dentin surfaces: a confocal laser scanning microscope analysis. Lasers Surg Med. 2010;42:662–70.
125. Cardoso MV, Coutinho E, Ermis RB, Poitevin A, Van Landuyt K, De Munck J, Carvalho RC, Lambrechts P, Van Meerbeek B. Influence of Er,Cr:YSGG laser treatment on the microtensile bond strength of adhesives to dentin. J Adhes Dent. 2008;10:25–33.
126. Adu-Arko AY, Sidhu SK, McCabe JF, Pashley DH. Effect of an Er,Cr:YSGG laser on water perfusion in human dentin. Eur J Oral Sci. 2010;118:483–8.
127. De Moor RJG, Delmé KIM. Laser-assisted cavity preparation and adhesion to erbium-lased tooth structure: part 2. Present-day adhesion to erbium-lased tooth structure in permanent teeth. J Adhes Dent. 2010;12:91–102.
128. Van Meerbeek B, Yoshibar K. Clinical recipe for durable dental bonding: why and how? J Adhes Dent. 2014;16:94.

129. Mithiborwala S, Chaugule V, Munshi AK, Patil V. A comparison of the resin tag penetration of the total etch and the self-etch dentin bonding systems in the primary teeth: an in vitro study. Contemp Clin Dent. 2012;3:158–63.
130. Nor JE, Feigal RJ, Dennison JB, Edwards CA. Dentin bonding: SEM comparison of the resin-dentin interface in primary and permanent teeth. J Dent Res. 1996;75(6):1396–403.
131. Oztas N, Olmec A. Effects of one versus two-layer applications of a self-etching adhesive to dentin of primary teeth: a SEM study. J Contemp Dent Pract. 2005;6(1):18–25.
132. Nakornchai S, Harnirattisai C, Surarit R, Thiradilok S. Microtensile bond strength of a total-etching versus self-etching adhesive to caries-affected and intact dentin in primary teeth. JADA. 2005;136(4):477–8.
133. Salim DA, Andia-Merlin RY, Arana V. Micromorphological analysis of the interaction between a one-bottle adhesive and mineralized primary dentin after superficial deproteination. Biomaterials. 2004;25(19):4521–7.
134. Rontani RM, Ducatti CH, Garcia-Godoy F, De Goes MF. Effect of etching agent on dentinal adhesive interface in primary teeth. J Clin Pediatr Dent. 2000;24(3):205–9.
135. Shafiei F, Jowkar Z, Fekrazad R, Khalafi-Nezhad A. Micromorphology analysis and bond strength of two adhesives to Er,Cr:YSGG laser-prepared vs. bur-prepared fluorosed enamel. Microsc Res Tech. 2014;77:779–84.
136. Celik EU, Ergücü Z, Türkün LS, Türkün M. Shear bond strength of different adhesives to Er:YAG laser-prepared dentin. J Adhes Dent. 2006;8:319–25.
137. Moura SK, Pelizzaro A, Dal Bianco K, de Goes MF, Loguercio AD, Reis A, Grande RH. Does the acidity of self-etching primers affect bond strength and surface morphology of enamel? J Adhes Dent. 2006;8:75–83.
138. Koshiro K, Inoue S, Niimi K, Koase K, Sano H. Bond strength and SEM observations of CO2 laser irradiated dentin, bonded with simplified-step adhesives. Oper Dent. 2005;30:170–9.
139. Oznurhan F, Olmez A. Morphological analysis of the resin-dentin interface in cavities prepared with Er,Cr:YSGG laser or bur in primary teeth. Photomed Laser Surg. 2013;31(8):386–91.
140. Monteiro Ramos T, Ramos-Oliveira TM, de Freitas PM, Azambuja N Jr, Esteves-Oliveira M, Gutknecht N, de Paula Eduardo C. Effects of Er:YAG and Er,Cr:YSGG laser irradiation on the adhesion to eroded dentin. Lasers Med Sci. 2015;30:17–26.
141. Olivi G, Margolis F, Genovese MD. Pediatric laser dentistry: a user's guide. Chicago: Quintessence Pub; 2011.
142. Dua D, Dua A, Anagnostaki E, Poli R, Parker S. Effect of different types of adhesive systems on the bond strength and marginal integrity of composite restorations in cavities prepared with the erbium laser—a systematic review. Lasers Med Sci. 2022;37:19–45.
143. Sanhadji El Haddar Y, Cetik S, Bahrami B, Atash R. A comparative study of microleakage on dental surfaces bonded with three self-etch adhesive systems treated with the Er:YAG laser and bur. Biomed Res Int. 2016;2016:2509757.
144. da Silva MP, Barceleiro MO, Dias KR, Zanin F. Shear bond strength of two adhesive systems bonded to Er:YAG laser-prepared dentin. Gen Dent. 2011;59(3):e96–100.
145. Burnett LH Jr, Conceicao EN, Pelinos JE, Eduardo CD. Comparative study of influence on tensile bond strength of a composite to dentin using Er:YAG laser, air-abrasion, or air turbine for preparation of cavities. J Clin Laser Med Surg. 2001;19:199–202.
146. Nishimoto Y, Otsuki M, Yamauti M, Eguchi T, Sato Y, Foxton RM, Tagami J. Effect of pulse duration of Er:YAG laser on dentin ablation. Dent Mater J. 2008;27:433–9.
147. Delme KI, Deman PJ, De Moor RJ. Microleakage of class V resin composite restorations after conventional and Er:YAG laser preparation. J Oral Rehabil. 2005;32(9):676–85.
148. Esteves-Oliveira M, Zezell DM, Apel C, Turbino ML, Aranha AC, Eduardo Cde P, Gutknecht N. Bond strength of self-etching primer to bur cut, Er, Cr:YSGG, and Er:YAG lased dental surfaces. Photomed Laser Surg. 2007;25:373–80.
149. Tachibana A, Marques MM, Soler JMP, Matos AB. Erbium, chromium:yttrium scandium gallium garnet laser for caries removal: influence on bonding of a self-etching adhesive system. Lasers Med Sci. 2008;23:435–41.
150. Kaptan A, Oznurhan F. Effects of Er:YAG and Er,Cr:YSGG laser irradiation and adhesive systems on microtensile bond strength of a self-adhering composite. Lasers Med Sci. 2023;38(1):41.
151. Korkmaz Y, Ozel E, Attar N, Bicer CO, Firatli E. Microleakage and scanning electron microscopy evaluation of all-in-one self-etch adhesives and their respective nanocomposites prepared by erbium:yttrium-aluminum-garnet laser and bur. Lasers Med Sci. 2010;25(4):493–502.
152. Ozel E, Tuna EB, Firatli S, Firatli E. Comparison of total-etch, self-etch, and selective etching techniques on class V composite restorations prepared by Er:YAG laser and bur: a scanning electron microscopy study. Microsc Res Tech. 2016;79(10):998–1004.
153. Ramos RP, Chinelatti MA, Chimello DT, Borsatto MC, Pécora JD, Palma-Dibb RG. Bonding of self-etching and total-etch systems to Er:YAG laser-irradiated dentin. Tensile bond strength and scanning electron microscopy. Braz Dent J. 2004;15:I9–I20.
154. Tay FR, Sano H, Carvalho R, Pashley E, Pashley DH. Ultrastructural study of the influence of acidity of self-etching primers and smear layer thickness on bonding to intact dentin. J Adhes Dent. 2000;2:83–98.
155. Aranha AC, Turbino ML, Powell GL, de Paula Eduardo C. Assessing microleakage of class V resin composite restorations after Er:YAG laser and bur preparation. Lasers Surg Med. 2005;37(2):172–7.
156. Yaman BC, Guray BE, Dorter C, Gomeç Y, Yazıcıoglu O, Erdilek D. Effect of the erbium:yttrium-aluminum-garnet laser or diamond bur cavity preparation on the marginal microleakage of class V cavities restored with different adhesives and composite systems. Lasers Med Sci. 2012;27(4):785–94.
157. Hashimoto M, Hirose N, Kitagawa H, Yamaguchi S, Imazato S. Improving the durability of resin-dentin bonds with an antibacterial monomer MDPB. Dent Mater J. 2018;37(4):620–7.
158. Phanombualert J, Chimtim P, Heebthamai T, Weera-Archakul W. Microleakage of self-etch adhesive system in class V cavities prepared by using Er:YAG laser with different pulse modes. Photomed Laser Surg. 2015;33(9):467–72.
159. Visuri SR, Walsh JT, Wigdor HA. Erbium laser ablation of dental hard tissue: effect of water cooling. Lasers Surg Med. 1996;18:294–300.
160. Obeidi A, McCracken MS, Liu PR, Litaker MS, Beck P, Rahemtulla F. Enhancement of bonding to enamel and dentin prepared by Er,Cr:YSGG laser. Lasers Surg Med. 2009;41(6):454–62.
161. Russel AD. Lethal effects of heat on bacterial physiology and structure. Sci Prog. 2003;86:115–37.
162. Türkün M, Türkün LS, Celik EU, Ates M. Bactericidal effect of Er,Cr:YSGG laser on Streptococcus mutans. Dent Mater J. 2006;25:81–6.
163. Mawhara S, Mordon S. Monitoring of bactericidal action of laser by in vivo imaging of bioluminescent E. coli in a cutaneous wound infection. Lasers Med Sci. 2006;21:153–9.

164. Moritz A. Oral laser application. Berlin: Quintessence; 2006. p. 258–77.
165. Hibst R, Stock K, Gall R, Keller U. Controlled tooth surface heating and sterilization by Er:YAG laser radiation. In: Altshuler GB, editor. Laser applications in medicine and dentistry, Proc SPIE, vol. 2922. SPIE; 1996. p. 119–61.
166. Valenti C, Pagano S, Bozza S, Ciurnella E, Lomurno G, Capobianco B, Coniglio M, Cianetti S, Marinucci L. Use of the Er:YAG laser in conservative dentistry: evaluation of the microbial population in carious lesions. Materials (Basel). 2021;14(9):2387.
167. Rizoiu I, Kohanghadosh F, Kimmel AI, Eversole LR. Pulpal thermal responses to an erbium,chromium:YSGG pulsed laser hydrokinetic system. Oral Surg Oral Med Oral Pathol Oral Radiol Endod. 1998;86(2):220–3.
168. Cavalcanti BN, Lage-Marques JL, Rode SM. Pulpal temperature increases with Er:YAG laser and high-speed handpieces. JPD. 2003;90(5):447–51.
169. Attrill DC. Thermal effects of the Er:YAG laser on a simulated dental pulp: a quantitative evaluation of the effects of a water spray. JOD. 2004;32(1):35–40.
170. Amasyalı M, Sabuncuoğlu FA, Ersahan Ş, Oktay EA. Comparison of the effects of various methods used to remove adhesive from tooth surfaces on surface roughness and temperature changes in the pulp chamber. Turk J Orthod. 2019;32(3):132–8.
171. Poli R, Parker S. Achieving dental analgesia with the erbium chromium yttrium scandium gallium garnet laser (2780 nm): a protocol for painless conservative treatment. Photomed Laser Surg. 2015;33(7):364–71.
172. Ayer WA Jr, Domoto PK, Gale EN, Joy ED Jr, Melamed BG. Overcoming dental fear: strategies for its prevention and management. JADA. 1983;107:18–27.
173. Bedi R, Sutcliffe P, Donnan PT, McConnachie J. The prevalence of dental anxiety in a group of 13- and 14-year-old Scottish children. Int J Paediatr Dent. 1992;2:17–24.
174. Caprioglio A, Mariani L, Tettamanti L. A pilot study about emotional experiences by using CFSS-DS in young patients. Eur J Paediatr Dent. 2009;10(3):121–4.
175. Genovese MD, Olivi G. Laser in paediatric dentistry: patient acceptance of hard and soft tissue therapy. Eur J Paediatr Dent. 2008;9(1):13–7.
176. Leal SC, de Menezes M, Abreu D, Frencken JE. Dental anxiety and pain related to ART. J Appl Oral Sci. 2009;17(Sp. issue):84–8.
177. Houpt MI, Limb R, Livingstone RL. Clinical effects of nitrous oxide conscious sedation in children. Pediatr Dent. 2004;26(1):29–36.
178. Ryding HA, Murphy HJ. Use of nitrous oxide and oxygen for conscious sedation to manage pain and anxiety. J Can Dent Assoc. 2007;73(8):711.
179. Zacny JP, Hurst RJ, Graham L, Janiszewski DJ. Preoperative dental anxiety and mood changes during nitrous oxide inhalation. JADA. 2002;133:82–8.
180. Holroyd I. Conscious sedation in pediatric dentistry. A short review of the current UK guidelines and the technique of inhalation sedation with nitrous oxide. Pediatr Anesth. 2008;18:13–7.
181. Chan A, Armati P, Moorthy AP. Pulsed Nd:YAG laser induces pulpal analgesia: a randomized clinical trial. J Dent Res. 2012;91(7 Suppl):79S–84S.
182. Matsumoto K, Nakamura Y, Mazeki K, Kimura Y. Clinical dental application of Er:YAG Laser for class V cavity preparation. J Clin Laser Med Surg. 1996;14(3):123–7.
183. Keller U, Hibst R. Effects of Er:YAG laser in caries treatment: a clinical pilot study. Lasers Surg Med. 1997;20(1):32–8.
184. Keller U, Hibst R, Geurtsen W, Schilke R, Heidemann D, Klaiber B, Raab WH. Erbium:YAG laser application in caries therapy. Evaluation of patient perception and acceptance. J Dent. 1998;26(8):649–56.
185. Matsumoto K, Hossain M, Hossain MM, Kawano H, Kimura Y. Clinical assessment of Er,Cr:YSGG laser application for cavity preparation. J Clin Laser Med Surg. 2002;20(1):17–21.
186. Boj J, Galofre N, Espana A, Espasa E. Pain perception in pediatric patients undergoing laser treatments. J Oral Laser Appl. 2005;5(2):85–9.
187. Liu JF, Lai YL, Shu WY, Lee SY. Acceptance and efficiency of Er:YAG laser for cavity preparation in children. Photomed Laser Surg. 2006;24(4):489–93.
188. Matsumoto K, Wang X, Zhang C, Kinoshita J. Effect of a novel Er:YAG laser in caries removal and cavity preparation: a clinical observation. Photomed Laser Surg. 2007;25(1):8–13.
189. Jacobson B, Asgari A. Restorative dentistry for children using a hard tissue laser. Alpha Omegan. 2008;101(3):133–9.
190. Olivi G, Genovese MD. Laser restorative dentistry in children and adolescents. Eur Arch Paediatr Dent. 2011;12(2):68–78.
191. Fulop MA, Dhimmer S, Deluca JR, Johanson DD, Lenz RV, Patel KB, Douris PC, Enwemeka CS. A meta-analysis of the efficacy of laser phototherapy on pain relief. Clin J Pain. 2010;26(8):729–36.
192. Whitters CJ, Hall A, Creanor SL, Moseley H, Gilmour WH, Strang R, Saunders WP, Orchardson R. A clinical study of pulsed Nd:YAG laser-induced pulpal analgesia. J Dent. 1995;23(3):145–50.
193. Hoke JA, Burkes EJ Jr, Gomnes ED, Wolbarsht ML. Erbium:YAG (2.94 μm) laser effects on dental tissues. J Laser Appl. 1990;2(3–4):61–5.
194. Walsh LJ. Laser analgesia with pulsed infrared lasers: theory and practice. J Oral Laser Appl. 2008;8:7–16.
195. Cavalcanti BN, Lage-Marques JL, Rode SL. Pulpal temperature increases with Er:YAG laser and high-speed handpieces. J Prosthet Dent. 2003;90(5):447–51.
196. Attrill DC, Davies RM, King TA, Dickinson MR, Blinkhorn AS. Thermal effects of the Er:YAG laser on a simulated dental pulp: a quantitative evaluation of the effects of a water spray. J Dent. 2004;32(1):35–40.
197. Doukas AG, FlotteTJ. Physical characteristics and biological effects of laser-induced stress waves. Ultrasound Med Biol. 1996;22:151–64.
198. Melzack R, Wall P. Pain mechanisms: a new theory. Science. 1965;150:171–9.
199. Testani E, Le Pera D, Del Percio C, Miliucci R, Brancucci A, Pazzaglia C, De Armas L, Babiloni C, Rossini PM, Valeriani M. Cortical inhibition of laser pain and laser-evoked potentials by non-nociceptive somatosensory input. Eur J Neurosci. 2015;42:2407–14.
200. Orchardson R, Peacock JM, Whitters CJ. Effect of pulsed Nd:YAG laser radiation on action potential conduction in isolated mammalian spinal nerves. Lasers Surg Med. 1997;21:142–8.
201. Orchardson R, Peacock JM, Whitters CJ. Effects of pulsed Nd:YAG laser radiation on action potential conduction in nerve fibres inside teeth in vitro. J Dent. 1998;26:421–6.
202. Orchardson R, Whitters CJ. Effect of HeNe and pulsed Nd:YAG laser irradiation on intradental nerve responses to mechanical stimulation of dentine. Lasers Surg Med. 2000;26:241–9.
203. Zeredo JL, Sasaki KM, Fujiyama R, Okada Y, Toda K. Effects of low power Er:YAG laser on the tooth pulp-evoked jaw-opening reflex. Lasers Surg Med. 2003;33:169–72.
204. Zeredo JL, Sasaki KM, Takeuchi Y, Toda K. Antinociceptive effect of Er:YAG laser irradiation in the orofacial formalin test. Brain Res. 2005;1032:149–53.

205. Zeredo JL, Sasaki KM, Yozgatian JH, Okada Y, Toda K. Comparison of jaw-opening reflexes evoked by Er:YAG laser versus scalpel incisions in rats. Oral Surg Oral Med Oral Pathol Oral Radiol Endod. 2005;100:31–5.
206. Zeredo JL, Sasaki KM, Kumei Y, Toda K. Hindlimb withdrawal reflexes evoked by Er:YAG laser and scalpel incisions in rats. Photomed Laser Surg. 2006;24:595–600.
207. Zeredo JL, Sasaki KM, Toda K. High-intensity laser for acupuncture-like stimulation. Lasers Med Sci. 2007;22:37–41.
208. Walsh LJ. Laser dentistry: membrane-based photoacoustic and biostimulatory applications in clinical practice. Australas Dent Pract. 2006;17:62–4.
209. Rice WJ, Young HS, Martin DW, Sachs JR, Stokes DL. Structure of Na+,K+-ATPase at 11-A resolution: comparison with Ca2+ATPase in E1 and E2 states. Biophys J. 2001;80:2187–97.
210. Pomfret AJ, Rice WJ, Stokes DL. Application of the iterative helical real-space reconstruction method to large membranous tubular crystals of P-type ATPases. J Struct Biol. 2007;157:106–16.
211. Snyder-Mackler L, Bork CE. Effect of helium-neon laser irradiation on peripheral sensory nerve latency. Phys Ther. 1988;68:223–5.
212. Maeda T. Morphological demonstration of low reactive-laser therapeutic pain attenuation effect of the gallium aluminium arsenide diode laser. Laser Ther. 1989;1:23–30.
213. Wesselman U, Lin S, Rymer W. Effects of Q-switched Nd:YAG laser irradiation on neural impulse propagation: I. Spinal cord. Physiol Chem Phys Med NMR. 1991;23:67–80.
214. Wakabayashi H, Hamba M, Matsumoto K, Tachibana H. Effect of irradiation by semiconductor laser on responses evoked in trigeminal caudal neurons by tooth pulp stimulation. Lasers Surg Med. 1993;13:605–10.
215. Kono T, Kasai S, Sakamoto T, Mito M. Cord dorm potentials suppressed by low power laser irradiation on a peripheral nerve in the cat. J Clin Laser Med Surg. 1993;11:115–8.
216. Sato T, Kawatani M, Takeshige C, Matsumoto I. Ga-Al-As laser irradiation neuronal activity associated with inflammation. Acup Electrother Res. 1994;19:141–51.
217. Tsuchiya D, Kawatani M, Takeshige C. Laser irradiation abates neuronal responses to nociceptive stimulation of rat-paw skin. Brain Res Bull. 1994;34:369–74.
218. Chow RT, David MA, Armati PJ. 830 nm laser irradiation induces varicosity formation, reduces mitochondrial membrane potential and blocks fast axonal flow in small and medium diameter rat dorsal root ganglion neurons: implications for the analgesic effects of 830 nm laser. J Periph Nerv Syst. 2007;12(1):28–39.
219. Chow RT, Armati PJ. Photobiomodulation: implications for anesthesia and pain relief. Photomed Laser Surg. 2016;34(12):599–609.
220. Elbay ÜŞ, Tak Ö, Elbay M, Uğurluel C, Kaya C. Efficacy of low-level laser therapy in the management of postoperative pain in children after primary teeth extraction: a randomized clinical trial. Photomed Laser Surg. 2016;34(4):171–7.
221. Bjordal JM, Johnson MI, Iversen V, Aimbire F, Lopes-Martins RA. Photoradiation in acute pain: a systematic review of possible mechanisms of action and clinical effects in randomized placebo-controlled trials. Photomed Laser Surg. 2006;24(2):158–68.
222. Angelieri F, Sousa MVDS, Kanashiro LK, Furquim Siqueira D, Ávila Maltagliati L. Effects of low intensity laser on pain sensitivity during orthodontic movement. Dent Press J Orthod. 2011;16(4):95–102.
223. Deana NF, Zaror C, Sandoval P, Alves N. Effectiveness of low-level laser therapy in reducing orthodontic pain: a systematic review and meta-analysis. Pain Res Man. 2017;2017:1–18.
224. Hamblin MR. The role of nitric oxide in low level light therapy. Proc SPIE. 2008;6846:1–14.
225. Farivar S, Malekshahabi T, Shiari R. Biological effects of low level laser therapy. J Lasers Med Sci. 2014;5:58–62.
226. Laasko E, Cramond T, Richardson C, Galligan J. Plasma ACTH and beta-endorphin levels in response to low-level laser therapy (LLLT) for myofascial trigger points. Laser Ther. 1994;6:33–42.
227. Hagiwara S, Iwasaka H, Hasegawa A, Noguchi T. Pre-irradiation of blood by gallium aluminium arsenide (830 nm) low-level laser enhances peripheral endogenous opioid analgesia in rats. Anesth Analg. 2008;107:1058–63.
228. Laasko E, Cabot PJ. Nociceptive scores and endorphin-containing cells reduced by low-level laser therapy (LLLT) in inflamed paws of Wistar rat. Photomed Laser Surg. 2005;23:32–5.
229. Bruehl S, Burns JW, Chung OY, Chont M. What do plasma beta-endorphin levels reveal about endogenous opioid analgesic function? Eur J Pain. 2012;16(3):370–80.
230. Moriyama Y, Nguyen J, Akens M, Moriyama EH, Lilge L. In vivo effects of low-level laser therapy on inducible nitric oxide synthase. Lasers Surg Med. 2009;41(3):227–331.
231. Cidral-Filho FJ, Mazzardo-Martins L, Martins DF, Santos AR. Light-emitting diode therapy induces analgesia in a mouse model of postoperative pain through activation of peripheral opioid receptors and the L-arginine/nitric oxide pathway. Lasers Med Sci. 2014;29:695–702.
232. Aimbire F, Albertini R, Pacheco MT, Castro-Faria-Neto HC, Leonardo PS, Iversen VV, Lopes-Martins RA, Bjordal JM. Low-level laser therapy indices dose-dependent reduction of TNFa levels in acute inflammation. Photomed Laser Surg. 2006;24:33–7.
233. Boschi ES, Leite CE, Saciura VC, Caberlon E, Lunardelli A, Bitencourt S, Melo DA, Oliveira JR. Anti-inflammatory effects of low-level laser therapy (660 nm) in the early phase in carrageenan-induced pleurisy in rat. Lasers Surg Med. 2008;40:500–8.
234. Poli R, Parker S, Anagnostaki E, Mylona V, Lynch E, Grootveld M. Laser analgesia associated with restorative dental care: a systematic review of the rationale, techniques, and energy dose considerations. Dent J (Basel). 2020;8(4):128.
235. Jacobsen T, Norlund A, Englund GS, Tranæus S. Application of laser technology for removal of caries: a systematic review of controlled clinical trials. Acta Odontol Scand. 2011;69:65–74.
236. Tantbirojn D, Walinski CJ, Ross JA, Taylor CR, Versluis A. Composite removal by means of erbium, chromium: yttrium-scandium-gallium-garnet laser compared with rotary instruments. J Am Dent Assoc. 2019;150(12):1040–7.
237. Ferrari M, Vichi A, Mannocci F, Mason PN. Retrospective study of the clinical performance of fiber posts. Am J Dent. 2000;13(Spec No):9–13B.
238. Pegoretti A, Fambri L, Zappini G, Bianchetti M. Finite element analysis of a glass fibre reinforced composite endodontic post. Biomaterials. 2002;23(13):2667–82.
239. Kirmali O, Kustarci A, Kaplan A, Er K. Effects of dentin surface treatments including Er,Cr:YSGG laser irradiation with different intensities on the push-out bond strength of the glass fiber posts to root dentin. Acta Odontol Scand. 2015;73:380–6.
240. Cekic-Nagas I, Sukuroglu E, Canay S. Does the surface treatment affect the bond strength of various fibre-post systems to resin-core materials? J Endod. 2011;39:171–9.
241. Valandro LF, Yoshiga S, de Melo RM, Galeano GA, Mallmann A, Marinho CP, Bottino MA. Microtensile bond strength between a quartz fiber post and a resin cement: effect of post surface conditioning. J Adhes Dent. 2006;8:105–11.

242. de Souza Menezes M, Queiroz EC, Soares PV, Faria-e-Silva AL, Soares CJ, Martins LR. Fiber post etching with hydrogen peroxide: effect of concentration and application time. J Endod. 2011;37:398–402.
243. Kurtulmus-Yilmaz S, Cengiz E, Ozan O, Ramoglu S, Yilmaz HG. The effect of Er,Cr:YSGG laser application on the micropush-out bond strength of fiber posts to resin core material. Photomed Laser Surg. 2014;32(10):574–81.
244. Elsaka SE. Influence of chemical surface treatments on adhesion of fiber posts to composite resin core materials. Dent Mater. 2013;29:550–8.
245. Staninec M, Darling CL, Goodis HE, Pierre D, Cox DP, Fan K, Larson M, Parisi R, Hsu D, Manesh SK, Ho C, Hosseini M, Fried D. Pulpal effects of enamel ablation with a microsecond pulsed μ=9.3-μm CO2 laser. Lasers Surg Med. 2009;41(4):256–63.
246. Assa S, Meyer S, Fried D. Ablation of dental hard tissues with a microsecond pulsed carbon dioxide laser operating at 9.3-μm with an integrated scanner. Proc SPIE Int Soc Opt Eng. 2008;6843:684308.
247. Fantarella D, Kotlow L. The 9.3-μm CO2 dental laser: technical development and early clinical experiences. J Lasers Dent. 2014;22(1):10–27.
248. Nguyen D, Chang K, Hedayatollahnajafi S, Staninec M, Chan K, Lee R, Fried D. High-speed scanning ablation of dental hard tissues with a λ = 9.3 μm CO2 laser: adhesion, mechanical strength, heat accumulation, and peripheral thermal damage. J Biomed Opt. 2011;16(7):071410-1–071410-419.
249. Chung LC, Tom H, Chan KH, Simon JC, Fried D, Darling CL. Image-guided removal of occlusal caries lesions with a λ= 9.3-μm CO2 laser using near-IR transillumination. Proc SPIE Int Soc Opt Eng. 2015;9306:93060N.
250. Fan K, Bell P, Fried D. Rapid and conservative ablation and modification of enamel, dentin, and alveolar bone using a high repetition rate transverse excited atmospheric pressure CO2 laser operating at lambda=9.3 micro. J Biomed Opt. 2006;11(6):064008.
251. Maung LH, Lee C, Fried D. Near-IR imaging of thermal changes in enamel during laser ablation. Proc SPIE Int Soc Opt Eng. 2010;7546(1):754902.
252. Dela Rosa AA, Sarna AV, Le CQ, Jones RS, Fried D. Peripheral thermal and mechanical damage to dentin with microsecond and sub-microsecond 9.6 μm, 2.79 μm, and 0.355 μm laser pulses. Lasers Surg Med. 2004;35:214–28.
253. Badreddine AH, Couitt S, Donovan J, Cantor-Balan R, Kerbage C, Rechmann P. Demineralization inhibition by high-speed scanning of 9.3 μm CO2 single laser pulses over enamel. Lasers Surg Med. 2021;53:703–12.
254. Fried D, Featherstone JD, Le CQ, Fan K. Dissolution studies of bovine dental enamel surfaces modified by high-speed scanning ablation with a λ=9.3μm TEA CO2 laser. Lasers Surg Med. 2006;38(9):837–45.
255. Fan K, Fried D. A high repetition rate TEA CO2 laser operating at λ=9.3-mm for the rapid and conservative ablation and modification of dental hard tissue. Proc SPIE Int Soc Opt Eng. 2006;6137 https://doi.org/10.1117/12.661794.
256. Can AM, Darling CL, Ho C, Fried D. Non-destructive assessment of inhibition of demineralization in dental enamel irradiated by a lambda=9.3-μm CO2 laser at ablative irradiation intensities with PS-OCT. Lasers Surg Med. 2008;40(5):342–9.
257. Chan KH, Hirasuna K, Fried D. Analysis of enamel surface damage after selective laser ablation of composite from tooth surfaces. Photon Lasers Med. 2014;3(1):37–45.
258. Chan KH, Hirasuna K, Fried D. Rapid and selective removal of composite from tooth surfaces with a 9.3 μm CO2 laser using spectral feedback. Lasers Surg Med. 2011;43(8):824–32.
259. Rechmann P, Kubitz M, Chaffee BW, Rechmann BMT. Fissure caries inhibition with a CO2 9.3-μm short-pulsed laser-a randomized, single-blind, split-mouth controlled, 1-year clinical trial. Clin Oral Investig. 2021;25(4):2055–68.
260. Tom H, Chan KH, Darling CL, Fried D. Near-IR image-guided laser ablation of demineralization on tooth occlusal surfaces. Lasers Surg Med. 2016;48(1):52–61.
261. Zach L, Cohen G. Pulp response to externally applied heat. Oral Surg Oral Med Oral Pathol. 1965;19:515–30.
262. Hallmo P, Naess O. Laryngeal papillomatosis with human papillomavirus DNA contracted by a laser surgeon. Eur Arch Otorhinolaryngol. 1991;248:425–7.
263. Freitag L, Chapman GA, Sielczak M, Ahmed A, Russin D. Laser smoke effect on the bronchial system. Lasers Surg Med. 1987;7:283–8.
264. McKinley IB Jr, Ludlow MO. Hazards of laser smoke during endodontic therapy. J Endod. 1994;20:558–9.
265. Badreddine A, Patter K, Kerbage C, Linden E. Decontamination of hard tissue using a scanning pulsed 9.3-μm CO laser. J Dent Sci. 2021;6(3):000304.
266. Karveli A, Tzoutzas IG, Raptis PI, Tzanakakis EC, Farmakis ETR, Helmis CG. Air quality in a dental clinic during Er:YAG laser usage for cavity preparation on human teeth-an ex-vivo study. Int J Environ Res Public Health. 2021;18(20):10920.

Laser-Assisted Endodontics

Roy George and Laurence J. Walsh

Contents

D. J. Coluzzi, S. P. A. Parker (eds.), *Lasers in Dentistry—Current Concepts*, Textbooks in Contemporary Dentistry,
https://doi.org/10.1007/978-3-031-43338-2_9

Core Message

Endodontics is a technically demanding aspect of dentistry because of the limited ability to see and instrument the root canals of teeth. Conventional endodontic treatment is painstakingly slow, and traditional hand and powered endodontic instruments do not contact all the walls of the root canal. The use of lasers can assist in providing enhanced detection of bacteria and microbial biofilms in the root canal to guide debridement approaches and help define endpoints for instrumentation. Fluorescence feedback can indicate where microbial deposits remain and where further treatment is needed. There are a range of ways that lasers can enhance biomechanical preparation of the root canal system, particularly through fluid agitation and inducing cavitation in water-based fluids, to remove debris and smear layers. Such actions can be optimized using modified fibers to deliver the laser energy in various side-firing patterns. Lasers can achieve disinfection of the root canal through both photothermal and photodynamic processes, reaching with laser energy areas that are difficult to access using instruments, to inactivate microbial pathogens. Additional applications of lasers in endodontics include the assessment of pulp vitality through laser Doppler flowmetry, photothermal and photodynamic bleaching of discolored sclerosed vital or stained non-vital teeth, pulp capping and pulpotomy, photobiomodulation and laser-induced analgesia, and endodontic surgical applications including periapical surgery and treatment of invasive cervical resorption. In each of these areas, the use of lasers can simplify treatment protocols and optimize clinical outcomes.

9

Table 9.1 Classification of uses of lasers in endodontics

Primary application	Examples
Diagnosis	Detection of pulp vitality • Doppler flowmetry • Low-level laser therapy (LLLT) • Laser fluorescence • Detection of bacteria
Pulp therapy	Pulp capping Pulpotomy
Canal preparation	Biomechanical preparation Removal of smear layer Sterilization of the root canal • High-level lasers—photothermal disinfection • Low-level lasers—photodynamic disinfection
Periapical surgery	Ablation of granulation tissue Bone cutting and root resection
Laser photobiomodulation	Laser-induced analgesia Accelerated healing after pulpotomy or periapical surgery
Other	Removal of root canal filling materials and fractured instrument Softening gutta-percha Removal of moisture/drying of canal

9.1 Introduction

The primary goal of root canal treatment is to eliminate microorganisms from the root canal system and the radicular dentin. Laser technology can assist in the diagnosis of microbial deposits to guide biomechanical treatments and can help to inactive organisms through a range of thermal, photodynamic, and photomechanical processes. Unlike mechanical instrumentation such as hand or rotary endodontic files, laser effects reach across the entire root canal system and penetrate to some extent into dentine tubules. It is well known that with conventional instrumentation, large parts of the root canal system are untouched. The use of lasers, in many cases combined with appropriate fluids, can assist in achieving the goals of three-dimensional cleaning and profound disinfection of the canal.

Supporting these primary goals of endodontics are other laser applications in the diagnostic and therapeutic categories, as shown in Table 9.1. The uses of various lasers used in endodontics and some of their more popular applications are listed in Table 9.2.

Table 9.2 Selected applications of lasers

Laser	Wavelength	Reported uses in endodontics
Short wave length	[a]445 nm–514.5 nm	Endodontic disinfection
	KTP 532 nm	Soft tissue surgery in endodontics, endodontic disinfection
	[a]He–Ne 633 nm Diode 635 nm	Doppler flowmetry, photoactivated disinfection of root canals
	Diode 810–980 nm	Soft tissue surgery in endodontics, endodontic disinfection, laser-induced analgesia, laser photobiomodulation
	Nd:YAG 1064 nm	Soft tissue surgery in endodontics, endodontic disinfection, biomechanical preparation
Long wave length	Er,Cr:YSGG 2780 nm	Tooth preparation, soft and hard tissue surgery in endodontics, endodontic disinfection, biomechanical preparation
	Er:YAG 2940 nm	Tooth preparation, soft and hard tissue surgery in endodontics, endodontic disinfection, biomechanical preparation
	Carbon dioxide 10,600 nm	Pulp capping; soft tissue surgery in endodontics

[a] Commercial availability for use in dentistry is limited due to the replacement with 635-nm diode lasers, so its inclusion here is purely for historical reasons

9.2 Diagnostic Laser Applications

9.2.1 Laser Doppler Flowmetry

During dental pulp sensibility testing, a pain response elicited to a hot or cold stimulus or to an electric pulp tester provides information about the dental pulpal sensory supply, but not about its blood supply. Although the sensitivity of these commonly used tests is high, false results can lead to unnecessary endodontic treatment. This is a particular problem when teeth have experienced dental trauma or are undergoing orthodontic tooth movement [1].

In laser Doppler flowmetry (LDF), laser light is transmitted through tooth structure to the dental pulp by means of a fiber-optic probe held in a reproducible position on the tooth surface. If the pulp is vital, there will be blood flow within the tissue. With movement of erythrocytes, the scattered light is frequency-shifted, while light reflected from static tissue is unshifted. The reflected light is analyzed for its frequency shift to give a noninvasive, objective, painless, semi-quantitative assessment of pulpal blood flow. LDF has been used to estimating pulpal vitality in both adults and children, particularly in teeth which have been affected by dental trauma, excessive occlusal forces, or orthodontic movement.

LDF can also assist in the recognition of non-vital teeth. For this application, LDF has been found to be particularly valuable for assessing the blood flow in luxated teeth, as the pattern of results over time can guide decisions around "at-risk" teeth so that loss of vitality can be recognized and can then trigger endodontic intervention [2, 3].

While LDF is regarded as a highly accurate method for diagnosing the state of pulpal health and indeed comes closest to serving as a "gold standard," it must be recognized that LDF readings are prone to interferences from environmental- and technique-related factors, including superimposed signals arising from blood flow in the periodontal tissues rather than in the pulp, the posture of the patient, and their heart rate. If the laser light reaches the periodontium, then the reflected signal will not be entirely of pulpal origin [4–7]. One technique to overcome this is to place the probe on dentine in the floor of a cavity in the tooth, rather than on the enamel surface, since this is closer to the dental pulp and improves the signal-to-noise ratio [8].

There has also been work to explore the applications of transmitted laser light, rather than reflected laser light as in LDF. It has been suggested that transmitted light would be useful for the assessment of tooth pulp vitality both because the blood flow signals do not include flow of non-pulpal (e.g., periodontal) origin and because the response to blood flow changes is more obvious [9].

With both LDF and the transmitted light approach, it must be borne in mind that transmission of laser light may be influenced to some extent by tooth shade as well as by the presence of dental caries and restorations. Light can however be conducted within irregular secondary dentine, so the presence of carious lesions or tooth-colored restorations in molar teeth does not always prevent laser light reaching the pulp space. Light will not, however, pass through amalgam restorations or gold crowns [10, 11].

Key Points for Laser Doppler Flowmetry

- Laser wavelengths must penetrate normal tooth structure.
- Visible red and near-infrared wavelengths are the most suitable.
- Finding a reproducible position for the tip is important as repeated measurements will be made
- LDF cannot be used when the tooth is covered with an opaque full coverage restoration.

9.2.2 Fluorescence Diagnosis of the Root Canal System

Traditional culture-based techniques for assessing the presence of microorganisms in planktonic form or in biofilms in root canal system are difficult to use and prone to error. Real-time assessment of the microbial status of the root canal system using laser fluorescence has been developed to overcome these limitations and provide information that can guide clinical decisions around treatment endpoints [12].

The proof of concept for this application used an existing laser fluorescence device, the DIAGNOdent (KaVo, Biberach, Germany), which utilizes visible red laser light (wavelength 655 nm) to elicit fluorescence emissions in the near-infrared range. Initially, this was used with a rigid sapphire tip to analyze the pulp chamber and coronal third of the root canal system in extracted teeth with infected and uninfected root canals. The fluorescence properties of bacterial cultures, mono-species biofilms in root canals, pulpal soft tissues, and sound dentin were also evaluated, together with extracted teeth with known endodontic pathology. The baseline for sound dentin and healthy pulpal soft tissue was established as an average fluorescence reading of 5 (on a scale of 100), whereas biofilms of *Enterococcus faecalis* and *Streptococcus mutans* established in root canals showed a progressive increase in fluorescence over time [13]. Fluorescence readings reduced to the "healthy" threshold reading of 5 when root canals were endodontically treated, and the experimentally created bacterial biofilms were removed completely. The mechanism of action for these observations is that components within bacteria, such as porphyrin derivatives, fluoresce when they are excited by ultraviolet or visible light [14, 15]. Even Gram-positive bacteria such as *Streptococcus mutans* and *Enterococcus faecalis* can be detected using fluorescence generated by 655-nm laser light [13, 16].

High fluorescence readings were recorded in the root canals and pulp chambers of extracted teeth with radiographic evidence of periapical pathology and scanning electron microscopy evidence of bacterial infection. This confirmed that a laser fluorescence diagnostic approach could be useful for assessing the status of the pulp chamber and root canal system [13].

Development of thin flexible fiber tips to gain greater penetration into middle and apical thirds of the root canal was necessary to evaluate the performance of optical fibers (◘ Fig. 9.1a, b). Fibers with either plain or conically modified ends, connected to a fluorescence diagnostic system, were also used to assess canals of extracted teeth with known periapical pathology. Diameter of fibers and their penetration into root canals with different curvatures were also tested. It was found that the fibers could reach the apical third of the root canal, unless the canals had distal curvatures greater than 15°. Penetration was greater for fiber optics with a conical/radial end design than for fibers with a plain/bare end design. The self-guiding action of the conical tip prevented frictional binding onto the canal walls and hence allowed for greater penetration. Fluorescence readings were significantly higher in infected canals (range, 19–99) than in noninfected canals and sound radicular dentin (range 2–8) [16].

To further enhance the ability to take fluorescence fiber-optic readings from the walls of the root canals, a cone-shaped tip with optimal properties for the lateral emission and collection of light was developed [19–21]. Commercial optical fibers were altered by tube etching with hydrofluoric acid, modified tube etching (after removing the protective polyimide coating), alumina abrasive particle beams, and etching and particle beams used in combination. Laser emissions both forward and laterally were measured and visibly traced using He-Ne lasers (632.8 nm) or InGaAsP diode lasers (635 and 670 nm). It was found that a particular etching/abrasion/etching combination gave a unique honeycomb surface configuration with grating-like properties. This had unique micro-patterns which were not seen on fibers which had been either etched or abraded. The honeycomb tips showed ideal radial emission and collection of light for fluorescence assessment of the root canal. This tip design was then used on fibers made of various materials and of different sizes [19–21].

The possibility now exists to combine fluorescence diagnostics with an endodontic treatment system (◘ Fig. 9.2). This has already been demonstrated for the removal of infected carious dentine and subgingival deposits of plaque, based on the fluorescence properties of the porphyrin compounds (contained in bacteria) [22–27]. For example, using the DIAGNOdent system, 655-nm visible red laser light elicits porphyrin fluorescence emissions over 780 nm in wavelength, which can readily be measured to give a quantitative relative fluorescence score ranging from 0 to 100. The healthy circumpulpal dentine, walls of a healthy uninfected root canal, and healthy dental pulp tissue all give fluorescence readings in the order of the 5–6 range using this

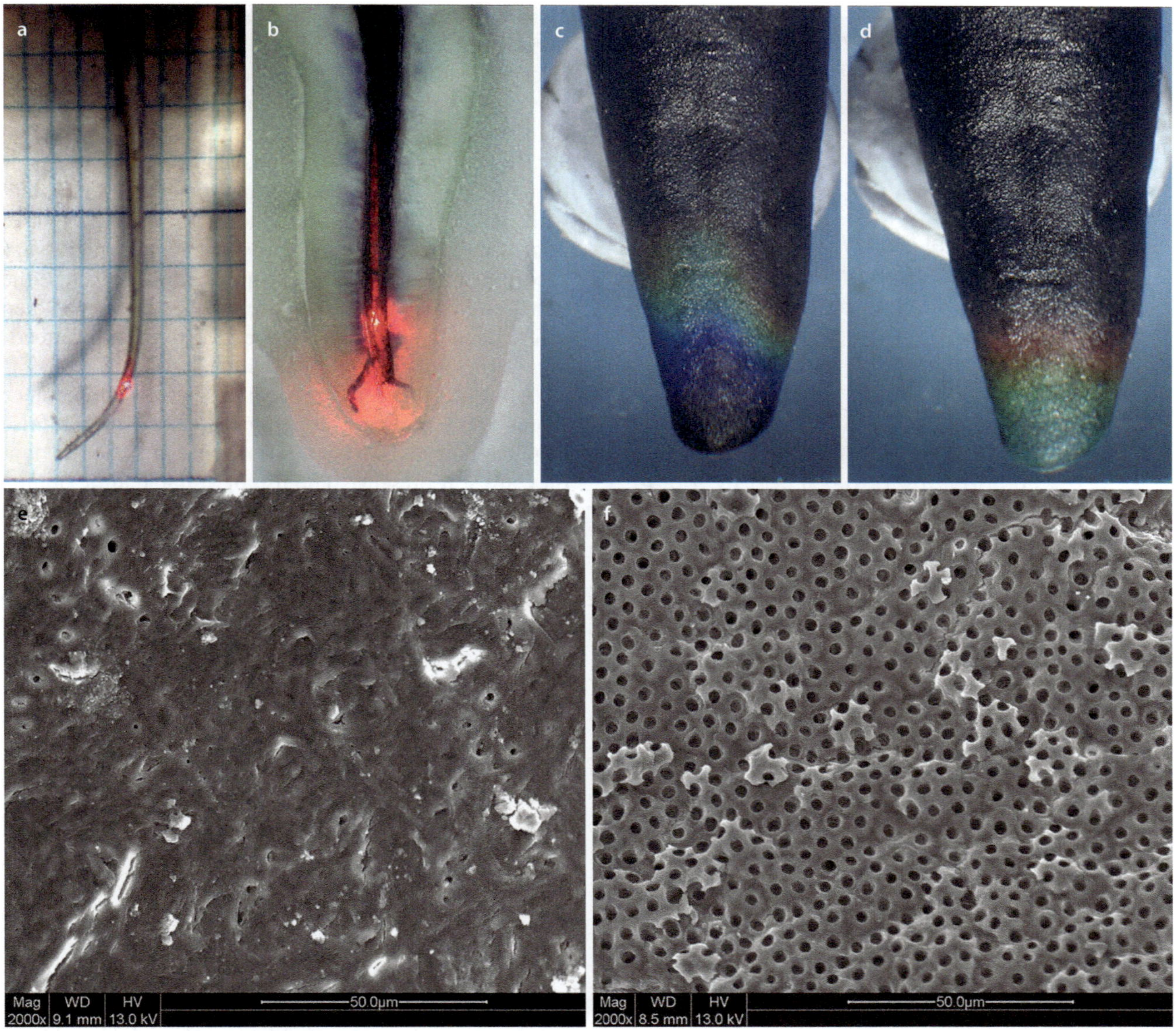

Fig. 9.1 Optical fibers and their applications in laser endodontics. **a** Conventional plain-ended fiber placed into an epoxy resin replica of the root canal showing forward emission of visible red laser energy. Using plain ended tips requires the fiber to be moved to achieve irradiation of the canal walls. **b** Honeycomb fiber placed into the root canal showing lateral emission of visible red laser energy. To enhance visibility, the canal was filled with ink prior to inserting the fiber. **c** Thermochromic (heat-sensitive) dye applied to the root surface showing subtle thermal changes during lasing of less than half a degree Celsius when the honeycomb tip is used to activate water-based fluid (500 mJ/pulse at 4 Hz). The point of greatest thermal change is the blue color change, followed by green and then by red. **d** The same tooth 5 s after lasing has stopped, showing dissipation of thermal changes at the root surface (for details, see Ref. [17]). **e** Smear layer present on the root canal walls when rotary instruments are used with water as the lubricant fluid. SEM magnification 2000×. **f** The same location after using a conical tip fiber to deliver 940-nm diode laser energy to activate EDTA irrigant. The laser was applied for ten cycles of 10-s duration using 80 mJ/pulse at 50 Hz (for details, see [18])

method [13, 16] (Fig. 9.3). There is also the possibility of using long wavelength ultraviolet light (380–400 nm) or violet light (405 nm) to elicit fluorescence emissions from bacterial deposits. In this instance, the emissions are in the visible red region [28–30].

For the successful use of fluorescence to guide laser-based treatment methods, it is important to recognize factors which could impair the fluorescence process, for example, by quenching fluorescence emissions (such as hydrogen peroxide or ozone).

A false-negative signal could affect the diagnostic value of fluorescence-assisted assessment of endpoint to chemo-mechanical preparation of the root canal system disinfection. In endodontic treatment, irrigating solutions such as sodium hypochlorite and hydrogen peroxide are used because of their chemical actions which

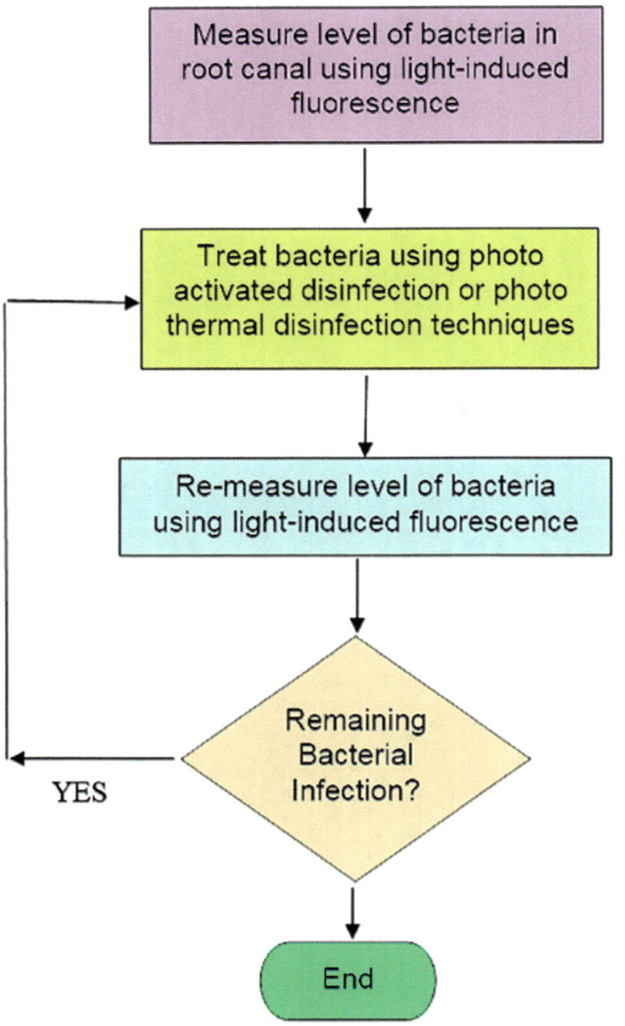

Fig. 9.2 Algorithm for using fluorescence to control laser-based debridement and disinfection. (Adapted from Ref. [19] and US patent 8,977,085)

assist the removal of debris and disinfection [31, 32]. Such irrigation solutions could potentially impair fluorescence readings through quenching. Sin et al. reported that NaOCl and H_2O_2 can quench fluorescence readings of human-uninfected and human-infected dentine [33, 34]. In addition, other solutions (e.g., 4% articaine containing 1:100,000 adrenaline) that may come in contact with dentine during endodontic treatment may reduce fluorescence readings [35]. These quenching effects vary according to concentration of quenching agents and the time that microbial deposits are exposed to them. When severe quenching occurs, recovery of fluorescence readings to the baseline may not occur even within 24 h.

After exposure to quenching agents, fluorescence scores are suppressed, creating the scenario of a false negative (suggesting bacteria are absent, when in fact they are still present). The use of suitable antioxidants and scavengers of oxygen free radicals can obviate such problems. Recently, it was reported that vitamin E oil and buffered 2% lignocaine with 1:80,000 adrenaline could act as suitable scavengers and were able to restore attenuated fluorescence readings back to baseline within approximately 10 min [36].

It is also important to recognize situations, which give rise to false-positive signals (i.e., when bacteria are not present), such as when tetracyclines are used in medicament pastes, and rapidly become incorporated into the dentine of the root canal walls. Moreover, Tsai et al. reported elevated dentine fluorescence readings following exposure to 0.2% chlorhexidine (CHX), 3% mepivacaine, chloroform, and Ledermix™ paste which contains demeclocycline hydrochloride [35].

It is important to note that not all solutions or medicaments used in endodontic practice have an effect on fluorescent reading. Tsai et al. (2022) reported that a 2-min application of distilled water, 15% EDTA,

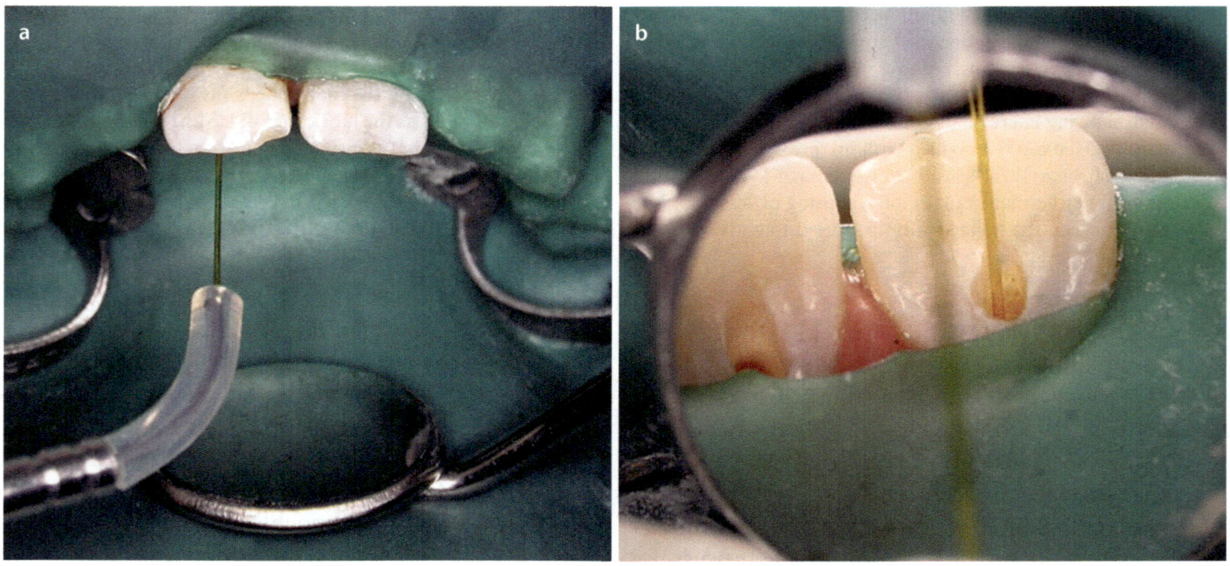

Fig. 9.3 Panels **a** and **b** show labual and palatal (mirror) views of the fiber entering the canal. Use of real-time fiber-optic detection of bacteria. A flexible fiber with a specially treated surface is inserted into the canal. The typical fiber diameter is 150–200 μm. The fiber is then linked to a fluorescence diagnostic system

Calmix™ calcium hydroxide paste, Odontopaste™ calcium hydroxide paste, 2% lignocaine with 1:80,000 adrenaline, and eucalyptus oil had no effect on dentine fluorescence readings [35].

Key Points for Laser Fluorescence Assessment

- False-positive fluorescence can occur with certain endodontic medicaments (e.g., tetracyclines).
- Oxidants can quench fluorescence signals.
- Fiber optics with lateral light collection properties are required.
- Low laser powers are used so there are no deleterious thermal effects.
- Suitable wavelengths include ultraviolet, visible, and near-infrared laser light.

9.2.3 Diagnosis of Symptomatic Cracks in Vital Teeth

Cracked tooth also known as cracked tooth syndrome (CTS) can be a difficult disorder to diagnose and manage [37, 38]. Failure to diagnose the condition early could result in crack progression and potential tooth loss, while early diagnosis may result in treatment with a conservative restoration. When cracks progress to communicate with the pulp, root canal treatment and a cuspal coverage restoration are often necessary [39, 40].

Lasers are of interest in the assessment of cracks because current diagnostic techniques (bite tests, transillumination, etc.) are limited in their ability to consistently determine the offending teeth and cusps, with most clinicians relying on the patient's reported symptoms in making a diagnosis. It is difficult to establish the location and direction of the crack responsible for symptoms. Moreover, a range of other pathologies can mimic the symptoms of CTS, making differential diagnosis challenging.

Sapra et al. in 2021 showed in an in vitro study that significantly more laser energy passes through teeth with cracks, compared to teeth without cracks [41]. In further work, they suggested that near-infrared lasers could be used as an adjunct for identifying cracked teeth in a clinical setting [42]. They used an 810-nm diode laser at a power of 1 W in a continuous wave mode, delivered through a 200-μm-diameter tip, for screening for cracks in vital teeth [42]. The surfaces of the suspect tooth were lased in a scanning motion across the surface. A crack was identified from the patient experiencing a transient sharp pain, which disappeared immediately as the scan moved past the crack [42]. This study provides a platform for further work to explore how a near-infrared diode laser could be an adjunctive tool for early detection and management of symptomatic cracks, facilitating early interventions with the aim of improving long-term survival of the teeth [42].

9.3 Laser-Assisted Widening of the Root Canal

One of the earliest explorations of the possible use of lasers to enlarge the root canal was the study of Levy [43], in which an Nd:YAG laser with water spray was used to widen root canals in the apical zone from ISO #20 to ISO #35, based on the fit of K files. The technique employed was a painting and sweeping action circumferentially, with lateral pressure on the canal walls during withdrawal of the fiber. The procedure took 60 s, using of 300 mJ and 1 Hz. Consistent with this, Matsuoka et al. [44] required approximately 2 min to enlarge root canals from 0.285 to 0.470 mm. Both Ali et al. [45] and Jahan et al. [46] took only 60 s of lasing time to prepare the root canal using a crown-down technique. This excludes the time required to change fiber-optic tips. It would be predicted that canals with larger tapers would be easier to prepare than those with narrower tapers.

With regard to the Ho:YAG laser, Cohen et al. [47] used a 245-μm-diameter optical fiber to enlarge canals. The fiber was inserted to the apex, energized, and then withdrawn slowly at 4 mm/s. Using this technique, canals with internal dimensions of ISO #25 were widened to an apical size of ISO #40. Using the same laser, Cohen et al. [48] employed a step-back technique with four different optical fiber tips (with diameters of 140, 245, 355, and 410 μm) to enlarge canals progressively, while Deutsch et al. [49] used six different sized optical fiber tips for enlarging the root canal with the Ho:YAG laser.

With regard to the Er,Cr:YSGG laser, Ali et al. [45] reported the use of fibers of various diameters to prepare root canals using a crown-down technique. While noting that this laser wavelength was useful for removal of smear layer and debris, the risk of ledging, zipping, perforation, or over-instrumentation of canals was noted. Matsuoka et al. [50] reported that the Er,Cr:YSGG laser could be used successfully to prepare root canals with curvatures up to 10°, using a step-back technique, with an average energy of 2 W, a pulse rate of 20 Hz, and an air and water spray. In contrast, Jahan et al. [46] reported that preparation of canals with a curvature above 5° could lead to zipping, ledge formation, or perforations. There is more limited information regarding the use of the Er:YAG laser for enlarging the canal. Matsuoka et al. [44] reported using the Er:YAG lasers to enlarge the root canal using three different size conventional optical fiber tips used sequentially, in line with the step-back approach.

Although several studies have shown the potential for lasers to widen the root canal, it is difficult to attain all the mechanical objectives of root canal preparation when laser energy is delivered with conventional optical fibers. This relates to their inability to deliver laser energy directly onto the walls of the root canal, as well as the operator challenge of maintaining a constant withdrawal rate. In 2006, Altundasar et al. [51] showed that delivery of laser energy onto the walls of the root canal using a conventional (plain) optical fiber to remove smear layer gives inconsistent ablation. From the standpoint of optics, a beam delivered from a plain fiber (and thus largely parallel to the walls of the root canal surface to be ablated) has low efficiency, and this has been demonstrated in the laboratory setting, by comparing the effects of parallel and perpendicular beams directed onto root canal dentine slices.

In an attempt to overcome some of these problems, fiber tips with sculpted polished ends and greater lateral emissions have been developed [49, 52–56]. Shoji et al. [52] employed a cone-shaped irradiation tip which could disperse laser energy in an annular pattern. This aluminum reinforced silicate tip was used to deliver Er:YAG laser energy to enlarge root canals. This tip design produced maximal enlargement when the laser was used at 30 mJ and 10 Hz.

Key Points for Laser-Assisted Widening of the Root Canal

- Laser wavelengths should ablate hard tissue for maximum effectiveness.
- Thermal side effects need to be controlled.
- Special tip designs improve safety and effectiveness.
- Suitable wavelengths are in the middle infrared.
- Laser energy must be pulsed to ensure thermal stresses are reduced.
- Concurrent irrigation assists cooling.

9.4 Removal of Smear Layer from Root Canal Walls

Many laser types have been reported to be useful in the removal of smear layer from root canal walls, including the argon fluoride (ArF) and other excimer lasers [57], argon ion lasers [58], KTP laser (532 nm) [59], diode lasers, Nd:YAG lasers [60, 61], HoYAG lasers [62], Er:YAG lasers [17, 63], Er,Cr:YSGG lasers [46, 64], and CO_2 lasers [65].

Diode lasers are cost-effective, compact, and portable devices. The near-infrared laser emissions from these devices (810–980 nm) have penetrating disinfecting actions, which are an additional advantage to being able to remove smear layer. Wang [66] used a 980-nm wavelength diode laser at 5 W for 7 s to remove smear layer; however, concerns remain in terms of generation and conduction of heat to the supporting apparatus if high irradiances are used [67].

Nd:YAG lasers are more effective for disinfecting the root canal, and relatively less effective for removing smear layer, compared to the erbium lasers [68]. Goya [69], who investigated the effect of the Nd:YAG laser on smear layer, found that black ink increases the removal of smear layer by enhancing absorption of laser energy. However, Wilder-Smith et al. [70] identified that thermal damage was a concern when using the Nd:YAG laser to remove smear layer.

The water-absorbing properties of the Er:YAG and Er,Cr:YSGG lasers make these useful both for disinfection of the root canal and for removal of smear layer [61, 64, 71]. Takeda et al. [61] undertook a comparative study of the argon ion laser (1 W, 50 mJ, 5 Hz), Nd:YAG laser (2 W, 200 mJ, 20 Hz), and Er:YAG laser (1 W, 100 mJ, 10 Hz) in terms of removal of smear layer from prepared root canal walls, compared to EDTA. All lasers achieved better smear layer removal than EDTA, and the Er:YAG laser was the most effective of the three lasers used. In a later study, Takeda et al. [65] reported that Er:YAG lasers were better than CO_2 lasers and three different acids in removal of smear layer. Ali et al. [45] reported less smear layer or debris when using an Er,Cr:YSGG laser, compared to the conventional root canal techniques; however, the mechanical quality of the canal preparation (smoothness, taper, etc.) was worse with the laser method. Biedma [72] also reported similar results of that of Ali et al. [45] but using the Er:YAG laser.

As already noted, several studies have reported inconsistent or inefficient removal of smear layer when using erbium lasers delivered using conventional optical fibers. Altundasar et al. [73] reported inconsistent smear layer removal of the walls of the root canal when the Er,Cr:YSGG laser (operated at 3 W and 20 Hz) was delivered using a conventional tip, while Anic et al. [51] reported greater efficiency of a perpendicular beam for ablation when compared to a parallel beam. Kimura et al. [74] stated that it was difficult to evenly irradiate root canal walls using a conventional fiber tip and advocated an improvement in the fiber tip design or method of irradiation to avoid obtaining an uneven surface.

To overcome such problems, several authors have employed sculptured fiber tips that have greater lateral delivery of laser energy [52–54]. Alves et al. [55] used the Er:YAG with forward-emitting sapphire tips and hollow fibers and compared these to modified tips which gave lateral emissions. Shoji et al. [52] used an Er:YAG laser delivered into a cone-shaped tip to enlarge artificial root canals in a block of bovine dentine using Er:YAG laser energy. The cone-shaped tip was faster for cavity preparation and smear layer removal, compared with conventional instruments. Likewise, Takeda et al. [65] used a conical tip with the CO_2 to remove smear layer from the root canal.

Stabholz et al. [75] designed an endodontic side-firing spiral tip (RCLase; Lumenis, OpusDent, Israel), which comprised a hollow waveguide with spiral slits along the length of the tube. The end of the tip was sealed to prevent the forward transmission of laser energy. The Er:YAG laser was used at 500 mJ and 12 Hz through this tip to remove smear layer successfully. However, such tips are too large and rigid to be used in narrow, curved root canals. Moreover, if the tip were to bend, more energy would be emitted across those slits that are in a straight line with the beam. To overcome such concerns, flexible fiber optics with various thicknesses have been used to gain better access to the full length of the root canal. Furthermore, fibers have been modified to allow for better delivery of laser energy that is directed onto the walls of the root canal [19, 21, 76]. For enhanced removal of smear layer from the root canal walls, water-absorbing middle-infrared laser wavelengths (e.g., Erbium:YAG) can be used. The laser energy that reaches the canal walls can directly ablate the dentine of the root canal walls and remove smear layer, especially when used with EDTA [77]. It is however important to note that the optimum energy settings need to be considered when using lasers for the removal of smear layer [78–81]. Operators need to also consider the suitability of the laser wavelength, the diameter and flexibility of the fiber-optic delivery system, the design of its tip, the dimensions of the root canal system, and the type of endodontic procedures being undertaken [19, 21, 76]. Understanding such factors ensures correct techniques are used, and this helps prevent carbonization, ledge formation, and apical extrusion of fluids [21, 76, 82–84].

Key Points for Removal of Smear Layer

- Laser wavelengths should absorb strongly in water to generate cavitation.
- The use of pulsed modes is essential.
- Laser pulse energy must be limited to prevent fluid extrusion from the apex and excessive projection of fluid from the root canal system.
- Most suitable laser wavelengths are in the middle infrared.
- Water-based irrigant fluids should be used; the procedure should never be done dry.
- Laser activation enhances the action of EDTA in smear layer removal.
- Lasers can enhance the actions of other water-based fluids such as sodium hypochlorite through agitation and warming of the fluid.
- Thermal side effects need to be controlled.
- Special tip designs improve safety and effectiveness.
- Tips degrade readily during use and this alters their emission characteristics.
- Laser energy must be pulsed to ensure thermal stresses are reduced.
- Concurrent irrigation assists cooling.

9.5 Disinfection

9.5.1 Photothermal Disinfection

Laser light can penetrate areas of canals where irrigating and disinfecting solutions cannot reach, such as fins, deltas, and lateral canals [85]. Selective photothermolysis occurs when laser energy is applied into the root canal system. For water-absorbing laser wavelengths, rapid expansion of water contained within microorganisms leads to their rupture, while for the visible and near-infrared wavelengths, primary absorption of laser energy into porphyrins, melanin, and other pigments occurs. The increase in temperature then denatures proteins, and this renders the organisms unviable [86–88].

When such methods are used, it is important to employ pulsed modes and rest periods to allow for cooling of the root structure so that there is no collateral injury to the periodontal ligament from thermal stress. Assessments of safety undertaken in laboratory conditions are based on threshold values around 5.5–7 °C as the limit of acceptable temperature increases on the root surface [89, 90].

Directing laser energy onto the walls of the root canal is essential for effective disinfection. To maximize this effect, different fiber modifications have been developed to increase lateral emission of laser energy, including designs with safe tips to reduce irradiation directed toward the root apex. Examples include conical tips, side-firing honeycomb tips, and honeycomb tips with silver-coated ends (safe-ended fibers) [76].

Photothermal laser disinfection is a useful supplement to existing protocols for canal disinfection as the properties of laser light may allow a bactericidal effect beyond 1 mm of dentine. It must be remembered that endodontic pathogens can be present not only in the canal but also extending into the dentine tubules for several hundred microns. This emphasizes the value of actions such as laser fluid agitation to enhance the efficacy of current irrigating protocols, which can increase the distance of the laser effect [91].

At the present time, the lasers used most commonly for photothermal disinfection are the Nd:YAG, KTP, and near-infrared diode lasers. All of these have been shown to have excellent antibacterial efficacy, with greater penetration of the disinfecting action than middle-infrared wavelengths [92].

Key Points for Photothermal Disinfection

- Laser energy must absorb into major chromophores (water, porphyrins, melanin, and other pigments) for bacterial inactivation to occur.
- This can be done with almost any laser system, but preferred lasers are Nd:YAG, KTP and near-infrared diode lasers. Middle-infrared lasers will show the lowest penetration (~0.5 mm).
- Lateral emitting/side-firing tips are preferred to ensure even irradiation is achieved.
- Disinfection can be achieved for microbial deposits deep within dentin which would not be reached by most medicaments placed into the canal.
- Penetration depths vary according to the laser wavelength used. Maximum penetration occurs with near-infrared laser energy.
- Pulsed modes must be used to lower thermal stress to the root and periodontium.
- Total dosimetry must be monitored so that the irradiation remains within safe limits.
- Movement of the fiber enhances coverage of the walls of the root canal.
- The fiber is moved from the apex in a coronal direction, tilting and rotating the fiber to help gain better exposure of the canal walls.
- The fiber must be kept in constant motion.
- Several passes one after the other are required to ensure that all parts of the root canal receive sufficient laser energy to inactivate microorganisms.

9

9.5.2 Photodynamic Disinfection

Photoactivated disinfection (PAD), also known as photoactivated chemotherapy (PACT), is based on the interaction of laser light with photosensitizers. These may be endogenous (such as porphyrins found in Gram-negative bacteria) or exogenous in the form of dyes such as tolonium chloride or methylene blue applied into the root canal, which then bind to microbial outer membranes (◘ Fig. 9.4). When the photosensitizer is exposed to laser light of the appropriate wavelength, reactive oxygen species are generated, which then damage the microbial cell membrane, leading to leakage of contents through it and denaturation of microbial proteins and DNA [88, 93].

Since LLP relies on the chromophore becoming electronically activated, it is essential to match the laser wavelength used with the chromophore, in exactly the same manner as is done for laser photodynamic therapy of oral lesions, where the laser energy activates otherwise nontoxic dyes producing reactive oxygen species that cause injury and death of tumor cells [94, 95].

PAD is a specific interaction, in that treatment with the laser alone (i.e., in the absence of the enhancing dye), or with the dye alone, produces much less microbial killing than the combination of dye with laser. For example, using visible red laser light, bactericidal effects can be achieved using a range of blue, purple, and green dyes within the phenylmethane family, all of which are strong absorbers of red light [96, 97]. Other photosensitizers of interest include indocyanine green (ICG) and curcumin, which are activated at 808 and 470 nm, respectively [98, 99]. PAD can be undertaken with LEDs as well as with lasers, since either will activate the photosensitizers [100].

Both in vitro and clinical studies of PAD have demonstrated its ability to kill photosensitized oral bacteria (such as *Enterococcus faecalis*). To date, 12 studies have reported PAD as being effective in eliminating *Enterococcus faecalis* from infected root canals [101].

While PAD can be undertaken as part of the routine disinfection of the root canal system, it also has a potential use for eradicating persistent endodontic infections for which conventional methods have been unsuccessful [102–104]. It does not cause significant thermal stress to the roots of teeth or the adjacent periodontal tissues [105].

There are many possible dyes which could be used for PAD. In its simplest form, the dye should undergo photodynamic activation and produce reactive oxygen species (ROS), which are the means by which microorganisms are inactivated. Dyes such as tolonium chloride and methylene blue are excellent producers of ROS in that regard. Using colored dyes which are activated by shorter wavelengths of light enhances effectiveness, since shorter wavelength light has a higher photon energy than longer wavelength light (such as in the near-infrared region of the spectrum).

There has been some confusion as to the mechanisms involved when green dyes such as indocyanine green (ICG) are exposed to near-infrared laser energy around 800–830 nm, which absorbs strongly in this material. The effect of the laser energy being absorbed is to heat the dye and therefore indirectly heat what the dye has become attached to. This is a type of photothermal disinfection process and is not a photodynamic process since the

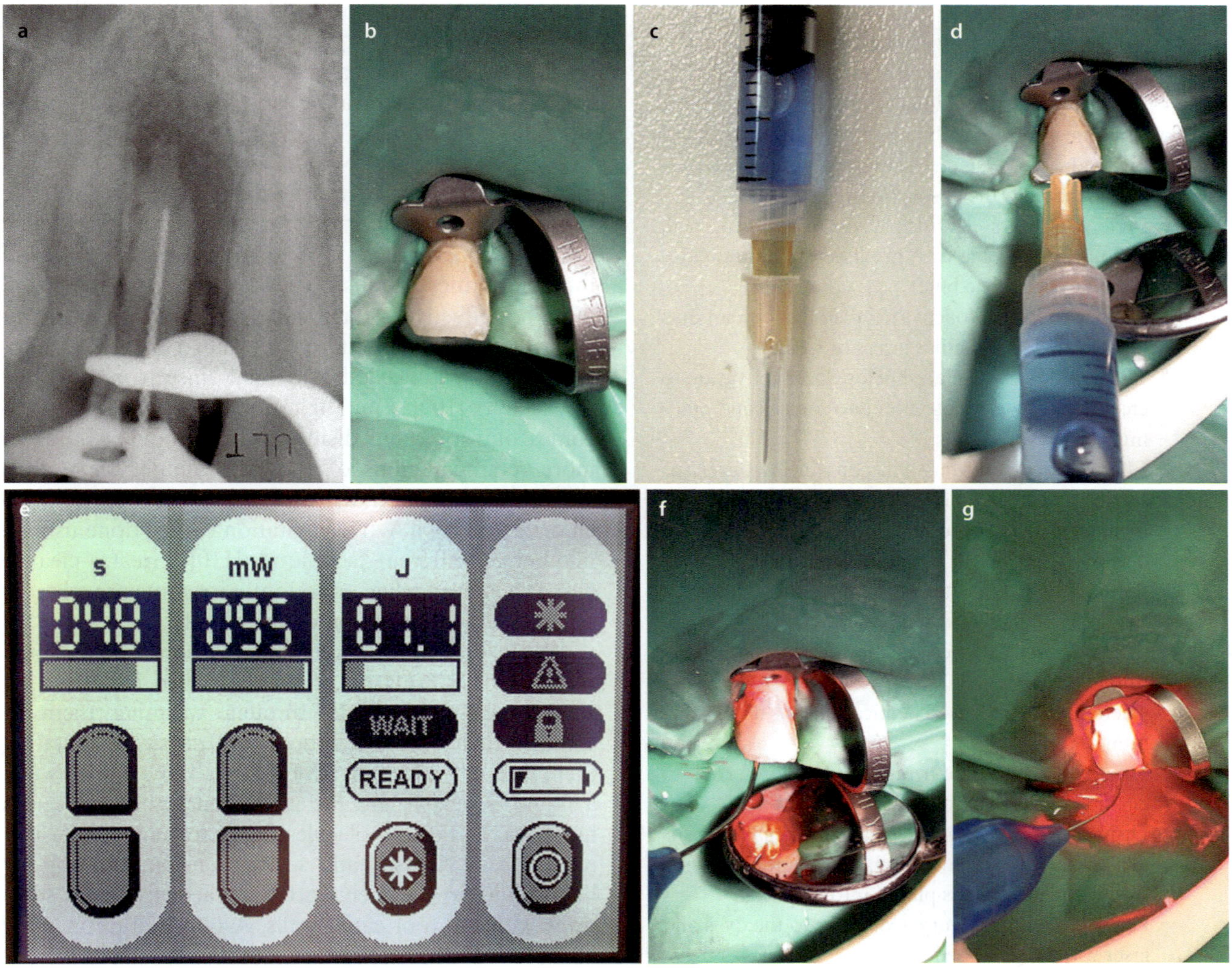

Fig. 9.4 Root canal photodynamic disinfection. This case presented with a large periapical lesion on the lateral incisor and apical root resorption. **a** Diagnostic file. **b** Isolated tooth. **c** Tolonium chloride dye solution. **d** Injection of dye solution into the root canal system. **e** Laser control panel for 635-nm diode laser midway during treatment showing delivered power 95-mW continuous wave mode. Typical irradiation is 60–90 s. **f**, **g** Transmission of visible red laser light through the coronal and radicular tooth structure. This activates the dye and provides biostimulation effects

action is mediated through heat rather than through the generation of ROS. This underpins the applications of ICG dye in laser-based tumor therapies. ICG dye can absorb between 600 nm in the visible red region all the way through to 900 nm, and it can emit fluorescence between 750 and 950 nm. ICG when exposed to 810-nm laser light will fluoresce, which is a major way ICG is used in medical diagnostics. Nevertheless, it is a simple fluorescence dye and is not a photosensitizer.

There has also been a level of confusion used in the terminology surrounding photodynamic disinfection, with terms such as photoactivated chemotherapy (PACT), photo-disinfection, and lethal laser photosensitization (LLP) all having been used to describe the effect with blue dyes; however, some studies using ICG also use these terms, which is incorrect.

For the use of photoactivated disinfection (PAD) for root canal disinfection, it is important to ensure that the dye is in contact with all the parts of root canal, even when its shape is complex. Agitation of dyes can enhance the contact of dyes with the root canal walls, giving better penetration into dentine tubules [106]. Some photosensitizer dyes are smaller molecules and so will have an inherently better penetration by diffusion than others with larger molecular sizes, and in any event, penetration of the dye can be enhanced further by agitation, such as by using LAI [106]. The dye photosensitizer solution must be in contact with the target for a short period of time, prior to its activation by light, to allow sufficient time for binding to microorganisms. PAD procedures should only be initiated after ensuring that the root canal system is cleared of all fluids that could dilute or inactivate the dye, such as saliva or blood or soft tissues. These may contain scavenging systems for reactive oxygen species which could impede the photosensitization process [107].

Key Points for Photoactivated Disinfection

- Laser energy must absorb into the photosensitizer for bacterial inactivation to occur.
- This can be done with almost any visible or near-infrared laser system, as long as the laser wavelength matches the absorption of the dye.
- Preferred lasers are visible red (633, 635, 660, 670 nm) when blue dyes are used (tolonium chloride and methylene blue)
- Dyes used in photoactivated disinfection can also be activated using either lasers or LEDs.
- The liquid must be placed before laser activation to ensure adequate penetration into tubules and binding to bacteria.
- Effective dye solutions will contain low levels of surfactants to enhance penetration and reduce the formation of vapor locks.
- The dye used should not permanently stain teeth.
- Some dyes will effectively kill bacteria in the dark before being activated with laser light.
- Thermal effects caused by photoactivated disinfection are minimal.
- There are no adverse chemical effects on normal human cells.
- Lateral emitting/side-firing tips are preferred to ensure even irradiation is achieved.
- Disinfection can be achieved for microbial deposits very deep within dentin which would not be reached by most medicaments placed into the canal.
- Penetration depths vary according to the dyes and laser wavelengths used.
- Longer irradiation times or multiple passes help ensure that all parts of the root canal receive sufficient laser energy to activate the dye to kill microorganisms.

9.6 Debridement of the Root Canal System

The use of lasers for debridement of the root canal systems offers several important advantages. Conventional instrumentation only touches some of the walls of the canal, since few canals are not perfectly round. Laser energy and laser-activated fluids, in contrast, can reach all the walls of the canal. In addition, the use of files results in both widening and alterations in canal curvature. This problem of transportation does not occur when lasers are employed since energy can be delivered into the root canal without significant ablation of the walls of the root canal [108].

Finally, conventional instruments produce a smear layer, which then requires additional work to remove, such as alternating rinses with sodium hypochlorite and then extended periods of flushing with ethylenediaminetetraacetic acid (EDTA). Lasers can remove smear layer created by rotary or hand files and do not generate a smear layer when they are used to cut into root dentine.

9.6.1 Fluid Agitation and Activation

Sodium hypochlorite is the main irrigating solution used in endodontics to dissolve organic matter and kill microbes effectively. High concentration sodium hypochlorite (4%) has a better effect than 1% and 2% solutions. EDTA is needed as a final rinse to remove the smear layer [109]. Fluid agitation can enhance the outcomes of root canal disinfection by moving fluids to more remote regions of the root canal system, thus enhancing the action of irrigants such as sodium hypochlorite and EDTA. This agitation can be done using a number of different techniques [110]. Greater cleanliness is achieved when endodontic irrigants are agitated during the final irrigation regimen [111].

Because of their strong water absorption, Er:YAG and Er,Cr:YSGG lasers are ideally suited for activating water-based fluids, both through warming them to enhance their chemical actions and physically agitating them through cavitation (◘ Fig. 9.5). Importantly, when a liquid is heated and the kinetic energy of molecules increases, as those molecules begin to move, there is a reduction in viscosity and surface tension, allowing for better fluid flow and enhanced contact of irrigant solutions with the root canal walls. A temperature increase also enhances the bactericidal action and tissue-dissolving capabilities of sodium hypochlorite solutions [112].

When using middle- and far-infrared lasers with water-based fluids in the canal, useful improvements can also be gained in the removal of debris and smear layer (◘ Fig. 9.1f). Using conical tips created using a tube etching process, both Er:YAG and Er,Cr:YSGG lasers have been shown to be able to remove extraordinarily thick smear layers that had been created intentionally to provide a challenge to the laser system. When the extent of smear layer was assessed from scanning electron microscopy images with an objective digital method, it was found that lasing improved the action of EDTA in removing smear layer. Conical fibers performed better than plain fibers [76, 80].

Since the description of laser fluid activation in 2008, studies have documented the benefits of laser fluid agitation (also known as laser activated irrigation (LAI)) for enhancing cleaning of the root canal system, using EDTA, peroxide, and sodium hypochlorite as the irrigant solutions. This shows enhanced antibacterial actions for sodium hypochlorite and improved biofilm removal.

One of the variants of this is known as photon-induced photoacoustic streaming (PIPS), which is typi-

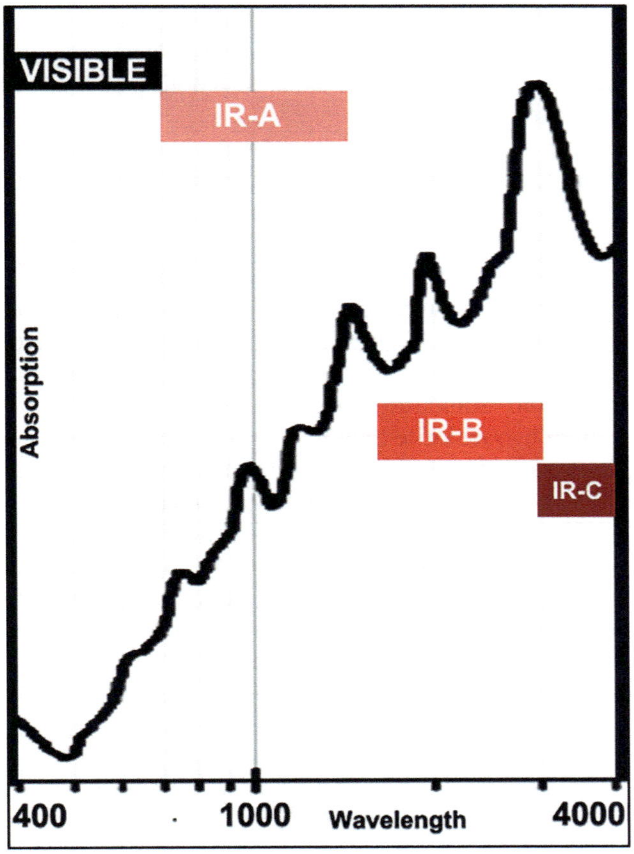

Fig. 9.5 Absorption of pure water in the visible region (400–700 nm) and adjacent infrared regions. The horizontal axis is wavelength in nanometers and the vertical axis is absorption. IR-A = 700–1400 nm; IR-B = 1400–3000 nm; IR-C = 3000 nm–1 mm

cally used with sodium hypochlorite. Laser-activated irrigation utilizing PIPS has been reported to enhance the disinfection of the root canal system [113–117]. However, when creating acoustic waves in narrow volume geometries of the dental root canal using the Er:YAG laser with very short duration pulses, there is evidence for superior effects by altering the pulse chain; hence, a further method, termed SWEEPS® (shock wave-enhanced emission photoacoustic streaming) was developed by Fotona (Slovenia, EU). PIPS uses single short pulse, while SWEEPS® delivers pulses in pairs, with the second pulse involved in accelerating the collapse of first bubble and the emergence of new bubbles. This process results in more shock waves within the root canal. A further evolution of this concept is Resonant SWEEPS (R-SWEEPS)®, which has the goal of enhancing the flushing action through increased pressure along the root canal walls, but without increasing the risk of apical extrusion of irrigation fluids.

Laser protocols which employ Er:YAG or Er,Cr:YSGG lasers with water-based fluids have been shown to cause minimal thermal stress to the root structure. Both plain and laterally emitting conical or honeycomb fiber tips can be used safely for intracanal irradiation without harmful thermal effects on the periodontium. When the irrigant fluid is refreshed between cycles of laser exposure, there is a strong beneficial effect on temperature, which attenuates completely the thermal effects of individual lasing cycles [83]. Laser-based protocols for fluid agitation may give better reductions in levels of bacteria than ultrasonic agitation delivered through an endodontic file, especially when the laser is used with NaOCl [118]. Enhancement of how well NaOCl can penetrate throughout the root canal system is an important clinical advantage, which adds to the direct photothermal disinfectant actions of lasers. As an example, one study has shown that LAI using the Er,Cr:YSGG laser for 60 s enhanced the distribution and action of 0.5% NaOCl to reach the same level of effectiveness as 2.5% NaOCl. This is of clinical interest, since working with lower concentrations of NaOCl may lower the risk of undesired effects [119]. In a similar way, laser agitation of fluid may enhance the removal of certain intracanal medicaments, such as water-based calcium hydroxide dressings that contain carboxymethylcellulose thickeners, making them difficult to remove by simple irrigation methods [120].

Key Points for Laser Activation of Fluids

- Water-based fluids such as EDTA are preferred.
- Optimal lasers are the middle-infrared lasers which have strong water absorption.
- Absorption leads to cavitation and thus to agitation, fluid movement, and shock waves.
- Fluid can be ejected from the root canal, and the canal must then be topped up with more fluid.
- Irrigation between lasing cycles reduces thermal stress.
- Excessively high-pulse energies can cause fluid extrusion through the apical foramen.

9.6.2 Cavitation

With conventional irrigant solutions, fluid motion is limited to the relatively passive flow of fluid into and outside the root canal system. The root canal has confined geometry, which through surface tension effects makes the dispersion of irrigant more difficult because of the absence of turbulence over much of the canal volume [121]. Finally, in root canals, the problems of bubble entrapment/vapor lock occur when using conventional irrigation approaches [122].

When lasers generate cavitation, the turbulence created agitates the fluids within the root canal. This can be done with the laser fiber stationary or being gently withdrawn. The laser tip does not have to be placed into the apical third of the root canal, while with conventional irrigation it is important to place the tip of the irrigation needle to within 1 mm of the working length to ensure adequate fluid exchange [122, 123]. The laser-generated agitation causes fluid motion, which overcomes the bubble entrapment effect. Fluid streaming, which is caused by the collapse of the laser-induced bubbles, is a major aspect of how laser-activated fluids clean the walls of the root canal [124, 125].

Cavitation and agitation generated by lasers in fluid-filled root canals create fluid movement and shear stresses along the root canal walls, enhancing removal of the smear layer and biofilm. Rapid fluid motion is caused by expansion and subsequent implosion of laser-induced bubbles [126, 127].

When used with sodium hypochlorite and EDTA, laser activation of aqueous fluids can increase the efficiency of debridement and disinfection of root canals [128–130]. Moreover, there is now direct evidence that pressure changes and shock waves which accompany cavitation may enhance the susceptibility of bacteria in biofilms to antimicrobial agents. The problem of biofilms in root canals has direct parallels to the tubing of medical catheters, where such shock wave approaches are now attracting interest [131, 132].

An obvious problem which arises is whether fluid movement in the root canal leads to greater extrusion of fluids beyond the root apex. Conventional needles used for irrigation create apical pressure and extrude some fluid [133]. Studies of fluid extrusion beyond the apical constriction using Er:YAG and Er,Cr:YSGG lasers with bare or conical fiber tips positioned at distances of 5 or 10 mm from the apex have shown that the extent of microdroplets of fluid displaced past the apex was no greater than that seen when conventional 25-gauge non-side-venting irrigation needles were used [82]. The PIPS technique aims to ensure fluid motion is along the root canal walls, rather than entirely toward the apical foramen, in order to reduce the volume that could be extruded. Several studies have shown that small volumes of fluid are extruded during PIPS and during other methods of LAI [134, 135].

When fibers with laterally emitting honeycomb patters are used, these generate agitation with fluid movement directed onto the walls of the canal, while both conventional plain fibers and tips with conical ends generate fluid movement largely in a forward direction. Having the laser energy directed laterally lowers the risk of fluid extrusion beyond the apex [79, 130].

Diode lasers in the 940–980-nm wavelength range can be used to generate cavitation, relying on their water absorption. Such lasers are used in pulsed modes, both to optimize the cavitation dynamics and to reduce collateral thermal effects on the roots (◘ Fig. 9.1c, d). Such lasers can then be used with water-based fluids to remove debris and smear layers from the walls of the root canal. For diode lasers, the cavitation effects can be enhanced by supplementing the water with hydrogen peroxide to a final concentration of 3%. Any thermal stresses at the cementum are reduced when irrigation fluids are replaced, which enhances cooling of the root structure [84, 136].

In a recent study which evaluated the efficiency of EDTAC activation for smear layer removal using a 940-nm diode laser operated in pulsed mode and delivered by plain fiber tips into 15% EDTAC or 3% hydrogen peroxide, lasing EDTAC was found to considerably improve smear layer removal to a greater extent than lasing into peroxide. Of interest, the diode laser protocol for smear layer removal was more effective than the clinical "gold standard" protocol using EDTAC with sodium hypochlorite (NaOCl). In addition, when using diode lasers, there are additional benefits gained through photothermal disinfection and biostimulation [18].

When using a diode laser versus an erbium laser, it must be remembered that the fluid agitation effects are less for a diode laser than for an erbium laser; however, both are a great improvement on irrigants which are simply held static in the root canal [137].

Key Points for Laser-Induced Cavitation

- Laser energy must absorb into water for cavitation to occur in a water-based fluid.
- A small volume of water will show greater cavitation than a large volume for the same laser pulse energy; this has relevance to the effects seen in small versus large diameter canals.
- Middle-infrared lasers (Er:YAG and Er,Cr:YSGG) will show the fastest cavitation (microseconds) versus 940–980-nm diode lasers (seconds) and will cause the fastest fluid motion in the canal.
- Lateral emitting/side-firing tips are preferred as this changes the direction of cavitation bubble formation and collapse.
- Pulsed modes must be used; shorter pulse durations will cause greater cavitation to occur for the same pulse energy but will increase the risk of fluid extrusion through the apex.
- More fluid extrusion occurs when the apical foramen is larger.

9.7 Laser-Enhanced Bleaching

Sclerosis following dental trauma and the severe forms of intrinsic staining due to the loss of vitality or endodontic treatment are challenging to manage. Some of these conditions are resistant to conventional bleaching treatments based on carbamide or hydrogen peroxide. Common factors in intrinsic staining include demeclocycline-containing tetracycline medicaments and bismuth oxide, an agent used to achieve radiopacity in some epoxy resin sealers and in mineral trioxide aggregate (MTA) [138–141].

The underlying chemistry which explains the patterns of discoloration with these different types of materials is quite complex. In the case of MTA, it is the formation of bismuth sulfide, which is black in color and therefore causes the tooth to appear gray. Iron released from hemoglobin following trauma can also form iron sulfide. Such sulfide compounds are very stable and not readily oxidized [142].

Removal of tetracycline medicaments from the root canal does not prevent later discoloration. In fact, studies have shown that current irrigation methods using plain needles, open-ended notched irrigation needles, or side-vented needles do not completely remove all traces of such medicaments [143]. Laser-activated irrigation is however significantly more effective for removing endodontic medicaments than any protocols based on needle irrigation [144].

Tetracyclines bind readily to tooth structure and then form a red-purple degradation tetracycline product (4 alpha, 12 alpha-anhydro-4-oxo-4--dedimethylaminotetracycline, known as AODTC) when moisture is present. AODTC is resistant to oxidation but can undergo photolysis when exposed to visible green light (530–535 nm), opening possibilities for laser therapy using a KTP laser (◘ Fig. 9.6) [145–147].

There are numerous ways that lasers can be used to enhance bleaching. These include photothermal effects (warming the gel to make hydrogen peroxide more active chemically), photochemical actions (such as the Fenton reaction), photocatalytic actions, and photodynamic actions, where the laser energy activates a suitable photosensitizer. There is also photooxidation, an effect which is essential for breaking up tetracyclines and AODTC [148].

Sclerosed discolored traumatized teeth, which have remained vital, can be treated using the KTP laser with photodynamic bleaching, employing rhodamine as the photosensitizer. The same technique can also be used successfully to treat using external bleaching applied in the office setting non-vital teeth with stains from endodontic treatment and teeth with tetracycline staining [149].

We undertook a clinical study of photodynamic bleaching for treating confirmed cases of tetracycline discoloration as a single-appointment procedure used the KTP (frequency-doubled Nd:YAG) laser (wavelength 532 nm) combined with a rhodamine-B photosensitizer gel (Smartbleach) applied to the teeth and activated for 30 s. Each tooth underwent four cycles of 30 s of laser exposure. Digital image analysis was undertaken in a blinded manner, and this showed a significant lightening effect was achieved in 78% of the teeth treated. An in-office KTP laser photodynamic bleaching treatment provides a useful option for improving tooth shade in teeth with tetracycline discoloration [150].

In later work, we showed that KTP laser photodynamic bleaching for tetracycline staining was more effective than using arrays of LEDs in the visible green range (535 nm) with the same photosensitizer or photo-Fenton bleaching using LED arrays in the visible blue (460 nm) [151, 152].

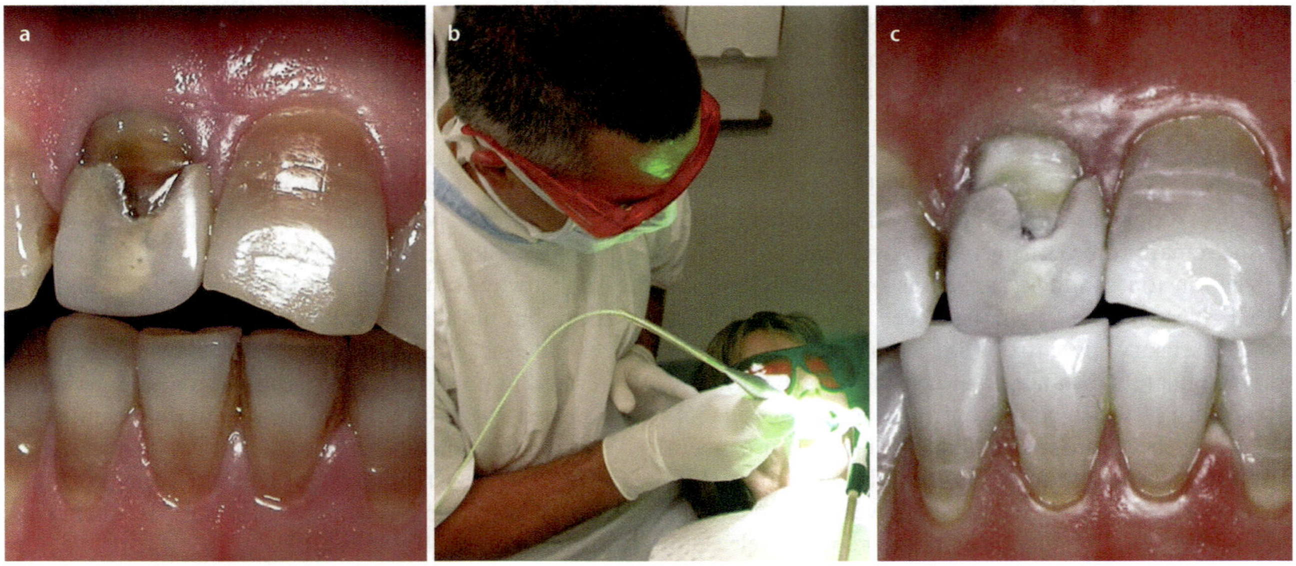

◘ **Fig. 9.6** KTP laser photodynamic bleaching of a discolored non-vital maxillary central incisor tooth (#11) from an external approach. **a** Preoperative view of the discolored tooth. All the teeth have tetracycline staining of developmental origin. Tooth #11 has undergone endodontic treatment. **b** Application of KTP laser onto a rhodamine photosensitizer in repeated cycles to treat all the anterior teeth. **c** Postoperative view at the end of the same appointment. There has been a useful improvement in the shade of the root filled #11 tooth as well as the adjacent teeth from the laser treatment. A tooth-colored restoration was subsequently placed 2 weeks later to restore the #11 tooth

Key Points for Laser-Assisted Bleaching

- Photothermal laser bleaching requires careful control of the irradiation protocol to limit heat stress to the dental pulp.
- Photodynamic laser bleaching is effective for more challenging intrinsic stains including tetracyclines deposited during tooth formation and sclerosed vital teeth.
- External bleaching approaches overcome problems of invasive cervical resorption associated with internal bleaching (walking bleach) methods where peroxides can come into contact with periodontal tissues.

9.8 Laser-Induced Analgesia and Photobiomodulation

Photobiomodulation (PBM) is a term used to describe a process by which photon interactions with atoms or molecules cause biological alterations. The term photobiomodulation is preferred over the previously used term LLLT (low-level laser therapy), as a vast majority of therapies can also be performed with light-emitting diodes (LEDs), which have also been investigated for their effectiveness, as an alternative to semiconductor diode lasers [153].

Mechanisms of PBM-induced dental analgesia have been explored in a range of in vitro studies [154–156]. With erbium lasers, the analgesic action is thought to occur from the use of low-energy density pulses with a frequency that coincides with the resonance frequency of Na/K ATPase ion channels. Altering this ion channel prevents impulse propagation in nociceptor nerves, thus leading to analgesia [157]. The exact mechanisms for laser analgesia effects seen with near-infrared lasers (such as of Nd:YAG and GaAlAs lasers) when applied to nociceptors (A delta and C fibers) located within dental pulp tissue remain to be conclusively proven but may include effects on ion channels as well as on axonal transport [153]. Near-infrared light has been found to decrease somatosensory potential (SSEP) amplitudes and to downregulate substance P in central dorsal horn neurons, both of which could contribute to analgesia [156].

It has been proposed that PBM can cause neural inhibition, with blockage of axonal conduction [158], due to effects on microtubules, lower neuronal mitochondrial membrane potential, and reduced release of ATP [158]. The various hypothesized mechanisms could explain why there are dose-related PBM effects that are either stimulatory (at low dose) or inhibitory to ATP synthesis. Lower doses, such as between 0.5 and 8 J/cm^2, are stimulatory, while doses of 10–12 J/cm^2 have inhibitory effects on nociceptors [158]. Similar dose-related variations have been found in the cortical neuronal responses, where reactive oxygen species (ROS) were elevated at low doses, but suppressed when irradiation was increased to 10 J/cm^2 [159].

Positive analgesic outcomes from in vitro studies may not translate directly to real-world clinical applications, due to beam attenuation and scattering as the beam passes through other structures to reach the target neurons [153]. Further, the presence of nerve-related structures, such as the presence or absence of myelin sheath, may alter how a neuron responds to PBM [160].

Studies conducted in the early 1990s using free-running pulsed Nd:YAG lasers showed that pulsed laser radiation which could penetrate dentine was responsible for a component of the desensitizing effect on sensitive cervical dentine. Later studies by Orchardson and Zeredo, using free-running pulsed Nd:YAG and Er:YAG lasers in rodents, showed conclusively that there was a dramatic blockage of neuronal activity and a corresponding increase in the pain threshold of teeth after laser irradiation. The effect had a clear dose response for its onset, declined after 15–20 min, and was also associated with blockade of late-phase neurogenic inflammation (which is driven by the effects of neuropeptides). These effects were identical to those noted in clinical practice when preparing cavities with erbium lasers (Er:YAG and Er,Cr:YSGG). The animal studies however removed all possibility of placebo effects and psychogenic influences and demonstrated that there was a fundamental reversible alteration occurring in the nociceptive response caused by the laser treatment, which suppressed nerve firing for a given level of stimulus [161–165].

These effects can be used therapeutically for analgesia associated with restorative dentistry, for oral surgery (including bone ablation procedures), and for endodontic procedures, including pulp capping and extirpation. Clinically, the blockade with shorter exposures is more selective for depolarization of A-delta fibers (rapid, sharp, well-localized pain) than for C fibers, which explains why some patients experience vibrational sensations but not discomfort [166].

Analgesic effects can be induced by diode lasers operated in pulsed or continuous wave mode, as well as by pulsed Nd:YAG and middle-infrared erbium lasers. With the former, the wavelength and irradiance are key variables in determining the potency of the effect, while with the latter the pulse energy and pulse frequency are critical variables [93, 167–170].

In a recent systematic review of the effectiveness of PBM for reducing pain and inducing dental analgesia, four out of the five included studies documented analgesia. Despite this, the strength of evidence remains low, as studies lack reporting and discussion of key non-laser variables such as preexisting pulpal disease and patient anxiety levels [153].

The laser dose that reaches the dental pulp can be affected by the enamel thickness and shade, the dentine thickness and shade, and tooth surface characteristics such as roughness, curvature, translucency, or opacity [153]. When laser incident beam enters the tooth, it will be scattered by enamel prisms and also by organic components such as collagen and pigments. In some cases, a greater than expected penetration depth could be achieved because enamel crystallites and dentinal tubules act as miniature waveguides, directing light to the pulpal tissue. On the other hand, natural variations in dentine, such as sclerosis, the presence of a smear layer, and the configuration of dentinal tubules, can all influence penetration of laser light applied to the tooth [171, 172]. Other factors that need to be considered for optimizing PBM-induced analgesia include the pulsing mode and the contact mode [173–177].

In PBM, the overall effect is photochemical in nature and arises from the action of visible red (633–635 nm) or near-infrared (810–1100 nm) light on the enzymes of the electron transport chain in mitochondria, resulting in a broad activation of normal cellular functions. LLLT effects underpin the beneficial effects of lasers when used for pulp capping and pulpotomy, where there is direct exposure of pulpal soft tissues. This explains why there is accelerated healing, nerve sprouting, and dentinogenesis after pulpotomy [178].

Bystander LLLT effects occur in the periodontal ligament and periapical bone when lasers are used for intracanal procedures such as disinfection and promote the resolution of inflammation and healing responses after infection [88, 179].

Key Points for Laser-Induced Analgesia

- Analgesic effects can be induced with near- or middle-infrared lasers.
- Irradiation parameters for analgesia with diode lasers are higher than those for enhancement of wound healing and other photobiomodulation treatments with the same lasers.
- Laser-induced analgesia effects occur when lasers are used to treat hypersensitive cervical dentine and contribute to the overall clinical effects seen.

9.9 Pulp Therapy and Pulpotomy

Pulpotomy techniques for primary teeth traditionally have used formocresol, but this is becoming less widely used because of its toxic effects on living tissues and mutagenic potential. Alternatives such as MTA are expensive, and this has led to interest in using lasers for pulpotomy procedures. The lasers used have included Nd:YAG, Er:YAG, carbon dioxide, and 632- or 980-nm diode lasers.

Several clinical studies support the use of lasers for pulpotomy. The reported advantages include better clinical as well as radiographic outcomes than ferric sulfate, MTA, or electrosurgery, as well as a shorter operating time, simpler procedure, and less postoperative pain. Effective photothermal disinfection combined with low-level laser effects likely account for the favorable outcomes seen clinically with laser pulpotomy [180–183]. Similarly, there is clinical trial data to support the effectiveness of direct pulp capping using lasers with carbon dioxide, 808-nm diode, and Er:YAG and Er,Cr:YSGG lasers (Fig. 9.7) [184–188].

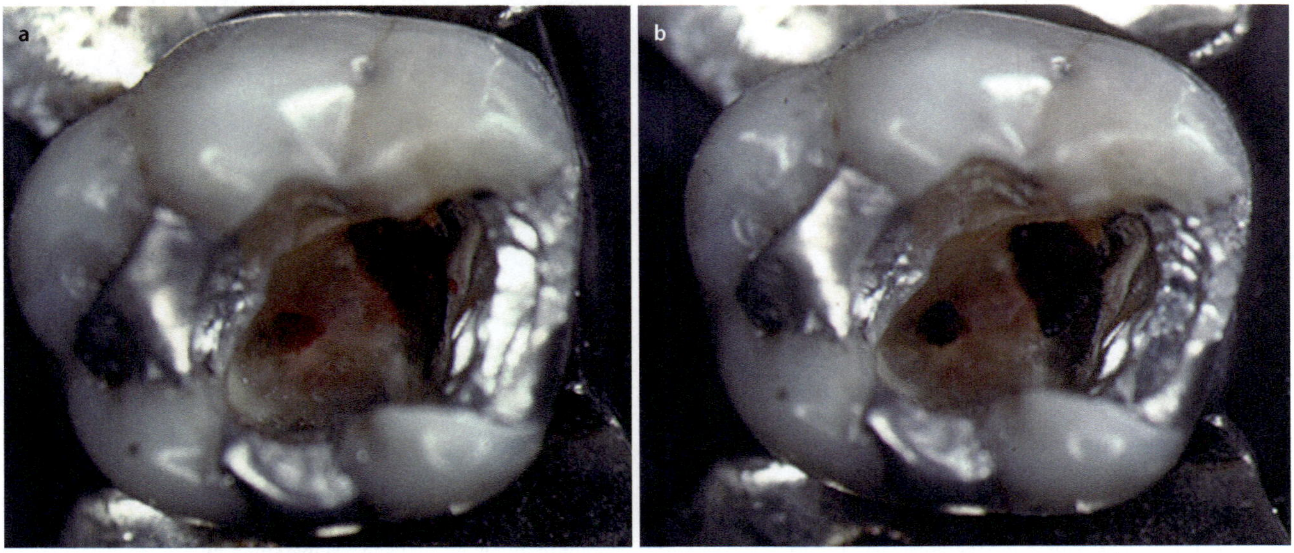

Fig. 9.7 Vital pulp capping. **a** Bleeding pulp at the base of the cavity preparation following an iatrogenic exposure of vital pulp tissue. **b** Immediately after firing several pulses of carbon dioxide laser radiation to seal the area and control bleeding. The cavity was then lined with glass ionomer cement and the tooth restored with amalgam. There was no loss of vitality over time

Key Points for Laser Pulpotomy

- Laser energy must absorb into major chromophores (water, porphyrins, melanin, and other pigments) for coagulation and bacterial inactivation to occur.
- This can be done with almost any laser system, but preferred lasers are Nd:YAG, KTP, and near-infrared diode lasers.
- If middle-infrared lasers are used, long pulse durations are needed to maximize coagulation.
- This typically employs very short exposure times.
- The techniques to treat the exposed pulp stumps are the same as for direct pulp capping.

9.10 Endodontic Surgery and Treatment of Resorption Lesions

Lesions of invasive cervical resorption can be treated by laser ablation, as an alternative to the traditional approach using trichloroacetic acid. The advantages of using lasers for this application include greater precision and less collateral injury to the tissues (◘ Fig. 9.8) [189].

For periapical surgery, lasers can be used to ablate granulation tissue and to sterilize the root apex, as well as for gaining access to the lesion by removing overlying bone. Bone is ablated readily by Er:YAG and Er,Cr:YSGG laser radiation, and in clinical practice, this is typically undertaken using an accompanying

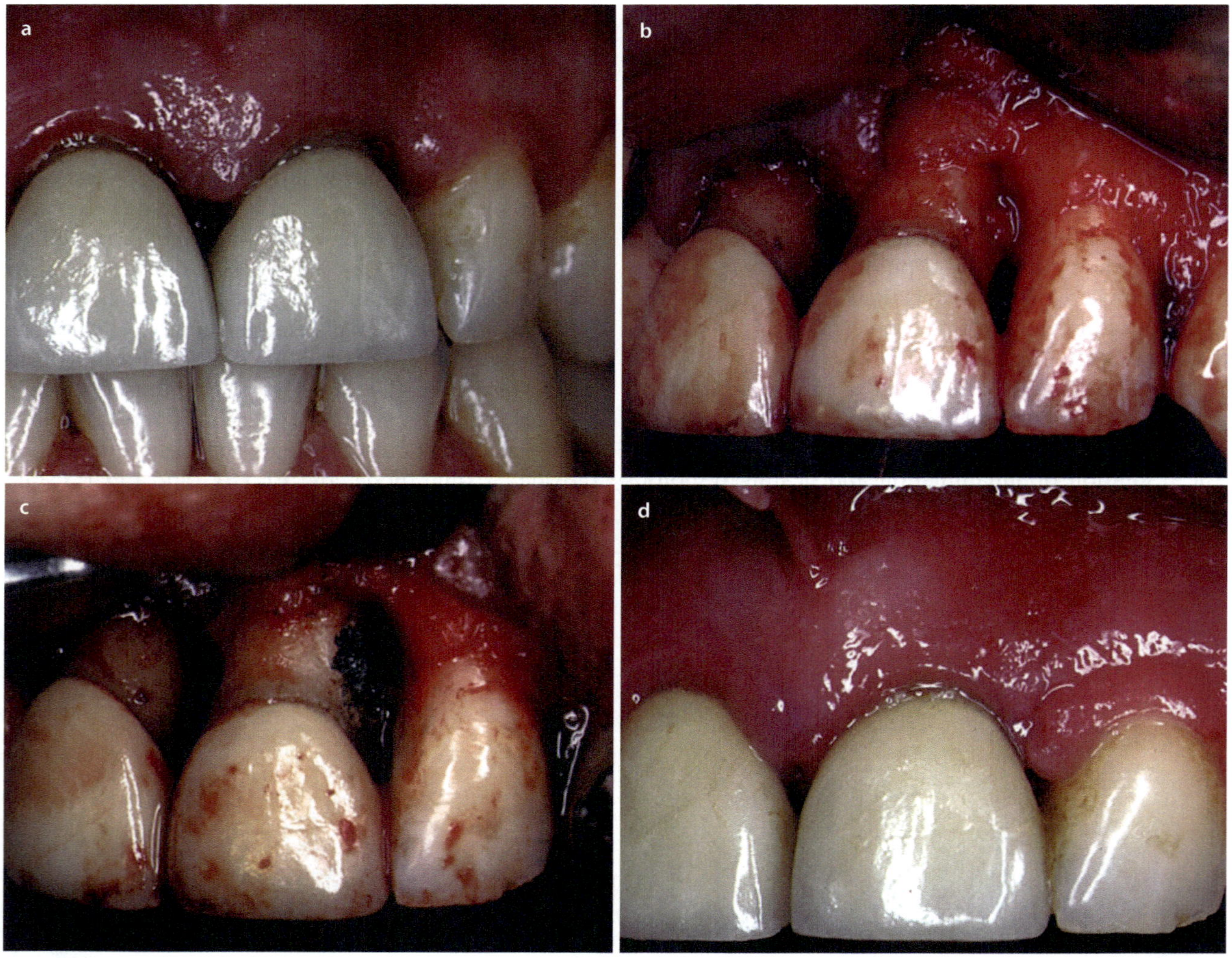

◘ **Fig. 9.8** Laser treatment of invasive cervical resorption. Tooth #21 (left maxillary central incisor) distal developed invasive resorption after internal "walking" bleaching with hydrogen peroxide. The bleaching occurred prior to PFM crowns being placed on both maxillary central incisor teeth. **a** Preoperative view. **b** With a flap raised, the granulation tissue filling the resorption defects on the root surface can be seen. **c** Pulses from a carbon dioxide laser were used to ablate the resorbing granulation tissue. After this, the root surface was conditioned and a glass ionomer cement restoration placed. **d** 12-month follow-up showing a stable situation water mist spray. Appropriate flow of water spray prevents desiccation of the bone, ensures cooling of the site to maintain bone viability, and irrigates the site to remove debris. These middle-infrared lasers give deep cuts with sharp edges which are free of charring. Similar benefits are found when these lasers are used for root resection procedures [190, 191].

Lasers have been used successfully for root end resection and root end cavity preparation during apical surgery [192]. The use of an Er:YAG or Er,Cr:YSGG laser with an operating microscope for periapical surgery has been shown to give significantly better results in terms of postoperative healing, in comparison with using conventional surgical approaches for apicoectomy. Such lasers can be used safely for root resections provided short-pulse duration is used and the water spray flow rate is sufficient [193, 194].

Key Points for Laser Endodontic Surgery

- Laser energy must absorb into major chromophores (water, porphyrins, melanin, and other pigments) for soft tissue ablation to occur.
- This can be done with almost any laser system, but preferred lasers are Nd:YAG, KTP, and near-infrared diode lasers. With the carbon dioxide laser, extreme care is needed to avoid deleterious thermal changes to tooth structure and the dental pulp.
- Hard tissue ablation (bone cutting, root end resection) requires a middle-infrared laser for high cutting efficiency.

9.11 Safety Issues Related to the Use of Lasers in Endodontics

Lasers can be used in conjunction with conventional endodontic equipment such as operating microscopes provided the appropriate considerations are made for eye safety, such as filters fitted to the objective of the microscope to match the laser wavelengths in use. With wavelengths longer than 2000 nm, this is not needed as the glass elements in the microscope provide sufficient attenuation.

9.11.1 Prevention of Transmission of Infection Through Contact

Laser endodontic fiber tips used within the root canal would be expected in many cases to encounter blood or other fluids which could be a source of patient-to-patient transmission of infection if the fibers are not appropriately disinfected. Disposable tips have become available for some laser systems; however, many fiber-optic systems are used where the fibers are cleaved after each use [195]. Appropriate disinfection and sterilization must be carried out for laser accessories and components that come into direct contact with oral soft and hard tissues. Other relevant recommendations include the following:

1. Fluid fed through a sleeve around the laser to cool it during surgery must be sterile.
2. Deposits of carbonized tissue residue can reduce the quantity and quality of the light emission. Therefore, it is necessary to wipe the tip after use. It may be necessary to calibrate the tip during the procedure.
3. Sapphire tips that come into contact with sterile tissue must be sterile and need to be cleaned and then sterilized after each use.

Piccione [196] further recommended that all controls of the laser should be disinfected or covered with a barrier, in a manner similar to other dental equipment, while smaller laser accessories such as handpiece should be steam sterilized.

9.11.2 Temperature Effects of Lasers on the Dental Pulp

In all endodontic applications using higher-powered lasers, care is needed to address thermal changes in the root structure, to preserve tissue vitality. Andersen [197] has demonstrated that in the human dental pulp, both cold and heat evoked a decrease in pulpal blood flow, when measured using a Doppler flowmetry. There is, therefore, a low potential of pulpal blood flow for cooling. The absorption coefficient and the reflectivity of the laser wavelength used are important in determining the pulpal reaction. Nyborg and Brannstrom [198] determined that a temperature of 150 °C on the enamel surface for 30 s could cause necrosis of the dental pulp. According to Zach and Cohen [89], an intra-pulpal temperature increase of approximately 5.5 °C can promote necrosis of the dental pulp in 15% of cases, while temperature increases of 11 and 17 °C will cause necrosis in 60% and 100% of cases [89, 199].

Pulpal damage can be avoided or minimized by a suitable choice of laser parameters and by appropriate use of irrigation or an air/water spray. Armengo [200] studied the effect of water spray on the temperature rise when using an Er:YAG or Nd:YAP laser. Water spray reduced the temperature rise associated with laser treatment and also helped to clear the ablation site of debris and keep it moist. The importance of air/water spray is exemplified in the study of Glockner et al. [201], which demonstrated that during coronal cavity preparation with the Er:YAG laser, a temperature reduction occurs after a few seconds from 37 to 25 °C, because of the cooling effect of the air/water spray.

9.11.3 Temperature Effect of Lasers on Periodontal Tissues

Maintaining the health of the periodontal apparatus is critical for the success or failure of endodontic treatment undertaken with lasers. Modern endodontic rotary instruments produce little or no increase in periradicular root surface temperature [202]. In contrast, several studies have shown that certain canal preparation techniques [203, 204] and obturation techniques [205–208] can transfer heat to the periodontal tissues. Er:YAG lasers cause evaporation and expansion of water within the crystals of hard tissue, and this evaporation can have a cooling action.

9

Several authors have studied the thermal effect of lasers on the periodontal ligament and surrounding bone [47, 48, 209, 210]. The supporting periodontal apparatus is known to be sensitive to temperatures of 47 °C, while temperatures of 60 °C and above will permanently stop blood flow and cause bone necrosis [211]. On the other hand, periodontal tissues are not damaged if the temperature increase is kept below 5 °C [212]. A threshold temperature increase of 7 °C is commonly considered as the highest thermal change which is biologically acceptable to avoid periodontal damage [90, 213–215].

Kimura et al. [216] using the Er:YAG laser noted that the root surface temperature increase was less than 6 °C at the apical third and 3 °C at the middle third. Similarly, Theodoro et al. [217] using the same laser reported temperature increases below 7 °C, while in the study of Machida et al. [213] where water spray was used the temperature increase at the apex was less than 2 °C. Thus, the use of air or water coolants in combination with lasers will help prevent adverse thermal effects on the periodontal ligament and surrounding bone [200, 201].

A further consideration is that of the thermal relaxation time (T_R), which is the amount of time required for heat to flow into adjacent regions or otherwise be dissipated [218]. The use of pulsed lasers with short-pulse durations will minimize the zone of thermal damage, by producing a thermal event that is shorter than the T_R of the tissue [219].

In case of root canal ablation, the conduction of heat from dentine to periodontal ligament and bone can be reduced by using a continuous stream of water during ablation. On the other hand, a dry root canal is devoid of fluid and will conduct energy similar to a solid body, which is uniformly in all directions. However, canals that are irrigated with fluids will benefit from transfer of heat into that fluid.

9.12 Conclusion

There are now several areas in endodontics where the use of laser technology offers superior outcomes for patients and simplification of techniques for the clinician. Pulsed near- or middle-infrared lasers combined with irrigants provide several advantages in terms of effective canal debridement as well as accompanying disinfection. Given the growing evidence in support of such applications, the integration of laser-based technologies into everyday clinical practice is likely to grow over the coming years. A wide range of lasers have been used successfully in endodontics, and this opens the pathway for laser systems which offer more than one wavelength, delivered through separate delivery systems or the one delivery system.

9.12.1 Future Aspects

The use of lasers in endodontics has entered a new phase with research over the past decade indicating that laser-based methods can provide not only equivalent but now superior results in terms of effective debridement of the root canal when compared to hand or powered conventional endodontic instruments. The potential in the future is to link systems for debridement and disinfection to approaches which also give accompanying analgesic effects and biostimulation, so that several therapeutic benefits are gained at the same time from a single laser irradiation protocol. There is considerable potential to incorporate feedback systems into endodontic laser systems so that fibers used within the root canal space for treatment can also support detection and diagnosis applications. This will enhance clinical efficiency and reduce the complexity of equipment which clinicians use. Further development of techniques for laser-induced analgesia will promote the development of endodontic and restorative treatments of teeth.

An analysis of metrics for publication activity (◘ Fig. 9.9) shows that the growth of published research in the field of lasers in endodontics has been spectacular over the past 20 years, with consistent gradients for the growth in the total number of publications on key topics within that field.

For endodontic diagnosis, growth in the number of publications in the use of laser Doppler flowmetry and in the use of laser-induced fluorescence (DiagnoDENT™) has been much lower than in topics

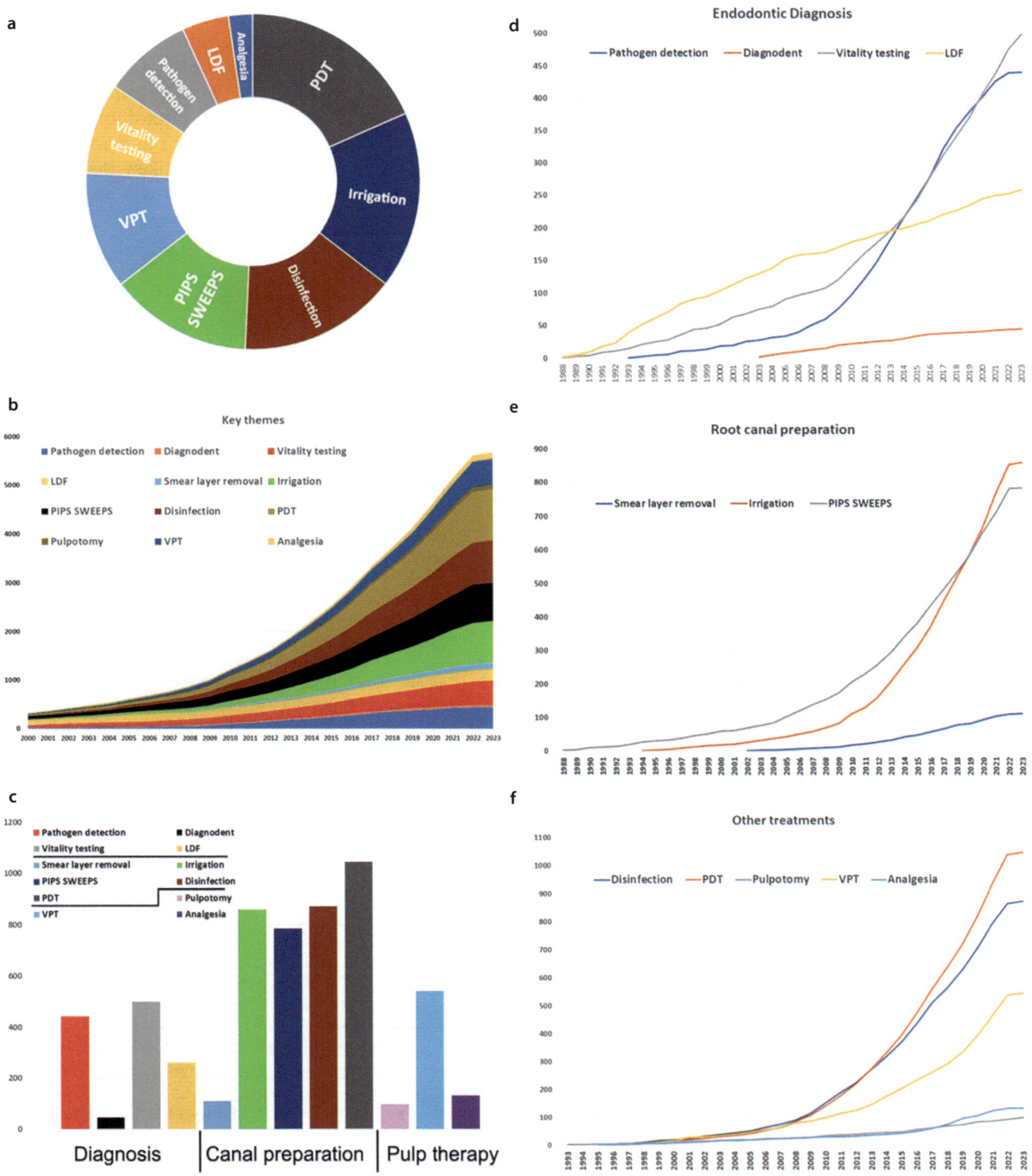

Fig. 9.9 An overview of the published literature on lasers in endodontics. Panel **a** (left top): this pie diagram shows the distribution of research topics on lasers in endodontics for the period from 2000 to 2023, based on search results from PUBMED for English language journals. The topics are arranged in descending order clockwise, for the nine top topics within lasers in endodontics. Panel **b** (left middle): this diagram shows that the relative distribution of publications on lasers in endodontics across the 12 most common topics has been consistent across from 2000 to 2023. Panel **c** (left bottom): this bar graph shows the total number of publications identified in a PubMed search for English language journals from 1965 to late January 2023, for topics within the three major themes of lasers in endodontics—diagnosis, canal preparation, and pulp therapy. Panels **d–f** show the cumulative number of publications growing over time for key topics within the three broad areas of diagnosis, root canal preparation, and vital pulp therapy and other treatments

relating to using laser technologies both to detect bacteria and other pathogens and also to assist in determining the vitality of the dental pulp. This indicates there is a strong ongoing interest in being able to use these various laser technologies to augment existing clinical diagnostic tools, to help in forming a diagnosis, and to assess the progress of treatment. Over time, it is likely that progress in the area of pathogen detection may lead to devices that can be employed widely both in general dental practice and in specialist endodontic practice for real-time assessment of the load of pathogens in the root canal system. If and when this occurs, it will be important to document how role introducing such technologies into everyday clinical practice alters the style of clinical practice and clinical efficiency, as well as influencing patient outcomes in both the short term and the long term.

Around the topic of root canal preparation, there continues to be a strong interest in the topics of using lasers for irrigation and fluid agitation and in the further evolution of specific protocols for using the middle-infrared laser wavelengths to achieve better debridement of the root canal system. The strong growth in publications relating to specific protocols, namely, PIPS and SWEEPS, shows the strong level of interest in the development of approaches that aim to improve the quality of debridement. Further refinement of such protocols can be anticipated in the coming years. Hopefully, the direction of this research will include not only laboratory studies but also investigations that track clinical outcomes. Such information can be important in bringing along large parts of the dental profession in what would be a considerable paradigms shift.

Interest in the use of laser technologies for photothermal and photodynamic disinfection treatments remains strong and has continued to grow over time. Once again, there is a need to assess how improved microbiological outcomes within the root canal system translate into short- and long-term clinical outcomes. Such studies will need to control for other factors that can influence short- and long-term endodontic outcomes, so that there can be a valid comparison between traditional endodontic treatments and those undertaken using laser disinfection.

An emerging area over the past decade has been the use of lasers for vital pulp therapy. This is an extension of past work using lasers for pulpotomy and for pulp capping, with a greater focus on maintaining sufficient viable pulp tissue in the root canal system that traditional anterograde root canal therapy can be avoided. By preserving the vitality of pulp tissue, adverse consequences of root canal treatment, such as a reduction in strength and discoloration, can be avoided.

An area which is still at a relatively early stage is the use of lasers for analgesia, including the reduction of discomfort following traditional endodontic treatment visits, as well as reducing the pain of dental injections. Treating painful conditions using photobiomodulation is a well-established clinical approach in many areas of medicine and allied health, and it is likely that over the coming decade, much greater interest will be placed on using lasers to improve the patient experience during endodontics and restorative dentistry.

References

1. Abd-Elmeguid A, Yu DC. Dental pulp neurophysiology: part 2. Current diagnostic tests to assess pulp vitality. J Can Dent Assoc. 2009;75(2):139–43.
2. Emshoff R, Moschen I, Strobl H. Use of laser Doppler flowmetry to predict vitality of luxated or avulsed permanent teeth. Oral Surg Oral Med Oral Pathol Oral Radiol Endod. 2004;98(6):750–5.
3. Roy E, Alliot-Licht B, Dajean-Trutaud S, Fraysse C, Jean A, Armengol V. Evaluation of the ability of laser Doppler flowmetry for the assessment of pulp vitality in general dental practice. Oral Surg Oral Med Oral Pathol Oral Radiol Endod. 2008;106(4):615–20.
4. Firestone AR, Wheatley AM, Thuer UW. Measurement of blood perfusion in the dental pulp with laser Doppler flowmetry. Int J Microcirc Clin Exp. 1997;17(6):298–304.
5. Jafarzadeh H. Laser Doppler flowmetry in endodontics: a review. Int Endod J. 2009;42(6):476–90.
6. Odor TM, Chandler NP, Watson TF, Ford TR, McDonald F. Laser light transmission in teeth: a study of the patterns in different species. Int Endod J. 1999;32(4):296–302.
7. Alghaithy RA, Qualtrough AJ. Pulp sensibility and vitality tests for diagnosing pulpal health in permanent teeth: a critical review. Int Endod J. 2017;50:135.
8. Banthitkhunanon P, Chintakanan S, Wanachantararak S, Vongsavan N, Matthews B. Effects of enamel and dentine thickness on laser Doppler blood-flow signals recorded from the underlying pulp cavity in human teeth in vitro. Arch Oral Biol. 2013;58(11):1692–5.
9. Sasano T, Nakajima I, Shoji N, Kuriwada S, Sanjo D, Ogino H, Miyahara T. Possible application of transmitted laser light for the assessment of human pulpal vitality. Endod Dent Traumatol. 1997;13(2):88–91.
10. Chandler NP, Pitt Ford TR, Watson TF. Pattern of transmission of laser light through carious molar teeth. Int Endod J. 2001;34(7):526–32.
11. Chandler NP, Pitt Ford TR, Monteith BD. Laser light passage through restored and carious posterior teeth. J Oral Rehabil. 2014;41(8):630–4.
12. Walsh LJ. Applications of laser fluorescence for diagnosis of bacterial infections in the root canal. Australas Dent Pract. 2010;21:54–6.
13. Sainsbury AL, Bird PS, Walsh LJ. DIAGNOdent laser fluorescence assessment of endodontic infection. J Endod. 2009;35(10):1404–7.
14. Hibst RP, R. Molecular basis of red excited caries fluorescence. Caries Res. 2000;34:323.
15. Koenig K, Schneckenburger H. Laser-induced autofluorescence for medical diagnosis. J Fluoresc. 1994;4(1):17–40.

16. Ho QV, George R, Sainsbury AL, Kahler WA, Walsh LJ. Laser fluorescence assessment of the root canal using plain and conical optical fibers. J Endod. 2010;36(1):119–22.
17. Kesler G, Gal R, Kesler A, Koren R. Histological and scanning electron microscope examination of root canal after preparation with Er:YAG laser microprobe: a preliminary in vitro study. J Clin Laser Med Surg. 2002;20(5):269–77.
18. Lagemann M, George R, Chai L, Walsh LJ. Activation of ethylenediaminetetraacetic acid by a 940 nm diode laser for enhanced removal of smear layer. Aust Endod J. 2014;40(2):72–5.
19. George R, Walsh LJ. Performance assessment of novel side firing flexible optical fibers for dental applications. Lasers Surg Med. 2009;41(3):214–21.
20. Shakibaie F, George R, LJ. W. Applications of laser-induced fluorescence in dentistry. Int J Dent Clin. 2011;3(3):38–44.
21. George R, Walsh LJ. Laser fiber-optic modifications and their role in endodontics. J Laser Dent. 2012;20:24–30.
22. Krause F, Braun A, Eberhard J, Jepsen S. Laser fluorescence measurements compared to electrical resistance of residual dentine in excavated cavities in vivo. Caries Res. 2007;41(2):135–40.
23. Clark J, Symons AL, Diklic S, Walsh LJ. Effectiveness of diagnosing residual caries with various methods during cavity preparation using conventional methods, chemo-mechanical caries removal, and Er:YAG laser. Aust Dent J. 2001;46:S20.
24. Yonemoto K, Eguro T, Maeda T, Tanaka H. Application of DIAGNOdent as a guide for removing carious dentin with Er:YAG laser. J Dent. 2006;34:269–76.
25. Eberhard J, Eisenbeiss AK, Braun A, Hedderich J, Jepsen S. Evaluation of selective caries removal by a fluorescence feedback-controlled Er:YAG laser in vitro. Caries Res. 2005;39(6):496–504.
26. Folwaczny M, Mehl A, Haffner C, Benz C, Hickel R. Root substance removal with Er:YAG laser radiation at different parameters using a new delivery system. J Periodontol. 2000;71(2):147–55.
27. Krause F, Braun A, Brede O, Eberhard J, Frentzen M, Jepsen S. Evaluation of selective calculus removal by a fluorescence feedback-controlled Er:YAG laser in vitro. J Clin Periodontol. 2007;34(1):66–71.
28. Walsh LJ, Mubarak S, A. M. Autopilot laser-based systems for guiding caries and calculus removal: from concept to clinical reality. Australas Dent Pract. 2007;18:122–8.
29. Coulthwaite L, Pretty IA, Smith PW, Higham SM, Verran J. The microbiological origin of fluorescence observed in plaque on dentures during QLF analysis. Caries Res. 2006;40(2):112–6.
30. Walsh LJ, Shakibaie F. Ultraviolet-induced fluorescence: shedding new light on dental biofilms and dental caries. Australas Dent Pract. 2007;18:52–6.
31. Johnson WT, Noblett WC. Cleaning and shaping in endodontics: principles and practice. 4th ed. Philadelphia, PA: Saunders; 2009.
32. Gomes BP, Vianna ME, Sena NT, Zaia AA, Ferraz CC, de Souza Filho FJ. In vitro evaluation of the antimicrobial activity of calcium hydroxide combined with chlorhexidine gel used as intracanal medicament. Oral Surg Oral Med Oral Pathol Oral Radiol Endod. 2006;102(4):544–50.
33. Sin JH, Hamlet S, Walsh LJ, Love RM, George R. Oxidising agents and its effect on human dentine fluorescence diagnostic measurements. Photodiagn Photodyn Ther. 2020;31:101950.
34. Sin JH, Ipe DS, Hamlet S, Walsh LJ, Love RM, George R. Fluorescence characteristics of E. faecalis in dentine following treatment with oxidizing endodontic irrigants. Photodiagn Photodyn Ther. 2021;35:102344.
35. Tsai AC, George R, Walsh LJ. Evaluation of the effect of various endodontic irrigants and medicaments on dentine fluorescence. Photodiagn Photodyn Ther. 2022;37:102651.
36. Tsai AC, George R, Teoh YY, Walsh LJ. Laser-fluorescence assessment of sodium hypochlorite quenching reversal agents on human dentine. Photodiagn Photodyn Ther. 2022;38:102791.
37. Mathew S, Thangavel B, Mathew CA, Kailasam S, Kumaravadivel K, Das A. Diagnosis of cracked tooth syndrome. J Pharm Bioallied Sci. 2012;4(Suppl 2):S242–4.
38. Sekhar KSSGPV. Cracked tooth syndrome (CTS)—a roller coaster ride—a review. Ann Essences Dent. 2012;IV(1):92–5.
39. Nguyen V, Palmer G. A review of the diagnosis and management of the cracked tooth. SADJ. 2010;65(9):396–8, 400–3.
40. Lubisich EB, Hilton TJ, Ferracane J. Cracked teeth: a review of the literature. Wiley Periodicals; 2010. p. 158–67.
41. Sapra A, Darbar A, George R. Near-infrared laser energy transmission through teeth with crack lines: an in-vitro study. Med Lasers. 2021;10(4):214–9.
42. Sapra A, Darbar A, George R. Laser-assisted diagnosis of symptomatic cracks in teeth with cracked tooth: a 4-year in-vivo follow-up study. Aust Endod J. 2020;46(2):197–203.
43. Levy G. Cleaning and shaping the root canal with a Nd:YAG laser beam: a comparative study. J Endod. 1992;18(3):123–7.
44. Matsuoka E, Yonaga K, Kinoshita J, Kimura Y, Matsumoto K. Morphological study on the capability of Er:YAG laser irradiation for root canal preparation. J Clin Laser Med Surg. 2000;18(4):215–9.
45. Ali MN, Hossain M, Nakamura Y, Matsuoka E, Kinoshita J, Matsumoto K. Efficacy of root canal preparation by Er,Cr:YSGG laser irradiation with crown-down technique in vitro. Photomed Laser Surg. 2005;23(2):196–201.
46. Jahan K, Hossain M, Nakamura Y, Yoshishige Y, Kinoshita J-I, Matsumoto K. An assessment following root canal preparation by Er,Cr:YSGG laser irradiation in straight and curved roots, in vitro. Lasers Med Sci. 2006;21(4):229–34.
47. Cohen BI, Deutsch AS, Musikant BL. Effect of power settings on temperature change at the root surface when using a Holmium YAG laser in enlarging the root canal. J Endod. 1996;22(11):596–9.
48. Cohen BI, Deutsch AS, Musikant BL, Pagnillo MK. Effect of power settings versus temperature change at the root surface when using multiple fiber sizes with a Holmium YAG laser while enlarging a root canal. J Endod. 1998;24(12):802–6.
49. Deutsch AS, Cohen BI, Musikant BL. Temperature change at the root surface when enlarging a root canal with a holmium: YAG (Ho:YAG) laser, using six different fiber-optic sizes. Gen Dent. 2004;52(3):222–7.
50. Matsuoka E, Jayawardena JA, Matsumoto K. Morphological study of the Er,Cr:YSGG laser for root canal preparation in mandibular incisors with curved root canals. Photomed Laser Surg. 2005;23(5):480–4.
51. Anic I, Segovic S, Katanec D, Prskalo K, Najzar-Fleger D. Scanning electron microscopic study of dentin lased with argon, CO2, and Nd:YAG laser. J Endod. 1998;24(2):77–81.
52. Shoji S, Hariu H, Horiuchi H. Canal enlargement by Er:YAG laser using a cone-shaped irradiation tip. J Endod. 2000;26(8):454–8.
53. Martin T. Comparison of the emission characteristics of three erbium laser systems—a physical case report. J Oral Laser Appl. 2004;4(4):263–9.
54. Verdaasdonkz RM, van Swol CFP. Laser light delivery systems for medical applications. Phys Med Biol. 1997;42:869–94.
55. Alves PR, Aranha N, Alfredo E, Marchesan MA, Brugnera Junior A, Sousa-Neto MD. Evaluation of hollow fiberoptic

tips for the conduction of Er:YAG laser. Photomed Laser Surg. 2005;23(4):410–5.
56. Shirk GJ, Gimpelson RJ, Krewer K. Comparison of tissue effects with sculptured fiberoptic cables and other Nd:YAG laser and argon laser treatments. Lasers Surg Med. 1991;11(6):563–8.
57. Stabholz A, Neev J, Liaw LH, Stabholz A, Khayat A, Torabinejad M. Effect of ArF-193 nm excimer laser on human dentinal tubules. A scanning electron microscopic study. Oral Surg Oral Med Oral Pathol. 1993;75(1):90–4.
58. Harashima T, Takeda FH, Zhang C, Kimura Y, Matsumoto K. Effect of argon laser irradiation on instrumented root canal walls. Endod Dent Traumatol. 1998;14(1):26–30.
59. Tewfik HM, Pashley DH, Horner JA, Sharawy MM. Structural and functional changes in root dentin following exposure to KTP/532 laser. J Endod. 1993;19(10):492–7.
60. Harashima T, Takeda FH, Kimura Y, Matsumoto K. Effect of Nd:YAG laser irradiation for removal of intracanal debris and smear layer in extracted human teeth. J Clin Laser Med Surg. 1997;15(3):131–5.
61. Takeda FH, Harashima T, Kimura Y, Matsumoto K. Comparative study about the removal of smear layer by three types of laser devices. J Clin Laser Med Surg. 1998;16(2):117–22.
62. Nuebler-Moritz M, Norbert G, Hermann FS, Peter H, Wilhelm P. Laboratory investigation of the efficacy of holmium:YAG laser irradiation in removing intracanal debris. In: Wigdor HA, Featherstone JD, Peter R, editors. Lasers in dentistry III, vol. 2973. SPIE; 1997. p. 150–6.
63. Takeda FH, Harashima T, Kimura Y, Matsumoto K. Efficacy of Er:YAG laser irradiation in removing debris and smear layer on root canal walls. J Endod. 1998;24(8):548–51.
64. Yamazaki R, Goya C, Yu DG, Kimura Y, Matsumoto K. Effects of erbium,chromium:YSGG laser irradiation on root canal walls: a scanning electron microscopic and thermographic study. J Endod. 2001;27(1):9–12.
65. Takeda FH, Harashima T, Kimura Y, Matsumoto K. A comparative study of the removal of smear layer by three endodontic irrigants and two types of laser. Int Endod J. 1999;32(1):32–9.
66. Wang X, Sun Y, Kimura Y, Kinoshita J, Ishizaki NT, Matsumoto K. Effects of diode laser irradiation on smear layer removal from root canal walls and apical leakage after obturation. Photomed Laser Surg. 2005;23(6):575–81.
67. da Costa RA, Nogueira GE, Antoniazzi JH, Moritz A, Zezell DM. Effects of diode laser (810 nm) irradiation on root canal walls: thermographic and morphological studies. J Endod. 2007;33(3):252–5.
68. Kaitsas V, Signore A, Fonzi L, Benedicenti S, Barone M. Effects of Nd: YAG laser irradiation on the root canal wall dentin of human teeth: a SEM study. Bull Group Int Rech Sci Stomatol Odontol. 2001;43(3):87–92.
69. Goya C, Yamazaki R, Tomita Y, Kimura Y, Matsumoto K. Effects of pulsed Nd:YAG laser irradiation on smear layer at the apical stop and apical leakage after obturation. Int Endod J. 2000;33(3):266–71.
70. Wilder-Smith P, Arrastia AM, Schell MJ, Liaw LH, Grill G, Berns MW. Effect of ND:YAG laser irradiation and root planning on the root surface: structural and thermal effects. J Periodontol. 1995;66(12):1032–9.
71. Takeda FH, Harashima T, Eto JN, Kimura Y, Matsumoto K. Effect of Er:YAG laser treatment on the root canal walls of human teeth: an SEM study. Endod Dent Traumatol. 1998;14(6):270–3.
72. Biedma BM, Varela Patino P, Park SA, Barciela Castro N, Magan Munoz F, Gonzalez Bahillo JD, Cantatore G. Comparative study of root canals instrumented manually and mechanically, with and without Er:YAG laser. Photomed Laser Surg. 2005;23(5):465–9.
73. Altundasar E, Ozcelik B, Cehreli ZC, Matsumoto K. Ultramorphological and histochemical changes after Er,Cr:YSGG laser irradiation and two different irrigation regimes. J Endod. 2006;32(5):465–8.
74. Kimura Y, Wilder-Smith P, Matsumoto K. Lasers in endodontics: a review. Int Endod J. 2000;33(3):173–85.
75. Stabholz A, Zeltser R, Sela M, Peretz B, Moshonov J, Ziskind D, Stabholz A. The use of lasers in dentistry: principles of operation and clinical applications. Compend Contin Educ Dent. 2003;24(12):935–48.
76. George R, Walsh LJ. Performance assessment of novel side firing safe tips for endodontic applications. J Biomed Opt. 2011;16(4):048004.
77. George R, Meyers IA, Walsh LJ. Laser activation of endodontic irrigants with improved conical laser fiber tips for removing smear layer in the apical third of the root canal. J Endod. 2008;34(12):1524–7.
78. Abduljalil M, Kalender A. Efficacy of Er,Cr:YSGG laser with different output powers on removing smear layer after retreatment of two different obturation techniques. Photobiomodul Photomed Laser Surg. 2020;38(2):84–90.
79. George R, Chan K. Modifying laser induced shock waves for use in clinical endodontics. In: Walsh LJ, editor. The 9th world association of laser therapy congress. Bologna: Medimond International Proceedings; 2013.
80. George R, Rutley EB, Walsh LJ. Evaluation of smear layer: a comparison of automated image analysis versus expert observers. J Endod. 2008;34(8):999–1002.
81. Anagnostaki E, Mylona V, Parker S, Lynch E, Grootveld M. Systematic review on the role of lasers in endodontic therapy: valuable adjunct treatment? Dent J (Basel). 2020;8(3):63.
82. George R, Walsh LJ. Apical extrusion of root canal irrigants when using Er:YAG and Er,Cr:YSGG lasers with optical fibers: an in vitro dye study. J Endod. 2008;34(6):706–8.
83. George R, Walsh LJ. Thermal effects from modified endodontic laser tips used in the apical third of root canals with erbium-doped yttrium aluminium garnet and erbium, chromium-doped yttrium scandium gallium garnet lasers. Photomed Laser Surg. 2010;28(2):161–5.
84. Hmud R, Kahler WA, George R, Walsh LJ. Cavitational effects in aqueous endodontic irrigants generated by near-infrared lasers. J Endod. 2010;36(2):275–8.
85. Asnaashari M, Safavi N. Disinfection of contaminated canals by different laser wavelengths, while performing root canal therapy. J Lasers Med Sci. 2013;4(1):8–16.
86. Whitters CJ, MacFarlane TW, MacKenzie D, Moseley H, Strang R. The bactericidal activity of pulsed Nd-YAG laser radiation in vitro. Lasers Med Sci. 1994;9:297–303.
87. Moshonov J, Orstavik D, Yamauchi S, Pettiette M, Trope M. Nd:YAG laser irradiation in root canal disinfection. Endod Dent Traumatol. 1995;11(5):220–4.
88. Walsh LJ. The current status of low level laser therapy in dentistry. Part 2. Hard tissue applications. Aust Dent J. 1997;42(5):302–6.
89. Zach L, Cohen G. Pulp response to externally applied heat. Oral Surg Oral Med Oral Pathol. 1965;19:515–30.
90. Nammour S, Kowaly K, Powell GL, Van Reck J, Rocca JP. External temperature during KTP-Nd:YAG laser irradiation in root canals: an in vitro study. Lasers Med Sci. 2004;19(1):27–32.
91. Bergmans L, Moisiadis P, Teughels W, Van Meerbeek B, Quirynen M, Lambrechts P. Bactericidal effect of Nd:YAG laser irradiation on some endodontic pathogens ex vivo. Int Endod J. 2006;39(7):547–57.

92. Mathew J, Emil J, Paulaian B, John B, Raja J, Mathew J. Viability and antibacterial efficacy of four root canal disinfection techniques evaluated using confocal laser scanning microscopy. J Conserv Dent. 2014;17(5):444–8.
93. Konopka K, Goslinski T. Photodynamic therapy in dentistry. J Dent Res. 2007;86(8):694–707.
94. Komerik N, MacRobert AJ. Photodynamic therapy as an alternative antimicrobial modality for oral infections. J Environ Pathol Toxicol Oncol. 2006;25(1–2):487–504.
95. Gursoy H, Ozcakir-Tomruk C, Tanalp J, Yilmaz S. Photodynamic therapy in dentistry: a literature review. Clin Oral Investig. 2013;17(4):1113–25.
96. Okamoto H, Iwase T, Morioka T. Dye-mediated bactericidal effect of He-Ne laser irradiation on oral microorganisms. Lasers Surg Med. 1992;12(4):450–8.
97. Dobson J, Wilson M. Sensitization of oral bacteria in biofilms to killing by light from a low-power laser. Arch Oral Biol. 1992;37(11):883–7.
98. Chiniforush N, Pourhajibagher M, Shahabi S, Bahador A. Clinical approach of high technology techniques for control and elimination of endodontic microbiota. J Lasers Med Sci. 2015;6(4):139–50.
99. Tennert C, Feldmann K, Haamann E, Al-Ahmad A, Follo M, Wrbas KT, Hellwig E, Altenburger MJ. Effect of photodynamic therapy (PDT) on Enterococcus faecalis biofilm in experimental primary and secondary endodontic infections. BMC Oral Health. 2014;14:132.
100. Rios A, He J, Glickman GN, Spears R, Schneiderman ED, Honeyman AL. Evaluation of photodynamic therapy using a light-emitting diode lamp against Enterococcus faecalis in extracted human teeth. J Endod. 2011;37(6):856–9.
101. Siddiqui SH, Awan KH, Javed F. Bactericidal efficacy of photodynamic therapy against Enterococcus faecalis in infected root canals: a systematic literature review. Photodiagn Photodyn Ther. 2013;10(4):632–43.
102. Lee MT, Bird PS, Walsh LJ. Photo-activated disinfection of the root canal: a new role for lasers in endodontics. Aust Endod J. 2004;30(3):93–8.
103. Bonsor SJ, Pearson GJ. Current clinical applications of photo-activated disinfection in restorative dentistry. Dent Update. 2006;33(3):143–4, 7–50, 53.
104. Bago I, Plecko V, Gabric Panduric D, Schauperl Z, Baraba A, Anic I. Antimicrobial efficacy of a high-power diode laser, photo-activated disinfection, conventional and sonic activated irrigation during root canal treatment. Int Endod J. 2013;46(4):339–47.
105. Dickers B, Lamard L, Peremans A, Geerts S, Lamy M, Limme M, Rompen E, De Moor RJ, Mahler P, Rocca JP, Nammour S. Temperature rise during photo-activated disinfection of root canals. Lasers Med Sci. 2009;24(1):81–5.
106. Kosarieh E, Bolhari B, Sanjari Pirayvatlou S, Kharazifard MJ, Sattari Khavas S, Jafarnia S, Saberi S. Effect of Er:YAG laser irradiation using SWEEPS and PIPS technique on dye penetration depth after root canal preparation. Photodiagn Photodyn Ther. 2021;33:102136.
107. Sin J, Walsh LJ, Figueredo CM, George R. Evaluation of effectiveness of photosensitizers used in laser endodontics disinfection: a systematic review. Transl Biophoton. 2021;3(1):e202000007.
108. Kuzekanani M, Walsh LJ, Yousefi MA. Cleaning and shaping curved root canals: Mtwo vs ProTaper instruments, a lab comparison. Indian J Dent Res. 2009;20(3):268–70.
109. Haapasalo M, Shen Y, Wang Z, Gao Y. Irrigation in endodontics. Br Dent J. 2014;216(6):299–303.
110. Gu LS, Kim JR, Ling J, Choi KK, Pashley DH, Tay FR. Review of contemporary irrigant agitation techniques and devices. J Endod. 2009;35(6):791–804.
111. Caron G, Nham K, Bronnec F, Machtou P. Effectiveness of different final irrigant activation protocols on smear layer removal in curved canals. J Endod. 2010;36(8):1361–6.
112. Bukiet F, Soler T, Guivarch M, Camps J, Tassery H, Cuisinier F, Candoni N. Factors affecting the viscosity of sodium hypochlorite and their effect on irrigant flow. Int Endod J. 2013;46(10):954–61.
113. Peters OA, Bardsley S, Fong J, Pandher G, Divito E. Disinfection of root canals with photon-initiated photoacoustic streaming. J Endod. 2011;37(7):1008–12.
114. Pedulla E, Genovese C, Campagna E, Tempera G, Rapisarda E. Decontamination efficacy of photon-initiated photoacoustic streaming (PIPS) of irrigants using low-energy laser settings: an ex vivo study. Int Endod J. 2012;45(9):865–70.
115. Zhu X, Yin X, Chang JW, Wang Y, Cheung GS, Zhang C. Comparison of the antibacterial effect and smear layer removal using photon-initiated photoacoustic streaming aided irrigation versus a conventional irrigation in single-rooted canals: an in vitro study. Photomed Laser Surg. 2013;31(8):371–7.
116. Al Shahrani M, DiVito E, Hughes CV, Nathanson D, Huang GT. Enhanced removal of Enterococcus faecalis biofilms in the root canal using sodium hypochlorite plus photon-induced photoacoustic streaming: an in vitro study. Photomed Laser Surg. 2014;32(5):260–6.
117. Balic M, Lucic R, Mehadzic K, Bago I, Anic I, Jakovljevic S, Plecko V. The efficacy of photon-initiated photoacoustic streaming and sonic-activated irrigation combined with QMiX solution or sodium hypochlorite against intracanal E. faecalis biofilm. Lasers Med Sci. 2016;31(2):335–42.
118. Betancourt P, Merlos A, Sierra JM, Arnabat-Dominguez J, Vinas M. Er,Cr:YSGG laser-activated irrigation and passive ultrasonic irrigation: comparison of two strategies for root canal disinfection. Photobiomodul Photomed Laser Surg. 2020;38(2):91–7.
119. Betancourt P, Sierra JM, Camps-Font O, Arnabat-Dominguez J, Vinas M. Er,Cr:YSGG laser-activation enhances antimicrobial and antibiofilm action of low concentrations of sodium hypochlorite in root canals. Antibiotics (Basel). 2019;8(4):232.
120. Laky M, Volmer M, Arslan M, Agis H, Moritz A, Cvikl B. Efficacy and safety of photon induced photoacoustic streaming for removal of calcium hydroxide in endodontic treatment. Biomed Res Int. 2018;2018:2845705.
121. Gulabivala K, Ng YL, Gilbertson M, Eames I. The fluid mechanics of root canal irrigation. Physiol Meas. 2010;31(12):R49–84.
122. Boutsioukis C, Kastrinakis E, Lambrianidis T, Verhaagen B, Versluis M, van der Sluis LW. Formation and removal of apical vapor lock during syringe irrigation: a combined experimental and computational fluid dynamics approach. Int Endod J. 2014;47(2):191–201.
123. Boutsioukis C, Lambrianidis T, Verhaagen B, Versluis M, Kastrinakis E, Wesselink PR, van der Sluis LW. The effect of needle-insertion depth on the irrigant flow in the root canal: evaluation using an unsteady computational fluid dynamics model. J Endod. 2010;36(10):1664–8.
124. Boutsioukis C, Lambrianidis T, Kastrinakis E. Irrigant flow within a prepared root canal using various flow rates: a computational fluid dynamics study. Int Endod J. 2009;42(2):144–55.

125. de Groot SD, Verhaagen B, Versluis M, Wu MK, Wesselink PR, van der Sluis LW. Laser-activated irrigation within root canals: cleaning efficacy and flow visualization. Int Endod J. 2009;42(12):1077–83.
126. Matsumoto H, Yoshimine Y, Akamine A. Visualization of irrigant flow and cavitation induced by Er:YAG laser within a root canal model. J Endod. 2011;37(6):839–43.
127. Verhaagen B, Fernandez RD. Measuring cavitation and its cleaning effect. Ultrason Sonochem. 2016;29:619–28.
128. Blanken J, De Moor RJ, Meire M, Verdaasdonk R. Laser induced explosive vapor and cavitation resulting in effective irrigation of the root canal. Part 1: a visualization study. Lasers Surg Med. 2009;41(7):514–9.
129. De Moor RJ, Blanken J, Meire M, Verdaasdonk R. Laser induced explosive vapor and cavitation resulting in effective irrigation of the root canal. Part 2: evaluation of the efficacy. Lasers Surg Med. 2009;41(7):520–3.
130. George R, Chan K, Walsh LJ. Laser-induced agitation and cavitation from proprietary honeycomb tips for endodontic applications. Lasers Med Sci. 2015;30(4):1203–8.
131. Wanner S, Gstottner M, Meirer R, Hausdorfer J, Fille M, Stockl B. Low-energy shock waves enhance the susceptibility of staphylococcal biofilms to antimicrobial agents in vitro. J Bone Joint Surg Br. 2011;93(6):824–7.
132. Gnanadhas DP, Elango M, Janardhanraj S, Srinandan CS, Datey A, Strugnell RA, Gopalan J, Chakravortty D. Successful treatment of biofilm infections using shock waves combined with antibiotic therapy. Sci Rep. 2015;5:17440.
133. Boutsioukis C, Verhaagen B, Versluis M, Kastrinakis E, Wesselink PR, van der Sluis LW. Evaluation of irrigant flow in the root canal using different needle types by an unsteady computational fluid dynamics model. J Endod. 2010;36(5):875–9.
134. Arslan H, Akcay M, Ertas H, Capar ID, Saygili G, Mese M. Effect of PIPS technique at different power settings on irrigating solution extrusion. Lasers Med Sci. 2015;30(6):1641–5.
135. Yost RA, Bergeron BE, Kirkpatrick TC, Roberts MD, Roberts HW, Himel VT, Sabey KA. Evaluation of 4 different irrigating systems for apical extrusion of sodium hypochlorite. J Endod. 2015;41(9):1530–4.
136. Hmud R, Kahler WA, Walsh LJ. Temperature changes accompanying near infrared diode laser endodontic treatment of wet canals. J Endod. 2010;36(5):908–11.
137. Deleu E, Meire MA, De Moor RJ. Efficacy of laser-based irrigant activation methods in removing debris from simulated root canal irregularities. Lasers Med Sci. 2015;30(2):831–5.
138. Moffitt JM, Cooley RO, Olsen NH, Hefferren JJ. Prediction of tetracycline-induced tooth discoloration. J Am Dent Assoc. 1974;88(3):547–52.
139. Kim ST, Abbott PV, McGinley P. The effects of Ledermix paste on discolouration of mature teeth. Int Endod J. 2000;33(3):227–32.
140. Walsh LJ, Athanassiadis B. Endodontic aesthetic iatrodontics. Australas Dent Pract. 2007;18:62–4.
141. Thomson AD, Athanassiadis B, Kahler B, Walsh L. Tooth discolouration: staining effects of various sealers and medicaments. Aust Endod J. 2012;38(1):2–9.
142. Walsh LJ, Verheyen P. Bleaching—accelerated with the laser. In: Moritz A, Beer F, Verheyen P, Wernisch J, Schoop U, Blum R, et al., editors. Oral laser application. Berlin: Quintessence; 2006.
143. Chou K, George R, Walsh LJ. Effectiveness of different intracanal irrigation techniques in removing intracanal paste medicaments. Aust Endod J. 2014;40(1):21–5.
144. Kustarci A, Er K, Siso SH, Aydin H, Harorli H, Arslan D, Kirmali O. Efficacy of laser-activated irrigants in calcium hydroxide removal from the artificial grooves in root canals: an ex vivo study. Photomed Laser Surg. 2016;34(5):205–10.
145. Chen BK, George R, Walsh LJ. Root discolouration following short-term application of steroid medicaments containing clindamycin, doxycycline or demeclocycline. Aust Endod J. 2012;38(3):124–8.
146. Chen BK, George R, Walsh LJ. Discoloration of roots caused by residual endodontic intracanal medicaments. Sci World J. 2014;2014:404676.
147. Bennett ZY, Walsh LJ. Factors affecting the rate of oxidation and resultant discolouration of tetracyclines contained in endodontic medicaments and irrigants. Int Endod J. 2015;48(4):373–9.
148. De Moor RJ, Verheyen J, Diachuk A, Verheyen P, Meire MA, De Coster PJ, Keulemans F, De Bruyne M, Walsh LJ. Insight in the chemistry of laser-activated dental bleaching. Sci World J. 2015;2015:650492.
149. Walsh LJ, Liu JY, P. V. Tooth discolouration and its treatment using KTP laser-assisted tooth whitening. J Oral Laser Appl. 2004;4:7–20.
150. Kuzekanani M, Walsh LJ. Quantitative analysis of KTP laser photodynamic bleaching of tetracycline-discolored teeth. Photomed Laser Surg. 2009;27(3):521–5.
151. Bennett ZY, Walsh LJ. Efficacy of LED versus KTP laser activation of photodynamic bleaching of tetracycline-stained dentine. Lasers Med Sci. 2015;30(7):1823–8.
152. Bennett ZY, Walsh LJ. Effect of photo-fenton bleaching on tetracycline-stained dentin in vitro. J Contemp Dent Pract. 2015;16(2):126–9.
153. Kulkarni S, George R, Love R, Ranjitkar S. Effectiveness of photobiomodulation in reducing pain and producing dental analgesia: a systematic review. Lasers Med Sci. 2022;37(7):3011–9.
154. Jimbo K, Noda K, Suzuki K, Yoda K. Suppressive effects of low-power laser irradiation on bradykinin evoked action potentials in cultured murine dorsal root ganglion cells. Neurosci Lett. 1998;240(2):93–6.
155. Yoda K. Suppressive effects of low-power laser on cytosolic calcium ion changes elicited by bradykinin in cultured murine dorsal root ganglion neurons. Pain Res. 2005;20:127–33.
156. Yan W, Chow R, Armati PJ. Inhibitory effects of visible 650-nm and infrared 808-nm laser irradiation on somatosensory and compound muscle action potentials in rat sciatic nerve: implications for laser-induced analgesia. J Peripher Nerv Syst. 2011;16(2):130–5.
157. Walsh LJ. Laser analgesia with pulsed infrared lasers: theory and practice. J Oral Laser Appl. 2008;8:7–16.
158. Yan W, Chow RT, Armati PJ. Effects of 650 nm and 808 nm laser irradiation on rat sciatic nerve: implications for the analgesic effects of LLLT. In: 8th Congress of the world association of laser therapy, Bergen, Norway, 2010.
159. Sharma SK, Kharkwal GB, Sajo M, Huang Y-Y, De Taboada L, McCarthy T, Hamblin MR. Dose response effects of 810 nm laser light on mouse primary cortical neurons. Lasers Surg Med. 2011;43(8):851–9.
160. Yachnev IL, Plakhova VB, Podzorova SA, Shelykh TN, Rogachevsky IV, Krylov BV. Mechanism of pain relief by low-power infrared irradiation: ATP is an IR-target molecule in nociceptive neurons. Med Chem. 2012;8(1):14–21.
161. Whitters CJ, Hall A, Creanor SL, Moseley H, Gilmour WH, Strang R, Saunders WP, Orchardson R. A clinical study of

pulsed Nd: YAG laser-induced pulpal analgesia. J Dent. 1995;23(3):145–50.
162. Orchardson R, Peacock JM, Whitters CJ. Effects of pulsed Nd:YAG laser radiation on action potential conduction in nerve fibres inside teeth in vitro. J Dent. 1998;26(5–6):421–6.
163. Orchardson R, Whitters CJ. Effect of HeNe and pulsed Nd:YAG laser irradiation on intradental nerve responses to mechanical stimulation of dentine. Lasers Surg Med. 2000;26(3):241–9.
164. Zeredo JL, Sasaki KM, Fujiyama R, Okada Y, Toda K. Effects of low power Er:YAG laser on the tooth pulp-evoked jaw-opening reflex. Lasers Surg Med. 2003;33(3):169–72.
165. Zeredo JL, Sasaki KM, Takeuchi Y, Toda K. Antinociceptive effect of Er:YAG laser irradiation in the orofacial formalin test. Brain Res. 2005;1032(1–2):149–53.
166. Walsh LJ. Membrane-based photoacoustic and biostimulatory applications in clinical practice. Australas Dent Pract. 2006;17:62–4.
167. Peres F, Felino A, Carvalho JF. [Analgesic effect of 904-nm laser radiation (IR) in oral surgery]. Rev Port Estomatol Cir Maxilofac. 1985;26(3):205–17.
168. Mezawa S, Iwata K, Naito K, Kamogawa H. The possible analgesic effect of soft-laser irradiation on heat nociceptors in the cat tongue. Arch Oral Biol. 1988;33(9):693–4.
169. Tsuchiya K, Kawatani M, Takeshige C, Sato T, Matsumoto I. Diode laser irradiation selectively diminishes slow component of axonal volleys to dorsal roots from the saphenous nerve in the rat. Neurosci Lett. 1993;161(1):65–8.
170. Baxter GD, Walsh DM, Allen JM, Lowe AS, Bell AJ. Effects of low intensity infrared laser irradiation upon conduction in the human median nerve in vivo. Exp Physiol. 1994;79: 227–34.
171. Cheong WF, Prahl SA, Welch AJ. A review of the optical properties of biological tissues. IEEE J Quant Electron. 1990;26:2166–85.
172. Lalla Y, Kulkarni S, Walsh LJ, George R. Does luminosity and smear layer influence 810nm laser energy transmission through human dentine? Photodiagn Photodyn Ther. 2023;42:103311.
173. Ebrahimi A, Marques MM, Miniello TG, Gutknecht N. Photobiomodulation therapy with 810-nm laser as an alternative to injection for anesthesia in dentistry. Lasers Dent Sci. 2021;5(2):117–23.
174. Chan A, Armati P, Moorthy AP. Pulsed Nd:YAG laser induces pulpal analgesia: a randomized clinical trial. J Dent Res. 2012;91(7 Suppl):79S–84S.
175. Liang R, George R, Walsh LJ. Pulpal response following photo-biomodulation with a 904-nm diode laser: a double-blind clinical study. Lasers Med Sci. 2016;31(9):1811–7.
176. Efthymiou A, Marques MM, Franzen R, Moreira MS, Gutknecht N. Acceptance and efficiency of anesthesia by photobiomodulation therapy during conventional cavity preparation in permanent teeth: a pilot randomized crossover clinical study. Lasers Dent Sci. 2017;1(2):65–71.
177. Al Bukhary R, Wassell R, Sidhu S, Al Naimi O, Meechan J. The local anaesthetic effect of a dental laser prior to cavity preparation: a pilot volunteer study. Oper Dent. 2015;40(2):129–33.
178. Kurumada F. A study on the application of Ga-As semiconductor laser to endodontics. The effects of laser irradiation on the activation of inflammatory cells and the vital pulpotomy. Ohu Daigaku Shigakushi. 1990;17:233–44.
179. Walsh LJ. The current status of low level laser therapy in dentistry. Part 1. Soft tissue applications. Aust Dent J. 1997;42(4):247–54.
180. Yadav P, Indushekar K, Saraf B, Sheoran N, Sardana D. Comparative evaluation of ferric sulfate, electrosurgical and diode laser on human primary molars pulpotomy: an "in-vivo" study. Laser Ther. 2014;23(1):41–7.
181. Marques NC, Neto NL, Rodini Cde O, Fernandes AP, Sakai VT, Machado MA, Oliveira TM. Low-level laser therapy as an alternative for pulpotomy in human primary teeth. Lasers Med Sci. 2015;30(7):1815–22.
182. Gupta G, Rana V, Srivastava N, Chandna P. Laser pulpotomy-an effective alternative to conventional techniques: a 12 months clinicoradiographic study. Int J Clin Pediatr Dent. 2015;8(1):18–21.
183. Uloopi KS, Vinay C, Ratnaditya A, Gopal AS, Mrudula KJ, Rao RC. Clinical evaluation of low level diode laser application for Primary teeth pulpotomy. J Clin Diagn Res. 2016;10(1):ZC67–70.
184. Nammour S, Tielemans M, Heysselaer D, Pilipili C, De Moor R, Nyssen-Behets C. [Comparative study on dogs between CO2 laser and conventional technique in direct pulp capping]. Rev Belg Med Dent (1984). 2009;64(2):81–6.
185. Walsh LJ. Clinical evaluation of dental hard tissue applications of carbon dioxide lasers. J Clin Laser Med Surg. 1994;12(1):11–5.
186. Yazdanfar I, Gutknecht N, Franzen R. Effects of diode laser on direct pulp capping treatment : a pilot study. Lasers Med Sci. 2015;30(4):1237–43.
187. Komabayashi T, Ebihara A, Aoki A. The use of lasers for direct pulp capping. J Oral Sci. 2015;57(4):277–86.
188. Cengiz E, Yilmaz HG. Efficacy of erbium, chromium-doped:yttrium, scandium, gallium, and garnet laser irradiation combined with resin-based tricalcium silicate and calcium hydroxide on direct pulp capping: a randomized clinical trial. J Endod. 2016;42(3):351–5.
189. Walsh LJ, Ryan PC. Management of external root resorption by carbon dioxide laser ablation and sealing. Aust Endod Newsl. 1992;18:15–7.
190. Buchelt M, Kutschera HP, Katterschafka T, Kiss H, Lang S, Beer R, Losert U. Erb:YAG and Hol:YAG laser osteotomy: the effect of laser ablation on bone healing. Lasers Surg Med. 1994;15(4):373–81.
191. Paghdiwala AF. Root resection of endodontically treated teeth by erbium: YAG laser radiation. J Endod. 1993;19(2):91–4.
192. Angiero F, Benedicenti S, Signore A, Parker S, Crippa R. Apicoectomies with the erbium laser: a complementary technique for retrograde endodontic treatment. Photomed Laser Surg. 2011;29(12):845–9.
193. Lietzau M, Smeets R, Hanken H, Heiland M, Apel C. Apicoectomy using Er:YAG laser in association with microscope: a comparative retrospective investigation. Photomed Laser Surg. 2013;31(3):110–5.
194. Bodrumlu E, Keskiner I, Sumer M, Sumer AP, Telcioglu NT. Temperature variation during apicectomy with Er:YAG laser. Photomed Laser Surg. 2012;30(8): 425–8.
195. Dutch Working party Infection Prevention (WIP). Infection prevention when using laser instruments. Dec 2003. Available www.wip.nl.
196. Piccione PJ. Dental laser safety. Dent Clin N Am. 2004;48(4):795–807, v.
197. Andersen E, Aars H, Brodin P. Effects of cooling and heating of the tooth on pulpal blood flow in man. Endod Dent Traumatol. 1994;10(6):256–9.
198. Nyborg H, Brannstrom M. Pulp reaction to heat. J Prosthet Dent. 1968;19(6):605–12.

199. Powell GL, Morton TH, Whisenant BK. Argon laser oral safety parameters for teeth. Lasers Surg Med. 1993;13(5):548–52.
200. Armengol V, Jean A, Marion D. Temperature rise during Er:YAG and Nd:YAP laser ablation of dentin. J Endod. 2000;26(3):138–41.
201. Glockner K, Rumpler J, Ebeleseder K, Stadtler P. Intrapulpal temperature during preparation with the Er:YAG laser compared to the conventional burr: an in vitro study. J Clin Laser Med Surg. 1998;16(3):153–7.
202. Capelli A, Guerisoli DM, Barbin EL, Spano JC, Pecora JD. In vitro evaluation of the thermal alterations on the root surface during preparation with different Ni-Ti rotary instruments. Braz Dent J. 2004;15(2):115–8.
203. Eriksson JH, Sundstrom F. Temperature rise during root canal preparation—a possible cause of damage to tooth and periodontal tissue. Swed Dent J. 1984;8(5):217–23.
204. Walters JD, Rawal SY. Severe periodontal damage by an ultrasonic endodontic device: a case report. Dent Traumatol. 2007;23(2):123–7.
205. Barkhordar RA, Goodis HE, Watanabe L, Koumdjian J. Evaluation of temperature rise on the outer surface of teeth during root canal obturation techniques. Quintessence Int. 1990;21(7):585–8.
206. Bailey GC, Cunnington SA, Ng YL, Gulabivala K, Setchell DJ. Ultrasonic condensation of gutta-percha: the effect of power setting and activation time on temperature rise at the root surface—an in vitro study. Int Endod J. 2004;37(7):447–54.
207. Lee FS, Van Cura JE, BeGole E. A comparison of root surface temperatures using different obturation heat sources. J Endod. 1998;24(9):617–20.
208. Weller RN, Koch KA. In vitro radicular temperatures produced by injectable thermoplasticized gutta-percha. Int Endod J. 1995;28(2):86–90.
209. Lee BS, Jeng JH, Lin CP, Shoji S, Lan WH. Thermal effect and morphological changes induced by Er:YAG laser with two kinds of fiber tips to enlarge the root canals. Photomed Laser Surg. 2004;22(3):191–7.
210. Schoop U, Barylyak A, Goharkhay K, Beer F, Wernisch J, Georgopoulos A, Sperr W, Moritz A. The impact of an erbium, chromium:yttrium-scandium-gallium-garnet laser with radial-firing tips on endodontic treatment. Lasers Med Sci. 2009;24:59.
211. Eriksson AR, Albrektsson T. Temperature threshold levels for heat-induced bone tissue injury: a vital-microscopic study in the rabbit. J Prosthet Dent. 1983;50(1):101–7.
212. Gutknecht N, Kaiser F, Hassan A, Lampert F. Long-term clinical evaluation of endodontically treated teeth by Nd:YAG lasers. J Clin Laser Med Surg. 1996;14(1):7–11.
213. Machida T, Mazeki K, Narushima K, Matsumoto K. Study on temperature raising in tooth structure at irradiating Er:YAG laser. J Jpn Endod Assoc. 1996;17:38–40.
214. Machida T, Wilder-Smith P, Arrastia AM, Liaw LH, Berns MW. Root canal preparation using the second harmonic KTP:YAG laser: a thermographic and scanning electron microscopic study. J Endod. 1995;21(2):88–91.
215. Sauk JJ, Norris K, Foster R, Moehring J, Somerman MJ. Expression of heat stress proteins by human periodontal ligament cells. J Oral Pathol. 1988;17(9–10):496–9.
216. Kimura Y, Yonaga K, Yokoyama K, Kinoshita J, Ogata Y, Matsumoto K. Root surface temperature increase during Er:YAG laser irradiation of root canals. J Endod. 2002;28(2):76–8.
217. Theodoro LH, Haypek P, Bachmann L, Garcia VG, Sampaio JE, Zezell DM, Eduardo CP. Effect of ER:YAG and diode laser irradiation on the root surface: morphological and thermal analysis. J Periodontol. 2003;74(6):838–43.
218. Kane SA. Introduction to physics in modern medicine. London, New York: Talyor & Francis; 2003.
219. Moriyama EH, Zangaro RA, Lobo PD, Villaverde AB, Pacheco MT, Watanabe IS, Vitkin A. Optothermal transfer simulation in laser-irradiated human dentin. J Biomed Opt. 2003;8(2):298–302.

Lasers in Oral Implantology

Robert J. Miller

Contents

D. J. Coluzzi, S. P. A. Parker (eds.), *Lasers in Dentistry—Current Concepts*, Textbooks in Contemporary Dentistry,
https://doi.org/10.1007/978-3-031-43338-2_10

Core Message

This chapter will describe the various uses of available laser wavelengths for dental implant procedures. Incisions and contouring of soft tissue, along with impression procedures for restorations, can all be accomplished with many different laser wavelengths. When the laser is used properly, hemostasis, precise tissue removal, and increased patient comfort are some of the immediate benefits. Certain lasers may also assist in the osseous procedures necessary to prepare the implant site. In areas of peri-implant disease, lasers can remove granulomatous tissue, decontaminate implant surfaces, and aid in establishing a more favorable healing environment.

10.1 Introduction

Implant Dentistry has become an important and predictable treatment alternative for the replacement of missing teeth and is a welcome addition to the standard of care for oral rehabilitation. Consequently, some new and very advanced treatment protocols have been developed to enable clinicians to incorporate these procedures into their practices [1].

Most dental implant procedures are performed utilizing traditional surgical instrumentation. The special armamentarium employed for dental implant procedures enables good precision, leading to acceptable clinical outcomes. With the introduction of dental lasers, we have the ability to not only manipulate both hard and soft tissue in a manner that is consistent with traditional protocols, but we can also now observe an additional benefit. Laser energy can be very biologically friendly and compatible, changing the way tissue heals both short and long term [2]. From hemostasis, to decontamination, to compression of the wound cascade, lasers are quickly becoming a vital instrument in shaping the clinical outcomes of our implant cases [3].

Lasers can be used in implant dentistry in a variety of ways, from pre-surgical tissue conditioning to post-operative photobiomodulation (PBM). Peri-surgical procedures include ablation, resection, and decontamination. Selection of the procedural base must be matched to the appropriate wavelength to ensure the best clinical outcome [4]. Each wavelength has a unique absorption characteristic and effect on peri-implant tissue [5]. Therefore, it is important to understand the characteristics of these wavelengths and their unique tissue interactions [6].

10.2 Laser Applications in Implant Dentistry

The currently available dental laser wavelengths, their tissue interaction, and optimum parameters have been discussed in ▸ Chaps. 3 and 4, and the reader should consult those for further information. The following is a more focused discussion with specific reference to implant procedures.

Laser Applications in Implant Dentistry

1. The presurgical therapy utilizing a laser includes:
   - Maintaining healthy attached gingiva or increasing the width of attached gingiva prior to implant placement with procedures such as a frenectomy, vestibuloplasty, gingival grafting, or apically repositioned flap
   - Preparation of surgical site prior to implant placement
2. The laser surgical procedures consist of:
   - Disinfection of the implant site in case of immediate extraction and implant placement
   - Incision and debridement using lasers
   - Laser-assisted osteotomy
   - Sinus lift used in the creation of the window in the direct sinus lift procedure
   - Preparation of the donor site prior to bone augmentation
3. Postsurgical treatment can employ lasers for:
   - Implant fixture exposure
   - Retraction and management of tissues prior to making impressions
   - Creation of attached gingiva prior to final restoration
   - PBM therapy during healing after the implant placement for pain relief and accelerated wound healing

10.2.1 Soft Tissue Lasers

While all currently available laser wavelengths can perform soft tissue surgery, the following section will describe those instruments whose only indication for use is on soft tissue. As a general rule, the clinician must use the reference manual of the laser employed to operate within the zone of safety for all procedures.

Diode Lasers

Diode lasers have a very small-sized footprint, resulting in good ergonomics and portability. The currently available dental surgical diode wavelengths range from 445 to 1064 nm, delivered in a contact mode [7]. Selection of the appropriate wavelength should be matched to the absorption characteristics of the target tissue. Most clinicians will use a diode laser purely as a cutting instrument. However, there are additional biologic benefits that accrue to the use of soft tissue lasers:

1. The Hematologic Response: selective photothermolysis, in the context of blood vessel coagulation, suggests that pulses of light with a wavelength targeted for absorption by hemoglobin with appropriate pulse duration and fluence can heat blood vessels to achieve coagulation and hemostasis without causing nonspecific damage to surrounding tissues [8].
2. The Neurogenic Response: C-fibers are small-diameter fibers acting as nociceptors from the periphery to the central nervous system. The diameter of these axons ranges from 0.2 to 1.2 μm and can reach up to 3 μm in some cases. The thermal effect of lasers will block depolarization of these afferent C-fibers, particularly in low velocity neural pathways such as non-myelinated afferent axons from nociceptors, preventing pain impulses from being transmitted [9].
3. The Lymphatic Response: lymphedema refers to tissue swelling caused by an accumulation of protein-rich fluid that is drained through the body's lymphatic system, most often caused by bacterial infection or tissue trauma. Laser decontamination, as well as thermolysis, can prevent lymphatic fluid buildup with concomitant soft tissue edema [10].

Diode lasers should be used with caution near osseous tissue since the photonic energy can scatter in soft tissue. However, when used judiciously with lower average power (approximately 1 W), they can aid in disinfecting the soft tissue site prior to grafting or implant placement. The reduction of bacterial pathogens in periodontal pockets after irradiation with a diode laser has been studied. A comparison between the initial and the final bacterial counts reveals that irradiation with the diode laser facilitates considerable bacterial reduction, especially of Actinobacillus actinomycetemcomitans, from periodontal pockets [11]. It follows that diode lasers can be used with the same result on soft tissue flaps around dental implants prior to implant placement.

A low average power output must be used for secondary implant recovery. If not, the cover screw could potentially become fused to the implant body, making it impossible to remove.

The Nd:YAG Laser

The Nd:YAG laser at 1064 nm is commonly used in periodontal therapy to incise and excise soft tissues as well as for curettage and disinfection of periodontal pockets [12]. The high peak power produced by a free-running pulsed mode laser can cause deep tissue penetration. This possible thermal effect of this laser on tissues lying below the irradiated area is a matter of concern during periodontal treatment [13]. Hence, caution must be exercised before using the Nd:YAG laser for decontamination of surgical sites or on the surface of implants [14, 15]. It is essential that the correct power densities be utilized, or the surface of the implant can be damaged or melted at power densities as low as 50 mJ/pulse at 20 Hz [16]. There can be deleterious effects on the soft tissue with burning, charring, or denaturation of the soft tissue complex, leading to tissue necrosis or a tissue dehiscence around the neck of the implant.

The 10,600 nm Carbon Dioxide (CO_2) Laser

The 10,600 nm Carbon Dioxide (CO_2) laser can also be used for decontamination of soft tissue tags and granulation tissue as well as hard tissue surgery. Studies using the carbon dioxide laser report a very high reduction of *S. sanguinis* and *P. gingivalis* bacteria using that wavelength [17, 18].

10.2.2 All Tissue Lasers

All tissue lasers are used to cut and shape hydroxyapatite (HA) containing tissues, including tooth structure and bone. They are optimized for absorption in water containing tissue and HA, resulting in precise cutting trajectories within a short period of time. However, they may also be employed to cut soft tissue with a much lower thermal coefficient. This combination effect makes all tissue lasers an excellent choice for dental implant procedures. These types of procedures can be categorized into two groups:

1. Ablation: considered to be refined removal of layers of tissue; it can be described as including gingivoplasty or osteoplasty.
2. Resection: considered to be removal of large segments of tissue *en masse*; it can be described as including gingivectomy or ostectomy.

Erbium-Based Lasers: Er,Cr:YSGG, 2780 nm; and Er:YAG, 2940 nm

The Erbium-based lasers are useful for implant procedures because they have the lowest thermal coefficient. This is particularly important in preventing overheating of bone where implants are to be placed, whether in shaping of the osteotomy, debridement, or decontamination of the extraction site.

Erbium lasers are antibacterial and can be used to remove both calculus and biofilm around tooth structure and implant surfaces [19, 20]. Er:YAG lasers possess suitable characteristics for both oral soft and hard tissue ablation [21]. Contouring and cutting of bone can be achieved with minimal damage and faster healing. In addition, irradiation with the Er:YAG laser has a bactericidal effect with reduction of lipopolysaccharides [22]. This is a major component of the outer membrane of Gram-negative bacteria and plays an active role in the pathogenesis of periodontal and peri-implant tissue breakdown. The properties of the photonic energy from the Er,Cr:YSGG laser also verifies its effectiveness for decontamination of hard and soft tissue. One advantage of the Er,Cr:YSGG laser is the deeper penetration of laser energy in water containing tissue, making this wavelength more effective for decontamination and osseous ablation in a shorter period of time.

Clinical studies have concluded that even at low energy densities, the Er:YAG laser has a high bactericidal potential on implant surfaces. Additionally, no excessive temperature elevations or morphological implant surface alterations were detected [23]. Other authors have found that Er,Cr:YSGG laser irradiation can be equally effective at decontaminating the surface of titanium implants [24].

The osteotomy within bony structures can be challenging since thin and fragile bony walls of the maxilla and mandible are prone to fracture from contact pressure and vibration caused by mechanical instruments. A laser osteotomy offers a viable and non-destructive alternative.

The 9300 nm Carbon Dioxide Laser

The 9300 nm Carbon Dioxide Laser is also indicated for osseous and soft tissue procedures, similar to the erbium wavelengths.

10.3 Flap Incisions Using Lasers

Traditionally, a scalpel is used to make incisions prior to flap reflection for implant placement with high precision. The major drawback of the use of a scalpel is lack of hemostasis and a high degree of post-operative pain because of the release of inflammatory modifiers. However, the use of scalpels is still indicated in the aesthetic zone for accurate coadaptation of flap margins and prevention of scarring. An exception to this rule may be in patients taking anticoagulants where excessive bleeding may interfere with surgical execution. Most of the commercially available dental lasers are effective in making incisions, almost replacing the scalpel. The advantages of using the laser rather than the scalpel are numerous. A laser incision cannot spread infection, and there is a dramatically reduced inflammatory cascade. Laser use also seals lymphatic and blood vessels, reducing post-operative edema. There is also a clinically measurable reduction in pain, and other potential post-operative sequelae.

The Er,Cr:YSGG laser can be used in contact mode with a chisel shaped tip on end which is directed at right angles to the tissue as shown in ◘ Fig. 10.1. The tip is used on edge in a pulling motion creating a sharp dissection down to periosteum with ideal hemostasis and no visible change in color of the soft or hard tissue (◘ Fig. 10.2). The choice of tip is important when used for flap design. A wide diameter end-firing tip will ablate a larger margin of tissue with lower accuracy, resulting in more challenging closure and potential scarring from second intention healing. Narrow end-cutting tips at angle to the tissue will be more precise with a more precise depth control.

If the patient is on anticoagulant medication or supplements that mimic anticoagulants, an Nd:YAG or any wavelength diode laser can be employed to create an incision with slightly less precise margins and a more pronounced thermalized zone. Deeper thermolytic zones can be more effective at sealing blood vessels for hemostasis in critical cases. Alternatively, a sapphire laser scalpel tip can be used to mimic the precision of an all-tissue laser. In ◘ Figs. 10.3 and 10.4, an InGaAsP diode with an emission wavelength of 1064 nm is shown utilizing that tip. An advantage to the sapphire scalpel is that it has an incision width similar to a conventional

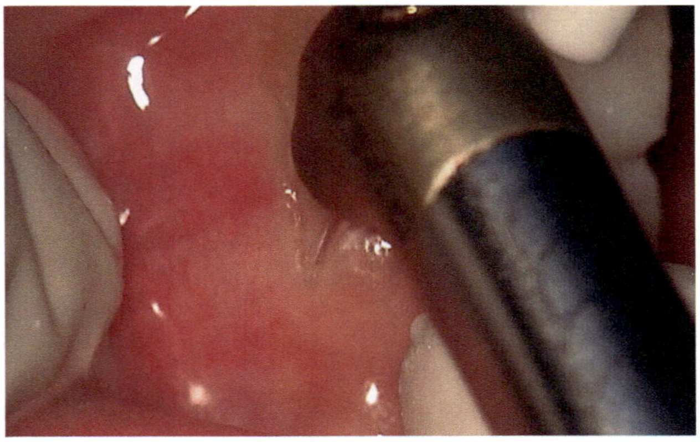

◘ **Fig. 10.1** An Er,Cr:YSGG laser and C3 chisel tip is used at 2.5 W, 25 Hz, 30% water and 30% air to perform an incision

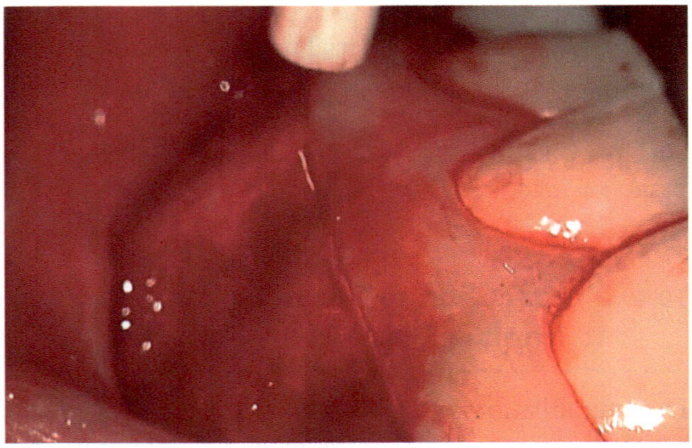

◘ **Fig. 10.2** The finished laser incision shows a very precise flap margin with excellent balance of cutting and hemostasis

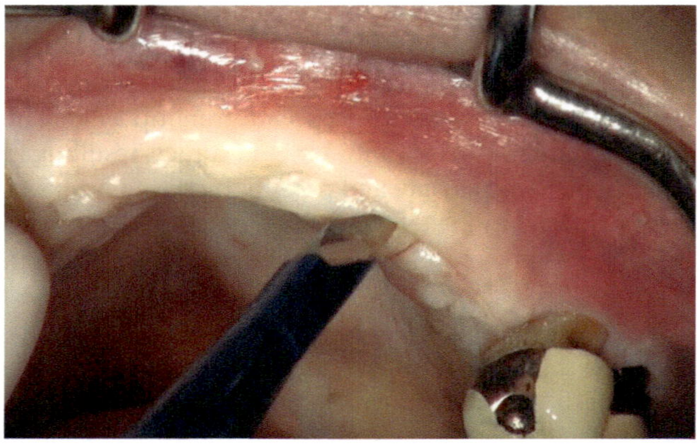

Fig. 10.3 An InGaAsP laser with an average power of 2 W and a sapphire scalpel makes an incision

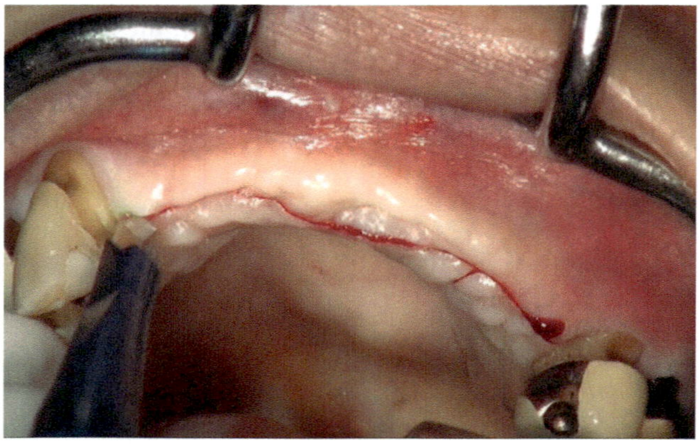

Fig. 10.4 The completion of a well-defined, precise incision, also with a good balance of cutting efficiency and bleeding control

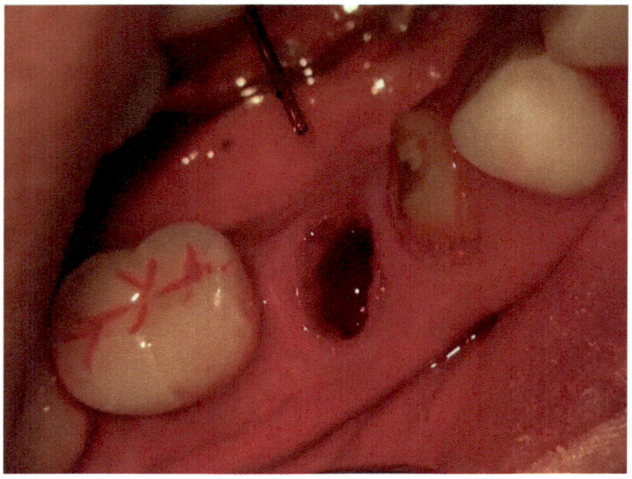

Fig. 10.5 Extraction site prior to laser decontamination

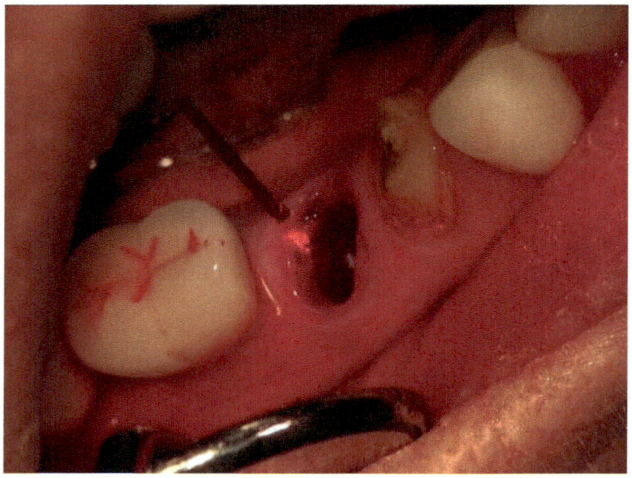

Fig. 10.6 Ablation of the gingival sulcus to gingival margin with Er,Cr:YSGG laser with 17 mm Z6 tip at 3.5 W, 25 Hz, 30% water, 30% air

bladed scalpel, while enjoying the benefits of hemostasis and post-op pain control [25].

In studies using laser-based instrumentation, there was a significant reduction of dark pigmented anaerobic Gram-negative rods, the most relevant being *Fusobacteria*, *Prevotella*, and *Porphyromonas* species [26]. These pathogens have been assigned a predominant role in the breakdown of the supporting periodontal tissue. Hence, a diode laser is ideal for bacterial reduction, decontamination of diseased soft tissue around implants, and hemostasis during implant surgery. They can be particularly effective when used for intrasulcular incisions where the pathogen load is relatively high.

10.4 Preparation of Surgical Site Prior to Implant Placement

Decontamination of the surgical site is essential for the successful integration of immediate dental implants. The goal prior to immediate implantation is to ensure that the post extraction surgical site is free of debris, granulation tissue, and pathogens.

One of the most critical aspects of extraction/immediate implant cases is peri-operative disinfection of the extraction site defect. A typical immediate post extraction view is shown in Fig. 10.5. This important step of degranulation must be completed prior to implant placement [21], and Fig. 10.6 shows the Er,Cr:YSGG laser's use for degranulation and decontamination of the site.

The photonic energy first removes gingival sulcus up to the coronal aspect of the gingival margin. This will eliminate periodontal pathogens associated with the infected tooth and allow a hemi-desmosomal attachment to the healing abutment or final abutment in a single stage surgery.

In a circular fashion, the laser tip is moved from coronal to apical, removing any remnants of periodontal ligament or apical pathology. This reduces the chances for crestal bone remodeling, and a condition known as retrograde peri-implantitis [27].

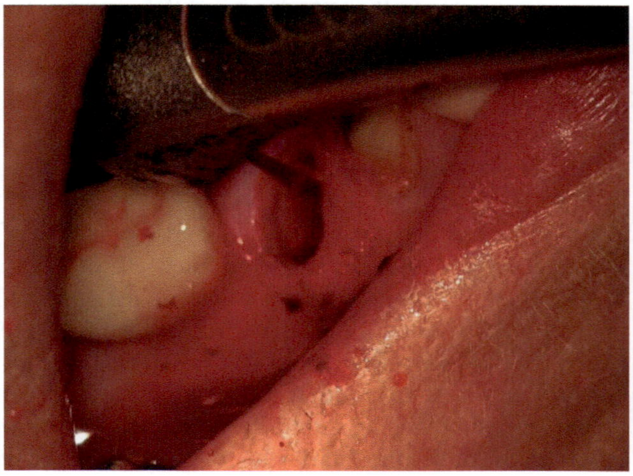

Fig. 10.7 Removal of the periodontal ligament and apical pathology with Er,Cr:YSGG laser with 17 mm Z6 tip at 3.5 W, 25 Hz, 30% water, 30% air

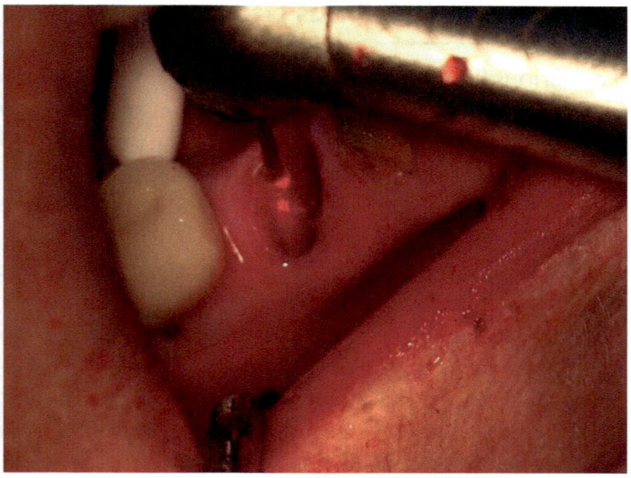

Fig. 10.8 Decortication of the dense lamina dura prior to implant and graft placement with Er,Cr:YSGG laser with 17 mm Z6 tip at 3.5 W, 15 Hz, 30% water, 30% air

Apical infections around implants are the result of failure to remove apical granulomas or cysts following extractions. While extraction of teeth will remove the offending organ, remaining granulation tissue may set up a histological response that is self-perpetuating [28]. In time, the lesion may expand resulting in a fenestration defect in the arch. It is imperative, therefore, that all granulomatous or cystic tissue be removed prior to implant or graft placement.

If a retrograde lesion does occur around the implant apex, surgical excision is necessary. In most cases, an envelope flap can be employed to expose the defect while not disturbing the crestal dentogingival complex. If the defect is solely on the facial of the apex, debridement is straightforward. However, in most cases the defect is circumferential in the apical zone. This requires complete debridement of all pathologic tissue on the inner aspect of the implant with both curettes and a laser with radially-firing or side-firing tips.

The failure rate of immediate implants is higher because of pre-existing disease in the teeth and periodontal tissue being replaced. Two examples are teeth with periapical infection or periodontal disease. The protocol would be to remove the gross amount of soft tissue with a curette and then use the laser to remove any visible tissue tags. The entire inner wall of the extraction socket can then be decontaminated as shown in Fig. 10.7. The laser is used in a circular motion with the tip parallel and in contact with the osteotomy walls. It is slowly moved apically to remove all remaining soft tissue. All laser wavelengths are antibacterial in nature and can be used to varying degrees to disinfect the surgical site [29].

A final evaluation of the debrided site should include moderate bleeding prior to implant placement. A thickened lamina dura can occur around some teeth, particularly when the tooth has been endodontically treated. To ensure proper healing of the implant and potential graft in the extraction site gap, the osteotomy should be decorticated to produce copious bleeding points. Bleeding is essential for early integration and the turnover of any graft materials that are placed in the extraction site gap. The laser is ideal for this procedure without causing thermal necrosis of the bone, provided the correct settings are used. Figure 10.8 depicts the Er,Cr:YSGG laser's use, producing a very biocompatible environment. Implant fixture placement, including an additional graft when necessary, can now be accomplished safely and predictably with the same clinical outcomes as in healed sites (Fig. 10.9). The blood clot that forms acts as a barrier to soft tissue invagination and bacterial colonization. A flared healing abutment or restorative abutment for immediate loading can now be placed.

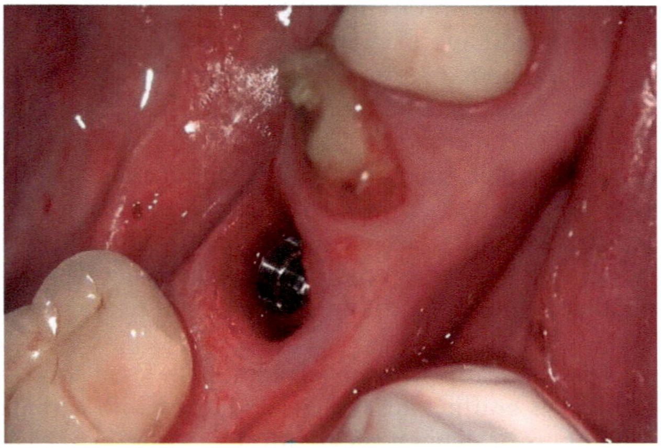

Fig. 10.9 Implant fixture placement with bleeding following complete laser ablation

10.5 Lateral Wall Sinus Lifts

Two serious limiting factors in the placement of implants in the posterior maxilla are the anatomical shape and location of the maxillary sinus and quantity of bone. To ensure successful implant placement in the posterior region, a minimum of 8–10 mm of sound bone structure is necessary.

Additionally, bone density in the posterior maxilla is often poor, which could lead to complications during implant placement such as lack of initial stability. To improve placement outcomes, maxillary sinus lift surgery was developed as a method to increase the amount of bone available for the implant initial stability. This has now become a routine procedure to address deficient maxillary posterior bone.

The lateral window sinus lift is a direct sinus lift procedure which allows for visualization of the sinus space and to facilitate graft placement. Also, tearing of the membrane can be more easily treated, minimizing contamination or extravasation of the graft during healing.

10

The lateral approach involves a modified Caldwell-Luc operation to gain access to the sinus cavity. A bony window is created in the lateral maxillary wall, the Schneiderian membrane is elevated, and bone grafting material which may consist of autogenous bone, allograft, alloplast, or xenograft is placed between the membrane and sinus cavity. A resorbable collagen membrane may be placed between the bony window and flap to prevent graft migration or soft tissue invagination.

The surgical instruments traditionally used to perform sinus grafting are rotary handpieces with carbide round burs or diamonds [30]. More recently, piezoelectric ultrasonic devices have replaced rotary instruments because of the reduced risk of membrane perforation [31, 32].

The development of hard tissue wavelengths has enabled bone ablation to be carried out with minimal adjacent damage. The use of Erbium-based lasers in dentoalveolar surgery represents a less traumatic experience for the patient when compared to the intense vibration of the slow-speed surgical bur. However, to prevent overheating of the bone, it is important to maintain a sufficient coaxial water spray. The water spray will contribute to lifting of the Schneiderian membrane off of the inner wall of the sinus cavity, thereby reducing the potential for membrane perforation.

In the maxilla, the speed of the laser ablation is comparable with that of the bur due to the cancellous structure of the osseous tissue. However, it is important to set appropriate power settings to minimize the potential for overheating the bone and to prevent perforation of the Schneiderian membrane.

10.5.1 Lateral Sinus Graft Procedure

A full-thickness flap must be raised, and the outline of the bony window must be predetermined. As shown in ◘ Fig. 10.10, after tracing the outline of the corticotomy, a bony window is created by placement of the Er,Cr:YSGG laser's beam at 30–45° to the cortical surface in a non-contact mode (with 1–3 mm distance from the target tissue). Lasing should be performed with slow movements until the darkness of the underlying Schneiderian sinus cavity is visualized. Ablation should be stopped after completing the predetermined window border decortication and the window can be rotated superiorly or removed. The remaining part of the surgical procedure is completed using conventional surgical instruments.

The Schneiderian membrane must be kept intact so it can help to contain the graft material as well as to prevent migration of the graft particles in the sinus cavity. If the membrane is perforated or torn, the graft material may become infected leading to a failed sinus graft procedure. ◘ Figure 10.11 shows the completed membrane reflection and the graft material placed. The radiographic information in ◘ Fig. 10.12 confirms the completed graft filling in that part of the sinus.

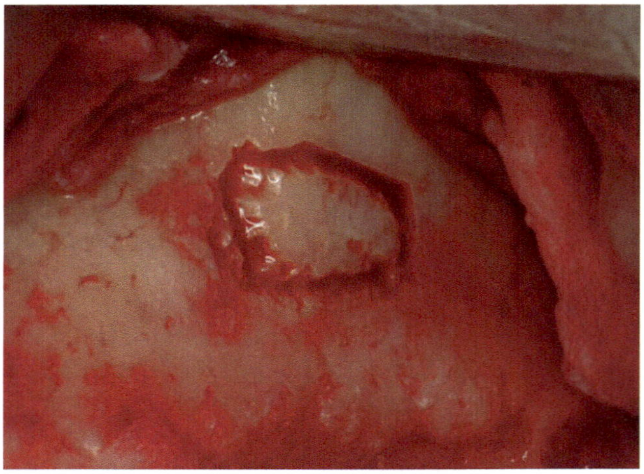

◘ **Fig. 10.10** Preparation of lateral wall window with an Er,Cr:YSGG laser without damage to Schneiderian membrane. Laser settings are 6 mm Z6 tip, 3.5 W, 20 Hz, 30% water, 30% air

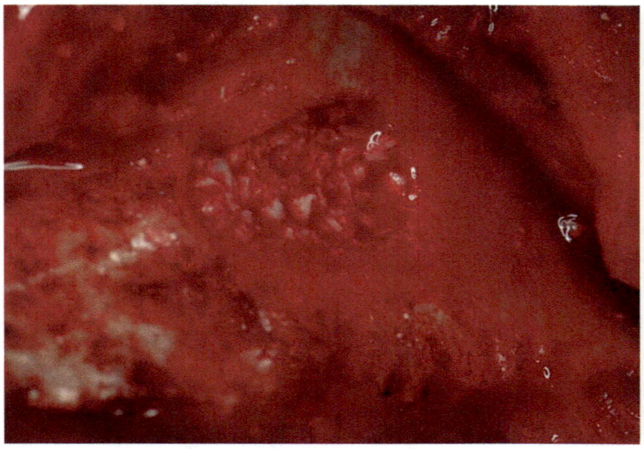

■ **Fig. 10.11** Reflection of sinus membrane to medial wall and graft placement

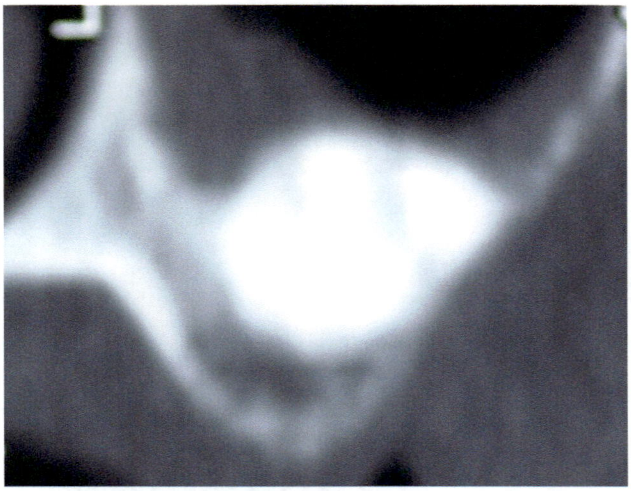

■ **Fig. 10.12** Radiograph demonstrating good fill of sinus compartment

10.6 Bone Harvesting for Ridge Augmentation

Autogenous bone harvesting for ridge augmentation has traditionally been performed using cross cut fissure burs, diamonds, diamond discs, or piezosurgery. A new technique base using ablative lasers has been developed

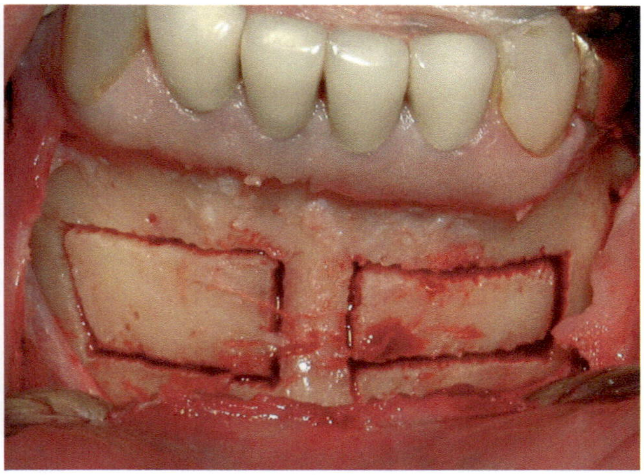

■ **Fig. 10.13** Precise laser preparation of donor site in anterior mandible. The Er,Cr:YSGG laser used a 6 mm Z6 tip, with settings of 3.5 W, 20 Hz, 30% water, 30% air

offering unique advantages. Lasers can limit the depth of cut more discreetly and allow a greater range of unique architectural shapes that fit more appropriately into the graft site. Additionally, they offer the same hemostatic and decontamination effects as in other procedures.

Using an ablative laser with an end-cutting tip, the bone cut is initiated. ■ Figure 10.13 shows the Er,Cr:YSGG laser, with a Z5 or Z6 tip, at 2.5 W, 20 Hz, 30% water, 30% air, has prepared the osseous graft sites. The tip should be brought to within 3 mm of the bone surface and moved slowly as the laser energy troughs the bone margin. Visual examination of the color of the bone should be maintained throughout the procedure. At the completion of the laser incisions, there should be bleeding from the bone margins which would confirm that the bone complex is not overly thermalized. Furthermore, there should be no discoloration of the bone margins which will result in bone necrosis. A histologic example of minimal thermal damage from this procedure is depicted in ■ Fig. 10.14.

The corticocancellous bone plates can be removed with bone chisels (■ Fig. 10.15) and then moved to the recipient site. The laser can also be used to decorticate the recipient site prior to placement of bone screws on the recipient site, shown in ■ Fig. 10.16.

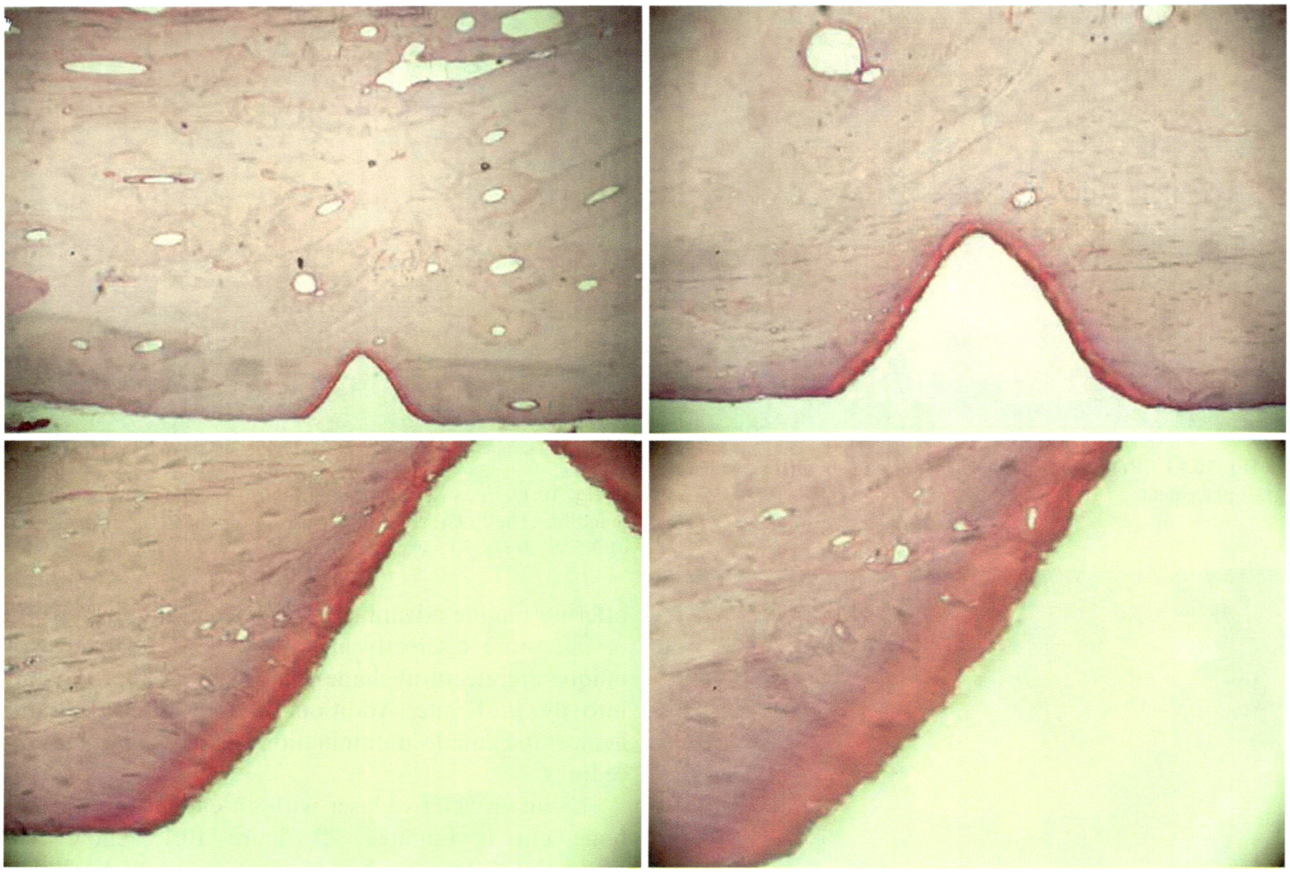

Fig. 10.14 A photomicrograph of a minor thermalized zone of 8–15 µm adjacent to an erbium-based laser osseous incision, depicted approximately between the two blue arrows, above. (Courtesy Dr. Harold Passow)

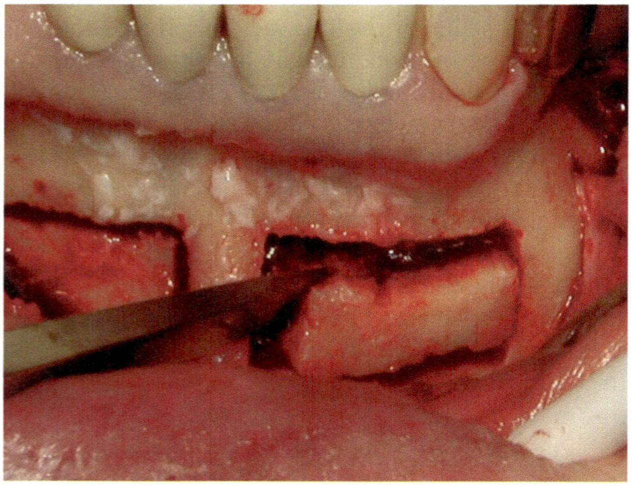

Fig. 10.15 Harvesting of cortical plates with bone chisels. Donor sites are filled with a graft material to prevent a post-surgery bone defect

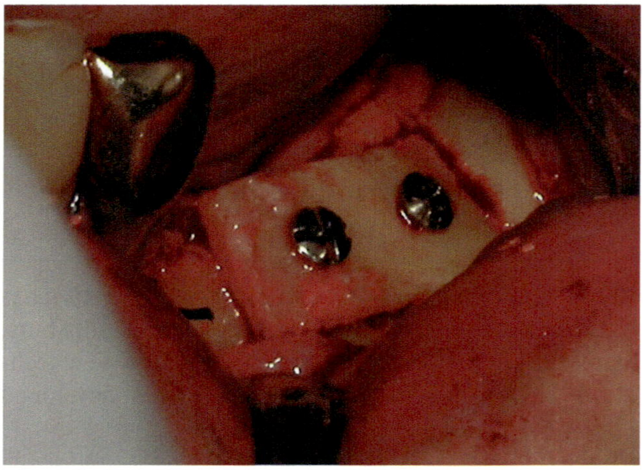

Fig. 10.16 Placement of cortical plate on recipient site in posterior mandible. Edges are faired with a particulate graft material to smooth edges of grafted area

10.7 Secondary Phase Implant Uncovering

There are various techniques to perform secondary implant recovery. These include a tissue punch, scalpel, diode laser, or hard tissue laser. There are advantages and disadvantages to each instrumentation technique and the choice should be based on the soft tissue architecture. If there is an inadequate keratinized zone, a flap should be raised and apically repositioned to allow the reformation of a keratinized tissue zone. If there is adequate keratinized tissue zone, a tissue punch can be used. However, this will remove the healed peri-implant tissue zone at crest. The epithelium will scar down into the surgical zone, violating biologic width. Crestal bone remodeling will then occur, potentially exposing the neck of the implant. This may then predispose the implant to bacterial colonization and future peri-implantitis. A third option is laser ablation of the keratinized tissue overlying the cover screw. It is important to note that the laser is never used beyond the radius of the cover screw. In this way, biologic width is not violated and we tend to not experience crestal bone remodeling as a consequence.

A small amount of local anesthetic is infiltrated over the position of the implant. The center of the cover screw is located using an explorer, periodontal probe, or laser tip as shown in ◘ Fig. 10.17. The Er,Cr:YSGG laser employs an end-firing tip with low average power to ablate the soft tissue while having no effect on peri-implant bone or the titanium surface. The laser is then used in a circular motion (◘ Fig. 10.18), widening the exposure of the cover screw until it can be removed (◘ Fig. 10.19). A healing abutment is then delivered and the tissue allowed to heal before impressions are taken (◘ Fig. 10.20).

The tissue is allowed to mature until three zones appear. Reformation of a gingival sulcus occurs at crest, followed by a junctional epithelium to protect the implant, and an area of hemi-desmosomal soft tissue attachment.

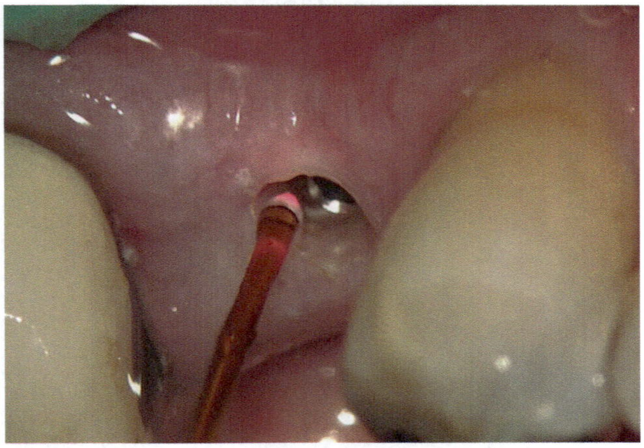

◘ **Fig. 10.18** Ablation of the tissue overlying the cover screw to the diameter of the implant without extension to the lateral tissue. The Er,Cr:YSGG laser is used with a 14–17 mm Z6 zirconia tip at 3.5 W, 25 Hz, 30% water, 30% air

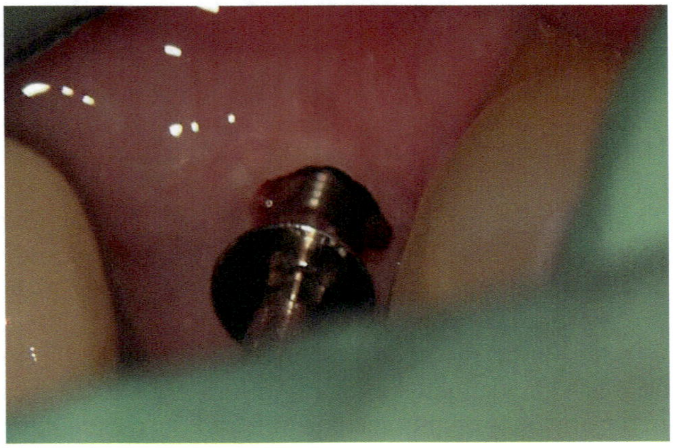

◘ **Fig. 10.19** Removal of the cover screw

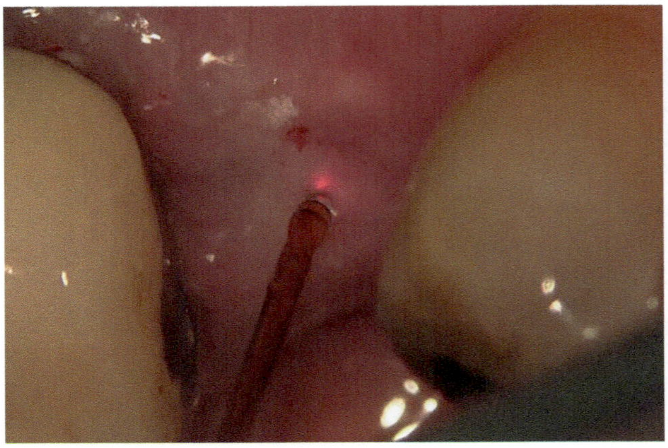

◘ **Fig. 10.17** Locating the center of the implant cover screw and implant recovery is initiated. A 14–17 mm Z6 zirconia tip of the Er,Cr:YSGG laser is used at 2.5 W, 25 Hz, 30% water, 30% air

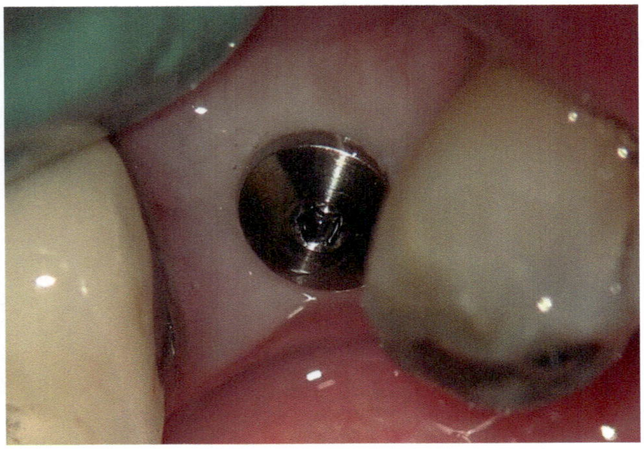

◘ **Fig. 10.20** Placement of a flared healing about to create the emergence profile

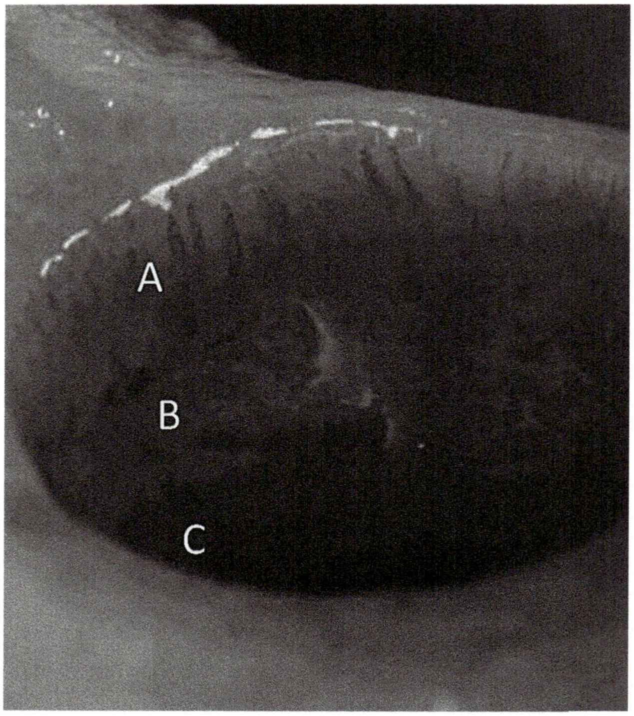

■ **Fig. 10.21** Maturation of the peri-implant soft tissue demonstrating three distinct zones: (A) implant sulcus, (B) junctional epithelium, (C) zone of hemi-desmosomal attachment

This mature tissue zone reestablishes biologic width and protects the crestal component of the implant body against trauma and bacterial colonization. ■ Figure 10.21 is a photomicrograph depicting the three zones.

Maturation of the soft tissue complex is essential to establish an adequate biologic width to prevent changes in the apico-coronal tissue position and the crestal bone position.

10.8 Laser Troughing Prior to Impression

In implant dentistry, the margin of the abutment shoulder may be sub-gingival. To prepare the gingival tissues for an accurate impression for the final restoration, a sulcular gap must be created; and the procedure is referred to as troughing. For non-implant restorative procedures, this is accomplished with retraction cords, a scalpel, or electrosurgery. Electrosurgery is contraindicated around implants because of the high thermal coefficient, but a retraction cord or scalpel are options. However, the retraction cord technique is time-consuming with the potential to traumatize the dentogingival complex or causing post-operative discomfort. The use of a scalpel may result in reduction in the height of keratinized tissue post-operatively.

In contrast to conventional techniques, laser troughing allows for clear visualization of gingival margins without affecting soft tissue height (■ Fig. 10.22). Most lasers are excellent coagulation devices with minimal to no bleeding. Lasers usually require 30–60 s to achieve retraction which does not rebound because lasers remove the internal epithelial lining of the gingival sulcus. Another advantage of laser troughing is that it promotes the ideal environment for current impression scanning devices by creating space between the tissue and abutment shoulder and establishing hemostasis. All of the currently available wavelengths can perform this procedure; however, as mentioned above, a pulsed Nd:YAG laser should be used with caution next to an implant fixture because of the higher heat buildup. ■ Figure 10.22 shows the troughing procedure using an Er,Cr:YSGG laser.

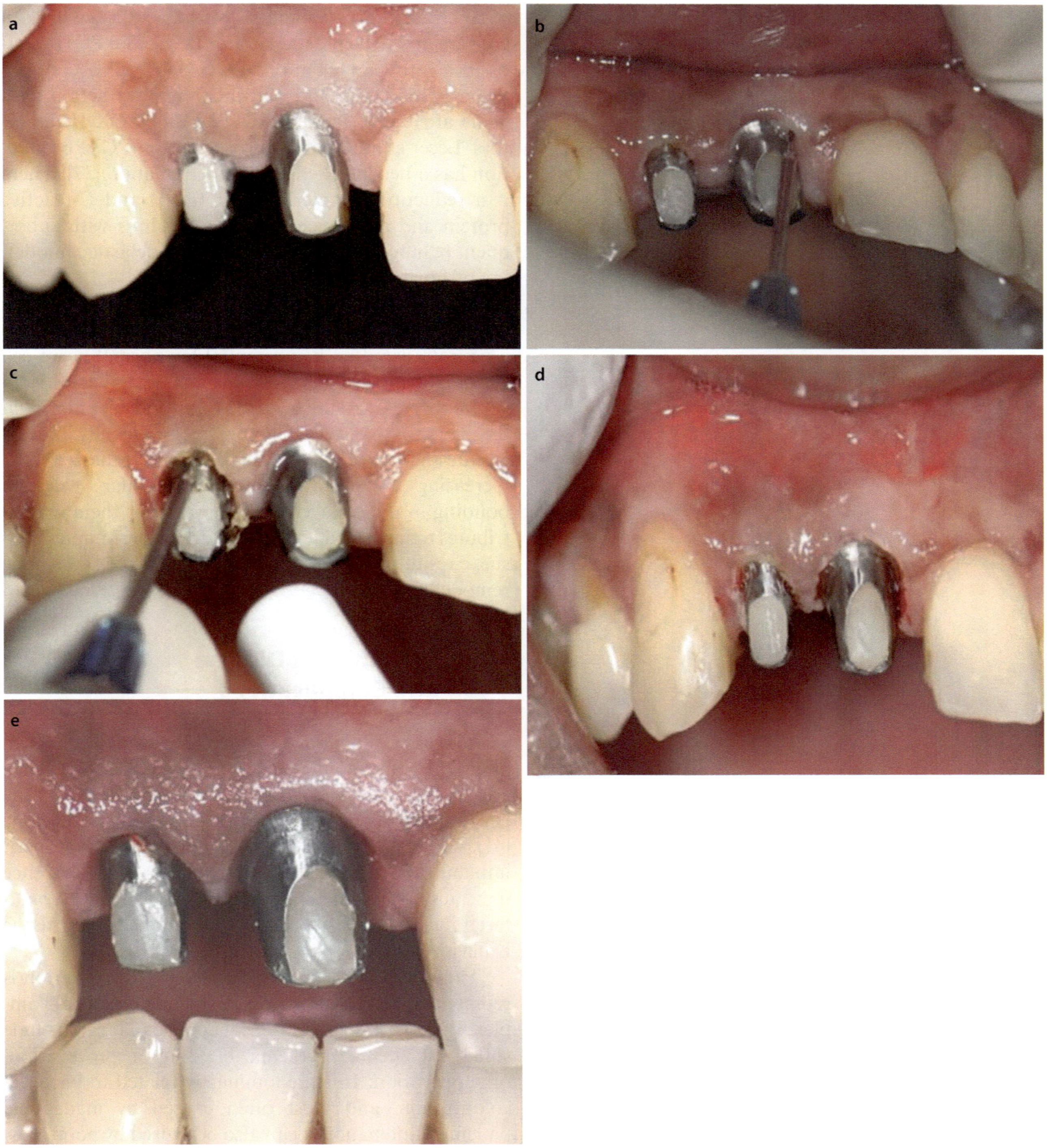

■ **Fig. 10.22** The upper right central and lateral incisors are extracted and replaced with immediately loaded implants and final abutments. **a** Is the pre-operative view of the healed, bulky tissue. **b**, **c** Show an Er,Cr:YSGG laser, used at an average power of 2 W, 100 Hz, H mode, air 10%, and water 10%, delivered with an MZ5 tip to trough the gingiva around the abutment and the prepared teeth. **d**, **e** Depict the immediate post-operative view with excellent tissue contours, which are ready for final impressions or digital scan

10.9 Photobiomodulation (PBM)

Diode lasers can be used to shape the metabolic activity of soft and hard tissue, both pre- and post-operatively [33]. Biologic pathways at the cellular level are very well understood. This understanding of how various laser wavelengths directly affect cellular metabolism can now be applied to compress treatment time following implant procedures. ■ Figures 10.23 and 10.24 show examples of using PBM photonic energy for healing and pain

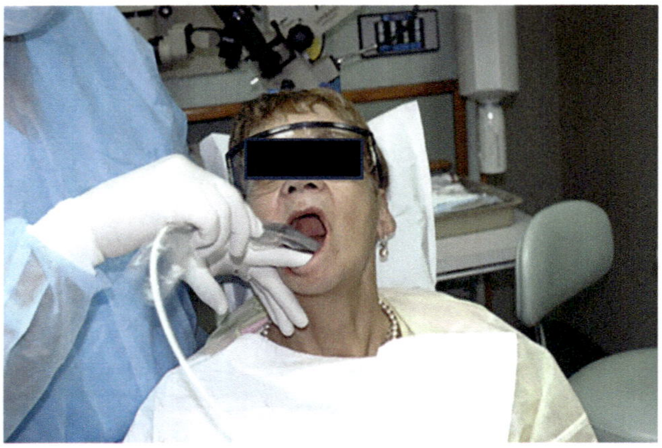

Fig. 10.23 Intraoral photobiomodulation to increase vascular perfusion

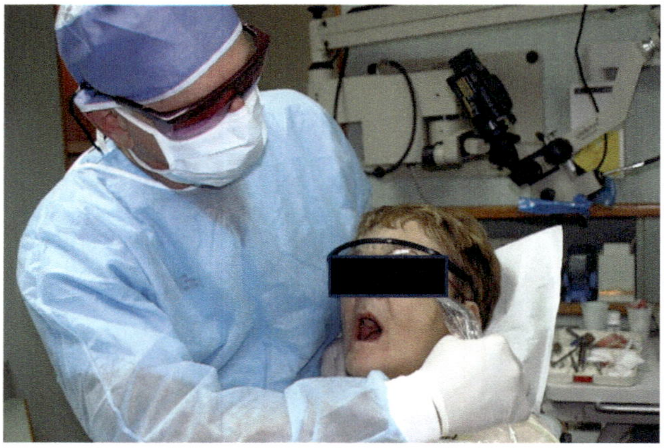

Fig. 10.24 Extraoral photobiomodulation to decrease post-operative joint discomfort

relief. For detailed information about PBM, please refer to ▶ Chap. 7.

Stabilization of cellular membrane: Ca^{2+}, Na^{+}, and K^{+} concentrations as well as the proton gradient over the mitochondria membrane are positively influenced by PBM. This is accomplished in part by the production of beneficial ROS (Reactive Oxygen Species) wherein triplet oxygen molecules absorb laser energy, improving Ca^{2+} uptake in the mitochondria [34]. ATP production and synthesis are significantly enhanced, contributing to cellular repair, reproduction, and functional ability. Vasodilation is stimulated via Histamine, Nitric Oxide (NO), and Serotonin increases, resulting in reduction of ischemia and improved perfusion. Beneficial acceleration of leukocytic activity results in enhanced removal of non-viable cellular and tissue components, allowing for a more rapid repair and regeneration process [35]. PBM post implant placement has also been shown to enhance the early biomechanical response of implants [36]. This can be particularly useful in immediate load cases.

Increased Prostaglandin synthesis particularly in conversion of the prostaglandins PGG_2 and PGH_2 perioxsides into prostaglandin PGI_2. PGI_2 (Prostacyclin) has a vasodilating and anti-inflammatory action with some attributes similar to COX-I and COX-II inhibitors. Laser stimulation of prostaglandin PGE_2 production has a beneficial effect on inflammation [37]. There is a reduction in Interleukin 1 (IL-1) and C-reactive protein and an increase in b-Endorphins with subsequent pain reduction and reduced osteoclastic remodeling of bone.

Nitric Oxide production is also increased; NO has both a direct and indirect impact on pain sensation. As a neurotransmitter, it is essential for normal nerve cell action potential in impulse transmission activity [38].

In soft tissue surgery, PBM can increase the final tensile strength of the healed tissue complex. By increasing the amount of collagen production/synthesis and by increasing the intra and inter-molecular hydrogen bonding in the collagen molecules. Laser therapy contributes to improved tensile strength of peri-implant tissue, resulting in less peri-implant crestal bone remodeling [39].

10.10 Peri-implant Disease and Complications

There are two potential post-operative sequalae that can affect the clinical outcomes of implant cases [40]. They are classified as disease processes and can affect both the peri-implant tissue and the surrounding bony envelope [41]. However, not all researchers believe that peri-implantitis is identical to periodontitis. New evidence indicates that cell metabolism may be affected by biomaterial interactions, altering the phenotype of soft tissue behavior [42].

The first is peri-implant mucositis. This condition is represented by inflammation of the dentogingival complex, with the absence of crestal bone loss. It can include soft tissue edema, erythema, bleeding on provocation, or suppuration. It is predominately related to bacterial colonization, and most often caused by inadequate patient maintenance. It can also be caused by poor prosthesis design preventing access to the transmucosal portion of the implant.

The second is peri-implantitis. This condition may include all of the conditions relative to mucositis, with additional evidence of crestal bone loss [43–45]. Early peri-implantitis may be treated in a manner consistent with mucositis. However, long standing peri-implantitis, with greater than 2–3 mm of bone loss, requires a more definitive approach. This would include open flap debridement with an ablative laser, followed by a graft procedure to regenerate the lost tissue volume.

10.10.1 Understanding the Mode of Failure

Prior to surgical intervention for peri-implant disease, it is imperative that the clinician understand the factors resulting in the biologic breakdown of the peri-implant tissue. Not all cases of peri-implantitis can be treated surgically; triage should be completed first to determine if regenerative surgery is warranted. Inappropriate axial positioning of the implant, an inadequately designed prosthesis, or problems related to the health of the patient are contraindications for surgical revision.

10.10.2 Removal of Pathology

Removal of all pathology is an absolute prerequisite for regeneration. While most gross pathology can be removed with surgical curettes, final debridement at the cellular level on the implant surface should be accomplished with an ablative laser [46]. The process of ablating the metallic oxide surface also has secondary effects. The first is decontamination. Removal of all pathogenic bacteria will prevent a recurrence of peri-implant infection. Lasers may also affect the electron state of the surface, rendering the implant surface more hydrophilic. This phenomenon will enhance early fibrin adhesion to the implant body, accelerating the attachment of reparative cells [47].

▶ Chapter 14 discusses laser-assisted therapy for peri-implant disease in great detail.

10.10.3 Treatment of the Contaminated Implant Surface

The treatment of peri-implantitis is predicated on the regeneration of the lost soft and hard tissue envelope [48]. Peri-implantitis is almost identical to periodontitis, with the same flora being responsible for both conditions [49]. The biologic imperative of preparation of the defect site also includes complete debridement of the implant surface to both decontaminate and surgically clean the implant surface to the condition that existed when the implant was placed [50]. Prior to the development of ablative lasers, chemotherapeutic agents were employed as a means to remove bioburden from the implant surface. These materials included tetracycline paste, citric acid, and EDTA. However, these agents are not very effective in removing calculus and other hardened material.

In the case of peri-implant mucositis, diode lasers are a common choice to perform bacterial reduction, an important step in non-surgical therapy. Elimination of pathogen load will dramatically reduce the production of bacterial endotoxins and lipopolysaccharides, the primary drivers of inflammation [51]. Home care instructions must be reviewed with the patient to reduce the probability of bacterial recolonization on the transmucosal portion of the implant. The initiated diode tip is brought to the depth of the sulcus and moved circumferentially to decontaminate the implant surface and inner lining of the sulcus. The power settings for diode lasers will be wavelength dependent but should be less than 1.0 W of average power. Other wavelengths can be used, and it is therefore essential to review the reference manual for the correct combinations of tip and power setting.

◘ Figure 10.25 shows SEM studies of erbium-based lasers demonstrate that bioburden can be completely removed from the implant surface to allow reintegration of the exposed implant body [52].

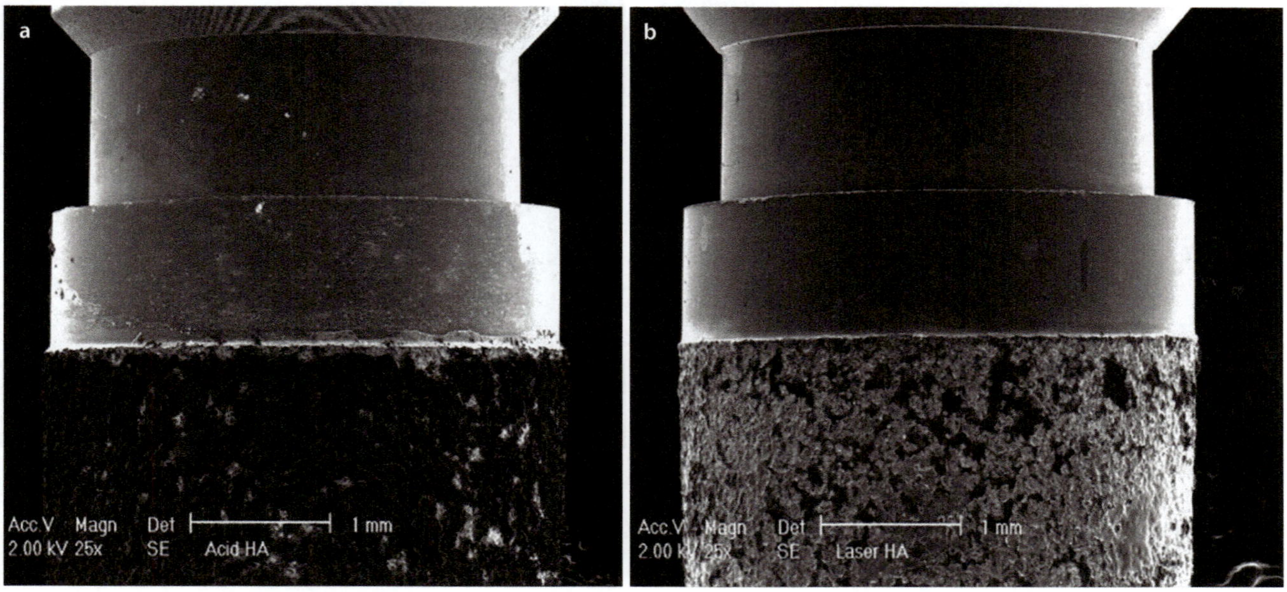

◘ **Fig. 10.25** Comparison of citric acid **a** and Erbium laser ablation **b** of hydroxyapatite on an implant surface [52]

10.10.4 Laser-Assisted Surgical Treatment of Peri-implantitis

The treatment protocol for peri-implantitis involves surgery to ensure proper visualization and access. ◘ Figure 10.26 depicts a full-thickness flap being raised to expose diseased tissue and the implant defect. The complete defect must be exposed to the margins of healthy bone. This will allow visualization of the extent of the defect and allow access for complete debridement of all pathology. Curettes are then employed to remove the gross volume of tissue followed by the Er,Cr:YSGG laser's ablation of remaining pathologic tissue as shown in ◘ Fig. 10.27.

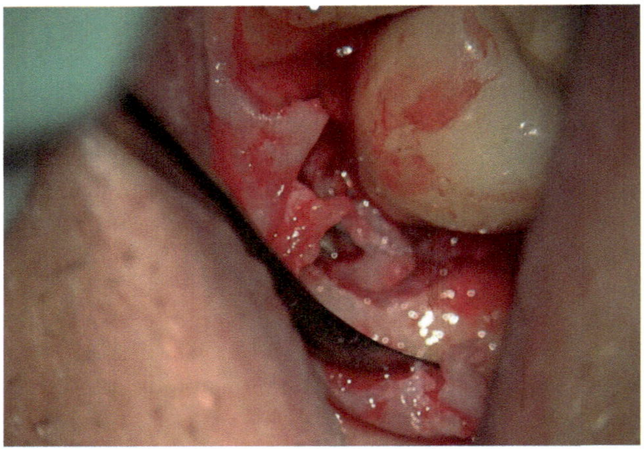

◘ **Fig. 10.26** A full-thickness flap is reflected, revealing the bioburden attached to implant surface and peri-implant granulation tissue

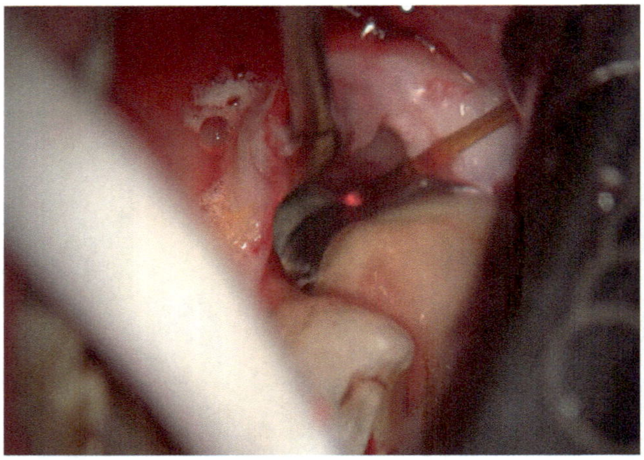

◘ **Fig. 10.27** Laser ablation of the implant surface and removal of infected tissue using an Er,Cr:YSGG laser. A 14–17 mm Z6 end-firing tip ablates granulation tissue and debrides the implant surface circumferentially. Approximate settings are 2.5 W, 20 Hz, 30% water, 30% air, and the tip is used in an oscillating motion so that ablation can be visualized with no heat buildup

Once the implant body has been debrided circumferentially to the bone level, a graft is replaced to the contours of the original bone volume (◘ Fig. 10.28). The choice of graft material should be designed to regenerate host bone and not become fibrous encapsulated.

◘ Figure 10.29 shows a failure of peri-implantitis treatment, with remnants of a non-resorbable graft extravasating from the bone defect. A full-thickness flap was raised, and the Er,Cr:YSGG laser, with the parameters employed in ◘ Fig. 10.27, removed all granulation tissue and bone graft remnants. A new resorbable graft was placed, primary closure around the transmucosal abutment was achieved and a post-op radiograph was taken. The graft material appears as red on a colorized digital radiograph (◘ Fig. 10.30). The 6-month postoperative colored radiograph of ◘ Fig. 10.31 shows autogenous bone fill in the previous defect with stable periodontium.

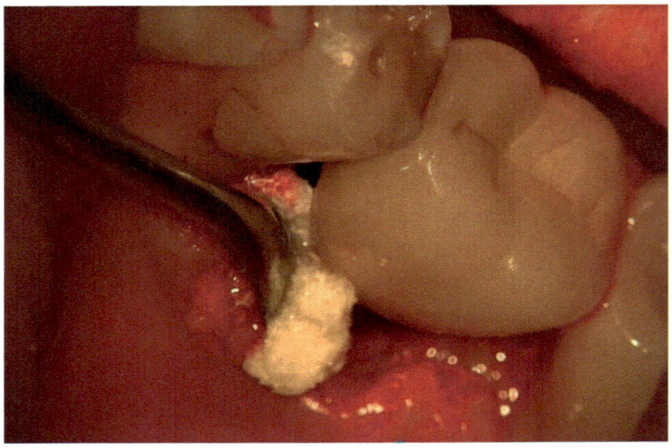

◘ **Fig. 10.28** Placement of a resorbable graft material hydrated with autologous plasma to restore lost tissue volume

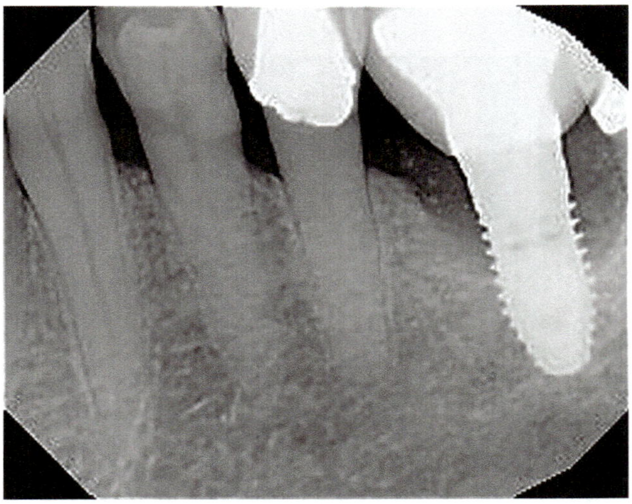

◘ **Fig. 10.29** Pre-operative condition of peri-implantitis with a failing bone graft and further bone loss

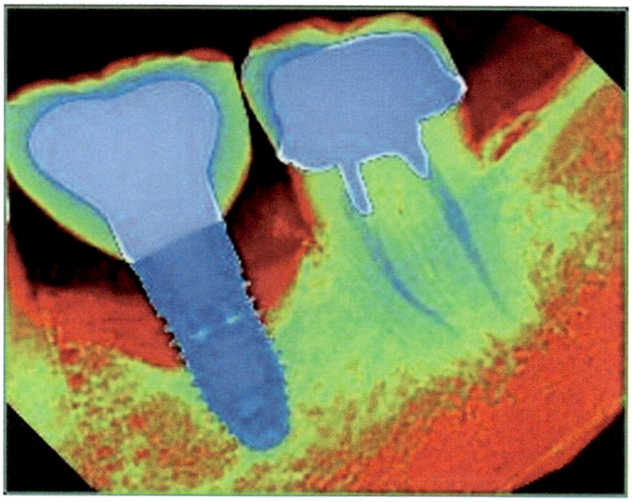

Fig. 10.30 Colored radiograph of the immediate post-operative condition. The area was debrided and ablated with the Er,Cr:YSGG laser and the graft material was placed. The red area represents grafted volume

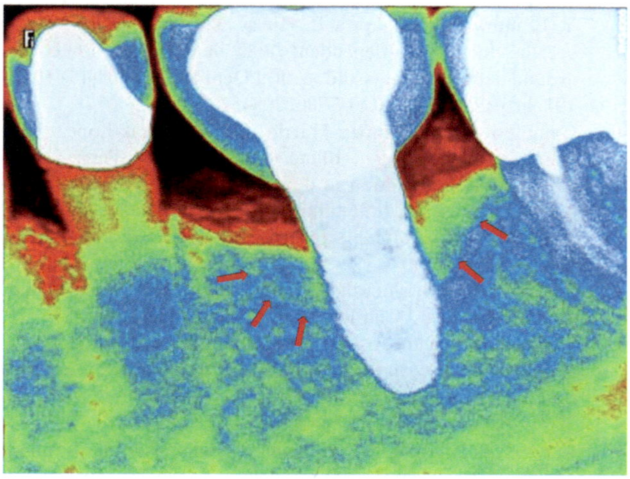

Fig. 10.31 A 6-month post-op radiograph demonstrating regeneration of hard tissue complex. Arrows represent the pre-surgical bone levels

It has been shown that ablation with erbium-based lasers facilitate osteoblast attachment to the titanium oxide surface [53]. Potentiating bone cell attachment and subsequent metabolism can speed up the process of early integration and potentially compress treatment time [54]. Laser irradiation can also potentiate the release of growth factors involved in the early production of bone. Autologous growth factors, such as platelet-derived growth factors (PDGF), are directly involved in early soft and hard tissue production [55].

10.11 Conclusions

Lasers are useful in all aspects of implant dentistry from pre-operative to post-operative applications. As more research and further long-term longitudinal studies are conducted, the results should confirm the effectiveness of various wavelengths.

Lasers in dentistry have significantly enhanced the concept of patient-centered care, and this is especially true of their use in implant dentistry. Laser-assisted treatments result in less pain, swelling, and inflammation, with greater efficacy as compared to conventional treatments. This results in greater post-operative comfort and increased patient acceptance. The clinician who desires excellent patient outcomes in implant dentistry should include lasers as part of the armamentarium.

References

1. Elani HW, Starr JR, et al. Trends in dental implant use in the U.S., 1999–2016, and projections to 2026. J Dent Res. 2018;97(13):1424–30. https://doi.org/10.1177/0022034518792567. Epub 2018 Aug 3.
2. Mizutani K, Aoki A, Coluzzi D, et al. Lasers in minimally invasive periodontal and peri-implant therapy. Periodontol 2000. 2016;71(1):185–212.
3. Schwarz F, Aoki A, Sculean A, et al. The impact of laser application on periodontal and peri-implant wound healing. Periodontol 2000. 2009;51:79–108.
4. Locke M. Clinical applications of dental lasers. Gen Dent. 2009;57(1):47–59.
5. Swick M. Laser-tissue interaction. I. J Laser Dent. 2009;17(1):28–32.
6. Romanos G, Nentwig GH. Diode laser (980 nm) in oral and maxillofacial surgical procedures: clinical observations based on clinical applications. J Clin Laser Med Surg. 1999;17(5):193–7.
7. Gokhale SR, Padhye AM, Byakod G, et al. A comparative evaluation of the efficacy of diode laser as an adjunct to mechanical debridement versus conventional mechanical debridement in periodontal flap surgery: a clinical and microbiological study. Photomed Laser Surg. 2012;30:598–603.
8. Katta N, Santos D, McElroy AB, et al. Laser coagulation and hemostasis of large diameter blood vessels: effect of shear stress and flow velocity. Sci Rep. 2022;12:8375. https://doi.org/10.1038/s41598-022-12128-1.
9. Dubin AE, Patapoutian A. Nociceptors: the sensors of the pain pathway. J Clin Invest. 2010;120(11):3760–72.
10. Scallan J, Huxley VH, Korthuis RJ. Chapter 3: The lymphatic vasculature. In: Capillary fluid exchange: regulation, functions, and pathology. San Rafael, CA: Morgan & Claypool Life Sciences; 2010. https://www.ncbi.nlm.nih.gov/books/NBK53448/.
11. Moritz A, Schoop U, Goharkhay K, et al. Treatment of periodontal pockets with a diode laser. Lasers Surg Med. 1998;22(5):302–11.

12. Romanos GE. Clinical applications of the Nd:YAG laser in oral soft tissue surgery and periodontology. J Clin Laser Med Surg. 1994;12(2):103–8.
13. White JM, Goodis HE, Rose CL. Use of the pulsed Nd:YAG laser for intraoral soft tissue surgery. Lasers Surg Med. 1991;11(5):455–361.
14. Chu RT, Watanbe L, White JM, et al. Temperature rise and surface modification of lased titanium cylinders. J Dent Res. 1992;71:144. Special issue abstract no 312.
15. Romanos GE, Everts H, Nentwig GH. Effects of diode and Nd:YAG laser irradiation on titanium discs: a scanning electron microscope examination. J Periodontol. 2000;71:810–5.
16. Fenolon T, Bakr M, et al. Effects of lasers on titanium surfaces: a narrative view. Lasers Dent Sci. 2022;6:152–67.
17. Kato T, Kusakari H, Hoshino E. Bactericidal efficacy of carbon dioxide laser against bacteria-contaminated titanium implant and subsequent cellular adhesion to irradiated area. Lasers Surg Med. 1998;23(5):299–309.
18. Wilder-Smith P, Dang J, Kurosaki T. Investigating the range of surgical effects on soft tissue produced by a carbon dioxide laser. J Am Dent Assoc. 1997;128(5):583–8.
19. Ishikawa I, Aoki A, Takasaki AA. Potential applications of Erbium:YAG laser in periodontics. J Periodontal Res. 2004;39(4):275–85.
20. Ting CC, Fukuda M, Watanabe T, et al. Effects of Er,Cr:YSGG laser irradiation on the root surface: morphologic analysis and efficiency of calculus removal. J Periodontol. 2007;78(11):2156–64.
21. Aoki A, Mizutani K, Schwarz F, et al. Periodontal and peri-implant wound healing following laser therapy. Periodontol 2000. 2015;68:217–69.
22. Kelbauskiene S, Baseviciene N, Goharkhay K, et al. One-year clinical results of Er,Cr:YSGG laser application in addition to scaling and root planning in patients with early to moderate periodontitis. Lasers Med Sci. 2011;26(4):445–52.
23. Kreisler M. Bactericidal effect of the Er:YAG laser on dental implant surfaces: an in vitro study. J Periodontol. 2002;73(11):1292–8.
24. Schwarz F, Nuesry E, Bieling K, et al. Influence of an erbium, chromium-doped yttrium, scandium, gallium, and garnet (Er:Cr:YSGG) laser on the reestablishment of the biocompatibility of contaminated titanium implant surfaces. J Periodontol. 2006;77(11):1820–7.
25. Kurlov VN, Shikunova AV, et al. Sapphire smart scalpel. Bull Russ Acad Sci Phys. 2009;73(2):1341.
26. Wang CW, Ashnagar S, Gianfilippo RD, Arnett M, Kinney J, Wang HL. Laser-assisted regenerative surgical therapy for peri-implantitis: a randomized controlled clinical trial. J Periodontol. 2021;92(3):378–88. https://doi.org/10.1002/JPER.20-0040.
27. Ramanauskaite A, et al. Apical/retrograde peri-implantitis/implant periapical lesion etiology, risk factors, and treatment options; a systematic review. Implant Dent. 2016;25(5):684–97. https://doi.org/10.1097/ID.0000000000000424.
28. Burdurlu MÇ, Dagasan VÇ, Tunç O, Güler N. Retrograde peri-implantitis: evaluation and treatment protocols of a rare lesion. Quintessence Int. 2021;52(2):112–21. https://doi.org/10.3290/j.qi.a45264.
29. Valderrama P, Wilson TG Jr. Detoxification of implant surfaces affected by peri-implant disease: an overview of surgical methods. Int J Dent. 2013;2013:740680. https://doi.org/10.1155/2013/740680.
30. Bhalla N, Dym H. Update on maxillary sinus augmentation. Dent Clin N Am. 2021;65(1):197–210. https://doi.org/10.1016/j.cden.2020.09.013.
31. Delilbasi C, Gurler G. Comparison of piezosurgery and conventional rotative instruments in direct sinus lifting. Implant Dent. 2013;22:662–5.
32. Jordi C, Mukaddam K, Lambrecht JT, Kühl S. Membrane perforation rate in lateral maxillary sinus floor augmentation using conventional rotating instruments and piezoelectric device-a meta-analysis. Int J Implant Dent. 2018;4(1):3. https://doi.org/10.1186/s40729-017-0114-2.
33. Tang E, Arany P. Photobiomodulation and implants: implications for dentistry. J Periodontol Implant Sci. 2013;43(6):262–8.
34. Martin R. Laser-accelerated inflammation/pain reduction and healing. Pract Pain Manag. 2003;3(6):20–5.
35. Romanos GE, Gutknecht N, Dieter S, et al. Laser wavelengths and oral implantology. Lasers Med Sci. 2009;24(6):961–70.
36. Boldrini C, de Almeida JM, Fernandes LA, et al. Biomechanical effect of one session of low-level laser on the bone-titanium implant interface. Lasers Med Sci. 2013;28(1):349–52.
37. Scher J, Pillimger M. The anti-inflammatory effects of prostaglandins. Investig Med. 2009;57(6):703–8. https://doi.org/10.2310/JIM.0b013e31819aaa76.
38. Cury Y, et al. Pain and analgesia: the dual effect of nitric oxide in the nociceptive system. Nitric Oxide. 2011;25(3):243–54. https://doi.org/10.1016/j.niox.2011.06.004.
39. Heyman O, Horev Y, Koren N, Barel O, Aizenbud I, Aizenbud Y, Brandwein M, Shapira L, Hovav AH, Wilensky A. Niche specific microbiota-dependent and independent bone loss around dental implants and teeth. J Dent Res. 2020;99(9):1092–101. https://doi.org/10.1177/0022034520920577.
40. Camolesi GCV, Somoza-Martín JM, Reboiras-López MD, Camacho-Alonso F, Blanco-Carrión A, Pérez-Sayáns M. Photobiomodulation in dental implant stability and post-surgical healing and inflammation. A randomised double-blind study. Clin Oral Implants Res. 2023;34(2):137–47. https://doi.org/10.1111/clr.14026.
41. Zitzmann NU, Berglundh T. Definition and prevalence of peri-implant diseases. J Clin Periodontol. 2008;35(8 Suppl):286–91.
42. Kotsakis GA, Olmedo DG. Peri-implantitis is not periodontitis: scientific discoveries shed light on microbiome-biomaterial interactions that may determine disease phenotype. Periodontol 2000. 2021;86(1):231–40. https://doi.org/10.1111/prd.12372.s.
43. Renvert S, Polyzois I. Treatment of pathologic peri-implant pockets. Periodontol 2000. 2018;76(1):180–90. https://doi.org/10.1111/prd.12149.
44. Figuero E, Graziani F, Sanz I, Herrera D, Sanz M. Management of peri-implant mucositis and peri-implantitis. Periodontol 2000. 2014;66(1):255–73. https://doi.org/10.1111/prd.12049.
45. Romanos GE, Javed F, Delgado-Ruiz RA, Calvo-Guirado JL. Peri-implant diseases: a review of treatment interventions. Dent Clin N Am. 2015;59(1):157–78. https://doi.org/10.1016/j.cden.2014.08.002.
46. Yao WL, Lin JCY, Salamanca E, Pan YH, Tsai PY, Leu SJ, Yang KC, Huang HM, Huang HY, Chang WJ. Er,Cr:YSGG laser performance improves biological response on titanium surfaces. Materials (Basel). 2020;13(3):756. https://doi.org/10.3390/ma13030756.
47. Funato A, Yamada M, Ogawa T. Success rate, healing time, and implant stability of photofunctionalized dental implants. Int J Oral Maxillofac Implants. 2013;28(5):1261–71. https://doi.org/10.11607/jomi.3263.
48. Roos-Jansaker A-M, Renvert S, Egelberg J. Treatment of peri-implant infections: a literature review. J Clin Periodontol. 2003;30(6):467–85.
49. Ohsugi Y, Aoki A, Mizutani K, Katagiri S, Komaki M, Noda M, Takagi T, Kakizaki S, Meinzer W, Izumi Y. Evaluation of

bone healing following Er:YAG laser ablation in rat calvaria compared with bur drilling. J Biophotonics. 2019;12(3):e201800245. https://doi.org/10.1002/jbio.201800245.
50. Gonçalves F, Zanetti AL, Zanetti RV, et al. Effectiveness of 980-mm diode and 1064-nm extra-long-pulse neodymium-doped yttrium aluminum garnet lasers in implant disinfection. Photomed Laser Surg. 2010;28:273–80.
51. Elsreti M, Smeo K, Gutknecht N. The effectiveness of diode lasers in detoxification of exposed implant surfaces in comparison with mechanical and chemical measures in the treatment of peri-implantitis: a literature review. Laser Dent Sci. 2022;6:1–14. https://doi.org/10.1007/s41547-021-00142-x.
52. Miller RJ. Treatment of the contaminated implant surface using the Er,Cr:YSGG laser. Implant Dent. 2004;13(2):165–70.
53. Romanos G, Crespi R, Barone A, et al. Osteoblast attachment on titanium disks after laser irradiation. Int J Oral Maxillofac Implants. 2006;21(2):232–6.
54. Pourzarandian A, Watanabe H, Aoki A, et al. Histological and TEM examination of early stages of bone healing after Er:YAG laser irradiation. Photomed Laser Surg. 2004;22(4):342–50.
55. Kesler G, Shvero DK, et al. Platelet derived growth factor secretion and bone healing after Er:YAG laser irradiation. J Oral Implantol. 2011;37:195–204.

Laser-Assisted Pediatric Dentistry

Konstantinos Arapostathis, Dimitrios Velonis, and Marianna Chala

Contents

D. J. Coluzzi, S. P. A. Parker (eds.), *Lasers in Dentistry—Current Concepts*, Textbooks in Contemporary Dentistry,
https://doi.org/10.1007/978-3-031-43338-2_11

Core Message

The progress of laser application in dentistry is continuous. There are many debates between researchers, clinicians, and scientists who try to carry on research within and with respect to clinical everyday dental practice. The American Academy of Pediatric Dentistry acknowledges using lasers as scientifically documented, alternative, and/or adjunctive treatment provision methods of soft and hard tissue management for infants, children, adolescents, and persons with disabilities. The aim of this chapter is to describe the indications for their use in various therapeutic procedures in pediatric dentistry and to analyze the advantages and disadvantages compared to traditional techniques. Together with the appropriate child's psychological management, proper presentation and approach with the laser are crucial. The technological evolution of dental lasers offers the possibility of completing several therapeutic procedures, such as removing carious dental tissue in permanent and deciduous teeth, usually with less or no anesthesia, and performing laser-assisted pulpotomy and pulpectomy, soft tissue interventions, dental trauma treatment procedures, etc. Depending on the treatment procedure and the targeted chromophores, all laser wavelengths could be used (e.g., KTP, diodes, Nd:YAG, erbium family lasers, CO_2).

11.1 Laser-Assisted Pediatric Dentistry

Pediatric dentistry is a demanding part of dentistry because of its nature to deal with children from birth through adolescence as well as with their parents' compliance. It requires from the clinician a high level of knowledge regarding the stomatognathic system conformation, the special anatomical figures, the prevention and cure, and the prognosis of dental pathologies found in children, but, above all, it requires expertise in treating the child itself. Pediatric dentistry practitioners are responsible not only for providing and promoting good oral and dental health for their patients but also to educate parents that oral health is an integral part of general health with continuous informative sources.

In general, the occurrence of oral diseases in children and adolescents includes dental caries, periodontal diseases (mainly in the form of gingival inflammation), developmental disturbances (morphological or numerical variations in both permanent and deciduous dentition), erosions, malocclusions, cranio-mandibular disorders, oral mucosal lesions (mainly aphthous ulcers, herpes simplex, and other virus infections or oral candidiasis), and, of course, dental trauma [1]. Over the past few years, traditional dentistry has been innovated with the embracement of more microinvasive techniques, moving from the era of "extension for prevention" to "prevention for extension" model of modern dentistry. In this technological-dental evolution with microabrasion, the application of topical fluoride and the use of sealants and the general adhesive techniques, laser technology has started to become more popular to the pediatric dental world. The widespread use of lasers in dentistry can be employed for both diagnosis and treatment, and as stated by the American Academy of Pediatric Dentistry (AAPD), "the use of lasers is an alternative and complementary method of providing soft and hard tissue dental procedures for infants, children, adolescents, and persons with special health care needs" [2].

11.2 Behavior Management and Laser Application

Dental specialists are trained to diagnose and treat dental diseases according to evidence-based dentistry with the behavior guidance to be the priority of the dental treatment. The dental practitioner interacts with the patient and their parents and through that procedure identifies appropriate or not appropriate behaviors, understands the emotional state of each person, and promotes empathy and compassion. The goal is to achieve communication and eliminate dental fear and anxiety in order to build a circle of trust between the child, the parent, the dentist, and the dental staff.

Earning child/parent's trust before managing to achieve high patient cooperation is the ultimate issue in pediatric dentistry. This is a difficult and demanding task, because many children perceive a visit to the dentist as stressful. This is an expected reaction, since an appointment includes several stresses—evoking components, such as strange sounds and tastes, having to lie down, meeting unfamiliar adult people and authority figures, discomfort, and even pain. Even though laser therapy sounds promising and well-accepted by the parents due to the possibility of better therapeutic results for their children and the assumption of no pain treatment (anesthesia may be necessary), the use by the dentist of the new technology still requires a degree of compliance by the child patient. Although a dental practice may have several modern and friendly devices, it remains an unknown and peculiar environment for the young child and may provoke negative emotions and stress during child's first visit. Therefore, the practitioner should choose and offer dental treatment with the appropriate methods and instruments that are suitable for each patient. Sometimes, laser treatment is preferable, especially for young children who refused the transitional dental treatment (◘ Fig. 11.1a–e). Laser treatment can be used to introduce dentistry, gain the

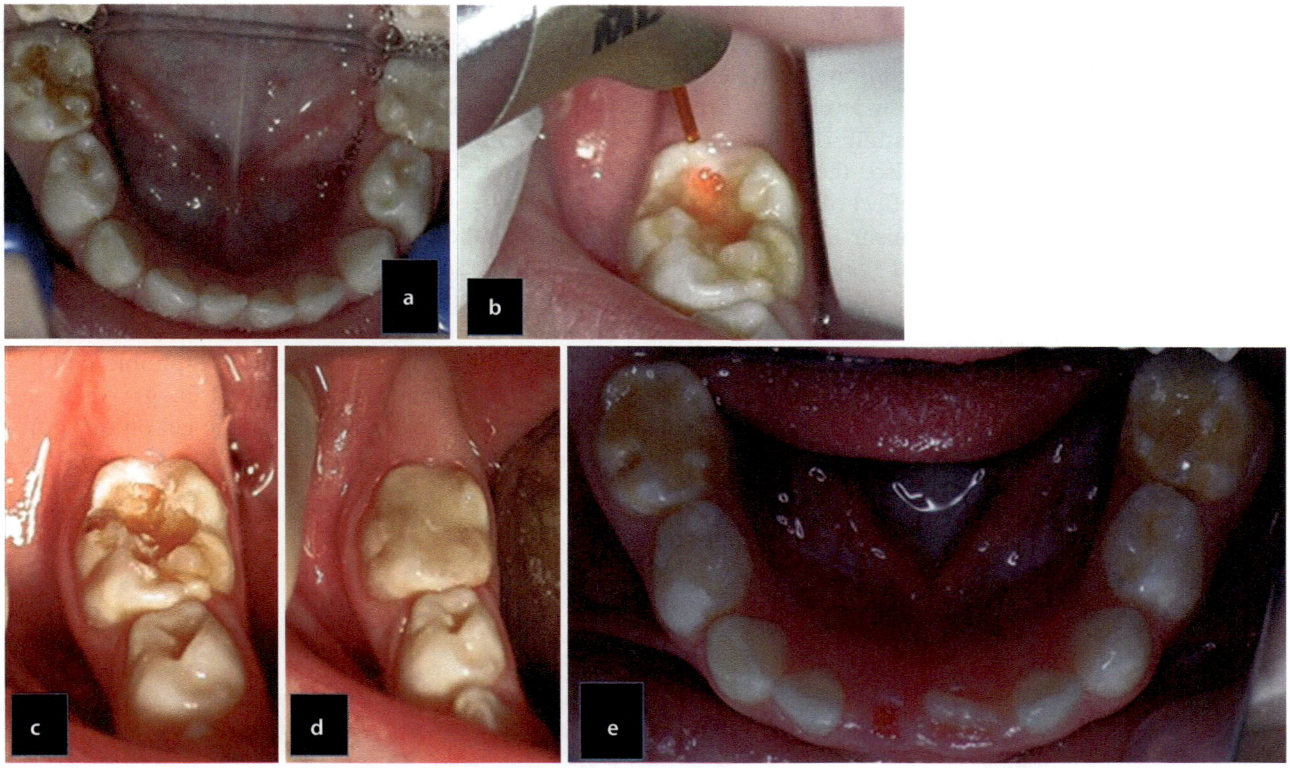

Fig. 11.1 Resin-modified glass ionomer (RMGI) restorations on the primary second molars, without the use of local anesthesia, of a 3-year-old uncooperative girl with primary molar hypomineralization (PMH). **a** Preoperative intraoral view. According to the transitional treatment recommendations, stainless steel crowns (SSC) should be placed on teeth #75 and #85 under local anesthesia. **b** Laser analgesia (starting with 50 mJ, 10 Hz, 82% water (16 mL/min), 70% air, distance 6–10 mm from the tooth, for 40–60 s and continuing with 80–100 mJ for 60 more seconds before tooth preparation) and cavity preparation (see Table 11.2 for energy parameters) by Er,Cr:YSGG (2780-nm gold handpiece, 0.6-mm MZ tip, H tissue mode). **c** After laser preparation. The cavity was well extended into dentin. The child revealed no pain and no complain and cooperation was relatively good. **d** Final RMGI restoration. **e** Intraoral view 24 months after treatment. Restorations are still in place and there are no caries lesions. The patient is now 5 years old and cooperative, and it is the practitioner's decision if and when will provide a more permanent rehabilitation

trust of the child, and perform needle-free and also no painful procedures. Through this, oral laser applications may also offer an alternative strategy in behavior management. A positive experience during dental treatment is of paramount significance for a lifelong confiding relationship between the child and the dentist, which may also lead to better oral health in the adulthood.

Either way, for its successfulness and the child's acceptance, a well-prepared presentation, training, and education on that have to be proceeded before use. The pediatric dentist may use some of the basic behavior techniques to introduce laser to the child. One of the most powerful techniques is "tell, show, do" in which the practitioner explains verbally the consecutive stages of the dental treatment (tell), demonstrates the equipment and shows the different tools/instruments on the hand/finger (show), and executes the procedure (do) [3]. Laser technology can be presented using friendly, familiar, easy-for-the-child-to-understand words like "special flashlight," "magic light," "colored light," etc. The sound of the laser could be like "making popcorn," "playing metal music," etc. The special glasses are going to make you look like "a ninja," "a princess," etc. In conjunction with the technique "tell, show, do," positive reinforcement (e.g., use of phrases like "great job" or a reward at the end of the session) and distraction techniques (e.g., television, movies, music) should be adopted. Children who do not cooperate or the mental status does not allow them to comply cannot be candidates for laser therapy.

11.3 Local Anesthesia and Laser Application

Local anesthesia is the basis in controlling pain during dental treatment but, at the same time, one of the most common and major fears for the patient. Traditionally, most of the dental treatment procedures need to be performed under local anesthesia. Laser analgesia provides an extra tool for the dentist to avoid or reduce the use of local anesthesia in some cases. It should be stated that analgesia is not really anesthesia but a way to reduce sensitivity, needing a more intensive stimulus for the patient to feel pain. Studies using infrared wavelengths (diode, Nd:YAG) conclude that low-level laser therapy (LLLT) can suppress the excitation of unmediated C-fiber afferents of the pulp. Also, there are studies regarding the potential analgesic effect of erbium family laser irradiation and the mechanism resulting in this effect. Many clinicians report that they have been successful in performing a variety of dental procedures, in pediatric dentistry too [4, 5].

In all clinical cases presented in this section, laser analgesia was applied using the Er,Cr:YSGG (2780 nm) laser with the following parameters: starting with 50 mJ, 10 Hz, (0.5 W), 82% water (16 mL/min), 70% air, and distance 6–10 mm from the tooth for 40–60 s and continuing with 80–100 mJ for 60 more seconds before tooth preparation (gold handpiece, 0.6-mm MZ tip, H tissue mode) (◘ Figs. 11.1a–e, 11.2a–g, and 11.3a–d). There are no studies reporting any analgesic effect of CO_2 wavelength. Theoretically, the ideal laser wavelength choice would be the one that has an analgesic effect and that can be used in all of those treatment procedures at the same time.

The performance of laser analgesia using erbium family lasers could be a useful tool to overcome behavioral problems, especially for needle-phobic children seeking dental treatment (◘ Fig. 11.2a–g). Also, only the application of topical anesthetic gel on dry gingival or mucosa for 3–5 min (e.g., EMLA cream (lido-

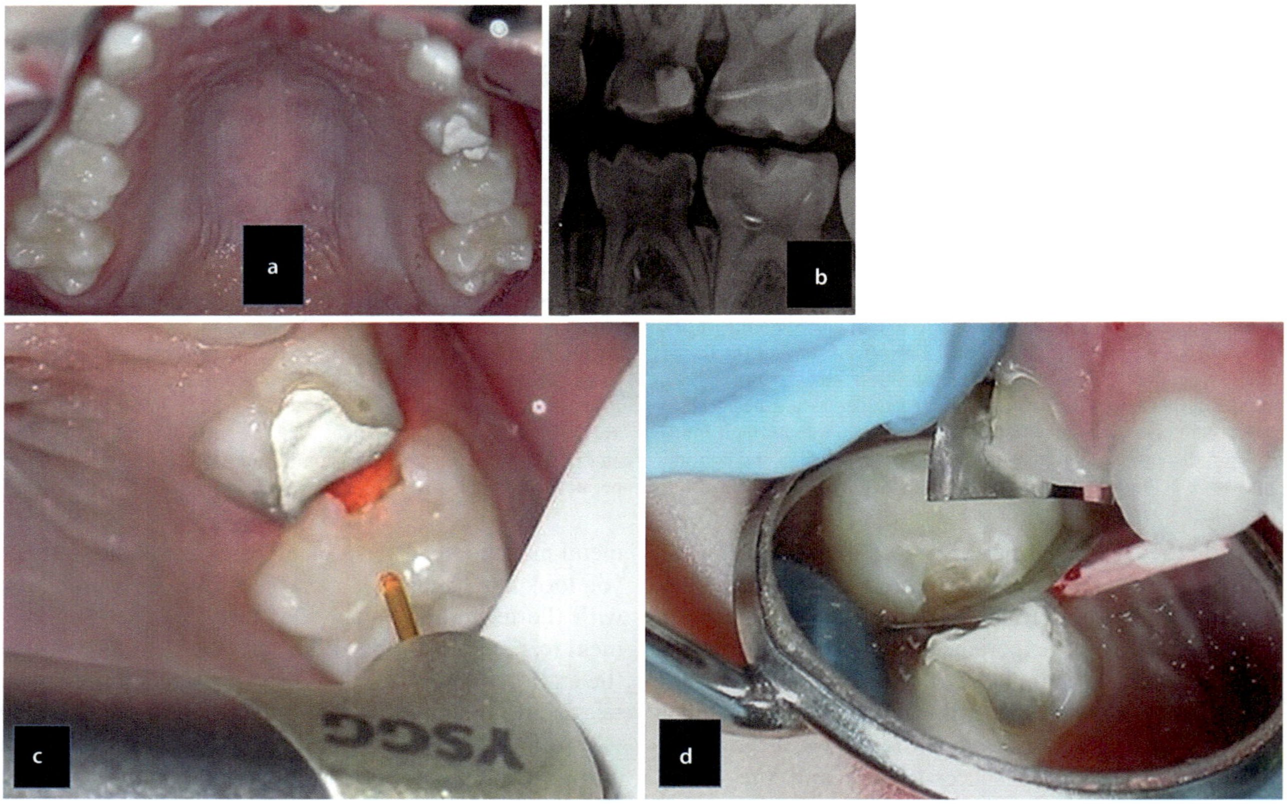

◘ **Fig. 11.2** Behavior shaping using laser for the completion of dental treatment of a referred 7.5-year-old needle-phobic girl with low cooperation at the dental office. Laser treatment used to introduce dentistry, gain the trust of the child, perform needle-free and also no painful procedures, and "desensitize" the patient through gradual exposure to dental treatment: perform first sealants, needle-free restorations (laser analgesia and preparation), and finally extraction. **a** Initial intraoral view of the upper arch. **b** Left bitewing radiograph. Tooth #64 had to be extracted due to abscess and root resorption. Note that caries was well extended into dentin on #65. **c** Cavity preparation on #65 (see ◘ Table 11.2 for energy parameters). No local anesthesia but laser analgesia (see text and ◘ Fig. 11.1b for laser parameters) and preparation by Er,Cr:YSGG (2780 nm, gold handpiece, 0.6-mm MZ tip, H tissue mode). **d** After laser preparation of #65. The child revealed no pain and no complaint and cooperation was good. **e** After RMGI placement on #65. **f** Intraoral view of the upper arch after 2 months. The patient presented for extraction of tooth #64 with the administration of local anesthesia (4% articaine, 1:200,000 epinephrine). Cooperation was excellent. **g** After 20 months. The permanent successor is erupting and space maintenance has to be removed. Restoration on #65 remains intact

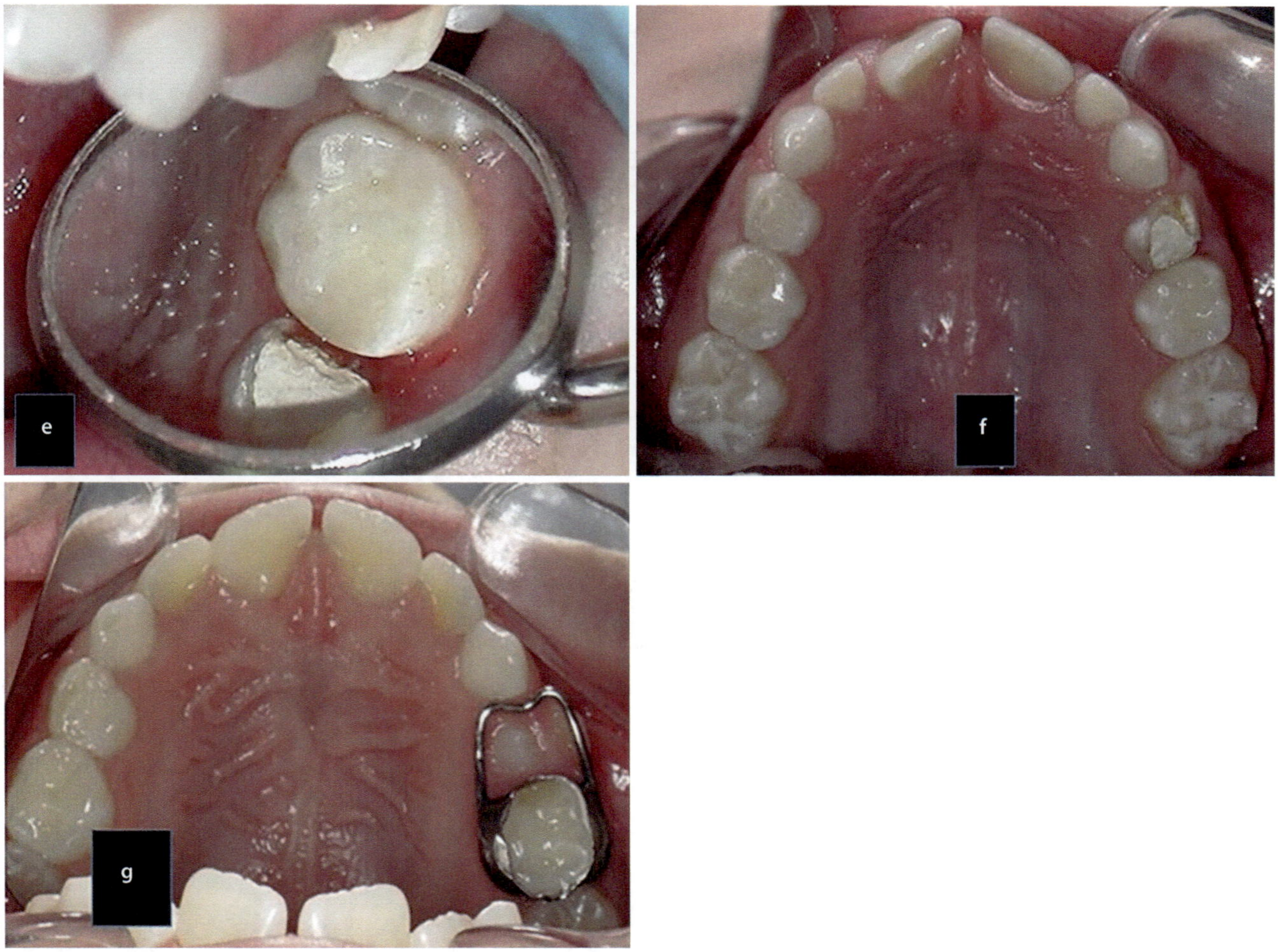

■ **Fig. 11.2** (continued)

caine 2.5% and prilocaine 2.5%); each gram of EMLA cream contains 25-mg lidocaine and 25-mg prilocaine), without the administration of injected local anesthesia, is efficient in performing minimal gingival interventions in several clinical cases by erbium family lasers (■ Figs. 11.3a–d and 11.4a–f).

It should be noticed that a prerequisite for achieving cooperation with the child and complete dental treatment is the minimization of disturbance and the absence of pain. Completion of dental treatment with children is directly related to the absence of pain. There is always a possibility of pain during dental treatment after laser analgesia, and in this case, laser energy parameters should be altered, or local anesthesia should be delivered. Adult patients can communicate their feelings with the dentist and may tolerate the pain to some extent and to remain cooperative, but children are frightened, lose trust to the dentist when their teeth ache, and then do not cooperate. It is the dentist's responsibility, after evaluating the child's maturity and providing adequate psychological preparation to reach a high degree of cooperation, to decide if local anesthesia should be administered before laser-assisted dental treatment. In general, if there is a possibility of pain, it is preferable to deliver local anesthesia before the start rather than during the dental treatment in children with low cooperation. Examples of such cases are shown at ■ Figs. 11.5a–e and 11.7a–g. These patients were not cooperative (one had extremely high gagging reflex which is very often associated with "hidden" dental anxiety) [6, 7]. Laser analgesia could be used, but it was decided that block anesthesia was more appropriate for these patients.

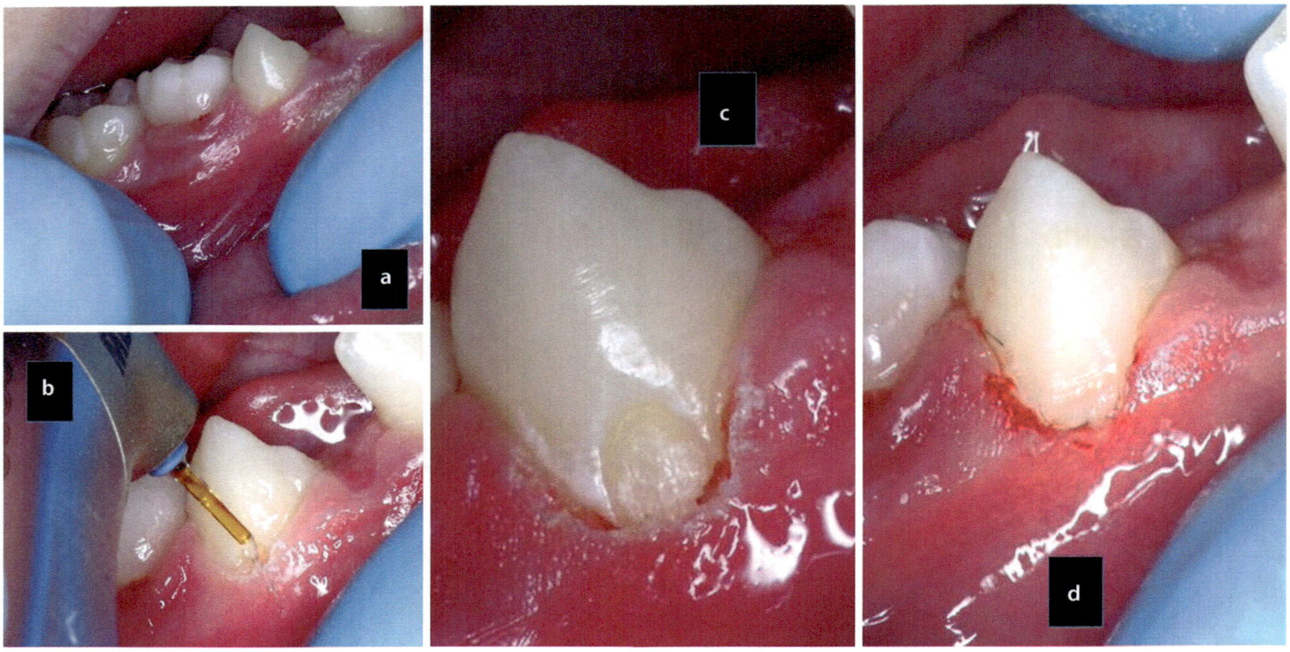

Fig. 11.3 Minimal gingivoplasty and subgingival composite resin restoration, in a single visit and without administration of local anesthesia, on tooth #83 of a 7-year-old boy. **a** Initial clinical view. Placement for 3 min only EMLA cream (lidocaine 2.5% and prilocaine 2.5%) on dry gingiva. **b** Minimal gingivoplasty using Er,Cr:YSGG (2780 nm) at 50 mJ, 20 Hz, (1.0 W), 30% water (6 mL/min), 70% air, tip distance 1 mm (close contact), tip parallel to the long axis of the tooth (gold handpiece, 0.6-mm MZ tip, H tissue mode). **c** After gingivoplasty and cavity preparation by Er,Cr:YSGG (2780 nm) (see Table 11.2 for energy parameters). **d** Final composite resin restoration

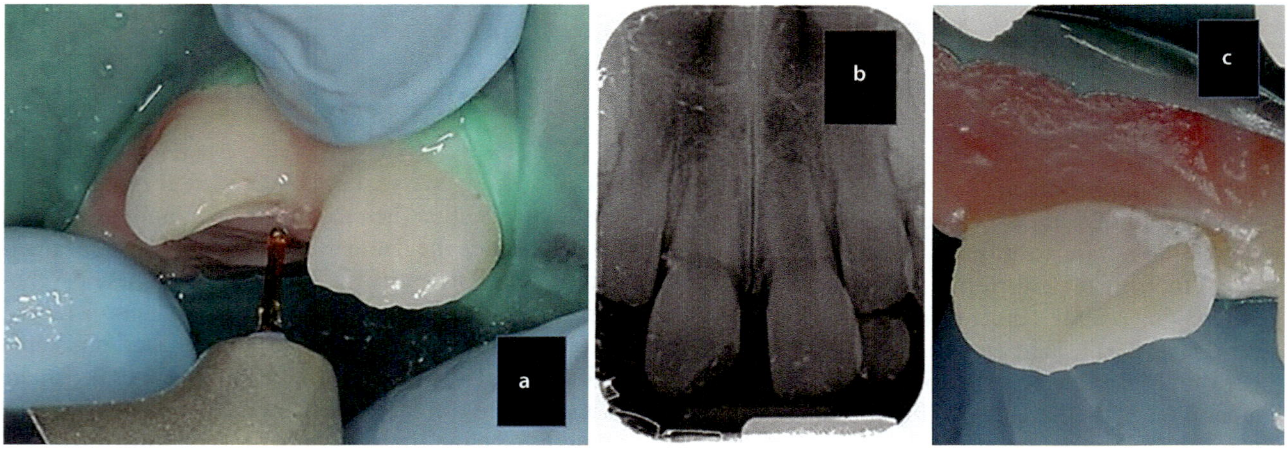

Fig. 11.4 Minimal gingivoplasty using Er,Cr:YSGG and reattachment of tooth fragment, in a single visit and without local anesthesia, following enamel-dentine subgingival (no pulp involvement) crown fracture on a permanent incisor of a 7.5-year-old boy. **a** Rubber dam placement, EMLA cream (lidocaine 2.5% and prilocaine 2.5%) for 3 min. Minimal gingivoplasty using Er,Cr:YSGG (2780 nm) (see laser energy parameters on Fig. 11.3b). **b** Initial radiographic image (no complete root formation). **c** After minimal gingivoplasty and before tooth fragment reattachment using composite resin. There is no gingival bleeding and tooth structures have been exposed. **d** Final restoration. **e**, **f** Clinical and radiographic views 30 months after treatment showing root formation

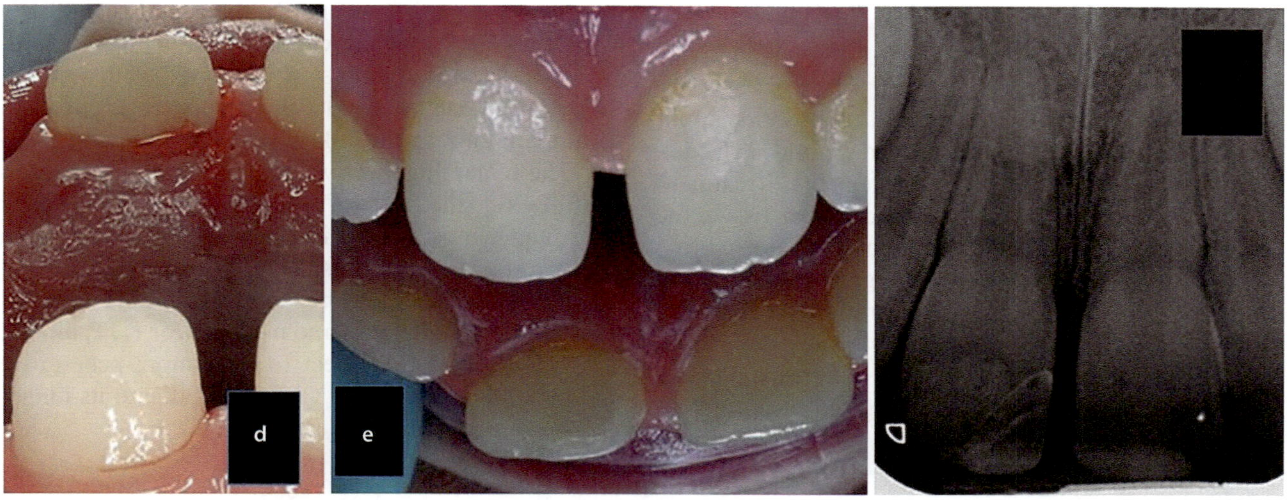

Fig. 11.4 (continued)

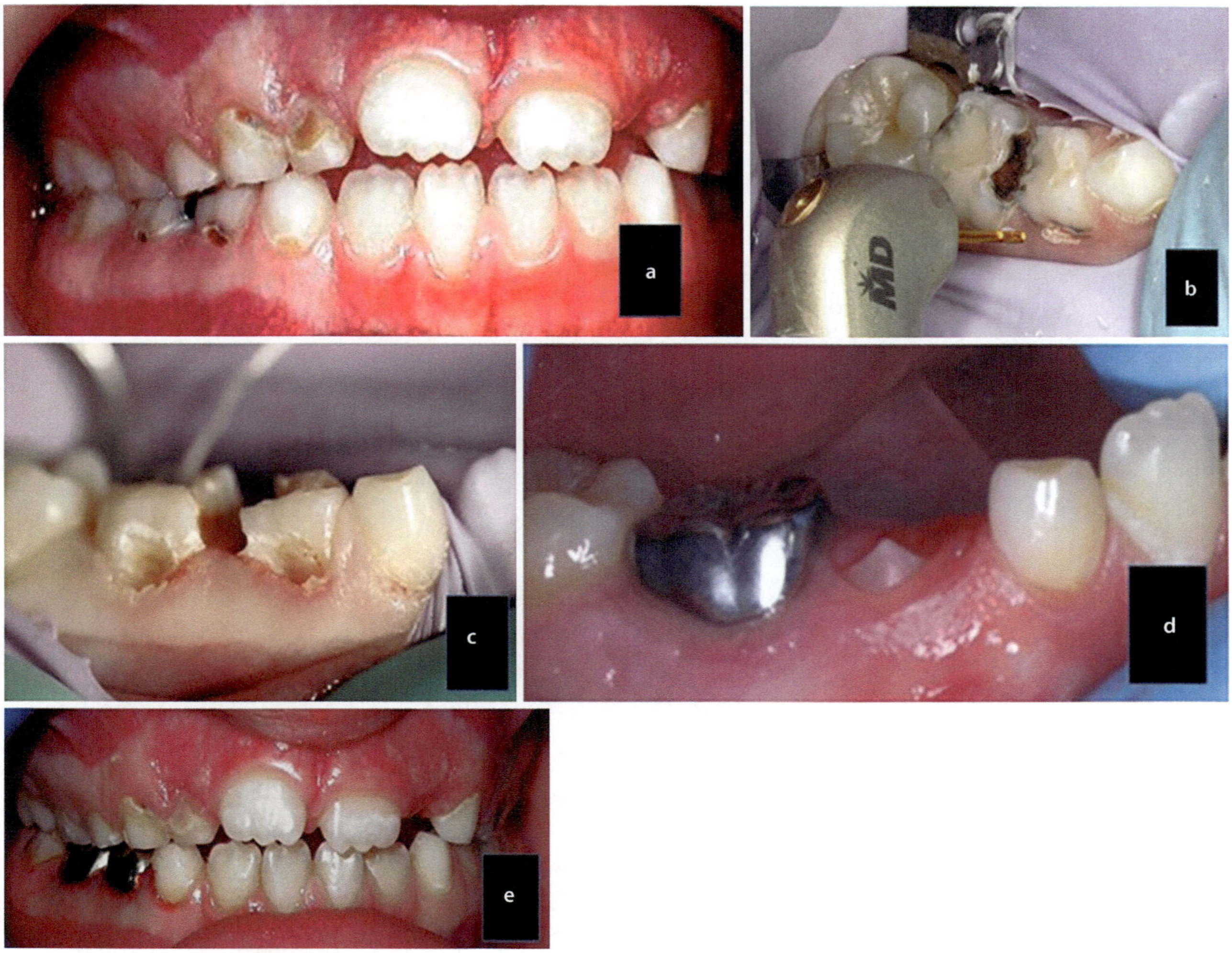

Fig. 11.5 Minimal gingivoplasty and treatment of subgingival caries, in a single visit, on teeth #83, #84, and #85 after preparation by Er,Cr:YSGG (2780 nm) of a 7.5-year-old girl. Also treatment of occlusal caries on #46. **a** Intraoral initial clinical view. **b** Block anesthesia (4% articaine, 1:200,000 epinephrine) and placement of rubber dam. Using Er,Cr:YSGG (2780 nm): (1) remove caries from #46 (enamel: 4 W, 20 Hz, 200 mJ, 82% (16 mL/min) H_2O, 70% air), (RC restoration and sealant) and (2) minimal gingivoplasty on teeth #83, #84, and #85 (see laser energy parameters on Fig. 11.3b). **c** Caries removal from teeth #83, #84, and #85 (see Table 11.2 for energy parameters for primary teeth). **d** Clinical view after 26 months. The girl is almost 10 years old, #83 is movable, and #84 has been normally exfoliated. Restoration on #83 is intact. **e** Placement of SSC on #84 and #85 and buccal RMGI restoration on #83

11.4 Types of Lasers Used in Pediatric Dentistry

Caries management includes prevention (fluoride application, dietary instructions, everyday oral hygiene), detection, and treatment management. Treatment includes the removal of the infected dental tissue, the cavity preparation, and, depending on the case severity, the indirect or direct pulp capping, pulpotomy, and pulpectomy, followed by tooth restoration. At this time, erbium family lasers are the ones that can be commonly used on both hard tissues, for caries removal and cavity preparation, and soft tissues. The targeted chromophore for this wavelength is primarily water and secondarily hydroxyapatite. This in combination with the mid-infrared wavelength (less penetrative compared to shorter wavelengths) results in its superficial effect on tissues, minimizing the risk for collateral thermal damage. The remaining laser wavelengths can be used successfully on the rest of the procedures, especially regarding hemostasis achievement in pulp or gingiva before restoration (◘ Figs. 11.6a–d and 11.7a–g) and decontamination, since the targeted chromophore in soft tissues is hemoglobin and melanin (for KTP, diodes, and Nd:YAG), with respect to their more penetrative wavelength (except for the CO_2 which, due to its longer

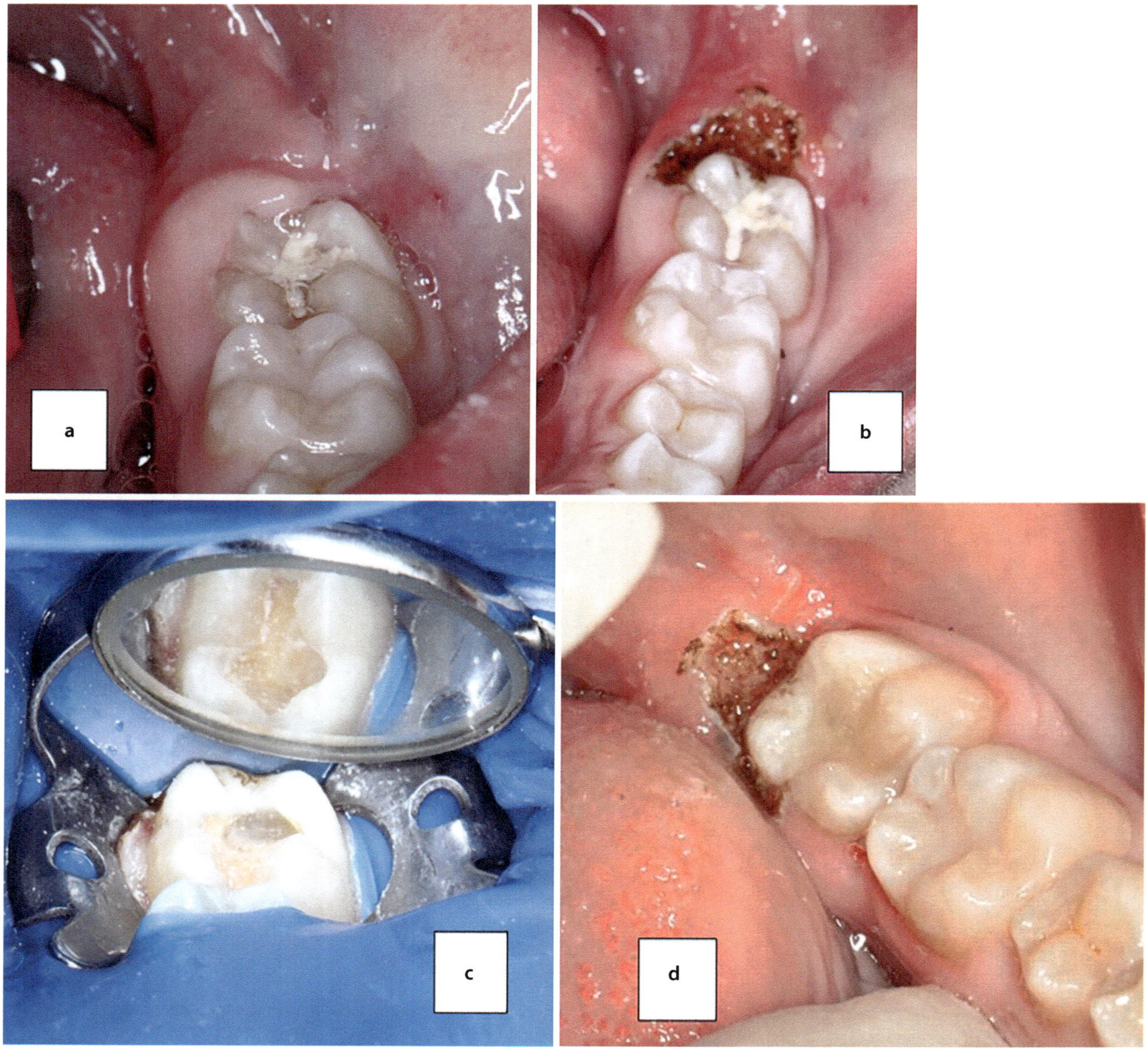

◘ **Fig. 11.6** Mimi, 12-year-old girl, clear medical history, caries on #37, under eruption, composite restoration, block anesthesia, gingivectomy with diode laser, Biolase, Epic X, 940-nm wavelength. **a** Intraoral initial clinical view. **b** After gingivectomy 1 W, continuous mode, 300-μm initiated fiber. **c** After cavity preparation. **d** Final intraoral view

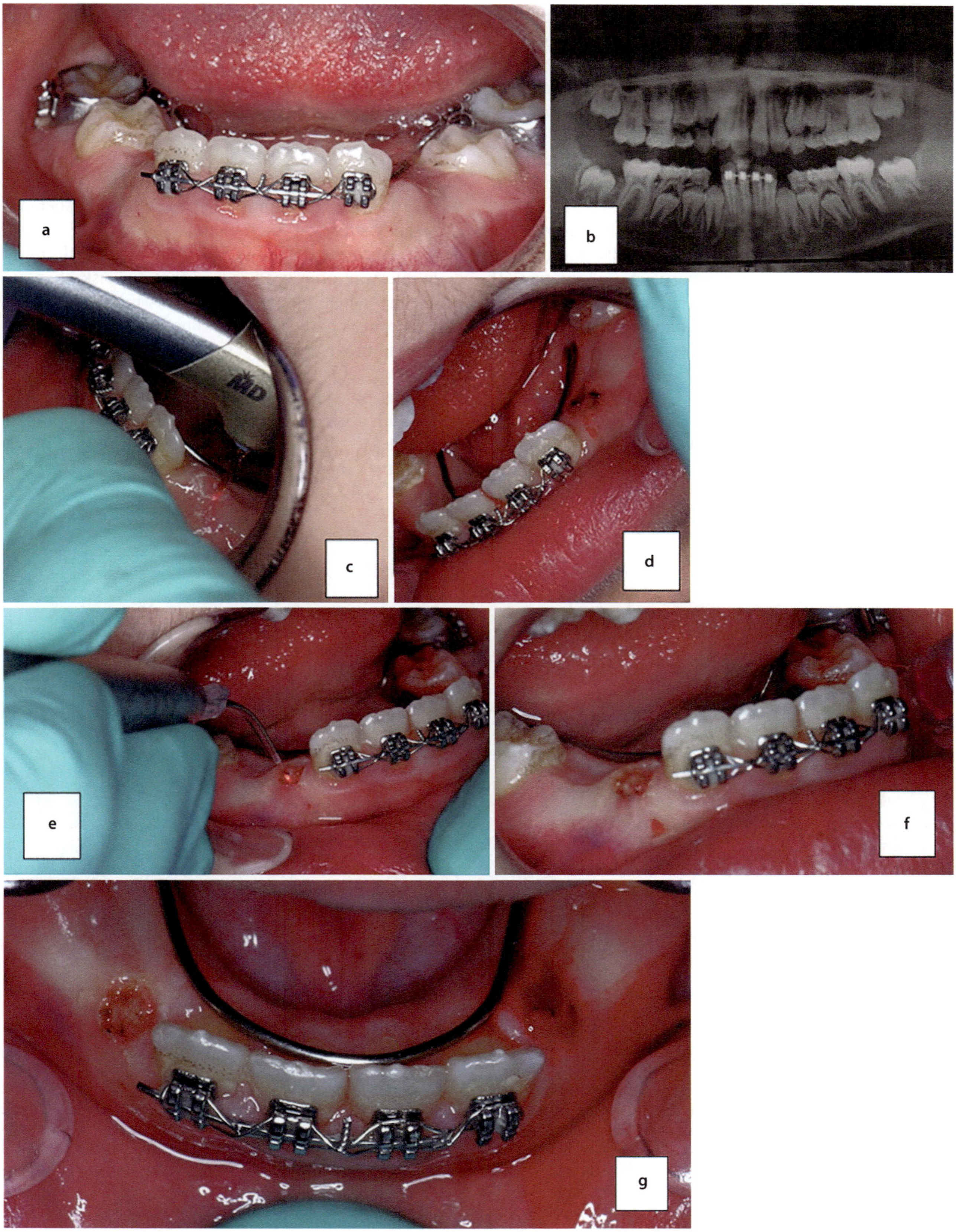

Fig. 11.7 Konstantinos, 11-year-old male, clear medical history, under orthodontic treatment, delay eruption of canines (#43, 33). **a**, **b** Intraoral initial clinical view and X-ray. **c**, **d** Crown exposure with Er,Cr:YSGG, 2.780-nm, Biolase (tooth #33). **e**, **f** Crown exposure with diode laser, Biolase, Epic X, 940-nm wavelength 1 W, continuous mode, 300-μm initiated fiber (tooth #43). **g** Postoperative intraoral view. Hemostasis was better and easier with diode laser

wavelength and high absorption in water, is the less penetrative of all) [8, 9]. Regarding caries prevention CO_2, erbium family lasers and Nd:YAG (due to their high power values emitted and ability to photo-thermally melt enamel) have been tested alone or in combination with fluoride, especially through in vitro studies. Infrared irradiation (diode lasers), due to its high penetration (and low absorption on hard tissue), is used widely in detecting caries.

Periodontal diseases in children usually include minimal severity gingivitis infections, usually due to poor everyday oral hygiene and hyperplastic gingivitis with the formation of pseudo-pockets (not completely erupted teeth). In addition, gingival and periodontal changes may be seen during or following orthodontic treatment, due to difficulties in maintaining good oral hygiene and/or the periodontal tissues following the teeth movement during the orthodontic treatment (◘ Fig. 11.8a–h). All laser wavelengths can be used in these instances for laser decontamination and if needed removal of hyperplastic gingival tissue. They simplify the surgical procedures by minimizing the use of flaps, provide excellent bleeding control without suturing, and result in a fast and less eventful healing (◘ Fig. 11.9a–e)

Apart from tooth decay, tooth injuries represent the most frequent pathology encountered in pediatric dentistry. Around 20% of children suffer a traumatic injury to their primary teeth and over 15% to their permanent teeth [10]. Dental trauma is a stressful and challenging emergency situation for the child, the parents, and the dentist. Accurate diagnosis in combination with immediate intervention is required, so that any risk of sequel problems or healing complications is minimized. Mid-infrared wavelength lasers could be used to reduce acute pain, to improve and speed up tissue healing (photobiostimulation effect), to provide decontamination and inflammation control, and to help control bleeding.

Among other advantages, the use of lasers can often make it easier for the dentist to perform several procedures in the same appointment (◘ Figs. 11.3a–d, 11.4a–f, and 11.5a–e).

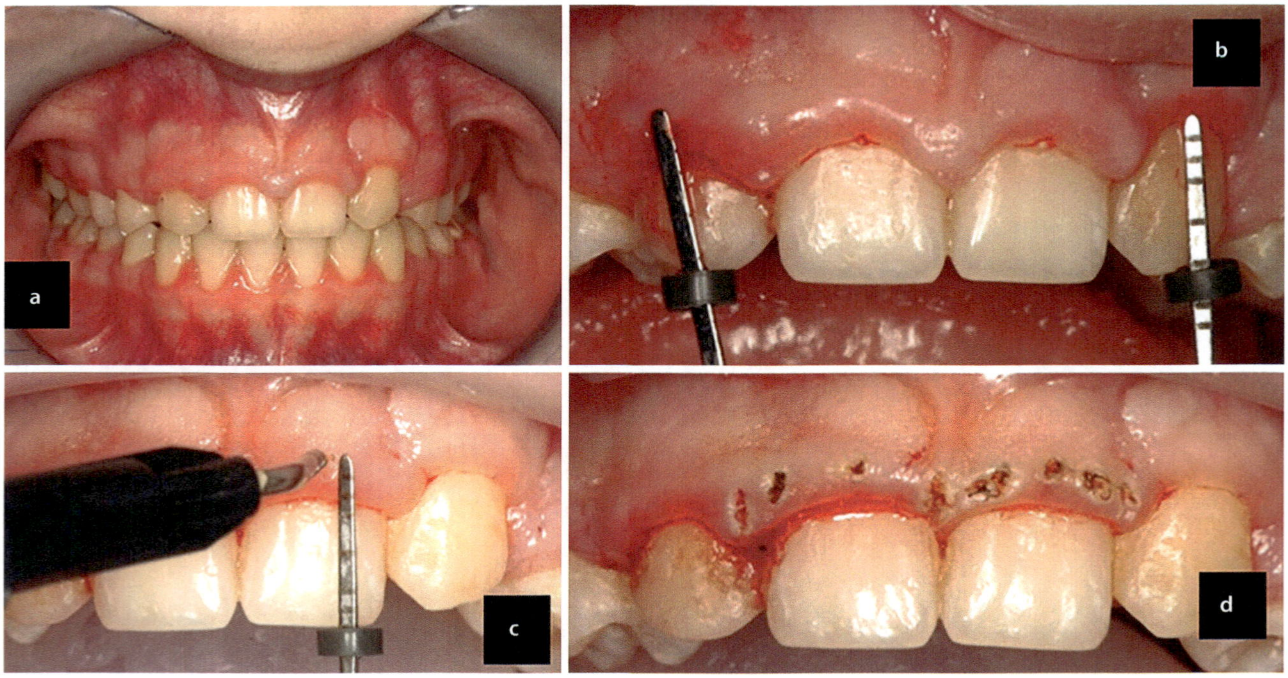

◘ **Fig. 11.8** Gingivectomy due to hyperplastic gingiva after orthodontic treatment and aesthetic reasons [canines replacing the missing second incisors] of a 16-year-old girl. **a** Initial situation, **b–d** measuring and comparing the size between the same teeth of the opposite sides, using a 1064-nm diode laser [1 W, continuous mode, 300-μm initiated fiber speed of movement 1 mm/s] to create dots and guide the soft tissue removal. **e** Measuring and comparing the size between the same teeth of the opposite sides, using a 1064-nm diode laser [1 W, continuous mode, 300-μm initiated fiber speed of movement 1 mm/s] to create dots and guide the soft tissue removal. **f** Immediate postoperative, **g–h** healing at 1 and 3 months postoperative

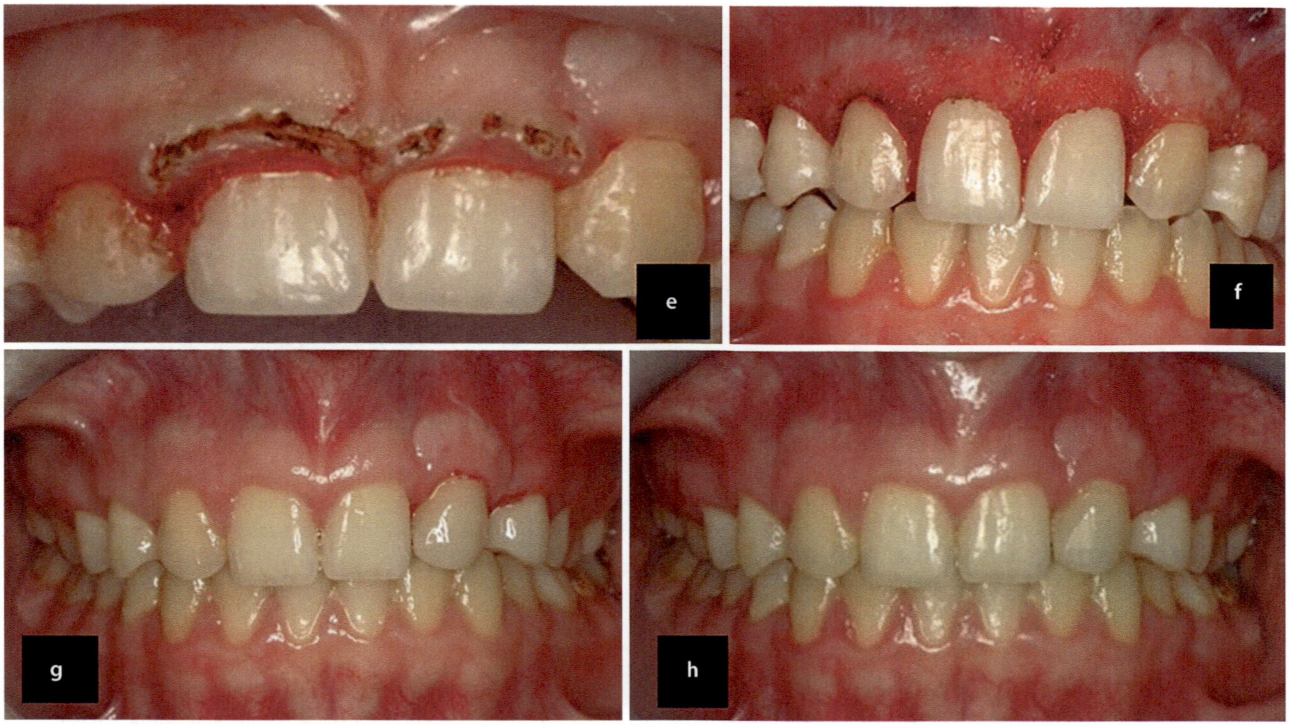

Fig. 11.8 (continued)

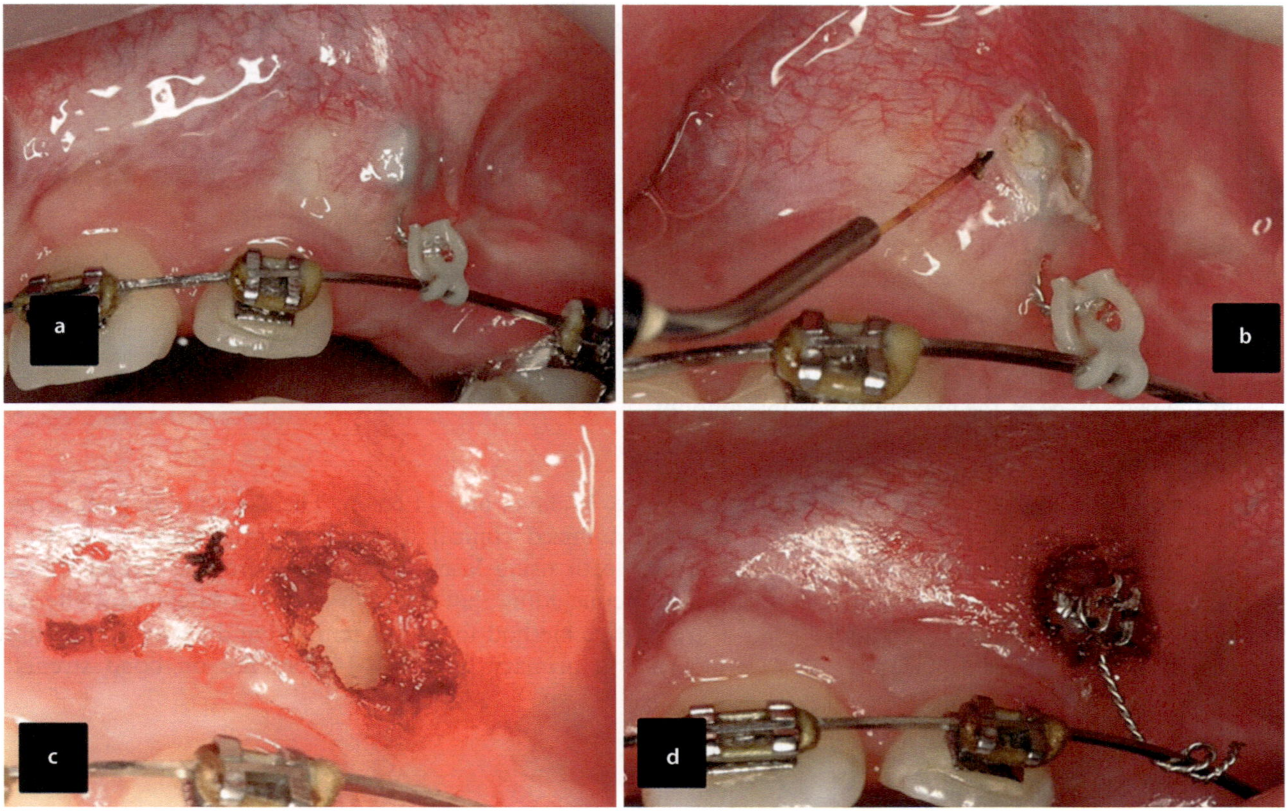

Fig. 11.9 Bracket recovery and re-bonding, 12-year-old boy. **a** Initial situation: the bracket is located under the alveolar mucosa and is no longer bonded on the canine. **b** Surgical procedure and bracket recovery using a 1064-nm diode [0.8 W, continuous mode, 300-μm initiated fiber, speed of movement 2 mm/s]. Extra caution and care to avoid extended thermal trauma [water cooling every 10 s, for 10 s] since the tissue is very thin and the metallic surface of the bracket accumulates heat rapidly. **c**, **d** Excellent hemostasis and bracket bonding. **e** Intraoral view after the end of the orthodontic treatment

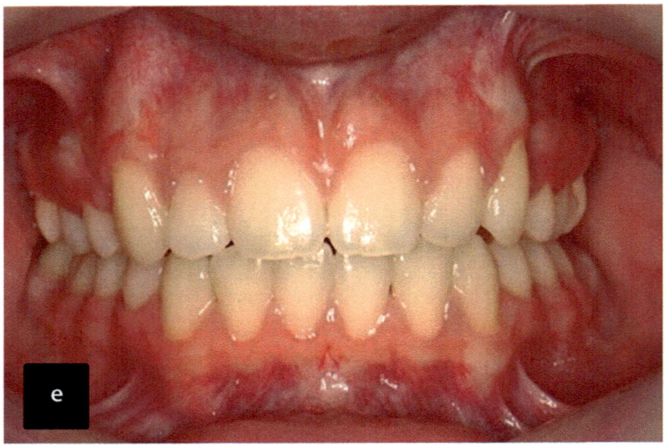

Fig. 11.9 (continued)

Table 11.1 Parameters for cavity preparation with Er:YAG laser (2940 nm) on primary teeth

Energy per pulse (mJ)	Average power (W)
(1) 160–200 mJ, 10 pps	(1) 1.6–2
(2) 80–100 mJ, 10 pps	(2) 0.8–1
(3) 40–60 mJ, 10 pps	(3) 0.4–0.6
(4) 35–50 mJ, 20 pps	(4) 0.70–1
(5) 50 mJ, 20 pps, defocus, for 15 s	(5) 1

Tip diameter 600 µm, 70% water (32 mL/min for Fotona LightWalker), 1-mm tip to tissue distance (1) enamel preparation, (2) dentine preparation, (3) dentine finishing-conditioning and removal of dental caries, (4) enamel finishing-conditioning, and (5) decontamination (based on Professor Selting laser parameter calculation sheet)

11.5 Restorations on Primary Teeth

Dental caries is one of the most common diseases in childhood, and several well-established restorative methods and materials have been used for replacing the carious dental tissues of primary teeth. Lasers can be used as alternative instruments to completely or partly substitute traditional instruments and techniques or to help and contribute to traditional dental treatment. The erbium family lasers are used for caries removal and cavity preparation on primary teeth. Enamel and dentine in primary teeth have compositional and structural differences from those of permanent teeth. Primary tooth enamel is less mineralized and more porous, and prisms do not have an orderly spatial organization. Primary tooth dentine has more water, less in number, and narrower dentinal tubules. Therefore, lower laser energy parameters than those for permanent teeth should be used for caries removal and cavity preparation on primary teeth (Tables 11.1 and 11.2). Water flow is given in both percentage and mL/min. The percentage of water given means the percentage of the maximum possible amount of water the specific laser unit could provide. For example, 70% (7 out of 10) for Fotona LightWalker (Er:YAG, 2940 nm) is water flow of 32 mL/min, while 82% for Er,Cr:YSGG (Waterlase MD, 2.780 nm) is 16 mL/min.

Table 11.2 Parameters for cavity preparation with Er,Cr:YSGG (Waterlase MD, 2.780 nm) on primary teeth

Average power (W)	Energy per pulse (mJ)
(1) 2.0 W, 10 pps	(1) 200
(2) 1.5 W, 10 pps	(2) 150
(3) 1 W, 10 pps	(3) 100
(4) 0.5 W, 10 pps	(4) 50
(5) 0.75 W, 20 pps	(5) 37.5
(6) 1 W, 20 pps, defocus, for 15 s	(6) 50

Tip diameter 600 µm, 82% water (16 mL/min), 1-mm tip to tissue distance (1) enamel preparation, (2) dentine preparation, (3) removal of dental caries, (4) dentine finishing-conditioning, (5) enamel finishing-conditioning, and (6) decontamination (based on Professor Selting laser parameter calculation sheet)

All dental restorative materials (composite resin (CR), compomers (C), resin-modified class ionomer (RMGI), glass ionomer (GI)) could be placed after laser cavity preparation on primary teeth (Figs. 11.1a–e, 11.3a–d, and 11.5a–e). There are no long-term randomized clinical trials about restoration of primary teeth using lasers. However, there are several studies concluding that laser abrasion is a safe, useful alternative method for caries removal and cavity preparation on primary teeth [11–14]. Studies on bond strength restorative materials after preparation of primary teeth by laser or traditional method showed lower or equal results [15–20]. The results on marginal microleakage are controversial, but most of the studies report good results (similar or better than the diamond bur) for both laser wavelengths of the erbium family. The restorative materials studied include several types of CR, C, RMGI, and GI. In the case of CR and C, several etching (total etch, self-etch) and adhesive systems (one-step adhesive, two-step adhesive, self-etching adhesive) are studied [21–29]. Also, a study showed no statistically significant difference on marginal microleakage between Er:YAG and Er,Cr:YSGG lasers for any of CR, RMGI, and GI restorations [30]. The main advantages of laser use in restorative pediatric dentistry are patient and parent's acceptance, the administration of no or less local anesthesia, the absence of vibration, the cavity decontamination effect, and the selectivity of dental caries.

Table 11.3 Advantages of laser use with minor surgical applications

Minor surgical applications using proper technique with a dental laser have several advantages
• Little local anesthesia
• No intraoperative or postoperative bleeding
• No suturing
• No postoperative pain and discomfort
• Normal function since day 1
• No antibiotic or analgesic medication use
• Accelerated healing
• No scarring
• Patient acceptance

11.6 Soft Tissue Applications

In any case a laser is used, it is imperative that the dentist/operator is fully familiar with the particular laser's settings and capabilities. It is recommended that a calculation spreadsheet is readily available, allowing the operator to instantly calculate values such as energy density (fluence), power density, and peak power, which are not available from the laser device monitor/dashboard.

11.6.1 Minor Surgical Applications (Table 11.3)

Labial Frenectomy

The frenum of the upper lip (maxillary labial frenum (MLF)) is a dense connective tissue structure with a high content of elastic fibers that tethers the upper lip to the maxilla. Based on the attachment of the fibers, the MLF has been classified by Mirko et al. [31] into four categories: (a) mucosal attachment (at the mucogingival junction), (b) gingival attachment (within the attached gingivae), (c) papillary attachment (extends into the interdental papilla), and (d) papilla penetrating attachment (crosses the alveolar ridge extending into the palatine papilla). Based on other characteristics, an MLF can also be described as simple frenum with a nodule, simple frenum with an appendix, bifid frenum, double frenum, or wider frenum.

The most common indication for frenectomy or frenum modification is when it causes or it is expected to cause an undesired diastema between the central incisors or when it causes periodontal problems, such as dehiscence, gum recession, or inflammation, or is easily traumatized during eating or oral hygiene. The optimal time or age to perform an upper labial frenectomy is under continuous debate. Most evidence points to the time when the upper canines have erupted or are erupting or as part of an orthodontic treatment plan in the mixed dentition. One of the arguments for closing a diastema before resecting the frenum has been that if performed before tooth movement the scar tissue may hinder orthodontic tooth movement and closing of the diastema. While the latter may be true for frenectomies performed traditionally with a scalpel and suturing, in the following section, it will become obvious that this is not true when the frenectomy or frenum modification is performed with a laser and with good technique. Several lasers can be and have been used to perform a maxillary labial frenectomy, and clinical examples are shown (Figs. 11.10a–h, 11.11a–e, 11.12a–k, 11.13a–e, 11.14a–e, 11.15a–d, and 11.16a–d).

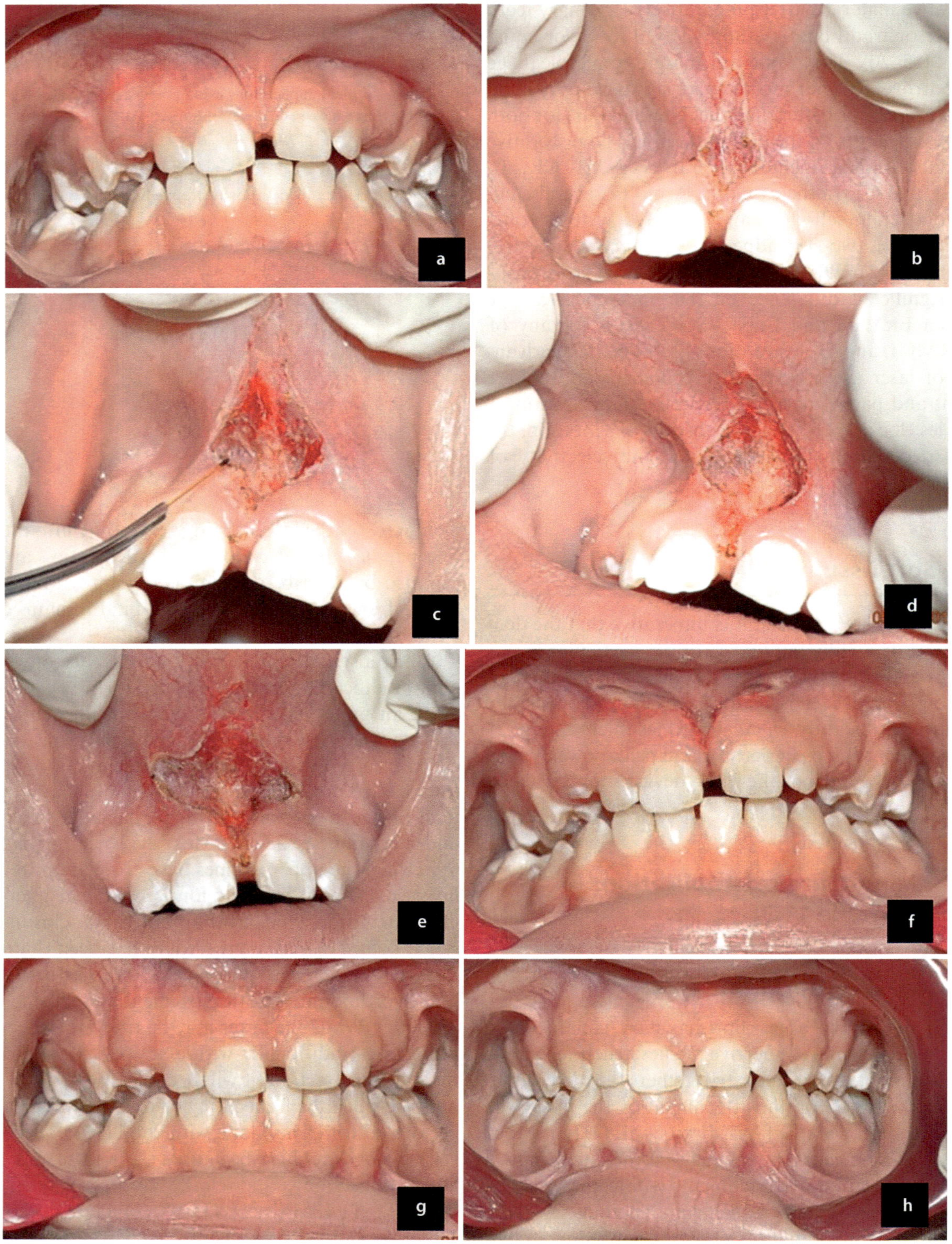

Fig. 11.10 Nine-year-and-5-month-old female (dental age ~ 11) with a short labial frenum and midline diastema prior to orthodontic treatment. Frenectomy performed under local anesthesia. *Laser parameters*: pulsed Nd:YAG initiated fiber 320 um, average power 3.2 W, pulse 40 mJ, repetition rate 80 Hz, pulse duration 100 μs. Treatment duration: total < 20 min, 30 s ON/30 s OFF. Mode: contact, parallel to teeth, stretched fibers, by manually pulling the upper lip. Progressive steps are shown. **a** Preop. **b** Step 1. **c** Step 2. Note the direction of the fiber parallel to the roots to minimize absorption of energy in the dental pulp and bone. **d** Step 3. **e** Final postop result. **f** One week F/U. **g** One month F/U. **h** Nine months F/U. Note no excellent coagulation, no carbonization, and no need for suturing. Instructions include rinsing with mild antiseptic solution during the first week of healing with chamomile or active oxygen solutions or drops (Unisept). Also, note excellent healing without scarring

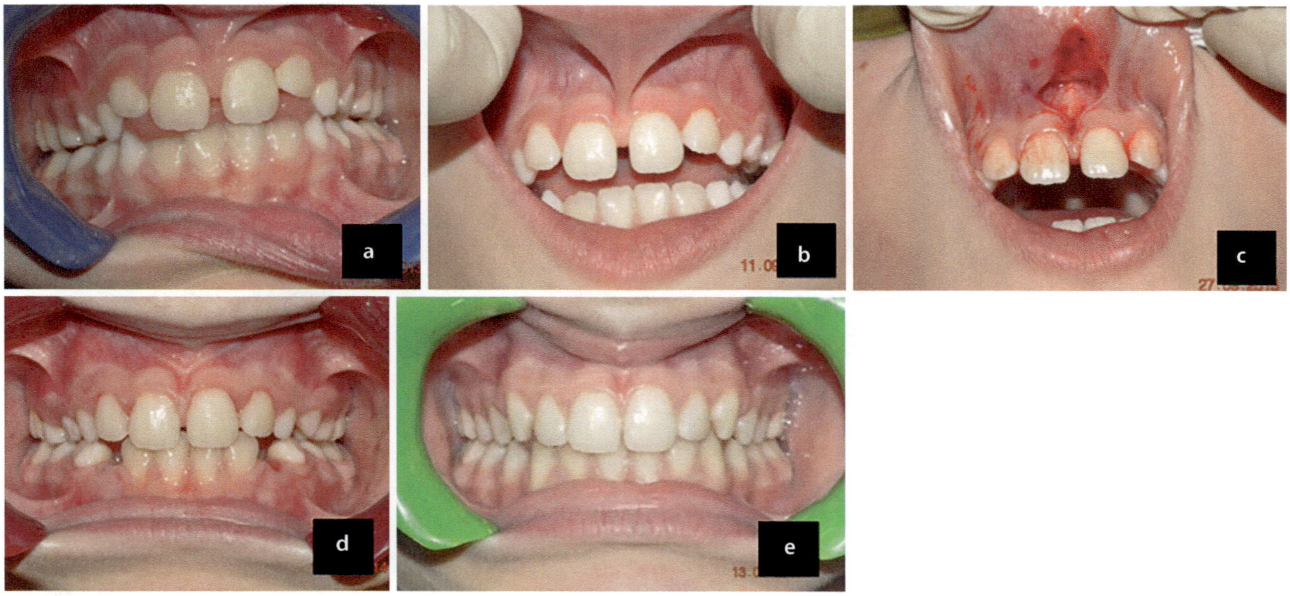

Fig. 11.11 Nd:YAG spontaneous diastema closing after laser frenectomy performed at the stage of upper lateral incisors' eruption in the mid-mixed dentition. No orthodontic movements were made. **a** Six months preop, **b** 2 weeks preop, **c** immediate postop, **d** 1-year F/U, **e** 3-year F/U. *Laser parameters*: pulsed Nd:YAG, average power 3.2 W, pulse 40 mJ, repetition rate 80 Hz, pulse duration 100 μs. Peak power 400 W, fiber diameter. Treatment duration: total < 20 min, 30 s ON/30 s OFF. Mode: contact, parallel to teeth, stretched fibers, by manually pulling the upper lip

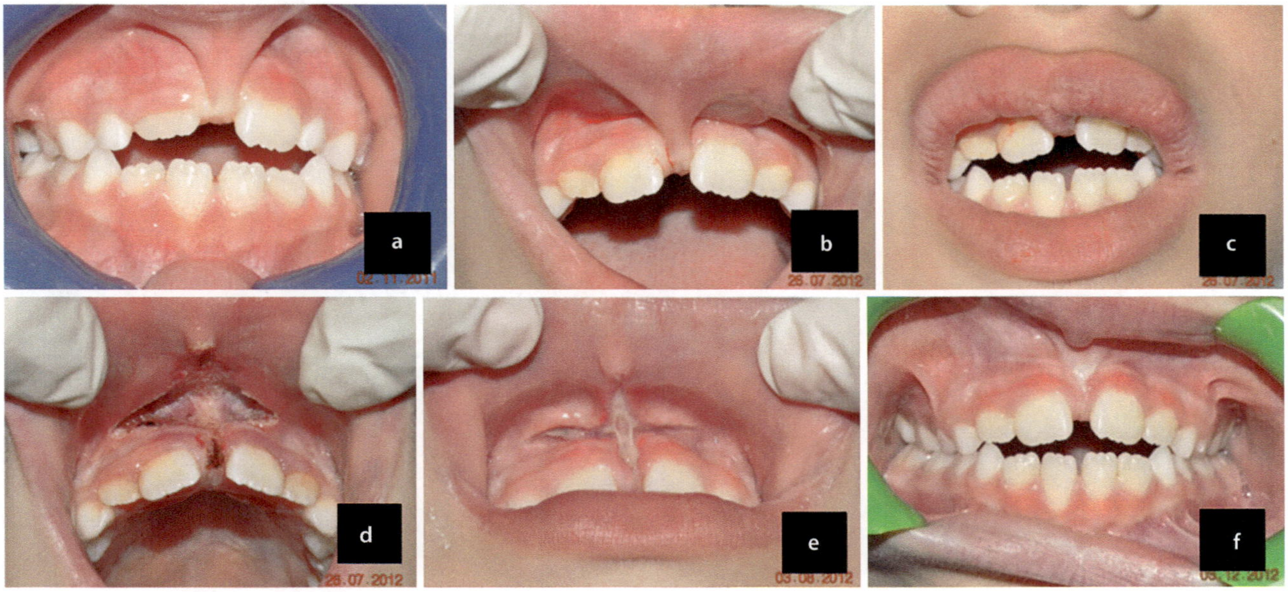

Fig. 11.12 Nd:YAG extreme case of upper labial frenum. Nine-year-old male with excessively thick and short frenum, combined with partial orthodontic treatment. **a** Ten months prior to laser frenectomy. **b**, **c** Preop. **d** Immediate postop. **e** One week F/U. **f** Four months F/U diastema closing spontaneously. **g** Sixteen months with partial orthodontic treatment of the lower anteriors and tongue thrust. **h** 22 months F/U with orthodontic treatment. **i** Four years F/U. Note good healing despite extensive modification. **j** Residual mucosal tag. **k** Laser aesthetic removal of the mucosal tag. This is best prevented at the time of the frenectomy by smoothening the end of the attachment of the frenum in the lip mucosa with the laser. *Laser parameters for frenectomy*: pulsed Nd:Yag, average power 4.0 W, pulse 40 mJ, repetition rate 100 Hz, pulse duration 100 μs. Fiber diameter 320 um. Treatment duration: total < 20 min, 30 s ON/30 s OFF. Mode: contact, parallel to teeth, stretched fibers, by manually pulling the upper lip. Laser parameters for the residual mucosal tag: repetition rate reduced to 80 Hz

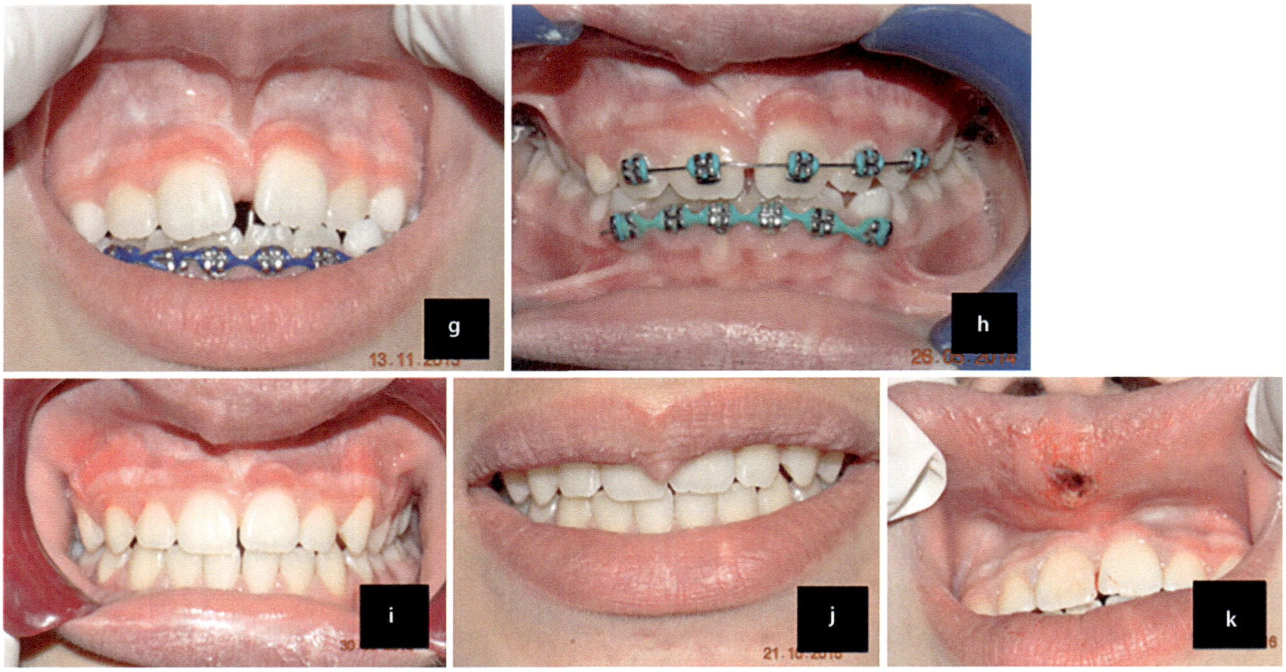

Fig. 11.12 (continued)

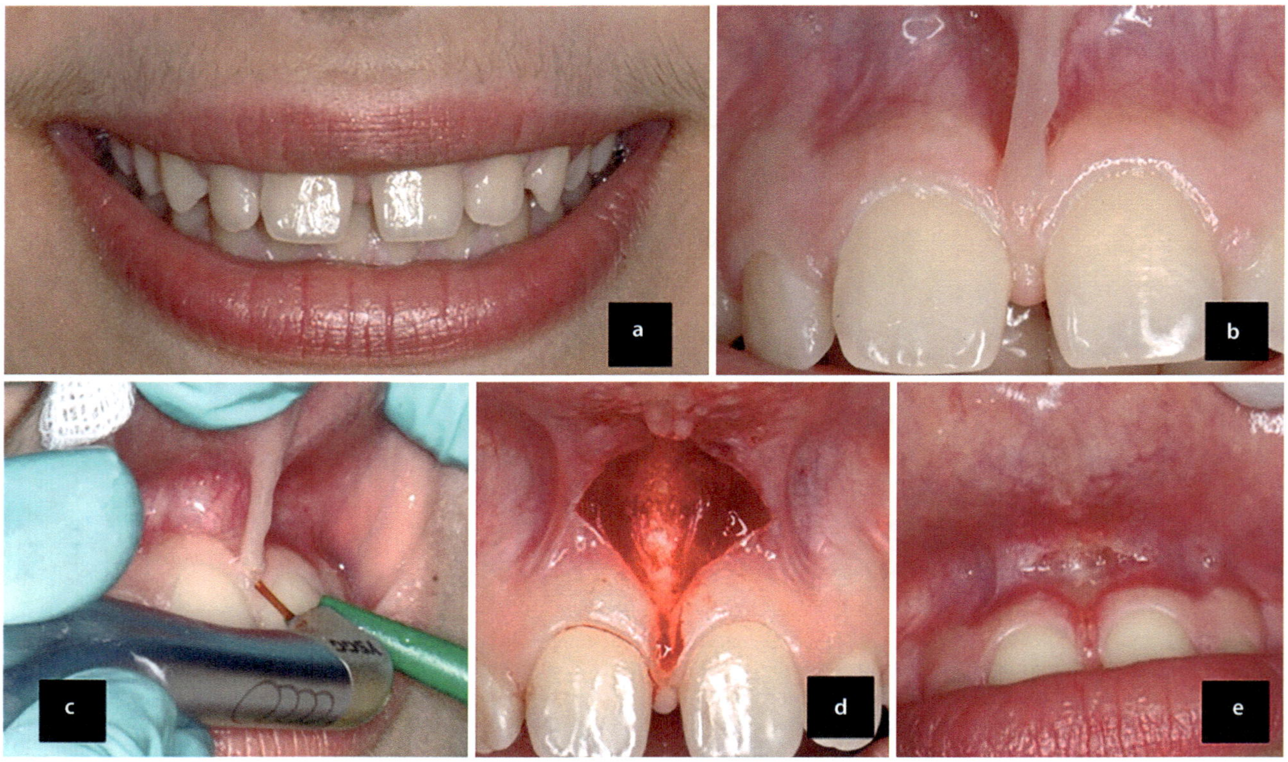

Fig. 11.13 Panagiotis, 12 years old, clear medical history, does not like bleeding (taste) and sutures. Er,Cr:YSGG (2780 nm), gold handpiece, 0.6-mm MZ tip. Distance: 1 mm, close contact, hard tissue mode. **a**, **b** Local anesthesia (4% articaine, 1:200,000 epinephrine). **c** Er,Cr:YSGG (2780 nm) was applied at 1.00–1.50 W, 20 Hz, 30% H_2O, 70% air, 6 mL/min H_2O. **d** Complete coagulation after laser application. **e** After 3 days

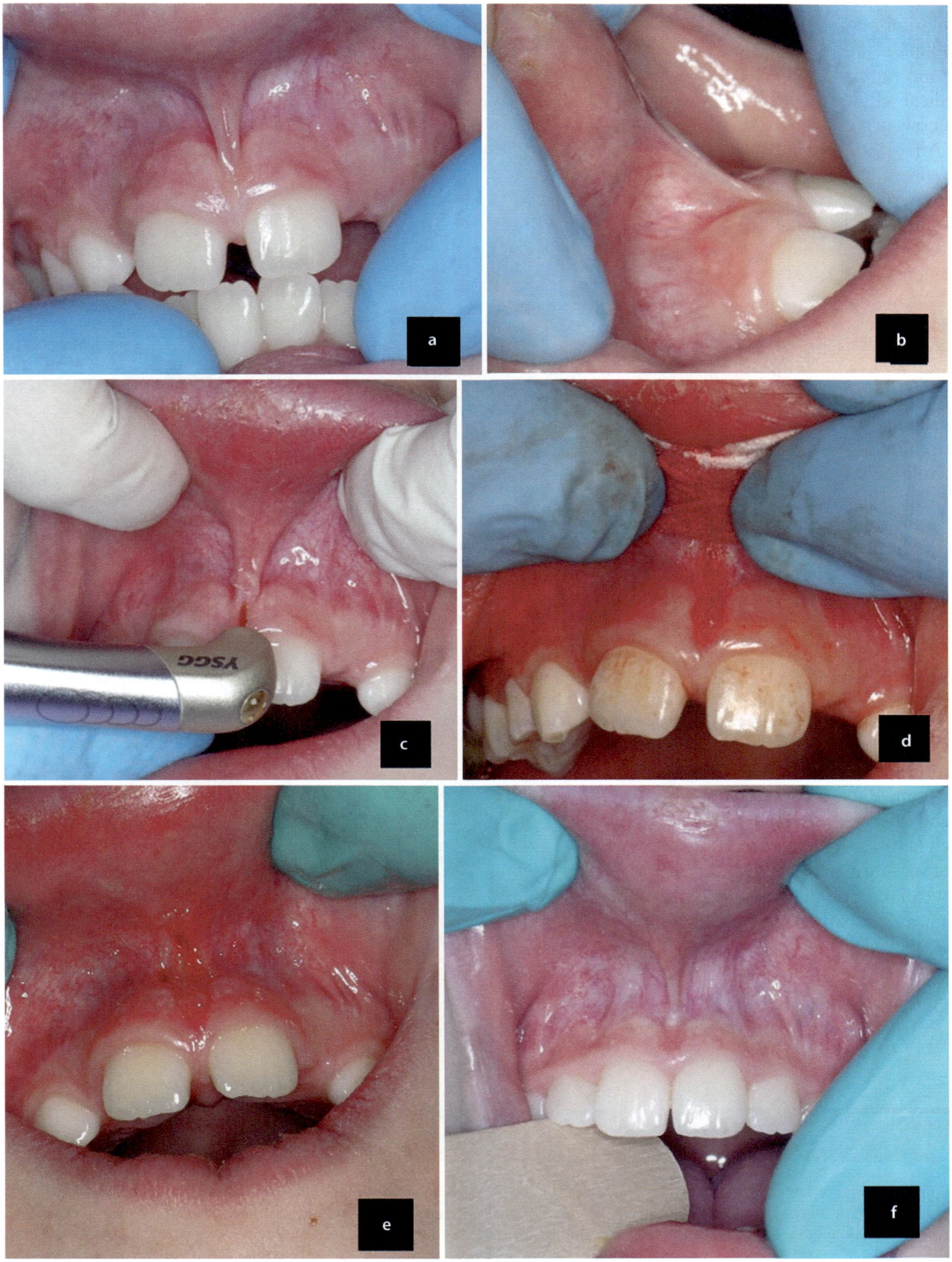

Fig. 11.14 Aggelos, 7.5 years old. **a**, **b** Clear medical history, space between #11 and #21 and no space for #22, not cooperative, does not like bleeding and sutures Er,Cr:YSGG (2780 nm), gold handpiece, 0.6-mm-MZ tip. Distance: 1 mm, close contact, hard tissue mode. **a**, **b** Local anesthesia (4% articaine, 1:200,000 epinephrine). **c** Er,Cr:YSGG (2780 nm) was applied at 1.00–1.50 W, 20 Hz, 30% H_2O, 70% air, 6 mL/min H_2O. **d** Complete coagulation after laser application. **e** After 2 days. **f** After 1 year. Tooth #22 is in place without any orthodontic treatment

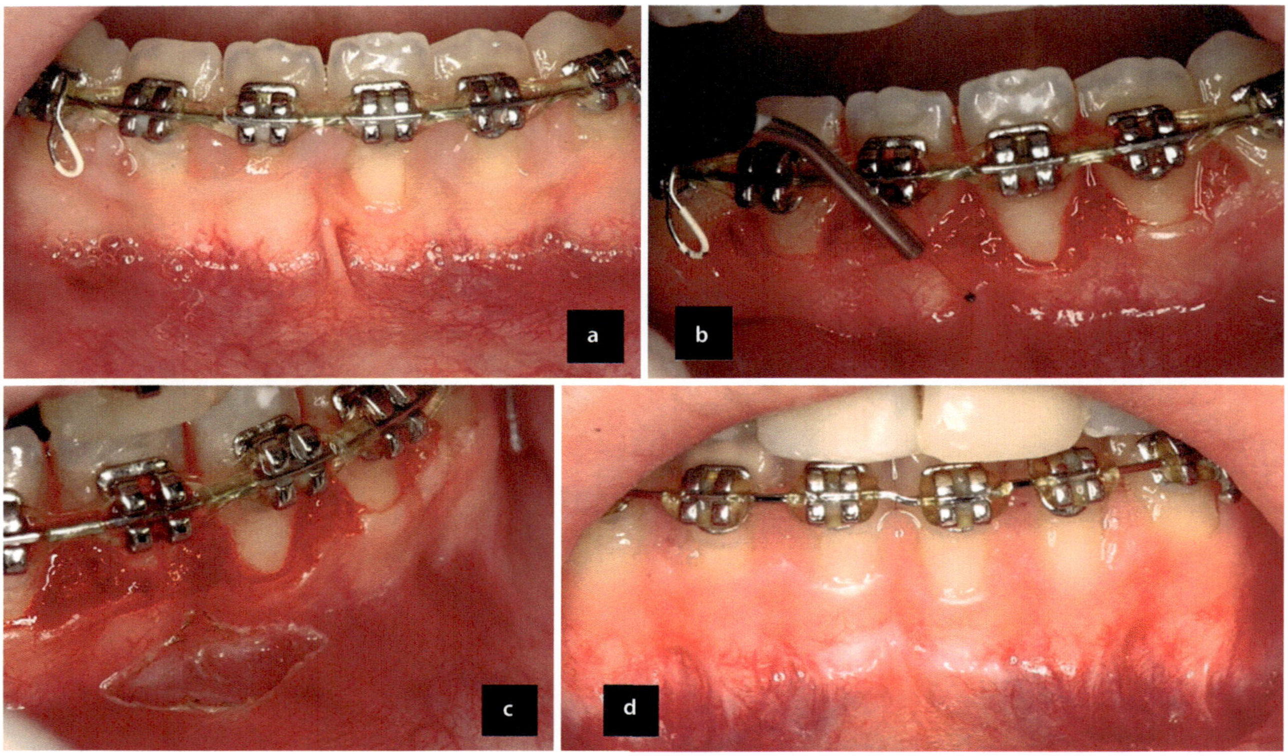

11

Fig. 11.15 Frenectomy and pseudo-pocket removal, 12-year-old girl. **a** Initial intraoral view. Hyperplastic tissue entirely covering the interdental spaces and in contact with the brackets and the orthodontic wire, compromising dental hygiene. Frenum attached too close to the attached gingiva [less than 2 mm], buccal position of the incisor roots and thin biotype, indicating frenectomy to prevent future gingival recession. **b** Using a 1064-nm diode [0.8 W, continuous mode, 300-μm initiated fiber, speed of movement 1-mm/s water cooling every 10 s for 5 s for thermal relaxation] for both surgical procedures. **c** Immediately postop. No carbonization can be observed. **d** Follow-up after 6 months. The distance between the frenum and the attached gingiva in more than 3 mm, stable result, no scar formation

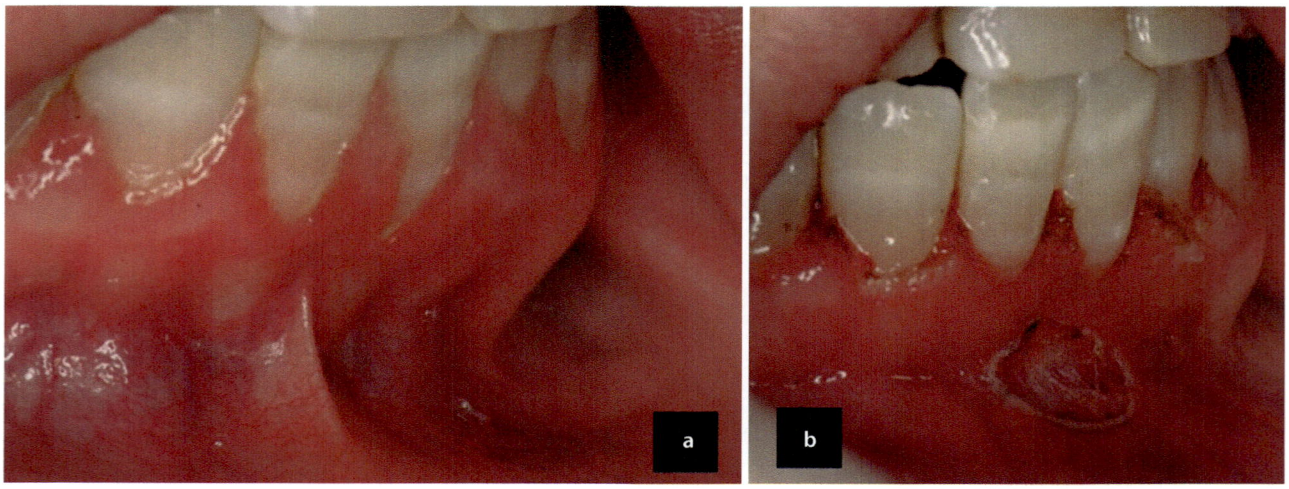

Fig. 11.16 Frenectomy and laser decontamination for gingivitis, 12-year-old girl. **a** Initial intraoral view: gingivitis due to pure hygiene and frenum attached too close to the attached gingiva [less than 2 mm], buccal position of the incisor roots, and thin biotype, indicating frenectomy to prevent future gingival recession. **b**, **c** Using a 1064-nm diode [0.5 W, continuous mode, 300-μm fiber, no initiation, speed of movement 1-mm/s water cooling every 10 s for 5 s for thermal relaxation] for decontamination and a CO_2 laser [2 W, continuous mode, 200-μm spot area at focus, cooling with water every 5 s for 2 s and charring removal with gauze] for the frenectomy. No carbonization and no bleeding can be observed. **d** Follow-up after 1 year. The distance between the frenum and the attached gingiva is more than 3 mm, stable result, no scar formation

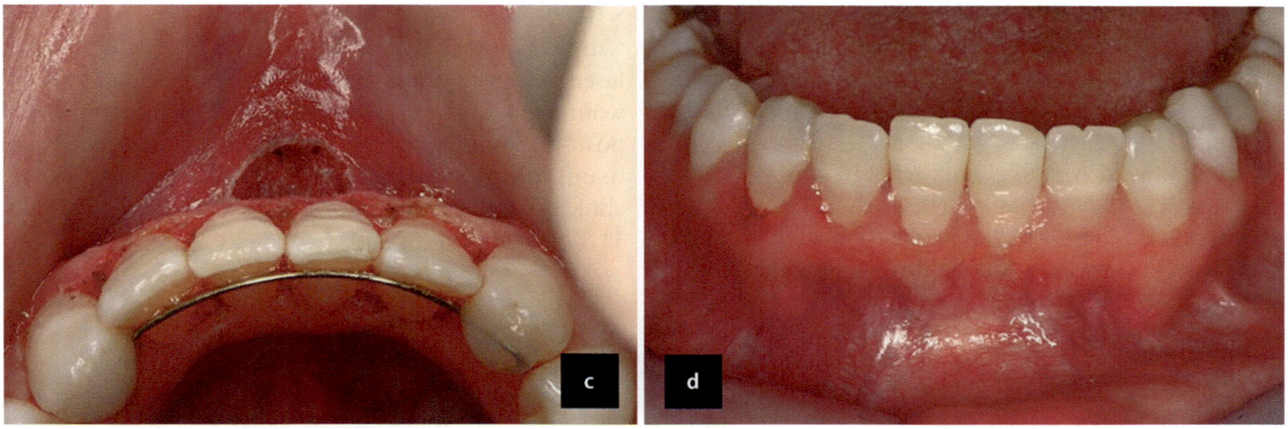

Fig. 11.16 (continued)

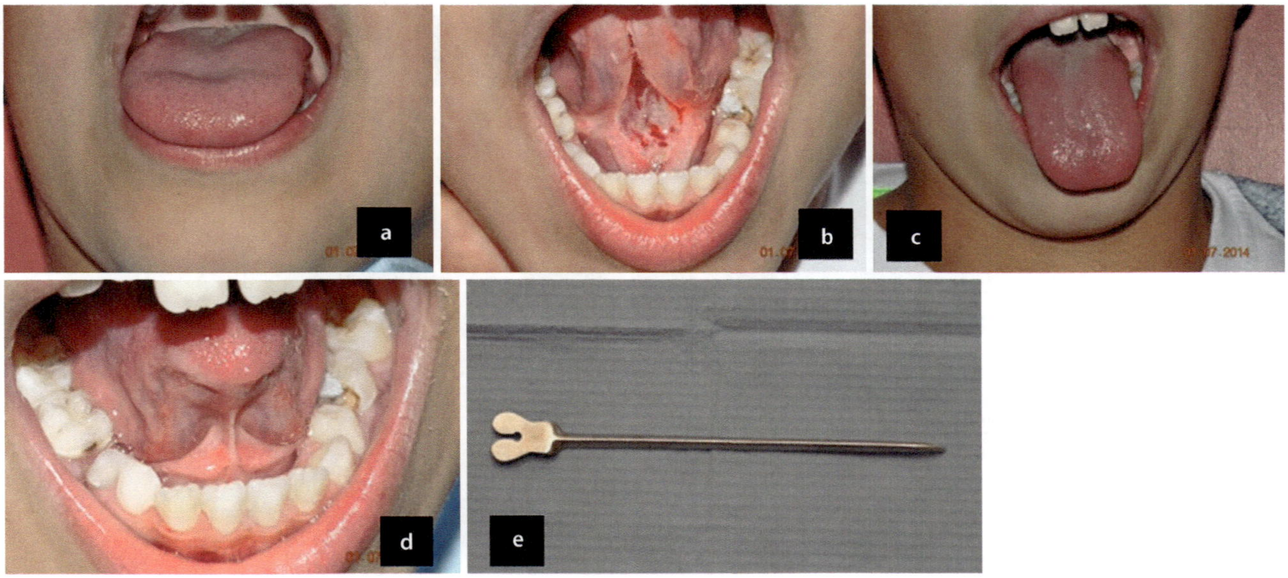

Fig. 11.17 Lingual frenectomy on a 9-year-old male. **a** Tongue mobility restrained to vermillion border. **b** Intraoral view of the lingual frenum. **c** Immediately postop. No suturing needed, no bleeding. Care must be taken to avoid and preserve the sublingual fold and carunculae. **d** Increase in outward tongue mobility immediately postop. **e** "Tongue-tie" instrument, or Sklar Director, or Groove director 5.5″ probe, to hold and stretch the lingual frenum while protecting adjacent structures. *Laser parameters*: pulsed Nd:YAG, average power 3.2 W, pulse 40 mJ, repetition rate 80 Hz, pulse duration 100 µs. Peak power 400 W, fiber diameter 300 µm

Lingual Frenectomy (Fig. 11.17a–e)

Ankyloglossia, or high lingual frenum attachment, or "tongue-tie" is a congenital variation or anomaly in which the tongue has restricted mobility secondary to a short lingual frenum, and cannot be protruded beyond the lower incisors, or the vermillion border of the lips. The restricted mobility of the tongue may also lead to difficulty in swallowing and pronouncing certain consonants that require the tip of the tongue to reach the hard palate [32].

There is no general consensus both on its diagnostic criteria and in the treatment protocols among practitioners of various specialties that examine or treat such patients, including pediatric dentists, oral and maxillofacial surgeons, ENT specialists, plastic surgeons, pediatric surgeons, pediatricians, speech therapists, and so on.

The most widely accepted and used clinical assessment tools are the Hazelbaker Assessment Tool for Lingual Frenulum Function (HATLFF) [33] and the Bristol Tongue Assessment Tool (BTAT) [34].

Ankyloglossia or tongue-tie is also classified into four classes by Kotlow [35] based on the length of the tongue from an insertion of lingual frenum at the base of the tongue to the tip of the tongue: (Normal length is 16 mm.)

- Class I: Mild Ankyloglossia—12–16 mm
- Class II: Moderate Ankyloglossia—8–11 mm
- Class III: Severe Ankyloglossia—3–7 mm
- Class IV: Complete Ankyloglossia—less than 3 mm

Crown Exposure

Crown exposure using proper technique with a dental laser has several advantages such as minimal surgical wound, no flaps, no sutures, disinfection, no or minimal postop pain/discomfort, quality-of-life improvement (pain, eating, speaking), and no antibiotic use. Several clinical examples are shown below (Figs. 11.18a–f, 11.19a–c, and 11.20a–g).

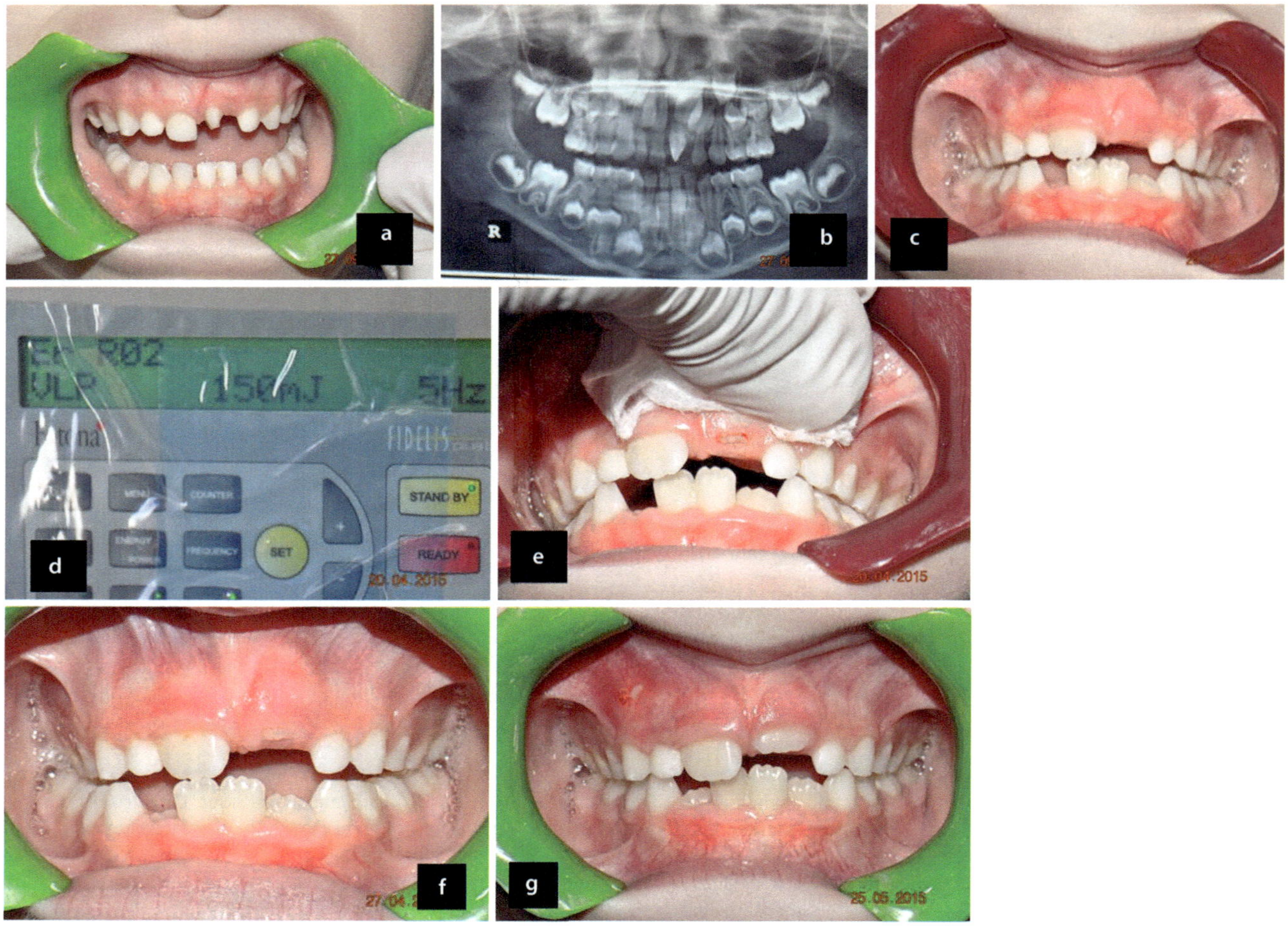

Fig. 11.18 Mesiodens **a**, **b** extracted at age 5 years and 10 months, causing delayed eruption of UL1, which has not erupted 1.5 years after extraction of the mesiodens **c**. The exposure was performed under no local anesthesia with an Er:YAG laser, at low repetition rate (5 Hz), 150 mJ, VLP (1000 μs) pulse de-focussed non-contact, spot size 600 um diameter **d**. A linear opening at the incisal ridge of the crown is sufficient **e**. One week **f** and 1 month F/U **g**

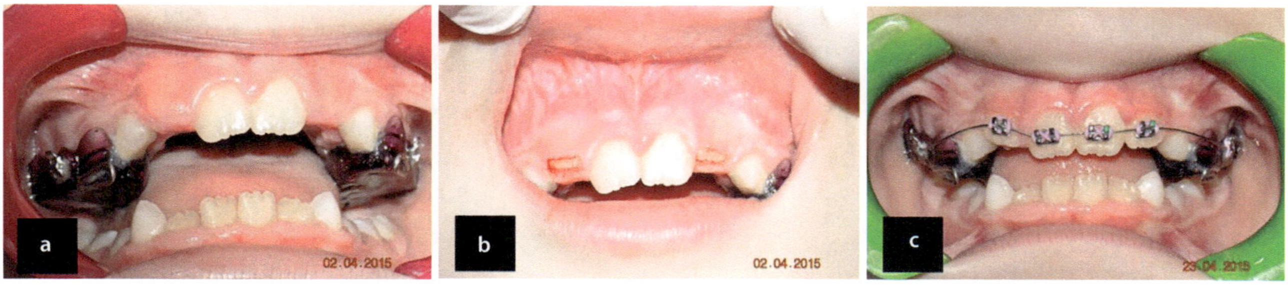

Fig. 11.19 Er:YAG exposure of upper lateral permanent incisors was requested by the orthodontist to accelerate orthodontic treatment **a**. With minimal amount of local anesthesia which can be applied either at the mucobuccal fold or only around the incision site, a higher repetition rate can be used, such as 20 Hz, 150 mJ, and very long pulse de-focussed non-contact, spot size 600 um diameter. Following a minimal incision that runs the length of the incisal edge of the crown **b**, within 3 weeks, the orthodontist was able to bond orthodontic brackets **c**

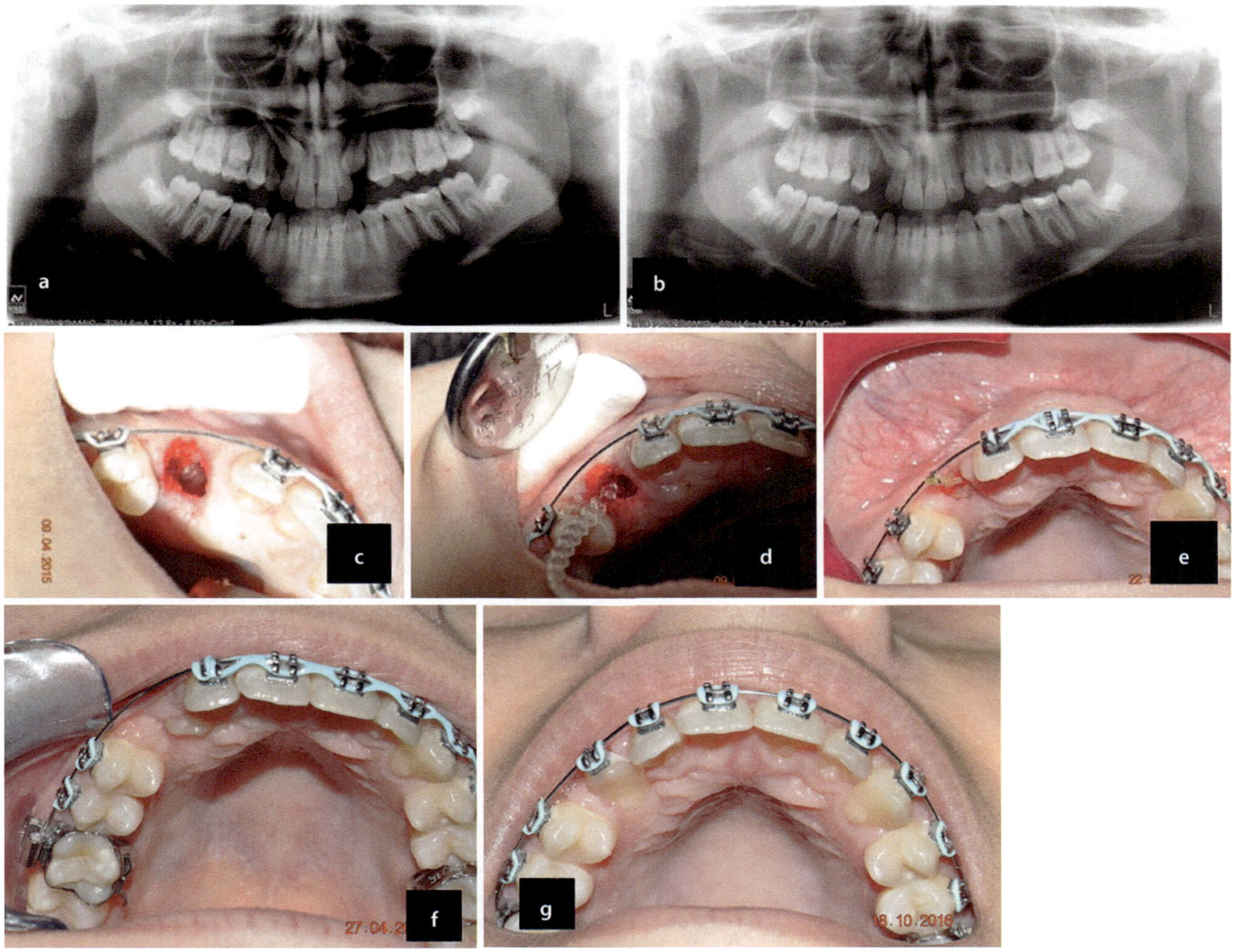

Fig. 11.20 Initial OPG before extraction of the upper right primary canine **a**. One year later, the upper right permanent canine is still impacted **b**. Using a simple SLOB (same lingual opposite buccal) method, it was determined that the distal aspect of the crown lies below the alveolar crest (not shown). All access was performed by an Er:YAG laser. Soft tissues were removed at 150 mJ, 20 Hz, and VLP (1000 µs), with and RO7 handpiece with a sapphire tip, in contact mode without water irrigation. Bone was removed with the same handpiece and tip, but at 200 mJ, 20 Hz, and VSP (100 µs) **c**. Note the minimally invasive surgical access and the dry field with good coagulation **c**. A bracket was bonded to the distal aspect of the crown with a pre-attached elastic chain **d**. In minimal access cases without a surgical flap, it is advisable that a chemically cured resin is used, as limited access does not always allow adequate light curing of the resin material under the orthodontic bracket or button. Note optimal healing in less than 2 weeks **e**; no analgesics were used. One year later **f**, the tooth emerges and is re-bracketed and brought to the arch in 1.5 years after the initial surgery **g**

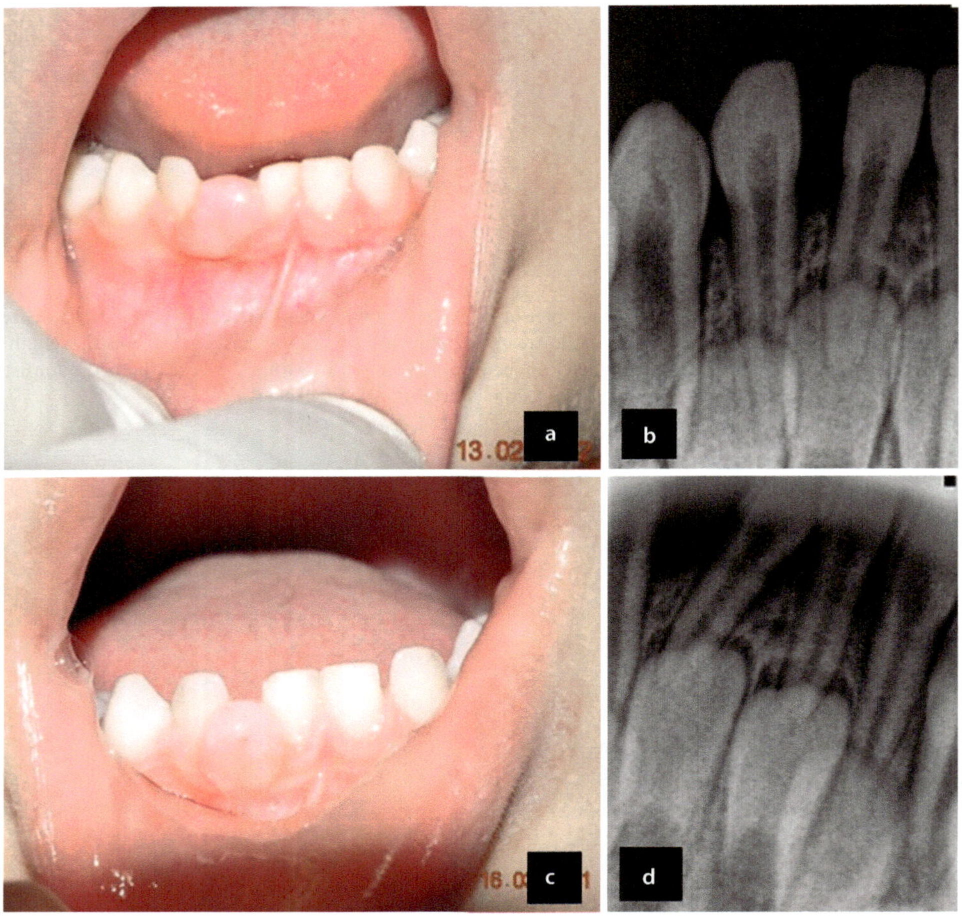

Fig. 11.21 Pyogenic granuloma—fibroma. Exophytic lesion of normal mucosal color on the gingivae between lower central and lateral. Using a pulsed Nd: YAG laser with a 300-μm fiber at 3.2 W, 80 Hz, 40 mJ in contact mode. Primary incisors, causing tooth displacement and bothersome to this 2-year-and-10-month-old boy **a**, **b**. Ten months later, the lesion is causing further displacement of the teeth, and parents agreed to have the lesion removed by laser in the dental office setting **c**, **d**

Exophytic Lesion Laser Excision

See Figures 11.21a–d, 11.22a–d, 11.23, 11.24a–c, 11.25a–c, 11.26a–c, and 11.27a–t.

11.6.2 Pain Management and Wound Healing

Aphthous Ulcers

Aphthous ulcer treatment using proper technique and protocols (Table 11.4) with a dental laser has several advantages such as immediate relief of symptoms, improves quality of life (pain, feeding, speech), accelerates wound healing mechanism, is fast and simple, and requires no use of local or systemic medications (Fig. 11.28a, b).

Soft Tissue Trauma (Figs. 11.29a–c and 11.30a, b)

An unfortunate but all too common interlude to everyday pediatric dental practice is the child who has sustained dental or facial trauma. For both patient and parent, the presenting features may often amount to an impression of greater damage than actually sustained, but, nevertheless, the ability to provide support and initial treatment to combine positive action with empathetic care will go some way to calm the anxious situation. Table 11.5 provides an overview of the advantages offered through the adjunctive use of a dental laser.

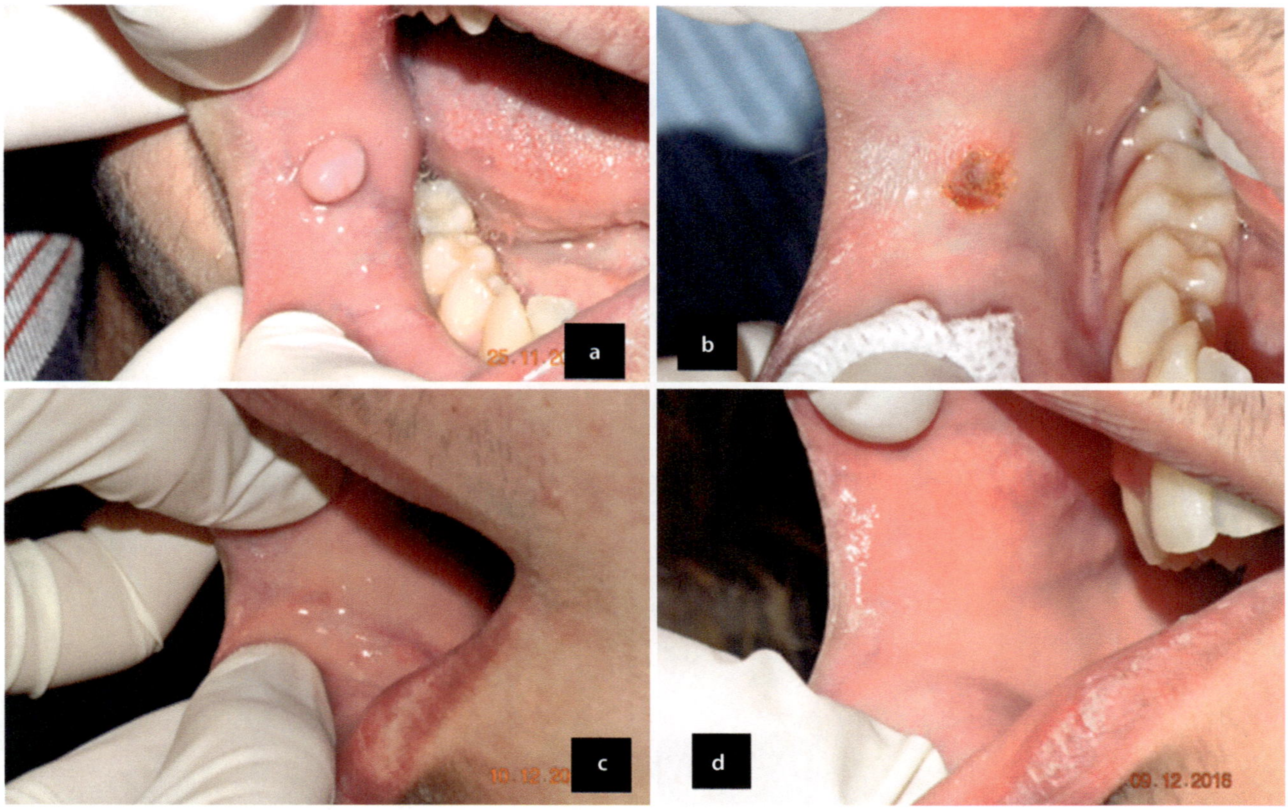

Fig. 11.22 Irritation fibroma removal with Nd:YAG laser in an 18-year-old male. Lesion **a** was excised using a pulsed Nd:YAG laser with a 300-μm fiber at 3.2 W, 80 Hz, 40 mJ in a contact mode holding the lesion by its base with forceps and placing the tip parallel to the tissue base. **b** Immediately postop. **c** Two weeks f/u. **d** One year f/u. Note the quality of healing and the absence of scar tissue

Dentin Hypersensitivity

Dentin hypersensitivity (DHS) is a common condition, described as pain or unpleasant sensation upon intake of cold food or liquids, even with breathing cold air. It occurs as the cervical dentin, and the root may be exposed to external stimuli secondary to periodontal inflammation, gingival recession, iatrogenic (scaling and root planning) causes, abfraction, aggressive toothbrushing, and/or consumption of soft, erosive drinks. It is a rather uncomfortable or debilitating condition that may influence chewing, drinking, and performing oral hygiene and may also have emotional consequences. Its management includes the use of desensitizing agents for office or home use (GLUMA desensitizer, fluoride varnish, HA paste, bioglass) and desensitizing toothpastes (Sensodyne, Emoform KNO_3, Colgate Sensitive Pro-Relief).

The use of Nd:YAG for DHS has more than 20 years of research. Er:YAG was also studied with some success, but Nd:YAG remains the laser of choice for DHS [36–38].

Its mode of action has been attributed to the following:

1. Surface modification and sealing of dentinal tubules to some extent (5–25%)
2. Blocking Aβ and C nerve fibers

Note The same protocol may be used for pulpal analgesia and for alleviating hypersensitivity in teeth with molar incisor hypomineralization (MIH) both for minimizing discomfort between dental appointments and for making dental work better tolerated by the young patient [39]. Teeth with MIH, owing to their porous structure, have been shown to be in a state of continuous mild pulpal inflammation [40].

An 8-year-old patient with MIH shares his experience: "I am Triantafyllos and I am 8 years old. My teeth hurt. I could not eat anything or drink water; neither could I wash my teeth. When I drank cold water, it hurts a lot; when I drank hot water, it hurts a little less, and when I ate food, it hurts even less. When I brushed my

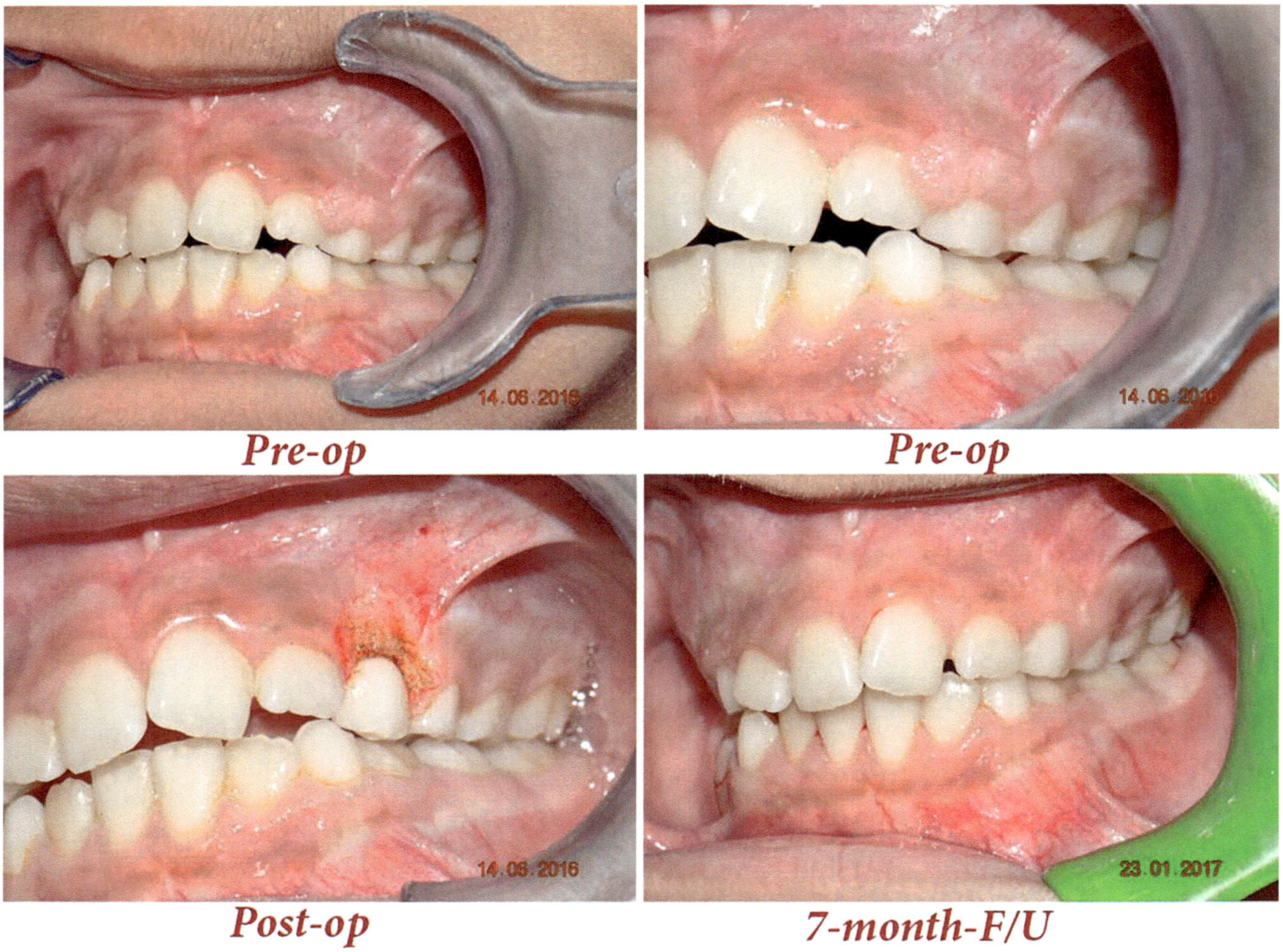

Fig. 11.23 (Published in LAHA (► www.laserandhealthacademy.com) 2017 as is). Oral warts: a 10-year-old girl noticed a "peculiar growth on her gums" that was growing during the last month and was not bleeding upon brushing. On intraoral examination, a 1-cm exophytic band-like lesion following the attached gingiva contour of the left upper primary canine. The lesion was of irregular surface, pedunculated on its base, and differential diagnosis consisted of oral papilloma (oral wart) and pyogenic granuloma. The lesion was removed under local anesthesia with 20-mg lidocaine and 0.11-µg epinephrine infiltration at the mucobuccal fold. The lesion was held with forceps and removed with an Nd:YAG laser (Fotona) at a setting of 4 W, 100 Hz, 40 mJ, VSP (100 µs), and 300-µm fiber from its base. Fluence was 56.6 J/cm^2; peak power was 400 W. There was no bleeding and no need for suturing, and no antibiotics or analgesics were prescribed. Healing was excellent without scarring, no pain or swelling was reported, and no recurrence was noted after 6 months F/U. The biopsy report confirmed the diagnosis of oral papilloma. Figure. Preoperative (Pre-op) and immediately postoperative (Post-op) views of the surgical site. Excellent healing at 7 months without scarring or recurrence

teeth though, I was crying because it hurts a lot when the toothbrush touched my teeth.

Dr. Dimitris did something to my teeth, and they immediately stopped hurting, and I am happy."

It is strongly discouraged to increase the power to 1.5 W or the time of application >1 min as this poses significant risk for thermal damage, microcracking, and carbonization [36].

Case A 15-year-and-8-month male complained of sensitivity only in the front lower teeth. On clinical examination, he showed calculus deposits (grade 1 scale) on upper first permanent molars and lower incisors with mild localized gingival inflammation. Air syringe test for sensitivity was negative on the upper molars but positive on all lower incisors. After running the following laser protocol, the air test was negative immediately after, and scaling was performed without any significant discomfort. The protocol used was with a pulsed Nd:YAG laser, at 100 mJ, 10 Hz, 1 W, 300-µm fiber, and 100-µs pulse duration, for 1 min in a scanning motion around the cervical area of the tooth, noncontact. It is very important to always move the fiber tip during laser emission, as there is evidence that keeping the probe still may result in irreversible damage to the pulp and ultimately necrosis. The sensitivity did not recur after 5 months follow-up. Because of the nature of such cases, the only evidence that can be provided is the patient's testimony, which is hereafter quoted in exact

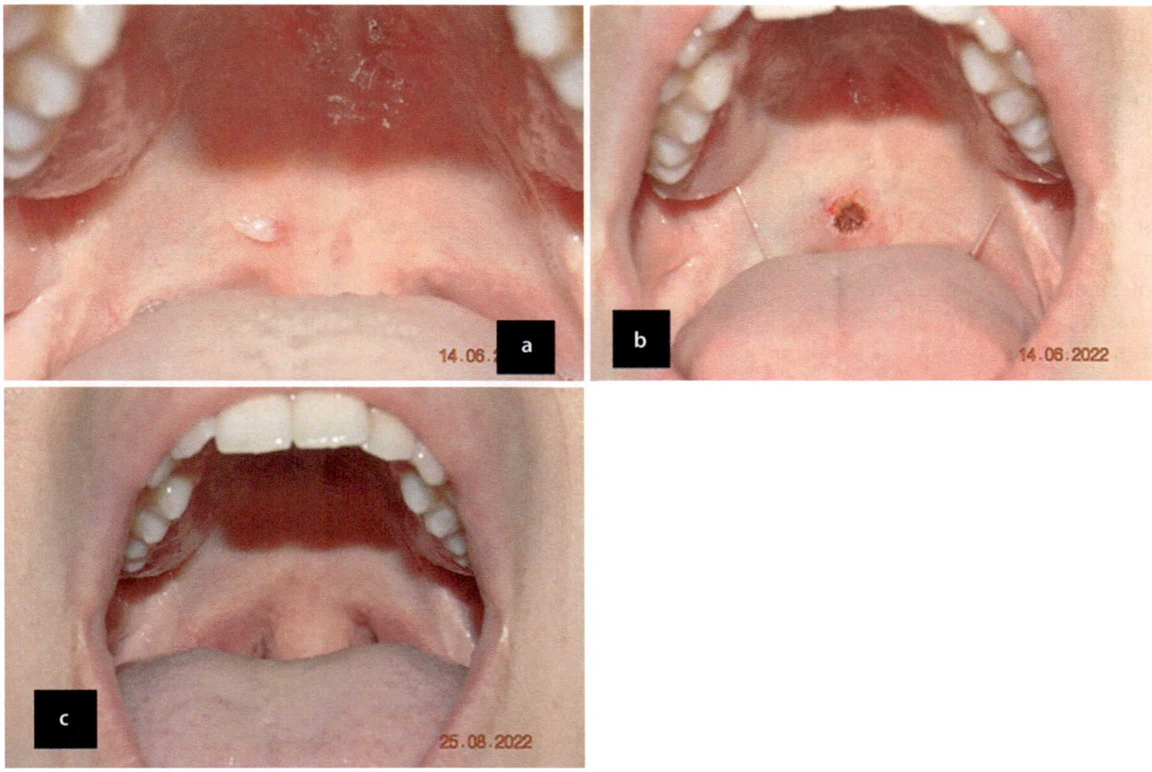

■ **Fig. 11.24** **a** Seventeen-year-old girl presented with an exophytic lesion on the soft palate mucosa. The lesion was pedunculated and whitish, and the clinical impression was oral wart. **b** The lesion was removed under local anesthesia and an Nd:YAG laser at 3.5 W, 100 Hz, 300 μm fiber, and 100 μs pulse duration from its base. No minor salivary gland was detected at the vicinity of the lesion, and the specimen was sent for histopathologic diagnosis, which confirmed the viral origin of the lesion. Postoperative instructions included rinses with mild antiseptic solutions, such as chamomile and Unisept (activated oxygen-based solution). **c** Healing after 9 days

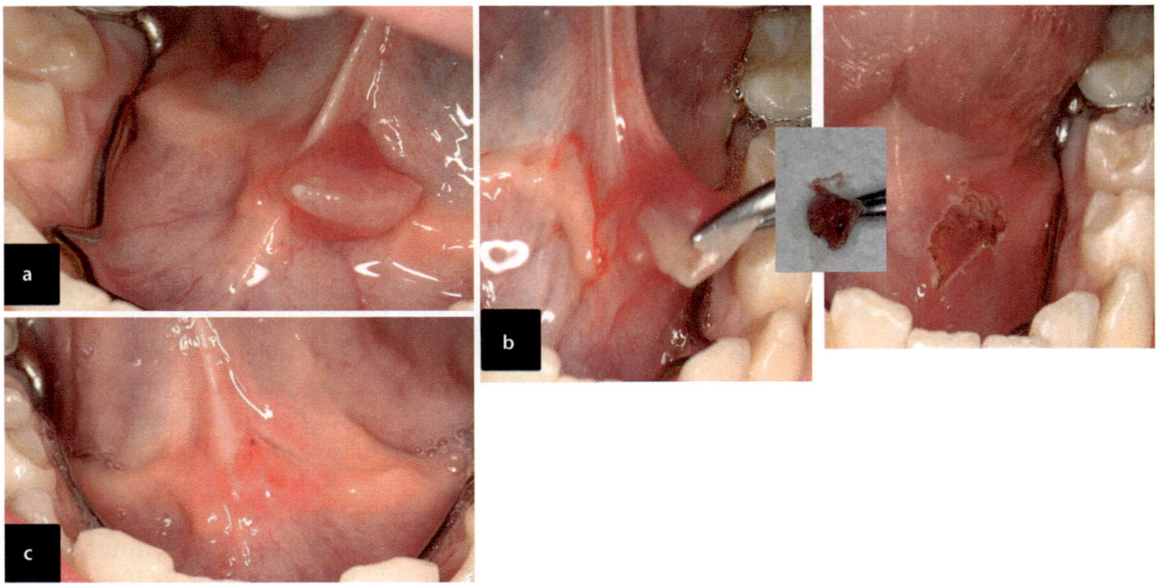

■ **Fig. 11.25** Traumatic fibroma attached to the left sublingual duct removal, 11-year-old boy. **a** Initial intraoral view. **b** Surgical excision using a CO_2 laser [3 W, continuous mode, 200-μm spot area at focus, speed of movement 1.5 mm/s, cooling with water every 10 s for 5 s, and charring removal with gauze]. Extra caution to avoid traumatizing the duct or causing postsurgical edema, which may result in blocking the sublingual duct. No carbonization and no bleeding can be observed. Thus, suturing can be avoided, which is an advantage, since sutures on this area may, too, block the duct. **c** Follow-up after 1 week. Uneventful healing and free sublingual duct

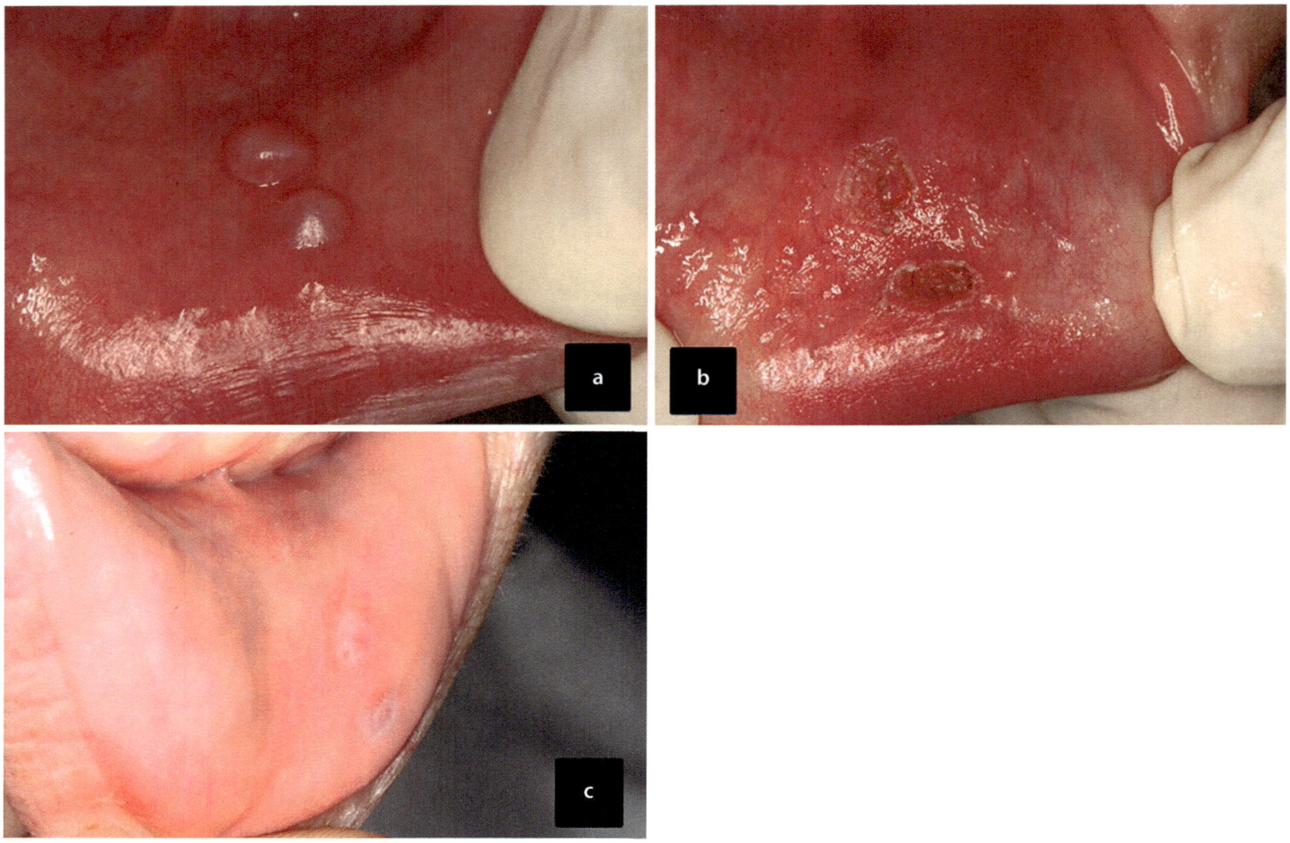

Fig. 11.26 Marsupialization of two mucoceles on the mucosa of the vestibular region of the left lip, 12-year-old boy. **a** Initial situation. Traumatic injury due to orthodontic brackets resulting in superficial blocking of the salivary glands (normal texture under palpation). Major concerns which exclude surgical excision: (a) the number of mucoceles [extended surgical area], (b) the use of sutures [can cause blocking on the surrounding small saliva glands], (c) an excessive postsurgical edema [may increase the traumatic effect of the brackets and result in relapse]. **b** Marsupialization of the two saliva submucosal cysts using the CO_2 laser 10,600 nm with minimal surgical trauma [2 W, CW, speed of movement 1.5 mm/s, at focus, 200 μm, total working time 1 min]. **c** Follow-up after 1 week. Uneventful healing and normal function of the small salivary glands

translation. "For a long period of time, I had big difficulty drinking cold beverages, because of my tooth sensitivity. However, with the use of lasers, Dr. Velonis made that sensitivity a distant memory" (*EM, 15-year-old*).

Dry Socket

In cases where a dry socket develops after a difficult tooth extraction, the Er:YAG can offer immediate relief from the pain by removing the necrotic tissue from the socket. The settings used are similar to those of dentin removal, i.e., 150–200 mJ, 10–15 Hz, and 100-μs pulse duration, with water irrigation and a contact handpiece. It does not need local anesthesia. Light bleeding occurring after removal of the necrotic tissue is beneficial for the healing process. Usually, a single application is adequate to manage a dry socket.

Incision and Drainage of Abscess and Fistula

Drainage of an abscess of dental/periodontal origin can be performed by a laser with minimal topical or local anesthesia and minimally invasive access.

With an Er:YAG laser, the opening can be performed with 80–120 mJ, 10–15 Hz, and 50–1000-μs pulse duration. If the abscess is under the periosteum, adequate local anesthesia is required, and the Er:YAG is the laser of choice to proceed with 150–200 mJ, 10–20 Hz, and 100-μs pulse duration with a contact or noncontact handpiece.

When using near-infrared laser, local anesthesia is required. With an Nd:YAG, laser settings range from 4 to 5 W, 70 to 100 Hz, and 100 μs to 300 μm fiber. With a diode 810/980-nm laser, use 4–6-W in continuous wave.

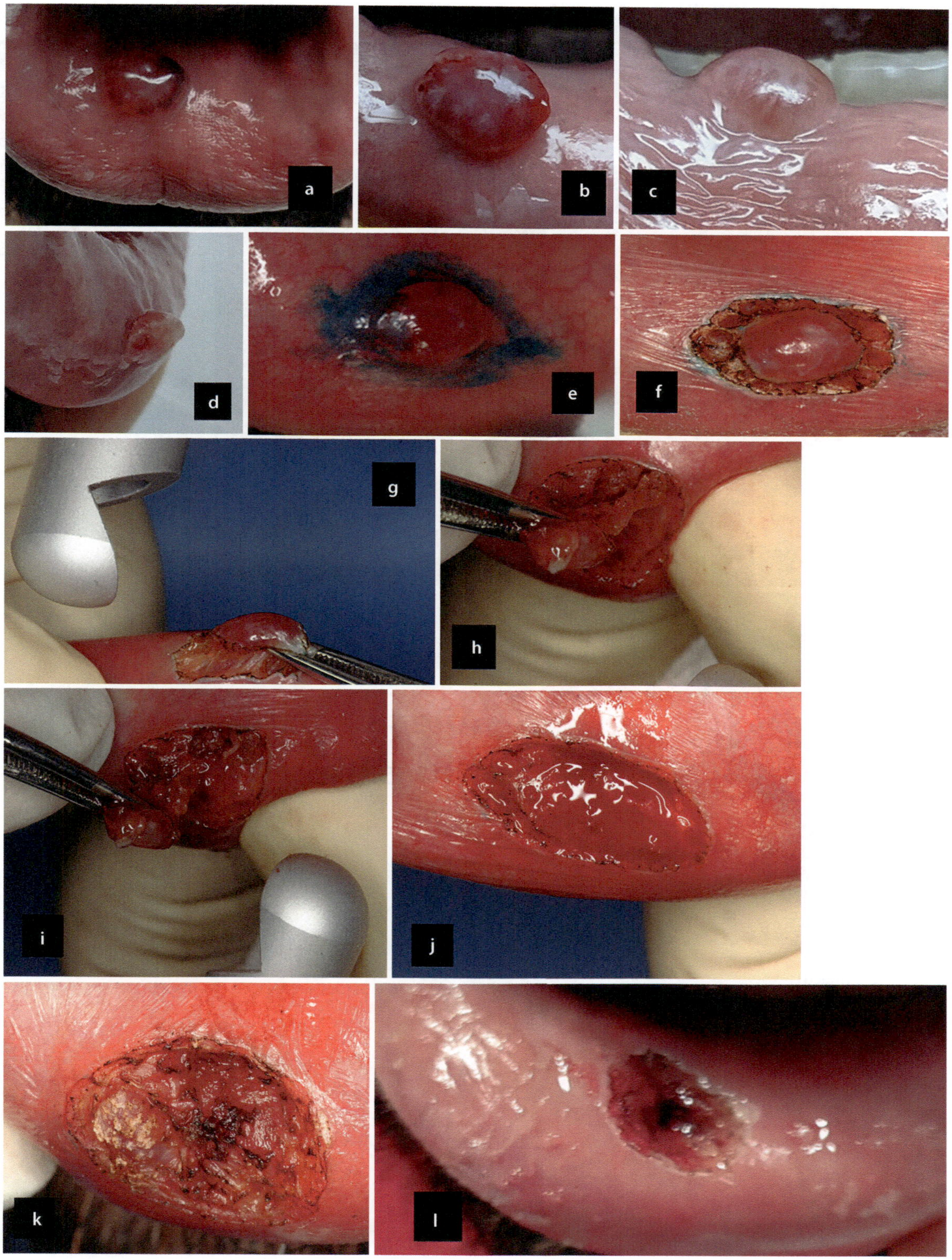

Fig. 11.27 Excision of one traumatic medium-size mucocele located superficially on the oral mucosa of the lower lip, 16-year-old boy. **a–d** Intraoral view. **e** Drawing the line of the incision. **f–j** Step-by-step surgical procedure: laser excision using a CO_2 laser 10,600 nm [2 W, CW, 200-μm spot area at focus, speed of movement 1.5 mm/s, total irradiation time 4 min], avoiding the direct irradiation of the mucocele. Every 20 s of irradiation, 40 s water cooling. **k** Coagulation for bleeding control using the CO_2 laser [same as above only on bleeding spots] and immediate postop view. **l–t** Day-by-day healing up to 3 months: 12 h **l**, 24 h **m**, 2 days **n**, 5 days **o**, 7 days **p**, 10 days **q**, 2 weeks **r**, 3 weeks **s**, and 3 months **t**. No scar formation

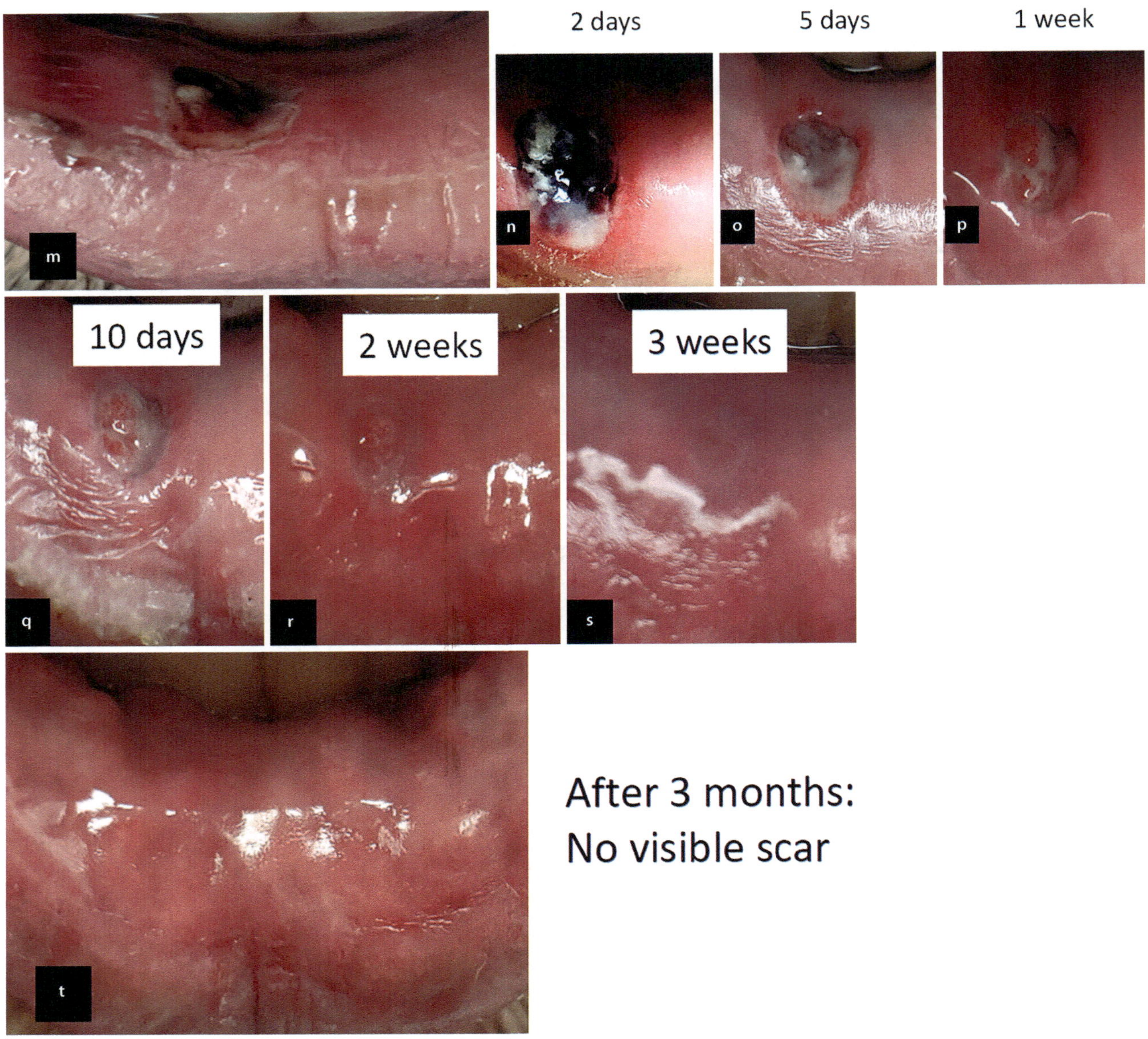

Fig. 11.27 (continued)

Table 11.4 Protocols of aphthous ulcer treatment

	Nd:YAG laser	Diode 810/940	Er:YAG, Er,Cr:YSGG
Power	2 W	Power: 1–2 W	Power: 0.45–0.75 W
Pulse E	100 mJ	CW (continuous wave)	30–50 mJ—no H_2O
Pulse frequency	20 Hz		15 Hz
Pulse duration	100 µs		
Fiber diameter	300 µm	Fiber diameter: 300 µm	Mirror or sapphire tip 0.8–1 mm
Mode	8–12 mm defocused	8–12 mm defocused	8–12 mm defocused
Movement	30 s/cm^2	30 s/cm^2	2–3×, white superficial layer

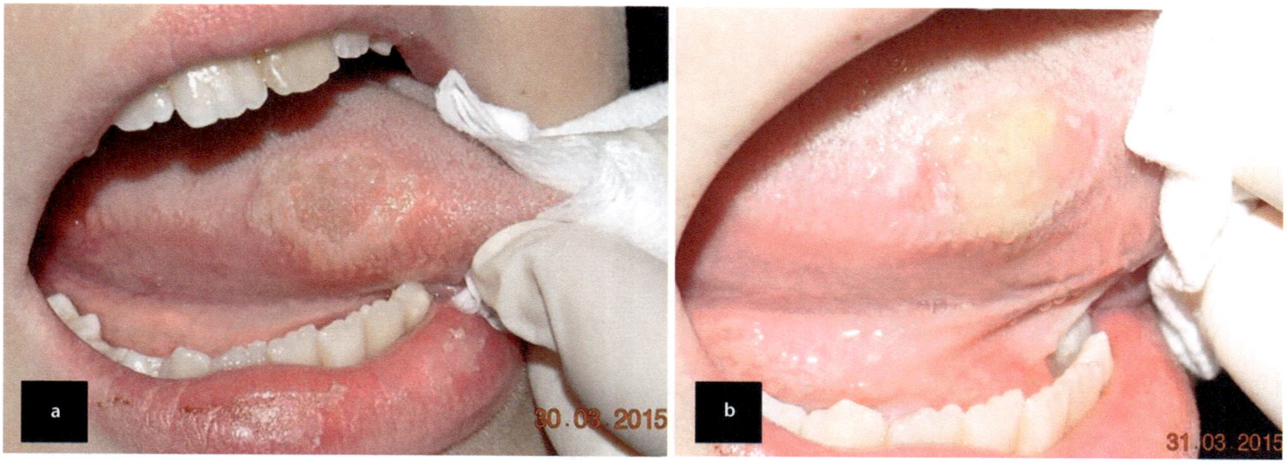

Fig. 11.28 Nd:YAG, before **a** and 1 day F/U **b**. Note beginning of fibrin formation to heal the ulceration

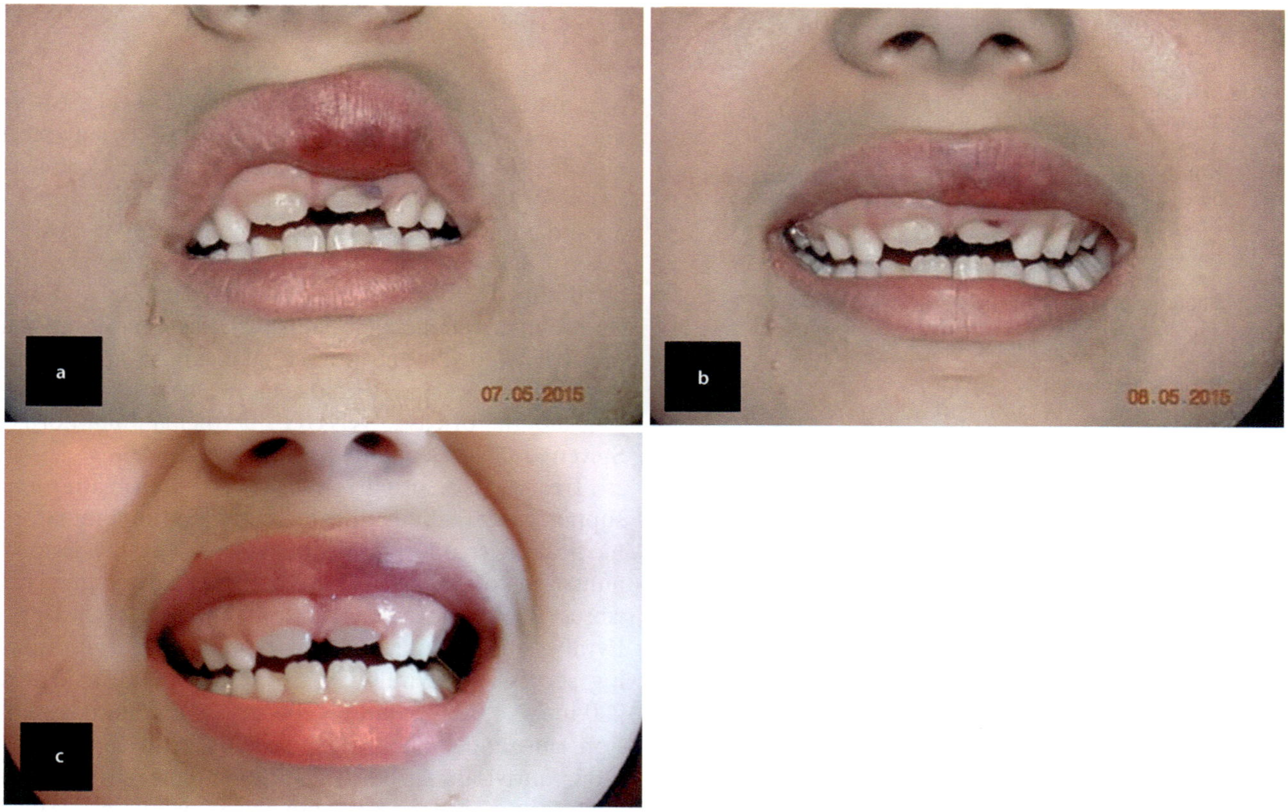

Fig. 11.29 Seven-year-old girl collided with schoolmate and sustained only a soft tissue injury. She was asked to not take any anti-inflammatories, or other topical ailments, to see the effect of photobiomodulation (PBM) treatment over the course of 3 days. A pulsed Nd:YAG laser was used with a fiber of 6-mm-diameter (R24 handpiece and fiber, Fotona). The settings were 75 mJ, 20 Hz, and 100-μs pulse duration. Average power was 1.5 W and peak power was 750 W, but energy density (fluence) was 0.27 J/cm^2 because of the large spot size of 6 mm. **a** Day 0, **b** day 1, **c** day 3

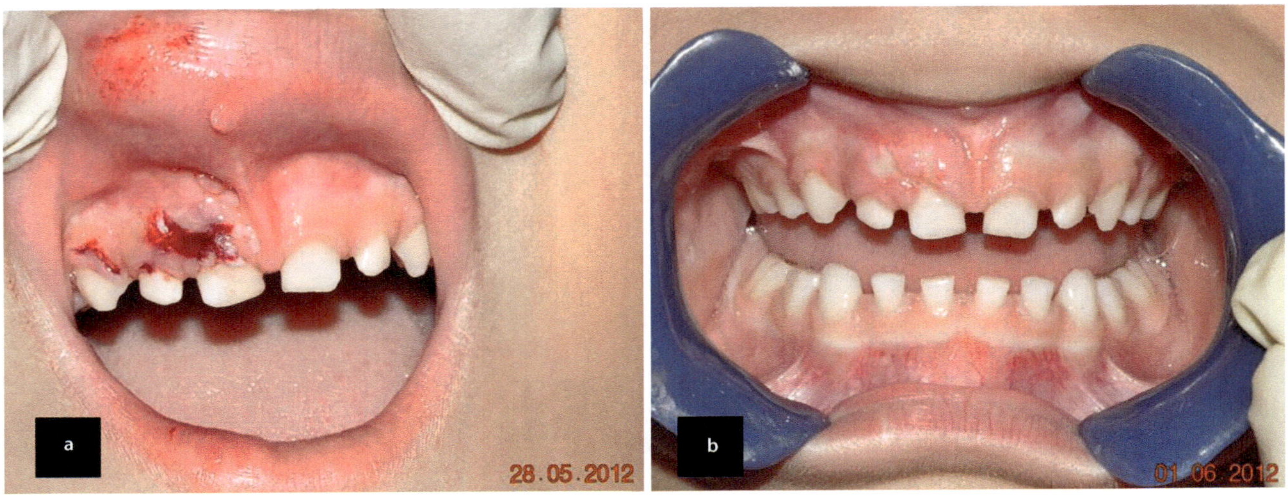

Fig. 11.30 a, b Five-year-old sustained intraoral lacerations at the swimming pool. Settings for aphthous ulcers were used, high-intensity laser treatment (HILT) to accelerate wound healing with a pulsed Nd:YAG laser, at 2 W, 20 Hz, with a 300-μm fiber, following the aphthous ulcer protocol described earlier in this chapter. Note the healing of the lacerations after 4 days

Table 11.5 Summary of laser advantages

Soft tissue trauma using proper technique with a dental laser has several advantages
• Immediate relief of symptoms
• Improves quality of life (pain, feeding, speech)
• Accelerates wound healing mechanisms
• Fast and simple
• No use of local or systemic medications
• Can be done with a number of less expensive diode lasers

11

11.6.3 Oral Infection Management

Herpetic Gingivostomatitis

Primary herpetic gingivostomatitis (PHGS) is caused primarily by HSV I and II, Herpesviridae that can cause latent or lytic infections. The first encounter with the virus usually occurs between ages 5 months and 6 years. It is common in children and 95% is caused by HSV 1 and 5% by HSV 2. While 90% of infections can have a subclinical course, 10% exhibit a symptomatic infection with 7–14-day duration and of varying severity.

The clinical characteristics are intense gingival erythema and bleeding; oral vesicles, which rupture leaving painful ulcers; and lip fissuring. Common locations are the gingivae, tongue, palate, lips, buccal mucosa, and floor of the mouth. General symptoms include fever (38–40 °C), lymphadenopathy, malaise, headache, weakness, oral malodor, anorexia, irritability, and difficulty eating and swallowing even liquids. As the lesions may start near the stomato-pharynx, it is often mistaken for pharyngitis or tonsillitis and often is mistreated with antibiotics and antifungals. As it is contagious, caution must be taken to avoid autoinoculation to the eye, cornea, and hand. Inoculation in the eye can be very painful and may lead to permanent ocular damage. Other conditions that may be included in the differential diagnosis are herpetiform ulcerations and aphthous ulcers, herpangina, hand-foot-mouth disease, ANUG, erythema multiforme, Stevens-Johnsons, and infectious mononucleosis.

Treatment is usually palliative, symptomatic, and supportive with fluids, anti-fever, and anti-inflammatory medication (paracetamol, ibuprofen). For the lesions, numerous topical agents have been used: dyclonine HCl rinse, viscous lidocaine, Mundisal, methylene blue (ink), Pyralvex, Tantum Verde (benzydamine HCl), with varying success. Antivirals per os are also used: aciclovir, 200 mg 4–6×/day, children 200 mg 4×/day until resolution, rinse and swallow, effective in the prodromal phase. Other antivirals, such as valaciclovir, femciclovir, penciclovir, and ganciclovir, are also being prescribed (Figs. 11.31 and 11.32).

After the initial infection, the virus enters a latency phase in 70–90% of cases in Gasser's ganglion to reemerge as herpes labialis ("cold sore" or "fever blister").

An alternate management/treatment of PHGS can be done with a laser. With an Nd:YAG laser (Table 11.6), the parameters and settings are shown.

Mode of Action Early in vitro studies have indicated that the laser energy may activate leukocytes to inactivate the

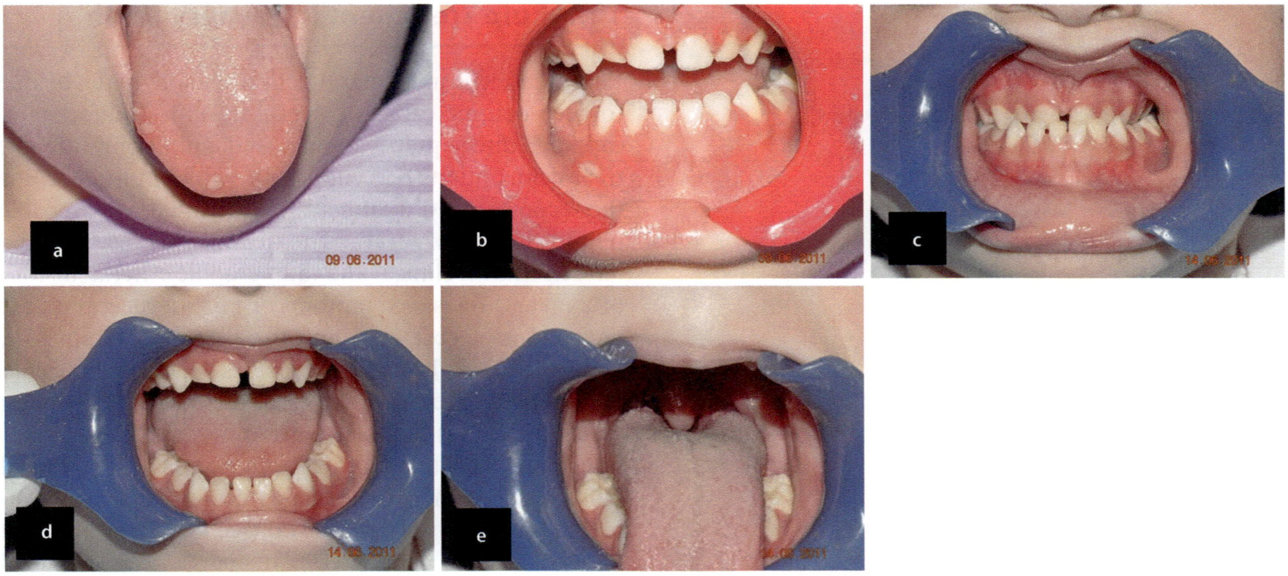

Fig. 11.31 Case #1. Four-year-and-2-month-old female presented to my private practice on September 6, 2011, with her mother. Chief complaint (CC) and HCC: her mouth and gums hurt for 2 days; she has fever 38.5 °C and does not want to eat or drink or brush her teeth. Her gums bleed easily and there is halitosis. Patient was referred from her pediatrician who had prescribed topical antifungals and antibiotics, which she did not yet take. PMHx and PDHx: no significant medical history. Diagnosis: clinically (not microbiologically or serologically) documented primary herpetic gingivostomatitis with appearance of painful vesicles and bullae that ulcerate in the labial mucosa, dorsum of the tongue, and swelling and bleeding of the gums. Mother had recently had an episode of recurrent herpes labialis. Treatment plan: laser (Nd: YAG) irradiation of oral lesions. No use of antibiotics or antifungals or antivirals. **a**, **b** Day 3 of primary herpetic gingivostomatitis, initial exam. **c**, **d** Day 8 of primary herpetic gingivostomatitis, 5 days post-laser irradiation. **e** Day 8 of primary herpetic gingivostomatitis, 5 days post-laser irradiation—only mild difficulty swallowing secondary to lesion in the palatal pillar which was not seen at the first visit. Irradiated this lesion at this appointment, and the patient felt immediate relief

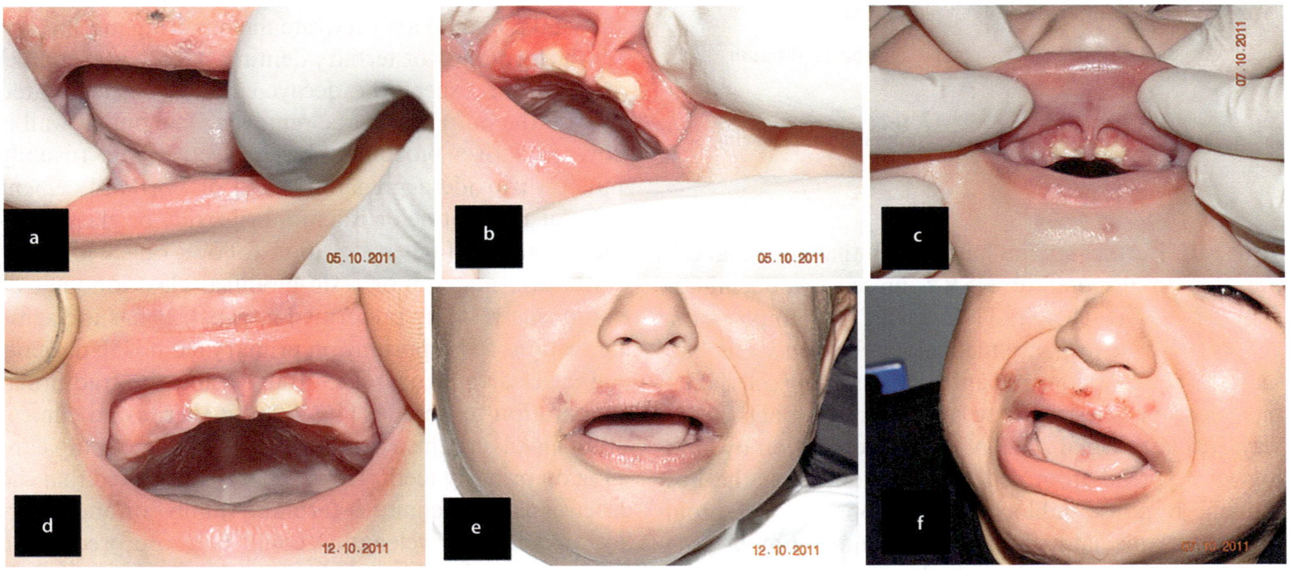

Fig. 11.32 Case #2. Primary herpetic gingivostomatitis treatment with Nd:YAG laser (DEKA). **a**, **b** Seven-month-old male presented on ca. Day 9 of infection. He does not drink water or milk and cries most of the time; parents and child do not sleep well; still fever >38 °C and was prescribed Daktarin gel. May 10, 2011: first laser treatment. Pre-cooperative (Frankl 1). Peak power 1500 W. **c**, **d** July 10, 2011: second laser treatment for few new blisters. Improving, parents reporting that he finally smiled on the night of the first treatment and slept well and started talking again. Difficult with water on that same day but drank water the next day. **e**, **f** December 10, 2011, F/U

Table 11.6 Laser (Nd:YAG 1064 nm) parameters for herpetic gingivostomatitis

Laser parameters for herpetic gingivostomatitis
• Operational mode: pulsed Nd:YAG
• Repetition rate: 20 Hz
• Pulse energy: 150 mJ
• Average power: 3.00 W
• Pulse duration: 100–350 μs
• Fiber diameter: 300 μm–6 mm
• Treatment duration: 30 s/cm^2
• Mode: noncontact, defocused at 8–10-mm distance

Table 11.7 Advantages of laser-assisted treatment of herpetic gingivostomatitis

Advantages of herpetic gingivostomatitis treatment using proper technique with a dental laser
• No medication use
• Immediate relief from pain
• Immediate resuming of eating and drinking
• Afebrile since day 1 of treatment
• Fast healing of lesions, no full-blown development
• Avoidance of dehydration and possible hospitalization
• Fast, painless, acceptable treatment
• Parents do not lose their sleep, family, and work benefits

virus [41]. Further studies have shown that epithelial cells are able to respond to HSV-1 presence inducing the expression of IL-6, IL-1, TNF-α, and IL-8, important in the acute-phase response mediation, chemotaxis, inflammatory cell activation, and antigen-presenting cells [42]. It has been hypothesized that laser irradiation acts in two ways, (a) in the final stage of HSV-1 replication by limiting viral spread from cell to cell and (b) on the host immune response unblocking the suppression of proinflammatory mediators induced by accumulation of progeny virus in infected epithelial cells (Table 11.7).

11.7 Pulp Treatment in Primary Teeth

Pulp treatment in primary teeth is usually required due to deep dentine caries or dental trauma. Indirect pulp capping, direct pulp capping, and pulpotomy are the treatment options for vital teeth, while pulpectomy is the recommended treatment for necrotic or irreversible pulpitis in primary teeth. The treatment choice is based on well-known clinical and radiographic criteria: history of pain and signs or symptoms of pulp degeneration are indications of necrotic pulp or irreversible pulpitis [43]. The use of the erbium family laser is beneficial for cavity preparation, especially in cases of teeth with deep dentin caries, due to (a) the selective and minimal tooth structure removal aiming to avoid unnecessary mechanical pulp exposure and (b) the facility of dentine decontamination and smear layer removal (see Tables 11.1 and 11.2 for energy parameters). In many cases, no local anesthesia is required when erbium family lasers are used. In addition, for all the above reasons, when interim therapeutic restorations (ITR) [44] is the choice of contemporary treatment in order to prevent the progression of dental caries on uncooperative patients, the use of lasers could be beneficial.

11.7.1 Indirect Pulp Capping

The goal of the technique is to preserve the integrity of the vital pulp and also activate the repairing mechanism for the formation of tertiary dentine. All decayed enamel and dentine except the decayed dental tissue located next to the pulp have to be removed. The pulpal wall is covered with a biocompatible protective base (usually mineral trioxide aggregate (MTA) or Portland cement (PC) or Biodentine or calcium hydroxide or glass ionomer), and the final restoration follows (glass ionomer restorative material or resin-modified glass ionomer or composite resin or preformed crowns). It has the same indications to pulpotomy on primary teeth [45], presenting success rates up to 83–100% using the traditional preparation techniques [46, 47], but there is no clinical study involving the use of laser at the indirect pulp capping on primary teeth. However, it is speculated that the laser-assisted technique (erbium family or/and near-infrared laser wavelengths) could be more predictable and successful due to decontamination of the cavity, the remaining carious dentine, and the positive effect on pulpal tissue healing and recovery [48]. See ▫ Tables 11.1 and 11.2 for laser wavelength parameters for deep dentine removal (erbium family) and decontamination.

11.7.2 Direct Pulp Capping

When the vital pulp is exposed because of mechanical caries removal or trauma, direct pulp capping could be performed. However, direct pulp capping is not recommended for primary teeth [43]. The success rate of the traditional techniques is 70–80%, using MTA or PC or Biodentine or calcium hydroxide as pulp dressing material, while usually there are acute edema and pain after 7–15 days in case of failure [33]. Therefore, there is a general recommendation to avoid direct pulp capping in primary teeth and perform pulpotomy, in case of any size of pulp exposure [43, 49]. Successful laser-assisted direct pulp capping cases have been reported [48], but there is no clinical study involving the use of any laser wavelength in such a treatment on primary teeth. Following cavity preparation using a diamond bur or a laser from the erbium family, the laser-assisted technique (erbium family, diode, Nd:YAG, CO_2) is introducing pulp tissue coagulation (erbium family: 50 mJ, 10 Hz, no water, 40% air, defocus for 5–10 s) along with decontamination before the placement of pulp dressing [48]. After laser-assisted direct pulp capping, it is expected that better pulpal healing occurs than with the transitional technique; the pulp will retain its vitality and perform the formation of tertiary dentine. See ▫ Tables 11.1 and 11.2 for laser wavelength parameters for decontamination.

11.7.3 Pulpotomy

The traditional technique of pulpotomy has clinical success rates up to 98–100% (MTA/PC/Biodentine or ferric sulfate (FS)) and is the most common technique performed after pulp exposure on vital primary teeth with deep carious dentine lesions. The technique involves the removal (amputation) of the coronal pulp tissue with burs and spoon excavator, achievement of hemostasis using sterile cotton pellets, and placement of MTA/PC/Biodentine or FS over the pulp stumps [43]. When the bleeding from the pulp stumps could not be controlled, it is an indication of irreversible pulpitis beside the absence of clinical and radiographical symptoms, and pulpectomy is indicated. Formocresol had been used for several years (before the wide use of FS) with great success, but its use is not currently recommended due to possible carcinogenic effect. The MTA/PC/Biodentine is covered by glass ionomer, while in the case of FS, a fast-setting zinc oxide and eugenol paste (IRM) is placed over the pulp stumps before the placement of the final restoration. FS forms a ferric ion and protein complex on contact with blood, providing a bridge between the vital root canal pulp tissues, and the paste contains eugenol (IRM), while the biocompatible and also bioinductive MTA/PC/Biodentine has to be in contact to the pulp tissue.

Alternatively, instead of using medicaments like FS, laser (erbium family, diode, Nd:YAG, CO_2) could be used for the pulp tissue coagulation over the pulp stumps before the placement of IRM (▫ Fig. 11.33a–g). Clinical studies show that either there is no significant difference in success rate (clinical or radiographically) between laser-assisted and traditional pulpotomy or the result is in favor for the laser-assisted method [50–57]. After coronal pulp was removed with burs and spoon excavator and hemorrhage was controlled, a type of laser [diode (five studies), Er:YAG (one study), Nd:YAG (three studies), CO_2 (one study)], using a variation in laser application parameters (power, frequency, exposure time) and capping materials (MTA, zinc oxide eugenol, IRM) report success rate of *laser-assisted pulpotomy* (follow-up period from 1 to 66 months) ranged from 71.4% to 100% clinically and 71.4% to 100% radiographically. The amputation through vaporization of the coronal pulp tissue using lasers (erbium family, diode, Nd:YAG, CO_2) is not recommended because they create coagulation and necrotic tissue which may camouflage possible inflammation or necrosis of the root canal pulp. See ▫ Tables 11.1 and 11.2 for laser wavelength parameters for decontamination (erbium family).

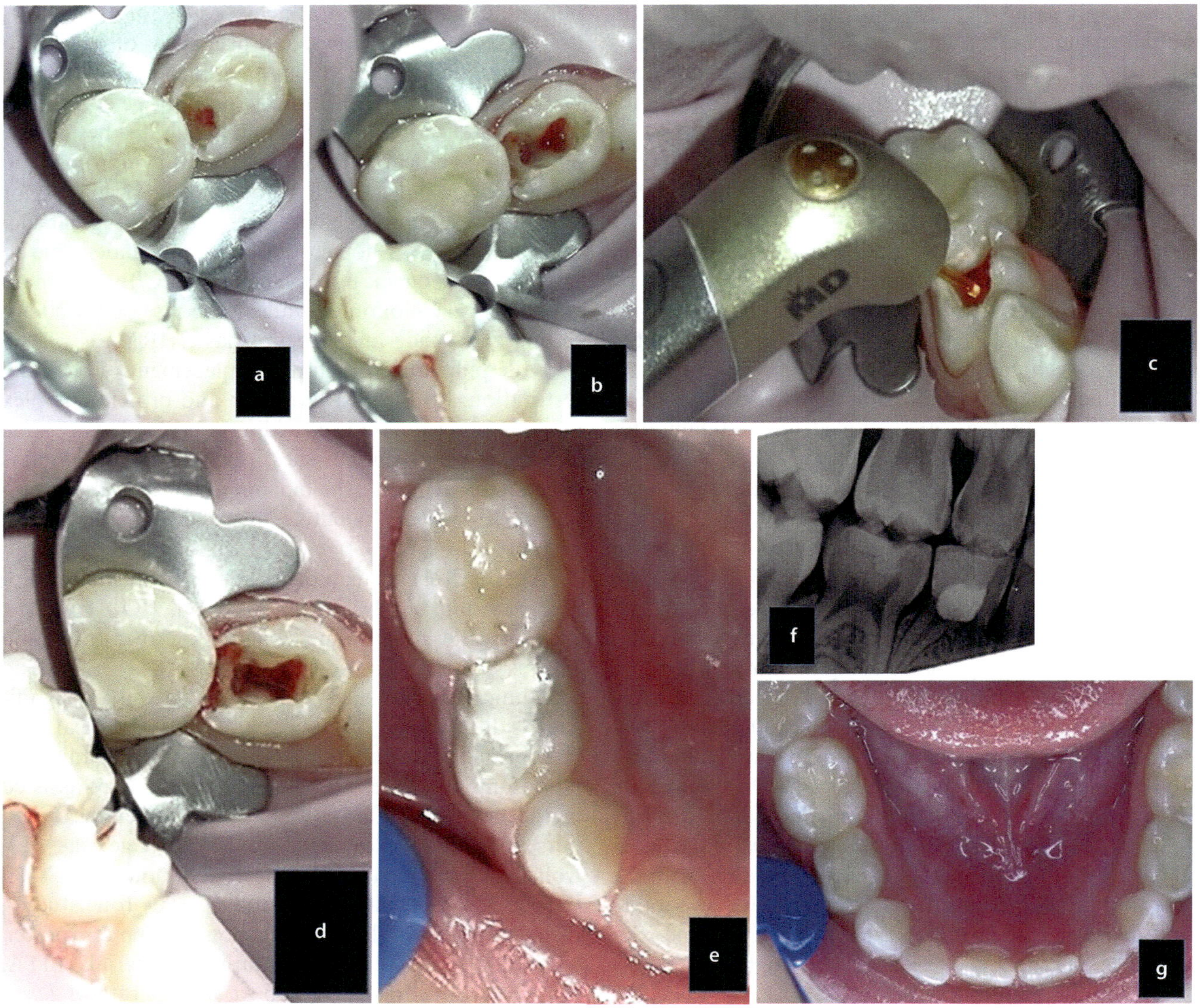

Fig. 11.33 Laser-assisted pulpotomy on a first primary molar of a 5.5-year-old girl. Block anesthesia (4% articaine, 1:200,000 epinephrine) and placement of rubber dam. **a** Pulp exposure during caries removal. **b** Removal of the coronal pulp tissue (diamond bur and spoon excavator). Hemorrhage was controlled. **c** Er,Cr:YSGG (2780 nm) was applied at 50 mJ, 10 Hz, (0.5 W), no water, 40% air, defocus for 5–10 s (gold handpiece, 0.6-mm MZ tip, S tissue mode) over the canal orifices **d**. Complete coagulation after laser application. **e** The cavity was filled up with IRM (fast-setting zinc oxide eugenol paste). Clinical view after 15 days. The placement of RMGI restoration was followed. **f**, **g** Radiographic and clinical pictures 16 months after treatment

11.7.4 Pulpectomy

It is the endodontic treatment for primary teeth and is indicated for teeth without or minimal pathological (internal or external) root resorption due to irreversible pulpitis or necrotic pulp. The traditional technique involves removing of all coronal and root pulp tissue, limited mechanical instrumentation, root canal disinfection using the appropriate irrigants, and filling the root canals with resorbable material (pure zinc oxide eugenol paste or iodoform-calcium hydroxide paste). Several protocols have been developed using lasers (erbium family, diode, Nd:YAG,) with great results for better decontamination of the main and the lateral canals on permanent teeth [58–62] (see ▶ Chap. 9). These same protocols for permanent teeth, using the same parameters, are also recommended for primary teeth, but there are only four studies (one in vivo, one in vitro, and two case reports) for deciduous teeth, all using photodynamic therapy, leading to satisfactory results [63–66]. In addition to the traditional technique, a laser-assisted disinfection method could be performed before the final conclusion of the endodontic treatment (Fig. 11.34a–i). Laser-assisted disinfec-

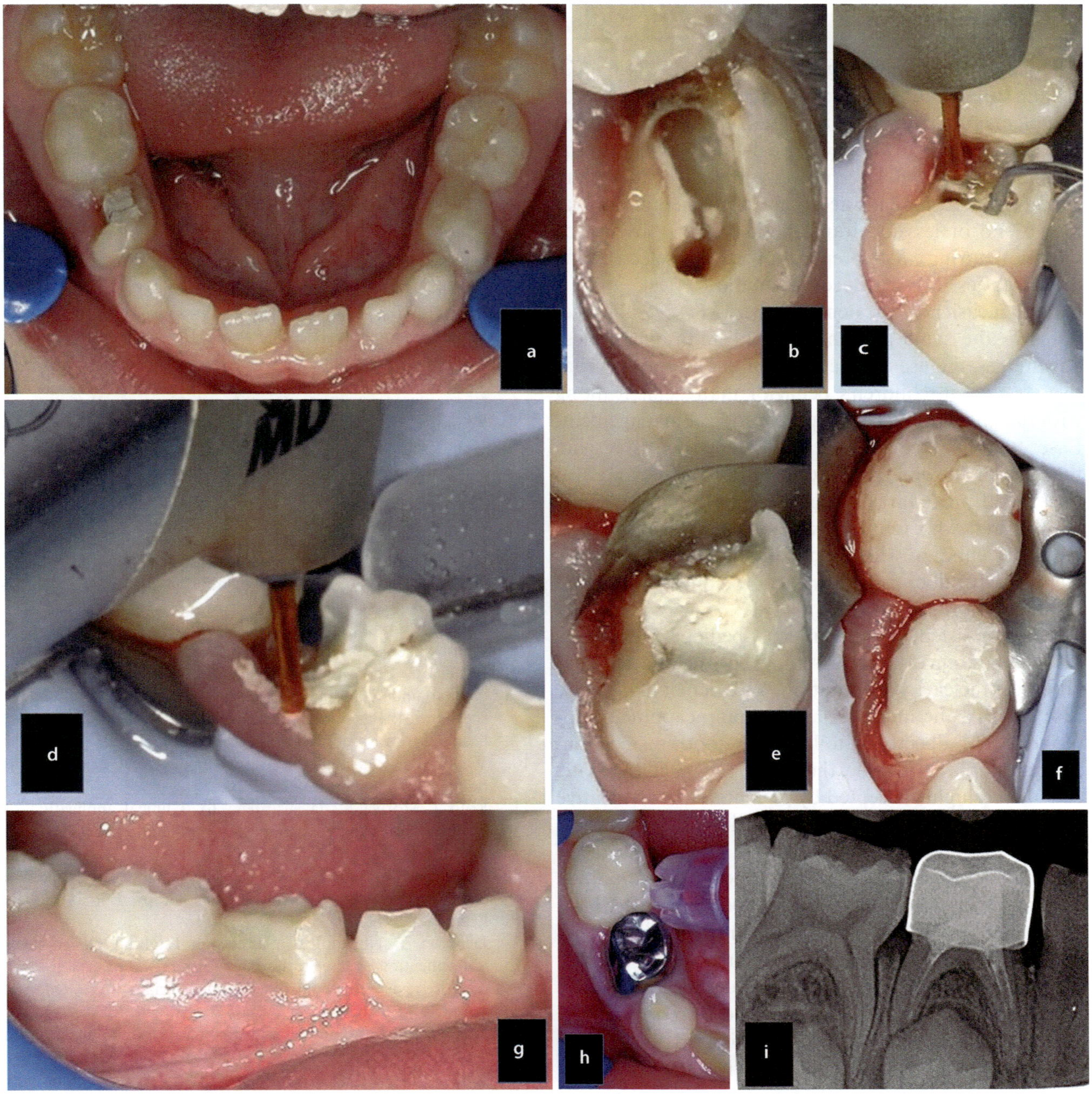

■ **Fig. 11.34** Laser-assisted pulpectomy and gingivoplasty on a first primary molar of a 6.5-year-old boy. Endodontic therapy of #84 due abscess. The patient returned 10 days after the initiation of treatment with a subgingival crown fracture. **a** Intraoral view of lower teeth. **b** Block anesthesia (4% articaine, 1:200,000 epinephrine) and placement of rubber dam. Occlusal view of #84. **c** Laser-activated irrigation (Er,Cr:YSGG 2780 nm) was applied at 33 mJ, 30 Hz, (1.0 W), no water, no air, tip inside the tooth chamber, saline solution for 5 s each root canal (gold handpiece, 0.6-mm MZ tip, H tissue mode). **d** Obstruction of the root canals with pure zinc oxide and eugenol paste and filling the tooth chamber with fast-setting zinc oxide and eugenol paste (IRM). Gingivoplasty (see laser energy parameters on ■ Fig. 11.3b) and cavity preparation followed (see ■ Table 11.2 for energy parameters). **e** After gingivoplasty and cavity preparation and decontamination. **f** Final restoration with RMGI. **g** Pulpectomy and restoration after 3 months. SSC was placed on #84 at that visit. **h**, **i** Periapical radiograph and clinical views 16 months after treatment

tion could have better results on primary teeth where there are more complex, with variable morphology, root canals making instrumentation and disinfection complicated. Irrigation with sodium hypochlorite should be avoided, especially when the laser-activated irrigation protocol is used, because if extruded from the open or resorbed root apex it could be irritant to the surrounding tissues.

11.8 Conclusion

All dental laser wavelengths (KTP, diode, Nd:YAG, erbium family, CO_2) could be used as alternative and/or complementary treatment methods of soft and hard tissue management for the pediatric dentistry patients. The main advantages of laser use in pediatric dentistry are (a) patient and parent's acceptance, (b) the administration of no or less local anesthesia, (c) the absence of vibration during cavity preparation, (d) the selectivity of dental caries, (e) the decontamination effect, and (f) making it easier for the dentist to perform several procedures in the same appointment. In addition to these advantages, the use of lasers can often offer an alternative strategy in children's behavior management along with the appropriate child's psychological management. Laser treatment can be used to introduce dentistry, gain the trust of the child, and perform needle-free and also no painful procedures using laser analgesia, especially for children who refused traditional dental treatment. However, children who do not finally cooperate or the mental status does not allow them to comply cannot be candidates for laser therapy.

Laser-assisted treatment in pediatric dentistry includes, among others, the removal of the infected dental tissue, the cavity preparation, and, depending on the case severity, the indirect or direct pulp capping, pulpotomy, and pulpectomy, followed by tooth restoration. Several studies conclude that laser abrasion is a safe, useful, and highly accepted by patients as an alternative method for caries removal and cavity preparation on primary teeth (erbium family). All dental restorative materials (composite resin, compomers, resin-modified class ionomer, glass ionomer) could be placed after laser cavity preparation on primary teeth revealing high success. Laser-assisted indirect and direct pulp capping techniques for primary teeth (erbium family or/and near-infrared laser wavelengths) could be more predictable and successful, than the transitional techniques, due to decontamination of the cavity, the remaining dentine, and the positive effect on pulpal tissue healing and recovery in order to form tertiary dentine. Instead of using medicaments (like ferric sulfate) during primary teeth pulpotomy, laser (erbium family, diode, Nd:YAG, CO_2) could be used, with great clinical and radiographical success, for the pulp tissue coagulation over the pulp stumps before the placement of the fast-setting zinc oxide and eugenol paste (IRM).

Laser-assisted disinfection, before the final root canal obstruction, could have better results on primary teeth pulpectomy where there are more complex, with variable morphology, root canals making instrumentation and disinfection complicated.

11

References

1. Koch G, Poulsen S. Pediatric dentistry. A clinical approach. 1st ed. Copenhagen: Munksgaard; 2001.
2. AAPD Oral health Policies. Policy on the use of lasers for pediatric dental patients. http://www.aapd.org/research/oral-health-policies--recommendations/.
3. AAPD Clinical Practice Guidelines on guideline on behavior guidance for the pediatric dental patient. http://www.aapd.org/policies/quidelines.
4. Poli R, Parker S. Achieving dental analgesia with the Erbium Chromium Yttrium Scandium Gallium Garnet laser (2780 nm): a protocol for painless conservative treatment. Photomed Laser Surg. 2015;33(7):364–71.
5. Olivi G, Magnolis FS, Genovese MD. Treatment considerations, Chapter 4. In: Pediatric laser dentistry. A user's guide. Quintessence Publishing Co, Inc; 2011.
6. Katsouda M, Tollili C, Coolidge T, Simos G, Kotsanos N, Arapostathis KN. Gagging prevalence and its association with dental fear in 4–12-year-old children in a dental setting. Int J Paediatr Dent. 2018;29:169. https://doi.org/10.1111/ipd.12445.
7. Katsouda M, Coolidge T, Simos G, Kotsanos N, Arapostathis KN. Gagging and cooperation in 4–12-year-old children over a series of dental appointments. Eur Arch Paediatr Dent. 2021;22(5):937–46. https://doi.org/10.1007/s40368-021--00654-x.
8. Moritz A. Oral laser application. Berlin: Quintessenz Verlags-GmbH; 2006.
9. Fisher JC. Photons, physiatrics, and physicians: a practical guide to understanding interaction of laser light with living tissue: Part II: Basic mechanisms of tissue destruction by laser beams. J Clin Laser Med Surg. 1993;11(6):291–303.
10. Vitale MC, Caprioglio C. Lasers in dentistry. Practical text book. Edizioni Martina s.r.l; 2010.
11. Kato J, Moriya K, Jayawardena JA. Clinical application of Er:YAG laser for cavity preparation in children. J Clin Laser Med Surg. 2003;21(3):151–5.
12. Genovese MD, Olivi G. Laser in paediatric dentistry: patient acceptance of hard and soft tissue therapy. Eur J Peadiatr Dent. 2008;9:13–7.
13. Jacobsen T, Norlund A, Sandborgh Englund G, Tranaeus S. Application of laser technology for removal of caries: a systematic review of controlled clinical trials. Acta Odontol Scand. 2011;69:65–74.
14. Martens LC. Laser physics and a review of laser applications in dentistry for children. Eur Arch Paediatr Dent. 2011;12(2):61–7.
15. Bahrololoomi Z, Kabudan M, Gholami L. Effect of Er:YAG laser on shear bond strength of composite to enamel and dentin of primary teeth. J Dent. 2015;12(3):163–70.
16. Monghini EM, Wanderley RL, Pecora JD, Palma-Dibb RG, Corona SAM, Borsatto MC. Shear bond strength to dentine of primary teeth irradiated with varying Er:YAG laser energies and SEM examination of the surface morphology. Lasers Surg Med. 2004;24:254–9.
17. Wanderley RL, Monghini EM, Pecora JD, Palma-Dibb RG, Borsatto MC. Shear bond strength to enamel of primary teeth irradiated with varying Er:YAG laser energies and SEM examination of the surface morphology: an in vitro study. Photomed Laser Surg. 2005;23(3):260–7.
18. Lessa FC, Mantovani CP, Barroso JM, Chinelatti MA, Palma-Dibb RG, Pécora JD, Borsatto MC. Shear bond strength to

primary enamel: influence of Er:YAG laser irradiation distance. J Dent Child (Chic). 2007;74(1):26–9.
19. Scatena C, Torres CP, Gomes-Silva JM, Contente MM, Pécora JD, Palma-Dibb RG, Borsatto MC. Shear strength of the bond to primary dentin: influence of Er:YAG laser irradiation distance. Lasers Med Sci. 2011;26(3):293–7.
20. Flury S, Koch T, Peutzfeldt A, Lussi A. Micromorphology and adhesive performance of Er:YAG laser-treated dentin of primary teeth. Lasers Med Sci. 2012;27(3):529–35.
21. Stiesch-Scholz M, Hannig M. In vitro study of enamel and dentin marginal integrity of composite and compomer restorations placed in primary teeth after diamond or Er:YAG laser cavity preparation. J Adhes Dent. 2000;2(3):213–22.
22. Hossain M, Nakamura Y, Yamada Y, Murakami Y, Matsumoto K. Microleakage of composite resin restoration in cavities prepared by Er,Cr:YSGG laser irradiation and etched bur cavities in primary teeth. J Clin Pediatr Dent. 2002;26(3):263–8.
23. Yamada Y, Hossain M, Nakamura Y, Murakami Y, Matsumoto K. Microleakage of composite resin restoration in cavities prepared by Er:YAG laser irradiation in primary teeth. Eur J Paediatr Dent. 2002;3(1):39–45.
24. Kohara EK, Hossain M, Kimura Y, Matsumoto K, Inoue M, Sasa R. Morphological and microleakage studies of the cavities prepared by Er:YAG laser irradiation in primary teeth. J Clin Laser Med Surg. 2002;20(3):141–7.
25. Borsatto MC, Corona SA, Chinelatti MA, Ramos RP, de Sá Rocha RA, Pecora JD, Palma-Dibb RG. Comparison of marginal microleakage of flowable composite restorations in primary molars prepared by high-speed carbide bur, Er:YAG laser, and air abrasion. J Dent Child (Chic). 2006;73(2):122–6.
26. Baygin O, Korkmaz FM, Arslan I. Effects of different types of adhesive systems on the microleakage of compomer restorations in Class V cavities prepared by Er,Cr:YSGG laser in primary teeth. Dent Mater J. 2012;31(2):206–14.
27. Ghandehari M, Mighani G, Shahabi S, Chiniforush N, Shirmohammadi Z. Comparison of microleakage of glass ionomer restoration in primary teeth prepared by Er:YAG laser and the conventional method. J Dent. 2012;9(3):215–20.
28. Baghalian A, Nakhjavani YB, Hooshmand T, Motahhary P, Bahramian H. Microleakage of Er:YAG laser and dental bur prepared cavities in primary teeth restored with different adhesive restorative materials. Lasers Med Sci. 2013;28:1453–60.
29. Bahrololoomi Z, Heydari E. Assessment of tooth preparation via Er:YAG laser and bur on microleakage of dentine adhesives. J Dent. 2014;11(2):172–8.
30. Arapostathis KN. An in vitro study comparing cavity preparation by Er:YAG, Er,Cr:YSGG lasers and diamond bur on primary molars: effect on microleakage of three different restorative materials and scanning electron microscopy examination of the cavity preparation. Thesis of Master of Science in Lasers Dentistry, Genoa, Italy; 2014.
31. Mirko P, Miroslav S, Lubor M. Significance of the labial frenum attachment in periodontal disease in man. Part 1. Classification and epidemiology of the labial frenum attachment. J Periodontol. 1974;45(12):891–4.
32. Walsh J, Links A, Boss E, Tunkel D. Ankyloglossia and lingual frenotomy: national trends in inpatient diagnosis and management in the United States, 1997–2012. Otolaryngol Head Neck Surg. 2017;156(4):735–40.
33. Hazelbaker AK. The assessment tool for lingual frenulum function (ATLFF): use in a lactation consultant private practice. California, Pacific Oaks College: Pasadena; 1993.
34. Ingram J, Johnson D, Copeland M, et al. The development of a tongue assessment tool to assist with tongue-tie identification. Arch Dis Child Fetal Neonatal Ed. 2015;100:F344–8.
35. Kotlow L. Diagnosis and treatment of ankyloglossia and ties maxillary frenum in infants using Er:YAG and 1064 diode lasers. Eur Arch Pediatr Dent. 2011;12(2):106–12.
36. Gutknecht N, Moritz A, Dercks HW, Lampert FJ. Treatment of hypersensitive teeth using neodymium:yttrium-aluminum-garnet lasers: a comparison of the use of various settings in an in vivo study. Clin Laser Med Surg. 1997;15(4):171–4. https://doi.org/10.1089/clm.1997.15.171.
37. Lopes AO, Aranha AC. Comparative evaluation of the effects of Nd:YAG laser and a desensitizer agent on the treatment of dentin hypersensitivity: a clinical study. Photomed Laser Surg. 2013;31(3):132–8. https://doi.org/10.1089/pho.2012.3386. Epub 2013 Feb 19.
38. Farmakis ET, Beer F, Kozyrakis K, Pantazis N, Moritz A. The influence of different power settings of Nd:YAG laser irradiation, bioglass and combination to the occlusion of dentinal tubules. Photomed Laser Surg. 2013;31(2):54–8. https://doi.org/10.1089/pho.2012.3333. Epub 2012 Dec 16. PMID: 23240877.
39. Orchardson R, Whitters CJ. Effect of HeNe and pulsed Nd:YAG laser irradiation on intradental nerve responses to mechanical stimulation of dentine. Lasers Surg Med. 2000;26(3):241–9. https://doi.org/10.1002/(sici)1096--9101(2000)26:3<241::aid-lsm1>3.0.co;2-i.
40. Lygidakis NA, Wong F, Jälevik B, Vierrou AM, Alaluusua S, Espelid I. Best clinical practice guidance for clinicians dealing with children presenting with Molar-Incisor-Hypomineralisation (MIH): an EAPD Policy Document. Eur Arch Paediatr Dent. 2010;11(2):75–81. https://doi.org/10.1007/BF03262716.
41. Körner R, Bahmer F, Wigand R. The effect of infrared laser rays on herpes simplex virus and the functions of human immunocompetent cells. Hautarzt. 1989;40(6):350–4.
42. Donnarumma G, De Gregorio V, Fusco A, Farina E, Baroni A, Esposito V, Contaldo M, Petruzzi M, Pannone G, Serpico R. Inhibition of HSV-1 replication by laser diode-irradiation: possible mechanism of action. Int J Immunopathol Pharmacol. 2010;23(4):1167–76.
43. AAPD Clinical Practice Guidelines on pulp therapy for primary and immature permanent teeth. http://www.aapd.org/policies/quidelines.
44. AAPD Oral health Policies on Policy on Interim Therapeutic Restorations (ITR). http://www.aapd.org/policies/quidelines.
45. AAPD Oral health Policies. Use of vital pulp therapies in primary teeth with deep caries lesions. http://www.aapd.org/policies/quidelines.
46. Farooq NS, Coll JA, Kuwabara A, Shelton P. Success rates of formocresol pulpotomy and indirect pulp therapy in the treatment of deep dentinal caries in primary teeth. Pediatr Dent. 2000;22(4):278–86.
47. Parisay I, Ghoddusi J, Forghani M. A review on vital pulp therapy in primary teeth. Iran Endod J. 2015;10(1):6–15. Epub 2014 Dec 24. Review.
48. Olivi G, Magnolis FS, Genovese MD. Endodontics, Chapter 8. In: Pediatric laser dentistry. A user's guide. Quintessence Publishing Co, Inc; 2011.
49. Konstanos N, Arapostathis KN, Arhakis A, Menexes G. Direct pulp capping of carious primary molars. A specialty practice based study. J Clin Pediatr Dent. 2014;38(4):307–12.
50. De Coster P, Rajasekharan S, Martens L. Laser-assisted pulpotomy in primary teeth: a systematic review. Int J Paediatr Dent. 2013;23(6):389–99. Epub 2012 Nov 22. Review.
51. Uloopi KS, Vinay C, Ratnaditya A, Gopal AS, Mrudula KJ, Rao RC. Clinical evaluation of low level diode laser application

for primary teeth pulpotomy. J Clin Diagn Res. 2016;10(1):ZC67–70.
52. Liu JF. Effects of Nd:YAG laser pulpotomy on human primary molars. J Endod. 2006;32:404–7.
53. Saltzman B, Sigal M, Clokie C, Rukavina J, Titley K, Kulkarni GV. Assessment of a novel alternative to conventional formocresol-zinc oxide eugenol pulpotomy for the treatment of pulpally involved human primary teeth: diode laser-mineral trioxide aggregate pulpotomy. Int J Paediatr Dent. 2005;15:437–47.
54. Gupta G, Rana V, Srivastava N, Chandna P. Laser pulpotomy—an effective alternative to conventional techniques: a 12 months clinicoradiographic study. Int J Clin Pediatr Dent. 2015;8(1):18–21.
55. Durmus B, Tanboga I. In vivo evaluation of the treatment outcome of pulpotomy in primary molars using diode laser, formocresol, and ferric sulphate. Photomed Laser Surg. 2014;32(5):289–95.
56. Yadav P, Indushekar K, Saraf B, Sheoran N, Sardana D. Comparative evaluation of ferric sulfate, electrosurgical and diode laser on human primary molars pulpotomy: an "in-vivo" study. Laser Ther. 2014;23(1):41–7.
57. Niranjani K, Prasad MG, Vasa AA, Divya G, Thakur MS, Saujanya K. Clinical evaluation of success of primary teeth pulpotomy using mineral trioxide aggregate(®), laser and Biodentine(TM)—an in vivo study. J Clin Diagn Res. 2015;9(4):ZC35–7.
58. Subbaiah R. Bacterial efficacy of Ca(OH)2 against E. faecalis compared with three dental lasers on root canal dentin—an In vitro study. J Clin Diag Res. 2014;8(11):ZC135–7.
59. Rebecca G, et al. Er:YAG 2,940-nm laser fiber in endodontic treatment: a help in removing smear layer. Lasers Med Sci. 2014;29(1):69–75.
60. Pedullà E, et al. Decontamination efficacy of photon-initiated photoacoustic streaming (PIPS) of irrigants using low-energy laser settings: an ex vivo study. Int Endod J. 2012;45:865–70.
61. Shoaib H. Bactericidal efficacy of photodynamic therapy against Enterococcus faecalis in infected root canals: a systematic literature review. Photodiagn Photodyn Ther. 2013;10(4):632.
62. Vahid Z. Antimicrobial efficacy of photodynamic therapy and sodium hypochlorite on monoculture biofilms of Enterococcus faecalis at different stages of development. Photomed Laser Surg. 2014;32(5):245–51.
63. Pinheiro SL, et al. Photodynamic therapy in endodontic treatment of deciduous teeth. Lasers Med Sci. 2009;24(4):521–6.
64. Pinheiro SL, et al. Manual and rotary instrumentation ability to reduce Enterococcus faecalis associated with photodynamic therapy in deciduous molars. Braz Dent J. 2014;25(6):502–7.
65. Giselle de Sant'Anna, Photodynamic therapy for the endodontic treatment of a traumatic primary tooth in a diabetic pediatric patient. J Dent Res Dent Clin Dent Prospect. 2014; 8(1): 56–60.
66. da Silva Barbosa P, et al. Photodynamic therapy in pediatric dentistry. Case Rep Dent. 2014;2014:217172.

Laser-Assisted Oral Soft Tissue Management

Contents

Laser Use in Muco-Gingival Surgical Orthodontics

Ali Borzabadi-Farahani

Contents

D. J. Coluzzi, S. P. A. Parker (eds.), *Lasers in Dentistry—Current Concepts*, Textbooks in Contemporary Dentistry,
https://doi.org/10.1007/978-3-031-43338-2_12

Core Message

Lasers are excellent adjuncts to the clinical discipline of surgical orthodontics. This chapter reviews the laser wavelengths available and discuss some lasers for soft tissue procedures. These include improved healing (photobiomodulation) and provision of a bloodless surgical site, laser gingivectomy to improve oral hygiene or bracket positioning, aesthetic laser gingival recontouring, and laser exposure of the superficially impacted teeth. Selected treated cases will be presented throughout. The emphasis was to provide more information on the use of diode lasers due to their popularity, smaller devices, and lower cost; however, treated cases with non-contact laser (CO_2 and erbium lasers) were also included. This chapter also reviews the available evidence that compares the diode laser surgery with conventional scalpel surgery for surgical exposure of impacted teeth or teeth with delayed eruption (both with no overlaying bone).

12.1 Soft Tissue Procedures Introduction

The healthy gingival margin is located 1–2 mm coronal to the cemento-enamel junction [1]. However, this gingival architecture does not always satisfy smile aesthetics during or after orthodontic treatment. Muco-gingival surgery, on the other hand, is a periodontal treatment to correct the defects in the morphology, position, and/or amount of soft tissue and underlying bone support around teeth and implants. Laser incision/excision has a proven place in modern muco-gingival surgery [2]. Compared with a scalpel, a laser beam or the delivery tip of a laser device can more easily cut, ablate and reshape the oral soft tissues in the oral cavity, with no or reduced bleeding and less pain, as well as with no or less need for suturing [2]. This represents a range of tissue interactions, such as tissue warming, welding, coagulation, protein denaturation, drying, and finally vaporization (ablation) and carbonization, where soft tissues are evaporated or incised [2–7]. This process also provides hemostasis, microbial inhibition, and destruction and photobiomodulation (PBM) [2–7]. In particular, there is increasing evidence that the appropriate use of lasers is associated with reduced intra-operative and post-operative pain and enhanced wound healing or tissue regeneration, compared to conventional use of scalpel or electrosurgery [2–4]. Electrosurgery can be used for incising soft tissues with good hemostasis [2–4] but comes with a risk of delayed wound healing due to unwanted thermal damage [2, 6] and necrosis of the underlying periosteum and alveolar bone.

12

12.1.1 Advantages of Laser Excision vs. Scalpel Surgery

Dental lasers have been widely used for soft tissue procedures, such as gingivectomy, gingivoplasty, and frenectomy, and, in particular, for aesthetic gingival procedures, such as recontouring or reshaping of gingiva, crown lengthening, and depigmentation [2]. For instance, compared to conventional scalpel surgery, the diode laser cut is more precise and more visible due to the laser ability to seal off blood vessels and lymphatics, leaving a clear dry field [2, 7]. The laser also contributes to significant pathogen reduction as it cuts; and residual bacteria are evaporated, destroyed, or denatured by laser irradiation [2]. Laser incision with high-level laser therapy (HLLT) excises (ablates) the diseased tissues, with simultaneous provision of low-level laser therapy (photobiomodulation, PBM) that penetrates or scatters into the surrounding tissues during high-level laser treatment and modulates tissues and cells without producing irreversible changes (◘ Fig. 12.1).

PBM promotes periodontal wound healing of the adjacent tissues as a desired effect [2, 8, 9] of tissues and cells following laser irradiation [2]. PBM generates an array of transient biochemical intermediates that result in cascading biological reactions in favor of tissue healing [10–12]. This process works by altering the cellular redox state [10] and production of reactive oxygen species (ROS) in mitochondria, such as superoxide ($O_2^{\bullet -}$) and hydrogen peroxide (H_2O_2), which mainly affect and stimulate cells in a low redox state [12]. Cells in a low redox state are acidic, but after laser irradiation, the cells become more alkaline and are able to perform optimally, inducing the activation of numerous intracellular signaling pathways [12]. Photoabsorption by mitochondrial chromophores, in particular *cytochrome c oxidase*, leads to dissociation of the binding between *nitric oxide* (NO) and *cytochrome c oxidase*, allowing mitochondria to increase ATP production and *nitric oxide* (NO) release [10–12]. The produced ATP modulates a wide range of biological responses, including activation or synthesis of DNA, RNA, enzymes, and other cellular components necessary for optimal performance and repair/regeneration of tissues [11]. The PBM-mediated NO release leads to vasodilatation, involving cGMP-mediated activation of Ca-sensitive K (Kc) channels [11], as well as promotes keratinocyte and tenocyte proliferation, endothelial migration and lumenization, macrophage function, angiogenesis in ischemic limb injuries, and stem cell differentiation [11]. Overall, PBM positively affects each of the four phases of wound healing [11, 13] (◘ Table 12.1).

Within the progressive stages of wound stabilization and healing, the many cellular and biochemical pathways are potential recipients of sub-ablative (low-level) laser photonic energy between approximately 600 and 1400 nm wavelength. Outside this range, similar induced effects may be attributable to low thermal rise and consequent tissue stimulation.

PBM can promote changes at the cell level and expression of cytokines that can collectively promote wound healing, by increasing collagen production, reduction of

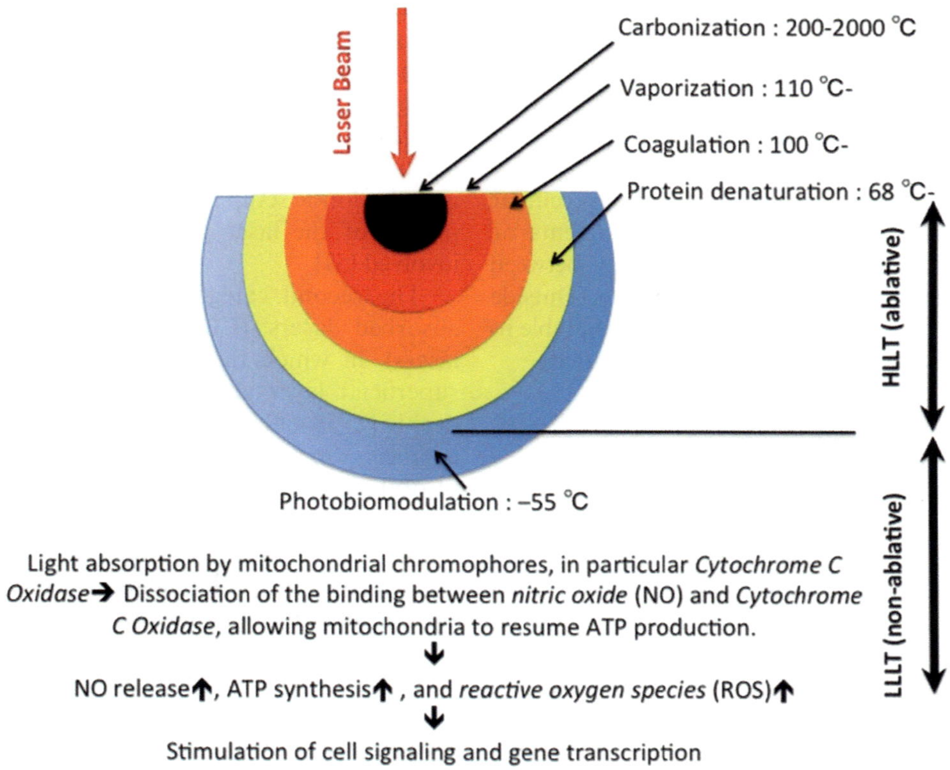

Fig. 12.1 Diagram showing the simultaneous work of high-level laser therapy (HLLT) and low-level laser therapy (PBM). HLLT initiates various thermal effects on tissues, such as carbonization, vaporization, coagulation, and ablation of soft tissue, as well as sometimes removal of the hard tissue (erbium lasers). Simultaneously, a low level of energy penetrates or scatters into the surrounding tissues during high-level laser treatment, which stimulates tissues and cells without producing irreversible thermal changes in the tissues, resulting in modulation of wound healing in the surrounding tissues. (Partially adopted from Aoki et al. [2])

Table 12.1 Effects of photobiomodulation (PBM) on four phases of wound healing [11, 13]

Four phases of wound healing	Effects of PBM on wound healing
Hemostatic phase	Promotes platelet aggregation and activation
Inflammatory phase	Promote proliferation and degranulation of mast cells
Proliferative phase	Promote proliferation of fibroblasts, keratinocytes, osteoblasts, and chondrocytes as well as induces matrix synthesis
Maturation phase	Improves reorganization and remodeling of wounds, aids improved tensile strength and restoring functional architecture of the repaired tissues

inflammation, and pain relief [2, 10–53]. PBM has been effective in pain reduction [15, 18], wound healing [19–24], bone repair and remodeling [25–33], nerve repair [34–39], angiogenesis [40, 41], as well as increased cell proliferation and biomodulation for cell lines such as fibroblasts [20, 42], keratinocytes [43], osteoblasts [44], and mesenchymal stem cells [45]. PBM predominantly stimulates macrophages and fibroblasts [46–48] and collectively modulates secretion of vascular endothelial growth factor (VEGF), platelet-derived growth factor (PDGF), fibroblast growth factor (FGF) and tumor necrosis factor alpha (TNF-α) by macrophages, neutrophils, endothelial cells and fibroblasts, stimulating cell proliferation, cell differentiation, and neoangiogenesis, as well as synthesis of extracellular matrix components such as types I and III collagen fibers [46–53]. Less wound contraction and edema also occur during mucosal healing; scars tend not to develop as less damage occurs to adjacent tissues, and there is rarely a need for periodontal dressing [7, 54]. This phenomenon can be attributed to the low-power (PBM) zones that surround the high-power surgical laser site [11].

These qualities allow faster or more favorable wound healing, needing less pain medication, as well as less post-operative discomfort, compared to usual scalpel surgery [2]. This can lead to reduction in the orthodontic treatment time, when there is a need for soft tissue procedures that otherwise need referral to other specialties such as periodontist or oral surgeon, in particular for fee-paying patients who demand optimal results with minimal effort as quickly as possible [55].

Table 12.1 outlines the progressive stages of wound stabilization and healing of the many cellular and biochemical pathways as potential recipients of sub-ablative (low-level) laser photonic energy between approximately 600 and 1400 nm wavelength. Outside this range, similar induced effects may be attributable to low thermal rise and consequent tissue stimulation.

12.2 Overview of Lasers Used for Soft Tissue Procedures

Various laser systems have been used for soft tissue procedures, which work by ablating, incising, and excising the soft tissue, as well as providing the much-needed coagulating effect. The frequently used soft tissue lasers include the carbon dioxide laser (10,600 nm), erbium lasers [erbium-doped yttrium-aluminum-garnet (Er:YAG) laser (2940 nm) and erbium chromium-doped yttrium-scandium-gallium-garnet (Er,Cr:YSGG) laser (2780 nm)], neodymium-doped yttrium-aluminum-garnet (Nd:YAG) laser (1064 nm), the diode group of lasers (800–980 nm), and the potassium, titanyl phosphate ($KTiOPO_4$, KTP-"green") laser (532 nm), together with the InGaN ("blue") 445 nm laser [2, 3].

12

12.2.1 The Shallow or Deeply Penetrating Lasers and Hemostasis

The soft tissue lasers can be categorized into two broadly acting types: the deeply penetrating-type lasers (visible and near-infrared spectrum, 532–1100 nm) that are essentially transmitted through water, showing a lower absorption coefficient in water [56] such as KTP green laser. This explains their deep penetration into healthy soft tissue, such as Nd:YAG and diode lasers, in which the laser light penetrates and scatters deeply into tissue [56]. However, they are selectively absorbed in areas of inflammation by blood components and tissue pigment [56], which makes them suitable for soft tissue procedures [56]. The Nd:YAG laser is often used in free-running pulsed mode, with very-short-duration pulses and an emission cycle (ratio of "on" time to total treatment time) of <1% and very high peak power per pulse (100–1000 W) [56]. The Nd:YAG laser is a deeply penetrating type of laser and produces a relatively thick coagulation layer on the lased soft tissue surface, exhibiting strong hemostasis. Therefore, the Nd:YAG laser is effective for ablation of potentially hemorrhagic soft tissue. Diode lasers represent a shallower penetration depth compared to Nd:YAG lasers and are less likely to cause pulpal damage after use [55]. The diode lasers can be used in a continuous-wave or gated-CW mode [7] and are the ideal choice for the use in orthodontic setup because of the smaller size ("footprint") of the laser device and relatively lower cost involved [57].

The second category highlights the superficially absorbed lasers (CO_2, Er:YAG and Er,Cr:YSGG lasers), in which the laser beam is absorbed in the superficial layer and does not penetrate or scatter deeply [2, 58–60]. These lasers have higher absorption coefficient in water, and due to the high-water content of oral mucosa (>90%), they are very effective soft tissue lasers [55]. The CO_2 laser beam is absorbed at the tissue surface with very little scatter or penetration [2] and is associated with relatively thin layer of coagulation around the ablated site. The ablation for CO_2 laser is basically caused by heat generation (photothermolysis) [2]. Erbium lasers have the highest absorption into water and target molecular water or the hydroxide ion as primary targets and mineral as a secondary target and therefore are used for ablation of both hard and soft tissues [56]. Erbium lasers provide the most rapid, favorable, and uneventful wound healing due to their precise ablation with minimal thermal effects as well as low inflammatory response [60]. However, hemostasis is less effective with the erbium lasers because of the minimal tissue denaturation, accentuated by the coaxial water spray which guarantees subsequent sufficient bleeding and blood clot formation in the ablated defects and thereby induces favorable wound healing [6]. Overall, erbium lasers provide the highest absorption into water, minimizing the thermal effects on the surrounding tissues during irradiation.

12.2.2 Tissue Ablation: Non-contact or Contact Cutting Mode

As has been seen elsewhere (▶ Chap. 3), laser-tissue interaction is the result of electromagnetic (photonic) energy being absorbed and converted into other (predominately thermal) energy. Three forms of energy transfer can be observed:

(a) Radiation—where the photon stream is delivered through a short airspace with no contact between delivery tip and target tissue. This may be commonly referred to a "non-contact mode."
(b) Conduction—where enhancement of the energy conversion can be achieved through direct contact

between the delivery tip and the tissue. This may be commonly referred to a "contact mode."

(c) Convection—transfer of energy within the body of the tissue through fluid movement or circulation. This may occur regardless of either contact or non-contact modes.

Most surgical lasers produce a photothermal effect on soft tissue, evaporating soft tissues through rapid thermal rise. The non-contact lasers such as CO_2 or erbium lasers (Er:YAG and Er,Cr:YSGG) directly and easily evaporate soft tissues by photothermal effects that vaporize interstitial water. However, the non-contact mode is associated with less precise cut and lack of proprioceptive feedback compared to contact mode lasers.

When lasers are used in contact mode to make an incision or excise soft tissue, they often need "initiation" of the laser tip end. In this process, part of the emitting light in the Nd:YAG and diode lasers is attenuated and converted into heat; the laser tip end gets initiated, creating a condition called "hot tip." This initiation produces secondary thermal effects at the heated tip end that can cut or incise soft tissue as well as offer coagulation of the tissue as a result of contact with the over-heated tip rather than by the laser energy itself [2, 3]. ◘ Figure 12.2 shows an initiated fiber-optic tip prior to laser exposure of an ectopic lower left canine. Diode and Nd:YAG lasers produce a relatively thicker coagulation layer on the treated surface than superficially absorbed lasers [4]. Diode lasers are considered to be suitable for daily practice of orthodontic soft tissue procedures owing to sufficient hemostasis and precise incision margins [61, 62]. For the purpose of this chapter, the use of diode lasers for soft tissue procedures will be discussed in more detail due to their ease of use in orthodontics.

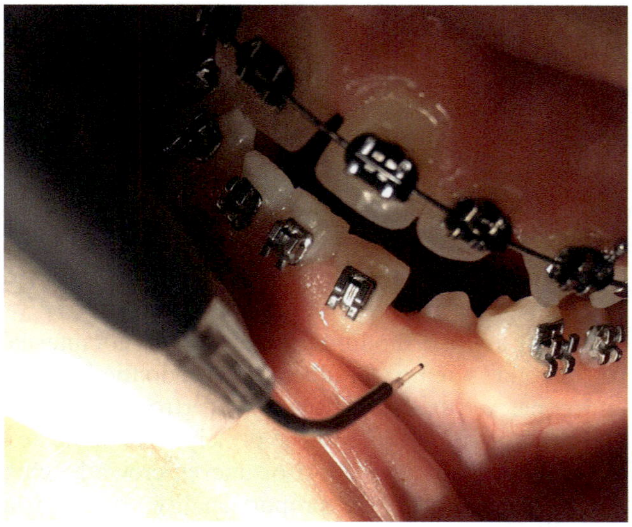

◘ **Fig. 12.2** A typical initiated fiber-optic laser tip will be used for laser excision

12.3 Soft Tissue Diode Lasers

Since their introduction in 1962 [63, 64], the diode laser family has grown considerably and diode lasers with wavelengths in the range of 445–2200 nm have been used for treatment of various medical conditions [55, 57, 65–67]. However, reports on the use of the 810–830, 940, 980, and 1064 nm wavelengths are more frequent in the literature [55, 57, 69]. They have high absorption coefficients in water and hemoglobin and particularly in oxy-hemoglobin, therefore rendering different soft tissue effects. However, diode laser light is poorly absorbed by the hydroxyapatite and enamel [54, 55]; however, it is an excellent soft tissue surgical laser for incising, excising, and coagulating gingiva and mucosa. The active media of semiconductor (diode) lasers are varied and can include aluminum (Al), gallium arsenide (GaAs), and occasionally, indium (In) [55, 68, 69]. Examples are gallium-aluminum-arsenic (Ga-Al-As), arsenic-gallium (As-Ga), and indium-gallium-aluminum-phosphorus (In-Ga-Al-P) lasers. The diode lasers are portable (<5 kg), small, relatively simple to use [68]. There is also a stable power output, long lifetime, and low installation and maintenance costs [68].

12.3.1 Fiber-Optic Tip Size, Power Output, and Continuous/Gated-CW Mode for Diode Lasers

The soft tissue diode lasers usually work in a "contact mode," and the laser beam is delivered by a fine glass optic fiber, with a fiber system tip that can be angled, so that the dentist holds it in a pencil-like holder for accurate manipulation of the areas that are difficult to handle [68]. For surgical incisions and excision, a 400-μm diameter fiber-optic tip is recommended, as smaller diameter fibers tend to be more friable and liable to fracture [69]. The fiber-optic tip needs initiation prior to performing surgical excision, often by tapping the initiated tip on a blue articulating paper, the use of black ink, a solid color in a magazine page or a piece of cork have been met with varying degrees of success [55, 70]. Diode lasers with power outputs of <500 mW are used in *low-level laser therapy* to provide photobiomodulation (PBM) and associated wound repair and pain relief [55]. However, for excision there is often a need for a continuous power output of 1.0–1.5 W [70], depending on the fibrotic nature of the tissue. In order to decrease the carbonization and thermal damage and allow for thermal recovery of the tissue, a gated-CW mode (with repetitive "on–off" cycles of varying length and frequency depending on the make of the laser) has been suggested and implemented in many contemporary diode laser units [69, 71].

12.3.2 Provision of Anesthesia and Basic Soft Tissue Guidelines

Lower pain sensation and less need for analgesia have been reported when diode laser with superpulsed mode [72] or with gated-CW of one millisecond pulse duration (on/off cycle of 50/50) [73] was used for soft tissue surgery, as compared to continuous wave diode laser. Diode laser soft tissue surgery is often performed using local infiltration (e.g., 2% lidocaine) approximately 5 min before procedure, but literature also reports using topical lignocaine anesthetic gel, applied for 3 min, particularly with the gated-CW mode [69, 74], or compound topical anesthetics such as TAC Alternate for 3 min (20% lidocaine, 4% tetracaine, and 2% phenylephrine) [69, 70, 75]. Given enough time, topical anesthetics often provide enough analgesia for laser exposure of buccally superficially impacted teeth; if enough analgesia is not achieved, additional topical dosage can be applied [69]. Palatal mucosa, however, is thicker and local infiltration is often necessary [69]. In order to confirm adequate anesthesia prior to laser soft tissue surgery, gently probing the soft tissue will confirm that the patient feels pressure only or feels anything sharp that indicates the need for added dose of local anesthesia.

During laser ablation, vaporized tissue, water, bacteria, and organic chemical residues are liberated; this is known as the "laser plume," and the use of a high-speed suction is recommended to remove this plume and objectionable charred odor, as well as provide a degree of safety against inhalation by patient and attending clinicians [69, 70]. Following the surgical excision, the soft tissue margins can appear dark and possibly charred, and the remnants of such tissue ablation at the surgical margins can be removed using sterile gauze dampened with saline [74] or a micro-applicator brush soaked in 3% hydrogen peroxide solution [70].

Various manufacturers present different arrangements for diode laser with respect to output power, diameter of fiber, and wavelength. Although these parameters may influence collateral tissue damage, there is currently lack of standardization in setting the best operating parameters of diode laser for orthodontic soft tissue procedures, which needs to be investigated in future studies [76].

Diode lasers are useful in recontouring the gingiva to gain access to the clinical crown, where there is gingival overgrowth or in case of partially erupted teeth, which prevent the proper positioning of a bracket. When planning laser soft tissue procedures, the general guideline is to leave at least 1.0 mm of pocket depth and to preserve at least 2.0 mm of keratinised tissue to avoid further soft tissue complications such as gingival recession [55]. The aforementioned guidelines are based on the "biologic width" concept, as measured from the free gingival margin to the crestal bone, which is approximately 3 mm, consisting of, on average, 1 mm of junctional epithelium, 1 mm of connective tissue attachment, as well as a gingival sulcus depth of approximately 1 mm [55, 69]. In order to decide between the conventional flap approach and laser gingivectomy, the gingivectomy location should be probed, and the amount of attached gingiva, the location of the crest of bone and the desired amount of crown lengthening should be looked into based on the limitations of the biologic width. In general, an average of 3 mm of soft tissue will rebound (regrow) coronal to the alveolar crest in about 3 months [77].

12.3.3 Laser Gingivectomy to Improve Oral Hygiene or Bracket Positioning

Difficulties in cleaning approximal tooth surfaces and reduction in aerobe/anaerobe ratio of sub- and supragingival flora [78, 79] may contribute to the gingival hyperplasia and pseudo-pocketing. This is common following bonding fixed orthodontic appliances and can be seen in about 10% of orthodontic patients [76, 80]. Gingival enlargement often impedes the maintenance of oral hygiene, causing aesthetic and functional problems, and has been reported to compromise orthodontic tooth movement [76, 81, 82]. Conventional treatment for gingival enlargement often includes oral hygiene instruction, scaling, root planning, and prophylaxis, but extreme and often fibrotic gingival enlargement compromises the self-care and may necessitate a few, interspersed gingivectomy sessions to maintain oral health [76, 83]. The adjunct use of laser-assisted gingivectomy can be less painful, producing less traumatic and faster improvement in gingival health of patient with gingival enlargement [76] (■ Fig. 12.3a, b).

In addition, laser gingivectomy can be performed to remove excess soft tissue and expose the crown of the partially erupted teeth, allowing brackets to be placed properly, ideally in the center of the teeth, allowing maintenance of an improved level of hygiene during treatment [7, 55].

12.3.4 Aesthetic Laser Gingival Recontouring

Following active orthodontic treatment, it is not unusual to "debond"—remove—adherent orthodontic brackets and come across unsightly and hypertrophic gingival margins not conforming to the principles of smile aesthetics, presenting with short or uneven crown heights, disproportionate tooth proportionality ratios and unaesthetic enlarged and fibrotic interdental papillae and gingival margins [7]. Aesthetic procedures such as

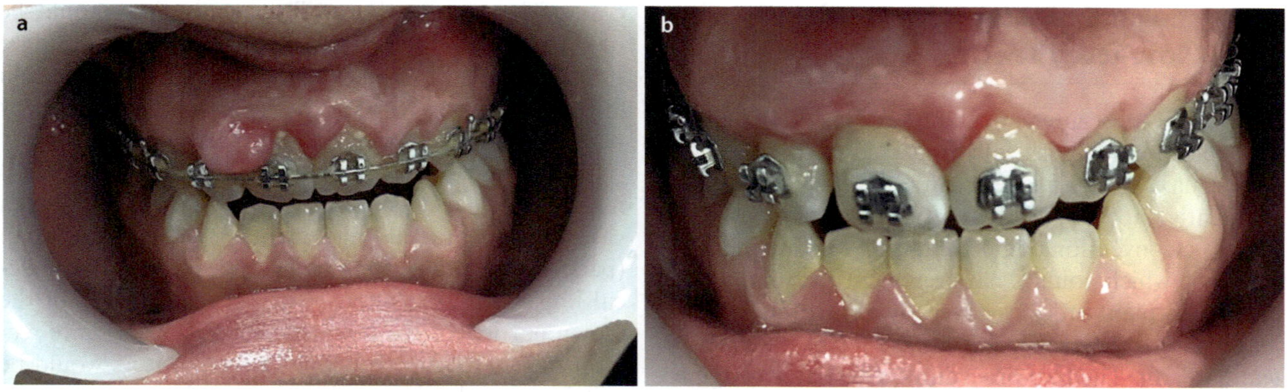

Fig. 12.3 **a** A pre-operative intra-oral frontal view of a patient with gingival hyperplasia of the maxillary anterior teeth, in particular in the maxillary right lateral incisor (UR2) region. **b** Post-operative appearance at 1 week. Laser operating parameters used were the continuous wave 940 nm diode (InGaAsP) laser (Epic 10, Biolase, Irvine, CA), with a 400-μm diameter fiber-optic tip, in a contact mode after initiation (power output = 1 W)

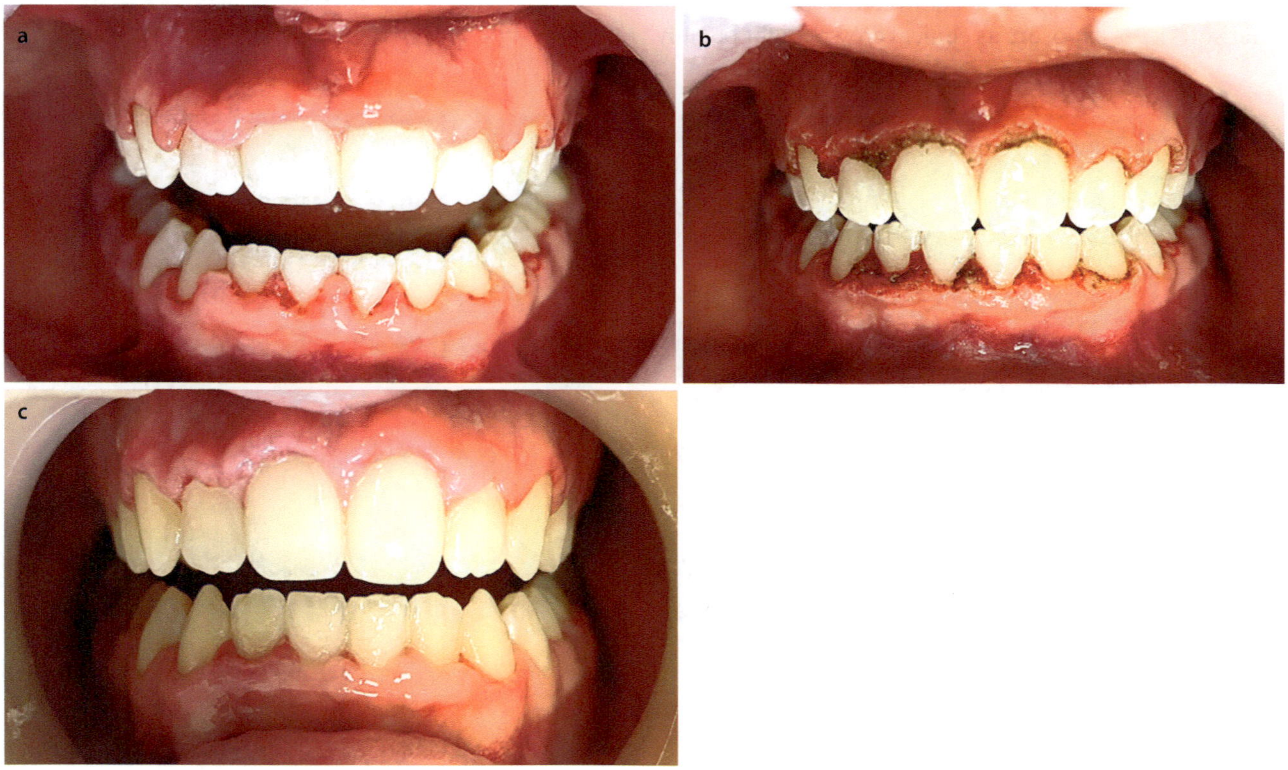

Fig. 12.4 **a** A pre-operative intra-oral frontal view of a patient with gingival hyperplasia of the maxillary and mandibular front teeth immediately after removal of fixed braces. This amount of gingival hyperplastic makes the provision of orthodontic retainers very challenging. **b** Immediate post-operative appearance following sulcular local infiltration (2% lidocaine) and removal and debulking of hypertrophic gingival tissue. Laser operating parameters used were the continuous wave 940 nm diode (InGaAsP) laser (Epic 10, Biolase, Irvine, CA), with a 400-μm diameter fiber-optic tip, in a contact mode after initiation (power output = 1 W). Some carbonization is evident at the laser gingivectomy sites. **c** Close-up view at 1-week post-operative follow-up after laser gingivectomy.

aesthetic crown lengthening or papilla flattening can be technically demanding tasks in that the gingival margins sometimes need very minor recontouring that needs a higher degree of precision than that achieved with a scalpel blade, regardless of the operator's skill level [84]. Diode lasers offer the precise incision control because of less bleeding and a clear dry field during surgery, offering an alternative to the scalpel for the aesthetic gingival recontouring [85–87]. Figure 12.4a–c show a patient, following removal of fixed braces, who has undergone gingival recontouring of the maxillary and mandibular front teeth.

12.3.5 Laser Exposure of Superficially Impacted Teeth

One of the most interesting applications of the laser is for exposure of superficially impacted teeth, in particular for maxillary permanent canines, which are the most frequently impacted teeth after third molars (0.92–4.3%) [88]. The conventional approach is to wait for the tooth to erupt, which could delay treatment for months and affect treatment efficacy adversely, or to refer for the placement of an apically positioned flap or mucoperiosteal flap [70]. The flap procedures are relatively aggressive in nature. Accurate localization of the impacted tooth prior to laser exposure is vital to establish if the impacted tooth is positioned superficially and not covered completely by bone or needs referral to an oral surgeon or periodontist for surgical exposure. The presence of a labial bulge does not guarantee access to crown after soft tissue exposure as clinical crown might be fully covered by alveolar bone. The localization should be based on both clinical (blanching of tissue with finger pressure) and, if in doubt, by radiographic examination [87]. Approximately, 85% of canine impactions occur palatally and 15% buccally [89–91].

Diode laser exposure is not applicable in cases of full impaction of teeth covered by cortical bone. In such cases, a conventional full-thickness mucoperiosteal flap (palatal impaction) or an apically positioned flap (buccal impaction) and removal of cortical bone until the crown portion of the retained tooth is exposed are recommended. Erbium and 9300 nm CO_2 lasers would be suitable for the latter situations. When superficially impacted teeth are present, it is recommended to create sufficient space before the surgical laser exposure to facilitate bonding an eyelet or bracket and apply orthodontic forces right after laser exposure.

Figure 12.5a–c demonstrate a male patient with a buccally impacted maxillary right canine, which has

12

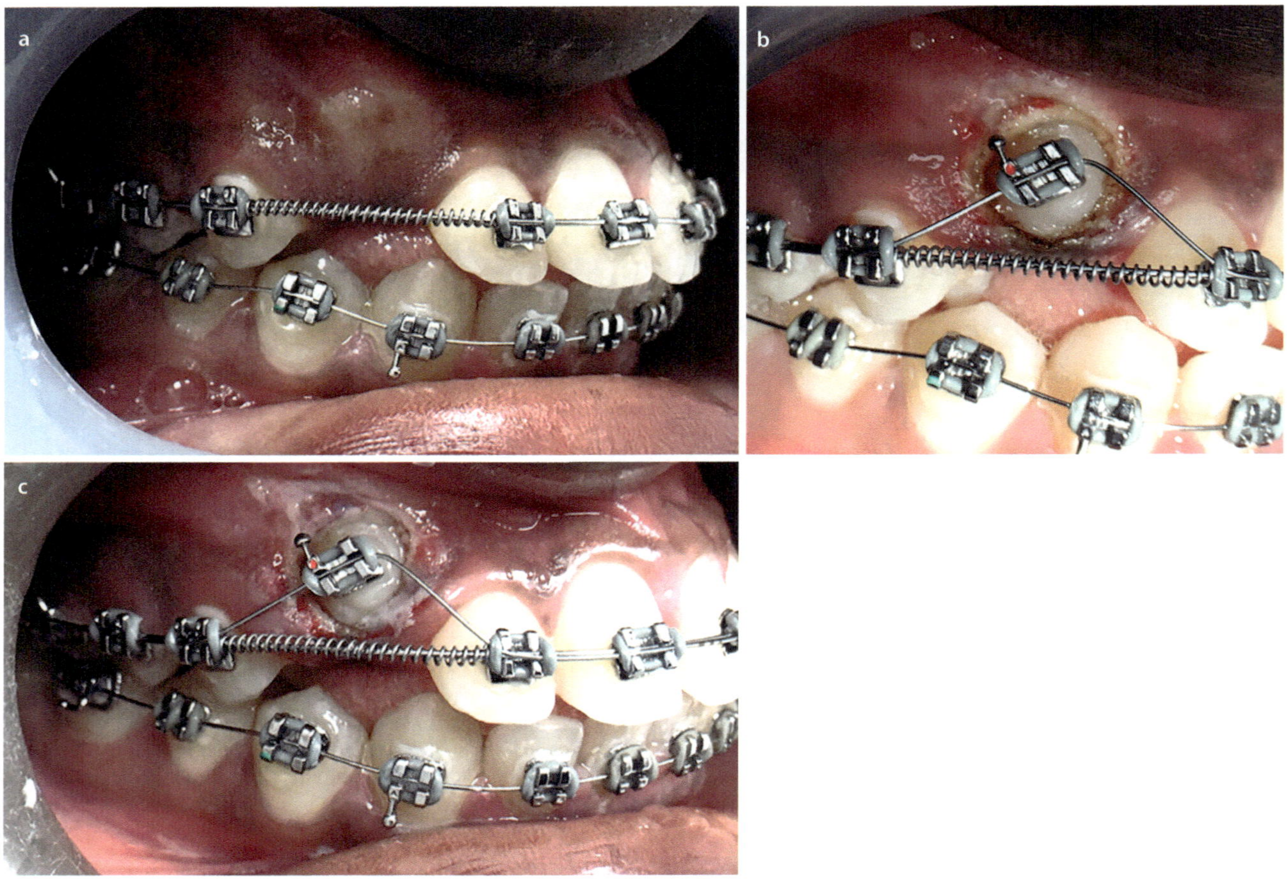

Fig. 12.5 **a** A male patient with a buccally impacted maxillary right canine. **b** Immediate post-operative view of laser-assisted exposure through soft tissue ablation. The bloodless field facilitates the bonding process and placement of orthodontic brackets. Laser operating parameters were the 940 nm diode laser (Epic 10, Biolase, Irvine, CA), with an initiated 400-μm diameter fiber-optic tip, in a contact mode (gated-CW mode, average power output = 1 W, pulse duration = 1 ms, time on/time off = 50%). Time spent for the laser exposure was approximately 10 min. **c** Appearance at 24 h follow-up. Note that the amount of inflammation is minimal. The patient reported very minimal pain and discomfort during the first 24 h

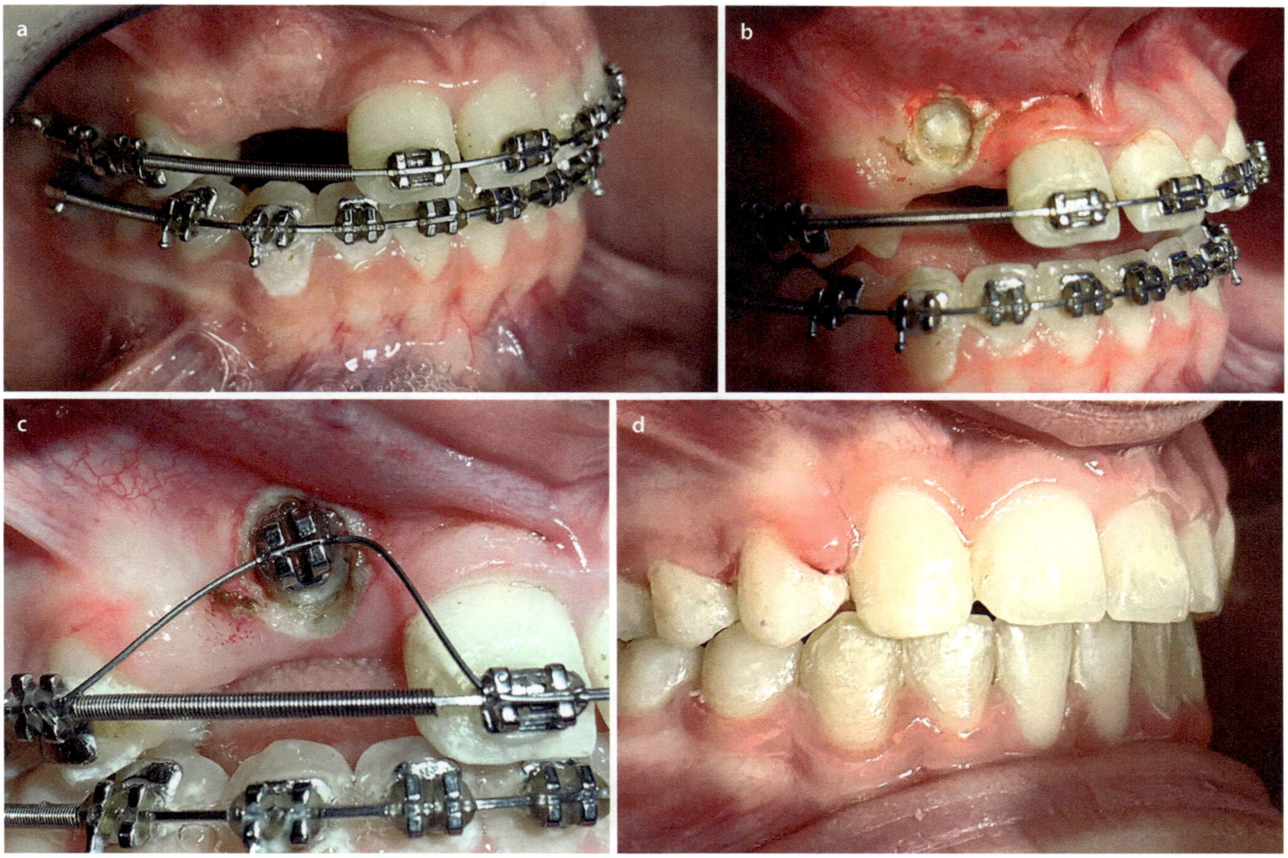

■ **Fig. 12.6** **a** A female patient with an impacted maxillary right canine. The maxillary right lateral incisor was extracted due to severe root resorption caused by the impacted canine. **b** Immediately after laser-assisted exposure. A 940 nm diode laser (Epic 10, Biolase, Irvine, CA) with an initiated 400-μm diameter fiber-optic tip, in a contact mode, was used (gated-CW mode, average power output = 1 W, pulse duration = 1 ms, time on/time off = 50%). Time spent for the laser exposure was approximately 10 min. In order to preserve the keratinised tissue, the starting exposure point should be the tip of the crown in buccal exposures. **c** After bonding a small size bracket, please note that the bloodless exposure site facilitates immediate placement of orthodontic bracket. Orthodontic force was applied with 014 niti wire immediately after exposure that reduces the treatment time. **d** At 6-month follow-up after canine substitution and at the time brace removal, note the healthy keratinised tissue at the maxillary right canine labial aspects

undergone laser exposure right after exposure and at 24 h follow-up. Note that the amount of post-operative inflammation is minimal.

■ Figure 12.6a shows a female patient with an impacted maxillary right canine that caused severe root resorption of upper right lateral incisor (which was extracted due to poor prognosis), right after laser exposure (■ Fig. 12.6b), bonding and applying orthodontic force (■ Fig. 12.6c) as well as at 6 months follow-up (■ Fig. 12.6d).

■ Figures 12.7a–d illustrate the remarkable healing process in a male patient with a palatally impacted maxillary right canine, after palatal laser exposure.

■ Figures 12.8a–j show another patient with a buccally impacted maxillary right canine and a partially erupted maxillary left canine. The gated-CW mode and pulse duration of 1 m second was used, which led to minimum post-operative discomfort and excellent healing at subsequent follow-ups. The diode laser in this case clearly provided bloodless site that allowed immediate orthodontic bonding of the maxillary canines and reducing the treatment time.

Compared to laser tooth exposure, a patient with a similar superficially palatally impacted canine is usually referred for full-thickness mucoperiosteal flap, which can be very aggressive and often needs placement of a protective dressing (pack) over the surgical site while it heals [92]. The use of scalpel usually involves suturing with stitches during surgical procedure that need to be removed 1–2 weeks post-operatively [93]. All demonstrated cases were performed using the 940 nm diode (InGaAsP) laser (maximum power output = 10 W, Epic 10, Biolase, Irvine, CA), using a 400-μm diameter fiber-optic tip, in a contact mode and after initiation.

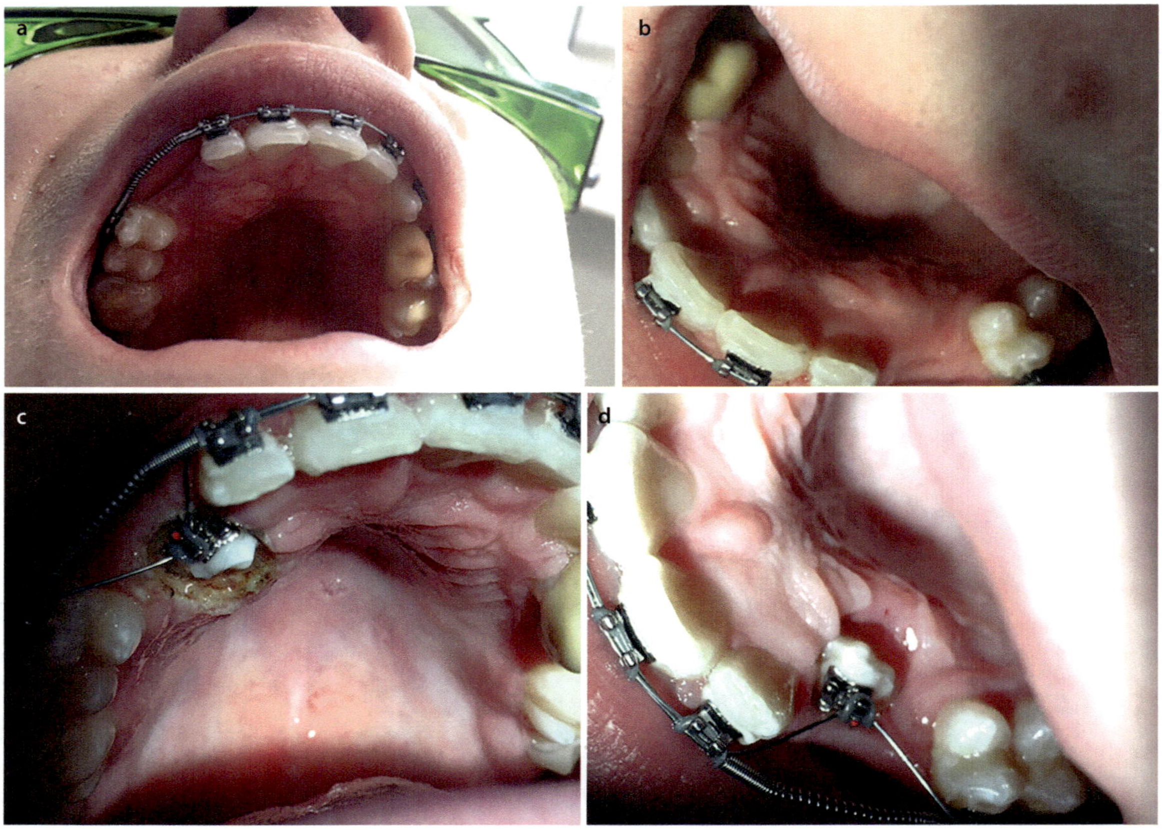

12

Fig. 12.7 **a** Pre-operative appearance with the patient wearing the safety protective glasses. The palatal canine bulge is clinically evident. **b** Palatal view. **c** Immediately after palatal laser exposure. Laser operating parameters were the 940 nm diode laser (Epic 10, Biolase, Irvine, CA), with an initiated 400-µm diameter fiber-optic tip, in a contact mode (gated-CW mode, average power output = 1 W, pulse duration = 1 ms, time on/time off = 50%). Time spent for the laser exposure was approximately 15 min. An aggressive conventional full-thickness mucoperiosteal flap often needs placement of a protective dressing (pack) over the surgical site while it heals and is associated with significant bleeding in the exposure site that can compromise the bonding process. **d** At 2 weeks follow-up. Please note the excellent healing without signs of inflammation

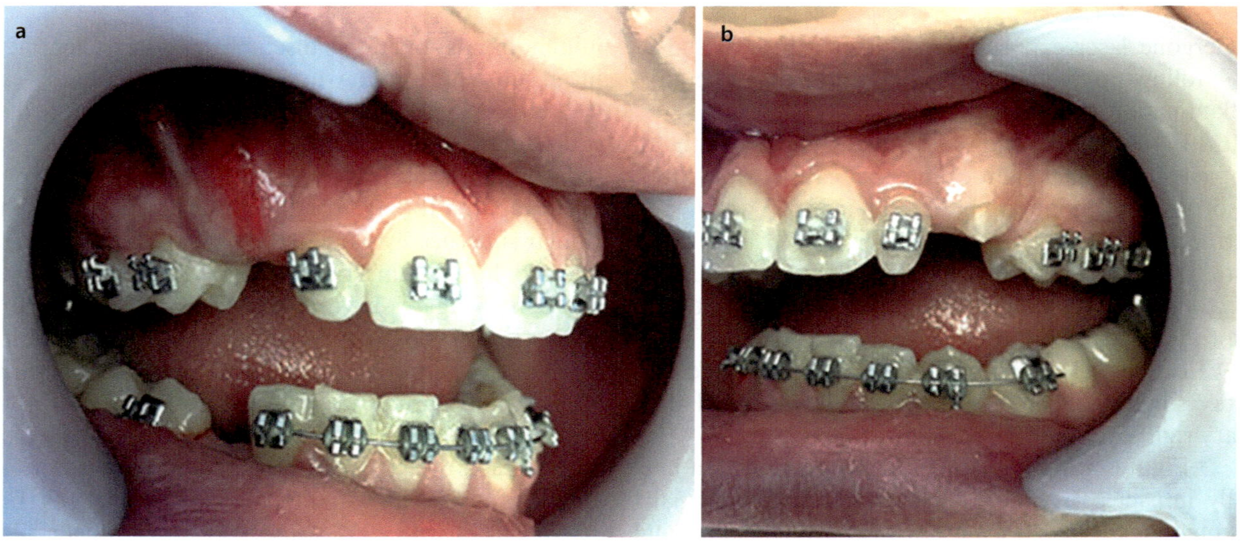

Fig. 12.8 **a**, **b** A male patient with a buccally impacted maxillary right canine and a partially erupted maxillary left canine, immediate post-operative view. **c**, **d** Laser operating parameters were the 940 nm diode laser (Epic 10, Biolase, Irvine, CA), with an initiated 400-µm diameter fiber-optic tip, in a contact mode (gated-CW mode, average power output = 1 W, pulse duration = 1 ms, time on/time off = 50%). Time spent for the laser exposure was approximately 15 min. **e**, **f** Following placement of orthodontic wire. **g**, **h** Appearance at 1 week. **i**, **j** Appearance at 4-month follow-up

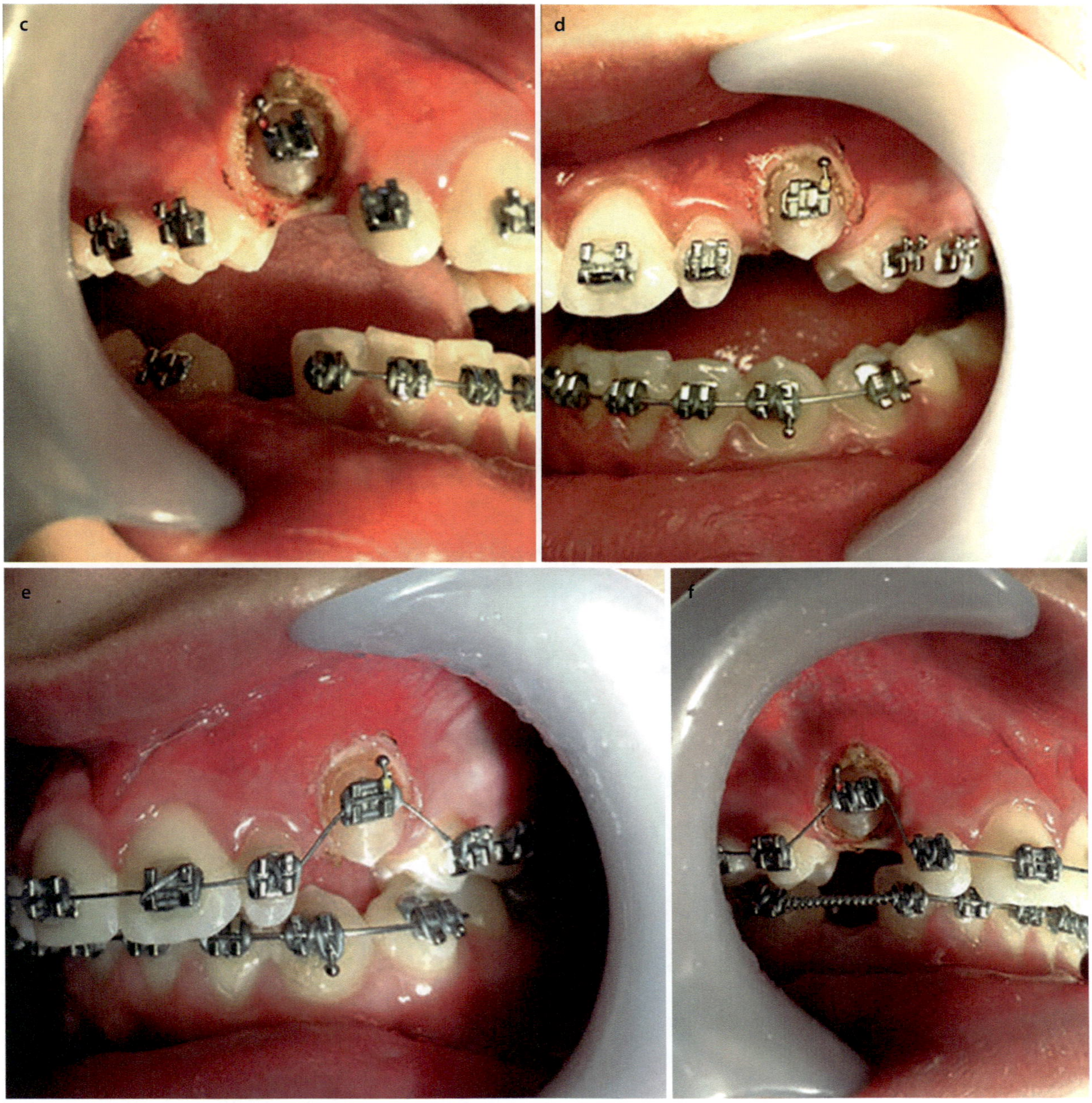

Fig. 12.8 (continued)

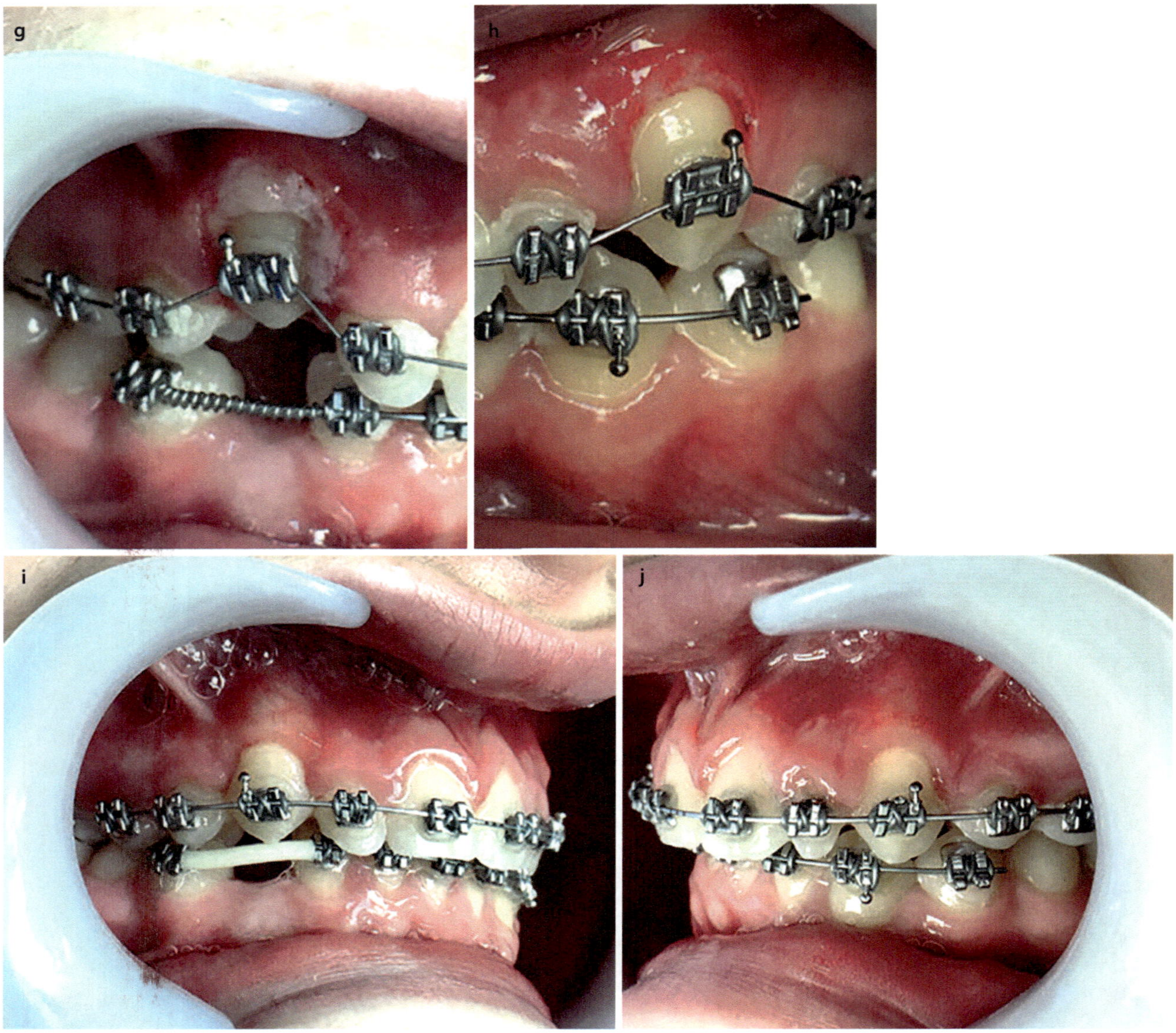

Fig. 12.8 (continued)

12.3.6 Comparison of the Efficacy of Diode Lasers and Scalpel Used for Minimally Invasive Exposure of Impacted Teeth or Teeth with Delayed Eruption

A recent scoping review [94] identified and analyzed the evidence evaluating the efficacy of the diode laser surgical exposure of impacted teeth or teeth with delayed eruption (both with no overlaying bone) vs. conventional scalpel surgical exposure. The literature search revealed no high-quality evidence. However, four prospective studies were identified [95–98]. Diode laser application was associated with less pain or analgesic need, minimal/no bleeding, and no need for suturing after surgery [94]. The laser wavelengths used were 808 nm, 810 nm, 935 nm, and 980 nm. This review concluded that comparison of the study outcomes was not possible; all the studies had methodological issues and their funding sources were not mentioned.

Therefore, adequately powered clinical trials are needed for comparing outcomes from diode laser surgical exposure vs. conventional methods, identifying the ideal laser characteristics, and assessing the long-term periodontal health of laser-exposed teeth and any potential risks.

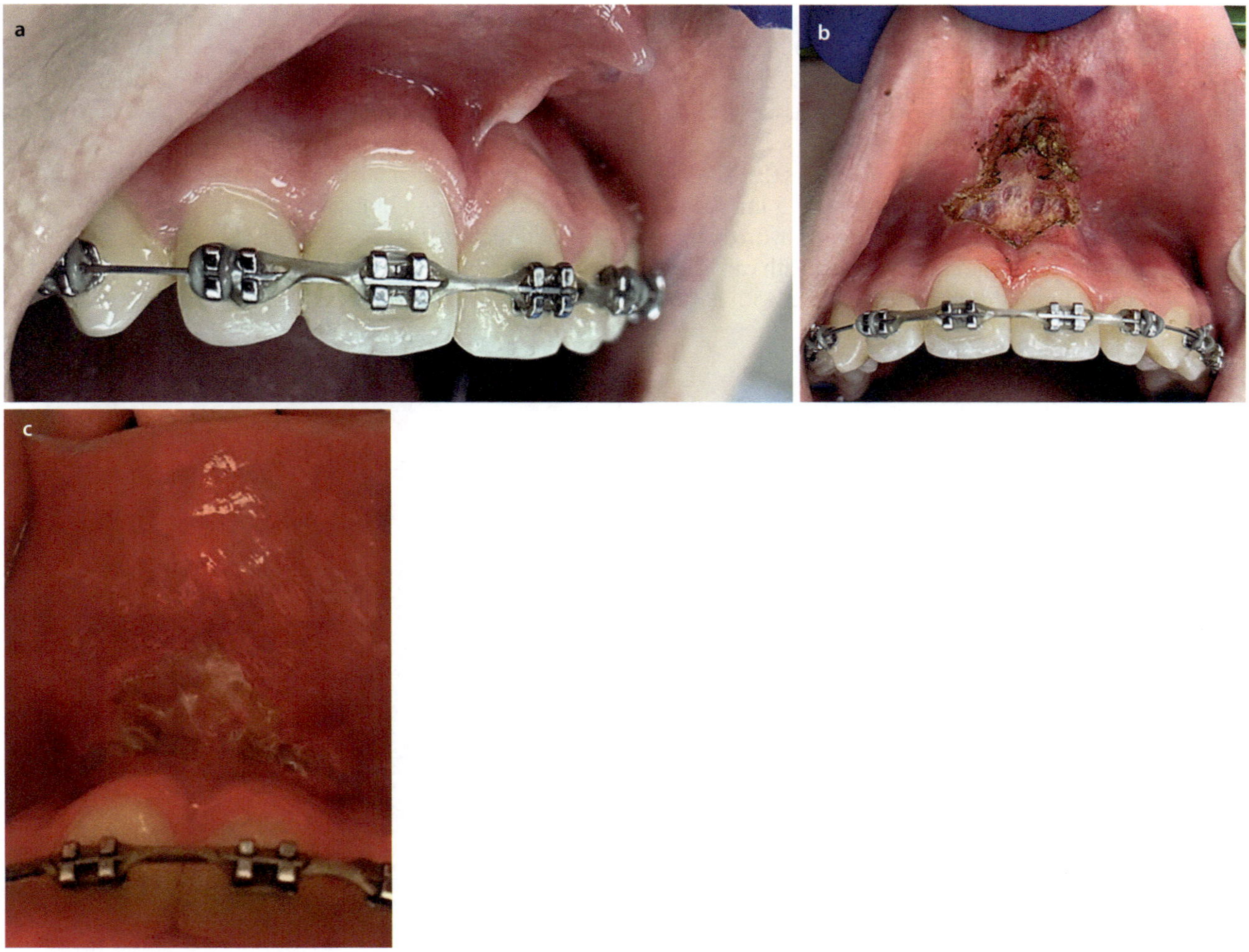

Fig. 12.9 **a** A female patient presenting with a labial frenum. **b** Following laser frenectomy with a 940 nm diode laser (Epic 10, Biolase, Irvine, CA). **c** At 1-week follow-up

12.3.7 Frenectomy and Other Applications of Diode Lasers for Soft Tissue Procedures in Orthodontics

Diode lasers have been used to uncover temporary anchorage devices (TADs), in frenectomy where highly attached frenum impedes tooth movement in diastema cases (Fig. 12.9a–c), circumferential supracrestal fiberotomy to prevent relapse of rotated teeth, removal of operculum on mandibular second molars that prevents banding, or to improve healing of minor aphthous ulceration following placement of fixed orthodontic braces [7, 55, 70, 99].

12.4 The Use of Non-contact Ablative Lasers (CO_2 and Erbium Lasers) for Frenectomy and Tooth Exposure

Traditionally ablative non-contact lasers (CO_2 and erbium lasers) have been used in the field of oral and maxillofacial surgery and aesthetic medicine [100–102]. Unfractionated CO_2 laser (10,600 nm) can't be delivered with a quartz fiber as it is absorbed by the quartz and instead is delivered using a system of mirrors and waveguides [102]. CO_2 laser has been used to treat wrinkles, skin scars tissues, premalignant lesions (oral leukoplakia), ▶ warts, and benign neoplasms [102, 103]. CO_2 and erbium lasers both target water, however the water absorption coefficient of Er:YAG (2940 nm) is about ten times that of a CO_2 laser, limiting its tissue penetration providing more precise ablation with minimal thermal damage of surrounding tissue [100–103]. Soft tissue penetration depth for CO_2 laser is approximately 0.2 mm [56, 60] and for erbium lasers (Er:YAG and Er,Cr:YSGG lasers) can be as shallow as 5 μm [56, 58]. CO_2 lasers have the highest absorption in hydroxyapatite and calcium phosphate and must be used with care during soft tissue procedures to avoid direct contact with hard tissue [56]. As mentioned earlier, erbium lasers target molecular water or the hydroxide ion as primary targets and mineral as a secondary target and can be used for both hard and soft tissue procedures [56]. However, erbium lasers are limited by their lack of ▶ hemostasis, which makes them unattractive for exposure and bonding of impacted teeth and limited oral surgery setup available in orthodontic practices.

Selected uses of CO_2 (◘ Figs. 12.10a–e and 12.11a–i) and erbium (◘ Fig. 12.12a–g) lasers in three patients

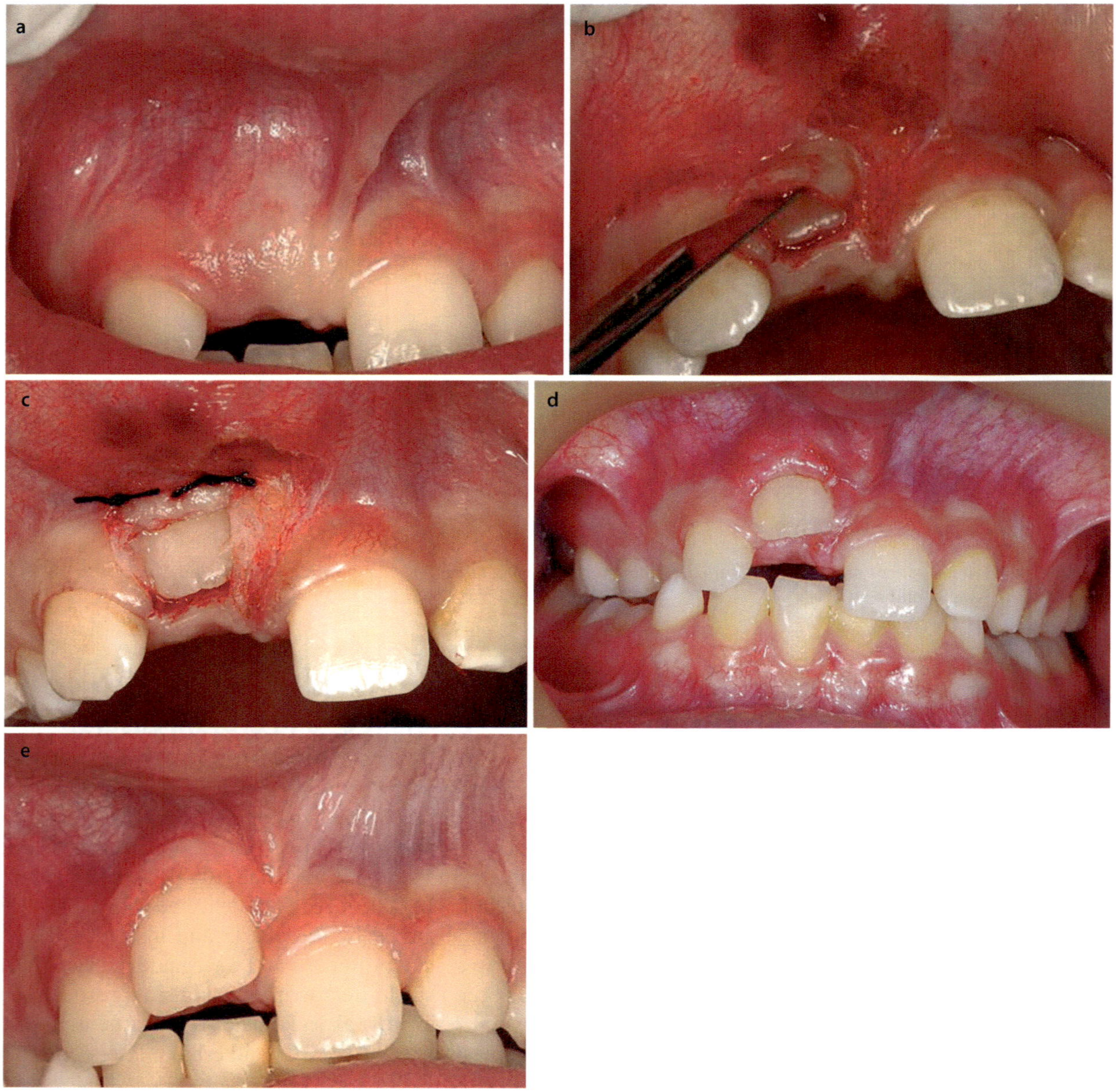

◘ **Fig. 12.10** **a** A 10-year-old patient with impacted upper right central incisor (UR1) and labial frenum. **b** Following the application of local anesthesia, simultaneous crown exposure of UR1 (apically positioned flap) and frenectomy carried out using a 10,600 nm CO_2 laser (average power: 2 W, 50% on, spot diameter: 200 μm, total irradiation time: 3 mm, Beam direction: parallel to the attached gingiva to avoid any damage on periosteum and leave enough tissue thickness, irradiation time: 20 s, cooling with water for 40 s), charring was removed with gauze. **c** Immediate post-operative appearance. **d** Post-operative appearance at 2 weeks. **e** At 3-month follow-up. Case courtesy Dr M. Chala

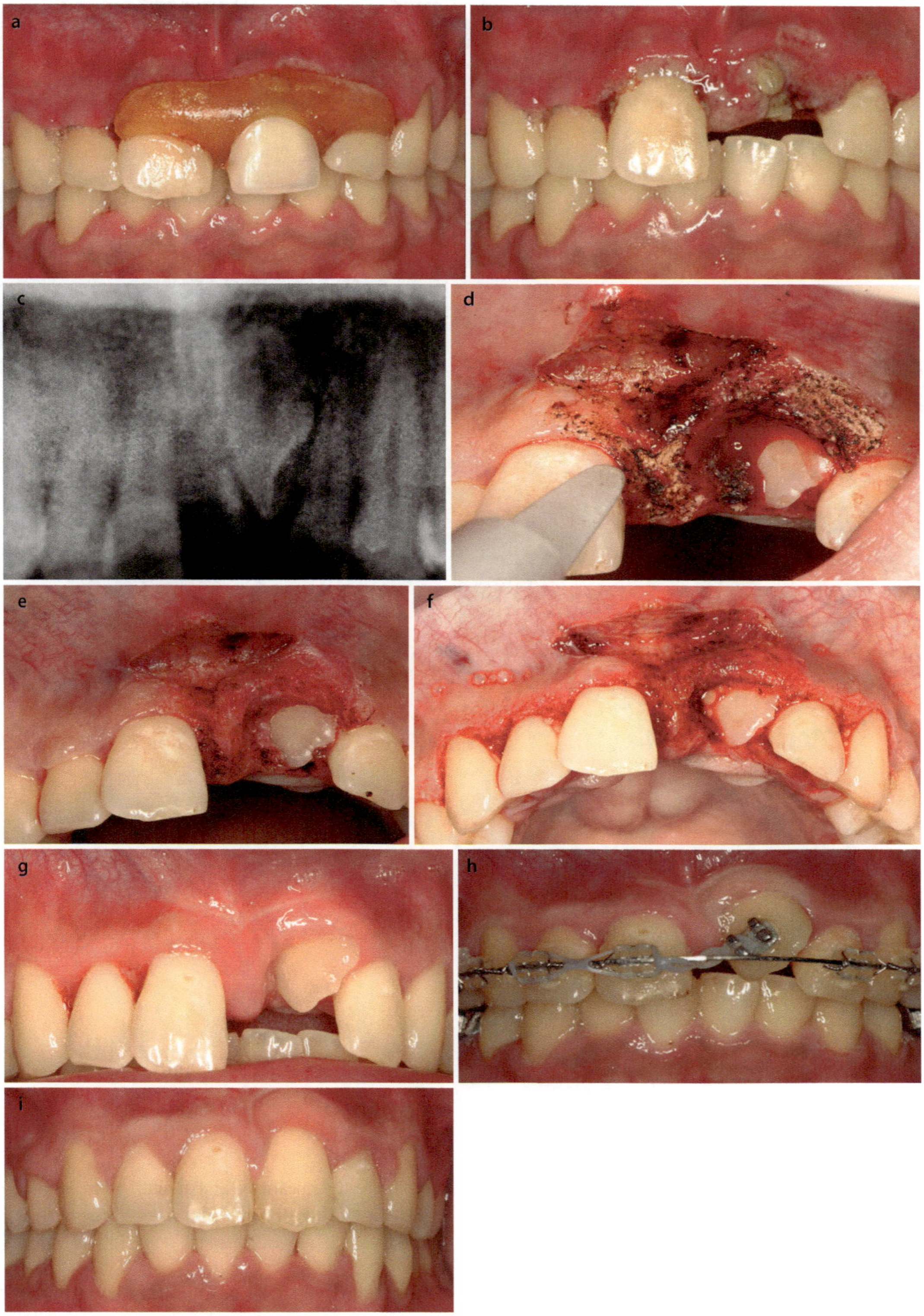

Fig. 12.11 **a** A 16-year-old patient with impacted upper left central incisor (UL1), remnant of URA, and labial frenum; the patient wore a provisional restoration in UL1 area for almost 8 years. **b** Appearance pre-operatively, demonstrating lack of attached gingiva in UL1 region. **c** Periapical radiograph showing rotated UL1 and fragment of dental origin, believed to be ULA. **d** Following extraction of URA remnant, frenectomy and vestibuloplasty with 10,600 nm CO_2 laser to remove the hyperplastic inflamed tissues and further exposure of UL1 crown. **e** The surgical site after removal of carbonized tissue. **f** Diode 1064 nm laser used for decontamination of the UL1 region to improve the gingival health. **g** The appearance at 1-month follow-up. **h** The appearance during orthodontic treatment. **i** Appearance after finishing orthodontic treatment at 18 years of age. Case courtesy Dr M. Chala

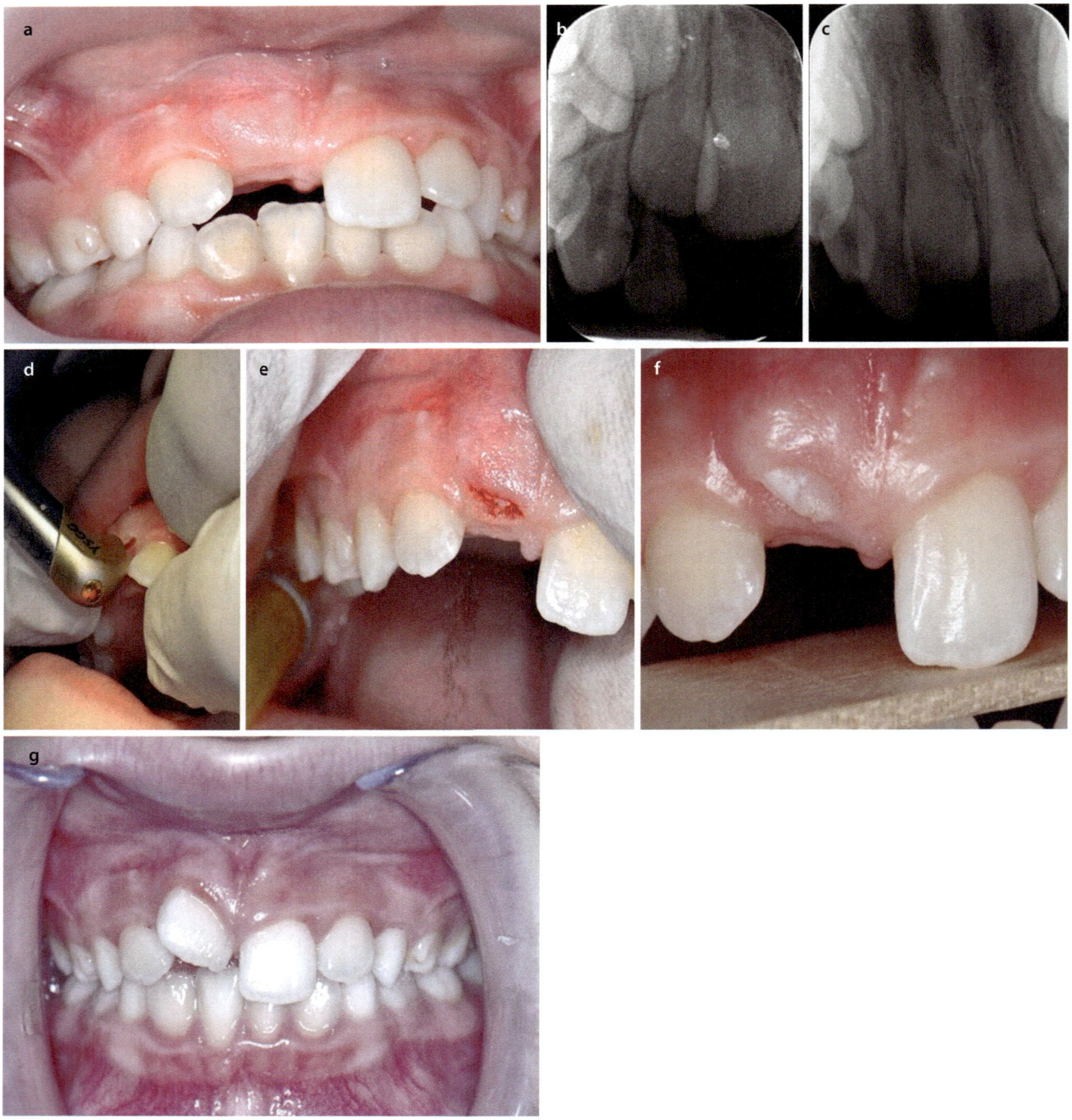

Fig. 12.12 **a** Intra-oral view showing delayed eruption of UR1. **b** Periapical X-Ray view. **c** Follow-up at 2 years. **d** Laser exposure of UR1 with a 2780 nm Er,Cr:YSGG laser. **e** Immediate post-operative appearance. **f** Appearance at 1 month. **g** Appearance at 2 months follow-up. Case courtesy Dr M. Chala

for frenectomy and exposure of impacted incisors are shown.

The case represented by Fig. 12.11a–i, is of a 16-year-old patient with impacted upper left central incisor (UL1); the patient wore an upper anterior acrylic removable prosthesis for almost 8 years. Lack of attached gingiva in UL1 region, traumatic fitting of the prosthesis, excessive plaque accumulation, are evident. The radiographic examination reveals UL1, impacted and rotated, with an associated fragment of dental origin. Treatment included extraction of URA remnant, frenectomy, and vestibuloplasty with 10,600 nm CO_2 laser to remove the hyperplastic inflamed tissues and further exposure of UL1 crown. The use of the diode 1064 nm laser was employed for decontamination of the UL1 region and to improve the gingival health. Ongoing orthodontic treatment and follow-up provided a finished result at 18 years of age.

In case images ◘ Fig. 12.12a–g, demonstration of a further laser-assisted soft tissue exposure of an unerupted upper central incisor (UR1). Intra-oral view of a 9-year-old boy presenting with delayed eruption of UR1 (a), X-Ray view (b) and follow-up at 2 years (c). Application of local anesthesia followed by laser exposure of UR1 with a 2780 nm Er,Cr:YSGG laser (d, e) (Biolase, Gold hand piece, tip: 0.6 mm MZ tip, Distance: 1 mm, Hard tissue close contact mode, power: 1–1.50 W, 20 Hz, 30% H_2O, 70% Air, 6 mL/min H_2O). Appearance at 1 month (f) and 2 months follow-up (g).

12.4.1 Post-operative Instructions

A review of published literature indicates suggestions such as keeping the area clean, using soft-bristle toothbrush (or cotton swab), rinsing the mouth with salt water three or four times daily for several days and removing any remaining tissue with a wet cotton swab [7, 55], rubbing vitamin E gel over the healing area (to aid in healing and keeping the treated area moist), as well as taking over-the-counter analgesics such as acetaminophen (500-mg tablet qid prn 3 3/7) that have been suggested for pain control [55, 76].

12.4.2 Summary

The use of lasers for soft tissue procedures introduced alternative adjuncts to scalpel conventional surgery. Laser-assisted gingivectomy in orthodontics is used to improve oral hygiene or bracket positioning, to enhance gingival aesthetics, and to surgically expose superficially impacted teeth reducing post-operative pain, and potentially decreasing orthodontic treatment time. Historically, non-contact ablative lasers (CO_2 and erbium lasers) have been used for soft tissue oral surgery. The diode lasers are the most commonly used in orthodontics for incision/excision, which is performed in a contact mode, and the use of a 400-μm diameter fiber-optic tip is recommended. Compared to scalpel surgery, diode lasers offer a clean and bloodless surgical site, with an added benefit of photobiomodulation that enhances the wound healing and reduces the patient discomfort. There is obviously a strong argument for laser safety that can be addressed with proper training. However, there is paucity of data regarding the most effective wavelength (810, 940, or 980 nm) for performing diode laser gingivectomy or tooth exposure, as well as other laser parameters such as continuous or pulsed/gated mode of delivery, and the optimum power output that requires further research.

Acknowledgment The author acknowledges the contribution of Dr. Marianna Chala (Private Practice, Athens, Greece) for illustrated patients in ◘ Figs. 12.10 and 12.11, as well as Dr. Konstantinos Arapostathis (Associate Professor, Paediatric Dentistry, School of Dentistry, Faculty of Health Sciences, Aristotle University of Thessaloniki) for illustrated patient in ◘ Fig. 12.12.

References

1. Zucchelli G, Mounssif I. Periodontal plastic surgery. Periodontol 2000. 2015;68(1):333–68.
2. Aoki A, Mizutani K, Schwarz F, et al. Periodontal and peri-implant wound healing following laser therapy. Periodontol 2000. 2015;68:217–69.
3. Pang P, Andreana S, Aoki A, et al. Laser energy in oral soft tissue applications. J Laser Dent. 2010;18:123–31.
4. Aoki A, Sasaki K, Watanabe H, Ishikawa I. Lasers in nonsurgical periodontal therapy. Periodontol 2000. 2004;36:59–97.
5. Chambrone L, Zadeh HH, Evidence-based rationale for the management of mucogingival deformities before or after orthodontic treatment, Seminars in Orthodontics, 2023, https://doi.org/10.1053/j.sodo.2023.09.002.
6. Sawabe M, Aoki A, Komaki M, et al. Gingival tissue healing following Er:YAG laser ablation compared to electrosurgery in rats. Lasers Med Sci. 2015;30:875–83.
7. Borzabadi-Farahani A, Cronshaw M. Lasers in orthodontics. In: Coluzzi D, Parker S, editors. Lasers in dentistry—current concepts. Textbooks in contemporary dentistry. Cham: Springer; 2017. https://doi.org/10.1007/978-3-319-51944-9_12.
8. Aoki A, Takasaki A, Pourzarandian A, et al. Photo-biomodulation laser strategies in periodontal therapy. In: Proceedings of Light-Activated Tissue Regeneration and Therapy II. New York: Springer; 2008. p. 181–90.
9. Izumi Y, Aoki A, Yamada Y, et al. Current and future periodontal tissue engineering. Periodontol 2000. 2011;56:166–87.
10. Avci P, Gupta A, Sadasivam M, et al. Low-level laser (light) therapy (LLLT) in skin: stimulating, healing, restoring. Semin Cutan Med Surg. 2013;32:41–52.
11. Khan I, Arany P. Biophysical approaches for oral wound healing: emphasis on photobiomodulation. Adv Wound Care (New Rochelle). 2015;4:724–37.
12. Tunér J, Beck-Kristensen PH, Ross G, Ross A. Photobiomodulation in dentistry. In: Convissar RA, editor. Principles and practice of laser dentistry. 2nd ed. St. Louis: Elsevier; 2016. p. 251–74.
13. Al Ghamdi KM, Kumar A, Moussa NA. Low-level laser therapy: a useful technique for enhancing the proliferation of various cultured cells. Lasers Med Sci. 2012;27:237–49.
14. Mester E, Spiry T, Szende B, Tota JG. Effect of laser rays on wound healing. Am J Surg. 1971;122:532–5.
15. Enwemeka CS, Parker JC, Dowdy DS, et al. The efficacy of low-power lasers in tissue repair and pain control: a meta-analysis study. Photomed Laser Surg. 2004;22:323–9.
16. Woodruff LD, Bounkeo JM, Brannon WM, et al. The efficacy of laser therapy in wound repair: a meta-analysis of the literature. Photomed Laser Surg. 2004;22:241–7.
17. Albertini R, Villaverde AB, Aimbire F, et al. Anti-inflammatory effects of low-level laser therapy (LLLT) with

two different red wavelengths (660 nm and 684 nm) in carrageenan-induced rat paw edema. J Photochem Photobiol B. 2007;89:50–5.
18. Bjordal JM, Johnson MI, Iversen V, et al. Low-level laser therapy in acute pain: a systematic review of possible mechanisms of action and clinical effects in randomized placebo controlled trials. Photomed Laser Surg. 2006;24:158–68.
19. Isman E, Aras MH, Cengiz B, et al. Effects of laser irradiation at different wavelengths (660, 810, 980, and 1064 nm) on transient receptor potential melastatin channels in an animal model of wound healing. Lasers Med Sci. 2015;30:1489–95.
20. Ogita M, Tsuchida S, Aoki A, et al. Increased cell proliferation and differential protein expression induced by low-level Er:YAG laser irradiation in human gingival fibroblasts: proteomic analysis. Lasers Med Sci. 2015;30:1855–66.
21. Dias SB, Fonseca MV, Dos Santos NC, et al. Effect of GaAIAs low-level laser therapy on the healing of human palate mucosa after connective tissue graft harvesting: randomized clinical trial. Lasers Med Sci. 2015;30:1695–702.
22. de Melo Rambo CS, Silva JA Jr, Serra AJ, et al. Comparative analysis of low-level laser therapy (660 nm) on inflammatory biomarker expression during the skin wound-repair process in young and aged rats. Lasers Med Sci. 2014;29:1723–33.
23. Liao X, Xie GH, Liu HW, et al. Helium-neon laser irradiation promotes the proliferation and migration of human epidermal stem cells in vitro: proposed mechanism for enhanced wound re-epithelialization. Photomed Laser Surg. 2014;32:219–25.
24. Fujimura T, Mitani A, Fukuda M, et al. Irradiation with a low-level diode laser induces the developmental endothelial locus-1 gene and reduces proinflammatory cytokines in epithelial cells. Lasers Med Sci. 2014;29:987–94.
25. Batista JD, Sargenti-Neto S, Dechichi P, et al. Low-level laser therapy on bone repair: is there any effect out-side the irradiated field? Lasers Med Sci. 2015;30:1569–74.
26. Pinheiro AL, Aciole GT, Ramos TA, et al. The efficacy of the use of IR laser phototherapy associated to biphasic ceramic graft and guided bone regeneration on surgical fractures treated with miniplates: a histological and histomorphometric study on rabbits. Lasers Med Sci. 2014;29:279–88.
27. de Vasconcellos LM, Barbara MA, Deco CP, et al. Healing of normal and osteopenic bone with titanium implant and low-level laser therapy (GaAlAs): a histomorphometric study in rats. Lasers Med Sci. 2014;29:575–80.
28. Tim CR, Pinto KN, Rossi BR, et al. Low-level laser therapy enhances the expression of osteogenic factors during bone repair in rats. Lasers Med Sci. 2014;29:147–56.
29. Pagin MT, de Oliveira FA, Oliveira RC, et al. Laser and light-emitting diode effects on pre-osteoblast growth and differentiation. Lasers Med Sci. 2014;29:55–9.
30. Peccin MS, de Oliveira F, Muniz Renno AC, et al. Helium-neon laser improves bone repair in rabbits: comparison at two anatomic sites. Lasers Med Sci. 2013;28:1125–30.
31. Boldrini C, de Almeida JM, Fernandes LA, et al. Biomechanical effect of one session of low-level laser on the bone-titanium implant interface. Lasers Med Sci. 2013;28:349–52.
32. Pinheiro AL, Soares LG, Barbosa AF, et al. Does LED phototherapy influence the repair of bone defects grafted with MTA, bone morphogenetic proteins, and guided bone regeneration? A description of the repair process on rodents. Lasers Med Sci. 2012;27:1013–24.
33. Rosa AP, de Sousa LG, Regalo SC, et al. Effects of the combination of low-level laser irradiation and recombinant human bone morphogenetic protein-2 in bone repair. Lasers Med Sci. 2012;27:971–7.
34. Anders JJ, Moges H, Wu X, et al. In vitro and in vivo optimization of infrared laser treatment for injured peripheral nerves. Lasers Surg Med. 2014;46:34–45.
35. Takhtfooladi MA, Jahanbakhsh F, Takhtfooladi HA, et al. Effect of low-level laser therapy (685 nm, 3 J/cm(2)) on functional recovery of the sciatic nerve in rats following crushing lesion. Lasers Med Sci. 2015;30:1047–52.
36. Masoumipoor M, Jameie SB, Janzadeh A, et al. Effects of 660- and 980-nm low-level laser therapy on neuropathic pain relief following chronic constriction injury in rat sciatic nerve. Lasers Med Sci. 2014;29:1593–8.
37. Gasperini G, de Siqueira IC, Costa LR. Lower-level laser therapy improves neurosensory disorders resulting from bilateral mandibular sagittal split osteotomy: a randomized crossover clinical trial. J Craniomaxillofac Surg. 2014;42:e130–3.
38. Akgul T, Gulsoy M, Gulcur HO. Effects of early and delayed laser application on nerve regeneration. Lasers Med Sci. 2014;29:351–7.
39. Lazovic M, Ilic-Stojanovic O, Kocic M, et al. Placebo-controlled investigation of low-level laser therapy to treat carpal tunnel syndrome. Photomed Laser Surg. 2014;32:336–44.
40. de Sousa AP, Silveira NT, de Souza J, et al. Laser and LED phototherapies on angiogenesis. Lasers Med Sci. 2013;28:981–7.
41. Góralczyk K, Szymańska J, Łukowicz M, et al. Effect of LLLT on endothelial cells culture. Lasers Med Sci. 2015;30:273–8.
42. Esmaeelinejad M, Bayat M, Darbandi H, et al. The effects of low-level laser irradiation on cellular viability and proliferation of human skin fibroblasts cultured in high glucose mediums. Lasers Med Sci. 2014;29:121–9.
43. Basso FG, Oliveira CF, Kurachi C, et al. Biostimulatory effect of low-level laser therapy on keratinocytes in vitro. Lasers Med Sci. 2013;28:367–74.
44. Fujihara NA, Hiraki KR, Marques MM. Irradiation at 780 nm increases proliferation rate of osteoblasts independently of dexamethasone presence. Lasers Surg Med. 2006;38:332–6.
45. Borzabadi-Farahani A. Effect of low-level laser irradiation on proliferation of human dental mesenchymal stem cells; a systemic review. J Photochem Photobiol B. 2016;162:577–82.
46. Gonçalves RV, Novaes RD, Matta SLP, et al. Comparative study of the effects of gallium-aluminum-arsenide laser photobiomodulation and healing oil on skin wounds in Wistar rats: a histomorphometric study. Photomed Laser Surg. 2010;28:597–602.
47. Xavier M, David DR, Souza RA, et al. Anti-inflammatory effects of low-level light emitting diode therapy on Achilles tendinitis in rats. Lasers Surg Med. 2010;42:553–8.
48. Novaes RD, Gonçalves RV, Cupertino MC, et al. The energy density of laser light differentially modulates the skin morphological reorganization in a murine model of healing by secondary intention. Int J Exp Pathol. 2014;95:138–46.
49. Posten W, Wrone DA, Dover JS, et al. Low-level laser therapy for wound healing: mechanism and efficacy. Dermatol Surg. 2005;31:334–40.
50. Houreld NN, Sekhejane PR, Abrahamse H. Irradiation at 830 nm stimulates nitric oxide production and inhibits pro-inflammatory cytokines in diabetic wounded fibroblast cells. Lasers Surg Med. 2010;42:494–502.
51. Pogrel MA, Chen JW, Zhang K. Effects of low-energy gallium-aluminum-arsenide laser irradiation on cultured fibroblasts and keratinocytes. Lasers Surg Med. 1997;20:426–32.
52. Martignago CC, Oliveira RF, Pires-Oliveira DA, et al. Effect of low-level laser therapy on the gene expression of collagen

and vascular endothelial growth factor in a culture of fibroblast cells in mice. Lasers Med Sci. 2015;30:203–8.
53. Gonçalves RV, Mezêncio JMS, Benevides GP, et al. Effect of gallium-arsenide laser, gallium-aluminum-arsenide laser and healing ointment on cutaneous wound healing in Wistar rats. Braz J Med Biol Res. 2010;43:350–5.
54. Kang Y, Rabie AB, Wong RW. A review of laser applications in orthodontics. Int J Orthod Milwaukee. 2014;25:47–56.
55. Chmura LG, Convissar RA. Lasers in orthodontics. In: Convissar RA, editor. Principles and practice of laser dentistry. 2nd ed. St. Louis: Elsevier; 2016. p. 203–19.
56. Mizutani K, Aoki A, Coluzzi D, et al. Lasers in minimally invasive periodontal and peri-implant therapy. Periodontol 2000. 2016;71:185–212.
57. Migliorati EKJ, de Almeida Rosa DS. Regenerative laser periodontal therapy. In: Convissar RA, editor. Principles and practice of laser dentistry. 2nd ed. St. Louis: Elsevier; 2016. p. 67–88.
58. Ishikawa I, Aoki A, Takasaki AA, et al. Application of lasers in periodontics: true innovation or myth? Periodontol 2000. 2009;50:90–126.
59. Pang P, Andreana S, Aoki A, et al. Laser energy in oral soft tissue applications. J Laser Dent. 2011;18:123–31.
60. Merigo E, Clini F, Fornaini C, et al. Laser-assisted surgery with different wavelengths: a preliminary ex vivo study on thermal increase and histological evaluation. Lasers Med Sci. 2013;28:497–504.
61. Goharkhay K, Moritz A, Wilder-Smith P, et al. Effects on oral soft tissue produced by a diode laser in vitro. Lasers Surg Med. 1999;25:401–6.
62. Jin JY, Lee SH, Yoon HJ. A comparative study of wound healing following incision with a scalpel, diode laser or Er,Cr:YSGG laser in guinea pig oral mucosa: a histological and immunohistochemical analysis. Acta Odontol Scand. 2010;68:232–8.
63. Hall RN, Fenner GE, Kingsley JD, et al. Coherent light emission from GaAs junctions. Phys Rev Lett. 1962;9:366–8.
64. Nathan MI, Dumke WP, Burns G, et al. Stimulated emission of radiation from GaAs pn junctions. Appl Phys Lett. 1962;1:62–4.
65. Nasim H, Jamil Y. Diode lasers: from laboratory to industry. Opt Laser Technol. 2014;56:211–22.
66. Fornaini C, Merigo E, Rocca JP, et al. 450 nm Blue laser and oral surgery: preliminary ex vivo study. J Contemp Dent Pract. 2016;17:795–800.
67. Braun A, Berthold M, Frankenberger R. The 445-nm semiconductor laser in dentistry—introduction of a new wavelength. Quintessenz. 2015;66:205–11.
68. Arroyo HH, Neri L, Fussuma CY, Imamura R. Diode laser for laryngeal surgery: a systematic review. Int Arch Otorhinolaryngol. 2016;20:172–9.
69. Borzabadi-Farahani A. The adjunctive soft-tissue diode laser in orthodontics. Compend Contin Educ Dent. 2017;38(eBook 5):e18–31.
70. Kravitz ND. The application of lasers in orthodontics. In: Krishnan V, Davidovitch Z, editors. Integrated clinical orthodontics. 1st ed. West Sussex: Wiley; 2012. p. 422–43.
71. Colluzzi DJ. Fundamentals of dental lasers: science and instruments. Dent Clin N Am. 2004;48:751–70.
72. Borchers R. Comparison of diode lasers in soft-tissue surgery using CW- and superpulsed mode: an in vivo study. Int J Laser Dent. 2001;1:17–27.
73. Al-Khatib AA, Al-Azzawi AS. Comparative study of diode laser 940 nm in performing frenectomy in both: continuous and pulsed modes: an in vivo study. J Dent Lasers. 2015;9:50–68.
74. Prabhu M, Ramesh A, Thomas B. Treatment of orthodontically induced gingival hyperplasia by diode laser—case report. Nitte Univ J Health Sci. 2015;5:66–8.
75. Kravitz ND, Graham JW, Nicozisis JL, Gill J. Compounded topical anesthetics in orthodontics. J Clin Orthod. 2015;49:371–7.
76. To TN, Rabie AB, Wong RW, McGrath CP. The adjunct effectiveness of diode laser gingivectomy in maintaining periodontal health during orthodontic treatment. Angle Orthod. 2013;83:43–7.
77. Hempton TJ, Dominici JT. Contemporary crown-lengthening therapy: a review. J Am Dent Assoc. 2010;141:647–55.
78. van Gastel J, Quirynen M, Teughels W, et al. Longitudinal changes in microbiology and clinical periodontal variables after placement of fixed orthodontic appliances. J Periodontol. 2008;79:2078–86.
79. Yáñez-Vico RM, Iglesias-Linares A, Ballesta-Mudarra S, et al. Short-term effect of removal of fixed orthodontic appliances on gingival health and subgingival microbiota: a prospective cohort study. Acta Odontol Scand. 2015;73:496–502.
80. Kloehn JS, Pfeifer JS. The effect of orthodontic treatment on the periodontium. Angle Orthod. 1974;44:127–34.
81. Palomo L, Palomo JM, Bissada NF. Salient periodontal issues for the modern biologic orthodontist. Semin Orthod. 2008;14:229–45.
82. Camargo PM, Melnick PR, Pirih FQ, et al. Treatment of drug-induced gingival enlargement: aesthetic and functional considerations. Periodontol 2000. 2001;27:131–8.
83. De Oliveira Guaré R, Costa SC, Baeder F, et al. Drug-induced gingival enlargement: biofilm control and surgical therapy with gallium–aluminum–arsenide (GaAlAs) diode laser—a 2-year follow-up. Spec Care Dentist. 2010;30:46–52.
84. Lee EA. Laser-assisted gingival tissue procedures in esthetic dentistry. Pract Proced Aesthet Dent. 2006;18(Suppl):2–6.
85. Sobouti F, Rakhshan V, Chiniforush N, Khatami M. Effects of laser-assisted cosmetic smile lift gingivectomy on postoperative bleeding and pain in fixed orthodontic patients: a controlled clinical trial. Prog Orthod. 2014;15:66.
86. Lione R, Pavoni C, Noviello A, et al. Conventional versus laser gingivectomy in the management of gingival enlargement during orthodontic treatment: a randomized controlled trial. Eur J Orthod. 2020;42:78–85.
87. Abdelhafez RS, Rawabdeh RN, Alhabashneh RA. The use of diode laser in esthetic crown lengthening: a randomized controlled clinical trial. Lasers Med Sci. 2022;37:2449–55.
88. Ericson S, Kurol J. Radiographic assessment of maxillary canine eruption in children with clinical signs of eruption disturbance. Eur J Orthod. 1986;8:133–40.
89. Wriedt S, Jaklin J, Al-Nawas B, Wehrbein H. Impacted upper canines: examination and treatment proposal based on 3D versus 2D diagnosis. J Orofac Orthop. 2011;73:28–40.
90. Counihan K, Al-Awadhi EA, Butler J. Guidelines for the assessment of the impacted maxillary canine. Dent Update. 2013;40(770–2):775–7.
91. Benson PE, Atwal A, Bazargani F, et al. Interventions for promoting the eruption of palatally displaced permanent canine teeth, without the need for surgical exposure, in children aged 9 to 14 years. Cochrane Database Syst Rev. 2021;12(12):CD012851.
92. Firestone AR, Scheurer PA, Bürgin WB. Patients' anticipation of pain and pain-related side effects, and their perception of pain as a result of orthodontic treatment with fixed appliances. Eur J Orthod. 1999;21:387–96.
93. Parkin N, Benson PE, Thind B, et al. Open versus closed surgical exposure of canine teeth that are displaced in the roof of the mouth. Cochrane Database Syst Rev. 2017;8(8):CD006966.

94. Borzabadi-Farahani A. A scoping review of the efficacy of diode lasers used for minimally invasive exposure of impacted teeth or teeth with delayed eruption. Photonics. 2022;9:265.
95. Migliario M, Rizzi M, Lucchina AG, Renò F. Diode laser clinical efficacy and mini-invasivity in surgical exposure of impacted teeth. J Craniofac Surg. 2016;27:e779–84.
96. Ize-Iyamu IN, Saheeb BD, Edetanlen BE. Comparing the 810 nm diode laser with conventional surgery in orthodontic soft tissue procedures. Ghana Med J. 2013;47:107–11.
97. Seifi M, Vahid-Dastjerdi E, Ameli N, et al. The 808 nm laser-assisted surgery as an adjunct to orthodontic treatment of delayed tooth eruption. J Lasers Med Sci. 2013;4:70–4.
98. Yossif RS, El-Destawy MT, El-Patal MA, Elbaiomy SY. Clinical outcome of diode laser usage versus conventional surgical technique in management of delayed erupted tooth. Al-Azhar J Dent Sci. 2017;20:201–8.
99. Miresmæili AF, Mollabashi V, Gholami L, et al. Comparison of conventional and laser-aided fiberotomy in relapse tendency of rotated tooth: a randomized controlled clinical trial. Int Orthod. 2019;17:103–13.
100. Renapurkar S, Strauss RA. Lasers in oral and maxillofacial surgery. In: Bonanthaya K, Panneerselvam E, Manuel S, Kumar VV, Rai A, editors. Oral and maxillofacial surgery for the clinician. Singapore: Springer; 2021. https://doi.org/10.1007/978-981-15-1346-639.
101. Waner M, Teresa O. Lasers and the treatment of congenital vascular lesions. J Oral Pathol Med. 2022;51:849–53.
102. Bhargava S, Goldust M, Singer H, et al. Evaluating resurfacing modalities in aesthetics. Clin Dermatol. 2022;40:274–82.
103. Vilar-Villanueva M, Somoza-Martín JM, Blanco-Carrión A, et al. Importance of the vaporization margin during CO laser treatment of oral leukoplakia: a survival study. Oral Dis. 2022; https://doi.org/10.1111/odi.14345.

Laser Use in Minor Oral Surgery

Omar Hamadah

Contents

D. J. Coluzzi, S. P. A. Parker (eds.), *Lasers in Dentistry—Current Concepts*, Textbooks in Contemporary Dentistry,
https://doi.org/10.1007/978-3-031-43338-2_13

Core Message

The initial experimentation and early clinical practice of laser use have been reported in the mid-1960s [1] and one of the first initial clinical surgeries was documented in 1977 [2]. Nowadays, adjunctive laser use is defining the standard of care for many oral surgeries, offering many advantages for both the surgeon and the patient. The purpose of this chapter is to explain those therapies and to demonstrate those benefits.

13.1 Introduction

13

In 1960, Maiman published his pioneering work on the ruby maser [3]. This was based on the theoretical statements of Einstein dated in 1917 [4] and the work of Schawlow and Townes 1958 [5]. Based on this breakthrough, many groups working on different laser types published their studies in the 1960s. Those articles led to the development of the main laser types in use today such as the CO_2 laser, the Nd:YAG laser, or the diode laser [6–11]. Within clinical dentistry, the long sought after goal was to replace the pain and vibration of the drill with a laser. This quest for the drill substitute held up the introduction of lasers in the dental field as it took until 1989 for the Er:YAG laser [12, 13] to be studied and then developed for practice a few years later. At this time, lasers were in widespread use in medicine working mainly on soft tissue. Beginning in 1990, laser therapy began to establish its place in dentistry.

Prior to lasers, there were three main methods of oral surgical treatment available:

- The conventional scalpel.
- Electro surgery [12].
- Cryotherapy (in use since the mid-1970s) [13].

For treatment of dental hard tissue, there are only a few wavelengths that may be used (Er:YAG, Er,Cr:YSGG, CO_2). However, in soft tissue surgery, up to ten different wavelengths are applicable. The oral/maxillofacial surgeon has a choice of a fiber-based delivery system (Nd:YAG, diode lasers) or a non-contact optical system (CO_2, erbium family).

- Diode lasers (445–1064 nm) are suitable for most oral soft tissue surgery procedures, especially on pigmented tissue. They are small, compact instruments whose portability can be an advantage for short procedures. They are delivered through a small diameter flexible optical fiber with an optional tip that can access small areas of tissue. Using a gated mode with digitally modulated pulse width and high output power diode laser have become very much more efficient than the early generations.
- Nd:YAG (1064 nm) has similar interaction with soft tissue although the free running pulse mode can produce very high peak powers for efficiency. Its delivery system is the same as the diode.
- The KTP laser emits at 532 nm and can be used similar to diode and Nd:YAG laser. Historically, the KTP was a frequency doubled FRP Nd:YAG. Now, the 532 nm "green" laser is a diode GaN semiconductor.
- Ho:YAG laser (2100 nm) is mostly used in soft tissue surgery. It utilizes an articulated delivery system and non-contact application.
- The argon laser (488 nm, 514 nm) has its indications in the use for pigmented lesions or vascular malformations.
- The carbon dioxide lasers (9300 and 10,600 nm) have been used traditionally in oral surgery; the latter wavelength having a very long history of clinical success. A higher average power than diode is readily available with very rapid tissue cutting speed, but there is a possibility of more tissue carbonization when used in continuous wave emission mode. A non-contact delivery mode and the use of an articulated arm makes the application of the 10,600 nm laser more demanding.
- The Er:YAG (2940 nm) and the Er,Cr:YSGG (2780 nm) lasers are readily absorbed in water. Their main indication is with hard tissue (osseous) surgery. Soft tissue surgery is also possible but much slower. Hemostasis is not as prominent due to poor absorption in pigmented chromophores and the coolant effect of co-axial air/water to assist in ablation site debridement. Depending on the particular laser, there is a choice of either contact or non-contact mode with the tissue, using delivery tips or a tip-less handpiece.

The main indications in selecting a laser for use in oral surgery:

- Hemostasis.
- Maintaining a decontaminated operation field.
- Controllable penetration depth of laser-tissue interaction.
- Minimal need for wound dressing/sutures.
- Less need for local anesthetic.
- Less postoperative pain.
- Less wound contraction and scarring.
- Uneventful wound healing.

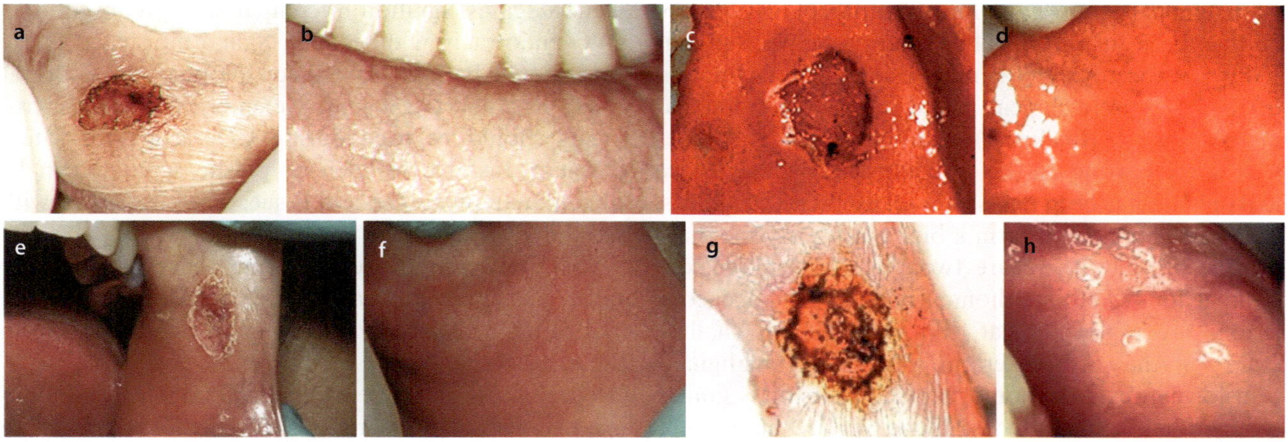

Fig. 13.1 **a** An immediate postoperative view of an excision of an irritation fibroma on the lower lip. An 810 nm diode was used with a 400 μm glass fiber in contact with the tissue at 1.2 W continuous wave. Fluence 149 J/cm^2. **b** A 3-week postoperative view of the healed area. **c** An immediate postoperative view of an excision of an irritation fibroma on the buccal mucosa. An Nd:YAG laser was used with an 320 μm glass fiber in contact with the tissue at an average power of 3 W (100 mJ, 30 Hz) Fluence 915 J/cm^2. **d** A 3-week postoperative view of the healed area. **e** An immediate postoperative view of an excision of an irritation fibroma on the inner lip mucosa. An Er:YAG laser was used with an 600 μm glass tip in contact with the tissue at an average power of 2.4 W (80 mJ, 30 Hz, without water spray.) Fluence 263 J/cm^2. **f** Two-week postoperative view of the healed area. **g** An immediate postoperative view of an excision of an irritation fibroma on the buccal mucosa. A 10,600 nm Carbon Dioxide laser was used with a 0.8 mm diameter hollow tubular tip that was focused on the lesion at 5 W continuous wave. Fluence 146 J/cm^2. **h** Three-week postoperative view of the healed area. (Courtesy of Dr. Donald Coluzzi)

Hemostasis is one of the most sought-after advantages that the laser provides. Depending on the target tissue, the laser wavelength and emission mode selected, an almost totally dry operation field is produced. This can also be achieved in patients with hemorrhagic diatheses [14]. The use of sufficiently matched energy for cutting will produce little carbonization leaving the wound unchanged in color and structure. This makes orientation during the procedures easier, even for the surgically inexperienced dentist. When using the laser not only is the wavelength important, but also parameters such as temporal emission mode (continuous wave or pulsed), pulse duration, emission cycle, exposure time, and the speed of the incision [15, 16]. All this in the right proportions will lead to a clean cut with little carbonization and contraction of proteins resulting in scarring [17–20].

As mentioned, all of the dental wavelengths currently available can be used for soft tissue oral surgery. As mentioned in ▶ Chap. 3, the laser-tissue interaction can vary because of the different absorption characteristics of the applied incident photonic energy. However, with careful technique and prudent choice of operating parameters, healing from laser surgery should be uneventful. Figure 13.1a–h depicts examples of an excisional procedure performed with four different wavelengths.

13.2 Benign Lesions and Tumors of the Oral Cavity

13.2.1 Leukoplakia

A color change from normally pale pink oral mucosa to white is one of the most often discovered abnormalities in the oral cavity. Failure to identify and recognize the cause of this alteration can be an omission with serious consequences since early squamous cell carcinomas may appear in early stages as a white lesion.

Clinically, the term leukoplakia has been used differently by many authors that it now presents a "white patch that cannot be rubbed away" [21]. As defined by the World Health Organization, leukoplakia is "a white patch or plaque that cannot be characterized clinically or pathologically as any other disease" [22, 23]. As such, leukoplakia should be used only as a clinical term; it has no specific histopathological connotation and should never be used as a microscopic diagnosis. Subsequently, the World Health Organization WHO (2005) changed the definition of leukoplakia to "A predominantly white plaque of questionable risk having excluded (other) known diseases or disorders that carry no increased risk for cancer" [24].

In clinical practice, the Malmö protocol of 1983 [25, 26] has been helpful in differentiating between potentially precancerous and benign lesions. The distinction of these is purely clinical, based on surface color and morphological (thickness) characteristics, and does have some bearing on the outcome or prognosis. Since leukoplakia has been and still is an exclusion diagnosis, it is mandatory to perform a biopsy to verify the diagnosis. Clinically, there are two types of leukoplakia: homogenous and nonhomogeneous. Homogeneous lesions are uniformly flat, thin, and exhibit shallow cracks of the surface keratin. The risk of malignant transformation in the homogenous type lesion is similar to any other normal mucosa tissue, whereas nonhomogeneous lesions exhibit oral cancer transformation rates ranging between 2.2% and 35%, especially when its biopsy reveled dysplastic entities [27–30]. In the therapeutic treatment protocol of leukoplakia, it is mandatory to excise all areas of the lesion that showed surface morphology that could warrant dysplasia. In conventional surgery, a flap procedure is usually employed, and the result can be scarring and prolonged and painful healing. Several treatment modalities have been suggested to manage this lesion including scalpel excision, electrosurgery, cryosurgery, laser surgery, and chemoprevention medications. A particularly difficult group of patients to manage are those with multiple oral, potentially premalignant lesions in whom extensive areas of mucosa may show signs of dysplastic change. Widespread leukoplakias have been shown to have higher rates of malignant transformation than more localized lesions [31]. The introduction of laser technology, for example, using the CO_2 laser, offers the option of vaporizing the lesion and leaving it to heal with no-to-minimal scar formation [32, 33]. However, the ability to analyze the submitted laser-excised specimen for pathology was soon questioned. Conventional excision produces a serial section of the whole specimen, and spot biopsies were criticized as insufficient. In the eyes of traditional surgeons, the pathologic evidence was being (literally) vaporized. Therefore, the standard of care before ablation of an oral potentially premalignant lesion is still to obtain an incisional biopsy for histopathological assessment.

The CO_2 laser manifests strong absorption ability in soft tissues, causing a superficial evaporation with minimal thermal damage to the surrounding tissue and maximum selectivity to remove targeted lesion. It works with a non-contact mode, using a handpiece attached to an articulating arm delivering the laser beam to the target tissue.

For procedures performed under general anesthesia, an endotracheal intubation with a cuffed ETT tube is recommended to be used as it provides significantly protection against ignition in comparison to conventional PVC endotracheal tubes [33]. Saline moistened gauze also should be used to protect the pharynx, endotracheal tube, and facial skin. Preferably, no metal instruments should be used during CO_2 lasing for safety measure. Instead, wooden or plastic spatulas can be used to manipulate the oral tissues when irradiated with CO_2 laser.

All staff within the operating theater must wear safety glasses for this laser wavelength. An evacuation system must be used to remove smoke and debris from the surgical site.

Oral potentially premalignant lesions can be managed under local or general anesthetic according to lesion size, location, and health status of the patient.

Oral leukoplakia without dysplastic histopathological features can be removed utilizing an ablation technique. The peripheral margins of the lesion are outlined with single mode pulse and vaporized with CO_2 laser in continuous mode at 4–6 W average power, with scanning mode covering the extension of the lesion. A second pass may be required to achieve a complete vaporization with adequate depth [34].

A complete resection of oral leukoplakia becomes mandatory when incisional biopsy of such lesion revealed dysplastic changes [34].

For lesion resection, the power levels vary between 5 and 15 W in continuous mode according to the thickness and extension of the lesion. For devices offering pulsed mode, the following parameters were suggested to facilitate resection of hyperkeratotic lesions: a power of 4.5 W on PW, 80 Hz, fluence of 44.78 J/cm^2, and a spot diameter of 400 μm [35].

Initially, a single pulse mode is used to outline the resection margins, situated at least 5–10 mm beyond the apparent clinical margin of the target lesion. Subsequently, the laser is adjusted to continuous mode and the peripheral marks are connected at the submucosal level. The whole specimen is then excised by undercutting at approximate 5 mm constant depth [34, 36]. However, when resecting mucosa overlying alveolar bone, the undercutting depth should be shallower.

It is recommended that the resected area and all peripheral margins should be vaporized using a defocused beam to eliminate residual disease and to facilitate hemostasis. The wound is left to heal by secondary intention, with no need for sutures, dressing, or grafting. Several studies demonstrated that the CO_2 laser offers precise cutting ability of the oral mucosa with less bleeding and scar formation in the surgery field.

All excised specimens should be preserved and fixed in 10% formalin and sent for histopathology examination. Large biopsies must be orientated by sited sutures to define the involved margin with the disease.

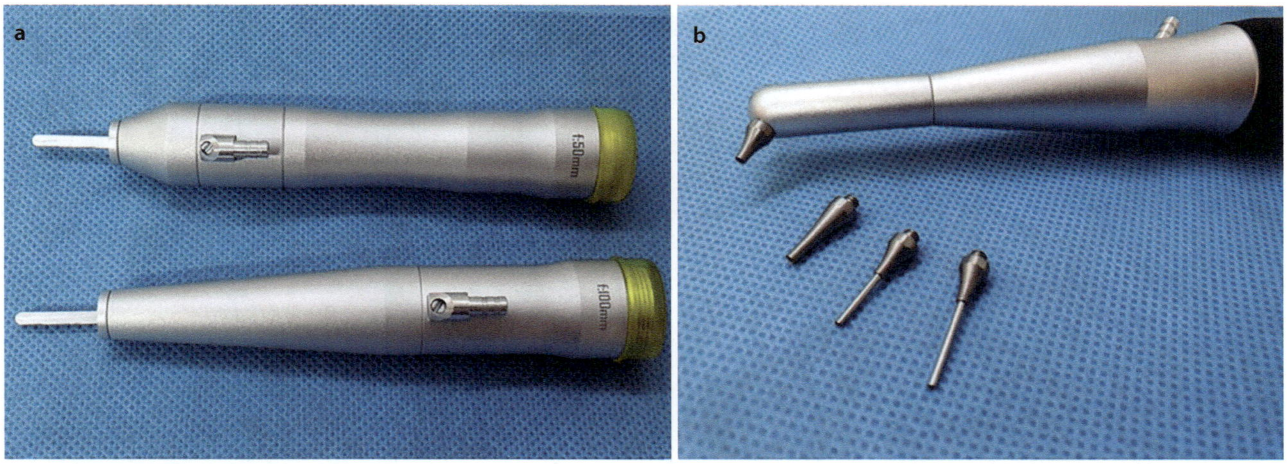

◼ **Fig. 13.2** Surgical and dental CO_2 laser handpieces used in the oral cavity. **a** Straight handpieces with different focal lenses (50/100 mm) used for resection of extended and bulk lesions. **b** Dental CO_2 laser handpiece with multiple tips facilitating treatment of small mucosal and gingival lesions within the oral cavity

As the wound will heal by secondary intention, postoperative instructions include using a mouth rinse such as chlorhexidine 0.12% and avoiding spicy and coarse food. Tranexamic acid 5% can be used as mouthwash if oozing happens. Augmentin antibiotic is prescribed as 1000 mg twice a day and paracetamol or ibuprofen for pain relief as required. Complete healing and mucosal re-epithelialization of the resected area is usually achieved within 4–6 weeks. Follow-up sessions usually are organized at 1, 3, and 6-month intervals thereafter.

Finally, although the CO_2 laser has been proven to be the treatment of choice of oral potential premalignant lesions, recurrence or malignant transformation is still unavoidable [31, 35].

Recurrence of a white lesion is a serious development. Studies show that the use of laser treatment has at least a similar if not better outcome than the conventional therapy [37–39]. In functionally sensitive regions as the lips, a laser excision demonstrated significant improvement in outcome [40].

Most of the studies regarding laser treatment of oral leukoplakia were performed with the 10,600 nm carbon dioxide wavelength. However, it is also possible in some small lesions, especially those located in the anterior part of the oral cavity to use a fiber-based delivery system of an Nd:YAG laser or a diode laser, depending on the accessibility of the lesion [41–44].

Two different types of handpieces used with CO_2 laser device in the oral cavity (◼ Fig. 13.2).

◼ Figure 13.3 shows a case of leukoplakia on the lateral border of the tongue. A 50-year-old female presented with an inconspicuous medical history except for smoking. An Nd:YAG was used to vaporize the lesion. Six months later, a small area has recurred.

Verrucous leukoplakia extending from the interior aspect of the upper left mucosa to the tuberosity (◼ Fig. 13.4). An incisional biopsy was performed which revealed the presence of severe dysplasia. Lesion resection was performed using the CO_2 laser. Exposed alveolar bone was covered by a platelet-rich fibrin PRF membrane which was stitched to resection margins. The healing was uneventful, and no recurrence was detected at 6 months postoperatively.

13.2.2 Lichen Planus

Lichen planus is a chronic inflammatory mucocutaneous disorder; it generally affects middle aged to elderly females and the lesions mostly appear in the buccal mucosa, tongue, gingiva, and vermillion border of the lower lip [45]. It is a T-cell mediated autoimmune disease in which the cytotoxic CD8+ T cells trigger apoptosis of the basal cells of the oral epithelium. Several antigen-specific and nonspecific inflammatory mechanisms have been proposed to explain the accumulation and homing of CD8+ T cells subepithelially, and the subsequent keratinocyte apoptosis [46].

Clinically, oral lichen planus (OLP) may appear in six forms: Reticular (Wickham striae), erosive, atrophic, plaque-like, papular, and bullous. The reticular shape is the most common type which is usually asymptomatic exhibiting localized or generalized extension with no need for treatment [47]. However, other clinical patterns such as erosive or atrophic forms may cause variable amounts of pain or discomfort, especially when the patient is consuming spicy or hot foods. Additionally, oral function and patient's quality of life may be affected [48].

The clinical diagnosis of OLP should be associated with a histopathological examination to rule out the presence of any associated sinister elements such as dysplasia or neoplasia [49]. The microscopic features of

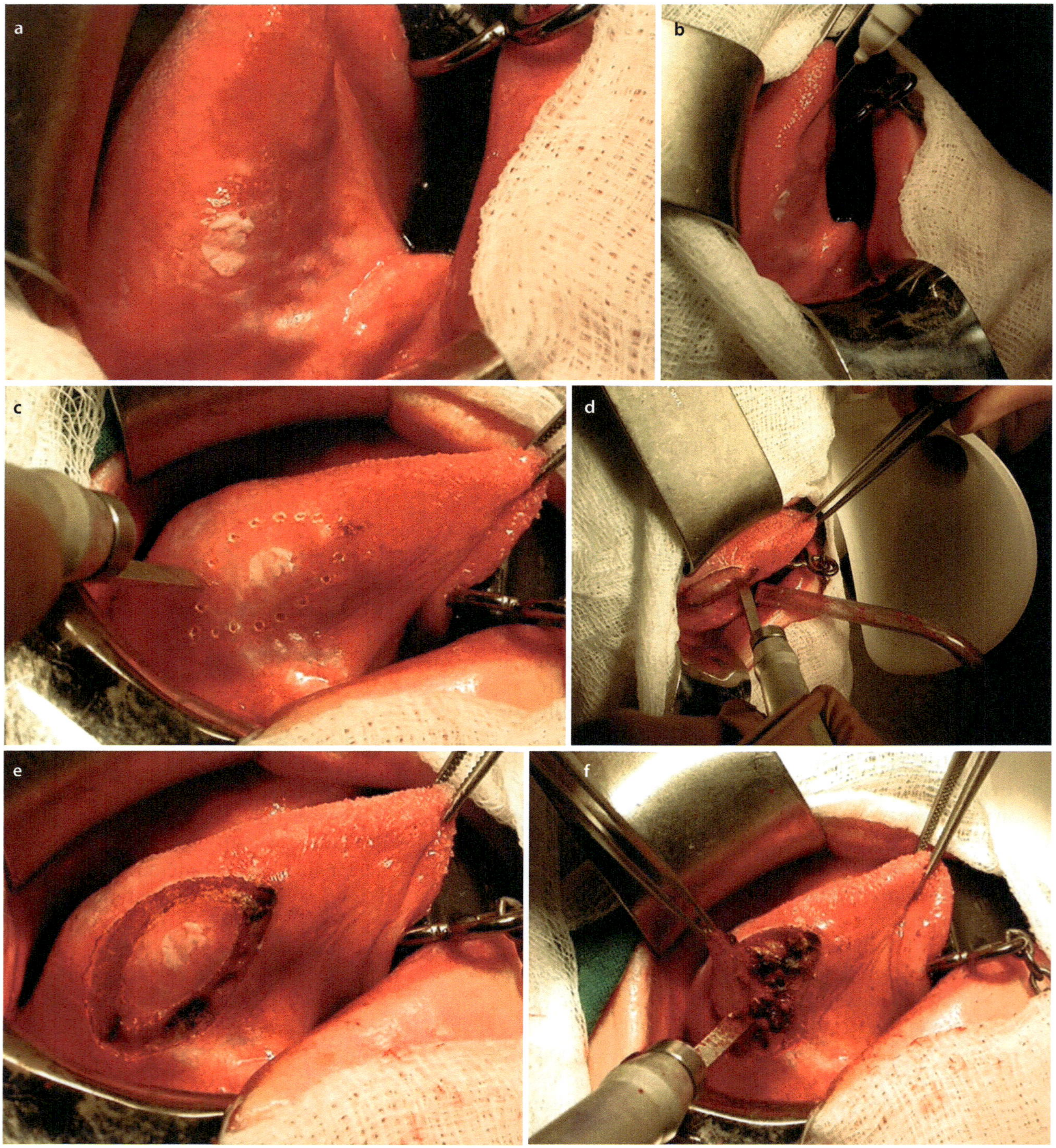

13

Fig. 13.3 **a** A 48-year-old male with severe dysplasia located on ventral tongue. **b** This procedure was performed under general anesthesia; however, local anesthesia was also provided to aid hemostasis during the resection. The CO_2 laser was used with power levels varying from 10 to 15 W **c**. **d**, **e** The outlined margins were first determined using single pulse mode which served as surgical boundaries, extending 0.5 cm beyond the apparent clinical margins of the lesion. Subsequently, the laser was set to continuous mode, and the peripheral marks were connected at the submucosal level. **f** The whole specimen was then resected by undercutting at approximate 5 mm constant depth. **g** The surgical bed and all peripheral margins were vaporized using a defocused beam to eliminate residual disease and to facilitate hemostasis. The wound surface was left, without closure or dressing, to heal by secondary intention. **h** The excisional specimen was oriented by placing multilength sutures at the margins and fixed in 10% formalin and transferred for histopathology evaluation. **i** Healing and resection field re-epithelialization was achieved within 4–6 weeks

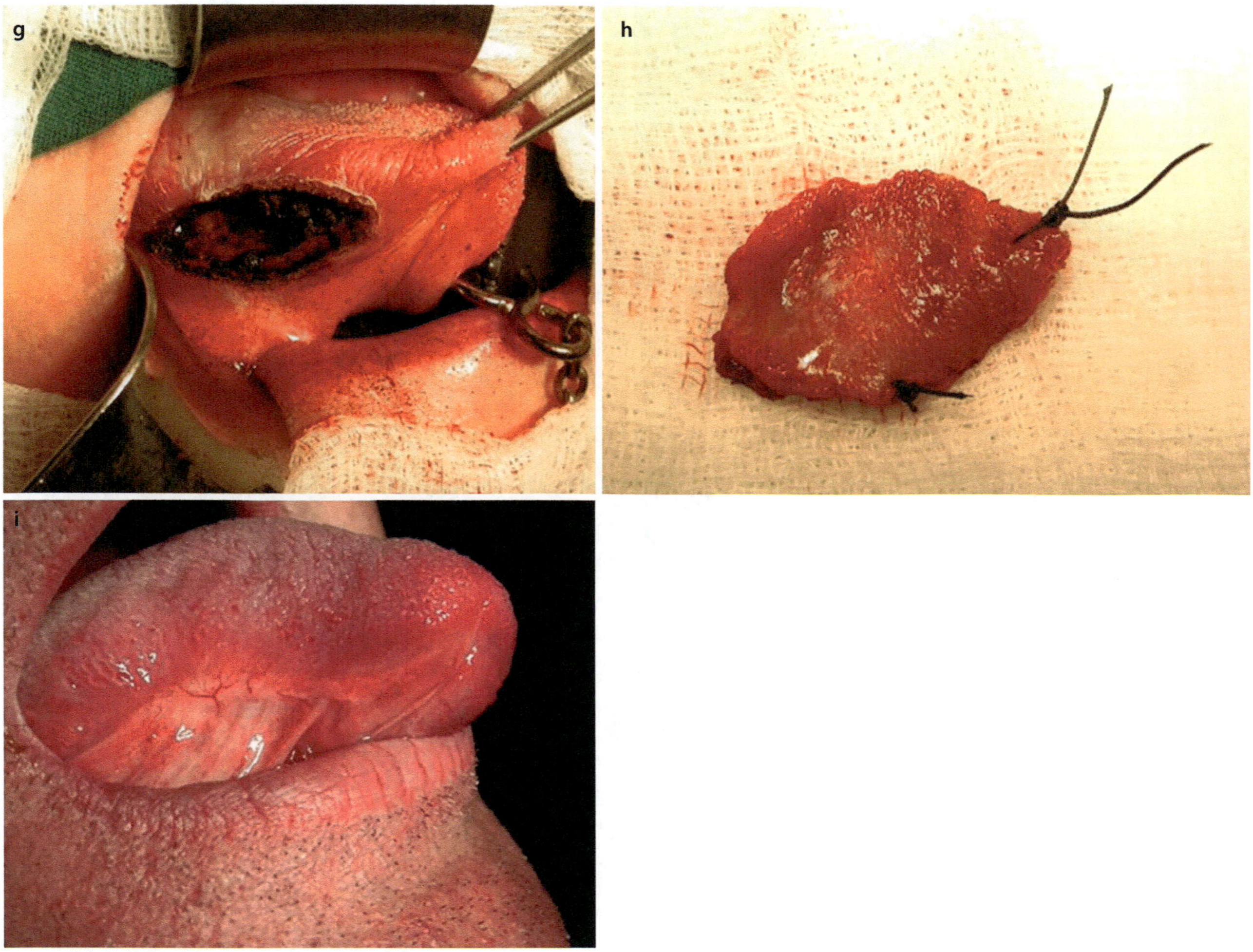

Fig. 13.3 (continued)

OLP usually include plasma cells, subepithelial band-like infiltration of lymphocytes and acanthosis [46, 50].

Bearing in mind that the risk of malignant transformation is higher in erosive and atrophic lesions, treatment of these types should be considered mandatory. Several treatment methods are available and topical corticosteroids are mostly considered the treatment of choice; in the light of side effects such as thinning of the oral mucosa and secondary candidiasis, it has been preferable to find alternative pharmacological treatments, including retinoids and immunosuppressive agents, especially cyclosporine and tacrolimus. Low level laser therapy (photobiomodulation) and high-power (surgical) laser may be considered as alternative tools for OPL treatment, especially for those unresponsive lesions to traditional pharmacological regimes [51, 52]. Surgical removal of affected mucosa may be performed either by a scalpel or high-power laser. Elimination of OLP with a laser offers complete resection of the lesion usually in one session with minimal postoperative complications such as pain, bleeding, or scar formation. Tarasenko et al. reported in their randomized controlled trail the superiority of Er:YAG and Nd:YAG on scalpel surgery in ablation of erosive lichen planus [52]. The Er:YAG laser was applied in a contact mode using reciprocal movements of 15 s per cm^2 with the following parameters: average power: 2 W, frequency: 10 Hz, pulse duration: 230 µs, spot size: 0.9 mm, water/air: on. The Nd:YAG laser was also applied in a contact mode using reciprocal movements of 15 s per cm^2 with the following parameters: average power: 1.5 W, frequency: 40 Hz, pulse duration: 350 µs, spot size: 300 µm. Coagulation was achieved by applying the two lasers with a non-contact mode using an average power of 3 W with circular movements of unfocused beam 1–3 mm far away from the treated area. The findings of this research showed that high-level laser intervention exhibits a superior clinical outcome compared to the scalpel excision for the surgical treatment of oral erosive lichen planus. Taking into consideration that the Er:YAG causes less thermal damage to the oral mucosal tissue, it has been considered as the most effective laser type in this study at the end of the first postoperative month [45].

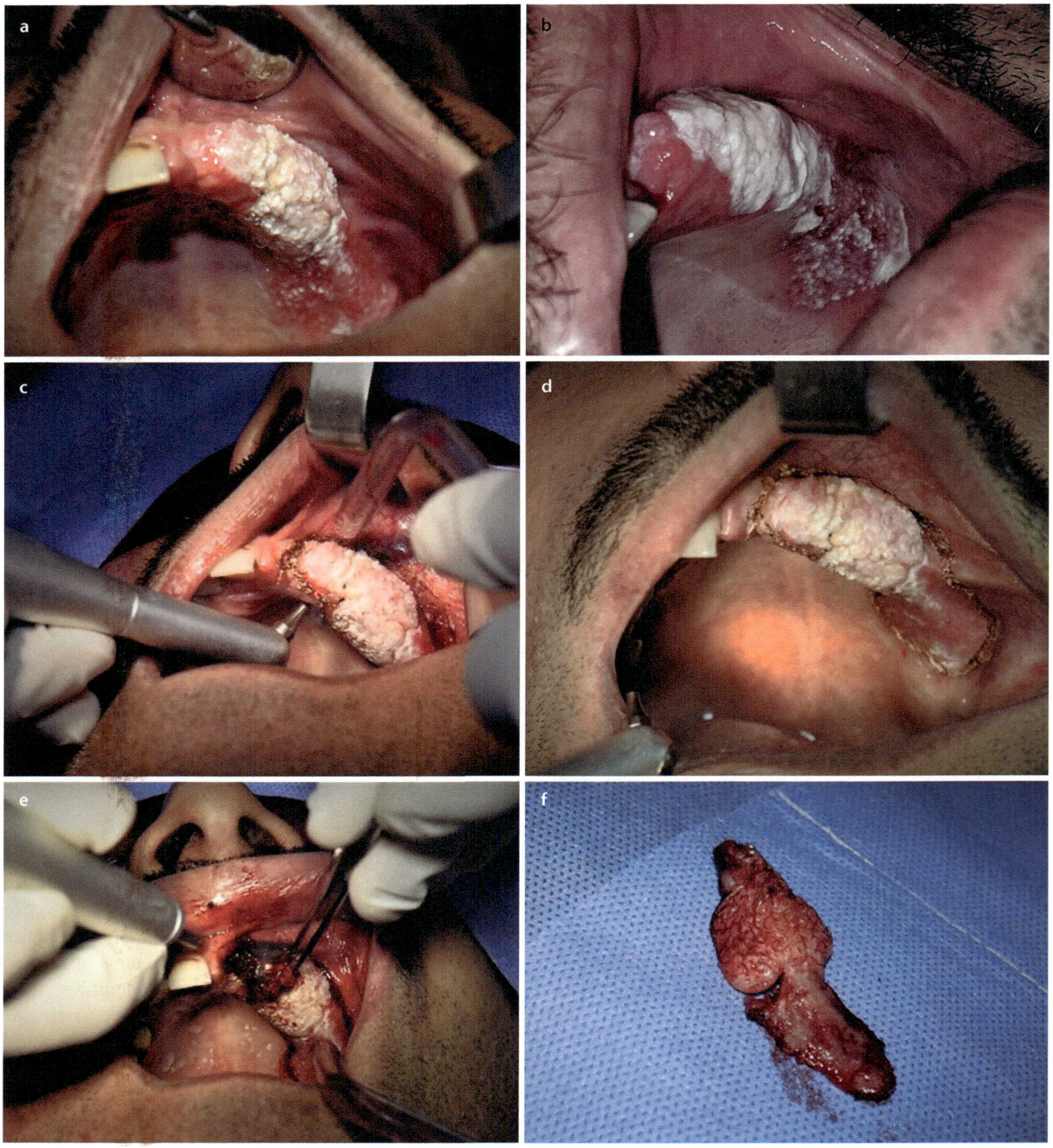

13

Fig. 13.4 **a**, **b** A 39-year-old male presented with a recurrent verrucous leukoplakia. The lesion extends clinically from the interior aspect of the upper left mucosa to the tuberosity. **c–f** Lesion peripheral outlining and resection were performed by applying CO_2 laser using both dental and straight handpieces. A periosteum elevator was used to facilitate the lesion dissecting. **g**, **h** The exposed upper alveolar maxillary bone was covered by a PRF membrane which was stitched to resection margins. A partial pressure denture was placed to support the membrane and to aid hemostasis control. The healing was uneventful, and no recurrence was detected 6 months post intervention

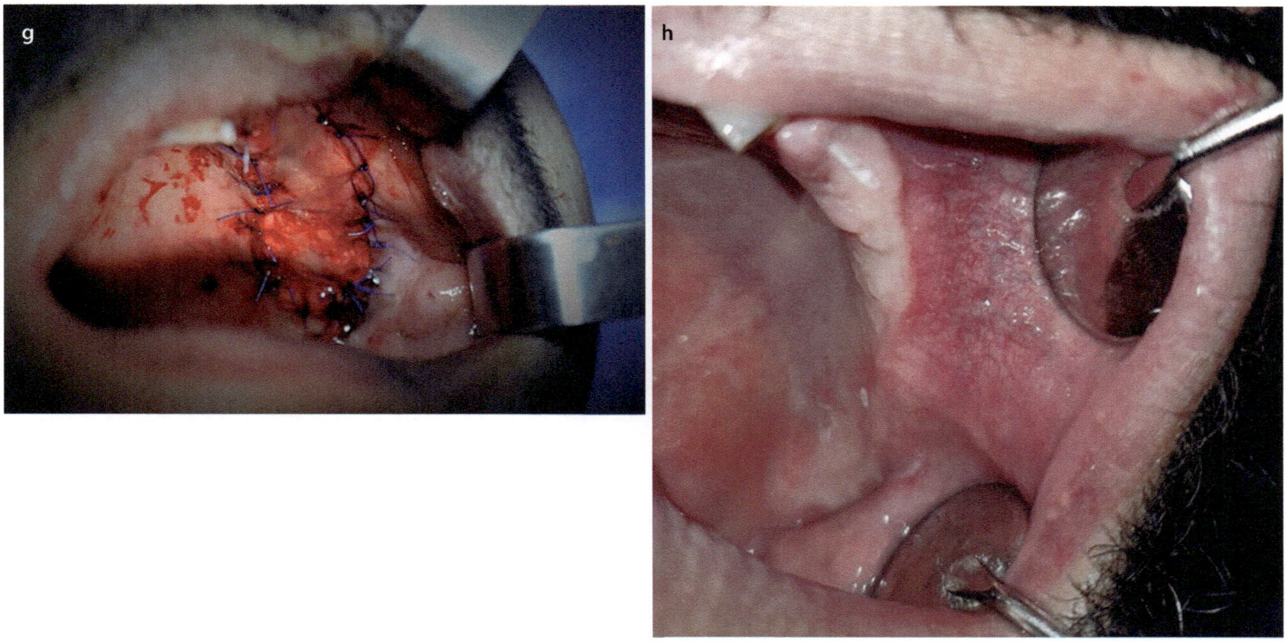

Fig. 13.4 (continued)

The CO_2 laser (10,600 nm) manifests a high absorption in the water component of oral mucosa which facilities a precise resection of oral mucosal lesions with post-surgery coagulation. It also can be used in a defocused mode enabling evaporation of superficial tissues even when the lesion exhibits a large extension thus the evaporated area will heal by secondary intention [53].

Treating of lichen planus using the CO_2 laser was reported in several studies with an advantage of reducing the lesion size and pain severity. The operation is performed after administration of local anesthesia using defocused continuous CO_2 laser (10,600 nm) with power range reported from 2 to 20 W. The lesion is irradiated to achieve superficial vaporization with a slow gliding motion, until the lesion is completely removed. The ablation should not reach the connective tissue level. Although a study suggested using CO_2 laser with defocused continuous mode, others recommended super pulsed mode [54–56].

OLP recurrence rate differs according to follow-up period, lesion extension, type of OLP, and technique of laser application and parameters used [54].

A 27-year-old male presented with pan-oral lichen planus (Fig. 13.5). Multiple incisional biopsies were performed, and the histopathological report confirmed the clinical entity of the lesions. A surgical CO_2 laser (λ = 10,600 nm) was used applying a continuous defocused mode with an output power of 3 W. The spot diameter was 0.5 mm, the speed of movement was approximately 3 mm/s, and the power density was 1527.8 W/cm^2. The procedure was conducted under local anesthesia.

The same patient presented with reticular and atrophic-shape lichen planus on the left buccal mucosa (Fig. 13.6). The lesion was vaporized using the CO_2 laser defocused beam with scanning mode covering the extension of the lesion. Similar laser parameters were used as in the abovementioned case.

13.2.3 Fibroma

The lesion most commonly found in the oral cavity is the fibroma. It occurs as a discrete, superficial, pedunculated mass commonly found on the buccal mucosa. It is usually of non-neoplastic nature and arises as a response to persistent mechanical irritation such as calculi, foreign bodies, chronic lip or check biting, overhanging margins restoration, sharp spicules of bones, and overextended borders of appliances [57, 58]. The irritation fibroma is also termed a traumatic fibroma, peripheral fibroma, fibrous nodule, fibroepithelial polyp, focal fibrous hyperplasia, and inflammatory fibrous hyperplasia [59]. It is composed of collagenous, fibrous connective tissue covered with keratinized or parakeratinized squamous epithelium. In the lesion, there can be myxomatous degeneration or pathological weakening of the connective tissue along with bone formation and in growth of fatty tissue. Affected patients often notice the lesion only after masticating on the area

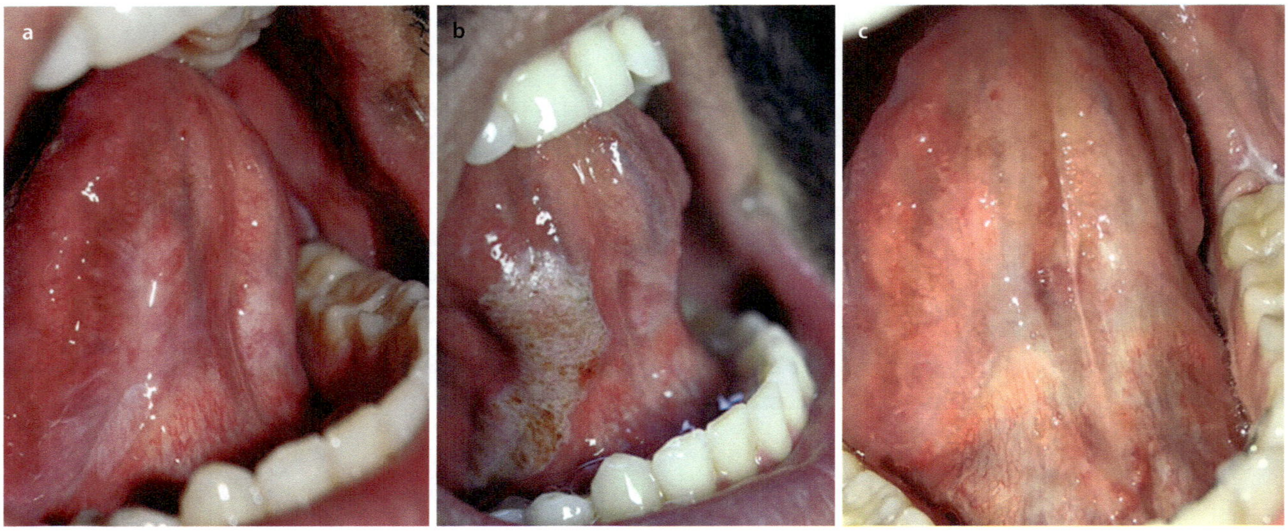

Fig. 13.5 **a** A 27-year-old male presented with a symptomatic lichenoid lesion on ventral tongue. **b** Immediately after CO_2 laser vaporization. **c** Four weeks postoperative view showing complete healing with no sign of recurrence

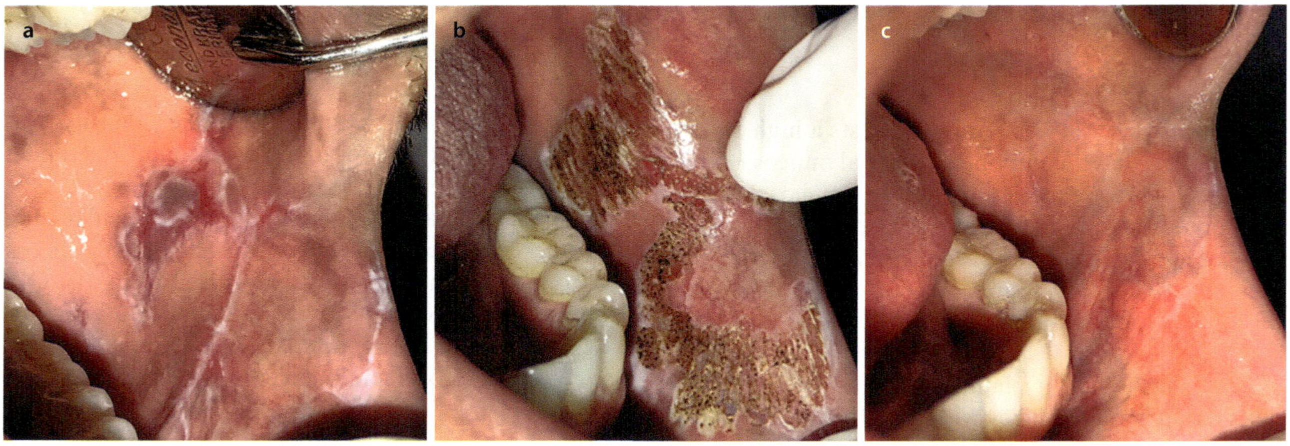

Fig. 13.6 **a** The same abovementioned patient exhibiting reticular and atrophic shape lichen planus lesions on the left buccal mucosa. **b** Lesion ablation using CO_2 laser with continuous mode. **c** Uneventful healing of the oral mucosa 3 weeks postoperatively

and then experiencing post traumatic pain and swelling. At this point, the fibroma will have grown to at least 2–3 mm in size depending on the region and would be visible in a mirror. The lesion will generally continue to grow, becoming a nuisance, and should be removed. When a scalpel is used, postoperative bleeding often makes the procedure difficult, especially if the lesion is inflamed with increased vascularization. Moreover, suturing is often necessary. Laser surgery allows both excision and hemostasis simultaneously [20, 57].

Treatment protocol includes injection of local anesthesia peripherally and underneath the lesion. A tissue forceps or a suture is used to grasp the lesion away from its base. Employing a contact mode, laser choice of Nd:YAG or diode laser with wavelength of 450, 810, 940, and 980 nm can be used to perform the resection [60]. Depending on the lesion size and location, it is outlined about 2 mm outside its border using an initiated 320–400 μm quartz optic fiber. As the tip contacts the base of the lesion, activate the laser to make an incision circumscribing the lesion, then move the tip in quick gentle 2–3 mm strokes to achieve complete resection of the lesion. It is important while performing the cutting to direct the tip at the base of the resected specimen and not toward the oral mucosa. This allows a precise resection without damaging underlying tissues. The area is left to heal in secondary intention with no need for dressing.

Figure 13.7 is a case of an asymptomatic fibroma present on a 45-year-old female patient, nonsmoker, with an inconspicuous medical history. A diode laser performed the excisional biopsy and the tissue healed without any scarring or residual lesion.

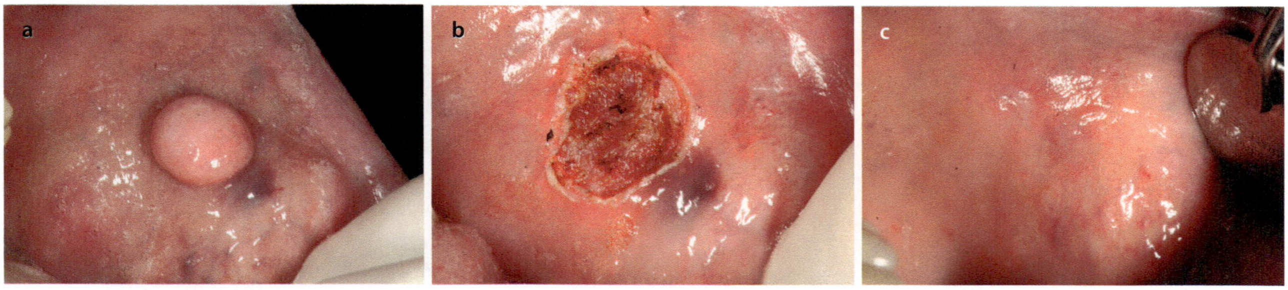

Fig. 13.7 **a** Pre-operative view of a fibroma in the buccal vestibule. **b** Immediate postoperative view of the excision. An 810 nm diode laser was used with a 200 μm glass fiber in contact with the tissue at 30 W, 12,500 Hz, and a 9 μs pulse duration, 200 μm glass fiber, contact mode. Average power 3.38 W. Fluence 554 J/cm^2. **c** A 4-week postoperative photo shows complete healing

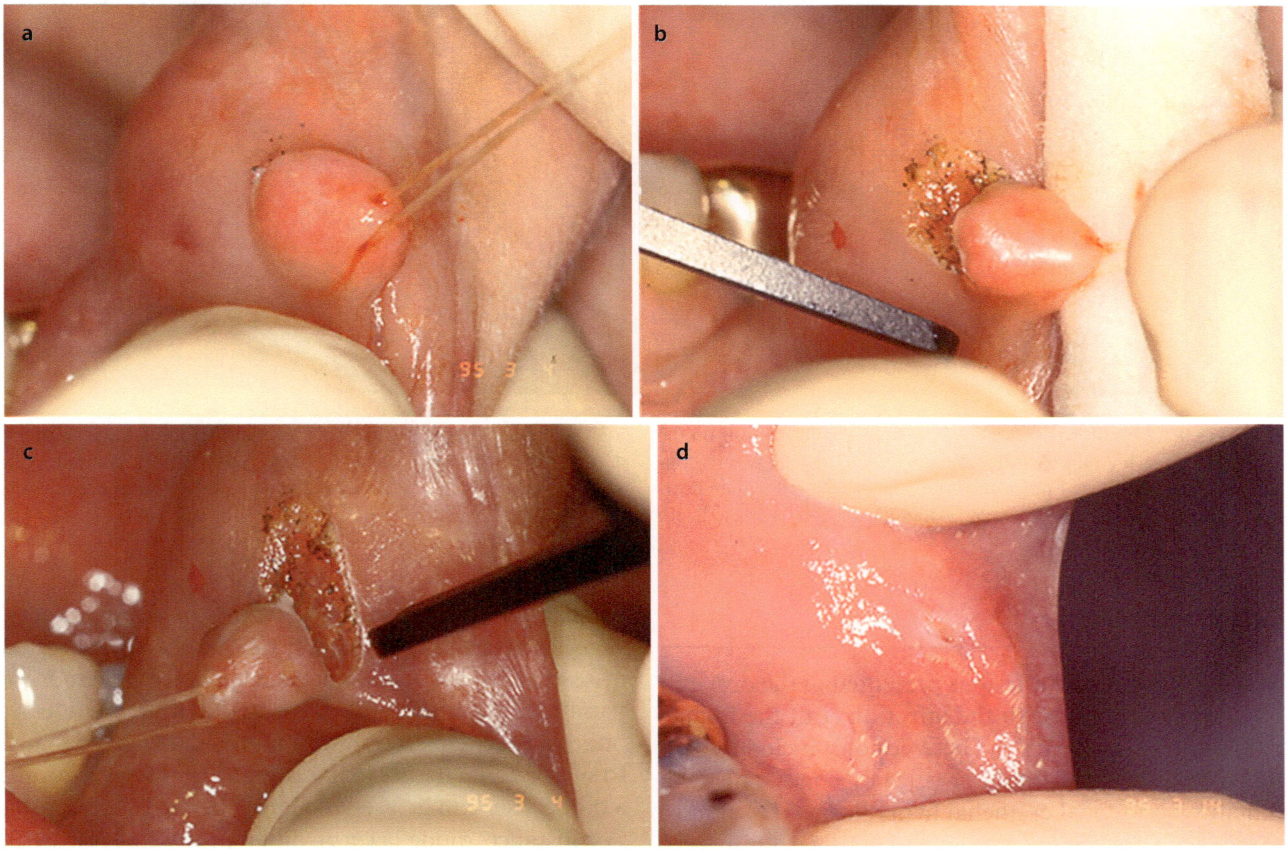

Fig. 13.8 Traumatic fibroma left buccal mucosa in a 50-year-old female. **a** Lesion secured with suture to stabilize the tissue and facilitate accurate estimation of the level of incision. **b**, **c** Using a CO_2 10,600 nm laser with non-contact, fixed focal length delivery: 1.0 W CW 800 μm beam diameter × 30 s. Incisional hemostasis achieved. **d** Postoperative uneventful early healing at 10 days. (Clinical case courtesy of Prof. Steven Parker)

Figure 13.8 depicts an irritation fibroma present inner mucosal lining of the lip of a (50-year-old female patient with a non-contributory medical history?) A superpulsed 10,600 nm carbon dioxide laser performed the excision, while a suture kept tension on the lesion. Clinical case courtesy of Prof. Steven Parker.

Figure 13.9 is that of a 38-year-old female, non-smoker, presented with multiple areas of fibromas. Her medical history included hypertension and cardiac dysrhythmia, which are drug controlled. She has a bruxism habit which could contribute to the presence of the lesions.

Figure 13.10 depicts a fibroepithelial polyp localized on the left buccal mucosa as a result of chronic irritation. The lesion was excised using the diode laser and complete re-epithelization was achieved after 3 weeks.

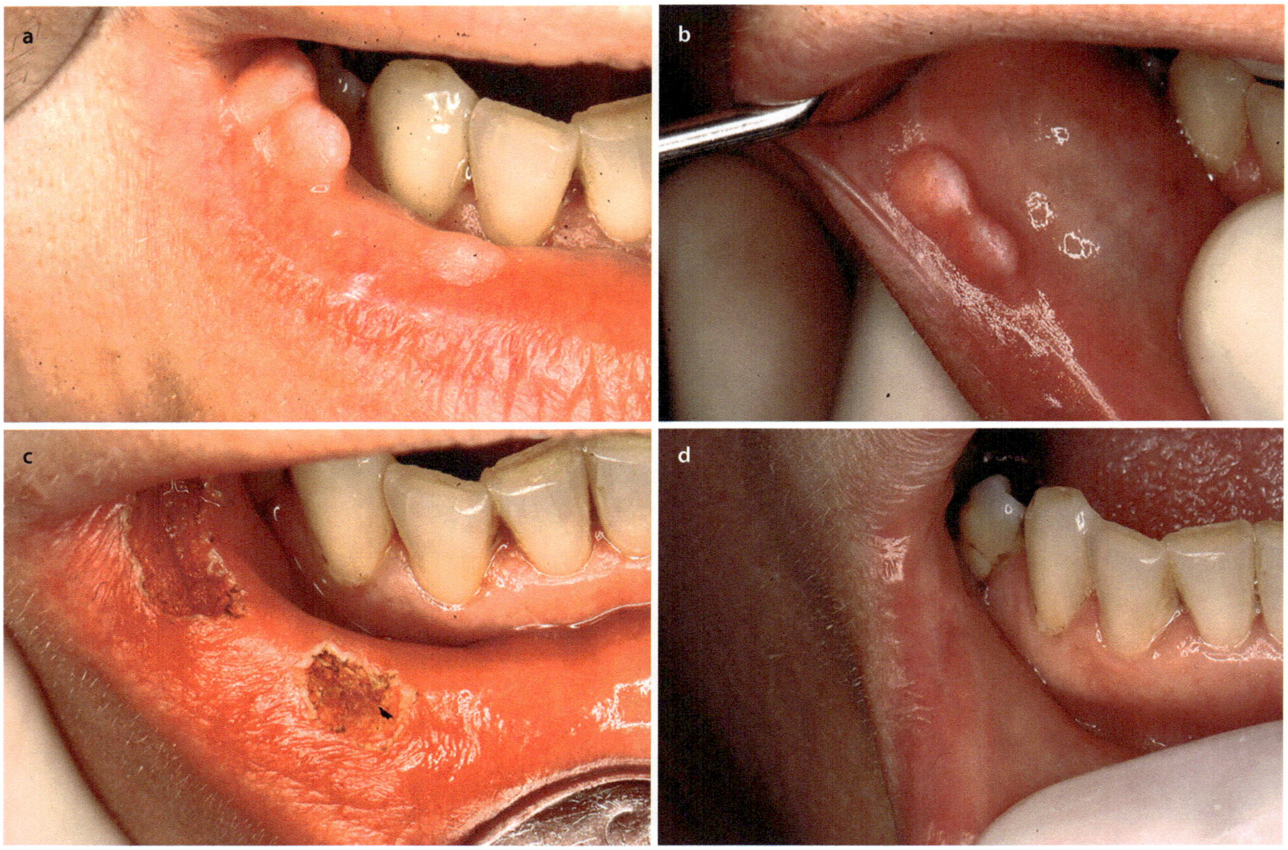

Fig. 13.9 **a** Pre-operative view of multiple fibroma lesions on the lower right lip. **b** A similar view showing that the lesions are contained in the keratinized portion of the tissue. **c** An immediate postoperative view of the excision areas. An Nd:YAG laser was used with a 320 µm glass fiber in contact with the tissue at an average power of 4 W (100 mJ, 40 Hz). Fluence 637 J/cm^2. **d** A 4-week postoperative photo showing complete healing with no residual lesion present

13.2.4 Papilloma

Several papillomatous lesions presented in the oral cavity are usually related to the infection with the human papilloma virus (HPV). These lesions include squamous cell papilloma, verruca vulgaris, condyloma acuminatum, and focal epithelial hyperplasia. HPV is also responsible for malignant lesions that occur in oropharyngeal cancer. A papilloma presents as an arborescent growth of numerous squamous epithelial finger-like projections. Each branch contains a well-vascularized fibrous connective tissue core. It can be seen throughout the oral cavity with a preference on the tongue and the periuvular region. The etiology of the lesion is generally viral and is thought to be induced by the human papilloma virus [61, 62]. Since human papilloma viruses infect basal cells that is the only dividing cells of the epithelium, it is necessary to remove the entire epithelial layer where the papilloma occurred. The treatment of papillomatous lesions comprises complete excision, including the base of the lesion with apparently free margins to avoid recurrence. This procedure can be performed with a conventional scalpel, an electrosurgery, or using a laser [63]. Diode and CO_2 lasers were reported in several studies as an acceptable choice to remove a papilloma from the oral cavity and aerodigestive tract offering precise resection with minimal thermal damage to surrounding tissues [64]. For small or solitary lesions, a diode laser can be used to achieve complete resection. For instance, diode laser of different wavelengths (450, 810–980 nm), with an average power of 2.1 W, in continuous wave mode, using 300–320 µm optical fibers has been recommended [65, 66]. However, for multiple and extended lesions ablation mode preferably using CO_2 laser is the treatment of choice to manage these lesions taking into consideration that several sessions be needed to achieve complete eradication of the lesion. A recurrence can occur even though laser treatment has shown lower reappearance rates than conventional therapy. However, any therapy could be limited since the latent behavior of infection of HPV in the apparently normal mucosa around papilloma which explains the frequent recurrence after surgical removal of a visible lesion [67, 68].

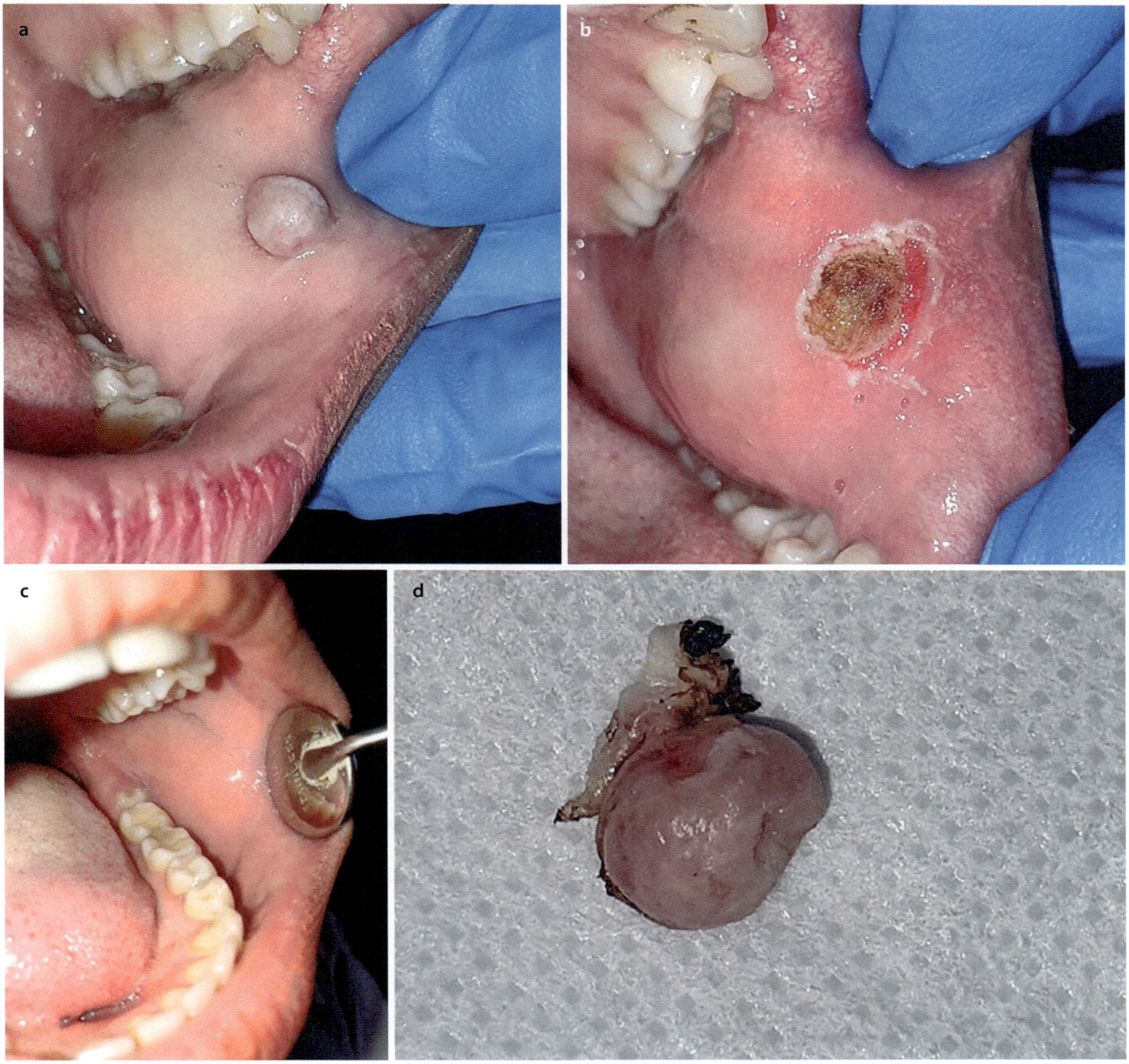

Fig. 13.10 **a** Fibroepithelial polyp seen on the left buccal mucosa caused by regular biting of the check. **b** The 980 diode laser with 3 W peak power and duty cycle 50% was used to excise the lesion. Complete hemostasis was achieved with no need for dressing or sutures. **c** Three weeks postoperative view showing complete healing. **d** The polyp after excision was fixed in 10% formalin and sent for histopathology examination

Figure 13.11 is a clinical case of an oral papilloma. The patient is a 36-year-old male, smoker with an inconspicuous medical history. A diode laser performed the excision and excellent healing is anticipated.

A 14-year-old female presented with extensively distributed nodules and papules, which appeared when she was 6 years old (Figs. 13.12 and 13.13). The size of the lesions ranges from 2 to 30 mm. The diagnosis of Heck's disease was confirmed depending on clinical characteristics, histopathological examination, and immunohistochemical staining. Depending on the extensive clinical situation, the decision was made to treat the patient using CO_2 laser at wavelength: 10,600 nm, continuous mode and power of 6 W. It showed good function and aesthetic results. Five treatment sessions were required to manage all regions in the oral cavity.

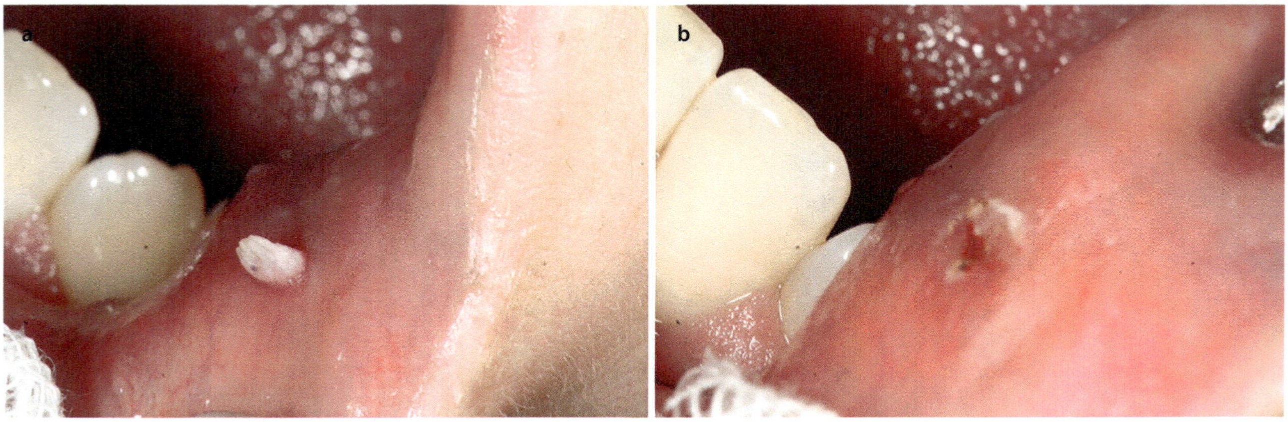

Fig. 13.11 **a** Pre-operative photo of a small oral papilloma on the buccal mucosa. **b** An immediate postoperative view of the excision of the lesion. An 810 nm diode laser was used with a 200 µm glass fiber in contact with the tissue at 30 W, 12,500 Hz, and a 9 µs pulse duration. Average power 3.38 W. Fluence 554 J/cm^2. Complete healing is expected

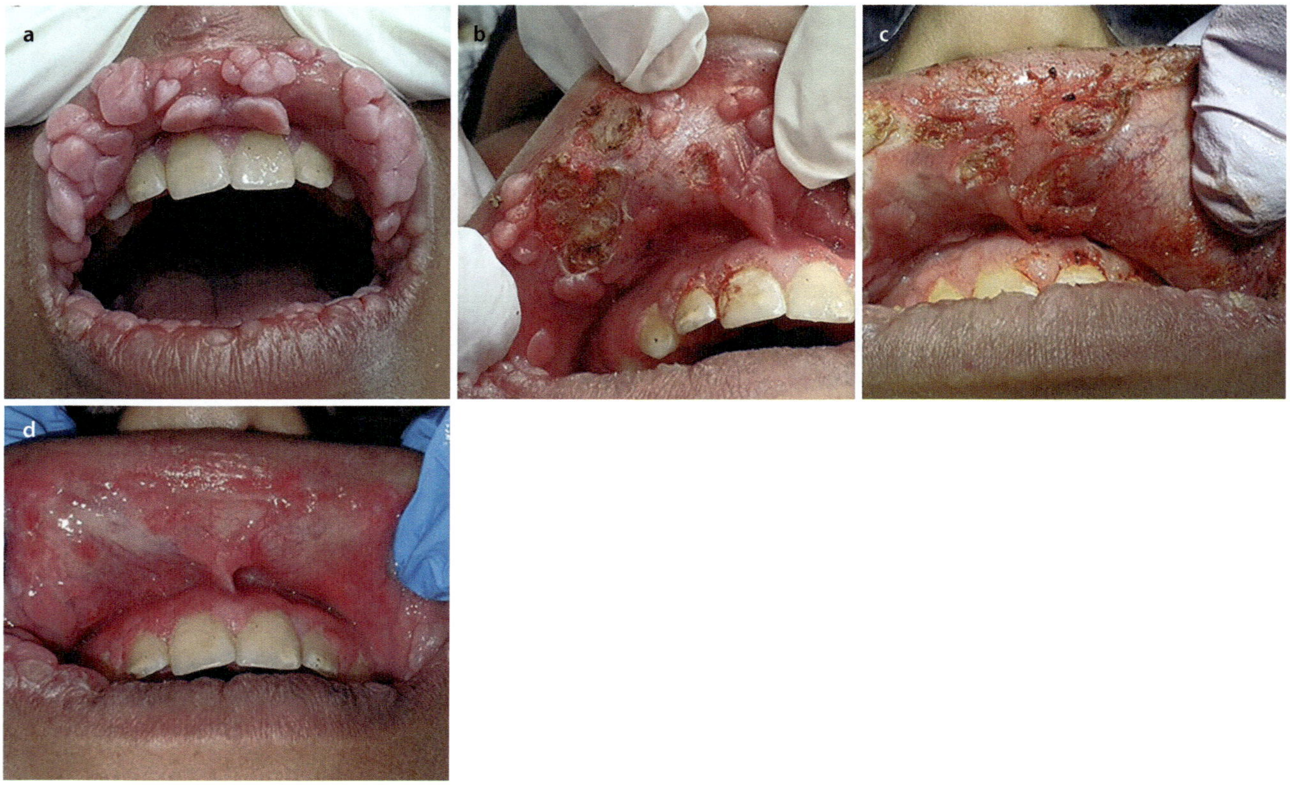

Fig. 13.12 **a** A 14-year-old female presented with an extensive case of pan-oral focal epithelial hyperplasia covering most of her lining oral mucosa. **b**, **c** Immediate postoperative view of the labial aspect of the upper lip after resection of lesions using the CO_2 laser with 6 W continues mode, which minimized the time of resection and reduced the thermal damage of the oral mucosa. **d** The labial aspect of the upper lip 1 month following CO_2 laser intervention with no sign of relapse or new primary lesions

13.2.5 Lipoma

Lipomas are the most common benign tumors of the human body. They can be solitary or in multiple clusters. They vary extensively in size from minute to large growths with weight in kilograms. They occur mostly between 40 and 50 years of age. In the oral cavity, they appear as soft, slightly elastic, and painless lesions and usually are noticed by the patients if they are of large dimension. They are normally found in the buccal, sub-

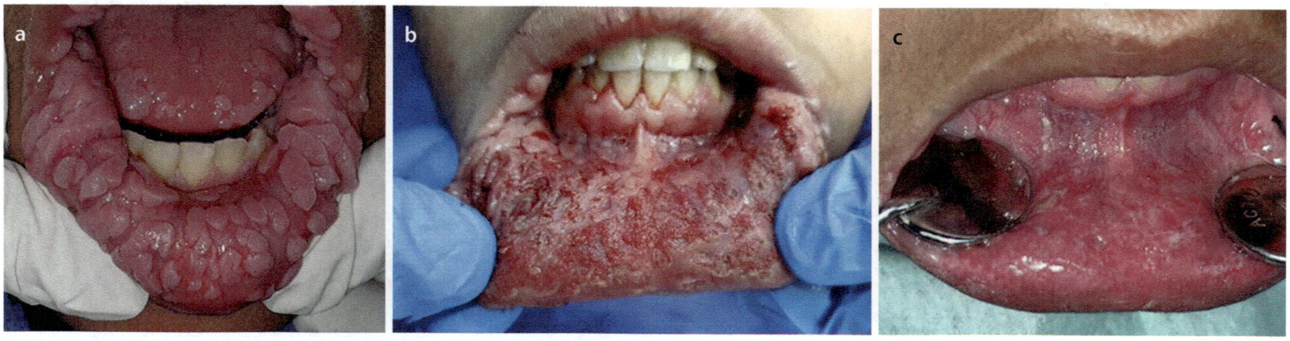

Fig. 13.13 **a** The same patient exhibiting multiple papillomatous lesions on the lining mucosa of the lower lip extending to the commissure in both sides. **b** Immediate postoperative view of the labial aspect of the lower lip after ablation of lesions using the CO_2 laser with continuous mode. **c** Uneventful healing of the oral mucosa 10 days postoperatively

mandibular, and vestibular region. The development of a lipoma is not necessarily hereditary although hereditary conditions, such as familial lipomatosis can stimulate its growth [69]. Genetic studies support prior epidemiologic data in humans showing a correlation between high mobility group proteins (HMG I-C) and mesenchymal tumors [70]. When the patient's complaint is the size of the lesion, or to verify the diagnosis, the treatment is excision. The laser's ability to achieve instant hemostasis offers a good view of the site, which is important in some areas of the mouth, for example, in direct proximity to the mental nerve [71]. While any laser wavelength will perform the surgery, a fiber delivered laser that has better maneuverability could be an advantage in the depth of the wound [72]. Depending on the dimension of the lesion, sutures would be placed to avoid food impaction. For bigger and subdermal lesions, irradiation with the Nd:YAG laser is performed by inserting a 600 μm optical fiber within a cannula through a 1 mm incision after applying local anesthesia underneath the tumor. The fiber is passed through the proximal end of the clamp, in order for the fiber to be a little longer than the cannula. The fiber is pouched no more than 2–3 mm beyond the end of the cannula while laser emission is activated. The recommended parameters to be used to treat subjects in a single session using Nd:YAG include average power 6 W, pulse repetition rate 30 Hz, energy per pulse 200 mJ, and pulse width 100 μs [73, 74].

Figure 13.14 is that of a 48-year-old male patient, nonsmoker with a non-contributory medical history presented with a slow growing mass in the mandibular vestibule. It was beginning to annoy the patient. A diode laser was used for an excisional biopsy and a diagnosis of lipoma was confirmed. Sutures were placed because of the large size of the excised tissue and the depth of the wound. Healing proceeded uneventfully.

Figure 13.15 is a clinical case of a lipoma on the lateral border of the tongue. The 58-year-old male patient presented because of his concern with the growing size of the tumor although it has not yet caused any functional impairment. His hypertension and diabetes are controlled with medication. A diode laser was used for the excision. A 15-year postoperative photo shows excellent healing and no recurrence.

13.2.6 Pyogenic Granuloma

Pyogenic granuloma is a vascular lesion of skin and oropharyngeal mucosa. It appears as reddish overgrowth due to mechanical, physical, chemical, or hormonal trauma [51]. It is generally located in the anterior of the maxilla and can be painful if it is constantly irritated. The lesion can grow rapidly and will often bleed profusely after little or no trauma, and it has the appearance of a highly vascular granulation tissue with inflammation on histologic examination. Special variants of the pyogenic granuloma are the epulis granulomatosum and the granuloma gravidarum. Treatment is not necessary except in cases of excessive bleeding or pain and ulceration. After pregnancy, the lesion normally regresses. [1, 52]. As expected, any surgical laser that achieves good control of bleeding may be used [45, 53]. A biopsy must be taken for histological verification of the lesion.

Pyogenic granuloma (PG) is one of the common inflammatory nonneoplastic vascular hyperplasias seen in the oral cavity. It occurs in response to many factors, for example, long-term irritation, traumas, or hormonal factors. It often arises in the second decade of life in young females, probably due to the effects of female hormones [75].

Clinically, oral PG is a smooth or lobulated pedunculated or sessile exophytic lesion which enlarge rapidly and easily bleeds. It is recognized as small pink to red to purple papules depending of the persistence of the

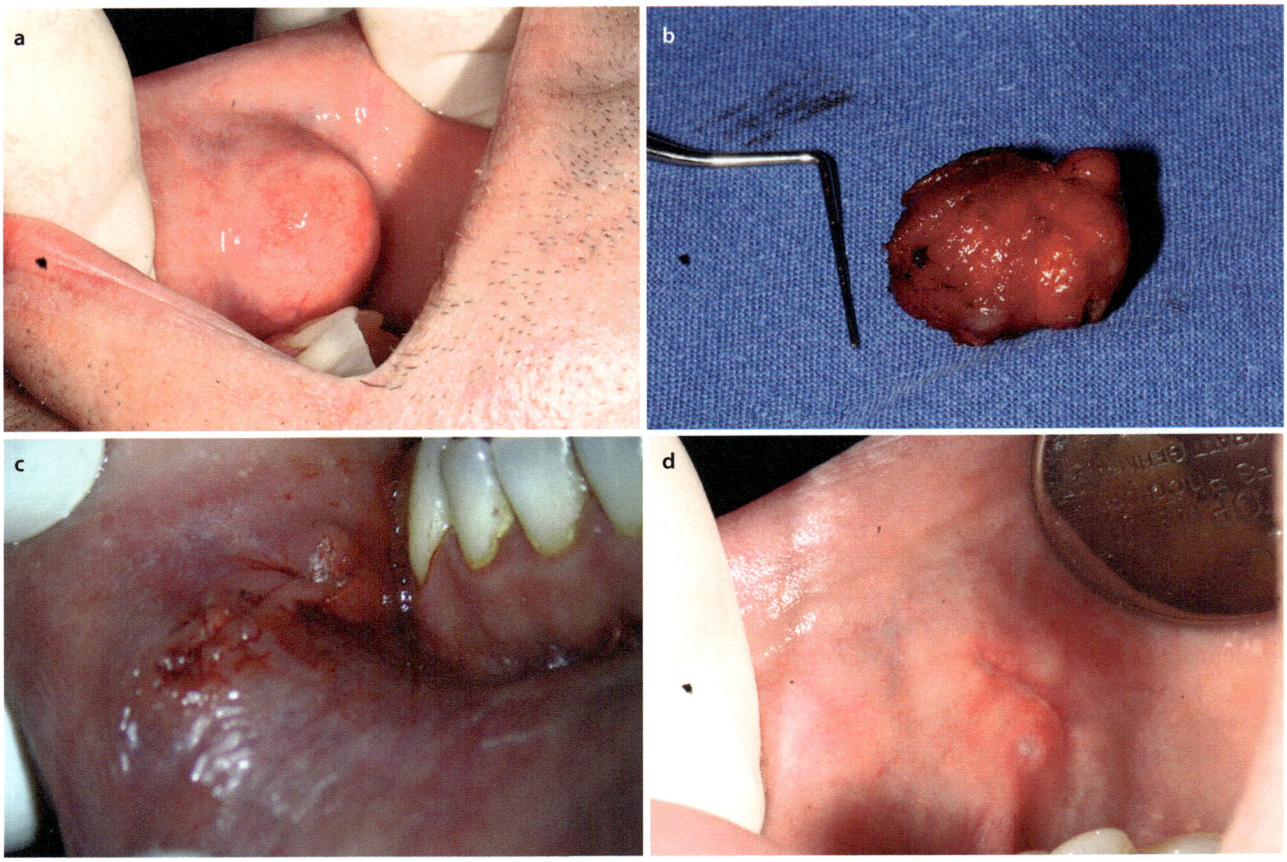

Fig. 13.14 **a** Pre-operative photo of a lipoma lesion in mandibular vestibule. It is soft and elastic, but firmly attached to the underlying and covering tissue. There is no sign of any inflammatory process. An 810 nm diode laser was used with a 300 μm glass fiber in contact at 30 W, 13,000 Hz, and a 9 μs pulse duration. Average power 2.93 W. Fluence 967 J/cm^2. **b** The excised specimen has an irregular lobular surface, and the histologic diagnosis was that of a common lipoma. **c** Immediate postoperative view showing closure of the deep wound with sutures. **d** One-week postoperative view depicts good healing

lesion. It is most likely to appear in the gingiva and less common in the palate, oral mucosa, tongue, and lips [76].

Histologically, PG comprises a highly vascular proliferation that resembles granulation tissue with a dense inflammatory infiltrate of neutrophils, plasma cells, and lymphocytes. The surface of chronic lesions becomes ulcerated and replaced by a thick fibrinopurulent membrane showing more fibrous entities. Long-standing PGs may progress to gingival fibromas as a result of fibrous maturation [77].

Differential diagnosis of PG includes peripheral giant cell granuloma, a peripheral odontogenic or ossifying fibroma, and less commonly Kaposi's sarcoma, bacillary angiomatosis, and non-Hodgkin's lymphoma. Thus, biopsy is the most decisive way to confirm the diagnosis. Radiographic findings are absent in most cases [77].

Scaling and curettage usually are the initial treatment, and the patient is instructed to improve his/her oral hygiene. In addition, all expected irritants like foreign bodies, calculus, and sharp restorations should be eliminated [78].

Complete surgical resection of PG is considered the treatment of choice. This could be performed using a scalpel as a conventional treatment or preferably using laser with indicated wavelengths [79]. The Nd:YAG, CO_2, Er:YAG, and high-power diode laser have been reported in several studies as a suitable tool for surgical removal of PG offering a precise excision with minimal bleeding [80, 81]. The wavelengths of diode 980 nm and Nd:YAG 1064 nm laser are well absorbed by oral lesions which exhibit a high amount of melanin and hemoglobin such PG. Therefore, they consider the ideal lasers for PG management [82].

The procedure is performed after infiltration of local anesthesia usually with a fiber lasing in a contact mode with average power ranging between 2 and 3.75 W. According to specifications of available laser device, continuous/interrupted or pulse mode can be used. The fiber tip (320–400 μm diameter) is pointed at an acute angle of about 15°, moving it circumferen-

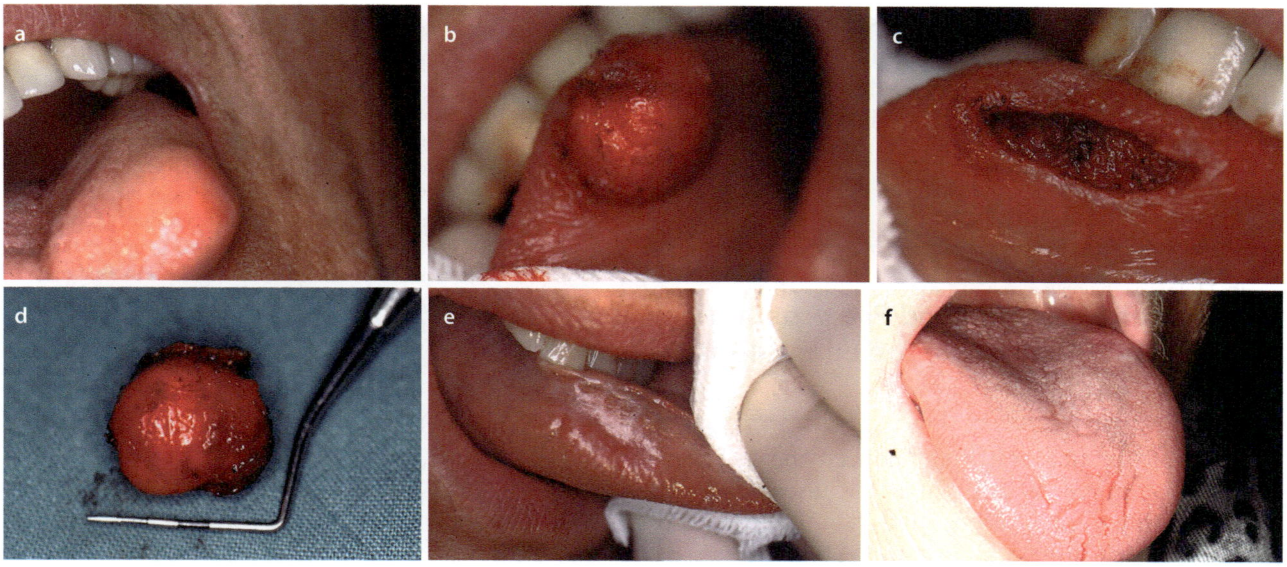

■ **Fig. 13.15** **a** A pre-operative view of the swelling in the left portion of tongue. **b** The tissue overlying the area is excised and the mass can be seen. It is adhered deeply in the underlying tissue. **c** An immediate postoperative view of the excision area. An 810 nm diode laser was used with a 400 μm glass fiber in contact with the tissue at 1.6 W continuous wave. Fluence 199 J/cm^2. The lesion does show some carbonization because of the emission mode, but the hemostasis is excellent. **d** A photo of the large sized lesion. **e** Three-week postoperative photo shows healing with a substantial tissue defect. **f** Fifteen-year postoperative view of the completely healed area with no sign of recurrence or a functional defect

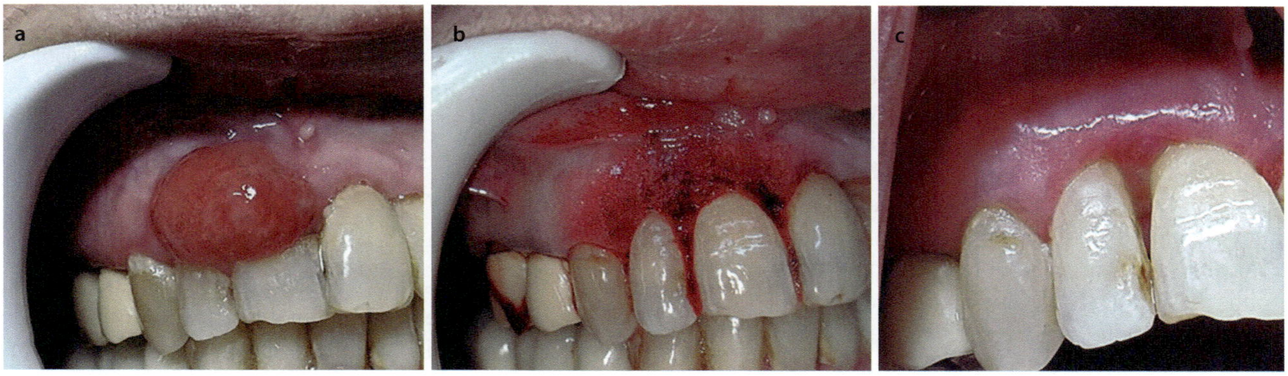

■ **Fig. 13.16** **a** A discrete nodular, erythematous, sessile overgrowth covering the cervical third of the crowns #11 and #12. **b** Complete removal of the lesion using 2 W super pulse CO_2 laser. The clinical diagnosis was a pyogenic granuloma which was confirmed histopathologically. **c** Three weeks postoperative view showed uneventful healing

tially around the base of the lesion. Whenever the lesion starts to detach from its adjacent tissues, the fiber tip is directed toward the base of lesion itself to avoid cutting deeply within the tissues underneath resected lesion. For PG located on gingiva and for aesthetic and functional request, the proper gingival contour is better to be achieved during laser resection and interdental papilla should be preserved when it is possible.

If PG is around an implant, then Er:YAG/Er,Cr:YSGG lasers have the advantage of removing the lesion with minimal thermal effects to the implant surface and surrounding bone [83]. The following parameters may apply safely in the field of intervention: Er:YAG laser 2940 nm with average power: 3 W, energy per pulse: 300 mJ, frequency: 10 Hz, short pulse, with contact headpiece, with 50% water and 50% air, Tip Ø = 1 mm) [84].

To avoid recurrence, the excision should extend down to the periosteum. Nonetheless, recurrence rate was reported in 8% of PGs resected with lasers in comparison to 15% for those undergoing conventional treatments [85].

■ Figure 13.16 is a healthy 48-year-old female complaining of a recurrent gingival overgrowth associated with the area of teeth 12/11. The patient manifested bad oral hygiene with interproximal open margin restorations. The lesion was about 1 cm in diameter with a sessile base. The clinical diagnosis displayed a pyogenic granuloma. It was asymptomatic and bled easily with light contact. The CO_2 laser was used to excise the lesion offering a precise cutting with good hemostasis.

13.3 Pre-prosthetic Surgery

13.3.1 Epulis Fissurata and Pre-prosthetic Vestibuloplasty of the Edentulous Patient

The epulis fissuratum or denture-induced fibrous hyperplasia is a trauma or inflammation caused lesion in patients with partial or full dentures [86]. Ill-fitting and overextended dentures can irritate the mucosal tissue and create a hyperplastic tissue flap that can develop signs of secondary inflammation [87, 88]. Due to constant irritation, the lesions can progressively grow, compromising the support for the denture, which in turn causes more irritation. Often these epulis are combined with alveolar ridge atrophy and diminished or missing vestibular depth [89]. During conventional surgery, hemostasis is difficult to achieve, especially for patients with bleeding dyscrasias. Furthermore, in most cases, suturing the incision can result in a diminished vestibular depth. Since laser surgical wounds show a reduced number of myofibroblasts, there will be a diminished contraction the tissue when healing by secondary intension [90, 91]. Secondary wound healing is ordinarily uneventful and significantly less painful than with a conventional surgical protocol. Therefore, excision of the epulis fissurata should be performed with a laser to achieve adequate hemostasis even in patients with bleeding disorder. If there are very large tissue flaps, a fiber-delivered laser can easily undercut them for excision [92].

Several dental lasers have been used with the following recommended protocols and parameters in the treatment of epulis fissuratum:

Laser	CO_2
Wavelength	10,600 nm
Power	5–6 W
Emission mode	Pulsed
Frequency	50 Hz
Spot size	0.9 mm
Local anesthesia	As needed
Application technique	Firstly, a gentle tension should be applied to pull the lesion away from its base and get clean cut with clear borders, then the clinician starts cutting the lesion with the focused mode and finally the mode is defocused to achieve the tissue vaporization with keeping the tip away from the tissue for about 6 mm during the operation, and finally the wound will stay exposed to heal in the second intention in the site, the excised mass should be sent to the pathological examination to confirm the diagnosis [93]

Laser	Diode
Wavelength	810 nm
Average power	2 W
Emission mode	Pulsed
Local anesthesia	As needed
Application technique	Firstly, the lesion is grasped with Adson forceps or any other helpful toothless forceps to fix the lesion and keep it under tension then the excision starts at the base of the mass in order not to over cut the tissues in the site with continuous removing and until the lesion is detached and until we reach a healthy connective tissue then the removed mass should be fixed in 10% formalin and sent to the histopathologist, neither sutures nor dressings are needed in the site [94]

Figure 13.17 is that of a 68-year-old female patient, nonsmoker, presents with an ill-fitting denture and a very painful epulis fissuratum lesion in the midline lingual vestibule. Her medical history is significant for hypertension, cardiac dysrhythmia, and Coumarin anticoagulant medication. A diode laser excised and contoured the tissue. Two weeks later, the area had healed, facilitating a new lower full denture, correctly extended.

Vestibular depth can be very challenging with elderly patients, and a vestibuloplasty can be planned. After losing their teeth at a young age, many of these patients show a profound progressive atrophy of the skeletal bone. This atrophy has anatomical limitations in the mandible due to the structure of the self-supporting bone; thus, the patient seldom presents with less than 8 mm of residual bone height. In the maxilla, the limitations are the nasal cavity with an apertura piriformis and the floor of the maxillary sinus. If the alveolar ridge is lost, the bony configuration of the maxillary sinus takes over to function as the ridge. In the anterior segment, the alveolar ridge can recede and disappear. Considering these anatomical situ-

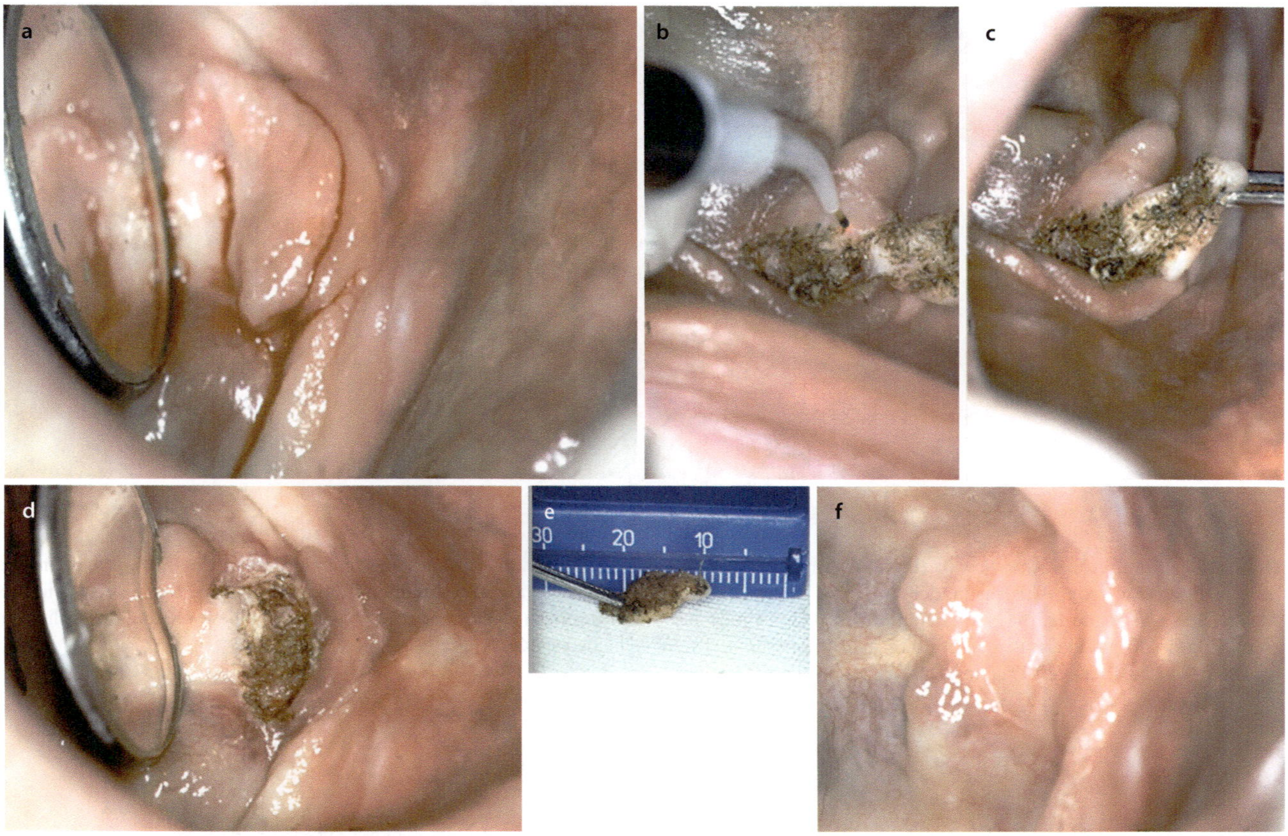

Fig. 13.17 **a** Pre-operative photo of an epulis fissuratum lesion in the midline lingual vestibule. The slight bleeding is due to administration of local anesthesia. A 980 nm diode laser was used with a 320 µm quartz fiber in contact at 1.5 W, CW. **b**, **c** The hypertrophic tissue is carefully dissected from the underlying tissue bed, with the advantage of incisional hemostasis. **d** Immediate postoperative view. **e** Excised tissue and histological confirmation of provisional diagnosis. **f** Healing at 2 weeks. (Case courtesy Prof. Steven Parker)

ations, experience has shown that, in a mandible with a residual height of approximately 10 mm, it is impossible to expect a satisfactory prosthodontic outcome. In the maxilla a vestibular extension in the premolar and molar region is possible. However, the vestibule can only extend to a certain height without compressing the residual soft tissue, so relapse can occur. It is possible to stabilize the wound healing by inserting a free gingival graft [95].

Figure 13.18 is that of a 68-year-old male with a medical history of cardiac insufficiency, hypertension, and nephrolithiasis, presents with insufficient vestibular depth. A diode laser was used for a vestibular excision and a free gingival graft was inserted. The border of the existing denture was extended apically to aid the graft in healing. Six months postoperatively, the vestibular depth is now adequate.

Atrophy of the alveolar ridge leads to the appearance of pre-existing frenula that can subsequently interfere with the denture flange or the periodontal tissue. Laser surgery offers benefits for frenum revision and vestibuloplasty and suturing, such as for conventional Z- or VY-plasty procedures, which are not necessary.

Gingival recession is normally located in the anterior mandible, due to a high insertion point of the mandibular frenulum and a mis-aligned mental muscle with mobility of the periodontal soft tissue under function. Treatment of this recession accompanied by a malalignment of the mental muscles should be carefully planned ahead of time so that patient can care for the healing edentulous area. The objective of the surgery is to divert the muscle movement away from the periodontium and establish a neutral zone in the affected anterior vestibule [96]. One option is called the Kazanjian technique. An incision line is made on the mucosa in the lip from canine to canine; or, it can be extended to the premolar region, depending on the patient's muscle activity zone [97]. A thin mucosal flap is prepared to the periosteum at the mucogingival border by using a laser with a 200 µm fiber. Then the mental muscle is excised away from the periosteum into the vestibule. When sufficient vestibular depth is achieved, the mucosal flap is placed back onto the periosteum, and fixed there with two or three resorbable sutures to spread out the flap. The muscle and wound in the lip are prone to secondary granulation. The postop-

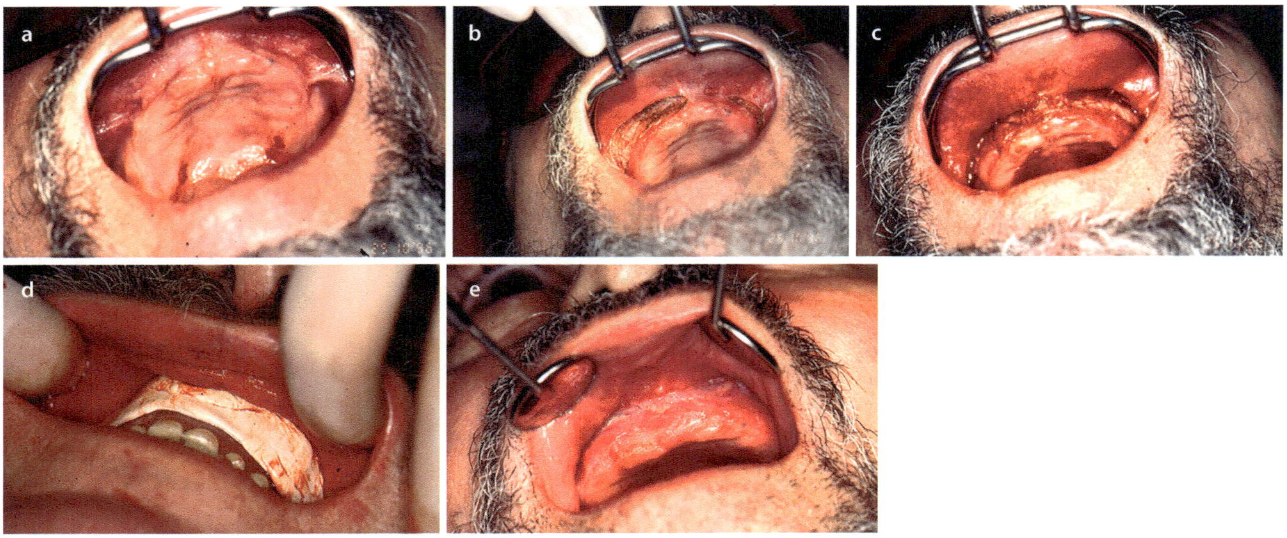

Fig. 13.18 **a** Pre-operative view of the very short maxillary anterior vestibule extension, which is inadequate for a stable denture. **b** An incision was made to the periosteum with 810 nm diode laser employing a 400 μm glass fiber in contact with the tissue at 1.6 W continuous wave. Fluence 199 J/cm^2. **c** A free gingival graft is placed, covering the periosteum. **d** An extension was placed on the denture border for a scaffold enabling the graft material to adapt and heal. **e** A 6-month postoperative view shows the new vestibular depth with healed tissue

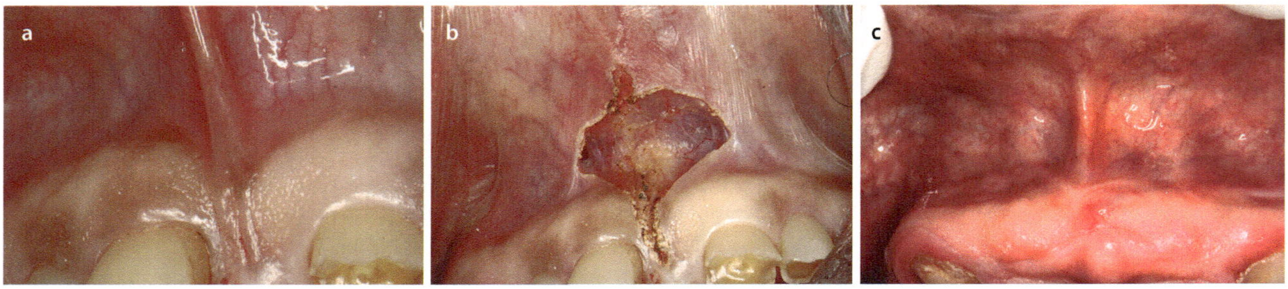

Fig. 13.19 **a** Pre-operative view of a maxillary anterior frenum whose attachment is embedded in the gingiva. **b** Immediate postoperative photo of the completed frenum revision. A 10,600 nm Carbon Dioxide laser was used in the SP mode at an average power of 1.5 W with an 800 μm beam diameter in non-contact mode. Fluence 135 J/cm^2. The incision's extension was well into the gingival tissue to eliminate the muscle pull. No sutures were placed. **c** Postoperative view of the revised frenum, showing completely healed tissue (Clinical case courtesy of Dr. Charles Hoopingarner)

13

erative outcome is generally uneventful, and patients report minimal bleeding with tolerable discomfort which is eased with mild oral pain medication. Some swelling could occur. Over time, a relapse of recession can occur; however, good oral hygiene measures can produce some stable reattachment. If the recession persists, a connective tissue graft can be placed to re-establish an adequate zone of attached gingiva.

Figure 13.19 shows the revision of the maxillary frenum of a 65-year-old male with no significant medical history. The maxillary anterior frenum attachment must be revised to allow improved periodontal health and further treatment planning for tooth restoration, which may involve a maxillary denture. Clearly, the revised frenum would also allow an adequate extension of the denture flange. A carbon dioxide laser in superpulsed mode was used for the surgery. A 2-month postoperative photo shows total healing and a stable position of the frenum's attachment.

A prosthetic treatment in an edentulous patient can be a real challenge to the dentist especially when the vestibular depth is shallow, also epulis fissuratum which is a soft tissue overgrowth related with a long-term irritation in the site or poorly adapted or over extended denture can make the case more complicated [98, 99].

Vestibuloplasty and removing the fibrous hyperplasia are mainly performed by conventional scalpel but recently, many lasers such as CO_2, Er:YAG, Nd:YAG, and diode lasers are considered as appropriate alternative tools because of their advantages in reducing the postoperative complications and acceleration the healing [98, 99].

Achieving adequate vestibular depth has an essential role to improve the retention of the complete denture and leads to a good oral hygiene; alternatively, inadequate vestibular depth will produce poor plaque control, gingival resection, and poor aesthetics, so that vestibuloplasty can be a necessary procedure in order to deepen the oral vestibule and increase the keratinized gingiva [100]. However, this may be achieved using a range of dental lasers following the instructions based on previous research as follows:

Laser	CO_2
Wavelength	10,600 nm
Average power	3 W
Emission mode	Pulsed
Focal spot size	0.25 mm
Local anesthesia	As needed
Application technique	The incision is horizontal across the muscle attachment by moving the hand parallel to the mucogingival junction line with pulling the lip outward to keep it under tension, the procedure is performed with an angled tipless handpiece which is 1–3 mm away from the tissue during the cutting procedure removing layer by layer in sweeping motions, moving it at the "hand-speed" for a few seconds until we reach the periosteum [101]

Laser	Er:YAG
Wavelength	2940 nm
Pulse energy	200 mJ
Frequency	2 Hz
Local anesthesia	As needed
Application technique	Ablating the strong attached tissues starting from the mucogingival junction and 2 mm from the bottom of the vestibule, removing the tissue by moving the laser tip until we reach the ideal depth and after finishing the procedure we can coagulate the surface with a very long pulse defocused mode, finally the wound will remain exposed to heal in the second intention without any sutures [98, 102]

Laser	Diode
Wavelength	940 nm
Average power	3 W
Emission mode	Pulsed
Local anesthesia	As needed
Application technique	Incising the tissues using an initiated 400 μm fiber tip extending all over the mucogingival junction with a horizontal stroke parallel to the bone relieving the muscle fibers from the periosteum until we reach the desired depth [103]

13.4 Frenulae Revision for Children and Adults

A frenulum attachment positioned within the attached gingiva can lead to gingival recession and accompanying periodontal problems. The revision of this attachment can be easily performed with a laser. The wound is usually not sutured and the secondary intention healing generally proceeds uneventfully [104]. This same treatment method can be performed on children at any age with no negative influence on the development of the mucogingival complex.

Similarly, lingual frenulum the so-called Tongue-Tie can be easily released using any available diode laser wavelength, CO_2 laser, or Er:YAG/Er,Cr:YSSG laser. Early recognition and management of an abnormal tongue frenulum attachment, especially Type III and IV frenula can help prevent, breast feeding difficulty, airway and sleep disturbance, snoring, speech, and orthodontic problems [105].

Midline diastema caused by a hypertrophic maxillary frenulum may interfere with orthodontic therapy and increase the possibility of post-orthodontic relapse. If the space between the upper ▶ central incisors is more than 2–3 mm, frenectomy will become essential to facilitate its closure, as it is less likely to close spontaneously [106].

Using laser to remove a prominent frenum is a simple, quick, and efficient method in both contact and non-contact laser application modes. A small amount of local anesthesia is infiltrated then the frenum is incised using a focused beam to lyse the fibrous band. This is accomplished by grasping the tissue with a fine ▶ hemostat or forceps at its inferior and superior attachment. The frenum is then excised by directing the laser along the hemostat outer levels. No sutures are need, and to

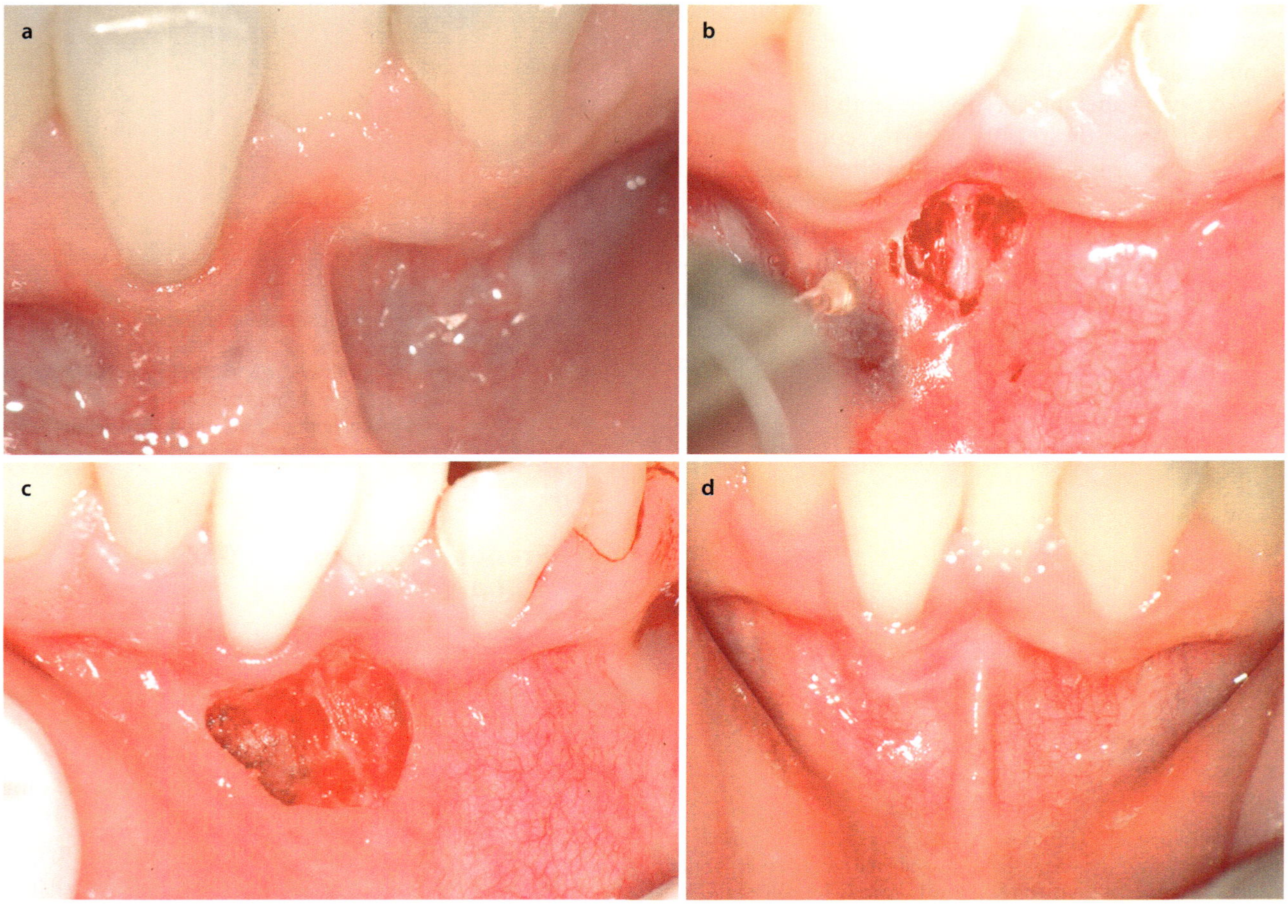

13

Fig. 13.20 **a** Pre-operative photo of a mandibular anterior frenum causing the beginning of clinical attachment loss. **b** A perioperative view. An Er:YAG laser (2940 nm) was used with a 400 μm contact tip at an average power of 2 W (40 mJ, 50 Hz) with a water spray for the incision. Fluence 57 J/cm^2. **c** Another peri-operative view showing the final incision dimensions. Subsequent to this photo, the same laser tip and parameters were used to score the underlying periosteum. No sutures were placed. **d** One-month postoperatively, the tissue is completely healed and the frenum is attached into the mucosal tissue. Case courtesy Dr D. Coluzzi

ensure hemostasis the bed of resected area is scanned by defocusing the beam [107].

Figure 13.20 depicts a mandibular anterior frenum of a 35-year-old female patient with no significant medical history. The frenum inserts at the base of the gingiva and is causing a developing loss of clinical attachment to the anterior teeth. One-month postoperatively, the frenum's attachment has been repositioned on the mucosa to eliminate the muscle pull on the gingival tissues.

Figure 13.21 demonstrates the advantage of laser soft tissue surgical management in the revision of a failed earlier upper labial frenectomy. Anecdotal evidence provides a potential for re-insertion of resected fibrous bands, where the underlying periosteum remains contiguous to the alveolar osseous tissue. Where indicated, the use of either scalpel or suitable laser wavelength to score the periosteum dissuades such complication and "finger-like" multiple frond reinsertion.

Figure 13.22 shows a clinical case of a 15-year-old female patient undergoing orthodontic treatment presented with the anterior impinging mandibular frenulum causing gingival recession. An incision for a mucosal flap procedure was performed with a diode laser and tissue was repositioned. One-year postoperatively, the attached tissue has regenerated, and periodontal health is restored.

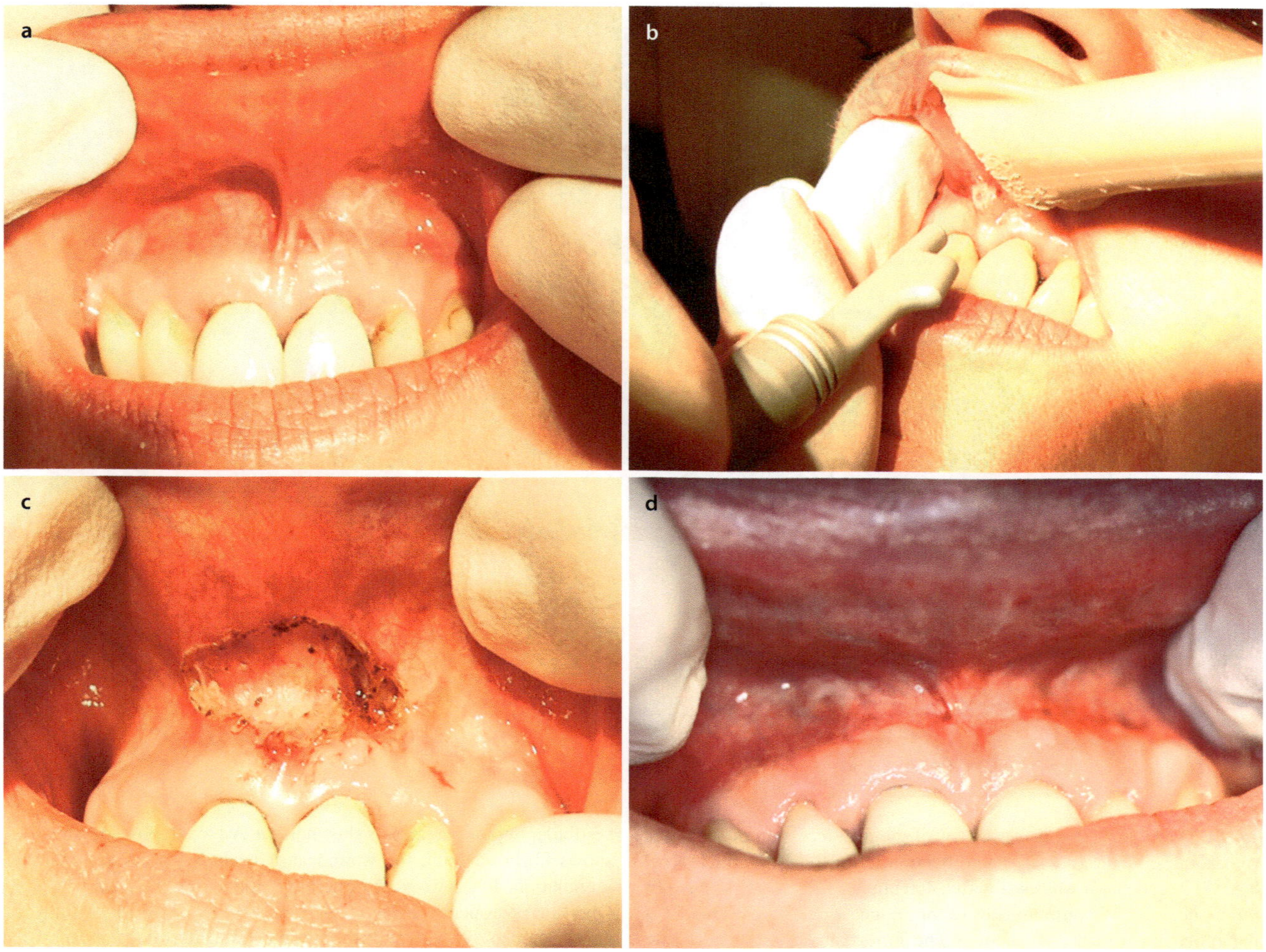

Fig. 13.21 **a** The case of a 33-year-old female patient who was referred by her general dentist for a developing periodontal problem with her frenum. Her history involved a previous scalpel frenectomy during teen years, that shows signs of fibrous reinsertion in the attached gingiva. Anecdotal reasons are offered that omission to score the periosteum during the surgical procedure may predispose to re-insertion. **b** Er:YAG 2940 laser was used to dissect all fibrous banding, using parameters: 1.5 W Average Power/150 mJ pp/10 Hz + water spray and air. Time taken 30 s. **c** Immediate post-operative appearance—a final pass with no water spray allows a tenacious plasma coagulum to protect the wound. No sutures required. **d** Early healing at 2 weeks. (Case courtesy Prof. Steven Parker)

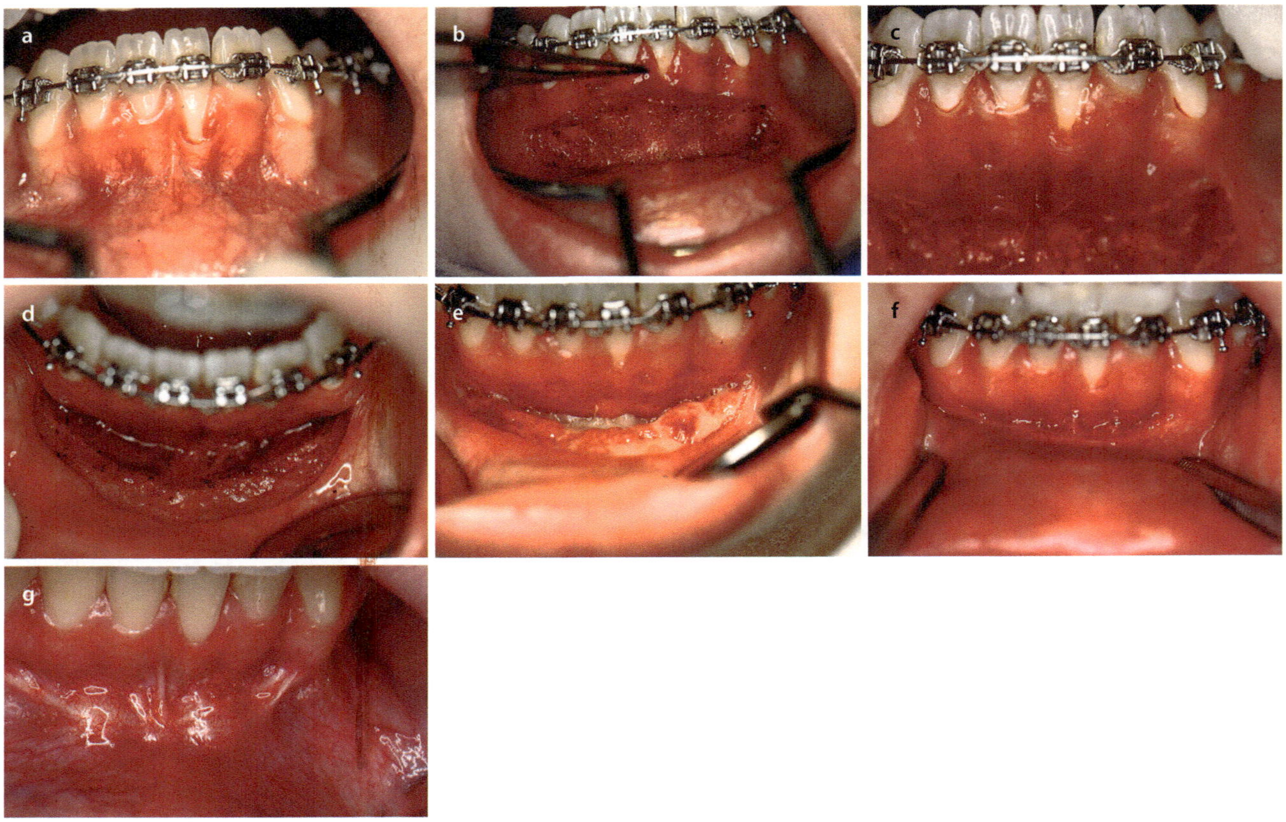

Fig. 13.22 **a** A pre-operative view of gingival recession of the lower central incisor caused by the mandibular anterior frenum's attachment. **b** An 810 diode laser was used with a 400 μm glass fiber in contact with the tissue at 1.6 W continuous wave for an incision in the mucosa for a flap procedure. Fluence 199 J/cm^2. **c** The flap is completed and adequate vestibular depth is created as shown in **d**. **e** A 1-week postoperative view shows fibrin covering the wound as it heals by secondary intention granulation. **f** A 3-week postoperative photo depicts good healing. **g** One-year postoperative view shows complete healing with a healthy frenal attachment revision and resolution of the gingival recession

13.5 Other Conditions

13.5.1 Vascular Lesions

Acknowledgment to Antoni J. España (Former Associate Professor at the Faculty of Medicine and Health Sciences. University of Barcelona) For his valuable contribution in this section.

Vascular Malformations

Vascular Lesions of the Oral Cavity

Vascular anomalies are a highly heterogeneous group of congenital disorders of the blood vessels. These types of lesions are classified as tumors and vascular malformations (which represent a localized defect in vascular morphogenesis) although each anomaly is characterized by a specific morphology, pathophysiology, and clinical behavior.

Most vascular tumors are malformations or hamartomas. Thus, they do not constitute a true neoplasm, but rather a congenital anomaly characterized by increased cell proliferation. Hemangiomas are the most common vascular tumors. Other tumors include hemangioendotheliomas, hemangiopericytomas, and less commonly angiosarcomas.

Vascular malformations, on the other hand, are the result of the abnormal development of vascular elements during embryogenesis and fetal life, these can be simple (capillary, arterial, lymphatic, or venous) or combined (arteriovenous malformations).

Hemangiomas

Hemangiomas are usually seen in newborns or at an early age. It is common for a regression to occur during puberty. There are also senile hemangiomas, usually secondary to a traumatic event. This pathology is more frequent in women. The most common locations are the lips, buccal mucosa, tongue, and palate. The size is highly variable, from a few millimeters to very extensive lesions, and they occur in approximately 10% of the population. Identifiable risk factors include female gender, prematurity, low birth weight, and fair skin

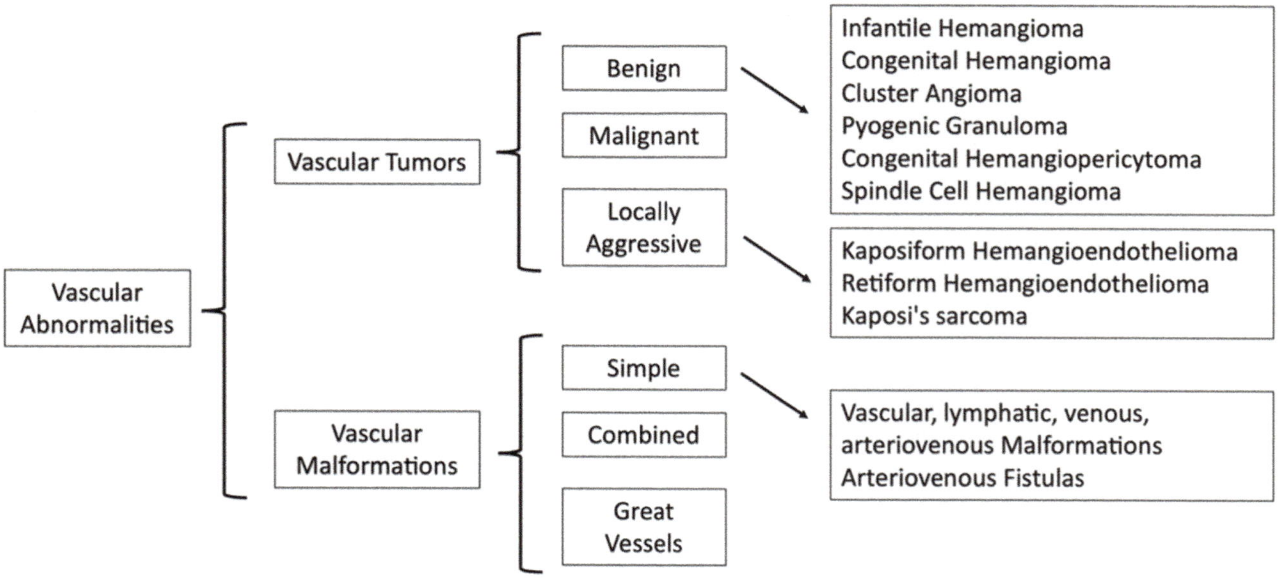

Fig. 13.23 Vascular abnormalities classification

[108]. Their growth is attributed to endothelial cell hyperplasia, they are classified as "children" or "congenital."

Another subclassification for hemangiomas is focal and segmental. Focal hemangiomas are localized and unilocular. Multifocal hemangiomatosis also exists, and newborns with more than five lesions should undergo a study to rule out visceral involvement [109]. Segmental hemangiomas are more diffuse in the form of a plaque and can lead to dysfunction and cosmetic alterations. The extremities and face are common places of appearance.

The current classification of vascular lesions can be consulted on the website of the International Society for the Study of Vascular Anomalies (ISSVA) [110–113]. As a summary, we can classify the vascular anomalies as seen in Fig. 13.23.

The presentation of hemangiomas is variable in terms of size, extent, and morphology. When there is superficial dermal involvement, the skin takes on a raised, firm appearance with a vivid crimson color. If the hemangioma is confined to the deeper dermis, subcutaneous tissue, or muscle, the overlying skin may be slightly raised and appear bluish in color. Hemangioma may present with a macular, telangiectatic appearance.

However, spontaneous involution may be incomplete, and ~15–20% of the residual lesion may remain. In cases of complete involution, scarring, replacement with fibrofatty tissue, tissue discoloration, and telangiectasia are generally observed. A total of 65.3% of affected patients are children. These lesions are more common among Caucasians, who have a prevalence of 10–12% [114].

Several treatment modalities were suggested for management of hemangiomas and vascular malformations. This will differ depending on the lesion's type, size, depth, and location [115]. Surgery has been the main form of treatment for these lesions although total removal is sometimes not possible due to the extent of the lesion, which may involve vital structures, significant deformity, prolonged pain, skin necrosis, nerve damage, systemic toxicity, and hemorrhagic phenomena [116].

Other possibilities include embolization, steroid therapy, cryosurgery, electrodessication, and laser therapy. High-power density laser treatment is an alternative, becoming one of the main treatment options for vascular lesions [117].

The mechanism of laser therapies in treating hemangiomas is achieved by targeting intravascular oxyhemoglobin, resulting in vascular injury. In general, laser irradiation is indictable for the treatment of early superficial hemangiomas or the superficial portion of a compound hemangioma, due to the limited penetration depth which is less than 5 mm [115]. However, Nd:YAG laser is more suitable for larger hemangiomas measuring up to 2 cm in depth. Percutaneous interstitial irradiation is another treatment modality that can be used for deep hemangiomas to avoid skin damage and to diminish lesions effectively. It is very frequent that vascular lesions are included between different planes, which makes it impossible to completely remove them due to the risk of injuring other structures. For this reason, when it is suspected that their anatomical relationship may pose problems, a detailed study of their location must be carried out. Nuclear magnetic resonance and selective arteriography, among others, are tests that can help us make the most appropriate therapeutic decision.

The use of different lasers for treatment of vascular lesions has been recommended in the literature. Any high-power density laser can be used, but it is convenient to know the effect the wavelength used will have, as well as the parameters applied on irradiated tissues. The most commonly used lasers are Nd:YAG and infrared diode lasers.

In the oral cavity, it is common to see isolated lesions that can be easily removed, but on some occasions their proximity to bone structures can hide a vascular shunt that could complicate the intervention and even jeopardize the patient's life.

Once the topographic scope of the lesion has been established, we can opt for different techniques; excisional biopsy, transmucosal thermocoagulation, or intralesional photocoagulation. However, before utilizing any interventional technique to manage a vascular lesion, a precise diagnosis should be obtained to determine the extent and flow pattern of the lesion. This will include having a thorough history of the present complaint and an accurate clinical examination. The extent and flow patterns of vascular malformations are usually confirmed depending on the findings of color Doppler ultrasound, MRI, or angiography.

13

For transmucosal thermocoagulation and intralesional photocoagulation techniques, it is recommended to infiltrate the bed of target lesions with local anesthesia without adrenalin to avoid vessel contraction. This will facilitate a homogeneous delivery of laser to the whole lesion.

Excisional Biopsy

In the case of excisional biopsies, any surgical laser that can emit a sufficient power density can be used. The most widely used lasers for this purpose are the CO_2 laser, Er:YAG, Er,Cr:YSGG and diode lasers. It is essential to consider that the fiber in infrared diode lasers must be activated to obtain the cutting effect. The emission parameters will vary depending on the laser which will be used. The lesion is removed by cutting through the healthy tissue that surrounds the visible lesion.

Unlike the other techniques—which will be discussed later, histological information can be obtained from an excisional biopsy.

Transmucosal Photothermal Coagulation

Another therapeutic option is transmucosal photothermal coagulation. This technique takes advantage of the mucosa's poor absorption of the applied laser wavelength, so that the energy accumulates in the hemoglobin contained in the vascular lesion.

This technique can be considered the least invasive of the three techniques proposed with the use of laser.

It is of high value to know the optimum wavelength absorbed by target tissue since this will be the basis of the success of the treatment. ◘ Figure 13.24 shows wavelength absorption peaks of different biological chromophores. It is important to note that it is a logarithmic graph and having a good understanding is paramount to anticipate the behavior of laser used.

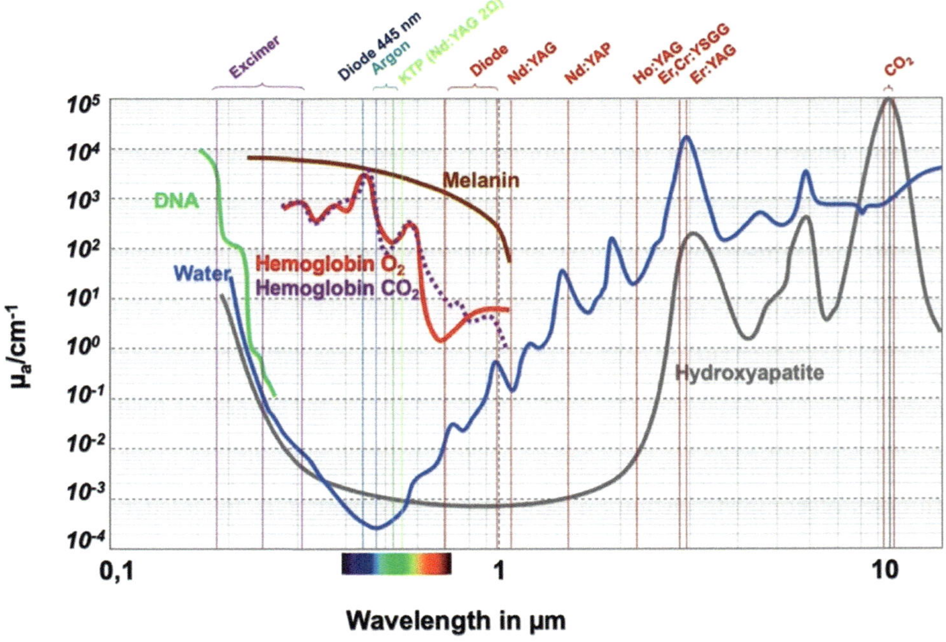

◘ **Fig. 13.24** Absorption graph according to laser wavelength for some tissue chromophores

The Nd:YAG laser is the laser of choice due to its high penetration depth of up to 5–6 mm in tissue, also being selectively absorbed by hemoglobin and poorly absorbed by water, which reduces the harmful effects on the surrounding connective tissue [118–124].

Near-infrared emitting diode lasers obtain similar absorption to Nd:YAG lasers and can also be used for the same purpose. Diode lasers (semiconductor lasers) have a range of wavelengths, which are in the visible and near-infrared spectrum. However, the 800–980 nm wavelengths are often used to treat superficial oral vascular lesions due to their good absorption by hemoglobin. The laser beam penetrates deeply into the tissue, down to a depth of 4–5 mm causing coagulation in the lesion that extends 7–10 mm because of the generated heat [125–128].

The penetration will be greater for the least absorbed wavelength, which will provide greater safety and a higher chance of completely eliminating the lesion. For this indication, the fiber should always be used without initiation.

Transmucosal photothermal coagulation (PHC) includes delivering the laser beam through a flexible quartz fiber, without contact between the fiber tip and the surface of the lesion (1–4 mm in distance). Also, a microscopic glass slide can be used to allow greater penetration of the laser beam into the lesion. Continuous wave mode can be used with a power of 2–6 W, or pulsed wave mode with a power of 6–12 W. Sweeping movements are applied from the circumference to the center of the lesion until blanching and shrinkage of the surface become clear [114, 129–133].

Transmucosal photothermal coagulation is indicated in the treatment of small to moderate vascular lesions in anticoagulated patients since the risk of bleeding is greatly reduced. By using this technique, no specimen will be available for histopathological evaluation. Another drawback is the lack of knowledge of the lesion being completely irradiated since the intraoperative appearance can be confusing, which would demand carrying out additional treatment sessions if the lesion has not disappeared in the following weeks.

Intralesional Photocoagulation

In some cases, the size and anatomical relationships of the lesion make excision or transmucosal photocoagulation unfeasible. Lesions that extend from the cutaneous plane to the mucosa, crossing subcutaneous tissue and muscle, or lesions of close proximity to the bone, can pose a risk to the patient's life. Accordingly, when a complete removal of large or extended vascular lesions is not achievable, partial removal or volume reduction of the lesion may help improve the patient's quality of life.

Intralesional photocoagulation (ILP) also called the "interstitial technique," depends on the direct puncture of the lesion by a needle connected to the end of the fiber. It should be inserted to cover the entire thickness of the lesion and moved in a radial pattern as photocoagulation is proceeded within the tissue. Continuous or pulsed wave modes can be used with a power of 9–13 W. The endpoint of treatment is obtaining pallor of the lesion, but the effectiveness of mentioned technique is limited when dealing with large and thick lesions [116]. The considerations in relation to the wavelength to be used are similar to wavelengths applied in transmucosal photothermal coagulation [130, 134–138].

In order to achieve optimal results, each clinical situation must be evaluated, as well as the wavelength and the parameters to be used. The clinician must select the best wavelength, from those available, to design the treatment strategy. In the same intervention, all the techniques described can be used when the injury requires it.

▪ Figure 13.25 depicts a clinical case of a large vascular formation that was present on the lip of a 72-year-old female patient with an inconspicuous medical history. The patient reported that the lesion had been slowly growing over a period of 10 years, but she was given medical advice to not treat it. A diode laser was used in a non-contact mode, and 4 weeks postoperatively, the lesion had disappeared.

▪ Figure 13.26 is that of a 78-year-old patient with a medical history consisting of hypertension, cardiac insufficiency, and cardiac dysrhythmia was taking Coumarin for anticoagulation therapy. A diode laser performed an excisional biopsy, and the tissue was determined to be a non-malignant vascular malformation. A 5-day postoperative photo shows the wound healing satisfactorily and that is expected to continue.

▪ Figure 13.27 shows a 65-year-old female patient with a medium size vascular lesion on the central part of her lower lip. Her medical history is non-contributory. The patient recalls biting her lip several months ago and she is concerned about the aesthetic appearance. A carbon dioxide laser was used for an excision. One month later, the tissue healed. Clinical case courtesy of Dr. Rick Kava.

▪ Figure 13.28 depicts a non-aesthetic hemangioma on the lower lip of a 72-year-old female, with a non-contributory medical history. The large lesion extends from the vermillion border to the mucosa. A diode laser was used in a non-contact mode to allow the laser radiant energy to penetrate into the lesion. A 3-month postoperative view shows complete healing. Clinical case courtesy of Dr. Giuseppe Iaria.

▪ Figure 13.29 depicts a clinical case of a large vascular formation that was seen on the upper lip of a 4-year-old male patient with no remarkable medical history. The parent reported that the lesion was observed since birth and had been slowly growing over the years. The patient was first prescribed Propranolol 3 mg/kg (beta-blocker) for 4 months before the laser application. This can effectively control the proliferation of heman-

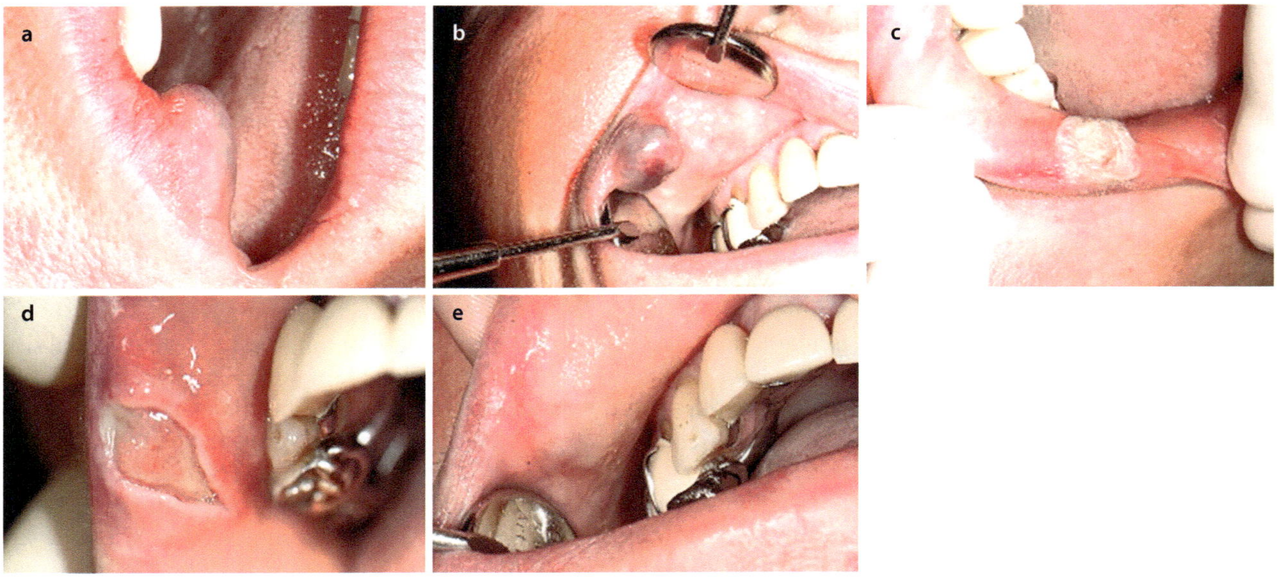

Fig. 13.25 **a**, **b** Two pre-operative views of a vascular malformation on the upper right lip. **c** Immediate postoperative view. An 810 nm diode laser was used with a non-initiated 600 μm glass fiber out of contact with the tissue at 2.5 W continuous wave. Fluence 128 J/cm^2. The laser was activated until the tissue appeared blanched, and then the laser was turned off. **d** A 1-week postoperative view shows granulation is proceeding. **e** Four-week postoperatively, the area is healed and the lesion disappeared

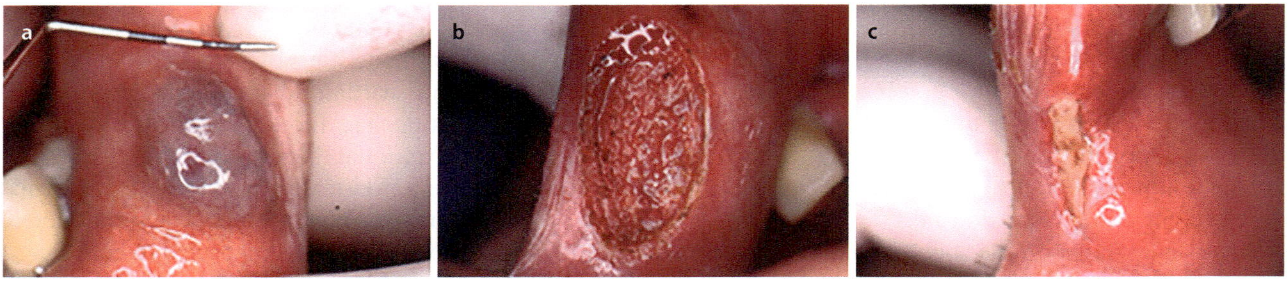

Fig. 13.26 **a** Pre-operative view of a large vascular malformation lesion on the lower lip. **b** An 810 nm diode laser was used with a 400 μm glass fiber in contact with 30 W, 12,500 Hz, and a 10 μs pulse duration. Average power 3.75 W. Fluence 928 J/cm^2. **c** A 5-day postoperative photo shows normal wound healing

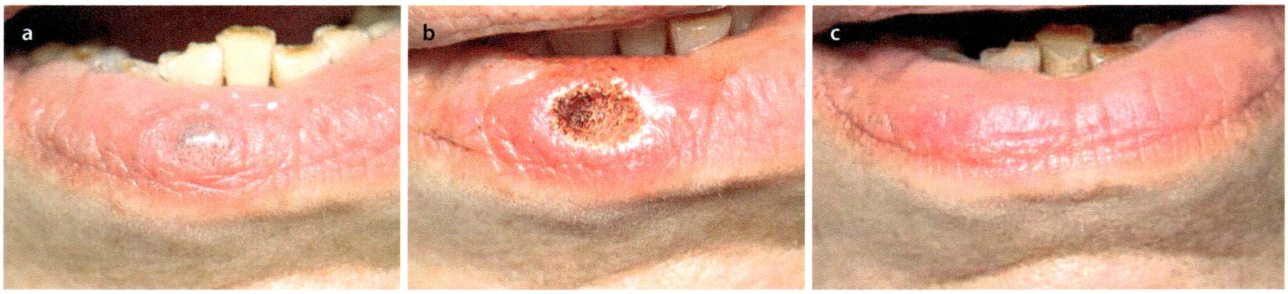

Fig. 13.27 **a** Pre-operative view of the hemangioma on the central area of the lower lip. **b** Immediate postoperative view of the lesion ablation. A 10,600 nm Carbon Dioxide laser was used with an average power of 2 W in SP mode with a 400 μm non-contact tip. Fluence 289 J/cm^2. **c** A 6-week postoperative photo showing complete resolution of the vascular lesion

gioma and promote its regression. The Nd:YAG laser was used in the transmucosal photothermal coagulation technique, in which the laser was applied in non-contact mode through a glass slide placed with added compression over the lesion. This will prevent the laser from penetrating the lining mucosa above the lesion and help in

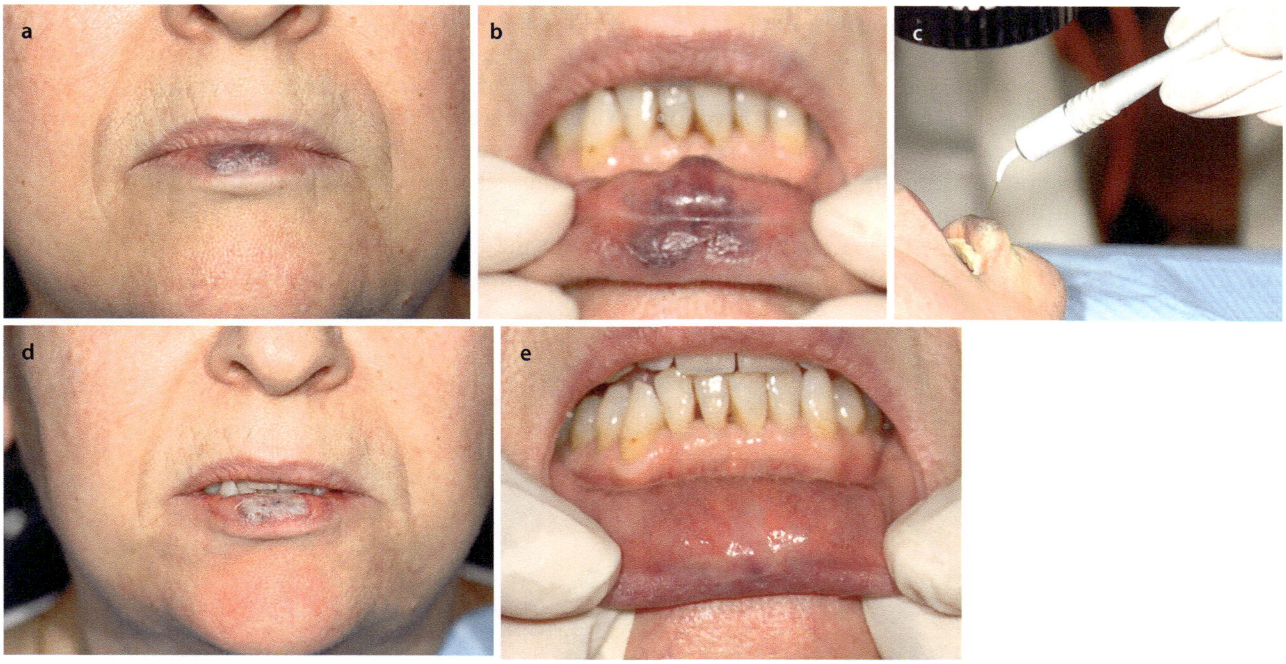

Fig. 13.28 **a** Pre-operative view of a hemangioma on the lower lip from the extraoral view. **b** The intraoral view showing the extent of the lesion to the inner mucosa. **c** An 808 nm diode laser was used with a 400 µm non-initiated fiber in non-contact, approximately 2 mm from the tissue surface. Average power 3.0 W CW. Fluence 154 J/cm^2. The handpiece was in constant movement with a circular motion, covering the extent of the lesion. No anesthesia was used, and the patient was comfortable. **d** The immediate postoperative view after 1 min shows the tissue is light gray color. Ice was applied for 2 min to reduce edema of the surrounding tissues. **e** Three-month postoperative view shows complete tissue healing. (Clinical case courtesy of Dr. Giuseppe Iaria)

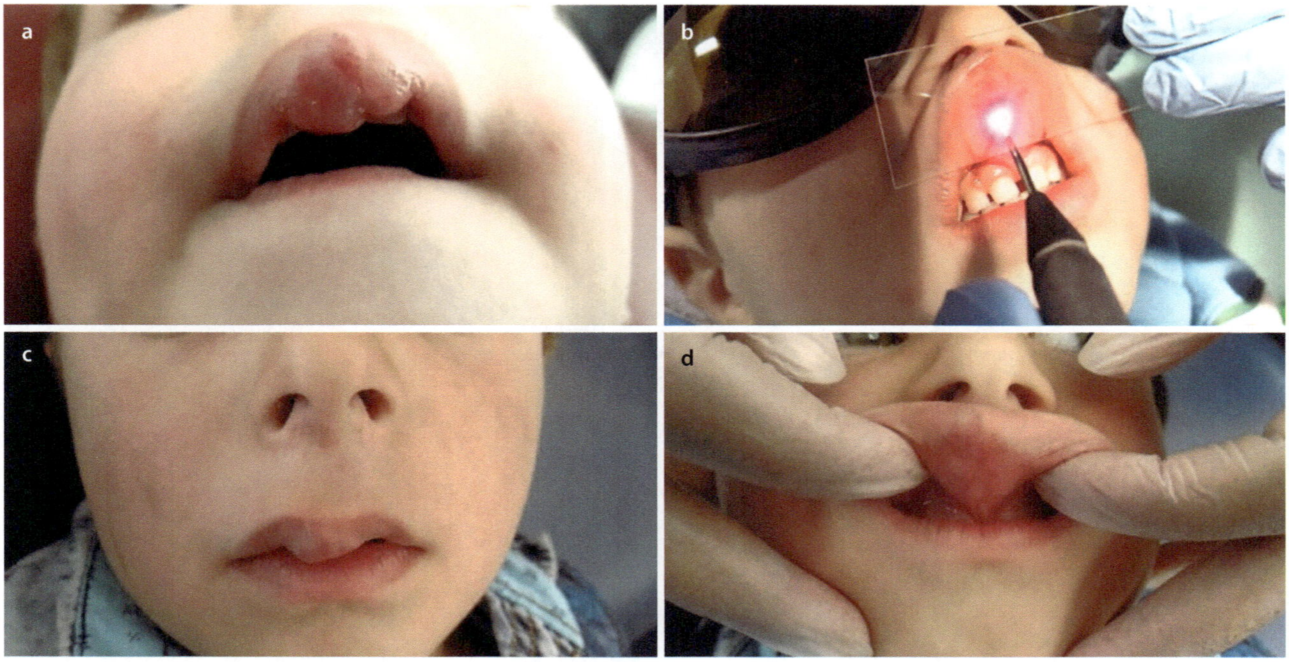

Fig. 13.29 **a** The initial clinical appearance of a hemangioma affecting the upper lip and philtrum. The lesion gradually increased in size since birth. **b** An irradiation session, in which the transmucosal photothermal coagulation technique was applied, used Nd:YAG laser with pulse mode to facilitate vessels shrinkage inside the lesion without damaging the superficial surface of the lesion. This was achieved by delivering the laser in non-contact mode through a glass slide placed with added compression over the lesion. **c**, **d** A complete resolution of the vascular lip lesion after five irradiation sessions. The patient is set to undergo a surgical intervention to correct the lip asymmetry after puberty

lasing the lump homogenously. Obviously, the Nd:YAG laser was applied in pulse mode with following operating parameters: 3 W, 60 Hz, and 50 mJ. Five sessions were required. The case requires cosmetic correction after puberty and complete development of the lip.

The clinical appearance of multiple vascular lesions noticed on the right buccal mucosa which were treated previously without complete resolving (Fig. 13.30). An Er,Cr:YSGG laser with the following parameters was used: 2 W (100 mJ per pulse), 20 Hz, pulse duration 140 μs with a fiber tip of 800 μm, 12% water, and 12% air for irrigation. No anesthesia was used in this procedure. All lesions were resected in the same session and left to heal by secondary intention. Complete re-epithelization was achieved after 3 weeks.

A 32-year-old woman who presented with a deformity in the lip/nose area on the right side of her face (Fig. 13.31). It was a congenital lesion that increased

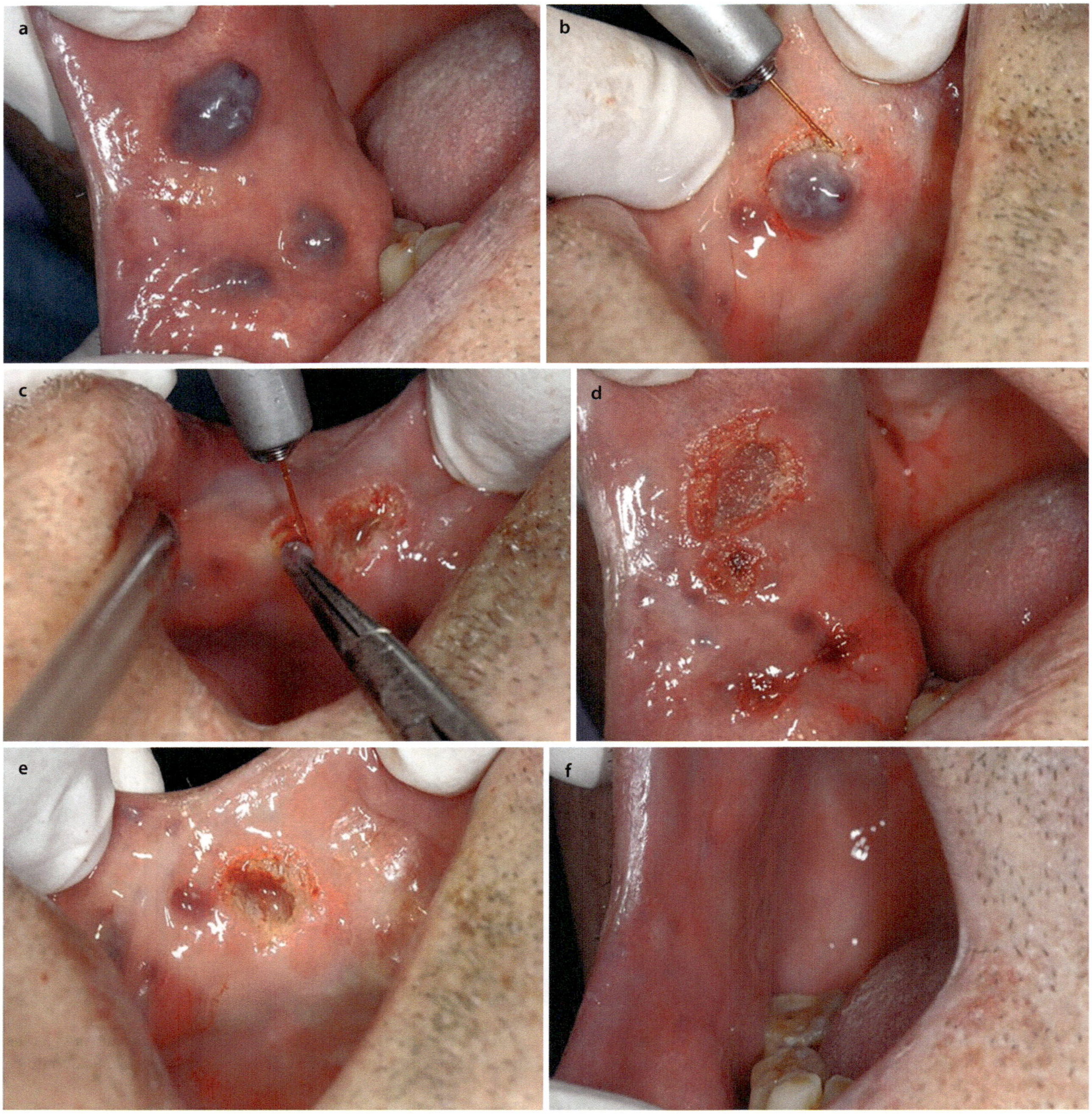

Fig. 13.30 **a** Multiple vascular lesions appearing on the buccal mucosa. **b**, **c** Lesions removal using Er:Cr:YSGG. Cutting was performed about 1–2 mm outward of the lesion margin to avoid lesion penetration and incidental bleeding. **d** Immediate postoperative. **e** Healing process after 1 week of treatment. **f** Three weeks post laser application view showing uneventful healing

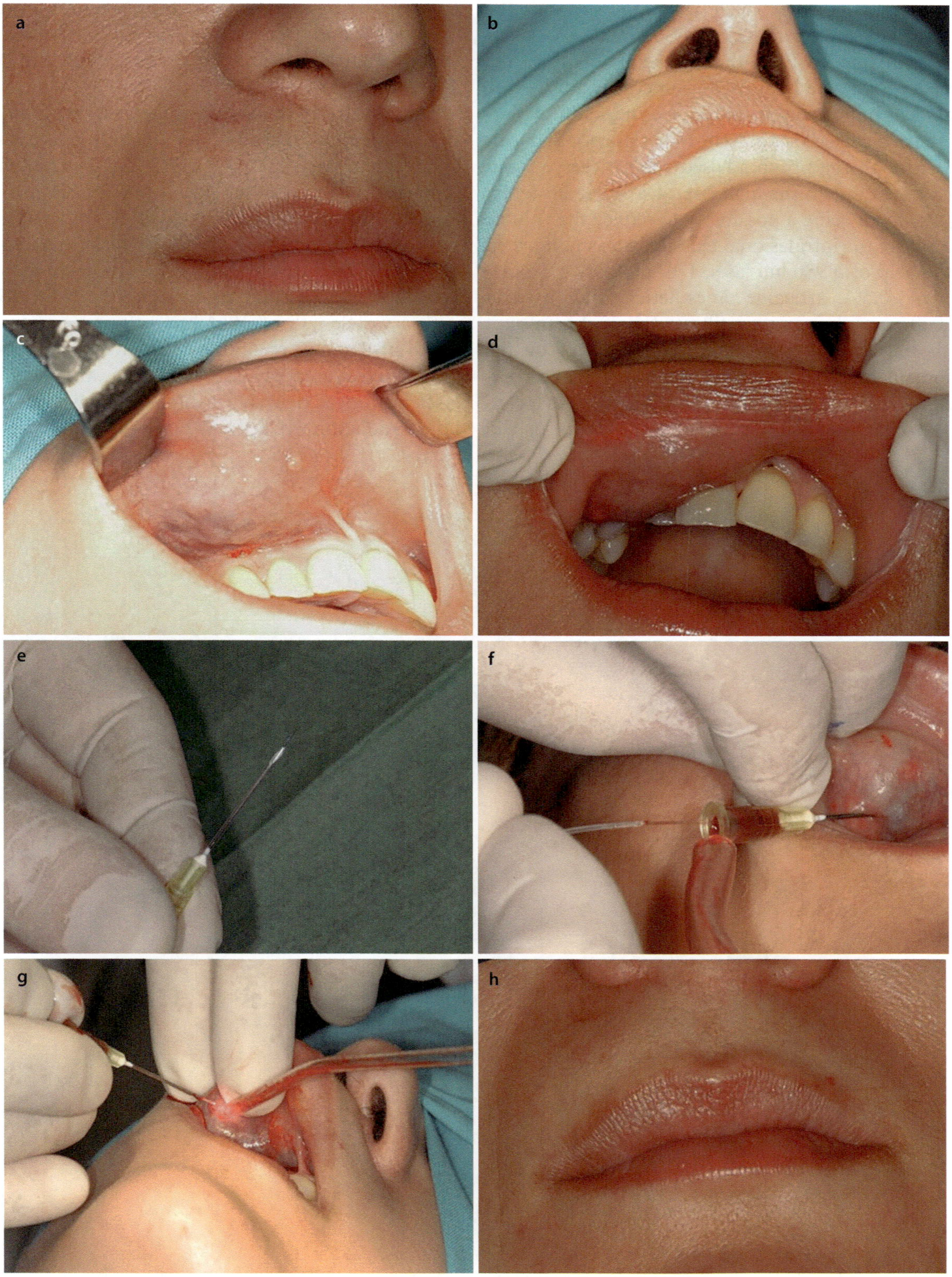

Fig. 13.31 **a, b** A vascular lesion affecting the right upper lip producing facial asymmetry. **c, d** Intraoral view of the lesion which measures about 2 cm in diameter. **e–g** The intralesional lasing technique was used by inserting a 320 μm optical fiber within a cannula through the lesion wall to the center of the lump, passing 2–3 mm beyond the end of the cannula, while laser emission is activated. The 980 nm diode laser, 1 W in continuous mode was used in this case. **h** Three weeks post laser application showing significant reduction in lesion size with no sign of scaring or discoloration

slowly in size over time, which was prominent over the last year. The pre-operative study showed a vascular lesion that extended from the cutaneous plane to the mucosa, crossing the levator labii superioris muscle, the zygomaticus minor muscle, the risorius muscle and, partially, the orbicularis oris muscle. In her previous consultation visits, she had been advised against carrying out any treatment due to the high risk of a worse unsightly result. As there being a possibility to reduce the volume of the lesion, the intralesional technique was performed with the help of a 980 nm laser, 1 W in a CW. One cartridge of 0.5% articaine anesthetic solution with 1:200,000 epinephrine was infiltrated, half the amount near the infraorbital nerve and the other half on the anterior superior alveolar nerve. The procedure was performed by inserting a 320 μm optical fiber within a cannula, through the lesion wall. The fiber was passed through the proximal end of the clamp no more than 2–3 mm beyond the end of the cannula while laser emission was activated. Clinical case courtesy of Prof Antoni J. España.

A 14-year-old female was diagnosed with an upper lip congenital hemangioma, exhibiting consistent enlargement overtime (Fig. 13.32). The patient presented complaining of her lip asymmetry. Surgical resection of the lesion was excluded as it would cause unfavorable deformity of the lip. The transmucosal photothermal coagulation technique was used to reduce the volume of the vascular lesion using the Nd:YAG laser with a wavelength of 1064 nm, offering a penetration depth of 4–6 mm by non-contact mode. The following parameters was applied: 3 W, 50 Hz, and 60 mJ. The laser beam was delivered using a 320 μm fiber tip and laser irradiation was applied 2 mm away from the visible

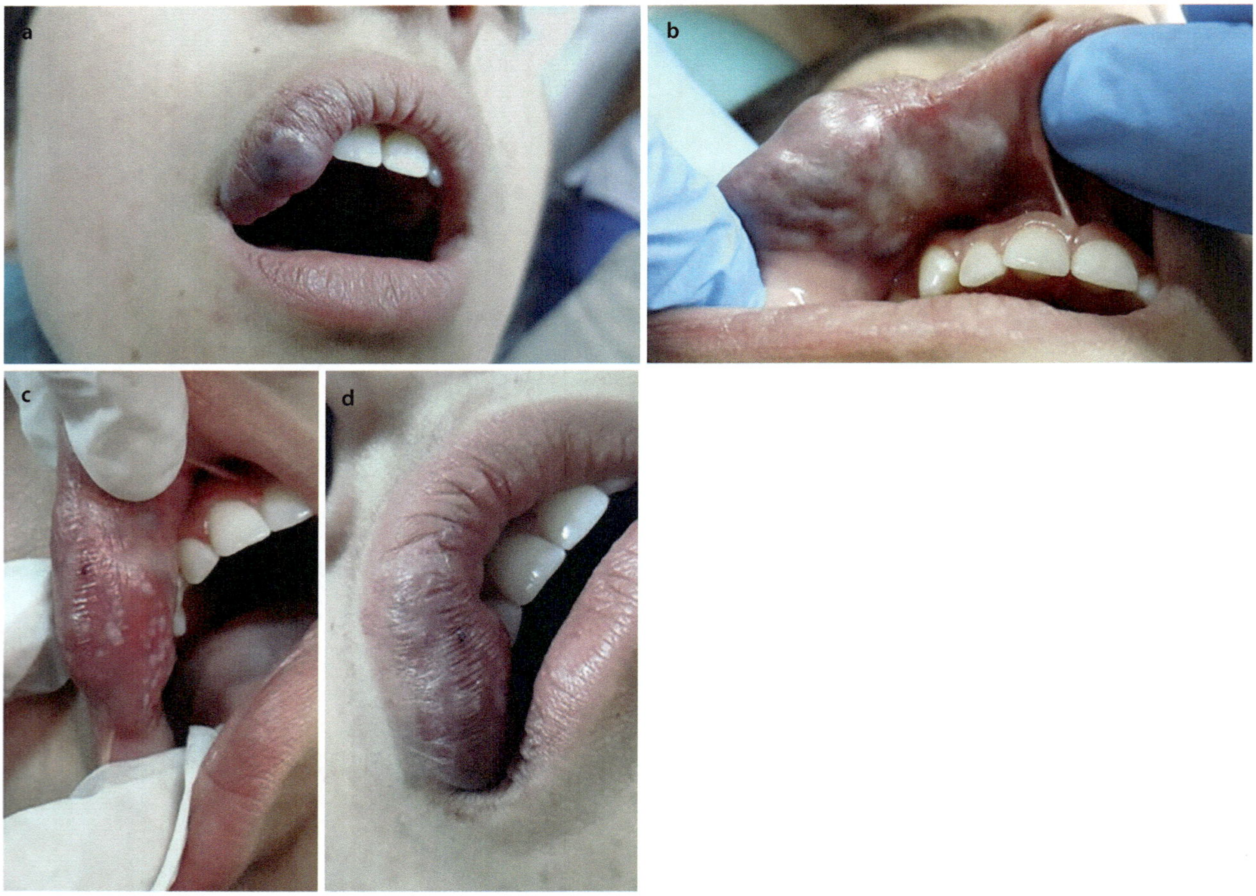

Fig. 13.32 **a**, **b** Initial clinical appearance of a hemangioma affecting the right upper lip. The lesion showed a persistent increase in size especially over the last year. MRI was organized to determine the extent and the flow pattern of the lesion. **c**, **d** First irradiation session. For aesthetic and function purposes, transmucosal photothermal coagulation technique using Nd:YAG laser with pulse mode was applied to facilitate vessels shrinkage inside the lesion without damaging the superficial surface of the lesion. Immediately after radiation, the treated area turned off-white, with a surrounding erythematous flare with a slight swelling. **e**, **f** A 2 weeks view post laser irradiation showing a good recovery with fibrin residues. **g**, **h** A complete resolution of the lip vascular lesion after seven irradiation sessions. **i**, **j** The patient underwent a surgical intervention to correct the lip asymmetry. **k**, **l** Six weeks post-surgery, whereby the final result was remarkable with no sign of scaring or discoloration

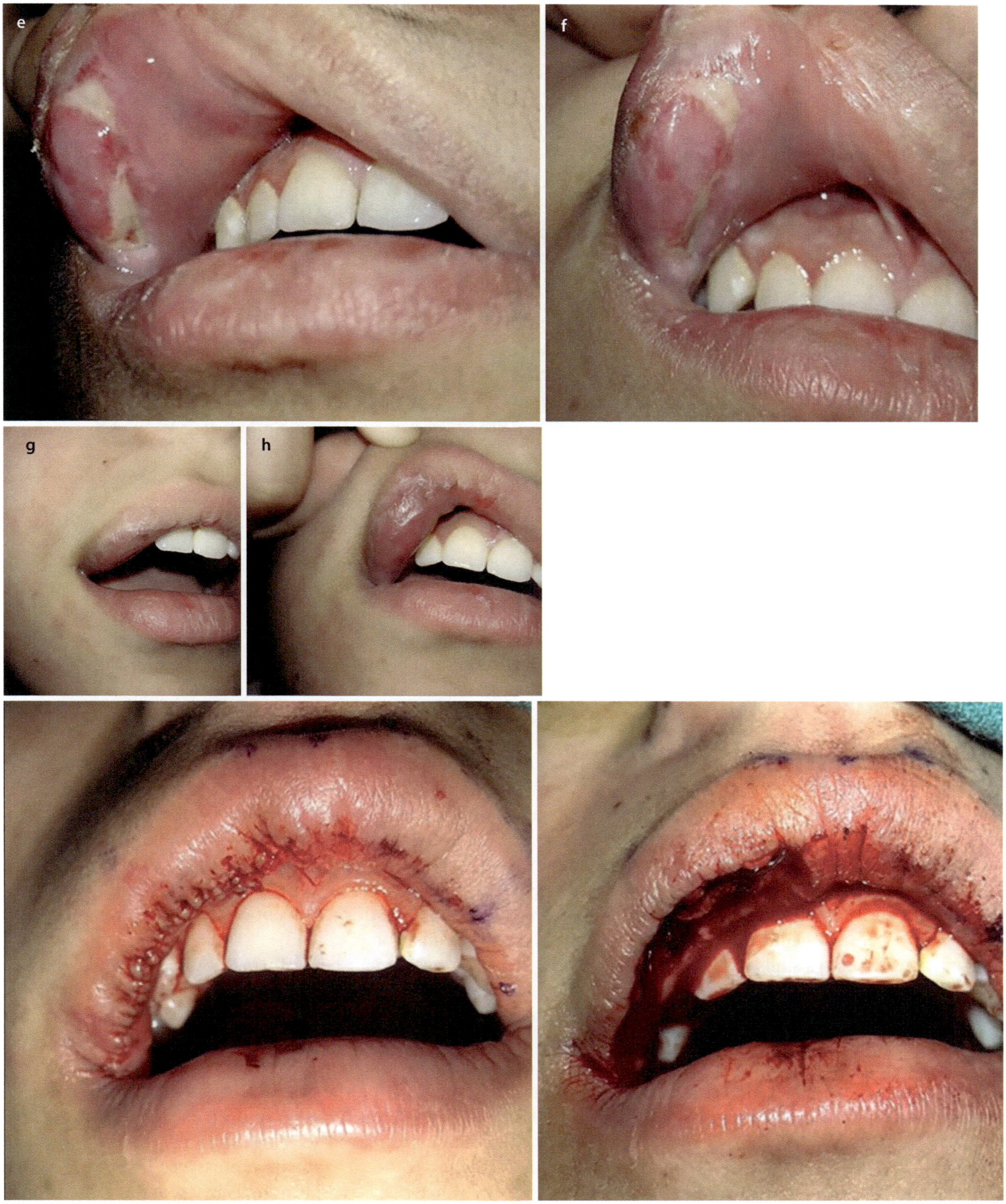

Fig. 13.32 (continued)

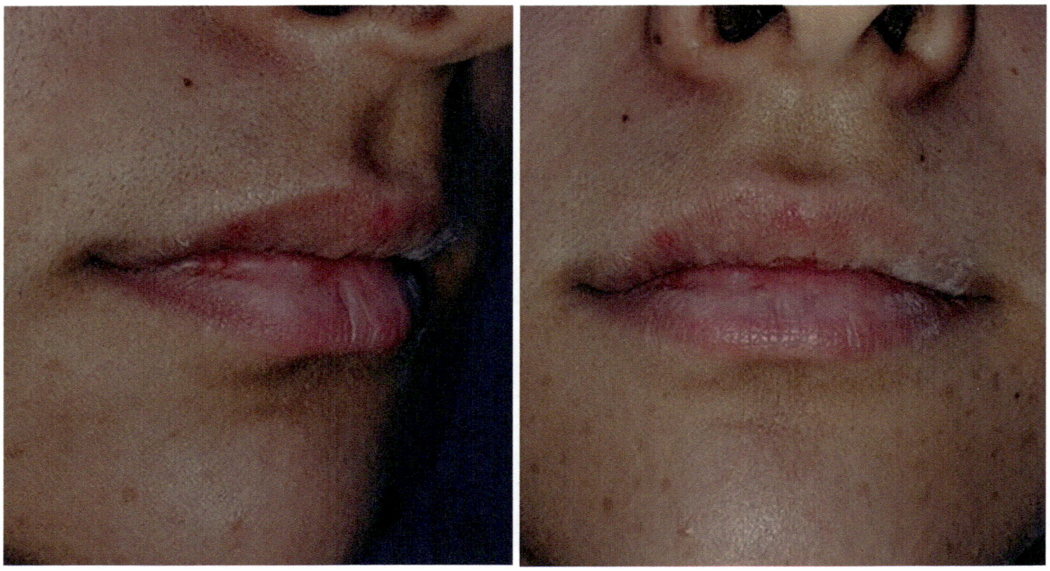

Fig. 13.32 (continued)

margins of the lesion toward its center in a spiral mode. Treatment was repeated at an interval of 4 weeks for seven treatment sessions. Subsequently, the patient underwent a surgical correction to re-contour the residual deformity and hypertrophied tissues in the lip, in order to improve cosmetics and competency.

13.5.2 Mucocele

A mucocele is considered the most common minor salivary gland disorder with a high incidence rate among older children or young adults in both genders, nonetheless, mucoceles have been reported in patients of all ages [139]. It is caused by accumulation of mucus following rupture of a salivary gland or a traumatic injury to its duct in which the saliva spills into the oral cavity's subepithelial tissue [77]. Clinically, it appears as a round, well-circumscribed, asymptomatic dome-like swelling that ranges from a few millimeters to several centimeters in size. Depending on the persistency and the amount of mucosa accumulated inside the lesion, the outer surface may exhibit deep blue to pinky-pale color [140]. Mucoceles are usually painless, however, in some persistent long-standing cases, the size of mucoceles can cause discomfort and interfere with speech, chewing and swallowing and even causes a concern to the patients from being sinister lesion [141]. Although the lower lip is the most common site for a mucocele, it is also present on the ventral tongue, the floor of the mouth, and check mucosa [142].

According to its cause, mucoceles are classified as retention cysts and extravasation accumulations. Extravasation mucoceles are considered pseudo-cysts as no epithelial lining is detected histopathologically. In contrast, retention mucoceles are true cysts with cubic or squamous cell epithelial linings [77]. They are less common but more frequent among major salivary gland ducts as a result of ductal obstruction by a calculus or scar that interferes with the normal salivary flow causing mucosal lump and ductal dilatation.

In some cases, extravasation mucoceles can regress spontaneously with no sign of recurrence. Treatment is only necessary when the patient's function is impaired. Several treatment modalities of mucoceles have been described in the literature including: cryosurgery, intralesion injection of corticosteroid, micro-marsupialization, conventional surgical removal, and laser ablation [142, 143].

Excision can be performed by scalpel which can be challenging since there is no true cystic capsule. Similarly, a laser can also lacerate the lesion; but, with a fiber delivery, it is possible to weld the injured and now overlapping wound edges together. This can conserve some of the form of the cyst and gland without blindly dissecting too much tissue. In deep excisions, two or three sutures are placed to avoid food impaction and allow healing to occur. Patients report very little pain or loss of function. Wound healing is by secondary intention.

Ramkumar et al. suggested using diode laser in wavelength of 940 nm, with fiber tip 400 µm at 1.5 W in continuous mode. The initial incision was performed on the uppermost site of the swelling and then the lesion was resected completely down to the muscular layer [144].

Choi et al. depending on available data of 164 patients found that oral mucocele recurrence is significantly higher on the ventral tongue (50%) than on the labial/buccal mucosa. In addition, patients aged under 30 years old exhibited a higher recurrence rate (16.0%), compared with older individuals (aged >30 years, 4.4%). With respect to intervention method, both surgical procedures using scalpels and those using laser showed similar recurrence rate [145].

Figure 13.33 is a 16-year-old female who has a recent history of biting her lower lip. A rounded swelling with semi-translucent appearance suggested the development of a mucocele. Under local anesthesia, the lesion was surgically dissected, taking advantage of laser hemostasis and using a diode 980 nm laser. The lesion was removed intact, to facilitate histological examination. Early (5 days) and later (1 month) indicate uneventful healing, attributable in part to wavelength-associated PBM.

Figure 13.34 is a 22-year-old patient who had a sublingual retention cyst for more than 2 months. The swelling appeared after a traumatic injury to the floor of the mouth, causing a scar in the salivary gland duct which obstructed saliva flow. The occlusal view was negative with no sign of calcified stones. The outer surface of the swelling was removed by a diode laser with a wavelength of 810 nm and power of 2.5 W. The patient showed a good recovery after 1 month postoperatively.

Figure 13.35 is a healthy, non-smoking 26-year-old female, was concerned about a lump on her lower lip that had been growing for 3 months, measuring about 5 mm in diameter. It was soft on palpation with a bluish color, nonetheless, blanching test was negative. The

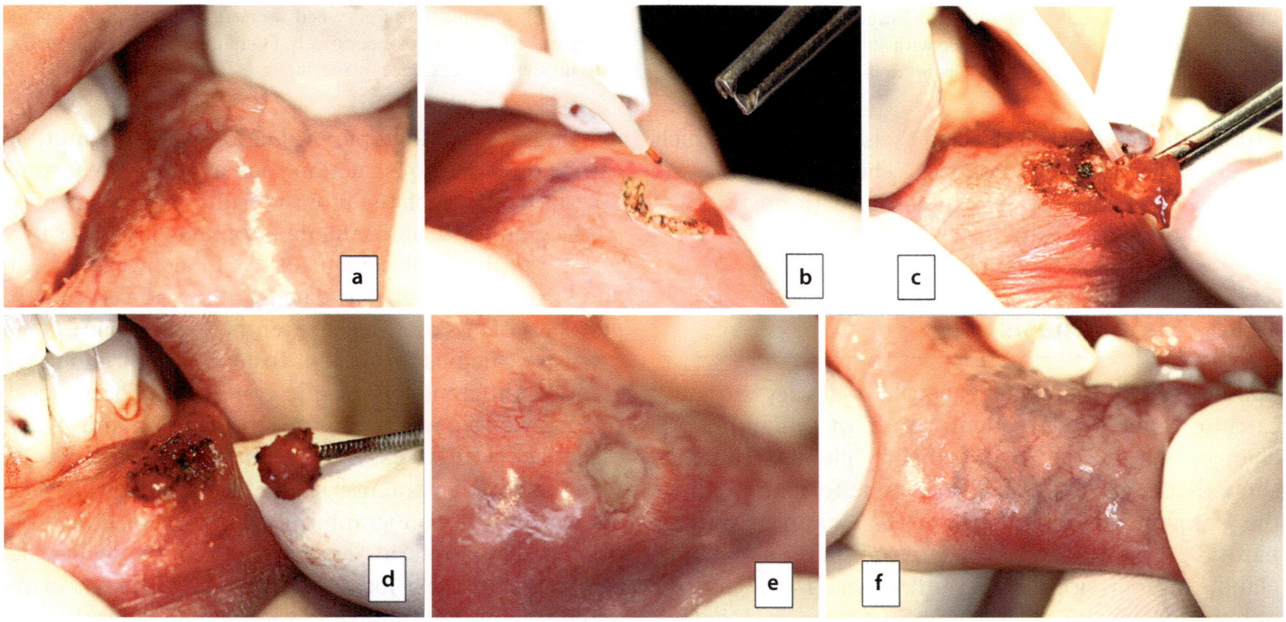

Fig. 13.33 **a** Pre-operative appearance of rounded translucent swelling lower lip in a 16-year-old female. History of recent biting trauma. **b, c** Using a 980 nm diode laser and 320 µm quartz fiber 1.2 W CW power parameters, together with incisional hemostasis, it was possible to dissect the lesion. **d** Immediate postoperative appearance. No sutures required. **e, f** Early (5 days) and later (1 month) uneventful healing and complete resolution of the surgical site. (Case courtesy Prof. Steven Parker)

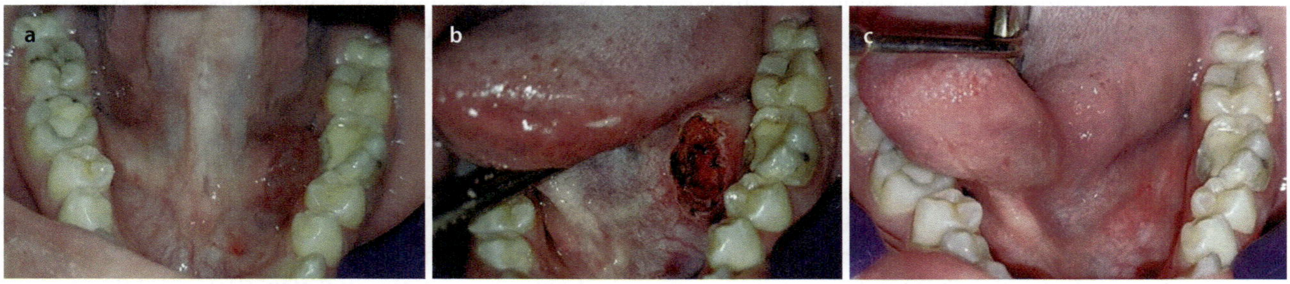

Fig. 13.34 **a** A well-defined retention cyst on the on the floor of the mouth expanding for more than a month. **b** The cyst was excised using the 810 nm diode laser. **c** No relapse was seen after 1 month and the operation bed showed good recovery

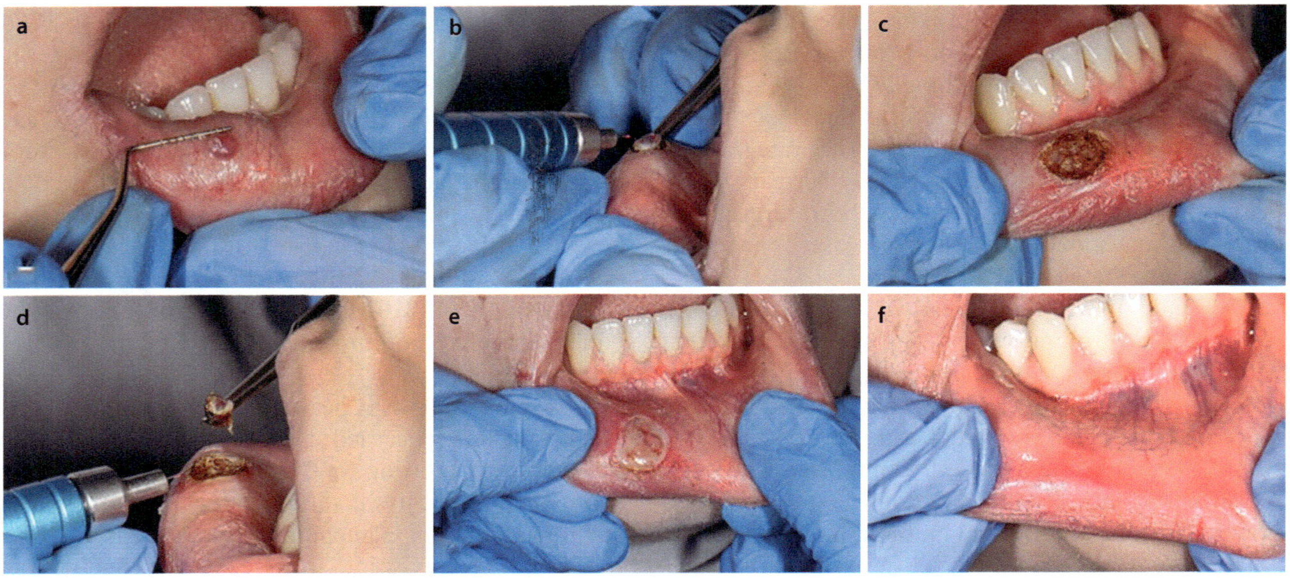

Fig. 13.35 **a** Mucocele of the right lower lip was recognized at 3 months. **b** The 810 nm diode laser with initiated 300 mic tip was used to remove the lesion. **c**, **d** Excision area extended down to the muscular layer and the surgery bed was left to heal with secondary intension. **e** Ten days postoperatively. **f** One-month post laser surgery with no sign of recurrence or scaring

810 nm diode laser was used to excise the lesion, which was identified as a retention cyst. The excision area was left to heal with secondary intention with no need for dressing or closure sutures.

13

13.5.3 Sialolithiasis

Sialolithasis is the occurrence of a salivary stone or calculus and is a frequent cause of inflammatory changes within the large salivary glands. Statistics show that over 78% of salivary stone formation affect the submandibular gland and 81% appear in the parotid gland; but is a rare occurrence in the sublingual gland, though not totally unknown [146]. Clinical practice detects signs of stone formation in 1 out of 10,000 patients. The incidence in male patients is 2–3 times higher than in females. The greatest presence can be found in the age range of 50–70 years, but sialiths have been reported in children. Some reports assume a correlation between sialolithiasis and other stone diseases as urinary or biliary stones, but large multicenter studies show no correlation at all [146, 147]. Major salivary gland stones are usually measured between 5 and 10 mm in size. Giant sialoliths is a term used to describe those stones exceeding 15 mm in any one dimension [148].

The typical symptoms include a painful swelling of the affected salivary gland which is intensified by chewing and at mealtimes. The diagnosis is based on the clinical symptoms, ultrasound examination, radiographic imaging, and endoscopic evaluation. Other symptoms can be due to an obstruction such as one in Wharton's duct in the floor of the mouth resulting in an infection of the gland. Chronic inflammation may lead over time to an atrophic replacement of the acinar cells by scar and fatty tissue [149]. An acute inflammation is often the first sign of a sialolithiasis even in the presence of an extensive calculus formation. Instead of extirpating the gland, a different therapeutic concept has been developed nowadays which analyzes the location and size of the stones [150]. In cases of normally small size parotid stones, a basket extraction shows promising results. Also extracorporeal sonographically controlled lithotripsy plays a major role in therapy. Endoscopic techniques can more precisely locate the stones. Although extra- and intracorporal lithotripsy show excellent results, that method is not often used due to accessibility and cost of the apparatus.

A trans oral approach has been established as an alternative; as a result, more than 90% of all submandibular stones can be removed preserving the gland. After sounding the duct to locate the stone, it is easy to incise the tissue and locate the stone which can be mobilized and removed. Following the incision using a 320–400 μm surgical fiber, a dissecting scissors is used to dilate the incision wound and expose the sialolith. After the exposure, the sialolith is grasped with the mosquito hemostat or forceps and removed from the duct. Once the stone is retrieved, the duct is marsupialized into the floor of the mouth with 5/0 vicryl sutures [151]. Checking duct patency is an essential step after sialolithotomy, in

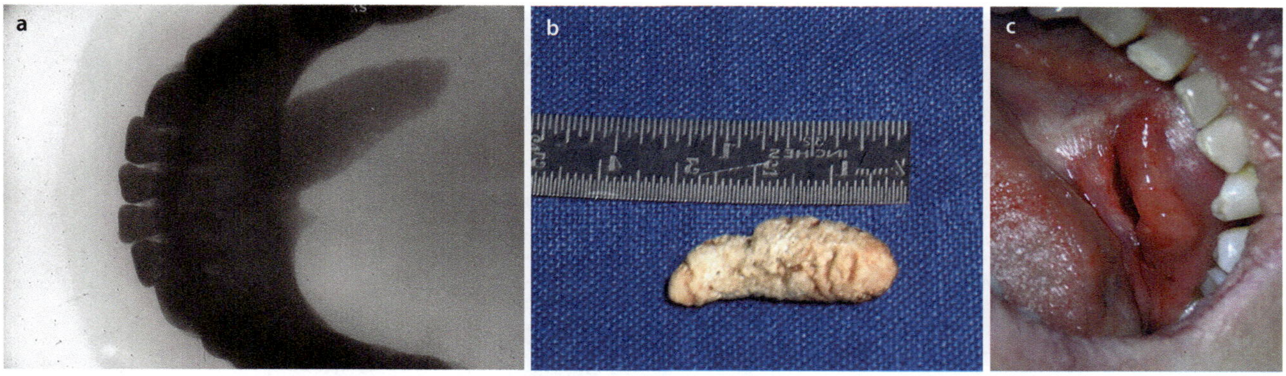

◻ **Fig. 13.36** **a** A radiograph of the area showing a large sialolith in the Warton's duct. X-ray shows a large sialolith in the Wharton's duct. **b** A photo of the excised sialolith. An 810 nm diode laser was used for the incision with a 200 µm glass fiber in contact with the tissue at 30 W, 12,500 Hz, and a 9 µs pulse duration. Average power 3.38 Watts. Fluence 554 J/cm^2. **c** Immediate postoperative view shows good hemostasis. No sutures were placed and healing is expected to be normal

some cases, a micro-drainage may be placed in the duct to avoid scar formation and blockage of the duct. Conventional instrumentation offers no control of bleeding and with the flow of saliva, visibility can be extremely difficult during the surgery. A laser insures good hemostasis and therefore a good overview of the operation site. Sutures are not usually placed, and the healing is uneventful healing with little scaring [152–154]. Accordingly, several laser wavelengths have been employed to treat sialolithiasis, including carbon dioxide, diode, Ho:YAG, and Nd:YAG lasers.

▶ Angiero et al. reported the efficacy of 810 nm diode laser as an alternative surgical tool to remove stones from the submandibular duct even those over 4.5 cm in size. The laser was used in continuous mode at 2.5 W power, and the beam was delivered using a 320-µm flexible fiber [152].

◻ Figure 13.36 shows a clinical case of a 48-year-old male smoker presented with a painful swelling submandibular which intensifies during chewing and eating. His medical history includes hypertension and hyperlipidemia. A radiograph was taken and the diagnosis was made as sialolithiasis. A diode laser performed the incision, and the stone was removed. No sutures were placed and the healing is expected to be normal.

◻ Figure 13.37 is a 56-year-old female complained of firm painful swelling in the right floor of the mouth. The clinical and radiological examination confirmed the presence of a 1 cm stone in the submandibular gland. The 980 nm diode laser was used to create an incision over the stone. A dissecting scissor was used to dilate the incision wound to facilitate the stone's removal. A micro-drainage was placed in the duct for 3 days after stone retrieval to avoid scar formation and blockage of the duct. The patient was seen in 1 month time with no sign of swelling or salivary flow disturbance.

13.5.4 Excessive Gingival Display (Gummy Smile)

A smile is the most essential expression on a person's face. It has a vital role in facial attractiveness, and it can positively affect the patient's social interactions. However, an ideal smile comes from a balance between three mutually related main factors: teeth, gingiva, and lips [155]. Increased interest has been shown to rectify excessive gingival display in growing numbers of individuals. Reports demonstrate a prevalence of 7% in males and 14% in females. Several factors may take part in producing this aesthetic issue. For instance, delayed tooth eruption, a short or hypermobile upper lip, gingival hyperplasia, and skeletal reasons [156]. Nevertheless, lip repositioning surgery had seen the light for the first time by Rubinstein and Kostianovsky in 1973, and it was considered as an effectual, safe, and predictable approach to manage gummy smiles with an average of 2.71 mm of exposed gingiva [157]. However, gummy smiles can be managed using several treatment approaches. For example, orthognathic surgery, orthodontic intrusion, crown lengthening, botulinum toxin, and lip repositioning. Thus, the underlying causing factor should be identified decisively in order to determine the appropriate treatment plan accurately [158]. The concept includes the removal of a partial-thickness mucosal strip from the vestibular alveolar mucosa of the internal upper lip, then suturing the apical edge of the exposed wound to the mucogingival margin aiming to shorten the vestibule and reduce lip movements. Whereas various modifications were dissimilar to the original technique, some of them described a full-thickness flap excluding the frenum by creating a V-shape in the middle of the flap [156].

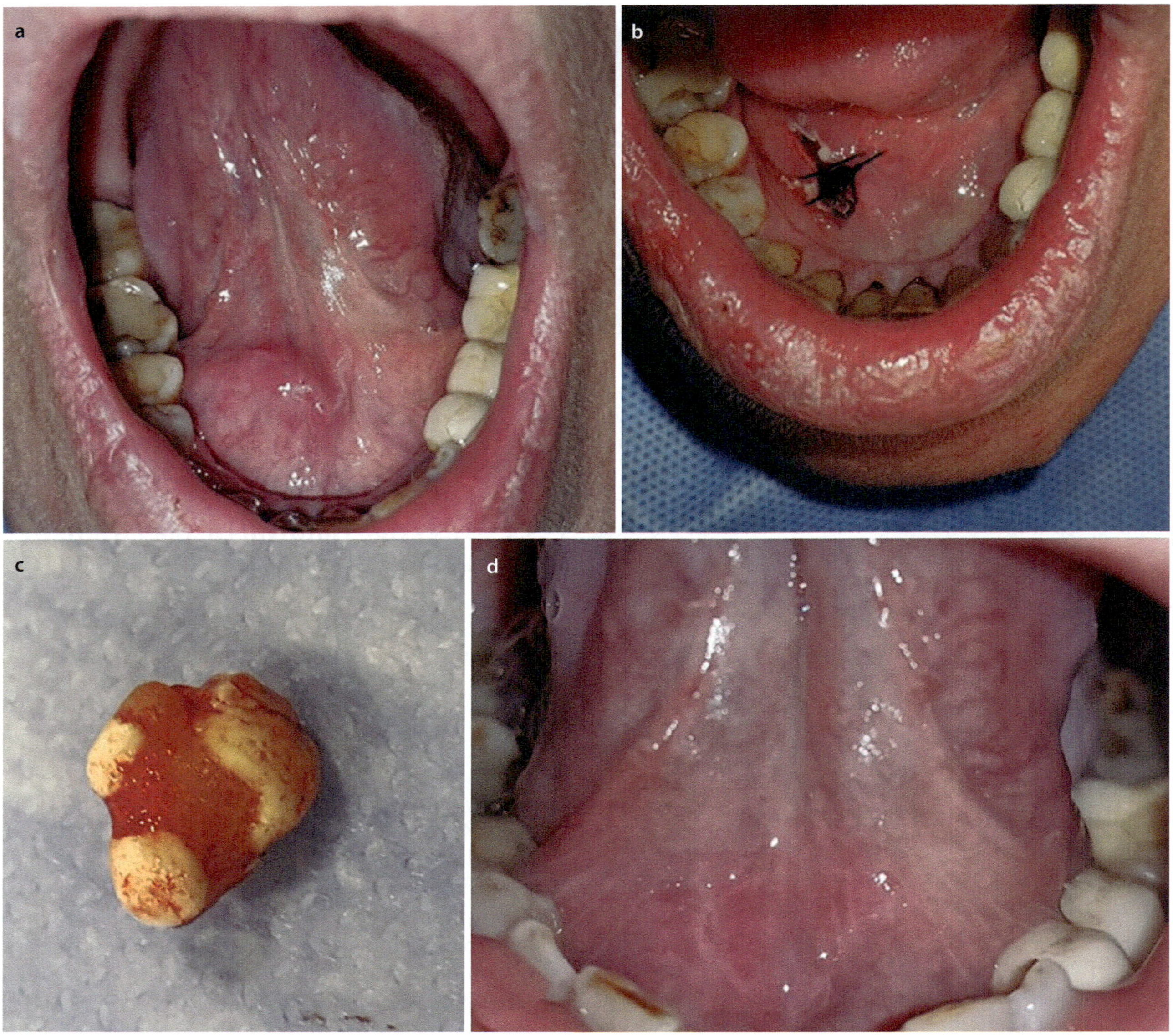

Fig. 13.37 **a** Painful swelling associated with the right submandibular gland duct. The location of the stone was confirmed radiographically. **b** Incision was performed using the 980 nm diode laser which helped in stone removal. Then a micro-drainage was placed inside the duct and sutured to adjacent mucosa to enhance saliva flow and prevent scar formation. **c** 1 cm sialolith was removed from the duct. **d** One-month view of the floor of the mouth showed uneventful healing with normal saliva flow from the salivary gland orifices

Laser surgery is considered as a proper alternative tool in soft tissue oral surgeries due to its recorded advantages in decreasing the postoperative complications and in accelerating the healing process. Putting in mind the fact that the CO_2 laser is the laser of choice in the majority of soft tissue oral surgeries, consequential to its efficient role in providing coagulation and an extraordinary visual surgical field during the procedure even in the extremely vascularized tissues [159, 160].

Lip repositioning using laser surgery is an effective, safe, and predictable approach to manage excessive gingival display caused by soft tissue disorders (short upper lip or hyperactive elevator muscles). The procedure is explained step by step in the related photos in Fig. 13.38.

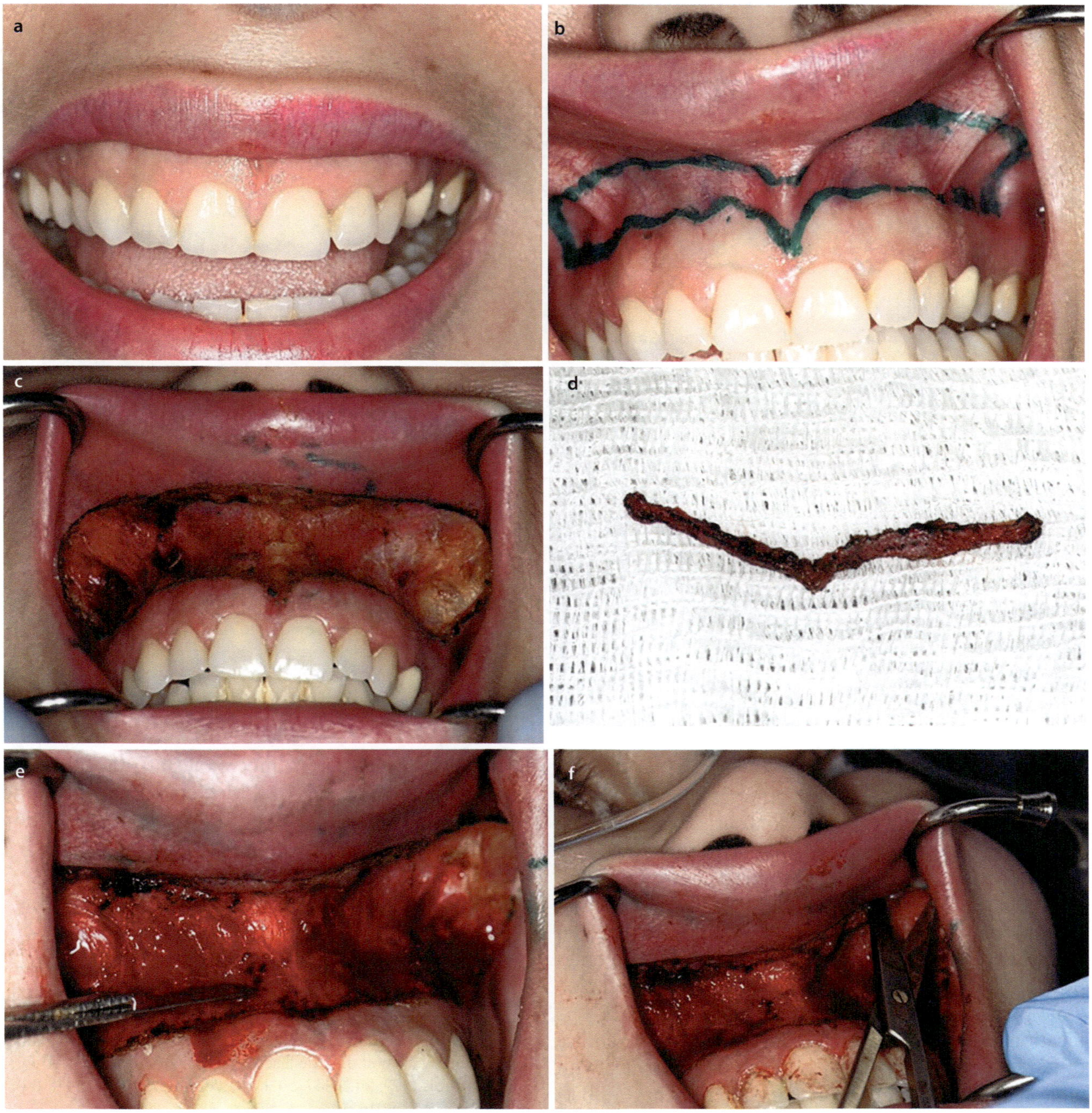

Fig. 13.38 **a** A 24-year-old female presented complaining of an excessive gingival display while smiling. **b** The tissue was dried prior to marking of the incision outlines using a sterile surgical marking pen, the margins were made 1 mm coronal to the mucogingival line in order to achieve accurate suturing from the distal line of the second premolar in both of the right and the left sides, in addition, a V-shape was made in the area of the upper lip frenum. Thus, the lip midline was punctually determined and the symmetry was well achieved. Then the second incision was made in the labial mucosa parallel to the first incision and 10–12 mm apical to it, the two incisions were connected. **c** A partial thickness epithelial strip was carefully removed utilizing 10,600 nm CO_2 laser with 4 W in continuous mode, this facilitated complete resection of the strip without bleeding. **d** A strip of mucosa removed by CO_2 laser demonstrates a partial-thickness flap removed without bleeding. **e**, **f** A 15 blade was used in peeling movements to encourage a slight bleeding which helps in accelerating the wound recovery, also the margins were dissected about 2 mm around to minimize tension during wound suturing. **g** Subsequently, the wound's borders were sutured with interrupted technique using 4/0 silk sutures. The first one was performed in the midline in order to ensure the symmetry of the lip. **h** A couple of indicative sutures in the midpoint of the canines areas in each side, so that two windows were made in order to compare the symmetry of the lip. Additional sutures were used in the rest of the two regions. **i** Follow-up appointment illustrates the punctual midline and crystal-clear lip symmetry

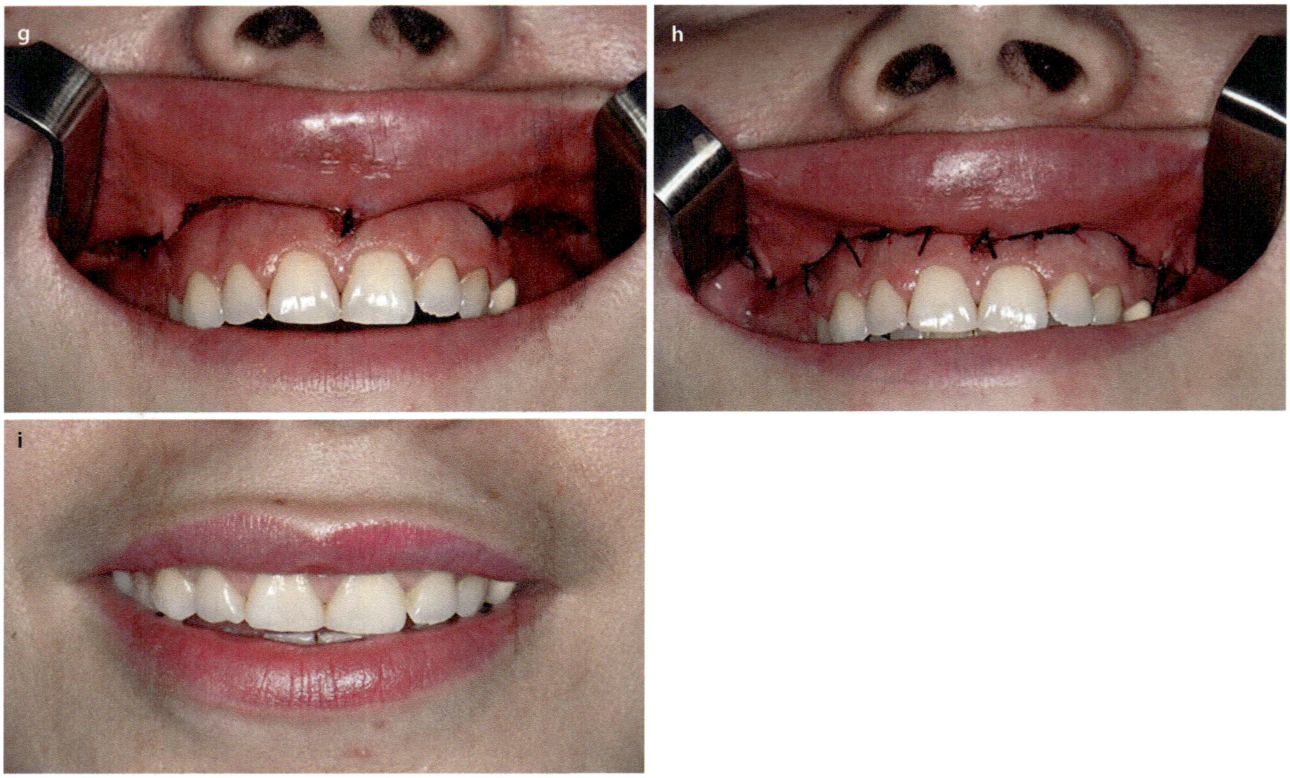

Fig. 13.38 (continued)

13.6 Conclusion

The use of lasers in oral soft tissue surgery is beneficial for patient and surgeon. The good hemostasis provided offers a better view of the operation site and can help to make the procedure more straightforward and even easier for both the accomplished and the novice surgeon. Treatment for patients with bleeding problems becomes possible. Wound healing is mostly uneventful. Conventional biopsies, so necessary for histological verification, can be easily taken. Depending on the choice of parameters, soft tissue surgery can be performed with all available wavelengths. On some instruments, a fiber-based delivery system facilitates accessibility in areas with challenging anatomy, such as undercuts. All told, for many indications laser treatment is superior to conventional therapy in many instances and can help to deliver safe and effective dental care.

Acknowledgment To Dr. Claus Neckel for his valuable contribution in the first edition of this chapter. The author would like to thank: Prof. Peter J. Thomson for the encouragement and advice he has provided throughout my time as his PhD student at Newcastle University, especially his advice on how to use the CO_2 laser in oral and maxillofacial surgery. I also wish to express my appreciation to all the members of staff and my postgraduate students in the Oral Medicine department and the Higher Institute of Laser Research and Application at Damascus University-Syria.

References

1. Gaspar L. The use of high-power lasers in oral surgery. J Clin Laser Med Surg. 1994;12(5):281–5. PMID: 10150673.
2. Shafir R, Slutzki S, Bornstein L. Excision of buccal hemangioma by carbon dioxide laser beam. Oral Surg Oral Med Oral Pathol. 1977;44(3):347–50. https://doi.org/10.1016/B978-0-08-013320-1.50016-6.
3. Maiman TH. Stimulated optical radiation in ruby. Nature. 1960;187:493–4. https://doi.org/10.1038/187493a0.
4. Einstein A. Zur quantentheorie der strahlung. [Article in German]. Phys Z. 1917;18:121–8.
5. Schawlow A, Townes C. Proposal of laser. Phys Rev. 1958;112(6):1940.
6. Snitzer E. Optical maser action of Nd^{+3} in a barium crown glass. Phys Rev. 1961;7(12):444. https://doi.org/10.1103/PhysRevLett.7.444.
7. Lobene RR, Fine S. Interaction of laser radiation with oral hard tissues. J Pros Dent. 1966;16(3):589–97. https://doi.org/10.1016/0022-3913(66)90066-7.
8. Hall RN, Fenner G, Kingsley J, Soltys T, Carlson R. Coherent light emission from GaAs junctions. Phys Rev. 1962;9(9):366. https://doi.org/10.1016/B978-0-08-013320-1.50022-1.

9. Nathan MI, Dumke WP, Burns G, Dill FH Jr, Lasher G. Stimulated emission of radiation from GaAs pn junctions. Appl Phys Lett. 1962;1(3):62–4. https://doi.org/10.1063/1.1777371.
10. Holonyak N Jr, Bevacqua S. Coherent (visible) light emission from Ga ($As_{1-x}P_x$) junctions. Appl Phys Lett. 1962;1(4):82–3. https://doi.org/10.1063/1.1753706.
11. Quist T, Rediker R, Keyes R, Krag W, Lax B, McWhorter AL, et al. Semiconductor maser of GaAs. Appl Phys Lett. 1962;1(4):91–2. https://doi.org/10.1063/1.1753710.
12. Hibst R, Keller U, Steiner R. Die wirkung gepulster Er:YAG Laserstrahlung auf zahngewebe. Laser Med Surg. 1988;4:163–5.
13. Hibst R, Keller U. Experimental studies of the application of the Er:YAG laser on dental hard substances: I. Measurement of the ablation rate. Lasers Surg Med. 1989;9:338–44. https://doi.org/10.1002/lsm.1900090405.
14. Das S, Mohammad S, Singh V, Gupta S. Neodymium:Yttrium aluminum garnet laser in the management of oral leukoplakia: a case series. Contemp Clin Dent. 2015;6(Suppl 1):S32–5. https://doi.org/10.4103/0976-237x.152934. PMID: 25821371.
15. Parker S, Cronshaw M, Grootveld M, George R, Anagnostaki E, Mylona V, Chala M, Walsh L. The influence of delivery power losses and full operating parametry on the effectiveness of diode visible-near infra-red (445–1064 nm) laser therapy in dentistry-a multi-centre investigation. Lasers Med Sci. 2022;37(4):2249–57. https://doi.org/10.1007/s10103-021-03491-y.
16. Farah C, Savage N. Cryotherapy for treatment of oral lesions. Aust Dent J. 2006;51(1):2. https://doi.org/10.1111/j.1834-7819.2006.tb00392.x.
17. Parker S, Cronshaw M, Anagnostaki E, Mylona V, Lynch E, Grootveld M. Current concepts of laser-oral tissue interaction. Dent J (Basel). 2020;8(3):61. https://doi.org/10.3390/dj8030061.
18. Parker S. Lasers and soft tissue: 'loose' soft tissue surgery. Br Dent J. 2007;202(4):185–91. https://doi.org/10.1038/bdj.2007.128.
19. Luomanen M, Meurman J, Lehto VP. Extracellular matrix in healing CO_2 laser incision wound. J Oral Pathol Med. 1987;16(6):322–31. https://doi.org/10.1111/j.1600-0714.1987.tb00702.x.
20. Zeinoun T, Nammour S, Dourov N, et al. Myofibroblasts in healing laser excision wounds. Lasers Surg Med. 2001;28(1):74–9. https://doi.org/10.1002/1096-9101(2001)28:1%3C74::aid-lsm1019%3E3.0.co;2-b. PMID: 11430446.
21. Gáspár L, Tóth J. Comparative study on wound healing in the oral cavity following experimental surgery using a scalpel, electrocauterization and CO_2 laser beam. Fogorv Sz. 1991;84(11):339–43. PMID: 1773835.
22. Neckel CP. Comparative study on cw mode versus pulsed mode in AlGaAs-diode lasers. In: Proc SPIE 4249, Lasers in dentistry, vol. VII. Bellingham, WA: BiOS 2001 The International Society for Optical Engineering; 2001. p. 44–9. https://doi.org/10.1117/12.424515.
23. Warnakulasuriya S, Johnson N, Van der Waal I. Nomenclature and classification of potentially malignant disorders of the oral mucosa. J Oral Pathol Med. 2007;36(10):575–80. PMID: 17944749.
24. Warnakulasuriya S, Kujan O, Aguirre-Urizar JM, et al. Oral potentially malignant disorders: a consensus report from an international seminar on nomenclature and classification, convened by the WHO Collaborating Centre for Oral Cancer. Oral Dis. 2021;27:1862–80. https://doi.org/10.1111/odi.13704.
25. Brouns E-R, Baart JA, Bloemena E, et al. The relevance of uniform reporting in oral leukoplakia: definition, certainty factor and staging based on experience with 275 patients. Med Oral Patol Oral Cir Bucal. 2013;18(1):19–26. https://doi.org/10.4317/medoral.18756.
26. Abidullah M, Kiran G, Gaddikeri K, et al. Leukoplakia—review of potentially malignant disorder. J Clin Diagn Res. 2014;8:ZE01–4. https://doi.org/10.7860/jcdr/2014/10214.4677.
27. Thomson PJ, Hamadah O. Cancerisation within the oral cavity: the use of 'field mapping biopsies' in clinical management. Oral Oncol. 2007;43(1):20–6. https://doi.org/10.1016/j.oraloncology.2005.12.019.
28. Axell T, Holmstrup P, Kramer I, Pindborg J, Shear M. International seminar on oral leukoplakia and associated lesions related to tobacco habits. Commun Dent Oral Epidemiol. 1984;12(3):145–54. https://doi.org/10.1111/j.1600-0528.1984.tb01428.x.
29. Odell E, Kujan O, Warnakulasuriya S, Sloan P. Oral epithelial dysplasia: recognition, grading and clinical significance. Oral Dis. 2021;27:1947–76. https://doi.org/10.1111/odi.13993.
30. Brouns ER, Evren I, Wils LJ, Poell JB, Brakenhoff RH, Bloemena E, de Visscher JGAM. Oral leukoplakia classification and staging system with incorporation of differentiated dysplasia. Oral Dis. 2022; https://doi.org/10.1111/odi.14295. Epub ahead of print. PMID: 35765231.
31. Hamadah O, Goodson M, Thomson PJ. Clinicopathological behaviour of multiple oral dysplastic lesions compared with that of single lesions. Br J Oral Maxillofac Surg. 2010;48(7):503–6. https://doi.org/10.1016/j.bjoms.2009.08.027.
32. Thomson PJ, Goodson ML, Cocks K, Turner JE. Interventional laser surgery for oral potentially malignant disorders: a longitudinal patient cohort study. Int J Oral Maxillofac Surg. 2017;46(3):337–42. https://doi.org/10.1016/j.ijom.2016.11.001.
33. Doroshenko M, Guerra A, Vu L. Airway for laser surgery. [Updated 2022 Oct 20]. In: StatPearls [Internet]. Treasure Island, FL: StatPearls Publishing; 2022.
34. Thomson PJ, Wylie J. Interventional laser surgery: an effective surgical and diagnostic tool in oral precancer management. Int J Oral Maxillofac Surg. 2002;31(2):145–53. https://doi.org/10.1054/ijom.2001.0189.
35. Romeo U, Mohsen M, Palaia G, Bellisario A, Del Vecchio A, Tenore G. CO_2 laser ablation of oral leukoplakia: with or without extension of margins? Clin Ter. 2020;171(3):e209–15. https://doi.org/10.7417/CT.2020.2215. PMID: 32323707.
36. Hamadah O, Thomson PJ. Factors affecting carbon dioxide laser treatment for oral precancer: a patient cohort study. Lasers Surg Med. 2009;41(1):17–25. https://doi.org/10.1002/lsm.20733.
37. Hazrati E, Horch H, Gerlach K, Schaffer H. CO_2 laser surgery of oral premalignant lesions. Plast Reconstruct Surg. 1988;82(2):378. https://doi.org/10.1016/s0300-9785(86)80006-0.
38. Van der Hem P, Nauta J, Van der Wal J, Roodenburg J. The results of CO_2 laser surgery in patients with oral leukoplakia: a 25 year follow up. Oral Oncol. 2005;41(1):31–7. https://doi.org/10.1016/j.oraloncology.2004.06.010.
39. Chandu A, Smith A. The use of CO_2 laser in the treatment of oral white patches: outcomes and factors affecting recurrence. Int J Oral Maxillofac Surg. 2005;34(4):396–400. https://doi.org/10.1016/j.ijom.2004.10.003.
40. Gooris P, Roodenburg J, Vermey A, Nauta J. Carbon dioxide laser evaporation of leukoplakia of the lower lip: a retrospective evaluation. Oral Oncol. 1999;35(5):490–5. https://doi.org/10.1016/s1368-8375(99)00022-6.

41. Vivek V, Jayasree RS, Balan A, Sreelatha KT, Gupta AK. Three-year follow-up of oral leukoplakia after neodymium:yttrium aluminum garnet (Nd:YAG) laser surgery. Lasers Med Sci. 2008;23(4):375–9. https://doi.org/10.1007/s10103-007-0500-8.
42. Montebugnoli L, Frini F, Gissi DB, Gabusi A, Cervellati F, Foschini MP, Marchetti C. Histological and immunohistochemical evaluation of new epithelium after removal of oral leukoplakia with Nd:YAG laser treatment. Lasers Med Sci. 2012;27(1):205–10. https://doi.org/10.1007/s10103-011-0941-y.
43. Sagalow ES, Kumar AT, Banoub RG, Xiao KB, Zhan T, Luginbuhl A, Curry JM. Recurrence of premalignant oral cavity and oropharynx lesions after pulsed diode laser treatment. Am J Otolaryngol. 2022;43(5):103556. https://doi.org/10.1016/j.amjoto.2022.103556. PMID: 35952526.
44. Bombeccari GP, Garagiola U, Candotto V, Pallotti F, Carinci F, Giannì AB, Spadari F. Diode laser surgery in the treatment of oral proliferative verrucous leukoplakia associated with HPV-16 infection. Maxillofac Plast Reconstr Surg. 2018;40(1):16. https://doi.org/10.1186/s40902-018-0156-2.
45. Tarasenko S, Stepanov M, Morozova E, Unkovskiy A. High-level laser therapy versus scalpel surgery in the treatment of oral lichen planus: a randomized control trial. Clin Oral Investig. 2021;25(10):5649–60. https://doi.org/10.1007/s00784-021-03867-y. PMCID: PMC8443509.
46. Lavanya N, Jayanthi P, Rao UK, Ranganathan K. Oral lichen planus: an update on pathogenesis and treatment. J Oral Maxillofac Pathol. 2011;15:127. https://doi.org/10.4103/0973-029x.84474.
47. Silverman S, Griffith M. Studies on oral lichen planus: II. Follow-up on 200 patients, clinical characteristics, and associated malignancy. Oral Surg Oral Med Oral Pathol. 1974;37(5):705–10. https://doi.org/10.1016/0030-4220(74)90135-2. PMID: 4524378.
48. Sonawane S, Sawane H, Pasalkar L, Khare V. Diode laser—a cure for obstinate oral lichen planus: a case series and review of the literature. Indian J Case Rep. 2022;8:120–2. https://doi.org/10.32677/ijcr.v8i5.3403.
49. Idrees M, Farah CS, Khurram SA, Firth N, Soluk-Tekkesin M, Kujan O. Observer agreement in the diagnosis of oral lichen planus using the proposed criteria of the American Academy of Oral and Maxillofacial Pathology. J Oral Pathol Med. 2021;50(5):520–7. https://doi.org/10.1111/jop.13170.
50. Rad M, Hashemipoor MA, Mojtahedi A, Zarei MR, Chamani G, Kakoei S, Izadi N. Correlation between clinical and histopathologic diagnoses of oral lichen planus based on modified WHO diagnostic criteria. Oral Surg Oral Med Oral Pathol Oral Radiol Endod. 2009;107(6):796–800. https://doi.org/10.1016/j.tripleo.2009.02.020.
51. Mutafchieva MZ, Draganova-Filipova MN, Zagorchev PI, Tomov GT. Effects of low level laser therapy on erosive-atrophic oral lichen planus. Folia Med. 2018;60(3):417–24. https://doi.org/10.2478/folmed-2018-0008.
52. Trehan M, Taylor CR. Low-dose excimer 308-nm laser for the treatment of oral lichen planus. Arch Dermatol. 2004;140(4):415–20. https://doi.org/10.1001/archderm.140.4.415.
53. Dalirsani Z, Seyyedi SA. Treatment of plaque-like oral lichen planus with CO_2 laser. Indian J Dermatol. 2021;66(6):698–703. https://doi.org/10.4103/ijd.ijd_1170_20.
54. Mücke T, Gentz I, Kanatas A, Ritschl LM, Mitchell DA, Wolff KD, Deppe H. Clinical trial analyzing the impact of continuous defocused CO_2 laser vaporisation on the malignant transformation of erosive oral lichen planus. J Craniomaxillofac Surg. 2015;43(8):1567–70. https://doi.org/10.1016/j.jcms.2015.06.044.
55. de Magalhaes-Junior EB, Aciole GT, Santos NR, dos Santos JN, Pinheiro AL. Removal of oral lichen planus by CO_2 laser. Braz Dent J. 2011;22(6):522–6. https://doi.org/10.1590/s0103-64402011000600014.
56. Mozaffari HR, Ziaei N, Nazari H, Amiri SM, Sharifi R. Oral lichen planus treatment by CO_2 laser: a systematic review. Asian J Sci Res. 2017;10:1–9. https://doi.org/10.3923/ajsr.2017.1.9.
57. Bakhtiari S, Taheri JB, Sehhatpour M, Asnaashari M, Attarbashi Moghadam S. Removal of an extra-large irritation fibroma with a combination of diode laser and scalpel. J Lasers Med Sci. 2015;6(4):182–4. https://doi.org/10.15171/jlms.2015.16.
58. Singh A, Vengal M, Patil N, Sachdeva SK. Traumatic fibroma—a saga of reaction against irritation. Dental Impact. 2012;4(1):49–52.
59. Halim DS, Pohchi A, Yi EEP. The prevalence of fibroma in oral mucosa among patient attending USM dental clinic year 2006–2010. Indonesian J Dent Res. 2010;1(1):61–6. https://doi.org/10.22146/theindjdentres.9991.
60. Sotoode SM, Azimi S, Taheri SA, Asnaashari M, Khalighi H, Rahmani S, Jafari S, Elmi Rankohi Z. Diode laser in minor oral surgery: a case series of laser removal of different benign exophytic lesions. J Lasers Med Sci. 2015;6(3):133–8. https://doi.org/10.15171/jlms.2015.08. PMID: 26464782.
61. Yamaguchi T, Shindoh M, Amemiya A, Inoue N, Kawamura M, Sakaoka H, et al. Detection of human papillomavirus type 2 related sequence in oral papilloma. Anal Cell Pathol. 1998;16(3):125–30. https://doi.org/10.1155/1998/705216. PMID: 9699941.
62. Eversole L. Papillary lesions of the oral cavity: relationship to human papillomaviruses. J Calif Dent Assoc. 2000;28(12):922–7. PMID: 11323946.
63. da Cunha W, Souza A, Pina P, Azevedo L. Efficacy of diode laser in treating oral papilloma: a case report. Open Dent J. 2021;15:262–5. https://doi.org/10.2174/1874210602115010262 .
64. Omi T, Numano K. The role of the CO_2 laser and fractional CO_2 laser in dermatology. Laser Ther. 2014;23(1):49–60. https://doi.org/10.5978/islsm.14-re-01.
65. Angiero F, Buccianti A, Parma L, Crippa R. Human papilloma virus lesions of the oral cavity: healing and relapse after treatment with 810–980 nm diode laser. Lasers Med Sci. 2015;30(2):747–51. https://doi.org/10.1007/s10103-013-1401-7.
66. Boj JA, Hernandez M, Espasa E, Poirier C. Laser treatment of an oral papilloma in the pediatric dental office: a case report. Quintessence Int. 2007;38(4):307–12. PMID: 17432786.
67. Shim J, Choi S, Jung K, Baek S. Treatment of laryngotracheal papilloma with 532-nm diode laser. Med Lasers. 2020;9(2):190–3. https://doi.org/10.25289/ML.2020.9.2.190.
68. Toledano-Serrabona J, López-Ramírez M, Sánchez-Torres A, España-Tost A, Gay-Escoda C. Recurrence rate of oral squamous cell papilloma after excision with surgical scalpel or laser therapy: a retrospective cohort study. Med Oral Patol Oral Cir Bucal. 2019;24(4):e433–7. https://doi.org/10.4317/medoral.22943.
69. Leffell DJ, Braverman IM. Familial multiple lipomatosis: report of a case and a review of the literature. J Am Acad Dermatol. 1986;15(2):275–9. PMID: 3745530.
70. Toy BR. Familial multiple lipomatosis. Dermatol Online J. 2003;9(4):29. PMID: 14594582.

71. Misir AF, Demiriz L, Barut F. Laser treatment of an oral squamous papilloma in a pediatric patient: a case report. J Indian Soc Pedodont Prevent Dent. 2013;31(4):279. https://doi.org/10.4103/0970-4388.121833. PMID: 24262405.
72. Capodiferro S, Maiorano E, Scarpelli F, Favia G. Fibrolipoma of the lip treated by diode laser surgery: a case report. J Med Case Rep. 2008;2:301. https://doi.org/10.1186/1752-1947-2-301.
73. Piccolo D, Mutlag MH, Pieri L, Fusco I, Conforti C, Crisman G, Bonan P. Lipoma management with a minimally invasive 1,444 nm Nd:YAG laser technique. Front Med (Lausanne). 2022;9:1011468. https://doi.org/10.3389/fmed.2022.1011468.
74. Lee SH, Jung JY, Roh MR, Chung KY. Treatment of lipomas using a subdermal 1,444-nm micropulsed neodymium-doped yttrium aluminum garnet laser. Dermatol Surg. 2011;37(9):1375–6. https://doi.org/10.1111/j.1524-4725.2011.02084.x. PMID: 22988996.
75. Kamal R, Dahiya P, Puri A. Oral pyogenic granuloma: various concepts of etiopathogenesis. J Oral Maxillofac Pathol. 2012;16(1):79–82. https://doi.org/10.4103/0973-029X.92978. PMID: 22434943.
76. Jafarzadeh H, Sanatkhani M, Mohtasham N. Oral pyogenic granuloma: a review. J Oral Sci. 2006;48(4):167–75. https://doi.org/10.2334/josnusd.48.167. PMID: 17220613.
77. Neville BW, Damm DD, Allen CM, Chi AC. Oral & maxillofacial pathology. 4th ed. Philadelphia, Pa: Saunders; 2016. p. 604–5.
78. Bugshan A, Patel H, Garber K, Meiller TF. Alternative therapeutic approach in the treatment of oral pyogenic granuloma. Case Rep Oncol. 2015;8(3):493–7. https://doi.org/10.1159/000441839.
79. Rai S, Kaur M, Bhatnagar P. Laser: a powerful tool for treatment of pyogenic granuloma. J Cutan Aesthet Surg. 2011;4(2):144–7. https://doi.org/10.4103/0974-2077.85044.
80. Raulin C, Greve B, Hammes S. The combined continuous-wave/pulsed carbon dioxide laser for treatment of pyogenic granuloma. Arch Dermatol. 2002;138(1):33–7. https://doi.org/10.1001/archderm.138.1.33.
81. Karandeep SA, Prabhpreet K, Kasper DH, Rupinder DK, Shrawan KS. Efficacy of Er,Cr:YSGG Laser in treatment of unusual presentation of Pyogenic Granuloma in a 9 year old girl. S Afr Dent J. 2016;71(5):218–21.
82. Khan MK, Jindal MK. Diode laser as minimal invasive treatment modality for oral pyogenic granuloma in a pediatric patient: a case report. J Pediatr Dent. 2021;7(3):178–83. https://doi.org/10.14744/JPD.2021.09_67.
83. Fekrazad R, Nokhbatolfoghahaei H, Khoei F, Kalhori KA. Pyogenic granuloma: surgical treatment with Er:YAG laser. J Lasers Med Sci. 2014;5(4):199–205. PMID: 25653822.
84. Burgess DK, Levi PA Jr, Kim DM. The management of an oral pyogenic granuloma around osseointegrated dental implants. Clin Adv Periodontics. 2022;13:50. https://doi.org/10.1002/cap.10219. PMID: 35908249.
85. Hasanoglu Erbasar GN, Senguven B, Gultekin SE, Cetiner S. Management of a recurrent pyogenic granuloma of the hard palate with diode laser: a case report. J Lasers Med Sci. 2016;7(1):56–61. https://doi.org/10.15171/jlms.2016.12.
86. Iinuma T, Arai Y, Abe Y, Takayama M, Fukumoto M, Fukui Y, Iwase T, Takebayashi T, Hirose N, Gionhaku N, Komiyama K. Denture wearing during sleep doubles the risk of pneumonia in the very elderly. J Dent Res. 2015;94(3 Suppl):28S–36S. https://doi.org/10.1177/0022034514552493. PMID: 25294364.
87. Mohan RP, Verma S, Singh U, Agarwal N. Epulis fissuratum: consequence of ill-fitting prosthesis. BMJ Case Rep. 2013;2013:bcr2013200054. https://doi.org/10.1136/bcr-2013-200054.
88. Ibrahim AH, Merzouk N, Abdelkoui A. Prosthetic and surgical management of a sizeable epulis fissuratum: a case report. Pan Afr Med J. 2022;41:49. https://doi.org/10.11604/pamj.2022.41.49.31339. PMID: 35317487.
89. Veena K, Jagadishchandra H, Sequria J, Hameed S, Chatra L, Shenai P. An extensive denture-induced hyperplasia of maxilla. Ann Med Health Sci Res. 2013;3(Suppl 1):S7–9. https://doi.org/10.4103/2141-9248.121208. PMID: 24349860.
90. de Freitas AC, Pinheiro AL, de Oliveira MG, Ramalho LM. Assessment of the behavior of myofibroblasts on scalpel and CO(2) laser wounds: an immunohistochemical study in rats. J Clin Laser Med Surg. 2002;20(4):221–5. https://doi.org/10.1089/104454702760230555.
91. Zeinoun T, Nammour S, Dourov N, Aftimos G, Luomanen M. Myofibroblasts in healing laser excision wounds. Lasers Surg Med. 2001;28(1):74–9. https://doi.org/10.1002/1096-9101(2001)28:1/74:AID-LSM1019>3.0.CO;2-B. PMID: 11430446.
92. Tamarit-Borrás M, Delgado-Molina E, Berini-Aytés L, Gay-Escoda C. Removal of hyperplastic lesions of the oral cavity. A retrospective study of 128 cases. Med Oral Patol Oral Cir Bucal. 2005;10:151–62. PMID: 15735548.
93. Monteiro LS, Mouzinho J, Azevedo A, Câmara MID, Martins MA, La Fuente JM. Treatment of epulis fissuratum with carbon dioxide laser in a patient with antithrombotic medication. Braz Dent J. 2012;23:77–81. https://doi.org/10.1590/s0103-64402012000100014.
94. Agrawal AA, Mahajan M, Mahajan A, Devhare S. Application of diode laser for excision of non-inflammatory vascular epulis fissuratum. Int J Case Rep Images. 2012;3(9):42–5. https://doi.org/10.5348/ijcri201209182CR12.
95. Kumar JN, Bhaskaran M. Denture-induced fibrous hyperplasia. Treatment with carbon dioxide laser and a two year follow-up. Indian J Dent Res. 2007;18(3):135. https://doi.org/10.4103/0970-9290.33791. PMID: 17687178.
96. Neckel CP. Vestibuloplasty: a retrospective study on conventional and laser operation techniques. In: Proc SPIE 3593, Lasers in dentistry, vol. V. Bellingham, WA: International Society for Optical Engineering; 1999. p. 76–80. https://doi.org/10.1117/12.348330.
97. Kazanjian VH. Surgical operations as related to satisfactory dentures. Dent Cosmos. 1924;66:387. https://doi.org/10.12691/ijdsr-3-6-3.
98. Kacarska M, Dimitrovski O, Popovic-Monevska D. Preprosthetic laser assisted mandibular vestibuloplasty. Balkan J Dent Med. 2016;20(3):182–5. https://doi.org/10.1515/bjdm-2016-0030.
99. de Arruda Paes-Junior TJ, Cavalcanti SCM, Nascimento DFF, Saavedra GDSFA, Kimpara ET, Borges ALS, Niccoli-Filho W, Komori PCDP. CO_2 laser surgery and prosthetic management for the treatment of epulis fissuratum. Int Scholar Res Notices. 2011; https://doi.org/10.5402/2011/282361.
100. Blessing EP, Adlin RS. Vestibuloplasty using diode laser—a case report. Int Org Sci Res J Dent Med Sci. 2019;18(8):43–7. https://doi.org/10.9790/0853-1808084347.
101. Levine R, Vitruk P. The use of a 10,600-nm CO_2 laser mandibular vestibular extension in a patient with a chromosomal disorder. Compend Contin Educ Dent. 2016;37(9):527–33. PMID: 27608196.
102. Pisevska SG, Simjanovska L, Markovska M, Petreska MP, Chadikovska E. ER:YAG LASER: minimal invasive tech-

nique for vestibuloplasty in the lower jaw. J Morphol Sci. 2019;2(2):37–42.
103. Bhullar SK, Goel V, Bhullar A, Goyal L, Mehta V, Nanda T, Sethi M. Comparative evaluation of pain in vestibular depth extension procedure using scalpel, electrocautery and diode laser. J Dent Specialities. 2017;5:148–51. https://doi.org/10.18231/2393-9834.2017.0033.
104. Kalakonda B, Farista S, Koppolu P, Baroudi K, Uppada U, Mishra A, Savarimath A, Lingam AS. Evaluation of patient perceptions after vestibuloplasty procedure: a comparison of diode laser and scalpel techniques. J Clin Diagn Res. 2016;10(5):ZC96–ZC100. https://doi.org/10.7860/jcdr/2016/17623.7820.
105. Fioravanti M, Zara F, Vozza I, Polimeni A, Sfasciotti GL. The efficacy of lingual laser frenectomy in pediatric OSAS: a randomized double-blinded and controlled clinical study. Int J Environ Res Public Health. 2021;18(11):6112. https://doi.org/10.3390/ijerph18116112.
106. Messner AH, Ha JF. Ankyloglossia and tight maxillary frenula. In: Lesperance MM, editor. Cummings pediatric otolaryngology. 2nd ed. Elsevier; 2021. p. 473–82. https://doi.org/10.1016/b978-0-323-69618-0.00033-0.
107. Fürchtgott N, Paterson BC, Bernard J, Costello BJ. The use of lasers in maxillofacial surgery. In: Fonseca RJ, editor. Oral and maxillofacial surgery, vol. 19. 3rd ed. Elsevier; 2018. p. 336–63.
108. Elias G, McMillan K, Monaghan A. Vascular lesions of the head and oral cavity—diagnosis and management. Dent Update. 2016;43:859–66. https://doi.org/10.12968/denu.2016.43.9.859.
109. Richter GT, Friedman AB. Hemangiomas and vascular malformations: current theory and management. Int J Pediatr. 2012;2012:645678. https://doi.org/10.1155/2012/645678.
110. Steiner JE, Drolet BA. Classification of vascular anomalies: an update. Semin Intervent Radiol. 2017;34:225–32. https://doi.org/10.1055/s-0037-1604295. PMID: 28955111.
111. Kunimoto K, Yamamoto Y, Jinnin M. ISSVA classification of vascular anomalies and molecular biology. Int J Mol Sci. 2022;23:2358. https://doi.org/10.3390/ijms23042358. PMID: 35216474.
112. Oomen KPQ, Wreesmann VB. Current classification of vascular anomalies of the head and neck. J Oral Pathol Med. 2022;51:830–6. https://doi.org/10.1111/jop.13353. PMID: 36066308.
113. ISSVA Classification. Accessed 9 Jan 2023. https://www.issva.org/classification.
114. Angiero F, Benedicenti S, Romanos GE, Crippa R. Treatment of hemangioma of the head and neck with diode laser and forced dehydration with induced photocoagulation. Photomed Laser Surg. 2008;26:113–8. https://doi.org/10.1089/pho.2007.2143.
115. Zheng JW, Zhang L, Zhou Q, Mai HM, Wang YA, Fan XD, Qin ZP, Wang XK, Zhao YF. A practical guide to treatment of infantile hemangiomas of the head and neck. Int J Clin Exp Med. 2013;6(10):851–60. PMID: 24260591.
116. Vesnaver A, Dovsak DA. Treatment of vascular lesions in the head and neck using Nd:YAG laser. J Craniomaxillofac Surg. 2006;34(1):17–24. https://doi.org/10.1016/j.jcms.2005.07.009. Epub 2005 Dec 13. PMID: 16352435.
117. Azma E, Razaghi M. Laser treatment of oral and maxillofacial hemangioma. J Lasers Med Sci. 2018;9:228–32. https://doi.org/10.15171/jlms.2018.41. PMID: 31119015.
118. Cadavid AMH, de Campos WG, Aranha ACC, Lemos-Junior CA. Efficacy of photocoagulation of vascular malformations in the oral mucosa using Nd:YAG. J Craniofac Surg. 2018;29:e614–7. https://doi.org/10.1097/scs.0000000000004676.
119. Bastos JT, Balassiano LKA, Mariano da Rocha CR, Freitas BMP, Bravo LG, Bravo BSF. Treatment of vascular lesions located in the lip and in the oral cavity with Nd:YAG laser. J Cosmet Laser Ther. 2017;19:256–8. https://doi.org/10.1080/14764172.2017.1314503.
120. Parisi D, Ciancio F, Cagiano L, Rucci M, Annacontini L, Portincasa A. Hemangioma of the oral and perioral region: Nd:YAG vs surgery. Ann Ital Chir. 2016;2016:2578. PMID: 27830673.
121. Medeiros R Jr, Silva IH, Carvalho AT, Leão JC, Gueiros LA. Nd:YAG laser photocoagulation of benign oral vascular lesions: a case series. Lasers Med Sci. 2015;30:2215–20. https://doi.org/10.1007/s10103-015-1764-z.
122. Asai T, Suzuki H, Takeuchi J, Komori T. Effectiveness of photocoagulation using an Nd:YAG laser for the treatment of vascular malformations in the oral region. Photomed Laser Surg. 2014;32:75–80. https://doi.org/10.1089/pho.2013.3594.
123. Yang HY, Zheng LW, Yang HJ, Luo J, Li SC, Zwahlen RA. Long-pulsed Nd:YAG laser treatment in vascular lesions of the oral cavity. J Craniofac Surg. 2009;20:1214–7. https://doi.org/10.1097/SCS.0b013e3181acdd9f. PMID: 19553836.
124. Vesnaver A, Dovsak DA. Treatment of large vascular lesions in the orofacial region with the Nd:YAG laser. J Craniomaxillofac Surg. 2009;37:191–5. https://doi.org/10.1016/j.jcms.2008.10.006.
125. Bacci C, Sacchetto L, Zanette G, Sivolella S. Diode laser to treat small oral vascular malformations: a prospective case series study. Lasers Surg Med. 2018;50:111–6. https://doi.org/10.1002/lsm.22737.
126. Voynov PP, Tomov GT, Mateva NG. Minimal invasive approach for lips venous lake treatment by 980 nm diode laser with emphasis on the aesthetic results. A clinical series. Folia Med. 2016;58:101–7. https://doi.org/10.1515/folmed-2016-0017. PMID: 27552786.
127. Dementieva N, Jones S. The treatment of problematic hemangiomas in children with propranolol and 940nm diode laser. J Pediatr Surg. 2016;51:863–8. https://doi.org/10.1016/j.jpedsurg.2016.02.038.
128. Favia G, Tempesta A, Limongelli L, Suppressa P, Sabbà C, Maiorano E. Diode laser treatment and clinical management of multiple oral lesions in patients with hereditary haemorrhagic telangiectasia. Br J Oral Maxillofac Surg. 2016;54:379–83. https://doi.org/10.1016/j.bjoms.2015.08.260.
129. Jasper J, Camilotti RS, Pagnoncelli RM, Poli VD, da Silveira Gerzson A, Gavin Zakszeski AM. Treatment of lip hemangioma using forced dehydration with induced photocoagulation via diode laser: report of three cases. Oral Surg Oral Med Oral Pathol Oral Radiol. 2015;119:89–94. https://doi.org/10.1016/j.oooo.2014.03.005.
130. Romeo U, Del Vecchio A, Russo C, Palaia G, Gaimari G, Arnabat-Dominguez J, España AJ. Laser treatment of 13 benign oral vascular lesions by three different surgical techniques. Med Oral Patol Oral Cir Bucal. 2013;18:e279–84. https://doi.org/10.4317/medoral.18156.
131. Fekrazad R, Am Kalhori K, Chiniforush N. Defocused irradiation mode of diode laser for conservative treatment of oral hemangioma. J Lasers Med Sci. 2013;4:147–50. PMID: 25606323.
132. Azevedo LH, Del Vechio A, Nakajima E, Galletto V, Migliari DA. Lip and oral venous varices treated by photocoagulation with high-intensity diode laser. Quintessence Int. 2013;44:171–4. https://doi.org/10.3290/j.qi.a28926. PMID: 23444183.

133. Azevedo LH, Galletta VC, Eduardo Cde P, Migliari DA. Venous lake of the lips treated using photocoagulation with high-intensity diode laser. Photomed Laser Surg. 2010;28:263–5. https://doi.org/10.1089/pho.2009.2564.
134. Abukawa H, Kono M, Hamada H, Okamoto A, Satomi T, Chikazu D. Indications of potassium titanyl phosphate laser therapy for slow-flow vascular malformations in oral region. J Craniofac Surg. 2017;28:771–4. https://doi.org/10.1097/scs.0000000000003445.
135. Miyazaki H, Kato J, Watanabe H, Harada H, Kakizaki H, Tetsumura A, Sato A, Omura K. Intralesional laser treatment of voluminous vascular lesions in the oral cavity. Oral Surg Oral Med Oral Pathol Oral Radiol Endod. 2009;107:164–72. https://doi.org/10.1016/j.tripleo.2008.08.004.
136. Derby LD, Low DW. Laser treatment of facial venous vascular malformations. Ann Plast Surg. 1997;38:371–8. https://doi.org/10.1097/00000637-199704000-00011.
137. Miyazaki H, Ohshiro T, Romeo U, Noguchi T, Maruoka Y, Gaimari G, Tomov G, Wada Y, Tanaka K, Ohshiro T, Asamura S. Retrospective study on laser treatment of oral vascular lesions using the "leopard technique": the multiple spot irradiation technique with a single-pulsed wave. Photomed Laser Surg. 2018;36:320–5. https://doi.org/10.1089/pho.2017.4410.
138. Álvarez-Camino JC, España-Tost AJ, Gay-Escoda C. Endoluminal sclerosis with diode laser in the treatment of orofacial venous malformations. Med Oral Patol Oral Cir Bucal. 2013;18:e486–90. https://doi.org/10.4317/medoral.18528.
139. Khandelwal S, Patil S. Oral mucoceles—review of the literature. Minerva Stomatol. 2012;61(3):91–9. PMID: 22402300.
140. More CB, Bhavsar K, Varma S, Tailor M. Oral mucocele: a clinical and histopathological study. J Oral Maxillofac Pathol. 2014;18(Suppl 1):S72–7. https://doi.org/10.4103/0973-029X.141370. PMID: 25364184.
141. Nagar SR, Fernandes G, Sinha A, Rajpari KN. Mucocele of the tongue: a case report and review of literature. J Oral Maxillofac Pathol. 2021;25(Suppl 1):S37–41. https://doi.org/10.4103/jomfp.jomfp_396_20. PMID: 34083968.
142. Bagher SM, Sulimany AM, Kaplan M, Loo CY. Treating mucocele in pediatric patients using a diode laser: three case reports. Dentistry J. 2018;6(2):13. https://doi.org/10.3390/dj6020013.
143. Sinha R, Sarkar S, Khaitan T, Kabiraj A, Maji A. Nonsurgical management of oral mucocele by intralesional corticosteroid therapy. Int J Dent. 2016;2016:2896748. https://doi.org/10.1155/2016/2896748.
144. Ramkumar S, Ramkumar L, Malathi N, Suganya R. Excision of mucocele using diode laser in lower lip. Case Rep Dentistry. 2016;2016:1746316, 4 pages. https://doi.org/10.1155/2016/1746316.
145. Choi YJ, Byun JS, Choi JK, Jung JK. Identification of predictive variables for the recurrence of oral mucocele. Med Oral Patol Oral Cir Bucal. 2019;24(2):e231–5. https://doi.org/10.4317/medoral.22690.
146. Zenk J, Constantinidis J, Kydles S, et al. Clinical and diagnostic findings of sialolithasis. HNO. 1999;47(11):963–9. https://doi.org/10.1007/s001060050476. PMID: 10602786.
147. Huoh KC, Eisele DW. Etiologic factors in sialolithiasis. Otolaryngol Head Neck Surg. 2011;145(6):935–9. https://doi.org/10.1177/0194599811415489. PMID: 21753035.
148. Gupta A, Rattan D, Gupta R. Giant sialoliths of submandibular gland duct: report of two cases with unusual shape. Contemp Clin Dentistry. 2013;4:78–80. https://doi.org/10.4103/0976-237X.111599. PMID: 23853458.
149. Iro H, Zenk J, Koch M. Stenosis and other non-sialolithiasis-related obstructions of the major salivary gland ducts. HNO. 2010;58(3):211–7. https://doi.org/10.1007/s00106-009-2076-z. PMID: 20213107.
150. Rzymska-Grala I, Stopa Z, Grala B, Gołębiowski M, Wanyura H, Zuchowska A, Sawicka M, Zmorzyński M. Salivary gland calculi—contemporary methods of imaging. Pol J Radiol. 2010;75(3):25–37. PMID: 22802788.
151. Kılınç Y, Çetiner S. Surgical removal of a giant sialolith by diode laser. Open J Stomatol. 2014;4:484–8. https://doi.org/10.4236/ojst.2014.410065.
152. Yang SW, Chen TA. Transoral carbon dioxide laser sialolithectomy with topical anaesthesia. A simple, effective, and minimally invasive method. Int J Oral Maxillofac Surg. 2011;40:169–72. https://doi.org/10.1016/j.ijom.2010.09.020.
153. Angiero F, Benedicenti S, Romanos GE, Crippa R. Sialolithiasis of the submandibular salivary gland treated with the 810- to 830-nm diode laser. Photomed Laser Surg. 2008;26:517–21. https://doi.org/10.1089/pho.2007.2226.
154. Martellucci S, Pagliuca G, de Vincentiis M, Greco A, Fusconi M, De Virgilio A, Gallipoli C, Gallo A. Ho:YAG laser for sialolithiasis of Wharton's duct. Otolaryngol Head Neck Surg. 2013;148:770–4. https://doi.org/10.1177/0194599812451438a34.
155. Younespour S, Yaghobee S, Aslroosta H, Moslemi N, Pourheydar E, Ghafary ES. Effectiveness of different modalities of lip repositioning surgery for management of patients complaining of excessive gingival display: a systematic review and meta-analysis. Biomed Res Int. 2021;2021:9476013. https://doi.org/10.1155/2021/9476013. PMID: 34660802; PMCID: PMC8516537.
156. Tatakis DN. Lip repositioning techniques and modifications. Dent Clin N Am. 2022;66(3):373–84. https://doi.org/10.1016/j.cden.2022.02.002. Epub 2022 Jun 1. PMID: 35738733.
157. Ardakani MT, Moscowchi A, Valian NK, Zakerzadeh E. Lip repositioning with or without myotomy: a systematic review. J Korean Assoc Oral Maxillofac Surg. 2021;47(1):3–14. https://doi.org/10.5125/jkaoms.2021.47.1.3. PMID: 33632971; PMCID: PMC7925163.
158. Alammar A, Heshmeh O, Mounajjed R, Goodson M, Hamadah O. A comparison between modified and conventional surgical techniques for surgical lip repositioning in the management of the gummy smile. J Esthet Restor Dent. 2018;30(6):523–31. https://doi.org/10.1111/jerd.12433. Epub 2018 Nov 9. PMID: 30412347.
159. Luke AM, Mathew S, Altawash MM, Madan BM. Lasers: a review with their applications in oral medicine. J Lasers Med Sci. 2019;10(4):324–9. https://doi.org/10.15171/jlms.2019.52. Epub 2019 Oct 1. PMID: 31875126; PMCID: PMC6885906.
160. Levine R, Vitruk P. Use of a 10,600-nm CO_2 laser mandibular vestibular extension in a patient with a chromosomal abnormality. Compend Contin Educ Dent. 2016;37(8):527–533; quiz 534. PMID: 27608196.

Laser-Assisted Oral Multi-tissue Management

Contents

Laser Treatment of Periodontal and Peri-implant Disease

Donald J. Coluzzi, Akira Aoki, and Nasim Chiniforush

Contents

D. J. Coluzzi, S. P. A. Parker (eds.), *Lasers in Dentistry—Current Concepts*, Textbooks in Contemporary Dentistry,
https://doi.org/10.1007/978-3-031-43338-2_14

Core Message

Therapy for periodontal and peri-implant disease continues to evolve with new methodologies, medications, and instrumentation added to the conventional armamentarium. Dental lasers have been used both adjunctively and alone in the protocol. Clinical studies and basic investigations have shown that laser photonic energy has been a useful addition to increase the effectiveness and outcomes of treatment of the disease.

14.1 Introduction

The periodontium is essential for optimal oral function and health. Any inflammation will affect both soft and hard tissues and could lead to loss of those structures. Periodontal disease is an infection whose primary etiologic factor is the oral pathogens existing in the plaque biofilm. Initially, the gingiva will become inflamed without attachment loss, and the disease is thus termed gingivitis. With increasing pathogen invasion, there will be loss of connective tissue attachment as well as apical migration of the epithelial tissue with subsequent infection and resorption of the alveolar bone. Periodontal disease is a worldwide public health burden. Indeed, studies show that various stages of periodontitis are one of the most commonly occurring diseases in middle-aged adults [1]. Moreover, periodontal disease has been linked with other systemic diseases. Oral pathogens can migrate through the inflamed and ulcerated gingival epithelium into the rest of the body. There are suggested clinically important associations between periodontal bacteria and conditions ranging from peripheral artery disease, liver cirrhosis, and chronic kidney disease to other systemic disorders including cardiovascular, respiratory, and osteoarticular problems. These connections highlight the importance of treating this disease because of its implication on general medical health. Interestingly, the reverse association is also important: the patient's age, smoking habits, and the presence of diabetes can worsen chronic periodontal inflammation. Other risk factors include genetics and lifestyle, and, clearly, this disease is multifactorial.

Putative periodontal pathogens, such as *Aggregatibacter actinomycetemcomitans* and *Porphyromonas gingivalis*, have long been considered the primary contributors to the disease. However, the red complex especially three species—*Porphyromonas gingivalis*, *Tannerella forsythia*, and *Treponema denticola*—are now regarded as the most pathogenic and are prevalent in biofilm. Moreover, these pathogens can reinfect the patient, diminishing the effect of therapy, especially with sporadic supportive maintenance care [2]. Unfortunately, determining which organisms are important can be a daunting task: there can be several species of pathogens in any one site; some can be more opportunistic than others and may proliferate after the initial inflammation rather than cause it; and the patient's immune response can vary.

Peri-implant disease shares the same etiology—microorganisms causing a plaque-associated pathologic condition—and the literature is beginning to report significant statistics that indicate many implant sites will develop the disease [3]. All of the abovementioned conditions and factors will contribute to the severity as well as to the success of any therapy. For clarity, soft tissue inflammation—bleeding on gentle probing—is termed peri-implant mucositis, whereas inflammation of the peri-implant mucosa and subsequent implant bone loss is termed peri-implantitis.

The recent periodontal and peri-implant disease classification scheme describes both the staging of the severity and complexity of the disease and the grading which can help assess the outcome and progression of the disease [4, 5]. Once the diagnosis is reached, a treatment plan can be developed.

The general understanding is that the gold standard for successful treatment of these diseases is gained in clinical attachment level. The root/implant surface should be restored to biocompatibility to re-establish that attachment without the presence of inflammation [6, 7]. There are however other clinical creditable endpoints such as lack of bleeding on probing, complete removal of accretions, and measurable regeneration of bone, periodontal ligament, and cementum. Of course, the patient's oral hygiene improvement and reduction of other risk factors can also be considered and are crucial for maintenance of a stable periodontium.

This chapter will be divided into three modalities of treatment utilizing lasers in nonsurgical, surgical, and antimicrobial photodynamic modalities. These methods are separate therapies but may be combined to produce the best result. A nonsurgical protocol is the first approach, but surgery may follow to help unresolved problems. Photoactivated medications can be a useful addition for either procedure. Since periodontal or peri-implant disease can have episodic progression, one or more of these treatments may be employed for the current stage of the disease.

14.2 Nonsurgical Periodontal and Peri-implant Disease Laser Therapy

The term "nonsurgical therapy" is defined as a protocol to remove as much calculus as possible, to disrupt or eliminate the biofilm and accompanying microbes, and to reduce inflammation contributing to periodontal and peri-implant disease as initial therapy. After this phase of treatment, the patient's periodontal condition will be evaluated. Two possibilities then exist: one, the patient will receive nonsurgical periodontal maintenance; two, a surgical procedure must be performed as a next step.

During initial nonsurgical therapy, it is essential that root/implant accretions be thoroughly removed [8]. Conventional periodontal treatment begins with calculus and biofilm removal, using scaling instruments on the tooth surfaces while using carbon fiber or plastic curettes on the implant fixture. For the patient with gingivitis or beginning peri-implant mucositis, that procedure is very straightforward with ease of access. As the severity of the disease increases, root/implant debridement becomes more difficult. Some erbium lasers feature a separate laser feedback system that indicates a calculus deposit. This is an adaptation of a new modality of a visible light diode laser to detect subgingival accretions [9]. Studies have shown that some calculus remains, despite careful root planing or implant debriding, and treatment outcomes may not always be successful with deeper pockets [10–12]. Thus, surgery would be necessary to access those areas, along with placing regenerative materials.

Another consideration is that conventional ultrasonic and sonic scalers used for subgingival debridement may not be effective to produce a bactericidal effect [13].

This initial therapy is usually performed in the general dentist's clinic if immediate referral to a periodontist is not indicated. Within that office setting, a dental hygienist may deliver all or part of the treatment in accordance with the scope of practice and other regulations governing his/her license. During any therapeutic session, the patient must be instructed in an effective oral hygiene regimen. Several appointments may be necessary to complete the initial nonsurgical protocol, and adequate evaluation periods will determine how successful the patient compliance and the practitioner's efforts have been.

14

14.2.1 Currently Available Laser Wavelengths

Dental lasers are generally used adjunctively for the above-described initial nonsurgical therapy [14, 15]. For purposes of this section, the laser instrument described will have a minimum output of approximately 0.5 W of average power. This is to distinguish it from other lasers used for photobiomodulation or for antimicrobial photodynamic therapy.

Any of the commercially available dental lasers can be utilized for nonsurgical periodontal or peri-implant disease therapy. At this date, the generic types and nominal emission wavelengths include diode (445, 532, 810, 940, 980, and 1064 nm), Nd:YAG (1064 nm), Er,Cr:YSGG (2780 nm), Er:YAG (2940 nm), and CO_2 (9300 and 10,600 nm). For treatment of periodontal and peri-implant diseases, all of the above wavelengths can be used for debridement of the soft tissue side of the periodontal pocket; both erbium wavelengths are also currently indicated for calculus removal on the tooth structure. With the exception of Nd:YAG, there are no general contraindications for use of these wavelengths around implant fixtures. Studies have shown that the high-peak-power emission of the Nd:YAG laser with microsecond pulses caused melting on sandblasted, acid-etched, and titanium plasma-sprayed surfaces of titanium implants [16]. The details are described in Table 14.1.

Table 14.1 Details of dental wavelengths used in adjunctive nonsurgical therapy

Laser type	Nominal wavelength in nm	Periodontal tissue target for laser photonic energy used in nonsurgical therapy	Precautions
Diode	445, 532, 810, 940, 980, 1064 nm	Debridement and detoxification of inflammatory tissue due to selective absorption in areas of inflammation by soft tissue pigments and blood components, including pigmented bacteria. Very good hemostasis of blood in the sulcus	1. For periodontitis, prolonged contact with dark colored calculus, root surface, and osseous tissue should be avoided 2. For peri-implant mucositis, no implant surface damage has been reported
Nd:YAG	1064	Same as diode	1. For periodontitis, prolonged contact with dark colored calculus, root surface, and osseous tissue should be avoided 2. For peri-implant mucositis, the beam should be placed parallel to the long axis of the implant fixture so that any interaction will be minimized

Table 14.1 (continued)

Laser type	Nominal wavelength in nm	Periodontal tissue target for laser photonic energy used in nonsurgical therapy	Precautions
Er, Cr:YSGG Er:YAG	2780 2940	Debridement of inflammatory soft tissue due to the absorption in water of sulcular fluid and organic components of soft tissue inflammation along with the cellular water of pathogens Removal of root accretions due to the primary absorption in the water component of dental calculus and secondary absorption in the mineral component. Good hemostasis of blood in the sulcus	1. For periodontitis, care should be used to avoid excessive removal of cementum during calculus removal. Water spray must be used 2. For peri-implant mucositis, low average power should be used
CO_2	9300 10,600	Debridement and detoxification of inflammatory soft tissue due to the absorption of water and organic components of sulcular fluid and soft tissue inflammation along with the cellular water of pathogens. Very good hemostasis of blood in the sulcus	1. For periodontitis, prolonged contact with tooth surface should be avoided[a] 2. For peri-implant mucositis, low average power should be used

[a] Note: the potential exists for 9300-nm CO_2 lasers to be used for calculus removal. Currently, there is no indication for use in this procedure

14.2.2 Adjunctive Laser Use

The general principle of adjunctive laser use for periodontal and peri-implant disease therapy is to supplement conventional instrumentation in removing or disrupting the biofilm and calcified deposits. Conventional mechanical therapy of periodontal pockets does not necessarily achieve complete removal of bacterial deposits and toxins. Employing a laser has the potential to improve therapeutic results [17, 18].

All dental lasers produce a temperature rise in the target tissue, which would affect the pathogens and the resulting inflammation. In general, most non-sporulating bacteria, including periodontopathic anaerobes, are readily deactivated at temperatures of 50 °C [19]. Both coagulation of the inflamed soft tissue wall of a periodontal pocket and hemostasis are achieved at a temperature of 60 °C [20]. It should be noted that surgical excision of soft tissue occurs at 100 °C; thus, using a laser at these lower temperatures defines a nonsurgical therapy. When erbium lasers are used for calculus removal, the primary interaction occurs when the photonic energy vaporizes the interstitial water of the mineralized matrix at a minimum temperature of 100 °C. However, the rapid pulsing of those lasers used with water spray minimizes any significant temperature rise in the surrounding tissues.

Considering the microbial component, it follows that laser irradiation would have significant potential as an adjunct to traditional scaling instrumentation used on teeth and implants. All of the lasers listed in Table 14.1 use the photothermal effect capable of strong bactericidal and detoxification effects [21]. As an example, a recent study confirmed a significant decrease in several pathogens 6 months posttreatment with the adjunctive use of a 980-nm diode [22]. In addition, the infected soft tissue in the pocket can be debrided; the lymphatic and blood vessels can also be coagulated to enable healing.

14.2.3 General Protocol [23]

Following the examination and diagnosis, the clinician should refer to the periodontal charting and perform initial nonsurgical therapy. A suggested protocol is as follows:

1. Prior to any other instrumentation, laser irradiation at low average power is used to reduce the microbial population in the sulcus [24]. This will lower the risk of bacteremia and reduce the aerosolized contaminants during conventional instrumentation. When using the diode, Nd:YAG, and CO_2 wavelengths, care should be taken to avoid prolonged laser contact with subgingival calculus and root surfaces. For

implant surfaces, care should be exercised with the Nd:YAG beam placement. When using erbium lasers, calculus removal is occasionally performed at the same time with the initial laser irradiation.
2. Appropriate conventional instrumentation is used to perform calculus removal of the tooth or implant surface. Erbium lasers can be used primarily or adjunctively. Better clinical results have been reported on laser decontamination of implants compared to mechanical treatment [25].
3. Decontamination of the pocket epithelium is performed with laser irradiation. The photonic energy interacts with different components of the inflamed soft tissue to disrupt the biofilm and microbial components. The parameters employed produce an average power below that is used for excisional surgery, and the clinician should refer to the laser's operating manual to verify the average power settings. The treatment objective is to aim the laser beam toward the soft tissue with overlapping strokes to ensure that the entire area of the pocket is irradiated. The time needed for this portion of the protocol depends on the pocket anatomy—its shape, depth, and width. Visible debris will accumulate on the contact tip of some lasers or will be flushed out of the pocket with others. Decontamination is complete when fresh bleeding emanates from the pocket.
4. To ensure coagulation and sealing of the blood capillaries and lymphatic vessels, laser energy is used. Generally, this occurs in a short time, and the last beam placement will be at the entrance to the pocket. In more shallow pockets, the decontamination procedure may produce the desired hemostasis without any additional irradiation. After the laser is turned off, digital pressure will help re-adaptation of the tissue to the tooth, especially in deeper pockets. In more shallow pockets, the decontamination procedure may help the initial healing.
5. The patient is given postoperative and oral hygiene instructions. There should be minimal tissue manipulation of the treated area so that the fibrin clot is not disrupted. Very gentle brushing and flossing should be performed for 2 days. Spicy and crunchy foods should be avoided for at least 1 day. Gentle rinsing with warm salt water three times a day should soothe the tissues in a short period of time, and only mild discomfort should be expected. Subgingival irrigation must be avoided.

14

14.2.4 Treatment Planning

The above protocol of initial therapy should be followed for every patient manifesting periodontal or peri-implant disease. The extent of the disease will be determined during the periodontal exam, charting, and diagnosis. When planning treatment, several points should be considered:

- The patient's physical limitations such as posture or temporomandibular joint disease
- The patient's pain sensitivity during the procedure and medications necessary to control it, ranging from topic and local anesthetics to sedation
- The patient's systemic health along with any risk factors that would affect the treatment outcome
- The number of pockets to be treated and the anatomy of each
- The amount and tenacity of debris and biofilm to be disrupted/removed
- Any restorations or occlusal problems that need attention and could compromise access or success of the therapy
- The patient's ability to continue adequate oral hygiene techniques

The severity of the disease will determine the appointment schedule both for therapy and for the patient's tolerance for treatment. Most importantly, the treatment plan must be customized for each patient. Some cases of gingivitis may only require full-mouth debridement and disinfection and can be accomplished in two appointments, including polishing. Other advanced conditions with excessive deposits and biofilm may necessitate that only a few teeth are to be treated in each visit.

The length of each appointment can also vary. For example, moderate generalized disease would be treated with hourly visits in each area of the disease. Some clinicians divide the mouth into quadrants for therapy; others choose to treat all the deeper pockets first. The latter approach has an advantage in that those pockets with more disease can be retreated with steps 3 and 4 on subsequent appointments, especially if some inflammation remains after the first session. To ensure those pockets receive maximum debridement, the laser can be used again during the other therapy visits.

Locally delivered chemotherapeutic agents, such as minocycline hydrochloride, doxycycline hyclate, and chlorhexidine gluconate may be placed in pockets to help biofilm suppression. They are most effective after the biofilm has been disrupted by the debridement procedure. As such, those additions should be performed after the last laser treatment. Antimicrobial photodynamic therapy should also be considered, as discussed in the next section of this chapter.

The patient's oral care skills must be continually assessed and reinforced in this protocol. If the presence of biofilm is not minimized, the intended healing will not progress. An assessment appointment should be scheduled approximately 4 weeks after the completion of the initial therapy.

Following the initial nonsurgical therapy, the next appointment 3 months later will assess both the patient's home care and the periodontal status. Expected outcomes are inflammation reduction or absence, healthier tissue tone, and reduced pocket depths without bleeding. Minimum force should be used during probing in this period, since the attachment apparatus is easily disrupted. Normal detailed probing can be performed at the 6-month post-therapy appointment. Reevaluation can continue at 3-month intervals, with careful assessment of how the disease is resolving. Supportive therapy to preserve the improved clinical attachment and minimum inflammation can continue. This will probably include additional debridement and laser decontamination, along with the patient's daily oral hygiene regimen.

14.2.5 Clinical Case Examples

▫ Figure 14.1 shows a diode laser used in a shallow inflamed periodontal pocket on a patient with Stage I grade A periodontitis (gingivitis). ▫ Figure 14.2 depicts the adjunctive use of an Nd:YAG laser during therapy of Stage III grade B periodontitis. ▫ Figure 14.3 shows the adjunctive use of an Er,Cr:YSGG laser for initial treatment of Stage IV grade B periodontitis. ▫ Figure 14.4 demonstrates the use of a diode laser for adjunctive treatment of peri-implant mucositis.

14.2.6 Considerations About Laser Use in Initial Nonsurgical Therapy

In any laser-tissue interaction, the absorption of the photonic energy depends on many factors, as discussed in ► Chap. 3. For periodontal and peri-implant therapy, those same factors are at work in a very limited space—the periodontal pocket and surrounding structures. The clinician should be mindful of the periodontal anatomy of each site so that both conventional and adjunctive laser therapies can proceed with as much precision and effectiveness as possible while minimizing trauma to the surrounding tissue. Therefore, the following points are important:

- Each wavelength will have different interaction on the various periodontal tissues and implant fixtures. For example, the near-infrared wavelengths are easily scattered and are only absorbed by inflammation. Their depth of penetration in sulcular fluid can be significant, which means that the energy could travel beyond the intended target tissue [26]. On the other hand, erbium lasers can be used efficiently to remove subgingival calculus, although it is possible to remove

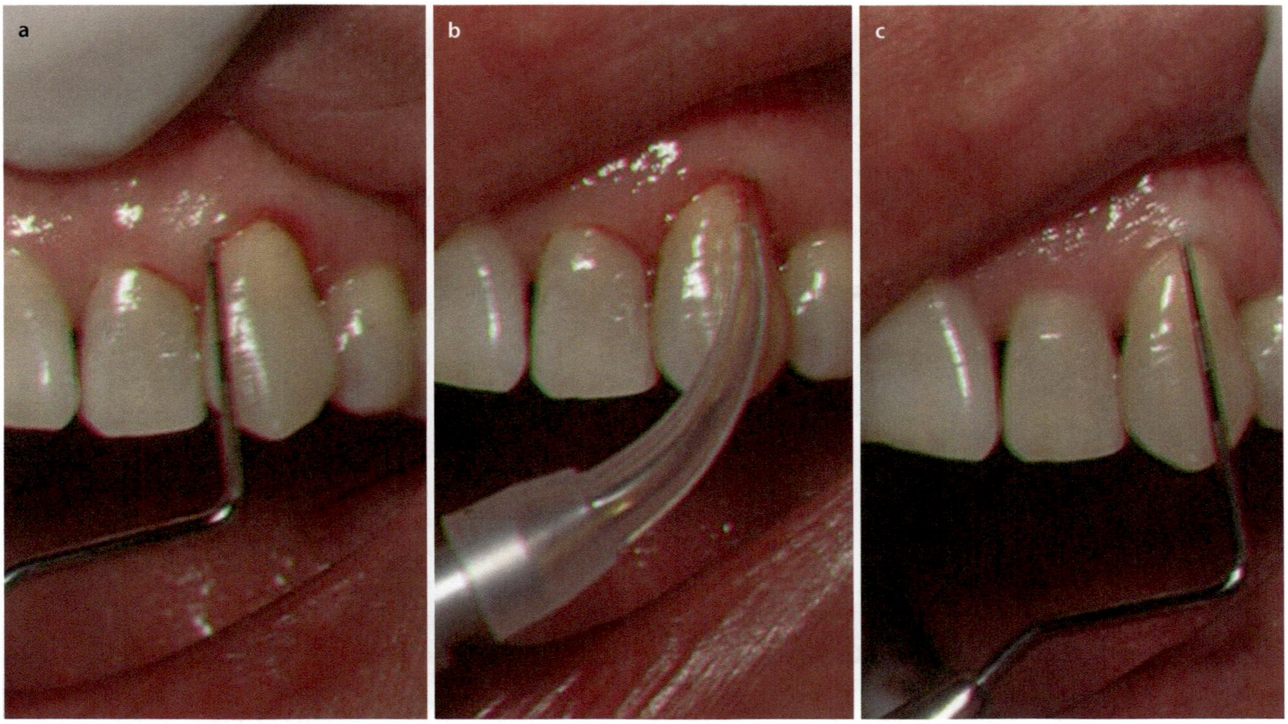

▫ **Fig. 14.1** **a** Preoperative view of an inflamed shallow gingival sulcus of the upper left cuspid. **b** An 810-nm diode laser with a 300-μm bare fiber and 0.4-W CW emission directed toward the soft tissue side of the sulcus. **c** Six-month postoperative view showing no inflammation

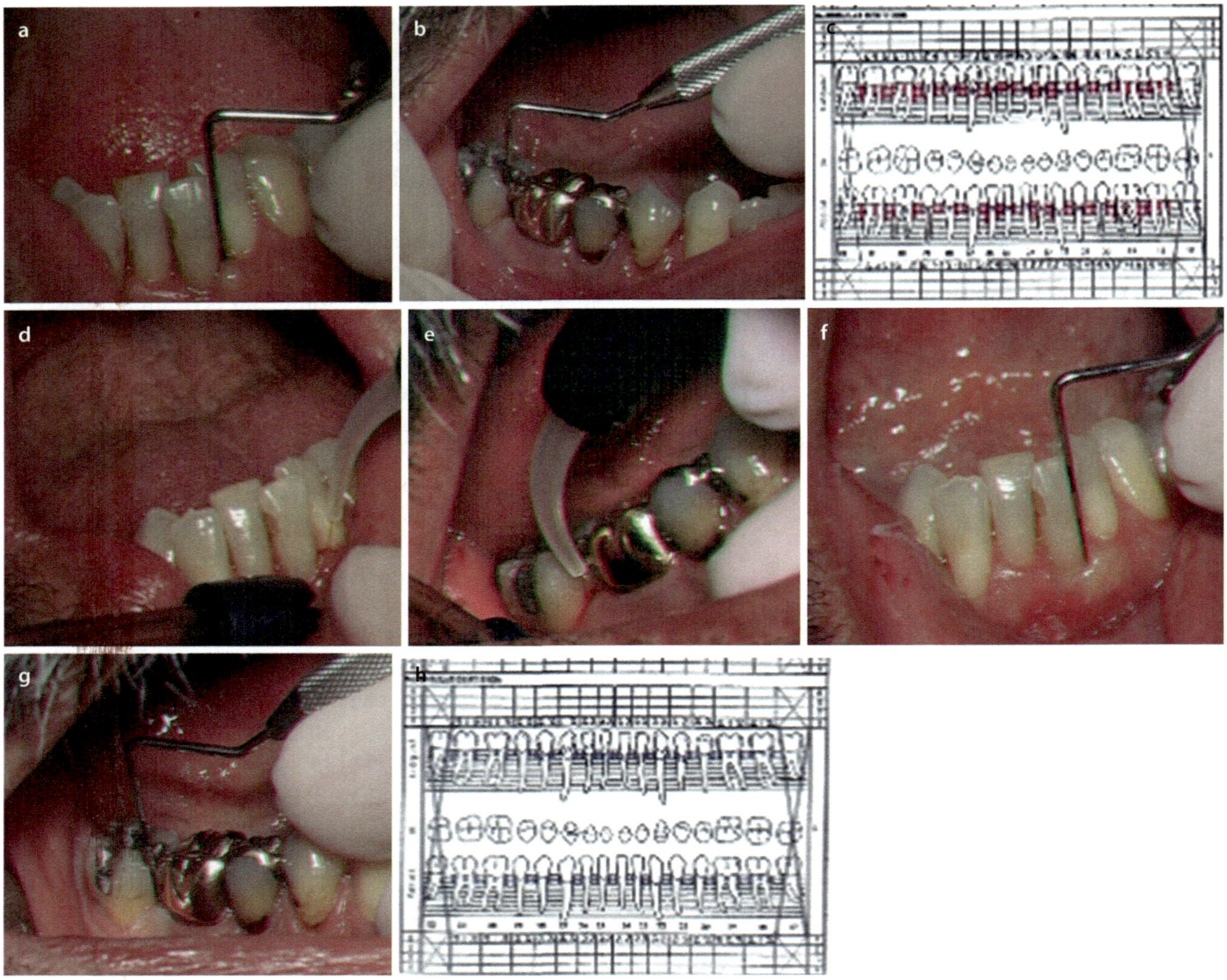

Fig. 14.2 **a** Preoperative view of a 5-mm pocket on the lower left central incisor. **b** Preoperative view of a 6-mm pocket on the lower right second molar. **c** Preoperative periodontal probe chart. **d**, **e** After hand and ultrasonic scaling, an Nd:YAG laser is used with a 400-µm fiber and an average power of 1.8 W (30 mJ/pulse and 60 Hz) and directed toward the soft tissue side of the pocket. **f**, **g** Six-month postoperative view showing pocket depth reduction and lack of inflammation. **h** Six-month postoperative periodontal probe chart

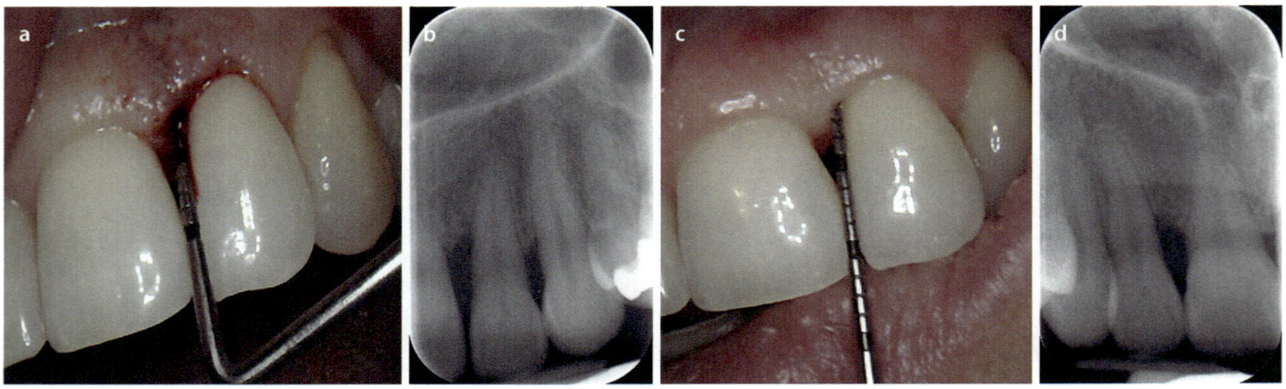

Fig. 14.3 **a** Preoperative view of an 8-mm pocket with bleeding on probing. **b** Preoperative radiograph of the pockets. After ultrasonic removal of the calculus, an Er,Cr:YSGG laser was used with a 500-µm-diameter radial firing tip at an average power of 1.5 W (50 mJ, 30 Hz) with a pulse duration of 60 µs for debridement. **c** Seven-month postoperative probing shows significant pocket depth reduction without bleeding on probing. **d** Seven-month postoperative radiograph depicts a more stable periodontium. (Clinical case courtesy of Dr. Rana Al-Falaki)

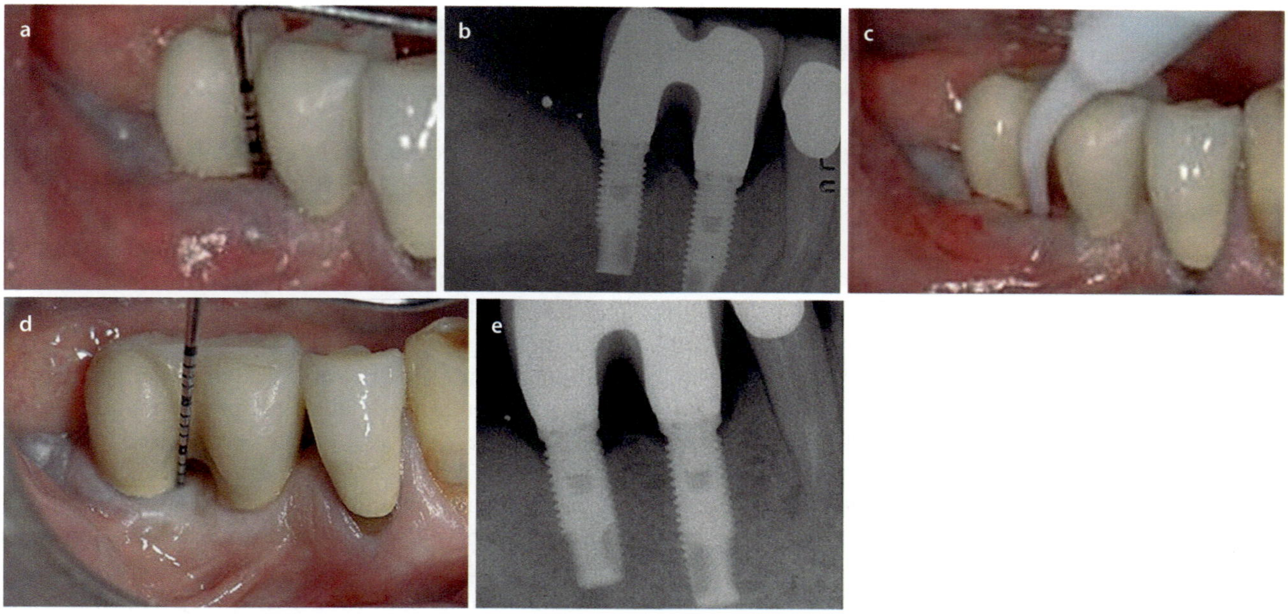

Fig. 14.4 **a** Preoperative view of a 7-mm pocket with bleeding on probing around an implant. **b** Preoperative radiograph. **c** After careful conventional debridement of any calculus, an 810-nm diode laser with a 400-μm tip was used with an average power of 0.4 CW emission directed toward the soft tissue and away from the implant fixture. **d** Six-month postoperative view showing pocket depth reduction and lack of inflammation. **e** Six-month postoperative radiograph

some cementum on the root surface [27]. Interestingly, a recent systematic review indicated that adjunctive use of erbium lasers with conventional instrumentation can be appropriate to remove residual debris from root surfaces [28].

- The current literature indicates that lasers are generally safe for treatment of peri-implant mucositis, but some precaution should be exercised [29, 30]. As mentioned previously, Nd:YAG lasers' usual emission mode produces very short pulse durations and a very high peak power per pulse. Those high powers have been known to damage titanium surfaces. Erbium lasers have the same pulse durations but produced no surface alterations with low energy density use. Understanding of those differences is necessary for the clinician to choose the appropriate wavelength for beneficial treatment.
- The laser parameters must produce low average power to minimize ablation of healthy tissue. Each laser instrument has specific operating instructions for this procedure with suggested settings, and these should be used as a guide to begin the therapy.
- Each laser has a specific handpiece and emission device—for example, an optical fiber tip or a small tube. The clinician should ensure that the laser beam is aimed as precisely as possible toward the intended target tissue. For example, diode photonic energy will be readily absorbed by dark calculus causing a significant temperature rise; so, the tip should be angled toward the soft tissue. Likewise, when using an erbium laser for calculus debridement, the tip should be as parallel to the tooth axis to avoid excessive cementum removal.
- As granulation tissue is removed, it may accumulate around the laser tip or tube. Those should be checked and cleaned often to avoid concentration of the energy in the debris.
- Proper case selection is important and continuing evaluation must be performed. If areas of disease do not respond to the nonsurgical approach, then subsequent surgical therapy will be necessary.

14.2.7 Acronyms for Nonsurgical Initial Periodontal and Peri-implant Therapy

Clinicians may find various acronyms in the operating manuals of different laser instruments or in scientific literature. The intent of these terms is the same—to provide the first phase of treatment. Such terms as LAD/LABR (laser-assisted decontamination/laser-assisted bacterial reduction), LAPT (laser-assisted periodontal therapy), and LCPT (laser-assisted comprehensive pocket treatment) can give specific additional details about the protocol. Some companies have legally protected their acronyms:

- REPaiR (Regenerative Er,Cr:YSGG Periodontitis Regimen) uses the company's Er,Cr:YSGG laser for sulcular debridement and root surface cleaning.
- WPT™ (Wavelength-Optimized Periodontal Therapy) uses the company's Nd:YAG and Er:YAG to remove the diseased epithelial lining and to debride the root surface calculus, respectively.

Whichever terminology or abbreviations are used, various laser wavelengths can add beneficial results for the treatment of periodontal and peri-implant disease.

14.2.8 Selected Literature Review for Lasers in Nonsurgical Therapy

The following is a sampling of the literature describing various wavelengths used adjunctively for nonsurgical therapy:

- Chambrone et al. in a best evidence review suggested that adjunctive laser use may promote statistically significant small improvements in pocket depth and clinical attachment [31].
- Celik et al., in a randomized clinical trial, used conventional scaling along with Er:YAG adjunctively and found significantly better pocket depth reduction and clinical attachment gain [32].
- Ciurescu et al., in a randomized clinical trial, employed both 940-nm diode and Er,Cr:YSGG lasers adjunctively and showed significant improvement in pocket depth reduction and clinical attachment gain at 6 months [33].
- Tenore et al., in a randomized clinical trial, showed added benefit with adjunctive 980-nm diode therapy for peri-implant mucositis [34].
- Estrin et al. in a systematic review found that a combination of Nd:YAG and Er:YAG lasers may lead to additional clinical improvements when used adjunctively [35].
- Yu et al., in a systematic review, showed that adjunctive use of a diode laser showed significant improvement in various periodontal health parameters at 6 months [36].

14

14.2.9 Photobiomodulation

Photobiomodulation (PBM) is explained in great detail in ▶ Chap. 7. It is also an adjunctive laser use both for nonsurgical procedures and surgical therapy. It can reduce inflammation and stimulate healing of the periodontium. Numerous manuscripts including systematic reviews report positive results when used for periodontal and peri-implant disease therapy [37, 38]. Such benefits include reducing clinical inflammatory indices or markers, increasing healthy periodontal parameters, improving bone healing, and providing analgesic effects. Those same studies report variations in protocol and irradiation parameters. Thus, more evidence-based guidelines are necessary.

In summary, the adjunctive use of lasers for initial, nonsurgical periodontal therapy must follow specific protocols, and the treatment must be continually evaluated to determine if surgery is necessary. This therapy is well accepted by patients, and it may contribute to their improved home care. Well-designed scientific studies are always required to support the evolving reported benefits of the procedure.

14.3 Surgical Therapy for Periodontal and Peri-implant Disease

After initial therapy is completed, periodic evaluations and maintenance appointments follow. There can be challenges to complete debridement and disinfection of periodontal or peri-implant pockets which range from anatomic complexity of the defects to the patient's inability to maintain good oral hygiene. Moreover, the clinical attachment level and pocket depth produced by initial therapy may still be inadequate for optimum health. Thus, some surgical intervention will be occasionally necessary.

◘ Table 14.2 lists the details of laser wavelengths that can be used for surgical therapy. Note that all wavelengths can be used for soft tissue surgery, but only the erbium and the 9.3 CO_2 instruments can be used for osseous procedures. The small diameter delivery system including curved tips can especially aid in access to infrabony and furcation pockets.

In this section, surgical therapy will be divided into two sections. The first is a flapless technique, and the second is the more conventional protocol where a flap is reflected and then repositioned.

Table 14.2 Details of wavelengths used in surgical therapy

Laser type	Nominal wavelength in nm	Periodontal tissue target for laser photonic energy used in surgical therapy	Precautions
Diode	445, 532, 810, 940, 980, 1064	Incision and excision of gingival tissue along with hemostasis. Good absorption in pigmented tissue and hemoglobin, likewise in areas of acute inflammation	1. For gingival surgery or debridement of granulation tissue, prolonged contact with dark colored calculus, root surface, and osseous tissue should be avoided 2. For peri-implantitis, no implant surface damage has been reported when low average power irradiation is used
Nd:YAG	1064	Same as diode	1. For gingival surgery and granulation tissue debridement, prolonged contact with dark colored calculus, root surface, and osseous tissue should be avoided 2. For peri-implantitis, the beam should be placed parallel to the long axis of the implant fixture so that any interaction will be minimized
Er, Cr:YSGG Er:YAG	2780 2940	Excellent incision and excision of soft tissue with minimal depth of cut due to the very high absorption in the water content of that tissue. Good hemostasis Excellent cutting and shaving of osseous tissue	1. For osseous surgery and calculus removal, water spray must be used 2. For peri-implantitis, low average power should be used near the implant fixture
CO_2	9300 10,600	Excellent incision and excision of soft tissue due to the high absorption in the water content of that tissue. Very good hemostasis Excellent cutting and shaving of osseous tissue (9300-nm wavelength *only*)	1. For osseous surgery with 9300 nm, water spray must be used 2. For peri-implantitis, low average power should be used near the implant fixture. Care should be taken to minimize reflection from the metal implant toward surrounding tissue

14.3.1 Flapless Periodontal and Peri-implant Surgery

Description of Flapless Surgery: There are two current flapless techniques that are considered to be surgical but without employing any open-flap technique for debridement. These fulfill the concept of a minimally invasive procedure but could have limitations because of limited access to the entire diseased area of the periodontium around the root or the implant. One, termed LANAP®, is an acronym for laser-assisted new attachment procedure and uses a proprietary Nd:YAG instrument. The other is termed laser-assisted comprehensive pocket treatment (LCPT) [39, 40] where any erbium laser can be employed.

The LANAP® protocol entails a specific step, single-session treatment, shown in Fig. 14.5. After verifying the pocket depth, the laser selectively removes the pocket's epithelial lining. The root surfaces are debrided with conventional scaling instruments, and then blunt dissection is performed at the osseous crest. The laser is then used to obtain hemostasis and to form a fibrin clot so that the loose gingival tissue can be approximated back to the tooth. Occlusal adjustments are performed and postoperative instructions given. This procedure has generated case report studies that offer histologic evidence of new connective tissue attachment, new cementum, and new alveolar bone [41, 42]. The Nd:YAG wavelength (1064 nm) is usually safe when using appropriate parameters for pocket irradiation. However, the photonic energy has a potential of deep tissue penetrability, so care must be taken to avoid thermal damage to the underlying tissues.

Figure 14.6 shows a clinical case of Stage III grade B periodontitis in the maxillary anterior sextant. The LANAP® protocol is used for successful treatment (clinical case courtesy of Dr. Raymond Yukna).

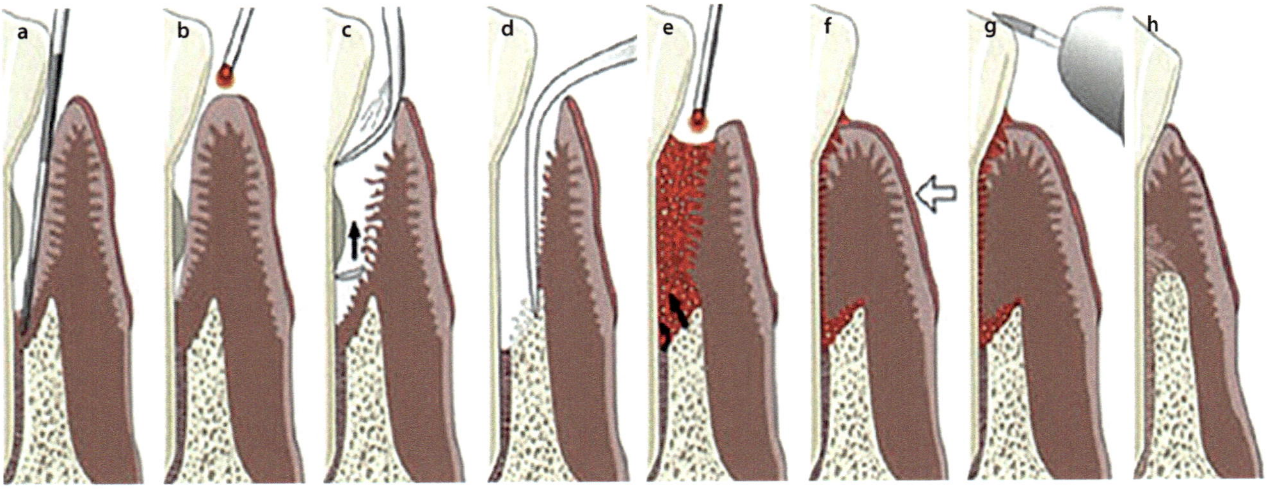

Fig. 14.5 Graphic depiction of LANAP® using the pulsed Nd:YAG laser. **a** Bone sounding to determine pocket depth. **b** Under local anesthesia, typically a 360-μm optic fiber delivers 3.6–4.0 W average power at a pulse duration of 100–150 μs to selectively remove the diseased epithelial lining of the pocket, denature pathologic proteins, and create bacterial antisepsis. **c** The root surface accretions are removed with piezo ultrasonics and conventional instruments. **d** Blunt dissection with a conventional dental instrument is used to modify the osseous contour at the alveolar crest and perform intramarrow penetration to gain access to stem cells and growth factors. **e** Using the same fiber but with a pulse duration of 550–650 μs, the laser energy performs hemostasis; establishes a thick, stable fibrin clot; activates growth factors; and upregulates gene expression. **f** The gingival tissue is pressed toward the tooth to secure it without sutures. **g** Occlusal adjustments are performed to eliminate improper contacts and to allow for passive eruption. **h** Showing anticipated healing in an environment conducive to true regeneration of new cementum, new periodontal ligament, and new alveolar bone. (LANAP® is a patented and registered trademark of Millennium Dental Technologies, Inc., Cerritos, Calif., USA.). (Graphic reproduced with permission from Millennium Dental Technologies)

14

The same company has an identical procedure for treatment of peri-implant disease, termed LAPIP™ (Laser-Assisted Peri-Implantitis Protocol.) The laser emission is reduced so that much less average power is applied around the implant structure. The fiber is aimed as parallel as possible to the long axis of the fixture to avoid the metal absorbing the energy and thus overheating as well as to minimize any reflected photons off the surface.

LCPT (laser-assisted comprehensive pocket therapy) uses erbium lasers with wavelength emissions of 2870 or 2940 nm [39, 40]. As noted previously, these lasers can be used for soft tissue and calculus removal. In addition, they are indicated for use in contouring osseous tissue. Thus, they can be useful for debridement of both granulation tissue and bone defects in moderate to deep periodontal pockets, depending on the accessibility. The procedural steps are shown in Fig. 14.7. After assessing the pocket, the laser and hand instrumentation is used for root surface debridement. That is followed by removal of the epithelial and diseased connective tissue of the lining of the gingival pocket as well as diseased osseous tissue. The treatment objective is thorough decontamination of the whole pocket as well as enhancement of bleeding from bone surface, including bone marrow-derived cells which are a major source of mesenchymal stem cells. The parameters used will not affect hemostasis in the bone; on the contrary, the procedure should enhance bleeding which would be advantageous for tissue regeneration. There may also be some biostimulatory effects in the surrounding tissues. The next step is laser ablation of the external gingival tissue at the pocket entrance. The epithelium and occasionally a layer of connective tissue are removed. The pocket depth is automatically reduced with this small dimension gingivectomy, and the exposure of the connective tissue will delay the migration of the epithelium into the pocket while the attachment is being re-established. That procedure will cause some gingival recession, but the primary benefit of pocket healing will be realized. The last step is to ensure adequate coagulation for a stable blood clot to seal the pocket entrance. The erbium laser is used in a noncontact mode without water spray to achieve this.

This procedure can also be used for the treatment of peri-implant mucositis or the initial stages of peri-implantitis.

Figure 14.8 is a clinical case of Er:YAG (2940 nm) laser-assisted LCPT for treatment of Stage III grade B periodontitis. There is a radiographic evidence of osseous healing and new attachment at 1 year postoperatively.

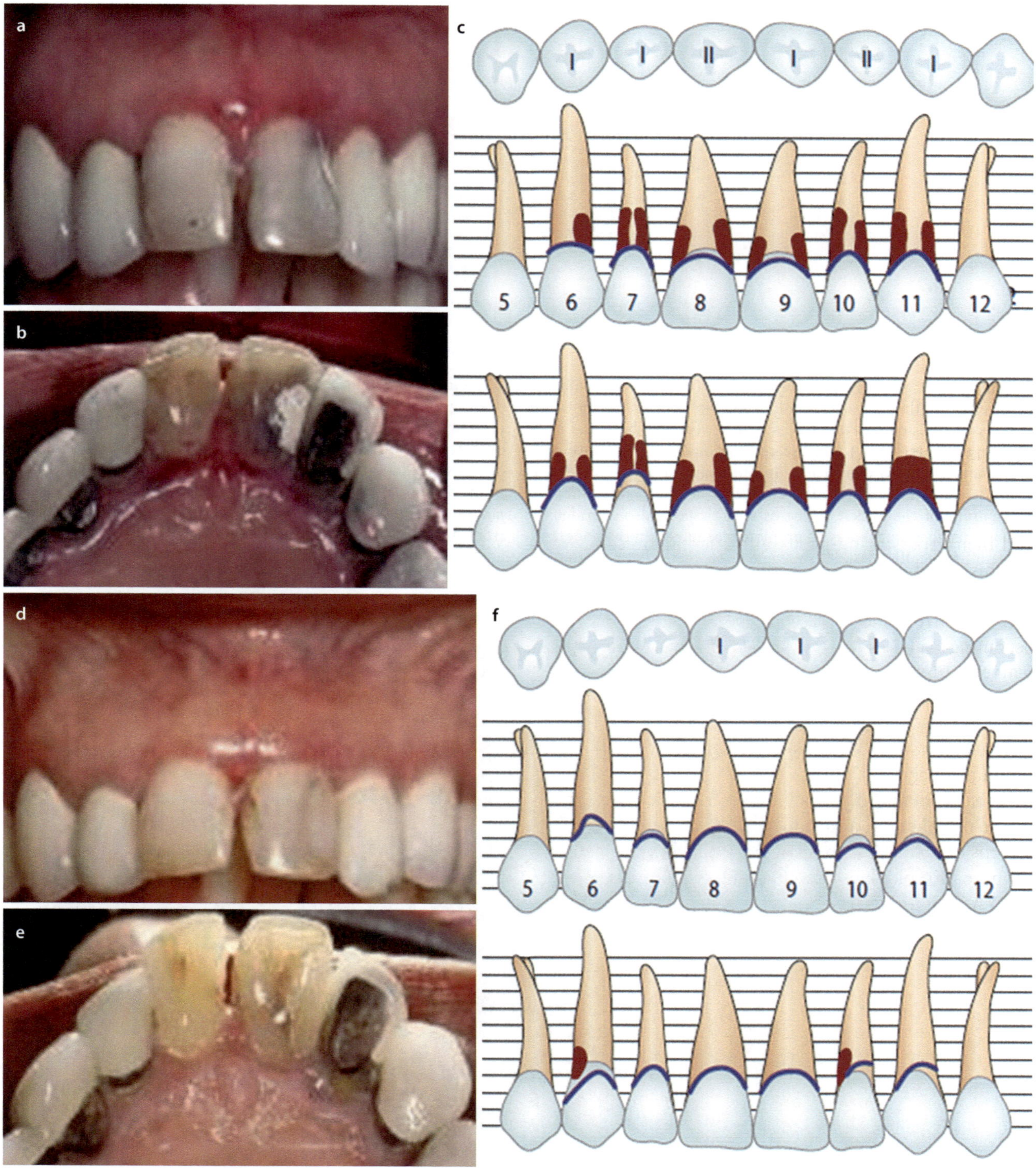

Fig. 14.6 **a** Clinical view of the anterior facial region. **b** Clinical view of the lingual anterior region. **c** Pre-treatment periodontal probe chart showing pockets and mobility on all the anterior teeth. Each horizontal line represents a 2-mm increment, and the red markings indicate bleeding on probing. Mobility is indicated with Roman numerals on the incisal view icon at the top of the chart. The patient was treated with the LANAP® protocol as described in Fig. 14.4. **d** Three-year facial postoperative view. **e** Three-year lingual postoperative view. Note the reduction in inflammation. **f** Three-year postoperative periodontal probe chart shows significant pocket depth and mobility decrease. (Clinical case courtesy Dr. Raymond Yukna)

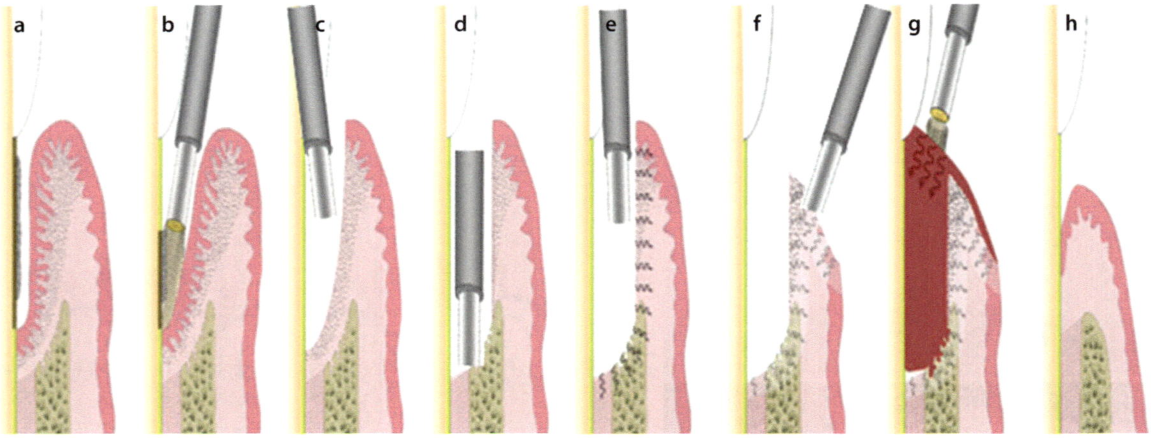

Fig. 14.7 A graphic depiction of LCPT (laser-assisted comprehensive pocket therapy) using an Er:YAG laser. **a** The pocket depth is assessed. **b** Subgingival calculus is removed with both the laser (utilizing a water spray) and the conventional instrumentation so that the diseased root surface is decontaminated and detoxified. **c** The laser is used to remove the diseased epithelial and connective tissue lining of the pocket. **d** The osseous tissue is also debrided by the laser using a water spray to promote bleeding from the bone, and the resulting healthy tissue is shown. **e** Some of the laser irradiation can offer biostimulation to the surrounding intrasulcular tissue. **f** The outer epithelium and some connective tissue are removed to delay epithelial migration into the healing pocket. This gingivectomy does produce some gingival recession. **g** The laser is used in a non-contact mode without a water spray to ensure hemostasis and to produce a stable blood clot to protect the entrance to the pocket, as well as to stimulate the outer surface of the periodontium. **h** Showing the new attachment and pocket depth reduction. (Graphic modified from Aoki et al. [39] with permission © copyright 2015 John Wiley and Sons A/S)

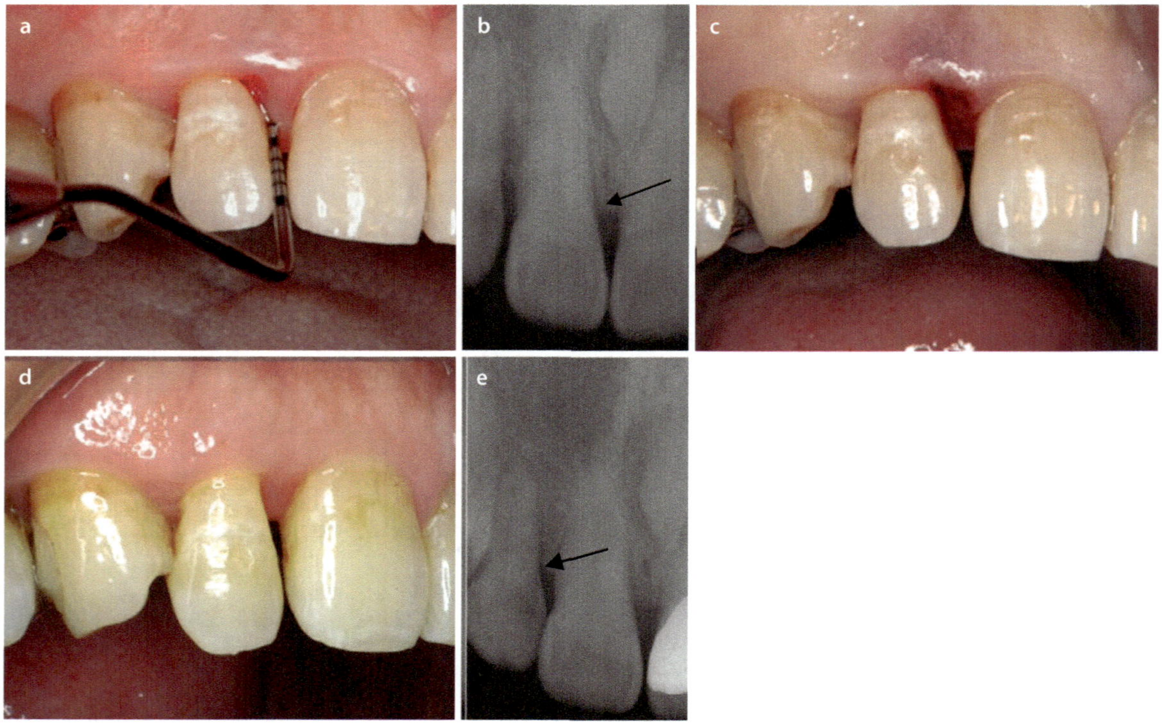

Fig. 14.8 **a** Deep pockets are present on the lateral incisor. The one measures 8 mm with bleeding on probing. **b** The radiograph shows the vertical bony defect (black arrow). **c** Immediate postoperative view showing a stable blood clot. The Er:YAG laser was used with a 600-μm curved tip at 1.0-W average power (50 mJ/pulse at 20 Hz) with a water spray to remove the inflamed soft tissue in the pocket and adjunctively with a curette to debride the root surface. Then the inner epithelial wall and the osseous defect were also debrided. The outer epithelium was recontoured to delay gingival downgrowth, and a stable clot was formed. The latter procedure was performed without a water spray. **d** One-year postoperative view shows good healing with a slight loss of the gingival papilla. **e** The radiograph confirms the osseous defect has filled in (black arrow). (Clinical case and details courtesy of Dr. Koji Mizutani and modified from citation [9] with permission © copyright 2016 John Wiley and Sons A/S)

14.3.2 Osseous Periodontal Surgery Employing a Flap

Osseous surgery, during which bone is removed, recontoured, and/or reshaped, is one of the major periodontal surgical procedures. Optimum bone anatomy will help establish and maintain clinical attachment, shallow pockets, and physiologic gingival architecture—all of which are critical for long-term stability of periodontal tissue.

At the time of writing, no studies are available for the 9300-nm wavelength, but there are manuscripts showing that both Er, Cr:YSGG (2780) and Er:YAG (2940 nm) [30, 31] wavelengths are effective in ablation of bone tissue with minimal thermal damage. In addition, the healing assessment of those lasers performing an osteotomy is at least comparable to conventional instrumentation [45] and may be advantageous for faster and improved outcomes [46, 47].

The correct laser wavelength delivered through a small diameter tip can offer more precision and better access than mechanical instruments. Conventional surgical instruments usually need a wider area of access compared to the laser with its irradiation confined to the end of the tip. Thus, more precision is possible.

Bone grafting procedures with appropriate membranes may be used in areas where the defect cannot be properly contoured. A laser produces minimal thermal damage resulting in a new osseous surface with good vascularity and a lack of smear layer, which should aid in successful bone augmentation [43, 44].

Figure 14.9 shows the use of an Er:YAG laser for open-flap surgery for Stage III grade C periodontitis on a 9-mm deep pocket on the distal of the mandibular right cuspid. This pocket remained after initial therapy. The flap was elevated, and the laser was used to remove the granulation tissue and debride the root surface. The osseous defect was also debrided, and no grafting material was placed. The inner surface of the flap was irradiated for debridement, and sutures were carefully placed. Eight years postoperatively, there were significant pocket depth reduction and clinical attachment gain (clinical case courtesy of Dr. Akira Aoki).

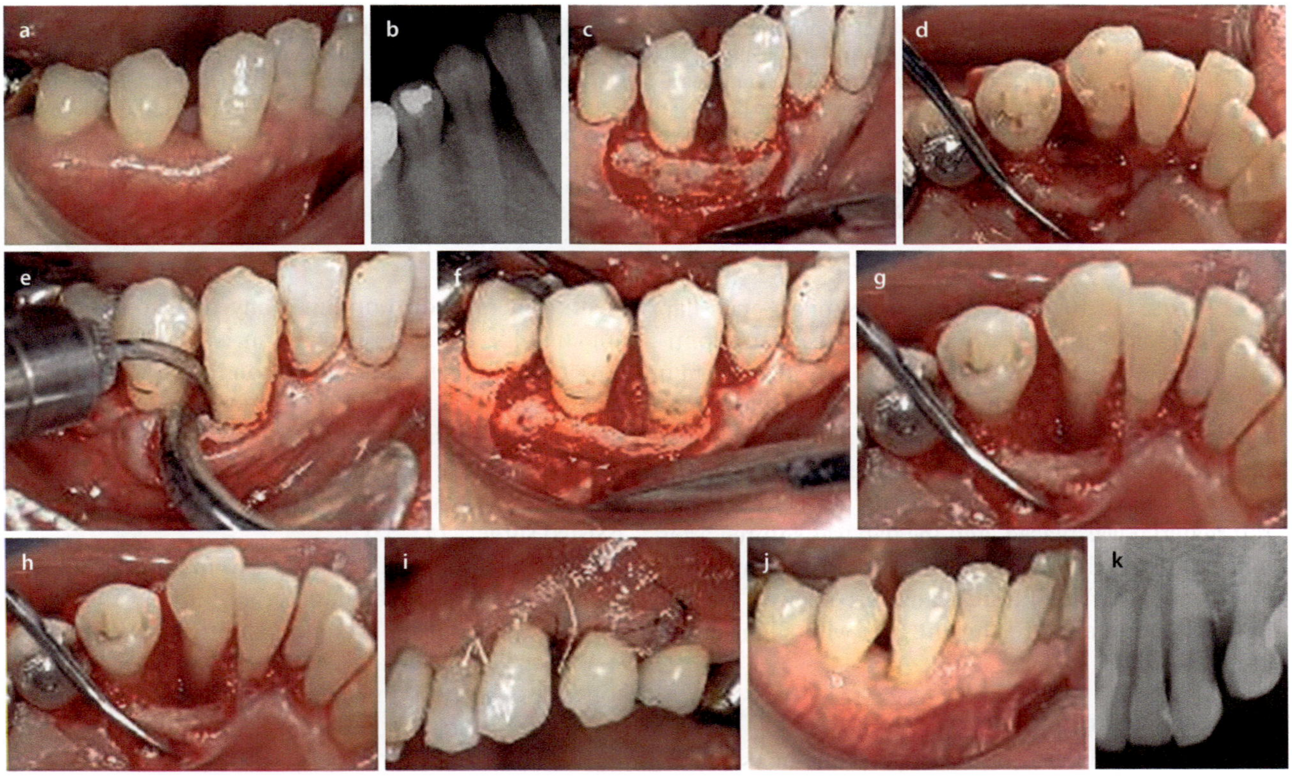

Fig. 14.9 **a**, **b** After initial therapy, a 9-mm pocket with bleeding on probing remains on the distal of the mandibular right cuspid shown clinically and radiographically. **c**, **d** A flap is elevated, and granulation tissue fills the pocket when viewed from the buccal and lingual aspects. **e** An Er:YAG laser is used with an 80° 400-μm tip at 1.2 W average power (40 mJ per pulse at 30 Hz) with a saline water spray for debridement. **f**, **g** Showing the clean vertical bone defect with no thermal damage from the laser energy. No bone augmentation material was placed. **h** The laser is used with the same parameters to decontaminate and stimulate the gingival flap tissue. **i** The flap is sutured in place. **j** Eight-year postoperative view showing healthy gingival tissue with some recession. In fact, there was 7 mm of pocket depth reduction and 5 mm of clinical attachment gain. **k** Eight-year postoperative radiograph. (Case photos and details modified from Aoki et al. [39] with permission © copyright 2015 John Wiley and Sons A/S)

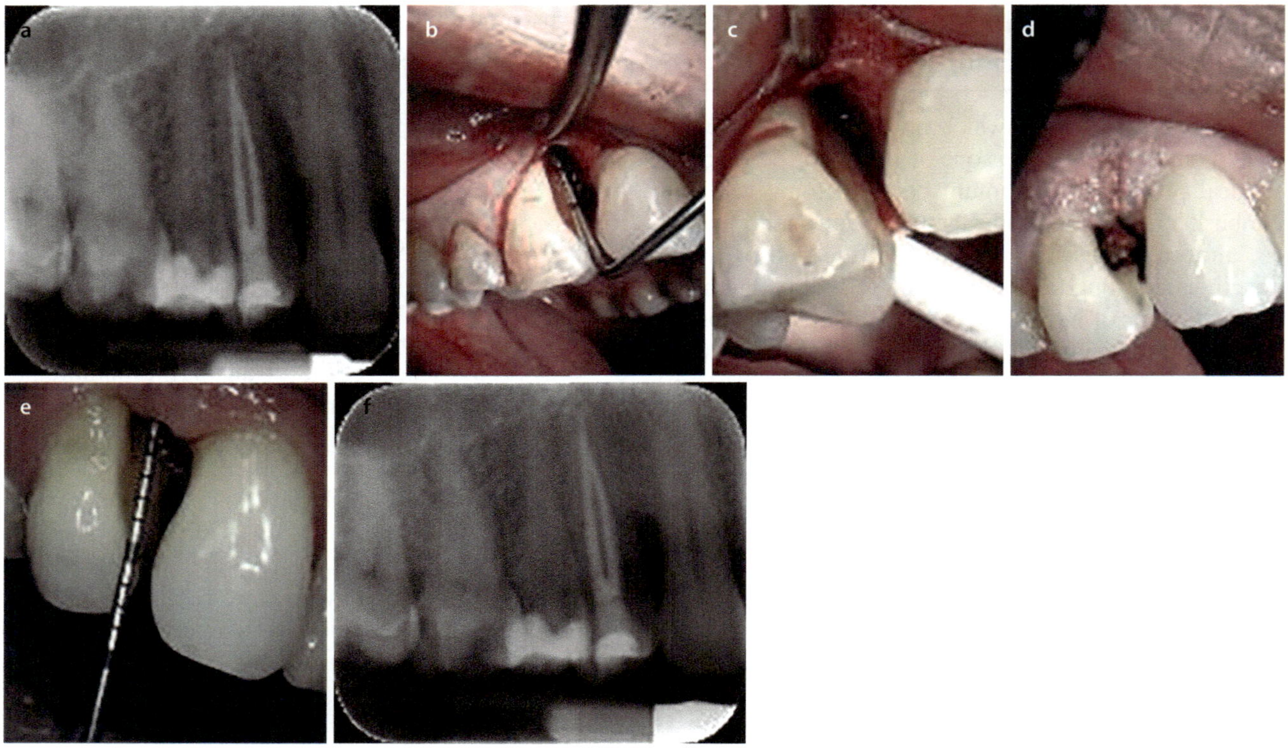

Fig. 14.10 **a** Radiograph showing 11-mm pocket on the mesial of the maxillary first premolar that remained after initial therapy. **b** After the subgingival calculus was removed with ultrasonic instrumentation, a flap is raised, and the osseous defect is explored with the periodontal probe. **c** Immediate postoperative view of the debrided infrabony pocket. An Er,Cr:YSGG laser was used with a 600-μm diameter contact tip. To remove the granulation tissue from the soft and hard tissue, an average power of 1.5 W (50 mJ, 30 Hz) was used with 50% air and 40% water spray in the short pulse mode. Subsequently, the smear layer was removed from the root surface and osseous tissue with an average power of 0.75 W (15 mJ, 50 Hz) with 50% water and 40% air spray. **d** View of the sutured flap. **e** Eight-month postoperative photo with periodontal probe, demonstrating good reattachment with slight gingival recession. **f** Eight-month postoperative radiograph showing a more stable periodontal condition with bone regeneration. (Clinical case courtesy of Dr. Rana Al-Falaki)

Figure 14.10 demonstrates a similar open-flap procedure for treatment of Stage IV grade B periodontitis, where the Er,Cr:YSGG laser was used on an 11-mm pocket on the mesial of the maxillary right first premolar that did not respond to initial therapy. After raising a flap, the laser was used to debride the root surface, the pocket epithelium, and the osseous defect. No bone graft material was placed, and the flap was sutured after confirming that the defect was filled with bone marrow-derived blood. An 8-month analysis showed pocket depth reduction and attachment gain with slight gingival recession.

Surgical Therapy for Peri-implantitis

Debridement and detoxification of the implant surfaces as well as the diseased tissues surrounding implant fixtures are the primary objectives for the treatment of peri-implantitis. Many laser wavelengths have been studied for their ability to efficiently debride implant surfaces. The diode, carbon dioxide, and erbium instruments generally do not cause any surface damage to implants [48–51]. The precaution is that high average power settings can generate heat on the peri-implant tissues and/or directly affect the titanium fixture. Thus, appropriate parameters and techniques must be employed during the surgical session.

Several recent studies report the benefits of laser-assisted surgery [52, 53].

After debridement of the surrounding tissues and decontamination of the implant itself, bone augmentation materials and appropriate membranes can be placed in the defect to enhance regeneration. Clearly, good bone vascularity is important to achieve, and proper laser parameters can accomplish this. As mentioned, lasers should allow for beneficial bone healing following surgery along with a biocompatible implant fixture.

Figure 14.11 demonstrates how an Er,Cr:YSGG laser for severe peri-implantitis therapy. After reflecting a flap, the osseous defect was filled with granulation tissue, and the laser debrided soft and hard tissues in the

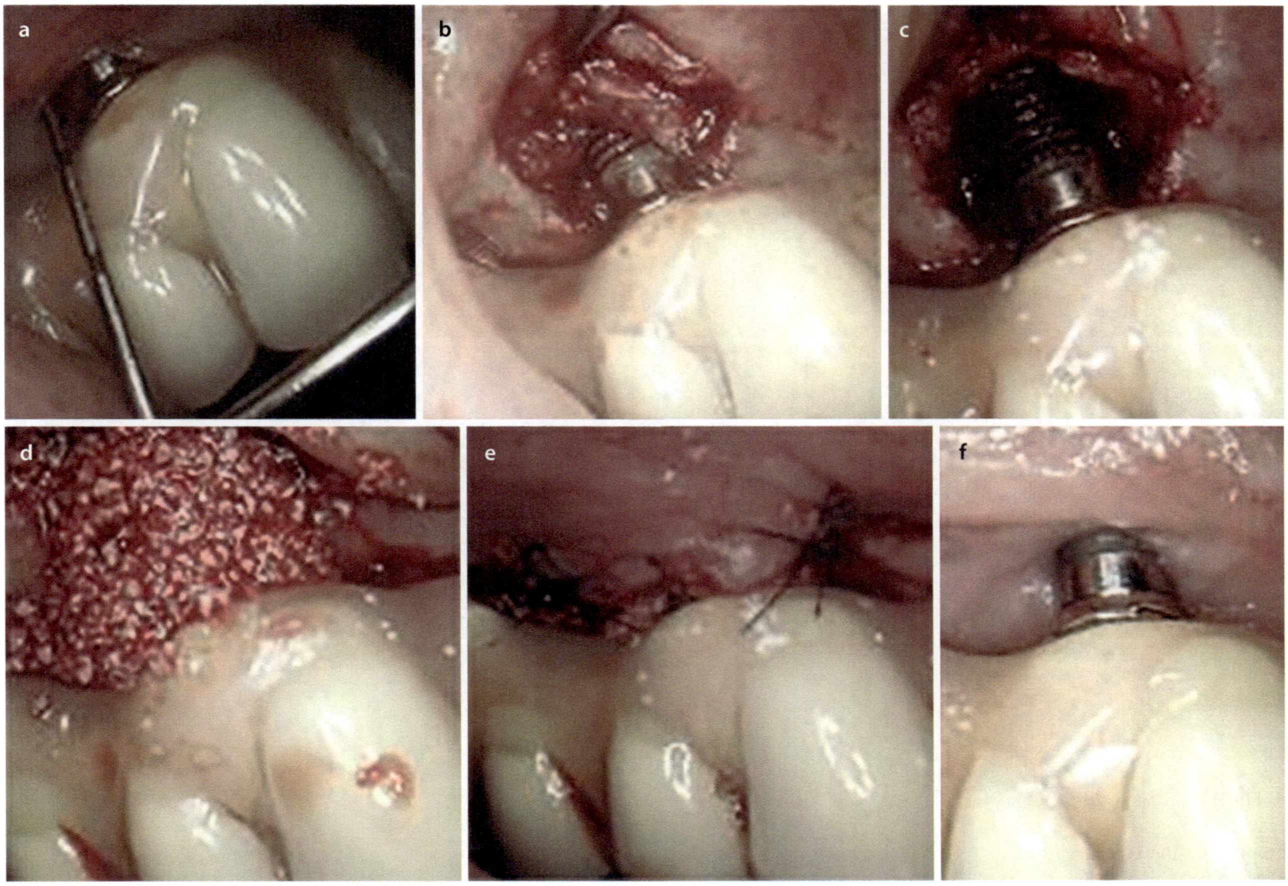

Fig. 14.11 **a** Preoperative view of peri-implantitis around a maxillary posterior implant with an 8-mm pocket. **b** After the flap is reflected, the extent of the defect, filled with granulation tissue, can be seen. **c** Immediate postoperative view of the debridement therapy. An Er,Cr:YSGG laser was used with a 600-μm-diameter contact tip. The granulation tissue was removed from the soft and hard tissue with an average power of 2.0 W (66 mJ, 30 Hz) with 70% water and 50% air, angling the tip away from the implant surface. Then the implant was debrided at an average power of 1.25 W (25 mJ, 50 Hz) with 70% water and 50% air. Lastly, the internal surface of the flap was decontaminated at any average power of 0.75 W (15 mJ, 50 Hz) with 50% water and 40% air. Note the good vascularity of the osseous tissue. **d** Bone grafting material is placed immediately to fill the area. **e** The flap is sutured in place. **f** Six-month postoperative view of the healed periodontium with no inflammation. (Clinical case courtesy of Dr. Rana Al-Falaki)

area. A bone graft and membrane were placed, and the flap was sutured in place. Six months later, the periodontal health was restored (clinical case courtesy of Dr. Rana Al-Falaki).

Figure 14.12 shows the use of an Er:YAG laser for the treatment of severe peri-implantitis. In this case, there is a large per-implant defect present in the maxillary left cuspid and bicuspid with recorded pocket depths of 8–12 mm with bleeding and suppuration. A surgical flap was raised conventionally, after which the area was debrided with the laser. Bone graft material was placed, hemostasis was achieved with the laser, and the flap was sutured. The 1-year clinical follow-up shows healthy periodontium, and subsequent radiographs of 1, 2, and 3 years confirm that the original osseous defect was successfully repaired (clinical case courtesy of Dr. Taichen Lin).

In summary, dental lasers can be used in a surgical approach for treatment of periodontal and peri-implant diseases with benefits such as precision and enhanced visibility during debridement. In addition, osseous tissue can be predictably ablated and contoured. Future research should emphasize proper power settings and describe the details of the protocol. Laser application for bone ablation is becoming a very useful modality for developing the various usages on periodontal and implant therapy. Preventing thermal damage following laser treatment is critical for optimal wound healing.

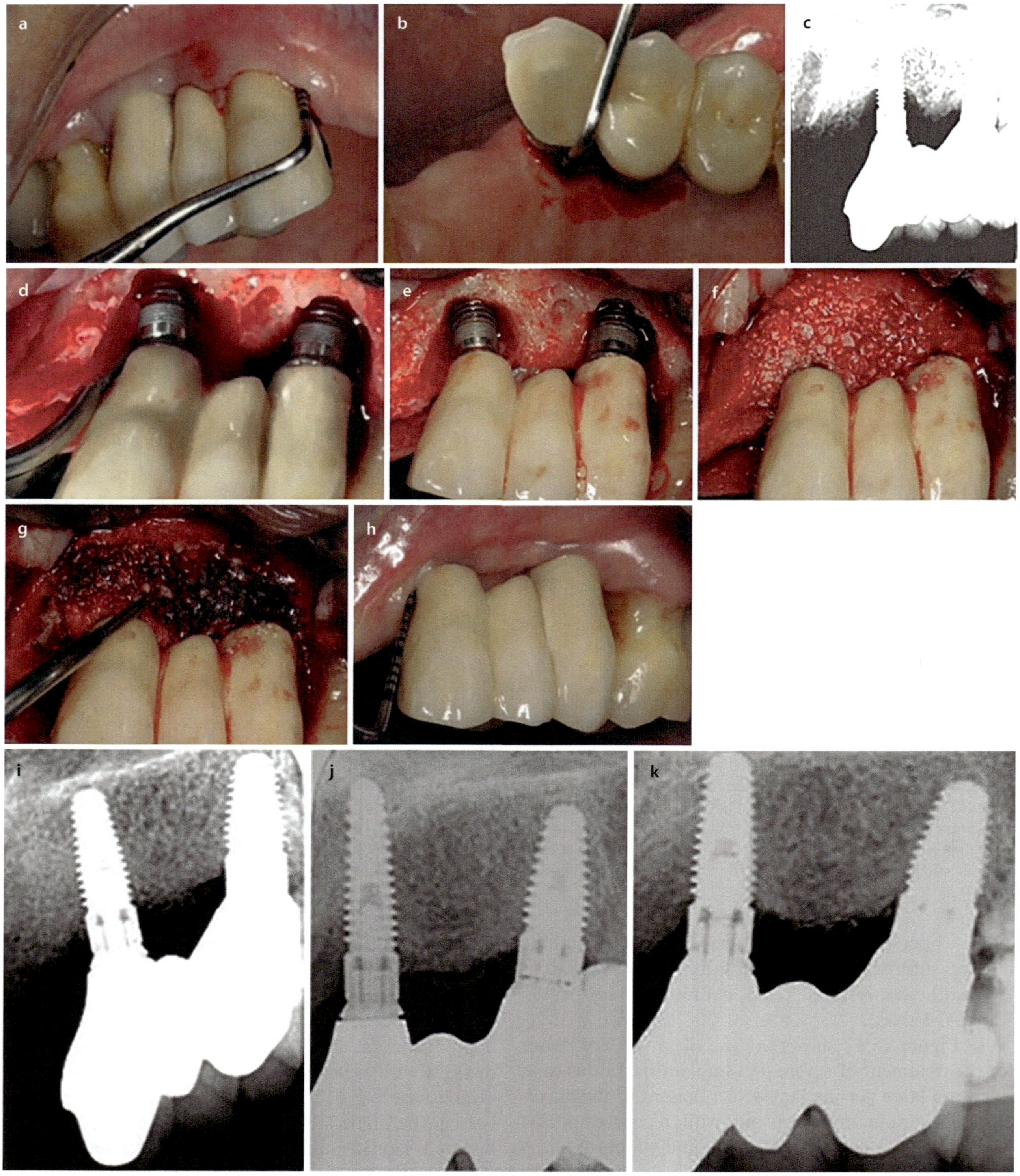

■ Fig. 14.12 a, b Preoperative view of 8–12-mm pocket depths around the maxillary left cuspid, part of a three-unit implant-supported fixed prosthesis. c Preoperative radiograph of the area shows a significant infrabony defect. d The Er:YAG laser was used at 20 Hz and 50 mJ/pulse with a water spray in near contact mode to remove granulation tissue and debride the implant surface. e No thermal damage was observed on the implant surface or the osseous structures. f Bovine bone graft material (InterOss) was grafted into the bony defects. g The laser was used without water spray in a non-contact mode to produce a clot, and the flap was sutured back into position. h One-year postoperative view shows some gingival recession but otherwise healthy tissue. i One-year postoperative radiograph shows the complete repair of the previous osseous defects. j Two-year postoperative radiograph. k Three-year postoperative radiograph. ((Clinical case courtesy of Dr. Taichen Lin.) Graphic modified from Lin et al. [54] with permission © copyright 2019 Elsevier A/S)

14.4 Antimicrobial Photodynamic Therapy in Management of Periodontal and Peri-implant Disease

14.4.1 Photodynamic Therapy

Photodynamic therapy (PDT) is a new approach in killing or eliminating pathogens and uses light of a specific wavelength to activate a nontoxic photoactive dye (photosensitizer) in the presence of oxygen to produce cytotoxic products [55, 56]. Various terms are used for PDT such as photoactivated chemotherapy (PACT), photodynamic disinfection (PDD), light-activated disinfection (LAD), photodynamic inactivation (PDI), photoactivated disinfection (PAD) and antimicrobial photodynamic therapy (aPDT) in different studies and literature. Among these terms, aPDT is the most accepted one for antimicrobial purposes [57–59].

The successful outcome of PDT critically depends on three elements: photosensitizer, light source, and oxygen (◘ Fig. 14.13).

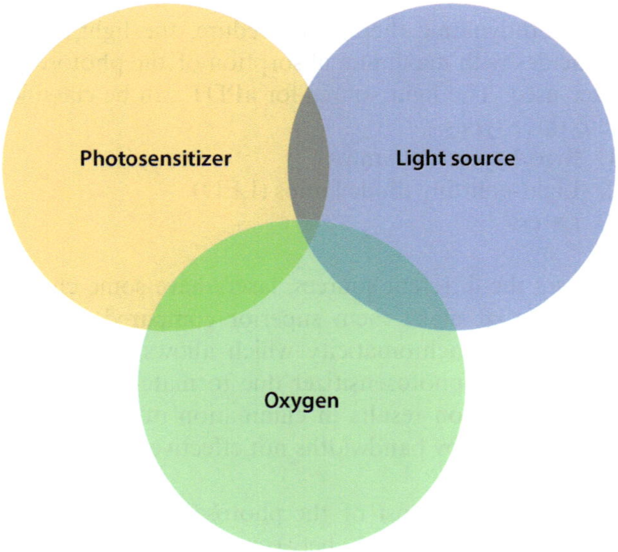

◘ **Fig. 14.13** The basic elements of PDT

14.5 Photosensitizer

A photosensitizer is a chemical compound, which when activated by an appropriate wavelength forms a highly reactive oxygen species which results in cell death.

The photosensitizers should have some characteristics including the following:

- Exist as nontoxic and chemically pure compound
- Have the ability to stain the target
- Be economical and easily available
- Possess a short interval between administration of the drug and peak accumulation in the tissue
- Have a short half-life
- Be rapidly eliminated from normal tissue
- Have activation at specific wavelength
- Possess the ability to produce the huge amount of cytotoxic products
- Have the ability to act on wide range of microorganism [60]

The most applicable photosensitizers which used in dentistry for antimicrobial procedures are described below.

14.5.1 Toluidine Blue O

Toluidine blue O (TBO) is a cationic blue coloring agent used for histological staining. It can also be applied for differential diagnosis between benign and malignant precancerous leukoplakia. It can be activated by wavelength of 635 nm. It can act on both gram-positive and gram-negative bacteria due to its physical and chemical properties and hydrophilic characteristics, and it showed attraction to the mitochondria which has negative charge. TBO can bind to LPS of the outer cell envelope in gram-negative bacteria and the teichuronic acid residues of the outer wall in gram-positive bacteria [61–63].

14.5.2 Methylene Blue

Methylene blue (MB) is used for selective coloring in histology. It is a hydrophilic compound with positive charge. This photosensitizer can be applied for both gram-positive and gram-negative bacteria. It can penetrate through the porin channels in the outer membrane of gram-negative bacteria and interacts with the anionic macromolecule lipopolysaccharide creating MB dimmers which have a role in the photosensitization process. It has a peak absorption at wavelength of 660 nm [64, 65].

14.5.3 Indocyanine Green

Indocyanine green (ICG), a green-coloring agent, has recently been introduced as photosensitizer. The mechanism of this photosensitizer is somehow different from other ones. The effect of ICG is mainly that of photothermal therapy (PTT) rather than photochemical reaction. This anionic photosensitizer can be activated by 810 nm, but its absorption critically depends on the dissolving medium, the chemical bonds of plasma proteins, and its concentration [60, 66].

14.5.4 Curcumin

Curcumin is a yellow-orange pigment isolated from *Curcuma longa* L. which is mostly used as a spice. It has some therapeutic effects on liver diseases, wounds, and inflamed joints, as well as for blood purification and microbial effects. Curcumin has shown no toxic effects on a number of cell cultures and animal studies. It has a broad absorption peak in the 300–500-nm range (maximum 430 nm) and produces strong phototoxic effects. Therefore, curcumin has the capability to be used as a photosensitizer. Easy handling, low cost, and efficacy make this photosensitizer more popular [67–69].

14.5.5 Chlorella

Chlorella is a green natural microalga which contains proteins, vitamins, and minerals and is used as a dietary compound. It has antimicrobial, anticancer, anti-inflammatory, and antioxidant wound-healing characteristics. It can be activated by red wavelength [70].

14.5.6 Phycocyanin

Phycocyanin is a blue-green pigment used as a coloring agent in the food industry. It is extracted from spirulina with antioxidant, immunomodulation, and antimicrobial effects, which makes it a potential photosensitizer in aPDT procedures. Under irradiation by an appropriate wavelength (red range), it can produce singlet oxygen [71].

14.5.7 Chlorophyll

Chlorophyll is a green pigment that is found in green plants activated by the range of blue and red wavelengths. The high potency of ROS production and structure modification by replanting metal ion complex make it suitable as a photosensitizer in aPDT procedures [72].

14.5.8 Riboflavin

Riboflavin (vitamin B2) is an efficient photosensitizer inducing oxidative damage when activated with visible light, especially blue light and LED lamps make it suitable for aPDT procedure [73].

14.6 Light Source

In photodynamic therapy procedure, the light source coincides with maximum absorption of the photosensitizer used. The light source for aPDT can be classified into three types:

1. Broad spectrum lamps
2. Light-emitting diode lamps (LED)
3. Lasers

Among the different sources, lasers have some characteristics that make them superior compared to other sources. Monochromaticity which allows the laser to interact with photosensitizer due to matching with its peak absorption results in elimination of unnecessary tissue heating by bandwidths not effective in PDT reaction [74, 75].

In dentistry, most of the photosensitizers are activated by wavelengths between 630 and 700 nm. Currently, with the introduction of new photosensitizer such as ICG, infrared wavelength like 810 nm is also used which has more penetration depths. On the other hand, blue light LED (400–500 nm) which coincidences with curcumin can be a suitable option due to its availability in all dental offices for curing of dental resin composites and capability in creating free radicals more efficiently compared to red light. LEDs are more cost-effective and compact in comparison to lasers [76, 77].

14.7 Mechanism of Photodynamic Therapy

When a photosensitizer is activated by an appropriate wavelength, electrons are transferred from a lower level of energy to a higher one which is called the triplet state. Then, the energy is transferred to a biomolecule or to oxygen which leads to the production of cytotoxic species. These products damage the cellular plasma membrane or DNA. Both consequences lead to cell death [78, 79].

The transfer of electrons in activated photosensitizer can be done in two pathways including transfer to the neighboring molecule (type 1 reaction) or to oxygen (type 2 reaction) to produce reactive oxygen species (ROS) like singlet oxygen and other radicals like hydroxyl radical. Although the two pathways can have a role on bacterial killing, type 2 by producing highly reactive singlet oxygen is detected as the main pathway in killing bacteria (◘ Figs. 14.14 and 14.15). This mechanism is totally different from that of antibiotics; hence, the resistance of bacterial strain is not likely, due to acting on multiple targets inside the bacteria [80, 81].

It's important to note that antioxidant enzymes produced by bacteria may protect against some oxygen radicals but not singlet oxygen which makes aPDT a more effective procedure. Singlet oxygen has a short lifetime in biological system (≤0.04 μs) and a short radius of action (0.02 μm) that make its action localized without affecting distant cells [82, 83].

One of the main concerns during photodynamic therapy is the photosensitivity of bacteria which seems mainly related to the charge of the photosensitizer used. The neutral or anionic sensitizer binds effectively to gram-positive bacteria. It also binds to some degree to the outer membrane of gram-negative bacteria [84]. The porous layer of peptidoglycan and lipoteichoic acid outside the cytoplasmic membrane of gram-positive species allows the photosensitizer to cross into the cell. On the other hand, gram-negative bacteria have an inner cytoplasmic membrane and an outer membrane separated by the peptidoglycan-containing periplasm which acts as a physical barrier between cells and its environment [85]. The binding of negatively charged photosensitizer to gram-negative bacteria may be improved by linking the photosensitizer to a cationic molecule [86].

The success rate of photodynamic therapies depends on the type, dose, incubation time, and localization of the photosensitizer, the availability of oxygen, the wavelength of light (nm), the light power density, and the light energy fluency. Limitations of this treatment which should be considered are the low-oxygenated environment and the diffusion ability of the photosensitizer and light to be used [87].

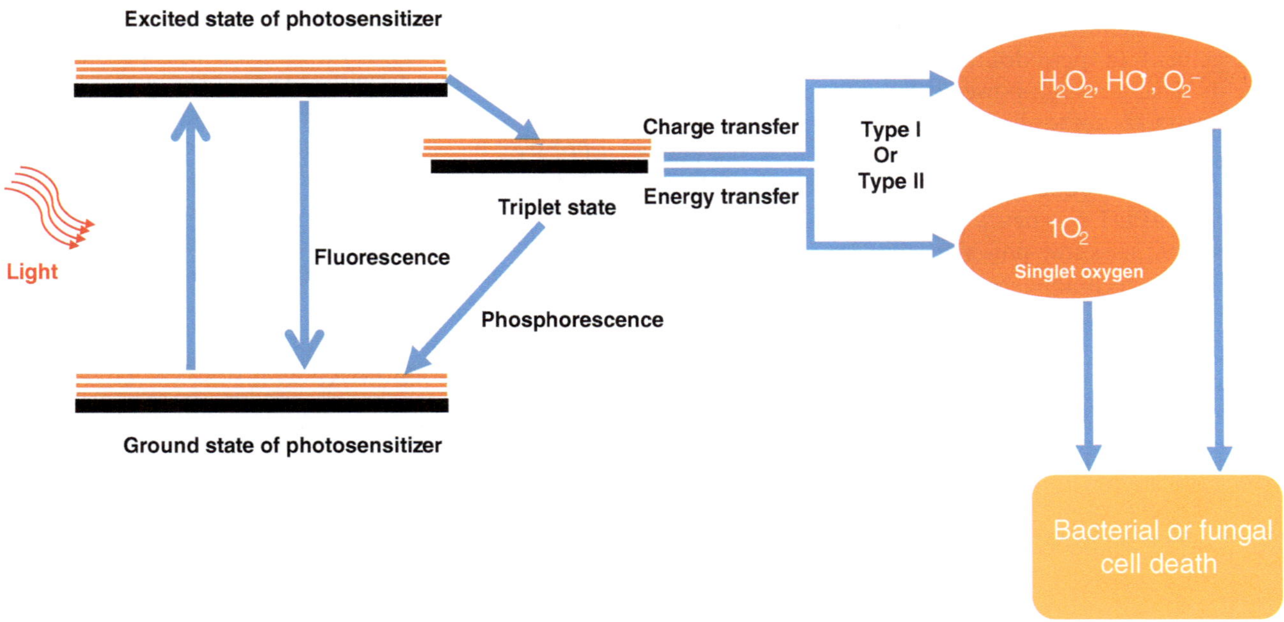

◘ **Fig. 14.14** The mechanism of PDT. The photosensitizer is raised to an excited state by absorbing photonic energy, and a triplet state is created. Its energy is transferred to create either a hydroxyl radical or a singlet oxygen. The latter is thought to be the main toxic agent for pathogens

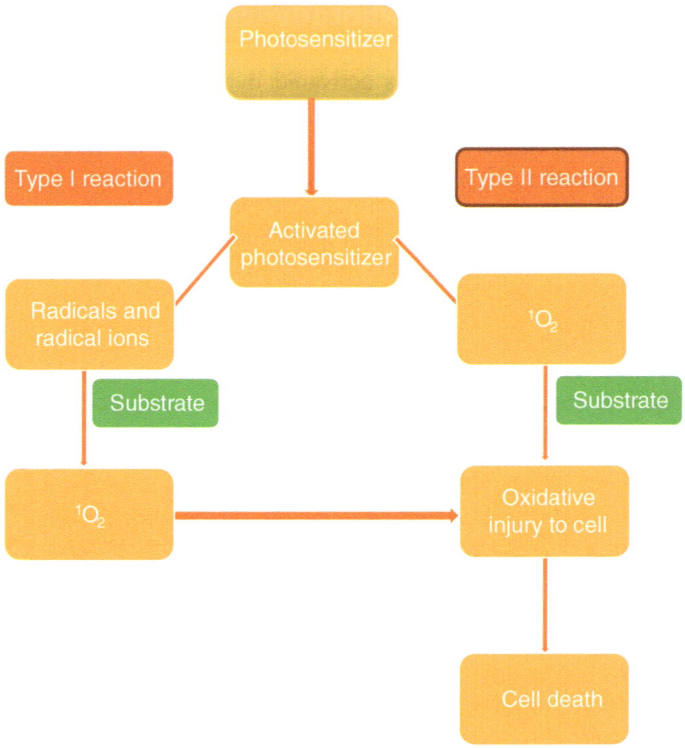

Fig. 14.15 Two pathways in function mechanism of PDT are shown in the flow chart

14.8 aPDT in Periodontal and Peri-implant Disease

It is now established that the application of lasers in management of periodontal diseases can offer some benefits. Using a high-powered laser for antimicrobial purposes raises some concerns like irreversible thermal damage to surrounding periodontal tissues, thermal coagulation, carbonization, and root necrosis [88]. Therefore, aPDT was developed, which is a noninvasive method that uses low power to overcome these limitations. In this technique, the bacteria are selectively targeted without damaging the neighboring tissue [89]. Cadore et al. reported that multiple sessions of adjunctive aPDT produced significantly improved clinical parameters in patients with chronic periodontitis at 90 days postoperatively [90].

14.8.1 Procedure

- Assessing periodontal clinical parameters

  The clinical parameters are collected before treatment. The record of parameters was as follows: (a) bleeding on probing (BOP), (b) clinical attachment level (CAL), (c) plaque index (PI), (d) probing pocket depth (PPD), (e) full-mouth plaque score (FMPS), and (f) full-mouth bleeding score (FMBS).

- Treatment

  After educating oral hygiene instructions, the patients receive full-mouth scaling and root planing (SRP). The photosensitizer solution is applied to the bottom of the periodontal or peri-implant pocket and gingival sulcus with the use of syringe. Following this, the pocket is exposed to the laser light due to protocol moving from bottom of pocket to coronal. Special safety glasses are provided to the patients, operator, and dental assistant to prevent possible eye damage by the laser irradiation. The procedure can be repeated in the same manner for following weeks due to treatment plan.

  Bacteria can penetrate epithelial cells and connective tissue during periodontal diseases. *P. gingivalis* and *A. actinomycetemcomitans* can infiltrate the epithelial barrier into periodontal tissues. In this case, aPDT can be a solution for eliminating them [91]. Sulcular epithelium has increased penetration of photosensitizer due to nonkeratinized pattern. The uptake of photosensitizer in epithelial cells is dependent on incubation time (the interval between applying photosensitizer and laser irradiation). So, there should be a few minutes waiting time after applying photosensitizer before starting laser irradiation [92, 93].

  Photodynamic therapy has some advantages like detoxification of endotoxins such as lipopolysaccharides which inhibit the production of pro-inflammatory cytokines. Also, it can reach to deep or limited access sites without the need for flap surgery in some cases; there is no need to anesthetize the area, and there is no need to prescribe antibiotics. In addition, there is a low risk of bacteremia, which is useful for at-risk patients (those with cardiovascular diseases, diabetes, and immunosuppression) [94].

Furthermore, aPDT increases tissue blood flow in microcirculatory system and reduces venous congestion in gingival tissues.

In assessing different studies, controversial results are obtained due to the different wavelengths of laser and the type of photosensitizer type used.

Bassir et al. assessed photoactivated disinfection using LED and TBO as an adjunct in the management of patients with moderate to severe chronic periodontitis. The study concluded that at 1 and 3 months, PDT showed significant improvements with regard to all clinical parameters compared to baseline but did not have additional effects on clinical parameters in patients [95].

On the other hand, Prasanth et al. in evaluation of aPDT by methylene blue and 655-nm diode laser in management of chronic periodontitis concluded that aPDT has an important role in improving clinical outcomes obtained by SRP and single application of aPDT resulted in effective gingival inflammation and pocket depth reduction over a period of 6 months. The group also suggested that the procedure be repeated at frequent intervals [96].

Monzavi et al. tried to test the efficacy of adjunctive aPDT with ICG compared with scaling and root planing (SRP) alone in chronic periodontitis treatment (◘ Fig. 14.5). The PDT group yielded higher improvements in bleeding on probing (BOP) and full-mouth bleeding score (FMBS) rather than the control group after 1- and 3-month follow-up examinations. After 3 months, the patients received PDT which showed 0% of BOP score, while the control group displays 48% of BOP-positive [97]. Boehm and Ciancio found rapid and significant uptake of ICG into periodontal pathogens that is activated by 810-nm diode laser, resulted in significant killing of *A. actinomycetemcomitans* and *Porphyromonas gingivalis* [98]. Nagahara et al. during an in vitro study also has shown the effectiveness of using ICG with an 810-nm diode laser can be considered as a promising candidate for adjunctive periodontal treatment [99].

The effect of ICG is mainly photothermal therapy rather than photochemical (80% photothermal and 20% photochemical). In addition, the peak absorption of ICG is close to available soft tissue diode lasers (808 nm), compared to the peak absorption of methylene blue and toluidine blue O which are at 660 nm and 635 nm, respectively. The higher penetration of 810-nm diode laser compared to other wavelengths with an easy insertion of the fiber-optic applicator allows an easier access into deep pockets. Besides, the photothermal effects of ICG accompanied by photochemical effects make this photosensitizer important for eradication of pathogens [100]. The implication is that treatment of deep periodontal pockets or non-reachable sites (i.e., furcation or invaginations) would be facilitated.

14.8.2 Clinical Cases

The clinical cases shown are examples that reflect the studies showing efficacy of aPDT [101, 102].

◘ Figure 14.16 illustrates the use of toluidine blue O for treatment of a Stage II grade A periodontal pocket. An LED source of photonic energy activates the chemical.

◘ Figure 14.17 depicts the use of indocyanine green for treatment of a Stage II grade A periodontal pocket. An 810-nm laser was used to activate the photosynthesizer.

◘ Figure 14.18 illustrates the use of methylene blue for the treatment of a Stage III grade B periodontal pocket. A proprietary visible red laser (632 nm) was used to activate the photosynthesizer (clinical case courtesy of Dr. Steven Parker).

◘ Figure 14.19 shows the application of antimicrobial photodynamic therapy for management of severe peri-implantitis in a 50-year-old woman. In this case, 7-mm pocket depth and bleeding on probing were observed. After manual debridement, the phenothiazine chloride as photosensitizer was applied in the implant sulcus. After 3 min, the excess photosensitizer was rinsed, and diode laser (670 nm) with output power of 75 mW was irradiated for 1 min in follow-up of 8 months, and pocket depth of 3 mm and absence of bleeding on probing were observed.

Angular bony defects at both mesial and distal parts were observed before treatment, but 8 and 19 months later, bone gains at both sides were detected, and 31 months later, maintenance of the bone level at mesial and distal aspects was approved (clinical case courtesy of Dr. Chen-Ying Wang).

◘ Figure 14.20 indicated the application of methylene blue as photosensitizer and 660-nm diode laser as an adjunct to scaling and root planing for a 45-year-old man. After 3 days following scaling and root planing, methylene blue at concentration of 100 µg/mL was applied inside the pockets. After 5 min, irradiation was done by laser light with photodynamic tip for 60 s in

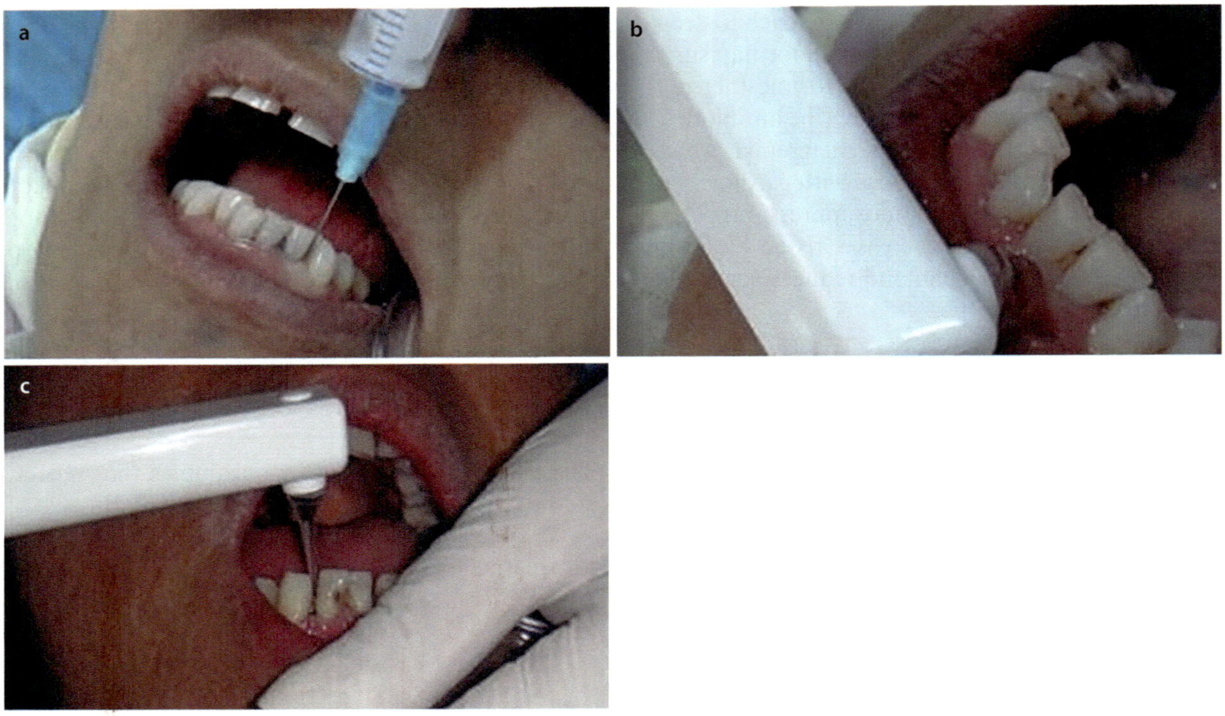

Fig. 14.16 **a** Application of TBO inside the pocket. **b** Irradiation of papilla by blunt tip of LED. **c** Irradiation of the pocket by intra-pocket tip of LED. (Clinical case courtesy of Dr. Chen Ying Wang)

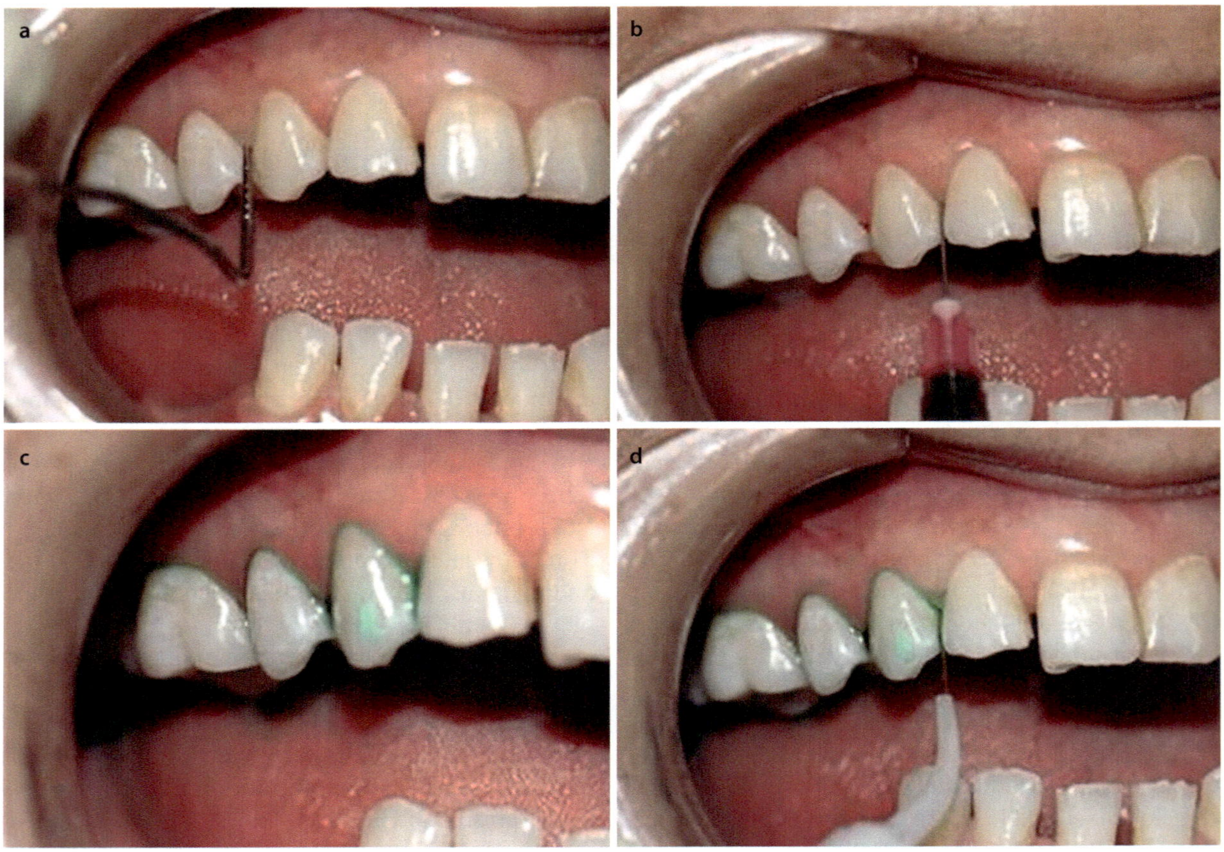

Fig. 14.17 **a** Clinical aspect of treatment site. **b** ICG was applied inside the pockets. **c** The view of treatment site after ICG application. **d** Irradiation of pockets by diode laser at wavelength of 808 nm

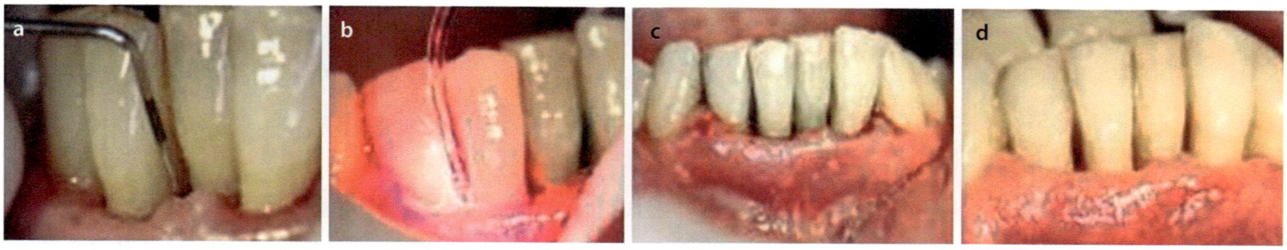

Fig. 14.18 **a** Facial view of the periodontitis condition of the lower anterior area. **b** After scaling of the pockets, the methylene blue solution is applied. A visible diode laser activates the photosensitizer. **c** Immediate posttreatment view of the site. **d** A 1-month posttreatment photo showing tissue health with no inflammation. (Clinical case courtesy of Dr. Steven Parker)

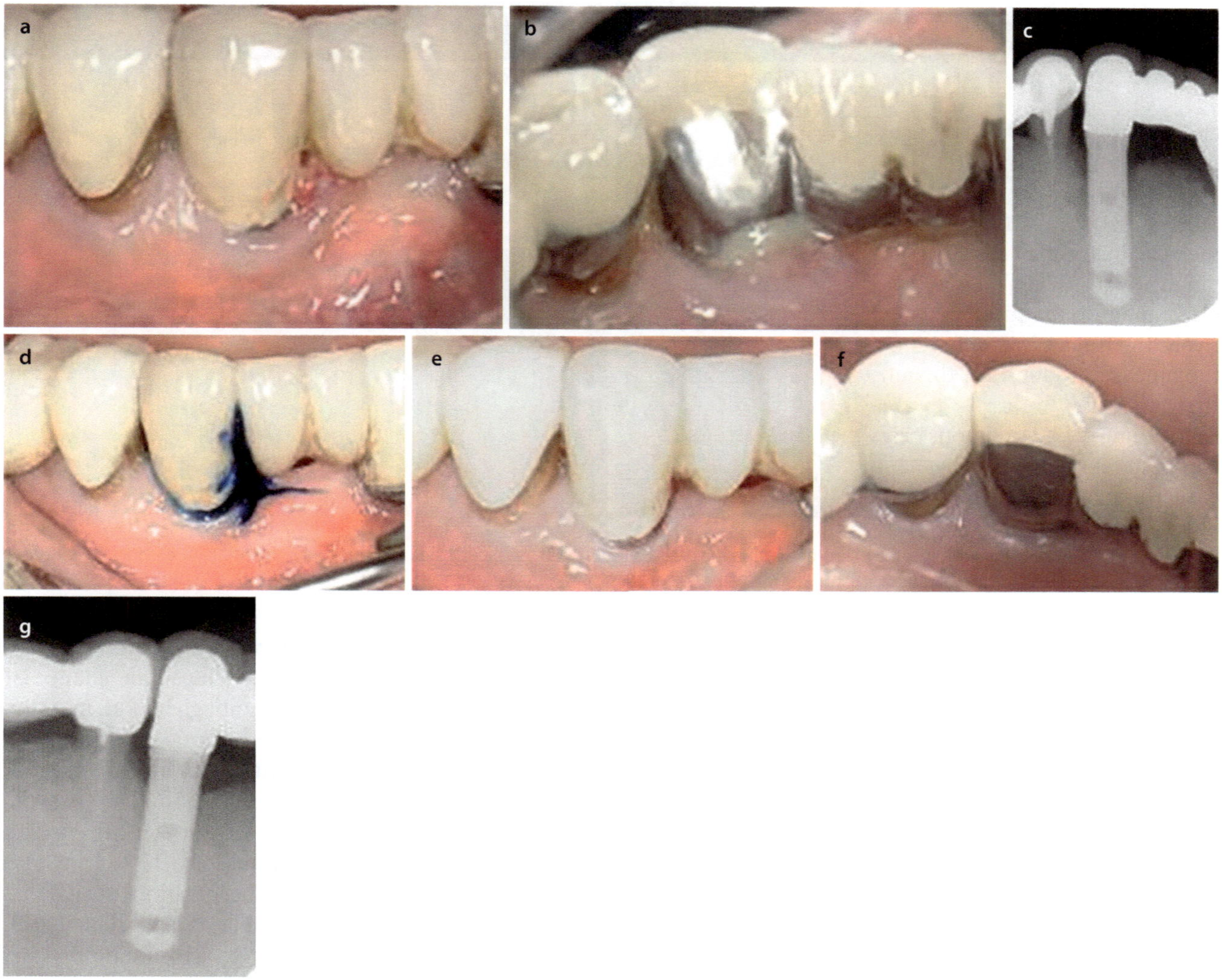

Fig. 14.19 **a** Facial view of peri-implantitis of the lower right cuspid. **b** The lingual view of the implant area with a discharge of exudates. **c** Preoperative radiograph showing the peri-implant defect. **d** The application of phenothiazine chloride inside the pocket. **e** Facial view 31 months later showing excellent peri-implant health. **f** Lingual view 31 months later of the healed area. **g** Radiograph of the treated implant, 31 months later. (Clinical case courtesy of Dr. Chen-Yeng Wang, modified from citation [14] with permission © copyright 2016 John Wiley and Sons A/S)

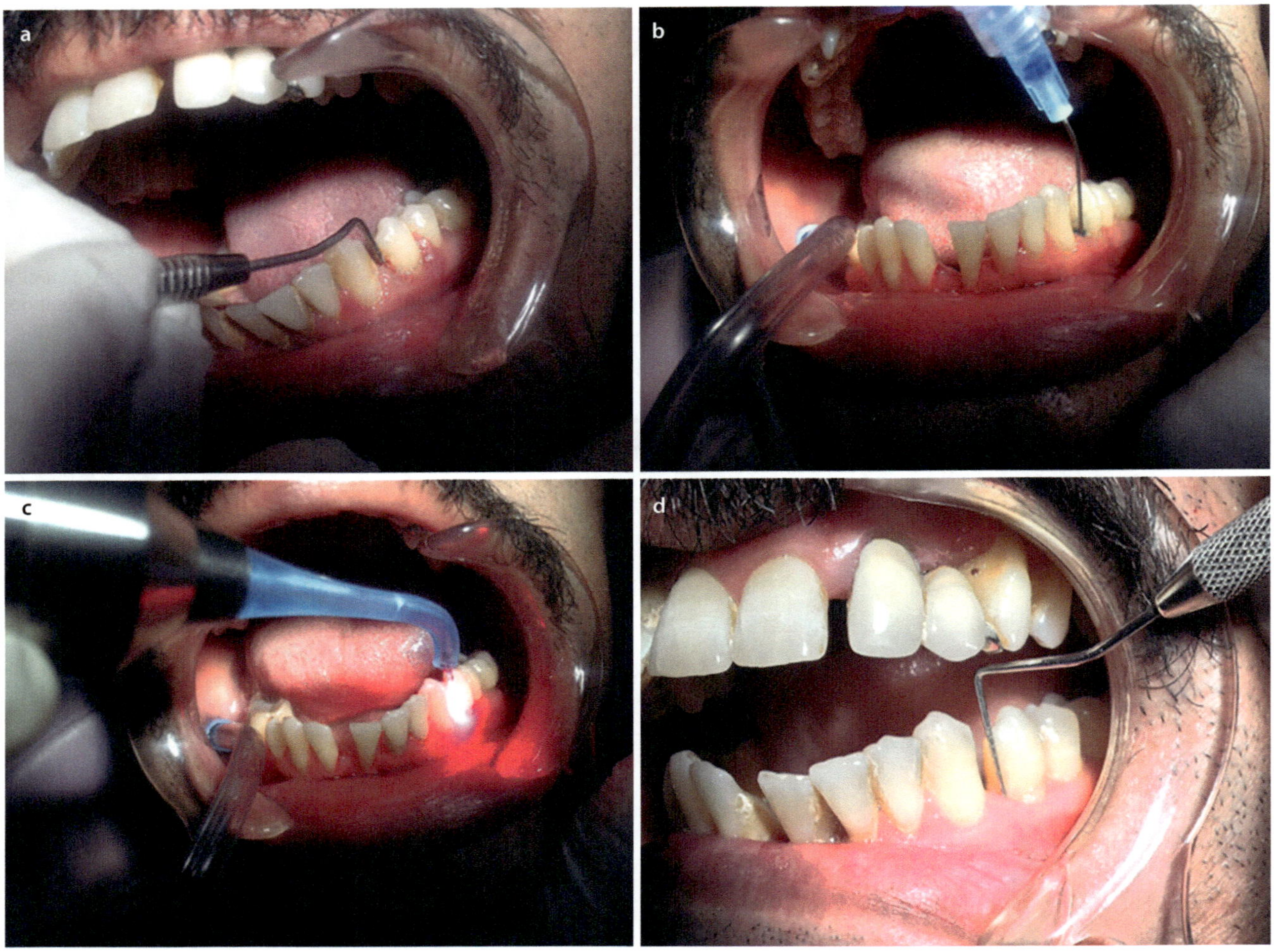

Fig. 14.20 **a** Clinical examination of treatment site. **b** Methylene blue was applied inside the pockets. **c** Irradiation of pockets by diode laser at wavelength of 660 nm. **d** Clinical examination after 6 months

each pocket moving apico-coronally both buccal and lingual parts of the tooth. The patient was followed after 6 weeks and 3 and 6 months.

14.9 Considerations During aPDT Therapy

The effective treatment of periodontal problems must include proper oral hygiene instructions, which consist of a combination of daily tooth brushing, interdental cleaning, and, when necessary, use of chemotherapeutic agents [e.g., mouthwash]. Thus, the patient's compliance with those instructions is fundamental for the success of the treatment of periodontal and peri-implant disease.

A low-oxygenated environment and the diffusion ability of the photosensitizer and light to be used are limitations of photodynamic therapy, which should be taken into consideration.

Moreover, the application of some photosynthesizers like methylene blue (MB) can stain the teeth if the appropriate concentration is not used. It was suggested that MB in concentrations below 100 μg/mL, activated by 660-nm diode laser for 60 s, reduces the chance of tooth discoloration [103]. Pourhajibagher et al. in evaluation of antimicrobial photodynamic therapy with indocyanine green and curcumin on human gingival fibroblast cells came to this conclusion that to avoid cytotoxicity the concentration of the sensitizers and laser irradiation time are essential for aPDT. They also observed that the optimum concentration of ICG as photosensitizer for aPDT should be at least 1000 μg/mL with 30- or 60-s irradiation time by 810-nm diode laser [104].

In a review for assessing the efficacy of aPDT for the treatment of periodontitis and peri-implantitis, Chambrone et al. [105] stated that aPDT provided similar clinical improvements as conventional therapy in periodontal and peri-implant patients. In one randomized controlled clinical trial, Martins et al. found significant improvement in clinical parameters with aPDT's adjunctive use in open-flap debridement of periodontitis [106]. In another clinical trial, Segarra-Vidal et al.

found that adjunctive aPDT use with SRP produced a significant reduction of *Aa* pathogens [107]. As a fourth example, Xue showed evidence of additional clinical improvement in the maintenance of residual pockets with aPDT [108]. Other studies reviewing both clinical and experimental protocols demonstrate some positive effect of adjunctive aPDT [109–111].

On the other hand, Sanz et al. [112] suggested as a clinical guideline to not use adjunctive aPDT at wavelength ranges of either 660–670 nm or 800–900 nm in patients with periodontitis.

Moreover, other technologies such as multifunctional nanoparticles in development have been shown to have strong anti-biofilm activity against periodontitis-related pathogens via aPDT [113].

Sufficient evidence to replace systemic antibiotic therapy by PDT in Stage IV periodontitis is not available. Also, limited evidence exists to consider PDT as an alternative to local antibiotics for the treatment of peri-implantitis [114].

14.10 Conclusion

Clinicians continue to search for and learn about novel methods to aid in the treatment of periodontal and peri-implant diseases. Various benefits such as pocket depth reduction, gain of clinical attachment, and improved wound healing are reported in the scientific studies. However, the use of the laser for these therapies generates controversial discussion in the literature. Due to the heterogeneity of photosensitizers and wavelengths of light sources, it can be difficult to decide how to apply aPDT as an adjunctive treatment for management of periodontitis and peri-implantitis.

Nonetheless, adjunctive or alternative use of dental lasers, both in direct minimally invasive nonsurgical or surgical procedures and in photochemical activation, is becoming part of the practitioner's armamentarium.

References

1. Wu L, Zhang SQ, Zhao L, Ren ZH, Hu CY. Global, regional, and national burden of periodontitis from 1990 to 2019: results from the Global Burden of Disease study 2019. J Periodontol. 2022;93(10):1445–54. https://doi.org/10.1002/JPER.21-0469.
2. Shiloah J, Patters M. Repopulation of periodontal pockets by microbial pathogens in the absence of supportive therapy. J Periodontal. 1996;67(2):130–9. https://doi.org/10.1902/jop.1996.67.2.130.
3. Sgolastra F, Petrucci A, Severino M, Gatto R, Monaco A. Periodontitis, implant loss, and peri-implantitis. A meta-analysis. Clin Oral Implants Res. 2015;26:e8–e16. https://doi.org/10.1111/clr.12319.
4. Papapanou P, Sanz M, Buduneli N, Dietrich T, Feres M, Fine DH, Flemmig TF, Garcia R, Giannobile WV, Graziani F, Greenwell H, Herrera D, Kao RT, Kebschull M, Kinane DF, Kirkwood KL, Kocher T, Kornman KS, Kumar PS, Loos BG, Machtei E, Meng H, Mombelli A, Needleman I, Offenbacher S, Seymour GJ, Teles R, Tonetti MS. Periodontitis: consensus report of workgroup 2 of the 2017 world workshop on the classification of periodontal and peri-implant diseases and conditions. J Periodontol. 2018;89(Suppl 1):S173–82. https://doi.org/10.1002/JPER.17-0721.
5. Renvert S, Person GR, Pirih F, Camargo P. Peri-implant health, peri-implant mucositis, and peri-implantitis: case definitions and diagnostic considerations. J Periodontol. 2018;89(Suppl 1):S304–12. https://doi.org/10.1002/JPER.17-0588.
6. Cobb CM. Non-surgical pocket therapy: mechanical. Ann Periodontol. 1996;1(1):443–90. https://doi.org/10.1902/annals.1996.1.1.443.
7. Cobb CM. Lasers in periodontics: a review of the literature. J Periodontol. 2006;77(4):545–64. https://doi.org/10.1902/jop.2006.050417.
8. Cobb C, Sottosanti J. A re-evaluation of scaling and root planing. J Periodontol. 2021;92:1370–8. https://doi.org/10.1002/JPER.20-0839.
9. Laky M, Laky B, Arslan M, Lettner S, Müller M, Haririan H, Husejnagic S, Rausch-Fan X, Wimmer G, Moritz A, Re Gregor R. Effectiveness of a 655-nm InGaAsP diode laser to detect subgingival calculus in patients with periodontal disease. J Periodontol. 2021;92:547–52. https://doi.org/10.1002/JPER.19-0663.
10. Kepic TJ, O'Leary TJ, Kafrawy AH. Total calculus removal: an attainable objective? J Periodontol. 1990;61(1):16–20. https://doi.org/10.1902/jop.1990.61.1.16.
11. Rabbani GM, Ash MM Jr, Caffesse RG. The effectiveness of subgingival scaling and root planing in calculus removal. J Periodontol. 1981;52(3):119–23. https://doi.org/10.1902/jop.1981.52.3.119.
12. Sherman PR, Hutchens LH Jr, Jewson LG. The effectiveness of subgingival scaling and root planing. II. Clinical responses related to residual calculus. J Periodontol. 1990;61(1):9–15. https://doi.org/10.1902/jop.1990.61.1.9.
13. Schenk G, Flemmig TF, Lob S, Ruckdeschel G, Hickel R. Lack of antimicrobial effect on periodontopathic bacteria by ultrasonic and sonic scalers in vitro. J Clin Periodontol. 2000;27(2):116–9. https://doi.org/10.1034/j.1600-051x.2000.027002116.x.
14. Mizutani K, Aoki A, Coluzzi D, Yukna R, Wang CY, Pavlic V, Izumi Y. Lasers in minimally invasive periodontal and peri-implant therapy. Periodontol 2000. 2016;71:185–212. https://doi.org/10.1111/prd.12123.
15. Low S, Mott A. Laser technology to manage periodontal disease: a valid concept? J Evid Based Dent Pract. 2014;14(Suppl):154–9. https://doi.org/10.1016/j/jebdp.2014.03.0.
16. Schwarz F, Aoki A, Sculean A, Becker J. The impact of laser application on periodontal and peri-implant wound healing. Periodontol 2000. 2009;51:79–108. https://doi.org/10.1111/j.1600-0757.2009.00301.x.
17. Ishikawa I, Aoki A, Takasaki AA, Mizutani K, Sasaki KM, Izumi Y. Application of lasers in periodontics: true innovation or myth? Periodontol 2000. 2009;50:90–126. https://doi.org/10.1111/j.1600-0757.2008.00283.x.
18. Coluzzi D, Anagnostaki E, Mylona V, Parker S, Lynch E. Do lasers have an adjunctive role in initial non-surgical periodon-

tal therapy? A systematic review. Dent J. 2020;8(3):93. https://doi.org/10.3390/dj8030093.
19. Russell AD. Lethal effects of heat on bacterial physiology and structure. Sci Prog. 2003;86(part 1–2):115–37. https://doi.org/10.3184/003685003783238699.
20. Knappe V, Frank F, Rohde E. Principles of lasers and biophotonic effects. Photomed Laser Surg. 2004;22(5):411–7. https://doi.org/10.1089/pho.2004.22.411.
21. Aoki A, Sasaki K, Watanabe H, Ishikawa I. Lasers in non surgical periodontal therapy. Periodontol 2000. 2004;36:59–7. https://doi.org/10.1111/j.1600-0757.2004.03679.x.
22. Nammour S, El Mobadder M, Maalouf E, Namour M, Namour A, Rey G, Matamba P, Matys J, Zeinoun T, Grzech-Lesniak K. Clinical evaluation of diode (980nm) laser assisted nonsurgical periodontal pocket therapy: a randomized comparative clinical trial and bacteriological study. Photobiomodul Photomed Laser Surg. 2021;39(1):10–22. https://doi.org/10.1089/photob.2020.4818.
23. Coluzzi DJ, Convissar RA. Laser periodontal therapy. In: Atlas of laser applications in dentistry. Hanover Park, IL: Quintessence Publishing Co, Inc; 2007. p. 25–9. ISBN-13: 978-0867154764.
24. Ciurescu C, Vanweersch L, Franzen R, Gutknecht N. The antibacterial effect of the combined Er,Cr:YSGG and 940 nm diode laser therapy in treatment of periodontitis: a pilot study. Laser Dent Sci. 2018;2:43–51. https://doi.org/10.1007/s41547-017-0018-8.
25. Renvert S, Polyzois I. Treatment of pathologic peri-implant pockets. Periodontol 2000. 2018;76(1):180–90. https://doi.org/10.1111/prd.12149.
26. Coluzzi DJ. Fundamentals of dental lasers: science and instruments. Dent Clin N Am. 2004;48(4):751–70. https://doi.org/10.1016/j.cden.2004.05.003.
27. Aoki A, Ando Y, Watanabe H, Ishikawa I. In vitro studies on laser scaling of subgingival calculus with an erbium:YAG laser. J Periodontol. 1994;65(12):1097–106. https://doi.org/10.1902/jop.1994.65.12.1097.
28. Alfergany MA, Nasher R, Gutknecht N. Calculus removal and root surface roughness when using the Er:YAG or Er,Cr:YSGG laser compared with conventional instrumentation method: a literature review. Photobiomodul Photomed Laser Surg. 2019;37(4):197–226. https://doi.org/10.1089/photob.2018.4465.
29. Kreisler M, Gotz H, Duschner H. Effect of Nd:YAG, Ho:YAG, Er:YAG, CO_2, and GaAIAs laser irradiation on surface properties of endosseous dental implants. Int J Oral Maxillofac Implants. 2002;17(2):202–11. PMID: 11958402
30. Chala M, Anagnostaki E, Mylona V, Chalas A, Parker S, Lynch E. Adjunctive use of lasers in peri-implant mucositis and peri-implantitis treatment: a systematic review. Dent J. 2020;8:68. https://doi.org/10.3390/dj8030068.
31. Chambrone L, Ramos U, Reynolds M. Infrared lasers for the treatment of moderate to severe periodontitis: an American Academy of Periodontology best evidence review. J Periodontol. 2018;89:743–65. https://doi.org/10.1902/jop.2017.160504.
32. Celik TZ, Saglam E, Ercan C, Akbas F, Nazaroglu K, Tunali M. Clinical and microbiological effects of the use of erbium:yttrium–aluminum–garnet laser on chronic periodontitis in addition to nonsurgical periodontal treatment: a randomized clinical trial—6 months follow-up. Photobiomodul Photomed Laser Surg. 2019;37(3):182–90. https://doi.org/10.1089/photob.2018.4510.
33. Ciurescu CE, Cosgarea R, Ciurescu D, Gheorghiu A, Popa D, Franzen R, Arweiler NB, Sculean A, Gutknecht N. Adjunctive use of InGaAsP and Er,Cr:YSGG lasers in nonsurgical periodontal therapy: a randomized controlled clinical study. Quintessence Int. 2019;50(6):436–47. https://doi.org/10.3290/j.qi.a42508.
34. Tenore G, Montori A, Mattarelli G, Palaia G, Romeo U. Evaluation of adjunctive efficacy of diode laser in the treatment of peri-implant mucositis: a randomized clinical trial. Lasers Med Sci. 2020;35(6):1411–7. https://doi.org/10.1007/s10103-020-03009-y.
35. Estrin NE, Moraschini V, Zhang Y, Romanos GE, Sculean A, Miron RJ. Combination of Nd:YAG and Er:YAG lasers in non-surgical periodontal therapy: a systematic review of randomized clinical studies. Lasers Med Sci. 2022;37(6):2737–43. https://doi.org/10.1007/s10103-022-03548-6.
36. Yu S, Zhao X, Zhang Y, Liu Y, Li A, Pei D. Clinical effectiveness of adjunctive diode laser on scaling and root planing in the treatment of periodontitis: is there an optimal combination of usage mode and application regimen? A systematic review and meta-analysis. Lasers Med Sci. 2022;37:759–69. https://doi.org/10.1007/s10103-021-03412-z.
37. Gholami L, Asefi S, Hooshyarfard A, Sculean A, Romanos GE, Aoki A, Fekrazad R. Photobiomodulation in periodontology and implant dentistry: Part 1. Photobiomodul Photomed Laser Surg. 2019;37(12):739–65. https://doi.org/10.1089/photob.2019.4710.
38. Gholami L, Asefi S, Hooshyarfard A, Sculean A, Romanos GE, Aoki A, Fekrazad R. Photobiomodulation in periodontology and implant dentistry: Part 2. Photobiomodul Photomed Laser Surg. 2019;37(12):766–83. https://doi.org/10.1089/photob.2019.4731.
39. Aoki A, Mizutani SF, Sculean A, Yukna RA, Takasaki AA, Romanos GE, Taniguchi Y, Sasaki KM, Zeredo JL, Koshy G, Coluzzi DJ, White JM, Abiko Y, Ishikawa I, Izumi Y. Periodontal and peri-implant wound healing following laser therapy. Periodontology 2000. 2015;68:217–69. https://doi.org/10.1111/prd.12080.
40. Aoki A, Mizutani K, Mikami R, Taniguchi Y, Ohsugi Y, Meinzer W, Izumi Y, Iwata T. Residual periodontal pocket treatment with Er:YAG laser-assisted comprehensive periodontal pocket therapy: a retrospective study. Clin Oral Investig. 2022;26:761–71.
41. Yukna RA, Carr RI, Evans GH. Histologic evaluation of an Nd:YAG laser-assisted new attachment procedure in humans. Int J Periodontics Restorative Dent. 2007;27(6):577–87. PMID: 18092452.
42. Nevins M, Kim SW, Camelo M, Martin IS, Kim D. A prospective 9-month human clinical evaluation of Laser-Assisted New Attachment Procedure (LANAP) therapy. Int J Periodontics Restorative Dent. 2014;34(1):21–7. https://doi.org/10.11607/prd.1848.
43. Kang HW, Oh J, Welch AJ. Investigations on laser hard tissue ablation under various environments. Phys Med Biol. 2008;53(12):3381–90. https://doi.org/10.1088/0031-9155/53/12/021.
44. Kimura K, Yu DG, Fujita A, Yamashita A, Murakami Y, Matsumoto K. Effects of erbium, chromium:YSGG laser irradiation on canine mandibular bone. J Periodontol. 2001;72(9):1178–82. https://doi.org/10.1902/jop.2000.72.9.1178.
45. Lewandrowski KU, Lorente C, Schomacker KT. Use of the Er:YAG laser for improved plating in maxillofacial surgery: comparison of bone healing in laser and drill osteotomies. Lasers Surg Med. 1996;19(1):40–5. https://doi.org/10.1002/(SICI)1096-9101(1996)19:1.
46. Pourzarandian A, Watanabe H, Aoki A, Ichinose S, Sasaki KM, Nitta H, Ishikawa I. Histological and TEM examination

of early stages of bone healing after Er:YAG laser irradiation. Photomed Laser Surg. 2004;22(4):355–63. https://doi.org/10.1089/pho.2004.22.342.
47. Mizutani K, Aoki A, Takasaki AA, Kinoshita A, Hayashi C, Oda S, Ishikawa I. Periodontal tissue healing following flap surgery using an Er:YAG laser in dogs. Lasers Surg Med. 2006;38(4):314–24. https://doi.org/10.1002/lsm.20299.
48. Tosun E, Tasar F, Strauss R, Kivanc DG, Ungor C. Comparative evaluation of antimicrobial effects of Er:YAG, diode, and CO_2 lasers on titanium discs: an experimental study. J Oral Maxillofac Surg. 2012;70(5):1064–9. https://doi.org/10.1016/j.joms.2011.11.021.
49. Romanos GE, Everts H, Nentwig GH. Effects of diode and Nd:YAG laser irradiation on titanium discs: a scanning electron microscope examination. J Periodontol. 2000;71(5):810–5. https://doi.org/10.1902/jop.2000.71.5.810.
50. Stubinger S, Etter C, Miskiewicz M, Homann F, Saldamli B, Wieland M, Sader R. Surface alterations of polished and sandblasted and acid-etched titanium implants after Er:YAG, carbon dioxide, and diode laser irradiation. Int J Oral Maxillofac Implants. 2010;25(1):104–11. PMID: 20209192.
51. Polak D, et al. The in vitro efficacy of biofilm removal from titanium surfaces using Er:YAG laser: comparison of treatment protocols and ablation parameters. J Periodontol. 2022;93:100–9. https://doi.org/10.1002/JPER.19-0574.
52. Clem D, Shani-Kdoshim S, Alias M, Shapira L, Stabholz A. Comparison of Er,Cr:YSGG laser to minimally invasive surgical technique in the treatment of intrabony defects: six-month results of a multicenter, randomized, controlled study. J Periodontol. 2021;92:496–506. https://doi.org/10.1002/JPER.20-0028.
53. Wang CW, Ashnagar S, Gianfilippo RD, Arnett M, Kinney J, Wang HL. Laser-assisted regenerative surgical therapy for peri-implantitis: a randomized controlled clinical trial. J Periodontol. 2021;92:378–88.
54. Lin T. Clinical evaluation of multiple peri-implant bony defect management by Er:YAG laser-assisted bone regenerative therapy. J Dent Sci. 2019;14:430–2. https://doi.org/10.1016/j.jds.2019.07.001.
55. Soukos NS, Goodson JM. Photodynamic therapy in the control of oral biofilms. Periodontol 2000. 2011;55(1):143–66. https://doi.org/10.1111/j.1600-0757.2010.00346.x.
56. Pandey RK, Zheng G. Porphyrins as photosensitizers in photodynamic therapy. In: Kadish KM, Smith KM, Guilard R, editors. The porphyrin handbook. Boston: Academic; 2000. p. 157–230. https://doi.org/10.1016/C2009-0-22719-X.
57. Parker S. The use of diffuse laser photonic energy and indocyanine green photosensitizer as an adjunct to periodontal therapy. Br Dent J. 2013;215(4):167–71. https://doi.org/10.1038/sj.bdj.2013.790.
58. Rajesh S, Koshi E, Philip K, Mohan A. Antimicrobial photodynamic therapy: an overview. J Indian Soc Periodontol. 2011;15(4):323–7. https://doi.org/10.4103/0972-124X.92563.
59. Konopka K, Goslinski T. Photodynamic therapy in dentistry. J Dent Res. 2007;86(8):694–707. https://doi.org/10.1177/154405910708600803.
60. Chiniforush N, Pourhajibagher M, Shahabi S, Bahador A. Clinical approach of high technology techniques for control and elimination of endodontic microbiota. J Lasers Med Sci. 2015;6(4):139–50. https://doi.org/10.15171/jlms.2015.09.
61. Seghatchian J, Struff WG, Reichenberg S. Main properties of the THERAFLEX MB-plasma system for pathogen reduction. Transfus Med Hemother. 2011;38(1):55–64. https://doi.org/10.1159/000323786.
62. Sridharan G, Shankar AA. Toluidine blue: a review of its chemistry and clinical utility. J Oral Maxillofac Pathol. 2012;16(2):251–5. https://doi.org/10.4103/0973-029X.99081.
63. Nikaido H. Molecular basis of bacterial outer membrane permeability revisited. Microbiol Mol Biol Rev. 2003;67(4):593–656. https://doi.org/10.1128/MMBR.67.4.593-656.2003.
64. Wainwright M, Giddens RM. Phenothiazinium photosensitisers: choices in synthesis and application. Dyes Pigments. 2003;57:245–57. https://doi.org/10.1002/chin.200343276.
65. Prento P. A contribution to the theory of biological staining based on the principles for structural organization of biological macromolecules. Biotech Histochem. 2001;76:137–61. PMID: 11475317.
66. George S, Hamblin MR, Kishen A. Uptake pathways of anionic and cationic photosensitizers into bacteria. Photochem Photobiol Sci. 2009;8(6):788–95. https://doi.org/10.1039/b809624d.
67. Aggarwal BB, Sundaram C, Malani N, Ichikawa H. Curcumin: the Indian solid gold. Adv Exp Med Biol. 2007;595:1–75. https://doi.org/10.1007/978-0-387-46401-5_1.
68. Dovigo LN, Pavarina AC, Ribeiro AP, Brunetti IL, Costa CA, Jacomassi DP, Bagnato VS, Kurachi C. Investigation of the photodynamic effects of curcumin against Candida albicans. Photochem Photobiol. 2011;87(4):895–903. https://doi.org/10.1111/j.1751-1097.2011.00937.x.
69. Araújo NC, Fontana CR, Bagnato VS, Gerbi ME. Photodynamic antimicrobial therapy of curcumin in biofilms and carious dentine. Lasers Med Sci. 2014;29(2):629–35. https://doi.org/10.1007/s10103-013-1369-3.
70. Afrasiabi S, Partoazar A, Chiniforush N, Goudarzi R. The potential application of natural photosensitizers used in antimicrobial photodynamic therapy against oral infections. Pharmaceuticals (Basel). 2022;15(6):767. https://doi.org/10.3390/ph15060767. PMID: 35745686; PMCID: PMC9227410.
71. Afrasiabi S, Barikani HR, Chiniforush N. Comparison of bacterial disinfection efficacy using blue and red lights on dental implants contaminated with Aggregatibacter actinomycetemcomitans. Photodiagn Photodyn Ther. 2022;40:103178. https://doi.org/10.1016/j.pdpdt.2022.103178. Epub 2022 Oct 29. PMID: 36602065.
72. Chiniforush N, Pourhajibagher M, Parker S, Benedicenti S, Bahador A, Sălăgean T, Bordea IR. The effect of antimicrobial photodynamic therapy using chlorophyllin–phycocyanin mixture on Enterococcus faecalis: the influence of different light sources. Appl Sci. 2020;10:4290. https://doi.org/10.3390/app10124290.
73. Etemadi A, Hamidain M, Parker S, Chiniforush N. Blue light photodynamic therapy with curcumin and riboflavin in the management of periodontitis: a systematic review. J Lasers Med Sci. 2021;12:e15. https://doi.org/10.34172/jlms.2021.15. PMID: 34733738; PMCID: PMC8558718.
74. Salva KA. Photodynamic therapy: unapproved uses, dosages or indications. Clin Dermatol. 2002;20(5):571–81. https://doi.org/10.1016/s0738-081x(02)00266-3.
75. Biel MA. Photodynamic therapy in head and neck cancer. Curr Oncol Rep. 2002;4:87–96. https://doi.org/10.1007/s11912-002-0053-8.
76. Grant WE, Hopper C, Speight PM, Bown SG. Photodynamic therapy, an effective, but non selective treatment for superficial cancers of the oral cavity. Int J Cancer. 1997;71:937–42. https://doi.org/10.1002/(sici)1097-0215(19970611)71:6.
77. Takasaki AA, Aoki A, Mizutani K, Schwarz F, Sculean A, Wang C-Y, Koshy G, Romanos G, Ishikawa I, Izumi Y. Application of

antimicrobial photodynamic therapy in periodontal and peri-implant diseases. Periodontol 2000. 2009;51:109–40. https://doi.org/10.1111/j.1600-0757.2009.00302.x.
78. Ochsner M. Photodynamic therapy in squamous cell carcinoma. J Photochem Photobial B. 2001;52:42–8.
79. Sharman WM, Allen CM, Van Lier JE. Photodynamic therapeutics: basic principles and clinical applications. Drug Discov Today. 1999;4(11):507–17. https://doi.org/10.1016/s1359-6446(99)01412-9.
80. Moan J, Berg K. The photodegradation of porphyrins in cells can be used to estimate the lifetime of singlet oxygen. Photochem Photobiol B. 1991;53:549–53. https://doi.org/10.1111/j.1751-1097.1991.tb03669.x.
81. Cieplik F, Tabenski L, Buchalla W, Maisch T. Antimicrobial photodynamic therapy for inactivation of biofilms formed by oral key pathogens. Front Microbiol. 2014;5:405. https://doi.org/10.3389/fmicb.2014.00405.
82. Ochsner M. Photophysical and photobiological processes in the photodynamic therapy of tumors. J Photochem Photobiol B. 1997;39(1):1–18. https://doi.org/10.1016/s1011-1344(96)07428-3.
83. Kishen A, Shi Z, Shrestha A, Neoh KG. An investigation on the antibacterial and antibiofilm efficacy of cationic nanoparticulates for root canal disinfection. J Endod. 2008;34(12):1515–20. https://doi.org/10.1016/j.joen.2008.08.035.
84. Kharkwal GB, Sharma SK, Huang YY, Dai T, Hamblin MR. Photodynamic therapy for infections: clinical applications. Lasers Surg Med. 2011;43(7):755–67. https://doi.org/10.1002/lsm.21080.
85. Minnock A, Vernon DI, Schofield J, Griffiths J, Parish JH, Brown SB. Mechanism of uptake of a cationic water-soluble pyridinium zinc phthalocyanine across the outer membrane of Escherichia coli. Antimicrob Agents Chemother. 2000;44(3):522–7. https://doi.org/10.1128/AAC.44.3.522-527.2000.
86. de Melo WC, Avci P, de Oliveira MN, Gupta A, Vecchio D, Sadasivam M, Chandran R, Huang Y-Y, Yin R, Perussi LR, Tegos GP, Perussi JR, Dai T, Hamblin MR. Photodynamic inactivation of biofilm: taking a lightly colored approach to stubborn infection. Expert Rev Anti Infect Ther. 2013;11(7):669–93. https://doi.org/10.1586/14787210.2013.811861.
87. Chrepa V, Kotsakis GA, Pagonis TC, Hargreaves KM. The effect of photodynamic therapy in root canal disinfection: a systematic review. J Endod. 2014;40:891–8. https://doi.org/10.1016/j.joen.2014.03.005.
88. Saxena S, Bhatia G, Garg B, Rajwar YC. Role of photodynamic therapy in periodontitis. Asian Pac J Health Sci. 2014;1(3):200–6.
89. Moslemi N, Soleiman-Zadeh Azar P, Bahador A, Rouzmeh N, Chiniforush N, Paknejad M, Fekrazad R. Inactivation of Aggregatibacter actinomycetemcomitans by two different modalities of photodynamic therapy using Toluidine blue O or Radachlorin as photosensitizers: an in vitro study. Lasers Med Sci. 2015;30(1):89–94. https://doi.org/10.1007/s10103-014-1621-5.
90. Cadore UB, Reis MBL, Martins SHL, Invernici MM, Novaes AB Jr, Mario Taba M, Palioto DB, Messora MR, Souza SLS. Multiple sessions of antimicrobial photodynamic therapy associated with surgical periodontal treatment in patients with chronic periodontitis. J Periodontol. 2019;90:339–49. https://doi.org/10.1002/JPER.18-0373.
91. Tribble GD, Lamont RJ. Bacterial invasion of epithelial cells and spreading in periodontal tissue. Periodontol 2000. 2010;52(1):68–83. https://doi.org/10.1111/j.1600-0757.2009.00323.x.
92. Soukos NS, Ximenez-Fyvie LA, Hamblin MR, Socranski SS, Hasan T. Targeted antimicrobial photochemotherapy. Antimicrob Agents Chemother. 1998;42(10):2595–601. https://doi.org/10.1128/AAC.42.10.2595.
93. Raghavendra M, Koregol A, Bhola S. Photodynamic therapy: a targeted therapy in periodontics. Aust Dent J. 2009;54(Suppl 1):S102–9. https://doi.org/10.1111/j.1834-7819.2009.01148.x.
94. Giannelli M, Pini A, Formigli L, Bani D. Comparative in vitro study among the effects of different laser and LED irradiation protocols and conventional chlorhexidine treatment for deactivation of bacterial lipopolysaccharide adherent to titanium surface. Photomed Laser Surg. 2011;29(8):573–80. https://doi.org/10.1089/pho.2010.2958.
95. Bassir SH, Moslemi N, Jamali R, Mashmouly S, Fekrazad R, Chiniforush N, Shamshiri AR, Nowzari H. Photoactivated disinfection using light-emitting diode as an adjunct in the management of chronic periodontitis: a pilot double-blind split-mouth randomized clinical trial. J Clin Periodontol. 2013;40(1):65–72. https://doi.org/10.1111/jcpe.12024.
96. Prasanth CS, Karunakaran SC, Paul AK, Kussovski V, Mantareva V, Ramaiah D, Selvaraj L, Angelov I, Avramov L, Nandakumar K, Subhash N. Antimicrobial photodynamic efficiency of novel cationic porphyrins towards periodontal Gram-positive and Gram-negative pathogenic bacteria. Photochem Photobiol. 2014;90(3):628–40. https://doi.org/10.1111/php.12198.
97. Monzavi A, Chinipardaz Z, Mousavi M, Fekrazad R, Moslemi N, Azaripour A, Bagherpasand O, Chiniforush N. Antimicrobial photodynamic therapy using diode laser activated indocyanine green as an adjunct in the treatment of chronic periodontitis: a randomized clinical trial. Photodiagn Photodyn Ther. 2016;14:93–7. https://doi.org/10.1016/j.pdpdt.2016.02.007.
98. Boehm TK, Ciancio SG. Diode laser activated indocyanine green selectively kills bacteria. J Int Acad Periodontol. 2011;13(2):58–63. PMID: 21913603.
99. Nagahara A, Mitani A, Fukuda M, Yamamoto H, Tahara K, Morita I, Ting CC, Watanabe T, Fujimura T, Osawa K, Sato S. Antimicrobial photodynamic therapy using a diode laser with a potential new photosensitized, indocyanine green-loaded nanospheres, may be effective for the clearance of Porphyromonas gingivalis. J Periodontal Res. 2013;48(5):591–9. https://doi.org/10.1111/jre.12042.
100. Chiniforush N, Pourhajibagher M, Shahabi S, Kosarieh E, Bahador A. Can antimicrobial photodynamic therapy (aPDT) enhance the endodontic treatment? J Lasers Med Sci. 2016;7(2):76–85. https://doi.org/10.15171/jlms.2016.14.
101. Chiniforush N, Coluzzi D, Aoki A. Antimicrobial photodynamic therapy for periodontal and peri-implant disease. Clin Dent Rev. 2022;6:9. https://doi.org/10.1007/s41894-022-00123-8.
102. Derikvand N, Ghasemi SS, Safiaghdam H, Piriaei H, Chiniforush N. Antimicrobial photodynamic therapy with diode laser and methylene blue as an adjunct to scaling and root planing: a clinical trial. Photodiagn Photodyn Ther. 2020;31:101818. https://doi.org/10.1016/j.pdpdt.2020.101818.
103. George S, Kishen A. Photophysical, photochemical, and photobiological characterization of methylene blue formulations for light-activated root canal disinfection. J Biomed Opt. 2007;12(3):034029. https://doi.org/10.1117/1.2745982.

104. Pourhajibagher M, Chiniforush N, Parker S, Shahabi S, Ghorbanzadeh R, Kharazifard MJ, Bahador A. Evaluation of antimicrobial photodynamic therapy with indocyanine green and curcumin on human gingival fibroblast cells: an in vitro photocytotoxicity investigation. Photodiagn Photodyn Ther. 2016;15(9):13–8. https://doi.org/10.1016/j.pdpdt.2016.05.003.

105. Chambrone L, Wang HL, Romanos GE. Antimicrobial photodynamic therapy for the treatment of periodontitis and peri-implantitis: an American Academy of Periodontology best evidence review. J Periodontol. 2018;89(7):783–803. https://doi.org/10.1902/jop.2017.170172.

106. Martins SHL, Novaes AB Jr, Taba M Jr, Palioto DB, Messora MR, Reino DM, Sérgio LS, Souza SLS. Effect of surgical periodontal treatment associated to antimicrobial photodynamic therapy on chronic periodontitis. A randomized controlled clinical trial. J Clin Periodontol. 2017;44:717–28. https://doi.org/10.1111/jcpe.12744.

107. Segarra-Vidal M, Guerra-Ojeda S, Vallés LS, López-Roldán A, Mauricio MD, Aldasoro M, Alpiste-Illueca F, Vila JM. Effects of photodynamic therapy in periodontal treatment: a randomized, controlled clinical trial. J Clin Periodontol. 2017;44:915–25. https://doi.org/10.1111/jcpe.12768.

108. Xue D, Zhao Y. Clinical effectiveness of adjunctive antimicrobial photodynamic therapy for residual pockets during supportive periodontal therapy: a systematic review and meta-analysis. Photodiagn Photodyn Ther. 2017;17:127–33. https://doi.org/10.1016/j.pdpdt.2016.11.011.

109. Peron D, Bergamo A, Prates R, Vieira SS, de Tarso Camillo de Carvalho P, Serra AJ. Photodynamic antimicrobial chemotherapy has an overt killing effect on periodontal pathogens? A systematic review of experimental studies. Lasers Med Sci. 2019;34:1527–34. https://doi.org/10.1007/s10103-019-02806.

110. Husenjnagic S, Lettner S, Laky M, Georgopoulos A, Moritz A, Rausch-Fan X. Photoactivated disinfection in periodontal treatment: a randomized controlled clinical split-mouth trial. J Periodontol. 2019;90:1260–9. https://doi.org/10.1002/JPER.18-0576.

111. Salvi GE, Stähli A, Schmidt JC, Ramseier CA, Sculean A, Walter C. Adjunctive laser or antimicrobial photodynamic therapy to nonsurgical mechanical instrumentation in patients with untreated periodontitis: a systematic review and meta-analysis. J Clin Periodontol. 2020;47:176–98. https://doi.org/10.1111/jcpe.13236.

112. Sanz M, Herrera D, Kebschull M, Chapple I, Jepsen S, Beglundh T, Sculean A, Tonetti MS. Treatment of stage I–III periodontitis—the EFP S3 level clinical practice guideline [published correction appears in J Clin Periodontol. 2021 Jan;48(1):163]. J Clin Periodontol. 2020;47(Suppl 22):4–60. https://doi.org/10.1111/jcpe.13290.

113. Sun X, Wang L, Lynch CD, Sun X, Li X, Qi M, Ma C, Li C, Dong B, Zhou Y, Xu HHK. Nanoparticles having amphilphilic silane containing Chlorin e6 with strong anti-biofilm activity against periodontitis-related pathogens. J Dent. 2019;81:70–84. https://doi.org/10.1016/j.jdent.2018.12.011.

114. Zhao Y, Pu R, Qian Y, Shi J, Si M. Antimicrobial photodynamic therapy versus antibiotics as an adjunct in the treatment of periodontitis and peri-implantitis. A systematic review and meta-analysis. Photodiagn Photodyn Ther. 2021;34:102231. https://doi.org/10.1016/j.pdpdt.2021.102231.

Laser-Assisted Multi-tissue Management During Aesthetic or Restorative Procedures

Donald J. Coluzzi, Mark Cronshaw, and Joshua Weintraub

Contents

D. J. Coluzzi, S. P. A. Parker (eds.), *Lasers in Dentistry—Current Concepts*, Textbooks in Contemporary Dentistry,
https://doi.org/10.1007/978-3-031-43338-2_15

Core Message

Whether treating new carious lesions or planning extensive prosthodontics, the dental clinician must consider how the restoration will harmonize with the periodontium so that the result will be both healthy and maintainable. Likewise, recontouring gingiva and the underlying supporting bone for improved aesthetics must entail the same deliberation. An important additional consideration is the patient's perception of the desired aesthetic result. Clearly, this can have a significant influence on the success of the planned treatment. Dental lasers can be used for modification of soft or hard supporting tissue. The appropriate wavelength and operating parameters must be chosen for the specific tissue, but the result can be very predictable and biologically compatible with the restoration.

15.1 Introduction

This chapter will describe the use of multiple laser wavelengths to alter and improve hard and soft dental tissues for improved aesthetics and successful restoration placement. While this chapter's intention is to not exhaustively review all aspects of periodontal surgery, fundamental concepts of soft tissue anatomy such as biotype and biologic width will be discussed. The basic principle is that after any alteration to soft or hard tissue, good physiologic contour must be restored. Predictable tissue management primarily relies on the clinician's choice of the proper wavelength to interact with the target tissue while using appropriate parameters and techniques to maximize the efficiency of tissue removal, establish proper contour, and minimize any collateral damage. ▶ Chapters 3 and 4 discuss these concepts in great detail.

15.2 Review of Laser Wavelengths and Tissue Interaction

15.2.1 Diode and Nd:YAG Lasers

These near-infrared wavelengths produce photonic energy that is generally scattered in soft tissue and is transmissive through water but will be absorbed by pigmented and/or inflamed areas. These lasers are for soft tissue only; they have virtually no interaction with healthy tooth structure and should not be used on the bone. They work well in well-vascularized tissue and provide excellent hemostasis. As noted previously, Nd:YAG instruments operate in a free-running pulsed mode, producing very high peak powers; some diode lasers also can operate with relatively short pulse durations and moderate peak power output. This pulsing modality can help to control collateral thermal damage.

15.2.2 Erbium Lasers

The two wavelengths of these mid-infrared instruments have the highest absorption by water compared to any other available lasers, with a smaller secondary interaction with the mineral components of hard tissue. These lasers employ a free-running pulsed emission, and their high peak power is primarily and rapidly absorbed into water. In soft tissue, this produces a shallow area of ablation. In hard tissue, there is superheating of the water content of the tooth or bone, resulting in an explosive expansion. This disrupts and ejects whole fragments of the calcified structure, resulting in a "cavity." The mineral remains unchanged. The term "all tissue laser" implies that an erbium instrument can perform soft tissue excisions, tooth preparation, and osseous procedures.

15.2.3 Carbon Dioxide Lasers

There are also two far-infrared wavelengths in this category. Both are highly absorbed by hard tissue and secondarily by water. Current developments in technology allow the 9300-nm machine to remove carious lesions and prepare teeth, contour the bone, and perform soft tissue surgery. The 10,600-nm device can only be used for soft tissue procedures in its present form.

15.2.4 With Any of the Above Wavelengths Employed for Soft Tissue Surgery, There Are Several Considerations

- Very fibrous gingival tissue surrounding chronic inflammation due to a margin discrepancy of a crown will be much more difficult to incise with the near-infrared wavelengths because of their photonic energy's preference for melanin or hemoglobin in the tissue. Moreover, these wavelengths can cause some conductive heat buildup in tissue distant to the surgical area with possible peripheral edema. Lasers with longer wavelengths, such as carbon dioxide, would be much better instruments for that type of tissue.

- Alternatively, acutely inflamed gingiva with its well-vascularized structure would be easily ablated by the same near-infrared lasers.
- A water spray can be used to control the tissue temperature during ablation. This irrigation can be emitted either from the laser (erbium and 9300-nm carbon dioxide instruments) or from other sources such as the operatory triplex syringe. The near-infrared wavelengths are generally transmitted through water, but the mid- and far-infrared photonic energy is highly interactive. While the water will cool the tissue, it would reduce the average power at the target tissue when using erbium or carbon dioxide wavelengths, since some of the laser energy will be actively absorbed.
- When using all tissue lasers, care must be exercised while removing soft tissue to avoid unintentional removal of tooth structure. The laser beam must be aimed as precisely as possible, and a suitable physical barrier (such as a matrix band or plastic instrument) could greatly aid in only ablating the target tissue. This precaution is especially important with some noncontact laser delivery systems.
- While performing a gingivectomy, the clinician must strive to match the healthy physiologic contour of the adjacent gingiva. The desired goal is that the healed site not only will be harmonious with the patient's periodontium but also will retain its shape with a healthy attachment. One benefit of using a laser for the procedure is that small areas of tissue can be treated in steps until the desired contour is achieved. This precision is easier to accomplish compared to tissue removed with a surgical blade.

15.2.5 For Procedures Involving Alveolar Bone, the Important Points to Remember Are Presented Below

- During osseous surgery, care must be taken to avoid overheating the bone and compromising its vascularity. The appropriate lasers for this procedure use a water spray, and the clinician should ensure that the irrigation is properly directed toward the target tissue.
- When performing osseous crown lengthening, an open-flap or closed-flap technique may be used. With the absence of visualization during a closed-flap procedure, the clinician must use maximum tactile sense while excising and contouring the bone and while simultaneously avoiding alterations to the healthy root surface.
- Whether an open or closed flap is chosen, it is recommended that the soft tissue modification proceeds first. It is much more difficult to excise and reshape the gingival tissue after the new bone form is established.
- Similar to soft tissue surgery, the laser beam's placement should be as precise as possible. The bony tissue must be properly contoured so that there are no remaining defects, troughs, or unusual anatomy. The underlying bone will determine the ultimate contour of the soft tissue covering it.

15.3 Gingival Biotype

Gingival anatomy has been generally described and categorized as either thin or thick [1, 2]. Variations of those terms sometimes appear in print as "thin-scalloped" or "thick-flat," and current terminology is phrased thick or thin biotype. Although somewhat difficult to determine visually, the disappearance of the tip of the periodontal probe into the sulcus usually indicates a thick tissue biotype. Other characteristics can aid in the classification:

- The thin biotype is generally less than 1.5 mm in thickness with a width of 3.5–5 mm and is characterized by a narrow zone of keratinized tissue with thin marginal bone surrounding teeth with triangular anatomic crowns. An example is shown in ▫ Fig. 15.1a.
- The thick biotype is at least 2 mm in thickness with a width of 5–6 mm and features a large amount of keratinized tissue, thick marginal bone, and bony plates surrounding square anatomic crowns. An example is shown in ▫ Fig. 15.1b.

The alveolar crest position and labial cortical plate thickness have a significant correlation with the gingival biotype [3]. It has been shown that patients with thin

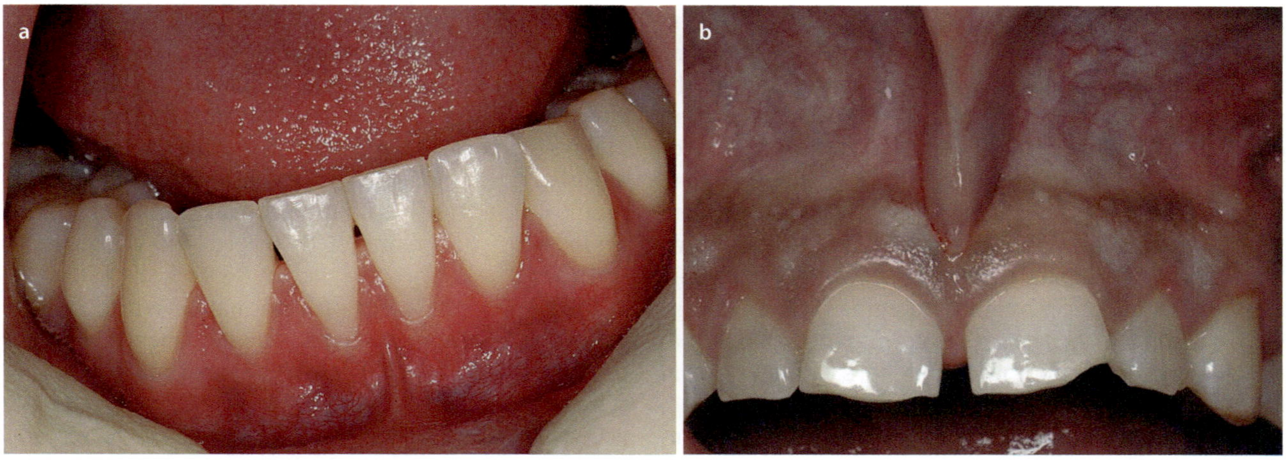

Fig. 15.1 **a** An example of thin biotype tissue in the mandibular incisor area. Note the narrow zone of keratinized tissue. **b** An example of thick biotype tissue with a large area of attached gingiva

biotype tissue had a great prevalence of gingival recession [4], whereas patients with thick biotype are less likely to experience those changes after surgical or restorative therapy [5]. Thus, the laser clinician should identify the gingival biotype before treatment and take special care with cases of thin anatomy.

15.4 Biologic Width and the Dentogingival Complex

This term is defined as the dimensions of the soft tissue attachments to the portion of the tooth coronal to the crestal alveolar bone. Based on measurements first published by Gargiulo [6], biologic width is generally stated as approximately 2 mm—the sum of the width of the epithelial and the connective tissue attachments. There are some variations in different studies of that 2-mm value, and clinicians also find the same variety. This can be due to many factors such as the position of the tooth in the alveolus, the anatomy of the roots, and especially the health of the periodontium [7]. Most practitioners generally use the term dentogingival complex which includes the gingival sulcular depth when discussing biologic width for ease of measurement. The literature states that 3.0 mm is the ideal distance from the free gingival margin to the alveolar crest on the facial aspect of anterior teeth and from 3.0 to 5.0 mm measured interproximally [8]. Thus, the apical aspect or bottom of the sulcus can be viewed as the top of the attachment. Therefore, the clinician can account for any variation in

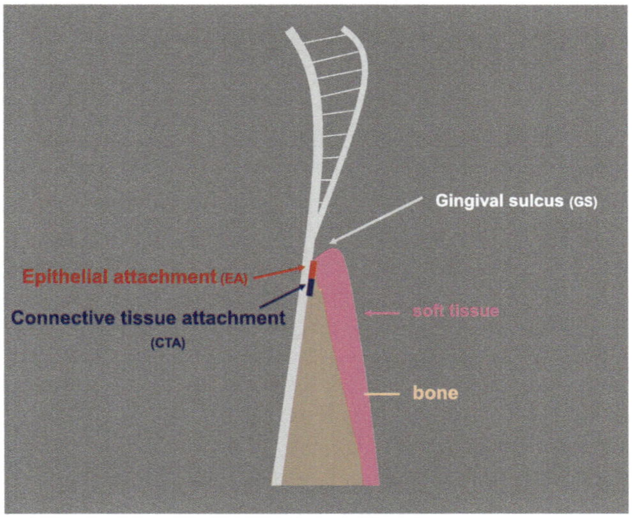

Fig. 15.2 A graphic depiction of the dentogingival complex (DGC) and the biologic width (BW). BW is composed of the epithelial attachment (EA) and the connective tissue attachment (CTA) and usually measured as a total of 2.0 mm. The DGC includes the gingival sulcus (GS) with a minimum depth of 1.0 mm

the attachment position by ensuring proper measurements. Figure 15.2 illustrates the ideal dentogingival complex and measurements.

As the clinician designs the restoration, this concept will guide the placement of the margin relative to the attachment and the bone to ensure optimal periodontal health [9]. This can be a critical decision in the aesthetic zone, where one of the treatment objectives is to mask the junction of the margin with the tooth. Other situations such as creating adequate resistance and retention form or to make significant alterations to the shape of the restoration will dictate the apical extension of the preparation. However, placement of the apical margin of the restoration within the biologic width can produce inflammatory periodontal disease [10]. This subgingival margin location can create the greatest biologic risk, and the best practice is to place that margin a maximum of 0.5 mm into the sulcus. This distance will minimize any chronic inflammation by not impinging on the biologic width. From another perspective, this would mean that the margin should be a minimum of 2 mm away from the alveolar crest.

For aesthetic procedures where only the periodontium is altered without placing any restorations, the new soft tissue must still retain an optimum biologic width so that there will be long-term stability.

Thus, any restorative or aesthetic procedure that alters hard or soft tissue must establish a new healthy biologic width and dentogingival complex.

15.5 Emergence Profile

The emergence profile is the portion of the clinical crown's contour extending from the base of the gingival sulcus to the proximal contacts and to the height of contour on the facial and lingual surfaces [11]. This circumferential shape of the tooth or restoration in relation to the surrounding soft tissue is crucial for both periodontal health and aesthetics. The emergence profile must be scrutinized on each axial surface depending on various clinical situations ranging from a diastema closure and height of contours of partial denture abutments to placement of interproximal contact areas and all subgingival margins. In all situations, the final restoration on an implant abutment or a pontic in a fixed bridge must harmonize with the rest of the patient's dentition.

The re-establishment of a normal embrasure with a restoration is particularly challenging when there is no papilla, as in pontic or implant spaces and some diastema areas [12]. Adding width to close a space generally necessitates a deeper subgingival margin of the restoration [13]. Of course, the final result must achieve periodontal health. Various techniques are necessary such as troughing the gingiva to add additional restorative material, contouring the edentulous ridge, and troughing around the implant fixture. All of these are ideal procedures that can be performed with a laser.

15.5.1 Clinical Cases Illustrating Emergence Profile

▪ Figure 15.3 shows three different clinical situations where a laser was used to create a new emergence profile. In each case, the soft tissue needed careful contouring so that the final restoration could be constructed with ideal axial surfaces to restore both function and health.

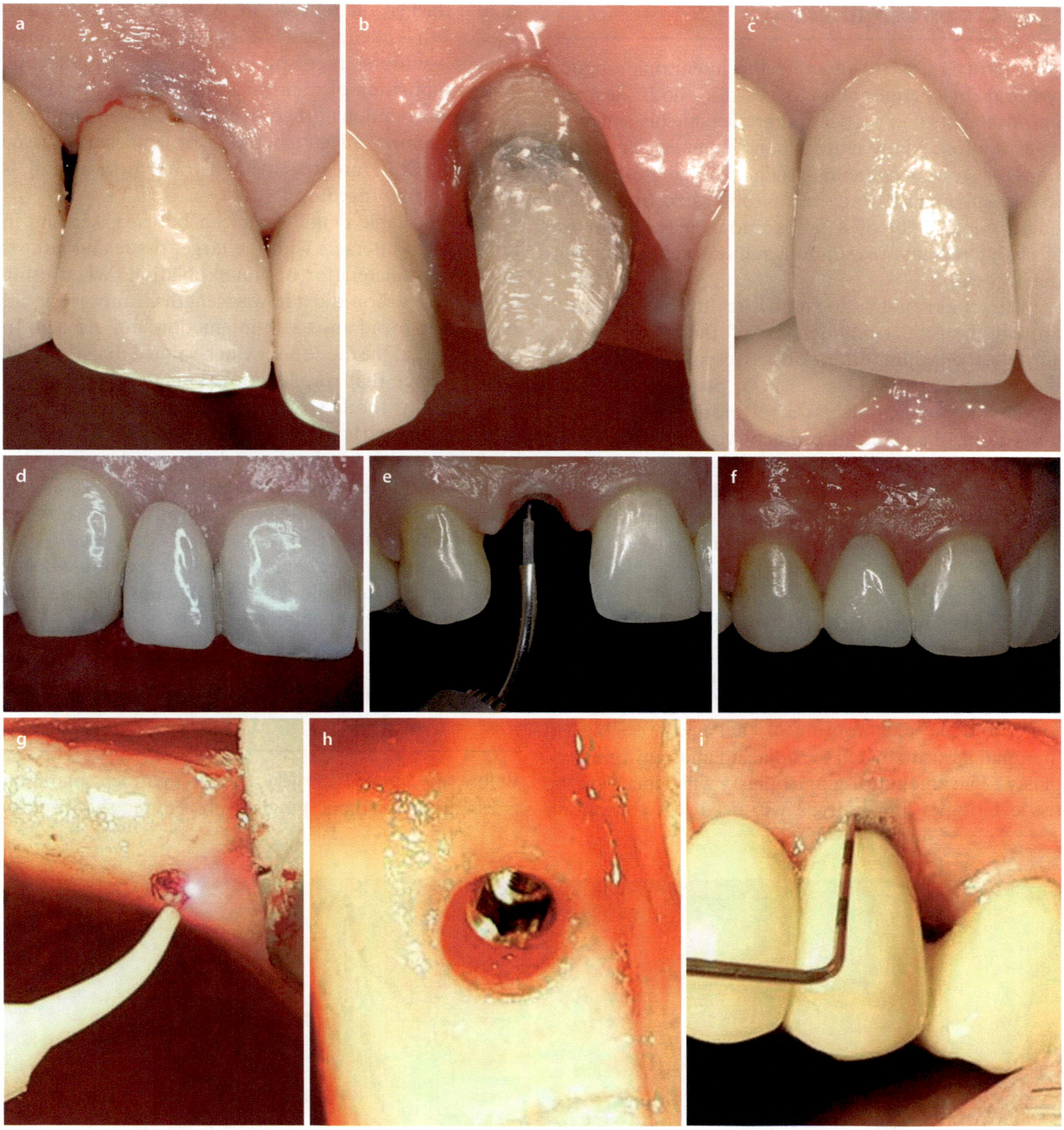

Fig. 15.3 **a** Preoperative view of a crown restoration with a recurrent carious lesion at the gingival margin. **b** A 2-week postoperative view showing the healed tissue after crown lengthening and troughing with an Nd:YAG laser using a 320-µm fiber and an average power of 2.0 W (100 mJ at 20 Hz). **c** Four-week postoperative view showing the crown delivery. Note the emergence profile. **d** Preoperative view of a bonded pontic replacing the maxillary lateral incisor. **e** An Er:YAG laser used with a 400-µm tip and an average power of 2.0 W (40 mJ per pulse at 50 Hz) without water spray to produce an ovate pontic concavity on the soft tissue. **f** Four-week postoperative view with new restoration in place. Note the much-improved gingival embrasures and papillae due to the improved emergence profile. **g** An 810-nm diode laser used with a 400-µm tip at 1.0-W continuous wave begins to uncover an implant fixture. **h** A 2-week postoperative view of the healed gingival contour. **i** A 6-week postoperative view of the restored implant. Note the healthy gingival tissue. (Implant case courtesy Dr. Steven Parker)

15.6 Crown Lengthening

This term is used to describe the intentional surgical removal of periodontal tissues for both aesthetic improvements and proper and predictable placement of a restoration. Many clinical conditions can be indications for crown lengthening, such as subgingival carious lesion, subgingival fracture of tooth structure, inadequate axial height of a preparation, unequal gingival levels, altered passive eruption, and short clinical crowns due to wear.

The primary goal is to attain a healthy biologic width around the total circumference of the tooth [14–16]. There are other important objectives such as achieving the proper aesthetic tooth form or providing sufficient tooth structure for a successful restoration. For aesthetic procedures, the clinician can achieve the desired result of a more pleasing smile by applying the principles of maintaining a healthy dentogingival complex. Crown lengthening can be limited to soft tissue only, or both soft and hard tissue can be contoured.

15.7 Soft Tissue Crown Lengthening

This surgery consists of two procedures—the excision of the gingival tissue to the desired height (gingivectomy) and the recontouring of that newly established marginal tissue to match the adjacent anatomy (gingivoplasty.) The amount of gingivoplasty will depend on the tissue biotype: thin biotype will need less contouring than a thicker anatomic form. After ensuring that biologic width is adequate, it is essential to restore the physiologic contours after soft tissue crown lengthening. This combination of removal and sculpting should produce a harmonious gingival outline segment and will also minimize any "rebound" or undesired tissue regrowth. ◘ Figure 15.4 demonstrates how a laser is used for both procedures.

Any available dental laser can be used, keeping in mind how it interacts with the target tissue and adjusting the parameters for optimum ablation. When using erbium or carbon dioxide wavelengths, caution should be taken to avoid any tooth interaction until needed. In ◘ Fig. 15.2a, the diode laser can be aimed directly at the enamel, since that wavelength has minimal interaction with healthy tooth structure. However, the beam of erbium and carbon dioxide wavelengths should be placed parallel to the enamel to avoid unintended removal of the enamel, as shown in ◘ Fig. 15.2b. As noted, biologic width must be considered both when planning this surgery and after it is completed. After soft tissue crown lengthening, the clinician should determine if adequate biologic width remains; if not, then osseous crown lengthening must be performed.

15.7.1 Soft Tissue Crown Lengthening for Aesthetics

Before any gingival surgery, proper treatment planning is essential. Aesthetic gingival procedures should con-

15

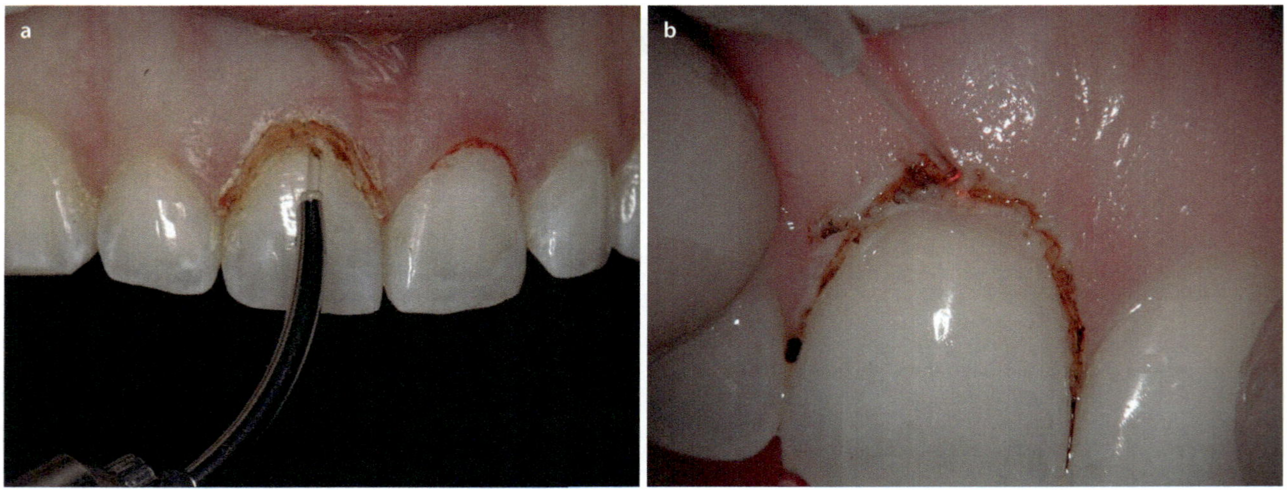

◘ **Fig. 15.4** **a** The diode laser is used for a gingivectomy and subsequent gingivoplasty on a maxillary central incisor. Note that the beam can be directed toward the tooth with minimum interaction or damage potential. **b** An Er:YAG laser is performing the gingivoplasty on the maxillary right central incisor. The gingivectomy was already performed on both central incisors. Since this wavelength can also be used for tooth preparation, the tip should not be aimed directly at the tooth surface during the soft tissue crown lengthening sider the overall design of the smile that exists and how the practitioner can change that form [17, 18]. Clearly, there are personal interpretations of aesthetics, and those can have a wide variance among patients both individually and culturally [19]. Moreover, the clinician may have specific opinions. In the end, the treatment objective is to produce a pleasing and healthy result for the patient [20].

The starting point for any smile design is the clinical crown profile of the maxillary central incisors and the corresponding gingival shape surrounding them [21]. If the patient desires some alteration, for example, a "gummy smile," then the biologic width must be located. After that, the surgical plan would be to create good symmetry on both sides of the midline. The zenith or apical most point of the gingival outline should ideally be the same height on the central and cuspid, while the lateral incisor's height can be 1–2-mm shorter [22].

The ideal gingival contour has a scalloped shape, and all of the interdental papillae fully occupy the interproximal embrasures. During gingival surgery, care should be taken to not produce a less scalloped, flatter gingival margin, since that could result in shorter interdental papilla and the opening of the embrasure spaces. The latter condition is sometimes referred to as "black triangles," and that would be a compromised aesthetic outcome. The most predictable gingival response will occur when the new postoperative outline follows the smile design principles as well as providing optimum periodontal health [23].

15.7.2 Soft Tissue Crown Lengthening for Restorative Dentistry

The traditional restorative requirements of adequate and sound tooth structure can be problematic when a carious lesion extends subgingivally. The clinician must be able to visualize and remove the diseased tooth structure while analyzing the periodontal condition. In addition, an acceptable emergence profile must be produced.

Retraction or removal of gingiva impinging on a lesion is essential for thorough caries removal. If biologic width is adequate after the preparation is complete, then the clinician must decide if the margin placement will aid or hinder the patient's ability to maintain oral hygiene to try to prevent another lesion [24]. In both cases, a laser can be used.

15.7.3 Clinical Cases of Soft Tissue Crown Lengthening

▫ Figure 15.5 shows the use of an erbium laser to improve gingival aesthetics. Excessive gingiva results in the appearance of short clinical crowns. After biologic width is measured, it was determined that soft tissue crown lengthening could proceed. A tissue marker provided a "layout" to guide the clinician for the procedure. Note that the laser is used parallel to the labial surface to avoid any interaction with the enamel. The immediate postoperative view shows good tissue contour.

▫ Figure 15.6 illustrates a case of porcelain veneers placed to improve aesthetics and to close the diastema of the maxillary incisors. The patient opted out of orthodontic treatment as a first step. A harmonious gingival architecture to produce a pleasing smile along with a good emergence profile for the restorations was meticulously planned. A diode laser was used.

▫ Figure 15.7 illustrates another case of aesthetic crown lengthening utilizing an Nd:YAG laser. Adequate biologic width was measured, and the laser performed a gingivectomy and gingivoplasty. The Nd:YAG wavelength produces a similar tissue interaction result to a diode laser, although the free-running pulse emission mode produces very short-duration pulses with a low emission cycle. The relatively long intervals of non-emission are periods of thermal relaxation for the tissue during the surgery, which is an advantage for thinner tissue biotype.

▫ Figure 15.8 shows the use of a diode laser to remove gingiva at an abfraction lesion in order to finish the apical extent of the preparation. The laser can easily recontour the tissue and maintain a dry field for placement of the restoration. As noted above, the diode wavelength has no interaction with the tooth structure. In addition, the new gingival level will facilitate the patient's oral hygiene in that area.

▫ Figure 15.9 depicts a recurrent carious lesion around an existing restoration on bicuspid. The inflamed marginal gingiva prevents total access to the lesion. A carbon dioxide laser was used to remove the tissue, repositioning it more apically so that a new composite could be placed. The 9300-nm instrument also removed the carious lesion, and that discussion can be found in ▶ Chap. 8 (clinical case courtesy of Dr. Joshua Weintraub).

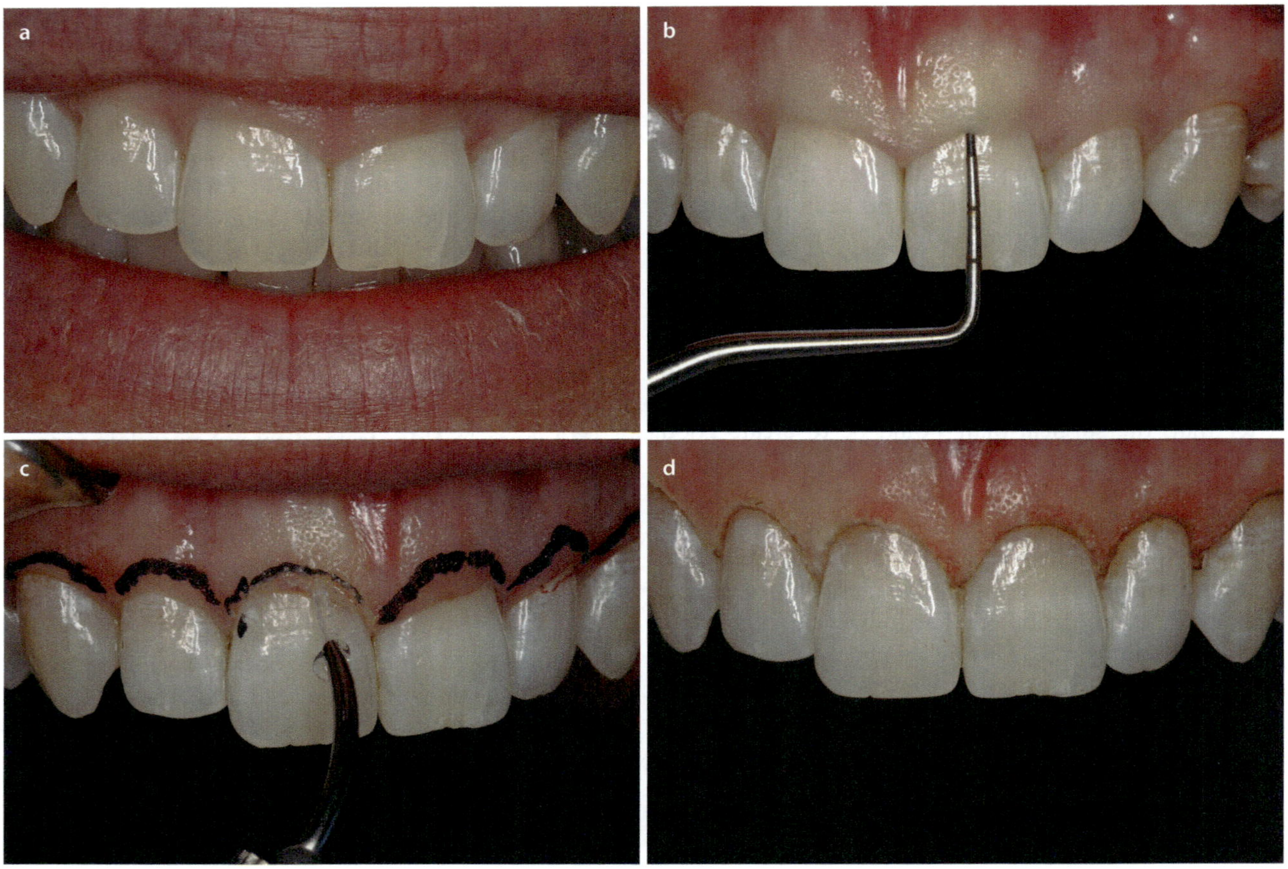

Fig. 15.5 **a** Preoperative view showing uneven gingival contour, with pronounced differences in the zeniths of the maxillary central incisors. **b** A periodontal probe is used for the determination of biologic width and the overall dimensions of the dentogingival complex. **c** After tissue is marked, an Er:YAG laser with a 600-μm tip is used with an average power of 2.0 W (40 mJ per pulse at 50 Hz) without a water spray to perform the gingivectomy and gingivoplasty on all six maxillary anterior teeth. **d** Immediate postoperative view showing excellent hemostasis and tissue contour. (Clinical case courtesy of Dr. David Hornbrook)

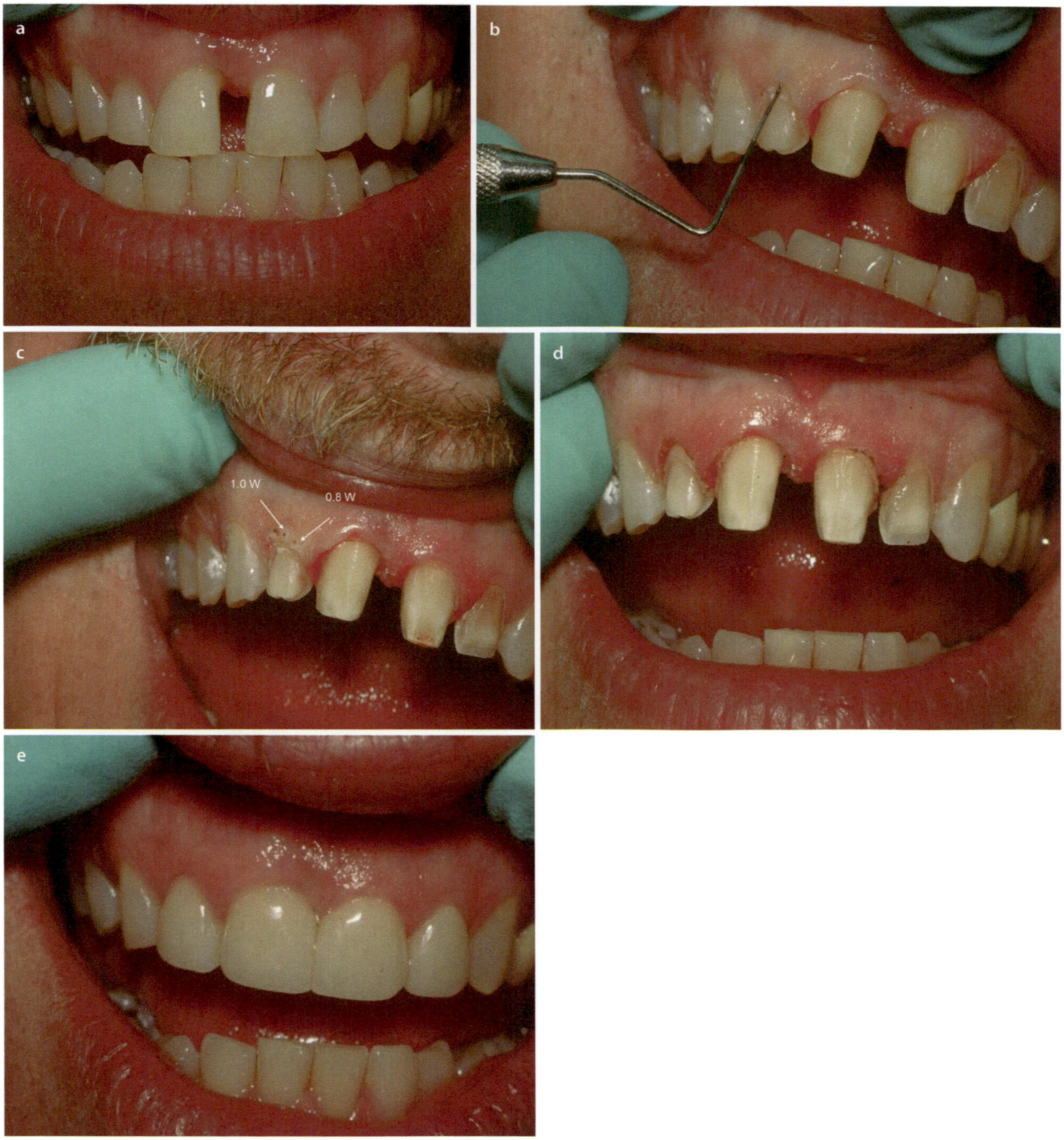

Fig. 15.6 **a** Preoperative view of the maxillary anterior segment with a large diastema between the central incisors and uneven gingival height of all incisors. **b** Biologic width determination and analysis of the dentogingival complex are performed with a periodontal probe. **c** After verifying the periodontal condition, soft tissue crown lengthening was performed with an 810-nm diode laser using a 400-μm bare fiber. In order to lay out the intended new gingival form, a 400-μm-diameter "dot" was placed at the new gingival zenith using a power of 1.0-W continuous wave. Careful inspection of that small area of ablation revealed slight carbonization, which indicates the tissue temperature was excessive. The laser parameter was adjusted to 0.8-W continuous wave and another dot was placed. That area showed normal ablation and that parameter was chosen to utilize for the surgery. **d** The immediate postoperative view shows the completed crown lengthening and finished preparations. Note that the central incisors were reduced on their mesial aspect and a subgingival trough was placed. Both of those procedures will enable new porcelain contours so that the diastema can be closed. **e** Six-month postoperative view demonstrates how the laser adjunctively fulfilled the principles of smile design and emergence profile while creating a healthy periodontal condition

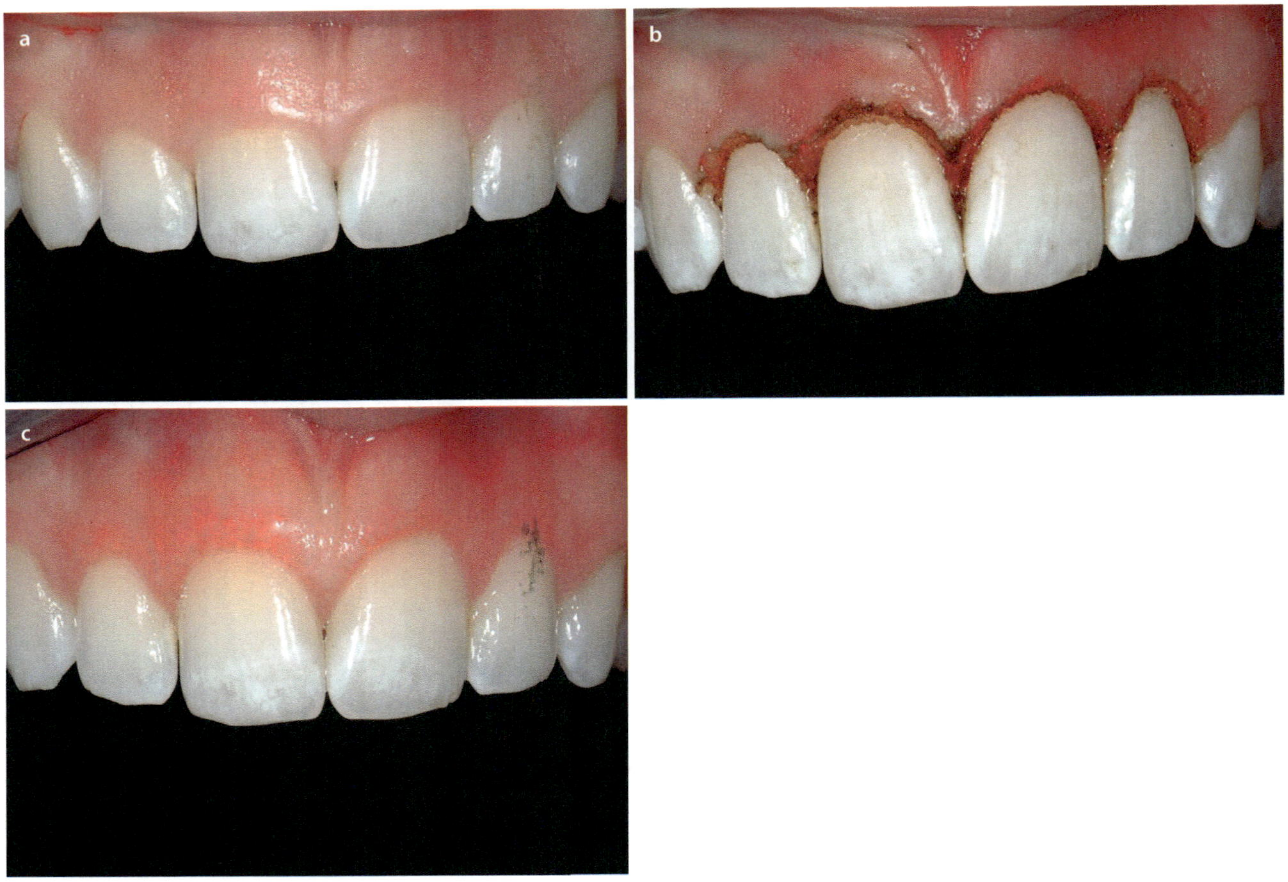

Fig. 15.7 **a** Preoperative view of the anterior maxillary sextant with asymmetrical gingival scalloping. Biologic width measurements revealed adequate tissue available for soft tissue crown lengthening. Note the thinner biotype on the later incisors compared to the central incisors. An Nd:YAG laser was used with a 320-µm fiber at an average power of 1.8 W (60 mJ per pulse at 30 Hz). **b** Immediate postoperative view. The laser's free-running pulse mode emission allows for thermal relaxation of the tissue, particularly on the lateral incisors' thinner biotype. Note the areas of gingivoplasty for new tissue form and outline. **c** Three-week postoperative view showing improved aesthetics with a more harmonious gingival scallop and embrasures

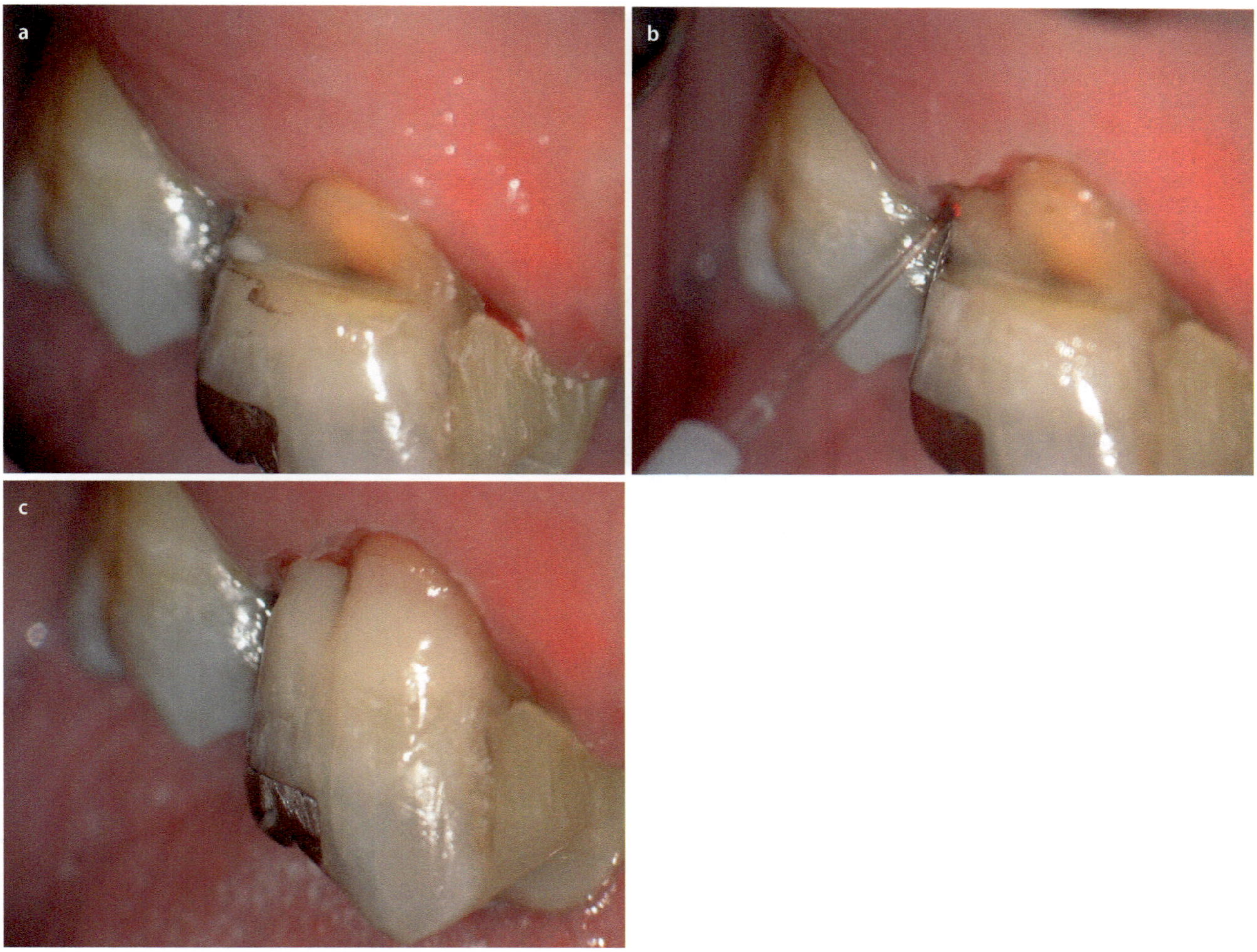

Fig. 15.8 **a** Preoperative view showing abfraction lesion on a maxillary molar. The gingival tissue has proliferated over the apical aspect of the lesion. **b** An 810-nm diode laser with a 400-μm bare fiber is used with an average power of 1.0 W to remove the gingival tissue and re-establish proper contour. The lesion can then be prepared and restored. **c** Immediate postoperative view of the final restoration. The laser created a dry field with lack of any bleeding from the tissue to aid in the placement of the restorative material. The final margin placement should allow easy patient access for maintenance

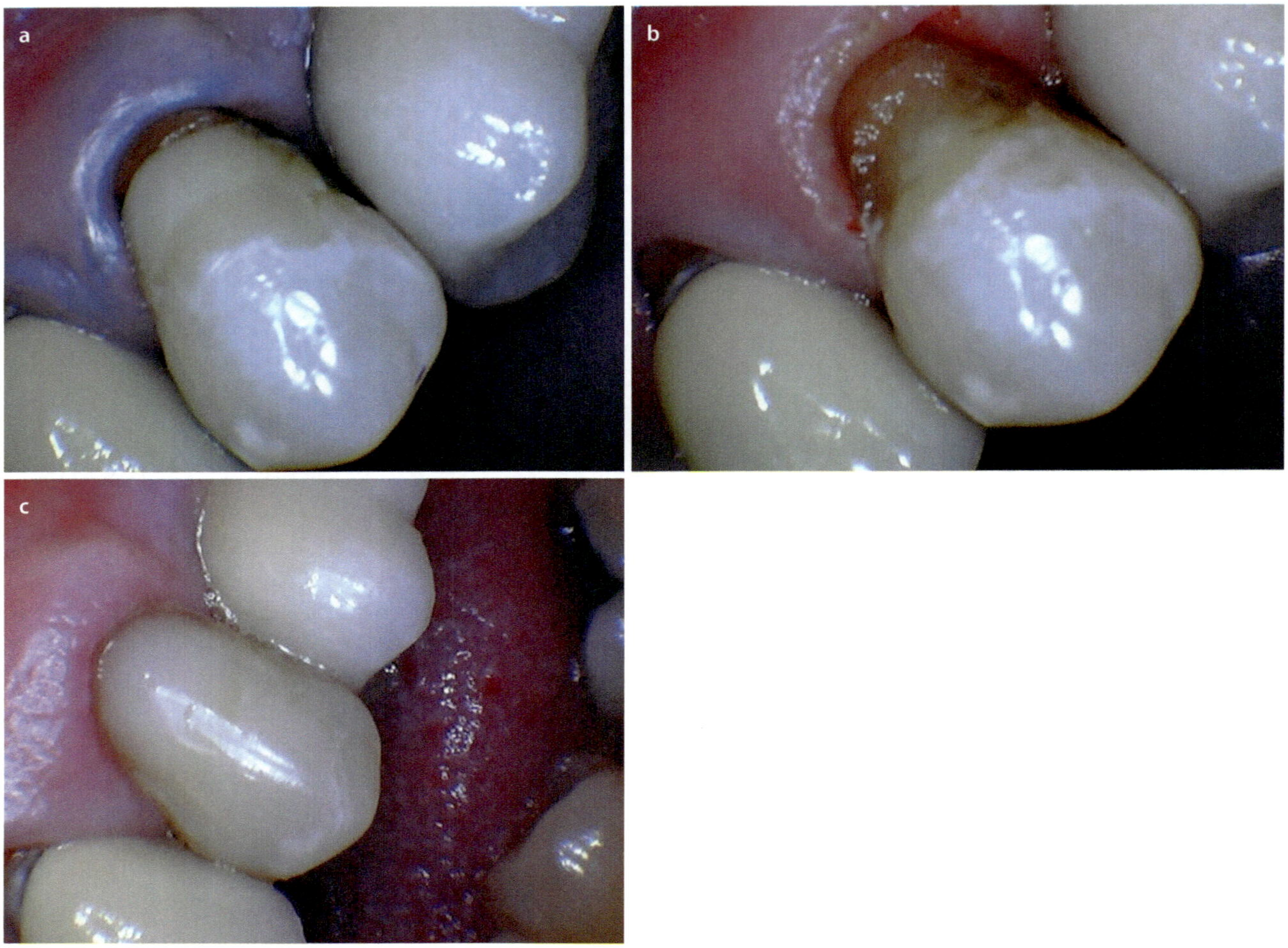

Fig. 15.9 **a** Preoperative view showing a recurrent carious lesion with inflamed gingival tissue. The carious lesion extends subgingivally. **b** The gingiva was recontoured with a 9300-nm carbon dioxide laser using a 0.25-mm spot size, a 65-μs pulse duration, and a cutting speed between 10% and 30% with minimal water spray. **c** The immediate postoperative view showing the new tissue contour with the margin of the new composite restoration placed at the free gingival margin. Note how the laser achieved good control of tissue fluids to aid in the composite placement

15.8 Osseous Crown Lengthening

If the intended crown lengthening will compromise the biologic width, an osseous procedure will be required. The desired goal is to shape the osseous crest to match the gingival scallop outline form, and both should parallel the restorative margin. As implied above, in general, soft tissue crown lengthening is performed first before the osseous procedure. The clinician must consider whether to proceed by raising a flap in an "open-flap" surgical approach or operate without elevating any soft tissue—the so-called closed-flap or flapless technique [25]. A contact laser tip can transmit tactile information to guide the clinician in the procedure; however, laser energy does not easily distinguish between bone and root surface cementum and/or dentin. Conventional flap reflection may be necessary to both visualize and properly contour bone, especially in large areas of missing tooth structure or in multiple adjacent sites. In a localized area, for example, with a subgingivally fractured cusp, a closed-flap osteotomy and osteoplasty can be performed [26]. In either case, the bone must be contoured as close to an ideal physiologic form as possible—without ledges, craters, or other deviations. Meticulous attention to creating proper anatomical form is much more challenging without flap access [27].

The overlying principle of biologic width dictates the amount of hard tissue removal along with the maintenance of adequate periodontal support.

The typical surgery begins with designing the new gingival outline and determining the initial biologic width. In this case, it is assumed that both soft and osseous crown lengthening will be performed. Next, the gingival tissue is excised and contoured to achieve that new sculpture. That may result in destruction of all or part of the soft tissue attachments. If possible, the existing osseous crest should be sounded. Then the clinician makes the decision about raising a flap. Removal of 2–3 mm of osseous resection is generally required to re-establish new biologic width [8]. Similar to soft tissue crown lengthening, an osteotomy and osteoplasty should be performed, resulting in a stable anatomic scaffold for the overlying gingiva. If an open-flap procedure was used, the soft tissue flap is usually apically repositioned and sutured. In a flapless technique, the clinician should ensure that the soft tissue is well approximated on the tooth surface [28].

15.8.1 Lasers for Osseous Crown Lengthening

As mentioned above, only the erbium family and the 9300-nm carbon dioxide lasers are currently indicated for bone procedures. The Er,Cr:YSGG (2780 nm) and the Er:YAG (2940 nm) instruments primarily target the water component in osseous tissue, whereas the 9300-nm carbon dioxide energy interacts the hydroxyapatite. All three wavelengths utilize free-running pulse emission with very short pulse durations. Each features a water spray to help minimize any overheated areas of ablation. To guide the beam, some instruments have contact tips, and others have small cylindrical guides used out of contact.

15.8.2 Osseous Crown Lengthening for Aesthetics

All of the concepts of smile design must be considered before any surgery begins. If any restorations will be placed, their gingival margin position should also be planned. Typically, multiple teeth are involved in aesthetic dentistry, and harmony among them must be achieved. A diagnostic wax up can certainly aid in visualization of the desired treatment outcome. In addition, approximate areas of laser excision and contouring can be simulated.

15.8.3 Osseous Crown Lengthening for Restorative Dentistry

As a tooth preparation extends deeply into the gingival sulcus, the clinician must evaluate how and where the bone will be repositioned. The restoration's margin and the surrounding periodontium will correspond to each other; therefore, the immediate postoperative tissue position, form, and contour will dictate the ultimate result.

15.8.4 Clinical Cases of Osseous Crown Lengthening

◘ Figure 15.10 shows an existing restoration of the upper left lateral incisor whose clinical crown's length was not harmonious with the other anterior teeth. Additionally, there was gingival pigmentation in the area (▶ Chap. 16 provides a detailed discussion of laser-assisted removal of gingival pigmentation). After depigmentation with the Nd:YAG laser, the Er:YAG laser performed both soft tissue crown lengthening and closed-flap osseous crown lengthening. A new restoration was placed (case courtesy of Dr. Shigeyuki Nagai).

◘ Figure 15.11 depicts a typical clinical dilemma where an existing crown restoration has fractured off with an inadequate amount of clinical crown remaining. Osseous crown lengthening was performed in an open-flap procedure using an Er:YAG laser. The successful procedure resulted in sufficient tooth structure for a new crown to be constructed.

◘ Figure 15.12 demonstrates an open-flap osseous crown lengthening procedure to restore a fractured lower right second bicuspid. The 9300-nm carbon dioxide laser was used, and a 5-month postoperative photo shows good healing with the re-establishment of biologic width (clinical case courtesy of Dr. Joshua Weintraub).

◘ Figure 15.13 depicts a fractured clinical crown on the upper left central incisor. The patient chose the option of endodontic treatment, crown lengthening, core buildup, and new restoration. The 9300-nm carbon dioxide laser completed the soft tissue and closed-flap osseous crown lengthening. The final restoration was delivered 3 weeks postoperatively, and the 1-year postoperative photo shows a healthy periodontium with a new restoration (clinical case courtesy of Dr. Joshua Weintraub).

◘ Figure 15.14 shows a case of varied gingival heights and contour of some maxillary teeth. Closed-

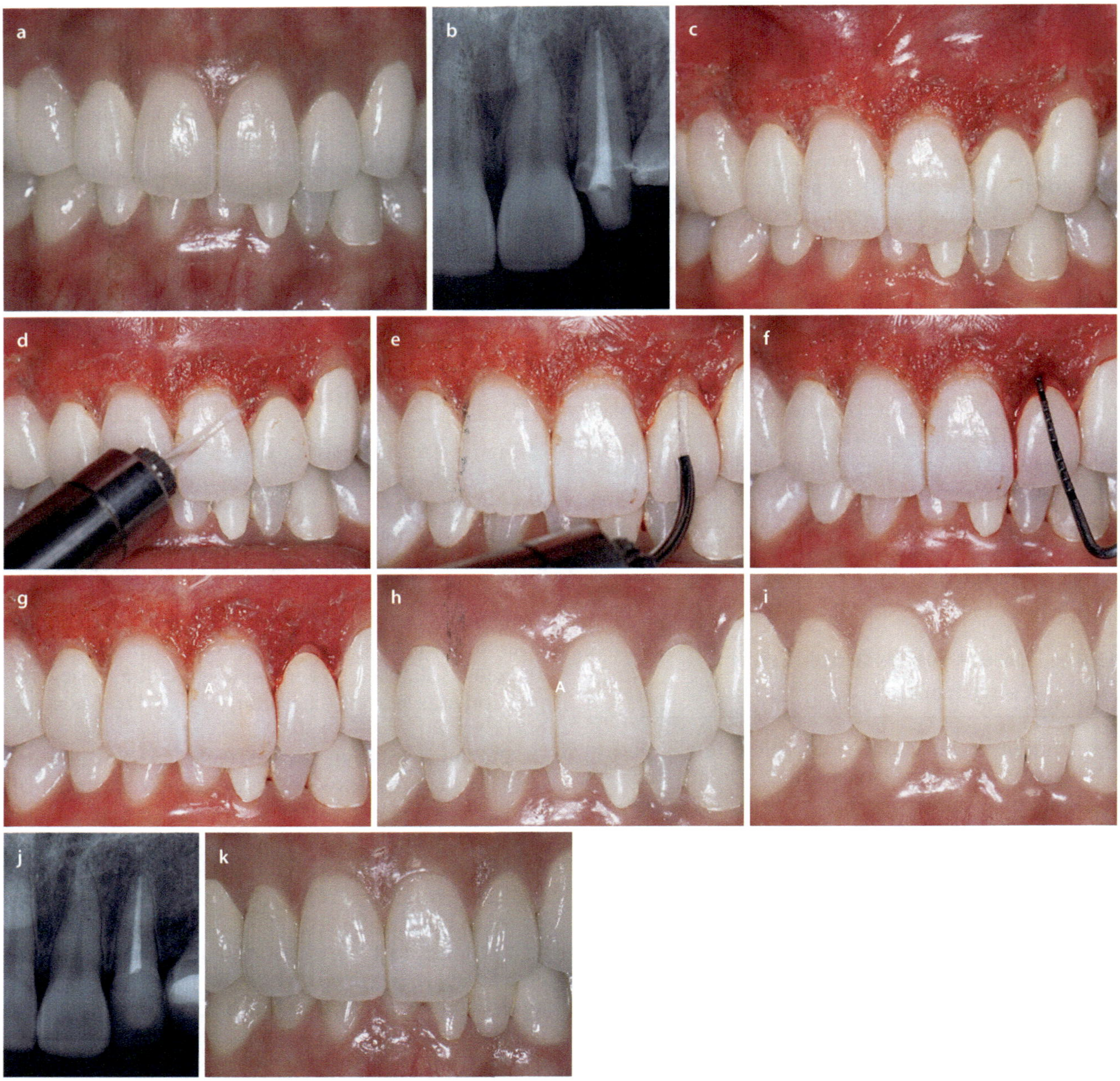

Fig. 15.10 **a** Patient presented with gingival pigmentation and uneven gingival architecture on the upper left maxillary lateral incisor, with a provisional restoration in place. **b** The radiograph shows the restoration margin very close to the osseous crest. The treatment plan is to remove the pigmentation and perform closed-flap gingival and osseous crown lengthening so that the new restoration could restore the aesthetics. **c** An Nd:YAG laser, delivered through a 320-μm quartz fiber, was used at 2.4 W, 80 mJ, 30 Hz, for 5 min to remove the epithelial pigmentation on the maxillary anterior tissue. The immediate postoperative result is depicted. **d** Immediately following, an Er:YAG laser with a pointed soft tissue tip (Fig. 15.4) was used at 1.5 W, 50 mJ, 30 Hz, without water for 12 s to remove the gingival tissue and to apically reposition the gingival margin. **e** The Er:YAG laser was then used with a 400-μm curved quartz tip for closed-flap osseous reduction. The average power of 1.75–3.0 W (70 mJ, 25 Hz–120 mJ, 25 Hz) was used to remove and contour the osseous crest with water spray. **f** Continuous periodontal probing determined the bone reduction and final contour. **g** The immediate postoperative result depicting additional tooth structure. **h** The healed tissue 2 weeks later. **i** The final restoration was placed 2 weeks later, which was 4 weeks after the two laser wavelength soft and hard tissue procedures. **j** A 14-month post procedure radiograph shows good periodontal health. **k** The 4-year postoperative view with excellent gingival architecture and tooth aesthetics with a slight recurrence of pigmentation

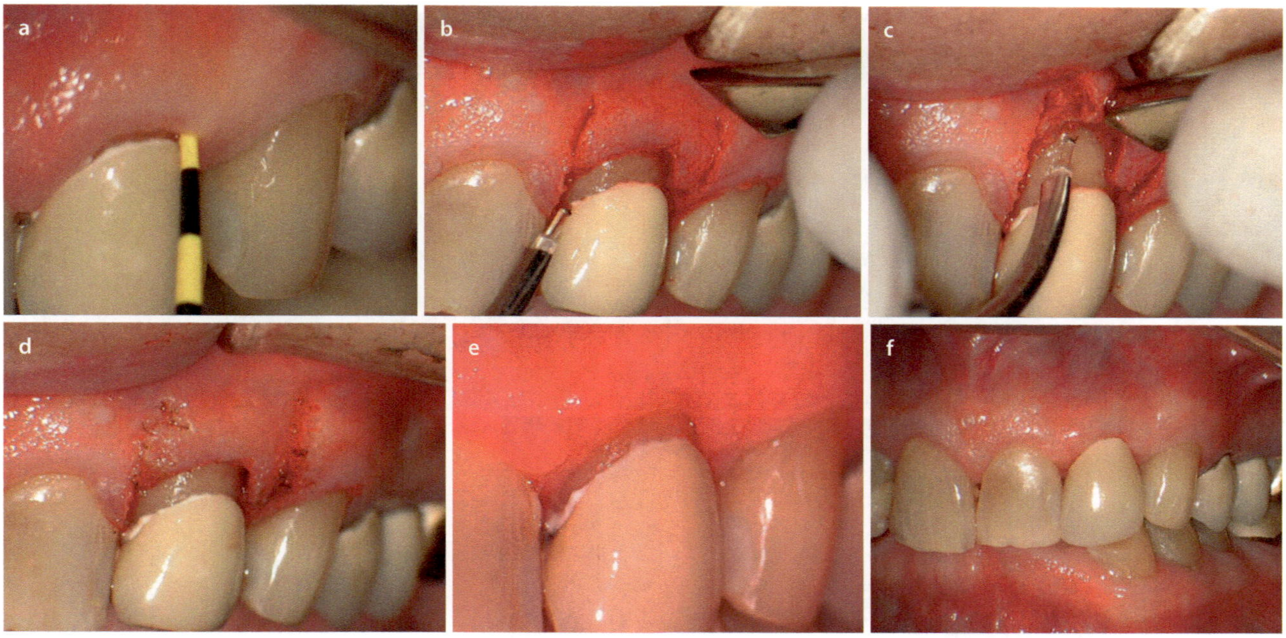

Fig. 15.11 **a** Preoperative view. The patient presented with a porcelain fused to metal crown that had become dislodged. The incisal one-third of the preparation had also fractured. The crown was recemented, and it was determined that osseous crown lengthening would be necessary because of the inadequate biologic width. **b** An Er:YAG laser was used with a 400-µm tip with an average power of 2.4 W (80 mJ at 30 Hz) with a copious water spray to apically reposition the gingival margin, achieving the soft tissue portion of the crown lengthening. **c** After raising a flap with conventional instruments, the same laser parameters were used to remove and reposition the osseous crest. Note that the tip is aimed at the bone, avoiding contact with the root. **d** The immediate postoperative view of the flap repositioned with new tooth form revealed. **e** One-month postoperative view shows the healed attachment and new gingival height. A new crown preparation can proceed with adequate ferule for good retention form. **f** Two weeks later, the crown is delivered, and the tissue should continue to heal for a successful restorative result

flap gingival and osseous crown lengthening were accomplished with an Er:YAG laser, and then porcelain veneers were placed. The clinical photos show a portion of that procedure on the cuspid. A 4-year postoperative view shows healthy periodontium and excellent aesthetics (clinical case courtesy of Dr. David Hornbrook).

Figure 15.15 shows a patient who presented with a desire to improve the aesthetics of the upper anterior teeth. The Er,Cr:YSGG laser with a 600 µm 9mm long tip assisting with flap surgery and removal of the apical portion of the root, and then performed gingival and osseous crown lengthening to provide adequate tooth structure for new restorations. . (Clinical case courtesy of Dr. Mark Cronshaw).

Figure 15.16 shows a patient who presented with worn and unaesthetic maxillary and mandibular anterior teeth. A treatment plan was developed that included increasing the vertical dimension, osseous crown lengthening for the upper right and left central incisors, a maxillary frenectomy, and restorations, and the Er,Cr:YSGG laser was utilized (clinical case courtesy of Dr. Mark Cronshaw).

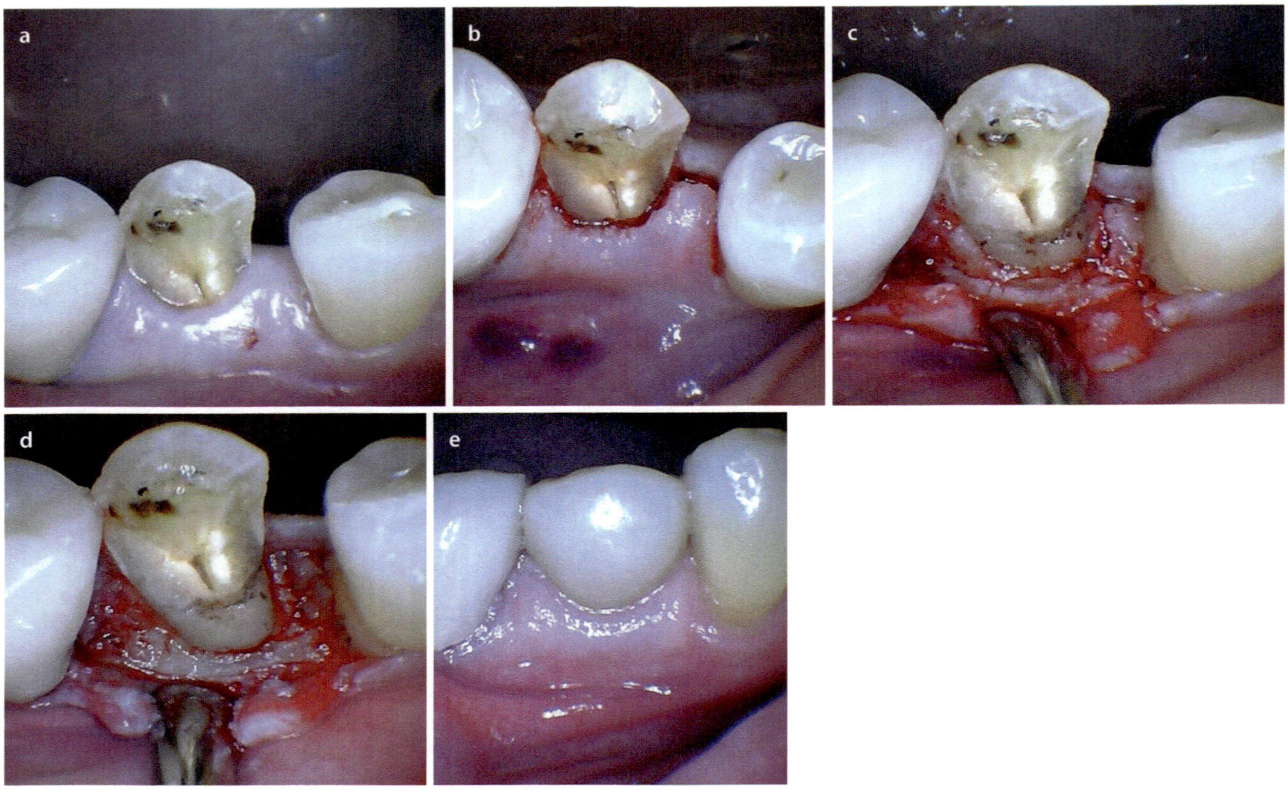

Fig. 15.12 **a** Preoperative view of a large fractured buccal cusp of the lower right second bicuspid. The patient considered all treatment options and chose crown lengthening, endodontic therapy, and a final restoration. **b** The 9300-nm carbon dioxide laser was used at a 1.0-mm spot size, 20–30% cutting speed, and 1% water mist spray to perform the soft tissue incision. **c** The buccal flap extended to the proximal surfaces of both adjacent teeth, showing that the cusp had fractured to the osseous crest. **d** The laser was then set at a 1.0-mm spot size, 40–50% cutting speed, and a 80% water mist spray for the alveolar bone ablation. Note the good vascularity of the ablated osseous structure. **e** A 5-month postoperative view shows the final restoration. (Clinical case courtesy of Dr. Joshua Weintraub)

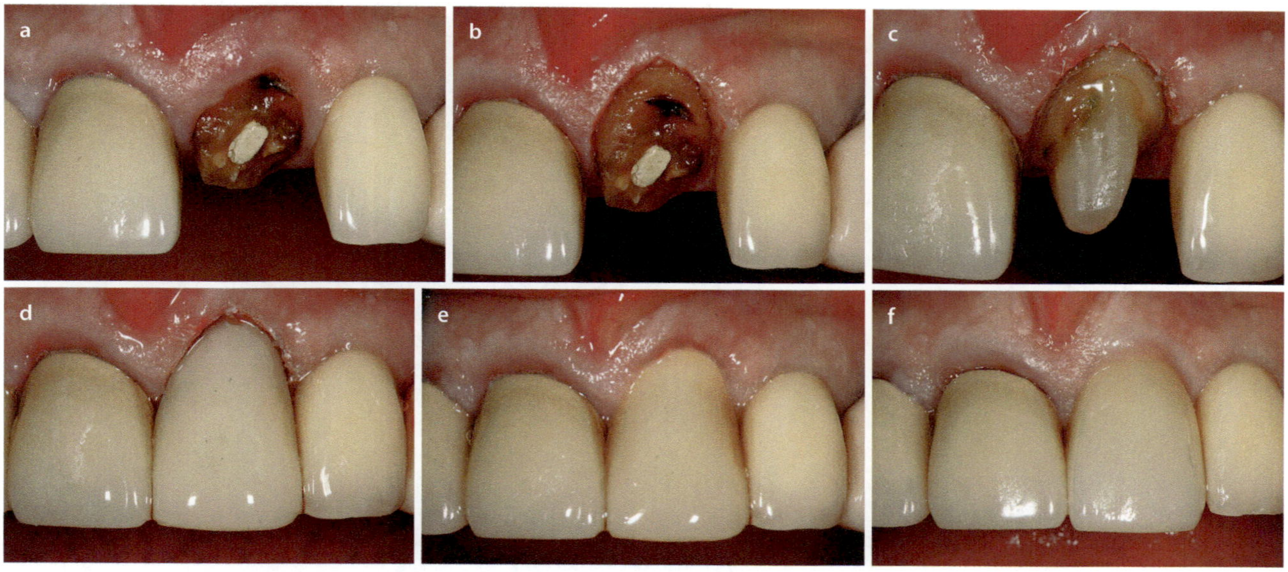

Fig. 15.13 **a** Preoperative view of a fractured clinical crown. **b** After endodontic treatment was completed, the soft tissue removed to expose the apical extent of the fracture. The 9300-nm carbon dioxide laser was used with a tipless contra angle handpiece and a 1.00-mm spot size, 1% water mist, a cutting speed between 20% and 30%. Control of the tissue removal was achieved by varying the working distance, hand speed, and cutting speed. **c** Showing the completed crown lengthening and core buildup. After sounding to the bone to determine the osseous crest, the same handpiece was used with a 0.5-mm spot size, 80% water mist, and a cutting speed between 60% and 80%. The result was to expose at least 2 mm of tooth structure coronal to the osseous crest. **d** Depicting the provisional crown in place. **e** Showing the 3-week postoperative view when the final restoration was delivered. **f** One-year postoperative view with healthy periodontium. (Clinical case courtesy of Dr. Joshua Weintraub)

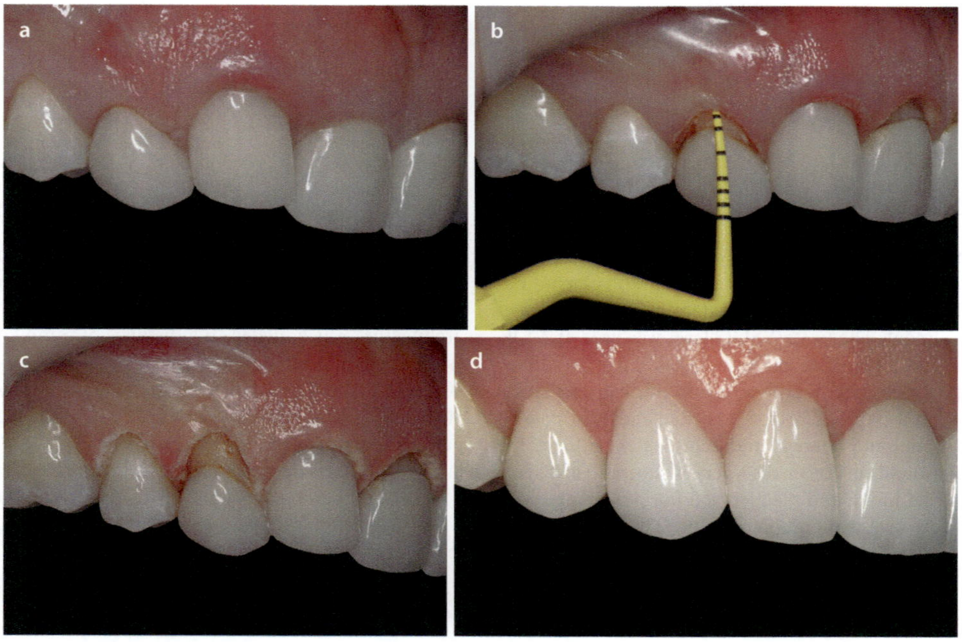

Fig. 15.14 **a** Preoperative view of varied gingival architecture around existing porcelain crowns. **b** An Er:YAG laser was used with a 400-μm tip with an average power of 2.0 W (40 mJ at 50 Hz) without water spray for the soft tissue removal. This photo shows that procedure in progress on the cuspid, and perio probing on that tooth shows that there will be a violation of biologic width in order to establish the intended new gingival outline and to place the margin of the restoration. Osseous crown lengthening is necessary on the labial surface, and a closed-flap technique was used. The Er:YAG was used with the same parameters—2.0 W (40 mJ 50 Hz) but with a water spray. **c** The crown lengthening is completed on all of the teeth. **d** Four-year postoperative view shows excellent periodontal health along with good smile design. (Clinical case courtesy of Dr. David Hornbrook)

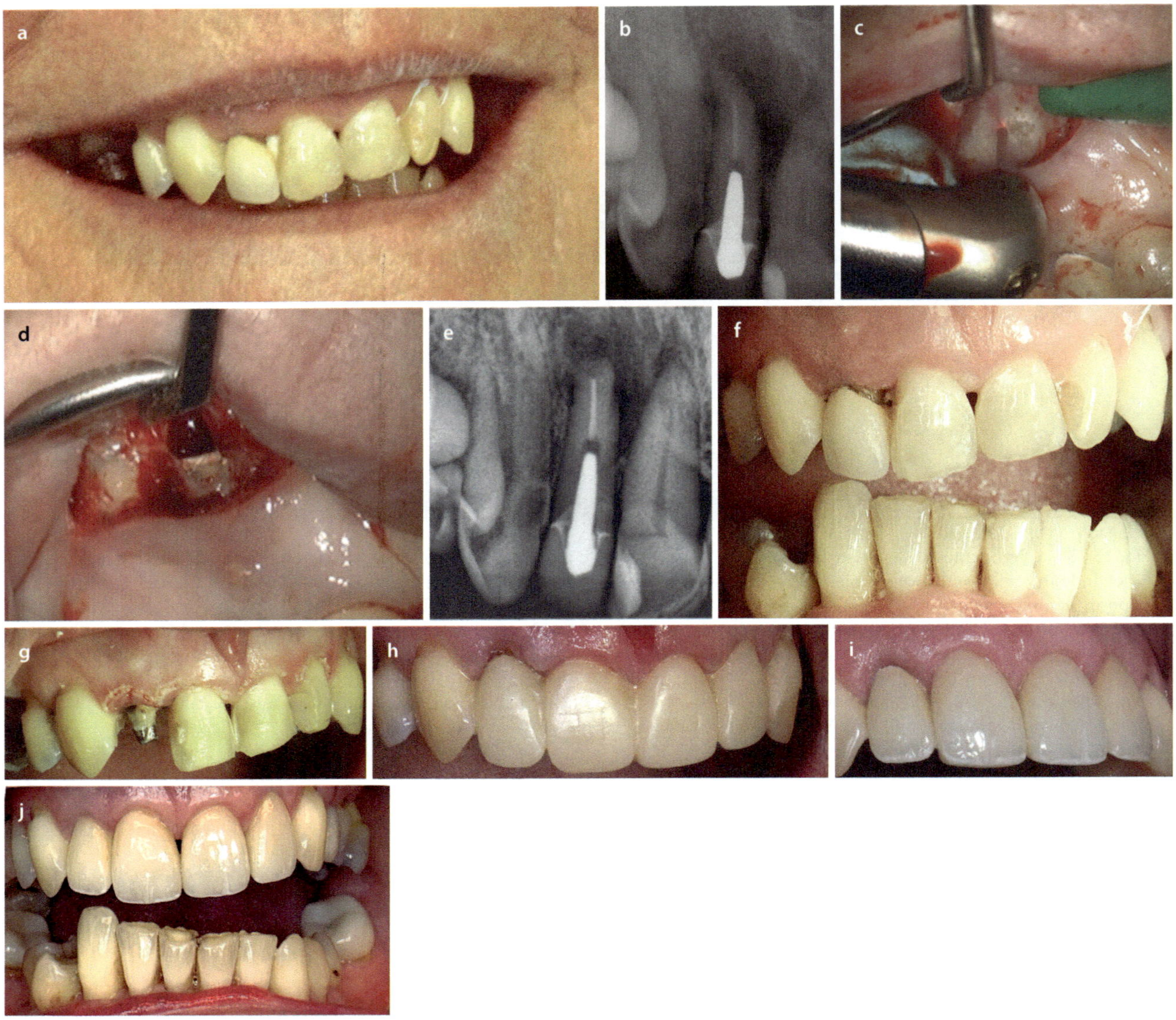

Fig. 15.15 **a** Showing the preoperative view of the patient who desired improved aesthetics. **b** Depicting the radiograph of continuing periapical pathology on the upper right lateral incisor with previous endodontic treatment. The Er,Cr:YSGG laser was used at 2.5 W, 25 Hz, 40% water, and 20% air to perform a semilunar soft tissue incision. **c** Depicting the osseous exposure, contouring, and apical resection performed at 3–3.5 W, 25 Hz, 85% water, and 20% air with the same tip. The granuloma removal and debridement utilized the soft tissue parameters, and the finished surgery is shown in **d**. Then a retrograde MTA apical filling and platelet-rich fibrin were placed. **e** Showing the postoperative radiographic result. The healed apical surgery site is shown in **f**. **g** Depicting the initial incision and the proposed contour for a closed-flap crown lengthening procedure. The Er,Cr:YSGG laser with a 600-μm 9-mm-long tip was used at 2.5 W, 40% water, and 20% air for the gingivectomy. The osseous architecture was adjusted to a distance of 3 mm apical to the newly established gingival margins. The same tip was used at 3.0 W, 30 Hz, 60% water, and 60% air. A photobiomodulation dose of 50 J/cm^2 for 100 s was then applied to the surgical sites. The teeth were then prepared for restoration and the provisionals were placed. Two weeks later, the tissue is healing well **h**. The final restorations were delivered 4 weeks later **i**, and the 4-year postoperative result is shown in **j**, with a healthy periodontium and excellent aesthetics. (Clinical case courtesy Dr. Mark Cronshaw)

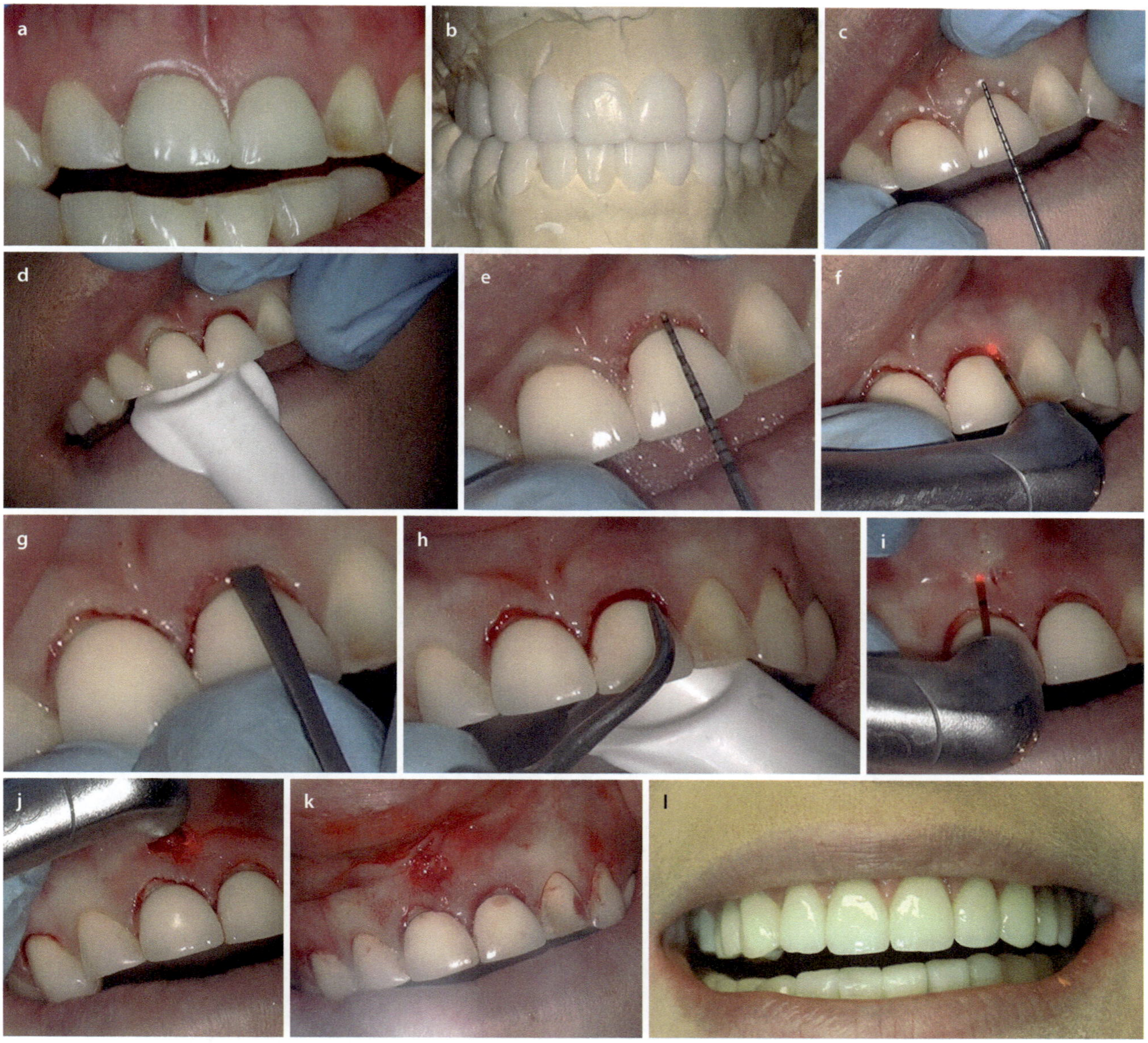

Fig. 15.16 **a** Preoperative view of patient presenting with worn and unaesthetic maxillary and mandibular anterior teeth. After studying all the findings, the treatment plan is to include increasing the vertical dimension of occlusion by 2 mm, lowering the incisal plane by 2 mm, osseous crown lengthening on UR and UL 1, frenectomy, and restorations. The diagnostic wax up is shown in **b**. Soft tissue crown lengthening was performed. The Er,Cr:YSGG laser was used with an MZ6 9-mm tip. The proposed gingival architecture was laid out with small dots on the tissue surface using 0.5 W without air or water. A probe measurement from the wax up verifies gingival margin placement, as shown in **c**. **d** Depicting the completed soft tissue crown lengthening. The same tip was used at 2.5 W, 25 Hz, 40% water, and 20% air to remove the soft tissue. The osseous crest was located with probing, as shown in **e**. The bone reduction must result in the crestal bone location 3 mm apical to the gingival margin. **f** Depicting closed-flap crown lengthening. The Er,Cr:YSGG laser utilized a 600-μm 9-mm long tip and 3 W, 30 Hz, 60% water, and 60% air. The tip enters the sulcus parallel to the root with a slow sweeping motion. The clinician should ensure good water flow in and out of the sulcus. To refine the osseous architecture, Wedelstaedt bone chisels are used **g**, and a blunt piezo ultrasonic tip removes any minor bony irregularities **h**. The frenectomy was next performed using the previous soft tissue parameters: MZ6 9 mm at 2.5 W, 25 Hz, 40% water, 20% air **i**. It is important to score the periosteum **j** to produce a scar which prevents frenal proliferation incisally. **k** Showing the completed soft and hard tissue surgery. All teeth were prepared and provisional restorations were placed 1 month postsurgery. **l** Showing the final restorations delivered, approximately 5 weeks later. (Clinical case courtesy Dr. Mark Cronshaw)

15.9 Soft Tissue Management for Placement of Direct or Indirect Restorations

Manipulation of soft tissue with various wavelengths and subgingival margin placement has been discussed earlier in this chapter. The restorative dentist must consider and apply all of those principles while preparing the tooth. Modern dental direct restorative materials generally require meticulous control of moisture and bleeding to ensure a good bonding environment. Successful fabrication of indirect restorations involves many factors. One important one is to duplicate the finished preparation with as much accuracy as possible. Whether for direct fillings, impressions, or optical scanning, any available dental laser can accomplish soft tissue management, moisture control, debridement, and hemostasis. Moreover, the laser can achieve excellent restorative results [29].

Figure 15.17 depicts a recurrent carious lesion on the upper right central incisor that extends into the gingival sulcus. The 9300-nm carbon dioxide laser recontoured the impinging gingival tissue, removed the carious lesion, and re-prepared the tooth for a new restoration. (Further discussion about lasers caries removal can be found in ► Chap. 8.)

Figure 15.18 shows how an Nd:YAG laser is used for troughing around two preparations during the construction of new crowns. The laser is gently placed in the sulcus and aimed toward the gingival soft tissue. It offers excellent tissue management which facilitates any impression technique.

Figure 15.19 depicts an Er,Cr:YSGG laser used for troughing around a molar. The rigid tip was aimed at

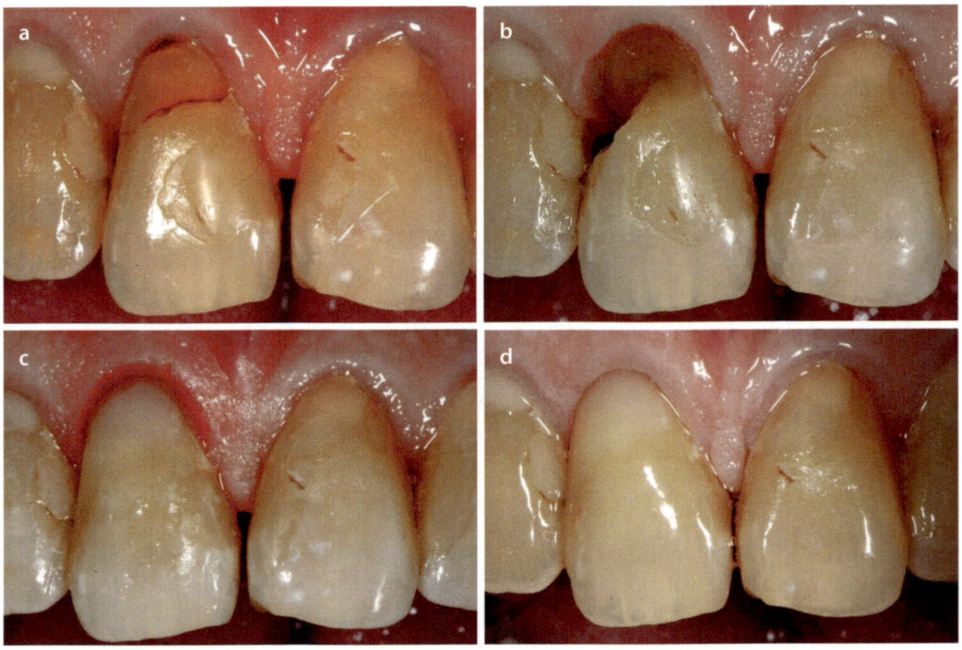

Fig. 15.17 **a** Preoperative view showing a carious lesion that extends subgingivally. **b** The gingiva was recontoured with a 9300-nm carbon dioxide laser using 1.00-mm spot size, 1% water mist spray, and a cutting speed of 20–30%. The existing composite was removed, and the tooth was prepared with the same settings except for increasing the water spray mist from 1% to 80%. Precise control of the speed of ablation was achieved by varying the working distance, the hand speed, and the cutting speed. **c** The immediate postoperative view shows the new restoration with good soft tissue contour. Note the good moisture control to aid in the composite placement. **d** A 6-day postoperative view showing the healed gingival tissue. (Clinical case courtesy of Dr. Josh Weintraub)

15

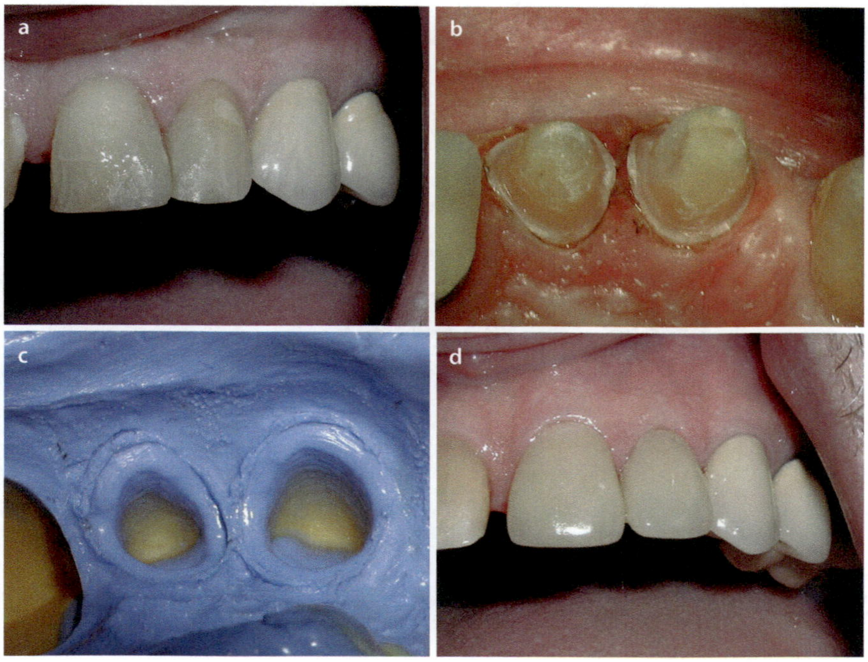

Fig. 15.18 **a** Preoperative view of the maxillary left central and lateral incisors which will be prepared for porcelain crowns. **b** Immediate postoperative view of the laser troughing. An Nd:YAG laser was used with a 320-μm bare fiber at an average power of 1.2 W (40 mJ at 30 Hz) in the sulcus, aimed at the soft tissue side of the pocket with gentle pressure. The fiber is used in short arcs of a circle, interacting with small segments of the tissue at a time. The goal was to gently retract and debride the tissue space along with controlling bleeding so that the impression material can accurately capture the subgingival margins. The interdental papilla was also slightly contoured. This average power is less than the parameters generally used for incisions since there is no need for any tissue removal. **c** The resulting impression shows accurate marginal detail. **d** Three-week postoperative and 1-week postdelivery view of the completed restorations showing an excellent tissue response

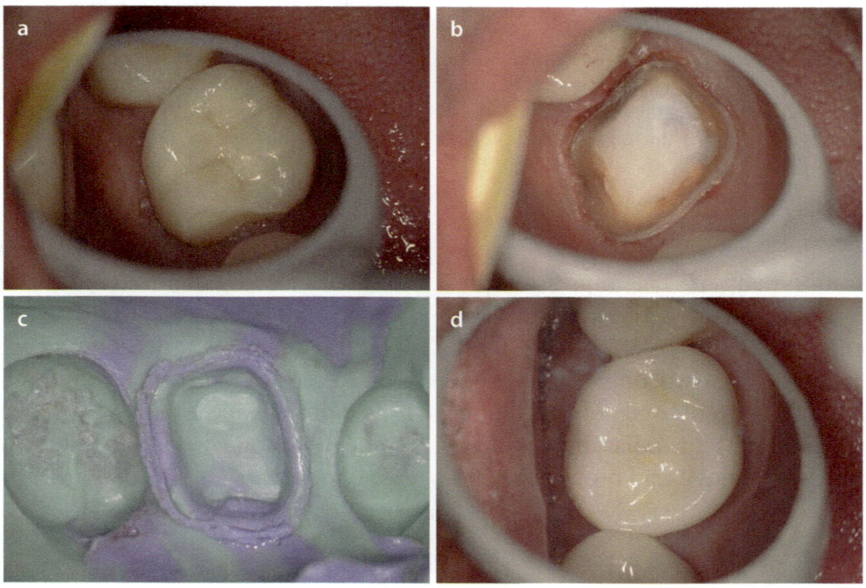

Fig. 15.19 **a** Preoperative mirror view of a maxillary molar defective crown restoration. A new crown will be constructed. **b** Immediate postoperative mirror view of the laser troughing. After the tooth was prepared, an Er,Cr:YSGG laser was used with an MZ 5 tip at an average power of 2.25 W (75 mJ per pulse at 30 Hz) in the sulcus with minimal water spray. The tip was moved around the tooth in short arcs of a circle to debride and widen the sulcular tissue so that the subgingival margins were revealed. (**c**) A photo of the final impression. Note the clear margin definition. **d** The final crown was delivered with excellent results including gingival health. (Clinical case courtesy of Dr. Glenn van As)

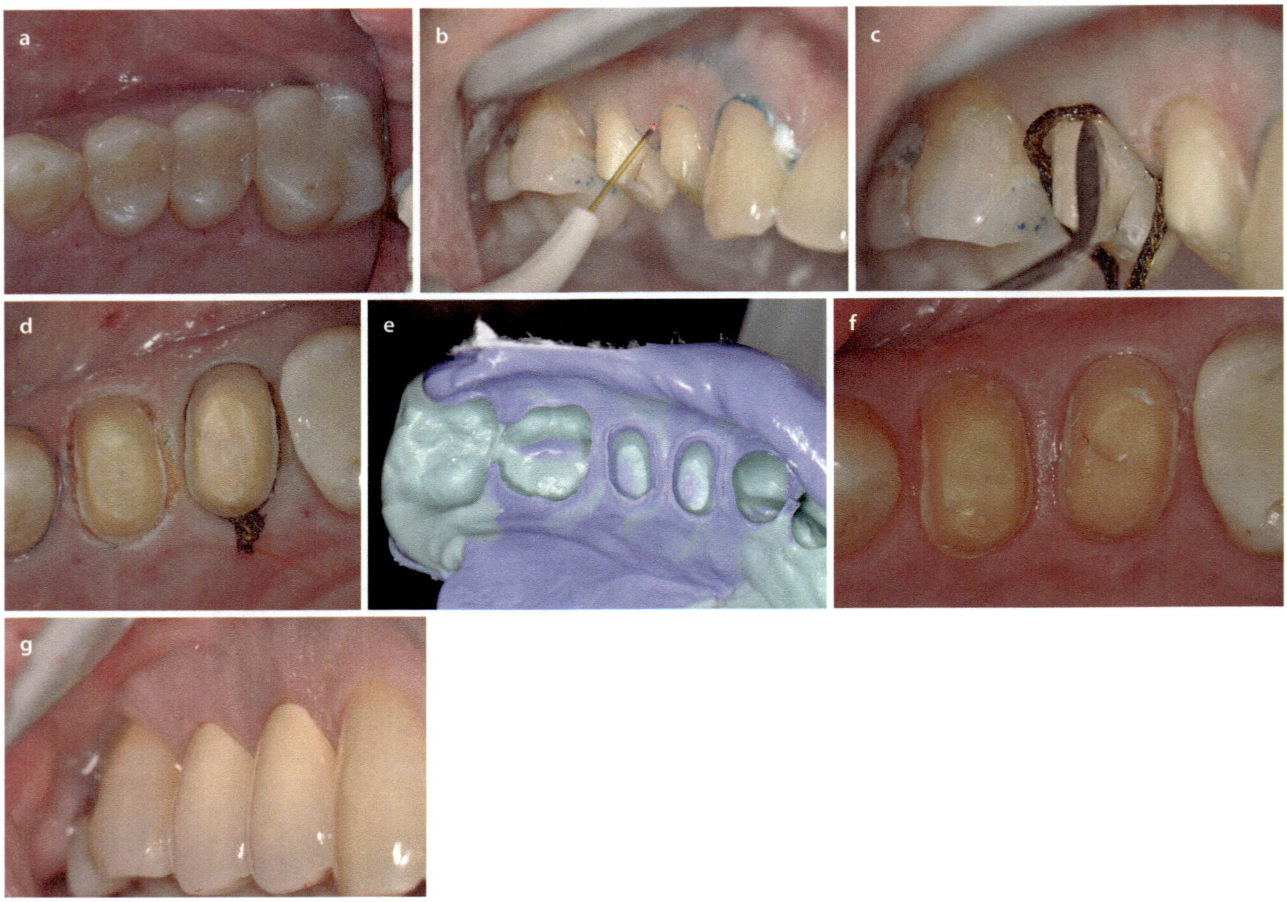

Fig. 15.20 **a** The mirror view of two maxillary premolar teeth that will receive full porcelain crown restorations. **b** A diode laser with a 400-μm bare fiber and an average power of 0.8 W is used parallel to the long axis of the preparation of the first premolar to expose the margins and debride the sulcus. The fiber is used in short arcs of a circle. **c** Retraction cord is placed in the sulcus of the second premolar. **d** The preparations are ready for the impression. Note the excellent control of bleeding and tissue retraction on the first premolar's gingiva after the laser use. **e** View of the final impression showing excellent capture of both preparations. **f** Two-week postoperative view of the preparations, after the provisional restorations are removed, showing adequate gingival health. **g** Three-week postoperative view of the final restorations showing an excellent result. (Clinical case courtesy of Dr. Glenn van As)

the soft tissue in the sulcus, being careful to avoid contact with the preparation margins. The impression precisely captured all the marginal detail (clinical case courtesy of Dr. Glenn van As).

Figure 15.20 shows a comparison between diode laser tissue retraction and conventional cord technique during full-crown preparations on adjacent maxillary premolars. Using careful technique and proper parameters, the laser easily reveals the subgingival margins with excellent bleeding control (clinical case courtesy of Dr. Glenn van As).

Figure 15.21 shows how a carbon dioxide laser is used to retract the tissue around a central incisor. The 10,600-nm wavelength is very effective in soft tissue removal while achieving excellent hemostasis (clinical case courtesy of Dr. Steven Parker).

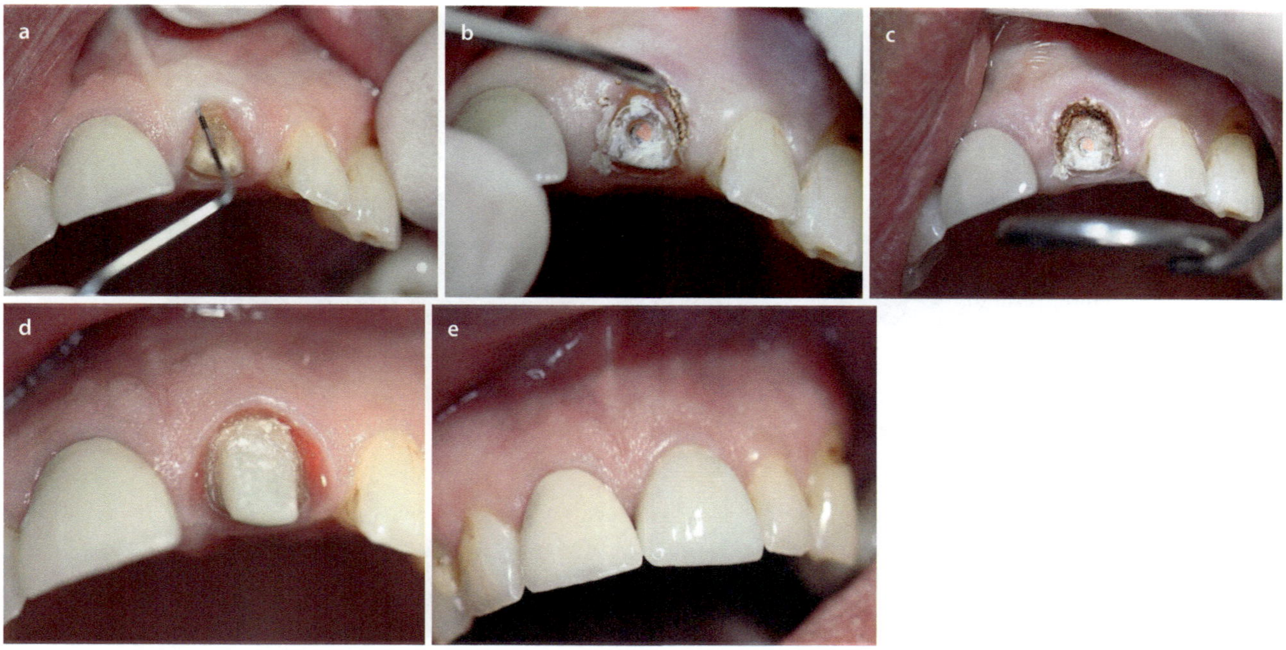

Fig. 15.21 **a** Preoperative view of a crown preparation in progress while determining the dentogingival complex measurement. **b** A 10,600-nm carbon dioxide laser used with a 600-μm beam diameter in noncontact at a power of 1.0-W continuous wave. The beam was aimed precisely at the soft tissue while avoiding interaction with any tooth structure. **c** Immediate postoperative view. The small area of tissue removal that appears carbonized will be rinsed away; however, the hemostasis is excellent. **d** Two-week postoperative view shows the healed tissue contour which will facilitate the good emergence profile of the restoration. **e** View with crown delivered. (Clinical Case courtesy of Dr. Steven Parker)

15.10 Tissue Preparation for a Fixed Prosthodontic Pontic Restoration

As discussed, the emergence profile ultimately determines the periodontal and aesthetic success of the dentition. Toward that end, the periodontal tissues can be manipulated with dental lasers to provide a stable and healthy foundation to guide the axial contours of the restoration or natural tooth. Crown lengthening can be used to perform these alterations. In the case of a fixed bridge pontic, the edentulous ridge can be prepared with a concave area so that the apical portion of the pontic can be made into a convex surface. This avoids a "ridge-lap" design which usually prevents the patient from adequate oral hygiene in the area. Instead, a more natural-appearing prosthodontic restoration can be fabricated.

Figure 15.22 shows the development of an ovate pontic area prior to replacing a fixed bridge. The tissue surface of the previous pontic was poorly contoured which resulted in chronically inflamed tissue. A diode laser was used to remove and reshape the soft tissue. The long-term (15-year) picture demonstrates how this procedure allowed the patient to maintain periodontal health (case courtesy of Dr. David Hornbrook).

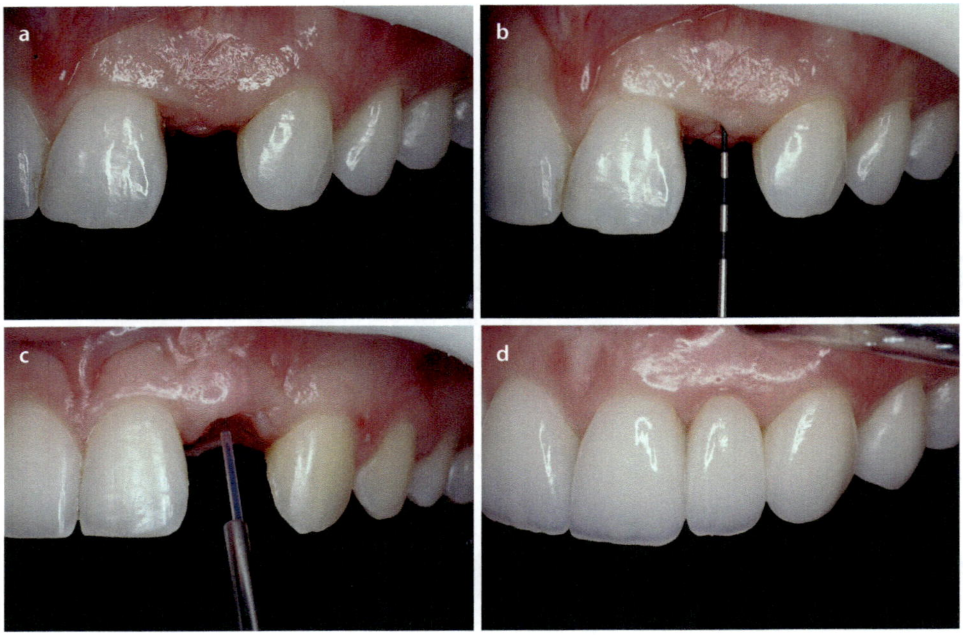

Fig. 15.22 **a** Preoperative view of edentulous ridge after removal of a bonded bridge. Note the chronic inflammation of the tissue due to the ridge lap design of the previous pontic, which prevented the patient from adequately cleaning the area. **b** A periodontal probe is used to measure the tissue thickness. There is sufficient tissue to allow removal for an ovate pontic design. **c** A diode laser is used with a bare 400-μm fiber at 1.0 W continuous wave emission to sculpt the concavity where the convex pontic will be positioned. The laser shaping can proceed with small amounts of tissue removal, but the clinician must be careful to leave at least 1.0 mm of tissue covering the bone. **d** A 15-year postoperative view showing excellent periodontal health with the new bridge restoration with its ovate pontic. The patient can easily maintain the pontic space and adjacent tissue. (Clinical case courtesy of Dr. David Hornbrook)

15.11 Conclusion

15

The purpose of this chapter is to demonstrate that utilization of the variety of dental wavelengths allows the clinician to manage the soft and hard tissue surrounding the teeth precisely and predictably. Any available dental surgical laser will incise and ablate soft tissue, although the interaction can vary among the emission wavelengths. For removal and contour of osseous tissue, the available choice of instruments is more limited to the erbium family and the 9300-nm carbon dioxide ones. These latter "all-tissue" lasers can facilitate treatment by allowing incremental removal of the tissues so that the target treatment section can harmonize with the adjacent areas. However, careful placement of the laser beam is essential to avoid unintended removal of one tissue while treating the other.

As always, thorough diagnosis and detailed treatment planning of a well-chosen case are highly important. It is equally important that the clinician be familiar with current periodontal surgical therapies and protocol, which can be found in any textbook [30]. The treatment phase must pay attention to several biologic principles so that the dentogingival complex and the tooth and/or restoration are harmonious and allow the patient to maintain good oral hygiene. Elective aesthetic procedures require the same principles along with elements of smile design and other dentofacial aesthetic details. In order to provide sufficient tooth structure for a successful restoration, the biologic width must be respected.

Thus, for successful placement of restorations and pleasing aesthetic procedures, a dental laser is a beneficial and significant addition to the clinician's armamentarium.

References

1. Ochsenbein C, Ross S. A re-evaluation of osseous surgery. Dent Clin N Am. 1969;13(1):87–102.
2. Seibert JL, Lindhe J. Esthetics and periodontal therapy. In: Lindhe J, editor. Textbook of clinical periodontology. 2nd ed. Copenhagen, Denmark: Munksgaard; 1989. p. 477–514.
3. Cook DR, Mealey BL, Verrett RG, Mills MP, Noujeim ME, Lasho DJ, Cronin RJ Jr. Relationship between clinical periodontal biotype and labial plate thickness: an in vivo study. Int J Periodontics Restorative Dent. 2011;31(4):345–54.

4. Serino G, Wennström JL, Lindhe J, Eneroth L. The prevalence and distribution of gingival recession in subjects with a high standard of oral hygiene. J Clin Periodontol. 1994;21(1):57–63.
5. Pontoriero R, Carnevale G. Surgical crown lengthening: a 12-month clinical wound healing study. J Periodontol. 2001;72(7):841–8.
6. Gargiulo AW, Wentz FM, Orban B. Dimensions and relations of the dentogingival junction in humans. J Periodontol. 1961;32(3):261–7.
7. Schmidt JC, Sahrmann P, Weiger R, Schmidlin PR, Walter C. Biologic width dimensions—a systematic review. J Clin Periodontol. 2013;40(5):493–504.
8. Lanning SK, Waldrop TC, Gunsolley JC, Maynard JG. Surgical crown lengthening: evaluation of biologic width. J Periodontol. 2003;74(4):468–74. https://doi.org/10.1902/jop.2003.74.4.468.
9. Kois JC. Altering gingival levels: the restorative connection. Part 1: Biologic variables. J Esthet Dent. 1994;6(1):3–9.
10. Gunay H, Seeger A, Tschernitschek H, Geurtsen W. Placement of the preparation line and periodontal health—a prospective two-year clinical study. Int J Periodontics Restorative Dent. 2000;20(2):171–81.
11. Croll BM. Emergence profiles in natural tooth contour. Part I: Photographic observations. J Prosthet Dent. 1989;62(1):4–10.
12. Tarnow D, Magner A, Fletcher P. The effect of the distance from the contact point to the crest of bone on the presence or absence of the interproximal dental papilla. J Periodontol. 1992;63(12):995–6.
13. Oquendo A, Brea L, David S. Diastema: correction of excessive spaces in the esthetic zone. Dent Clin N Am. 2011;55(2):265–81.
14. Hempton TJ, Dominici JT. Contemporary crown lengthening therapy: a review. J Am Dent Assoc. 2010;141(6):647–55. https://doi.org/10.14219/jada.archive.2010.0252.
15. Parwani SR, Parwani RN. Surgical crown lengthening: a periodontal and restorative interdisciplinary approach. Gen Dent. 2014;62(6):15–9.
16. Pilalas I, Tsalikis L, Tatakis DN. Pre-restorative crown lengthening surgery outcomes: a systematic review. J Clin Periodontolol. 2016;43(12):1094–108. https://doi.org/10.1111/jcpe.12617.
17. Marzadori M, Stefanini M, Sangiorgi M, Mounssif I, Monaco C, Zucchelli G. Crown lengthening and restorative procedures in the esthetic zone. Periodontol 2000. 2018;77(1):84–92. https://doi.org/10.1111/prd.12208.
18. Aroni MAT, Pigossi SC, Pichotano EC, de Oliveira GJPL, Marcantonio RAC. Esthetic crown lengthening in the treatment of a gummy smile. Int J Esthet Dent. 2019;14(4):370–82.
19. Kao R, Dault S, Frangadakis K, Salehieh JJ. Esthetic crown lengthening: appropriate diagnosis for achieving gingival balance. J Calif Dent Assoc. 2008;36(3):187–91.
20. Camargo P, Melnick P, Camargo L. Clinical crown lengthening in the esthetic zone. J Calif Dent Assoc. 2007;35(7):487–98.
21. Okuda W. Smile design 2.0: evolving from our past to be successful in treating the modern cosmetic patient. Gen Dent. 2016;64(1):10–3.
22. Garber D, Salama M. The aesthetic smile: diagnosis and treatment. Periodontol 2000. 1996;11:18–28.
23. Zucchelli G, Sharma P, Mounssif I. Esthetics in periodontics and implantology. Periodontol 2000. 2018;77(1):7–18. https://doi.org/10.1111/prd.12207.
24. Kois J. The restorative-periodontal interface: biological parameters. Periodontol 2000. 1996;11:29–38.
25. Altayeb W, Arnabat-Dominguez J, Low S, Abdullah A, Romanos GE. Laser-assisted esthetic crown lengthening: open-flap versus flapless. Int J Periodontics Restorative Dent. 2022;42(1):53–62. https://doi.org/10.11607/prd.5335.
26. Flax H, Radz G. Closed-flap laser-assisted esthetic dentistry using Er:YSGG technology. Compendium. 2004;25(8):622–8.
27. Cobb C. Lasers in periodontics: a review of the literature. J Periodontol. 2006;77:545–64.
28. McGuire M, Scheyer ET. Laser-assisted flapless crown lengthening: a case series. Int J Periodontics Restorative Dent. 2011;31:357–64.
29. Beyza UD, Bayza KN, Alperen D. Evaluation of gingival displacement methods in terms of periodontal health at crown restorations produced by digital scan: 1 year clinical follow-up. Lasers Med Sci. 2021;36:1323–35. https://doi.org/10.1007/s10103-021-03266-5.
30. Newman MG, Takei HH, Klokkevold PR, Carranza FA. Carranza's clinical periodontology. 11th ed. St Louis, MO: Elsevier Saunders; 2012.

Impact of Laser Dentistry in Management of Color in Aesthetic Zone

Kenneth Luk and Eugenia Anagnostaki

Contents

D. J. Coluzzi, S. P. A. Parker (eds.), *Lasers in Dentistry—Current Concepts*, Textbooks in Contemporary Dentistry,
https://doi.org/10.1007/978-3-031-43338-2_16

Core Message

A pleasingly attractive smile is composed of a harmonious balance of a well-aligned set of teeth with healthy anatomic contour of the gingiva and lips. Patients' demand for aesthetic dentistry can be met with the practitioner's choice of procedures and materials. Dental lasers can certainly be integrated into the treatment plan, and this chapter will provide details about modifying the color of the dental soft and hard tissues to help attain the desired outcome. The first section will describe on treatment for pigmentation of nonneoplastic origin. The second section will provide details about whitening.

16.1 Gingival and Lip Pigmentation Management

Kenneth Luk

16.1.1 Introduction

A well-aligned set of teeth not only improves function and oral hygiene maintenance but also most importantly produces a confident and attractive smile. Gingival contour, emergence profile, and teeth proportion all contribute to the smile profile [1]. However, dark pigmentation (Fig. 16.1) of the gingiva and lip may deflect the attention of a perfect smile [2]. Melanin pigmentation of the gingiva and lip is most commonly noticed in ethnic groups [2–4], smokers [3, 5, 6], patients under medication such as anti-malarial agents, tricyclic anti-depressant, and minocycline [7, 8] and hormonal disorder [9, 10]. Recent studies have also reported the correlation between passive smoking in women and children with hyperpigmentation [10–12]. Amalgam tattoo is a condition that is commonly observed in the gingiva [13]. Post-inflammatory hyperpigmentation (PIH) is a sequela of cutaneous disorder described in dermatology. PIH on the lip is a common cause by trauma (e.g., burns, lip biting, and trauma by pen and pencil).

Classification of Pigmentation

Pigmentation can be classified by Dummett index [14] and Takashi index [11].

The Dummett oral pigmentation index (DOPI) describes the color intensity of pigmentation into four scores:

1. No clinical pigmentation (pink-colored gingiva)
2. Mild clinical pigmentation (mild light brown color)
3. Moderate clinical pigmentation (medium brown or mixed pink and brown color)
4. Heavy clinical pigmentation (deep brown or bluish black color)

Takashi index describes the extension of melanin pigmentation into three scores:

1. Score 0: No pigmentation
2. Score 1: Solitary unit(s) of pigmentation in papillary gingiva without extension between neighboring solitary units
3. Score 2: Formation of continuous ribbon extending from neighboring solitary units

Fig. 16.1 Colours in oral pigmentation. Adapted from [15]

Color	Endogenous	Exogenous
Blue, red, purple	**Hemoglobin** – Varix, Hemangioma	-
Brown	**Hemosidrin** – Nevus, Drug induced pigmentation	-
Brown, black, grey	Melanin	**Chromogenic bacteria** – superficial colonization
Grey, black	-	**Silver amalgam, graphite** Amalgam tattoo, Nevus, trauma
Grey	-	**Lead, mercury, bismuth** Ingestion of paint or medicinals

Various Treatments for Depigmentation

The use of chemical peel and cryosurgery has been used for melanin depigmentation. Treatment with scalpel and diamond bur [16] has also been used and compared with lasers [17–21]. In a split mouth clinical trial, the use of lasers has been shown with less relapse 3 months postop when compared with electrosurgery [22].

16.1.2 Laser Ablation

For over 10 years, CO_2, Er:YAG, Er,Cr:YSGG, Nd:YAG, and diode lasers (◘ Fig. 16.2) have been reported with good results in melanin depigmentation [23–28]. The procedure requires ablation (vaporization) of the surface gingiva or mucosa to the basal layer where the melanocytes are located.

Erbium and CO_2 wavelengths (◘ Fig. 16.2) are well absorbed in water and thus will remove shallow layers of soft tissue and eventually reach the layer of melanin. Sapphire tips or focusing window are used in noncontact mode. The use of erbium lasers has the advantage of using water spray not only to keep the ablation front from dehydration but also to irrigate the debris to give a clear view during the procedure. Although this reduces the hemostatic effect, bleeding would be a good warning signal that the depth of ablation is beyond the basal layer. Water spray is not used with 10600nm CO_2 irradiation, but carbonization effect can be reduced with the short pulse durations that those CO_2 lasers offer.

9300nm CO_2 laser is the most recently developed CO_2 laser for both hard and soft tissue treatments. This laser delivers high output power and short pulse duration with the water spray. Precise control of ablation differs from accustomed laser parameters by selection of percentage of power and rheostatically controlled cutting speed. The powers and pulse durations were measured and converted to parameters generally discussed and illustrated by the author in ◘ Fig. 16.3 [29].

Visible light, near-infrared diode, and Nd:YAG (445–1064 nm) wavelengths are poorly absorbed in water but well absorbed in pigment. Initiated 320-μm fibers and similar-sized quartz tips are commonly used to ablate the layers in contact mode. Apart from direct surface absorption, the laser photons can also penetrate to the basal layer and capillaries to be absorbed (◘ Fig. 16.4). The tissue tag on the treatment site should be removed regularly to view the color improvement. For diode lasers, it has been demonstrated that water irrigation during ablation can be used to reduce the collateral thermal damage [30, 31].

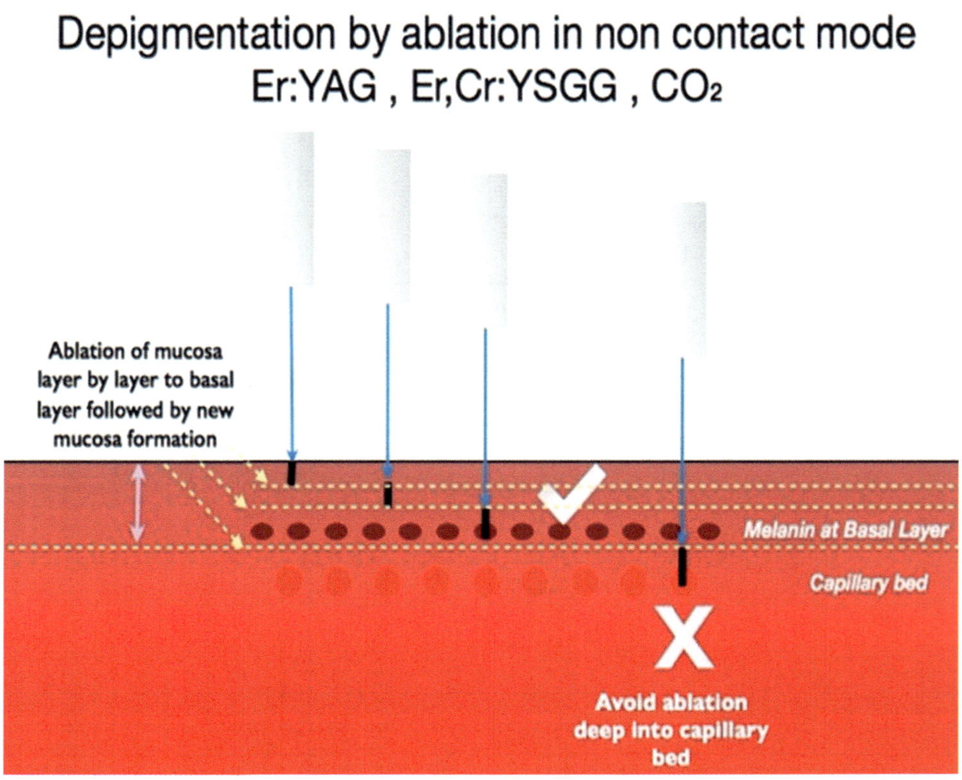

◘ **Fig. 16.2** Er:YAG, Er,Cr:YSGG, and CO_2 laser depigmentation by ablation in noncontact mode

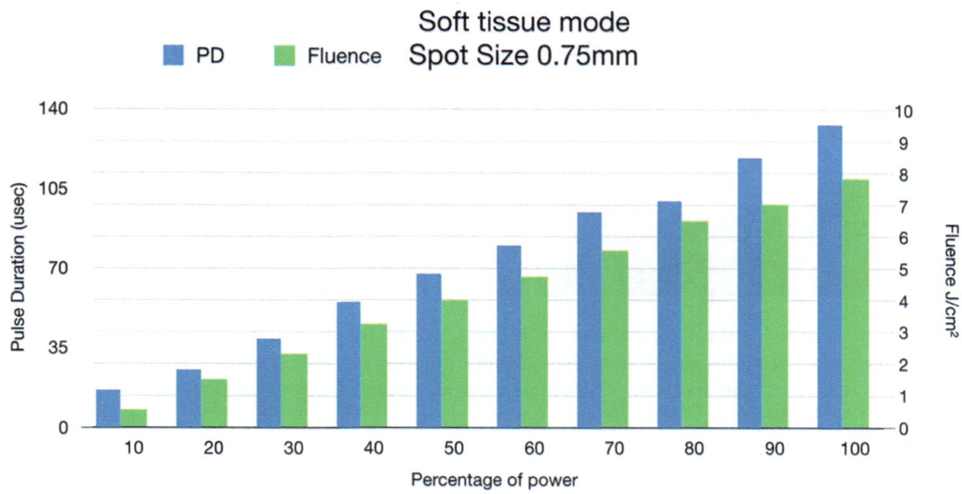

Fig. 16.3 Pulse duration and fluence in relation to the power percentage of 9300nm CO_2 laser in soft tissue mode with a repetition rate of 187 Hz. Peak power measured: 150 W at 10%, 260 W at 20–100% [29]

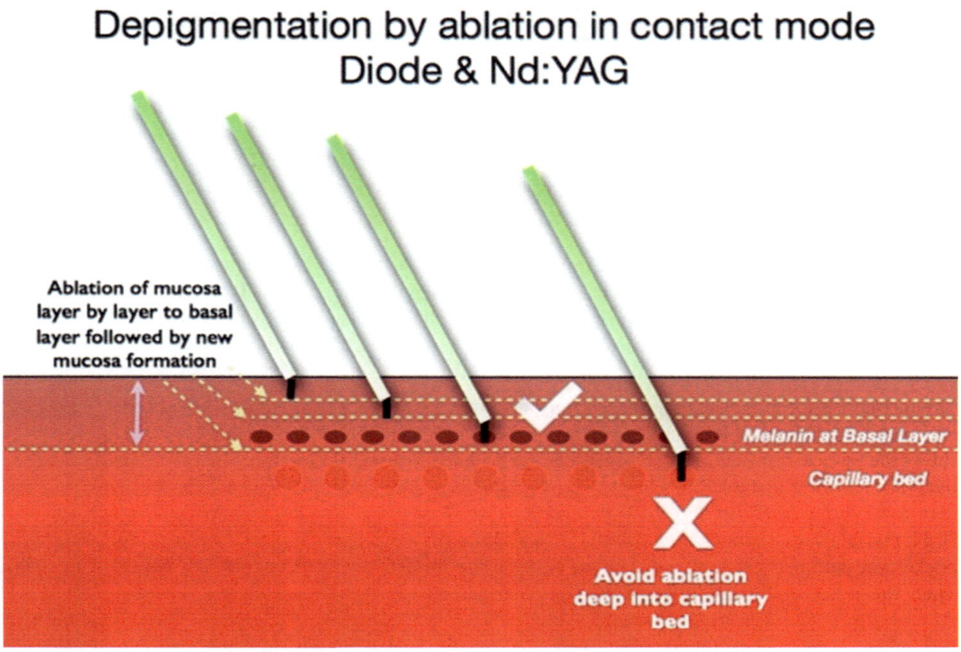

Fig. 16.4 Diode and Nd:YAG lasers depigmentation by ablation in contact mode

16.2 Clinical Cases of Ablative Depigmentation

A 19-year-old female requested for single-tooth gingivectomy on tooth #8. Patient also complained about her dark gingiva. The clinical objective was to remove hypertrophic tissue, expose tooth structure, and remove melanin for a more aesthetic smile with 9300nm CO_2 laser. Topical anesthetic (TAC 20 alternative) was used. The 2-week follow-up photo shows excellent healing and an aesthetically pleasing result (Fig. 16.5).

Figure 16.6 shows a case of split mouth depigmentation using an Er:YAG (2940 nm) and Nd:YAG (1064 nm) laser. The upper right quadrant was ablated using Er:YAG laser. This free-running pulse laser is well absorbed by water. Ablation of soft tissue is very effective as soft tissue contains high percentage of water by volume. Water spray was used not only to keep the surface cool but more importantly to prevent dehydration of the surface layer and to sustain the efficacy of ablation. The upper left quadrant was ablated by Nd:YAG laser which is effective in ablating the melanocytes. The free-running pulse emission can help to control the tissue temperature and avoid thermal damage to the underlying periodontium.

Figure 16.7 shows depigmentation with 980-nm diode laser. Under local anesthesia, an initiated 320-μm fiber was used to ablate the pigmented area with a sweeping motion. 1.9-W average power with 1000 Hz and a 30% emission cycle were set, and a peak power of 6.3 W

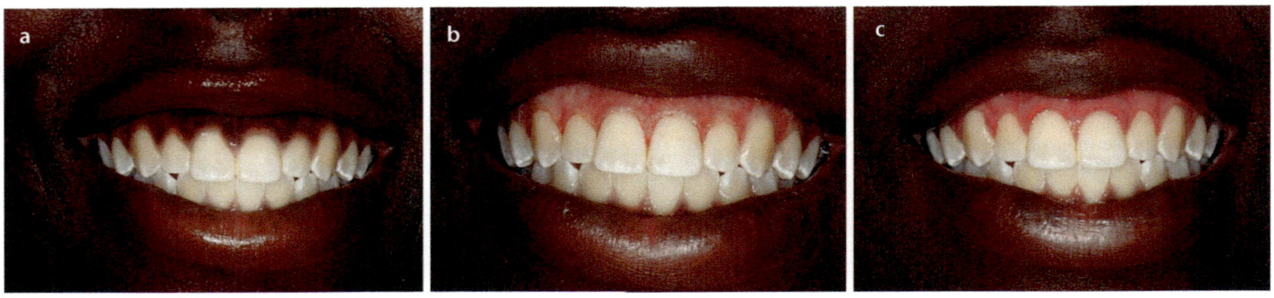

■ **Fig. 16.5** **a** Showing the preoperative view. **b** Immediate postoperative view. Gingivectomy on tooth #8 by 9300nm CO_2 laser was performed utilizing 1 mm spot size, 1% mist and cutting speed of ~30–40%. For the depigmentation, the same laser employed a 1.25-mm spot size, 1% mist, and cutting speed between ~20% and 30%. The procedure was performed utilizing the tipless contra-angle handpiece. The total procedure time was approximately 20 min. **c** Showing the 2-week postoperative view. (Clinical case courtesy of Dr Joshua P. Weintraub)

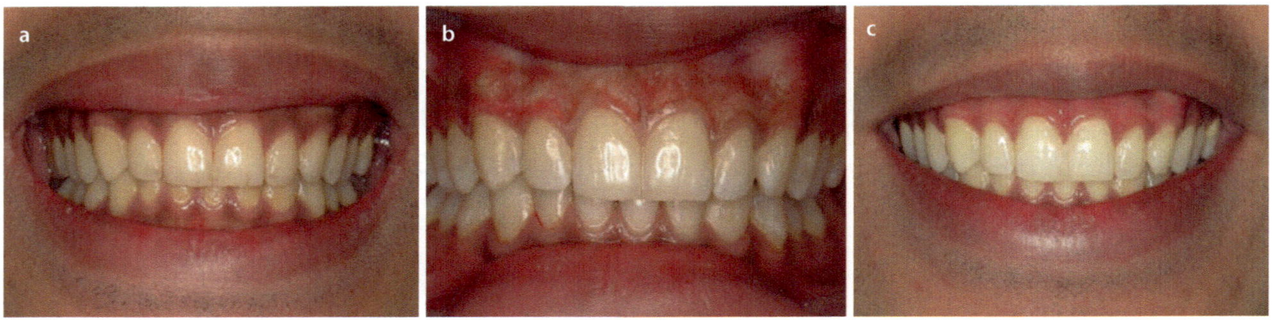

■ **Fig. 16.6** **a** Preoperative view of melanin pigmentation. **b** On the right quadrant, the Er:YAG laser was used with 800-μm sapphire tip at an average power of 1.8 W (120 mJ, 15 Hz) and a 1000-μs pulse duration. The Nd:YAG laser was used with a 320-μm bare fiber and an average power of 4 W (MSP mode and 20 Hz) on the left quadrant. Treatment time for each side was 10 min. **c** Six-month postoperative view shows stable tissue color and good tissue tone performed by both lasers.

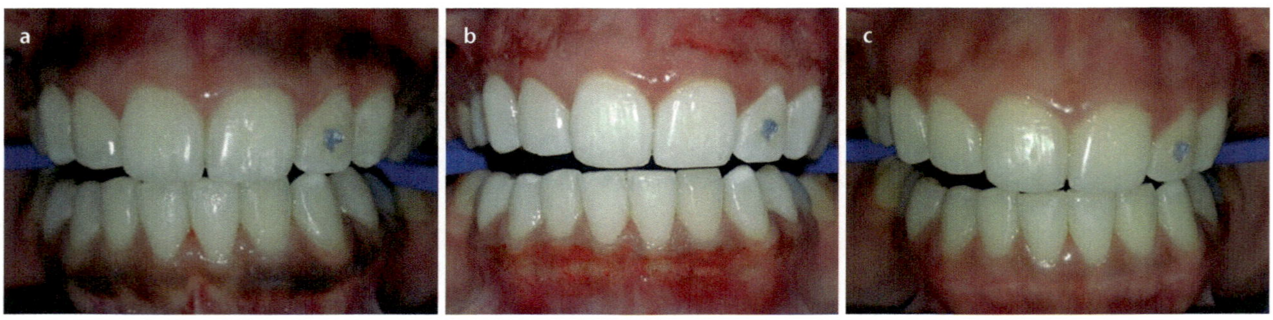

■ **Fig. 16.7** **a** Preoperative view of melanin pigmentation. **b** Immediate postoperative view of upper and lower anterior segments. **c** One week postoperative view showed mucosa with good maturation. Although some areas still showed pigmentation [11, 12], the overall aesthetic appearance was very pleasing. (Clinical case courtesy of Dr. Mudasser Iqbal)

was calculated. The treatment time took 12 min to complete the upper and lower anterior segments (clinical case courtesy of Dr. Mudasser Iqbal).

A recent study by Altayeb et al. [32] compared ablative gingival depigmentation on 60 patients using diode laser (940 nm) and Er,Cr:YSGG (2780 nm) lasers. A 2-year follow-up in efficiency of depigmentation, patient perceptions, and recurrence rates was evaluated. Treatment was significantly faster with Er,Cr:YSGG laser and required no anesthesia, with faster healing and less postoperative discomfort after 1 week of treatment, compared to the diode laser treatment. It was concluded that both lasers efficiently removed gingival pigments with comparable in clinical outcomes and overall positive patient experience.

16.2.1 Non-ablative Laser Procedures

The word non-ablative is frequently used in dermatology on skin resurfacing techniques. Unlike an ablative laser procedure which vaporizes the skin surface, a non-ablative procedure penetrates the surface and coagulates the subsurface layer of the skin while keeping the surface layer in place. In depigmentation, the author uses the description of "non-ablative" to describe this effect using near-infrared and visible light diode lasers. The surface mucosa will peel off at day 1 or 2 after treatment.

The Fundamental Interaction Is Photocoagulation

Photocoagulation, in this case laser coagulation, describes a laser-tissue interaction where the target tissue component is well absorbed by the laser wavelength and there is a coagulation effect when the temperature reaches 60 °C. Hemoglobin is very well absorbed in Nd:YAG, visible light, and near infra-diode lasers. Thus, photocoagulation is most commonly applied for hemostasis of a surgical wound such as extraction socket or a periodontal pocket. Treatment of vascular lesions such as varix and hemangioma can also benefit by photocoagulation.

Near-Infrared Diode Lasers

A non-ablative technique using a high-power 810-nm diode laser (30 W, 20 kHz, 16-μs pulse duration) was reported by the author [33, 34]. Since the 810-nm wavelength is well absorbed by melanin and hemoglobin, the use of a high-power laser can concentrate the thermal energy on the surface with a very short pulse (Fig. 16.8). A continuous sweeping motion should be used, and re-irradiation of the already coagulated area should be avoided. This technique resulted in a treatment time of 2 min for single arch between first premolars. Using a low-power diode laser with long pulse to continuous wave usually allows thermal conduction deeper into the tissue [30, 31]. A similar principle called laser-patterned microcoagulation using 20-W 980-nm diode laser was reported on depigmentation of one papilla [35].

16.2.2 Clinical Cases of Non-ablative Technique with an Infrared Diode

Figure 16.9 shows a case of melanin depigmentation using an 810-nm very short pulsed high peak power diode laser.

Figure 16.10 shows a case of lip depigmentation using an 810-nm very short pulsed high peak power diode laser. The patient, a young lady, had tried to mask the pigment with dark colored lipstick but was unable.

Figure 16.11 shows a case of photocoagulation of a lip hemangioma using an 810-nm diode. Three laser treatments were performed at a monthly interval. The patient wanted the lesion removed for her wedding photos.

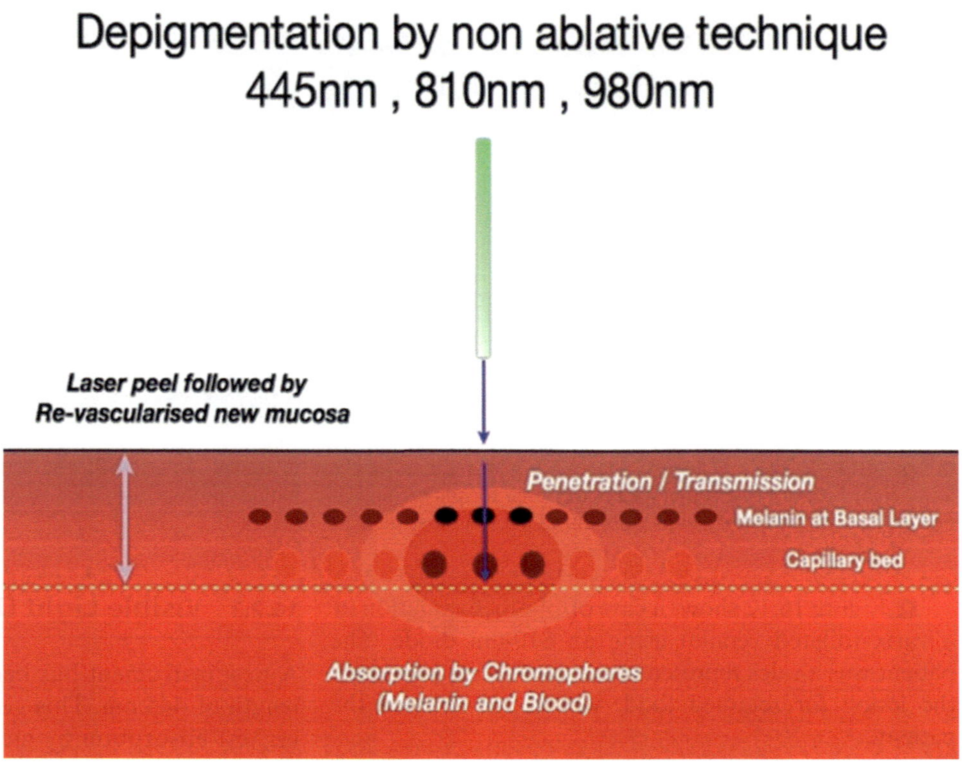

Fig. 16.8 445-, 810, and 980-nm laser depigmentation by non-ablative technique

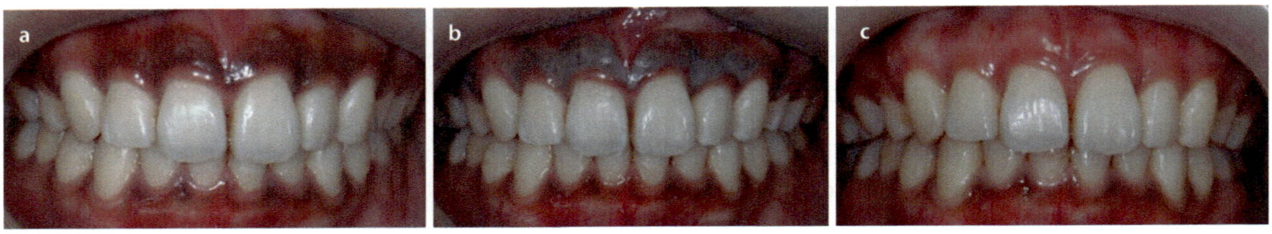

Fig. 16.9 **a** Preoperative photo of gingival melanin pigmentation. An 810-nm diode laser was used with a 600-μm bare fiber with 30 W of peak power, a 16-μs pulse duration, at 20 kHz. Under local anesthesia, the treatment time was 80 s on the maxillary arch only. **b** Immediate postop showing non-ablative mucosa. **c** The 6-week postoperative photo is shown and the pigment is not present

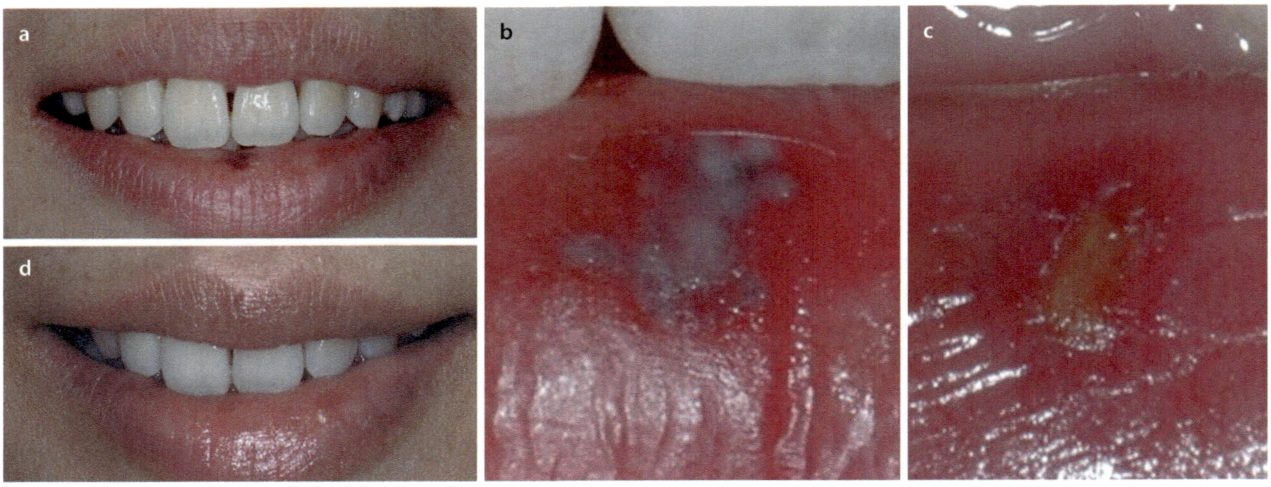

Fig. 16.10 **a** The preoperative view of a pigmented area on the lower lip. **b** The immediate postoperative view. Under local anesthesia, an 810-nm diode laser was used with a 600-μm fiber with 30 W of peak power, 16-μs pulse duration, at 20,000 Hz for an exposure time of 8 s with non-ablative technique. **c** Two days postoperative view. **d** A 5.5-year postoperative view showing no relapse of pigmentation

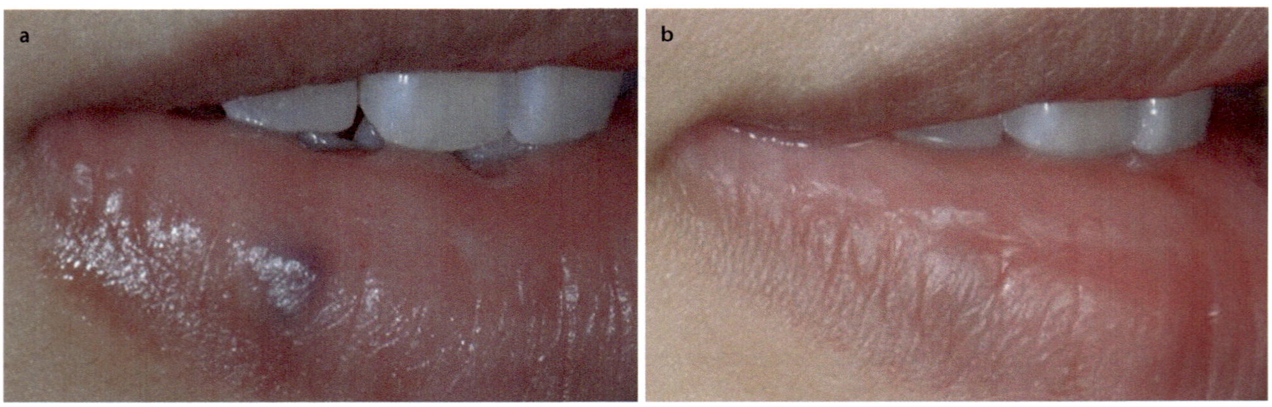

Fig. 16.11 **a** The preoperative view of a hemangioma on the lower right lip. An 810-nm diode laser was used for photocoagulation with 30 W, 20,000 Hz, 16-μs pulse duration (3 s in first visit, 3 s in second visit, and 10 s in third visit) without anesthesia. **b** Three months postoperative view showing normal lip tissue

Figure 16.12 shows a case of photocoagulation of a varix (dilated venule) using an 810-nm diode. The patient was a sales representative whose clients mistook the lesion for retained food, which embarrassed the patient.

16.2.3 Visible Light Diode

A new instrument has become available with a wavelength of 445 nm. This visible blue diode laser has the highest absorption by melanin and hemoglobin com-

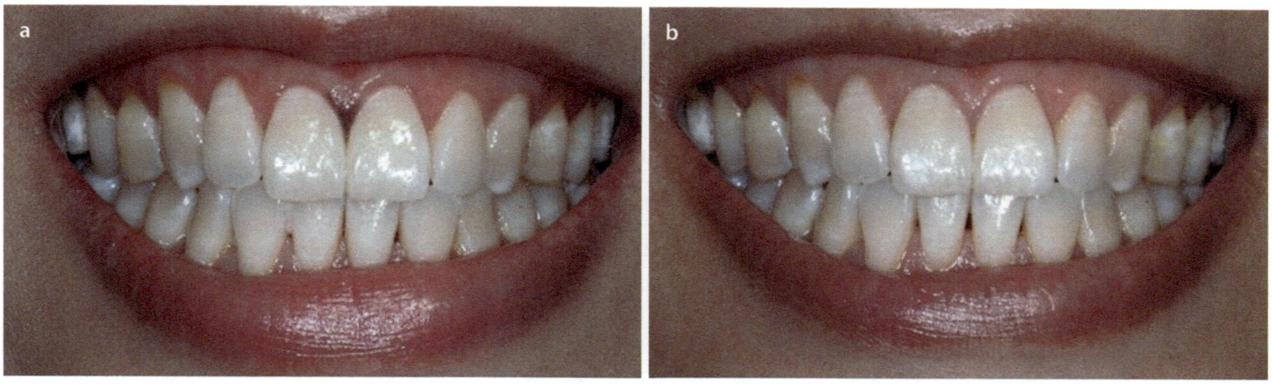

Fig. 16.12 **a** Preoperative view of a varix (venous blood vessel) on the interproximal papilla of the maxillary central incisors. An 810-nm diode laser was used for photocoagulation with 30 W, 20,000 Hz, 16-μs pulse duration for two visits. (Two seconds on labial in first visit, 2 s labial and 2 s palatal in second visit). **b** Eight months postoperative view

Table 16.1 A comparison of 445 and 810 nm

Wavelength (nm)	Parameters used	Power density	Treatment time	Comments
445	1 W, continuous wave, average power = 1.0 W	88 W/cm^2 2 mm defocused from tissue	40 s	Absorption in melanin ten times higher than 810 nm Absorption in hemoglobin 100 times higher than 810 nm Novel wavelength with no complications yet reported; however, the absorption and significantly lower power density should produce fewer complications
810	30-W peak power Pulse duration 16 μs, 20,000 pulses per second Average power 9.9 W	1697 W/cm^2 2 mm defocused from tissue	40 s to 3 min for a comparable area	Complications can be gingival recession and bone necrosis

pared with other dental laser wavelengths. In ablative technique, Taher Agha and Polenik [36] evaluated 30 patients using 445-, 940-, 2780-nm wavelengths. It was concluded that all three wavelengths were fast, effective in pigmentation removal, and well tolerated by patients.

Visible blue diode laser (445 nm) can also be used with non-ablative technique. The author showed 1-W continuous wave being comparable in achieving the same result and speed as 30-W 810-nm diode laser [37]. This much lower-power density produces the same effect and can minimize any unwanted thermal damage to the tissue. Table 16.1 shows some further details in comparing the two wavelengths.

16.2.4 Clinical Case of Visible Light Diode for Non-ablative Depigmentation

Figure 16.13 shows a depigmentation case by 445-nm wavelength.

16.2.5 Pigmentation of Exogenous Origin

Metal tattoo such as amalgam is not uncommon in pigmentation of the gingiva. Pigmentation lip is regularly caused by pencil (lead) and pen.

Clinical Cases of Metal Tattoo

Figure 16.14 shows a patient presented with amalgam tattoo at the gingival margin and papilla on the upper central incisors [38]. Under microscope and local anesthesia, the Er:YAG laser (40 mJ at 30 Hz) with water spray was used. The metal debris embedded in the connective tissue was effectively removed, without producing major thermal damage such as carbonization and coagulation (b). At one week of postoperative view (c) and one year of post treatment (d), favorable wound healing was achieved without any gingival tissue defects or recession. The gingival color recovered a natural aesthetic appearance. (Ishikawa I et al. Potential applications of Erbium:YAG laser in periodontics. J Periodont Res; 2004; 275–285.; with permission. Journal of Periodontal

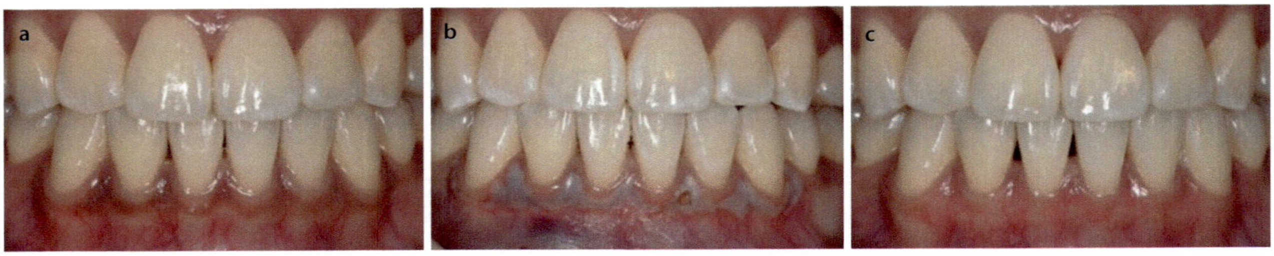

Fig. 16.13 **a** The preoperative view of the lower anterior gingiva with melanin pigmentation. **b** The 445-nm diode laser with a 320-µm bare fiber, defocused 2 mm from the tissue, was used at 1-W CW for 40 s under local anesthesia. Immediate postop with non-ablative technique. (Gingiva between the lower left central and lateral incisors shows an ablated site indicating over-irradiation beyond coagulation temperature. **c** The 5-month postoperative view shows normal tissue color and contour

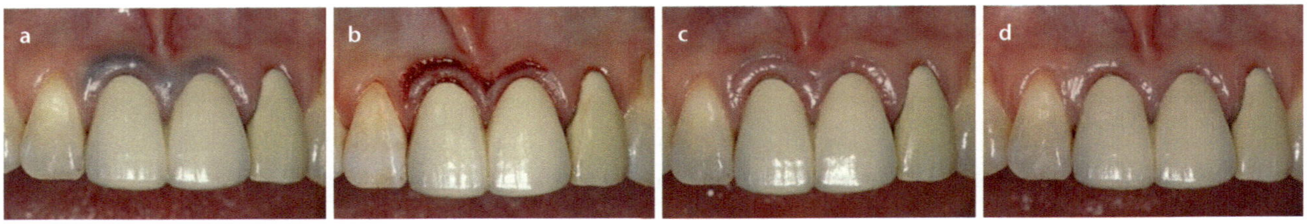

Fig. 16.14 A 73-year-old female presented with an amalgam tattoo on the attached gingiva on the distal buccal of the implant site. The previously extracted #4 had an existing large DO amalgam with radiographic evidence of the tattoo apical to the gingival floor of the restoration. The 9300nm CO laser and topical anesthetic were used. The laser does not selectively ablate pigment but removes the affected tissue and pigment (Fig. 16.15). (Clinical case courtesy of Dr Joshua P. Weintraub)

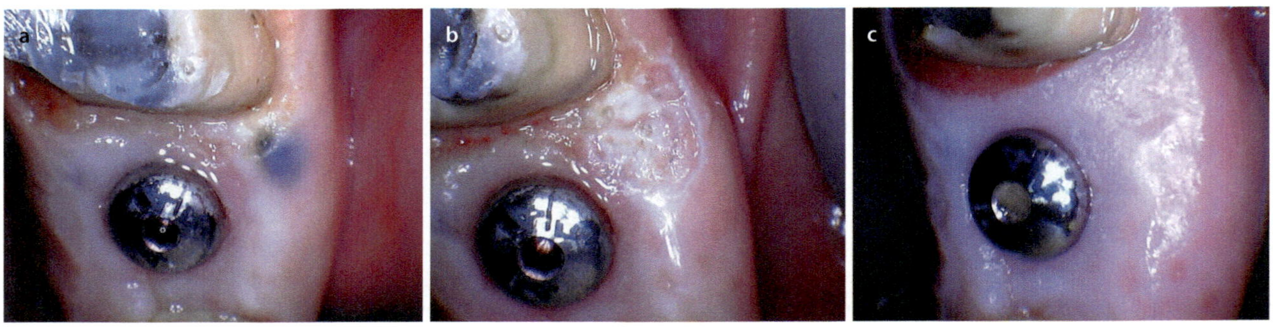

Fig. 16.15 **a** Preoperative photo. **b** The amalgam tattoo was ablated utilizing a 9300nm CO_2 laser with a 1.00-mm spot size, 1% mist, and cutting speed between ~20% and 30%. Tissue was carefully removed in layers until the tattoo was gone. The total procedure time was approximately 3 min. **c** A 6-week postoperative view, showing excellent healing

16

Research ©Copyright (2004) Blackwell Munksgaard, Inc., Pictures are courtesy of Dr. Akira Aoki).

A 73-year-old female presented with an amalgam tattoo on the attached gingiva on the distal buccal of the implant site. The previously extracted #4 had an existing large DO amalgam with radiographic evidence of the tattoo apical to the gingival floor of the restoration. The 9300nm CO_2 laser and topical anesthetic were used. The laser does not selectively ablate pigment but removes the affected tissue and pigment (Fig. 16.15). (Clinical case courtesy of Dr Joshua P. Weintraub)

The 6-week follow-up photo shows pink keratinized tissue.

Clinical Cases of Post-inflammatory Hyperpigmentation (PIH)

Figure 16.16 depicts a 25-year-old female reported pigmentation on lower lip traumatized by pencil at childhood. Such pigment is more likely to be a result of PIH. The procedure was accomplished with an 810-nm diode laser.

Figure 16.17 shows a 40-year-old male complaining of pigmentation on the lower lip. It was revealed that he had a habit of biting the lower lip at his younger age showing the result of PIH. An 810-nm diode was used in a non-ablative mode for the procedure.

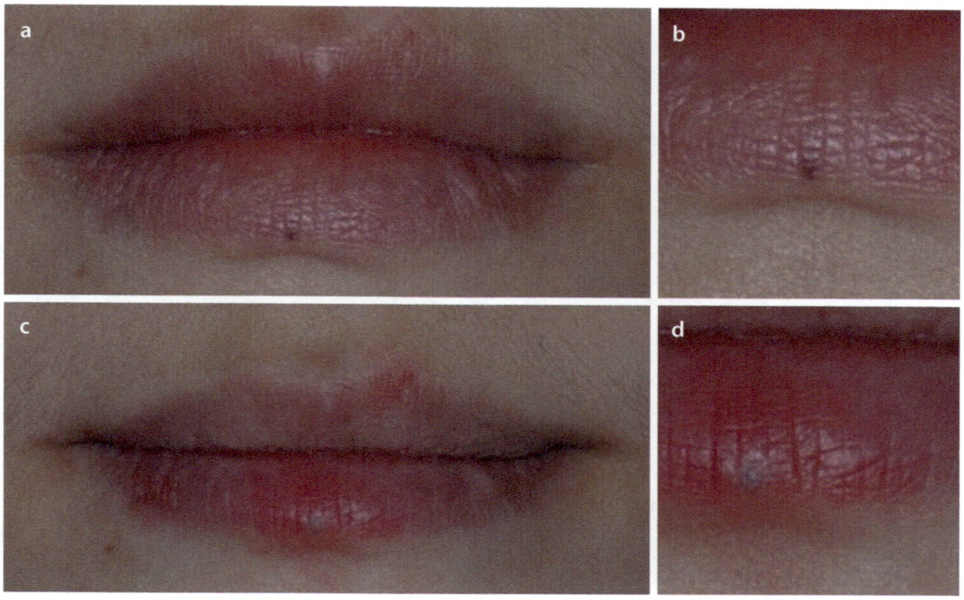

■ **Fig. 16.16** **a**, **b** Preoperative photo. **c** Immediate postoperative view where the 810-nm diode laser set at 30 W, 16 μs, and 20,000 Hz was used in noncontact mode. The pigment was ablated in less than 1 s without local anesthesia. **d** Five-year postoperative photo shows no recurrence

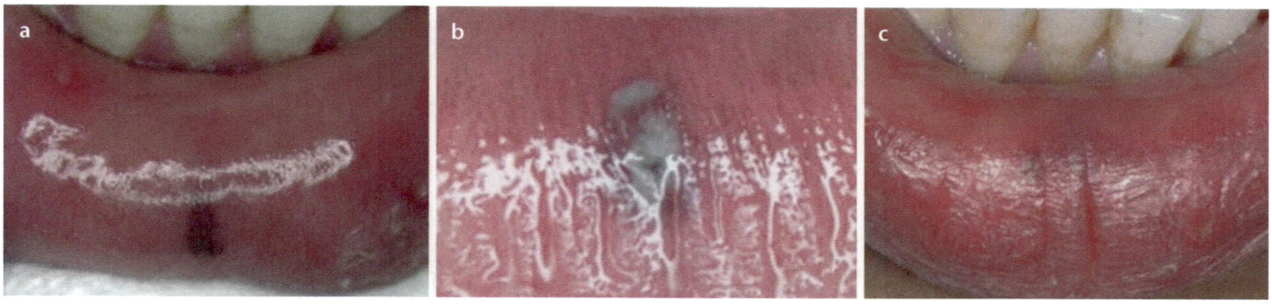

■ **Fig. 16.17** **a** Preoperative view. **b** Immediate postoperative view after an 810-nm laser exposure of 1 s with 30 W, 20,000 Hz, 16-μs pulse duration using a non-ablative technique with anesthesia. **c** One-year postoperative view

16.2.6 Postoperative Care

The patient should be advised to avoid smoking, alcohol, and acidic and spicy food. Sashimi is best avoided for the first few days. Gentle tooth brushing around the gingival margin and warm salt mouth rinse is recommended. In general, no analgesic is required. However, there has been report from mild discomfort to pain with hot food 1 day after treatment [39, 40].

16.2.7 Recurrence of Melanin Pigmentation

Although depigmentation with laser is an effective procedure, recurrence of pigment should be considered. The causes of pigmentation have been described in ▶ Sect. 16.1. The mixed reports of recurrence are listed in ■ Table 16.2 [24, 25, 28, 32, 40–46]. Depigmentation of gingiva was performed on the anterior segments of the maxilla and mandible. Ablative technique was used in all articles. Two of the four reports cited reported recurrence after 1 year [40] and at 2 years [42] after treatment. Recently, evaluation of 940-nm and 2780-nm depigmentation by Altayeb et al. [32] reported long-term stability of gingival color over the 2 years was better with diode laser. The re-pigmentation intensity and extensity were higher significantly in the Er,Cr:YSGG group than in the diode group at 1 year and 2 years. Patients should be aware of the recurrence of pigmentation, but this can be retreated readily. Cessation of medication and giving up on smoking habit reduce the likelihood of relapse.

16.2.8 Treatment Duration

El Shenawy [47] reported 15 cases of depigmentation on the maxilla and mandible using 3-W continuous wave 980-nm diode laser. The cases were completed in contact mode within 20–25 min. Berk [41] reported two cases using Er,Cr:YSGG laser completing the procedure on upper and lower anterior segments in 30 min. In con-

Table 16.2 Details of published articles about recurrence of pigmentation

Recurrence of pigmentation Author	Wavelength	Re-evaluation period (months)	No. of cases	Location	No. of Recurrence
Atsawasuwan J Periodontal 2000 [24]	Nd:YAG (1064 nm)	11–13	4	Max	0
Tal H. J Periodontal 2003 [25]	Er:YAG (2940 nm)	6	10	Max	0
Doshi Y Int J Laser Dent 2012 [28]	940 nm	12	1	Max	1
Altayeb et al. Clin Oral Invest 2021 [32]	940 nm (diode) Er,Cr:YSGG (2780 m)	24 24	30 30	Max	11 25
Ozbayrak Oral Surg Oral Med Oral Path 2000 [40]	CO_2 10,600 nm	18	8	Max	0
Berk G. J Oral Laser Appl 2005 [41]	Er,Cr:YSGG	6	2	Max and man	0
E. Esen Oral Surg Oral Med, Oral Path, Oral Rad, Endod 2004 [42]	980 nm (diode)	12–24	10	Max	2
Nakamura Lasers Surg Med 1999 [43]	Super pulsed CO_2 10,600 nm	12–24	7	Max	0/4
Gupta J Cutan Aesthet Surg 2011 [44]	980 nm	15	1	Max and man	0
Rosa DS J Periodontal 2007 [45]	Er:YAG	3	5	Max	1
Hedge R J Periodontol 2013 [46]	Er:YAG and 10,600-nm CO_2	6	35 on 140 sites	Split mouth	More with Er:YAG than CO_2

trast, when using the non-ablative procedure, the author reported a treatment time of 1–2 min to complete one dental arch [33, 34].

16

16.2.9 Ablative vs. Non-ablative Technique

The ablative technique can be performed by all Class IV surgical lasers and is most commonly used. The operator is able the observe color change layer by layer during ablation. Carbonization during treatment should be avoided, and any tissue tags should be removed during the procedure to have a clear view of the ablation front.

The non-ablative technique relies on the optical properties, biophysical properties (laser-tissue interaction), and laser parameters described above. Therefore, not all Class IV lasers can use this technique. Although the non-ablative technique can be completed much faster, the operator should have a clear understanding of the principles before performing this technique.

There have not been reports of any major clinical complications; however, there is a possibility of gingival recession and bone necrosis which can be caused due to excessive ablation, over-irradiation, and deep tissue heat conduction. Prophylaxis and good oral hygiene are essential to stabilize gingival tissue prior to depigmentation. This can avoid any dispute with gingival recession after depigmentation. The clinician should carry out the procedure with good understanding of the optical property of the laser he/she is using and the laser-tissue interaction involved. On the other hand, a clinician is accustomed to completing many procedures in one appointment. The thought of "touching up" a pigmented area should not be construed as poor operative technique; rather, it should be viewed as an opportunity to further improve the aesthetic result.

16.2.10 Summary

In the author's experience, many patients are troubled by the pigmentation, even one spot of pigment on the lip, for example, but not aware of possible treatment. Pigment on the lip is more of concern than gingiva. Furthermore, ethnic communities in the Middle Eastern countries are more concerned with gingival pigmentation.

For those who are made aware of their pigmentation, acceptance of treatment is high. They are usually more appreciative of our examination skills and will have confidence in our techniques. In addition, treatment acceptance is good for both men and women at any age group.

When discussing the procedure, the patients must be made aware of possibilities of relapse, depending on the cause of pigmentation. One example is shown in ◘ Fig. 16.5c, and the patient should be informed that a second procedure will be necessary. As an aside, this is an opportune time to discuss smoking cessation. After the laser treatment of the pigmentation, the shade of the teeth may look darker as their gingiva returns to a pinker color, since there will be less contrast between the enamel and the soft tissue.

16.3 Laser-Assisted Dental Bleaching

Eugenia Anagnostaki

16.3.1 Introduction

A smile creates an immediate visual first impression. A bright tooth shade is the most important factor for making a smile attractive, according to Dunn [48].

During recent years, aesthetic dentistry has become a very important field of dentistry. But it is also among the achievements of preventive dentistry, that people are taking more care of their teeth in recent times. The most popular and the least invasive procedure within aesthetic dentistry is dental bleaching.

Dental bleaching was known and performed from ancient years, but in the scientific literature, bleaching as a procedure was first described in 1951 by Pearson [49], night guard vital bleaching in 1989 by Haywood and Heymann [50], and laser-assisted tooth bleaching in 1998 by Reyto [51].

It is important to clarify the often interchangeable terms "bleaching" and "whitening": According to the FDA, the difference between bleaching and whitening is that the term "bleaching" should only be used when the teeth can be whitened beyond their natural color and when the products used contain bleach. "Whitening" refers to restoring a tooth's surface color by removing superficial staining and debris.

In order to keep bleaching noninvasive, we have to ensure that the materials applied and the methods performed will not harm the enamel, the dentin, the pulp of the teeth, and the surrounding tissues.

16.3.2 Tooth Structure

Enamel

Enamel is the outer layer of the tooth and the hardest tissue in the body.

Mature dental enamel has a complex organized structure, mainly containing inorganic minerals (96%) and a small amount of organic material and water (4%), but no cells or collagen [52]. The basic inorganic structural blocks are hydroxyapatite crystals. Ionic exchange can occur, resulting in ionic substitutions, which have key effects on the hydroxyapatite physical and chemical properties. As an example, the fluoride incorporation in enamel was shown to increase its resistance to demineralization [53]. Mineral gain and loss are a dynamic physicochemical process on the tooth surface. Light scattering by the mineral crystals plays a considerable role in light scattering processes.

The organic component contains non-collagenous proteins (60%) and lipids (40%) and can function similarly to a semipermeable membrane, allowing small molecules to penetrate through [54]. According to a study of Eimar et al. [55], this is a possible explanation for the mechanism of bleaching.

Dentin

Dentin contains about 48% of mineral, 28% of organic material, and 24% of water. Dentinal tubules extend from the dentin-enamel junction up to the edge of the pulp. The tubules are where the majority of scattering occurs, whereas collagen fibrils play a minor role; scattering by the mineral crystals in dentin is negligible [56].

16.3.3 Natural Tooth Color

The main determinant for the tooth shade is dentin with its yellow to brown shades. At the same time, enamel properties, like thickness, chemical and physical composition, as well as the hydroxyapatite crystal size, affect the scattering of the light. Smaller crystals scatter more light and the tooth appears brighter [56]. The enamel can have blue, green, and pink tints.

The color perception depends on the observer, the object, and the light source. It is a complex procedure resulting from reflection, absorption, and scattering of the light to the object and returning to the eye of the observer [57].

As visual examination is a subjective method, there are instruments available in dentistry—colorimeters and spectrophotometers—which provide unbiased information on the tooth color.

16.3.4 Discoloration

Extrinsic Staining

The exposed surface of the teeth is covered by a protein-polysaccharide coating, the pellicle. The pellicle is easily stained by exogenous colorants from food, drinks, mouth rinses, or oral medications but also from chromogenic bacteria, dental materials, or industrial exposure to metal dust. Chromogens either bind and directly stain the teeth or bind and then stain and darken with time, or pre-chromogens bind and subsequently undergo a chemical reaction to cause staining [58].

The extrinsic staining either may stain the pellicle, and can easily be removed by means of a thorough cleaning, or may be retained on the tooth surface forming a stain-enamel complex through ion interaction. Accumulation of extrinsic stains is affected by the oral hygiene habits, saliva composition and flow, and enamel surface roughness.

Intrinsic Staining

This staining is distributed throughout the internal structure of the teeth and may either develop preeruptive during odontogenesis or appear at any time after eruption. The staining can be localized to few teeth or be present on all of them depending on the period of development when it took effect. It can be caused by changes in the structure of the hard tissue itself or by incorporation of chromogenic molecules inside the hard tissue [59].

Conditions such as alkaptonuria, congenital erythropoietic porphyria, congenital hyperbilirubinemia, amelogenesis imperfecta, dentinogenesis imperfecta, tetracycline intake, fluorosis, and enamel hypoplasia contribute to preeruptive intrinsic staining.

Hemorrhagic products in the pulp, root resorption, and aging may cause posteruptive staining.

16.3.5 Chemistry of Bleaching Materials

The bleaching gel applied for home bleaching may contain carbamide peroxide as an active substance, in concentrations from 5% up to 20%, or hydrogen peroxide, in concentrations of maximal 10%. Some countries have specific regulations for the active ingredients. An example is the 2011/84/EU European Directive [60].

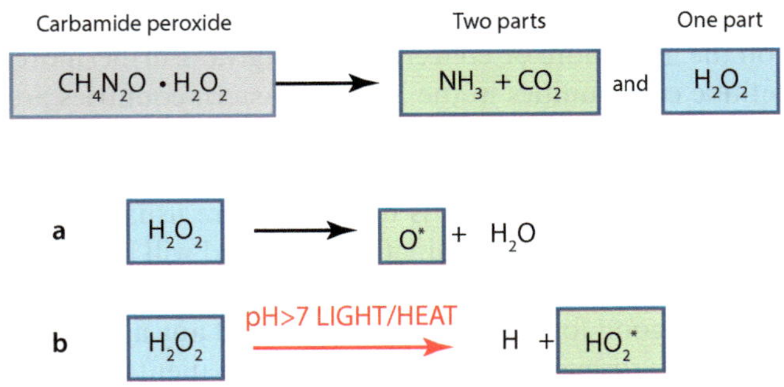

Fig. 16.18 Chemical reaction mechanisms of carbamide peroxide and hydrogen peroxide. The first phase is the reduction of carbamide peroxide into two parts of ammonia and carbon dioxide and one part hydrogen peroxide. Hydrogen peroxide can be further reduced (A) into an oxygen radical and water or (B) under certain conditions into a perhydroxyl radical and a hydrogen ion

Chemically, a carbamide peroxide concentration of 16% is equivalent to 5.76% of hydrogen peroxide, since carbamide peroxide is being reduced to approximately two parts of ammonia and carbon dioxide and one part of hydrogen peroxide.

For in-office use, available bleaching gels contain from 25% up to 40% hydrogen peroxide. The action of the gel can be accelerated chemically or by means of a light source such as plasma, LED, halogen lamps, or lasers.

The chemical reaction is a reduction of hydrogen peroxide into oxygen and water and preferably into perhydroxyl and hydrogen. The perhydroxyl ($HO_2^{\bullet}$) free radical belongs to the most reactive species in tooth bleaching, the formation of which is favored by high (alkaline) pH, but this is rarely the situation as the product shelf life is adversely affected under these conditions [61]. Figure 16.18 shows a graphic representation of the reduction and subsequent reaction of carbamide peroxide and hydrogen peroxide during activation.

16.3.6 Mechanisms of Bleaching

The mechanism of the action of either carbamide peroxide or hydrogen peroxide is not completely investigated. Until now, studies have shown that the unstable oxygen or perhydroxyl radicals are diffusing through the organic matrix of the enamel and dentine and are causing a breakdown of ring structures of stain molecules followed by a breakdown of long molecular chains in the organic tooth matrix into shorter ones, which absorb less and therefore reflect more light. It is important to know that there is a saturation point during the bleaching process, where there are only hydrophilic pigment-free structures and the lightening of the teeth stops dramatically. At this point, bleaching should be stopped immediately, or the next step would be a rapid loss of enamel [62].

However, this "breakdown of stain molecules" hypothesis might be weak since (1) the organic chromophore concentration, if they exist, in tooth enamel is extremely low (below the detection limit of many spectroscopy techniques) and (2) several studies have shown that following tooth bleaching the translucency of tooth enamel decreased significantly, making it more opaque.

Studies have shown that the tooth enamel organic matrix is mainly composed of amide groups that represent enamel proteins. Eimar et al. [55] found that hydrogen peroxide does modify neither the organic nor the inorganic relative contents of dental enamel but oxidizes their enamel organic matrix. The conclusion states that oxidation of enamel protein and the increase in enamel opacity following peroxide treatment seem to indicate that the peroxide whitens teeth by oxidizing its transparent organic matrix into an opaque whiter material. This is a more comprehensive theory of the mechanism by which peroxide might whiten teeth.

16.3.7 Bleaching Methods

There are two main approaches to vital tooth bleaching: home bleaching (night guard bleaching) and in-office bleaching. The two techniques may be combined in order to enhance the result. According to Auschill [63], 7 days of home bleaching result to a similar effect of one session of in-office bleaching.

Night/Day Guard Vital Bleaching or Home Bleaching

As first described by Haywood and Heymann [50], this is the most popular bleaching method. Customized trays, occasionally referred to as "night guards," have to be used for several hours a day or overnight, depending on the concentration of the material used.

The material mainly used is carbamide peroxide in concentrations up to 20%. Several companies also offer prefabricated-preloaded trays for home bleaching:

- Indications for home bleaching (according to So-Ran Kwon [62]):
  - Generalized yellow, orange, or light brown discoloration
  - Age-related yellow discoloration
  - Mild tetracycline staining
  - Superficial brown fluorosis stains
  - Discoloration due to smoking or chromogenic foods or drinks
  - Genetically yellow or gray teeth
  - Patient wishing shade improvement with minimally invasive treatment
  - Yellow discoloration of single vital teeth
- Contraindications for home bleaching (according to So-Ran Kwon [62]):
  - Amelogenesis or dentinogenesis imperfecta
  - Severe tetracycline staining
  - Discoloration due to restorative materials
  - Pregnancy/lactation
  - Severe surface damage due to attrition, abrasion, and erosion
  - Untreated dentinal hypersensitivity
  - Lack of compliance

In the case of home bleaching, the outcome depends on a good supervision and motivation by the dentist, which will lead to a good compliance from the side of the patient.

In-Office Bleaching or Power Bleaching

The procedure is performed chairside, and highly concentrated bleaching gels are applied on the buccal tooth surface. The appointment time can be lengthy, but the result is immediately visible and is even enhanced within the next 1–2 days. There are disadvantages: the costs for such treatment are higher, the postop sensitivity might be more intense, and there can be some reversible but moderately painful alteration to the soft tissue surrounding the teeth.

According to Walsh [64], the indications for bleaching are:

- Discoloration related to genetic disorders (dentinogenesis imperfecta)
- Permanent acquired discoloration: fluorosis staining, tetracycline staining, lead poisoning
- Posttraumatic discoloration (vital tooth with sclerotic dentin)
- Moderate tooth discoloration
- Discoloration related to the physiological aging of teeth
- Improvement of the teeth's natural color before the placement of a prosthetic restoration in the adjacent area

Furthermore, bleaching is contraindicated in cases of:

- Severe untreated tooth sensitivity due to exposed cervical dentine from dental erosion, gingival recession, or gingival pathology. In this case, the sensitivity problem should be resolved first.
- Unrealistic expectations of the treatment result.
- Inability of the patient to sit still in the dental chair during the procedure and tolerate the required soft tissue isolation devices.
- Inability of the patient to follow (at least for a while) changes needed to prevent reformation of extrinsic stains (smokers).
- Inability to have the restorations in the teeth to be bleached changed, after a necessary delay of 2 weeks after the whitening treatment [64].

Procedure

After a thorough cleaning with oil- and glycerine-free pumice, hydrogen peroxide gels in concentrations between 25% and 40% are applied on the teeth to be whitened. The soft tissue must be protected from the caustic action of the bleaching gel. Depending on the material used, activation is achieved chemically (by thorough mixing of the gel components usually contained in syringes) and/or by non-coherent light, heat, or a laser. The different light sources are used to enhance the chemical decomposition of hydrogen peroxide and with this to enhance the whitening efficacy. When heat is applied, a temperature elevation in the gel of 10 °C is reported to double the speed of hydrogen peroxide decomposition [65]. For a laser-activated application, the gel should contain a proper photocatalyst, which is an absorber that matches the spectrum of the light used to activate or accelerate the gel action. In addition, special filtering substances are used in order to keep the energy inside the gel, and not allow a temperature rise inside the pulp [66].

Studies have shown that the hydrogen peroxide penetrates deeper if activated by laser or other light sources [61, 67]. Furthermore, Kwon [68] found that light-activated bleaching is not more dangerous to the pulp by the stronger oxidizing action.

The difference between activation by ordinary light or by laser is that ordinary light sources emit a broad spectrum of photonic energy with increased possibility of thermal damage [69]. Using an appropriate laser, it is possible to shorten the interaction time of the material on the tooth surface, thus avoiding possible superficial damages.

Laser-Assisted In-Office Bleaching

During an in-office bleaching procedure, it is possible to activate the bleaching material by a laser. Figure 16.19 shows the laser wavelengths and their basic mechanism of action, which is photo-thermal catalysis in cases of diodes and Nd:YAG and photochemical in the case of Argon and KTP lasers. Understanding these differences will allow for the most effective bleaching outcome by producing the maximum number of powerful free radicals from the bleaching gel.

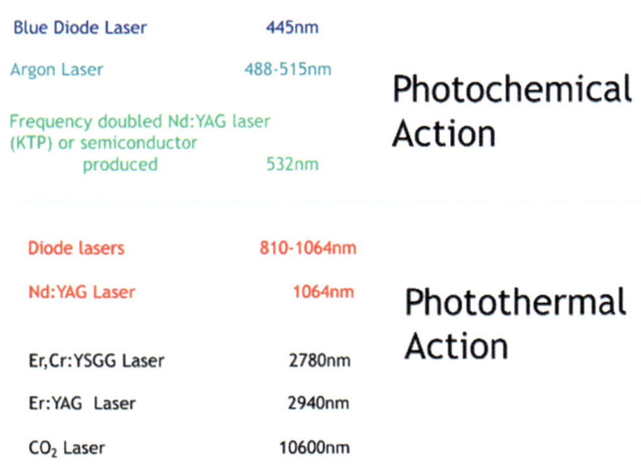

Fig. 16.19 A listing of generically named dental lasers with their corresponding wavelengths and the two mechanisms of action in laser-assisted bleaching

The application of KTP, argon, and diode lasers is widely supported by the literature [66, 68, 70]. Nd:YAG lasers possibly cause overheating due to their high peak power and are under investigation for their safety in bleaching [67, 71, 72]. Newly introduced diode lasers operating at a wavelength of 445 nm have gained approval for dental applications. However, their clinical safety in dental bleaching remains an area that requires further investigation.

CO_2 lasers were the first lasers indicated for use during bleaching procedures, but their application is no longer supported, since Luk et al. [73] showed overheating on tooth surface and pulp. Additionally, there have not been any controlled clinical studies for that wavelength.

CO_2 and erbium lasers show a photo-thermal action since they are absorbed in water contained in the bleaching gel. Caution must be given due to the absorption of these lasers in hydroxyapatite and due to the high peak power of erbium lasers [74]. It is essential to stay far below the ablation threshold for enamel and always keep the teeth covered by bleaching gel. As explained in ► Chap. 3, these wavelengths are immediately absorbed in the water within the gel, causing it to rapidly dehydrate. Thus, attention is necessary so as not to activate the laser on a dry layer of gel.

The application of erbium lasers (Er:YAG and Er,Cr:YSGG) in dental bleaching is still under investigation. In a study, Nguyen et al. [75] showed that laser-assisted bleaching (KTP and Er:YAG) gives similar results in a shorter time compared to nonactivated bleaching, but they concluded that data on mechanisms of action of the Er:YAG laser on bleaching gel and dental tissues are still limited and additional studies are needed to assess the contribution of the Er:YAG laser in tooth bleaching.

In the only available randomized controlled study comparing color outcome and sensitivity among conventional in-office bleaching, diode laser-activated and Er,Cr:YSGG laser-activated bleaching, the authors concluded that there were no significant intergroup differences in color outcome and that no significant hypersensitivity was observed in the Er,Cr:YSGG-laser-activated group [76].

The risks of damaging the underlying tissue are depending on the wavelength of the laser to be used: Erbium family and CO_2 lasers are well absorbed in water and may be absorbed in the tooth surface, but shorter wavelengths in the red to near-infrared, as well as in the visible spectrum, are more likely to penetrate into the pulp.

Therefore, it is essential to always use the appropriate bleaching gel with a particular laser wavelength, to have a high absorption of the energy inside the gel, and to minimize any increase of the temperature in the pulpal tissues. To achieve this high absorption, bleaching gels for the blue and green wavelengths contain orange-red dyes like rhodamine, and bleaching gels for red to infrared wavelengths contain bluish-purple dyes. Titanium-dioxide particles also act as a broadband absorber in a gel offered by many laser manufacturers [74].

The action of the argon and KTP wavelength is not only photo-thermal and photochemical but photolytic as well. This relies upon specific absorption of a narrow spectral range of green light (510–540 nm) not only in rhodamine but also into chelate compounds formed between apatite, porphyrin, and tetracycline. Through this, argon and KTP lasers are capable to achieve a good bleaching result even in teeth not responding to other techniques (tetracycline-stained teeth) [77].

In addition, the KTP laser induces a photochemical reaction in the special formulated bleaching gel, providing a higher free radical outcome than during a photothermal action [75]. Through buffering to an alkaline pH of 9 and high energy, the produced free radicals were shown to be more reactive [64].

16.3.8 Safety Concerns

Studies show varying results concerning the safety of the bleaching methods. One main point is the toxicity of hydrogen peroxide as a strong oxidative; another is the possible corrosiveness of the gels used. Additionally, activation by any light source might overheat the target and underlying tissues:

1. The exposure of the organism to the hydrogen peroxide, according to Li and Greenwall [78], is minimal when used properly, either in-office or at home. Enzymes in the saliva are capable to neutralize up to eight times the amount of hydrogen peroxide used in a home bleaching session. During in-office bleaching, with a good isolation of the soft tissue, the level of chemical is not detected systemically.
2. Local soft tissue effects usually manifest as chemical burns. These can appear due to either poor isolation or extended contact time with the material and poor fitting trays, in the case of home bleaching. Vitamin E used locally on the defects is supported by the literature [79].
3. Tooth sensitivity is another side effect and might be an indication of a pulp response to the oxygen and perhydroxyl free radicals [80] (produced from the breakdown of hydrogen peroxide; see ◘ Fig. 16.18). It usually appears for the first 24 h to 3 days, and this has to be differentiated from the post-bleaching sensitivity which appears like a sudden shooting stimulus and is possibly related with a direct activation of neuronal receptors by the hydrogen peroxide and its products [81]. In a study by Schulte [82], the sensitivity was severe enough to cause 14% of the patients to discontinue the bleaching treatment.

For sensitivity treatment, potassium nitrate gels or casein phosphopeptide-amorphous calcium phosphate (CPP-ACP) pastes are used immediately after the bleaching session with an effective reduction of sensitivity. Agents such as potassium salts, which depress nerve excitability, are claimed to be more effective than tubule-occluding agents in reducing sensitivity according to Markowitz [81], but according to a newer randomized controlled clinical trial, there is no significant difference between the above-mentioned agents [112].

In a study by Moosavi et al. [83], photobiomodulation therapy with an infrared diode laser could be recommended as a suitable strategy to reduce the intensity of tooth sensitivity after in-office bleaching. The parameters applied with an 810-nm diode laser were 3 J of energy, energy density of 12 J/cm^2, and power density of 800 mW/cm^2 on the cervical area of the tooth.

4. Penetration of the bleaching material into the pulp has been shown in several in vitro studies to take place. However, penetration is relatively low due to a positive pulpal pressure and the presence of dentinal fluid. The affected pulp tissue can protect itself from damage by hydrogen peroxide through the enzymatic breakdown of the molecule by peroxidase and catalase [84]. These cellular enzymatic systems eliminate the excess oxygen, but still it is not known how much hydrogen peroxide can be tolerated by the pulp tissue [85]. Additionally, an in vivo study demonstrated that bleaching with or without light activation did not cause damage on pulp tissue of young sound premolar teeth [86]. Another in vivo study showed that there was no increase in the levels of cell damage biomarkers in serum and saliva among the groups tested but found a significant increase in biomarkers in gingival crevicular fluid immediately after treatment [87].
5. Overheating of the pulp in the case of in-office bleaching needs to be avoided, through the choice of a material which needs only a short time activation and contains the proper filtering absorber to block the heat inside the gel. According to Zach and Cohen [66], the pulp can only tolerate a temperature rise up to 5.5 °C. Eriksson and Alberktsson [88] found that 42 °C might be a critical temperature to the pulp when sustained for 1 min. On the other hand, Baldissara et al. [89] reported that an intrapulpal temperature rise of 8.9–14.7 °C in humans does not induce pulpal pathology. The values of temperature rise obtained in this study were not critical for pulp health; however, this was a preliminary study.

6. Enamel surface morphology. The effect of bleaching materials (either home or in-office) on the enamel surface is controversially documented in the literature.

The studies that have been performed on the adverse effects of bleaching on chemical and physical characteristics of the enamel use several different methods: SEM (direct or indirect), profilometry, microhardness, calcium loss, and infrared spectroscopy. Out of these, the microhardness seems to be the preferred choice of the researchers [90]. However, those methods are destructive, which means that it is impossible to follow up the same samples during a period of time. But the processes in the mouth are dynamic, so there is the need of examining what is happening in vivo, over time. Out of the numerous studies investigating the impact of bleaching procedures on enamel, only a few are performed in vivo conditions.

Clinical studies found no statistically significant loss of surface enamel hardness or loss of calcium and phosphorus as minerals out of the enamel [91]. In contrary, many in vitro studies show microhardness reduction after bleaching. Attin et al. [92] reviewed the published literature on the effect of bleaching on enamel microhardness. They conclude that if in the different (in vitro) studies the intraoral conditions are simulated as close as possible (e.g., with the use of artificial or human saliva as a storage medium or re-fluoridation after bleaching), then the study outcome shows a lower risk of enamel microhardness reduction due to the bleaching treatment, compared to the remaining studies.

The most severe alterations in enamel in vitro have been described when acidic bleaching gels were used [93]. The hydrogen peroxide per se is acidic, and the gels are usually kept in acidic pH in order to increase shelf life. Gels produced for use with the KTP laser contain a special buffering system and set the pH to around 9.5 [66].

16.3.9 Long-Term Effectiveness and Stability of Tooth Whitening

A number of authors have investigated the effectiveness of tooth whitening by comparing different techniques and different peroxide concentrations. The evidence on effectiveness suggests that all of them are effective and reach similar results when their respective protocols are followed.

The stability of outcome of various whitening products has been widely evaluated, and significant shade improvement can be predicted when products such as carbamide peroxide gel, or hydrogen peroxide, with or without light activation are used according to a study of Sulieman [94].

Unfortunately, only short-term results on bleaching efficacy have been reported by most studies [68], while only a few of them report results over the long term [63, 95]. It has been shown that although higher peroxide concentrations produce a quicker shift than lower concentrations, the result on whitening is the same for all concentrations at the end of the whitening process. However, the differences in effectiveness of tooth whitening appear to be significant when results are evaluated over a longer period.

In a retrospective case series study, Leonard [96] had shown that long-term shade retention was reported by 82% of the participants 4 years post-bleaching. Later, the group of Boushell and Leonard [97] showed that satisfactory retention of the shade change without re-treatment can be expected in at least 43% at 10 years posttreatment.

Two long-term studies done by Grobler et al. [95] assessing the effectiveness of 10% carbamide peroxide showed that the majority of patients maintained whitening improvement for up to 6 months. From an initial shade change of 5, there was between 18% and 26% relapse. However, significant relapse was noticed when the patients were evaluated 14 months after tooth whitening, mostly for the whiteness/brightness parameter. It is suggested that re-whitening should be done at about 14 months post-whitening in the case of home bleaching.

Clinical studies evaluating the outcome of laser-assisted bleaching are controversial: Strobl et al. [71] concluded after a split mouth study in 20 patients that the Nd:YAG laser did not enhance the bleaching success; furthermore, the laser-activated sites have been more sensitive after treatment than the nonactivated sites. Gurgan et al. [70] could show that the use of a diode laser results in spectrophotometric measured better outcome, with less gingival and tooth sensitivity, and it might be preferred among in-office bleaching systems. Surmelioglou et al. [76, 98] concluded that application of lasers to induce the bleaching reaction does not provide better aesthetic or long-term color stability. Nevertheless, the treatment time was significantly less for the laser-activated groups.

Concerning tetracycline-discolored teeth, Kuzekanani and Walsh [98] found in vivo, through a quantitative analysis of digital pre- and post-bleaching digital images, that KTP laser photodynamic bleaching provides a clinically useful improvement in tooth shade.

Regarding the relapse in the color after tooth whitening, it is known that the oxygen within the tooth from the oxidative process initially alters the optical properties of the tooth. Then oxygen dissipates over the following week(s), and the tooth takes on the actual lightened shade [99].

Is it the protein molecules re-bonding the double bonds, is it aging or the oxidized organic matrix which is turning less opaque and more transparent again with time, or finally is it the enamel changes caused by the bleaching materials, guiding the teeth to a new discoloration by leakage from extrinsic stains? Investigations will continue, especially in comparing color relapse after laser-activated bleaching, with the traditional methods.

Unfortunately, the available studies concerning laser-assisted bleaching do not provide clear conclusions. The details of the protocols, laser wavelengths, and gels used are often not fully described. Long-term outcome studies are absolutely necessary in order to evaluate the advantages of this method.

16.3.10 Patient Selection

Further to the indications and contraindications mentioned in ▶ Sect. 16.3.7, before starting a bleaching procedure, there are several items to be considered:

- Informed consent must be obtained from the patients.
- An initial detailed examination is performed, including medical and dental history, occlusion and TMJ exam, and radiographic, soft, and hard tissue status with tooth vitality, percussion, and mobility testing.
- After this, a professional cleaning and polishing of all teeth must be performed, and hygiene instructions are given to the patient.
- In case of combined in-office/home bleaching, alginate impressions of both upper and lower arches have to be taken, in order to fabricate customized bleaching trays.

In case of poor general or dental health, untreated periodontal disease, pregnancy/lactation, inability to quit smoking during the bleaching and post-bleaching examination period, and carious or fractured teeth to be bleached, issues must be resolved before initiating any bleaching procedure.

16.3.11 Shade Evaluation

In order to verify that the bleaching procedure is successful, a shade evaluation has to be periodically performed. The initial shade inspection, as well as every shade evaluation—e.g., immediately after the procedure, 1 week later, and additionally every 6 months after the usual dental hygiene appointments—is advised to be performed with a spectrophotometer, as it has been found to give the most repeatable results [100]. ◘ Figure 16.20 shows the result of one such analysis.

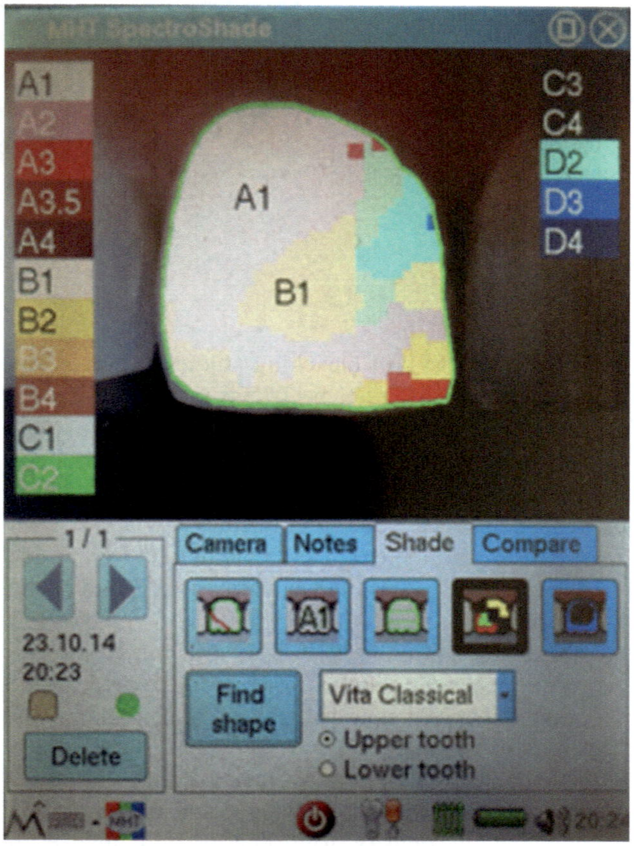

◘ Fig. 16.20 Screenshot of handheld spectrophotometer with tooth color map. SpectroShade MHT (MHT Optic Research AG, Niederhasli, Switzerland)

The color detected by the spectrophotometer can be confirmed visually by a value-oriented scale or shade guide. The color is expressed in "shade guide units" according to Paravina [101], and the difference in shade guide units shows the color improvement. A photographic documentation with a digital camera is mandatory. It is advantageous to obtain pictures with the corresponding shade tab placed on the tooth. ◘ Figure 16.21 shows a typical shade guide. An additional "bleach guide" might be useful for colors lighter than B1.

16.3.12 Laser Parameter Calculation and Reporting

Several recent papers have attempted to address the issue of laser application parametry during various treatment procedures [102–105]. Specifically, in an experimental spectrophotometric study on different bleaching gels [106], the authors could suggest a theoretical safe fluence range applicable during laser-activated dental bleaching. A fluence of 45 J/cm^2 is in accordance with fluences applied in several in vivo [107]

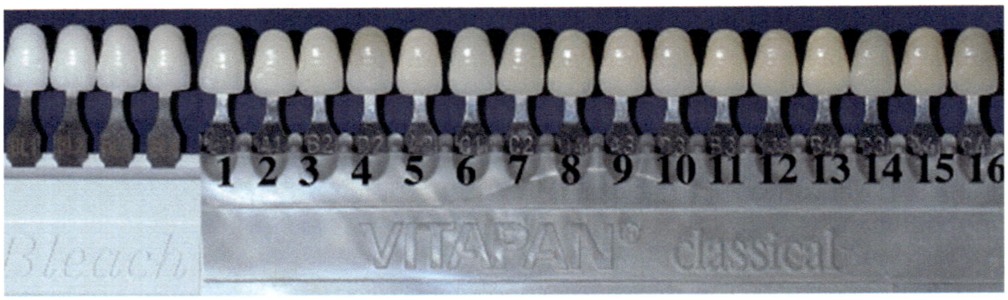

Fig. 16.21 Value-oriented shade scale (Vita, Zahnfabrick, Germany). The black numbers superimposed on the scale show the ascending order of darkness of the shade, with number 1 being the higher (brighter) value

and in vitro [108–110] bleaching studies. Nevertheless, fluences reported in other peer-reviewed studies ranged between 14 and 180 J/cm^2.

As mentioned in the literature [111], a complete parameter report is essential to ensure that a bleaching procedure produces consistent results:

- Intrinsic parameters: Device manufacturer, model and type, delivery system, emission mode, and energy distribution.
- Adjustable parameters: Pulse width, repetition rate or frequency, tip diameter or handpiece area, tip-to-tissue distance, beam divergence, tissue cooling, and length of treatment.
- Calculated parameters: Average power, peak power, spot diameter and spot area at tissue, power density, pulse energy density, total energy, and fluence are indispensable details for every case, in order to obtain a repeatability of the procedures.

Verification of the accuracy of the laser's display panel with a power meter is advantageous. Often, the power indicated on that panel does not correspond to the power at the exit of the handpiece/fiber [105].

Ideally, the energy distribution out of the handpiece used for bleaching should be "flattop." If not, care must be taken for not creating "hot spots" on the teeth which result from a Gaussian distribution of the beam. An option is to keep the handpiece/fiber continuously moving in the defined distance from the target.

16

16.3.13 Clinical Examples of Laser Bleaching Materials and Methods of Use

LaserWhite 20

LaserWhite 20 (Biolase Inc., Irvine, CA, USA) was developed for the company's 810- or 940-nm lasers. The gel is produced by mixing the base and activator gel contained in two syringes, and the resulting hydrogen peroxide concentration is 38%.

The clinical protocol is as follows: The initial color is documented. The teeth to be whitened are cleaned with a glycerine- and oil-free pumice and dried with air, and a gingival barrier is applied to protect the gingiva. Soft tissue is protected with cheek retractors and cotton rolls. The gel is applied on each tooth in a uniform layer, approximately 1 mm thick, directly from the syringe using a brush tip applicator. Figure 16.22 shows the gingival barrier in place and the gel applied to the teeth.

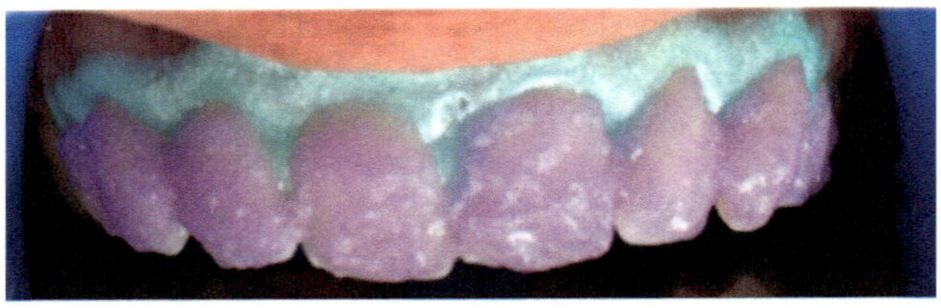

Fig. 16.22 LaserWhite 20 applied with gingival barrier in place

The patient, staff, and operator will wear the appropriate protective eyewear, and all other laser safety measures have to be followed.

The laser (Epic X, Biolase Inc., Irvine CA, USA), with an emission wavelength of 940 nm, is used with a power of 7-W continuous wave, delivered into a quadrant whitening handpiece with a tip area of 2.8 cm^2. The same protocol may be applied for the 810- and 980-nm diode lasers when using a similar quadrant handpiece. The application time is 30 s per quadrant with the handpiece approximately 1–2 mm away from the gel. After the 30-s exposure of all quadrants, the handpiece is moved back to the first quadrant, and the photonic energy is reapplied for 30 s to it and to each of the other quadrants. After this second exposure, the laser is deactivated for 5 min. Thus, the total laser activation time per quadrant is 60 s with an additional waiting time of 5 min. The resulting fluence for these parameters is approximately 70 J/cm^2 during each exposure.

The bleaching gel is removed with the high-volume suction and rinsed off carefully to avoid damage of the gingival and soft tissue barriers. Then a new layer of gel is applied on the teeth, and the above technique is repeated on all quadrants.

After completing the procedure, the protective barriers are removed, and the final color is documented.

If any bleaching gel leakage affected the soft tissue, vitamin E oil should be applied to the area.

In case of hypersensitivity or pain, a non-staining desensitizing gel can be applied.

The total contact time of the bleaching gel with the teeth is 18 min, with 2 min of laser exposure per quadrant. As an aside, the entire chairside appointment time needed is not included in the calculation. The barrier application time will vary, and the application time for all bleaching gels is approximately 2 min. The clinician should remember how the selected laser parameters will affect the tissue interaction. ► Chap. 4 has a detailed discussion about how to calculate those parameters.

Heydent JW Power Bleaching Gel

JW Power Bleaching Gel (Heydent, Kaufering-Germany) can be activated with any wavelength from 450 to 1064 nm since it contains a multiwavelength absorber. A gel is produced by mixing the kit contents of liquid hydrogen peroxide fluid with the powder, which contains titanium dioxide. The resulting hydrogen peroxide concentration is 35%.

The clinical protocol is as follows: The initial color is documented. The teeth to be whitened are cleaned with a glycerine-free pumice and dried with air, and a gingival barrier is applied to protect the gingiva. Soft tissue is protected with cheek retractors and cotton rolls if necessary. The gel is applied in a layer of approximately 2 mm on each tooth, using a plastic spatula. ◘ Figure 16.23 shows the gingival barrier in place and the gel applied to the teeth.

The patient, staff, and operator must wear the appropriate protective eyewear, and all other laser safety measures must be followed.

The laser (Fox 1064, ARC Lasers Nürnberg, Germany) with an emission wavelength of 1064 nm is used at 1.5-W continuous wave and delivered into a collimated handpiece of 6 mm diameter that is placed 10 mm from the gel. Each tooth is exposed for 30 s. The suggested fluence is 45 J/cm^2.

After having activated the bleaching gel on all teeth to be bleached, the gel is removed with high-volume suction and is rinsed off carefully to prevent damage of the gingival and soft tissue barriers, tooth by tooth every 30 s, starting with the tooth which was activated first. In this way, the resting and contact time of the gel on each tooth is approximately the same.

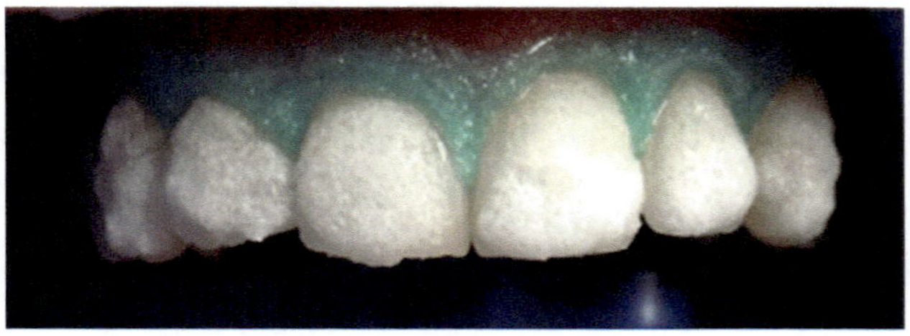

◘ **Fig. 16.23** JW power bleaching gel applied with the gingival barrier in place

Then a new layer of gel is applied on the teeth and the procedure is repeated twice.

After completing the procedure, the protective barriers are removed. The final color is documented.

If any bleaching gel leakage affected the soft tissue, vitamin E oil should be applied to the area.

In case of hypersensitivity or pain, a non-staining desensitizing gel can be applied.

Total contact time of the bleaching gel with the teeth is approximately 30 min, and each tooth was exposed to the laser photonic energy for 90 s.

Smartbleach Gel 36%

Smartbleach KTP Gel (SBI Dental-Herzele, Belgium) is activated by the 532-nm wavelength. The gel is produced by mixing the kit's contents of hydrogen peroxide and bleaching powder. The resulting hydrogen peroxide concentration is 36%.

The clinical protocol is as follows: The initial color is documented. The teeth to be whitened are cleaned with a glycerine-free pumice and dried with air, and a gingival barrier is applied to protect the gingiva. Soft tissue is protected with cheek retractors and cotton rolls if necessary. The gel is applied in a layer of approximately 1 mm on each tooth, using a plastic spatula. ◘ Figure 16.24 shows the gingival barrier in place and the gel applied.

The patient, staff, and operator must wear the appropriate protective eyewear, and all other laser safety measures have to be followed.

The laser (DEKA SmartLite, DEKA M.E.L.A. Srl.—Firenze, Italy) with an emission wavelength of 532 nm is used at 1-W continuous wave and delivered into a handpiece that is placed at a distance of approximately 40 mm to the gel. Each tooth is exposed to the laser energy for 30 s, and then the activated gel is allowed to interact with the tooth for a total of 10 min. The suggested fluence is 30 J/cm^2.

After this, the gel is removed with the high-speed suction, tooth by tooth every 30 s, starting with the

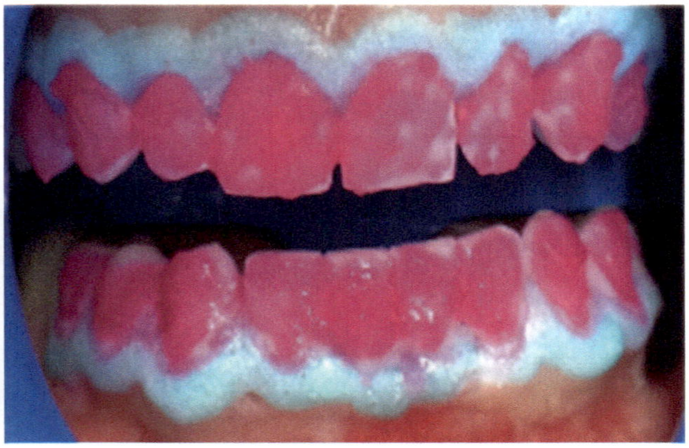

◘ **Fig. 16.24** Gingival barrier in place and the Smartbleach Gel applied to the teeth

tooth which was activated first, second, third, etc., and rinsed off carefully to avoid damage the gingival and soft tissue barriers. In this way, the resting and contact time of the gel on each tooth is approximately the same.

Then a new layer of gel is applied on the teeth, and the procedure outlined above is repeated twice. After the third cycle, the bleaching gel is removed tooth by tooth again, with the high-speed suction, and rinsed off, and then the protective barriers are removed. The final color is documented.

If any bleaching gel leakage affected the soft tissue, vitamin E oil should be applied to the area.

In case of hypersensitivity or pain, a non-staining desensitizing gel can be applied.

The total contact time of the gel on the teeth is therefore approximately 30 min, and the laser exposure time is 90 s per tooth.

A typical appointment for laser-assisted bleaching includes the following steps:

1. Obtain the patient's informed consent.
2. Obtain the initial shade using shade tabs and/or spectrophotometer, and document it with a clinical photograph.
3. Apply cheek retractors.
4. Polish the teeth with pumice or glycerine-/oil-free paste, and avoid any contact with saliva.
5. Remove bleaching gel from refrigerated storage and mix, following manufacturer's instructions.
6. Air-dry teeth to be bleached and apply gingival protection barrier.
7. Apply bleaching gel in an appropriate thickness.
8. Irradiate tooth by tooth or per quadrant following appropriate laser protocols.
9. Adhere to laser safety principles.
10. After recommended interaction time, remove gel, and rinse teeth with water.
11. Air-dry teeth, check gingival barriers, reapply gel if required, and repeat steps 8 and 9.
12. If necessary, apply desensitizing gel or other similar materials.
13. Give post-bleaching instructions—for example, avoid dark colored food and liquids, and continue good oral hygiene.
14. Evaluate shade change.

16.3.14 Clinical Cases

Two clinical cases of laser-assisted bleaching are shown. Case 1 was performed with the 532-nm KTP laser and Smartbleach, and case 2 was performed with a 1064-nm diode laser using JW Power Bleaching Gel.

▣ Figure 16.25 is a clinical case of laser-assisted with the KTP 532-nm device and Smartbleach. The 6-month post-bleaching photo shows good color stability.

▣ Figure 16.26 depicts bleaching using a 1064-nm diode with JW Power Bleaching Gel.

16.3.15 Summary

In summary, the use of lasers is a safe and effective way to enhance in-office bleaching procedures.

Thorough knowledge of laser-tissue interaction and laser safety is indispensable. In such a way, a well-trained clinician can perform a laser-assisted bleaching with no side effects. Precise evidence-based clinical protocols need to be established and followed to standardize the procedures and expect predictable outcomes.

16.4 Conclusion

The clinician's first goal for a dental patient is to help maintain a healthy dentition. Many patients also desire a smile that fits their own aesthetic criteria. Of course, the interpretation of those concepts will vary widely based on an individual's culture, philosophy, and personality, to name just a few. Thus, practitioners should be aware of techniques to help fulfill the patient's aesthetic demands. This chapter has demonstrated how dental lasers are one instrument that can be used for hard and soft tissue enhancement.

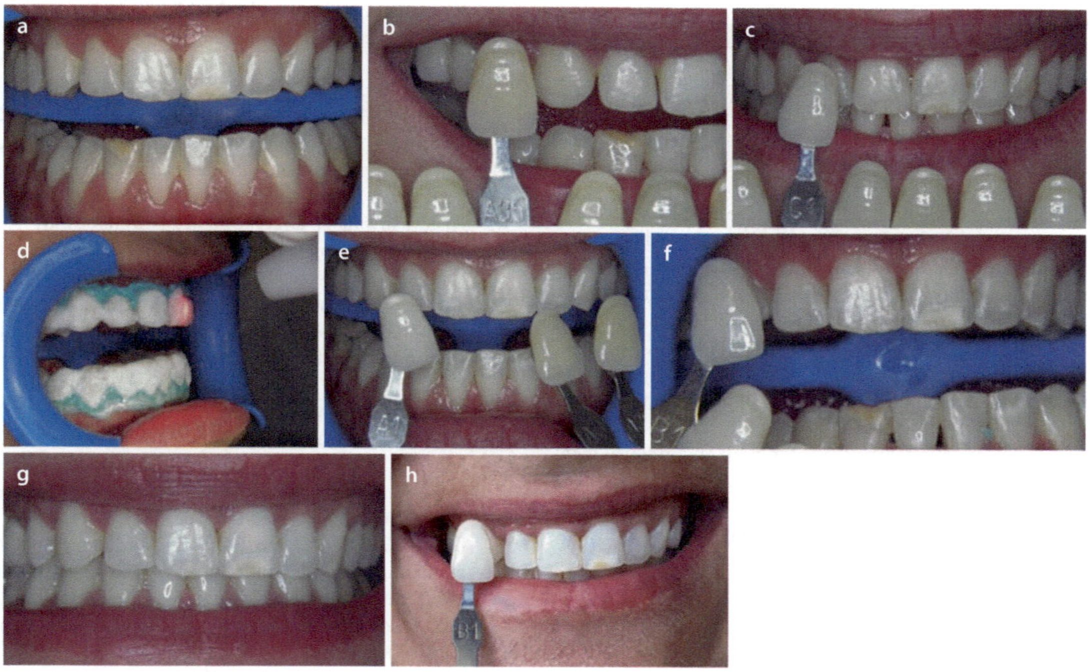

Fig. 16.26 **a** Pre-bleaching view. **b** Pre-bleaching shade match of the canines is A 3.5. **c** Pre-bleaching shade match of the central incisors is C-1. **d** After retractors and tissue protection are in place, the bleaching gel is placed on the teeth. The laser is placed approximately 2 cm from the teeth, and the gel is activated. **e** Immediately after the first bleaching session, the canines are matched to A-3 and the centrals are A-1. **f** Immediately after the second session, the canines are now A-2 and the centrals are B-1. **g** One month after bleaching, treatment of the color is stable. **h** Six years after treatment, the color is very stable and remains bright

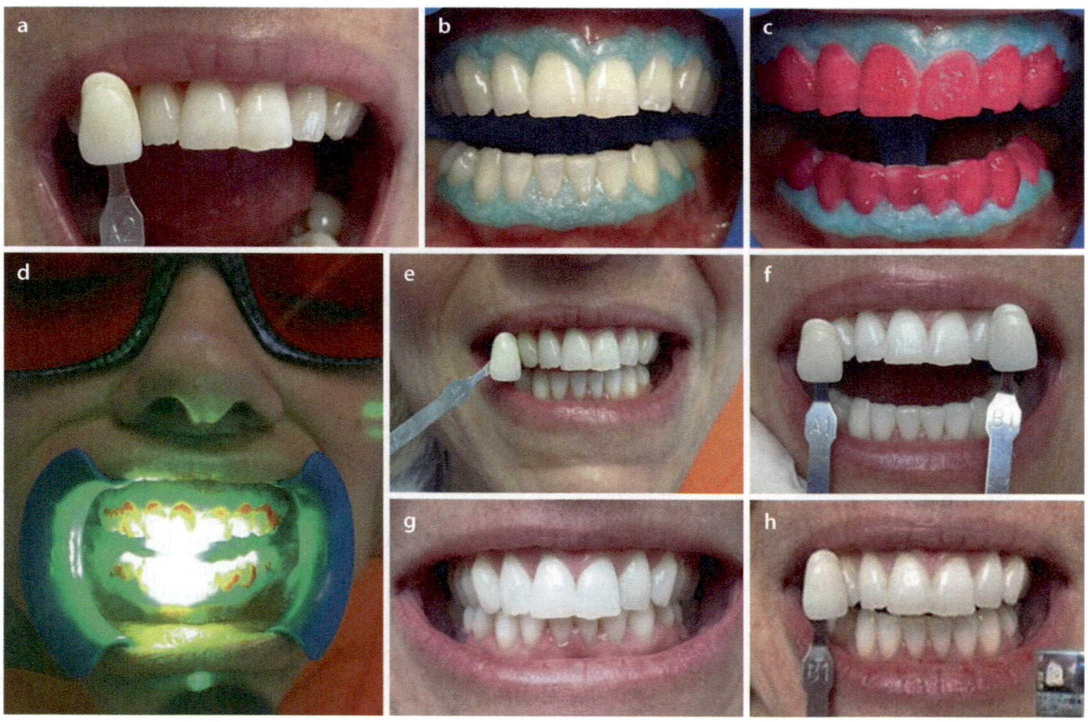

Fig. 16.25 **a** Pre-bleaching shade selection—A 2. **b** Retractors and gingival protection in place. **c** Bleaching gel applied. **d** Activation with the KTP (532 nm) laser. **e** Immediately after bleaching. Note some blanching on papilla between the central incisors due to leakage of the gel. **f** Five days after bleaching. Papilla has a normal appearance. Shade B1 has been achieved. **g** Six months after bleaching. **h** Two years after bleaching

References

1. Liebart M, Fouque-Deruelle C, Santini A, Dillier F, Monnet-Corti V, Gilse J, Borghetti A. Smile line and periodontium visibility. Medicine. 2004;1(1):17–25.
2. Eisen D. Disorders of pigmentation in the oral cavity. Clin Dermatol. 2000;18(5):579–87.
3. Unsal E, Paksoy C, Soykan E, Elhan AH, Sahin M. Oral melanin pigmentation related to smoking in a Turkish population. Community Dent Oral Epidemiol. 2001;29(4):272–7.
4. Babaee N, Nooribayat S. Frequency of oral pigmentation in patients referred to Babol Dental School (2008–2009). J Qazvin Univ Med Sci. 2011;15(3):87–90.
5. Haresaku S, Hanioka T, Tsutsui A. Association of lip pigmentation with smoking and gingival melanin pigmentation. Oral Dis. 2007;13(1):71–6.
6. Multani S. Interrelationship of smoking, lip and gingival melanin pigmentation, and periodontal status. Addict Health. 2013;5(1–2):57–65.
7. Granstien R, Sober A. Drug and heavy metal induced hyperpigmentation. J Am Acad Dermatol. 1981;5(1):1–6.
8. La Porta V, Nikitakis N, Sindler A, Reynolds M. Minocycline-associated intra-oral soft-tissue pigmentation: clinicopathologic correlations and review. J Clin Periodontol. 2005;32(2):119–22.
9. Ciçek Y, Ertaş U. The normal and pathological pigmentation of oral mucous membrane: a review. J Contemp Dent Pract. 2003;4(3):76–86.
10. Hajifattahi F, Azarshab M, Haghgoo R, Lesan S. Evaluation of the relationship between passive smoking and oral pigmentation in children. J Dent (Tehran). 2010;7(3):119–23.
11. Hanioka T, Tanaka K, Ojima M, Yuuki K. Association of melanin pigmentation in the gingiva of children with parents who smoke. Pediatrics. 2005;116(2):e186–90.
12. Moravej-Salehi E, Moravej-Salehi E, Hajifattahi F. Relationship of gingival pigmentation with passive smoking in women. Tanaffos. 2015;14(2):107–14.
13. Yilmaz H, Bayindir H, Kusakci-Seker B, Tasar S, Kurtulmus-Yilmaz S. Treatment of amalgam tattoo with an Er,Cr:YSGG laser. J Investig Clin Dent. 2010;1(1):50–4.
14. Dummett CO, Gupta OP. Estimating the epidemiology of gingival pigmentation. J Natl Med Assoc. 1964;56:419–20.
15. Eversole LE. Chapter 6: Pigmented lesions of the oral mucosa. In: Burket LW, Greenberg MS, Glick M, editors. Burket's oral medicine – diagnosis & treatment. 10th ed. Hamilton: BC Decker; 2003
16. Deepak P, Sunil S, Mishra R. Treatment of gingival pigmentation: a case series. Indian J Dent Res. 2005;16:171–6.
17. Lagdive S, Doshi Y, Marawarr P. Management of gingival hyperpigmentation using surgical blade and diode laser therapy: a comparative study. J Oral Laser Appl. 2009;9(1):41–7.
18. Mani A, Mani S, Saumil S. Management of gingival hyperpigmentation using surgical blade, diamond bur and diode laser therapy: a case report. J Oral Laser Appl. 2009;9:227–32.
19. Murthy MB, Kaur J, Das R. Treatment of gingival hyperpigmentation with rotary abrasive, scalpel, and laser techniques: a case series. J Indian Soc Periodontol. 2012;16(4):614–9.
20. Lee K, Lee D, Shin S, Kwon YH, Chung JH, Herr Y. A comparison of different gingival depigmentation techniques: ablation by erbium:yttrium-aluminum-garnet laser and abrasion by rotary instruments. J Periodontal Implant Sci. 2011;41(4):201–7.
21. Kaushik N, Gaurav V. Efficacy of different techniques of gingival depigmentation: a comparative evaluation with a case report. Int J Laser Dent. 2013;3(2):68–72.
22. Elavarasu S, Thangavelu A, Alex S. Comparative evaluation of depigmentation techniques in split-mouth design with electrocautery and laser. J Pharm Bioallied Sci. 2015;7(Suppl 2):S786–90.
23. Yousuf A, Hossain M, Nakamura Y, Yamada Y, Kinoshita J, Matsumoto K. Removal of gingival melanin pigmentation with the semiconductor diode laser: a case report. J Clin Laser Med Surg. 2000;18(5):263–6.
24. Atsawasuwan P, Greethong K, Nimmanon V. Treatment of gingival hyperpigmentation for esthetic purposes Nd:YAG laser: report of 4 cases. J Periodontal. 2000;71(2):315–21.
25. Tal H, Oegiesser D, Tal M. Gingival depigmentation by erbium:YAG laser: clinical observations and patient responses. J Periodontal. 2003;74(11):1660–7.
26. Esen E, Haytac MC, Oz IA, Erdogan O, Karsli ED. Gingival melanin pigmentation and its treatment with the CO_2 laser. Oral Surg Oral Med Oral Pathol Oral Radiol Endod. 2004;98(5):522–7.
27. Nagai S. Esthetic treatment of gingival melanin depigmentation with Er:YAG laser. J Laser Dent. 2011;19(1):172–5.
28. Doshi Y, Khandge N, Byakod G. Management of gingival pigmentation with diode laser: is it a predictive tool? Int J Laser Dent. 2012;2(1):29–32.
29. Luk K, Zhao IS, Gutknecht N, Chu CH. Use of carbon dioxide lasers in dentistry. Lasers Dent Sci. 2019;3:1–9. https://doi.org/10.1007/s41547-018-0047-y.
30. Swick MD. Cosmetic diode laser gingivectomy with frenectomy. Wavelengths. 2000;8(4):19.
31. Swick MD. A char-free technique for the Ceralas D15 diode laser. Wavelengths. 2000;8(4):20.
32. Altayeb W, Hamadah O, Alhaffar BA, et al. Gingival depigmentation with diode and Er,Cr:YSGG laser: evaluating repigmentation rate and patient perceptions. Clin Oral Invest. 2021;25:5351–61. https://doi.org/10.1007/s00784-021-03843-6.
33. Luk K. Clinical application of a digital pulsed diode laser in depigmentation therapy. J Acad Laser Dent. 2005;13(4):18–21.
34. Luk K. Depigmentation of gingivae and lip with digital pulsed diode laser—an integral part of cosmetic dentistry. Laser J. 2009;1(2):31–3.
35. Allen EP, Gladkova ND, Fomina YV, Karabut M, Kiseleva E, Feldchtein F, Altshuler G. Successful gingival depigmentation with laser-patterned microcoagulation: a case report. Clin Adv Periodont. 2011;1(3):210–4.
36. Taher Agha M, Polenik P. Laser treatment for melanin gingival pigmentations: a comparison study for 3 laser wavelengths 2780, 940, and 445 nm. Int J Dent. 2020;9:3896386. https://doi.org/10.1155/2020/3896386.
37. Luk K. Non-ablative melanin depigmentation of gingiva. Laser J. 2016;1(8):24–7.
38. Ishikawa I, Aoki A, Takasaki AA. Potential applications of erbium:YAG laser in periodontics. J Periodontal Res. 2004;39(4):275–85.
39. Azzeh M. Treatment of gingival hyperpigmentation by erbium doped:yttrium, aluminum, and garnet laser for esthetic purposes. J Periodontol. 2007;78(1):177–84.
40. Ozbayrak S, Dumlu A, Ercalik-Yalcinkaya S. Treatment of melanin-pigmented gingiva and oral mucosa by CO_2 laser. Oral Surg Oral Med Oral Pathol Oral Radiol Endod. 2000;90(1):14–5.
41. Berk G, Atici K, Berk N. Treatment of gingival pigmentation with Er,Cr:YSGG laser. J Oral Laser Appl. 2005;5:249–53.
42. Esen E, Haytac M, Öz IA, Erdogan O, Karsli ED. Gingival melanin pigmentation and its treatment with the CO_2 laser. Oral Surg Oral Med Oral Pathol. 2004;98:522–7.

16

43. Nakamura Y, Hossain M, Hirayama K, Matsumoto K. A clinical study on the removal of gingival melanin pigmentation with the CO_2 laser. Lasers Surg Med. 1999;25(2):140–7.
44. Gupta G. Management of gingival hyperpigmentation by semiconductor diode laser. J Cutan Aesthet Surg. 2011;4(3):208–10.
45. Rosa DS, Aranha AC, Eduardo Cde P, Aoki A. Esthetic treatment of gingival melanin hyperpigmentation with Er:YAG laser: short term clinical observations and patient follow-up. J Periodontol. 2007;78(10):2018–25.
46. Hegde R, Padhye A, Sumanth S, Jain AS, Thukral N. Comparison of surgical stripping: erbium-doped:yttrium, aluminum, and garnet laser; and carbon dioxide laser techniques for gingival depigmentation: a clinical and histologic study. J Periodontol. 2013;84(6):738–48.
47. El Shenawy H, Nasry S, Zaky A, Quriba MA. Treatment of gingival hyperpigmentation by diode laser for esthetical purposes. Open Access Maced J Med Sci. 2015;3(3):447–54.
48. Dunn W, Murchison D, Broome J. Esthetics: patients' perceptions of dental attractiveness. J Prosthodont. 1996;5(3):166–71.
49. Pearson H. Successful bleaching without secondary discolouration. J Can Dent Assoc. 1951;17(4):200–1.
50. Haywood V, Heymann H. Nightguard vital bleaching. Quintessence Int. 1989;20:173–6.
51. Reyto R. Laser tooth whitening. Dent Clin N Am. 1998;42(4):755–62.
52. Simmer J, Hu J. Dental enamel formation and its impact on clinical dentistry. J Dent Educ. 2001;65(9):896–905.
53. Attin T, Kielbassa A, Schwanenberg M, Hellwig E. Effect of fluoride treatment on remineralization of bleached enamel. J Oral Rehabil. 1997;24(4):282–6.
54. Girija V, Stephen H. Characterization of lipid in mature enamel using confocal laser scanning microscopy. J Dent. 2003;31(5):303–11.
55. Eimar H, Siciliano R, Abdallah M, Nader SA, Amin WM, Martinez PP, Celemin A, Cerruti M, Tamimi F. Hydrogen peroxide whitens teeth by oxidizing the organic structure. J Dent. 2012;40:25–33.
56. Ten Bosch J, Coops J. Tooth color and reflectance as related to light scattering and enamel hardness. J Dent Res. 1995;74(1):374–80.
57. Joiner A. Tooth colour: a review of the literature. J Dent. 2004;32:3–12.
58. Nathoo S. The chemistry and mechanisms of extrinsic and intrinsic discoloration. J Am Dent Assoc. 1997;128(1):6–10.
59. Hattab F, Qudeimat M, Al-Rimawi H. Dental discoloration: an overview. J Esthet Dent. 1999;11(6):291–310.
60. http://eurlex.europa.eu/LexUriServ/LexUriServ.do?uri=OJ:L:2011:283:0036:0038:en:PDF.
61. Moritz A, Beer F, Blum R. Orale Lasertherapie. Berlin: Quintessenz Verlags-GmbH; 2006. p. 422–3.
62. Kwon S, Ko S, Greenwall L, Goldstein R. Tooth whitening in esthetic dentistry. 1st ed. London: Quintessence; 2009.
63. Auschill T, Hellwig E, Schmidale S, Sculean A, Arweiler NB. Efficacy, side-effects and patients' acceptance of different bleaching techniques (OTC, in-office, at-home). Oper Dent. 2005;30(2):156–63.
64. Walsh L, Liu J, Verheyen P. Tooth discolouration and its treatment using KTP laser-assisted tooth whitening. J Oral Laser Appl. 2004;4:7–21.
65. Goldstein R, Garber D. Complete dental bleaching. Chicago: Quintessence Pub. Co.; 1995.
66. Zach L, Cohen G. Pulp response to externally applied heat. Oral Surg Oral Med Oral Pathol. 1965;19(4):515–30.
67. Camargo S, Cardoso P, Valera M, de Araujo M. Penetration of 35% hydrogen peroxide into the pulp chamber in bovine teeth after LED or Nd:YAG laser activation. Eur J Esthet Dent. 2009;4:82–9.
68. Kwon S, Oyoyo U, Li Y. Effect of light activation on tooth whitening efficacy and hydrogen peroxide penetration: an in vitro study. J Dent. 2013;41:39–45.
69. De Moor R, Vanderstricht K. The use of the KTP laser, an added value for tooth bleaching. J Oral Laser Applic. 2009;9(4):219.
70. Gurgan S, Cakir F, Yazici E. Different light-activated in-office bleaching systems: a clinical evaluation. Lasers Med Sci. 2009;25(6):817–22.
71. Strobl A, Gutknecht N, Franzen R, Hilgers R, Lampert F, Meister J. Laser-assisted in-office bleaching using a neodymium:yttrium–aluminum–garnet laser: an in vivo study. Lasers Med Sci. 2009;25(4):503–9.
72. Domínguez A, García JA, Costela A, Gómez C. Influence of the light source and bleaching gel on the efficacy of the tooth whitening process. Photomed Laser Surg. 2011;29(1):53–9.
73. Luk K, Tam L, Hubert M. Effect of light energy on peroxide tooth bleaching. J Am Dent Assoc. 2004;135(2):194–201.
74. De Moor RJG, Verheyen J, Verheyen P, Diachuk A, Meire MA, De Coster PJ, De Bruyne M, Keulemans F. Laser teeth bleaching: evaluation of eventual side effects on enamel and the pulp and the efficiency in vitro and in vivo. Sci World J. 2015;835405:1–12. https://doi.org/10.1155/2015/835405.
75. Nguyen C, Augros C, Rocca J, Lagori G, Fornaini C. KTP and Er:YAG laser dental bleaching comparison: a spectrophotometric, thermal and morphologic analysis. Lasers Med Sci. 2015;30(8):2157–64.
76. Surmelioglu D, Usumez A. Effectiveness of different laser-assisted in-office bleaching techniques: 1-year follow-up. Photobiomodul Photomed Laser Surg. 2020;38(10):632–9.
77. Bennett ZY, Walsh LJ. Efficacy of LED versus KTP laser activation of photodynamic bleaching of tetracycline-stained dentine. Lasers Med Sci. 2014;30(7):1823–8.
78. Li Y, Greenwall L. Safety issues of tooth whitening using peroxide-based materials. Br Dent J. 2013;215(1):29–34.
79. Furukawa M, K-Kaneyama J, Yamada M, Senda A, Manabe A, Miyazaki A. Cytotoxic effects of hydrogen peroxide on human gingival fibroblasts in vitro. Oper Dent. 2015;40(4):430–9.
80. Li Y, Lee S, Cartwright S, Wilson A. Comparison of clinical efficacy and safety of three professional at-home tooth whitening systems. Compend Contin Educ Dent. 2003;24:357–64.
81. Markowitz K. Pretty painful: why does tooth bleaching hurt? Med Hypotheses. 2010;74(5):835–40.
82. Schulte J, Morrissette D, Gasior E, Czajewski MV. The effects of bleaching application time on the dental pulp. J Am Dent Assoc. 1994;125(10):1330–5.
83. Moosavi H, Arjmand N, Ahrari F, Zakeri M, Maleknejad F. Effect of low-level laser therapy on tooth sensitivity induced by in-office bleaching. Lasers Med Sci. 2016;31(4):713–9.
84. Abbasi M, Pordel E, Chiniforush N, Firuzjaee SG, Omrani LR. Hydrogen peroxide penetration into the pulp chamber during conventional in-office bleaching and diode laser-assisted bleaching with three different wavelengths. Laser Ther. 2019;28(4):285–90.
85. Li Y. Tooth bleaching using peroxide-containing agents: current status of safety issues of gel in tray whitening. Compend Contin Educ Dent. 1998;19:783–6.
86. Kina JF, Huck C, Riehl H, Martinez TC, Sacono NT, Ribeiro AP, Costa CA. Response of human pulps after professionally applied vital tooth bleaching. Int Endod J. 2010;43(7):572–80.

87. Sürmelioğlu D, Gündoğar H, Taysi S, Bağiş YH. Effect of different bleaching techniques on DNA damage biomarkers in serum, saliva, and GCF. Hum Exp Toxicol. 2021;40(8):1332–41.
88. Eriksson AR, Albrektsson T. Temperature threshold levels for heat-induced bone tissue injury: a vital—microscopic study in the rabbit. J Pros Dent. 1983;50(1):101–7.
89. Baldissara P, Catapano S, Scotti R. Clinical and histological evaluation of thermal injury thresholds in human teeth: a preliminary study. J Oral Rehabil. 1997;24(11):791–801.
90. Metz M, Cochran M, Matis B, Gonzalez C, Platt J, Lund M. Clinical evaluation of 15% carbamide peroxide on the surface microhardness and shear bond strength of human enamel. Oper Dent. 2007;32(5):427–36.
91. Brunton P, Aminian A, Pretty I. Vital tooth bleaching in dental practice: 2. Novel bleaching systems. Dent Update. 2005;33(6):357–8.
92. Attin T, Schmidlin P, Wegehaupt F, Wiegand A. Influence of study design on the impact of bleaching agents on dental enamel microhardness: a review. Dent Mater. 2009;25(2):143–57.
93. Shannon H, Spencer P, Gross K, Tira D. Characterization of enamel exposed to 10% carbamide peroxide bleaching agents. Quintessence Int. 1993;24(1):39–44.
94. Sulieman M, Rees J, Addy M. Surface and pulp chamber temperature rises during tooth bleaching using a diode laser: a study in vitro. Br Dent J. 2006;200(11):631–4.
95. Grobler S, Majeed A, Hayward R, et al. A clinical study of the effectiveness of two different 10% carbamide peroxide bleaching products: a 6-month followup. Int J Dent. 2011;2011:167525.
96. Leonard R, Bentley C, Eagle JC, Garland GE, Knight MC, Phillips C. Nightguard vital bleaching: a long-term study on its efficacy, shade retention, side effects, and patients' perceptions. J Esthet Restor Dent. 2001;13:357–69.
97. Boushell L, Ritter A, Garland G, Tiwana K, Smith L, Broome A, Leonard R. Nightguard vital bleaching: side effects and patient satisfaction 10 to 17 years post-treatment. J Esthet Restor Dent. 2012;24(3):211–9.
98. Kuzekanani M, Walsh L. Quantitative analysis of KTP laser photodynamic bleaching of tetracycline-discolored teeth. Photomed Laser Surg. 2009;27(3):521–5.
99. do Amaral F, Sasaki R, da Silva T, Francca F, Florio F, Basting R. The effects of home-use and in-office bleaching treatments on calcium and phosphorus concentrations in tooth enamel: an in vivo study. J Am Dent Assoc. 2012;143(6):580–6.
100. Khurana R, Tredwin C, Weisbloom M, Moles D. A clinical evaluation of the individual repeatability of three commercially available colour measuring devices. Br Dent J. 2007;203(12):675–80.
101. Paravina R, Johnston W, Powers J. New shade guide for evaluation of tooth whitening—colorimetric study. J Esthet Restor Dent. 2007;19(5):276–83.
102. Hadis MA, Zainal SA, Holder MJ, Carroll JD, Cooper PR, Milward MR, Palin W. The dark art of light measurement: accurate radiometry for low-level light therapy. Lasers Med Sci. 2016;31(4):789–809.
103. Cronshaw M, Parker S, Anagnostaki E, Mylona V, Lynch E, Grootveld M. Photobiomodulation dose parameters in dentistry: a systematic review and meta-analysis. Dent J. 2020;8(4):114.
104. Parker S, Cronshaw M, Grootveld M. Photobiomodulation delivery parameters in dentistry: an evidence-based approach. Photobiomodul Photomed Laser Surg. 2022;40(1):42–50.
105. Parker S, Cronshaw M, Grootveld M, George R, Anagnostaki E, Mylona V, Chala M, Walsh L. The influence of delivery power losses and full operating parametry on the effectiveness of diode visible–near infra-red (445–1064 nm) laser therapy in dentistry—a multi-centre investigation. Lasers Med Sci. 2022;37(4):2249–57.
106. Anagnostaki E, Mylona V, Kosma K, Parker S, Chala M, Cronshaw M, Dimitriou V, Tatarakis M, Papadogiannis N, Lynch E, Grootveld M. A spectrophotometric study on light attenuation properties of dental bleaching gels: potential relevance to irradiation parameters. Dent J. 2020;8(4):137.
107. Gurgan S, Cakir FY, Yazici E. Different light-activated in-office bleaching systems: a clinical evaluation. Lasers Med Sci. 2010;25(6):817–22.
108. Moncada G, Sepúlveda D, Elphick K, Contente M, Estay J, Bahamondes V, Fernandez E, Oliveria OB, Martin J. Effects of light activation, agent concentration, and tooth thickness on dental sensitivity after bleaching. Oper Dent. 2013;38(5):467–76.
109. Kiomars N, Azarpour P, Mirzaei M, Kamangar SSH, Kharazifard MJ, Chiniforush N. Evaluation of the diode laser (810 nm, 980 nm) on color change of teeth after external bleaching. Laser Ther. 2016;25(4):267–72.
110. Mirzaie M, Yassini E, Ganji S, Moradi Z, Chiniforush N. A comparative study of enamel surface roughness after bleaching with diode laser and Nd:YAG laser. J Lasers Med Sci. 2016;7(3):197–200.
111. Shahabi S, Assadian H, Nahavandi AM, Nokhbatolfoghahaei H. Comparison of tooth color change after bleaching with conventional and different light-activated methods. J Lasers Med Sci. 2018;9(1):27–31.
112. Nanjundasetty J, Ashrafulla M. Efficacy of desensitizing agents on postoperative sensitivity following an in-office vital tooth bleaching: A randomized controlled clinical trial. J Conserv Dent. 2016;19(3):207–11.

The Way Forward?

Contents

Current Research and Future Dreams for Dental Lasers

Peter Rechmann

Contents

D. J. Coluzzi, S. P. A. Parker (eds.), *Lasers in Dentistry—Current Concepts*, Textbooks in Contemporary Dentistry,
https://doi.org/10.1007/978-3-031-43338-2_17

Core Message

While all commercially available dental lasers can be clinically used for soft tissue procedures, the quest continues for the ideal instrument to safely and efficiently interact with dental hard tissue. This chapter describes several years of research that has resulted in the second generation of hard tissue lasers. The interaction of carbon dioxide laser photonic energy with tooth enamel and dentin has been studied for several decades, with first publications in the 1960s. Carbon dioxide lasers at 9.3 and 9.6 µm wavelength show the highest absorption of all dental lasers in dental hard tissues. Laboratory studies have shown that these short-pulsed CO_2 lasers can efficiently be used to render enamel caries resistant by transforming the originally carbonated apatite into the much acid less soluble hydroxyapatite. Adding fluoride after the laser treatment additionally reduces the acid solubility of enamel and creates the desired least acid-soluble fluorapatite. Irradiation with the 9.3 µm laser wavelength can reduce mineral loss by 55% over untreated enamel. Safety studies have shown that without harmful effects to the pulpal tissue, these lasers can efficiently and safely be used on vital teeth. The first in vivo clinical study engaging an orthodontic bracket model showed over 4 weeks a 46% and over 12 weeks a 87% reduction in mineral loss. The first in vivo occlusal caries prevention by pulsed CO_2 laser and additional fluoride varnish application proof-of-concept pilot study demonstrated that a microsecond pulsed 9.6 µm CO_2 laser with additional fluoride varnish applications significantly inhibited the formation of carious lesions in fissures of molars in vivo in comparison to a non-irradiated control tooth in the same arch over a 1-year observation interval. These positive results were confirmed in a main clinical study with 60 subjects similarly designed as the pilot study. Using ICDAS and SOPROLIFE daylight and fluorescence assessment tools proved the reduction in caries. Moreover, the 9.3 and 9.6 µm CO_2 µs-short-pulsed lasers are very efficient in cutting dental hard and soft tissue. Results of shear bond strength testing with multiple bonding agents to such laser cuts are promising.

17.1 Rendering Enamel Caries Resistant: Laboratory Work

During the creation of tooth mineral, a pure hydroxyapatite $[Ca_{10}(PO_4)_6(OH)_2]$ is actually not formed. In fact, the mineral portion of the enamel and dentin is best called a highly substituted carbonated apatite [1]. The mineral is closely related to hydroxyapatite, but in acid, it is much more soluble. Carbonated apatite is deficient in calcium (sodium, magnesium, zinc, etc. replace the calcium) and contains between 3% and 6% carbonate by weight. In the crystal lattice, carbonate mostly replaces phosphate ions [2–4]. The mineral of enamel and dentin can be described by the formula of carbonated hydroxyapatite $[Ca_{10-x}(Na)_x(PO_4)_{6-y}(CO_3)_z(OH)_{2-u}(F)_u]$.

Numerous laboratory studies in the past have shown that increasing resistance to enamel demineralization may be achieved by microsecond-pulsed CO_2 laser irradiation [5, 6]. The most strongly absorbed wavelengths in dental enamel are the 9.3 and 9.6 µm CO_2 laser wavelengths [7, 8] (◘ Fig. 17.1).

At 9.3 and 9.6 µm wavelengths, the enamel absorption coefficient is ten times higher compared to the 10.6 µm CO_2 laser wavelength [7, 8]. This is demonstrated in ◘ Table 17.1.

Due to the irradiation heat, the carbonate phase loss from the enamel crystals is responsible for the reduced dissolution of enamel in acid [10] due to transformation of the carbonated hydroxyapatite into the more acid-resistant hydroxyapatite. If at this point in time fluoride is added, fluorapatite is created, which is even less acid soluble than hydroxyapatite [11].

Fluoride works predominantly via topical mechanisms, including (1) demineralization inhibition at the crystal surfaces inside the enamel, (2) remineralization enhancement at the surfaces of the crystal (the newly created remineralized layer is very resistant to acid attack), and (3) finally inhibition of bacterial enzymes [12].

Topical fluoride in solution in the oral cavity boosts remineralization by accelerating the growth of a new surface on top of the partially demineralized subsurface crystals in the carious tooth structure. The newly formed veneer-like surface layer on top of the crystal is like flu-

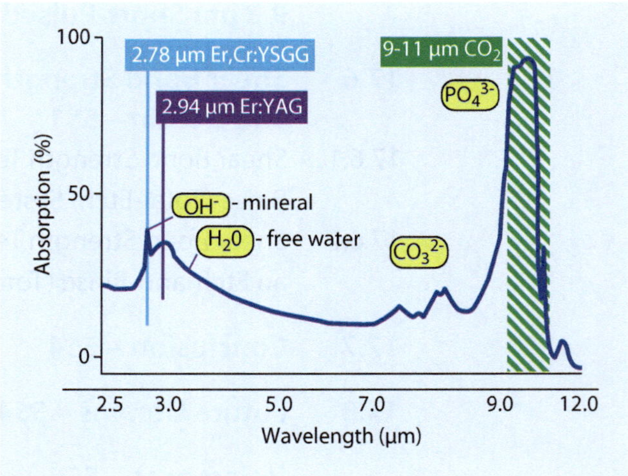

◘ **Fig. 17.1** Absorption of human dental enamel in the infrared (IR) spectral region showing the position of the primary absorbers, namely phosphate (PO_4^{3-}), carbonate (CO_3^{2-}), hydroxyl (OH^-), and water (H_2O), overlapped by the positions of the Er,Cr:YSGG, Er:YAG, and carbon dioxide (9.3, 9.6, 10.3, and 10.6 µm) emission wavelengths. (The curve is a simplified transformation from an infrared transmission spectrum of dental enamel; significantly modified from Refs. [5, 8, 9])

■ **Table 17.1** Selected optical properties of enamel and dentin, tabulated from references mentioned as sources

Absorption coefficients (μ_a) of light in dental enamel			
Wavelength		**μ_a (cm^{-1})**	**Source**
Visible			
450–700 nm		3–4	Fried et al./Ten Bosch
Near IR			
Nd:YAG	1.06 μm	<1	Fried et al.
Mid IR			
Ho:YAG	2.10 μm	<20	Estimate
Er:YSGG	2.79 μm	450	Zuerlein et al.
Er:YAG	2.94 μm	770	Zuerlein et al.
CO_2	9.3 μm	5500	Zuerlcin et al.
	9.6 μm	8000	
	10.3 μm	1125	
	10.6 μm	825	

From Ref. [9]

orapatite, exhibiting much lower acid solubility than the original carbonated apatite mineral [13, 14].

Initial investigations postulated that melting of enamel was required to accomplish caries resistance. Surface enamel melts at about 1100 °C and fuses at or above the hydroxyapatite melting point of about 1280 °C [15]. The goal in basic research was to determine parameters that will selectively melt and/or chemically alter crystals near the surface to a depth that will provide the greatest efficacy for caries prevention [16]. Consequently, McCormack et al. in 1995 irradiated bovine and human enamel by a tunable, pulsed CO_2 laser (■ Fig. 17.2) [16]. This specific laser prototype could be tuned to the wavelengths of 9.3, 9.6, 10.3, and 10.6 μm. To irradiate the samples, 5, 25, or 100 pulses were used, at fluences of 2, 5, 10, or 20 J/cm^2 with pulse widths of 50, 100, 200, and 500 μs, respectively. The authors observed crystal fusion at fluences of 5 J/cm^2 with the 9.3, 9.6, and 10.3 μm wavelength but never with the 10.6 μm wavelength [16].

Fried et al. in 1996 showed when using a 9.6 μm CO_2 laser at 100 μs pulsed duration and a fluence of 4 J/cm^2, they achieved an 800 °C enamel surface temperature, which was just not melting enamel. Surface temperatures of 800 °C and up to 1200 °C, achieved at 6 and 8 J/cm^2, respectively, caused the mineral to melt [17]. Applying 10 J/cm^2 resulted in vaporization of the enamel (■ Fig. 17.3).

As a consequence of irradiation with a 9.6 μm carbon dioxide laser with 100 μs pulse duration, Featherstone et al. in 1997 showed that using fluences of 0–3 J/cm^2 did not or only slightly reduced the carbonate content of enamel. In contrast, irradiation with fluences of 4, 5, or 6 J/cm^2 eliminated the carbonate from enamel surfaces. Measurements were done by Fourier transform infrared reflectance (FTIR) spectroscopy (■ Fig. 17.4) [9, 18].

Lately, a short-pulsed carbon dioxide laser emitting at a wavelength of 9.3 μm became available on the US market for use in dental offices (Solea, Convergent Dental, Inc., Natick, MA). The CO_2 gas of the laser medium is "radio-frequency excited," and thus the direct-pulsed laser can emit extremely short laser pulse durations as short as a 3 μs minimum pulse duration.

In order to test the caries preventive potential of the 9.3 μm short-pulsed CO_2 laser, five different pulse durations between 3 and 7 μs were used irradiating enamel samples in a laboratory study. The consequently delivered pulse energies ranged from 1.49 mJ/pulse and up to 2.9 mJ/pulse, resulting in fluences between 3.0 and 5.9 J/cm^2. Nonirradiated samples served as control in this study. In addition, a series of samples received additional fluoride treatment. After a 9-day pH cycling period, when using cross-sectional microhardness testing, this study showed that by laser treatment without additional fluoride, the average mineral loss of the test samples was already significantly reduced by 53% ± 11% (■ Fig. 17.5a). When additional fluoride applications were used without any laser treatment, the mineral loss of these controls was already reduced in average by more than 50%. Adding laser irradiation, the average mineral loss was significantly reduced by 55% ± 9% (■ Fig. 17.5b) [19] on top of the already gained resistance when using fluoride alone.

As mentioned above, it had been reported that enamel surface temperatures of 800 °C and above caused the mineral melting and the mineral transformation into less acid-soluble mineral after cooling [15, 20, 21]. Other work has established that temperatures of 400 °C and above are necessary to decompose the carbonate inclusions in the enamel and transform the carbonated hydroxyapatite to the much less acid-soluble hydroxyapatite [21, 22]. As seen below, the scanning electron microscope of the Rechmann et al. study in 2016 revealed that the lowest applied energies (pulse durations) did not produce very noticeable surface modifications, besides some minor areas with insignificant melting (■ Fig. 17.6). Nevertheless, the cross-sectional microhardness testing after simulating caries demineralization and remineralization in the pH cycling model revealed significantly reduced mineral loss in the laser treatment group; thus, one could conclude that enamel

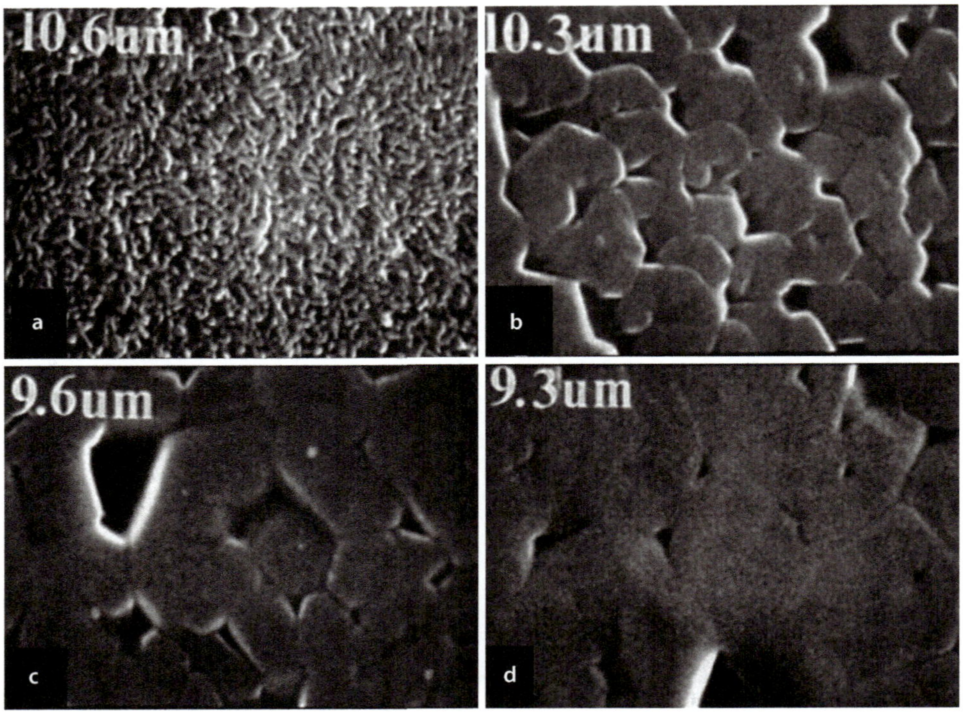

Fig. 17.2 SEM pictures of unpolished enamel after irradiation with CO_2 laser wavelength: **a** 10.6 μm, **b** 10.3 μm, **c** 9.6 μm, and **d** 9.3 μm, all 50 μs pulse duration; melting of the enamel prisms occurs at 10.3 μm, 9.6 μm, and 9.3 μm but not at 10.6 μm. (Photos from Ref. [16]. Used with permission)

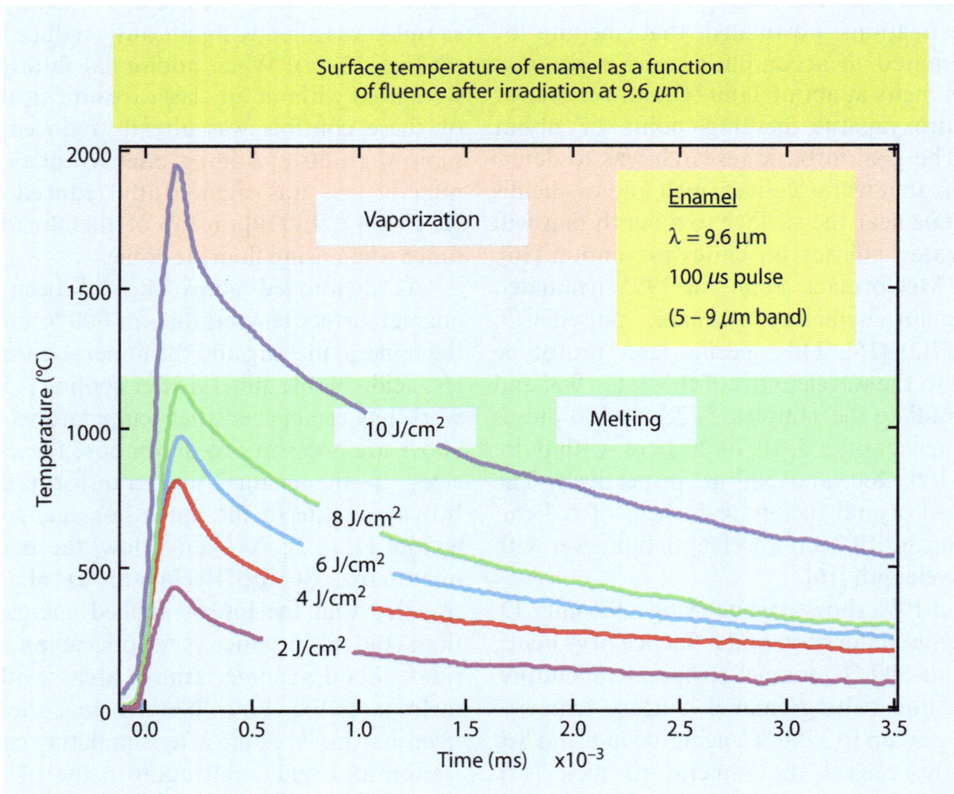

Fig. 17.3 Plot of temperature at the surface of dental enamel versus time following irradiation by a carbon dioxide laser at 9.6 μm, over a range of fluences, and with a pulse duration of 100 μs. (Adapted from Ref. [17])

17

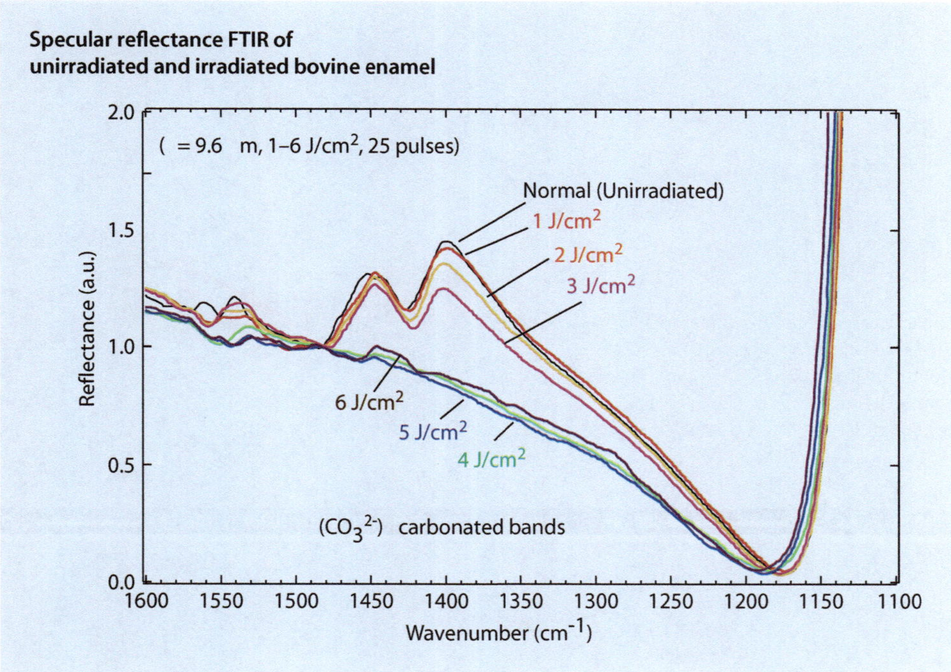

Fig. 17.4 Fourier transform infrared reflectance (FTIR) spectrum showing the carbonate bands following surface treatment of dental enamel after applying a range of fluences with a 9.6 μm CO_2 laser. (Adapted from Ref. [18]. Used with permission: SPIE, Bellingham, WA)

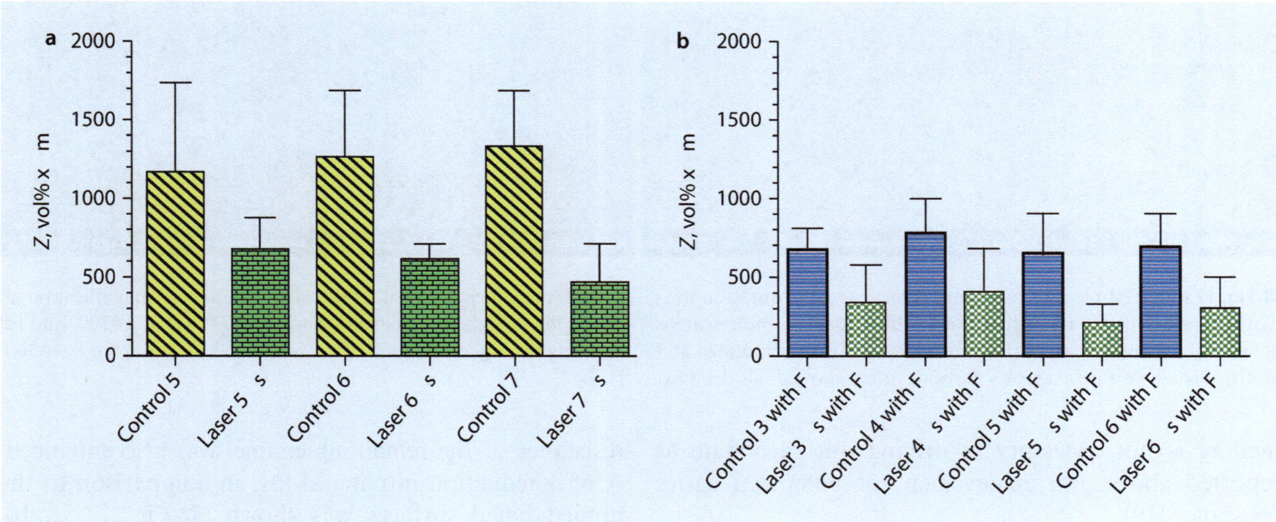

Fig. 17.5 **a** Mean relative mineral loss DZ for the laser-treated enamel and for the control groups, three different laser energies, after 9 days of pH cycling with no additional fluoride use resulting on average in 53% reduced mineral loss for the laser-treated teeth (statistically significant reduced mineral loss with $P < 0.0001$; error bars represent standard deviations). **b** Mean relative mineral loss DZ for the laser-treated and control groups, four different laser energies, after 9 days of pH cycling with additional fluoride, showing on average 55% reduced mineral loss for the laser-treated teeth (statistically significant reduced mineral loss with $P < 0.0001$; error bars represent standard deviations). (Adapted from Ref. [19])

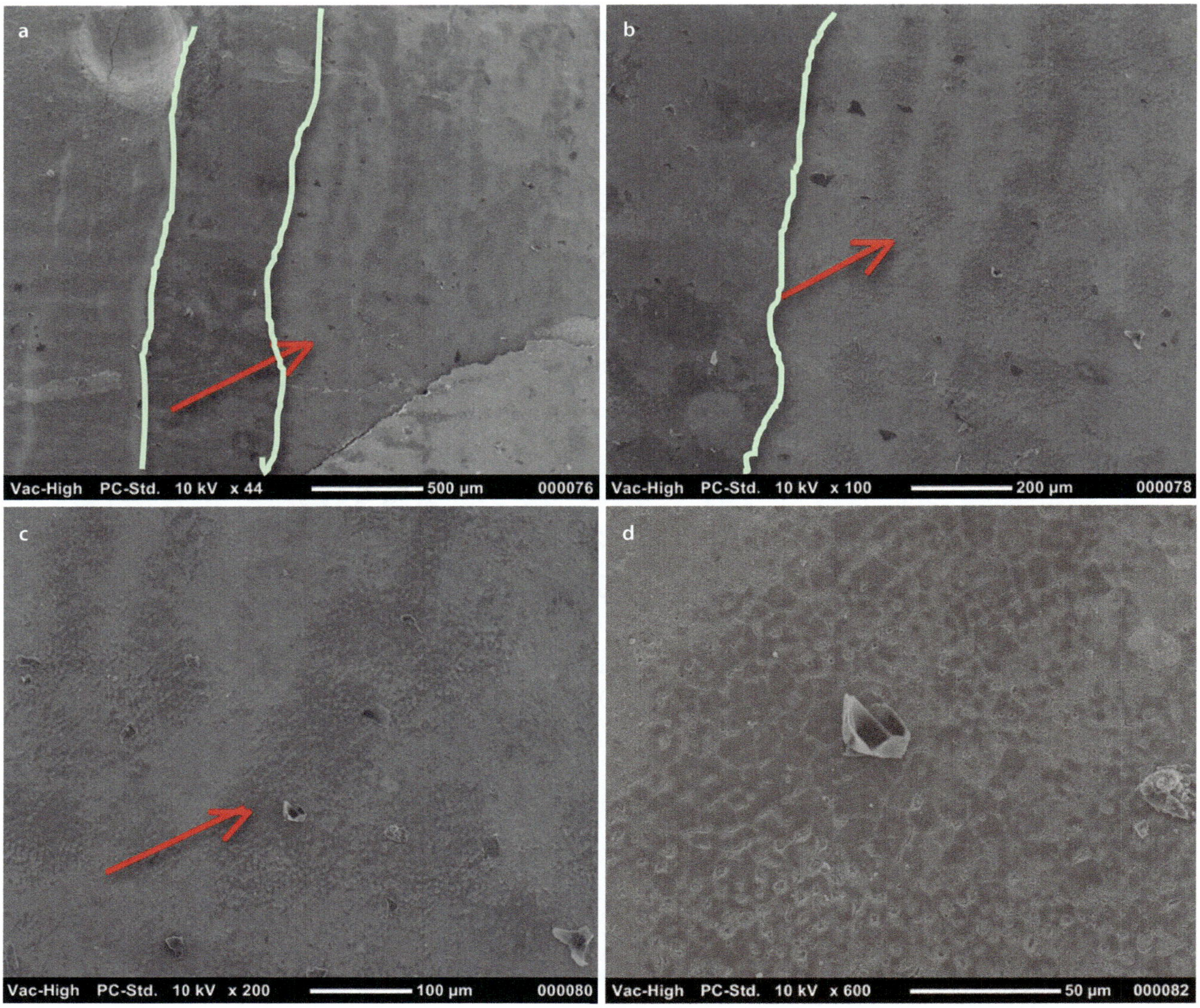

Fig. 17.6 SEM pictures of enamel after 3 µs pulse duration irradiation; only minor or no changes are visible with a few molten areas at the highest magnification (lines are drawn between irradiated and nonirradiated surfaces; arrows indicate area showed at the next higher magnification, with (**a**) being the lowest magnification at x=44, (**b**) at higher magnification with x=100, (**c**) at x=200, and (**d**) showing the highest magnification with x=600)). (Adapted from Ref. [19])

17

melting is not necessary in driving out carbonate as reported above and achievement of enhanced caries resistance [19].

It had been established that ablation of enamel occurs at temperatures above 1200 °C [17]. In the laboratory study presented here [19], when 9.3 µm CO_2 laser short-pulsed laser energies were applied, causing ablation of enamel for cutting teeth (Fig. 17.7), caries resistance of the remaining enamel was also enhanced. A 65% reduction in mineral loss in comparison to the nonirradiated surfaces was shown (Fig. 17.5a, b). This effect is an advantage when cavities are drilled with a 9.3 µm CO_2 short-pulsed laser and a restoration is then placed. The restoration margins will be better protected against recurrent caries. A failure of the restoration will be more unlikely.

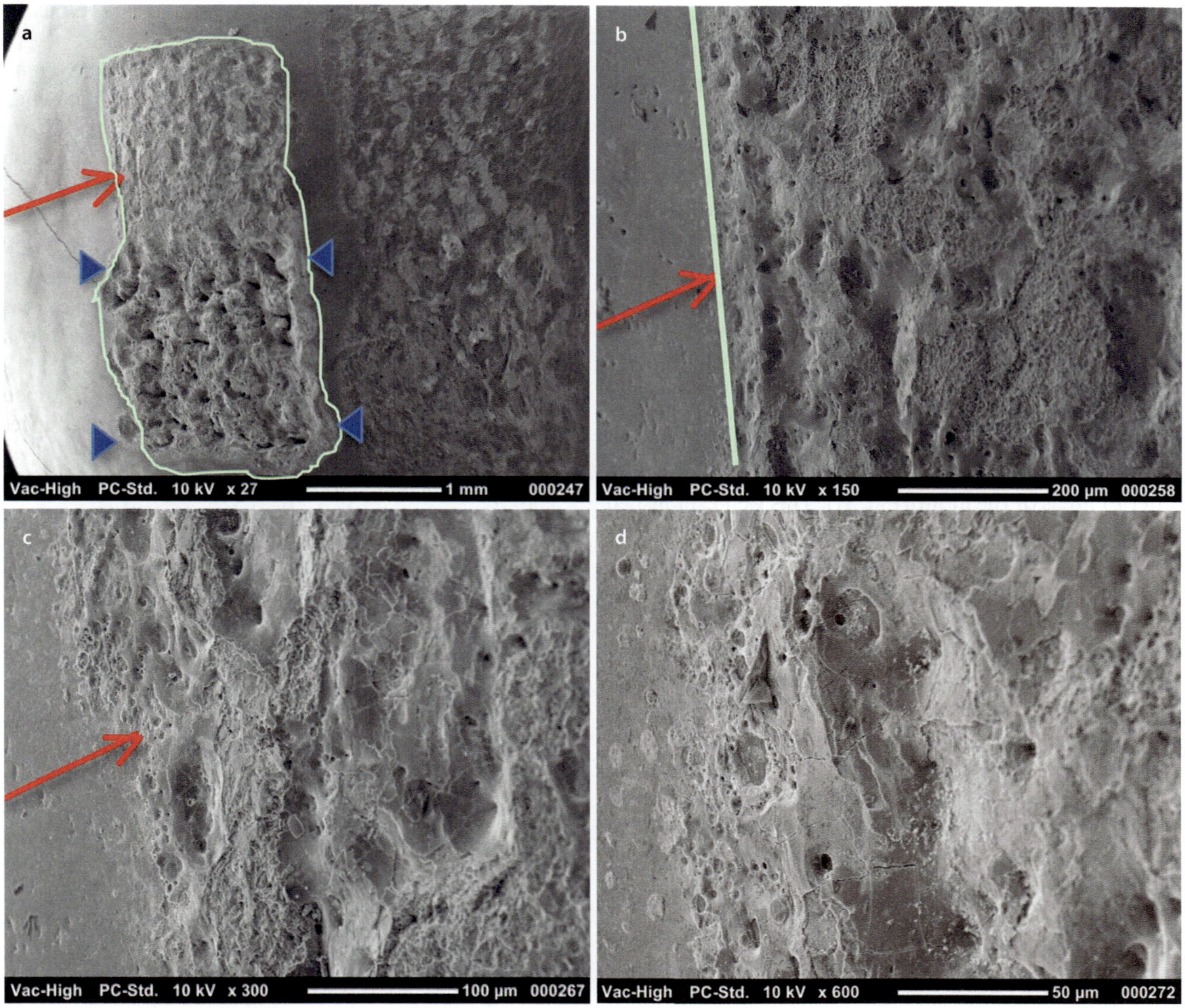

Fig. 17.7 SEM pictures of enamel after 6 and 7 μs pulse duration irradiation; rough surface morphology with slight ablation of the enamel occurs at 6 μs pulses; 7 μs pulses (in a, between triangles) result in noticeable ablation of the enamel (lines are drawn between irradiated and nonirradiated surfaces; arrows indicate area showed at the next higher magnification, with (**a**) being the lowest magnification at x=27, (**b**) at higher magnification with x=150, (**c**) at x=300, and (**d**) showing the highest magnification with x=600). (Adapted from Refs. [19, 22])

17.2 Pulpal Safety Study

Before proving that both short-pulsed CO_2 9.6 and 9.3 μm laser irradiation clinically render enamel more caries resistant, a pulpal safety study had to be performed. The intention of such a study was to show that the laser treatment would not harm the dental pulp. A clinical study using third molars was completed performing pulp histology after laser treatment and sham dental procedures, respectively. The conclusion was that the 9.6 μm CO_2 laser, pulsed with 5–8 μs pulse width, can be safely used to alter enamel surfaces to render them more resistant to caries without permanently damaging the dental pulp. Histological examination of all teeth disclosed no indication of an inflammatory response in the pulp tissue at any time point. All histological sections appeared normal with no changes seen in the normal pulpal morphology. Permanent or serious damaging of the dental pulp was not observed [23]. Lately, a second pulpal safety study performed with a 9.3 μm, 15 μs pulsed CO_2 laser confirmed that this laser can ablate enamel safely without harming the pulp [24].

17.3 Inhibition of Caries in Vital Teeth by CO_2 Laser Treatment: First In Vivo Study Using the Orthodontic Bracket Model

To test the caries preventive potential of specific CO_2 laser irradiation for the first time in vivo, an orthodontic model was used [25]. Typically, orthodontic treatment is associated with rapidly enhanced demineralization of the enamel due to the increased plaque accumulation around the brackets [26]. This also includes a change to a more cariogenic bacterial milieu [27, 28]. After bracket bonding in orthodontic patients, demineralization takes place at the gingival and middle thirds of the facial surfaces [29]. Demineralization switches from typically the interproximal areas to the facial as well as from posterior to anterior areas of the mouth [30, 31]. This well-established caries pattern was used as a model system [25, 32] to determine whether the laser treatment prevents demineralization and/or even enhances remineralization in vital human teeth at those more caries-prone regions.

Twenty-four orthodontic patients with a need for extraction of bicuspids within the planned orthodontic treatment gave their consent and were enrolled into the study. Brackets were bonded with a conventional light-cured composite resin (Transbond XT, 3M ESPE, St. Paul, MN) [25] onto the buccal surface of the bicuspid scheduled for extraction. An enamel area directly next to the bracket at the cervical area of the tooth (◘ Fig. 17.8) was treated according to the laser treatment protocol. The two pictures in ◘ Fig. 17.9 demonstrate the clinical aspect of the laser-treated surfaces; due to the biofilm removal, the irradiated enamel is very shiny. Nevertheless, after 1 week of home care, typically the shiny aspect faded off.

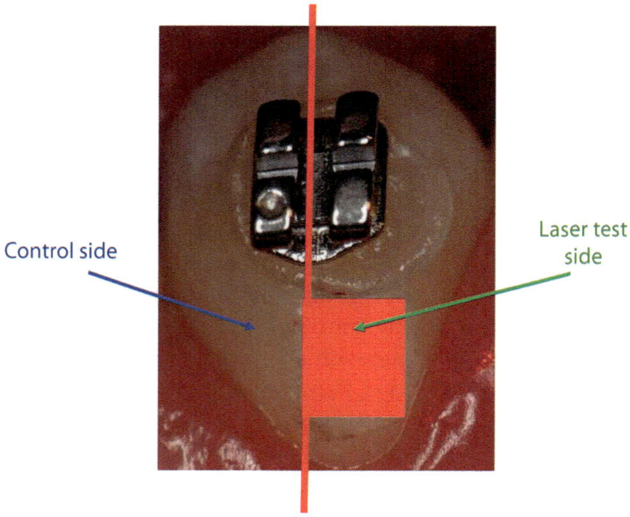

◘ **Fig. 17.8** Orthodontic bracket placed on a study bicuspid using an abundant amount of composite to create a microbial plaque trap; the irradiated area cervical to the bracket is marked red in this picture, and the other side served as control

The laser employed was a CO_2 laser, Pulse System, Inc. (PSI) (Model #LPS-500, Los Alamos, NM), which had been designed for dermatology indications (9.6 μm wavelength, 20 μs pulse duration, 20 Hz pulse repetition rate, 1100 μm beam diameter). To achieve caries preventive changes, each irradiated spot had to receive 20 laser pulses; the laser fluence per pulse averaged 4.1 ± 0.3 J/cm^2 (for more details, see [33]). An area cervical to the bracket (◘ Fig. 17.8) on one side of the bracket was irradiated, while the opposite site on the same tooth served as the control side later on.

The patients were instructed to brush for 1 timed minute twice daily with a provided over-the-counter (OTC) toothpaste (1100 ppm fluoride as NaF).

The study teeth were extracted 4 or 12 weeks after irradiation, respectively. In the laboratory, they were cut into halves through the bracket so that the laser-treated area was on one half and the non-laser control area on the other half. Afterwards, they were embedded in acrylic and prepared for the cross-section microhardness testing. ◘ Figure 17.10 shows a microscopic picture of a typical cross section with enamel, dentin, and orthodontic bracket with the composite and the microindents. During the cross-section microhardness testing procedure, a pyramid-shaped diamond tip is pushed into the enamel with a defined weight. The softer the enamel is due to the occurred demineralization, the wider the indent will be. The width of the indent is measured and the actual mineral loss can be calculated.

In ◘ Fig. 17.11, the mineral loss of enamel (volume percentage) is plotted versus the depth into the tooth presenting a mineral loss profile for the 4-week study arm teeth. The green line with the triangle symbols represents the average mean volume % mineral at each depth for the laser-treated teeth and the blue line with the square symbols for the non-laser-treated controls. At 15 μm depth, the control teeth (square symbols) show a very high loss of mineral with a remaining only 40% average volume % mineral, increasing to an average of 82% at a depth of 45 μm. In contrast, the CO_2 9.6 μm laser-irradiated enamel (triangle symbols) shows 62% volume % mineral at the 15 μm depth. Further into the tooth, the volume % mineral increases at 45 μm depth to 85%, which represents the typical volume % mineral content of sound enamel.

To compare the "mineral loss" between groups, the mean relative mineral loss, ΔZ (vol% × μm), can be calculated. After the 4-week observation, the mean relative mineral loss ΔZ for the laser-treated enamel was 402 ± 85 (mean ± standard error [SE]), while the controls showed an almost doubled mineral loss ΔZ of 737 ± 131. The laser treatment resulted in a significant 46% reduction of demineralization around the orth-

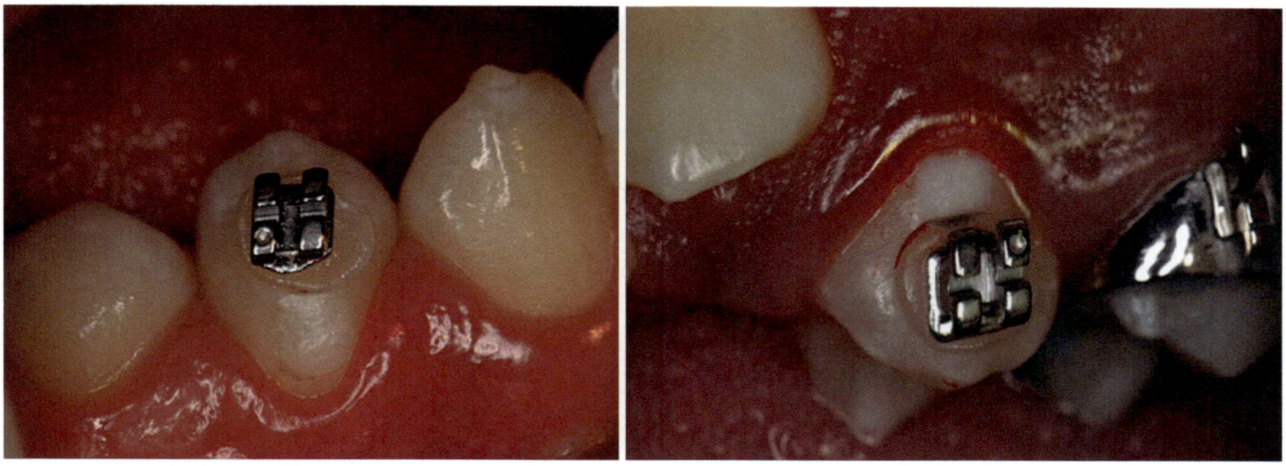

Fig. 17.9 Orthodontic bracket placed on a lower right (left picture) and on an upper left (right picture) study bicuspid showing an irradiated area beyond the bracket. Obviously, the biofilm is removed by the laser irradiation, and the area appears white and shiny, but after a period of a week of home care, the colors matched again

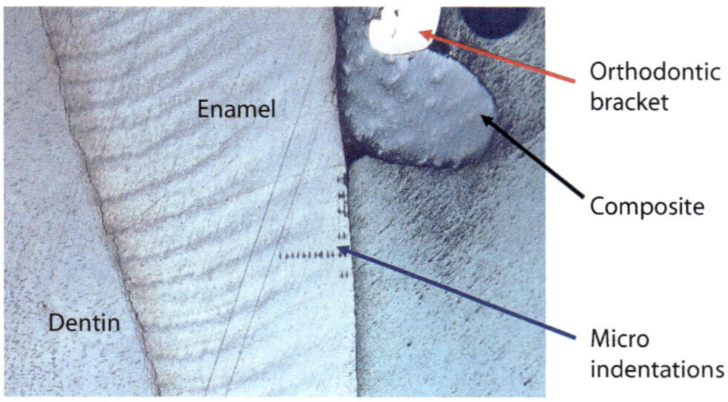

Fig. 17.10 Microscopical picture demonstrating the cross section microhardness testing; the bicuspid histological cross section shows dentin, enamel, and composite, used to bond the orthodontic bracket onto the enamel; the micro-indentations were placed right below the enamel surface following an elaborated distribution pattern; the indents are located right below the area where the orthodontic bracket was bonded to the enamel; this area represents the microbial plaque challenge, and consequently demineralization occurs here. (Adapted from Ref. [33])

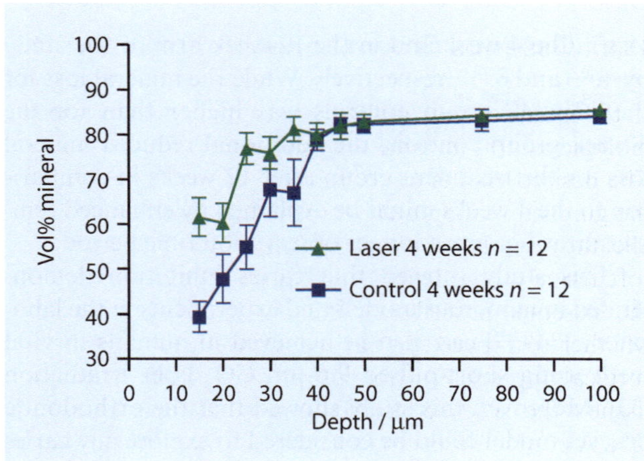

Fig. 17.11 A depth profile of volume % mineral loss for the laser-irradiated areas (green line with triangle symbols) and for the controls (blue line with square symbols) for the bicuspids at the 4-week time point (mean ± standard error). (Adapted from Ref. [33])

odontic brackets in comparison to the controls (Fig. 17.12) (significance level $P = 0.04$).

The reduction in mineral loss was even more impressive after 12 weeks (Fig. 17.12). The control areas showed a fairly high relative mineral loss ΔZ of 1067 ± 254. In contrast, for the CO_2 laser-treated area, the mean relative mineral loss was much lower (ΔZ 135 ± 98) and even lower than for the 4-week mineral loss. The difference in mineral loss between laser and control for the 12-week observation was significant (significance level $P = 0.002$). For the 12-week group, the laser treatment produced a noticeable 87% inhibition of enamel demineralization.

Previous studies had demonstrated that the orthodontic bracket model employed in this study here presents a high caries demineralization challenge to the enamel. In addition, this demineralization challenge cannot simply be overcome by using an OTC 1100 ppm fluoride toothpaste [25]. Gorton et al. in 2003 reported using the orthodontic bracket model that the mean mineral loss value ΔZ in their control group was 805 ± 78 (mean ± SE) vol% × µm. This establishes a considerable, measurable, demineralization in only 1 month even when fluoride toothpaste is used. In this CO_2 laser orthodontic bracket model study, the participants in the control regions around the brackets had a high mineral loss of ΔZ of 737 ± 131 vol% × µm at the 4-week and even higher one of 1067 ± 254 vol% × µm at the 12-week arm, similar to the mineral loss Gorton et al. stated [25].

However, the application of the CO_2 9.6 µm short-pulsed laser irradiation significantly reduced the mineral

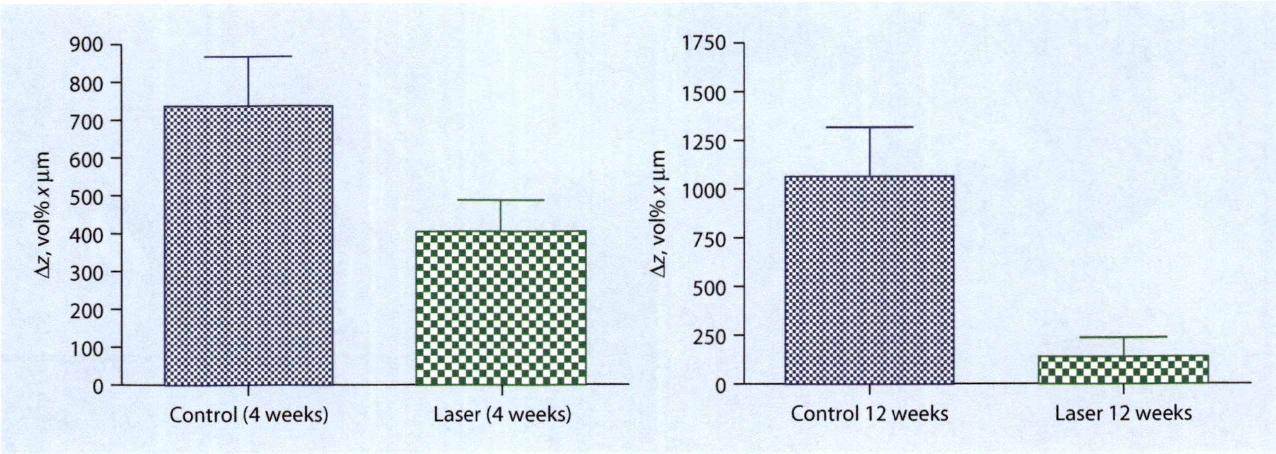

Fig. 17.12 Relative mineral loss ΔZ (vol% × μm) for the laser-treated enamel and for the non-laser-treated controls ($n = 12$ for both groups, mean ± SE) (4 weeks left and 12 weeks right graphs) after treatment. The differences in relative mineral loss between laser-treated and control groups are statistically significant with a significance level of $P = 0.04$ for the 4-week and $P = 0.002$ for the 12-week observations. (Adapted from Ref. [33])

loss in the 4-week and in the 12-week arm of the study by 46% and 87%, respectively. While the mineral loss for the 12-week group controls was higher than for the 4-week group controls, the additional reduced mineral loss for the treatment group after 12 weeks in comparison to the 4 weeks might be explained by enhanced remineralization over a longer observation time period.

This study showed that caries inhibition demonstrated in numerous models and experiments in the laboratory [34–37] can also be achieved in humans in vital teeth using short-pulsed 9.6 μm CO_2 laser irradiation [33]. Moreover, this study showed that the orthodontic bracket model could be considered to explore any caries inhibition agents or tools in living teeth in humans.

17.4 In Vivo Occlusal Caries Prevention by Pulsed CO_2 Laser and Fluoride Varnish Treatment: A Proof-of-Concept Pilot Study

17

In the orthodontic bracket model study, the test teeth were extracted to perform the cross-sectional microhardness testing. A subsequent in vivo pilot study was designed with the goal to determine the caries prevention effects of a short-pulsed CO_2 laser without the need to extract the study teeth for analytical assessments [38].

In this proof-of-concept study, for the first time, in vivo occlusal pits and fissures of second molars in the oral cavity were irradiated. The change in demineralization and remineralization was assessed by three visual methods: (1) ICDAS—the International Caries Detection and Assessment System; (2) SOPROLIFE in daylight and in blue fluorescence mode; and (3) DIAGNOdent fluorescence tool. The study's intention was to show caries inhibition in fissures of molars [38]. The significant challenge was to reach normal configurated as well as deep pits and fissures with the laser irradiation. In order to facilitate adequate laser interaction with the walls of deep fissures, we designed and used a custom contra-angle laser handpiece especially made for this study.

Twenty subjects were recruited and consented for the study. Their age ranged between 10 and 15 years; they were at high caries risk according to CAMBRA rules [39–41] and presented at least two fully erupted second molars in the same arch (contralateral) with untreated, non-carious (non-cavitated) occlusal surfaces. One molar was randomly selected for the laser irradiation, and the tooth on the opposite site in the same jaw functioned as control.

At baseline, right after laser treatment and at the 6- and 12-month recall visit, the occlusal surfaces of the study molars were visually judged for decalcification applying the ICDAS II criteria (International Caries Detection and Assessment System) and the SOPROLIFE Light Induced Fluorescence Evaluator system (SOPRO, ACTEON Group, La Ciotat, France).

17.4.1 Caries Assessment Methods Applied in the In Vivo Occlusal Caries Prevention by Pulsed CO_2 Laser Studies

The International Caries Detection and Assessment System provides a standardized method of lesion detection and assessment, leading to a caries diagnosis [42, 43]. ICDAS scores range from code 0 to 6, with code 0 as no mineral loss, code 1 and 2 as precavitated lesions, and code 3 showing the first physical, localized enamel breakdown (without clinical visual signs of dentinal involvement) while higher codes include enamel breakdown with up to 50% (code 5) and more than 50% of the tooth surface cavitated (code 6). ICDAS criteria are built on the visual enamel properties of translucency and microporosity. After increasing the number of demineralization attacks, the microporosity of the subsurface of enamel intensifies. Consequently, a change in translucency and light refraction of the enamel surface represents the first sign of a carious alteration. If the demineralization process continues, the enamel microporosity increases, which results in an even further decrease in the refractive index of enamel [44].

Ekstrand et al. [45–47] validated ICDAS by demonstrating an association between the lesions' histological depth and the severity of caries lesions (as described by ICDAS codes). Other authors confirmed this close relationship between ICDAS scoring and the histological depth of the caries lesion, especially in precavitated (ICDAS codes 1 and 2) but also in slightly cavitated stages (ICDAS code 3 and above) [48, 49].

Fluorescence is a property of specific materials to absorb energy at shorter wavelengths and emit light at longer wavelengths. Several caries detection methods use fluorescence. The SOPROLIFE system basically combines the benefits of a visual inspection method (high specificity) by employing a high-magnification oral camera with a laser fluorescence instrument (high reproducibility and discrimination) [50]. The system uses four white LEDs in the daylight mode and four blue LEDs in the blue fluorescence mode. Bacteria and their byproducts trigger the fluorescence signal and expression. The blue light transmits through healthy enamel and evokes a green fluorescence of the dentin core. The green fluorescence light coming back from the dentin core then leads to a red fluorescence from bacteria and bacterial byproducts like porphyrins [50–52]. ◘ Figure 17.13 shows a graphic representation of the SOPROLIFE system, and ◘ Fig. 17.14 depicts both the visible light and SOPROLIFE images of an ICDAS code 3 occlusal lesion.

◘ **Fig. 17.13** The SOPROLIFE light-induced fluorescence evaluator system uses in the blue fluorescence mode four blue LEDs emitting light at 450 nm. The light is transmitted through the enamel and then induces green fluorescence from the dentin body. If green light on the way back hits porphyrins in caries, the emitted fluorescence is red

17.4.2 Laser Application and Results of In Vivo Occlusal Caries Prevention by Pulsed CO_2 Laser and Fluoride Varnish Treatment: A Proof-of-Concept Pilot Study

The laser was the same instrument that was used in the inhibition of caries in vital teeth by CO_2 laser treatment orthodontic bracket model study (CO_2 laser, Pulse System, Inc. (PSI) Los Alamos, NM), with a 9.6 μm wavelength, 20 μs pulse duration, and 20 Hz pulse repetition rate, with an 800 μm beam diameter delivered through a custom-made contra-angle handpiece. Each irradiated spot received 20 laser pulses; the laser fluence per pulse averaged 4.5 ± 0.5 J/cm^2.

All subjects received fluoride varnish applications to all teeth (Omni Vanish fluoride varnish, Omni Preventive Care, West Palm Beach, FL) at baseline and 6-month recall (for more details, see Ref. [38]).

The ICDAS scores ranged at baseline from code 0 to code 2. ◘ Figure 17.15 shows the average change of ICDAS scores between baseline and 3-month, baseline and 6-month, and baseline and 12-month recall (mean ± SE) for laser-treated and control teeth. While for the controls the average ICDAS increased over time, a decrease in ICDAS could be observed for the laser-treated fissures. The differences in average change in ICDAS scores over time were statistically significant between laser and control at 6-month and again between laser and control at the 12-month recalls (significance level $P = 0.001$ and $P < 0.0001$, respectively).

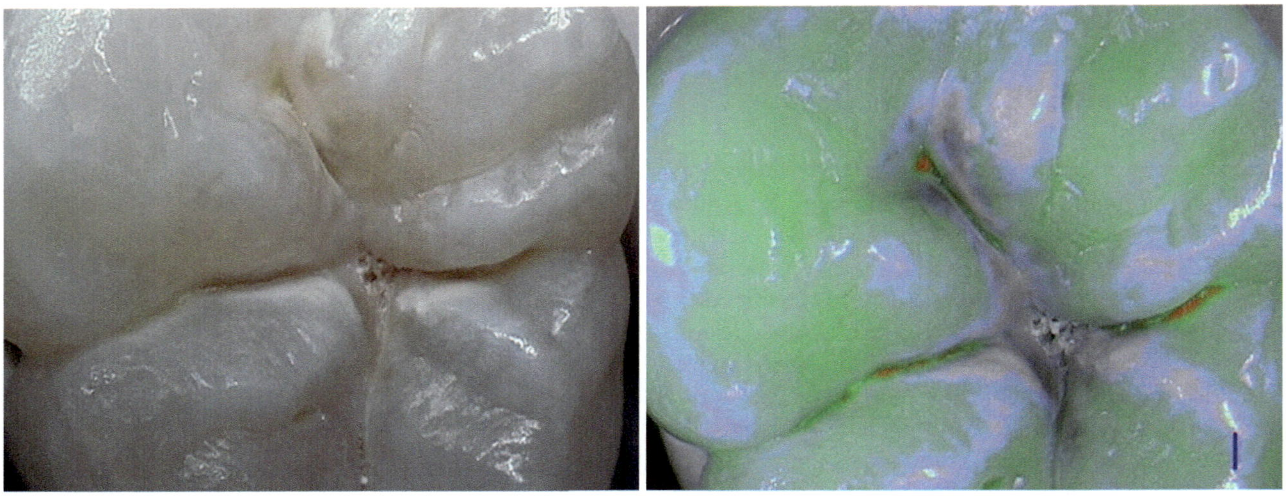

Fig. 17.14 The left picture of the occlusal surface of a molar is taken with SOPROLIFE in daylight mode (white LED illumination) and the right taken in blue fluorescence mode (blue LED illumination). The tooth shows an ICDAS 3 code, already exhibiting physical enamel breakdown in the central grove. The enamel breakdown became clearly visible

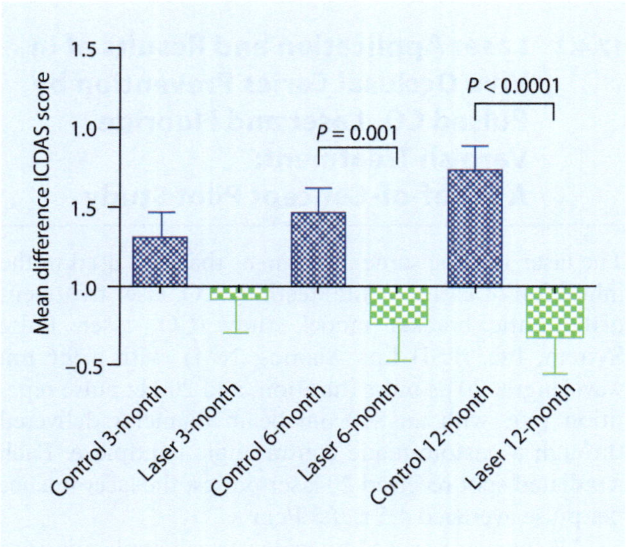

Fig. 17.15 Average change of ICDAS scores between baseline and 3-month, baseline and 6-month, and baseline and 12-month recall (mean ± SE) for laser-treated and control teeth; statistically significant differences between laser and control at 6- and 12-month recalls ($P = 0.001$ and $P < 0.0001$, respectively)

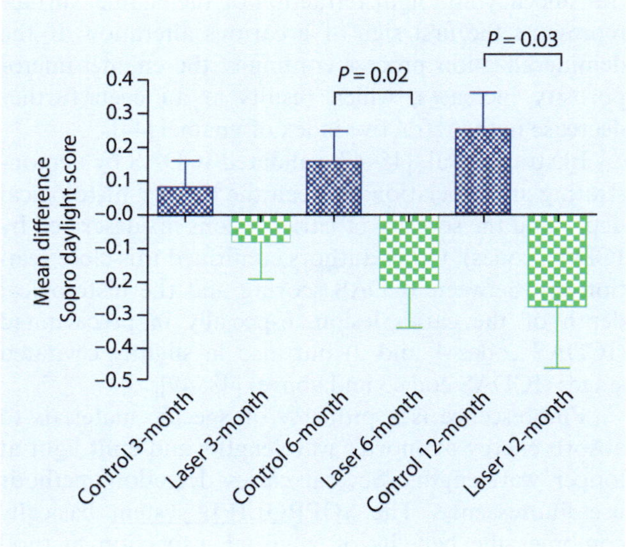

Fig. 17.16 Average changes of SOPROLIFE daylight scores for laser-treated and control teeth between baseline and the 3-, 6-, and 12-month recall (mean ± SE) showing statistically significant differences at the 6- and 12-month interval (significance level $P = 0.02$ and $P = 0.03$, respectively). (Graph adapted from Ref. [38])

In addition to the ICDAS scoring, the study teeth were also evaluated with the SOPROLIFE system. They were scored with a recently presented scoring system, which was developed for the SOPROLIFE Light Induced Fluorescence Evaluator for daylight and for the blue fluorescence mode [51, 52]. For the control as well as the laser-treated teeth, the SOPROLIFE scores ranged between 0 and 3 at baseline with no statistically significant differences between study and control group.

Calculating the changes in SOPROLIFE daylight scores between baseline and each recall time point (Fig. 17.16) revealed an increasing daylight score for the controls and a decreasing score for the laser-treated fissures similar to the ICDAS scoring. The differences between the average changes were again like the ICDAS scores statistically significant between baseline and 6-month and baseline and 12-month recall.

Figures 17.17 and 17.18 show the occlusal surface of a control and a laser-treated tooth, respectively, with the pictures taken with the SOPROLIFE camera in daylight mode and in the blue fluorescence mode. The pictures demonstrate the obvious differences over time

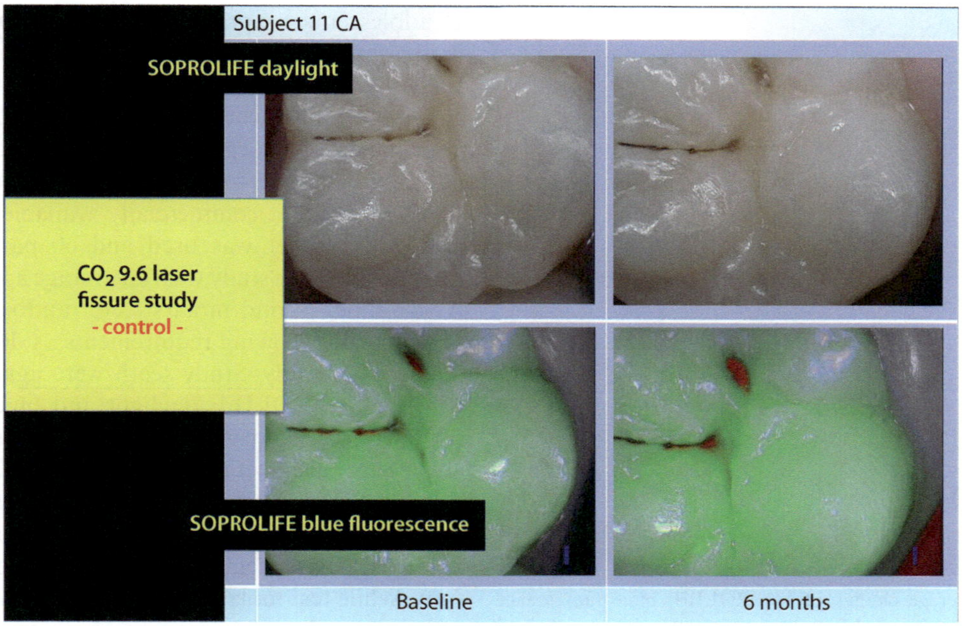

Fig. 17.17 Control tooth of a subject at baseline and at the 6-month recall, pictures taken in daylight and fluorescence mode, respectively. The area of demineralization appears wider in daylight mode and in fluorescence mode after 6 months (upper row daylight, lower row blue fluorescence mode; baseline left 6-month recall right). (Photos from Ref. [38]. Used with permission: John Wiley & Sons)

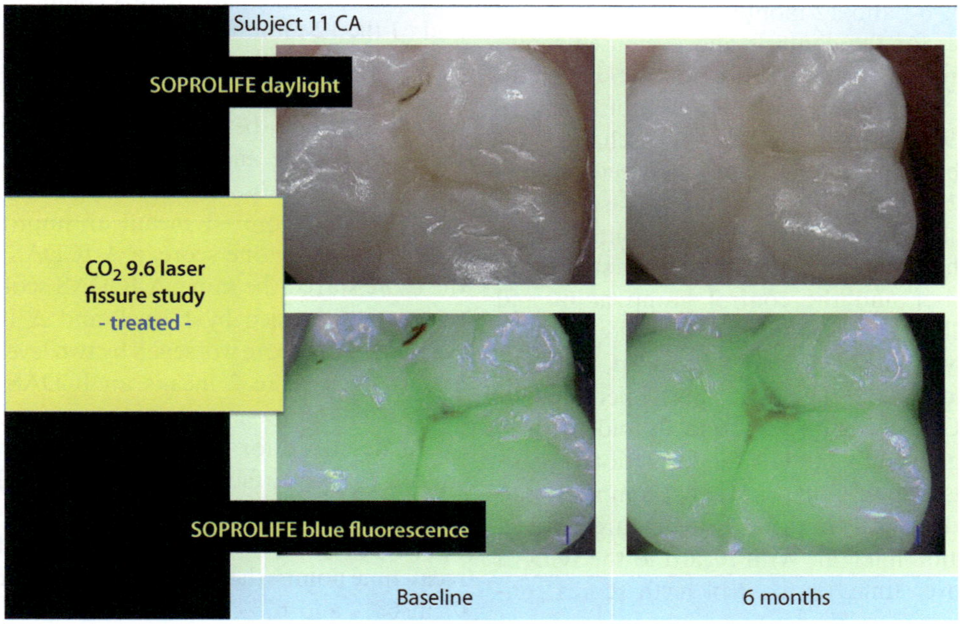

Fig. 17.18 Laser-irradiated tooth of a subject at baseline and at the 6-month recall, pictures taken in daylight and fluorescence mode, respectively. The area of demineralization in the distal fossa of the laser-treated tooth in daylight mode and fluorescence mode, respectively, is not visible anymore (upper row daylight, lower row blue fluorescence mode; baseline left 6-month recall right). (Photos from Ref. [38]. Used with permission: John Wiley & Sons)

between the baseline and the 6-month recall. Both molars show noticeable changes in the fissure system, being very distinct in the distal groove. While the area of demineralization in the control tooth appears to become more extended in both daylight mode and fluorescence mode, after 6 months, the demineralization zone and the red fluorescence width and intensity, respectively, disappeared on the laser-irradiated tooth.

Similar to the SOPROLIFE daylight scores, the SOPROLIFE blue fluorescence average scores for base-

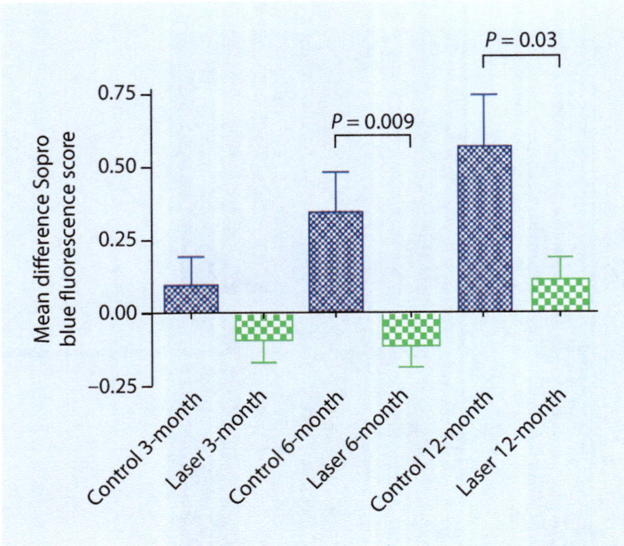

Fig. 17.19 Average change of SOPROLIFE blue fluorescence scores between baseline and 3-month, baseline and 6-month, and baseline and 12-month recall (mean ± SE) for laser-treated and control teeth; statistically significant differences between laser and control at 6- and 12-month recalls ($P = 0.009$ and $P = 0.03$, respectively). (Adapted from Ref. [38])

line and the 3-, 6-, and 12-month recall demonstrated increased scores for the controls and decreased scores for the laser-treated fissure (except for the 12-month recall for the laser group, which did not change). Figure 17.19 demonstrates the average changes for SOPROLIFE blue fluorescence scores between baseline and 3-month, baseline and 6-month, and baseline and 12-month recall for the laser-treated and control teeth, respectively. The differences are statistically significant between the laser and the control group at 6- and 12-month recalls.

In summary, this single-blind, controlled, randomized clinical pilot trial showed that using a microsecond-pulsed 9.6 μm CO_2 laser with additional fluoride varnish applications significantly inhibits the formation of carious lesions in fissures of molars in vivo in comparison to a nonirradiated control tooth in the same arch over a 1-year observation interval. With regard to the ICDAS score change over time, the control teeth scores constantly increased describing more severe mineral loss while the control teeth showed constantly decreasing ICDAS scores, demonstrating a certain mineral gain.

17

17.4.3 In Vivo Occlusal Caries Prevention by Pulsed CO_2 Laser: Main Study

Encouraged by the aforementioned positive results of the in vivo occlusal caries prevention by pulsed 9.6 μm wavelength CO_2 laser proof-of-principle pilot study in 20 adolescents, we performed the additional main study with a similar design. The objective of this randomized, single-blind, split-mouth-controlled, clinical trial was again to evaluate whether the use of a short-pulsed CO_2 laser increases the caries resistance of occlusal pit and fissures in addition to fluoride therapy over 12 months.

In this trial, a commercially available 9.3 μm short-pulsed CO_2 laser was used and 60 participants were recruited into the study with an average age of 13.1 years. At baseline, second molars were randomized into test and control following requirements as described before in the pilot study. Study teeth were again assessed by ICDAS, SOPROLIFE daylight and blue fluorescence, and DIAGNOdent. An independent investigator irradiated test molars with a CO_2 laser (Convergent Dental, Needham, MA), wavelength 9.3 μm, pulse duration 4 μs, pulse repetition rate 43 Hz, beam diameter 250 μm, average fluence 3.9 J/cm^2, 20 laser pulses per irradiation spot. While test molars received laser and fluoride treatment, control teeth received only fluoride application. Fluoride varnish (Vanish, 3M Oral Care, Saint Paul, MN) was applied at baseline and at 6 months. After 6 and 12 months, all study teeth were again assessed.

At the start of the study, 60 patients were recruited into this study; 57 participants completed the 6-month and 51 the 12-month recall.

In this main study, we looked into changes of the ICDAS score in a slightly different way than in the pilot study described before. Changes in ICDAS scores between baseline and 6- and 12-month recall, respectively, were expressed in a delta (Δ) ICDAS score. A delta ICDAS score −1 meant an improvement of the ICDAS score by one score, a Δ ICDAS score 0 meant the score stayed the same, Δ ICDAS score 1 meant that the score worsened by 1 level, and Δ ICDAS score 2 meant that the score worsened by two levels (for instance, a Δ ICDAS score 2 means an ICDAS 0 changed to ICDAS 2 at follow-up).

As described for laser-irradiated teeth before, laser-treated fissure surfaces showed slight ICDAS improvements over time with an ICDAS change −1 in 11% and 8% at the 6- and 12-month recalls, respectively. At these recall time points, there were no changes (ICDAS change 0) in 68% and 67%; slightly worsened scores (ICDAS change 1) in 19% and 24%, respectively; and the scores worsened by two scores in 2% at both recall time points.

In contrast, control teeth showed significantly higher ICDAS increases over time. At the 6- and 12-month recall, 47% and 25% showed no ICDAS changes, but ICDAS change 1 occurred in 49% and 55%, and ICDAS change 2 in 4% and 20% at 6- and 12-month recall, respectively. The differences in ICDAS changes between the groups were statistically significant ($P = 0.0002$ and $P < 0.0001$; Wilcoxon signed-rank test).

Table 17.2 2 × 2 contingency table, laser-treated versus control and ICDAS 3 score Yes or No

	ICDAS score 3 Yes	ICDAS score 3 No	Total
Control	13	44	44
Laser treated	0	57	57
Total	13	101	114

As reported from the previous proof-of-principle pilot study, 25% of the participants had developed an occlusal ICDAS 3 lesion, a first small cavity in enamel. All of them occurred in the control teeth [38]. In the pilot study, ICDAS 3 lesions occurred at 6 months in 7% and at 12 months in additional 18% of the control teeth ($P < 0.0001$). In this main study, almost the same percentage of participants (22%) developed an ICDAS 3 lesion, and again all ICDAS 3 lesions occurred in the control teeth (Table 17.2).

Comparable to the ICDAS scores, SOPROLIFE daylight scoring showed increasing scores over time for the control teeth and relatively unchanged scores for the laser-treated fissures [53].

Given that ICDAS score 3 refers to a physical loss of enamel and not just a reversible mineral loss as in demineralization, it describes the first visible cavity. Since none of the laser-treated molar fissures showed an ICDAS 3 score, obviously, application of CO_2 9.3 μm short-pulsed laser irradiation with additional fluoride use prevented development of first visible cavities over a 1-year period compared to non-laser-irradiated control teeth in the same mouth. By preventing a first cavity, a first "drilling action" is avoided. In fact, "drilling and filling" start the "repeat restoration cycle" process that ends with each restoration being less prophylactic and more iatrogenic than the previous one [54].

Fisher's exact test reveals that differences in ICDAS score 3 occurrence rate are highly significant, and the two-tailed P value equals 0.0001.

From both the orthodontic bracket study and the occlusal caries prevention studies, it can be reasonably concluded that the CO_2 short-pulsed laser irradiation drives out the carbonated phase from the enamel crystal and decreases the demineralization of the modified hydroxyapatite in an acid environment. The specific laser irradiation prevents over at least 1 year the occurrence of ICDAS 3 score lesions. Specifically, when fluoride is present, the transformed hydroxyapatite appears to be prone to higher remineralization. The phenomenon of increased remineralization proven by ICDAS and SOPROLIFE daylight and fluorescence assessments was observed in these studies over a period of 12 months.

17.5 Cavity Preparation and Soft Tissue Cutting with the CO_2 9.3 μm Short-Pulsed Laser

The previously mentioned 9.3 μm wavelength short-pulsed carbon dioxide laser (Solea, Convergent Dental, Inc., Needham, MA) has been available on the US market for soft and hard tissue procedures. The instrument offers a wide range of beam diameters; the basic beam diameter is actually 0.25 mm but by using computer-controlled galvo mirrors, the beam can cover up to a diameter of 1.25 mm by using a spiral movement of the focus point. The focus distance is relatively long and is set between 10 and 19 mm. The emitted laser energy is controlled by the pulse width, varying between 10 and 130 μs, controlled by the foot pedal.

In addition to preventive procedures, this laser can perform carious lesion removal, tooth preparation, and osseous surgery. Figure 17.20 shows the first patient receiving a Solea laser cavity preparation in 2013. The green pilot laser helps guiding the laser beam; the computerized laser control facilitates easy cavity preparation. Clinical case examples will be found in ► Chap. 8.

Figure 17.21 shows that, for the first time, true minimal invasive treatment becomes a reality. This fissure on a molar is prepared with a 0.25 mm laser beam. The preparation borders are completely smooth. The scanning electron microscopic picture (Fig. 17.22) shows a similar smooth cut with a diameter of roughly 0.75 mm, done by the author, in 1997 with an experimental carbon dioxide laser used to produce a carbon-13 (^{13}C) isotope. Thus, 20 years ago, the search for a second-generation hard tissue cutting laser had begun.

Figure 17.23 depicts a distal occlusal cavity preparation with the Solea laser. In addition, the overgrown gingival tissue was quickly removed with the laser without any bleeding. (Clinical case courtesy of Dr. Joshua Weintraub)

The CO_2 9.3 μm short-pulsed laser offers, when it is already too late for caries prevention, a wide potential for cutting hard tissues as well as soft tissue.

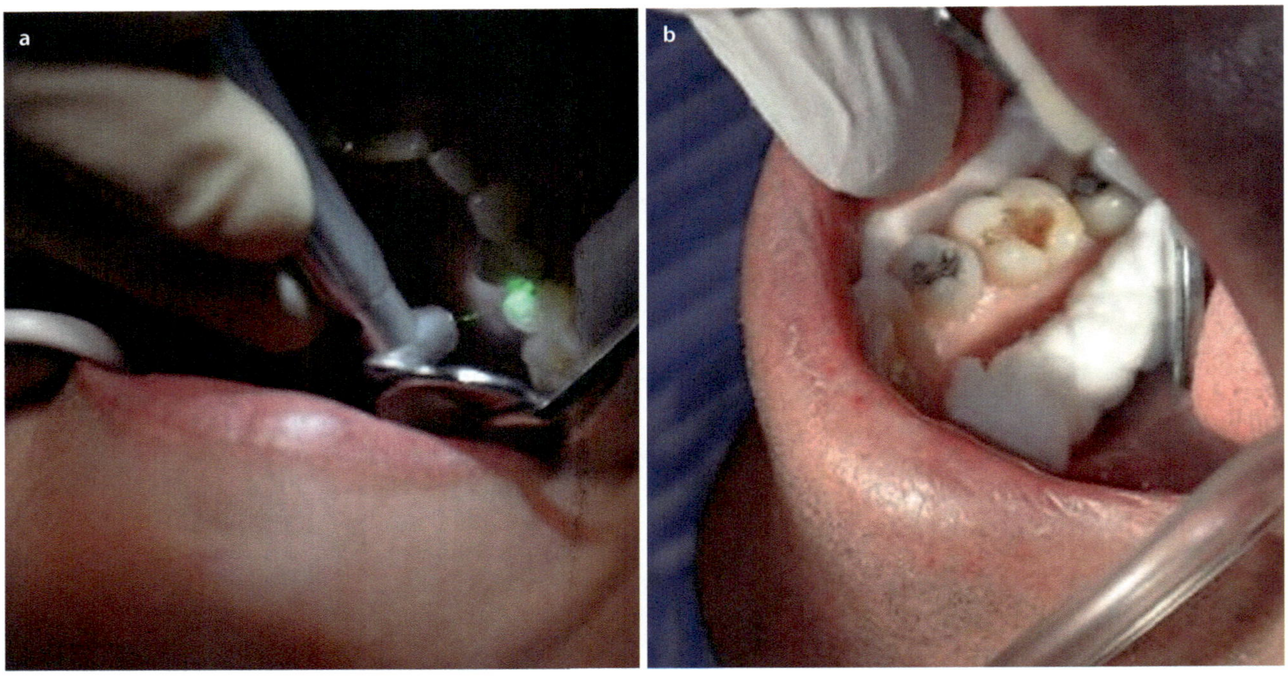

Fig. 17.20 First Solea cavity preparations: **a** Upper left jaw with the green pilot laser visible; **b** lower right jaw after the clean laser preparation was finished. (Clinical photos courtesy of Dr. Mark Mizner)

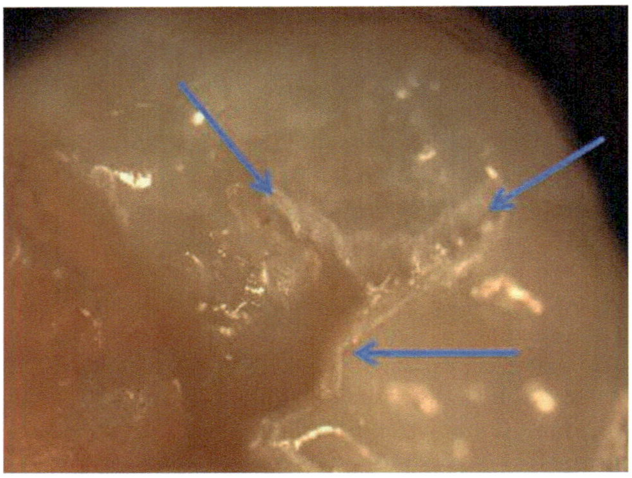

Fig. 17.21 The Solea was used for a minimal invasive preparation of an occlusal fissure. Note the very smooth preparation in enamel

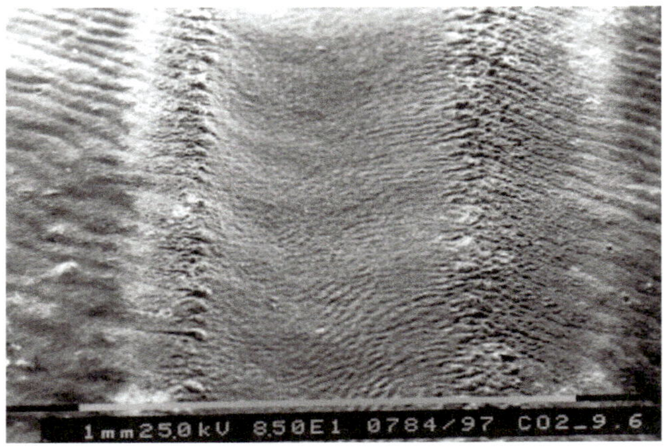

Fig. 17.22 Scanning electron microscope picture showing a similar smooth surface cut in enamel as in Fig. 17.21. Here, an experimental 9.6 μm short-pulsed CO_2 laser (described above) was used

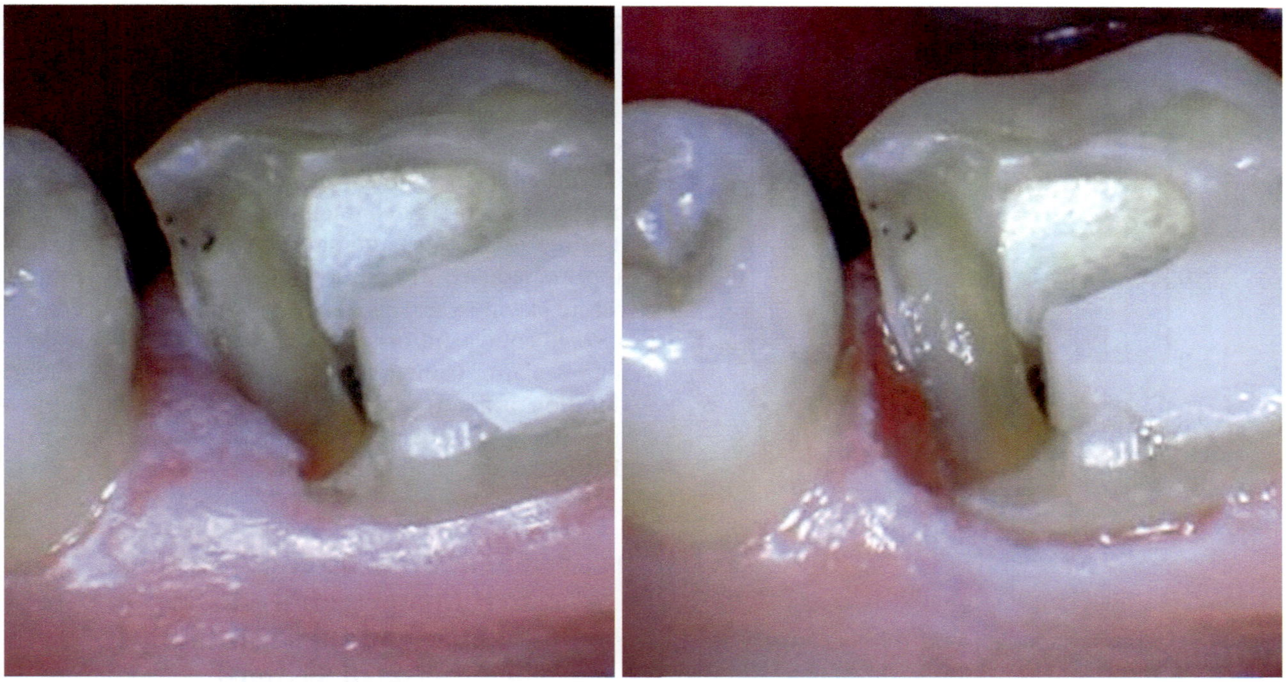

Fig. 17.23 Solea cavity preparation—left picture after laser preparation with overgrown gingiva; right picture after gingivectomy was performed without anesthesia, no bleeding occurs due to the high hemostatic effect when cutting soft tissue with a CO_2 laser. (Clinical photos courtesy of Dr. Joshua Weintraub)

17.6 Shear Bond Strength Testing to Human Enamel and Dentin

A series of bond strength testing was conducted to answer the question whether enamel, already rendered caries resistant with the CO_2 9.3 μm short-pulsed laser, would still allow sufficient bonding of composites. The additional question of how bond strength to dentin might be influenced by the laser irradiation was also of interest. Consequently, enamel and dentin samples were irradiated with a wide range of laser parameters. Different total etch and self-etch bonding systems of the fourth- and fifth-generation bonding materials were tested. The adhesive bonding strength was determined by performing a single-plane shear bond test.

17.6.1 Shear Bond Strength Testing to Human Enamel Using an Etch-and-Rinse (Total-Etch) System

The objective of this in vitro study was to evaluate whether irradiation of enamel with a CO_2 9.3 μm short-pulsed laser using energies that enhance caries resistance influences the shear bond strength of composite resin sealants to enamel.

Seventy bovine and 240 human enamel samples were irradiated with a 9.3 μm carbon dioxide laser (Solea, Convergent Dental, Inc., Needham, MA) with four different laser energies [55]. These energies had been shown to enhance caries resistance or were high enough to ablate enamel. The pulse durations used started as low as 3 μs at 1.6 mJ/pulse and went up to 43 μs at 14.9 mJ/pulse with resulting fluences between 3.3 and 30.4 J/cm^2. The pulse repetition rate varied between 4.1 and 41.3 Hz. The beam diameter was chosen as original beam of 0.25 mm diameter or was moved with the galvo elements in a 1 mm spiral pattern. The laser was sufficiently focused if a focus distance of 4 up to 15 mm was maintained. To achieve "real-world" conditions, the irradiation was performed "freehand." In addition, irradiation occurred while using a computerized, motor-driven stage.

Total etch was performed with 37% phosphoric acid (Scotchbond Universal etchant, 3M ESPE, St. Paul, MN). As a bonding agent, Adper Single Bond Plus was used before Z250 Filtek supreme flowable composite resin (both 3M ESPE) was placed. After 24-h water storage, the single-plane shear bond test was performed with an UltraTester (Ultradent Products, Inc., South Jordan, UT) [55].

All laser-irradiated samples showed equal or higher bond strength than non-laser-treated controls (Fig. 17.24). The highest shear bond strength values

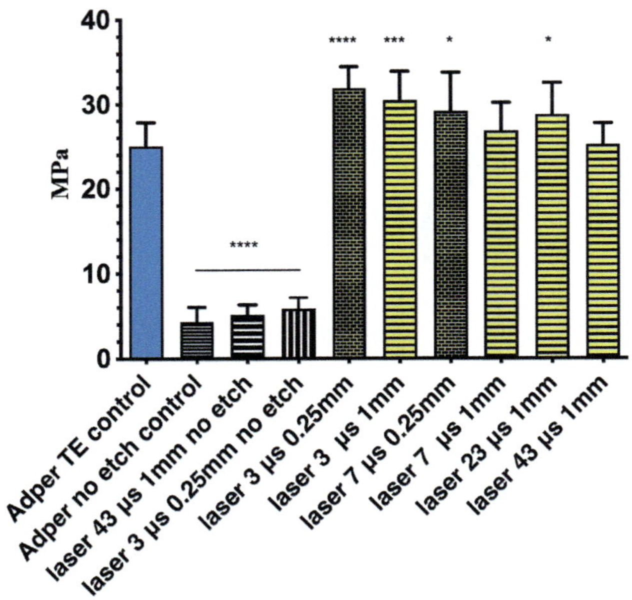

Fig. 17.24 Human enamel—hand irradiation no acid etch and Adper Single Bond Plus TE (with acid etching) (mean ± SD); statistically significant differences to the control are marked ($*P \leq 0.05$; $***P \leq 0.001$; $****P \leq 0.0001$)

of 31.90 ± 2.50 MPa (mean ± standard deviation [SD]) were observed with the 3 µs pulse duration/0.25 mm laser pattern. This represents a 27.4% bond strength increase over the controls, which showed a typical bond strength value of 25.04 ± 2.80 MPa. The difference in bond strength between the control samples and the laser-irradiated enamel samples was highly significant ($P \leq 0.0001$). Two other caries preventive irradiation settings (3 µs/1 mm and 7 µs/0.25 mm) and one ablative pattern (7 µs/0.25 mm) also achieved significantly increased bond strength compared to the controls (Fig. 17.24).

Figure 17.24 also shows the situation if acid etching of the enamel is not performed before composite bonding. When it is erroneously assumed that laser treatment on its own would increase enamel bond strength, shear bond tests showed that the bond strength to non-acid-etched but laser-treated enamel ranged only between 5.1 and 5.9 MPa (control without etching 3.8 MPa). These bond strength values are far below acceptable bond strength values for bonding composite to enamel. Acceptable bonding can only be achieved (with and without laser application) if acid etching of the laser-treated enamel is performed before bonding of the composite takes place.

The bond strength tests also revealed that when bovine enamel was used as substrate, this type of enamel also showed in all test groups increased shear bond strength over the controls. Furthermore, computerized motor-driven stage irradiation did not show superior bond strength values over the clinically more relevant freehand irradiation [55].

In summary, enamel that is rendered caries resistant by CO_2 9.3 µm short-pulsed laser irradiation showed at least equal but more important even significantly higher shear bond strength to pits-and-fissure sealants than non-laser-irradiated enamel. The risk of a sealant failure due to CO_2 9.3 µm short-pulsed laser irradiation appears reduced. If additional laser ablation is required before placing a sealant, the CO_2 9.3 µm laser-cut enamel showed equivalent or superior bond strength to a flowable sealant. Consequently, the success and lifetime duration of fissure sealants might be improved by applying caries-preventive CO_2 9.3 µm short-pulsed laser irradiation before sealant placement.

17.6.2 Shear Bond Strength Testing to Human Enamel and Dentin Using an Etch-and-Rinse (Total-Etch) and Self-Etch Systems

In this second bond strength study, the goal was to look again at enamel bonding as well as dentin bonding using not only etch-and-rinse systems but also popular self-etch systems [56]. Consequently, the influence of CO_2 9.3 µm short-pulsed laser irradiation on the shear bond strength of composite resin to enamel and dentin was to be tested combined with the different composite systems.

Two hundred enamel and 210 dentin samples were irradiated with the abovementioned 9.3 µm carbon dioxide laser with energies which again either enhanced caries resistance or were known to be effective for ablation. All samples were "freehand" irradiated.

Two fifth-generation etch-and-rinse systems were tested, the OptiBond Solo Plus (Kerr Corporation, Orange, CA) and the Peak Universal Bond light-cured adhesive (Ultradent Products, South Jordan, UT). As self-etching systems, Scotchbond Universal (3M ESPE, St. Paul, MN) and Peak SE self-etching primer with Peak Universal Bond light-cured adhesive (Ultradent Products) were investigated. Clearfil APX (Kuraray, New York, NY) was bonded to the samples. After 24 h, a single-plane shear bond test was performed with an UltraTester (Ultradent Products, Inc., South Jordan, UT).

Enamel Bonding

Using the caries-preventive setting on enamel resulted in increased shear bond strength values for all bonding agents except for the self-etch Peak SE with only 26.34 ± 5.95 MPa versus the control showing 31.46 ± 6.12 MPa (mean ± standard deviation [SD]).

The highest overall bond strength to enamel rendered caries resistant with the CO_2 9.3 µm laser was seen with the etch-and-rinse system Peak TE (41.29 ± 6.04 MPa, control 38.74 ± 6.44 MPa). This confirms the finding from the first bond strength study reported above.

To ablated enamel, all etch-and-rinse systems achieved higher bond strength values than the self-etch systems did. Peak TE showed the highest shear bond strength with 35.22 ± 4.40 MPa, achieving 90.9% of the bond strength control value.

OptiBondTE reached 93.8% of its control value (25.72 ± 3.08 versus control 27.43 ± 4.45 MPa). The self-etch system Peak SE presented significantly lower bond strength (16.95 ± 2.93 MPa versus control 31.46 ± 6.12 MPa; 53.9% of the control).

Dentin Bonding

Overall, the shear bond strength to ablated dentin ranged material depended widely from 19.15 ± 3.49 MPa for OptiBondTE as lowest bond strength value and up to 43.94 ± 6.47 MPa for Peak SE as the highest observed bond strength value.

Figures 17.25 and 17.26 represent the shear bond strength test results for laser-ablated dentin after applying total etch bonding systems. The Peak TE control sample achieved the highest bond strength value for dentin bonding (63.02 ± 7.56 MPa (mean ± SD)) of all in this study tested systems. OptiBond Solo Plus showed roughly half the bond strength (31.58 ± 6.02 MPa). For both bonding systems, Peak TE and OptiBond Solo Plus TE, the bond strength values of the irradiated dentin significantly decreased compared to the controls ($P \leq 0.0001$). Still, the bond strength values achieved between 57.2% and 65% of the control bond strength values. For Peak TE, they were always greater than 36 MPa, and for OptiBond Solo Plus, they were still around 19 MPa.

Figure 17.27 shows that the dentin bonding self-etch system Peak SE achieved the second highest bond strength value to regular untreated dentin (control 58.79 ± 10.94 MPa) for all dentin samples. Figure 17.28 shows the situation for Scotchbond Universal. The

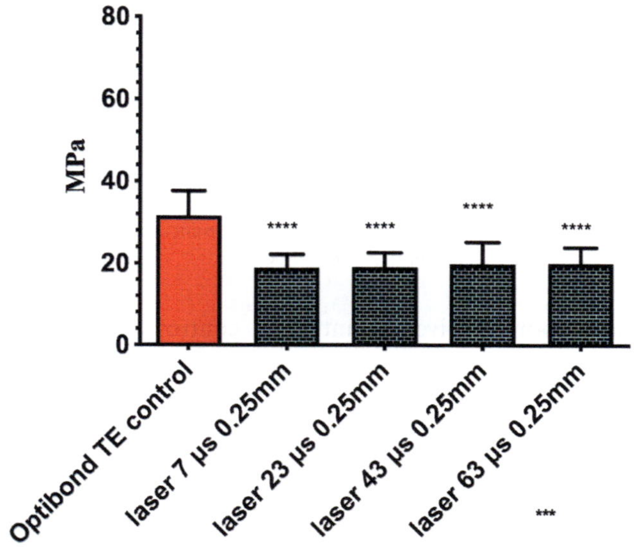

Fig. 17.26 Human dentin—hand irradiation shear bond strength OptiBond Solo Plus TE, (mean ± SD); statistically significant differences to the controls are marked (****$P \leq 0.0001$)

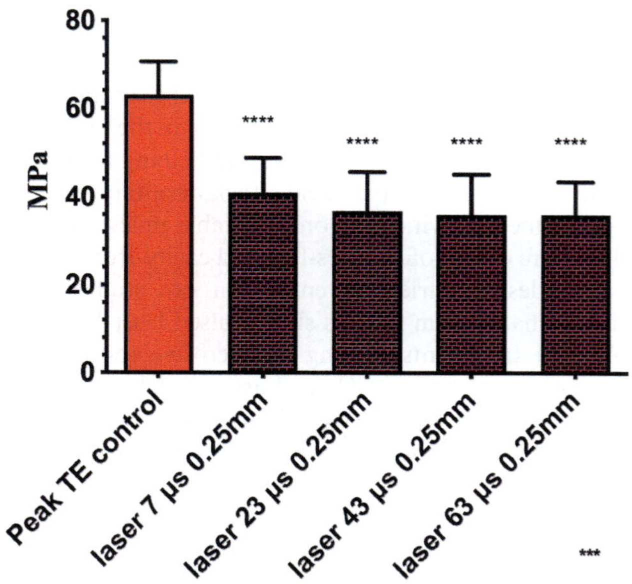

Fig. 17.25 Human dentin—hand irradiation shear bond strength Peak TE (mean ± SD); statistically significant differences to the controls are marked (****$P \leq 0.0001$)

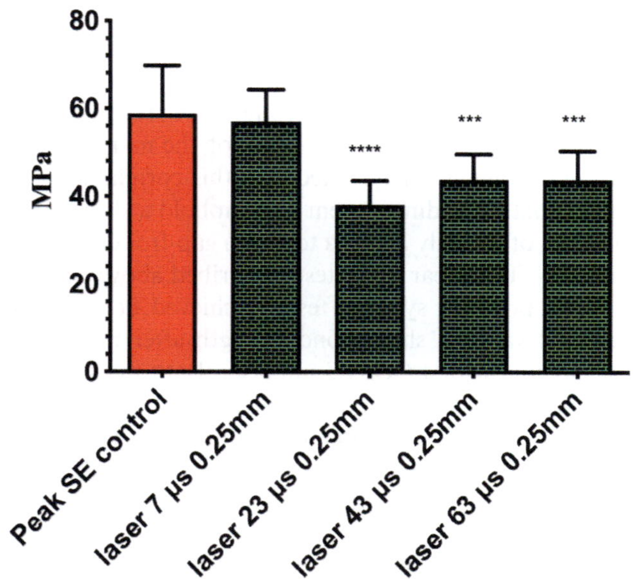

Fig. 17.27 Human dentin—hand irradiation shear bond strength Peak SE (mean ± SD); statistically significant differences to the controls are marked (***$P \leq 0.001$; ****$P \leq 0.0001$)

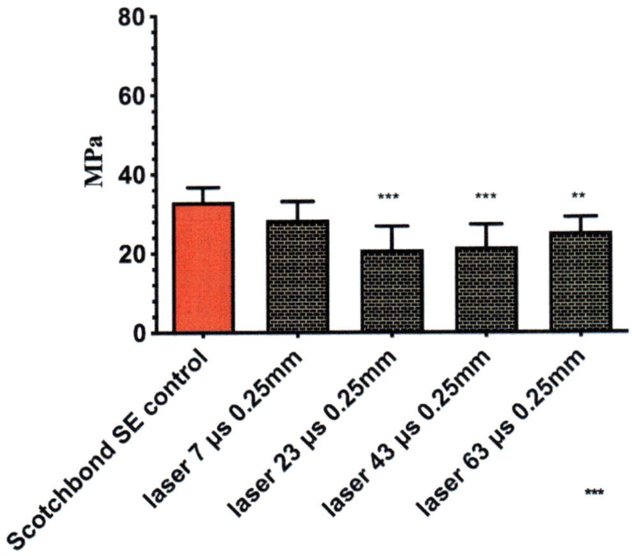

Fig. 17.28 Human dentin—hand irradiation shear bond strength Scotchbond Universal SE (mean ± SD); differences are statistically not significant (**$P \leq 0.01$; ***$P \leq 0.001$)

Scotchbond Universal untreated control achieved an average shear bond strength of 33.15 ± 3.55 MPa. Nevertheless, samples irradiated with ablative energy (23, 43, and 63 µs pulse duration) showed significantly lower bond strength values compared to the controls, but the bond strength values still reached between 62.8% and 75.4% of the control values. For Peak SE, even under these ablation conditions, the bond strength values were still around 40 MPa, while Scotchbond Universal still demonstrated dentin shear bond strength values of around 20 MPa and higher.

When composite resin polymerizes and shrinks [57], the bond between the dentin bonding system and the tooth should be strong enough to prevent the resin from withdrawing from the cavity surface [57]. Munksgaard and Asmussen described a strong correlation between shear bond strength and the width of the marginal gap of dentin fillings. They stated that this correlation predicted that a bonding system must uphold a shear bond strength of roughly 17 MPa to attain gap-free dentin fillings [58]. The shear bond tests described above showed that all bonding systems tested achieved at least the required value in shear bond strength after the dentin was ablated with a CO_2 9.3 µm laser. Both the self-etch system Peak SE and the etch-and-rinse Peak TE achieved more than double the minimum required bond strength value (36.0 ± 7.4 and 43.9 ± 6.5 MPa, respectively).

In summary, etch-and-rinse systems resulted consistently in higher bond strength to CO_2 9.3 µm laser-ablated enamel than self-etch systems. Using the maximum recommended energy for dentin ablation, the self-etch system Peak SE reached the highest bond strength (43.9 ± 6.5 MPa) to laser-ablated dentin.

17

17.7 Conclusion

9.3 and 9.6 µm CO_2 µs short-pulsed lasers are the most highly absorbed laser wavelengths in dental hard tissues. They can safely and efficiently be used to render enamel caries resistant by transforming carbonated apatite into the much less soluble hydroxyapatite. Adding fluoride after the laser treatment additionally reduces the acid solubility of enamel and creates the desired least acid-soluble fluorapatite. These 9.3 and 9.6 µm CO_2 µs short-pulsed lasers are very efficient in cutting dental hard and soft tissue. Results of shear bond strength testing with multiple bonding agents to such laser cuts are promising. Acid etching of enamel can still be performed after the enamel was rendered more caries resistant with the 9.3 µm CO_2 µs short-pulsed laser. The resulting shear bond strength to such treated enamel is even significantly enhanced and thus might result clinically in superior conditions and longevity for composite fillings and even fissure sealants.

17.8 Future Dreams

Erbium and µs short-pulsed carbon dioxide lasers have been shown to efficiently ablate dental hard tissue and while doing that being safe for the pulpal tissue. CO_2 µs short-pulsed lasers are most advantageous for this task since the 9.3 and 9.6 µm wavelengths are the highest absorbed wavelength in enamel as well as dentin. Moreover, these two lasers offer enormous caries preventive advantages as side effects while cutting enamel for a cavity preparation. In addition, caries prevention can be achieved before even one cavity occurs. In a dream for the future, enamel surfaces of freshly erupted teeth may be irradiated very early to prevent any loss of mineral. The lately achieved FDA 510(k) clearance for a 9.3 µm CO_2 µs short-pulsed laser aiding in the reduction of mineral loss in dental enamel and helping to protect teeth against the effects of acid attack, resulting in caries development (cavities), is one desirable milestone into the future dream of a caries-free and cavity-free world.

Besides, in caries prevention, in situ studies have shown that 9.3 µm CO_2 µs short-pulsed laser has their place in treatments dealing with erosive tooth wear. Application of 9.3 µm CO_2 µs short-pulsed laser irradiation combined with specific fluoride products can be extremely efficient in preventing as well as in controlling erosive tooth wear progression in human enamel. These lasers showed to act even with greater efficiency when the irradiation was combined with AmF/NaF/$SnCl_2$ solution application [59, 60]. It may be predicted that in the future, especially when prices for these lasers might have dropped significantly and even tabletop 9.3 µm wavelength laser models might be available, each child and young adult entering a dental office might be treated for caries prevention as well as erosive tooth wear prevention.

Finally, even for adults, caries-preventive irradiation with CO_2 µs short-pulsed lasers might be added in the future as a routine armamentarium to CAMBRA—Caries Management By Risk Assessment. CAMBRA represents an evidence-based approach to preventing, reversing, and treating dental caries. The CAMBRA protocol consists of analyzing the patient's profile by assessing the risk and protective factors and assigning a risk level to the patient to allow an individualized treatment plan, which combines restorative treatments with a preventive chemical therapy [61–63]. As protective factors, sufficient saliva flow and its components, fluoride calcium and phosphate for remineralization, and antibacterials like chlorhexidine and xylitol are listed, and protection against demineralization by CO_2 µs short-pulsed laser may be routinely added in the future.

Furthermore, to enhance the dentists' armamentarium against recurrent caries, new bactericidal composites should be the routine clinical future for filling cavities. Dental composite with quaternary ammonium silica dioxide (QASi) nanoparticles incorporated with other fillers into the restorative material, which by now are FDA cleared, demonstrated antibacterial activity by reducing enamel demineralization in an in situ gap model [64].

Combining big technology like lasers with small technology like new bactericidal composites and adding CAMBRA as prevention philosophy will be the dream for a successful future resulting in healthy, caries-free patients.

References

1. LeGeros RZ. Properties of osteoconductive biomaterials: calcium phosphates. Clin Orthop Relat Res. 2002;395:81–98.
2. Featherstone JD, Mayer I, Driessens FC, Verbeeck RM, Heijligers HJ. Synthetic apatites containing Na, Mg, and CO_3 and their comparison with tooth enamel mineral. Calcif Tissue Int. 1983;35(2):169–71.
3. Featherstone JD, Pearson S, LeGeros RZ. An infrared method for quantification of carbonate in carbonated apatites. Caries Res. 1984;18(1):63–6.
4. Budz JA, Lore M, Nancollas GH. Hydroxyapatite and carbonated apatite as models for the dissolution behavior of human dental enamel. Adv Dent Res. 1987;1:314–21.
5. Featherstone JD, Nelson DG. Laser effects on dental hard tissues. Adv Dent Res. 1987;1(1):21–6.
6. Featherstone JD, Barrett-Vespone NA, Fried D, Kantorowitz Z, Seka W. CO_2 laser inhibitor of artificial caries-like lesion progression in dental enamel. J Dent Res. 1998;77(6):1397–403.
7. Fried D, Zuerlein MJ, Le CQ, Featherstone JD. Thermal and chemical modification of dentin by 9–11-microm CO_2 laser pulses of 5–100-micros duration. Lasers Surg Med. 2002;31(4):275–82.
8. Fried D, Glena RE, Featherstone JD, Seka W. Nature of light scattering in dental enamel and dentin at visible and near-infrared wavelengths. Appl Opt. 1995;34(7):1278–85.
9. Featherstone JDB, Fried D. Fundamental interactions of lasers with dental hard tissues. Med Laser Appl. 2001;16(3):181–94.
10. Featherstone JD, Nelson DG. Recent uses of electron microscopy in the study of physico-chemical processes affecting the reactivity of synthetic and biological apatites. Scanning Microsc. 1989;3(3):815–27; discussion 27–8.
11. Takagi S, Liao H, Chow LC. Effect of tooth-bound fluoride on enamel demineralization/remineralization in vitro. Caries Res. 2000;34(4):281–8.
12. Featherstone JD. Prevention and reversal of dental caries: role of low level fluoride. Community Dent Oral Epidemiol. 1999;27(1):31–40.
13. Featherstone JD, Glena R, Shariati M, Shields CP. Dependence of in vitro demineralization of apatite and remineralization of dental enamel on fluoride concentration. J Dent Res. 1990;69 Spec No:620–5; discussion 34–6.
14. ten Cate JM, Featherstone JD. Mechanistic aspects of the interactions between fluoride and dental enamel. Crit Rev Oral Biol Med. 1991;2(3):283–96.
15. Kuroda S, Fowler BO. Compositional, structural, and phase changes in vitro laser-irradiated human tooth enamel. Calcif Tissue Int. 1984;36:361–9.
16. McCormack SM, Fried D, Featherstone JD, Glena RE, Seka W. Scanning electron microscope observations of CO_2 laser effects on dental enamel. J Dent Res. 1995;74(10):1702–8.
17. Fried D, Seka W, Glena RE, Featherstone JDB. Thermal response of hard dental tissues to 9- through 11-µm CO_2-laser irradiation. Opt Eng. 1996;35(7):1976–84.
18. Featherstone JDB, Fried D, Bitten ER. Mechanism of laser induced solubility reduction of dental enamel. Lasers Dent III SPIE Proc. 1997;2973:112–6.
19. Rechmann P, Rechmann BM, Groves WH Jr, Le CQ, Rapozo-Hilo ML, Kinsel R, et al. Caries inhibition with a CO_2 9.3 µm laser: an in vitro study. Lasers Surg Med. 2016;48(5):546–54.
20. Fried D, Glena RE, Featherstone JD, Seka W. Permanent and transient changes in the reflectance of CO_2 laser-irradiated dental hard tissues at lambda = 9.3, 9.6, 10.3, and 10.6 microns and at fluences of 1–20 J/cm_2. Lasers Surg Med. 1997;20(1):22–31.
21. Fowler BO, Kuroda S. Changes in heated and in laser-irradiated human tooth enamel and their probable effects on solubility. Calcif Tissue Int. 1986;38:197–208.
22. Legeros RZ. Calcium phosphates in enamel, dentin and bone. In: Myers HM, editor. Calcium phosphates in oral biology and medicine, vol. 15. Basel: Karger; 1991. p. 108–29.
23. Goodis HE, Fried D, Gansky S, Rechmann P, Featherstone JD. Pulpal safety of 9.6 microm TEA CO_2 laser used for caries prevention. Lasers Surg Med. 2004;35(2):104–10.
24. Staninec M, Darling CL, Goodis HE, Pierre D, Cox DP, Fan K, et al. Pulpal effects of enamel ablation with a microsecond pulsed lambda = 9.3-microm CO_2 laser. Lasers Surg Med. 2009;41(4):256–63.
25. Gorton J, Featherstone JD. In vivo inhibition of demineralization around orthodontic brackets. Am J Orthod Dentofac Orthop. 2003;123(1):10–4.
26. Gwinnet J, Ceen F. Plaque distribution on bonded brackets. Am J Orthod. 1979;75:667–77.
27. Corbett JA, Brown LR, Keene HJ, Horton IM. Comparison of Streptococcus mutans concentrations in non-banded and banded orthodontic patients. J Dent Res. 1981;60(12):1936–42.

28. Mattingly JA, Sauer GJ, Yancey JM, Arnold RR. Enhancement of Streptococcus mutans colonization by direct bonded orthodontic appliances. J Dent Res. 1983;62(12):1209–11.
29. Mizrahi E. Enamel demineralization following orthodontic treatment. Am J Orthod. 1982;82(1):62–7.
30. Zachrisson BU, Zachrisson S. Caries incidence and oral hygiene during orthodontic treatment. Scand J Dent Res. 1971;79(6):394–401.
31. Zachrisson BU. Fluoride application procedures in orthodontic practice, current concepts. Angle Orthod. 1975;45(1):72–81.
32. Ogaard B, Rolla G. The in vivo orthodontic banding model for vital teeth and the in situ orthodontic banding model for hard-tissue slabs. J Dent Res. 1992;71 Spec No:832–5.
33. Rechmann P, Fried D, Le CQ, Nelson G, Rapozo-Hilo M, Rechmann BM, et al. Caries inhibition in vital teeth using 9.6-μm CO_2-laser irradiation. J Biomed Opt. 2011;16(7):071405.
34. Featherstone JDB, Barrett-Vespone NA, Fried D, Kantorowitz Z, Lofthouse J, Seka WD, editors. Rational choice of laser conditions for inhibition of caries progression. Washington: SPIE; 1995.
35. Kantorowitz Z, Featherstone JD, Fried D. Caries prevention by CO_2 laser treatment: dependency on the number of pulses used. J Am Dent Assoc. 1998;129(5):585–91.
36. Featherstone JD. Lasers in dentistry 3. The use of lasers for the prevention of dental caries. Ned Tijdschr Tandheelkd. 2002;109(5):162–7.
37. Fried D, Ragadio J, Akrivou M, Featherstone JD, Murray MW, Dickenson KM. Dental hard tissue modification and removal using sealed transverse excited atmospheric-pressure lasers operating at lambda = 9.6 and 10.6 microm. J Biomed Opt. 2001;6(2):231–8.
38. Rechmann P, Charland DA, Rechmann BM, Le CQ, Featherstone JD. In-vivo occlusal caries prevention by pulsed CO_2-laser and fluoride varnish treatment—a clinical pilot study. Lasers Surg Med. 2013;45(5):302–10.
39. Domejean S, White JM, Featherstone JD. Validation of the CDA CAMBRA caries risk assessment—a six-year retrospective study. J Calif Dent Assoc. 2011;39(10):709–15.
40. Rechmann P, Featherstone JD. Quality assurance study of caries risk assessment performance by clinical faculty members in a school of dentistry. J Dent Educ. 2014;78(9):1331–8.
41. Young DA, Featherstone JD. Caries management by risk assessment. Community Dent Oral Epidemiol. 2013;41(1):e53–63.
42. ICDAS-Foundation. International Caries Detection & Assessment System. www.icdasorg.
43. Pitts N. "ICDAS"—an international system for caries detection and assessment being developed to facilitate caries epidemiology, research and appropriate clinical management. Community Dent Health. 2004;21(3):193.
44. ICDAS. Rationale and evidence for the International Caries Detection and Assessment System (ICDAS II). International Caries Detection and Assessment System (ICDAS) Coordinating Committee; 2005.
45. Ekstrand KR, Kuzmina I, Bjorndal L, Thylstrup A. Relationship between external and histologic features of progressive stages of caries in the occlusal fossa. Caries Res. 1995;29(4):243–50.
46. Ekstrand KR, Ricketts DN, Kidd EA. Reproducibility and accuracy of three methods for assessment of demineralization depth of the occlusal surface: an in vitro examination. Caries Res. 1997;31(3):224–31.
47. Tranaeus S, Lindgren LE, Karlsson L, Angmar-Mansson B. In vivo validity and reliability of IR fluorescence measurements for caries detection and quantification. Swedish Dent J. 2004;28(4):173–82.
48. Mendes FM, Siqueira WL, Mazzitelli JF, Pinheiro SL, Bengtson AL. Performance of DIAGNOdent for detection and quantification of smooth-surface caries in primary teeth. J Dent. 2005;33(1):79–84.
49. Astvaldsdottir A, Holbrook WP, Tranaeus S. Consistency of DIAGNOdent instruments for clinical assessment of fissure caries. Acta Odontol Scand. 2004;62(4):193–8.
50. Tassery H, Levallois B, Terrer E, Manton DJ, Otsuki M, Koubi S, et al. Use of new minimum intervention dentistry technologies in caries management. Aust Dent J. 2013;58(Suppl 1):40–59.
51. Rechmann P, Charland D, Rechmann BM, Featherstone JD. Performance of laser fluorescence devices and visual examination for the detection of occlusal caries in permanent molars. J Biomed Opt. 2012;17(3):036006.
52. Rechmann P, Rechmann BM, Featherstone JD. Caries detection using light-based diagnostic tools. Compend Contin Educ Dent. 2012;33(8):582–4, 6, 8–93; quiz 94, 96.
53. Rechmann P, Kubitz M, Chaffee BW, Rechmann BMT. Fissure caries inhibition with a CO(2) 9.3-μm short-pulsed laser-a randomized, single-blind, split-mouth controlled, 1-year clinical trial. Clin Oral Investig. 2021;25(4):2055–68.
54. Brantley CF, Bader JD, Shugars DA, Nesbit SP. Does the cycle of rerestoration lead to larger restorations? J Am Dent Assoc. 1995;126(10):1407–13.
55. Rechmann P, Sherathiya K, Kinsel R, Vaderhobli R, Rechmann BMT. Influence of irradiation by a novel CO(2) 9.3-μm short-pulsed laser on sealant bond strength. Lasers Med Sci. 2017;32(3):609–20.
56. Rechmann P, Bartolome N, Kinsel R, Vaderhobli R, Rechmann BMT. Bond strength of etch-and-rinse and self-etch adhesive systems to enamel and dentin irradiated with a novel CO_2 9.3 μm short-pulsed laser for dental restorative procedures. Lasers Med Sci. 2017;32(9):1981–93.
57. Ferracane JL, Hilton TJ. Polymerization stress—is it clinically meaningful? Dent Mater. 2016;32(1):1–10.
58. Munksgaard EC, Irie M, Asmussen E. Dentin-polymer bond promoted by Gluma and various resins. J Dent Res. 1985;64(12):1409–11.
59. Tavares JP, da Silva CV, Engel Y, de Freitas PM, Rechmann P. In situ effect of CO_2 laser (9.3 μm) irradiation combined with $AmF/NaF/SnCl_2$ solution in prevention and control of erosive tooth wear in human enamel. Caries Res. 2021;55(6):617–28.
60. Silva CV, Mantilla TF, Engel Y, Tavares JP, Freitas PM, Rechmann P. The effect of CO_2 9.3 μm short-pulsed laser irradiation in enamel erosion reduction with and without fluoride applications—a randomized, controlled in vitro study. Lasers Med Sci. 2020;35(5):1213–22.
61. Rechmann P, Kinsel R, Featherstone JDB. Integrating caries management by risk assessment (CAMBRA) and prevention strategies into the contemporary dental practice. Compend Contin Educ Dent. 2018;39(4):226–33; quiz 34.
62. Featherstone JDB, Alston P, Chaffee BW, Rechmann P. Caries management by risk assessment (CAMBRA): an update for use in clinical practice for patients aged 6 through adult. J Calif Dent Assoc. 2019;47(1):25–36.
63. Featherstone JDB, Crystal YO, Alston P, Chaffee BW, Domejean S, Rechmann P, et al. Evidence-based caries management for all ages-practical guidelines. Front Oral Health. 2021;2:657518.
64. Rechmann P, Le CQ, Chaffee BW, Rechmann BMT. Demineralization prevention with a new antibacterial restorative composite containing QASi nanoparticles: an in situ study. Clin Oral Investig. 2021;25(9):5293–305.

Lasers in General Dental Practice: Is There a Place for Laser Science in Everyday Dental Practice? Evidence-Based Laser Use, Laser Education—Medico-Legal Aspects of Laser Use

Steven P. A. Parker

Contents

D. J. Coluzzi, S. P. A. Parker (eds.), *Lasers in Dentistry—Current Concepts*, Textbooks in Contemporary Dentistry,
https://doi.org/10.1007/978-3-031-43338-2_18

Core Message

"Science is the pursuit and application of knowledge and understanding (*of the natural and social world*) following a systematic methodology based on evidence"—*source:* ▶ www.sciencecouncil.org. Within such a generalized statement, there is much to commend the acquisition of knowledge and understanding when using a dental laser of any description and preferably *before* using a dental laser. The author is acknowledged with the statement "to know is to use; to understand is to empower," in connection with laser use. The key to the understanding and correct application of laser photonic energy must encompass awareness of the basics of light physics, laser–tissue interaction, and dosimetry of the laser being chosen for a given procedure. Evidence-based research and peer-reviewed clinical investigations provide the bedrock of a growing interdependent structure to enable the dental professional to use laser technology expressly to the benefit of the patient and within a safe environment, free of risk.

18.1 Introduction

The following extract is taken from: Raymond J. Lanzafame, Editor-In-Chief, *Photomedicine and Laser Surgery* Volume 26, Number 4, 2008—"Over the years, the clarity of the positive results of laser use in dentistry has been clouded by gizmo idolatry, occasional sales hype and clinical anecdote. Whatever the reasons, we must continue to educate ourselves, our colleagues and our patients about the science, the risks, the benefits … as well as the limitations of phototherapy."

There are two protagonists in the debate as to who benefits from laser use in dentistry. Primarily, it would seem that the patient, anxious to receive cavity preparation without the noise, vibration and pain of the "drill," should demand of their dental clinician that they should embrace this technology and therapy, but a deeper conviction is that it should always be the clinician who not only should decide what is an appropriate therapy, but should also assume responsibility for employing techniques that are evidence-based and deliver maximal benefit with minimal risk.

To those of us old enough to have witnessed at first-hand the progress of dental care during the past 35 years, two areas of primary dental care appear to have emerged and evolved outside the undergraduate teaching structure—dental implants and laser dentistry. Both modalities derived their theoretical and primary application outside clinical dentistry and certainly in the case of implantology, has led to a wide and far-reaching expansion of techniques and application that benefits clinician and patient alike.

By way of specific example of the latter, the consideration of dental implants within a post-extraction treatment plan is now considered mandatory by many dental licensing authorities. With the review of an ever-expanding time-base over nearly 40 years, the initial near-certainty of success with implants has now been supplanted by a realization that a new "disease" has emerged, with ever-increasing prevalence among those aging patients with long-term implant-supported prostheses. In 1989, Dr Carl Misch is recorded as saying "Peri-implantitis—if you haven't had a case of PI, you're either lying or you haven't done enough cases" (International Congress of Oral Implantologists London Congress, 1989).

In response to this emerging picture of possible permanent damage and deterioration, the whole emphasis of investigation into implant design, placement techniques, restoration and long-term survivability of implants has concentrated on a recognized science and evidence-base, built upon, and sustained through peer-reviewed rigorous studies.

Another significant area of concern to dental professionals lies within the near-epidemic of "cracked tooth syndrome" among the baby-boomer and younger generation. At the time of a strong belief in the principles of restorative dentistry that had spanned nearly 150 years, undergraduate teaching of cavity design and the use of amalgam formed a benchmark of understanding, sustained by near dogma postulated through G. V. Black.

With the emergence of failing dentition through fractures within the clinical crown, an understanding of the causative factors emerged through scientific investigation. "Micro-fractures adjacent to cavity margins, damage to teeth adjacent to inter-proximal cavity design, thermal injury to the pulp caused by rotary instrumentation" [1, 2] and "Restorative procedures …() … can all induce stresses onto the residual tooth structure culminating in a possible fracture" [3]. Even the underlying philosophy of university education became open to examination through publication: "the placement of poor-quality dental amalgam alloys, the contamination of freshly placed dental amalgam by moisture and excessive condensation pressures when placing amalgam may also induce fractures" [4].

The purpose of this discussion in what purports to be a textbook of laser use in dentistry, serves to highlight the precarious nature of our profession; the employment of ever-increasing sophisticated therapies applied to address dental and oral disease, for which the patient undertakes to pay and through which a legal contract of service and responsibility exists.

Laser dentistry?—isn't that some fringe activity that "amuses" a few dentists who believe in its poorly defined application in dentistry and convinces patients that it's a "magical" and painless alternative to the drill? How common is such a statement and over the years this has been the response of the cynics who question the application of lasers within the many disciplines of clinical dental practice.

Having written a great deal on the integration of laser use into dentistry, the author is at once compelled to defend its use as being as important to modern dentistry as implantology and restorative therapy, but to concede that he once coined the phrase "lasers are just looking for a home in dentistry." During personal association with lasers spanning 34 years of their availability in clinical dentistry and a growing passion over the expanding opportunity for integration, there was disappointment with the early machines, to the extent that laser use in dentistry was simply justified because they enjoyed success in other areas of medicine, so why not dentistry?

If there is a place for lasers in general within dentistry, there must be a place for laser science and much more ....

There is also an emerging shift in treatment patterns that every dentist will recognize. Emphasis on prevention, early interceptive treatment of caries and preservation of healthy tooth structure, maintenance of sound periodontal support for natural teeth and an over-riding patient-driven quest for an aesthetically pleasing smile, have all contributed to modern dentistry. The demands of the once-sound biologic and mechanical principles of Black's cavity design—outline form, retention form, extension for prevention, have been transformed by the near-ubiquitous use of micro-retentive direct composite resin restoratives. It is through this revolution that has growing acceptance through evidence-based protocols, that dentists can offer supportive oral treatment that at least has a potential of being acceptable to the patient.

Within this framework, laser dentistry—to the cynic to be "just looking for a home in dentistry"—might establish itself as an ideal philosophy and treatment tool and this might be best seen in the following clinical scenario:

A 9-year-old patient attends as a new patient with her mother. Both are anxious through past experience of dental treatment. The patient has several deciduous cavities, evidence of malocclusion, an upper permanent central incisor has failed to erupt and remains sub-gingival, and there is a high insertion of a fibrous maxillary midline frenum that may prevent approximation of the permanent central incisors. Few of us have the blessed sanctity of the pediatric dentist and the hope of a successful long-term happy relationship is slowly ebbing with the realization of what treatment is required. Equipped with a suitable laser, there is a possibility of enjoyment for patient, mother and dentist. Pain control in such situations can often be achieved through the use of topical anesthetics and the removal of caries and cavity preparation easily performed with an Erbium YAG or Erbium, Chromium YSGG laser; indeed, with the adoption of a gradual ascending irradiation of the target tooth a level of analgesia may be achieved to obviate the use of even topical analgesia. During cavity preparation, it is discovered that there is a vital pulp exposure and gingival in-growth interproximally; here, the soft tissue correction and effective pulpotomy can be achieved without bleeding, using either erbium lasers or possibly a near-infrared diode or Nd:YAG or far infrared CO_2 laser. In addressing any possible bacterial contamination, it is possible to use appropriate sub-ablative laser fluence to activate a photo-chemical liquid placed in the cavity, prior to placement of the restoration. The exposure of the un-erupted incisor using a suitable surgical laser can follow and labial frenectomy achieved almost imperceptibly without bleeding or the need for sutures or dressings. Fanciful or a detailed breakdown of justified laser-assisted therapies in everyday general dental practice?

Our challenge is not to dwell on the near-certainty of laser-assisted therapy, as considered above but to establish the evidence base that defines the use of laser technology as the prime consideration of instrumentation, in delivering such treatment.

18.2 Wisdom and Knowledge: Are They Mutually Exclusive and Beneficial?

> *Wisdom is the reward for surviving our own stupidity.* (Brian Rathbone, author).

The author first "knew" of lasers was in 1989 when encouraged to purchase a newly arriving model in the United Kingdom of the first dental laser—the dLase 300 Nd:YAG American Dental Laser. It was considered a positive vehicle in the provision of dental care under private contract and the encouragement received from the seller was that such a machine would lead to greater patient uptake. As the promotional material relating to this laser proclaimed:

> *....Imagine a beam of light so powerful it vaporizes tooth decay, cuts dentin and cementum, etches enamel, desensitizes teeth, incises soft tissue .... A beam of light so intense that it kills bacteria.*
>
> *So quick that pain is virtually eliminated ... what you have imagined is a dental laser ......*

The personal understanding of lasers at that moment was zero and initial personal instruction in its use amounted to a 1-h demonstration, once the laser was purchased and delivered. Frustration at the total lack of understanding as to what was exactly happening (the laser operating beam was invisible, adding to the "magical" effects on tissue) was compounded by the fact that this first laser—a near-infrared Nd:YAG—was ideally suited—indeed originally developed—for soft tissue ablation; indeed, any possible use on tooth tissue was limited to a slow and erratic ablation of accessible pigmented caries. Hardly an instrument that would deliver pain-free cavity preparation which was what the purchase was expected to provide.

Undeterred and within a belief that the interaction of this laser wavelength with soft tissue would deliver predictable outcomes, it rapidly emerged that something was drastically wrong and mostly it was the lack of operator understanding (◘ Fig. 18.1). In this clinical case, the fibrous nature of the frenum was recognized and high output laser power was chosen to ablate this tissue. Unfortunately, the power levels—with the benefit of hindsight, far too high for such delicate tissue, together with an incorrect angle of approach with the delivery fiber, led to thermal damage to the soft tissue, the underlying periosteum and necrosis of the underlying bone. As a result, the early post-operative period was very painful with a delayed period of sloughing of tissue and a resulting gingival fenestration. The patient was later referred to a periodontist for corrective gingival surgery.

There was no pride associated with such failure and the combination of personal shame and litigation fuelled an anger to overcome the lack of knowledge as to what was exactly happening and how the effects of laser use on oral tissue might be better controlled. Much of the early personal learning about laser use was through overcoming mishaps in wavelength choice and dose.

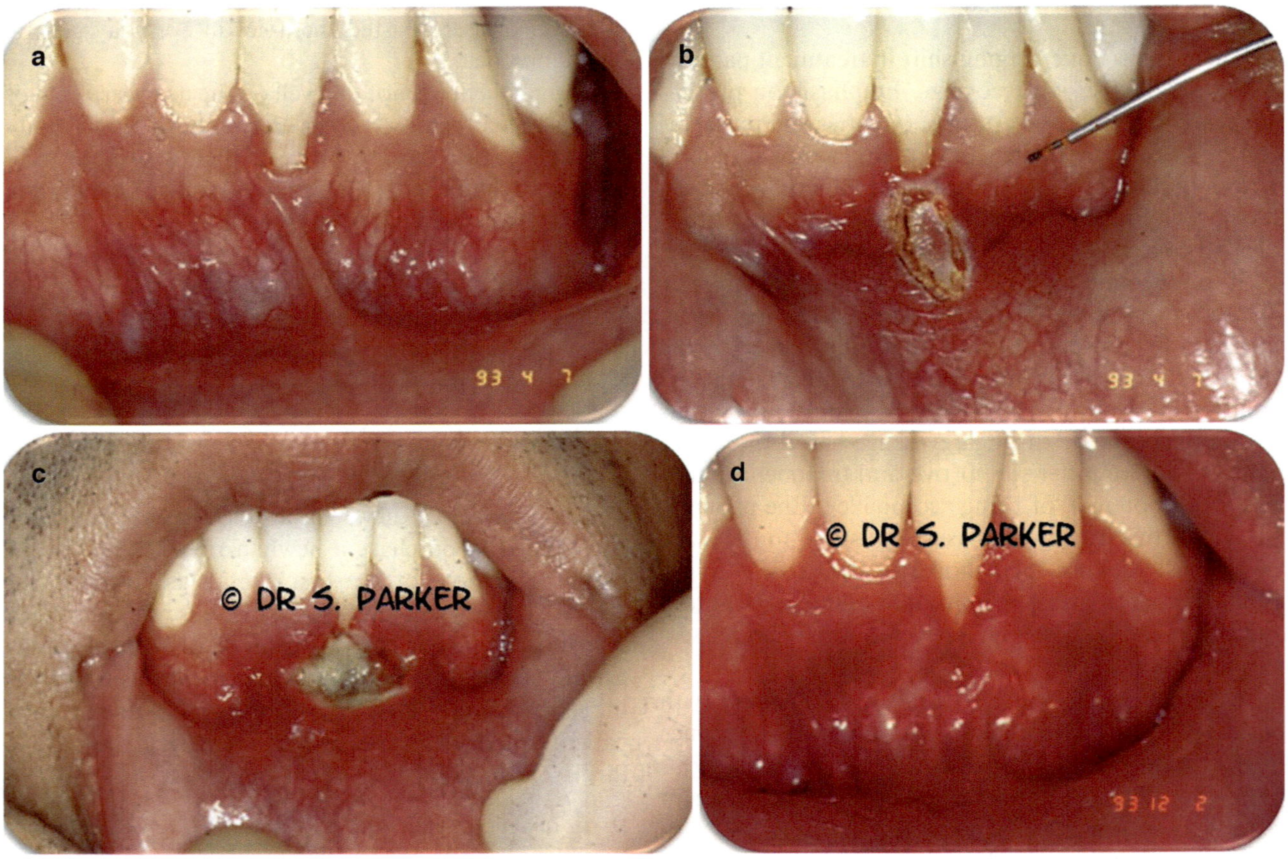

◘ **Fig. 18.1** The use of the Nd:YAG 1064 nm laser in a case of lower labial frenectomy **a**. **b** Initial laser surgery at excessive power and lack of consideration of anatomical structures, leading to deeper tissue damage. **c** Appearance at 4 weeks and **d** 8 months, necessitating onward specialist referral. Laser operating parameters: 320 μm fiber in contact mode. 250 mJ per pulse/15 Hz. Average power 3.74 W. Fluence 1162 J/cm^2. Time taken 30 s. Total energy delivered 113 J. Pulse width 100 μs. Peak power density 3,108,495 W/cm^2

During the following 4 years, a great deal of work was made available with several laser manufacturers together with extensive travel; in addition, other laser wavelengths such as the carbon dioxide and some early "diode" wavelengths were experienced through company affiliation. A fundamental example of lack of understanding of a new laser wavelength—the carbon dioxide 10,600 nm—was highlighted through the belief that the examples of relationship between laser wavelength and absorptive target tissue elements (essentially, tissue pigments, proteins, water and hard tissue mineral) would alone provide a safe guide as to the choice of laser (◘ Fig. 18.2).

The oral cavity is a complex mixture of tissue elements and each element—protein, melanin, water, hemoglobin, and carbonated hydroxyapatite—exhibits differing levels of absorption with any of the laser wavelengths available in dentistry. Essentially, shorter laser wavelengths, those within the blue visible and near-infrared wavelength regions of the electromagnetic spectrum enjoy positive absorption by pigmented protein and hemoglobin. Longer wavelengths (mid to far infrared) are absorbed by water and hard tissue mineral. The phenomenon of absorption leads to a conversion of incident photonic energy into thermal energy within the target tissue element, and this can be predictably configured to deliver a desired change in the tissue. With high-powered surgical lasers, the rise in temperature will lead to structural change, which will cause ablation.

The supposed high absorption in hydroxyapatite led to the belief that here was an answer to the quest to own a laser suitable for tooth cavity preparation. As shown in ◘ Fig. 18.2, the result of using incorrect and inappropriate laser operating parameters continued to expose a personal lack of understanding of the relevant funda-

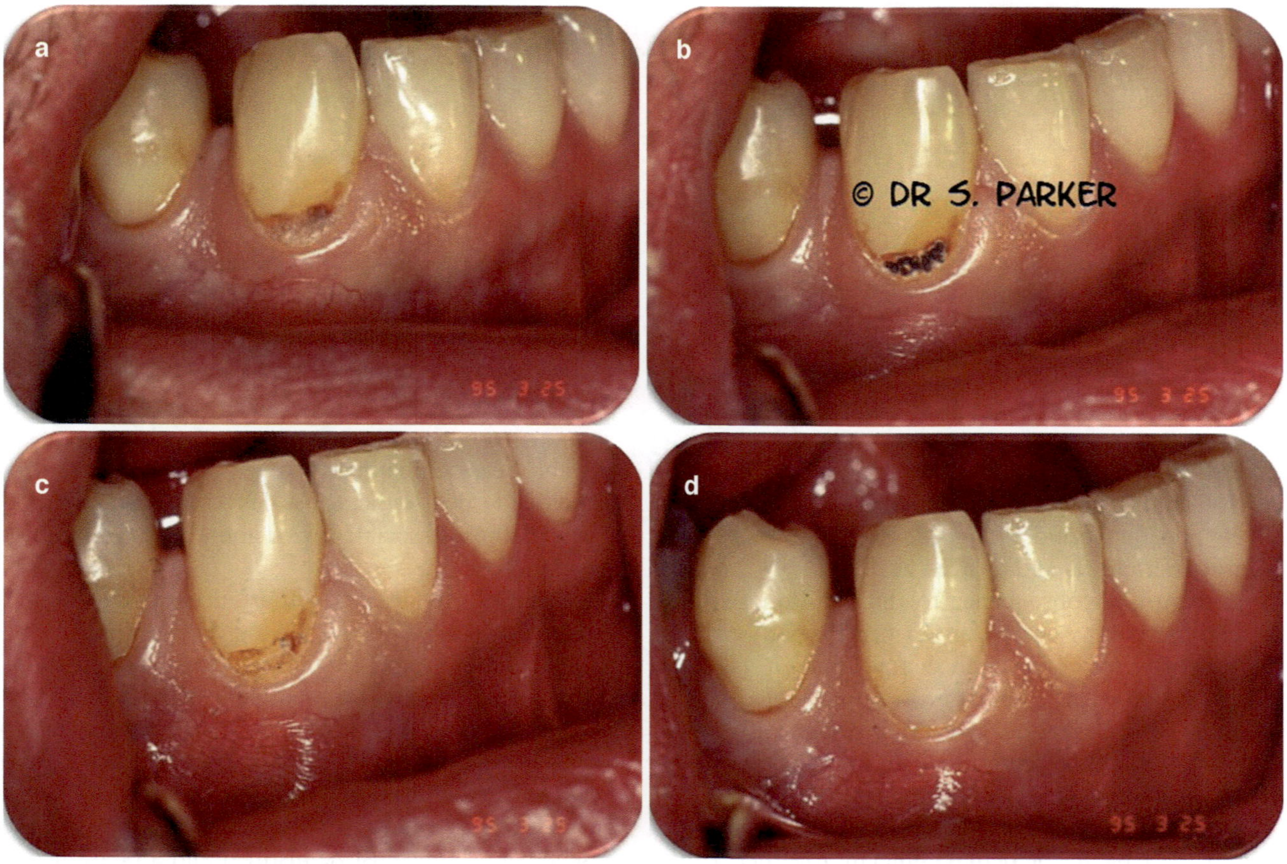

◘ **Fig. 18.2** The use of the CO_2 10,600 nm laser in an attempted Class V cavity preparation, having provided assurance to the patient that no rotary instruments would be used **a**. Relatively low average power delivered (1.0 W) was not the prime issue in the complications incurred **b**—carbon deposits after initial laser irradiance), but the emission mode (CW) and lack of water, leading to over-heating of the tissue. **c** Following attempts to remove the carbonized dentine and residual caries, the restoration was completed **d**. Fortunately, no pulpal damage followed. Average power 1.0 W. Continuous wave emission. Non-contact delivery/beam size 600 µm. Fluence 111 J/cm^2. Time taken 5 s. Power density 164 W/cm^2

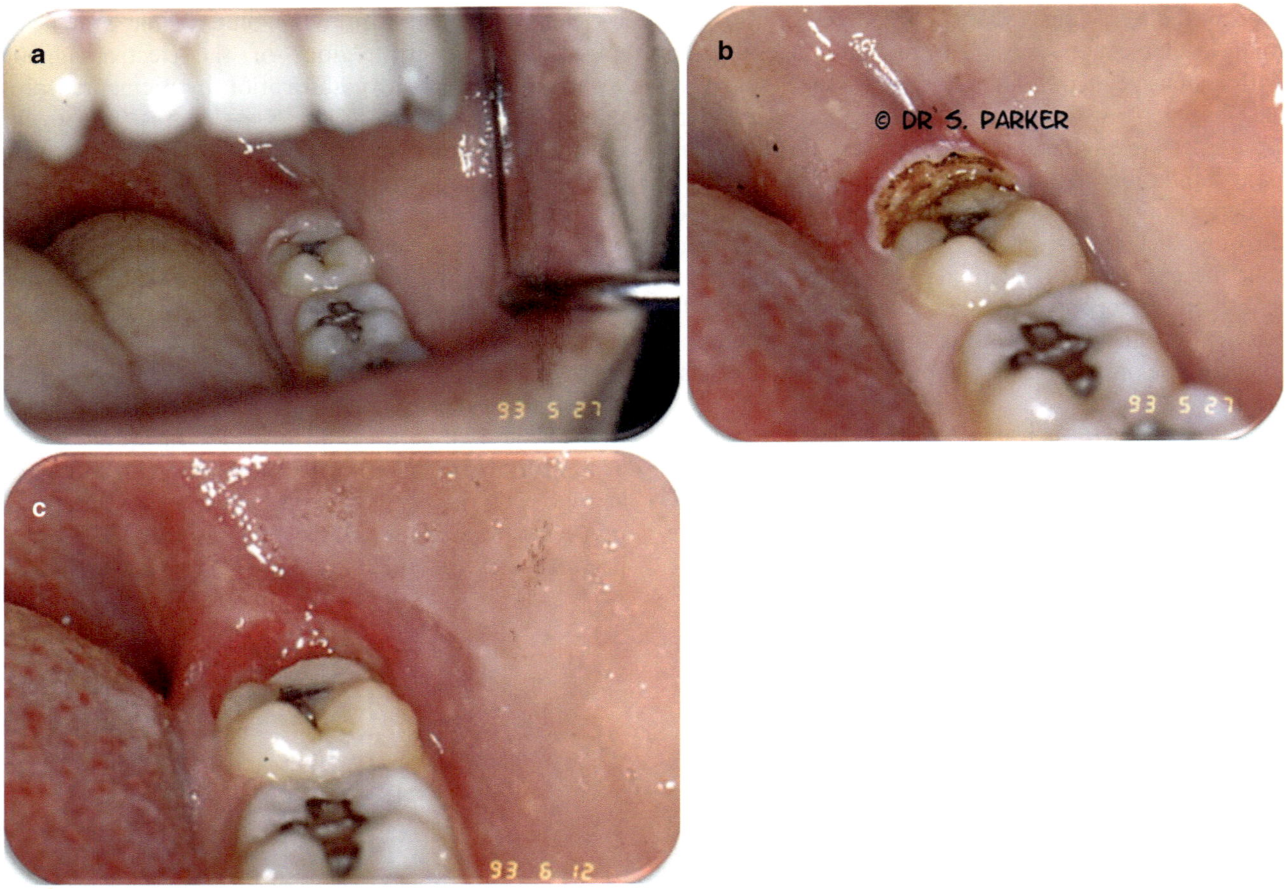

Fig. 18.3 The use of the 810 nm diode laser to treat pericoronitis at lower left wisdom tooth. **a** Shows inflamed operculum associated with functional wisdom tooth. **b** Immediately post-laser resection of soft tissue. Care must always be given to the equal irradiation of keratinised and non-keratinised epithelium and in the case of excessive power, the possibility exists for deeper edema to spread within "loose" retromolar anatomical structures and spaces. **c** Sixteen days post-operative presentation. The patient had experienced severe pain, discharge, halitosis and trismus. Healing eventually achieved. Laser operating parameters: Average power 2.0 W. Continuous wave emission. Contact delivery/beam size 320 μm. Fluence 310 J/cm^2. Time taken 25 s. Total energy 50 J

mentals of laser science and biophysics. Once again, a lack of control of rapid over-heating led to damaging effects and furthermore, as in Fig. 18.3, this led to painful after-effects and delayed healing.

18

If there is a consequence to this self-imposed and dangerous learning curve that formed the early part of the author's association with dental lasers, it was that clinical experience should be complemented by a grounding in relevant science, applied clinical training and development of prescriptive analytical skill in case of selection and laser choice.

The rapid expansion in recent years in laser wavelengths available to the dental clinician, together with a rationalization toward machine size, sophistication and in many instances, cost has allowed a full range of clinical treatment areas to be further explored and managed with the help of laser photonic energy. Figure 18.4 provides a (non-exhaustive) list of those clinical procedures or areas of treatment that can be carried out with the help of lasers—either those delivering tissue ablation as well as those associated with non-ablative photobiomodulation.

What is of fundamental importance is that no matter what the level of adjunctive use a given laser may bring, the clinician is first and foremost a dentist and it remains a duty to the patient that treatment offered should be based on evidence and within the capability or regulation of the dentist to provide it. Evidence demands science and furthermore enforces the belief that like all other aspects of clinical procedures and techniques, the presence of a laser must be justified and all parameters of its use for a given procedure defined through evidence-base.

This philosophy and approach can be simply demonstrated through the adoption of the "MIMO" concept—Minimum Input …… Maximum Outcome.

Surgery
Gingivectomy, Gingivoplasty
Crown lengthening, Operculectomy
Frenectomy
Tumour Removal (fibroma, epulis)
Dainage and incision of abscess
Osseous sugery

Restorative
Cavity preparation
Caries removal

Endodontology
Decontamination of the Root Canal
Granuloma
Pulpectomy
Pulpotomy

Therapy
Apthae, Herpes, Angular Cheilitis
Bio-stimulation – TMJ, etc.
Leucoplakia. Lichen Planus

Orthodontics
Tooth exposure
PBM accelerated tooth movement

Periodontology
Parodontitis / treatment of periodontal pockets
Curettage
Antimicrobial photodynamic therapy
Gingival grafting

Implantology
Uncovering of implants
Sterilisation and decontamination of implants
Peri-implantitis

Prosthetics
Ovate pontic development

Aesthetics
Laser activation of bleaching materials for teeth whitening

Photobiomodulation
PBM-mediated treatment of inflammation, pain, healing

Fig. 18.4 Specific areas of therapy and clinical procedures where laser-assisted treatment may be carried out. Each procedure has received peer-reviewed publication as listed in PubMed (▸ www.ncbi.nlm.nih.gov/pubmed)

18.3 "MIMO"

The predominance among laser manufacturers to safeguard against untutored and inexperienced use of a laser, may be seen through the availability of treatment "menus"—dial-up options with associated pre-set laser operating parameters. These are far from a "safety net" and in many cases provide the only means of selecting operating values for a chosen clinical procedure. In Fig. 18.5, a simple algorithm provides contrast to reliance on manufacturer's setting, as opposed to individual assessment of patient needs, condition and laser settings.

In all cases, for a proposed treatment, the exact nature of a chosen target tissue is unique; the physical structure and optical parameters relative to incident laser irradiation will define the extent to which the electromagnetic energy is absorbed or scattered/transmitted and thus allow conversion to thermal energy with attendant effect. An example may be seen in a procedure such as gingivectomy, where the rate of target ablation may differ significantly between a north European and an Afro-Caribbean patient through the degree of melanin pigmentation present.

Total reliance on manufacturer's settings may lead the unwary or untrained to encounter unforeseen outcomes, whereas a fully developed and qualified experience and training pathway will allow the clinician to anticipate the demands of the case and adjust operating parameters accordingly to protect the patient. Minimal "input" amounts to operating parameters (power density variants) sufficiently low to produce a desired outcome and avoid unwanted tissue damage.

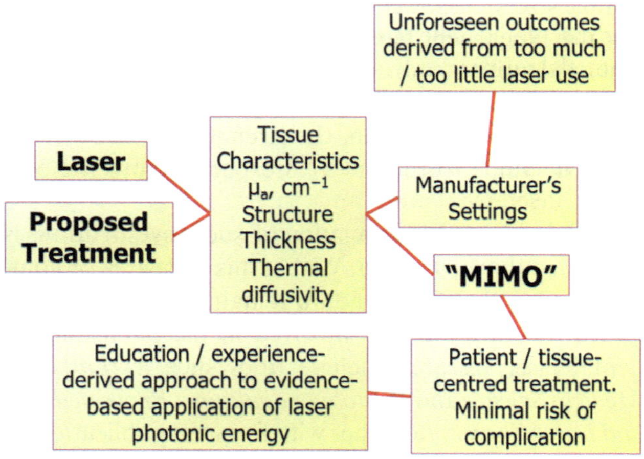

Fig. 18.5 "MIMO" algorithm. Minimum Input–Maximum Outcome. The use of a minimal level of total laser photonic energy, relative to irradiated tissue volume, in order to deliver a desired outcome and avoid collateral damage events

18.4 Evidence-Based Pathways in Laser Use in Dentistry

A suitable explanation of the meaning and extent of evidence base may be seen as follows:

> *the conscientious, explicit and judicious use of current best evidence in making decisions about the care of the individual patient. It means integrating individual clinical expertise with the best available external clinical evidence from systematic research* [5].

Within the field of laser dentistry, at the time of the launch of the first Nd:YAG laser in 1989, the extent of dedicated research as applied to dentistry was virtually non-existent. Unfortunately, the early investigations to establish scientific rigor to claims of clinical benefits with this laser were to some extent open to criticism and claims of bias and limited direct significance to possible use within clinical dental practice.

A pathway to research might explore available science and define a hypothesis as to examination of variables within a system. Accuracy in the so-called materials and method must be ideally underpinned by degrees of bias limitation or "blinding." In essence, this would be when one investigator defines the subject matter and sample size(s), followed by another investigator carrying out the experiment but not aware of the sampling process or identity of subjects. Results might be collected by another "blinded" worker, adding further to the objectivity of the investigation, the strength of the outcome and true value to the progress expected from the outcome.

The method of propagation of such investigations is through published media. Within this wide area of communication, the gold standard remains high-value peer-review. Quality journals involved in the publishing of dental laser research include titles such as *Lasers in Medical Science* and *Photobiomodulation, Photomedicine and Laser Surgery*, together with frequent publication of laser-related studies and articles in journals of periodontology, endodontics, implantology, and restorative dentistry. Key to the value of such publications must always surround a declaration of strict peer-review, where each article submission is subject to review by more than one referee who is considered expert in the given subject matter. Additional guidance as to sustainability of rigor is the Impact Factor, a computation for an individual journal as a measure of the frequency with which the "average article" in a journal has been cited in a particular year or period. The impact factor is useful in clarifying the significance of absolute (or total) citation frequencies. It eliminates some of the bias of such counts which favor large journals over small ones, or frequently issued journals over less frequently issued ones, and of older journals over newer ones. Particularly in the latter case such journals have a larger citable body of literature than smaller or younger journals. All things being equal, the larger the number of previously published articles, the more often a journal will be cited.

18.5 *"Apple Pie"* Philosophy and Sustainability of Scientific Investigation

Imagine if you will, the following relating to a published case of laser-assisted frenectomy:

> diode laser used. Power 1.5 Watts. Frenectomy completed ....

Maybe simplistic but examples abound of similar accounts of laser use. Even the laser wavelength is not recorded and operating technique absent.

Of course, of more concern might be the materials and methods report of a clinical investigation:

> patient group C treated with the laser. Laser used: Nd:YAG. Average power 1.5 Watts. Pocket debridement carried out ......

If there is a basic, profound flaw in each of these examples, it is the abject lack of subsequent opportunity to reproduce the laser action and outcome by others. The former case presentation might give rise to risk as similar to that outlined in Fig. 18.1; the latter, if part of a study into the use of the Nd:YAG laser in the treatment of periodontal disease, provides grave danger of compounding the risk to an indeterminate number of patients, were such a study to be cited or implemented in further clinical investigation.

In ▶ Chap. 4, it was shown that important parameters relating to the laser and the output represent a bare minimum.

In cooking an apple pie, it would be necessary to consult and follow a recipe. To add to the attraction of what awaits, a tempting image of the finished pie would add to the excitement. But—what if some stages of the recipe were missing; or maybe instead of "2 eggs" or "100 g sugar," the recipe merely stated "eggs" and "sugar." How confident would we be of success—and first-time success! In the same way, a published article or investigative study highlighting laser use should detail exactly the laser, its emission wavelength and mode, beam/spot size and average power. From there, computed values such as fluence, irradiance, possible peak power can be recorded, together with co-axial water/air if use and total time taken. When such more extensive parameters are applied to our two "basic recipe" examples above, a much greater degree of certainty emerges—reproducibility that enables successive use of each laser in similar clinical scenario to meet the basic "apple pie" rule!

The situation is changing; for so long the level of reproducibility among published material was low, often with merely laser control panel readings of laser output. Recognition of possible power losses—*vis a vis* between control panel data and hand-piece emission—along a delivery system can affect the actual applied fluence dose by as much as 30% [6].

Through precise and comprehensive recording of all standing and computed laser parameters, the rigor of science and research can be sufficiently strong to allow truly representative progress in the broadening application of laser technology to develop.

The risk of "bias" is considered greater if the article is company sponsored, either through direct support or through investigators with declared interest. The type and extent of criteria under investigation—the so-called inclusion and exclusion criteria may have a profound effect on the predictability of outcome and/or the statistical significance of the results.

Figure 18.6 provides an overview of the relationship of published work, the type of investigation, the type or format of published media and associated distorting factors that might affect any of the preceding factors. Each of these will have relevance to the "objectivity" of the paper or article, the scientific rigor, and the reproducibility ("apple pie" factor).

Distorting factors and influence of bias through predetermination of material or funding support, together with "manipulation" of data or materials and method, have been discussed above.

Within the Science Publication Sequence, there is a hierarchical relationship between type of medium and lack of subjectivity. Apart from occasional national blanket controls on public dissemination of sensitive data, the internet is limitless in terms of breadth of

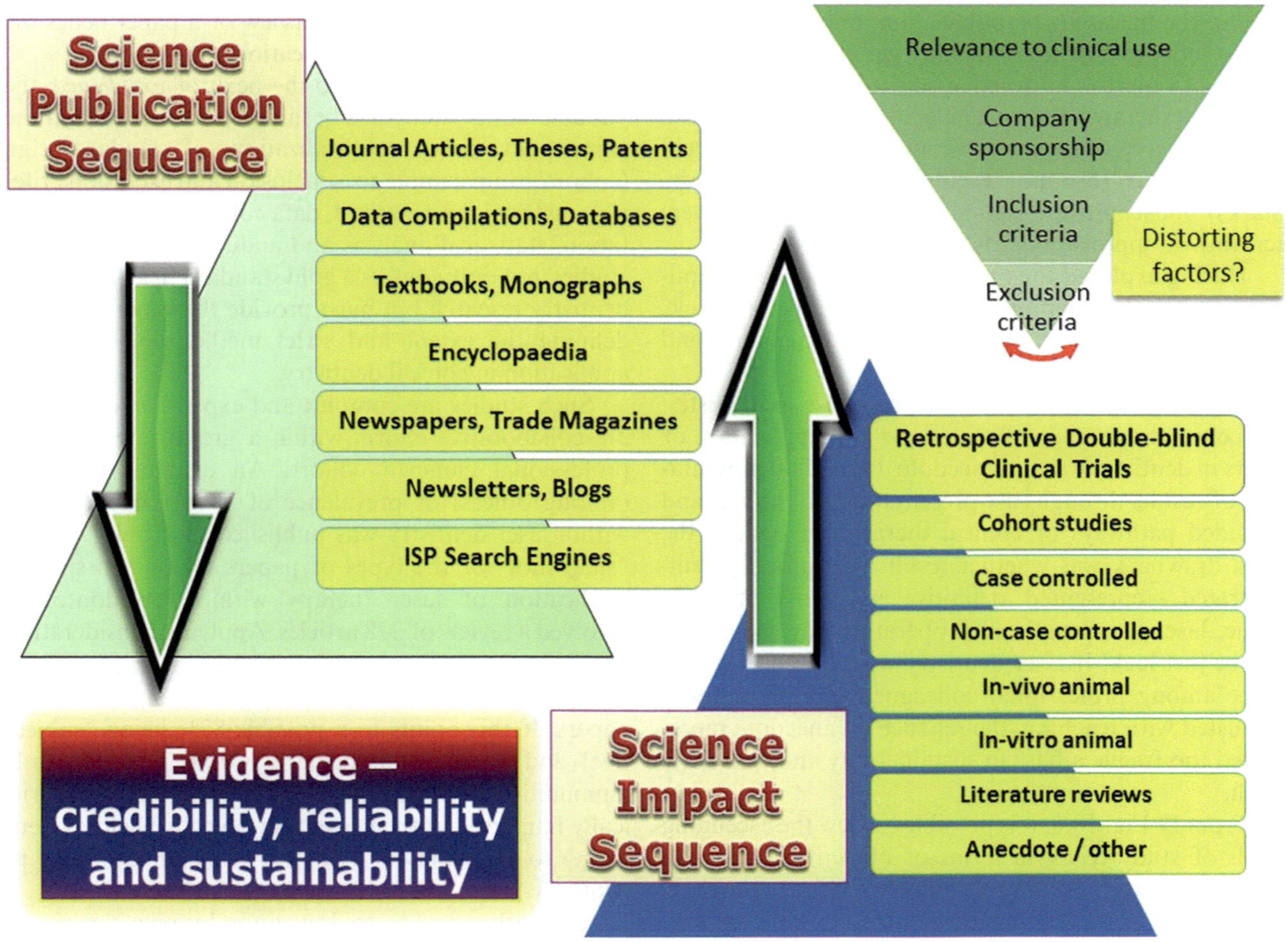

Fig. 18.6 Hierarchical relationships of publication media, study design and distorting factors as they relate to the sustainability of science and evidence in laser dentistry

knowledge but also position within a spectrum of truth. Anyone can upload anything and claim intellectual property rights; whether a statement is true or not or even if the material is original or not, it is possible to both host a dubious source of knowledge and manipulate its truthfulness. Of course, such a viewpoint is offensive to those who post honest and scientifically strong material (clinical and non-clinical) relating to laser action within dentistry. Nonetheless, the internet as a generic source may be seen as least robust or dependable.

From there, online newsletters, blogs together with corporation-led "advertorial" and "infomercial" (yes—there is a difference!) publicity of a clinical team, company, technique, or laser instrument, while striving to uphold the corporate or personal ethics of those involved may be seen as being open to claims of subjectivity and massaging of truth to some extent.

The so-called reference texts, of which this book is an example, are often seen as totally credible but lacking in peer-review; hence, there is only the reputation of individual authors and overall publisher's approach to scrutiny that provides reinforcement of truth and ethics. Within the hierarchy of publication, only high-impact-factor independent journals may get close to a perfect source of objective knowledge, science, and applied clinical laser therapy. Public university degree theses will enjoy a degree of peer-review, and instrument or technique patents (but not exclusively, registered trademarks) must prove originality and pass the severe scrutiny of approval boards.

Having explored the medium of knowledge dissemination, the other hierarchy relates to the type of article, study or investigation in our quest for originality and objectivity within laser dentistry.

Certainly, in the early days of laser use in dentistry, the common currency of technique and application of lasers in dentistry was the anecdote. Individual presentation of results through case presentation led the way and provided pathways of clinical therapy. Of course, the great drawback was whether results—so proudly demonstrated—represented definitive protocols of technique, laser wavelength, and operating parameters, or maybe just luck! If laser dentistry was to receive acceptance among professional colleagues, groups or those entrusted with teaching, the reliance on anecdote represented too fragile a base in sustainability and (possibly) truth.

From ◘ Fig. 18.6, it is possible to view the ascending levels of study types, in terms of objectivity, scientific rigor, reproducibility and ultimately, clinical relevance. Literature reviews are sometimes suspect—what control exists to mandate the selection of published material? The risk would lie with individual reviews that declare their own chosen agenda. Certainly, "super-objective" filters such as the Cochrane Database of Systematic Reviews represent the leading resource for systematic reviews in health care. The following is taken from their website (▶ www.cochrane.org):

> *The Cochrane editorial process follows a consistent and structured path. It is unique in two ways: (1) to monitor the process of review development throughout the editorial life cycle, beginning with registration of a title, through preparation and publication of the protocol and completed review; (2) Cochrane Reviews are updated to take account of emerging evidence, to provide the best and most current evidence to guide decision-making.*

However, sometimes the restrictive nature of literature review might result in (for instance) several hundred published papers being considered, but with application of inclusion criteria only a handful are further examined. Certainly, the reader is encouraged to look at the exact nature of review and if this is a personal enquiry into a given laser technique or study, care, and caution should be exercised in total reliance on article abstract material, as often a brief overview of a paper belies the complete impact of the publication.

Further upwards and to the peak of *excellence*; the "double-blind" study has been described earlier and the component compartmentalization of study design (cohort/group/sample size, inclusion and exclusion criteria), materials and method, data collection and interpretation, statistical analysis and audit through successive studies, not only confers a gold standard in all aspects of dentistry research but must provide the way forward in defining the extent and strict methodology of laser application in clinical dentistry.

Such studies are complex and expensive; often they are collaborative efforts within a group, university or professional clinician cohort. An example of audit (among others) of prevalence of such refined studies within laser dentistry was published by Cobb [7]. This "snap-shot" of the types of papers relating to specific application of laser therapy within periodontology involved a review of 278 articles. Applying consideration that has been explored above, only 12.6% of papers were classified as "objective, controlled," amounting to 35 papers. If this points to a possible 87% being "subjective" and possibly unreliable, it would provide ample ammunition to those non-laser users (possibly hypocritically immersed in their own non-objective field of dentistry) to brand lasers as of no sustainable benefit to dentistry.

The author prefers to draw upon his own experience during the past 15 years, not only as a clinician but within university teaching and research, journal editorial duties and personal publication, to bear witness that the general standard of objective research and clinical application of laser use in dentistry has progressed enormously and with it the quality and sustainability of objective publication.

18.6 Education and Qualification Pathways in Laser Dentistry

So why is laser use in dentistry so sporadic? There may be several reasons: financial, lack of awareness among dentists or knowledge of what lasers can perform, but perhaps the greater one is the lack of grounding and integrated teaching of laser therapy at undergraduate level. If a student is introduced to tooth cavity preparation using a rotary drill, such treatment modality forms the basis of the concept of restorative dentistry; equally, a soft tissue incision performed with a scalpel defines an approach to surgery that will be carried into one's professional career. The early learning in undergraduate life forms a bed-rock foundation to clinical opinion and techniques and this, together with the bonds established between student and professor is part of the excellence and integrity of clinical dentistry seen all over the world.

Early adopters of laser dentistry were not able to access structured educational pathways; rather, those who wished to consider laser use would attend an introductory course, often provided by the company selling the machine. Such courses would prove invaluable in providing the novice with basics. Many such courses had a drawback in that they would often limit presentation material to the extent of the particular laser and what could be achieved with it. Notwithstanding, introductory experiences continue to provide initiation into the world of lasers.

Awareness of different laser wavelengths and clinical opportunities will drive the clinician to explore wider education avenues. Courses at varying levels of intensity, subject matter and complexity are now frequently promoted—often as break-out sessions at a laser scientific conference.

With both these types of courses, verifiable continuing education credits underline the significance of learning and awareness to benefit clinician, patient and professional licencing body. As such they will be quantifiable in terms of course content and expectation. Many of such courses draw upon accepted core curriculum structures an example of which is set out in a paper by White et al. on curriculum guidelines for dental laser education [8].

Often, however the value of such courses is thwarted through the non-linear pathway that such attendance implies. Course X given by organization Y may have integrity and convey accurate contemporary opinion, presented by an accredited experienced practitioner. Taking such credit to a higher level, such as post-graduate degree application may suffer from non-compliance with the entry criteria of that university. In consequence, so much of education in laser use in dentistry has suffered because of lack of structure and progressive pathway.

This is not a new phenomenon; in 2005 when installed as President of one laser organization, the author was invited to join the editorial board of another laser organization, located in a different continent. Despite best efforts to effect change, neither organization would forego their own exclusivity and allow reciprocation of (otherwise complementary) education achievement levels in laser use. What frustration! Those having progressed through one organization and wishing to join another, would need to begin once more at the bottom. Examples of such perceived selfishness and blinkered ambition continue to the present time in dissuading dental professionals from joining and benefitting from the forum of like-minded colleagues in laser use.

In many regions of the world, public university-based education has supplanted some historically popular laser organization-centered course pathways with their own. Accredited diploma and post-graduate MSc and PhD degree opportunities are now common—certainly in Europe, with the benefit to the busy clinician that many are part-time and distance-learning modular courses. Of particular benefit is the International recognition of such education and graduates of such courses are becoming the opinion-makers within an emerging generation of specialists and experts. In turn, those universities that have developed degree opportunities have themselves witnessed expansion in accredited personnel, over and complementary to existing faculty staff—ambassadors who in turn develop the outreach of evidence-based education. The appeal of these courses is high and in the author's experience as a university faculty member, students have traveled thousands of miles across continents to take advantage of accredited learning pathways.

At the beginning of this chapter, I wrote "to know is to use; to understand is to empower." The fundamental elements to support such a statement must be the acknowledgment of a scientific base to all laser–tissue interaction, together with an accredited grasp of biophysical principles that govern the degree of laser–tissue interaction, by laser type, laser wavelength, and target tissue composition.

18.7 Regulation and Medico-Legal Aspects of Laser Use

It is beyond the scope of this chapter to provide definitive regulatory requirements as they might apply in every country. However, there is scope for discussion of core principles in both regulation and medico-legal aspects that might impact on the use of lasers.

18.7.1 Regulation as It Might Impact on Laser Use in Dentistry

A common and logical question that might be asked by those wishing to use a laser in clinical practice is "why?" Is there regulation governing the use of a high-speed air rotor; or a scalpel? Surely, as a dentist or hygienist, the issue or re-issue of a licence or practicing certificate defines capability in using any instrument of choice in the delivery of primary dental care?

Of course, as dental professional there is obligation to provide specific treatment only inasmuch as defined within the scope of practice. There are different regulations of practice in many areas of the world. For example, it is outside the remit of hygienists in the United Kingdom to provide treatment that "consists of or is implicit in making an incision in the periodontal tissues" (▶ www.gdc-uk.org). Equally, US dentists may be prohibited from providing treatment that is peri-oral and outside the vermillion border. To some, such regulation may appear illogical relative to their own geographical location and licencing authority. In general, however the basic duty of care of the clinical dentist or allied professional is only to provide treatment for which they are licenced, only to provide treatment that the dentist feels able or experienced in providing, and to only provide treatment that the patient understands and is willing to undergo.

Added to such basic general licencing regulation, there is the impact of owning a device emitting electromagnetic radiation. All dental professionals must be aware of and abide by Ionising Radiation Regulations as they impact on the use of X-Ray equipment and use. At present, the spectral range of current commercial lasers in dentistry is from the visible blue to far infra-red non-ionizing band of the electromagnetic spectrum. No inherent risk posed by photonic energy within this range to cause ionization of DNA, but sufficient risk by virtue of exposure of the unprotected eye exists, aspects of which have been covered extensively in ▶ Chap. 5.

18

There is therefore considerable legislation surrounding the safety aspects of laser use in general and many of these statutes have been adapted or simply adopted to the use of lasers in dentistry. The International Electrotechnical Commission provides exacting standards for the construction and use of lasers in clinical practice, and these are implemented as a whole, or through National or Federal agencies in individual countries. Cross-over legislation governing safety at work and statutory instrument specification (e.g., OSHA and ANSI within the USA and CE accreditation within the EU) will have implication for the clinician in housing and using equipment, provision of laser therapy and safety protocols. Objective resources exist to provide the dentist or hygienist with reference material as to how laser use impacts their duty of care as a provider of dental care, and the reader is recommended to investigate their position within this framework and how that is implemented through their individual country or licencing authority.

It is acknowledged that laser dentistry, like dental implants as modalities have supplanted long-standing therapies within the armamentarium of the practicing dental clinician. As a result, some "quirky" legislative outcomes have provided confusion and possible incredulity! Within the United Kingdom, a growing use of early lasers in dermatology during the late 1970s and early 1980s—use that outside the licence to practice of the doctors concerned was completely un-regulated. At that time a popular National television consumer program championed the claims of patients who were disfigured through supposed malpractice using these lasers. So strong was the level of concern that the UK government sought to legislate clinical laser use and thought it expedient to "piggy back" onto a developing Act of Parliament—the 1985 Nursing Homes Act. As such, overnight the duty of medical practitioners (and some years later the first dentists to use lasers) was to register their premises as a nursing home and to abide by all aspects of the act in treating patients. As a laser dentist in 1989, the author was required to register with the local health authority, seek the services and report of a laser physicist appointed as a laser protection advisor, as well as complete a yearly audit on quite unrelated aspects as care of the dying, number of beds, registration and training of nursing staff, etc.

That early legislation was supplanted by the UK Care Standards Act in 2000 and later versions, with added obligation to undergo fitness to practice medical examination, financial liquidity disclosure and even criminal records check for the dentist and all staff.

Hopefully, the reader may be excused such levels of suffocation in examining their obligation in using Class III and IV lasers!

18.7.2 Medico-Legal Aspects of Laser Use

Those lasers powerful enough to ablate tissue carry additional risk of damage through breach of duty of care and causation. Simple reference to the case in ◘ Fig. 18.1 provides an example of breach of duty of care in that inappropriate levels of power were used without thought to underlying tissue and collateral gingival damage. It may be alleged that *on the balance of probability* a breach of duty of care on the part of the dentist occurred. Furthermore, the outcome of such breach—termed causation—would suggest that *on the balance of probability* either the need for reparative treatment or permanent damage was linked directly to the excessive laser power or inadequate technique.

Very few patients attend for dental treatment; most patients are required or even compelled to attend by virtue of the need for treatment of a dental condition, and it is the duty of the clinician to diagnose, treat, and maintain the function and health of the oral cavity. Aristotle, perhaps himself not a dental patient is attributed with the phrase "the object of the wise is not to secure pleasure, but to avoid pain." This is surely the mantra of the dental patient and the objective of the successful dentist. The growth in consumerism in general and the liberalization of choice has rendered the provision of dental treatment to that of a "customer-centered service experience." Most of the therapeutic instrumentation that formed the basis of our undergraduate training is quite abhorrent to the anxious patient; the noise, vibration and perception of pain associated with the high-speed drill and the bleeding, post-operative swelling inflammation and associated sutures and dressings that accompany intra-oral soft tissue surgical procedures, serve to interfere with all aspects of oral function for the patient during what may be a long period of healing. This doesn't suggest that dental treatment is intrinsically wrong or justifiably avoided by the patient, but the opportunity to address these disadvantages that are subjective to the patient experience and also to deliver high-quality dental treatment must surely represent a gold standard. It also represents an extremely potent marketing tool that compliments that essential business base of every private dental clinic.

However, the choice of a laser to deliver prescribed treatment or therapy should be made by the dental professional. Lack of proper training and development of core knowledge and education (both quantitative and qualitative) may be viewed as serious examples of negligent action.

It should be acknowledged that few cases of litigation have reached a level of seriousness or constituted sufficient breach of licencing protocols to be the subject of dental registering and governing bodies. However, were such events to occur, it is highly likely that the relative lack of precedence (case history) within dentistry would result in such breach being examined against the wider health and safety legislation in the workplace, together with IEC standards on laser use.

18.8 Summary

Of course, the myths persist that laser dentistry is somehow "magical," delivering "painless dentistry" and that the laser dentist never uses a drill or scalpel. Such perceptions are not just in the domain of the patient as he or she seeks therapeutic absolution of their dental sins. The author is guilty of seeking a marketing advantage over local colleagues in 1989 when purchasing his first laser. As explained above, it was quickly learnt that it was not possible to deliver what patients demanded as treatment, with that laser. Internal marketing—the quiet dissemination of evidence-based techniques that are targeted at the needs of individual dental patients, provides a base for practice growth that, although less spectacular than the glittering advertising campaign, promotes a value-added loyal patient list. Laser dentistry is not painless: it is certainly *less* painful, when compared to cavity preparation with an air rotor, as it is possible to target caries selectively using a non-tactile interaction of light energy with target structural elements. Equally, with the enormous amount of research into the local tissue and wider effects of PBM, laser-assisted therapy—either stand-alone or adjunctive to laser surgery, has helped to provide understanding of the benefits of PBM in reducing inflammation, promoting uneventful healing and—reducing pain! As such, the determination of patient acceptance, itself often shrouded in a personal history of a painful experience is much more a process of comparative evaluation and gradual co-operation, and it is my opinion that it is negligent to promote laser dentistry as a panacea that will transform the patient experience as if by "magic." Equally, no laser will cut metal restorations or prepare self-retentive crown abutments. Early investigations into the suitability of laser energy applied to oral and dental tissue suffered from allegations of subjective supportive inclusion criteria, company sponsorship or anecdotal experience or, at best the application of incorrect laser wavelengths to subject tissue. With the tremendous growth in wavelengths and machines that are dentistry-specific, the objectivity of investigation through peer-reviewed retrospective cross-over studies has led to an acceptance not only of the suitability of lasers in dentistry but also the evidence-based protocols for their supportive role in treatment.

References

1. Lenters M. Iatrogenic damage to the adjacent surfaces of primary molars, in three different ways of cavity preparation. Eur Arch Paediatr Dent. 2006;7(1):6–10.
2. Baldissara P, Catapano S. Clinical and histological evaluation of thermal injury thresholds in human teeth: a preliminary study. J Oral Rehabil. 1997;24(11):791–801.
3. Turp C, Gobetti J. The cracked tooth syndrome: an elusive diagnosis. J Am Dent Assoc. 1996;127:1502–7.
4. Rosen H. Cracked tooth syndrome. J Prosthet Dent. 1982;47:36–43.
5. Sackett DL, Rosenberg WM, Gray JA, Haynes RB, Richardson WS. Evidence based medicine: what it is and what it isn't. BMJ. 1996;312(7023):71–2.
6. Parker S, Cronshaw M, Grootveld M, Anagnostaki E, Mylona V, Chala M, George R, Walsh L. The influence of delivery power losses and full operating parametry on the effectiveness of diode visible–near-infrared (445–1064nm) laser therapy in dentistry—a multi-center investigation. Lasers Med Sci. 2022;37(4):2249–57. https://doi.org/10.1007/s10103-021-03491-y.
7. Cobb CM. Lasers in periodontics: a review of the literature. J Periodontol. 2006;77:545–64.
8. White JM, et al. Curriculum guidelines and standards for dental laser education. Lasers in dentistry V. SPIE Int Soc Opt Eng. 1999;3593:110–22.

Supplementary Information

D. J. Coluzzi, S. P. A. Parker (eds.), *Lasers in Dentistry—Current Concepts*, Textbooks in Contemporary Dentistry, https://doi.org/10.1007/978-3-031-43338-2

Glossary

Ablation Removal of a segment of tissue or (dental restorative) material using photothermal energy. When applied to laser-tissue interaction, it describes the irreversible disruption of the physical structure of the target. It is sometimes termed vaporization, although that is not technically correct.

Absorption The transfer of electromagnetic radiation into the target tissue resulting in a change in that tissue. For available dental lasers, that transfer is primarily light into heat. The true opposite of absorption is transmission.

Absorption coefficient The change in energy as the wave passes through a layer is a constant of the material for a given wavelength and is called its absorption coefficient.

Active medium The material within the optical cavity that, when stimulated and amplified into a population inversion, will emit laser energy. This medium may be an ion, molecule, crystal, semiconductor wafer, or a combination of gases. A synonymous term is lasant.

Amplification A process that occurs within the optical resonator of the laser whereby stimulated emission produces a population inversion. The amount of amplification is known as gain.

Articulated arm One type of laser delivery system that uses segments of a hollow tube that are coupled with right angle mirrors that allow propagation of the laser beam along its length. Commonly found as part of the delivery system in mid- and far-infra red wavelength dental lasers.

Attenuation The observed decline in energy as a beam passes through an absorbing or scattering medium.

Average power An expression of the amount of laser photonic energy delivered over a unit of time. In continuous wave emission, it is the total power delivered; with free-running or gated pulse emission, it is an expression of the product of the peak power multiplied by the emission cycle.

Beam A directional stream of photons; any collection of radiant electromagnetic rays that may be divergent, convergent, or collimated.

Carbon dioxide laser A laser whose active medium is composed primarily of helium, with carbon dioxide, nitrogen, and small amounts of hydrogen and helium. The laser is pumped with an electrical discharge, and the emission is due to the population inversion of the carbon dioxide molecules. Current available wavelengths range from 9300 to 10,600 nm and are in the far-infrared thermal portion of the electromagnetic spectrum.

Chopped pulse See gated pulse mode.

Chromium A transition metal element used as a dopant for laser active medium crystals, such as yttrium, scandium, gallium, and garnet (see YSGG).

Chromophore A compound or molecule normally occurring in tissues that is an absorber of specific wavelengths of visible light.

Cladding A thin coating that surrounds the core of glass in a fiber-optic delivery system. The cladding maintains the propagation of the laser beam along the glass. The cladding is surrounded by a thicker jacket to aid in flexibility.

Coagulation For thermal tissue interaction, an observed denaturation of soft tissue proteins that occurs at approximately 60 °C. The term is usually applied to blood when it changes from a fluid to semisolid congealed mass.

Coherency A term that describes radiant waves traveling in phase both temporally and spatially.

Collimation The state in which all electromagnetic rays are parallel with virtually no divergence.

Contact mode The direct touching of the laser delivery system to the target tissue.

Continuous wave mode A manner of applying laser energy in which beam power density remains constant over time; also abbreviated CW. This mode is commonly used clinically.

Delivery system The manner in which laser energy is transferred to the target tissue. For dental lasers, there are fiber-optic, hollow waveguide, and articulated arm systems. The distal end can employ additional tips, and the beam may be used in a contact or noncontact mode.

Diffraction The bending of a light ray as the light passes through a medium, also known as refraction.

Diode laser A laser whose active medium consists of an array of semiconductor wafers in a double-heterostructure arrangement. These lasers are pumped with electrical current, and the resulting light is collected and focused into a beam. Various elements, such as aluminum, arsenide, gallium, indium, nitrogen, and phosphorous, are part of the composition of the wafer. The emission wavelengths can range from the visible into the near-infrared thermal portion of the electromagnetic spectrum.

Divergence An observed amount of the spread of the laser beam as it increases its distance from the emission aperture or focal point; the opposite of collimation. It is measured in degrees of an angle.

Doping The addition of an element to the laser active medium crystal, resulting in a specific emission of energy. Those elements are known as dopants and in most cases; they are either rare earth ions or transition metal ions. An example is doping an yttrium aluminum garnet crystal with the element of erbium.

Electromagnetic radiation Flow of energy consisting of oscillating electric and magnetic fields lying transverse to the direction of the wave's propagation.

Electromagnetic spectrum The entire range of all forms of radiant energy from gamma rays to radio waves and is usually depicted with increasing wavelength and/or decreasing frequency.

Emission cycle A ratio of the emission on time to the on plus off time, expressed as a percentage. An alternative term is duty cycle, although that term usually refers to the limit of operation of a machine.

Energy The ability to perform work, measured in a unit known as a Joule, abbreviated J. For dental lasers, a common value is a millijoule (mJ) which is one-one thousandth of a Joule.

Energy density The measurement of energy per unit area, usually expressed as Joules/square centimeter, also known as fluence/radiant exposure.

Erbium A rare earth element that is used to dope a crystal of yttrium, aluminum, garnet or yttrium, scandium, gallium, and garnet (see YAG or YSGG).

Excited state An atom or molecule with electron orbit(s) in an energized or higher level than the resting state.

Extinction length The distance into a material where the power density has dropped to 37% of its original value. A synonymous term is attenuation length.

Fiber optics A laser delivery system composed of a glass fiber, which can contain multiple strands and is used to propagate the laser beam along its length. The glass is surrounded by cladding and a jacket or layers of jackets. The bare fiber can be used for laser procedures, or additional quartz or sapphire tips can be added to the distal end of the fiber optics.

Fluence See energy density.

Focal length The distance between the focusing lens and the focal point, which is the place where the laser beam's power and energy are delivered at maximum density. It is usually measured in millimeters. In bare fiber-optically delivered lasers, there is no focusing lens so focal length does not exist. The greatest emission is at the end of the fiber. In other delivery systems, such as an articulated arm, the focal point is usually several millimeters from the end of the delivery system.

Free-running pulse mode A primary (inherent) laser operating mode where the emission is truly pulsed and not gated. The pumping process only sustains lasing conditions for a very short time. Flashlamps, radio frequency electron signals, and diode lasers are examples of the pumping mechanism used. The resulting laser emission consists of very short pulse durations in the microsecond range, and peak powers of thousands of Watts are possible. A laser operating in this mode cannot be operated in continuous wave. This mode is commonly used clinically for dental hard tissue procedures.

Frequency The number of oscillations or cycles of a wave of electromagnetic radiation per second, usually expressed as Hertz and abbreviated Hz. Frequency is the inverse of wavelength.

Gated pulse mode A laser operating mode where the emission is a repetitive on and off cycle. The laser beam is actually emitted continuously, but a mechanical shutter or electronic controls "chop" the laser beam into pulses. This term is synonymous with chopped pulse mode, and this mode is commonly used clinically. When to "on time" and the "off time" can be controlled separately, it is termed variable gated continuous wave mode.

Gaussian curve A graphic depiction of normal distribution of an entity. For lasers, it illustrates the cross section of the power density during a certain time period, usually a pulse.

Handpiece An instrument attached to the distal portion of the delivery system that contains the focusing lens system. In some cases, an additional tip is attached to the handpiece to complete the assembly.

Hertz A term of frequency of an electromagnetic wave. The term is sometimes used for the number of pulses per second or repetition rate of a pulsed dental laser, with some confusion.

Hollow wave guide A laser delivery system that uses a flexible hollow tube with a mirrored inner surface to propagate the laser beam along its length.

Infrared spectrum That portion of the invisible, nonionizing electromagnetic spectrum radiation whose wavelengths range from the red border of the visible spectrum at 700 nm up to 10,000 nm (0.07–1.0 μm). Commonly understood, but not universally agreed to, subdivisions are (1) near-infrared wavelengths from 700 to 1400 nm, (2) mid-infrared 1400–3000 nm, and (3) long infrared 3000–10,000 nm.

Intensity See power density.

Irradiance See power density.

Joule A unit of expression of energy.

Lasant See active medium.

Laser An acronym of light amplification by stimulated emission of radiation. The basic components of the device are the active medium, external energy source or pumping mechanism, optical resonator, and the focusing and delivery systems.

Meter A unit of measurement and, for electromagnetic waves, used to describe the wavelength. For dental lasers, it is divided by a million and termed a micron symbolized by μm or divided by a billion and termed a nanometer, abbreviated nm.

Mode-locked laser This laser uses a group of techniques of modulation or absorption in the resonator to produce ultrashort pulses in the picosecond or femtosecond range. These lasers are currently used experimentally.

Monochromatic The characteristic of a laser beam where only one wavelength is present.

Neodymium A rare earth element used to dope an active medium laser crystal, such as yttrium aluminum garnet (see YAG).

Noncontact mode The delivery system is used without touching the target tissue.

Optical resonator (optical cavity) The component of a laser containing the active medium in which the population inversion occurs. At each end of the resonator, there are reflective surfaces or mirrors which produce amplification and coherency. The distal mirror is partially transmissive; when there is sufficient energy, the beam can exit through that mirror.

Peak power The measurement of the maximum power in each pulse. It is the result of dividing the energy per pulse by the pulse duration.

Photon A unit or quantum of radiant energy.

Plume Essentially the smoke produced from aerosolization of by-products due to the laser-tissue interaction. It is composed of particulate matter, cellular debris, carbonaceous and inorganic materials, and potentially biohazardous products.

Population inversion A state within the laser cavity in which the quantity of excited species of the active medium exceeds that of the unexcited species (those at the resting, stable state).

Power The amount of work performed per unit time, expressed in Watts (W) or thousandth of a Watt (mW). A Watt is one Joule per second.

Power density The measurement of power per unit area, usually expressed as Watts/square centimeter, also known as intensity, irradiance, and radiance.

Pulse duration A measurement of the total amount of time that the pulse is emitted, also known as pulse width.

Pulse rate The number of pulses per second produced by the laser, sometimes known as "hertz," although that is technically incorrect.

Pulse width See pulse duration

Pumping The process of energy transfer from an external energy source into the gain medium of the laser. This pumping provides for the excitation and stimulation of the active medium. Some examples of pumping mechanisms are electricity, electrical fields, radio frequency signals, or a flashlamp.

Q-switched laser This laser uses Q switching to produce pulse durations in the nanosecond range. The Q (quality) factor of resonator losses is modulated, and the result is that the power can build up quickly. These lasers are currently used in the laboratory.

Radiant energy Energy transferred by an electromagnetic wave, also called radiation.

Reflection The returning of electromagnetic radiation by surfaces upon which it is incident. The two general types are specular, which is created from a smooth polished surface, and diffuse, which emanates from a rough surface.

Refraction See diffraction.

Repetition rate See hertz.

Selective photothermolysis A precise laser-tissue interaction in which the radiation is well absorbed and the pulse duration is shorter than the thermal relaxation time which minimizes tissue damage.

Spontaneous emission The release of energy (a photon) as the previously excited particle level returns to its resting, stable state.

Stimulated emission The release of energy (a photon) from an already excited particle by interaction with a particle of identical energy, producing two coherent particles. This process was theorized by Albert Einstein in 1916 and is the basis for laser operation.

Superpulse A variation of gated pulsed mode in which the pulse durations are very short, often producing high peak power, also termed very short pulse.

Thermal effect For lasers, the absorption of the radiant energy by tissue producing an increase in temperature.

Thermal relaxation time The amount of time required for temperature of the tissue that was raised by absorbed laser radiation to cool down to 37% of that value immediately after the laser pulse.

Transmission The passage of electromagnetic radiation through any medium, the opposite of absorption.

Vaporization The physical process of converting a solid or liquid into a gas; for dental procedures, it describes conversion of liquid water in the target into steam.

Watt A unit of power.

Wavelength The distance between any two similar points on a wave; for example, from peak to peak, measured in meters.

YAG An acronym describing a solid crystal of yttrium, aluminum, and garnet that can be doped with various rare earth elements and is used as an active medium for some lasers.

YAP An acronym describing a solid crystal of yttrium, aluminum, and perovskite (a calcium titanium oxide mineral) that can be doped with various rare earth elements and is used as an active medium for some lasers.

YSGG An acronym describing a sold crystal of yttrium, scandium, gallium, and garnet that can be doped with various rare earth elements and is used as an active medium for some lasers.

Index

E

F

G

H

I

K

L

M

N

O

Q

R

V

W

GPSR Compliance

The European Union's (EU) General Product Safety Regulation (GPSR) is a set of rules that requires consumer products to be safe and our obligations to ensure this.

If you have any concerns about our products, you can contact us on ProductSafety@springernature.com

In case Publisher is established outside the EU, the EU authorized representative is:

Springer Nature Customer Service Center GmbH
Europaplatz 3
69115 Heidelberg, Germany

Batch number: 10731661

Printed by Printforce, the Netherlands